Student Solutions Manual

Chemistry & Chemical Reactivity

ELEVENTH EDITION

John C. Kotz

Paul M. Treichel

John R. Townsend

David A. Treichel

Prepared by

Charles Atwood

Australia • Brazil • Canada • Mexico • Singapore • United Kingdom • United States

Copyright © 2024 Cengage Learning, Inc.

Unless otherwise noted, all content is © Cengage

ALL RIGHTS RESERVED. No part of this work covered by the copyright herein may be reproduced or distributed in any form or by any means, except as permitted by U.S. copyright law, without the prior written permission of the copyright owner, except as may be permitted by the license terms below.

The names of all products mentioned herein are used for identification purposes only and may be trademarks or registered trademarks of their respective owners. Cengage Learning disclaims any affiliation, association, connection with, sponsorship, or endorsement by such owners.

> For product information and technology assistance, contact us at Cengage Customer & Sales Support, 1-800-354-9706 or **support.cengage.com**.
>
> For permission to use material from this text or product, submit all requests online at **www.copyright.com**..

ISBN: 978-0-357-85183-8

Cengage
200 Pier 4 Boulevard
Boston, MA 02210
USA

Cengage is a leading provider of customized learning solutions. Our employees reside in nearly 40 different countries and serve digital learners in 165 countries around the world. Find your local representative at **www.cengage.com**.

To learn more about Cengage platforms and services, register or access your online learning solution, or purchase materials for your course, visit **www.cengage.com**.

Printed at CLDPC, USA, 08-23

Table of Contents

Chapter 1: Basic Concepts of Chemistry 1

Chapter 1R: Let's Review..15

Chapter 2: Atoms, Molecules, and Ions..36

Chapter 3: Chemical Reactions..88

Chapter 4: Stoichiometry: Quantitative Information about Chemical Reactions 119

Chapter 5: Principles of Chemical Reactivity: Energy and Chemical Reactions 184

Chapter 6: The Structure of Atoms 231

Chapter 7: The Structure of Atoms and Periodic Trends...256

Chapter 8: Bonding and Molecular Structure 283

Chapter 9: Bonding and Molecular Structure: Orbital Hybridization and Molecular Orbitals ...328

Chapter 10: Gases & Their Properties 356

Chapter 11: Intermolecular Forces and Liquids...398

Chapter 12: The Solid State 418

Chapter 13: Solutions and Their Behavior...448

Chapter 14: Chemical Kinetics The Rates of Chemical Reactions..492

Chapter 15: Principles of Chemical Reactivity: Equilibria 533

Chapter 16: Principles of Chemical Reactivity: The Chemistry of Acids and Bases................569

Chapter 17: Principles of Chemical Reactivity: Other Aspects of Aqueous Equilibria 612

Chapter 18: Thermodynamics-Entropy and Free Energy 671

Chapter 19: Principles of Chemical Reactivity: Electron Transfer Reactions..........................713

Chapter 20: Nuclear Chemistry 763

Chapter 21: The Chemistry of the Main Group Elements 789

Chapter 22: The Chemistry of the Transition Elements 833

Chapter 23 Carbon: Not Just Another Element 861

Chapter 24 Biochemistry 901

Chapter 25 Environmental Chemistry Earth's Environment, Energy, and Sustainability..........919

Solution and Answer Guide

Kotz Treichel Townsend Treichel, Chemistry and Chemical Reactivity 11e, 978-0-357-85140-1, Chapter 1: Basic Concepts of Chemistry

TABLE OF CONTENTS

Applying Chemical Principles 1
Practicing Skills 1
General Questions 7

Applying Chemical Principles

CO_2 in the Oceans

1.1.1. Name of CO_2: carbon dioxide

1.1.2. Symbols for metals mentioned in the article:

    calcium, Ca;   copper, Cu;   manganese, Mn;   iron, Fe

1.1.3. Most dense metal: Cu (8960 kg/m³)   Least dense metal: Ca (1550 kg/m³)
Data taken from www.ptable.com

1.1.4. $CaCO_3$ (calcium carbonate) contains Ca (calcium), C (carbon), and O (oxygen).

Practicing Skills

Nature of Science

1.1. (a) Proposal that pressure increases with decreased volume—hypothesis

    (b) Over time experiments indicate that pressure and volume are inversely proportional—law

    (c) Proposal that more molecules colliding per given area results in increased pressure—theory

1.2. Categorize as hypothesis, theory, or law: Hypothesis--a tentative explanation or prediction in accord with current knowledge.

Green Chemistry

1.3. Sustainable development means meeting today's needs while ensuring that future generations will be able to meet theirs.

1.4. Green chemistry refers to practices that reduce waste products during chemical processes, use materials wisely, use renewable materials, generate substances with the lowest possible toxicity, and conserve energy as well as materials.

1.5. Practices of Green Chemistry described:
- Synthetic methods to maximize incorporation of all materials
- Synthetic methods to generate substances with little or no toxicity
- Raw materials should be renewable
- Energy requirements recognized for environmental and economic impact minimized

1.6. Practices of Green Chemistry described:
- Raw materials (plant-based materials) renewable
- Energy saved
- Synthesis uses products with low or no toxicity (water and plant-based materials)
- Synthetic methods to generate substances with little or no toxicity

Matter: Elements and Atoms, Compounds and Molecules

1.7. The names of each of the elements:

(a)	N	nitrogen	(c)	Br	bromine	(e)	Li	lithium
(b)	Ca	calcium	(d)	I	iodine	(f)	Fe	iron – from the Latin *ferrium*

1.8. The names of each of the elements:

(a)	Cr	chromium	(c)	Mg	Magnesium – often confused with manganese	(e)	Ar	argon
(b)	Ni	nickel	(d)	Cl	chlorine	(f)	Ti	titanium

1.9. The symbol for each of the elements:

(a)	strontium	Sr	(c)	cobalt	Co	(e)	selenium	Se
(b)	cadmium	Cd	(d)	mercury	Hg	(f)	bismuth	Bi

1.10. The symbol for each of the elements:

(a)	platinum	Pt	(c)	uranium	U	(e)	tungsten	W
(b)	gallium	Ga	(d)	thallium	Tl	(f)	xenon	Xe

1.11. In each of the pairs, decide which is an element and which is a compound:

[HINT: If the isolated symbol is on the periodic table, it's an element!]

(a) Na or NaCl—Sodium(Na) is an element, and sodium chloride(NaCl) is a compound.

(b) Sugar or carbon—Sugar, composed of C, H, and O, is a compound, and carbon(C) is an element.

(c) Gold or gold(III) chloride—Gold(Au) is an element, and gold(III) chloride (AuCl$_3$) is a compound.

1.12. In each of the pairs, decide which is an element and which is a compound:

[HINT: If the isolated symbol is on the periodic table, it's an element!]

(a) Pt(NH$_3$)$_2$Cl$_2$ is a compound; Pt is an element

(b) Copper is an element; copper(II) oxide is a compound

(c) Silicon is an element; SiO$_2$ is a compound

1.13. Masses of hydrogen and oxygen gases prepared from 27 g of water?

An 18 g sample of water contains 2 g of hydrogen gas and 16 g of oxygen gas. A 27 g sample will contain the same proportion of hydrogen and oxygen.

$$\frac{2 \text{ g hydrogen}}{18 \text{ g water}} = \frac{x}{27 \text{ g water}}$$

$$x = \frac{2 \times 27}{18} = 3$$

The amount of oxygen would be 27 − 3 or 24 g oxygen. We could have used the ratio of oxygen to water to solve for the amount of oxygen in 27 g water.

The Law of Constant Composition (or the Law of Definite Proportions) is used.

1.14. 60. g of magnesium produces 100. g of magnesium oxide. A ratio will tell us the amount of oxide formed when 30. g of magnesium are used (An example of The Law of Constant Composition or the Law of Definite Proportions).

$$\frac{60. \text{ g magnesium}}{40. \text{ g oxygen}} = \frac{30. \text{ g magnesium}}{x}$$

$$x = \frac{30. \times 40.}{60.} = 20. \text{ g oxygen}$$

Physical and Chemical Properties

1.15. Determine if the property is a physical or chemical property for the following:

(a) color — a physical property

(b) transformed into rust — a chemical property

(c) explode — a chemical property

(d) density — a physical property

(e) melts — a physical property

(f) green — a physical property (as in (a))

Physical properties are those that can be observed or measured without changing the composition of the substance. Exploding or transforming into rust results in substances that are **different** from the original substances—and represent chemical properties.

1.16. Determine if the property is a physical or chemical property for the following:

[HINT: Physical changes do not alter the chemical composition of the substances whereas chemical changes produce new substances with different chemical compositions.]

(a) physical property—electrical conductivity does not affect the chemical state of copper

(b) physical property—viscosity is changed upon heating but it does not change the chemical state of olive oil (or other liquids)

(c) physical property—density is one of the basic physical properties of substances

(d) chemical property—fermentation is a chemical process that changes sugar into alcohol

(e) physical property—color of a substance is also a basic physical property

(f) chemical property—burning (combustion) converts gasoline to carbon dioxide and water substances that are different from gasoline

1.17. Determine if the change is a physical or chemical change for the following:

[HINT: Physical changes do not alter the chemical composition of the substances whereas chemical changes produce new substances with different chemical compositions.]

(a) physical change—changes of state, from solid to liquid to gas and so forth, are physical changes

(b) chemical change—milk souring converts the milk sugars into acids

(c) chemical change—a chemical reaction occurs when baking soda and vinegar combine making a new product, carbon dioxide gas

(d) physical change—after sugar is dissolved in water we can evaporate the water then recover the sugar

(e) chemical change—color changes are frequently associated with chemical changes, in this case as the tomato ripens several chemical compounds change structure to produce a red, ripe tomato

(f) chemical change—colors produced by fireworks are the result of chemicals burning in air producing new chemical compounds and releasing energy in the form of light

1.18. Determine if the change is a physical or chemical change for the following:

[HINT: Physical changes do not alter the chemical composition of the substances whereas chemical changes produce new substances with different chemical compositions.]

(a) chemical change—color changes are frequently associated with chemical changes, in this case bleach reacts with the T-shirt dye changing its chemical structure

(b) physical change— changes of state, from solid to liquid to gas and so forth, are physical changes, in this case your exhaled breath is initially a gas that converts to a liquid in the cold air.

(c) chemical change—when one chemical compound, carbon dioxide, is converts to another chemical compound, sugar, a chemical reaction occurs

(d) physical change— changes of state, from solid to liquid to gas and so forth, are physical changes, in this case melting butter changes from a solid to a liquid

1.19. Which are physical properties or chemical properties:

[HINT: While a substance's color is a physical property a change in color is almost always a chemical change.]

(a) color and physical state are physical properties, reactivity is a chemical property

chlorine's color, yellow-green, and physical state, gas, are physical properties

chlorine's reaction with sodium is a chemical property

(b) color and physical state are physical properties, reactivity is a chemical property

sodium bicarbonate's color, white, and physical state, solid, are physical properties

sodium bicarbonate's reaction with an acid is a chemical property

1.20. Which are physical properties or chemical properties:

[HINT: While a substance's color is a physical property a change in color is almost always a chemical change.]

(a) color, physical state, and density are physical properties, reactivity is a chemical property

copper(II) sulfide's color, black, density, 2.71 g/cm³, and physical state, solid, are physical properties

copper(II) sulfide's reaction with an acid is a chemical property

(b) color and physical state are physical properties, reactivity is a chemical property

magnesium's color, silver, and physical state, metal, are physical properties

magnesium's reaction with oxygen is a chemical property

Energy

1.21. To move the lever, one uses mechanical energy. The energy resulting is manifest in electrical energy (which produces light); thermal (radiant) energy would be released as the bulb in the flashlight glows.

1.22. Mechanical energy propels the car, electrical energy recharges the batteries, (thermal) radiant energy is released as the sun shines on the solar panels.

1.23. Which represents potential energy and which represents kinetic energy:

(a) acoustic energy represents matter in motion (molecules in air are moving) — kinetic

(b) thermal energy represents matter in motion—kinetic

(c) gravitational energy represents the attraction of the earth for an object—and therefore energy due to position—potential

(d) chemical energy represents the energy stored in fuels—potential

(e) electrostatic energy represents the energy of separated charges—and therefore potential energy.

1.24. Kinetic to Potential or vice versa:

(a) Potential → kinetic as water falls

(b) Potential → kinetic as natural gas burns, air inside house is heated

(c) Kinetic → potential as electrons are stored during battery charge

(d) Potential → kinetic the potential energy of the chemicals is converted into kinetic energy of the moving piston

1.25. Since 1500 J of energy is lost by the metal, the water must gain 1500 J of energy, as dictated by the Law of Conservation of Energy.

1.26. The energy lost by the falling book is gained by the floor (which typically doesn't move owing to a larger mass). Some of the energy is gained by surrounding air molecules in the form of sound.

General Questions

1.27. For the gemstone turquoise:

(a) Qualitative: blue-green color, solid state   Quantitative: density; mass

(b) Extensive: Mass   Intensive: Density; Color; Physical state

(c) Volume: $2.5 \text{ g} \left(\dfrac{1 \text{ cm}^3}{2.65 \text{ g}} \right) = 0.94 \text{ cm}^3$

1.28. Qualitative vs Quantitative observations; Extensive vs Intensive observations:

(a) Qualitative: shiny golden metallic appearance, crystals in form of perfect cubes

Quantitative: length of 0.40 cm on a side, mass of 0.32 g

(b) Extensive: Mass and length; Intensive: color, luster, and crystalline form

(c) Density = Mass/Volume = 0.32 g/(0.40 cm)³ = 0.064 g/0.064 cm³ = 5.0 g/cm³
   = 5.0 g/mL

1.29. Of the observations below, those which identify chemical properties:

[Chemical properties, in general, are those observed during a chemical change—as opposed to during a physical change.]

(a) Vinegar and sodium bicarbonate reaction –Chemical

(b) Sugar soluble in water--Physical

(c) Water boils at 100°C--Physical

(d) Ultraviolet light converts O_3 to O_2--Chemical

(e) Ice is less dense than water—Physical

1.30. Of the observations below, those which identify chemical properties:

[Chemical properties, in general, are those observed during a chemical change—as opposed to during a physical change.]

(a) Iron rusting—a chemical property as iron reacts with oxygen

(b) Sodium metal reacts—a chemical property as sodium metal and water react

(c) Octane combustion—a chemical property as C_8H_{18} forms CO_2 and H_2O

(d) Chlorine is a yellow-green gas—a physical property (observable without a chemical reaction)

(e) Ice melting from heat—a physical property (observable without a chemical reaction)

1.31. Regarding fluorite:

(a) The symbols for the elements in fluorite: Ca (calcium) and F (fluorine);

(b) Shape of the crystals: cubic
Arrangement of ions in the crystal indicates that the fluoride ions are arranged around the calcium ions in the lattice in such a way as to form a cubic crystal.

1.32. Regarding azurite:

(a) Symbols of the elements: Copper, Cu; Carbon, C; Oxygen, O

(b) Oxygen is a gas, while copper, carbon, and azurite are solids at room temperature. Oxygen is colorless, while copper has a reddish color and carbon is gray/black. The gemstone is a bluish color clearly different from its component elements.

1.33. A solution is a mixture, so the components can be separated using a physical technique. If one heats the NaCl solution to dryness, evaporating all the water, the NaCl solid remains behind. Hence the physical property of boiling points is useful in this separation.

1.34. The non-uniform appearance of the mixture indicates that samples taken from different regions of the mixture would be different—a characteristic of a heterogeneous mixture. Recalling that iron is attracted to a magnetic field while sand is generally not attracted suggests that passing a magnet through the mixture would separate the sand and iron.

1.35. Identify physical or chemical changes:

(a) As there is no change in the composition of the carbon dioxide in the sublimation process, this represents a physical change.

(b) A change in density as a function of temperature does not reflect a change in the composition of the substance (mercury), so this phenomenon represents a physical change.

(c) The combustion of methane represents a change in the substance present as methane is converted to the oxides of hydrogen and carbon that we call water and carbon dioxide—a chemical change.

(d) Dissolving NaCl in water represents a physical change as the solid NaCl ion pairs are separated by the solvent, water. This same phenomenon, the separation of ions, also occurs during melting.

1.36. Identify physical or chemical changes:

(a) The desalination of sea water represents a physical change—as the salts and solvent (water) are separated.

(b) The formation of SO_2 as sulfur-containing coal is burned represents a combination of sulfur and oxygen—a chemical change.

(c) The tarnishing of silver represents a chemical change as silver compounds form on the exterior of the silver object.

(d) Iron is heated to red heat. Changing the temperature of an object is a physical change.

1.37. A segment of Figure 1.2 is shown here:

The macroscopic view is the large crystal in the lower left of the figure, and the particulate view is the representation in the upper right. If one imagines reproducing the particulate (sometimes called submicroscopic) in all three dimensions—imagine a molecular duplicating machine—the macroscopic view results.

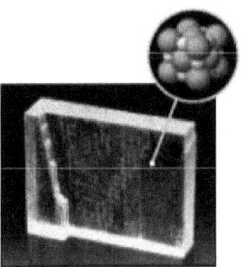

1.38. The orange solid and liquid in the bowl (top right) and the orange liquid and gas in the round-bottom flask (bottom right) represent the macroscopic views. The spheres (left column) represent the particulate view.

The particulate view of the solid displays molecules of bromine tightly packed to produce the solid.

The liquid view displays molecules of Br_2 that are still close together but not arranged in an orderly fashion and can move past one another.

The gas view displays molecules of Br_2, with the molecules being farther apart than in the liquid view.

1.39. A substance will float in any liquid whose density is greater than its own, and sink in any liquid whose density is less. The piece of plastic soda bottle (d =1.37 g/cm^3) will float in liquid CCl_4 and the piece of aluminum (d = 2.70 g/cm^3) will sink in the liquid CCl_4.

1.40. Liquids: mercury and water;   Solid: copper
Of the substances shown, mercury is most dense and water is least dense.

1.41. Categorize each as an element, a compound, or a mixture:

(a) Iron pyrite (fool's gold)—is a compound composed of iron and sulfur

(b) Carbonated mineral water is a mixture. It certainly contains the compound water AND carbon dioxide. The term "mineral" implies that other dissolved materials are present.

(c) Molybdenum—an element

(d) Sucrose—a compound, with formula $C_{12}H_{22}O_{11}$

1.42. Categorize each as an element, a compound, or a mixture:

(a) Sea water is a mixture of water with many dissolved salts and gases.

(b) Sodium chloride is a compound of sodium and chlorine.

(c) Bronze is a mixture of copper and tin.

(d) Gold (24-carat) is elemental gold refined to be at least 99.9% pure gold.

1.43. Indicate the relative arrangements of the particles in each of the following:

(a) solid iron (b) liquid water (c) gaseous water

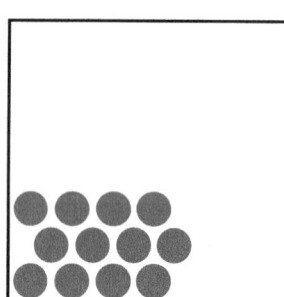 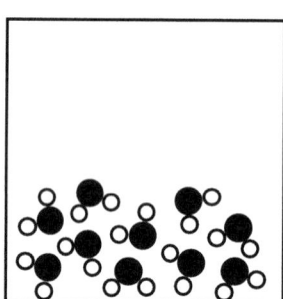 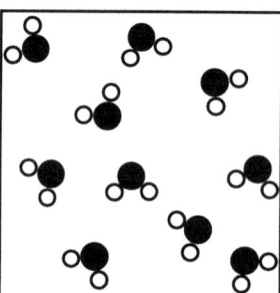

1.44. Indicate the relative arrangements of the particles in each of the following:

(a) $H_2O(g)$ & $He(g)$ (b) $H_2O(\ell)$ & $Al(s)$ (c) $Cu(s)$ & $Zn(s)$

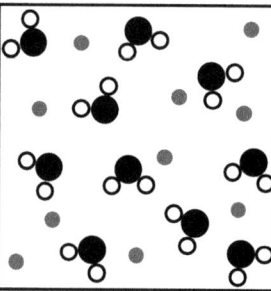 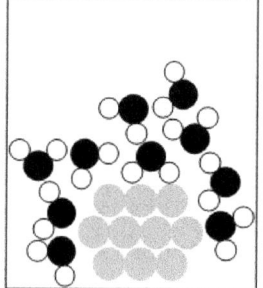 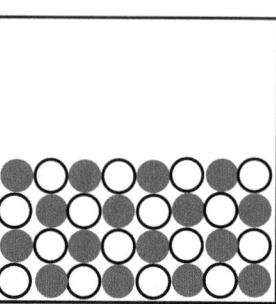

1.45. When the three liquids are placed into the graduated cylinder, they will "assemble" in layers with increasingly smaller densities (from the bottom to the top) in the cylinder.

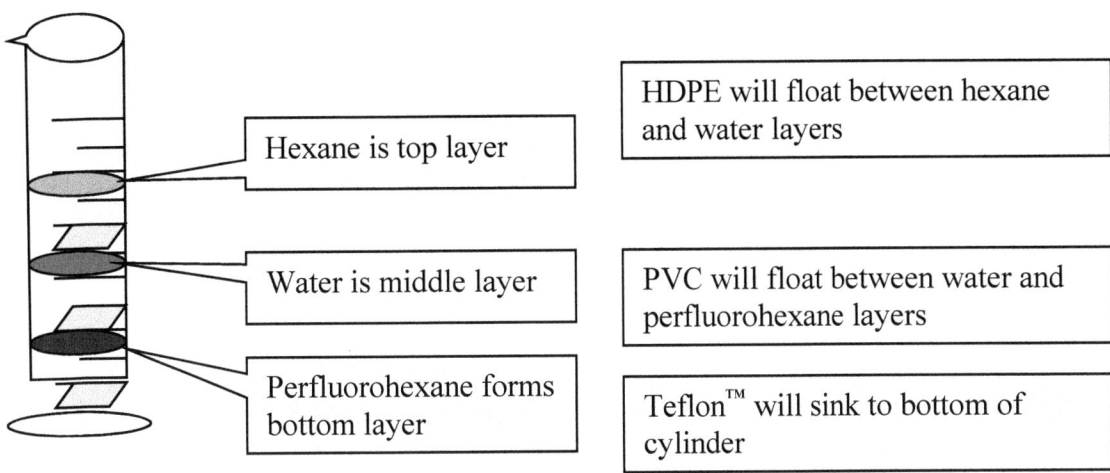

[Shadings are added to the top of each layer to provide clarity only (NOT to indicate the color of the liquids). Similarly the parallelogram symbols indicating the plastic samples are shaded—only to provide clarity in locating them—and not to imply any specific colors.]

1.46. There are several properties that could be used. One is the melting point. Sugar melts around 160-186 °C while table salt melts about 800 °C.

1.47. HDPE with a density of 0.97 g/cm^3 will float in any liquid whose density is greater than 0.97 g/cm^3 and sink in any liquid whose density is less than that of HDPE. Of the liquids listed, HDPE should float in ethylene glycol, water, acetic acid, and glycerol.

1.48. Measure the density and melting point of the silvery metal and compare to published data points for silver.

1.49. Water, a large component of milk, expands as it is converted to the solid state—as the density of solid water is less than that of liquid water. The expanded solid escapes via the avenue of least resistance—the cap.

1.50. First, weigh the object. Then immerse it in a liquid (e.g. in a graduated cylinder) in which it sinks. If you measure the volume of liquid before and after you immerse the metal, the difference in volume is the volume of the object. Calculate the density of the object by dividing the mass by its volume.

1.51. If one excretes too much sugar, the concentration of the sugar "solution" left in the body would decrease, resulting in urine with a higher density. If one excretes too much water, the concentration of the sugar "solution" in the body would increase, with a concomitant decrease in density of the urine.

1.52. To determine the identity, use the physical properties of water: measure the density of the unknown liquid. Freeze and boil the liquid. If the density of the liquid is about 1g/cm^3, if the liquid freezes about 0 °C and boils about 100 °C, the liquid is probably water. To test for the presence of salt, use a conductivity device. Water containing dissolved salts will conduct an electric current while pure water will not.

1.53. For the reaction of elemental potassium reacting with water:

(a) States of matter involved: **Solid** potassium reacts with **liquid** water to produce **gaseous** hydrogen and **aqueous** potassium hydroxide solution (a homogenous mixture).

(b) The observed change is **chemical**. The products (hydrogen and potassium hydroxide) are quite different from elemental potassium and water. Litmus paper would also provide the information that while the original water was neither acidic nor basic, the solution produced would be basic. (The color of red litmus paper would change to blue.)

(c) The **reactants**: potassium and water
The **products**: hydrogen, potassium hydroxide solution, heat, and light

(d) Potassium reacts **vigorously** with water. Potassium is less dense than water, and floats atop the surface of the water. The reaction produces enough heat to ignite the hydrogen gas evolved. The flame observed is typically violet-purple in color. The potassium hydroxide formed is soluble in water (and therefore not visible).

1.54. The three liquids with the least dense liquid at the top and most dense liquid at bottom:

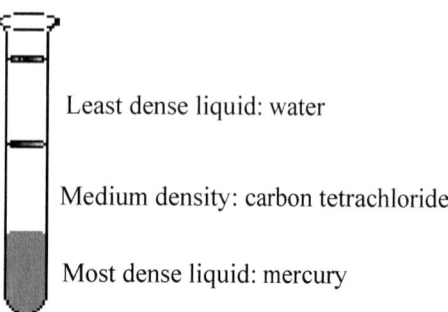

Least dense liquid: water

Medium density: carbon tetrachloride

Most dense liquid: mercury

1.55. Since gases rise to an area with a similar density as their own, balloons with hydrogen and methane—with densities less than that of dry air--will float (and, if untethered, float away), while the balloons containing chlorine and propane will "sink"—to the lowest nearby surface.

1.56. You cannot say that the metal is definitely copper based on the fact that it conducts electricity. Use the physical properties of copper. One could measure the density of the metal and compare it to the published density of copper. Other options are to determine the metal's boiling and melting points and then compare them to the published values.

1.57. The dissolution of iodine in ethanol (to make a solution) is a **physical** change, with iodine being the solute and ethanol the solvent.

1.58. For the mixture:

(a) Density of the mixture:

Calculate the mass of each substance (multiply density by volume):

Mass of $CHCl_3$: 10.0 mL · 1.492 g/mL = 14.92 g

Mass of $CHBr_3$: 5.0 mL · 2.890 g/mL = 14.45 g

The mass of the solution is: 14.92 g + 14.45 g or 29.37 g (29 to 2 significant figures) and a volume of 15.0 mL

The density of the mixture is then: 29 g/15.0 mL = 2.0 g/mL (to 2 significant figures).

(b) Density of yellow crystal:

Mass of solution = 2.07 g/mL · 20.0 mL = 41.4 g

Mass of solution = mass $CHCl_3$ + mass $CHBr_3$

Let x = volume $CHCl_3$ and (20.0 mL − x) = volume $CHBr_3$

Then, mass of solution = (1.492 g/mL)x + (2.890 g/mL)(20.0 mL − x) = 41.4 g

x = 11.7 mL $CHCl_3$ and (20.0 − x) = 8.3 mL $CHBr_3$

So to obtain 20.0 mL of solution with d = 2.07 g/cm^3, mix 11.7 mL $CHCl_3$ and 8.3 mL $CHBr_3$.

1.59. For the wedding band in question:

(a) Gold- Au; Copper- Cu; Silver- Ag

(b) The reported density of Osmium (according to www.ptable.com) is 22.59 g/cm^3, making it the densest element.

(c) As the band is 18-carat (or 75% gold), we can multiply the mass of the ring by 75% written as a fraction.

$$5.58 \text{ g ring} \left(\frac{75 \text{ g Au}}{100 \text{ g ring}} \right) = 4.185 \text{ g Au or } 4.2 \text{ g Au to 2 significant figures}$$

(d) Mass of lost gold:

$$112 \times 10^6 \text{ rings} \left(\frac{6.15 \times 10^{-3} \text{ g Au lost}}{1 \text{ ring}} \right) = 6.89 \times 10^5 \text{g Au lost} (3 \text{ significant figures})$$

Value of lost gold:

$$6.89 \times 10^5 \text{g Au lost} \left(\frac{1 \text{troy oz}}{31.1 \text{ g Au}} \right) \left(\frac{\$1815}{1 \text{troy oz}} \right) = \$4.02 \times 10^7 \text{ or } \$40.2 \text{ million (3 significant figures)}$$

Note the assumption in the first calculation that each married couple would have 2 rings.

Solution and Answer Guide

Kotz Treichel Townsend Treichel, Chemistry and Chemical Reactivity 11e, 978-0-357-85140-1, Chapter 1R: Let's Review

TABLE OF CONTENTS

Applying Chemical Principles 15
Practicing Skills 16
General Questions 25
In The Laboratory 33

Applying Chemical Principles

Out of Gas!

1R.1. Fuel density in kg/L: $\dfrac{1.77 \text{ lb}}{1 \text{ L}} \cdot \dfrac{0.4536 \text{ kg}}{1 \text{ lb}} = 0.803 \text{ kg/L}$

1R.2. Mass and volume of fuel that should have been loaded:

Mass of fuel already in tank: $7682 \text{ L} \times \dfrac{0.803 \text{ kg}}{\text{L}} = 6170 \text{ kg}$

Mass of fuel needed: $22{,}300 \text{ kg} - 6170 \text{ kg} = 16{,}100 \text{ kg}$

Volume of fuel needed: $16{,}100 \text{ kg} \times \dfrac{1 \text{ L}}{0.803 \text{ kg}} = 20{,}100 \text{ L}$

Ties in Swimming and Significant Figures

1R.1. Distance traveled in 0.001 s (at world record rate of 20.91 s for 50-m):

$\dfrac{50 \text{ m}}{20.91 \text{ s}} \cdot \dfrac{0.001 \text{ s}}{1} \cdot \dfrac{1000 \text{ mm}}{1 \text{ m}} = 2.4 \text{ mm (2sf)}$

1R.2. Time to travel 3.0 cm at world record rate:

$\dfrac{20.91 \text{ s}}{50 \text{ m}} \cdot \dfrac{1 \text{ m}}{100 \text{ cm}} \cdot \dfrac{3.0 \text{ cm}}{1} = 0.0125 \text{ s or } 0.013 \text{ s (2sf)}$

1R.3. Percent error if lane is 3 cm longer than 50.00 m

Actual length of lane is 50.03 m (3 cm + 50.00 m converted to units of m)

Error = 50.03 m − 50.00 m = 0.03 m

% error = $\dfrac{\text{error}}{\text{accepted value}} \times 100\% = \dfrac{0.03 \text{ m}}{50.00} \times 100\% = 0.06\%$

Practicing Skills

Temperature Scales

1R.1. Express 25 °C in kelvins:
$T(K) = (25\ °C + 273.15\ °C)\dfrac{1\ K}{1\ °C} = 298\ K$

1R.2. Express 5.5×10^3 °C in kelvins: $(5.5 \times 10^3\ °C + 273.15\ °C)\dfrac{1\ K}{1\ °C} = 5.8 \times 10^3\ K$

1R.3. Make the following temperature conversions:

°C	K
(a) 16	16 + 273.15 = 289
(b) 370 − 273 or 97	370
(c) 40	40 + 273.15 = 310

Note no decimal point after 40

1R.4. Make the following temperature conversions:

°C	K
(a) 77 − 273.15 = −196	77
(b) 63	63 + 273.15 = 336
(c) 1450 − 273.15 = 1177	1450

Length, Volume, Mass, and Density

1R.5. The distance of a marathon (42.195 km) in meters; in miles:

$\dfrac{42.195\ km}{1} \cdot \dfrac{1000\ m}{1\ km} = 42195\ m \qquad \dfrac{42.195\ km}{1} \cdot \dfrac{0.62137\ miles}{1\ km} = 26.219\ miles$

The factor (0.62137 mi/km) is found inside the back cover of the text.

1R.6. Length of a 19 cm pencil in mm; in m:

$$\frac{19 \text{ cm}}{1} \cdot \frac{10 \text{ mm}}{1 \text{ cm}} = 190 \text{ mm} \qquad \frac{19 \text{ cm}}{1} \cdot \frac{1 \text{ m}}{100 \text{ cm}} = 0.19 \text{ m}$$

1R.7. Express the area of a 2.5 cm × 2.1 cm stamp in cm^2; in m^2:

$$2.5 \text{ cm} \cdot 2.1 \text{ cm} = 5.3 \text{ cm}^2$$

$$\frac{5.3 \text{ cm}^2}{1} \cdot \left(\frac{1 \text{ m}}{100 \text{ cm}}\right)^2 = 5.3 \times 10^{-4} \text{ m}^2$$

1R.8. Surface area of a cat food can's lid in cm^2 and m^2:

The radius of the lid is half the diameter.

$$\text{Area} = \pi r^2 = \pi \left(\frac{8.3 \text{ cm}}{2}\right)^2 = 54 \text{ cm}^2$$

$$8.3 \text{ cm} \cdot \frac{1 \text{ m}}{100 \text{ cm}} = 0.083 \text{ m}$$

$$\text{Area} = \pi r^2 = \pi \left(\frac{0.083 \text{ m}}{2}\right)^2 = 0.0054 \text{ m}^2$$

1R.9. Express volume of 250. mL beaker in cm^3; in liters (L); in m^3; in dm^3:

$$\frac{250. \text{ mL}}{1 \text{ beaker}} \cdot \frac{1 \text{ cm}^3}{1 \text{ mL}} = \frac{250. \text{ cm}^3}{1 \text{ beaker}}$$

$$\frac{250. \text{ cm}^3}{1 \text{ beaker}} \cdot \frac{1 \text{ L}}{1000 \text{ cm}^3} = \frac{0.250 \text{ L}}{1 \text{ beaker}}$$

$$\frac{250. \text{ cm}^3}{1 \text{ beaker}} \cdot \frac{1 \text{ m}^3}{1 \times 10^6 \text{ cm}^3} = \frac{2.50 \times 10^{-4} \text{ m}^3}{1 \text{ beaker}}$$

$$\frac{250. \text{ cm}^3}{1 \text{ beaker}} \cdot \frac{1 \text{ L}}{1000 \text{ cm}^3} \cdot \frac{1 \text{ dm}^3}{1 \text{ L}} = \frac{0.250 \text{ dm}^3}{1 \text{ beaker}}$$

1R.10. Volume of bottle in mL; in cm^3; in dm^3:

$$1.5 \text{ L} \cdot \frac{10^3 \text{ mL}}{1 \text{ L}} = 1.5 \times 10^3 \text{ mL}$$

$$1.5 \text{ L} \cdot \frac{10^3 \text{ mL}}{1 \text{ L}} \cdot \frac{1 \text{ cm}^3}{1 \text{ mL}} = 1.5 \times 10^3 \text{ cm}^3$$

$$1.5 \text{ L} \cdot \frac{10^3 \text{ mL}}{1 \text{ L}} \cdot \frac{1 \text{ cm}^3}{1 \text{ mL}} \cdot \left(\frac{1 \text{ dm}}{10 \text{ cm}}\right)^3 = 1.5 \text{ dm}^3$$

1R.11. Convert book's mass of 2.52 kg into grams:
$$\frac{2.52 \text{ kg}}{1 \text{ book}} \cdot \frac{1 \times 10^3 \text{ g}}{1 \text{ kg}} = \frac{2.52 \times 10^3 \text{ g}}{1 \text{ book}}$$

1R.12. Mass of 2.265 g in kg; in mg:
$$2.265 \text{ g} \cdot \frac{1 \text{ kg}}{1 \times 10^3 \text{ g}} = 2.265 \times 10^{-3} \text{ kg}$$
$$2.265 \text{ g} \cdot \frac{1 \times 10^3 \text{ mg}}{1 \text{ g}} = 2.265 \times 10^3 \text{ mg}$$

1R.13. What mass of ethylene glycol (in grams) possesses a volume of 500. mL of the liquid?
$$\frac{500. \text{ mL}}{1} \cdot \frac{1 \text{ cm}^3}{1 \text{ mL}} \cdot \frac{1.11 \text{ g}}{\text{cm}^3} = 555 \text{ g}$$

1R.14. Mass of 100. mL of acetone:
Remember that 1 mL = 1 cm^3
$$100. \text{ cm}^3 \left(\frac{0.7845 \text{ g}}{1 \text{ cm}^3}\right) = 78.5 \text{ g}$$

1R.15. Volume of 2.365 g of silver: $2.365 \text{ g} \cdot \frac{1 \text{ cm}^3}{10.5 \text{ g}} = 0.225 \text{ cm}^3$

1R.16. Volume of 50.0 g of lead: $50.0 \text{ g} \cdot \frac{1 \text{ cm}^3}{11.35 \text{ g}} = 4.41 \text{ cm}^3$

1R.17. To determine the density, given the data, one must first convert each length to units of cm and then calculate the volume:
$$\left[\frac{1.05 \text{ mm}}{1} \cdot \frac{1 \text{ cm}}{10 \text{ mm}}\right] \cdot \frac{2.35 \text{ cm}}{1} \cdot \frac{1.34 \text{ cm}}{1} = 0.3306 \text{ cm}^3 \; (0.331 \text{ to 3sf})$$
Density is Mass/Volume or $\frac{2.361 \text{ g}}{0.3306 \text{ cm}^3} = 7.14 \frac{\text{g}}{\text{cm}^3}$.

Given the selection of metals, the identity of the metal is **zinc**.

1R.18. Larger volume, 600 g of water or 600 g of lead:
$$600 \text{ g H}_2\text{O} \cdot \frac{1 \text{ cm}^3}{0.995 \text{ g}} = 600 \text{ cm}^3 \quad 600 \text{ g lead} \cdot \frac{1 \text{ cm}^3}{11.35 \text{ g}} = 50 \text{ cm}^3$$

Energy Units

1R.19. Express the energy of a 1200 Calories/day diet in joules:

$$\frac{1200 \text{ Cal}}{1 \text{ day}} \cdot \frac{1000 \text{ calorie}}{1 \text{ Cal}} \cdot \frac{4.184 \text{ J}}{1 \text{ cal}} = 5.0 \times 10^6 \text{ Joules/day}$$

1R.20. Express 1670 kJ in dietary Calories: $1670 \text{ kJ} \cdot \frac{10^3 \text{J}}{1 \text{ kJ}} \cdot \frac{1 \text{ cal}}{4.184 \text{ J}} \cdot \frac{1 \text{ Cal}}{10^3 \text{ cal}} = 399 \text{ Cal}$

1R.21. Compare 170 kcal/serving and 280 kJ/serving.

$$\frac{170 \text{ kcal}}{1 \text{ serving}} \cdot \frac{1000 \text{ calorie}}{1 \text{ kcal}} \cdot \frac{4.184 \text{ J}}{1 \text{ cal}} \cdot \frac{1 \text{ kJ}}{1000 \text{ J}} = 710 \text{ kJoules/serving}$$

So, 170 kcal/serving has a greater energy content.

1R.22. Greater energy per mL, berry juice or soft drink:

Soft drink: $130 \text{ Cal} \cdot \frac{10^3 \text{ cal}}{1 \text{ Cal}} \cdot \frac{4.184 \text{ J}}{1 \text{ cal}} = 5.4 \times 10^5 \text{ J}$ or 540 kJ

With the berry juice providing 630 kJ, the JUICE provides more energy.

$$\frac{540 \text{ kJ}}{335 \text{ mL}} = \frac{1.6 \text{ kJ}}{1 \text{ mL}} \qquad \frac{630 \text{ kJ}}{295 \text{ mL}} = \frac{2.1 \text{ kJ}}{1 \text{ mL}}$$

The juice also provides more energy per milliliter.

Accuracy, Precision, Error, and Standard Deviation

1R.23. Using the data provided, the averages and their deviations are as follows:

Data point	Method A	deviation	Method B	deviation
1	2.2	−0.2	2.703	0.000
2	2.3	−0.1	2.701	−0.002
3	2.7	0.3	2.705	0.002
4	2.4	0.0	2.703	0.000
Averages:	2.4		2.703	

Note that the deviations for both methods are calculated by first determining the average of the four data points, and then subtracting the individual data points from the average.

(a) The average density for method A is 2.4 g/cm³ while the average density for method B is 2.703 g/cm³

(b) The percent error for each method:

Error = experimental value - accepted value

From Method A error = (2.4 − 2.702) = −0.3 g/cm³

From Method B error = (2.703 − 2.702) = 0.001 g/cm³

and the percent error is then:

$$(\text{Method A}) = \frac{-0.3}{2.702} \cdot 100\% = -10\% \text{ (1 sf)}$$

$$(\text{Method B}) = \frac{0.001}{2.702} \cdot 100\% = 0.04\% \text{ (1 sf)}$$

(c) The standard deviation for each method:

Method A: $\sqrt{\frac{(-0.2)^2 + (-0.1)^2 + (0.3)^2 + (0.0)^2}{3}} = \sqrt{\frac{0.14}{3}} = 0.2 \text{ g/cm}^3$ (1 sf)

Method B:
$\sqrt{\frac{(0.000)^2 + (-0.002)^2 + (0.002)^2 + (0.000)^2}{3}} = \sqrt{\frac{0.000008}{3}} = 0.002 \text{ g/cm}^3$ (1 sf)

(d) Method B offers both *better accuracy* (average closer to the accepted value) and *better precision* (since the value has a lower standard deviation).

1R.24. Using the data provided, the averages and their deviations are as follows:

Data point	Your Data	deviation	Partner's Data	deviation
1	321	−4	327	−0.5
2	326	1	329	1.5
3	324	−1	328	0.5
4	329	4	326	−1.5
Averages:	325		327.5	

Note that the deviations for both methods are calculated by first determining the average of the four data points, and then subtracting the individual data points from the average.

(a) The average mass of aspirin for method A is 325 mg while the average mass of aspirin for method B is 327.5 mg = 328 mg (three sf).

(b) Your data is more accurate than your partner's data as your average mass coincides exactly with the manufacturer's accepted mass. Percent errors shown below.

Error = experimental value − accepted value

From Method A   error = (325 − 325) = 0 mg

From Method B   error = (327.5 − 325) = 2.5 mg

and the percent error is then:

$$(\text{Method A}) = \frac{0}{325} \times 100\% = 0\%$$

$$(\text{Method B}) = \frac{2.5}{325} \times 100\% = 0.8\% \text{ (1 sf)}$$

(c) Your partner's data is more precise. The data for your partner are closer to each other.

(d) The standard deviation for you and your partner:

Your data:

$$\sqrt{\frac{(-4)^2 + (1)^2 + (-1)^2 + (4)^2}{3}} = \sqrt{\frac{34}{3}} = 3 \text{ mg (1 sf)}$$

Your partner's data:

$$\sqrt{\frac{(-0.5)^2 + (1.5)^2 + (0.5)^2 + (-1.5)^2}{3}} = \sqrt{\frac{5.0}{3}} = 1 \text{ mg (1 sf)}$$

The smaller standard deviation for your partner's data confirms the prediction.

Exponential Notation and Significant Figures

1R.25. Express the following numbers in exponential (or scientific) notation:

(a) 0.0830 g = 8.30 × 10^{-2} g To locate the decimal behind the first non-zero digit, we move the decimal place to the right by 2 spaces (–2); 3 significant figures

(b) 136 g = 1.36 × 10^2 g To locate the decimal behind the first non-zero digit, we move the decimal place to the left by 2 spaces (+2); 3 significant figures

(c) 0.00602 g = 6.02 × 10^{-3} g To locate the decimal behind the first non-zero digit, we move the decimal place to the right by 3 spaces (–3); 3 significant figures

(d) 3000 mL = 3 × 10^3 mL To locate the decimal behind the first non-zero digit, we move the decimal place to the left by 3 spaces (+3); 1 significant figure

1R.26. Express the following numbers in exponential (or scientific) notation:

(a) 1356 mL = 1.356 × 10^3 mL To locate the decimal behind the first non-zero digit, we move the decimal place to the left by 3 spaces (+3); 4 significant figures

(b) 0.03042 L = 3.042 × 10^{-2} L To locate the decimal behind the first non-zero digit, we move the decimal place to the right by 3 spaces (–3); 4 significant figures

(c) 250.0 = 2.500 × 10^2 To locate the decimal behind the first non-zero digit, we move the decimal place to the left by 2 spaces (+2); 4 significant figures

(d) 120 = 1.2 × 10^2 To locate the decimal behind the first non-zero digit, we move the decimal place to the left by 2 spaces (+2); 2 significant figures

1R.27. Express the following without using scientific notation:

(a) 5.43 × 10^2 m = 543 m (3 significant figures)

(b) 4.306 × 10^{-2} L = 0.04306 L (4 significant figures)

(c) 6.20×10^{-4} L = 0.000620 L (3 significant figures) (zero after decimal point is significant)

(d) 8.42×10^3 mL = 8420 mL (3 significant figures)

1R.28. Express the following without using scientific notation:

(a) 3.25×10^{-2} m = 0.0325 m (3 significant figures)

(b) 4.02×10^{-3} mL = 0.00402 mL (3 significant figures)

(c) 4.2×10^3 mL = 4200 mL (2 significant figures)

(d) 9.305×10^4 g = 93050 g (4 significant figures)

1R.29. Perform operations and report answers to proper number of sf:

(a) $(1.52)(6.21 \times 10^{-3}) = 9.44 \times 10^{-3}$ (3 sf - each term in the product has 3)

(b) $(6.217 \times 10^3) - (5.23 \times 10^2) = 5.694 \times 10^3 = 5694$ [Convert 5.23×10^2 to 0.523×10^3 and subtract, leaving 5.694×10^3. With 3 decimal places to the right of the decimal place in both numbers, we can express the difference with 3 decimal places.

(c) $(6.217 \times 10^3) \div (5.23 \times 10^2) = 11.887$ or 11.9 (3 sf)

Recall that in multiplication and division, the result should have the same number of significant figures as the **term** with the fewest significant figures.

(d) $(0.0546)(16.0000)\left[\dfrac{7.779}{55.85}\right] = 0.121678$ or 0.122 (3 sf)

the same rule applies, as in part (c) above: The first term has 3 sf, the second term 6 sf—yes the zeroes count, and the third term 4 sf.

Dividing the two terms in the last quotient gives an answer with 4 sf. So with 3, 6, and 4 sf in the terms, the answer should have **no more than** 3 sf

1R.30. Provide result of calculation with proper number of significant figures:

In all of these calculations, the answer has 3 sf

(a) 2.44×10^8 (c) 0.133

(b) 4.85×10^{-2} (d) 0.0286

Graphing

1R.31. Plot the data for number of kernels of popcorn versus mass (in grams):

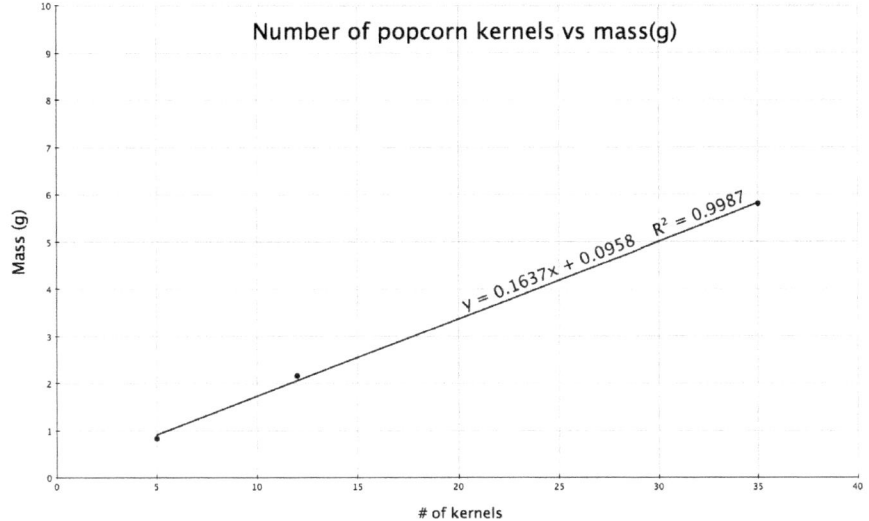

The best straight line has the equation, $y = 0.1637x + 0.096$, with a slope of 0.1637. This slope indicates that the mass increases by a factor of 0.1637 grams with each kernel of popcorn. The mass of 20 kernels would be:

mass = (0.1637)(20) + 0.096 or 3.370 grams.

To determine the number of kernels (x) with a mass of 20.88 grams, substitute 20.88 for mass (i.e., y) and solve for the number of kernels.

20.88 g = 0.1637(x) + 0.096; (20.88 − 0.096) = 0.1637x and dividing by the slope:

$$x = \frac{20.88 - 0.096}{0.1637} = 127 \text{ kernals}$$

1R.32. Using the provided graph:

(a) The value of x when $y = 4.0$ is 0.21

(b) The value of y when x is 0.30 is 5.6

(c) slope = $\frac{5.6 - 4.0}{0.30 - 0.21}$ = 18; y-intercept = 0.20

(d) The value of y when $x = 1.0$: $y = 18x + 0.20 = (18)(1.0) + 0.20 = 18$

1R.33. Using the graph shown, determine the values of the equation of the line:

(a) Using the points (0.00, 20.00) and (5.00, 0.00), we calculate the slope (rise/run):

$$\frac{20.00 - 0.00}{0.00 - 5.00} = -4.00$$

The y-intercept (b) is the y value when the x value is zero (0).

We can read the value of *b* from the graph (20.00).

The equation for the line is: $y = -4.00x + 20.00$

(b) The value of *y* when $x = 6.0$ is: $y = (-4.00)(6.0) + 20.00 = -4.00$.

1R.34. (a) The data (and the reciprocals) are as follows:

Amount of H_2O_2	1/Amount	Reaction Speed	1/Speed
1.96	0.5102	4.75×10^{-5}	21050
1.31	0.7634	4.03×10^{-5}	24810
0.98	1.02	3.51×10^{-5}	28490
0.65	1.54	2.52×10^{-5}	39680
0.33	3.03	1.44×10^{-5}	69440
0.16	6.25	0.585×10^{-5}	170900

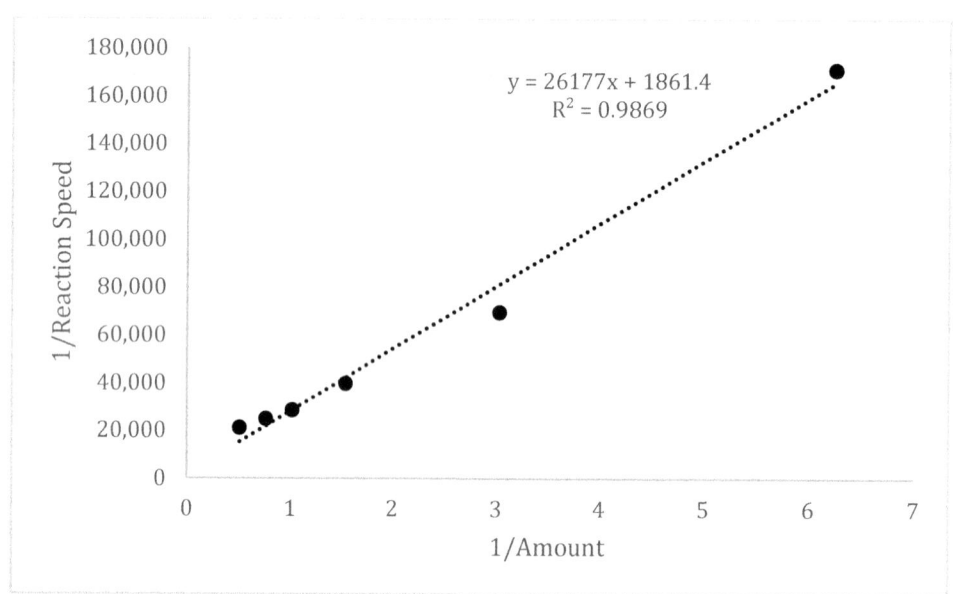

(b) $y = 2.62 \times 10^4 x + 1900$ with a slope of 2.62×10^4 (3 sf) and *y*-intercept = 1900 (rounded to the nearest hundred).

Solving Equations

1R.35. Solving the equation for "*C*":

$(0.502)(123) = (750.)C$ and rearranging the equation by dividing by 750. gives $C = \dfrac{(0.502)(123)}{750.} = 0.0823$ (3 sf)

1R.36. Solving the equation for *n*:

$(1.0)(22.4) = n(0.082057)(273.15)$ Rearranging: $n = \dfrac{(1.0)(22.4)}{(0.082057)(273.15)} = 1.0$ (2 sf)

1R.37. Solve the following equation for T:

$(4.184)(244)(T - 292.0) + (0.449)(88.5)(T - 369.0) = 0$

$020.9T - 298101.6 + 39.74T - 14663 = 0$

$1060.64T - 312764 = 0$; Solving for T:

$T = \dfrac{312764}{1060.64} = 295$ (3 sf)

1R.38. Solving the equation for n:

$-246.0 = 1312\left[\dfrac{1}{2^2} - \dfrac{1}{n^2}\right]$ Simplify by dividing by 1312: $\dfrac{-246.0}{1312} = \left[\dfrac{1}{2^2} - \dfrac{1}{n^2}\right]$

$-0.1875 = \left[\dfrac{1}{4} - \dfrac{1}{n^2}\right]$ Substituting 0.25 for 1/4, $-0.1875 = 0.25 - \dfrac{1}{n^2}$

$\dfrac{1}{n^2} = 0.4375$; Multiply by n^2, divide by 0.4375 and take the square root of both sides to

obtain: $n = \sqrt{\dfrac{1}{0.4375}} = 2$ (1 sf)

General Questions

1R.39. Express the length 1.97 Angstroms in nanometers; In picometers;

$\dfrac{1.97 \text{ Å}}{1} \cdot \dfrac{1 \times 10^{-10} \text{ m}}{1 \text{ Å}} \cdot \dfrac{1 \times 10^{9} \text{ nm}}{1 \text{ m}} = 0.197$ nm

$\dfrac{1.97 \text{ Å}}{1} \cdot \dfrac{1 \times 10^{-10} \text{ m}}{1 \text{ Å}} \cdot \dfrac{1 \times 10^{12} \text{ pm}}{1 \text{ m}} = 197$ pm

1R.40. Separation between C atoms in diamond is 0.154 nm.

(a) In meters: $\dfrac{0.154 \text{ nm}}{1} \cdot \dfrac{1 \text{ m}}{1 \times 10^{9} \text{ nm}} = 1.54 \times 10^{-10}$ m

(b) In picometers: 1.54×10^{-10} m $\cdot \dfrac{1 \text{ pm}}{1 \times 10^{-12} \text{ m}} = 154$ pm

(c) In Angstroms: 1.54×10^{-10} m $\cdot \dfrac{1 \text{ Å}}{1 \times 10^{-10} \text{ m}} = 1.54$ Å

1R.41. Diameter of red blood cell = 7.5 μm

(a) In meters: $\dfrac{7.5\ \mu m}{1} \cdot \dfrac{1\ m}{1 \times 10^6\ \mu m} = 7.5 \times 10^{-6}\ m$

(b) In nanometers: $\dfrac{7.5\ \mu m}{1} \cdot \dfrac{1\ m}{1 \times 10^6\ \mu m} \cdot \dfrac{1 \times 10^9\ nm}{1\ m} = 7.5 \times 10^3\ nm$

(c) In picometers: $\dfrac{7.5\ \mu m}{1} \cdot \dfrac{1\ m}{1 \times 10^6\ \mu m} \cdot \dfrac{1 \times 10^{12}\ m}{1\ m} = 7.5 \times 10^6\ pm$

1R.42. $1.53\ g\ cisplatin \cdot \dfrac{65.0\ g\ Pt}{100.0\ g\ cisplatin} = 0.995\ g\ Pt$

1R.43. Mass of procaine hydrochloride (in mg) in 0.50 mL of solution

$\dfrac{0.50\ mL}{1} \cdot \dfrac{1.0\ g}{1\ mL} \cdot \dfrac{10.\ g\ procaine\ HCl}{100\ g\ solution} \cdot \dfrac{10^3\ mg\ procaine\ HCl}{1\ g\ procaine\ HCl} = 50.\ mg\ procaine\ HCl$

1R.44. Length of a cube of Al to have a mass of 7.6 grams:

$7.6\ g \cdot \dfrac{1\ cm^3}{2.698\ g} = 2.8\ cm^3$; The cube has dimensions of the length³.

The cube edge = $\sqrt[3]{2.8\ cm^3}$ = 1.4 cm

1R.45. Average density of a marble: $\dfrac{95.2\ g}{(99-61)\ mL} \cdot \dfrac{1\ mL}{1\ cm^3} = 2.5\ g/cm^3$

1R.46. Calculate the density of the solid and identify the compound:

$\dfrac{18.82\ g}{(15.3-8.5)\ mL} \cdot \dfrac{1\ mL}{1\ cm^3} = 2.8\ g/cm^3$

The white solid's density matches that of (c) KBr.

1R.47. For the sodium chloride unit cell:

(a) The volume of the unit cell is the (edge length)³

With an edge length of 0.563 nm, the volume is (0.563 nm)³ or 0.178 nm³. The volume in cm³ is calculated by first expressing the edge length in cm: ,

$\dfrac{0.563\ nm}{1} \cdot \dfrac{1 \times 10^2\ cm}{1 \times 10^9\ nm} = 5.63 \times 10^{-8}\ cm$, so the volume is $1.78 \times 10^{-22}\ cm^3$

(b) The mass of the unit cell is:

M = V · D = $1.78 \times 10^{-22}\ cm^3 \cdot 2.17\ g/cm^3 = 3.87 \times 10^{-22}\ g$

(c) Given the unit cell contains 4 NaCl formula units, the mass of one formula unit is:
$$\frac{3.87 \times 10^{-22} \text{ g per unit cell}}{4 \text{ formula units per unit cell}} = 9.68 \times 10^{-23} \text{ g/formula unit}$$

1R.48. Volume of a 1.50-carat diamond:
$$1.50 \text{ carat} \cdot \frac{0.200 \text{ g}}{1 \text{ carat}} \cdot \frac{1 \text{ cm}^3}{3.513 \text{ g}} = 0.0854 \text{ cm}^3$$

1R.49. The accepted value for a normal human temperature is 98.6 °F. On the Celsius scale this corresponds to: $°C = \frac{5 \, °C}{9 \, °F}(98.6 \, °F - 32 \, °F) = 37 \, °C$

Since the melting point of gallium is 29.8 °C, the gallium should melt in your hand.

1R.50. A good estimate for the density would be between that at 15 °C and that at 25 °C. If one assumes it is halfway between those densities, a better estimate would be 0.99810 g/cm³. It should be noted, however, that the decrease in density with increasing temperature is not linear.

1R.51. The heating of popcorn causes the loss of water.
(a) The percentage of mass lost upon popping:
$$\frac{(0.125 \text{ g} - 0.106 \text{ g})}{0.125 \text{ g}} \times 100\% = 15\%$$
(b) With an average mass of 0.125 g, the number of kernels in a pound of popcorn:
$$\frac{1 \text{ kernel}}{0.125 \text{ g}}\left(\frac{453.6 \text{ g}}{1 \text{ lb}}\right) = 3630 \text{ kernels}$$

1R.52. What is the thickness of the aluminum foil in millimeters?
$$\text{Volume} = \frac{12 \text{ oz}}{1} \cdot \frac{28.4 \text{ g}}{1 \text{ oz}} \cdot \frac{1 \text{ cm}^3}{2.70 \text{ g}} = 126 \text{ cm}^3$$
$$\text{Area} = 75 \text{ ft}^2 \left(\frac{12 \text{ in}}{1 \text{ ft}}\right)^2 \left(\frac{2.54 \text{ cm}}{1 \text{ in}}\right)^2 = 7.0 \times 10^4 \text{ cm}^2$$
$$\text{Thickness} = \frac{\text{volume}}{\text{area}} = \frac{126 \text{ cm}^3}{7.0 \times 10^4 \text{ cm}^2} = 1.8 \times 10^{-3} \text{ cm} = 1.8 \times 10^{-2} \text{ mm}$$

1R.53. The mass of NaF needed for 150,000 people for a year:

One way to do this problem is to begin with a factor that contains the units of the "answer" (mass NaF in kg). Since NaF is 45% fluoride (and 55 %Na), we can write the factor:

$\dfrac{100.0 \text{ kg NaF}}{45.0 \text{ kg F}^-}$. Note that the expression of kg/kg has the same value as the expression g/g—and provides the "desired" units of our answer.

A concentration of 1 ppm can be expressed as: $\dfrac{1.00 \text{ kg F}^-}{1.00 \times 10^6 \text{ kg H}_2\text{O}}$ [We could use the fraction with the masses expressed in **grams**, but we would have to convert grams to kg. Note this factor can be derived from the factor using grams if you multiply BOTH numerator and denominator by 1000.]. Using the data provided in the problem, plus conversion factors (found in the rear cover of your textbook)

$$150{,}000 \text{ people} \cdot \dfrac{660 \text{ L H}_2\text{O}}{1 \text{ person} \cdot \text{day}} \cdot \dfrac{365 \text{ days}}{\text{year}} \cdot \dfrac{1.00 \text{ kg H}_2\text{O}}{1.00 \text{ L H}_2\text{O}} \cdot \dfrac{1.00 \text{ F}^-}{1.00 \times 10^6 \text{ kg H}_2\text{O}} \cdot \dfrac{100 \text{ kg NaF}}{45.0 \text{ kg F}^-} = 8.0 \times 10^4 \text{ kg NaF/day}$$

Note that 660 L of water per day limits the answer to 2 sf.

1R.54. Area in cm²: $0.5 \text{ acre} \cdot \dfrac{1.0 \times 10^4 \text{ m}^2}{2.47 \text{ acres}} \left(\dfrac{100 \text{ cm}}{1 \text{ m}} \right)^2 = 2 \times 10^7 \text{ cm}^2$

thickness = $\dfrac{\text{volume}}{\text{area}} = \dfrac{5 \text{ cm}^3}{2 \times 10^7 \text{ cm}^2} = 2 \times 10^{-7}$ cm

This is likely related to the "length" of oil molecules.

1R.55. Mass of sulfuric acid in 750. mL (or 750. cm³) solution.

$750. \text{ cm}^3 \text{ solution} \cdot \dfrac{1.285 \text{ g solution}}{1 \text{ cm}^3 \text{ solution}} \cdot \dfrac{38.08 \text{ g sulfuric acid}}{100 \text{ g solution}} = 367 \text{ g sulfuric acid (3 sf)}$

1R.56. Mass of hydrochloric acid in 50. mL of 37% by mass solution.

$50. \text{ mL HCl solution} \left(\dfrac{1.2 \text{ g HCl solution}}{1 \text{ mL solution}} \right) = 60. \text{ g HCl solution}$

$60. \text{ g HCl solution} \left(\dfrac{37 \text{ g HCl}}{100 \text{ g HCl solution}} \right) = 22 \text{ g HCl (2sf)}$

1R.57. Volume of 5.25% bleach to contain 8.0 g of NaClO.

$8.0 \text{ g NaClO} \cdot \dfrac{100 \text{ g bleach}}{5.25 \text{ g NaClO}} \cdot \dfrac{1 \text{ mL bleach}}{1.1 \text{ g bleach}} = 140 \text{ mL bleach (2 sf)}$

1R.58. The volume is: $\dfrac{279 \text{ kg}}{1} \cdot \dfrac{10^3 \text{ g}}{1 \text{ kg}} \cdot \dfrac{1 \text{ cm}^3}{19.3 \text{ g}} = 1.45 \times 10^4 \text{ cm}^3$

volume = (area)(thickness)

$1.45 \times 10^4 \text{ cm}^3 = (\text{area}) \left(0.0015 \text{ mm} \times \dfrac{1 \text{ cm}}{10 \text{ mm}} \right)$

Solve for area:

$\text{area} = \dfrac{1.45 \times 10^4 \text{ cm}^3}{\left(0.0015 \text{ mm} \cdot \dfrac{1 \text{ cm}}{10 \text{ mm}} \right)} = 9.6 \times 10^7 \text{ cm}^2$

Converting to m² : $9.6 \times 10^7 \text{ cm}^2 \left(\dfrac{1 \text{ m}}{100 \text{ cm}} \right)^2 = 9.6 \times 10^3 \text{ m}^2$

1R.59. Density of water changes with state:

(a) Volume of solid water at –10°C when a 250. mL can is filled with liquid water at 25°C:

The volume of liquid water at 25 °C is 250. mL (or cm³). The mass of that water is:

$\dfrac{0.997 \text{ g water}}{1.000 \text{ cm}^3 \text{ water}} \cdot \dfrac{250. \text{ cm}^3}{1} = 249.25$ g water (249 to 3 sf)

That mass of water at the lower temperature will occupy:

$\dfrac{1.000 \text{ cm}^3 \text{ water}}{0.917 \text{ g water}} \cdot \dfrac{249.25 \text{ g water}}{1} = 271.81$ (or 272 cm³ to 3 sf)

(b) With the can being filled to 250. mL at room temperature, the expansion (an additional 22 mL) cannot be contained in the can. (Get out the sponge—there's a mess to clean up.)

1R.60. Volume of room and mass of air:

(a) Volume of room = (length)(width)(height)

$= (18 \text{ ft})(15 \text{ ft})(8.5 \text{ ft}) \left(\dfrac{12 \text{ in}}{1 \text{ ft}} \right)^3 \left(\dfrac{2.54 \text{ cm}}{1 \text{ in}} \right)^3 \left(\dfrac{1 \text{ m}}{100 \text{ cm}} \right)^3 = 65 \text{ m}^3$

and $65 \text{ m}^3 \cdot \dfrac{1 \text{ L}}{10^{-3} \text{ m}^3} = 6.5 \times 10^4$ L

(b) Mass of air in kg: $\dfrac{6.5 \times 10^4 \text{ L}}{1} \cdot \dfrac{1.2 \text{ g}}{1 \text{ L}} \cdot \dfrac{1 \text{ kg}}{10^3 \text{ g}} = 78$ kg

and in pounds: $\dfrac{78 \text{ kg}}{1} \cdot \dfrac{10^3 \text{ g}}{1 \text{ kg}} \cdot \dfrac{1 \text{ lb}}{454 \text{ g}} = 170$ lb

1R.61. Calculate the density of steel if a steel sphere of diameter 9.40 mm has a mass of 3.475 g:

The radius of the sphere is (0.5)(9.40 mm) or 4.70 mm. Since density is usually expressed in cm^3, express the radius in cm (0.470 cm) and substitute into the volume equation:

$$V = \frac{4}{3}\pi r^3 = \frac{4}{3}(3.1416)(0.470 \text{ cm})^3 = 0.435 \text{ cm}^3 \text{ (to 3 sf)}$$

The density is 3.475 g / 0.435 cm^3 = 7.99 g/cm^3.

1R.62. Identify the liquid:

(a) density = $\dfrac{16.08 \text{ g} - 12.20 \text{ g}}{3.50 \text{ mL}} \cdot \dfrac{1 \text{ mL}}{1 \text{ cm}^3}$ = 1.11 g/cm^3.

Of the liquids supplied, the liquid is most likely ethylene glycol.

(b) With 2 sf, the calculated density would be 1.1 g/cm^3. While this value still suggests that the liquid is ethylene glycol, it is close to the values for acetic acid and water. Additional testing would be needed to uniquely identify the liquid.

1R.63. (a) Calculate the density of an irregularly shaped piece of metal:

$$\text{Density} = \frac{m}{V} = \frac{74.122 \text{ g}}{36.7 \text{ cm}^3 - 28.2 \text{ cm}^3} = \frac{74.122 \text{ g}}{8.5 \text{ cm}^3} = 8.7 \text{ g/cm}^3$$

Note that the subtraction of volumes leaves only 2 sf, limiting the density to 2 sf

(b) From the list of metals provided, one would surmise that the metal is **cadmium**. Since the major uncertainty is in the volume, one can substitute 8.4 and 8.6 cm^3 as the volume and calculate the density (resulting in 8.82 and 8.62 g/cm^3 respectively). The hypothesis that the metal is cadmium is reasonably sound, but the calculated density is close to that of cobalt, nickel, and copper. Further testing should be done on the metal.

1R.64. (a) Density = $\dfrac{3.2745 \text{ g}}{5.0 \text{ mL}}$ = 0.65 g/mL

(b) Using volumes of 4.9 mL and 5.1 mL, corresponding to ±0.1 mL of the recorded volume, the range of densities possible is 0.66 g/mL to 0.64 g/mL. No, all five hydrocarbon density values fall within the range of possible values for the liquid.

(c) No, all five hydrocarbon density values fall within the range of possible values.

(d) Using a volume of (4.93 ± 0.01) mL, the calculated density is 0.664 g/mL, and maximum and minimum density values are 0.666 g/mL and 0.663 g/mL. The liquid is 2-methylpentane.

1R.65. Mass of Hg in the capillary:

Mass of capillary with Hg	3.416 g
Mass of capillary without Hg	3.263 g
Mass of Hg	0.153 g

To determine the volume of the capillary, calculate the volume of Hg that is filling it.

$$0.153 \text{ g Hg} \cdot \frac{1 \text{ cm}^3}{13.546 \text{ g Hg}} = 1.13 \times 10^{-2} \text{ cm}^3 \text{ (3 sf)}$$

Now that we know the volume of the capillary, and the length of the tubing (given as 16.75 mm—or 1.675 cm), we can calculate the radius of the capillary using the equation: Volume = $\pi r^2 l$.

1.13×10^{-2} cm^3 = (3.1416) r^2(1.675 cm), and solving for r^2:

$$r^2 = \frac{1.13 \times 10^{-2} \text{ cm}^3}{(3.1416)(1.675 \text{ cm})} = 2.15 \times 10^{-3} \text{ cm}^2 \text{ and } r = 4.63 \times 10^{-2} \text{ cm.}$$

Diameter = $2r$ = 2(4.63 × 10^{-2} cm) = 0.0927 cm

1R.66. Volume of this amount of copper: $57 \text{ kg} \cdot \frac{10^3 \text{ g}}{1 \text{ kg}} \cdot \frac{1 \text{ cm}^3}{8.96 \text{ g}} = 6.4 \times 10^3 \text{ cm}^3$ copper

The length of wire = $\frac{\text{volume}}{(\pi)(\text{radius})^2} = \frac{6.4 \times 10^3 \text{ cm}^3}{(\pi)(0.950 \text{ cm}/2)^2} = 9.0 \times 10^3$ cm

and expressed in meters: $9.0 \times 10^3 \text{ cm} \cdot \frac{1 \text{ m}}{100 \text{ cm}} = 90.$ M

1R.67. Regarding a cube of Cu:

(a) The number of Cu atoms in a cube whose mass is 0.1206 g.

$$\frac{0.1206 \text{ g}}{\text{cube}} \cdot \frac{1 \text{ atom Cu}}{1.055 \times 10^{-22} \text{g}} = 1.143 \times 10^{21} \text{ atoms Cu}$$

Fraction of the lattice that contains Cu atoms:

Given the radius of a Cu atom is 128 pm, and the number of Cu atoms, the total volume occupied by the Cu atoms is the volume occupied by ONE atom (4/3πr^3) multiplied by the total number of atoms:

Volume of one atom: $4/3 \cdot 3.1416 \cdot (128 \text{ pm})^3 = 8.78 \times 10^6 \text{ pm}^3$

Total volume: $(8.78 \times 10^6 \text{ pm}^3/\text{Cu atom})(1.143 \times 10^{21} \text{ atoms Cu})$
= $1.00 \times 10^{28} \text{ pm}^3$.

The lattice cube has a volume of (0.236 cm)³ or (2.36 × 10⁹ pm)³ or 1.31 × 10²⁸ pm³.

The fraction occupied is: total volume of Cu atoms/total volume of lattice cube:

$$\frac{1.00 \times 10^{28} \text{ pm}^3}{1.31 \times 10^{28} \text{ pm}^3} = 0.764 \text{ or } 76.4\% \text{ occupied (3 sf)}$$

The empty space in a lattice is due to the inability of spherical atoms to totally fill a given volume. A macroscopic example of this phenomenon is visible if you place four marbles in a square arrangement. At the center of the square there are voids. In a cube, there are similar repeating incidents.

(b) Estimate the number of Cu atoms in the smallest repeating unit:

Since we know the length of the smallest repeating unit (the unit cell), let's calculate the volume (first converting the length to units of centimeters:

L = 361.47 pm so L = 361.47 pm · $\frac{1 \times 10^2 \text{ cm}}{1 \times 10^{12} \text{ pm}}$ = 3.6147 × 10⁻⁸ cm

V = L³ = (3.6147 × 10⁻⁸ cm)³ = 4.7230 × 10⁻²³ cm³

Since we know the density, we can calculate the mass of one unit cell:

D × V = 8.960 g/cm³ × 4.7230 × 10⁻²³ cm³ = 4.232 × 10⁻²² g

Knowing the mass of one copper atom (1.055 × 10⁻²² g) we can calculate the number of Cu atoms in that mass: $\frac{4.232 \times 10^{-22} \text{ g}}{1.055 \times 10^{-22} \text{ g/Cu atom}}$ = 4.011 Cu atoms

Four atoms are present in this smallest repeating unit. As you will learn later, the number of atoms for a face-centered cubic lattice is 4.

1R.68. Determine the density of lead, average density, percent error, and standard deviation using the provided data:

Data	Density (g/cm³)
1	11.6
2	11.8
3	11.5
4	12.0
Average	11.7

Percent error = $\frac{\text{error in measurement}}{\text{accepted value}}$ × 100% = $\frac{11.7 \text{ g/cm}^3 - 11.3 \text{ g/cm}^3}{11.3 \text{ g/cm}^3}$ × 100% = 4% (1 sf)

Standard Deviation is calculated as usual:

Data	Density (g/cm³)	Difference (Measurement – Average)	Square of Difference
1	11.6	– 0.1	1×10^{-2}
2	11.8	0.1	1×10^{-2}
3	11.5	– 0.2	4×10^{-2}
4	12.0	0.3	9×10^{-2}

The sum of squares is then 0.15 and square root of 0.15/3 = 0.2 g/cm³

In The Laboratory

1R.69. The metal will displace a volume of water that is equal to the volume of the metal.

The difference in volumes of water (20.2 – 6.9) corresponds to the volume of metal. Since 1 mL = 1 cm³, the density of the metal is then:

$$\frac{\text{Mass}}{\text{Volume}} = \frac{37.5 \text{ g}}{13.3 \text{ cm}^3} = \text{ or } 2.82 \frac{\text{g}}{\text{cm}^3}$$

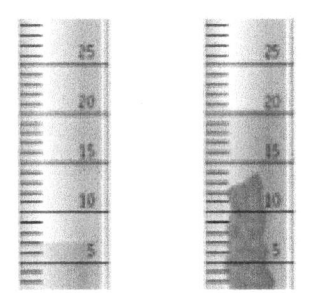

The metal is (d) aluminum.

1R.70. Calculate the density of the sample: $\frac{23.5 \text{ g}}{(52.2 - 47.5) \text{ mL}} \cdot \frac{1 \text{ mL}}{1 \text{ cm}^3} = 5.0 \text{ g/cm}^3$

The sample's density matches that of fool's gold.

1R.71. Plot of Absorbance vs mass of compound (g/L):

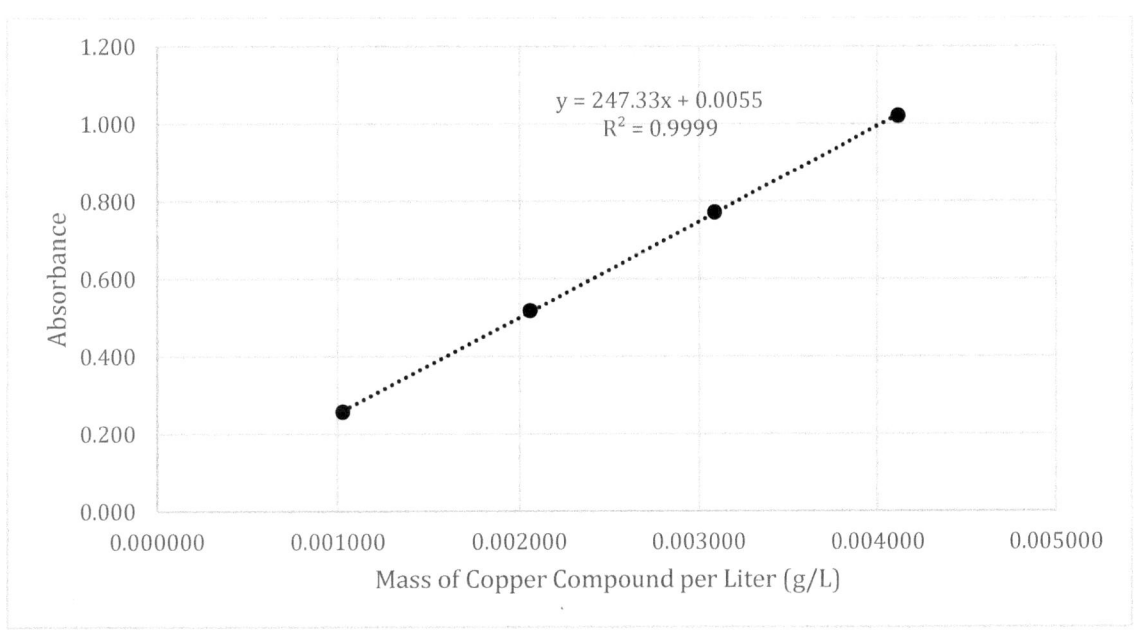

y = 247.33x + 0.0055
R² = 0.9999

The equation for the best straight line is $y = 247x + 0.006$

The best straight line indicates that the slope is 247 and the intercept is 0.006

If Absorbance is 0.635, what is the mass of Cu in g/L and mg/mL:

Substituting into the equation for the line:

$0.635 = (247)(\text{concentration}) + 0.006$; concentration $= 2.55 \times 10^{-3}$ g/L

$$\frac{2.55 \times 10^{-3} \text{ g}}{\text{L}} \cdot \frac{1 \text{ L}}{10^3 \text{ mL}} \cdot \frac{10^3 \text{ mg}}{1 \text{ g}} = 2.55 \times 10^{-3} \text{ mg/mL}$$

1R.72. Use data to determine % isooctane:

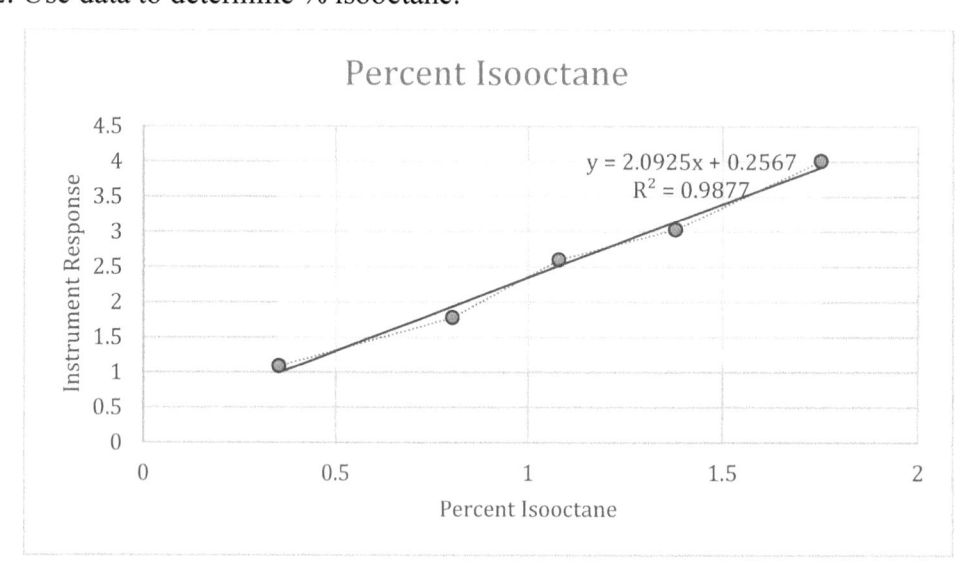

The best straight line indicates that the slope is 2.09 and the intercept is 0.26.

Substituting into the equation for the line with instrument response = 2.75:

$2.75 = 2.09(\% \text{ isooctane}) + 0.26$ and solving for (% isooctane)
% isooctane = 1.19%

1R.73. Insert the data in a spreadsheet (here, Excel is used), to obtain the results:

Student	% Acetic Acid
1	5.22
2	5.28
3	5.22
4	5.30
5	5.19
6	5.23
7	5.33
8	5.26
9	5.15
10	5.22
Average	5.24
St.Dev	0.05

Only the values 5.30%, 5.33%, and 5.15% fall *outside* the Average ± St.Dev, so seven points fall within the range $5.19 \leq x \leq 5.29$.

Solution and Answer Guide

Kotz Treichel Townsend Treichel, Chemistry and Chemical Reactivity 11e, 978-0-357-85140-1, Chapter 2: Atoms, Molecules, and Ions

TABLE OF CONTENTS

Applying Chemical Principles 36
Practicing Skills..39
General Questions..66
In the Laboratory..85
Summary and Conceptual Questions..87

Applying Chemical Principles:
Using Isotopes

2.1.1. Neutrons in ^{18}O; Neutrons in the two Pb isotopes:

Neutrons in ^{18}O: atomic mass (18) – proton number (8) = 18 – 8 = 10 neutrons

Neutrons in ^{206}Pb: atomic mass (206) – proton number (82) = 206 – 82 = 124 neutrons

Neutrons in ^{204}Pb: atomic mass (204) – proton number (82) = 204 – 82 = 122 neutrons

2.1.2. Using the abundance and mass of the three stable isotopes of O, calculate the atomic weight:

Atomic weight is a weighted average of the various isotopes. Add the product of (percent abundance) (isotopic mass) for all the isotopes to obtain the atomic weight:

Isotopic Mass	% Abundance (converted to a fraction)	Product
15.9949	0.99757	15.956
16.9991	0.00038	0.0065
17.9992	0.00205	0.0369
	SUM:	15.999

© 2024 Cengage Learning, Inc. All Rights Reserved. May not be scanned, copied or duplicated, or posted to a publicly accessible website, in whole or in part.

Arsenic, Medicine, and the Formula of Compound 606

2.2.1. What is the empirical formula of the compound enargite?

The percentages total to 100%, so think of the percent of each element in 100 g of the compound: Mass As: 19.024 g; Mass Cu: 48.407 g; Mass S: 32.569 g.
Now convert the masses to moles of each element:

$$19.024 \text{ g As} \left(\frac{1 \text{ mol As}}{74.922 \text{ g As}} \right) = 0.25392 \text{ mol As}$$

$$48.407 \text{ g Cu} \left(\frac{1 \text{ mol Cu}}{63.546 \text{ g Cu}} \right) = 0.76176 \text{ mol Cu}$$

$$32.569 \text{ g S} \left(\frac{1 \text{ mol S}}{32.06 \text{ g S}} \right) = 1.016 \text{ mol S}$$

Calculate the mole: mole ratio:

[Hint: divide the amounts (moles) by the smallest amount.]

$$\frac{1.016 \text{ mol S}}{0.25392 \text{ mol As}} = 4.001 \quad \frac{0.76176 \text{ mol Cu}}{0.25392 \text{ mol As}} = 3.0000$$

The ratio of the three elements is: $Cu_3As_1S_4$ or usually written: Cu_3AsS_4.

2.2.2. Molecular formulas of the compounds composing Salvarsan:

First determine empirical formulas for the substances:

$$39.37 \text{ g C} \left(\frac{1 \text{ mol C}}{12.011 \text{ g C}} \right) = 3.278 \text{ mol C}$$

$$3.304 \text{ g H} \left(\frac{1 \text{ mol H}}{1.008 \text{ g H}} \right) = 3.278 \text{ mol H}$$

$$8.741 \text{ g O} \left(\frac{1 \text{ mol O}}{15.999 \text{ g O}} \right) = 0.5463 \text{ mol O}$$

$$7.652 \text{ g N} \left(\frac{1 \text{ mol N}}{14.007 \text{ g N}} \right) = 0.5463 \text{ mol N}$$

$$40.932 \text{ g As} \left(\frac{1 \text{ mol As}}{74.922 \text{ g As}} \right) = 0.54633 \text{ mol As}$$

Determine mole:mole ratios:

Using the smallest # of mol (either for O, N, or As), determine ratios of C and H:

$$\frac{3.278 \text{ mol C}}{0.5463 \text{ mol N}} = \frac{6.000 \text{ mol C}}{1 \text{ mol N}} \text{ or 6 mol C/mol N}$$

Similarly, the ratio for H/N is also 6 mol H/mol N.

These ratios imply an empirical formula that is C₆H₆AsNO.

Determine the empirical formula mass.

$$6 \text{ mol C}\left(\frac{12.011 \text{ g C}}{1 \text{ mol C}}\right) + 6 \text{ mol H}\left(\frac{1.008 \text{ g H}}{1 \text{ mol H}}\right) + 1 \text{ mol N}\left(\frac{14.007 \text{ g N}}{1 \text{ mol N}}\right)$$

$$+ 1 \text{ mol O}\left(\frac{15.999 \text{ g O}}{1 \text{ mol O}}\right) + 1 \text{ mol As}\left(\frac{74.922 \text{ g As}}{1 \text{ mol As}}\right) = 183.04 \text{ g}/\text{empirical formula}$$

We are told that one of the compounds has a molar mass of 549 g/mole.

$$\frac{549 \text{ g/mol compound}}{183.04 \text{ g/mol empirical formula}} = 3.00 \text{ mol empirical formula/mol compound}$$

So, this compound has a molecular formula that is 3 empirical formulas/molecular compound,

C₁₈H₁₈As₃N₃O₃.

The other compound has a molar mass of 915 g/mol. Using a similar procedure to that above, there are 5 empirical formulas/molecular compound, or a molecular formula of C₃₀H₃₀As₅N₅O₅.

Argon-An Amazing Discovery

2.3.1. Volume of globe:

$$0.20389 \text{ g N}\left(\frac{1 \text{ L}}{1.25718 \text{ g N}}\right) = 0.16218 \text{ L or } 162.18 \text{ mL or } 162.18 \text{ cm}^3$$

2.3.2. Determine the density of argon:
Using the prescribed procedure of summing the products of (fractional volume)(density):
(0.2096)(1.42952 g/L) + (0.7811)(1.25092 g/L) + (0.00930)x = 1.000(1.29327 g/L)
and x = 1.78 g/L

2.3.3. From the periodic table, the atomic weight of Argon = 39.95. Now we deduce the abundance of ⁴⁰Ar. 100% − 0.334% − 0.063% = 99.603%

Now that we know the abundance of ⁴⁰Ar, we can determine the weighted average of the three isotopes: (0.00334)(35.967545) + (0.00063)(37.962732) + (0.99603)x = 39.95 and x = 39.96

2.3.4. Number of Ar atoms in a room 4.0 m × 5.0 m × 2.4 m:
Volume of room: $4.0 \text{ m} \cdot 5.0 \text{ m} \cdot 2.4 \text{ m} \cdot \dfrac{1 \text{ L}}{1.00 \times 10^{-3} \text{ m}^3} = 4.8 \times 10^4 \text{ L}$

Use this volume and the density of argon:
$4.8 \times 10^4 \text{ L} \left(\dfrac{1.78 \text{ g Ar}}{1 \text{ L}}\right)\left(\dfrac{1 \text{ mol Ar}}{39.95 \text{ g Ar}}\right)\left(\dfrac{6.022 \times 10^{23} \text{ atoms Ar}}{1 \text{ mol Ar}}\right) = 1.3 \times 10^{27}$ atoms Ar

---

Practicing Skills

Atoms: Their Composition and Structure

2.1.

Fundamental Particles	Protons	Electrons	Neutrons
Electrical Charges	+1	−1	0
Present in nucleus	Yes	No	Yes
Relative Masses	1.007	**0.00055**	1.009

Note that an electron is 1830 times less massive than a proton or neutron.

2.2. Mass number is the sum of the number of protons and number of neutrons for an atom. Atomic mass is the mass of an atom. Mass number is a simple TOTAL, while atomic mass conveys the sum of the weight of the individual particles.

2.3. Begin by expressing the diameter of the nucleus and the electron cloud in the same units.
2mm (diameter of nucleus) = 2×10^{-3} m (since 1 m = 10^3 mm, 1 mm = 10^{-3} m)

The ratio of diameters: $\dfrac{\text{electron cloud}}{\text{nucleus}} = \dfrac{200 \text{ m}}{2 \times 10^{-3} \text{ m}}$. So we set the actual diameters in the same ratio: $\dfrac{200 \text{ m}}{2 \times 10^{-3} \text{ m}} = \dfrac{1 \times 10^{-8} \text{ cm}}{x}$ and solving for *x*:
(200 m) *x* = (2 × 10⁻³ m)(1 × 10⁻⁸ cm) and *x* = 1 × 10⁻¹³ cm.

Note that we left the actual diameter of the electron cloud in units of centimeters, so the **ratio** would be the same as if we had changed the units to **meters**.

2.4. Each gold atom has a diameter of 2 × 145 pm = 290. pm

$$36 \text{ cm}\left(\frac{1 \text{ m}}{100 \text{ cm}}\right)\left(\frac{1 \times 10^{12} \text{ pm}}{1 \text{ m}}\right) = 3.6 \times 10^{11} \text{ pm}$$

$$3.6 \times 10^{11} \text{ pm}\left(\frac{1 \text{ Au atom}}{290. \text{ pm}}\right) = 1.2 \times 10^{9} \text{ Au atoms}$$

2.5. Isotopic symbol for:

(a) K (at. no. 19) with 22 neutrons: 19 + 22 = 41 $^{41}_{19}\text{K}$

(b) V (at. no. 23) with 28 neutrons: 23 + 28 = 51 $^{51}_{23}\text{V}$

(c) Ga (at. no. 31) with 38 neutrons: 31 + 38 = 69 $^{69}_{31}\text{Ga}$

The mass number represents the SUM of the protons + neutrons in the nucleus of an atom. The atomic number represents the # of protons, so (atomic no. + # neutrons) = mass number

2.6. Isotopic symbol for:

(a) Fe (at. no. 26) with 30 neutrons: 26 + 30 = 56 $^{56}_{26}\text{Fe}$

(b) U (at. no. 92) with 146 neutrons: 92 + 146 = 238 $^{238}_{92}\text{U}$

(c) Pb (at. no. 82) with 125 neutrons: 82 + 125 = 207 $^{207}_{82}\text{Pb}$

2.7.

substance	protons	neutrons	electrons
(a) magnesium-24	12	12	12
(b) tin-119	50	69	50
(c) thorium-232	90	142	90
(d) carbon-13	6	7	6
(e) copper-63	29	34	29
(f) bismuth-205	83	122	83

Note that the number of protons and electrons are **equal** for any **neutral atom**. The number of protons is **always** equal to the atomic number. The mass number equals the sum of the numbers of protons and neutrons.

2.8. (a) Number of protons = number of electrons = 43; number of neutrons = 56
 (b) Number of protons = number of electrons = 95; number of neutrons = 146

2.9. Cobalt has an atomic number of 27 that equals the proton and electron number. Remember the mass number is the sum of the protons and neutrons. The three mass numbers are:

27 + 30 = 57, 27 + 31 = 58, 27 + 33 = 60

Thus, we have the following three complete symbols $^{57}_{27}Co$, $^{58}_{27}Co$, $^{60}_{27}Co$

2.10. Atomic number of Ag is 47; both isotopes have 47 protons and 47 electrons.

^{107}Ag 107 – 47 = 60 neutrons   and ^{109}Ag 109 – 47 = 62 neutrons

2.11. Hydrogen has three isotopes:

Name	# Protons	# Neutrons	# Electrons
Protium	1	0	1
Deuterium	1	1	1
Tritium	1	2	1

The ONLY difference between the isotopes of an element is in the **number** of neutrons.

2.12. Recalling that isotopes of an element must all have the same number of protons, the following are isotopes of element X: $^{32}_{16}X$ and $^{36}_{16}X$

Key Experiments Developing Atomic Structure

2.13. The accepted mass of a proton is 1.672622×10^{-24} g, while that for the electron is 9.109384×10^{-28} g. The ratio of these two masses is:

$$\frac{\text{mass of proton}}{\text{mass of electron}} = \frac{1.672622 \times 10^{-24} \text{ g}}{9.109384 \times 10^{-28} \text{ g}} = 1836$$

So, Thomson's estimate was "off" by a factor of 2.

2.14. Negatively charged electrons in the cathode-ray tube collide with He atoms, splitting the atom into an electron and a He$^+$ cation.

2.15. One can observe from the accompanying graphic, that the β particles are attracted to the "+" plate, while the α particles are attracted to the "–" one. The gamma rays are attracted to neither.

From this information, we know that α particles are **positively charged**, and that β particles are **negatively charged**. With the α particle being essentially the nucleus of a He atom, it is the heavier of the two particles.

2.16. Atoms are not solid, hard, or impenetrable. We know, thanks to the Kinetic-Molecular Theory, that the atoms are in rapid motion at all temperatures above absolute zero. They have mass (an important aspect of Dalton's hypothesis). The Rutherford "gold foil" experiment proved that atoms were *not* impenetrable, solid or hard, as alpha particles penetrated the atom and revealed a better picture of the atom.

Atomic Weight

2.17. Thallium has two stable isotopes ^{203}Tl and ^{205}Tl. The more abundant isotope is:___?___

The atomic weight of thallium is 204.4 u. The fact that this weight is closer to 205 than 203 indicates that the **205 isotope is the more abundant** isotope. Recall that the atomic weight is the "weighted average" of all the isotopes of each element. Hence the more abundant isotope will have a "greater contribution" to the atomic weight than the less abundant one.

2.18. Strontium has an atomic weight of 87.62, nearly 88, so ^{88}Sr is the most abundant.

2.19. The atomic mass of lithium is: $(0.076)(6.015123) + (0.924)(7.016003) = 6.94$

Recall that the atomic mass is a weighted average of all isotopes of an element, and is obtained by **adding** the *product* of (relative abundance × mass) for all isotopes.

2.20. (^{24}Mg mass)(% abundance/100) + (^{25}Mg mass)(% abundance/100) + (^{26}Mg mass)(% abundance/100) = atomic weight of Mg
$(23.985042)(0.7899) + (24.985837)(0.1000) + (25.982593)(0.1101) = 24.31$

2.21. The average atomic weight of gallium is 69.723 (from the periodic table). If we let x represent the abundance of the lighter isotope, and $(1 - x)$ the abundance of the heavier isotope, the expression to calculate the atomic weight of gallium may be written:

$(x)(68.9256) + (1 - x)(70.9247) = 69.723$

[Note that the sum of all the isotopic abundances must add to 100% -- or 1 (in decimal notation).]

Simplifying the equation gives:

$68.9256\,x + 70.9247 - 70.9247\,x = 69.723$

$-1.9991\,x = (69.723 - 70.9247)$

$-1.9992\,x = -1.2017$

$x = 0.6011$

So, the relative abundance of gallium-69 is 60.11 % and that of gallium-71 is 39.89 %.

2.22. Let x represent the abundance of ^{151}Eu and $(1 - x)$ represent the abundance of ^{153}Eu.
$(x)(150.9199) + (1 - x)(152.9212) = 151.96$, and solving for x yields
$x = 0.4803$; ^{151}Eu abundance is 48% (2 sf), ^{153}Eu abundance is 52%

The Periodic Table

2.23. Comparison of Titanium and Thallium:

Name	Symbol	Atomic #	Atomic Weight	Group #	Period #	Metal, Metalloid, or Nonmetal
Titanium	Ti	22	47.867	4B (4)	4	Metal
Thallium	Tl	81	204.38	3A (13)	6	Metal

2.24. Elements with symbols beginning with S:

	silicon	tin	antimony	sulfur	selenium
Symbol	Si	Sn	Sb	S	Se
At. number	14	50	51	16	34
Period	3	5	5	3	4
Group	4A (14)	4A (14)	5A (15)	6A (16)	6A (16)
	metalloid	metal	metalloid	nonmetal	nonmetal

2.25. Periods with 8 elements: **2**; Periods 2 (at. no. 3-10) and 3 (at. no. 11-18)

Periods with 18 elements: **2**; Periods 4 (at. no 19-36) and 5 (at. no. 37-54)

Periods with 32 elements: **2**; Periods 6 (at. no. 55-86) and 7 (at. no. 87-118)

2.26. There are 32 elements in the seventh period, many of them are called the Actinides, most are metals, and many are human-made elements.

2.27. Elements fitting the following descriptions:

	Description	Elements
(a)	Nonmetals	C, Cl
(b)	Main group elements	C, Ca, Cl, Cs
(c)	Lanthanides	Ce
(d)	Transition elements	Cr, Co, Cd, Cu, Ce, Cf, Cm
(e)	Actinides	Cf, Cm
(f)	Gases	Cl

2.28. Name and chemical symbol for:

(a) C, carbon; N, nitrogen; O, oxygen; F, fluorine; Ne, neon

(b) Rb, rubidium

(c) Cl, chlorine

(d) H, hydrogen; He, helium; N, nitrogen; O, oxygen; F, fluorine; Ne, neon; Cl, chlorine; Ar, argon; Kr, krypton; Xe, xenon; Rn, radon are all gases at 20°C and 1 atmosphere pressure.

2.29. Classify the elements as metals, metalloids, or nonmetals:

	Metals	Metalloids	Nonmetals
N			X
Na	X		
Ni	X		
Ne			X
Np	X		

2.30. Match symbol with description:

(a) A radioactive element   Bk
(b) A liquid at room temperature   Br

(c) A metalloid   B
(d) An alkaline earth metal   Ba
(e) A group 5A (15) element   Bi

Molecules: Formulas, Models and Names

2.31. The molecular formula for nitric acid is HNO$_3$.

The structural formula is shown below (without lone-pair electrons). The structure around the N is flat. The oxygens are in a trigonal-planar arrangement, and the H–O–N atoms reside in bent arrangement.

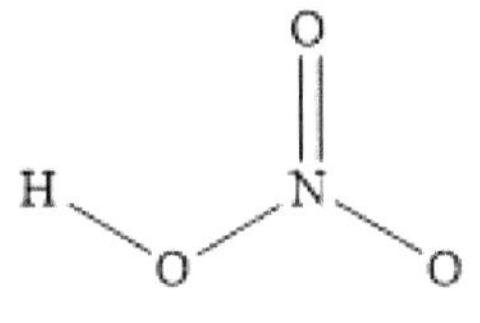

Nitric acid

2.32. Molecular formula for asparagine (an amino acid) is C$_4$H$_8$N$_2$O$_3$. The structural formula is:

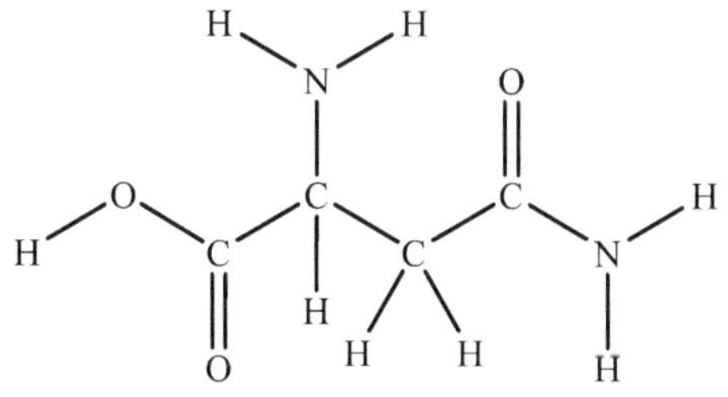

2.33. Names of binary molecular compounds

(a) NF$_3$ nitrogen trifluoride

(b) HI   hydrogen iodide

(c) BI$_3$ boron triiodide

(d) PF$_5$ phosphorus pentafluoride

2.34. Names of binary molecular compounds

(a) N$_2$O$_5$ dinitrogen pentaoxide

(b) P$_4$S$_3$ tetraphosphorus trisulfide

(c) OF$_2$ oxygen difluoride

(d) XeF$_4$ xenon tetrafluoride

2.35. Formulas for:

    (a) sulfur dichloride SCl_2

    (b) dinitrogen pentaoxide N_2O_5

    (c) silicon tetrachloride $SiCl_4$

    (d) diboron trioxide B_2O_3

2.36. Formulas for:

    (a) bromine trifluoride BrF_3

    (b) xenon difluoride XeF_2

    (c) hydrazine N_2H_4

    (d) diphosphorus tetrafluoride P_2F_4

    (e) butane C_4H_{10}

Ions and Ion Charges

2.37. Most commonly observed charge of monatomic ion for:

    (a) Magnesium: 2+ —like all the alkaline earth metals

    (b) Zinc: 2+

    (c) Nickel: 2+

    (d) Gallium: 3+ —an analog of aluminum

2.38. Most commonly observed charge of monatomic ion for:

    (a) Sodium: 1+ —like all the alkali metals

    (b) Aluminum: 3+

    (c) Iron: 2+ and 3+ (transition metals frequently form multiple cations)

    (d) Copper: 1+ and 2+ (transition metals frequently form multiple cations)

2.39. Most commonly observed charge of monatomic ion for:

    (a) Selenium: 2–

    (b) Fluorine: 1–

    (c) Oxygen: 2–

    (d) Nitrogen: 3–

2.40. Most commonly observed charge of monatomic ion for:

(a) Bromine: 1–

(b) Sulfur: 2–

(c) Phosphorus: 3–

(d) Chlorine: 1–

2.41. The symbol and charge for the following ions:

(a)	barium	Ba^{2+}	(e)	sulfide	S^{2-}
(b)	titanium(IV)	Ti^{4+}	(f)	perchlorate	ClO_4^-
(c)	phosphate	PO_4^{3-}	(g)	cobalt(II)	Co^{2+}
(d)	hydrogen carbonate	HCO_3^-	(h)	sulfate	SO_4^{2-}

2.42. Symbol and charge for the ions:

(a)	permanganate	MnO_4^-	(d)	ammonium	NH_4^+
(b)	nitrite	NO_2^-	(e)	phosphate	PO_4^{3-}
(c)	dihydrogen phosphate	$H_2PO_4^-$	(f)	sulfite	SO_3^{2-}

2.43. When potassium becomes a monatomic ion, potassium—like all alkali metals—**loses 1 electron.** The noble gas atom with the same number of electrons as the potassium ion is **argon.**

2.44. Both atoms gain two electrons. O^{2-} has the same number of electrons as Ne and S^{2-} has the same number of electrons as Ar.

2.45. Number of protons and electrons in these ions.

Ion	# of protons	# of electrons
Na^+	11	10
Mg^{2+}	12	10
F^-	9	10
O^{2-}	8	10

2.46. Number of protons and electrons in these ions.

Ion	# of protons	# of electrons
Ca^{2+}	20	18
Cu^{+}	29	28
Cu^{2+}	29	27
Cl^{-}	17	18

Ionic Compounds

2.47. Barium is in Group 2A (2) and is expected to form a 2+ ion while bromine is in group 7A (17) and expected to form a 1– ion. Since the compound would have to have an **equal amount** of negative and positive charges, the formula would be $BaBr_2$.

2.48. Cobalt is a transition metal. The Roman numeral indicates that this is a 3+ ion. Fluorine is in group 7A (17) and expected to form a 1– ion. Since the compound must have an **equal amount** of negative and positive charges, the formula would be CoF_3.

2.49. Formula, Charge, Number, and Name of ions in:

		cation & name	# of	anion & name	# of
(a)	K_2S	K^+ potassium	2	S^{2-} sulfide	1
(b)	$CoSO_4$	Co^{2+} cobalt(II)	1	SO_4^{2-} sulfate	1
(c)	$KMnO_4$	K^+ potassium	1	MnO_4^- permanganate	1
(d)	$(NH_4)_3PO_4$	NH_4^+ ammonium	3	PO_4^{3-} phosphate	1
(e)	$Ca(ClO)_2$	Ca^{2+} calcium	1	ClO^- hypochlorite	2
(f)	$NaCH_3CO_2$	Na^+ sodium	1	$CH_3CO_2^-$ acetate	1

2.50. Formula, Charge, Number, and Name of ions in:

		cation & name	# of	anion & name	# of
(a)	Mg(CH$_3$CO$_2$)$_2$	Mg^{2+} magnesium	1	CH$_3$CO$_2^-$ acetate	2
(b)	Al(OH)$_3$	Al^{3+} aluminum	1	OH$^-$ hydroxide	3
(c)	CuCO$_3$	Cu^{2+} copper(II)	1	CO$_3^{2-}$ carbonate	1
(d)	Ti(SO$_4$)$_2$	Ti^{4+} titanium(IV)	1	SO$_4^{2-}$ sulfate	2
(e)	KH$_2$PO$_4$	K$^+$ potassium	1	H$_2$PO$_4^-$ dihydrogen phosphate	1
(f)	CaHPO$_4$	Ca^{2+} calcium	1	HPO$_4^{2-}$ hydrogen phosphate	1

2.51. Regarding cobalt oxides: cobalt(II) oxide, CoO   cobalt ion : Co^{2+}

cobalt(III) oxide, Co$_2$O$_3$ cobalt ion : Co^{3+}

2.52. Platinum (II and IV) ions with chloride and sulfide ions:

(a) Pt^{2+}: PtCl$_2$ Pt^{4+}: PtCl$_4$
(b) Pt^{2+}: PtS   Pt^{4+}: PtS$_2$

2.53. Provide correct formulas for compounds:

(a) AlCl$_2$ Incorrect; Correct is AlCl$_3$.

(b) KF$_2$ Incorrect; Correct is KF.

(c) Ga$_2$O$_3$ Correct.

(d) MgS   Correct.

2.54. Provide correct formulas for compounds if given formula is incorrect:

   (a) Ca$_2$O   Incorrect; Correct is CaO.

   (b) SrBr$_2$ Correct as is.

   (c) Fe$_2$O$_5$ Incorrect; Correct is either Fe$_2$O$_3$ for iron(III) oxide or FeO for iron(II) oxide

   (d) Li$_2$O   Correct as is.

2.55. Names for the ionic compounds

   (a) K$_2$S   potassium sulfide

   (b) CoSO$_4$ cobalt(II) sulfate

   (c) (NH$_4$)$_3$PO$_4$ ammonium phosphate

   (d) Ca(ClO)$_2$ calcium hypochlorite

2.56. Name the following:

   (a) Ca(CH$_3$CO$_2$)$_2$, calcium acetate

   (b) Ni$_3$(PO$_4$)$_2$, nickel(II) phosphate

   (c) Al(OH)$_3$, aluminum hydroxide

   (d) KH$_2$PO$_4$, potassium dihydrogen phosphate

2.57. Names of the ionic compounds:

   (a) Na$_2$S, sodium sulfide

   (b) Na$_2$SO$_3$, sodium sulfite

   (c) NaHSO$_4$, sodium hydrogen sulfate

   (d) Na$_2$SO$_4$, sodium sulfate

2.58. Names of the ionic compounds:

   (a) Li$_3$N, lithium nitride

   (b) LiNO$_2$, lithium nitrite

   (c) LiNO$_3$, lithium nitrate

2.59. Formulas for the ionic compounds

   (a) ammonium carbonate (NH$_4$)$_2$CO$_3$

   (b) calcium iodide   CaI$_2$

   (c) copper(II) bromide   CuBr$_2$

(d) aluminum phosphate   AlPO$_4$

(e) silver(I) acetate   AgCH$_3$CO$_2$

2.60. Formula for the following:

(a) calcium hydrogen carbonate   Ca(HCO$_3$)$_2$

(b) potassium permanganate   KMnO$_4$

(c) magnesium perchlorate   Mg(ClO$_4$)$_2$

(d) potassium hydrogen phosphate   K$_2$HPO$_4$

(e) sodium sulfite   Na$_2$SO$_3$

2.61. Names and formulas for ionic compounds:

	anion	anion
cation	**CO$_3^{2-}$**	**I$^-$**
Na$^+$	Na$_2$CO$_3$ sodium carbonate	NaI sodium iodide
Ba^{2+}	BaCO$_3$ barium carbonate	BaI$_2$ barium iodide

2.62. Write formulas and name:

	anion	anion
cation	**NO$_3^-$**	**PO$_4^{3-}$**
Mg^{2+}	Mg(NO$_3$)$_2$ magnesium nitrate	Mg$_3$(PO$_4$)$_2$ magnesium phosphate
Fe^{3+}	Fe(NO$_3$)$_3$ iron(III) nitrate	FePO$_4$ iron(III) phosphate

Coulomb's Law

2.63. The fluoride ion has a smaller radius than the iodide ion. Hence the distance between the sodium and fluoride ions will be less than the comparable distance between sodium and iodide. Coulomb's Law indicates that the attractive force becomes greater as the distance between the charges grows smaller—hence NaF will have stronger forces of attraction.

2.64. The attractive forces are stronger in CaO because the ion charges are greater (+2/–2 in CaO and +1/–1 in NaCl).

Atoms and the Mole

2.65. The mass, in grams of:

(a) $3.2 \text{ mol Al} \left(\dfrac{26.982 \text{ g Al}}{1 \text{ mol Al}} \right) = 86 \text{ g Al (2 sf)}$

(b) $2.35 \times 10^{-3} \text{ mol Fe} \left(\dfrac{55.845 \text{ g Fe}}{1 \text{ mol Fe}} \right) = 0.131 \text{ g Fe (3 sf)}$

(c) $0.12 \text{ mol Ca} \left(\dfrac{40.078 \text{ g Ca}}{1 \text{ mol Ca}} \right) = 4.8 \text{ g Ca (2 sf)}$

(d) $23.0 \text{ mol Ne} \left(\dfrac{20.180 \text{ g Ne}}{1 \text{ mol Ne}} \right) = 464 \text{ g Ne (3 sf)}$

Note that, whenever possible, one should use a molar mass of the substance that contains at least **one more** significant figure than the data, to reduce round-off error. Correct number of significant figures indicated by sf in parentheses.

2.66. The mass, in grams of:

(a) $1.24 \text{ mol Au} \left(\dfrac{196.97 \text{ g Au}}{1 \text{ mol Au}} \right) = 244 \text{ g Au (3 sf)}$

(b) $14.8 \text{ mol He} \left(\dfrac{4.0026 \text{ g He}}{1 \text{ mol He}} \right) = 59.2 \text{ g He (3 sf)}$

(c) $0.43 \text{ mol Pt} \left(\dfrac{195.08 \text{ g Pt}}{1 \text{ mol Pt}} \right) = 84 \text{ g Pt (2 sf)}$

(d) $2.42 \times 10^{-4} \text{ mol Rh} \left(\dfrac{102.91 \text{ g Rh}}{1 \text{ mol Rh}} \right) = 0.0249 \text{ g Rh (3 sf)}$

2.67. The amount (moles) of substance represented by:

(a) $87.21 \text{ g Cu} \left(\dfrac{1 \text{ mol Cu}}{63.546 \text{ g}} \right) = 1.372 \text{ mol Cu (4 sf)}$

(b) $0.024 \text{ g Na} \left(\dfrac{1 \text{ mol Na}}{22.990 \text{ g Na}} \right) = 0.0010 \text{ mol Na (2 sf)}$

(c) $2.0 \text{ mg Ir} \left(\dfrac{1 \text{ g Ir}}{1000 \text{ mg Ir}}\right)\left(\dfrac{1 \text{ mol Ir}}{192.22 \text{ g Ir}}\right) = 1.0 \times 10^{-5}$ mol Ir (2 sf)

(d) $6.75 \text{ g Au}\left(\dfrac{1 \text{ mol Au}}{196.97 \text{ g Au}}\right) = 3.43 \times 10^{-2}$ mol Au (3 sf)

2.68. The amount (moles) of substance represented by:

(a) $9.4 \text{ g Li}\left(\dfrac{1 \text{ mol Li}}{6.94 \text{ g Li}}\right) = 1.4$ mol Li (2 sf)

(b) $0.942 \text{ g Sn}\left(\dfrac{1 \text{ mol Sn}}{118.71 \text{ g Sn}}\right) = 7.94 \times 10^{-3}$ mol Sn (3 sf)

(c) $0.037 \text{ g Pt}\left(\dfrac{1 \text{ mol Pt}}{195.08 \text{ g Pt}}\right) = 1.9 \times 10^{-4}$ mol Pt (2 sf)

(d) $1.84 \text{ g Xe}\left(\dfrac{1 \text{ mol Xe}}{131.29 \text{ g Xe}}\right) = 1.40 \times 10^{-2}$ mol Xe (3 sf)

2.69. The average mass of an atom (in grams) of each of these elements:

(a) He $\left(\dfrac{4.0026 \text{ g He}}{1 \text{ mol He}}\right)\left(\dfrac{1 \text{ mol He}}{6.02214 \times 10^{23} \text{ atoms He}}\right) = 6.6465 \times 10^{-24}$ g/He atom (5 sf)

(b) F $\left(\dfrac{18.998 \text{ g F}}{1 \text{ mol F}}\right)\left(\dfrac{1 \text{ mol F}}{6.02214 \times 10^{23} \text{ atoms F}}\right) = 3.1547 \times 10^{-23}$ g/F atom (5 sf)

(c) Sc $\left(\dfrac{44.956 \text{ g Sc}}{1 \text{ mol Sc}}\right)\left(\dfrac{1 \text{ mol Sc}}{6.02214 \times 10^{23} \text{ atoms Sc}}\right) = 7.4651 \times 10^{-23}$ g/Sc atom (5 sf)

(d) Bi $\left(\dfrac{208.98 \text{ g Bi}}{1 \text{ mol Bi}}\right)\left(\dfrac{1 \text{ mol Bi}}{6.02214 \times 10^{23} \text{ atoms Bi}}\right) = 3.4702 \times 10^{-22}$ g/Bi atom (5 sf)

2.70. The average mass of an atom (in grams) of each of these elements:

(a) Be $\left(\dfrac{9.0122 \text{ g Be}}{1 \text{ mol Be}}\right)\left(\dfrac{1 \text{ mol Be}}{6.02214 \times 10^{23} \text{ atoms Be}}\right) = 1.4965 \times 10^{-23}$ g/Be atom (5 sf)

(b) Al $\left(\dfrac{26.982 \text{ g Al}}{1 \text{ mol Al}}\right)\left(\dfrac{1 \text{ mol Al}}{6.02214 \times 10^{23} \text{ atoms Al}}\right) = 4.4805 \times 10^{-23}$ g/Al atom (5 sf)

(c) Kr $\left(\dfrac{83.798 \text{ g Kr}}{1 \text{ mol Kr}}\right)\left(\dfrac{1 \text{ mol Kr}}{6.02214 \times 10^{23} \text{ atoms Kr}}\right) = 1.3915 \times 10^{-22}$ g/Kr atom (5 sf)

(d) Hg $\left(\dfrac{200.59 \text{ g Hg}}{1 \text{ mol Hg}}\right)\left(\dfrac{1 \text{ mol Hg}}{6.02214 \times 10^{23} \text{ atoms Hg}}\right) = 3.3309 \times 10^{-22}$ g/Hg atom (5 sf)

2.71. 1.0-gram samples of He, Fe, Li, Si, C:

Which sample contains the **largest number** of atoms? …the **smallest number** of atoms?

If we calculate the number of atoms of any one of these elements, say He, the process is:

$1.0 \text{ g He}\left(\dfrac{1 \text{ mol He}}{4.0026 \text{ g He}}\right)\left(\dfrac{6.022 \times 10^{23} \text{ He atoms}}{1 \text{ mol He}}\right) = 1.5 \times 10^{23}$ He atoms (2 sf)

All the calculations proceed analogously, with the ONLY numerical difference attributable to the molar mass of the element. Therefore, the element with the **smallest** molar mass (He) will have the **largest number** of atoms, while the element with the **largest** molar mass (Fe) will have the **smallest number** of atoms. This is a great question to answer by **thinking** rather than by calculating.

2.72. List samples in increasing amounts:

$0.10 \text{ g K} \cdot \dfrac{1 \text{ mol K}}{39.098 \text{ g K}} = 0.0026 \text{ mol K}$

$0.10 \text{ g Mo} \cdot \dfrac{1 \text{ mol Mo}}{95.95 \text{ g Mo}} = 0.0010 \text{ mol Mo}$

$0.10 \text{ g Cr} \cdot \dfrac{1 \text{ mol Cr}}{51.996 \text{ g Cr}} = 0.0019 \text{ mol Cr}$

$0.10 \text{ g Al} \cdot \dfrac{1 \text{ mol Al}}{26.982 \text{ g Al}} = 0.0037 \text{ mol Al}$

0.0010 mol Mo < 0.0019 mol Cr < 0.0026 mol K < 0.0037 mol Al

Because there are equal masses of each sample the element with the largest molar mass will have the smallest number of moles. The element with the smallest molar mass will have the largest number of moles.

2.73. Analysis of a 10.0-g sample of apatite contained 3.99g Ca, 1.85g P, 4.14g O, and 0.020g H. Calculating moles of each element gives:

$3.99 \text{ g Ca} \cdot \dfrac{1 \text{ mol Ca}}{40.08 \text{ g Ca}} = 0.0996 \text{ mol Ca}$ and $1.85 \text{ g P} \cdot \dfrac{1 \text{ mol P}}{30.97 \text{ g P}} = 0.0597 \text{ mol P}$

$4.14 \text{ g O} \cdot \dfrac{1 \text{ mol O}}{16.00 \text{ g O}} = 0.259 \text{ mol O}$ and $0.020 \text{ g H} \cdot \dfrac{1 \text{ mol H}}{1.01 \text{ g H}} = 0.020 \text{ mol H}$

From smallest to largest # moles: 0.020 mol H < 0.0597 mol P < 0.0996 mol Ca < 0.259 mol O

Unlike SQ 2.72, each sample has a different mass. Consequently, we must calculate each one to determine the number of moles.

2.74. Semiconducting material contains 52 g Ga, 9.5 g Al, 112 g As. Element with largest number of atoms:

$52 \text{ g Ga} \left(\dfrac{1 \text{ mol Ga}}{69.7 \text{ g Ga}} \right) \left(\dfrac{6.022 \times 10^{23} \text{ atoms Ga}}{1 \text{ mol Ga}} \right) = 4.5 \times 10^{23} \text{ atoms Ga}$

$9.5 \text{ g Al} \left(\dfrac{1 \text{ mol Al}}{27.0 \text{ g Al}} \right) \left(\dfrac{6.022 \times 10^{23} \text{ atoms Al}}{1 \text{ mol Al}} \right) = 2.1 \times 10^{23} \text{ atoms Al}$

$$112 \text{ g As} \left(\frac{1 \text{ mol As}}{74.92 \text{ g As}}\right)\left(\frac{6.022 \times 10^{23} \text{ atoms As}}{1 \text{ mol As}}\right) = 9.00 \times 10^{23} \text{ atoms As}$$

Arsenic has the largest number of atoms in the mixture.

Molecules, Compounds, and the Mole

2.75. Moles of carbon in 1.0 mol and 0.20 mol of glucose:

The molecular formula of glucose, $C_6H_{12}O_6$, tells us the number of moles of each element in one mole of the compound. In one mole of glucose, there are 6 mol C, 12 mol H, and 6 mol O. Thus, in 1.0 mol of glucose, there are 6.0 mol C.

In 0.20 mol of glucose, there are $(0.20 \text{ mol glucose})\left(\dfrac{6 \text{ mol C}}{1 \text{ mol glucose}}\right) = 1.2 \text{ mol C}$

2.76. Moles of hydrogen and oxygen in 1.0 mol and 4.0 mol of water:

The molecular formula of water, H_2O, tells us the number of moles of each element in one mole of the compound. In 1.0 mol of water, there are 2.0 mol H and 1.0 mol O.

In 4.0 mol of water, there are $4.0 \text{ mol water}\left(\dfrac{2 \text{ mol H}}{1 \text{ mol water}}\right) = 8.0 \text{ mol H}$

In 4.0 mol of water, there are $4.0 \text{ mol water}\left(\dfrac{1 \text{ mol O}}{1 \text{ mol water}}\right) = 4.0 \text{ mol O}$

2.77. Molar mass of the following: (with atomic weights expressed to 4 significant figures)

    (a) Fe_2O_3 (2)(55.85) + (3)(16.00) = 159.70 g/mol

    (b) BCl_3 (1)(10.81) + (3)(35.45) = 117.16 g/mol

    (c) $C_6H_8O_6$ (6)(12.01) + (8)(1.008) + (6)(16.00) = 176.12 g/mol

    (d) $Mg(NO_3)_2$ (1)(24.31) + (2)(14.01) + (6)(16.00) = 148.33 g/mol

2.78. Molar mass of the following: (with atomic weights expressed to 4 significant figures)

    (a) $CaCO_3$ (1)(40.08) + 1(12.01) + 3(16.00) = 100.09 g/mol

    (b) $Fe(C_6H_{11}O_7)_2$ (1)(55.85) + (12)(12.01) + (22)(1.008) + (14)(16.00) = 446.15 g/mol

    (c) $CH_3CH_2CH_2CH_2SH$ (4)(12.01) + (10)(1.008) + (1)(32.06) = 90.18 g/mol

    (d) $C_{20}H_{24}N_2O_2$ (20)(12.01) + (24)(1.008) + (2)(14.01) + (2)(16.00) = 324.41 g/mol

2.79. Molar mass of the following: (with atomic weights expressed to 4 significant figures)

(a) Ni(NO$_3$)$_2$ · 6 H$_2$O (1)(58.69) + (2)(14.01) + 6(16.00) + (12)(1.008) + (6)(16.00)
$$= 290.81 \text{ g/mol}$$

(b) CuSO$_4$ · 5 H$_2$O (1)(63.55) + (1)(32.06) + 4(16.00) + (10)(1.008) + (5)(16.00)
$$= 249.69 \text{ g/mol}$$

2.80. Molar mass of the following:

(a) H$_2$C$_2$O$_4$·2 H$_2$O (6)(1.008) + (2)(12.01) + (6)(16.00) = 126.07 g/mol
(b) MgSO$_4$·7 H$_2$O (1)(24.31) + (1)(32.06) + (11)(16.00) + (14)(1.008) = 246.48 g/mol

2.81. Mass represented by 0.0255 moles of the following compounds:

Molar masses are calculated as before. To determine the mass represented by 0.0255 moles, recall that 1 mol of a substance has a mass equal to the molar mass expressed in units of grams.

We calculate for (a) the mass represented by 0.0255 moles:

$$0.0255 \text{ mol C}_3\text{H}_7\text{OH} \left(\frac{60.10 \text{ g C}_3\text{H}_7\text{OH}}{1 \text{ mol C}_3\text{H}_7\text{OH}} \right) = 1.53 \text{ g C}_3\text{H}_7\text{OH}$$

Compound	Molar mass	Mass of 0.0255 moles
(a) C$_3$H$_7$OH	60.10	1.53 g
(b) C$_{11}$H$_{16}$O$_2$	180.2	4.60 g
(c) C$_9$H$_8$O$_4$	180.2	4.60 g
(d) C$_3$H$_6$O	58.08	1.48 g

Masses are expressed to 3 sf, since the # of moles has 3.

2.82. Mass of 0.123 mol of each of the following:

(a) $0.123 \text{ mol C}_{14}\text{H}_{10}\text{O}_4 \left(\frac{242.2 \text{ g C}_{14}\text{H}_{10}\text{O}_4}{1 \text{ mol C}_{14}\text{H}_{10}\text{O}_4} \right) = 29.8 \text{ g C}_{14}\text{H}_{10}\text{O}_4$

(b) $0.123 \text{ mol C}_4\text{H}_8\text{N}_2\text{O}_2 \left(\frac{116.1 \text{ g C}_4\text{H}_8\text{N}_2\text{O}_2}{1 \text{ mol C}_4\text{H}_8\text{N}_2\text{O}_2} \right) = 14.3 \text{ g C}_4\text{H}_8\text{N}_2\text{O}_2$

(c) $0.123 \text{ mol C}_5\text{H}_{10}\text{S} \left(\frac{102.2 \text{ g C}_5\text{H}_{10}\text{S}}{1 \text{ mol C}_5\text{H}_{10}\text{S}} \right) = 12.6 \text{ g C}_5\text{H}_{10}\text{S}$

(d) $0.123 \text{ mol C}_{12}\text{H}_{17}\text{NO} \left(\frac{191.3 \text{ g C}_{12}\text{H}_{17}\text{NO}}{1 \text{ mol C}_{12}\text{H}_{17}\text{NO}} \right) = 23.5 \text{ g C}_{12}\text{H}_{17}\text{NO}$

2.83. Regarding sulfur trioxide:

1. Amount of SO₃ in 1.00 kg: $1.00 \times 10^3 \text{ g SO}_3 \left(\dfrac{1 \text{ mol SO}_3}{80.06 \text{ g SO}_3} \right) = 12.5 \text{ mol SO}_3$

2. Number of SO₃ molecules:

$12.5 \text{ mol SO}_3 \left(\dfrac{6.022 \times 10^{23} \text{ molecules SO}_3}{1 \text{ mol SO}_3} \right) = 7.52 \times 10^{24} \text{ molecules SO}_3$

3. Number of S atoms: With 1 S atom per SO₃ molecule—

$7.52 \times 10^{24} \text{ molecules SO}_3 \left(\dfrac{1 \text{ atom S}}{1 \text{ molecule SO}_3} \right) = 7.52 \times 10^{24} \text{ atoms S}$

4. Number of O atoms: With 3 O atoms per SO₃ molecule—

$7.52 \times 10^{24} \text{ molecules SO}_3 \left(\dfrac{3 \text{ atoms O}}{1 \text{ molecule SO}_3} \right) = 2.26 \times 10^{25} \text{ atoms O}$

2.84. Number of ammonium and sulfate ions present:

$0.20 \text{ mol (NH}_4)_2\text{SO}_4 \left(\dfrac{2 \text{ mol NH}_4^+}{1 \text{ mol (NH}_4)_2\text{SO}_4} \right) \left(\dfrac{6.022 \times 10^{23} \text{ NH}_4^+ \text{ ions}}{1 \text{ mol NH}_4^+} \right) = 2.4 \times 10^{23} \text{ NH}_4^+ \text{ ions}$

$0.20 \text{ mol (NH}_4)_2\text{SO}_4 \left(\dfrac{1 \text{ mol SO}_4^{2-}}{1 \text{ mol (NH}_4)_2\text{SO}_4} \right) \left(\dfrac{6.022 \times 10^{23} \text{ SO}_4^{2-} \text{ ions}}{1 \text{ mol SO}_4^{2-}} \right) = 1.2 \times 10^{23} \text{ SO}_4^{2-} \text{ ions}$

$0.20 \text{ mol (NH}_4)_2\text{SO}_4 \left(\dfrac{2 \text{ mol N}}{1 \text{ mol (NH}_4)_2\text{SO}_4} \right) \left(\dfrac{6.022 \times 10^{23} \text{ N atoms}}{1 \text{ mol N}} \right) = 2.4 \times 10^{23} \text{ N atoms}$

$0.20 \text{ mol (NH}_4)_2\text{SO}_4 \left(\dfrac{8 \text{ mol H}}{1 \text{ mol (NH}_4)_2\text{SO}_4} \right) \left(\dfrac{6.022 \times 10^{23} \text{ H atoms}}{1 \text{ mol H}} \right) = 9.6 \times 10^{23} \text{ H atoms}$

$0.20 \text{ mol (NH}_4)_2\text{SO}_4 \left(\dfrac{1 \text{ mol S}}{1 \text{ mol (NH}_4)_2\text{SO}_4} \right) \left(\dfrac{6.022 \times 10^{23} \text{ S atoms}}{1 \text{ mol S}} \right) = 1.2 \times 10^{23} \text{ S atoms}$

$0.20 \text{ mol (NH}_4)_2\text{SO}_4 \left(\dfrac{4 \text{ mol O}}{1 \text{ mol (NH}_4)_2\text{SO}_4} \right) \left(\dfrac{6.022 \times 10^{23} \text{ O atoms}}{1 \text{ mol O}} \right) = 4.8 \times 10^{23} \text{ O atoms}$

2.85. C₈H₉NO₂, the molecular formula for acetaminophen has a molar mass of 151.17 g/mol.

$2 \text{ caplets} \left(\dfrac{500 \text{ mg C}_8\text{H}_9\text{NO}_2}{1 \text{ caplet}} \right) \left(\dfrac{1 \text{ g}}{1000 \text{ mg}} \right) \left(\dfrac{1 \text{ mol C}_8\text{H}_9\text{NO}_2}{151.17 \text{ g C}_8\text{H}_9\text{NO}_2} \right) \left(\dfrac{6.022 \times 10^{23} \text{ molecules C}_8\text{H}_9\text{NO}_2}{1 \text{ mol C}_8\text{H}_9\text{NO}_2} \right)$

$= 4 \times 10^{21} \text{ molecules C}_8\text{H}_9\text{NO}_2 \text{ (1 sf)}$

2.86. In an Alka-Seltzer tablet:

(a) $325 \text{ mg C}_9\text{H}_8\text{O}_4 \left(\dfrac{1 \text{ g}}{10^3 \text{ mg}} \right) \left(\dfrac{1 \text{ mol C}_9\text{H}_8\text{O}_4}{180.2 \text{ g C}_9\text{H}_8\text{O}_4} \right) = 1.80 \times 10^{-3} \text{ mol C}_9\text{H}_8\text{O}_4$

$1916 \text{ mg NaHCO}_3 \left(\dfrac{1 \text{ g}}{10^3 \text{ mg}} \right) \left(\dfrac{1 \text{ mol NaHCO}_3}{84.006 \text{ g NaHCO}_3} \right) = 2.281 \times 10^{-3} \text{ mol NaHCO}_3$

$1000. \text{ mg C}_6\text{H}_8\text{O}_7 \left(\dfrac{1 \text{ g}}{10^3 \text{ mg}} \right) \left(\dfrac{1 \text{ mol C}_6\text{H}_8\text{O}_7}{191.12 \text{ g C}_6\text{H}_8\text{O}_7} \right) = 5.232 \times 10^{-3} \text{ mol C}_6\text{H}_8\text{O}_7$

(b) $1.80 \times 10^{-3} \text{ mol C}_9\text{H}_8\text{O}_4 \left(\dfrac{6.022 \times 10^{23} \text{ molecules C}_9\text{H}_8\text{O}_4}{1 \text{ mol C}_9\text{H}_8\text{O}_4} \right) = 1.09 \times 10^{21} \text{ molecules C}_9\text{H}_8\text{O}_4$

Percent Composition

2.87. Mass percent for: [4 significant figures]

(a) PbS: Molar mass = (1)(207.2) + (1)(32.06) = 239.3 g/mol

$\%\text{Pb} = \dfrac{207.2 \text{ g Pb}}{239.3 \text{ g PbS}} \times 100\% = 86.60\%$

$\%\text{S} = 100.00\% - 86.60\% = 13.40\%$

(b) C_3H_8: Molar mass = 44.097 g/mol

$\%\text{C} = \dfrac{36.03 \text{ g C}}{44.097 \text{ g C}_3\text{H}_8} \times 100\% = 81.71\%$

$\%\text{H} = 100.00\% - 81.71\% = 18.29\%$

(c) $C_{10}H_{14}O$: Molar mass = 150.221 g/mol

$\%\text{C} = \dfrac{120.11 \text{ g C}}{150.221 \text{ g C}_{10}\text{H}_{14}\text{O}} \times 100\% = 79.956\% = 79.96\%$ (4 sf)

$\%\text{H} = \dfrac{14.112 \text{ g H}}{150.221 \text{ g C}_{10}\text{H}_{14}\text{O}} \times 100\% = 9.394\%$

$\%\text{O} = 100.00\% - (79.96\% + 9.394\%) = 10.65\%$

2.88. Mass percent of elements in the compounds:

(a) $\dfrac{(8)(12.01) \text{ g C}}{166.18 \text{ C}_8\text{H}_{10}\text{N}_2\text{O}_2} \cdot 100\% = 57.82\% \text{ C}$ $\qquad \dfrac{(10)(1.008) \text{ g H}}{166.18 \text{ C}_8\text{H}_{10}\text{N}_2\text{O}_2} \cdot 100\% = 6.066\% \text{ H}$

$\dfrac{(2)(14.01) \text{ g N}}{166.18 \text{ C}_8\text{H}_{10}\text{N}_2\text{O}_2} \cdot 100\% = 16.86\% \text{ N}$ $\qquad \dfrac{(2)(16.00) \text{ g O}}{166.18 \text{ C}_8\text{H}_{10}\text{N}_2\text{O}_2} \cdot 100\% = 19.26\% \text{ O}$

(b) $\dfrac{(10)(12.01) \text{ g C}}{156.26 \text{ g C}_{10}\text{H}_{20}\text{O}} \cdot 100\% = 76.86\%$ C $\dfrac{(20)(1.008) \text{ g H}}{156.26 \text{ g C}_{10}\text{H}_{20}\text{O}} \cdot 100\% = 12.90\%$ H

$\dfrac{(1)(16.00) \text{ g O}}{156.26 \text{ g C}_{10}\text{H}_{20}\text{O}} \cdot 100\% = 10.24\%$ O

(c) $\dfrac{(1)(58.93) \text{ g Co}}{237.92 \text{ g CoCl}_2 \cdot 6\text{ H}_2\text{O}} \cdot 100\% = 24.77\%$ Co

$\dfrac{(2)(35.45) \text{ g Cl}}{237.92 \text{ g CoCl}_2 \cdot 6\text{ H}_2\text{O}} \cdot 100\% = 29.80\%$ Cl

$\dfrac{(12)(1.008) \text{ g H}}{237.92 \text{ g CoCl}_2 \cdot 6\text{ H}_2\text{O}} \cdot 100\% = 5.084\%$ H

$\dfrac{(6)(16.00) \text{ g O}}{237.92 \text{ g CoCl}_2 \cdot 6\text{ H}_2\text{O}} \cdot 100\% = 40.35\%$ O

2.89. Mass of CuS to provide 10.0 g of Cu:

To calculate the weight percent of Cu in CuS, we need the respective atomic weights:

Cu = 63.546; S = 32.06   adding CuS gives a molecular weight of 95.61

The % of Cu in CuS is then: $\left(\dfrac{63.546 \text{ g Cu}}{95.61 \text{ g CuS}}\right) \times 100 = 66.47\%$

Now with this fraction (inverted) calculate the mass of CuS that will provide 10.0 g of Cu:

$10.0 \text{ g Cu}\left(\dfrac{95.61 \text{ g CuS}}{63.546 \text{ g Cu}}\right) = 15.0 \text{ g CuS}$

2.90. Mass percent of Ti in ilmenite; Mass of ilmenite to obtain 750 g of Ti:

$\dfrac{47.87 \text{ g Ti}}{151.71 \text{ g FeTiO}_3} \cdot 100\% = 31.55\%$ Ti

$750 \text{ g Ti} \cdot \dfrac{100 \text{ g FeTiO}_3}{31.55 \text{ g Ti}} = 2.4 \times 10^3 \text{ g FeTiO}_3$

Empirical and Molecular Formulas

2.91. The empirical formula ($C_2H_3O_2$) would have a molar mass of 59.04 g/mol.

Since the molar mass is 118.1 g/mol we can write

$$\frac{1 \text{ empirical formula}}{59.04 \text{ g succinic acid}} \cdot \frac{118.1 \text{ g succinic acid}}{1 \text{ mol succinic acid}} = \frac{2.0 \text{ empirical formulas}}{1 \text{ mol succinic acid}}$$

So, the molecular formula contains 2 empirical formulas (2 × $C_2H_3O_2$) or $C_4H_6O_4$.

2.92. Molecular formula for a compound:

Empirical formula mass = 58.06 g/mol; $\frac{116.1 \text{ g/mol molecular formula}}{58.06 \text{ g/mol empirical formula}}$ = 2 empirical formulas/molecular formula. The molecular formula is $C_4H_8N_2O_2$.

2.93. Provide the empirical or molecular formula for the following, as requested:

	Empirical Formula	Molar Mass (g/mol)	Molecular Formula
(a)	CH	26.0	**C_2H_2**
(b)	CHO	116.1	**$C_4H_4O_4$**
(c)	**CH_2**	**112.2**	C_8H_{16}

Note that we can calculate the mass of an empirical formula by adding the respective atomic weights (13.0 for CH, for example). The molar mass (26.0 for part (a)) is twice that of an empirical formula, so the molecular formula is 2 × empirical formula (or **C_2H_2** in part (a)).

2.94. Provide the empirical or molecular formula for the following, as requested:

	Empirical Formula	Molar Mass (g/mol)	Molecular Formula
(a)	$C_2H_3O_3$	150.1	**$C_4H_6O_6$**
(b)	C_3H_8	44.1	**C_3H_8**
(c)	**B_2H_5**	**53.3**	B_4H_{10}

2.95. Calculate the empirical formula of acetylene by calculating the mole ratios of carbon and hydrogen in 100 g of the compound.

$$92.26 \text{ g C} \cdot \frac{1 \text{ mol C}}{12.011 \text{ g C}} = 7.681 \text{ mol C} \quad \text{and} \quad 7.74 \text{ g H} \cdot \frac{1 \text{ mol H}}{1.008 \text{ g H}} = 7.68 \text{ mol H}$$

Calculate the mole ratio: $\dfrac{7.681 \text{ mol C}}{7.68 \text{ mol H}} = \dfrac{1 \text{ mol C}}{1 \text{ mol H}}$

The mole ratio indicates that there is 1 C atom for 1 H atom (1:1). The **empirical formula is then CH**. The formula mass is 13.02. Given that the molar mass of the compound is 26.02 g/mol, there are two formula units per molecular unit, hence the **molecular formula for acetylene is C₂H₂**.

2.96. Empirical formula of a B–H compound:

The compound is 88.5% B and 100% – 88.5% = 11.5% H. Assume 100.0 g of compound.

$88.5 \text{ g B}\left(\dfrac{1 \text{ mol B}}{10.81 \text{ g B}}\right) = 8.19 \text{ mol B}$ $\quad 11.5 \text{ g H}\left(\dfrac{1 \text{ mol H}}{1.008 \text{ g H}}\right) = 11.4 \text{ mol H}$

$\dfrac{11.4 \text{ mol H}}{8.19 \text{ mol B}} = \dfrac{1.39 \text{ mol H}}{1 \text{ mol B}} = \dfrac{7/5 \text{ mol H}}{1 \text{ mol B}} = \dfrac{7 \text{ mol H}}{5 \text{ mol B}}$ So the empirical formula is B₅H₇.

2.97. Determine the empirical and molecular formulas of cumene:

Assume 100.0 g of compound.

The percentage composition of cumene is 89.94% C and 100.00% – 89.94% = 10.06%H. Calculate the ratio of mol C: mol H as done in SQ95.

$89.94 \text{ g C} \cdot \dfrac{1 \text{ mol C}}{12.011 \text{ g C}} = 7.488 \text{ mol C}$

$10.06 \text{ g H} \cdot \dfrac{1 \text{ mol H}}{1.008 \text{ g H}} = 9.980 \text{ mol H}$

Calculating the mole ratio:

$\dfrac{9.980 \text{ mol H}}{7.488 \text{ mol C}} = \dfrac{1.333 \text{ mol H}}{1 \text{ mol C}}$ or a ratio of 3C : 4H

So, the empirical formula for cumene is C₃H₄, with a formula mass of 40.07.

If the molar mass is 120.2 g/mol, then dividing the "empirical formula mass" into the molar mass gives: 120.2/40.07 or 3 empirical formulas **per** molar mass. The **molecular formula** is then 3 × C₃H₄ or C₉H₁₂.

2.98. Empirical and Molecular formula for sulflower:

Assume 100.0 g of compound.

$57.17 \text{ g S}\left(\dfrac{1 \text{ mol S}}{32.06 \text{ g S}}\right) = 1.783 \text{ mol S}$ $\quad 42.83 \text{ mol C}\left(\dfrac{1 \text{ mol C}}{12.011 \text{ g C}}\right) = 3.566 \text{ mol C}$

$\dfrac{3.566 \text{ mol C}}{1.783 \text{ mol S}} = \dfrac{2.000 \text{ mol C}}{1 \text{ mol S}}$

So the empirical formula is C₂S. The molar mass of a C₂S entity is 56.08 g/mol, so divide that into the molar mass:

$$\frac{448.70 \text{ g/mol}}{56.08 \text{ g/mol}} = 8.$$ The molecular formula is C₁₆S₈

2.99. Empirical and Molecular formula for mandelic acid:

Assume 100.0 g of compound.

$$63.15 \text{ g C} \cdot \frac{1 \text{ mol C}}{12.011 \text{ g C}} = 5.258 \text{ mol C} \qquad 5.30 \text{ g H} \cdot \frac{1 \text{ mol H}}{1.008 \text{ g H}} = 5.26 \text{ mol H}$$

$$31.55 \text{ g O} \cdot \frac{1 \text{ mol O}}{15.999 \text{ g O}} = 1.972 \text{ mol O}$$

Notice that the amounts (moles) of C and H are the same. Dividing by the smallest amount, we calculate the mole ratio for C and O:

$$\frac{5.258 \text{ mol C}}{1.972 \text{ mol O}} = \frac{2.666 \text{ mol C}}{1 \text{ mol O}} \text{ or } \frac{2\,^2/_3 \text{ mol C}}{1 \text{ mol O}} \text{ or } \frac{^8/_3 \text{ mol C}}{1 \text{ mol O}}$$

So 3 mol O combine with 8 mol C and 8 mol H and the empirical formula is C₈H₈O₃.

The formula mass of C₈H₈O₃ is 152.15. Given the data that the molar mass of the compound is 152.15 g/mL, the molecular formula for mandelic acid is C₈H₈O₃.

2.100. Empirical and Molecular formula for nicotine:

Assume 100.0 g of compound.

$$74.0 \text{ g C}\left(\frac{1 \text{ mol C}}{12.01 \text{ g C}}\right) = 6.16 \text{ mol C} \qquad 8.65 \text{ g H}\left(\frac{1 \text{ mol H}}{1.008 \text{ g H}}\right) = 8.58 \text{ mol H}$$

$$17.35 \text{ g N}\left(\frac{1 \text{ mol N}}{14.007 \text{ g N}}\right) = 1.239 \text{ mol N}$$

Now determine the mole:mole ratios:

$$\frac{6.16 \text{ mol C}}{1.239 \text{ mol N}} = \frac{4.97 \text{ mol C}}{1 \text{ mol N}} = \frac{5 \text{ mol C}}{1 \text{ mol N}} \qquad \frac{8.58 \text{ mol H}}{1.239 \text{ mol N}} = \frac{6.93 \text{ mol H}}{1 \text{ mol N}} = \frac{7 \text{ mol H}}{1 \text{ mol N}}$$

and the empirical formula is C₅H₇N.

Compare the mass of the empirical formula to that of the molecular formula: $\frac{116 \text{ g/mol}}{81.1 \text{ g/mol}} = 2$

so the molecular formula is C₁₀H₁₄N₂

Determining Formulas from Mass Data

2.101. Given the masses of xenon involved, we can calculate the number of moles of the element:

$$0.674 \text{ g Xe} \cdot \frac{1 \text{ mol Xe}}{131.29 \text{ g Xe}} = 0.00513 \text{ mol Xe}$$

The mass of fluorine present is: 0.869 g compound – 0.674 g Xe = 0.195 g F

$$0.195 \text{ g F} \cdot \frac{1 \text{ mol F}}{19.00 \text{ g F}} = 0.0103 \text{ mol F}$$

Calculating atomic ratios:

$$\frac{0.0103 \text{ mol F}}{0.00513 \text{ mol Xe}} = \frac{2 \text{ mol F}}{1 \text{ mol Xe}} \text{ indicating that the empirical formula is } XeF_2.$$

2.102. Value of x in the formula SF_x:

6.70 g compound – 1.47 g S = 5.23 g F; Determine mol of S and mol of F:

$$1.47 \text{ g S}\left(\frac{1 \text{ mol S}}{32.06 \text{ g S}}\right) = 0.0459 \text{ mol S} \quad 5.23 \text{ g F}\left(\frac{1 \text{ mol F}}{18.998 \text{ g F}}\right) = 0.275 \text{ mol F}$$

$$\frac{0.275 \text{ mol F}}{0.0459 \text{ mol S}} = \frac{6.00 \text{ mol F}}{1 \text{ mol S}}, \text{ so the empirical formula is } SF_6; x =$$

2.103. Value of x in the formula $BaCl_2 \cdot x\, H_2O$:

Molar mass of $BaCl_2$ = 208.23 g/mol

Molar mass of H_2O = 18.02 g/mol

Mass of H_2O evaporated by heating = 1.523 g – 1.298 g = 0.225 g H_2O evaporated

$$1.298 \text{ g BaCl}_2\left(\frac{1 \text{ mol BaCl}_2}{208.23 \text{ g BaCl}_2}\right) = 0.006233 \text{ mol BaCl}_2$$

$$0.225 \text{ g H}_2\text{O}\left(\frac{1 \text{ mol H}_2\text{O}}{18.02 \text{ g H}_2\text{O}}\right) = 0.0125 \text{ mol H}_2\text{O}$$

Mole ratio of H_2O to $BaCl_2 = \dfrac{0.0125 \text{ mol H}_2\text{O}}{0.006233 \text{ mol BaCl}_2} = \dfrac{2.00 \text{ mol H}_2\text{O}}{1 \text{ mol BaCl}_2}$

Thus, there are 2 moles of H_2O for every 1 mole of $BaCl_2$ indicating that x has a value of 2 and the compound's formula is $BaCl_2 \cdot 2\, H_2O$.

The name of this compound is barium chloride dihydrate.

2.104. Value of x in the formula $Na_2SO_4 \cdot x\ H_2O$:

Molar mass of Na_2SO_4 = 142.0 g/mol

Molar mass of H_2O = 18.02 g/mol

Mass of H_2O evaporated by heating = 2.343 g – 1.033 g = 1.310 g H_2O evaporated

$$1.033\ g\ Na_2SO_4 \left(\frac{1\ mol\ Na_2SO_4}{142.0\ g\ Na_2SO_4} \right) = 0.007275\ mol\ Na_2SO_4$$

$$1.310\ g\ H_2O \left(\frac{1\ mol\ H_2O}{18.02\ g\ H_2O} \right) = 0.07270\ mol\ H_2O$$

Mole ratio of H_2O to Na_2SO_4 = $\frac{0.07270\ mol\ H_2O}{0.007275\ mol\ Na_2SO_4} = \frac{9.993\ mol\ H_2O}{1\ mol\ Na_2SO_4} \approx \frac{10\ mol\ H_2O}{1\ mol\ Na_2SO_4}$

Thus, there are 10 moles of H_2O for every 1 mole of Na_2SO_4 indicating that x has a value of 10 and the compound's formula is $Na_2SO_4 \cdot 10\ H_2O$.

This compound is named sodium sulfate decahydrate.

2.105. Value of x in the formula $MgSO_4 \cdot x\ H_2O$:

Molar mass of $MgSO_4$ = 120.4 g/mol

Molar mass of H_2O = 18.02 g/mol

Molar mass of $MgSO_4 \cdot 7\ H_2O$ = 120.4 g + 7(18.02 g) = 246.5 g/mol $MgSO_4 \cdot 7\ H_2O$

In 1.394 g of $MgSO_4 \cdot 7\ H_2O$ there are

$$1.394\ g\ MgSO_4 \cdot 7\ H_2O \left(\frac{120.4\ g\ MgSO_4}{246.5\ g\ MgSO_4 \cdot 7\ H_2O} \right) = 0.6809\ g\ MgSO_4$$

In 0.885 g of $MgSO_4 \cdot x\ H_2O$ there are 0.6809 g $MgSO_4$ and 0.885 g – 0.6809 g = 0.204 g H_2O

$$0.6809\ g\ MgSO_4 \left(\frac{1\ mol\ MgSO_4}{120.4\ g\ MgSO_4} \right) = 0.005655\ mol\ MgSO_4$$

$$0.204\ g\ H_2O \left(\frac{1\ mol\ H_2O}{18.02\ g\ H_2O} \right) = 0.0113\ mol\ H_2O$$

Mole ratio of H_2O to $MgSO_4$ = $\frac{0.0113\ mol\ H_2O}{0.005655\ mol\ MgSO_4} = \frac{2.00\ mol\ H_2O}{1\ mol\ MgSO_4}$

Thus, in the partially hydrated sample there are 2 moles of H_2O for every 1 mole of $MgSO_4$ indicating that x has a value of 2 and the compound's formula is $MgSO_4 \cdot 2\ H_2O$.

2.106. Formula for Ge$_x$Cl$_y$:

3.69 g product − 1.25 g Ge = 2.44 g Cl; Determine # mol of Cl and Ge:

$1.25 \text{ g Ge}\left(\dfrac{1 \text{ mol Ge}}{72.63 \text{ g Ge}}\right) = 0.0172 \text{ mol Ge}$ $\qquad$ $2.44 \text{ g Cl}\left(\dfrac{1 \text{ mol Cl}}{35.45 \text{ g Cl}}\right) = 0.0688 \text{ mol Cl}$

Calculate the mole ratio: $\dfrac{0.0688 \text{ mol Cl}}{0.0172 \text{ mol Ge}} = \dfrac{4.00 \text{ mol Cl}}{1 \text{ mol Ge}}$, so the empirical formula is GeCl$_4$.

Mass Spectrometry

2.107. Regarding the mass spectrum for NO$_2$:

(a) The four cations are: 14(N$^+$); 16(O$^+$); 30(NO$^+$); 46(NO$_2^+$). Note the assignments are based upon the masses of the atoms involved.

(b) The mass spectrum does provide evidence for NO bonds as opposed to OO bonds. An OO bond configuration would have produced a peak at 32 *m/Z*.

2.108. Regarding the mass spectrum for POF$_3$:

(a) Fragment at *m/Z* ratio of 85: The peak at 104 is the molecular ion, POF$_3^+$, so the fragment at 85 represents (104 − 19) or POF$_2^+$

(b) Fragment at *m/Z* ratio of 69: This fragment is PF$_2^+$, indicating a further loss of O (85 − 16).

(c) Fragment at *m/Z* ratio of 47 indicates a PO$^+$ fragment, and the fragment at *m/Z* ratio of 50 indicates a PF$^+$ fragment. The lack of a peak at 35, indicates an absence of OF$^+$.

2.109. Regarding the mass spectrum of CH$_3$Cl:

(a) Species responsible for lines at 50 and 52:

^{12}CH$_3$^{35}Cl$^+$ gives rise to the line at *m/Z* = 50 (12 + 3 + 35).

^{12}CH$_3$^{37}Cl$^+$ gives rise to the line at *m/Z* = 52 (12 + 3 + 37).

The differential in sizes of the lines is due to the relative distribution of isotopes of Cl, with the abundance of Cl-37 being about 1/3 that of Cl-35.

(b) The line at *m/Z* = 51 is due to ^{13}CH$_3$^{35}Cl$^+$ with the accompanying smaller size of the peak. There is also a small contribution to this peak from ^{12}C^{2}H^{1}H$_2$^{35}Cl$^+$.

2.110. Consider the peaks in the Br$_2$ mass spectrum at 158, 160, and 162:

(a) The *m/Z* peak at 158 is: ^{79}Br$_2^+$; while the peak at *m/Z* 160 is: ^{79}Br^{81}Br$^+$. Finally, the peak at *m/Z* 162 represents ^{81}Br$_2^+$.

(b) The abundances are close enough to assume an equal abundance of ^{79}Br and ^{81}Br. Two atoms from the two isotopes can be combined in four different manners to form Br$_2$: ^{79}Br$_2$, ^{79}Br^{81}Br, ^{81}Br^{79}Br, and ^{81}Br$_2$. Thus, the peak at *m/Z* 160 should have twice the intensity of the peaks at *m/Z* 158 and 162.

General Questions

2.111.

Symbol	^{58}Ni	^{33}S	^{20}Ne	^{55}Mn
Number of protons	28	16	10	25
Number of neutrons	30	17	10	30
Number of electrons in the neutral atom	28	16	10	25
Name of element	nickel	sulfur	neon	manganese

2.112. As the atomic weight of potassium is 39.098 (remember this is a weighted average), the lighter isotope, ^{39}K is more abundant than ^{41}K.

2.113. Crossword puzzle: Clues:

Horizontal

1–2 A metal used in ancient times: tin (Sn)

3–4 A metal that burns in air and is found in Group 5A(15): bismuth (Bi)

Vertical

1–3 A metalloid: antimony (Sb)

2–4 A metal used in U.S. coins: nickel (Ni)

Single squares:

1. A colorful nonmetal: sulfur (S)

2. A colorless gaseous nonmetal: nitrogen (N)

3. An element that makes fireworks green: boron (B)

4. An element that has medicinal uses: iodine (I)

Diagonal:

1–4 An element used in electronics: silicon (Si)

2–3 A metal used with Zr to make wires for superconducting magnets: niobium (Nb)

Using these solutions, the following letters fit in the boxes

1	2
S	N
3	4
B	I

2.114. Regarding the abundance of elements:

(a) Mg is the most abundant main group metal.
(b) H is the most abundant nonmetal.
(c) Si is the most abundant metalloid.
(d) Fe is the most abundant transition element.
(e) F and Cl are the halogens included, and of these Cl is the most abundant.

2.115. Copper atoms:

(a) The average mass of one copper atom:

One mole of copper (with a mass of 63.546 g) contains 6.0221×10^{23} atoms. So, the average mass of **one** copper atom is: $\dfrac{63.546 \text{ g Cu/mol}}{6.02214 \times 10^{23} \text{ atoms Cu/mol}}$

$= 1.0552 \times 10^{-22}$ g/Cu atom

(b) Given the cost data: $80.10 for 7.0 g and the mass of a Cu atom (from part (a)), the cost of one Cu atom is:

$\dfrac{\$80.10}{7.0 \text{ g sample}} \left(\dfrac{100 \text{ g sample}}{99.999 \text{ g Cu}}\right)\left(\dfrac{1.0552 \times 10^{-22} \text{ g Cu}}{1 \text{ Cu atom}}\right) = 1.2 \times 10^{-21}$ dollars/Cu atom

2.116. Identify which of the following is possible:

(a) silver foil that is 1.2×10^{-4} m thick (thick enough for an atom of Ag—possible)

(b) a sample of potassium that contains 1.784×10^{24} atoms (about 3 moles of K—possible)

(c) a gold coin of mass 1.23×10^{-3} kg (massive enough for more than one Au atom—possible)

(d) 3.43×10^{-27} mol S_8 (amount is less than one molecule of S_8—impossible)

2.117. Identify the element that:

(a) Is in Group 2A (2) and the 5th period: Strontium

(b) Is in the 5th period and Group 4B (4): Zirconium

(c) Is in the second period in Group 4A (14): Carbon

(d) Is an element in the 4th period of Group 5A (15): Arsenic

(e) Is a halogen (Group 7A [17]) in the 5th period: Iodine

(f) Is an alkaline earth element (Group 2A [2]) in the 3rd period: Magnesium

(g) Is a noble gas (Group 8A [18]) in the 4th period: Krypton

(h) Is a nonmetal in Group 6A (16) and the 3rd period: Sulfur

(i) Is a metalloid in the 4th period: Germanium or Arsenic

2.118. **Carbon** has three allotropes. Graphite consists of flat sheets of carbon atoms, diamond has carbon atoms attached to four other others in a tetrahedron, and buckminsterfullerene is a 60-atom cage of carbon atoms. **Oxygen** has two allotropes. Diatomic oxygen consists of molecules containing two oxygen atoms and ozone consists of molecules containing three oxygen atoms. Other nonmetallic elements with allotropes include phosphorus (white, black and red), sulfur (more than 15 allotropic forms), and selenium (red, gray, and black).

2.119. Which of the following has the greater mass:

(a) 0.5 mol Na, 0.5 mol Si, 0.25 mol U

You can determine this without doing the entire calculation. Examine the molar mass of each element. 0.5 mol of any element has a mass that is one-half the molar mass. One quarter mol of U (molar mass approximately 238 g) will have the greatest mass of these three.

(b) 9.0 g of Na, 0.50 mol Na, 1.2×10^{22} atoms Na:

9.0 g Na is already in a unit of mass.

$$0.50 \text{ mol Na} \left(\frac{23.0 \text{ g Na}}{1 \text{ mol Na}} \right) = 12 \text{ g Na (2 sf)}$$

$$1.2 \times 10^{22} \text{ atoms Na} \left(\frac{1 \text{ mol Na}}{6.022 \times 10^{23} \text{ atoms Na}} \right) \left(\frac{23.0 \text{ g Na}}{1 \text{ mol Na}} \right) = 0.46 \text{ g Na}$$

0.50 mol Na will have the greatest mass of these three choices.

(c) 10 atoms of Fe or 10 atoms of K

As in (a), the full calculation is not needed. Fe has a greater atomic weight, so 10 atoms of Fe would have a greater mass than 10 atoms of K.

2.120. Regarding the RDA of iron for women:

$$18 \text{ mg Fe}\left(\frac{1 \text{ g}}{1000 \text{ mg}}\right)\left(\frac{1 \text{ mol Fe}}{55.85 \text{ g Fe}}\right) = 3.2 \times 10^{-4} \text{ mol Fe}$$

$$3.2 \times 10^{-4} \text{ mol Fe}\left(\frac{6.02 \times 10^{23} \text{ atoms Fe}}{1 \text{ mol Fe}}\right) = 1.9 \times 10^{20} \text{ atoms Fe}$$

2.121. Arrange the elements from least massive to most massive:

Select a common unit by which to compare the substances (say grams?)

(a) 3.79×10^{24} atoms Fe $\left(\dfrac{1 \text{ mol Fe}}{6.022 \times 10^{23} \text{ atoms Fe}}\right)\left(\dfrac{55.85 \text{ g Fe}}{1 \text{ mol Fe}}\right) = 351$ g Fe

(b) 19.92 mol $H_2 \left(\dfrac{2.016 \text{ g } H_2}{1 \text{ mol } H_2}\right) = 40.16$ g H_2

(c) 8.576 mol C $\left(\dfrac{12.011 \text{ g C}}{1 \text{ mol C}}\right) = 103.0$ g C

(d) 7.40 mol Si $\left(\dfrac{28.085 \text{ g Si}}{1 \text{ mol Si}}\right) = 208$ g Si

(e) 9.221 mol Na $\left(\dfrac{22.990 \text{ g Na}}{1 \text{ mol Na}}\right) = 212.0$ g Na

(f) 4.07×10^{24} atoms Al $\left(\dfrac{1 \text{ mol Al}}{6.022 \times 10^{23} \text{ atoms Al}}\right)\left(\dfrac{26.982 \text{ g Al}}{1 \text{ mol Al}}\right) = 182$ g Al

(g) 9.2 mol $Cl_2 \left(\dfrac{70.90 \text{ g } Cl_2}{1 \text{ mol } Cl_2}\right) = 6.5 \times 10^2$ g Cl_2

In ascending order of mass: H₂, C, Al, Si, Na, Fe, Cl₂

2.122. Regarding the atomic weights of P and O:

0.744 g phosphorus combined with (1.704 g – 0.744 g) = 0.960 g O

Expressing the mass per atom, divide P by 4 and O by 10: $\dfrac{(0.744/4) \text{ g P}}{(0.960/10) \text{ g O}} = \dfrac{1.94 \text{ g P}}{1 \text{ g O}}$

telling us that a P atom is 1.94 times more massive than a O atom. Using this ratio, we can determine the mass of a P atom if we define O as 16.00:

$16.00 \text{ O}\left(\dfrac{1.94 \text{ g P}}{1 \text{ g O}}\right) = 31.0 \text{ P}$

2.123. Regarding the atomic weights of C and O:

0.876 g carbon combined with (3.210 g – 0.876 g) = 2.334 g O

Expressing the mass per atom, divide C by 1 and O by 2: $\frac{(2.334/2) \text{ g O}}{(0.876/1) \text{ g C}} = \frac{1.33 \text{ g O}}{1 \text{ g C}}$ telling us that an O atom is 1.33 times more massive than a C atom. Using this ratio, we can determine the mass of a O atom if we define C as 12.01: $12.01 \text{ C} \left(\frac{1.33 \text{ g O}}{1 \text{ g C}} \right) = 16.0 \text{ O}$

2.124. Percent of K in a Na-K alloy:

For every 100 atoms of the alloy, there are 68 atoms of K and 32 atoms of Na.

$68 \text{ atoms K} \left(\frac{1 \text{ mol K}}{6.022 \times 10^{23} \text{ atoms K}} \right) \left(\frac{39.1 \text{ g K}}{1 \text{ mol K}} \right) = 4.4 \times 10^{-21} \text{ g K}$

$32 \text{ atoms Na} \left(\frac{1 \text{ mol Na}}{6.02 \times 10^{23} \text{ atoms Na}} \right) \left(\frac{23.0 \text{ g Na}}{1 \text{ mol Na}} \right) = 1.2 \times 10^{-21} \text{ g Na}$

mass % K = $\frac{4.4 \times 10^{-21} \text{ g K}}{4.4 \times 10^{-21} \text{ g K} + 1.2 \times 10^{-21} \text{ g Na}} \times 100\% = 78\% \text{ K}$

2.125. Possible compounds from ions:

	CO_3^{2-}	PO_4^{3-}
NH_4^+	$(NH_4)_2CO_3$	$(NH_4)_3PO_4$
Ni^{2+}	$NiCO_3$	$Ni_3(PO_4)_2$

Compounds are electrically neutral—hence the total positive charge contributed by the cations (+ ion) must equal the total negative charge contributed by the anions (– ion). Since carbonate is di-negative and phosphate is tri-negative, two ammonium ions are required for carbonate and three for phosphate. For the nickel(II) ion one carbonate ion is needed while three Ni(II) ions and two phosphate ions are necessary for neutral compounds.

2.126. A strontium atom has 38 electrons (same as atomic number). When an atom of strontium forms an ion, it loses two electrons, forming an ion having the same number of electrons as the noble gas krypton (with 36 electrons).

2.127. Compound from the list with the highest mass percent of Cl:
One way to answer this question is to calculate the %Cl in each of the five compounds. An observation that each compound has the same number of Cl atoms provides a "non-calculator" approach to answering the question.

Since 3 Cl atoms will contribute the same TOTAL mass of Cl to the formula weights, the compound with the highest weight percent of Cl will also have the **lowest** weight percent of the other atom. Examining the atomic weights of the other atoms:

B	As	Ga	Al	P
10.81	74.92	69.72	26.98	30.97

B contributes the smallest mass of these five atoms, hence the smallest contribution to the molar masses of the five compounds—so BCl₃ has the highest weight percent of Cl.

2.128. Sample with largest number of ions:
These calculations require (1) converting mass to amount (moles) of compound, (2) converting amount of compound into amount of ions, and (3) converting amount of ions into numbers of ions. Each of the calculations is identical simply by substituting the appropriate molar mass of the compound and the number of ions formed per molecule. Note especially for (d) and (e), that the polyatomic ions remain intact—giving ONE ion PER polyatomic ion (e.g. CO_3^{2-} is 1 ion, not 4).

(a) $1.0 \text{ g CaCl}_2 \left(\dfrac{1 \text{ mol CaCl}_2}{111 \text{ g CaCl}_2} \right) \left(\dfrac{3 \text{ mol ions}}{1 \text{ mol CaCl}_2} \right) \left(\dfrac{6.02 \times 10^{23} \text{ ions}}{1 \text{ mol ions}} \right) = 1.6 \times 10^{22}$ ions

(b) $1.0 \text{ g MgCl}_2 \left(\dfrac{1 \text{ mol MgCl}_2}{95.2 \text{ g MgCl}_2} \right) \left(\dfrac{3 \text{ mol ions}}{1 \text{ mol MgCl}_2} \right) \left(\dfrac{6.02 \times 10^{23} \text{ ions}}{1 \text{ mol ions}} \right) = 1.9 \times 10^{22}$ ions

(c) $1.0 \text{ g CaS} \left(\dfrac{1 \text{ mol CaS}}{72.1 \text{ g CaS}} \right) \left(\dfrac{2 \text{ mol ions}}{1 \text{ mol CaS}} \right) \left(\dfrac{6.02 \times 10^{23} \text{ ions}}{1 \text{ mol ions}} \right) = 1.7 \times 10^{22}$ ions

(d) $1.0 \text{ g SrCO}_3 \left(\dfrac{1 \text{ mol SrCO}_3}{148 \text{ g SrCO}_3} \right) \left(\dfrac{2 \text{ mol ions}}{1 \text{ mol SrCO}_3} \right) \left(\dfrac{6.02 \times 10^{23} \text{ ions}}{1 \text{ mol ions}} \right) = 8.1 \times 10^{21}$ ions

(e) $1.0 \text{ g BaSO}_4 \left(\dfrac{1 \text{ mol BaSO}_4}{233 \text{ g BaSO}_4} \right) \left(\dfrac{2 \text{ mol ions}}{1 \text{ mol BaSO}_4} \right) \left(\dfrac{6.02 \times 10^{23} \text{ ions}}{1 \text{ mol ions}} \right) = 5.2 \times 10^{21}$ ions

MgCl₂ sample has the largest number of ions.

2.129. To determine the greater mass, let's first ask the question about the molar mass of adenine.
The formula for adenine is: $C_5H_5N_5$ with a molar mass of 135.13 g. The number of molecules requested is approximately 1/2 mole of adenine molecules. So, 1/2 mol of adenine molecules would have a mass of 1/2(135.13g) or 67.57 g. Thus, 3.0×10^{23} molecules of adenine has a greater mass than 40.0 g of adenine.

2.130. Regarding BaF₂, SiCl₄, and NiBr₂:

(a) BaF₂: barium fluoride   SiCl₄: silicon tetrachloride   NiBr₂: nickel(II) bromide

(b) BaF₂ and NiBr₂ are ionic; SiCl₄ is molecular

(c) $0.50 \text{ mol BaF}_2 \left(\dfrac{175 \text{ g BaF}_2}{1 \text{ mol BaF}_2} \right) = 88 \text{ g BaF}_2$

$0.50 \text{ mol SiCl}_4 \left(\dfrac{170. \text{ g SiCl}_4}{1 \text{ mol SiCl}_4} \right) = 85 \text{ g SiCl}_4$

$1.0 \text{ mol NiBr}_2 \left(\dfrac{219 \text{ g NiBr}_2}{1 \text{ mol NiBr}_2} \right) = 2.2 \times 10^2 \text{ g NiBr}_2$

1.0 mol NiBr₂ has the largest mass

2.131. A drop of water has a volume of 0.050 mL. Assuming the density of water is 1.00 g/cm³, the number of molecules of water may be calculated by first determining the mass of water present: $0.050 \text{ mL H}_2\text{O} \left(\dfrac{1 \text{ cm}^3}{1 \text{ mL}} \right) \left(\dfrac{1.00 \text{ g H}_2\text{O}}{1 \text{ cm}^3 \text{ H}_2\text{O}} \right) = 0.050 \text{ g H}_2\text{O}$

The molar mass of water is 18.02 g/mol.

$0.050 \text{ g H}_2\text{O} \left(\dfrac{1 \text{ mol H}_2\text{O}}{18.02 \text{ g H}_2\text{O}} \right) \left(\dfrac{6.02 \times 10^{23} \text{ molecules H}_2\text{O}}{1 \text{ mol H}_2\text{O}} \right) = 1.7 \times 10^{21} \text{ molecules H}_2\text{O}$

2.132. Regarding capsaicin:

(a) Molar mass = (18)(12.011) + (27)(1.008) + (1)(14.007) + (3)(15.999) = 305.418 g/mol

(b) $55 \text{ mg C}_{18}\text{H}_{27}\text{NO}_3 \left(\dfrac{1 \text{ g}}{10^3 \text{ mg}} \right) \left(\dfrac{1 \text{ mol C}_{18}\text{H}_{27}\text{NO}_3}{305.418 \text{ g C}_{18}\text{H}_{27}\text{NO}_3} \right) = 1.8 \times 10^{-4} \text{ mol C}_{18}\text{H}_{27}\text{NO}_3$

(c) $\dfrac{(18)(12.011) \text{ g C}}{305.418 \text{ g C}_{18}\text{H}_{27}\text{NO}_3} \cdot 100\% = 70.788\% \text{ C}$

$\dfrac{(27)(1.008) \text{ g H}}{305.418 \text{ g C}_{18}\text{H}_{27}\text{NO}_3} \cdot 100\% = 8.911\% \text{ H}$

$\dfrac{(1)(14.007) \text{ g N}}{305.418 \text{ g C}_{18}\text{H}_{27}\text{NO}_3} \cdot 100\% = 4.586\% \text{ N}$

$\dfrac{(3)(15.999) \text{ g O}}{305.418 \text{ g C}_{18}\text{H}_{27}\text{NO}_3} \cdot 100\% = 15.715\% \text{ O}$

(d) $55 \text{ mg C}_{18}\text{H}_{27}\text{NO}_3 \left(\dfrac{70.788 \text{ mg C}}{100 \text{ mg C}_{18}\text{H}_{27}\text{NO}_3} \right) = 39 \text{ mg C}$

2.133. Molar mass and mass percent of the elements in Cu(NH$_3$)$_4$SO$_4$·H$_2$O:

Molar Mass: (1)(Cu) + (4)(N) + 12(H) + (1)(S) + (4)(O) + (2)(H) + (1)(O).
Combining the hydrogens and oxygen from water with the compound:

(1)(Cu) + (4)(N) + 14(H) + (1)(S) + (5)(O) =
(1)(63.546) + (4)(14.007) + 14(1.008) + (1)(32.06) + (5)(15.999) = 245.74 g/mol

The mass percentages are:

Cu: (63.546/245.74) × 100 = 25.86% Cu

N: (56.028/245.74) × 100 = 22.80% N

H: (14.112/245.74) × 100 = 5.743% H

S: (32.06/245.74) × 100 = 13.05% S

O: (79.995/245.74) × 100 = 32.55% O

The mass of copper and of water in 10.5 g of the compound:

For Copper: $10.5 \text{ g compound} \left(\dfrac{25.86 \text{ g Cu}}{100 \text{ g compound}} \right) = 2.72 \text{ g Cu}$

For Water: $10.5 \text{ g compound} \left(\dfrac{18.02 \text{ g H}_2\text{O}}{245.74 \text{ g compound}} \right) = 0.770 \text{ g H}_2\text{O}$

2.134. Molecular formula and molar mass for the compounds:

(a) Ethylene glycol   C$_2$H$_6$O$_2$ Molar mass = 62.07 g/mol

$\dfrac{(2)(12.01) \text{ g C}}{62.07 \text{ g C}_2\text{H}_6\text{O}_2} \cdot 100\% = 38.70\% \text{ C}$ $\dfrac{(2)(16.00) \text{ g O}}{62.07 \text{ g C}_2\text{H}_6\text{O}_2} \cdot 100\% = 51.55\% \text{ O}$

(b) Dihydroxyacetone   C$_3$H$_6$O$_3$ Molar mass = 90.08 g/mol

$\dfrac{(3)(12.01) \text{ g C}}{90.08 \text{ g C}_3\text{H}_6\text{O}_3} \cdot 100\% = 40.00\% \text{ C}$ $\dfrac{(3)(16.00) \text{ g O}}{90.08 \text{ g C}_3\text{H}_6\text{O}_3} \cdot 100\% = 53.29\% \text{ O}$

(c) Ascorbic acid   C$_6$H$_8$O$_6$ Molar mass = 176.12 g/mol

$\dfrac{(6)(12.01) \text{ g C}}{176.12 \text{ g C}_6\text{H}_8\text{O}_6} \cdot 100\% = 40.92\% \text{ C}$ $\dfrac{(6)(16.00) \text{ g C}}{176.12 \text{ g C}_6\text{H}_8\text{O}_6} \cdot 100\% = 54.51\% \text{ O}$

Ascorbic acid has the largest percentage of carbon **and** of oxygen.

2.135. The empirical formula of malic acid, if the ratio is: C$_1$H$_{1.50}$O$_{1.25}$

Since we prefer all subscripts to be integers, we ask what "multiplier" we can use to convert each of these subscripts to integers **while** retaining the given ratio of C:H:O. Multiplying each subscript by 4 (we need to convert the 0.25 to an integer) gives a ratio of C$_4$H$_6$O$_5$.

2.136. Substance delivering greater number of Fe atoms:

$$100. \text{ mg FeSO}_4 \left(\frac{1 \text{ g}}{1\times10^3 \text{ mg}}\right)\left(\frac{1 \text{ mol FeSO}_4}{151.90 \text{ g FeSO}_4}\right)\left(\frac{6.022\times10^{23} \text{ atoms Fe}}{1 \text{ mol FeSO}_4}\right) = 3.96\times10^{20} \text{ atoms Fe}$$

$$100. \text{ mg Fe}(C_6H_{11}O_7)_2 \left(\frac{1 \text{ g}}{1\times10^3 \text{ mg}}\right)\left(\frac{1 \text{ mol Fe}(C_6H_{11}O_7)_2}{446.14 \text{ g Fe}(C_6H_{11}O_7)_2}\right)\left(\frac{6.022\times10^{23} \text{ atoms Fe}}{1 \text{ mol Fe}(C_6H_{11}O_7)_2}\right) = 1.35\times10^{20} \text{ atoms Fe}$$

The tablet containing FeSO$_4$ will deliver more atoms of iron.

2.137. A compound Fe$_x$(CO)$_y$ is 30.70% Fe: This implies that the balance of the mass (69.30%) is attributable to the CO molecules. One approach is to envision CO as **one** unit, with a molar mass of (12.011 + 15.999) or 28.010 g. Assuming we have 100 grams, the ratios of masses are then:

$$30.70 \text{ g Fe}\left(\frac{1 \text{ mol Fe}}{55.845 \text{ g Fe}}\right) = 0.5497 \text{ mol Fe}$$

$$69.30 \text{ g CO}\left(\frac{1 \text{ mol CO}}{28.010 \text{ g CO}}\right) = 2.474 \text{ mol CO}$$

Dividing by the smaller amount gives

$$\frac{2.474 \text{ mol CO}}{0.5497 \text{ mol Fe}} = \frac{4.501 \text{ mol CO}}{1 \text{ mol Fe}} = \frac{4\frac{1}{2} \text{ mol CO}}{1 \text{ mol Fe}} = \frac{9 \text{ mol CO}}{2 \text{ mol Fe}}$$

The empirical formula is Fe$_2$(CO)$_9$.

2.138. Regarding ephedrine:

(a) C$_{10}$H$_{15}$NO   Molar mass = 165.236 g/mol

(b) $\dfrac{(10)(12.011) \text{ g C}}{165.236 \text{ g C}_{10}\text{H}_{15}\text{NO}} \cdot 100\% = 72.690\%$ C

(c) $0.125 \text{ g C}_{10}\text{H}_{15}\text{NO}\left(\dfrac{1 \text{ mol C}_{10}\text{H}_{15}\text{NO}}{165.236 \text{ g C}_{10}\text{H}_{15}\text{NO}}\right) = 7.56 \times 10^{-4} \text{ mol C}_{10}\text{H}_{15}\text{NO}$

(d)

$$7.56 \times 10^{-4} \text{ mol C}_{10}\text{H}_{15}\text{NO}\left(\dfrac{6.022 \times 10^{23} \text{ molecules C}_{10}\text{H}_{15}\text{NO}}{1 \text{ mol C}_{10}\text{H}_{15}\text{NO}}\right) = 4.56 \times 10^{20} \text{ molecules C}_{10}\text{H}_{15}\text{NO}$$

$$4.56 \times 10^{20} \text{ molecules C}_{10}\text{H}_{15}\text{NO}\left(\dfrac{10 \text{ C atoms}}{1 \text{ molecule C}_{10}\text{H}_{15}\text{NO}}\right) = 4.56 \times 10^{21} \text{ C atoms}$$

2.139. For the molecule saccharin, structural formula shown:

(a) The formula is C₇H₅NO₃S

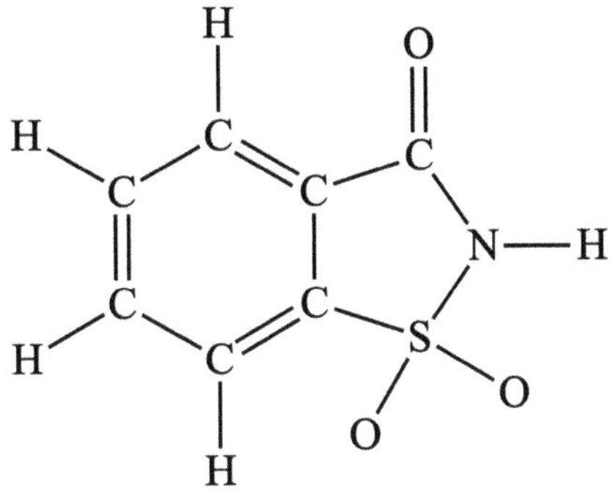

saccharin

(b) Mol of saccharin associated with 125 mg:

$$125 \text{ mg saccharin}\left(\frac{1 \text{ g}}{1000 \text{ mg}}\right)\left(\frac{1 \text{ mol saccharin}}{183.18 \text{ g saccharin}}\right) = 6.82 \times 10^{-4} \text{ mol saccharin}$$

(c) Mass of S in 125 mg saccharin:

$$125 \text{ mg saccharin}\left(\frac{32.06 \text{ mg S}}{138.18 \text{ mg saccharin}}\right) = 21.9 \text{ mg S}$$

2.140. Name the compounds and indicate which are ionic:

Ionic compounds are in **bold** font.

(a) chlorine trifluoride
(b) nitrogen trichloride
(c) **strontium sulfate**
(d) **calcium nitrate**
(e) xenon tetrafluoride
(f) oxygen difluoride
(g) **potassium iodide**
(h) **aluminum sulfide**
(i) phosphorus trichloride
(j) **potassium phosphate**

2.141. Formulas for compounds; identify the ionic compounds

Ionic compounds are in **bold** font.

(a) sodium hypochlorite   **NaClO**
(b) boron triiodide   BI₃
(c) aluminum perchlorate   **Al(ClO₄)₃**
(d) calcium acetate   **Ca(CH₃CO₂)₂**
(e) potassium permanganate   **KMnO₄**

(f) ammonium sulfite   **(NH₄)₂SO₃**

(g) potassium dihydrogen phosphate   **KH₂PO₄**

(h) disulfur dichloride   S₂Cl₂

(i) chlorine trifluoride   ClF₃

(j) phosphorus trifluoride   PF₃

The ionic compounds are identified by looking for a *metal*.

2.142. Supply the indicated information:

Cation	Anion	Name	Formula
NH_4^+	Br^-	ammonium bromide	NH_4Br
Ba^{2+}	S^{2-}	barium sulfide	BaS
Fe^{2+}	Cl^-	iron(II) chloride	$FeCl_2$
Pb^{2+}	F^-	lead(II) fluoride	PbF_2
Al^{3+}	CO_3^{2-}	aluminum carbonate	$Al_2(CO_3)_3$
Fe^{3+}	O^{2-}	iron(III) oxide	Fe_2O_3

2.143. Empirical and molecular formulas:
(a) For fluorocarbonyl hypofluorite:

In a 100.00 g sample there are:

$14.6 \text{ g C} \left(\dfrac{1 \text{ mol C}}{12.011 \text{ g C}} \right) = 1.22 \text{ mol C}$

$39.0 \text{ g O} \left(\dfrac{1 \text{ mol O}}{15.999 \text{ g O}} \right) = 2.44 \text{ mol O}$

$46.3 \text{ g F} \left(\dfrac{1 \text{ mol F}}{18.998 \text{ g F}} \right) = 2.44 \text{ mol F}$

Dividing all three terms by 1.22 gives a ratio of O:C of 2:1. Likewise F:C is 2:1
The empirical formula would be CF₂O₂ with an "empirical mass" of 82.0 g/mol Since the molar mass is also 82.0 g/mol, the molecular formula is also CF₂O₂.

(b) For azulene:

Given the information that azulene is a hydrocarbon, if it is 93.71 % C, it is also (100.00 – 93.71) or 6.29 % H.

In a 100.00 g sample of azulene there are

$$93.71 \text{ g C} \left(\frac{1 \text{ mol C}}{12.011 \text{ g C}} \right) = 7.802 \text{ mol C}$$

$$6.29 \text{ g H} \left(\frac{1 \text{ mol H}}{1.008 \text{ g H}} \right) = 6.24 \text{ mol H}$$

The ratio of C to H atoms is: 1.25 mol C : 1 mol H or a ratio of 5 mol C:4 mol H (C$_5$H$_4$). The mass of such an empirical formula is ~64. Given that the molar mass is ~128 g/mol, the molecular formula for azulene is C$_{10}$H$_8$.

2.144. Determine the empirical formula:

Assume 100.00 g of compound. Calculate moles of C, H, and As:

$$22.88 \text{ g C} \left(\frac{1 \text{ mol C}}{12.011 \text{ g C}} \right) = 1.905 \text{ mol C} \qquad 5.76 \text{ g H} \left(\frac{1 \text{ mol H}}{1.008 \text{ g H}} \right) = 5.71 \text{ mol H}$$

$$71.36 \text{ g As} \left(\frac{1 \text{ mol As}}{74.922 \text{ g As}} \right) = 0.9525 \text{ mol As}$$

Determine the mole ratios: $\dfrac{1.905 \text{ mol C}}{0.9525 \text{ mol As}} = \dfrac{2.000 \text{ mol C}}{1 \text{ mol As}}$ and

$\dfrac{5.71 \text{ mol H}}{0.9525 \text{ mol As}} = \dfrac{6.00 \text{ mol H}}{1 \text{ mol As}}$

The empirical formula is C$_2$H$_6$As. $\dfrac{210 \text{ g/mol}}{105.0 \text{ g/mol}} = 2.0$

With a molar mass of 210 g/mol, the molecular formula is C$_4$H$_{12}$As$_2$

2.145. Molecular formula of cadaverine:

Calculate the amount of each element in the compound (assuming you have 100 g)

$$58.77 \text{ g C} \left(\frac{1 \text{ mol C}}{12.011 \text{ g C}} \right) = 4.893 \text{ mol C}$$

$$13.81 \text{ g H} \left(\frac{1 \text{ mol H}}{1.008 \text{ g H}} \right) = 13.70 \text{ mol H}$$

$$27.40 \text{ g N} \left(\frac{1 \text{ mol N}}{14.007 \text{ g N}} \right) = 1.956 \text{ mol N}$$

The ratio of C:H:N can be found by dividing each by the smallest amount (1.956): to give C$_{2.50}$H$_7$N$_1$ and converting each subscript to an integer (multiplying by 2) C$_5$H$_{14}$N$_2$. The molar mass of this empirical formula would be approximately 102, hence the molecular formula is also C$_5$H$_{14}$N$_2$.

2.146. For Ni(CO)$_x$, determine the value of *x*:

The mass of CO is: 0.364 g Ni(CO)$_x$ – 0.125 g Ni = 0.239 g CO

Determine moles of each:

$0.239 \text{ g CO} \left(\dfrac{1 \text{ mol CO}}{28.01 \text{ g CO}} \right) = 0.00853 \text{ mol CO}$; $0.125 \text{ g Ni} \left(\dfrac{1 \text{ mol Ni}}{58.69 \text{ g Ni}} \right) = 0.00213 \text{ mol Ni}$

The ratio of these is: $\dfrac{0.00853 \text{ mol CO}}{0.00213 \text{ mol Ni}} = \dfrac{4.01 \text{ mol CO}}{1 \text{ mol Ni}}$. The compound formula is Ni(CO)$_4$.

2.147. The empirical formula for MMT:

Moles of each atom present in 100. g of MMT:

$49.5 \text{ g C} \left(\dfrac{1 \text{ mol C}}{12.011 \text{ g C}} \right) = 4.12 \text{ mol C}$

$3.2 \text{ g H} \left(\dfrac{1 \text{ mol H}}{1.008 \text{ g H}} \right) = 3.2 \text{ mol H}$

$22.0 \text{ g O} \left(\dfrac{1 \text{ mol O}}{15.999 \text{ g O}} \right) = 1.38 \text{ mol O}$

$25.2 \text{ g Mn} \left(\dfrac{1 \text{ mol Mn}}{54.938 \text{ g Mn}} \right) = 0.459 \text{ mol Mn}$

The ratio of C:H:Mn:O can be found by dividing each by the smallest amount (0.459): to give C$_9$H$_7$MnO$_3$.

2.148. Mass % of P; Mass of calcium phosphate to produce 15.0 kg P:

Molar mass of Ca$_3$(PO$_4$)$_2$ = 310.17 g/mol

Mass % of P in Ca$_3$(PO$_4$)$_2$: $\dfrac{2(30.97) \text{ g P}}{310.17 \text{ g Ca}_3(\text{PO}_4)_2} \times 100\% = 19.97\% \text{ P}$

With this percentage, we calculate the amount of the phosphate to produce the requested amount of P: $15.0 \text{ kg P} \cdot \dfrac{100 \text{ kg Ca}_3(\text{PO}_4)_2}{19.97 \text{ kg P}} = 75.1 \text{ kg Ca}_3(\text{PO}_4)_2$

2.149. Chromium(III) oxide has the formula Cr$_2$O$_3$.

The weight percent of Cr in Cr$_2$O$_3$ is: $\dfrac{(2 \cdot 52.00)}{(2 \cdot 52.00) + (3 \cdot 16.00)} \cdot 100\% = 68.42\% \text{ Cr}$.

[The numerator is the sum of the mass of 2 atoms of Cr, while the denominator is the sum of the mass of 2 atoms of Cr and 3 atoms of O.]

The weight of Cr$_2$O$_3$ necessary to produce 850 kg Cr:

$$850 \text{ kg Cr}\left(\frac{100 \text{ kg Cr}_2\text{O}_3}{68.42 \text{ kg Cr}}\right) = 1200 \text{ kg Cr}_2\text{O}_3 \text{ (2 sf)}$$

The second fraction represents the %Cr in the oxide.

2.150. Mass percent of Sb in the sulfide; Mass of stibnite in 1.00 kg of ore:

Percent of Sb: $\dfrac{(2)(121.76) \text{ g Sb}}{339.7 \text{ g Sb}_2\text{S}_3} \times 100\% = 71.69\%$ Sb

Mass of Stibnite in ore: $1.00 \text{ kg ore}\left(\dfrac{1000 \text{ g}}{1 \text{ kg}}\right)\left(\dfrac{10.6 \text{ g Sb}}{100 \text{ g ore}}\right)\left(\dfrac{100 \text{ g Sb}_2\text{S}_3}{71.69 \text{ g Sb}}\right) = 148 \text{ g Sb}_2\text{S}_3$

2.151. Empirical and Molecular formula for I$_x$Cl$_y$:

I$_2$	+	Cl$_2$	→	I$_x$Cl$_y$
0.678 g		(1.246 – 0.678)		1.246 g
		= 0.568 g		

Calculate the ratio of I:Cl atoms

$0.678 \text{ g I} \cdot \dfrac{1 \text{ mol I}}{126.9 \text{ g I}} = 5.34 \times 10^{-3}$ mol I atoms

$0.568 \text{ g Cl} \cdot \dfrac{1 \text{ mol Cl}}{35.45 \text{ g Cl}} = 1.60 \times 10^{-2}$ mol Cl atoms

The ratio of Cl:I is: $\dfrac{1.60 \times 10^{-2} \text{ mol Cl atoms}}{5.34 \times 10^{-3} \text{ mol I atoms}} = \dfrac{3.00 \text{ mol Cl atoms}}{1 \text{ mol I atoms}}$

The empirical formula is ICl$_3$ (FW = 233.3)
Given that the molar mass of I$_x$Cl$_y$ was 467 g/mol, we can calculate the number of empirical formulas per mole:

$\dfrac{467 \text{ g/mol}}{233.3 \text{ g/empirical formula}} = 2 \dfrac{\text{empirical formulas}}{\text{mol}}$ for a molecular formula of I$_2$Cl$_6$.

2.152. Empirical formula of the vanadium and sulfur compound:

Calculate the moles of each element:

$2.04 \text{ g V}\left(\dfrac{1 \text{ mol V}}{50.94 \text{ g V}}\right) = 0.0400$ mol V

$$1.93 \text{ g S} \left(\frac{1 \text{ mol S}}{32.06 \text{ g S}} \right) = 0.0602 \text{ mol S}$$

Now we can calculate the ratios of V:S

$$\frac{0.0602 \text{ mol S}}{0.0400 \text{ mol V}} = \frac{1.50 \text{ mol S}}{1 \text{ mol V}} = \frac{3 \text{ mol S}}{2 \text{ mol V}}$$

The empirical formula is V_2S_3.

2.153. Mass of Fe in 15.8 kg of FeS_2:

$$\% \text{ Fe in FeS}_2 = \frac{55.85 \text{ g Fe}}{119.97 \text{ g FeS}_2} \times 100\% = 46.55 \% \text{ Fe}$$

and in 15.8 kg FeS_2: $15.8 \text{ kg FeS}_2 \cdot \frac{46.55 \text{ kg Fe}}{100.00 \text{ kg FeS}_2} = 7.36 \text{ kg Fe}$

2.154. Statements about octane that are not true:

(a) True. $0.500 \text{ mol C}_8H_{18} \left(\frac{114.2 \text{ g C}_8H_{18}}{1 \text{ mol C}_8H_{18}} \right) = 57.1 \text{ g C}_8H_{18}$

(b) True. $\frac{(8)(12.01) \text{ g C}}{114.2 \text{ g C}_8H_{18}} \cdot 100\% = 84.1\% \text{ C}$

(c) True. Dividing the molecular formula by 2 gives the simplest integral ratio of C and H.

(d) False. $57.1 \text{ g C}_8H_{18} \cdot \frac{(18)(1.008) \text{ g H}}{114.2 \text{ g C}_8H_{18}} = 9.07 \text{ g H}$

2.155. The formula of barium molybdate is $BaMoO_4$. What is the formula for sodium molybdate? This question is answered by observing that the compound indicates ONE barium ion. Since the barium ion has a 2+ charge, 2 Na^+ cations would be needed, making the formula for sodium molybdate Na_2MoO_4 or choice (d).

2.156. Identity of the metal that forms MCl_4: The compound is 74.75% Cl, and we know that this mass represents 4 Cl atoms, so we determine the molar mass of the compound:

$$\frac{74.75 \text{ g Cl}}{100 \text{ g MCl}_4} = \frac{(4)(35.45) \text{ g Cl}}{\text{molar mass MCl}_4} ; \text{ Solving for molar mass: } 189.7 \text{ g/mol.}$$

Determine the mass of the metal by difference: $189.7 \text{ g MCl}_4 - (4)(35.45) \text{ g Cl} = 47.9 \text{ g}$.

M is Ti, titanium.

2.157. Mass of Bi in two tablets of Pepto-Bismol™ ($C_{21}H_{15}Bi_3O_{12}$):

Moles of the active ingredient:

$$2 \text{ tablets}\left(\frac{262 \text{ mg } C_{21}H_{15}Bi_3O_{12}}{1 \text{ tablet}}\right)\left(\frac{1 \text{ g}}{1000 \text{ mg}}\right)\left(\frac{1 \text{ mol } C_{21}H_{15}Bi_3O_{12}}{1086.3 \text{ } C_{21}H_{15}Bi_3O_{12}}\right) = 4.82 \times 10^{-4} \text{ mol } C_{21}H_{15}Bi_3O_{12}$$

Mass of Bi:

$$4.82 \times 10^{-4} \text{ mol } C_{21}H_{15}Bi_3O_{12}\left(\frac{3 \text{ mol Bi}}{1 \text{ mol } C_{21}H_{15}Bi_3O_{12}}\right)\left(\frac{208.98 \text{ g Bi}}{1 \text{ mol Bi}}\right) = 0.302 \text{ g Bi}$$

2.158. Molar mass of MO_2; Possible elements for M:

We know that the compound is 15.2% O, and the molecule has 2 O atoms. Calculate the molar mass: $\frac{15.2 \text{ g O}}{100 \text{ g } MO_2} = \frac{(2)(16.00) \text{ g O}}{\text{molar mass } MO_2}$. The molar mass for MO_2 = 211 g/mol.

Determine the mass of the metal by difference: 211 g MO_2 – (2)(16.00) g O = 179 g. An examination of the periodic table lets us know that M is Hf, hafnium.

2.159. What is the molar mass of ECl_4 and the identity of E?

2.50 mol of ECl_4 has a mass of 385 grams. The molar mass of ECl_4 would be:
$\frac{385 \text{ g } ECl_4}{2.50 \text{ mol } ECl_4} = 154$ g/mol ECl_4.

Since the molar mass is 154, and we know that there are 4 chlorine atoms per mole of the compound, we can subtract the mass of 4 chlorine atoms to determine the mass of E. 154 – 4(35.5) = 12. The element with an atomic mass of 12 g/mol is **carbon**.

2.160. Atomic weights of A and Z:

First, calculate the molar masses of A_2Z_3 and AZ_2 using the mass (g) and amount (moles) of each:

$\frac{22.48 \text{ g } A_2Z_3}{0.15 \text{ mol } A_2Z_3} = 150$ g/mol A_2Z_3

$\frac{12.44 \text{ g } AZ_2}{0.15 \text{ mol } AZ_2} = 83$ g/mol AZ_2

For AZ_2: (atomic weight A) + (2)(atomic weight Z) = 83; and

For A_2Z_3: (2)(atomic weight A) + (3)(atomic weight Z) = 150

Rearrange the AZ_2 equation to solve for (atomic weight A):
atomic weight A = 83 – 2(atomic weight Z)

Substitute this into the equation for A_2Z_3: (2)[83 – (2)(atomic weight Z)] + (3)(atomic weight Z) = 150.

2.161. For what value of *n*, will Br compose 0.105% of the mass of the polymer, Br₃C₆H₃(C₈H₈)ₙ?

Knowing that the 3Br atoms comprises 0.105% of the formula weight of the polymer, we can write the fraction:

$$\frac{3 \cdot Br}{\text{formula weight}} = 0.00105$$ (where 3·Br) is 3 times the atomic weight of Br. Substituting we get $\frac{239.7}{\text{formula weight}} = 0.00105$ or $\frac{239.7}{0.00105} = $ formula weight $= 2.283 \times 10^5$ g.

Noting that the "fixed" part of the formula contains 3 Br atoms, 6 C atoms, and 3 H atoms, we can calculate the mass associated with this part of the molecule (3(79.904) + 6(12.011) + 3(1.008) = 314.802). Subtracting this mass from the total (2.283 × 10⁵) gives 2.280 × 10⁵ g as the mass corresponding to the "C₈H₈" units. Since each such unit has a mass of 104.15 (8C + 8H), we can divide the mass of *one* C₈H₈ unit into the 2.280 × 10⁵ mass remaining:

$$\frac{2.280 \times 10^5 \text{g}}{104.15 \text{g/C}_8\text{H}_8} = 2189 \text{ C}_8\text{H}_8 \text{ units to give a value for } n \text{ of } 2.19 \times 10^3 \text{ (3 sf)}.$$

2.162. The molar mass of hemoglobin represents 100% of the mass of the molecule, while Fe represents 0.335% of the molecule. Set up a ratio of mass of Fe: mass of hemoglobin, which parallels this percentage of Fe in the molecule:

$$\frac{(4)(55.85) \text{ g Fe}}{\text{molar mass hemoglobin}} = \frac{0.335 \text{ g Fe}}{100 \text{ g hemoglobin}}$$ and solving for (molar mass hemoglobin) gives 6.67×10^4 g/mol

2.163. For the ⁶⁴Zn atom:
(a) Calculate the density of the nucleus:

Assume the nucleus of the Zn atom is a sphere (whose volume is 4/3 πr^3.) Given the desired units, express the radius in units of cm and calculate nuclear volume:

$$4.8 \times 10^{-6} \text{ nm} \left(\frac{100 \text{ cm}}{1 \times 10^9 \text{ nm}}\right) = 4.8 \times 10^{-13} \text{ cm}$$

$$V \text{ of nucleus} = \frac{4}{3}\pi(4.8 \times 10^{-13} \text{ cm})^3 = 4.6 \times 10^{-37} \text{ cm}^3$$

Then determine nuclear density

$$\text{density of nucleus} = \frac{1.06 \times 10^{-22} \text{ g}}{4.6 \times 10^{-37} \text{ cm}^3} = 2.3 \times 10^{14} \text{ g/cm}^3$$

(b) The density of the space occupied by the electrons: Express the radius of the atom in cm, as in part (a): 0.125 nm = 1.25 × 10⁻⁸ cm

Calculate the volume of that sphere: $V = \frac{4}{3}\pi(1.25 \times 10^{-8} \text{ cm})^3 = 8.18 \times 10^{-24} \text{ cm}^3$

Mass of 30 electrons (the atomic number of zinc is 30): 30 electrons × 9.11 × 10⁻²⁸ g = 2.73 × 10⁻²⁶ g Density of space occupied by electrons =

$\dfrac{2.73 \times 10^{-26} \text{ g}}{8.18 \times 10^{-24} \text{ cm}^3} = 3.34 \times 10^{-3} \text{ g/cm}^3$

(c) As we've learned, the mass of the atom is concentrated in the nucleus—borne out by these densities.

2.164. Estimate the radius of a Pb atom:

(a) Volume of cube = (1.000 cm)³ = 1.000 cm³

$1.000 \text{ cm}^3 \text{ Pb} \left(\dfrac{11.35 \text{ g Pb}}{1 \text{ cm}^3}\right)\left(\dfrac{1 \text{ mol Pb}}{207.2 \text{ g Pb}}\right)\left(\dfrac{6.022 \times 10^{23} \text{ atoms Pb}}{1 \text{ mol Pb}}\right) = 3.299 \times 10^{22} \text{ atoms Pb}$

(b) Volume of one lead atom = $\dfrac{0.60 \times 1.000 \text{ cm}^3}{3.299 \times 10^{22} \text{ atoms Pb}} = 1.8 \times 10^{-23} \text{ cm}^3$

1.8 × 10⁻²³ cm³ = (⁴/₃)(π)(Pb radius)³ and solving, Pb radius = 1.6 × 10⁻⁸ cm

2.165. Calculate:

(a) moles of nickel—found by density **once** the volume of foil is calculated.

0.550 mm = 0.0550 cm
$V = 1.25 \text{ cm} \times 1.25 \text{ cm} \times 0.0550 \text{ cm} = 8.59 \times 10^{-2} \text{ cm}^3$

Mass = $8.59 \times 10^{-2} \text{ cm}^3 \text{ Ni} \left(\dfrac{8.902 \text{ g Ni}}{1 \text{ cm}^3 \text{ Ni}}\right) = 0.765 \text{ g Ni}$

$0.765 \text{ g Ni} \cdot \dfrac{1 \text{ mol Ni}}{58.69 \text{ g Ni}} = 1.30 \times 10^{-2} \text{ mol Ni}$

(b) Formula for the fluoride salt:

Mass F = (1.261 g salt − 0.765 g Ni) = 0.496 g F

Moles F = $0.496 \text{ g F} \cdot \dfrac{1 \text{ mol F}}{19.00 \text{ g F}} = 2.61 \times 10^{-2}$ mol F, so 1.30 × 10⁻² mol Ni combines with 2.61 × 10⁻² mol F, indicating a formula of NiF₂

(c) Name: nickel(II) fluoride

2.166. Concerning uranium:

(a) 0.199 g U$_x$O$_y$ – 0.169 g U = 0.030 g O

$$0.169 \text{ g U} \left(\frac{1 \text{ mol U}}{238.03 \text{ g U}} \right) = 7.10 \times 10^{-4} \text{ mol U}$$

$$0.030 \text{ g O} \left(\frac{1 \text{ mol O}}{16.0 \text{ g O}} \right) = 1.9 \times 10^{-3} \text{ mol O}$$

$$\frac{1.9 \times 10^{-3} \text{ mol O}}{7.10 \times 10^{-4} \text{ mol U}} = \frac{2.64 \text{ mol O}}{1 \text{ mol U}} \approx \frac{2\frac{2}{3} \text{ mol O}}{1 \text{ mol U}} = \frac{8 \text{ mol O}}{3 \text{ mol U}}$$

so the empirical formula is U$_3$O$_8$.

$$7.10 \times 10^{-4} \text{ mol U} \left(\frac{1 \text{ mol U}_3\text{O}_8}{3 \text{ mol U}} \right) = 2.37 \times 10^{-4} \text{ mol U}_3\text{O}_8$$

(b) The atomic weight of U is 238.03 u, implying that the isotope ^{238}U is the most abundant.

(c) 0.865 g – 0.679 g = 0.186 g H$_2$O lost upon heating

$$0.186 \text{ g H}_2\text{O} \left(\frac{1 \text{ mol H}_2\text{O}}{18.02 \text{ g H}_2\text{O}} \right) = 0.0103 \text{ mol H}_2\text{O}$$

$$0.679 \text{ g UO}_2(\text{NO}_3)_2 \left(\frac{1 \text{ mol UO}_2(\text{NO}_3)_2}{394.0 \text{ g UO}_2(\text{NO}_3)_2} \right) = 0.00172 \text{ mol UO}_2(\text{NO}_3)_2$$

$$\frac{0.0103 \text{ mol H}_2\text{O}}{0.00172 \text{ mol UO}_2(\text{NO}_3)_2} = \frac{5.99 \text{ mol H}_2\text{O}}{1 \text{ mol UO}_2(\text{NO}_3)_2}$$

The formula of the hydrated compound is UO$_2$(NO$_3$)$_2$ · 6 H$_2$O.

2.167. The volume of a cube of Na containing 0.125 mol Na:

First, we need to know the **mass** of 0.125 mol Na:

$$0.125 \text{ mol Na} \cdot \frac{22.99 \text{ g Na}}{1 \text{ mol Na}} = 2.87 \text{ g Na} \quad (3 \text{ sf})$$

Now we can calculate the volume that contains 2.87 g Na: (using the density given)

$$2.87 \text{ g Na} \cdot \frac{1 \text{ cm}^3}{0.97 \text{ g Na}} = 3.0 \text{ cm}^3$$

To determine the length of one edge of a perfect cube we take the cube root of the volume.

$$\sqrt[3]{3.0 \text{ cm}^3} = 1.4 \text{ cm}$$

2.168. Calculate the atomic weight:

Convert Abundances to percentages first, for example 0.185% = 0.00185.

Determine the contribution of each isotope by multiplying the abundance × isotopic mass:

(0.00185)(135.9071) + (0.00251)(137.9060) + (0.8845)(139.9054) + (0.1111)(141.9093) = 140.1 which is the atomic weight of the element cerium, Ce.

In the Laboratory

2.169. Molecules of water per formula unit of $MgSO_4$:

From 1.687 g of the hydrate, only 0.824 g of the magnesium sulfate remain.

The mass of water contained in the solid is: (1.687 – 0.824) or 0.863 grams
Use the molar masses of the solid and water to calculate the number of moles of each substance present:

$$0.824 \text{ g MgSO}_4 \left(\frac{1 \text{ mol MgSO}_4}{120.36 \text{ g MgSO}_4} \right) = 6.85 \times 10^{-3} \text{ mol MgSO}_4$$

$$0.863 \text{ g H}_2\text{O} \left(\frac{1 \text{ mol H}_2\text{O}}{18.02 \text{ g H}_2\text{O}} \right) = 4.79 \times 10^{-2} \text{ mol H}_2\text{O}$$

The ratio of H_2O to $MgSO_4$ is: $\dfrac{4.79 \times 10^{-2} \text{ mol H}_2\text{O}}{6.85 \times 10^{-3} \text{ mol MgSO}_4} = \dfrac{7.00 \text{ mol H}_2\text{O}}{1 \text{ mol MgSO}_4}$

So we write the formula $MgSO_4 \cdot 7\ H_2O$.

2.170. Number of waters of hydration in alum:

Moles of anhydrous alum: 4.74 g hydrated compound – 2.16 g H_2O = 2.58 g $KAl(SO_4)_2$

$$2.58 \text{ g KAl(SO}_4)_2 \left(\frac{1 \text{ mol KAl(SO}_4)_2}{258.2 \text{ g KAl(SO}_4)_2} \right) = 0.00999 \text{ mol KAl(SO}_4)_2$$

Moles of water: $2.16 \text{ g H}_2\text{O} \left(\dfrac{1 \text{ mol H}_2\text{O}}{18.02 \text{ g H}_2\text{O}} \right) = 0.120 \text{ mol H}_2\text{O}$. Now we determine the

number of moles of water per mol of alum: $\dfrac{0.120 \text{ mol H}_2\text{O}}{0.00999 \text{ mol KAl(SO}_4)_2} = \dfrac{12.0 \text{ mol H}_2\text{O}}{1 \text{ mol KAl(SO}_4)_2}$

There are 12 water molecules per formula unit of $KAl(SO_4)_2$; $x = 12$

2.171. We can calculate the mass of Sn used in the compound by the **difference** of masses in the original mixture and the mass recovered after reaction: = 1.056 g – 0.601 g = 0.455 g Sn

The # mol of Sn is: $0.455 \text{ g Sn} \left(\dfrac{1 \text{ mol Sn}}{118.71 \text{ g Sn}} \right) = 0.00383 \text{ mol Sn}$

Similarly for iodine, the mass of iodine can be converted into moles of **atomic iodine.**

$$1.947 \text{ g I} \left(\frac{1 \text{ mol I}}{126.90 \text{ g I}} \right) = 0.01534 \text{ mol I}$$

The empirical formula is determined by calculating the RATIO of **iodine** to **tin**.

$$\frac{0.01534 \text{ mol I}}{0.00383 \text{ mol Sn}} = \frac{4.00 \text{ mol I}}{\text{mol Sn}}$$

The empirical formula for the compound: Sn₁I₄, or more commonly expressed SnI₄.

2.172. Determine the empirical formula:

Assume 100.0 g of sample.

$$54.0 \text{ g C} \left(\frac{1 \text{ mol C}}{12.01 \text{ g}} \right) = 4.50 \text{ mol C} \qquad 6.00 \text{ g H} \left(\frac{1 \text{ mol H}}{1.008 \text{ g H}} \right) = 5.95 \text{ mol H}$$

$$40.0 \text{ g O} \left(\frac{1 \text{ mol O}}{16.00 \text{ g O}} \right) = 2.50 \text{ mol O}$$

Determine the mol ratio for C:O $\quad \dfrac{4.50 \text{ mol C}}{2.50 \text{ mol O}} = \dfrac{1.80 \text{ mol C}}{1 \text{ mol O}} = \dfrac{1\frac{4}{5} \text{ mol C}}{1 \text{ mol O}} = \dfrac{9 \text{ mol C}}{5 \text{ mol O}}$

Determine the mol ratio for H:O $\quad \dfrac{5.95 \text{ mol H}}{2.50 \text{ mol O}} = \dfrac{2.38 \text{ mol H}}{1 \text{ mol O}} \approx \dfrac{2\frac{2}{5} \text{ mol H}}{1 \text{ mol O}} = \dfrac{12 \text{ mol H}}{5 \text{ mol O}}$

Answer (d) C₉H₁₂O₅ is correct. Several reasons exist for an incorrect empirical formula. The students apparently did not correctly calculate the number of moles of material in 100.0 g or they improperly calculated the ratio of those moles in determining their empirical formula.

2.173. Using the student data, calculate the number of moles of CaCl₂ and moles of H₂O:

$$0.739 \text{ g CaCl}_2 \cdot \frac{1 \text{ mol CaCl}_2}{111.0 \text{ g CaCl}_2} = 0.00666 \text{ mol CaCl}_2$$

$$(0.832 \text{ g} - 0.739 \text{ g}) \text{ or } 0.093 \text{ g H}_2\text{O} \cdot \frac{1 \text{ mol H}_2\text{O}}{18.02 \text{ g H}_2\text{O}} = 0.0052 \text{ mol H}_2\text{O}$$

The number of moles of water/mol of calcium chloride is $\dfrac{0.0052 \text{ mol H}_2\text{O}}{0.00666 \text{ mol CaCl}_2} = 0.78$

This is a strong indication that they should **(c) heat the crucible again, and then reweigh it**.

2.174. Empirical formula of tin oxide: 14.710 g crucible & Sn – 13.457 g crucible = 1.253 g Sn

$$1.253 \text{ g Sn} \left(\frac{1 \text{ mol Sn}}{118.71 \text{ g Sn}} \right) = 0.01056 \text{ mol Sn}$$

The O present is: 15.048 g (crucible & Sn & O) – 14.710 g (crucible & Sn) = 0.338 g O

$$0.338 \text{ g O} \left(\frac{1 \text{ mol O}}{15.999 \text{ g O}} \right) = 0.0211 \text{ mol O}$$

Now the ratio of O/Sn is: 0.0211 mol O/0.01056 mol Sn = 2.00 mol O/1 mol Sn

The formula is SnO_2.

Summary and Conceptual Questions

2.175. Necessary information to calculate the number of atoms in one cm³ of iron:

A sample calculation to arrive at an exact value is shown below:

$$1.00 \text{ cm}^3 \left(\frac{7.87 \text{ g Fe}}{1 \text{ cm}^3} \right) \left(\frac{1 \text{ mol Fe}}{55.85 \text{ g Fe}} \right) \left(\frac{6.022 \times 10^{23} \text{ atoms Fe}}{1 \text{ mol Fe}} \right) = 8.49 \times 10^{22} \text{ atoms Fe}$$

Note that we needed: (d) density of iron, (b) molar mass of Fe, and (c) Avogadro's number.

2.176. Relationship between abundance and atomic number:

Element abundance generally decreases with increasing atomic number (with exceptions at Li–B and Sc–Fe). Elements with an even atomic number appear to be slightly more abundant than those with an odd atomic number.

2.177. Reactivity of Mg, Ca, and Ba:

(a) Given the greater reactivity of Ca over Mg with water, one would anticipate that Ba would be even more reactive than Ca or Mg—with a more vigorous release of hydrogen gas.

(b) Mg is in period 3, Ca in period 4, and Ba in period 6. Reactivity of these metals increases down the group. (This trend is noted for Group 1A (1) as well.)

2.178. One possible method involves the following steps:

(1) weigh a representative sample of jelly beans (say 10 or so) in order to determine the average mass of a jelly bean;

(2) weigh the jelly beans in the jar (subtract the mass of the empty jar from the mass of the jar filled with jelly beans);

(3) use the average mass per jelly bean and the total mass of the jelly beans in the jar to determine the approximate number of jelly beans in the jar.

Solution and Answer Guide

Kotz Treichel Townsend Treichel, Chemistry and Chemical Reactivity 11e, 978-0-357-85140-1, Chapter 3: Chemical Reactions

TABLE OF CONTENTS

Applying Chemical Principles 88
Practicing Skills 90
General Questions 109
In the Laboratory 115
Summary and Conceptual Questions 116

Applying Chemical Principles

Superconductors

3.1.1. What is the value of x in a sample of $La_{2-x}Ba_xCuO_4$?

Given: %La = 63.43, %Ba = 5.085, %Cu = 15.69, and %O = 15.80.

Noting that the percentages supplied total to 100%, we can calculate the moles of each element present, assuming we have 100.g of the substance:

$$\frac{63.43 \text{ g La}}{1} \cdot \frac{1 \text{ mol La}}{138.91 \text{ g La}} = 0.4566 \text{ mol La}; \quad \frac{5.085 \text{ g Ba}}{1} \cdot \frac{1 \text{ mol Ba}}{137.33 \text{ g Ba}} = 0.03703 \text{ mol Ba}$$

$$\frac{15.69 \text{ g Cu}}{1} \cdot \frac{1 \text{ mol Cu}}{63.546 \text{ g Cu}} = 0.2469 \text{ mol Cu}; \quad \frac{15.80 \text{ g O}}{1} \cdot \frac{1 \text{ mol O}}{15.999 \text{ g O}} = 0.9876 \text{ mol O}$$

Now unlike our usual procedure of dividing each number of moles by the **smallest**, we can exercise a bit of reasoning here. Note that the formula of the superconductor shows the subscript of Cu as 1. So let's force our calculations to have 1 has the subscript by dividing each of the amounts of La, Ba, and O by the moles of Cu:

$$\frac{0.4566 \text{ mol La}}{0.2469 \text{ mol Cu}} = 1.849 \text{ mol La/mol Cu} \quad \frac{0.03703 \text{ mol Ba}}{0.2469 \text{ mol Cu}} = 0.1500 \text{ mol Ba/mol Cu}$$

$$\frac{0.9875 \text{ mol O}}{0.2469 \text{ mol Cu}} = 4.000 \text{ mol O/mol Cu}$$

Note that this ratio simply **confirms** our formula for the superconductor. Since Ba is x, and we found $x = 0.1500$, then the subscript for La is (2-x) or 1.85. The formula for the semiconductor is then $La_{1.85}Ba_{0.15}CuO_4$.

© 2024 Cengage Learning, Inc. All Rights Reserved. May not be scanned, copied or duplicated, or posted to a publicly accessible website, in whole or in part.

3.1.2. Percent by mass of the elements in $YBa_2Cu_3O_{6.93}$ if $x = 0.07$:
Determine mass of each of the elements:

$1.000 \text{ mol Y} \cdot \dfrac{88.906 \text{g Y}}{1 \text{ mol Y}} = 88.906 \text{ g Y}$ $2.000 \text{ mol Ba} \cdot \dfrac{137.33 \text{ g Ba}}{1 \text{ mol Ba}} = 274.7 \text{ g Ba}$

$3.000 \text{ mol Cu} \cdot \dfrac{63.546 \text{ g Cu}}{1 \text{ mol Cu}} = 190.6 \text{ g Cu}$ $6.930 \text{ mol O} \cdot \dfrac{15.999 \text{ g O}}{1 \text{ mol O}} = 110.9 \text{ g O}$

The sum of these masses = 665.1 g. Now determine the % of each element:

$\dfrac{88.906 \text{ g Y}}{665.1 \text{ g}} \cdot 100\% = 13.4\% \text{ Y}$ $\dfrac{274.7 \text{ g Ba}}{665.1 \text{ g}} \cdot 100\% = 41.3\% \text{ Ba}$

$\dfrac{190.6 \text{ g Cu}}{665.1 \text{ g}} \cdot 100\% = 28.7\% \text{ Cu}$ $\dfrac{110.9 \text{g O}}{665.1 \text{ g}} \cdot 100\% = 16.7\% \text{ O}$

3.1.3. Charges present on the copper ions in $YBa_2Cu_3O_7$?
Since all the charges must add to 0, and since we're confident that each O will have a charge of 2−, we can add the charges for Y, Ba, and O to obtain (3+) + 2(2+) + 7(2−) = 7−. The three Cu ions must have charges that add to 7+. So, if we have TWO Cu ions at 2+ and ONE Cu ion at 3+, we accomplish that requirement.

3.1.4. A balanced chemical equation for the requested reaction:
$Y_2O_3(s) + 4 BaCO_3(s) + 6 CuO(s) \rightarrow 2 YBa_2Cu_3O_{6.5}(s) + 4 CO_2(g)$
Given the amount of Cu per mol of the product compound, the value of x is 0.5.

3.1.5. What mass of oxygen is required to convert 1.00 g $YBa_2Cu_3O_{6.50}$ to $YBa_2Cu_3O_{6.93}$?
Molar mass of YBCO: 1(Y) + 2(Ba) + 3(Cu) + 6.50(O)

= 1(88.906) + 2(137.33) + 3(63.546) + 6.50(15.999) = 658.20 g/mol

$\dfrac{1.00 \text{ g YBCO}}{1} \cdot \dfrac{1 \text{ mol YBCO}}{658.20 \text{ g YBCO}} = 0.00152 \text{ mol YBCO}$.

Adding (6.93 − 6.50) or 0.43 mol O₂ requires:
$\dfrac{32.00 \text{ g O}_2}{1 \text{mol O}_2} \cdot \dfrac{1 \text{ mol O}_2}{2 \text{mol O}} \cdot \dfrac{0.43 \text{ mol O}}{1 \text{ mol YBCO}} \cdot \dfrac{0.00152 \text{ mol YBCO}}{1} = 0.010 \text{ g O}_2$

Sequestering Carbon Dioxide

3.2.1. Balanced net ionic equation for the reaction of Ca^{2+} with H_2CO_3:

$Ca^{2+}(aq) + H_2CO_3(aq) + 2 H_2O(\ell) \rightarrow CaCO_3(s) + 2 H_3O^+(aq)$

3.2.2. Protons, electrons, and neutrons in a C-14 atom:

Protons = 6; Electrons = 6 (it is neutral); Neutrons = (14 − 6) = 8

Black Smokers and Volcanoes

3.3.1. Formulas and names of ions making up compounds:

$CaSO_4$	Calcium ion, Ca^{2+}; sulfate ion, SO_4^{2-}
MnS	Manganese(II) ion, Mn^{2+}; sulfide ion, S^{2-}
FeS	Iron(II) ion, Fe^{2+}; sulfide ion, S^{2-}
NiS	Nickel(II) ion, Ni^{2+}; sulfide ion, S^{2-}

3.3.2. Oxidation number for S in sulfide and sulfate ions:

S^{2-}: oxidation number = charge on ion for a monatomic ion, so −2

SO_4^{2-}: (S) + 4(O) = −2 (polyatomic ions have oxidation numbers that sum to charge on ions) so (S) + 4(−2) = −2. A little algebra yields +6.

Practicing Skills

Introduction to Chemical Equations

3.1. $P_4(s) + 5\ O_2(g) \rightarrow P_4O_{10}(s)$
Reactants: $P_4(s)$, $O_2(g)$ Products: $P_4O_{10}(s)$
Stoichiometric coefficients: 1, 5, 1 (note that the 1's are implicit, and not typically written)
Designations s and g refer to the physical states of solid and gas respectively.

3.2. Equation from the description: $4\ NH_3(g) + 7\ O_2(g) \rightarrow 4\ NO_2(g) + 6\ H_2O(\ell)$

3.3. For the reaction: $P_4(s) + 6\ Cl_2(g) \rightarrow 4\ PCl_3(\ell)$ stoichiometric coefficients for the reactants are in a 1:6 ratio, so 8000 molecules of P_4 will require 48000 molecules of Cl_2.

3.4. For the reaction: $2\ Al(s) + 3\ Br_2(\ell) \rightarrow Al_2Br_6(s)$, stoichiometric coefficients for the reactants are in the ratio 2 Al: 3 Br_2. This ratio indicates that 6.0×10^{23} molecules of Br_2 will require 2/3 the number of Al atoms, or $\dfrac{2\ \text{atoms Al}}{3\ \text{molecules Br}_2} \cdot 6.0 \times 10^{23} \text{molecules Br}_2 = 4.0 \times 10^{23} \text{atoms Al}$

3.5. Balanced equation for the oxidation of CO: $2\ CO(g) + O_2(g) \rightarrow 2\ CO_2(g)$
What mass of oxygen is required to convert 1.00 g of CO to 1.57 g CO$_2$? Law of Conservation of Matter indicates that 0.57 g of O$_2$ are needed to convert 1.00 g to 1.57 g.

3.6. The oxidation of Mg can be represented by the equation: $2\ Mg(s) + O_2(g) \rightarrow 2\ MgO(s)$.
If 0.20 mol of Mg form 0.20 mol of MgO, it will require HALF that amount of O$_2$, or 0.10 mol.

Balancing Equations

Balancing equations can be a matter of "running in circles" if a reasonable methodology is not employed. While there isn't one "right place" to begin, generally you will suffer fewer complications if you begin the balancing process using a substance that contains the **greatest number** of elements **or** the **largest subscript** values. Noting that you must have at least that many atoms of each element involved, coefficients can be used to increase the "atomic inventory". In the next few questions, you will see one **emboldened** substance in each equation. This emboldened substance is the one that I judge to be a "good" starting place. One last hint--modify the coefficients of uncombined elements, i.e., those not in compounds, after you modify the coefficients for compounds containing those elements -- not before!

3.7. Balanced chemical equations for these reactions:

(a) $2\ Al(s) + Fe_2O_3(s) \rightarrow 2\ Fe(\ell) + Al_2O_3(s)$

1. Note the need for at least 2 Fe and 3 O atoms as products.
2. A coefficient of 2 for Fe (in products) balances the Fe inventory.
3. Al$_2$O$_3$ (as product) takes care of the O inventory, and mandates 2 Al atoms.
4. 2 Al would give 2 Al atoms on both sides of the equation.

(b) $H_2O(g) + C(s) \rightarrow CO(g) + H_2(g)$

Note this equation has 2 H atoms both left and right; 1 O atom on both sides and 1 C atom on both sides. Hooray—it's already balanced.

(c) $SiCl_4(\ell) + 2\ Mg(s) \rightarrow Si(s) + 2\ MgCl_2(s)$

1. Each SiCl$_4$ (reactant) will require 4 Cl atoms (as product). Since MgCl$_2$ has 2 Cl atoms per unit, we'll need 2 MgCl$_2$ units.
2. A coefficient of 2 for MgCl$_2$ mandates 2 Mg atoms (as reactants).
3. A coefficient of 2 for elemental Mg (in reactants) balances the Mg inventory.

3.8. Balanced chemical equations for these reactions:

(a) $N_2(g) + 3\ H_2(g) \rightarrow 2\ NH_3(g)$

(b) $2\ H_2(g) + CO(g) \rightarrow CH_3OH(\ell)$

(c) $2\ S(s) + 3\ O_2(g) + 2\ H_2O(\ell) \rightarrow 2\ H_2SO_4(\ell)$

3.9. Balanced chemical equations for these reactions:

(a) $4\ Cr(s) + 3\ O_2(g) \rightarrow 2\ Cr_2O_3(s)$

1. Note the need for at least 2 Cr and 3 O atoms.
2. Oxygen is diatomic -- we'll need an even number of oxygen atoms, so try: 2 Cr_2O_3.
3. 3 O_2 would give 6 O atoms on both sides of the equation.
4. 4 Cr would give 4 Cr atoms on both sides of the equation.

(b) $Cu_2S(s) + O_2(g) \rightarrow 2\ Cu(s) + SO_2(g)$

1. A minimum of 2 O in SO_2 is required, and is provided with one molecule of elemental oxygen.
2. 2 Cu atoms (on the right) indicates 2 Cu (on the left).

(c) $C_6H_5CH_3(\ell) + 9\ O_2(g) \rightarrow 4\ H_2O(\ell) + 7\ CO_2(g)$

1. A minimum of 7 C and 8 H is required.
2. 7 CO_2 furnishes 7 C and 4 H_2O furnishes 8 H atoms.
3. 4 H_2O and 7 CO_2 furnish a total of 18 O atoms, making the coefficient of O_2 = 9.

3.10. Balance the following equations:

(a) $2\ Cr(s) + 3\ Cl_2(g) \rightarrow 2\ CrCl_3(s)$

(b) $SiO_2(s) + 2\ C(s) \rightarrow Si(s) + 2\ CO(g)$

(c) $3\ Fe(s) + 4\ H_2O(g) \rightarrow Fe_3O_4(s) + 4\ H_2(g)$

3.11. Balance and name the reactants and products:

(a) $Fe_2O_3(s) + 3\ Mg(s) \rightarrow 3\ MgO(s) + 2\ Fe(s)$

1. Note the need for at least 2 Fe and 3 O atoms.
2. 2 Fe atoms would provide the proper iron atom inventory.
3. 3 MgO would give 3 O atoms on both sides of the equation.
4. 3 Mg would give 3 Mg atoms on both sides of the equation.

Reactants: iron(III) oxide and magnesium

Products: magnesium oxide and iron

(b) AlCl$_3$(s) + 3 NaOH(aq) → Al(OH)$_3$(s) + 3 NaCl(aq)

1. Note the need for <u>at least</u> 1 Al and 3 Cl atoms.
2. 3 NaCl molecules would provide the proper Cl atom inventory.
3. 3 NaCl would require 3 Na atoms on the left side—a coefficient of 3 for NaOH is needed.
4. 3 OH groups (from Al(OH)$_3$) would give 3 OH groups needed on both sides of the equation—so a coefficient of 3 for NaOH is needed to provide that balance.

    Reactants: aluminum chloride and sodium hydroxide

    Products: aluminum hydroxide and sodium chloride.

(c) Ba(NO$_3$)$_2$(s) + H$_2$SO$_4$(aq) → BaSO$_4$(s) + 2 HNO$_3$(aq)

1. Note the need for <u>at least</u> 1 Ba, 2 N, 1 S and 10 O atoms.
2. One Ba(NO$_3$)$_2$ will provide the proper Ba atom inventory.
3. The subscript 2 on Ba(NO$_3$)$_2$ requires a coefficient of 2 for HNO$_3$—providing a balance for N atoms.
4. The implied coefficient of 1 for BaSO$_4$ suggests a similar coefficient for H$_2$SO$_4$—to balance the S atom inventory.
5. O atom inventory is done "automatically" when we balanced N and S inventories.

    Reactants: barium nitrate and sulfuric acid
    Products: barium sulfate and nitric acid

(d) NiCO$_3$(s) + 2 HNO$_3$(aq) → Ni(NO$_3$)$_2$(aq) + CO$_2$(g) + H$_2$O(ℓ)

1. Note the need for <u>at least</u> 1 Ni atom on both sides. This inventory will mandate 2 NO$_3$ groups on the right—and also on the left. Since these come from HNO$_3$ molecules, we'll need 2 HNO$_3$ on the left.
2. The 2 H from the acid and the CO$_3$ from nickel carbonate, provide 2H, 1 C and 3 O atoms. 1 H$_2$O takes care of the 2H, and **one** of the O atoms, 1 CO$_2$ consumes the 1 C and the remaining 2 O atoms.

    Reactants: nickel(II) carbonate and nitric acid

    Products: nickel(II) nitrate, carbon dioxide, and water

3.12. Balance equations and name reactants and products:

(a) SF$_4$(g) + 2 H$_2$O(ℓ) → SO$_2$(g) + 4 HF(aq)

    sulfur tetrafluoride, water, sulfur dioxide, hydrofluoric acid

(b) 4 NH$_3$(aq) + 5 O$_2$(g) → 4 NO(g) + 6 H$_2$O(ℓ)

    ammonia, oxygen, nitrogen monoxide, water

(c) BF$_3$(g) + 3 H$_2$O(ℓ) → 3 HF(aq) + H$_3$BO$_3$(aq)

boron trifluoride, water, hydrogen fluoride, hydrogen borate (boric acid)

Chemical Equilibrium

3.13. Identify as True or False:

(a) At equilibrium the rates of the forward and reverse reactions are equal. TRUE. This is the definition of equilibrium.
(b) When a reaction reaches equilibrium the forward and reverse reactions cease to occur. FALSE. Occurring at the same rates, the opposing reactions give the appearance that nothing is happening, but equilibrium is a dynamic condition.
(c) Chemical reactions always proceed toward equilibrium. TRUE. Equilibrium is the "lowest-energy" state, so reactions proceed toward that state.

3.14 Identify as True or False:

(a) All chemical reactions are product-favored at equilibrium. FALSE.

(b) There is no observable change in a chemical system at equilibrium. TRUE. At equilibrium both forward and reverse reactions are proceeding at the same rate—hence the appearance of no change.

(c) An equilibrium involving a weak acid in water is product favored. FALSE. A weak acid solution contains mostly reactant species and is hence reactant-favored.

3.15. The greater electrical conductivity of the HCO$_2$H solution at equilibrium indicates a greater concentration of ions (H$_3$O$^+$ and HCO$_2^-$), indicating that the HCO$_2$H solution is more product-favored at equilibrium than the CH$_3$CO$_2$H solution.

3.16. Equations to describe the equilibrium in each solution:

H$_3$BO$_3$(aq) + H$_2$O(ℓ) → H$_3$O$^+$(aq) + H$_2$BO$_3^-$(aq)

H$_3$PO$_4$(aq) + H$_2$O(ℓ) → H$_3$O$^+$(aq) + H$_2$PO$_4^-$(aq)

The phosphoric acid ionizes to a greater extent than boric acid, producing more ions in solutions and the greater current.

Ions and Molecules in Aqueous Solution

3.17. What is an electrolyte? What are experimental means for discriminating between weak and strong electrolytes?
An electrolyte is a substance whose aqueous solution conducts an electric current.
As to experimental means for discriminating between weak and strong electrolytes, refer to the apparatus in Figure 3.8. NaCl is a strong electrolyte and would cause the bulb to glow brightly—reflecting many ions in solution while aqueous ammonia or vinegar (an aqueous solution of acetic acid) would cause the bulb to glow only dimly—indicating a smaller number of ions in solution.

3.18. Hydrochloric acid, HCl, and nitric acid, HNO_3, are strong electrolytes. Acetic acid, CH_3CO_2H, is a weak electrolyte. Sodium hydroxide, NaOH, and potassium hydroxide, KOH, are strong electrolytes. Ammonia, NH_3, is a weak electrolyte.

3.19. Predict water solubility:

(a) $CuCl_2$ is expected to be soluble, while CuO and $FeCO_3$ are not. Chlorides are generally water soluble, while oxides and carbonates are not.

(b) $AgNO_3$ is soluble. AgI and Ag_3PO_4 are not soluble. Nitrate salts are soluble. Phosphate salts are generally insoluble. While halides are generally soluble, those of Ag^+ are not.

(c) K_2CO_3, KI and $KMnO_4$ are soluble. In general, salts of the alkali metals are soluble.

3.20. Water soluble groups:

(a) $Ba(NO_3)_2$ (b) Na_2SO_4, $NaClO_4$, $NaCH_3CO_2$ (c) KBr, Al_2Br_6

3.21. Ions produced when the compounds dissolve in water:

Compound	Cation	Anion
(a) KOH	K^+	OH^-
(b) K_2SO_4	K^+	SO_4^{2-}
(c) $LiNO_3$	Li^+	NO_3^-
(d) $(NH_4)_2SO_4$	NH_4^+	SO_4^{2-}

3.22. Ions produced when the compounds dissolve in water:

(a) K^+ and I^- ions   (c) K^+ and HPO_4^{2-} ions

(b) Mg^{2+} and $CH_3CO_2^-$ ions   (d) Na^+ and CN^- ions

3.23.

Compound	Water Soluble	Cation	Anion
(a) Na_2CO_3	yes	Na^+	CO_3^{2-}
(b) $CuSO_4$	yes	Cu^{2+}	SO_4^{2-}
(c) NiS	no		
(d) $BaBr_2$	yes	Ba^{2+}	$2\ Br^-$

3.24. Identify ions formed if compound is soluble:

(a) soluble; Ni^{2+} and Cl^- ions   (c) soluble; Pb^{2+} and NO_3^- ions

(b) soluble; Cr^{3+} and NO_3^- ions   (d) $BaSO_4$ is insoluble

Precipitation Reactions and Net Ionic Equations

3.25. $ZnBr_2(aq) + 2\ NaOH(aq) \rightarrow Zn(OH)_2(s) + 2\ NaBr(aq)$

Net ionic equation: $Zn^{2+}(aq) + 2\ OH^-(aq) \rightarrow Zn(OH)_2(s)$

3.26. $K_2CO_3(aq) + Fe(ClO_4)_2(aq) \rightarrow FeCO_3(s) + 2\ KClO_4(aq)$

Net ionic equation: $Fe^{2+}(aq) + CO_3^{2-}(aq) \rightarrow FeCO_3(s)$

3.27. Balanced equations for precipitation reactions:

(a) $NiCl_2(aq) + (NH_4)_2S(aq) \rightarrow NiS(s) + 2\ NH_4Cl(aq)$

Net ionic equation: $Ni^{2+}(aq) + S^{2-}(aq) \rightarrow NiS(s)$

(b) $3\ Mn(NO_3)_2(aq) + 2\ Na_3PO_4(aq) \rightarrow Mn_3(PO_4)_2(s) + 6\ NaNO_3(aq)$

Net ionic equation: $3\ Mn^{2+}(aq) + 2\ PO_4^{3-}(aq) \rightarrow Mn_3(PO_4)_2(s)$

3.28. Balanced equations for precipitation reactions:

(a) $Pb(NO_3)_2(aq) + 2\ KBr(aq) \rightarrow PbBr_2(s) + 2\ KNO_3(aq)$

Net ionic equation: $Pb^{2+}(aq) + 2\ Br^-(aq) \rightarrow PbBr_2(s)$

(b) $Ca(NO_3)_2(aq) + 2\ KF(aq) \rightarrow CaF_2(s) + 2\ KNO_3(aq)$

Net ionic equation: $Ca^{2+}(aq) + 2\ F^-(aq) \rightarrow CaF_2(s)$

(c) $Ca(NO_3)_2(aq) + Na_2C_2O_4(aq) \rightarrow CaC_2O_4(s) + 2\ NaNO_3(aq)$

Net ionic equation: $Ca^{2+}(aq) + C_2O_4^{2-}(aq) \rightarrow CaC_2O_4(s)$

Reactions of Acids and Bases with Water

3.29. Names and strengths of four acids.

Formula	Name	Strong or weak
(a) HNO_3	Nitric acid	Strong
(b) HNO_2	Nitrous acid	Weak
(c) $HClO_4$	Perchloric acid	Strong
(d) $HClO_2$	Chlorous acid	Weak

3.30. Names and strengths of four acids.

Formula	Name	Strong or weak
(a) H_2SO_3	Sulfurous acid	Weak
(b) H_2SO_4	Sulfuric acid	Strong
(c) $HClO_3$	Chloric acid	Strong
(d) $HClO$	Hypochlorous acid	Weak

3.31. $HNO_3(aq) + H_2O(\ell) \rightarrow H_3O^+(aq) + NO_3^-(aq)$

3.32. $HClO_4(aq) + H_2O(\ell) \rightarrow H_3O^+(aq) + ClO_4^-(aq)$

3.33. Balanced equations for oxalic acid ionizations.

$H_2C_2O_4(aq) + H_2O(\ell) \rightleftarrows HC_2O_4^-(aq) + H_3O^+(aq)$

$HC_2O_4^-(aq) + H_2O(\ell) \rightleftarrows C_2O_4^{2-}(aq) + H_3O^+(aq)$

3.34. Balanced equations for phosphoric acid ionizations.

$H_3PO_4(aq) + H_2O(\ell) \rightleftarrows H_2PO_4^-(aq) + H_3O^+(aq)$

$H_2PO_4^-(aq) + H_2O(\ell) \rightleftarrows HPO_4^{2-}(aq) + H_3O^+(aq)$

$HPO_4^{2-}(aq) + H_2O(\ell) \rightleftarrows PO_4^{3-}(aq) + H_3O^+(aq)$

3.35. Balanced equation for methylamine ionization.

$CH_3NH_2(aq) + H_2O(\ell) \rightleftarrows CH_3NH_3^+(aq) + OH^-(aq)$

3.36. Balanced equation for formate ion ionization.

$HCO_2^-(aq) + H_2O(\ell) \rightleftarrows HCO_2H(aq) + OH^-(aq)$

3.37. Metal oxide reacts with water to form a base: $MgO(s) + H_2O(\ell) \rightarrow Mg(OH)_2(s)$

3.38. Metal oxide reacts with water to form a base: $Li_2O(s) + H_2O(\ell) \rightarrow 2\ LiOH(aq)$

3.39. Nonmetal oxide reacts with water to form an acid: $SO_3(g) + H_2O(\ell) \rightarrow H_2SO_4(aq)$

3.40. Nonmetal oxide reacts with water to form an acid: $P_4O_{10}(s) + 6\ H_2O(\ell) \rightarrow 4\ H_3PO_4(aq)$

3.41. Equilibrium when nitric acid dissolves in water:

$HNO_3(aq) + H_2O(\ell) \rightarrow H_3O^+(aq) + NO_3^-(aq)$

   BA   BB   BA   BB

BA = Brönsted acid;   BB= Brönsted base

Nitric acid is a STRONG acid, so the equilibrium favors the PRODUCTS.

3.42. Equilibrium when benzoic acid dissolves in water:

$C_6H_5CO_2H(aq) + H_2O(\ell) \rightarrow C_6H_5CO_2^-(aq) + H_3O^+(aq)$

   BA   BB   BB   BA

Benzoic acid is a WEAK acid, so the equilibrium favors the REACTANTS.

3.43. Show H₂O reacting (with HBr) as a Brönsted base:

HBr(aq) + H₂O(ℓ) → H₃O⁺(aq) + Br⁻(aq)   (water accepts a proton from HBr)

(BA)   (BB)

Show H₂O reacting (with NH₃) as a Brönsted acid: (water donates a proton to NH₃)

NH₃(aq) + H₂O(ℓ) → NH₄⁺(aq) + OH⁻(aq)

(BB)   (BA)

3.44. Show H₂PO₄⁻ reacting as an acid and HPO₄⁻ as a base

H₂PO₄⁻(aq) + CO₃²⁻(aq) → HPO₄⁻(aq) + HCO₃⁻(aq)

HPO₄⁻(aq) + CH₃CO₂H(aq) → H₂PO₄⁻(aq) + CH₃CO₂⁻(aq)

Reactions of Acids and Bases

3.45. Complete and balance:

(a)   2 CH₃CO₂H(aq) + Mg(OH)₂(s) → Mg(CH₃CO₂)₂(aq) + 2 H₂O(ℓ)

   acetic acid   magnesium   magnesium   water
                    hydroxide   acetate

(b)   HClO₄(aq) + NH₃(aq) → NH₄ClO₄(aq)
   perchloric   ammonia   ammonium
   acid   perchlorate

3.46. Complete and balance:

(a) H₃PO₄(aq) + 3 KOH(aq) → K₃PO₄(aq) + 3 H₂O(ℓ)

   phosphoric   potassium   potassium   water
   acid   hydroxide   phosphate

(b) H₂C₂O₄(aq) + Ca(OH)₂(s) → CaC₂O₄(s) + 2 H₂O(ℓ)

   oxalic   calcium   calcium   water
   acid   hydroxide   oxalate

3.47. Write and balance the equation for barium hydroxide reacting with nitric acid:

Ba(OH)$_2$(aq) + 2 HNO$_3$(aq) → Ba(NO$_3$)$_2$(aq) + 2 H$_2$O(ℓ)
barium   nitric   barium   water
hydroxide   acid   nitrate

3.48. Write and balance the equation for aluminum hydroxide reacting with sulfuric acid:
2 Al(OH)$_3$(s) + 3 H$_2$SO$_4$(aq) → Al$_2$(SO$_4$)$_3$(aq) + 6 H$_2$O(ℓ)

Writing Net Ionic Equations

3.49. Write and balance the equation for iron(II) carbonate reacting with nitric acid:

FeCO$_3$(s) + 2 HNO$_3$(aq) → Fe(NO$_3$)$_2$(aq) + H$_2$O(ℓ) + CO$_2$(g)

iron(II)   nitric   iron(II)   water   carbon
carbonate   acid   nitrate   dioxide

3.50. Balance equation and name the products:

MnCO$_3$(s) + 2 HCl(aq) → MnCl$_2$(aq) + CO$_2$(g) + H$_2$O(ℓ)
                          manganese(II) carbon   water
                          chloride   dioxide

3.51. Balanced equation for reaction of (NH$_4$)$_2$S with HBr:

(NH$_4$)$_2$S(aq) + 2 HBr(aq) → 2 NH$_4$Br(aq) + H$_2$S(g)
ammonium   hydrobromic   ammonium   hydrogen
sulfide   acid   bromide   sulfide

3.52. Balanced equation for reaction of Na$_2$SO$_3$ with CH$_3$CO$_2$H:

Na$_2$SO$_3$(aq) + 2 CH$_3$CO$_2$H(aq) → 2 NaCH$_3$CO$_2$(aq) + SO$_2$(g) + H$_2$O(ℓ)
sodium   acetic   sodium   sulfur   water
sulfite   acid   acetate   dioxide

3.53. Balance and write net ionic equations:

(a) (NH$_4$)$_2$CO$_3$(aq) + Cu(NO$_3$)$_2$(aq) → CuCO$_3$(s) + 2 NH$_4$NO$_3$(aq)

(net) CO$_3^{2-}$(aq) + Cu^{2+}(aq) → CuCO$_3$(s)

(b) Pb(OH)$_2$(s) + 2 HCl(aq) → PbCl$_2$(s) + 2 H$_2$O(ℓ)

(net) Pb(OH)$_2$(s) + 2 H$_3$O$^+$(aq) + 2 Cl$^-$(aq) → PbCl$_2$(s) + 4 H$_2$O(ℓ)

(c) BaCO$_3$(s) + 2 HCl(aq) → BaCl$_2$(aq) + H$_2$O(ℓ) + CO$_2$(g)

(net) BaCO$_3$(s) + 2 H$_3$O$^+$(aq) → Ba^{2+}(aq) + 3 H$_2$O(ℓ) + CO$_2$(g)

(d) 2 CH$_3$CO$_2$H(aq) + Ni(OH)$_2$(s) → Ni(CH$_3$CO$_2$)$_2$(aq) + 2 H$_2$O(ℓ)

(net) 2 CH$_3$CO$_2$H(aq) + Ni(OH)$_2$(s) → Ni^{2+}(aq) + 2 CH$_3$CO$_2^-$(aq) + 2 H$_2$O(ℓ)

3.54. Balance and write net ionic equations:

(a) Zn(s) + 2 HCl(aq) → H$_2$(g) + ZnCl$_2$(aq)
net: Zn(s) + 2 H$_3$O$^+$(aq) → H$_2$(g) + Zn^{2+}(aq) + 2 H$_2$O(ℓ)

(b) Mg(OH)$_2$(s) + 2 HCl(aq) → MgCl$_2$(aq) + 2 H$_2$O(ℓ)
net: Mg(OH)$_2$(s) + 2 H$_3$O$^+$(aq) → Mg^{2+}(aq) + 4 H$_2$O(ℓ)

(c) 2 HNO$_3$(aq) + CaCO$_3$(s) → Ca(NO$_3$)$_2$(aq) + H$_2$O(ℓ) + CO$_2$(g)
net: 2 H$_3$O$^+$(aq) + CaCO$_3$(s) → Ca^{2+}(aq) + 3 H$_2$O(ℓ) + CO$_2$(g)

(d) (NH$_4$)$_2$S(aq) + FeCl$_2$(aq) → 2 NH$_4$Cl(aq) + FeS(s)
net: S^{2-}(aq) + Fe^{2+}(aq) → FeS(s)

3.55. Balance and write net ionic equations:

(a) AgNO$_3$(aq) + KI(aq) → AgI(s) + KNO$_3$(aq)
net: Ag$^+$(aq) + I$^-$(aq) → AgI(s)

(b) Ba(OH)$_2$(aq) + 2 HNO$_3$(aq) → 2 H$_2$O(ℓ) + Ba(NO$_3$)$_2$(aq)
net: OH$^-$(aq) + H$_3$O$^+$(aq) → 2 H$_2$O(ℓ)

(c) 2 Na$_3$PO$_4$(aq) + 3 Ni(NO$_3$)$_2$(aq) → Ni$_3$(PO$_4$)$_2$(s) + 6 NaNO$_3$(aq)
net: 2 PO$_4^{3-}$(aq) + 3 Ni^{2+}(aq) → Ni$_3$(PO$_4$)$_2$(s)

3.56. Balance and write net ionic equations:

(a) 2 NaOH(aq) + FeCl$_2$(aq) → Fe(OH)$_2$(s) + 2 NaCl(aq)
2 OH$^-$(aq) + Fe^{2+}(aq) → Fe(OH)$_2$(s)

(b) BaCl$_2$(aq) + Na$_2$CO$_3$(aq) → BaCO$_3$(s) + 2 NaCl(aq)

Ba^{2+}(aq) + CO$_3^{2-}$(aq) → BaCO$_3$(s)

(c) 3 NH$_3$(aq) + H$_3$PO$_4$(aq) → (NH$_4$)$_3$PO$_4$(aq)

3 NH$_3$(aq) + H$_3$PO$_4$(aq) → 3 NH$_4^+$(aq) + PO$_4^{3-}$(aq)

3.57. Write balanced net ionic equations:

(a) HNO$_2$(aq) + NaOH(aq) → H$_2$O(ℓ) + NaNO$_2$(aq)

net: HNO$_2$(aq) + OH$^-$(aq) → H$_2$O(ℓ) + NO$_2^-$(aq)

(b) Ca(OH)$_2$(s) + 2 HCl(aq) → 2 H$_2$O(ℓ) + CaCl$_2$(aq)

net: Ca(OH)$_2$(s) + 2 H$_3$O$^+$(aq) → Ca^{2+}(aq) + 4 H$_2$O(ℓ)

3.58. Write balanced net ionic equations:

(a) Ag$^+$(aq) + I$^-$(aq) → AgI(s)

(b) Ba^{2+}(aq) + CO$_3^{2-}$(aq) → BaCO$_3$(s)

3.59. Net ionic equation for reaction of (NH$_4$)$_2$S with HCl:

S^{2-}(aq) + 2 H$_3$O$^+$(aq) → H$_2$S(g) + 2 H$_2$O(ℓ)

3.60. Net ionic equation for reaction of NH$_4$NO$_3$ with KOH:

NH$_4^+$(aq) + OH$^-$(aq) → H$_2$O(ℓ) + NH$_3$(g)

Oxidation Numbers

3.61. For questions on oxidation number, read the symbol (x) as "the oxidation number of x."

(a) BrO$_3^-$ (Br) + 3(O)   = −1

Since oxygen almost always has an oxidation number of −2, we can substitute this value and solve for the oxidation number of Br.

(Br) + 3(−2) = −1 and solving (Br) = +5

(b) C$_2$O$_4^{2-}$

$$2(C)+4(O)=-2$$
$$2(C)+4(-2)=-2$$
$$2(C)+(-8)=-2$$
$$2(C)=6$$
$$(C)=3$$

(c) F⁻ The oxidation number for any monatomic ion is the charge on the ion, so (F) = −1

(d) CaH₂ in hydrides H has a −1 oxidation number
$$(Ca)+2(H)=0$$
$$(Ca)+2(-1)=0$$
$$(Ca)=+2$$

(e) H₄SiO₄
$$4(H)+(Si)+4(O)=0$$
$$4(+1)+(Si)+4(-2)=0$$
$$(Si)=+4$$

(f) HSO₄⁻
$$(H)+2(S)+4(O)=-1$$
$$(+1)+2(S)+4(-2)=-1$$
$$(S)=+6$$

3.62. Oxidation number of each element in the following:

(a) PF₆⁻
$$(P)+6(F)=-1$$
$$(P)+6(-1)=-1$$
$$(P)=+5$$

(b) H₂AsO₄⁻
$$2(H)+(As)+4(O)=-1$$
$$2(+1)+(As)+4(-2)=-1$$
$$(As)=+5$$

(c) UO₂⁺

$$(U) + 2(O) = +1$$
$$(U) + 2(-2) = +1$$
$$(U) = +5$$

(d) N₂O₅

$$2(N) + 5(O) = 0$$
$$2(N) + 5(-2) = 0$$
$$(N) = +5$$

(e) POCl₃

$$(P) + (O) + 3(Cl) = 0$$
$$(P) + (-2) + 3(-1) = 0$$
$$(P) = +5$$

(f) XeO₄²⁻

$$(Xe) + 4(O) = -2$$
$$(Xe) + 4(-2) = -2$$
$$(Xe) = +6$$

Oxidation-Reduction Reactions

3.63. Classify as oxidation-reduction or acid-base:

(a) Oxidation-Reduction:

Zn(s) has an oxidation number of 0, while Zn^{2+}(aq) has an oxidation number of +2. Hence, Zn is oxidized. N in NO_3^- has an oxidation number of +5, while N in NO_2 has an oxidation number of +4. Hence, N is reduced.

(b) Acid-Base reaction:

There is no change in oxidation number for any of the elements in this reaction—hence it is NOT an oxidation-reduction reaction. H_2SO_4 is an acid, and $Zn(OH)_2$ acts as a base.

(c) Oxidation-Reduction:

Ca(s) has an oxidation number of 0, while Ca^{2+}(aq) has an oxidation number of +2. Hence, Ca is oxidized. H in H_2O has an oxidation number of +1, while H in H_2 has an oxidation number of 0. Hence, H is reduced.

3.64. Identify oxidation-reduction reactions, classify others:

(a) precipitation reaction, as a solid is formed

(b) oxidation-reduction reaction, as the oxidation number of Ca changes from 0 to +2, while that of O changes from 0 to –2

(c) oxidation-reduction reaction, as the oxidation number of Fe changes from +2 to +3, while that of O changes from 0 to –2

3.65. Determine which reactant is oxidized and which is reduced:

(a) $C_2H_4(g) + 3\ O_2(g) \rightarrow 2\ CO_2(g) + 2\ H_2O(\ell)$

Species	Ox. number before	Ox. number after	Experiences	Functions as
C	–2	+4	oxidation	(C_2H_4) reducing agent
H	+1	+1	no change	
O	0	–2	reduction	(O_2) oxidizing agent

(b) $Si(s) + 2\ Cl_2(g) \rightarrow SiCl_4(\ell)$

Species	Ox. number before	Ox. number after	Experiences	Functions as
Si	0	+4	oxidation	(Si) reducing agent
Cl	0	–1	reduction	(Cl_2) oxidizing agent

3.66. Determine which reactant is oxidized and which is reduced:

(a) $Cr_2O_7^{2-}(aq) + 3\ Sn^{2+}(aq) + 14\ H_3O^+(aq) \rightarrow 2\ Cr^{3+}(aq) + 3\ Sn^{4+}(aq) + 21\ H_2O(\ell)$

Species	Ox. number before	Ox. number after	Experiences	Functions as
Cr	+6	+3	reduction	($Cr_2O_7^{2-}$) oxidizing agent
H	+1	+1		
O	–2	–2	no change	
Sn	+2	+4	oxidation	(Sn^{2+}) reducing agent

(b) FeS(s) + 3 NO$_3^-$(aq) + 4 H$_3$O$^+$(aq) → 3 NO(g) + SO$_4^{2-}$(aq) + Fe^{3+}(aq) + 6 H$_2$O(ℓ)

Species	Ox. number before	Ox. number after	Experiences	Functions as
Fe	+2	+3	oxidation	(Fe) reducing agent
N	+5	+2	reduction	(NO$_3^-$) oxidizing agent
H	+1	+1	no change	
O	−2	−2	no change	
S	−2	−2	no change	

Types of Reactions in Aqueous Solution

3.67. Precipitation (PR), Acid-Base (AB), or Gas-Forming Acid-Base (GFAB)

(a) Ba(OH)$_2$(aq) + 2 HCl(aq) → BaCl$_2$ (aq) + 2 H$_2$O(ℓ)   AB

(b) 2 HNO$_3$(aq) + CoCO$_3$(s) → Co(NO$_3$)$_2$(aq) + H$_2$O(ℓ) + CO$_2$(g)   GFAB

(c) 2 Na$_3$PO$_4$(aq) + 3 Cu(NO$_3$)$_2$(aq) → Cu$_3$(PO$_4$)$_2$(s) + 6 NaNO$_3$(aq)   PR

3.68. Precipitation (PR), Acid-Base (AB), or Gas-Forming Acid-Base (GFAB)
(a) K$_2$CO$_3$(aq) + Cu(NO$_3$)$_2$(aq) → CuCO$_3$(s) + 2 KNO$_3$(aq)   PR
(b) Pb(NO$_3$)$_2$(aq) + 2 HCl(aq) → PbCl$_2$(s) + 2 HNO$_3$(aq)   PR
(c) MgCO$_3$(s) + 2 HCl(aq) → MgCl$_2$(aq) + H$_2$O(ℓ) + CO$_2$(g)   GFAB

3.69. Precipitation (PR), Acid-Base (AB), or Gas-Forming Acid-Base (GFAB)

(a) MnCl$_2$(aq) + Na$_2$S(aq) → MnS(s) + 2 NaCl(aq)   PR

net: Mn^{2+}(aq) + S^{2-}(aq) → MnS(s)

(b) K$_2$CO$_3$(aq) + ZnCl$_2$(aq) → ZnCO$_3$(s) + 2 KCl(aq)   PR

net: CO$_3^{2-}$(aq) + Zn^{2+}(aq) → ZnCO$_3$(s)

3.70. Precipitation (PR), Acid-Base (AB), or Gas-Forming Acid-Base (GFAB)

(a) Fe(OH)$_3$(s) + 3 HNO$_3$(aq) → Fe(NO$_3$)$_3$(aq) + 3 H$_2$O(ℓ)   AB

net: Fe(OH)$_3$(s) + 3 H$_3$O$^+$(aq) → Fe^{3+}(aq) + 6 H$_2$O(ℓ)

(b) FeCO$_3$(s) + 2 HNO$_3$(aq) → Fe(NO$_3$)$_2$(aq) + CO$_2$(g) + H$_2$O(ℓ) GFAB

net: FeCO$_3$(s) + 2 H$_3$O$^+$(aq) → Fe^{2+}(aq) + CO$_2$(g) + 3 H$_2$O(ℓ)

3.71. Balance the following and classify them as PR, AB, GFAB, or OR:
(a) CuCl$_2$(aq) + H$_2$S(aq) → CuS(s) + 2 HCl(aq)   PR
(b) H$_3$PO$_4$(aq) + 3 KOH(aq) → 3 H$_2$O(ℓ) + K$_3$PO$_4$(aq)   AB
(c) Ca(s) + 2 HBr(aq) → H$_2$(g) + CaBr$_2$(aq)   OR
(d) MgCl$_2$(aq) + 2 NaOH(aq) → Mg(OH)$_2$(s) + 2 NaCl(aq)   PR

3.72. Balance the following and classify them as PR, AB, GFAB, or OR:
(a) NiCO$_3$(s) + H$_2$SO$_4$(aq) → NiSO$_4$(aq) + H$_2$O(ℓ) + CO$_2$(g)   GFAB
(b) Co(OH)$_2$(s) + 2 HBr(aq) → CoBr$_2$(aq) + 2 H$_2$O(ℓ)   AB
(c) AgCH$_3$CO$_2$(aq) + NaCl(aq) → AgCl(s) + NaCH$_3$CO$_2$(aq)   PR
(d) NiO(s) + CO(g) → Ni(s) + CO$_2$(g)   OR

3.73. Identify the reactants (x and y) and write the complete balanced equation for each.
(a) x + y → H$_2$O(ℓ) + CaBr$_2$(aq)

(complete) Ca(OH)$_2$(s) + 2 HBr(aq) → 2 H$_2$O(ℓ) + CaBr$_2$(aq)

(b) x + y → Mg(NO$_3$)$_2$(aq) + CO$_2$(g) + H$_2$O(ℓ)

(complete) MgCO$_3$(aq) + 2 HNO$_3$(aq) → Mg(NO$_3$)$_2$(aq) + CO$_2$(g) + H$_2$O(ℓ)

(c) x + y → BaSO$_4$(s) + NaCl(aq)

(complete) BaCl$_2$(aq) + Na$_2$SO$_4$(aq) → BaSO$_4$(s) + 2 NaCl(aq)

(d) x + y → NH$_4^+$(aq) + OH$^-$(aq)

(complete) NH$_3$(g) + H$_2$O(ℓ) → NH$_4^+$(aq) + OH$^-$(aq)

3.74. Identify the reactants (x and y) and write the complete balanced equation for each.
(a) x + y → (NH$_4$)$_2$SO$_4$(aq)

(complete) 2 NH$_3$(aq) + H$_2$SO$_4$(aq) → (NH$_4$)$_2$SO$_4$(aq)

(b) $x + y \rightarrow CaCl_2(aq) + CO_2(g) + H_2O(\ell)$

(complete) $CaCO_3(s) + 2\ HCl(aq) \rightarrow CaCl_2(aq) + CO_2(g) + H_2O(\ell)$

(c) $x + y \rightarrow Ba(NO_3)_2\ (aq) + 2\ AgCl\ (s)$

(complete) $2\ AgNO_3(aq) + BaCl_2(aq) \rightarrow Ba(NO_3)_2(aq) + 2\ AgCl(s)$

(d) $x + y \rightarrow H_3O^+(aq) + ClO_4^-(aq)$

(complete) $HClO_4(aq) + H_2O(\ell) \rightarrow H_3O^+(aq) + ClO_4^-(aq)$

3.75. Identify the reactants (x and y) and write the complete balanced equation for each.

(a) $x + y \rightarrow H_2O(\ell) + NaNO_3(aq)$

(complete) $HNO_3(aq) + NaOH(aq) \rightarrow H_2O(\ell) + NaNO_3(aq)$

(net) $H_3O^+(aq) + OH^-(aq) \rightarrow 2\ H_2O(\ell)$

(b) $x + y \rightarrow CaCO_3(s) + NaCl(aq)$

(complete) $CaCl_2(aq) + Na_2CO_3(aq) \rightarrow CaCO_3(s) + 2\ NaCl(aq)$

(net) $Ca^{2+}(aq) + CO_3^{2-}(aq) \rightarrow CaCO_3(s)$

(c) $x + y \rightarrow Sr(NO_3)_2\ (aq) + CO_2(g) + H_2O(\ell)$

(complete) $2\ HNO_3(aq) + SrCO_3(s) \rightarrow Sr(NO_3)_2\ (aq) + CO_2(g) + H_2O(\ell)$

(net) $2\ H_3O^+(aq) + CO_3^{2-}(s) \rightarrow CO_2(g) + 3\ H_2O(\ell)$

(d) $x + y \rightarrow Zn_3(PO_4)_2(s) + NaCl(aq)$

(complete) $2\ Na_3PO_4(aq) + 3\ ZnCl_2(aq) \rightarrow Zn_3(PO_4)_2(s) + 6\ NaCl(aq)$

(net) $2\ PO_4^{3-}(aq) + 3\ Zn^{2+}(aq) \rightarrow Zn_3(PO_4)_2(s)$

3.76. Identify the reactants (x and y) and write the complete balanced equation for each.

(a) $x + y \rightarrow H_2O(\ell) + NaF(aq)$

(complete) $HF(aq) + NaOH(aq) \rightarrow H_2O(\ell) + NaF(aq)$

(net) $H_3O^+(aq) + OH^-(aq) \rightarrow 2\ H_2O(\ell)$

(b) $x + y \rightarrow SrSO_4(s) + KNO_3(aq)$

(complete) $K_2SO_4(aq) + Sr(NO_3)_2(aq) \rightarrow SrSO_4(s) + 2\ KNO_3(aq)$

(net) $SO_4^{2-}(aq) + Sr^{2+}(aq) \rightarrow SrSO_4(s)$

(c) $x + y \rightarrow$ KCl(aq) + H$_2$S(g)

(complete) 2 HCl(aq) + K$_2$S(aq) $\rightarrow$ 2 KCl(aq) + H$_2$S(g)

(net) 2 H$_3$O$^+$(aq) + S^{2-}(aq) $\rightarrow$ H$_2$S(g) + 2 H$_2$O(ℓ)

(d) $x + y \rightarrow$ Fe(OH)$_2$(s) + NaI(aq)

(complete) FeI$_2$(aq) + 2 NaOH(aq) $\rightarrow$ Fe(OH)$_2$(s) + 2 NaI(aq)

(net) Fe^{2+}(aq) + 2 OH$^-$(aq) $\rightarrow$ Fe(OH)$_2$(s)

General Questions

3.77. Balance:

(a) Synthesis of urea:

CO$_2$(g) + 2 NH$_3$(g) $\rightarrow$ NH$_2$CONH$_2$(s) + H$_2$O(ℓ)

1. Note the need for two NH$_3$ in each molecule of urea, so multiply NH$_3$ by 2.
2. 2 NH$_3$ provides the two H atoms for a molecule of H$_2$O.
3. Each CO$_2$ provides the O atom for a molecule of H$_2$O.

(b) synthesis of uranium(VI) fluoride

UO$_2$(s) + 4 HF(aq) $\rightarrow$ UF$_4$(s) + 2 H$_2$O(ℓ)

UF$_4$(s) + F$_2$(g) $\rightarrow$ UF$_6$(s)

1. The 4 F atoms in UF$_4$ require 4 F atoms from HF. (equation 1)
2. The H atoms in HF produce 2 molecules of H$_2$O. (equation 1)
3. The 1:1 stoichiometry of UF$_6$: UF$_4$ provides a simple balance. (equation 2)

(c) synthesis of titanium metal from TiO$_2$:

TiO$_2$(s) + 2 Cl$_2$(g) + 2 C(s) $\rightarrow$ TiCl$_4$(ℓ) + 2 CO(g)

TiCl$_4$(ℓ) + 2 Mg(s) $\rightarrow$ Ti(s) + 2 MgCl$_2$(s)

1. The O balance mandates 2 CO for each TiO$_2$. (equation 1)
2. A coefficient of 2 for C provides C balance. (equation 1)
3. The Ti balance (TiO$_2$:TiCl$_4$) requires 4 Cl atoms, hence 2 Cl$_2$ (equation 1)
4. The Cl balance requires 2 MgCl$_2$, hence 2 Mg. (equation 2)

3.78. Balance the following equations:

(a) Ca$_3$(PO$_4$)$_2$(s) + 2 H$_2$SO$_4$(aq) → Ca(H$_2$PO$_4$)$_2$(aq) + 2 CaSO$_4$(s)

(b) 2 NaBH$_4$(s) + H$_2$SO$_4$(aq) → B$_2$H$_6$(g) + 2 H$_2$(g) + Na$_2$SO$_4$(aq)

(c) WO$_3$(s) + 3 H$_2$(g) → W(s) + 3 H$_2$O(ℓ)

(d) (NH$_4$)$_2$Cr$_2$O$_7$(s) → N$_2$(g) + 4 H$_2$O(ℓ) + Cr$_2$O$_3$(s)

3.79. Formula for the following compounds:

(a) soluble compound with Br⁻ ion: almost any bromide compound with the exception of Ag$^+$, Hg$_2$$^{2+}$ and Pb^{2+} for example KBr

(b) insoluble hydroxide: almost any hydroxide except salts of NH$_4$$^+$ and the alkali metal ions for example, Fe(OH)$_2$

(c) insoluble carbonate: almost any carbonate except salts of NH$_4$$^+$ and the alkali metal ions for example, CaCO$_3$

(d) soluble nitrate-containing compound: all nitrate-containing compounds are soluble for example NaNO$_3$

The listing of soluble and insoluble compounds in your text will provide general guidelines for predicting the solubility of compounds.

(e) a weak Bronsted acid: the carboxylic acids are weak acids: or example CH$_3$CO$_2$H (acetic)

3.80. Formula for each of the following:

(a) A soluble acetate: one possible answer is NaCH$_3$CO$_2$

(b) An insoluble sulfide: one possible answer is NiS

(c) A soluble hydroxide: one possible answer is NaOH

(d) An insoluble chloride: one possible answer is PbCl$_2$

(e) A strong Bronsted base: one possible answer is NaOH

3.81. For the following copper salts:

Water soluble: Cu(NO$_3$)$_2$, CuCl$_2$ — nitrates and chlorides are soluble
Water insoluble: CuCO$_3$, Cu$_3$(PO$_4$)$_2$ — carbonates and phosphates are insoluble

3.82. Two of the possible anions that combine with Al^{3+} to make water-soluble compounds: NO$_3$⁻, nitrate ion, and ClO$_4$⁻, perchlorate ion

3.83. **Spectator ions** in the following equation and the net ionic equation:

2 H₃O⁺(aq) + 2 **NO₃⁻(aq)** + Mg(OH)₂(s) → 4 H₂O(ℓ) + Mg²⁺(aq) + 2 **NO₃⁻(aq)**

The emboldened nitrate ions are the spectator ions. The net ionic equation would be the first equation shown above without the spectator ions:

2 H₃O⁺(aq) + Mg(OH)₂(s) → 4 H₂O(ℓ) + Mg²⁺(aq)   [An *acid-base* exchange]

3.84. Name water insoluble products:

(a) CuS(s) copper(II) sulfide; Cu²⁺(aq) + S²⁻(aq) → CuS(s)

(b) CaCO₃(s) calcium carbonate; Ca²⁺(aq) + CO₃²⁻(aq) → CaCO₃(s)

(c) AgI(s) silver iodide; Ag⁺(aq) + I⁻(aq) → AgI(s)

3.85. For the reaction of chlorine with NaBr: Cl₂(g) + 2 NaBr(aq) → 2 NaCl(aq) + Br₂(ℓ)

(a) Oxidized: **bromine's** oxidation number is changed from −1 to 0
    Reduced: **chlorine's** oxidation number is changed from 0 to −1
(b) Oxidizing agent: **Cl₂** removes the electrons from NaBr
    Reducing agent: **NaBr** provides the electrons to the chlorine.

3.86. Oxidizing agents: HNO₃, Cl₂, O₂, KMnO₄   Reducing agent: Na

3.87. Reaction: MgCO₃(s) + 2 HCl(aq) → CO₂(g) + MgCl₂(aq) + H₂O(ℓ)

(a) The net ionic equation: MgCO₃(s) + 2 H₃O⁺(aq) → CO₂(g) + Mg²⁺(aq) + 3 H₂O(ℓ)

The spectator ion is the chloride ion (Cl⁻).

(b) The production of CO₂(g) makes this a gas-forming acid-base reaction.

3.88. Reaction of ammonium sulfide with mercury (II) nitrate:

(a) (NH₄)₂S(aq) + Hg(NO₃)₂(aq) → HgS(s) + 2 NH₄NO₃(aq)

(b) ammonium sulfide, mercury(II) nitrate, mercury(II) sulfide, ammonium nitrate

(c) precipitation reaction

3.89. Species present in aqueous solutions of:

compound	types of species	species present
(a) NH$_3$	molecules (weak base)	NH$_3$, NH$_4^+$, OH$^-$
(b) CH$_3$CO$_2$H	molecules (weak acid)	CH$_3$CO$_2$H, CH$_3$CO$_2^-$, H$^+$
(c) NaOH	ions (strong base)	Na$^+$ and OH$^-$
(d) HBr	ions (strong acid)	H$_3$O$^+$ and Br$^-$

In every case, H$_2$O will be present (but omitted in this list)

3.90. Some possible answers:

(a) Water soluble: CuCl$_2$, copper(II) chloride   Cu(NO$_3$)$_2$, copper(II) nitrate

Water insoluble: CuCO$_3$, copper(II) carbonate   CuS, copper(II) sulfide

(b) Water soluble: BaBr$_2$, barium bromide   Ba(CH$_3$CO$_2$)$_2$, barium acetate

Water insoluble: BaSO$_4$, barium sulfate   BaCrO$_4$, barium chromate

3.91. Balance and classify each as PR, AB, GFAB

(a) K$_2$CO$_3$(aq) + 2 HClO$_4$(aq) → 2 KClO$_4$(aq) + CO$_2$(g) + H$_2$O(ℓ)   GFAB

Reactants: potassium carbonate, perchloric acid (respectively)

Products: potassium perchlorate, carbon dioxide, water (respectively)

Net ionic equation: CO$_3^{2-}$(aq) + 2 H$_3$O$^+$(aq) → CO$_2$(g) + 3 H$_2$O(ℓ)

(b) FeCl$_2$(aq) + (NH$_4$)$_2$S(aq) → FeS(s) + 2 NH$_4$Cl(aq)   PR

Reactants: iron(II) chloride, ammonium sulfide (respectively)

Products: iron(II) sulfide, ammonium chloride (respectively)

Net ionic equation: Fe^{2+}(aq) + S^{2-}(aq) → FeS(s)

(c) Fe(NO$_3$)$_2$(aq) + Na$_2$CO$_3$(aq) → FeCO$_3$(s) + 2 NaNO$_3$(aq)   PR

Reactants: iron(II) nitrate, sodium carbonate (respectively)

Products: iron(II) carbonate, sodium nitrate (respectively)

Net ionic equation: Fe^{2+}(aq) + CO$_3^{2-}$(aq) → FeCO$_3$(s)

(d) 3 NaOH(aq) + FeCl$_3$(aq) → 3 NaCl(aq) + Fe(OH)$_3$(s)

Reactants: sodium hydroxide, iron(III) chloride (respectively)   PR

Products: sodium chloride, iron(III) hydroxide (respectively)

Net ionic equation: Fe^{3+}(aq) + 3 OH$^-$(aq) → Fe(OH)$_3$(s)

3.92. Overall balanced equation and net ionic equation:

(a) Pb(NO$_3$)$_2$(aq) + 2 KOH(aq) → Pb(OH)$_2$(s) + 2 KNO$_3$(aq)
Pb^{2+}(aq) + 2 OH$^-$(aq) → Pb(OH)$_2$(s)

(b) Cu(NO$_3$)$_2$(aq) + Na$_2$CO$_3$(aq) → CuCO$_3$(s) + 2 NaNO$_3$(aq)
Cu^{2+}(aq) + CO$_3^{2-}$(aq) → CuCO$_3$(s)

3.93. Which of the following compounds in each pair could be separated by stirring with water?

(a) NaOH dissolves in water, with Ca(OH)$_2$ remaining as undissolved.

(b) MgCl$_2$ will dissolve in water, while MgF$_2$ will remain undissolved.

(c) KI dissolves—as do all the salts of Group IA (1) cations, while AgI will remain undissolved.

(d) NH$_4$Cl dissolves readily—as a salt containing BOTH ammonium ions AND chloride ions, while PbCl$_2$ is an insoluble chloride.

3.94. Identify compounds that dissolve to give a solution that conducts electricity:

(a) CuCl$_2$. The others are not soluble.

(b) HCl, HNO$_3$, and H$_2$SO$_4$. Hypochlorous, HClO, will dissolve but is a weak electrolyte.

3.95. Compound(s) in each set that creates a solution that is **only** a weak conductor:

(a) NH$_3$ reacts only slightly with water to produce a weakly conducting solution. The sodium and barium hydroxides are strong bases, providing a strong conductor. CuSO$_4$ also dissolves appreciably so it will produce a strong conducting solution.

(b) Both CH$_3$CO$_2$H and HF are weak acids and will produce only a weakly conducting solution when dissolved in water. Na$_3$PO$_4$ will dissolve to a great extent—providing a strongly conducting solution, and HNO$_3$ is a strong acid—producing a strongly conducting solution.

3.96. Net ionic equations:

(a) CH$_3$CO$_2$H(aq) + OH$^-$(aq) → H$_2$O(ℓ) + CH$_3$CO$_2^-$(aq)

(b) Zn(s) + 2 H$_3$O$^+$(aq) → Zn^{2+}(aq) + H$_2$(g) + H$_2$O(ℓ)

3.97. Na$_2$S was treated with acid, producing a gas:

S^{2-}(aq) + 2 H$_3$O$^+$(aq) → H$_2$S(g) + 2 H$_2$O(ℓ)

A black precipitate forms when the gas is bubbled into lead nitrate solution:

Pb^{2+}(aq) + H$_2$S(g) + 2 H$_2$O(ℓ) → PbS(s) + 2 H$_3$O$^+$(aq)

3.98. 2 HI ⇌ H$_2$(g) + I$_2$(g)

The HI(g) decomposes to produce H$_2$(g) and I$_2$(g). The two product gases in turn react with each other to produce the initial reactant HI(g). The amount of HI decreases until just enough H$_2$ and I$_2$ is produced so that they react to generate HI at the same rate the HI decomposes. The concentrations of the three gases then remain constant although the forward and reverse reactions keep proceeding, indicating that the system has reached equilibrium.

3.99 Sulfurous acid has the formula H$_2$SO$_3$. It is commonly produced by the reaction of SO$_2$ and H$_2$O. Furthermore, the combustion (roasting) of sulfur in air produces SO$_2$. Thus, we can conclude that:

(a) formula of sulfur oxide: SO$_2$

(b) combustion reaction: S(s) + O$_2$(g) → SO$_2$(g)

(c) reaction of oxide with water: SO$_2$(g) + H$_2$O(ℓ) → H$_2$SO$_3$(aq)

3.100. White phosphorus combustion reaction.

(a) Formula of phosphorus oxide

Mass of P$_2$O$_5$ is $(2 \times 30.974 \text{ g/mol}) + (5 \times 15.999 \text{ g/mol}) = 141.943 \text{ g/mol}$

Molar mass of gas is 284 g/mol

Divide molar mass by mass of empirical formula to determine molecular formula.

$$\frac{284 \text{ g/mol}}{141.943 \text{ g/mol}} \approx 2$$

Thus, there are 2 empirical formulas per molecular formula.

Molecular formula is $2(P_2O_5)$ or P$_4$O$_{10}$.

(b) Combustion reaction equation

P$_4$(s) + 5 O$_2$(g) → P$_4$O$_{10}$(s)

(c) Phosphoric acid production equation

P$_4$O$_{10}$(s) + 6 H$_2$O(ℓ) → 4 H$_3$PO$_4$

In the Laboratory

3.101. For the reaction:

2 NaI(s) + 2 H$_2$SO$_4$(aq) + MnO$_2$(s) → Na$_2$SO$_4$(aq) + MnSO$_4$(aq) + I$_2$ (g) + 2 H$_2$O(ℓ)

(a) Oxidation number of each atom in the equation: (ox. numbers shown in order)
Reactants: NaI (+1, −1)   H$_2$SO$_4$ (+1, +6, -2)   MnO$_2$ (+4,−2)
Products: Na$_2$SO$_4$ (+1,+6,−2)   MnSO$_4$ (+2,+6,−2)   I$_2$(0)   H$_2$O(+1,−2)

(b) Oxidizing agent: MnO$_2$ Oxidized: I in NaI
Reducing agent: NaI   Reduced: Mn (in MnO$_2$)

(c) The formation of gaseous iodine "drives" the process—product-favored
(d) Names of reactants and products:

NaI	H$_2$SO$_4$	MnO$_2$	Na$_2$SO$_4$	MnSO$_4$	I$_2$	H$_2$O
sodium iodide	sulfuric acid	manganese(IV) oxide	sodium sulfate	manganese(II) sulfate	iodine	water

3.102. The first reaction is an oxidation-reduction reaction. The second reaction is a gas-forming acid-base reaction.

Acids: H$_2$S   Bases: NaHCO$_3$

Oxidizing agents: Ag$_2$S   Reducing agents: Al

3.103. Another way to prepare MgCl$_2$: Given the reactivity of both elemental magnesium and chlorine, one can bring the two elements into contact (carefully!)

Mg(s) + Cl$_2$(g) → MgCl$_2$(s)

Or another possibility is to use a gas-forming acid-base reaction, shown below, then evaporate the solution to recover the MgCl$_2$.

2 HCl(aq) + MgS(s) → MgCl$_2$(s) + H$_2$S(g)

3.104. One possibility for preparing barium phosphate:

3 Ba(OH)$_2$(aq) + 2 H$_3$PO$_4$(aq) → Ba$_3$(PO$_4$)$_2$(s) + 6 H$_2$O(ℓ)

Use an acid-base reaction to form barium phosphate and water. Filter the solid to isolate barium phosphate.

3.105. In the reaction:

$C_6H_{12}O_6(aq) + 2\ Ag^+(aq) + 2\ OH^-(aq) \rightarrow C_6H_{12}O_7(aq) + 2\ Ag(s) + H_2O(\ell)$

Oxidized: $C_6H_{12}O_6$ is oxidized to $C_6H_{12}O_7$ (simple observation—note that O is added)
Reduced: $Ag^+(aq)$ is reduced to $Ag(s)$ (oxidation number changes from +1 to 0)

Oxidizing agent: $Ag^+(aq)$ oxidizes the sugar

Reducing agent: $C_6H_{12}O_6$ reduces Ag^+

Summary and Conceptual Questions

3.106. Measure the conductivity of the solution and compare it to that of a soluble compound such as sodium chloride. A conductivity apparatus that uses a light bulb will glow only dimly if the solid dissolves to a small extent but the light bulb will glow brightly if the solid dissolves to a great extent. The dissolving process for $PbCl_2$ is reactant-favored.

3.107. A simple experiment to prove that lactic acid is a weak acid (ionizing to a small extent) is to test the conductivity of the solution. A conductivity apparatus (e.g., a light bulb) will indicate only a small current flow (a light bulb will glow only dimly). To prove that the establishment of equilibrium is reversible, add strong acid (H_3O^+). The shift of equilibrium to the left should result in the molecular acid precipitating from solution.

3.108. One possible answer:

$BaCO_3(s) + 2\ HCl(aq) \rightarrow BaCl_2(aq) + H_2O(\ell) + CO_2(g)$

3.109. Using the reagents: $BaCl_2$, $BaCO_3$, $Ba(OH)_2$, H_2SO_4, Na_2SO_4,

Prepare barium sulfate by:

a precipitation reaction

The reaction of $BaCl_2$ with Na_2SO_4 will perform this task:

$BaCl_2(aq) + Na_2SO_4(aq) \rightarrow BaSO_4(s) + 2\ NaCl(aq)$

a gas-forming acid-base reaction

$BaCO_3(s) + H_2SO_4(\ell) \rightarrow BaSO_4(s) + H_2O(\ell) + CO_2(g)$

3.110. Possible answers:

(a) acid-base reaction: $Zn(OH)_2(s) + 2\ HCl(aq) \rightarrow ZnCl_2(aq) + 2\ H_2O(\ell)$

(b) gas-forming acid-base reaction: $ZnCO_3(s) + 2\ HCl(aq) \rightarrow ZnCl_2(aq) + H_2O(\ell) + CO_2(g)$

(c) oxidation-reduction reaction: $Zn(s) + Cl_2(g) \rightarrow ZnCl_2(s)$

3.111. Begin by treating the percentages as mass (out of 100 g of the compound), and calculate the number of moles of each atom:

$$\frac{20.32 \text{ g Ni}}{1} \cdot \frac{1 \text{ mol Ni}}{58.693 \text{ g Ni}} = 0.3462 \text{ mol Ni}$$

$$\frac{33.26 \text{ g C}}{1} \cdot \frac{1 \text{ mol C}}{12.011 \text{ g C}} = 2.769 \text{ mol C}$$

$$\frac{4.88 \text{ g H}}{1} \cdot \frac{1 \text{ mol H}}{1.008 \text{ g H}} = 4.84 \text{ mol H}$$

$$\frac{22.15 \text{ g O}}{1} \cdot \frac{1 \text{ mol O}}{15.999 \text{ g O}} = 1.384 \text{ mol O}$$

$$\frac{19.39 \text{ g N}}{1} \cdot \frac{1 \text{ mol N}}{14.007 \text{ g N}} = 1.384 \text{ mol N}$$

Using the # of moles of nickel, calculate ratios for mole element X/mol Ni

$$\frac{2.769 \text{ mol C}}{0.3462 \text{ mol Ni}} = 8.00 \text{ mol C}/\text{mol Ni}$$

$$\frac{4.84 \text{ mol H}}{0.3462 \text{ mol Ni}} = 14.0 \text{ mol H}/\text{mol Ni}$$

$$\frac{1.384 \text{ mol O}}{0.3462 \text{ mol Ni}} = 4.00 \text{ mol O}/\text{mol Ni} \quad \text{and}$$

$$\frac{1.384 \text{ mol N}}{0.3462 \text{ mol Ni}} = 4.00 \text{ mol N}/\text{mol Ni} \quad \text{so the empirical formula is:} \quad NiC_8H_{14}N_4O_4$$

3.112. Formula of terbium oxide:

(73.95 g Tb)(1 mol/158.925 g Tb) = 0.4653 mol Tb

100.00 g Tb$_x$O$_y$ − 73.95 g Tb = 26.05 g O

(26.055 g O)(1 mol/15.999 mol O) = 1.628 mol O

Mole: mole ratio: 1.628 mol O/0.4653 mol Tb = 3.499 O:Tb

A formula would be: Tb$_2$O$_7$

Balanced equation to form this oxide: 4 Tb(s) + 7 O$_2$(g) → 2 Tb$_2$O$_7$(s)

3.113. Concerning arsenic compounds:

(a) Oxidation states for As,S,N in the reaction:

Reactants: As$_2$S$_3$(s): As = +3, S = −2; HNO$_3$: N = +5

Products: H$_3$AsO$_4$: As = +5 ; NO: N = +2 ; S = 0

(b) Formula for Ag$_x$AsO$_y$:

Using the composition, As, 16.20% and Ag, 69.96%, we know that the remaining percent belongs to O, so 100.000 – (16.20 + 69.96) = 13.84% O. Express the mass of each element (out of 100g), and calculate the # moles of As, Ag, and O.

$$\frac{16.20 \text{ g As}}{1} \cdot \frac{1 \text{ mol As}}{74.922 \text{g As}} = 0.2162 \text{ mol As}$$

$$\frac{69.96 \text{ g Ag}}{1} \cdot \frac{1 \text{ mol Ag}}{107.87 \text{g Ag}} = 0.6486 \text{ mol Ag}$$

$$\frac{13.84 \text{ g O}}{1} \cdot \frac{1 \text{ mol O}}{15.999 \text{ g O}} = 0.8651 \text{ mol O}$$

Express these as mol of element X/mol As:

$$\frac{0.6486 \text{ mol Ag}}{0.2162 \text{ mol As}} = 3.000 \text{ mol Ag}/\text{mol As} \text{ and } \frac{0.8651 \text{ mol O}}{0.2162 \text{ mol As}} = 4.001 \text{ mol O}/\text{mol As}$$

The compound has the formula: Ag$_3$AsO$_4$.

3.114. Regarding the reaction of barium hydroxide with sulfuric acid:

(a) Ba(OH)$_2$(aq) + H$_2$SO$_4$(aq) → BaSO$_4$(s) + 2 H$_2$O(ℓ)

(b) Ba^{2+}(aq) + 2 OH$^-$(aq) + H$_3$O$^+$(aq) + HSO$_4^-$(aq) → BaSO$_4$(s) + 3 H$_2$O(ℓ)

(c) diagram (b). The conductivity drops as sulfuric acid is added initially, as ions are removed to form BaSO$_4$ and H$_2$O. After the reaction is complete, the conductivity rises as excess sulfuric acid, a strong acid, ionizes to give H$_3$O$^+$ and HSO$_4^-$ ions.

Solution and Answer Guide

Kotz Treichel Townsend Treichel, Chemistry and Chemical Reactivity 11e, 978-0-357-85140-1, Chapter 4: Stoichiometry: Quantitative Information about Chemical Reactions

TABLE OF CONTENTS

Applying Chemical Principles 119
Practicing Skills 122
General Questions 153
In The Laboratory 169
Summary And Conceptual Questions 176

Applying Chemical Principles

Atom Economy

4.1.1. Calculate the atom economy to produce methyl methacrylate:

Sum the masses of reactants incorporated into MMA and product MMA:

Reactants: $C_2H_4 + CH_3OH + CO + CH_2O + H_2 + ½ O_2$

$(28) + (32) + (28) + (30) + 2 + 1/2 (32) = 136$

Product: $CH_3C(CH_2)CO_2CH_3$

$(5C + 8H + 2O) = (60 + 8 + 32) = 100$

% Atom Economy = Mass Product/Mass Reactants × 100% = (100/136) × 100 = 73.5%

Forensic Chemistry—Food Tampering

4.2.1. Bleach titration question

(a) Mol NaClO present in 25.0 mL sample of dilute bleach

$$\frac{0.151 \text{ mol Na}_2\text{S}_2\text{O}_3}{\text{L}} \left(29.34 \text{ mL} \cdot \frac{1 \text{ L}}{1000 \text{ mL}} \right) = 0.00443 \text{ mol Na}_2\text{S}_2\text{O}_3$$

$$0.00443 \text{ mol Na}_2\text{S}_2\text{O}_3 \left(\frac{1 \text{ mol I}_2}{2 \text{ mol Na}_2\text{S}_2\text{O}_3} \right) \left(\frac{1 \text{ mol HClO}}{1 \text{ mol I}_2} \right) \left(\frac{1 \text{ mol NaClO}}{1 \text{ mol HClO}} \right) = 0.00222 \text{ mol NaClO}$$

(b) Molarity NaClO in dilute solution

$$25.0 \text{ mL} \left(\frac{1 \text{ L}}{1000 \text{ mL}} \right) = 0.0250 \text{ L}$$

© 2024 Cengage Learning, Inc. All Rights Reserved. May not be scanned, copied or duplicated, or posted to a publicly accessible website, in whole or in part.

$$\frac{0.00222 \text{ mol NaClO}}{0.0250 \text{ L}} = 0.0886 \text{ M NaClO}$$

(c) Molarity of NaClO in original bleach solution

$$0.0886 \text{ M NaClO in dilute solution}\left(\frac{100.0 \text{ mL bleach}}{10.0 \text{ mL dilute solution}}\right) = 0.886 \text{ M NaClO in bleach}$$

(d) Mass of NaClO in 10.0 mL sample

molar mass of NaClO = 74.44 g/mol

$$\frac{0.886 \text{ mol NaClO}}{\text{L}}\left(10.0 \text{ mL} \cdot \frac{1 \text{ L}}{1000 \text{ mL}}\right) = 0.00886 \text{ mol NaClO}$$

$$0.00886 \text{ mol NaClO}\left(\frac{74.44 \text{ g NaClO}}{1 \text{ mol NaClO}}\right) = 0.660 \text{ g NaClO}$$

(e) Mass of 10.0 mL bleach solution

$$10.0 \text{ mL solution}\left(\frac{1.1 \text{ g}}{1 \text{ mL solution}}\right) = 11 \text{ g solution}$$

(f) Mass percent of NaClO

$$\text{mass percent} = \frac{0.660 \text{ g NaClO}}{11 \text{ g solution}} \times 100\% = 6.0 \% \text{ NaClO}$$

How Much Salt is There in Seawater

4.3.1. Determine the amount of chloride ion in the diluted sample:

$$26.25 \text{ mL}\left(\frac{1 \text{ L}}{1000 \text{ mL}}\right)\left(\frac{0.100 \text{ mol AgNO}_3}{1 \text{ L}}\right)\left(\frac{1 \text{ mol Ag}^+}{1 \text{ mol AgNO}_3}\right)\left(\frac{1 \text{ mol Cl}^-}{1 \text{ mol Ag}^+}\right) = 0.02625 \text{ mol Cl}^-$$

$$= 0.0263 \text{ mol Cl}^- \text{ (3 sf)}$$

This amount of Cl$^-$ was contained in the 50.00 mL (= 0.05000 L) sample, so the concentration of Cl$^-$ in this solution is

$$\frac{0.02625 \text{ mol Cl}^-}{0.05000 \text{ L}} = 0.0525 \text{ M Cl}^- \text{ (3 sf)}$$

This solution was created by two 10-fold dilutions. EACH dilution reduces the concentration by a factor of 10. So, the initial concentration would be 100 times the concentration of the diluted solution, 5.25 M.

The Martian

4.4.1. Unfortunately, Mark Watney has confused two chemical principles. While a certain mass, for example, a certain number, of O₂ molecules, can create twice that number of molecules containing single O atoms, that does not necessarily relate to volumes (liters). The density of liquid O₂, 1.14 g/mL, is different from the density of water, 1.00 g/mL. Consequently, equal volumes of these two species contain different numbers of molecules and atoms. This logic is also true of liquid hydrazine, with a density of 1.02 g/mL. The logic is flawed, and this is a **poor** assumption.

4.4.2. Calculate the volume of water using the step-wise procedure:

Vol O₂ → Mass O₂ → Mol O₂ → Mol H₂O → Mass H₂O → Volume H₂O

$$50.\text{ L O}_2 \left(\frac{1000 \text{ mL}}{1 \text{ L}}\right)\left(\frac{1.14 \text{ g O}_2}{1 \text{ mL O}_2}\right)\left(\frac{1 \text{ mol O}_2}{32.00 \text{ g O}_2}\right)\left(\frac{2 \text{ mol H}_2\text{O}}{1 \text{ mol O}_2}\right)\left(\frac{18.02 \text{ g H}_2\text{O}}{1 \text{ mol H}_2\text{O}}\right)$$

$$\left(\frac{1 \text{ mL H}_2\text{O}}{1.00 \text{ g H}_2\text{O}}\right)\left(\frac{1 \text{ L}}{1000 \text{ mL}}\right) = 64 \text{ L H}_2\text{O}$$

4.4.3. Volume of water obtained from 1.0 L of N₂H₄(ℓ):

$$1.0 \text{ L N}_2\text{H}_4 \left(\frac{1000 \text{ mL}}{1 \text{ L}}\right)\left(\frac{1.02 \text{ g N}_2\text{H}_4}{1 \text{ mL N}_2\text{H}_4}\right)\left(\frac{1 \text{ mol N}_2\text{H}_4}{32.05 \text{ g N}_2\text{H}_4}\right)\left(\frac{2 \text{ mol H}_2\text{O}}{1 \text{ mol N}_2\text{H}_4}\right)\left(\frac{18.02 \text{ g H}_2\text{O}}{1 \text{ mol H}_2\text{O}}\right)$$

$$\left(\frac{1 \text{ mL H}_2\text{O}}{1.00 \text{ g H}_2\text{O}}\right)\left(\frac{1 \text{ L}}{1000 \text{ mL}}\right) = 1.1 \text{ L H}_2\text{O}$$

4.4.4. Notice that 50. L of O₂ would make 64 L of H₂O, not 100 L. Similarly, 1.0 L of hydrazine would make 1.1 L of H₂O, not 2 L. Watney's predictions are incorrect.

4.4.5. Decomposition of hydrazine:

(a) Balanced chemical equation: N₂H₄(ℓ) → N₂(g) + 2 H₂(g)

(b) Percent yield if 125 g N₂H₄ produces 12.7 g H₂:

First, calculate the mass of H₂ that could be produced from 125 g N₂H₄:

$$\text{Theoretical yield} = 125 \text{ g N}_2\text{H}_4 \left(\frac{1 \text{ mol N}_2\text{H}_4}{32.05 \text{ g N}_2\text{H}_4}\right)\left(\frac{2 \text{ mol H}_2}{1 \text{ mol N}_2\text{H}_4}\right)\left(\frac{2.016 \text{ g H}_2}{1 \text{ mol H}_2}\right) = 15.7 \text{ g H}_2$$

$$\text{Percent yield} = \frac{\text{actual yield}}{\text{theoretical yield}} \cdot 100\%$$

$$= \frac{12.7 \text{ g}}{15.7 \text{ g}} \cdot 100\%$$

$$= 80.8\%$$

4.4.6 MOXIE

(a) Balanced chemical equation: $2\ CO_2(g) \rightarrow 2\ CO(g) + O_2(g)$

(b) Oxidation numbers: C in CO_2 = +4; O in CO_2 = –2; C in CO = +2, O in CO = –2; O in O_2 = 0.

The oxidation number of some of the oxygen atoms increases from –2 to 0, so oxygen is oxidized.

The oxidation number of carbon decreases from +4 to +2, so the carbon atom is reduced.

(c) Time needed to produce 750 g of O_2:

$$750\ g\ O_2 \left(\frac{32\ \text{minutes}}{5.4\ g\ O_2} \right) = 4.4 \times 10^3\ \text{minutes}\ (= 74\ \text{hours} = 3.1\ \text{days})$$

This rate of production is not sufficient to sustain an astronaut.

(d) Theoretical yield of O_2 from 1200 g CO_2:

$$1200\ g\ CO_2 \left(\frac{1\ mol\ CO_2}{44.01\ g\ CO_2} \right) \left(\frac{1\ mol\ O_2}{2\ mol\ CO_2} \right) \left(\frac{32.00\ g\ O_2}{1\ mol\ O_2} \right) = 4.4 \times 10^2\ g\ O_2$$

Practicing Skills

Mass Relationships in Chemical Reactions: Basic Stoichiometry

4.1. Moles of aluminum needed to react with 3.0 mol of Fe_2O_3:

$Fe_2O_3(s) + 2\ Al(s) \rightarrow 2\ Fe(\ell) + Al_2O_3(s)$

$$3.0\ mol\ Fe_2O_3 \left(\frac{2\ mol\ Al}{1\ mol\ Fe_2O_3} \right) = 6.0\ mol\ Al;$$

What mass of Fe should be produced?

$$3.0\ mol\ Fe_2O_3 \left(\frac{2\ mol\ Fe}{1\ mol\ Fe_2O_3} \right) \left(\frac{55.845\ g\ Fe}{1\ mol\ Fe} \right) = 340\ g\ Fe\ (2\ sf)$$

4.2. Potassium chlorate decomposition

$2\ KClO_3(s) \rightarrow 2\ KCl(s) + 3\ O_2(g)$

$$1.0\ mol\ KClO_3 \left(\frac{3\ mol\ O_2}{2\ mol\ KClO_3} \right) = 1.5\ mol\ O_2$$

$$1.5 \text{ mol O}_2 \left(\frac{32.00 \text{ g O}_2}{1 \text{ mol O}_2} \right) = 48 \text{ g O}_2 \text{ (2 sf)}$$

4.3. Octane combustion

$$2 \text{ C}_8\text{H}_{18}(\ell) + 25 \text{ O}_2(g) \rightarrow 16 \text{ CO}_2(g) + 18 \text{ H}_2\text{O}(g)$$

$$5.00 \text{ g C}_8\text{H}_{18} \left(\frac{1 \text{ mol C}_8\text{H}_{18}}{114.2 \text{ g C}_8\text{H}_{18}} \right) \left(\frac{25 \text{ mol O}_2}{2 \text{ mol C}_8\text{H}_{18}} \right) \left(\frac{32.00 \text{ g O}_2}{1 \text{ mol O}_2} \right) = 17.5 \text{ g O}_2$$

$$5.00 \text{ g C}_8\text{H}_{18} \left(\frac{1 \text{ mol C}_8\text{H}_{18}}{114.2 \text{ g C}_8\text{H}_{18}} \right) \left(\frac{16 \text{ mol CO}_2}{2 \text{ mol C}_8\text{H}_{18}} \right) \left(\frac{44.01 \text{ g CO}_2}{1 \text{ mol CO}_2} \right) = 15.4 \text{ g CO}_2$$

$$5.00 \text{ g C}_8\text{H}_{18} \left(\frac{1 \text{ mol C}_8\text{H}_{18}}{114.2 \text{ g C}_8\text{H}_{18}} \right) \left(\frac{18 \text{ mol H}_2\text{O}}{2 \text{ mol C}_8\text{H}_{18}} \right) \left(\frac{18.02 \text{ g H}_2\text{O}}{1 \text{ mol H}_2\text{O}} \right) = 7.10 \text{ g H}_2\text{O}$$

4.4. Mass of HCl to react with 0.750 g of Al(OH)$_3$:

$$0.750 \text{ g Al(OH)}_3 \left(\frac{1 \text{ mol Al(OH)}_3}{78.00 \text{ g Al(OH)}_3} \right) \left(\frac{3 \text{ mol HCl}}{1 \text{ mol Al(OH)}_3} \right) \left(\frac{36.46 \text{ g HCl}}{1 \text{ mol HCl}} \right) = 1.05 \text{ g HCl}$$

Mass of water produced:

$$0.750 \text{ g Al(OH)}_3 \left(\frac{1 \text{ mol Al(OH)}_3}{78.00 \text{ g Al(OH)}_3} \right) \left(\frac{3 \text{ mol H}_2\text{O}}{1 \text{ mol Al(OH)}_3} \right) \left(\frac{18.02 \text{ g H}_2\text{O}}{1 \text{ mol H}_2\text{O}} \right) = 0.520 \text{ g H}_2\text{O}$$

4.5. Quantity of Br$_2$ to react with 2.56 g of Al:
According to the balanced equation, 2 mol of Al react with 3 mol of Br$_2$.
Calculate the # of moles of Al, then multiply by 3/2 to obtain # mol of Br$_2$ required.

$$2.56 \text{ g Al} \left(\frac{1 \text{ mol Al}}{26.98 \text{ g Al}} \right) \left(\frac{3 \text{ mol Br}_2}{2 \text{ mol Al}} \right) \left(\frac{159.8 \text{ g Br}_2}{1 \text{ mol Br}_2} \right) = 22.7 \text{ g Br}_2$$

Mass of Al$_2$Br$_6$ expected:
This could be solved in several ways. The simplest is to recognize that—according to the Law of Conservation of Matter—mass is conserved in a reaction. If 22.7 g of bromine react with exactly 2.56 g of aluminum, the total products would also have a mass of (22.7 g + 2.56 g) or 25.3 g Al$_2$Br$_6$.

4.6. Mass of Fe in 454 g Fe₂O₃; Mass of CO to react with this mass of Fe₂O₃:

(a) $454 \text{ g Fe}_2\text{O}_3 \left(\dfrac{1 \text{ mol Fe}_2\text{O}_3}{159.7 \text{ g Fe}_2\text{O}_3}\right)\left(\dfrac{2 \text{ mol Fe}}{1 \text{ mol Fe}_2\text{O}_3}\right)\left(\dfrac{55.85 \text{ g Fe}}{1 \text{ mol Fe}}\right) = 318 \text{ g Fe}$

(b) $454 \text{ g Fe}_2\text{O}_3 \left(\dfrac{1 \text{ mol Fe}_2\text{O}_3}{159.7 \text{ g Fe}_2\text{O}_3}\right)\left(\dfrac{3 \text{ mol CO}}{1 \text{ mol Fe}_2\text{O}_3}\right)\left(\dfrac{28.01 \text{ g CO}}{1 \text{ mol CO}}\right) = 239 \text{ g CO}$

4.7. For the reaction of methane burning in oxygen:

(a) The products of the reaction are the oxides of carbon and of hydrogen: carbon dioxide (CO₂) and water (H₂O).

(b) The balanced equation requires 4 hydrogen atoms (so 2 water molecules), and 2 water molecules and 1 carbon dioxide mandate 2 oxygen molecules:

CH₄(g) + 2 O₂(g) → CO₂(g) + 2 H₂O(ℓ)

(c) The mass of O₂ required for complete combustion of 25.5 g of methane:

$25.5 \text{ g CH}_4 \left(\dfrac{1 \text{ mol CH}_4}{16.04 \text{ g CH}_4}\right)\left(\dfrac{2 \text{ mol O}_2}{1 \text{ mol CH}_4}\right)\left(\dfrac{32.00 \text{ g O}_2}{1 \text{ mol O}_2}\right) = 102 \text{ g O}_2$

(d) As in SQ 4.5, the simplest method to determine the total mass of products expected is to recognize that mass is conserved in a reaction. If 25.5 g of methane reacts with exactly 102 g of oxygen, the total products would have a mass of (25.5 g + 102 g) or 127.5 g (or 128 g to 3 sf).

4.8. For the reaction of barium chloride with silver nitrate:

(a) BaCl₂(aq) + 2 AgNO₃(aq) → 2 AgCl(s) + Ba(NO₃)₂(aq)

(b) Mass of AgNO₃ for complete reaction:

$0.156 \text{ g BaCl}_2 \left(\dfrac{1 \text{ mol BaCl}_2}{208.2 \text{ g BaCl}_2}\right)\left(\dfrac{2 \text{ mol AgNO}_3}{1 \text{ mol BaCl}_2}\right)\left(\dfrac{169.9 \text{ g AgNO}_3}{1 \text{ mol AgNO}_3}\right) = 0.255 \text{ g AgNO}_3$

$0.156 \text{ g BaCl}_2 \left(\dfrac{1 \text{ mol BaCl}_2}{208.2 \text{ g BaCl}_2}\right)\left(\dfrac{2 \text{ mol AgCl}}{1 \text{ mol BaCl}_2}\right)\left(\dfrac{143.3 \text{ g AgCl}}{1 \text{ mol AgCl}}\right) = 0.215 \text{ g AgCl}$

Amounts Tables and Chemical Stoichiometry

4.9. Emboldened quantities indicate given data:

Equation	2 PbS(s) +	3 O$_2$(g) →	2 PbO(s) +	2 SO$_2$(g)
Initial amount (mol)	**2.50**	3.75	0	0
Change in amount upon reaction (mol)	–2.50	–3/2(2.50) = –3.75	+2.50	+2.50
Amount after complete reaction (mol)	0	0	2.50	2.50

The amount of O$_2$ required is 1.5 times the amount of PbS (note the 3:2 ratio). With the required amount of oxygen present, the amount of PbO and SO$_2$ formed is equal to the amount of PbS consumed (note the ratios of 2:2 for both PbS:PbO and PbS:SO$_2$)

4.10. Amount of C required for 6.2 mol Fe$_2$O$_3$:

Equation	2 Fe$_2$O$_3$(s) +	3 C(s) →	4 Fe(s) +	3 CO$_2$(g)
Initial amount (mol)	**6.2**	9.3	0	0
Change in amount upon reaction (mol)	–6.2	–3/2(6.2) = –9.3	+4/2(6.2) = 12	+3/2(6.2) = 9.3
Amount after complete reaction (mol)	0	0	12	9.3

$$6.2 \text{ mol Fe}_2\text{O}_3 \cdot \frac{3 \text{ mol C}}{2 \text{ mol Fe}_2\text{O}_3} = 9.3 \text{ mol C}$$

Amount of Fe produced: $6.2 \text{ mol Fe}_2\text{O}_3 \cdot \dfrac{4 \text{ mol Fe}}{2 \text{ mol Fe}_2\text{O}_3} = 12 \text{ mol Fe}$

Amount of CO$_2$ produced: $6.2 \text{ mol Fe}_2\text{O}_3 \cdot \dfrac{3 \text{ mol CO}_2}{2 \text{ mol Fe}_2\text{O}_3} = 9.3 \text{ mol CO}_2$

4.11. For the reaction of elemental Cr with oxygen to produce chromium(III) oxide:

(a) Equation	4 Cr(s)	+ 3 O$_2$(g) →	2 Cr$_2$O$_3$(s)
Initial mass (g)	**0.175**	0.0808	0
(b) Initial amount (mol)	3.37 × 10^{-3}	2.52 × 10^{-4}	0
Change in amount upon reaction (mol)	–3.37 × 10^{-3}	–3/4(3.37 × 10^{-3}) = –2.52 ×10^{-4}	+2/4(3.37 × 10^{-3}) = 1.68 × 10^{-3}
Amount after complete reaction (mol)	0	0	+1.68 × 10^{-3}
Mass after complete reaction (g)	0	0	0.256

$$0.175 \text{ g Cr}\left(\frac{1 \text{ mol Cr}}{52.00 \text{ g Cr}}\right) = 3.37 \times 10^{-3} \text{ mol Cr}$$

The balanced equation indicates that for each mol of elemental Cr, 1/2 mol of Cr_2O_3 is produced (the 4:2 ratio), so 3.37×10^{-3} mol Cr produces 1.68×10^{-3} mol Cr_2O_3

$$1.68 \times 10^{-3} \text{ mol Cr}_2O_3\left(\frac{152.0 \text{ g Cr}_2O_3}{1 \text{ mol Cr}_2O_3}\right) = 0.255 \text{ g Cr}_2O_3$$

(c) The amount of oxygen required is 3/4 of the amount of elemental Cr (4:3 ratio), with each mole of oxygen having a mass of 31.9 g.

$$3.37 \times 10^{-3} \text{ mol Cr}\left(\frac{3 \text{ mol O}_2}{4 \text{ mol Cr}}\right)\left(\frac{32.00 \text{ g O}_2}{1 \text{ mol O}_2}\right) = 0.0808 \text{ g O}_2$$

4.12. For the burning of ethane:

(a) Combustion of hydrocarbons always yields CO_2, carbon dioxide, and H_2O, water.

(b) Balanced equation: $2 \text{ C}_2\text{H}_6(g) + 7 \text{ O}_2(g) \rightarrow 4 \text{ CO}_2(g) + 6 \text{ H}_2\text{O}(g)$

(c) $13.6 \text{ g C}_2\text{H}_6\left(\frac{1 \text{ mol C}_2\text{H}_6}{30.07 \text{ g C}_2\text{H}_6}\right)\left(\frac{7 \text{ mol O}_2}{2 \text{ mol C}_2\text{H}_6}\right)\left(\frac{32.00 \text{ g O}_2}{1 \text{ mol O}_2}\right) = 50.7 \text{ g O}_2$

(d) Mass of products: 13.6 g C_2H_6 + 50.7 g O_2 = 64.3 g reactants = 64.3 g products

Equation	2 C₂H₆(g) +	7 O₂(g) →	4 CO₂(g) +	6 H₂O(g)
Initial amount (mol)	0.452	1.58	0	0
Change (mol)	−0.452	−1.58	+0.905	+1.36
Amount after reaction (mol)	0	0	0.905	1.36

Amount of ethane: $13.6 \text{ g C}_2\text{H}_6 \cdot \frac{1 \text{ mol C}_2\text{H}_6}{30.07 \text{ g}} = 0.452 \text{ mol C}_2\text{H}_6$

Amount of oxygen needed: $0.452 \text{ mol C}_2\text{H}_6\left(\frac{7 \text{ mol O}_2}{2 \text{ mol C}_2\text{H}_6}\right) = 1.58 \text{ mol O}_2$

Amount of carbon dioxide produced: $0.452 \text{ mol C}_2\text{H}_6\left(\frac{4 \text{ mol CO}_2}{2 \text{ mol C}_2\text{H}_6}\right) = 0.905 \text{ mol CO}_2$

Amount of water produced: $0.452 \text{ mol C}_2\text{H}_6\left(\frac{6 \text{ mol H}_2\text{O}}{2 \text{ mol C}_2\text{H}_6}\right) = 1.36 \text{ mol H}_2\text{O}$

Limiting Reactants

4.13. The reaction to produce sodium sulfide: Na₂SO₄(s) + 4 C(s) → Na₂S(s) + 4 CO(g)

With 25.0 g of Na₂SO₄ and 12.5 g of C, what is the limiting reactant and what mass of Na₂S can be produced?

First, calculate the mass of Na₂S that could be produced if each reactant were the limiting reactant:

$$25.0 \text{ g Na}_2\text{SO}_4 \left(\frac{1 \text{ mol Na}_2\text{SO}_4}{142.0 \text{ g Na}_2\text{SO}_4}\right)\left(\frac{1 \text{ mol Na}_2\text{S}}{1 \text{ mol Na}_2\text{SO}_4}\right)\left(\frac{78.04 \text{ g Na}_2\text{S}}{1 \text{ mol Na}_2\text{S}}\right) = 13.7 \text{ g Na}_2\text{S}$$

$$12.5 \text{ g C} \left(\frac{1 \text{ mol C}}{12.01 \text{ g C}}\right)\left(\frac{1 \text{ mol Na}_2\text{S}}{4 \text{ mol C}}\right)\left(\frac{78.04 \text{ g Na}_2\text{S}}{1 \text{ mol Na}_2\text{S}}\right) = 20.3 \text{ g Na}_2\text{S}$$

A smaller mass of Na₂S can be produced from 25.0 g Na₂SO₄ than from 12.5 g C, so Na₂SO₄ is the limiting reactant. The mass of Na₂S that can be produced is 13.7 g.

4.14. For the reaction of CaO with NH₄Cl:

CaO(s) + 2 NH₄Cl(s) → 2 NH₃(g) + H₂O(g) + CaCl₂(s)

First, calculate the mass of NH₃ that could be obtained if each reactant were the limiting reactant:

$$112 \text{ g CaO} \left(\frac{1 \text{ mol CaO}}{56.08 \text{ g CaO}}\right)\left(\frac{2 \text{ mol NH}_3}{1 \text{ mol CaO}}\right)\left(\frac{17.03 \text{ g NH}_3}{1 \text{ mol NH}_3}\right) = 68.0 \text{ g NH}_3$$

$$224 \text{ g NH}_4\text{Cl} \left(\frac{1 \text{ mol NH}_4\text{Cl}}{53.49 \text{ g NH}_4\text{Cl}}\right)\left(\frac{2 \text{ mol NH}_3}{2 \text{ mol NH}_4\text{Cl}}\right)\left(\frac{17.03 \text{ g NH}_3}{1 \text{ mol NH}_3}\right) = 71.3 \text{ g NH}_3$$

A smaller mass of NH₃ can be produced from 112 g CaO than from 224 g NH₄Cl, so CaO is the limiting reactant. The mass of NH₃ that can be produced is 68.0 g.

4.15. For the reaction of fluorine with sulfur:

The balanced equation is: S₈(s) + 24 F₂(g) → 8 SF₆(g)

Determine the amount of SF₆ that could be obtained if each reactant were the limiting reactant:

$$1.6 \text{ mol S}_8 \left(\frac{8 \text{ mol SF}_6}{1 \text{ mol S}_8}\right) = 13 \text{ mol SF}_6$$

$$35 \text{ mol F}_2 \left(\frac{8 \text{ mol SF}_6}{24 \text{ mol F}_2}\right) = 12 \text{ mol SF}_6$$

(a) Identify the limiting reagent when 1.6 mol of S₈ and 35 mol of F₂ react:

A smaller amount of SF₆ can be produced from 35 mol F₂ than from 1.6 mol S₈, so F₂ is the limiting reactant.

(b) What amount of SF₆ is produced?

The amount of SF₆ that can be produced is 12 mol.

4.16. For the reaction of sulfur with chlorine:

$$S_8(\ell) + 4\,Cl_2(g) \rightarrow 4\,S_2Cl_2(\ell)$$

$$32.0\text{ g S}_8 \left(\frac{1\text{ mol S}_8}{256.5\text{ g S}_8}\right)\left(\frac{4\text{ mol S}_2Cl_2}{1\text{ mol S}_8}\right)\left(\frac{135.0\text{ g S}_2Cl_2}{1\text{ mol S}_2Cl_2}\right) = 67.4\text{ g S}_2Cl_2$$

$$71.0\text{ g Cl}_2 \left(\frac{1\text{ mol Cl}_2}{70.90\text{ g Cl}_2}\right)\left(\frac{4\text{ mol S}_2Cl_2}{4\text{ mol Cl}_2}\right)\left(\frac{135.0\text{ g S}_2Cl_2}{1\text{ mol S}_2Cl_2}\right) = 135\text{ g S}_2Cl_2$$

(a) A smaller mass of S₂Cl₂ can be produced from 32.0 g S₈ than from 71.0 g Cl₂, so S₈ is the limiting reactant.

(b) The theoretical yield is the mass produced by the limiting reactant, 67.4 g S₂Cl₂.

(c) Mass excess reactant:

First, determine the mass of the excess reactant (Cl2) required to react with all of the limiting reactant.

$$32.0\text{ g S}_8 \left(\frac{1\text{ mol S}_8}{256.5\text{ g S}_8}\right)\left(\frac{4\text{ mol Cl}_2}{1\text{ mol S}_8}\right)\left(\frac{70.90\text{ g Cl}_2}{1\text{ mol Cl}_2}\right) = 35.4\text{ g Cl}_2 \text{ required}$$

The mass remaining is the difference between the mass available and the mass required:
71.0 g Cl₂ available – 35.4 g Cl₂ required = 35.6 g Cl₂ remains

4.17. For the reaction of methane with water:

$$CH_4(g) + H_2O(g) \rightarrow CO(g) + 3\,H_2(g)$$

$$995\text{ g CH}_4 \left(\frac{1\text{ mol CH}_4}{16.04\text{ g CH}_4}\right)\left(\frac{3\text{ mol H}_2}{1\text{ mol CH}_4}\right)\left(\frac{2.016\text{ g H}_2}{1\text{ mol H}_2}\right) = 375\text{ g H}_2$$

$$2510\text{ g H}_2O \left(\frac{1\text{ mol H}_2O}{18.02\text{ g H}_2O}\right)\left(\frac{3\text{ mol H}_2}{1\text{ mol H}_2O}\right)\left(\frac{2.016\text{ g H}_2}{1\text{ mol H}_2}\right) = 842\text{ g H}_2$$

(a) A smaller mass of H₂ can be produced from 996 g CH₄ than from 2510 g H₂O, so CH₄ is the limiting reactant.

(b) The maximum mass of H₂ that can be prepared is 375 g.

(c) Mass of water remaining:

First, determine the mass of water that is required to react with all the limiting reactant, CH$_4$:

$$995 \text{ g CH}_4 \left(\frac{1 \text{ mol CH}_4}{16.04 \text{ g CH}_4}\right)\left(\frac{1 \text{ mol H}_2\text{O}}{1 \text{ mol CH}_4}\right)\left(\frac{18.02 \text{ g H}_2\text{O}}{1 \text{ mol H}_2\text{O}}\right) = 1120 \text{ g H}_2\text{O required (3 sf)}$$

The mass remaining is the difference between the mass present and the mass required:

2510 g H$_2$O present − 1120 g H$_2$O required = 1390 g H$_2$O remains

4.18. For the formation of aluminum chloride:

2 Al(s) + 3 Cl$_2$(g) → 2 AlCl$_3$(s)

$$2.70 \text{ g Al}\left(\frac{1 \text{ mol Al}}{26.98 \text{ g Al}}\right) = 0.100 \text{ mol Al}$$

$$0.100 \text{ mol Al}\left(\frac{2 \text{ mol AlCl}_3}{2 \text{ mol Al}}\right)\left(\frac{133.3 \text{ g AlCl}_3}{1 \text{ mol AlCl}_3}\right) = 13.3 \text{ g AlCl}_3$$

$$4.05 \text{ g Cl}_2\left(\frac{1 \text{ mol Cl}_2}{70.90 \text{ g Cl}_2}\right) = 0.0571 \text{ mol Cl}_2$$

$$0.0571 \text{ mol Cl}_2\left(\frac{2 \text{ mol AlCl}_3}{3 \text{ mol Cl}_2}\right)\left(\frac{133.3 \text{ g AlCl}_3}{1 \text{ mol AlCl}_3}\right) = 5.08 \text{ g AlCl}_3$$

(a) Less AlCl$_3$ can be produced from 4.05 g Cl$_2$ than from 2.70 g AlCl$_3$, so Cl$_2$ is the limiting reactant.

(b) The mass of AlCl$_3$ that can be produced is 5.08 g.

(c) Mass excess of reactant required:

$$4.05 \text{ g Cl}_2\left(\frac{1 \text{ mol Cl}_2}{70.90 \text{ g Cl}_2}\right)\left(\frac{2 \text{ mol Al}}{3 \text{ mol Cl}_2}\right)\left(\frac{26.98 \text{ g Al}}{1 \text{ mol Al}}\right) = 1.03 \text{ g Al required}$$

Remaining: 2.70 g Al available − 1.03 g Al used = 1.67 g Al remains

(d) Amounts table:

Reaction:	2 Al(s) +	3 Cl$_2$(g) →	2 AlCl$_3$(s)
Initial amount (mol)	0.100	0.0571	0
Change (mol)	−(2/3)(0.0571) = −0.0381	−0.0571	+(2/3)(0.0571) = +0.0381
Amount after reaction (mol)	0.062	0	0.0381

4.19. Iron(III) oxide reacts with aluminum to give Fe and Al_2O_3:

$$Fe_2O_3(s) + 2\ Al(s) \rightarrow 2\ Fe(\ell) + Al_2O_3(s)$$

$$15.0\ g\ Fe_2O_3 \left(\frac{1\ mol\ Fe_2O_3}{159.7\ g\ Fe_2O_3} \right) = 0.0939\ mol\ Fe_2O_3$$

$$0.0939\ mol\ Fe_2O_3 \left(\frac{2\ mol\ Fe}{1\ mol\ Fe_2O_3} \right)\left(\frac{55.85\ g\ Fe}{1\ mol\ Fe} \right) = 10.5\ g\ Fe$$

$$30.0\ g\ Al \left(\frac{1\ mol\ Al}{26.98\ g\ Al} \right) = 1.11\ mol\ Al$$

$$1.11\ mol\ Al \left(\frac{2\ mol\ Fe}{2\ mol\ Al} \right)\left(\frac{55.85\ g\ Fe}{1\ mol\ Fe} \right) = 62.1\ g\ Fe$$

(a) Which reactant is limiting?

Less Fe can be produced from 15.0 g Fe_2O_3 than from 30.0 g Al, so Fe_2O_3 is the limiting reactant.

(b) What mass of Fe can be produced in the reaction?

The mass of Fe that can be produced is 10.5 g.

(c) What mass of the excess reactant remains?

$$15.0\ g\ Fe_2O_3 \left(\frac{1\ mol\ Fe_2O_3}{159.7\ g\ Fe_2O_3} \right)\left(\frac{2\ mol\ Al}{1\ mol\ Fe_2O_3} \right)\left(\frac{26.98\ g\ Al}{1\ mol\ Al} \right) = 5.07\ g\ Al\ required$$

30.0 g Al available – 5.07 g Al required = 24.9 g Al remains

(d) The amounts table:

Equation	Fe_2O_3 (s) +	2 Al (s) →	2 Fe (ℓ)	+ Al_2O_3 (s)
Initial mass (g)	**15.0**	**30.0**	0	0
Initial amount (mol)	0.0939	1.11	0	0
Change in amount upon reaction (mol)	−0.0939	−2(0.0939)	+2(0.0939)	+0.0939
Amount after complete reaction (mol)	0	0.92	0.188	0.0939

4.20. Maximum mass of aspirin possible:

$C_7H_6O_3(s)$	+	$C_4H_6O_3(\ell)$	→	$C_9H_8O_4(s)$	+	$CH_3CO_2H(\ell)$
salicylic acid		acetic anhydride		aspirin		acetic acid

$$100.\text{ g C}_7\text{H}_6\text{O}_3\left(\frac{1\text{ mol C}_7\text{H}_6\text{O}_3}{138.1\text{ g C}_7\text{H}_6\text{O}_3}\right)\left(\frac{1\text{ mol C}_9\text{H}_8\text{O}_4}{1\text{ mol C}_7\text{H}_6\text{O}_3}\right)\left(\frac{180.2\text{ g C}_9\text{H}_8\text{O}_4}{1\text{ mol C}_9\text{H}_9\text{O}_4}\right) = 1.30 \times 10^2 \text{ g C}_9\text{H}_8\text{O}_4 \text{ (aspirin)}$$

$$100.\text{ g C}_4\text{H}_6\text{O}_3\left(\frac{1\text{ mol C}_4\text{H}_6\text{O}_3}{102.1\text{ g C}_4\text{H}_6\text{O}_3}\right)\left(\frac{1\text{ mol C}_9\text{H}_8\text{O}_4}{1\text{ mol C}_7\text{H}_6\text{O}_3}\right)\left(\frac{180.2\text{ g C}_9\text{H}_8\text{O}_4}{1\text{ mol C}_9\text{H}_9\text{O}_4}\right) = 176 \text{ g C}_9\text{H}_8\text{O}_4 \text{ (aspirin)}$$

Less aspirin can be produced from 100. g of salicylic acid than from 100. g of acetic anhydride, so salicylic acid is the limiting reactant and the maximum mass of aspirin that can be produced is 1.30×10^2 g of aspirin.

Percent Yield

4.21. Percent yield of CH_3OH:

$$\frac{\text{actual yield}}{\text{theoretical yield}} \cdot 100\% = \frac{332\text{ g CH}_3\text{OH}}{407\text{ g CH}_3\text{OH}} \cdot 100\% = 81.6\% \text{ yield}$$

4.22. Percent yield of NH_3:
From Study Question 4.14, the theoretical yield of NH_3 is 68.0 g of NH_3. The percent yield is then $\frac{16.3\text{ g}}{68.0\text{ g}} \cdot 100\% = 24.0\%$ yield

4.23. In the formation of $Cu(NH_3)_4SO_4$:

(a) The theoretical yield of $Cu(NH_3)_4SO_4$ from 10.0 g of $CuSO_4$:

$$10.0\text{ g CuSO}_4 \cdot \frac{1\text{ mol CuSO}_4}{159.6\text{ g CuSO}_4} \cdot \frac{1\text{ mol Cu(NH}_3)_4\text{SO}_4}{1\text{ mol CuSO}_4}$$

$$\cdot \frac{227.7\text{ g Cu(NH}_3)_4\text{SO}_4}{1\text{ mol Cu(NH}_3)_4\text{SO}_4} = 14.3\text{ g Cu(NH}_3)_4\text{SO}_4$$

(b) Percentage yield of the compound:

With an actual yield of 12.6 g, the percent yield is:

$$\frac{12.6\text{ g Cu(NH}_3)_4\text{SO}_4}{14.3\text{ g Cu(NH}_3)_4\text{SO}_4} \times 100\% = 88.1\%$$

4.24. For the reaction of CH_3SH with CO:

(a) $10.0\text{ g CH}_3\text{SH}\left(\frac{1\text{ mol CH}_3\text{SH}}{48.10\text{ g CH}_3\text{SH}}\right)\left(\frac{1\text{ mol CH}_3\text{COSCH}_3}{2\text{ mol CH}_3\text{SH}}\right)$

$$\left(\frac{90.14\text{ g CH}_3\text{COSCH}_3}{1\text{ mol CH}_3\text{COSCH}_3}\right) = 9.37\text{ g CH}_3\text{COSCH}_3$$

(b) The percent yield is then: $\dfrac{8.65 \text{ g}}{9.37 \text{ g}} \cdot 100\% = 92.3 \%$ yield

4.25. For the reaction of AgNO$_3$ with K$_2$CrO$_4$:

$$2 \text{ AgNO}_3(\text{aq}) + \text{K}_2\text{CrO}_4(\text{aq}) \rightarrow \text{Ag}_2\text{CrO}_4(\text{s}) + 2 \text{ KNO}_3(\text{aq})$$

(a) $2.00 \text{ g AgNO}_3 \left(\dfrac{1 \text{ mol AgNO}_3}{169.9 \text{ g AgNO}_3}\right)\left(\dfrac{1 \text{ mol Ag}_2\text{CrO}_4}{2 \text{ mol AgNO}_3}\right)\left(\dfrac{331.7 \text{ g Ag}_2\text{CrO}_4}{1 \text{ mol Ag}_2\text{CrO}_4}\right)$

$= 1.95 \text{ g Ag}_2\text{CrO}_4$

$2.00 \text{ g K}_2\text{CrO}_4 \left(\dfrac{1 \text{ mol K}_2\text{CrO}_4}{194.2 \text{ g K}_2\text{CrO}_4}\right)\left(\dfrac{1 \text{ mol Ag}_2\text{CrO}_4}{1 \text{ mol K}_2\text{CrO}_4}\right)\left(\dfrac{331.7 \text{ g Ag}_2\text{CrO}_4}{1 \text{ mol Ag}_2\text{CrO}_4}\right)$

$= 3.42 \text{ g Ag}_2\text{CrO}_4$

Less Ag$_2$CrO$_4$ can be formed from 2.00 g AgNO$_3$ than from 2.00 g K$_2$CrO$_4$, so AgNO$_3$ is the limiting reactant, and the theoretical yield of Ag$_2$CrO$_4$ is 1.95 g.

(b) Finally, the percent yield is:

$\dfrac{1.79 \text{ g}}{1.95 \text{ g}} \cdot 100\% = 91.7\%$

4.26. For the reaction of lead(II) nitrate and sodium chloride:

$$\text{Pb(NO}_3)_2(\text{aq}) + 2 \text{ NaCl(aq)} \rightarrow \text{PbCl}_2(\text{s}) + 2 \text{ NaNO}_3(\text{aq})$$

(a) $1.50 \text{ g Pb(NO}_3)_2 \left(\dfrac{1 \text{ mol Pb(NO}_3)_2}{331.2 \text{ g Pb(NO}_3)_2}\right)\left(\dfrac{1 \text{ mol PbCl}_2}{1 \text{ mol Pb(NO}_3)_2}\right)\left(\dfrac{278.1 \text{ g PbCl}_2}{1 \text{ mol PbCl}_2}\right)$

$= 1.26 \text{ g PbCl}_2$

$1.00 \text{ g NaCl} \left(\dfrac{1 \text{ mol NaCl}}{58.44 \text{ g NaCl}}\right)\left(\dfrac{1 \text{ mol PbCl}_2}{2 \text{ mol NaCl}}\right)\left(\dfrac{278.1 \text{ g PbCl}_2}{1 \text{ mol PbCl}_2}\right) = 2.38 \text{ g PbCl}_2$

Less PbCl$_2$ can be formed from 1.50 g Pb(NO$_3$)$_2$ than from 1.00 g NaCl, so Pb(NO$_3$)$_2$ is the limiting reactant, and the theoretical yield of PbCl$_2$ is 1.26 g.

(b) The percent yield is:

$\dfrac{1.05 \text{ g}}{1.26 \text{ g}} \cdot 100\% = 83.4\%$

4.27. The percent yield is the ratio of actual/theoretical × 100. Calculate the **theoretical** yield of hydrogen if the **actual yield** is 15 g.

% yield $= \dfrac{\text{actual}}{\text{theoretical}} = \dfrac{15 \text{ g H}_2}{x} = 0.37$

[Note that the " × 100" factor was omitted because 37% was converted into its decimal equivalent.]

Solving for x: $\dfrac{15 \text{ g H}_2}{0.37} = 40.5 \text{ g H}_2$. To **actually** obtain 15 g of H₂, you must **make** 40.5 g of the gas. So, the question is "What mass of methane is needed to make 40.5 g of hydrogen?"

$$\dfrac{40.5 \text{ g H}_2}{1} \cdot \dfrac{1 \text{ mol H}_2}{2.016 \text{ g H}_2} \cdot \dfrac{1 \text{ mol CH}_4}{3 \text{ mol H}_2} \cdot \dfrac{16.0 \text{ g CH}_4}{1 \text{ mol CH}_4} = 107 \text{ g CH}_4 \text{ (or 110 g CH}_4 \text{ to 2 sf)}$$

4.28. Mass of H₂ to produce 1.0 L CH₃OH:

$$1.0 \text{ L CH}_3\text{OH}\left(\dfrac{1000 \text{ mL}}{1 \text{ L}}\right)\left(\dfrac{0.791 \text{ g CH}_3\text{OH}}{1 \text{ mL CH}_3\text{OH}}\right) = 7.9 \times 10^2 \text{ g CH}_3\text{OH}$$

$$7.9 \times 10^2 \text{ g CH}_3\text{OH}\left(\dfrac{1 \text{ mol CH}_3\text{OH}}{32.042 \text{ g CH}_3\text{OH}}\right)\left(\dfrac{2 \text{ mol H}_2}{1 \text{ mol CH}_3\text{OH}}\right)\left(\dfrac{2.016 \text{ g H}_2}{1 \text{ mol H}_2}\right) = 1.0 \times 10^2 \text{ g H}_2$$

This is the mass that would be needed if there was a 100% yield, but there is only a 74% yield.

1.0×10^2 g H₂ × (100%/74%) = 1.3×10^2 g H₂ is needed

Analysis of Mixtures

4.29. Mass percent of CuSO₄ • 5 H₂O in the mixture:

Mass of H₂O = 1.245 g − 0.832 g = 0.413 g H₂O

Since this water was a part of the hydrated salt, calculate the mass of that salt present:

In 1 mol of CuSO₄ • 5 H₂O there are 90.08 g H₂O and 159.60 g CuSO₄ or 249.68 g CuSO₄ • 5 H₂O. These masses correspond to the molar masses of anhydrous CuSO₄ and 5 mol H₂O. So:

$$0.413 \text{ g H}_2\text{O}\left(\dfrac{249.68 \text{ g CuSO}_4 \cdot 5 \text{ H}_2\text{O}}{90.08 \text{ g H}_2\text{O}}\right) = 1.14 \text{ g CuSO}_4 \cdot 5 \text{ H}_2\text{O}$$

% hydrated salt = $\dfrac{1.14 \text{ g hydrated salt}}{1.245 \text{ g mixture}} \cdot 100\% = 91.9\%$

4.30. Mass percent of dihydrate in sample:

2.634 g mixture − 2.125 g after heating = 0.509 g H₂O lost

$$0.509 \text{ g H}_2\text{O}\left(\frac{1 \text{ mol H}_2\text{O}}{18.02 \text{ g H}_2\text{O}}\right)\left(\frac{1 \text{ mol CuCl}_2 \cdot 2\text{H}_2\text{O}}{2 \text{ mol H}_2\text{O}}\right)\left(\frac{170.5 \text{ g CuCl}_2 \cdot 2\text{H}_2\text{O}}{1 \text{ mol CuCl}_2 \cdot 2\text{H}_2\text{O}}\right) = 2.41 \text{ g CuCl}_2 \cdot 2\text{H}_2\text{O}$$

$$\frac{2.41 \text{ g CuCl}_2 \cdot 2\text{H}_2\text{O}}{2.634 \text{ g mixture}} \cdot 100\% = 91.4\% \text{ CuCl}_2 \cdot 2\text{H}_2\text{O}$$

4.31. Mass percent of CaCO₃ in a limestone sample:

Note that the ratio of carbon dioxide to calcium carbonate is 1:1 (from balanced equation). Calculate the amount of carbon dioxide represented by 0.638 g CO₂.

$$0.638 \text{ g CO}_2\left(\frac{1 \text{ mol CO}_2}{44.01 \text{ g CO}_2}\right) = 0.0145 \text{ mol CO}_2$$

Since there would have been 0.0145 mol of CaCO₃, the mass is:

$$0.0145 \text{ mol CaCO}_3\left(\frac{100.08 \text{ g CaCO}_3}{1 \text{ mol CaCO}_3}\right) = 1.45 \text{ g CaCO}_3$$

The percent of CaCO₃ in the sample is: $\frac{1.45 \text{ g CaCO}_3}{1.624 \text{ g sample}} \times 100\% = 89.3\% \text{ CaCO}_3$

4.32. Mass percent of NaHCO₃:

$$0.196 \text{ g CO}_2 \cdot \frac{1 \text{ mol CO}_2}{44.01 \text{ g}} \cdot \frac{2 \text{ mol NaHCO}_3}{1 \text{ mol CO}_2} \cdot \frac{84.01 \text{ g}}{1 \text{ mol NaHCO}_3} = 0.748 \text{ g NaHCO}_3$$

$$\frac{0.748 \text{ g}}{1.7184 \text{ g}} \cdot 100\% = 43.5\% \text{ NaHCO}_3$$

4.33. "Backtrack" through the compounds containing nickel to solve this problem! Calculate the amount (moles) of the Ni(DMG)₂ complex (last equation). That number of moles can be used to determine the amount of nickel(II) nitrate. The amount of nickel(II) nitrate can be related to the amount of NiS (from the first equation). From this, you can calculate the mass of NiS. To simplify the calculation, DMG is used to represent C₄H₈N₂O₂. So the chemical formula for Ni(DMG)₂ is Ni(C₄H₇N₂O₂)₂, with a molar mass of 288.92 g.

$$0.206 \text{ g Ni(DMG)}_2\left(\frac{1 \text{ mol Ni(DMG)}_2}{288.92 \text{ g Ni(DMG)}_2}\right)\left(\frac{1 \text{ mol Ni(NO}_3)_2}{1 \text{ mol Ni(DMG)}_2}\right)\left(\frac{1 \text{ mol NiS}}{1 \text{ mol Ni(NO}_3)_2}\right)\left(\frac{90.75 \text{ g NiS}}{1 \text{ mol NiS}}\right)$$
$$= 0.0647 \text{ g NiS}$$

Now calculate the mass percent of NiS in the sample:

$$\frac{0.0647 \text{ g NiS}}{0.468 \text{ g millerite}} \cdot 100\% = 13.8\% \text{ NiS}$$

4.34. Mass percent of Al in the sample:

$$0.127 \text{ g Al}_2\text{O}_3 \left(\frac{1 \text{ mol Al}_2\text{O}_3}{102.0 \text{ g Al}_2\text{O}_3}\right)\left(\frac{2 \text{ mol Al}}{1 \text{ mol Al}_2\text{O}_3}\right)\left(\frac{26.98 \text{ g Al}}{1 \text{ mol Al}}\right) = 0.0672 \text{ g Al}$$

$$\frac{0.0672 \text{ g}}{0.764 \text{ g}} \cdot 100\% = 8.79 \% \text{ Al}$$

Using Stoichiometry to Determine Empirical and Molecular Formulas

4.35. Empirical formula of styrene:

Calculate the amount of C in the sample of styrene:

$$1.481 \text{ g CO}_2 \left(\frac{1 \text{ mol CO}_2}{44.009 \text{ g CO}_2}\right)\left(\frac{1 \text{ mol C}}{1 \text{ mol CO}_2}\right) = 0.03365 \text{ mol C}$$

Calculate the amount of H in the sample of styrene:

$$0.303 \text{ g H}_2\text{O} \left(\frac{1 \text{ mol H}_2\text{O}}{18.02 \text{ g H}_2\text{O}}\right)\left(\frac{2 \text{ mol H}}{1 \text{ mol H}_2\text{O}}\right) = 0.0336 \text{ mol H}$$

Mole ratio = $\frac{0.03365 \text{ mol C}}{0.0336 \text{ mol H}} = 1.00 \frac{\text{mol C}}{\text{mol H}}$

The empirical formula for styrene is CH.

4.36. Empirical formula of mesitylene:

Determine # mol of C and H: $0.379 \text{ g CO}_2 \cdot \frac{1 \text{ mol CO}_2}{44.01 \text{ g}} \cdot \frac{1 \text{ mol C}}{1 \text{ mol CO}_2} = 0.00861 \text{ mol C}$

and for H: $0.1035 \text{ g H}_2\text{O} \cdot \frac{1 \text{ mol H}_2\text{O}}{18.015 \text{ g}} \cdot \frac{2 \text{ mol H}}{1 \text{ mol H}_2\text{O}} = 0.01149 \text{ mol H}$

Ratio of H:C is: $\frac{0.01149 \text{ mol H}}{0.00861 \text{ mol C}} = \frac{1.33 \text{ mol H}}{1 \text{ mol C}} = \frac{4 \text{ mol H}}{3 \text{ mol C}}$ and the empirical formula: C_3H_4

4.37. Combustion of Naphthalene:

Calculate the amount of C in naphthalene:

$$1.0620 \text{ g CO}_2 \left(\frac{1 \text{ mol CO}_2}{44.009 \text{ g CO}_2}\right)\left(\frac{1 \text{ mol C}}{1 \text{ mol CO}_2}\right) = 0.024131 \text{ mol C}$$

Similarly, the amount of H in naphthalene:

$$0.1739 \text{ g H}_2\text{O}\left(\frac{1 \text{ mol H}_2\text{O}}{18.015 \text{ g H}_2\text{O}}\right)\left(\frac{2 \text{ mol H}}{1 \text{ mol H}_2\text{O}}\right) = 0.01931 \text{ mol H}$$

(a) The mole ratio is then: $\dfrac{0.024131 \text{ mol C}}{0.01931 \text{ mol H}} = \dfrac{1.250 \text{ mol C}}{\text{mol H}} = \dfrac{5 \text{ mol C}}{4 \text{ mol H}}$

The empirical formula is C₅H₄.

(b) If the molar mass is 128.2 g/mol, the molecular formula is:

Since the molecular formula represents some **multiple** of the empirical formula, calculate the "empirical formula mass".

For C₅H₄ that mass is 64.1 g/empirical formula [(5 × 12.011) + (4 × 1.008)]

Calculate the # of empirical formulas in a molecular formula:

So: $\dfrac{128.2 \text{ g/molecular formula}}{64.1 \text{ g empirical formula}} = \dfrac{2 \text{ empirical formulas}}{\text{molecular formula}}$, giving a molecular formula of C₁₀H₈.

4.38. Concerning azulene:

(a) Calculate mol of C and H: $0.364 \text{ g CO}_2 \cdot \dfrac{1 \text{ mol CO}_2}{44.01 \text{ g}} \cdot \dfrac{1 \text{ mol C}}{1 \text{ mol CO}_2} = 0.00827 \text{ mol C}$

$0.0596 \text{ g H}_2\text{O} \cdot \dfrac{1 \text{ mol H}_2\text{O}}{18.02 \text{ g}} \cdot \dfrac{2 \text{ mol H}}{1 \text{ mol H}_2\text{O}} = 0.00661 \text{ mol H}$

The ratio of C:H is $\dfrac{0.00827 \text{ mol C}}{0.00661 \text{ mol H}} = \dfrac{1.25 \text{ mol C}}{1 \text{ mol H}} = \dfrac{5 \text{ mol C}}{4 \text{ mol H}}$

The empirical formula is C₅H₄.

(b) The number of empirical formulas/ molecular formula is:

$\dfrac{128.2 \text{ g/mol}}{64.09 \text{ g/mol}} = 2$, and the molecular formula is C₁₀H₈

4.39. An unknown compound has the formula $C_xH_yO_z$.

Calculate the amount (moles) of C in the sample:

$$0.1356 \text{ g CO}_2 \left(\frac{1 \text{ mol CO}_2}{44.009 \text{ g CO}_2}\right)\left(\frac{1 \text{ mol C}}{1 \text{ mol CO}_2}\right) = 0.003081 \text{ mol C}$$

and the mass of C:

$$0.003081 \text{ mol C} \left(\frac{12.011 \text{ g C}}{1 \text{ mol C}}\right) = 0.03701 \text{ g C}$$

Calculate the amount (moles) of H in the sample:

$$0.0833 \text{ g H}_2\text{O}\left(\frac{1 \text{ mol H}_2\text{O}}{18.02 \text{ g H}_2\text{O}}\right)\left(\frac{2 \text{ mol H}}{1 \text{ mol H}_2\text{O}}\right) = 0.00925 \text{ mol H}$$

and the mass of H:

$$0.00925 \text{ mol H}\left(\frac{1.008 \text{ g H}}{1 \text{ mol H}}\right) = 0.00932 \text{ g H}$$

To determine the mass of O in the sample, subtract the masses of C and H from the mass of the sample: 0.0956 g sample – 0.03701 g C – 0.00932 g H = 0.0493 g O

Calculate the amount (moles) corresponding to this mass:

$$0.0493 \text{ g O}\left(\frac{1 \text{ mol O}}{16.00 \text{ g O}}\right) = 0.00308 \text{ mol O}$$

Establish a small whole-number ratio of the amounts of C, H, O:
Note that C and O are present in equimolar amounts (0.00308 mol). Dividing the number of moles of H by the moles of C (or O), gives a ratio of 3 H: 1 C or 3 H:1 O—for an empirical formula of $C_1H_3O_1$, more commonly written CH_3O,
If the molar mass is 62.1 g/mol, what is the molecular formula?
Adding the atomic weights of 1 C, 3 H, and 1 O gives approximately (12.0 + 3.0 + 16.0) = 31.0 for an "empirical formula" weight.
The molecular formula is then :

$$\frac{62.1 \text{ g/mol molecular formula}}{31.0 \text{ g/mol empirical formula}} = 2.00 \text{ empirical formulas/molecular formula}$$

So, the molecular formula is $C_2H_6O_2$.

4.40. What are empirical and molecular formulas?

Calculate mole of C: $0.3718 \text{ g CO}_2\left(\frac{1 \text{ mol CO}_2}{44.009 \text{ g CO}_2}\right)\left(\frac{1 \text{ mol C}}{1 \text{ mol CO}_2}\right) = 0.008448 \text{ mol C}$

and the mass of C: $0.008448 \text{ mol C} \cdot \frac{12.011 \text{ g}}{1 \text{ mol C}} = 0.1015 \text{ g C}$

Calculate mole of H and mass of H:

$$0.1522 \text{ g H}_2\text{O} \cdot \frac{1 \text{ mol H}_2\text{O}}{18.015 \text{ g}} \cdot \frac{2 \text{ mol H}}{1 \text{ mol H}_2\text{O}} = 0.01690 \text{ mol H}$$

$$0.01690 \text{ mol H} \cdot \frac{1.008 \text{ g}}{1 \text{ mol H}} = 0.01703 \text{ g H}$$

Now determine the mass of O in the sample:

mass of O = sample mass – mass of C – mass of H

$$= 0.1523 \text{ g} - 0.1015 \text{ g C} - 0.01704 \text{ g H}$$

$$= 0.0338 \text{ g O}$$

$$0.0338 \text{ g O} \cdot \frac{1 \text{ mol O}}{16.00 \text{ g}} = 0.00211 \text{ mol O}$$

$$\frac{0.008448 \text{ mol C}}{0.00211 \text{ mol O}} = \frac{4 \text{ mol C}}{1 \text{ mol O}} \qquad \frac{0.01690 \text{ mol H}}{0.00211 \text{ mol O}} = \frac{8 \text{ mol H}}{1 \text{ mol O}}$$

Using these ratios, the empirical formula is C₄H₈O. The empirical formula mass is equal to the molar mass, so the molecular formula is also C₄H₈O.

4.41 What are empirical and molecular formulas?

Calculate mole of C: $0.5517 \text{ g CO}_2 \left(\frac{1 \text{ mol CO}_2}{44.009 \text{ g CO}_2} \right) \left(\frac{1 \text{ mol C}}{1 \text{ mol CO}_2} \right) = 0.01254 \text{ mol C}$

and the mass of C: $0.01254 \text{ mol C} \cdot \frac{12.011 \text{ g}}{1 \text{ mol C}} = 0.1506 \text{ g C}$

Calculate mole of H and mass of H:

$$0.2258 \text{ g H}_2\text{O} \left(\frac{1 \text{ mol H}_2\text{O}}{18.015 \text{ g H}_2\text{O}} \right) \left(\frac{2 \text{ mol H}}{1 \text{ mol H}_2\text{O}} \right) = 0.02507 \text{ mol H}$$

$$0.02507 \text{ mol H} \cdot \frac{1.008 \text{ g}}{1 \text{ mol H}} = 0.02527 \text{ g H}$$

Now determine the mass of O in the sample:

mass of O = sample mass – mass of C – mass of H

$$= 0.2427 \text{ g} - 0.1506 \text{ g C} - 0.02527 \text{ g H}$$

$$= 0.0669 \text{ g O}$$

$$0.0669 \text{ g O} \cdot \frac{1 \text{ mol O}}{16.00 \text{ g O}} = 0.00418 \text{ mol O}$$

$$\frac{0.01254 \text{ mol C}}{0.00418 \text{ mol O}} = \frac{3.00 \text{ mol C}}{1 \text{ mol O}} \qquad \frac{0.02507 \text{ mol H}}{0.00418 \text{ mol O}} = \frac{6.00 \text{ mol H}}{1 \text{ mol O}}$$

Using these ratios, the empirical formula is C_3H_6O. The empirical formula mass is equal to the molar mass, so the molecular formula is also C_3H_6O.

4.42 What are empirical and molecular formulas?

Calculate mole of C: $0.2089 \text{ g CO}_2 \left(\dfrac{1 \text{ mol CO}_2}{44.009 \text{ g CO}_2} \right) \left(\dfrac{1 \text{ mol C}}{1 \text{ mol CO}_2} \right) = 0.004747 \text{ mol C}$

and the mass of C: $0.004747 \text{ mol C} \cdot \dfrac{12.011 \text{ g}}{1 \text{ mol C}} = 0.05701 \text{ g C}$

Calculate mole of H and mass of H:

$0.0855 \text{ g H}_2\text{O} \left(\dfrac{1 \text{ mol H}_2\text{O}}{18.015 \text{ g H}_2\text{O}} \right) \left(\dfrac{2 \text{ mol H}}{1 \text{ mol H}_2\text{O}} \right) = 0.00949 \text{ mol H}$

$0.00949 \text{ mol H} \cdot \dfrac{1.008 \text{ g}}{1 \text{ mol H}} = 0.00957 \text{ g H}$

Now determine the mass of O in the sample:

mass of O = sample mass – mass of C – mass of H

$= 0.1425 \text{ g} - 0.05701 \text{ g C} - 0.00957 \text{ g H}$

$= 0.0759 \text{ g O}$

$0.0759 \text{ g O} \cdot \dfrac{1 \text{ mol O}}{16.00 \text{ g O}} = 0.00474 \text{ mol O}$

$$\frac{0.004747 \text{ mol C}}{0.00474 \text{ mol O}} = \frac{1 \text{ mol C}}{1 \text{ mol O}} \qquad \frac{0.00949 \text{ mol H}}{0.00474 \text{ mol O}} = \frac{2 \text{ mol H}}{1 \text{ mol O}}$$

Using these ratios, the empirical formula is CH_2O which has a mas of 30.03 g/mol.

$$\frac{150.1 \text{ g/molecular formula}}{30.03 \text{ g/empirical formula}} = \frac{5 \text{ empirical formulas}}{\text{molecular formula}}$$

Molecular formula is $C_5H_{10}O_5$.

4.43. Formula of the carbonyl compound formed with nickel:

$Ni_x(CO)_y(s) + O_2(g) \rightarrow x \text{ NiO}(s) + y \text{ CO}_2(g)$

Given the mass of NiO formed, calculate the mass of Ni in NiO (and in the nickel carbonyl compound).

Quantity of Ni:

0.0426 g NiO $\cdot \dfrac{1 \text{ mol NiO}}{74.692 \text{ g NiO}} = 5.70 \times 10^{-4}$ mol NiO (and an equal number of moles of nickel since the oxide has 1mol Ni:1mol NiO). That number of moles of Ni would have a mass of: 5.70×10^{-4} mol Ni $\cdot \dfrac{58.693 \text{ g Ni}}{1 \text{ mol Ni}} = 0.0335$ g Ni.

Quantity of CO contained in the nickel carbonyl compound:

0.0973 g compound – 0.0335 g Ni = 0.0638 g CO

Amount of CO contained in the compound:

0.0638 g CO $\left(\dfrac{1 \text{ mol CO}}{28.010 \text{ g CO}}\right) = 2.28 \times 10^{-3}$ mol CO

The ratio of Ni:CO is: $\dfrac{2.28 \times 10^{-3} \text{ mol CO}}{5.70 \times 10^{-4} \text{ mol Ni}} = \dfrac{4.00 \text{ mol CO}}{1 \text{ mol Ni}}$ and the empirical formula is Ni(CO)₄.

4.44. Empirical formula of Fe$_x$(CO)$_y$:

Calculate mol of Fe and of CO: 0.799 g Fe₂O₃ $\cdot \dfrac{1 \text{ mol Fe}_2\text{O}_3}{159.7 \text{ g Fe}_2\text{O}_3} \cdot \dfrac{2 \text{ mol Fe}}{1 \text{ mol Fe}_2\text{O}_3} = 0.0100$ mol Fe

2.200 g CO₂ $\cdot \dfrac{1 \text{ mol CO}_2}{44.009 \text{ g CO}_2} \cdot \dfrac{1 \text{ mol CO}}{1 \text{ mol CO}_2} = 0.04999$ mol CO

Ratio of mol Fe: mol CO: $\dfrac{0.04999 \text{ mol CO}}{0.0100 \text{ mol Fe}} = \dfrac{5 \text{ mol CO}}{1 \text{ mol Fe}}$, so the empirical formula is Fe(CO)₅.

Solution Concentration

4.45. Molarity of Na₂CO₃ solution:

6.73 g Na₂CO₃ $\cdot \dfrac{1 \text{ mol Na}_2\text{CO}_3}{106.0 \text{ g Na}_2\text{CO}_3} = 0.0635$ mol Na₂CO₃

Molarity $\equiv \dfrac{\text{\# mol}}{\text{Liter}} = \dfrac{0.0635 \text{ mol Na}_2\text{CO}_3}{0.250 \text{ L}} = 0.254$ M Na₂CO₃

Concentration of Na⁺ and CO₃²⁻ ions:

$\dfrac{0.254 \text{ mol Na}_2\text{CO}_3}{\text{L}} \cdot \dfrac{2 \text{ mol Na}^+}{1 \text{ mol Na}_2\text{CO}_3} = 0.508$ M Na⁺

$$\frac{0.254 \text{ mol Na}_2\text{CO}_3}{\text{L}} \cdot \frac{1 \text{ mol CO}_3^{2-}}{1 \text{ mol Na}_2\text{CO}_3} = 0.254 \text{ M CO}_3^{2-}$$

4.46. Concentration of K_2SO_4, and concentration of the ions:

$$2.335 \text{ g K}_2\text{SO}_4 \cdot \frac{1 \text{ mol K}_2\text{SO}_4}{174.25 \text{ g K}_2\text{SO}_4} = 0.01340 \text{ mol K}_2\text{SO}_4$$

$$\frac{0.01340 \text{ mol K}_2\text{SO}_4}{0.500 \text{ L}} = 0.0268 \text{ M K}_2\text{SO}_4$$

$[K^+] = 2 \times [K_2SO_4] = 0.0536$ M $[SO_4^{2-}] = [K_2SO_4] = 0.0268$ M

4.47. Moles and mass of HCl

500. mL = 0.500 L

$$0.500 \text{ L} \left(\frac{0.200 \text{ mol HCl}}{1 \text{ L HCl}} \right) = 0.100 \text{ mol HCl}$$

$$0.100 \text{ mol HCl} \left(\frac{36.46 \text{ g HCl}}{1 \text{ mol HCl}} \right) = 3.65 \text{ g HCl}$$

4.48. Moles and mass of acetone

250. mL = 0.250 L

$$0.250 \text{ L} \left(\frac{4.00 \text{ mol CH}_3\text{COCH}_3}{1 \text{ L CH}_3\text{COCH}_3} \right) = 1.00 \text{ mol CH}_3\text{COCH}_3$$

$$1.00 \text{ mol CH}_3\text{COCH}_3 \left(\frac{58.08 \text{ g CH}_3\text{COCH}_3}{1 \text{ mol CH}_3\text{COCH}_3} \right) = 58.08 \text{ g CH}_3\text{COCH}_3$$

4.49. Mass of $KMnO_4$: $0.250 \text{ L} \left(\frac{0.0125 \text{ mol KMnO}_4}{1 \text{ L}} \right) \left(\frac{158.0 \text{ g KMnO}_4}{1 \text{ mol KMnO}_4} \right) = 0.494 \text{ g KMnO}_4$

4.50. Mass of Na_3PO_4, and concentrations of sodium and phosphate ions:

$$0.125 \text{ L} \cdot \frac{1.023 \times 10^{-3} \text{ mol Na}_3\text{PO}_4}{1 \text{ L}} \cdot \frac{163.9 \text{ g Na}_3\text{PO}_4}{1 \text{ mol Na}_3\text{PO}_4} = 0.0210 \text{ g Na}_3\text{PO}_4$$

$[Na^+] = 3 \times [Na_3PO_4] = 3.069 \times 10^{-3}$ M $[PO_4^{3-}] = [Na_3PO_4] = 1.023 \times 10^{-3}$ M

4.51. Volume of 0.123 M NaOH to contain 25.0 g NaOH:

Calculate moles of NaOH in 25.0 g:

$$\frac{25.0 \text{ g NaOH}}{1} \cdot \frac{1 \text{ mol NaOH}}{40.00 \text{ g NaOH}} = 0.625 \text{ mol NaOH}$$

The volume of 0.123 M NaOH that contains 0.625 mol NaOH:

$$0.625 \text{ mol NaOH} \cdot \frac{1 \text{ L}}{0.123 \text{ mol NaOH}} \cdot \frac{1000 \text{ mL}}{1 \text{ L}} = 5.08 \times 10^3 \text{ mL}$$

4.52. Volume of solution containing 322 g of solute:

$$322 \text{ g KMnO}_4 \cdot \frac{1 \text{ mol KMnO}_4}{158.0 \text{ g KMnO}_4} \cdot \frac{1 \text{ L}}{2.06 \text{ mol KMnO}_4} = 0.989 \text{ L solution}$$

4.53. Identity and concentration of ions in each of the following solutions:
(a) 0.25 M $(NH_4)_2SO_4$ gives rise to (2 × 0.25 M) = 0.50 M NH_4^+ ions and 0.25 M SO_4^{2-} ions

(b) 0.123 M Na_2CO_3 gives rise to (2 × 0.123 M) = 0.246 M Na^+ ions and 0.123 M CO_3^{2-}.

(c) 0.056 M HNO_3 gives rise to 0.056 M H_3O^+ ions and 0.056 M NO_3^- ions.

4.54. Identity and concentration of ions in each of the following solutions:
(a) 0.12 M $BaCl_2$ gives rise to 0.12 M Ba^{2+} and 0.24 M Cl^-

(b) 0.0125 M $CuSO_4$ gives rise to 0.0125 M Cu^{2+} and 0.0125 M SO_4^{2-}

(c) 0.500 M $K_2Cr_2O_7$ gives rise to 1.00 M K^+ and 0.500 M $Cr_2O_7^{2-}$

Preparing Solutions

4.55. Prepare 500. mL of 0.0200 M solution of Na_2CO_3:

Amount (# moles) of sodium carbonate needed:

$$\frac{0.0200 \text{ M Na}_2\text{CO}_3}{1 \text{ L}} \cdot \frac{0.500 \text{ L}}{1} = 0.0100 \text{ mol Na}_2\text{CO}_3$$

Mass of 0.0100 mol sodium carbonate?
Molar mass = 106.0 g Na_2CO_3/mol Na_2CO_3 and using 0.0100 mol would require:
0.0100 mol Na_2CO_3 · 106.0 g Na_2CO_3/mol Na_2CO_3 or 1.06 g Na_2CO_3.

To prepare the solution, take 1.06 g of Na_2CO_3, and transfer it to the volumetric flask. Rinse all the solid into the flask. Add a bit of distilled water and stir carefully until all the solid

Na₂CO₃ has dissolved. Once the solid has dissolved, add distilled water to the calibrated mark on the neck of the volumetric flask. Stopper and stir to assure complete mixing.

4.56. Mass of oxalic acid to prepare 250. mL of 0.15 M H₂C₂O₄:

$$0.250 \text{ L} \cdot \frac{0.15 \text{ mol H}_2\text{C}_2\text{O}_4}{1 \text{ L}} \cdot \frac{90.03 \text{ g H}_2\text{C}_2\text{O}_4}{1 \text{ mol H}_2\text{C}_2\text{O}_4} = 3.4 \text{ g H}_2\text{C}_2\text{O}_4$$

4.57. Molarity of HCl in the diluted solution: You can calculate the molarity if you know the number of moles of HCl in the 25.0 mL solution.

1. Moles of HCl in 25.0 mL of 1.50 M HCl:

$$c \times V = \frac{1.50 \text{ mol HCl}}{\text{L}} \cdot \frac{25.0 \times 10^{-3} \text{ L}}{1} = 0.0375 \text{ mol HCl}$$

2. When that number of moles is distributed to 500. mL: $\frac{0.0375 \text{ mol HCl}}{0.500 \text{ L}} = 0.0750 \text{ M HCl}$

Another way to solve this problem is to note the number of moles (found by multiplying the original molarity times the volume) is distributed in a given volume, resulting in the diluted molarity. Mathematically: $c_c \times V_c = c_d \times V_d$

4.58. Molar concentration of CuSO₄ when 4.00 mL is diluted to 10.0 mL. Because there is 1 Cu²⁺ ion per formula unit of CuSO₄, [Cu²⁺] = [CuSO₄]

$$c_d = c_c \times \frac{V_c}{V_d} = 0.0250 \text{ M} \cdot \frac{4.00 \text{ mL}}{10.0 \text{ mL}} = 0.0100 \text{ M CuSO}_4 = 0.0100 \text{ M Cu}^{2+}$$

4.59. Volume of 6.0 M NaOH to prepare 500. mL of 0.10 M NaOH

$$V_c = \frac{c_d V_d}{c_c} = \frac{(0.10 \text{ M})(500. \text{ mL})}{6.0 \text{ M}} = 8.3 \text{ mL}$$

4.60 Volume of concentrated HCl to make 250. mL of 3.0 M HCl

$$V_c = \frac{c_d V_d}{c_c} = \frac{(3.0 \text{ M})(250. \text{ mL})}{12 \text{ M}} = 63 \text{ mL}$$

4.61. Method to prepare 1.00 L of 0.125 M H$_2$SO$_4$

(a) Dilute 20.8 mL of 6.00 M H$_2$SO$_4$ to 1.00 L.

$$c_d = \frac{c_c V_c}{V_d} = \frac{(6.00 \text{ M})(20.8 \text{ mL})}{1.00 \times 10^3 \text{ mL}} = 0.125 \text{ M}$$

Method (a) is correct.

(b) Add 950. mL of water to 50.0 mL of 3.00 M H$_2$SO$_4$.

$$c_d = \frac{c_c V_c}{V_d} = \frac{(3.00 \text{ M})(50.0 \text{ mL})}{(950. \text{ mL} + 50.0 \text{ mL})} = 0.150 \text{ M}$$

Method (b) is not correct.

4.62. Method to prepare 300. mL of 0.500 M K$_2$Cr$_2$O$_7$:

(a) Add 30.0 mL of 1.50 M K$_2$Cr$_2$O$_7$ to 270. mL of water.

$$c_d = \frac{c_c V_c}{V_d} = \frac{(1.50 \text{ M})(30.0 \text{ mL})}{(30.0 \text{ mL} + 270. \text{ mL})} = 0.150 \text{ M}$$

Method (a) is not correct.

(b) Dilute 250. mL of 0.600 M K$_2$Cr$_2$O$_7$ to a volume of 300. mL.

$$c_d = \frac{c_c V_c}{V_d} = \frac{(0.600 \text{ M})(250. \text{ mL})}{(300. \text{ mL})} = 0.500 \text{ M}$$

Method (b) is correct.

Serial Dilutions

4.63. Concentration of HCl after carrying out two dilutions:

The dilution equation is $c_c V_c = c_d V_d$

First dilution: 25.00 mL of 0.136 M HCl is diluted to 100.00 mL.

$$c_d = \frac{c_c V_c}{V_d} = \frac{(0.136 \text{ M})(25.00 \text{ mL})}{100.00 \text{ mL}} = 0.0340 \text{ M}$$

Second dilution: 10.00 mL of 0.0340 M HCl is diluted to 100.00 mL.

$$c_d = \frac{c_c V_c}{V_d} = \frac{(0.0340 \text{ M})(10.00 \text{ mL})}{100.00 \text{ mL}} = 0.00340 \text{ M}$$

The concentration of HCl in the final solution is 0.00340 M.

4.64. Calculate concentration of dye in original solution if its concentration after two dilutions is known:

In the second dilution, 5.00 mL of the first dilution was diluted to 100.00 mL, and the concentration of this final solution was 0.000158 M.

$$c_c = \frac{c_d V_d}{V_c} = \frac{(0.000158 \text{ M})(100.00 \text{ mL})}{5.00 \text{ mL}} = 0.00316 \text{ M}$$

In the first dilution, 2.00 mL of the original solution was diluted to 100.00 mL.

$$c_c = \frac{c_d V_d}{V_c} = \frac{(0.00316 \text{ M})(100.00 \text{ mL})}{2.00 \text{ mL}} = 0.158 \text{ M}$$

The concentration of the dye in the original solution was 0.158 M.

Calculating and Using pH

4.65. The hydrogen ion concentration of a wine whose pH = 3.40:
Since pH is defined as $-\log[H_3O^+]$;

$$[H_3O^+] = 10^{-pH} = 10^{-3.40} = 4.0 \times 10^{-4} \text{ M } H_3O^+$$

The solution has a pH < 7.00, so it is **acidic**.

4.66. Hydronium ion concentration of $Mg(OH)_2$ solution:

$[H_3O^+] = 10^{-pH} = 10^{-10.5} = 3 \times 10^{-11}$ M   The solution is basic (pH > 7).

4.67. The $[H_3O^+]$ and pH of a solution of 0.0013 M HNO_3:

Since nitric acid is a strong acid (and strong electrolyte), we can state that:
$[H_3O^+] = [HNO_3] = 0.0013$ M; the pH = –log (0.0013) or 2.89

4.68. The $[H_3O^+]$ and pH of a solution of 1.2×10^{-4} M $HClO_4$:
Since perchloric acid is a strong acid $[H_3O^+] = [HClO_4] = 1.2 \times 10^{-4}$ M;
the pH = $-\log[H_3O^+] = -\log(1.2 \times 10^{-4}) = 3.92$

4.69. Make the interconversions and decide if the solution is acidic or basic:

	pH	[H₃O⁺]	Acidic/Basic
(a)	**1.00**	0.10	Acidic
(b)	10.50	3.2 × 10⁻¹¹ M	Basic
(c)	4.89	**1.3 × 10⁻⁵ M**	Acidic
(d)	7.64	**2.3 × 10⁻⁸ M**	Basic

pH values less than 7 indicate an acidic solution while those greater than 7 indicate a basic solution. Similarly, solutions for which the [H₃O⁺] is greater than 1.0×10^{-7} M are acidic while those with [H₃O⁺] LESS THAN 1.0×10^{-7} M are basic.

4.70. Make the interconversions and decide if the solution is acidic or basic:

	pH	[H₃O⁺]	Acidic/Basic
(a)	9.17	**6.7 × 10⁻¹⁰ M**	Basic
(b)	5.66	**2.2 × 10⁻⁶ M**	Acidic
(c)	**5.25**	5.6 × 10⁻⁶ M	Acidic
(d)	1.60	**2.5 × 10⁻² M**	Acidic

See 4.69 for explanation of acidic/basic nature.

Stoichiometry of Reactions in Solution

4.71. Volume of 0.109 M HNO₃ to react with 2.50 g of Ba(OH)₂:

Need several steps:

1. Calculate mol of barium hydroxide in 2.50 g: $2.50 \text{ g Ba(OH)}_2 \cdot \dfrac{1 \text{ mol Ba(OH)}_2}{171.3 \text{ g Ba(OH)}_2}$

2. Calculate mol of HNO₃ needed to react with that # of mol of barium hydroxide

$$2.50 \text{ g Ba(OH)}_2 \cdot \dfrac{1 \text{ mol Ba(OH)}_2}{171.3 \text{ g Ba(OH)}_2} \cdot \dfrac{2 \text{ mol HNO}_3}{1 \text{ mol Ba(OH)}_2}$$

3. Calculate volume (in L) of 0.109 M HNO₃ that contains that # of mol of nitric acid and convert to mL.

$$2.50 \text{ g Ba(OH)}_2 \cdot \dfrac{1 \text{ mol Ba(OH)}_2}{171.3 \text{ g Ba(OH)}_2} \cdot \dfrac{2 \text{ mol HNO}_3}{1 \text{ mol Ba(OH)}_2} \cdot \dfrac{1 \text{ L}}{0.109 \text{ mol HNO}_3} \cdot \dfrac{1000 \text{ mL}}{1 \text{ L}}$$

$$= 268 \text{ mL}$$

4.72. Mass of Na₂CO₃ to react with 50.0 mL of 0.125 M HNO₃

$$50.0 \text{ mL} \cdot \dfrac{1 \text{ L}}{10^3 \text{ mL}} \cdot \dfrac{0.125 \text{ mol HNO}_3}{1 \text{ L}} \cdot \dfrac{1 \text{ mol Na}_2\text{CO}_3}{2 \text{ mol HNO}_3} \cdot \dfrac{106.0 \text{ g Na}_2\text{CO}_3}{1 \text{ mol Na}_2\text{CO}_3} = 0.331 \text{ g Na}_2\text{CO}_3$$

4.73. Mass of NaOH formed from 15.0 L of 0.35 M NaCl:

$$\frac{0.35 \text{ mol NaCl}}{1 \text{ L}} \cdot \frac{15.0 \text{ L}}{1} \cdot \frac{2 \text{ mol NaOH}}{2 \text{ mol NaCl}} \cdot \frac{40.0 \text{ g NaOH}}{1 \text{ mol NaOH}} = 210 \text{ g NaOH (2 sf)}$$

Mass of Cl$_2$ obtainable:

$$\frac{0.35 \text{ mol NaCl}}{1 \text{ L}} \cdot \frac{15.0 \text{ L}}{1} \cdot \frac{1 \text{ mol Cl}_2}{2 \text{ mol NaCl}} \cdot \frac{70.9 \text{ g Cl}_2}{1 \text{ mol Cl}_2} = 190 \text{ g Cl}_2 \text{ (2 sf)}$$

4.74. Mass of hydrazine to react with 250. mL of 0.146 M H$_2$SO$_4$:

$$250. \text{ mL} \cdot \frac{1 \text{ L}}{10^3 \text{ mL}} \cdot \frac{0.146 \text{ mol H}_2\text{SO}_4}{1 \text{ L}} \cdot \frac{2 \text{ mol N}_2\text{H}_4}{1 \text{ mol H}_2\text{SO}_4} \cdot \frac{32.05 \text{ g N}_2\text{H}_4}{1 \text{ mol N}_2\text{H}_4} = 2.34 \text{ g N}_2\text{H}_4$$

4.75. Volume of 0.0138M Na$_2$S$_2$O$_3$ to dissolve 0.225 g of AgBr:
Calculate: 1. mol of AgBr

2. mol of Na$_2$S$_2$O$_3$ needed to react (balanced equation)

3. volume of 0.0138 M Na$_2$S$_2$O$_3$ containing that number of moles.

$$0.225 \text{ g AgBr} \cdot \frac{1 \text{ mol AgBr}}{187.8 \text{ g AgBr}} \cdot \frac{2 \text{ mol Na}_2\text{S}_2\text{O}_3}{1 \text{ mol AgBr}} \cdot \frac{1 \text{ L}}{0.0138 \text{ mol Na}_2\text{S}_2\text{O}_3} \cdot \frac{1000 \text{ mL}}{1 \text{ L}}$$

$$= 174 \text{ mL of 0.0138 M Na}_2\text{S}_2\text{O}_3$$

4.76. Mass of Al remaining; Mass of KAl(OH)$_4$ formed:

Mass of KAl(OH)$_4$ formed from 2.05 g Al:

$$2.05 \text{ g Al} \cdot \frac{1 \text{ mol Al}}{26.98 \text{ g Al}} \cdot \frac{2 \text{ mol KAl(OH)}_4}{2 \text{ mol Al}} \cdot \frac{134.1 \text{ g KAl(OH)}_4}{1 \text{ mol KAl(OH)}_4} = 10.2 \text{ KAl(OH)}_4$$

Mass of alum formed from 0.185 L of 1.35 M KOH:

$$185 \text{ mL} \cdot \frac{1 \text{ L}}{10^3 \text{ mL}} \cdot \frac{1.35 \text{ mol KOH}}{1 \text{ L}} \cdot \frac{2 \text{ mol KAl(OH)}_4}{2 \text{ mol KOH}} \cdot \frac{134.1 \text{ g KAl(OH)}_4}{1 \text{ mol KAl(OH)}_4} = 33.5 \text{ g KAl(OH)}_4$$

Less KAl(OH)$_4$ can be produced from 2.05 g Al than from 185 mL of 1.35 M KOH, so Al is the limiting reactant. Because it is the limiting reactant, no Al will remain.
The mass of KAl(OH)$_4$ produced in the reaction is 10.2 g.

4.77. The balanced equation:

Pb(NO$_3$)$_2$(aq) + 2 NaCl(aq) → PbCl$_2$(s) + 2 NaNO$_3$(aq)

Volume of 0.750 M Pb(NO$_3$)$_2$ needed to react with 1.00 L of 2.25 M NaCl:

$$\frac{2.25 \text{ mol NaCl}}{1 \text{ L}} \cdot \frac{1.00 \text{ L}}{1} \cdot \frac{1 \text{ mol Pb(NO}_3)_2}{2 \text{ mol NaCl}} \cdot \frac{1 \text{ L}}{0.750 \text{ mol Pb(NO}_3)_2} \cdot \frac{1000 \text{ mL}}{1 \text{ L}}$$

$$= 1500 \text{ mL or } 1.50 \times 10^3 \text{ mL}$$

4.78. Volume of 0.125 M oxalic acid to react with 35.2 mL of 0.546 M NaOH?

Calculate moles of NaOH. Note that 1 mol oxalic reacts with 2 mol NaOH; then calculate volume of oxalic acid solution containing that # of mol.

$$35.2 \text{ mL} \cdot \frac{1 \text{ L}}{10^3 \text{ mL}} \cdot \frac{0.546 \text{ mol NaOH}}{1 \text{ L}} \cdot \frac{1 \text{ mol H}_2\text{C}_2\text{O}_4}{2 \text{ mol NaOH}} \cdot \frac{1 \text{ L}}{0.125 \text{ mol H}_2\text{C}_2\text{O}_4} = 0.0769 \text{ L}$$

4.79. Reaction of lead(II) nitrate and potassium iodide

(a) Balanced chemical equation

$Pb(NO_3)_2(aq) + 2 \text{ KI}(aq) \rightarrow PbI_2(s) + 2 \text{ KNO}_3(aq)$

(b) Mass of lead(II) iodide formed in reaction

This is a limiting reactant situation, so determine the mass of PbI₂ that could be formed from each reactant.

$$50.0 \text{ mL}\left(\frac{1 \text{ L}}{1000 \text{ mL}}\right)\left(\frac{0.0500 \text{ mol Pb(NO}_3)_2}{1 \text{ L}}\right)\left(\frac{1 \text{ mol PbI}_2}{1 \text{ mol Pb(NO}_3)_2}\right)\left(\frac{461.0 \text{ g PbI}_2}{1 \text{ mol PbI}_2}\right) = 1.15 \text{ g PbI}_2$$

$$50.0 \text{ mL}\left(\frac{1 \text{ L}}{1000 \text{ mL}}\right)\left(\frac{0.150 \text{ mol KI}}{1 \text{ L}}\right)\left(\frac{1 \text{ mol PbI}_2}{2 \text{ mol KI}}\right)\left(\frac{461.0 \text{ g PbI}_2}{1 \text{ mol PbI}_2}\right) = 1.73 \text{ g PbI}_2$$

Less PbI₂ can be formed from 50.0 mL of 0.0500 M Pb(NO₃)₂ than from 50.0 mL of 0.150 M KI, so Pb(NO₃)₂ is the limiting reactant, and 1.15 g of PbI₂ can be formed.

4.80. Reaction of silver nitrate and copper(II) chloride:

(a) $2 \text{ AgNO}_3(aq) + \text{CuCl}_2(aq) \rightarrow 2 \text{ AgCl}(s) + \text{Cu(NO}_3)_2(aq)$

(b)

$$100. \text{ mL}\left(\frac{1 \text{ L}}{1000 \text{ mL}}\right)\left(\frac{0.100 \text{ mol AgNO}_3}{1 \text{ L}}\right)\left(\frac{2 \text{ mol AgCl}}{2 \text{ mol AgNO}_3}\right)\left(\frac{143.32 \text{ g AgCl}}{1 \text{ mol AgCl}}\right) = 1.43 \text{ g AgCl}$$

$$100. \text{ mL}\left(\frac{1 \text{ L}}{1000 \text{ mL}}\right)\left(\frac{0.100 \text{ mol CuCl}_2}{1 \text{ L}}\right)\left(\frac{2 \text{ mol AgCl}}{1 \text{ mol CuCl}_2}\right)\left(\frac{143.32 \text{ g AgCl}}{1 \text{ mol AgCl}}\right) = 2.87 \text{ g AgCl}$$

Less AgCl can be prepared from 100. mL of 0.100 M AgNO₃ than from 100. mL of 0.100 M CuCl₂, so AgNO₃ is the limiting reactant, and 1.43 g of AgCl can be formed.

4.81. Carbon dioxide produced from reaction of sodium carbonate and hydrochloric acid:

$Na_2CO_3(s) + 2\ HCl(aq) \rightarrow 2\ NaCl(aq) + CO_2(g) + H_2O(\ell)$

$$1.00\ g\ Na_2CO_3 \left(\frac{1\ mol\ Na_2CO_3}{105.99\ g\ Na_2CO_3}\right)\left(\frac{1\ mol\ CO_2}{1\ mol\ Na_2CO_3}\right)\left(\frac{44.01\ g\ CO_2}{1\ mol\ CO_2}\right) = 0.415\ g\ CO_2$$

$$150.\ mL\left(\frac{1\ L}{1000\ mL}\right)\left(\frac{0.750\ mol\ HCl}{1\ L}\right)\left(\frac{1\ mol\ CO_2}{2\ mol\ HCl}\right)\left(\frac{44.01\ g\ CO_2}{1\ mol\ CO_2}\right) = 2.48\ g\ CO_2$$

Less CO_2 can be produced from 1.00 g of Na_2CO_3 than from 150. mL of 0.750 M HCl, so Na_2CO_3 is the limiting reactant, and 0.415 g of CO_2 can be formed.

4.82. Carbon dioxide produced from reaction of calcium carbonate and nitric acid:

$CaCO_3(s) + 2\ HNO_3(aq) \rightarrow Ca(NO_3)_2(aq) + CO_2(g) + H_2O(\ell)$

$$1.50\ g\ CaCO_3\left(\frac{1\ mol\ CaCO_3}{100.09\ g\ CaCO_3}\right)\left(\frac{1\ mol\ CO_2}{1\ mol\ CaCO_3}\right)\left(\frac{44.01\ g\ CO_2}{1\ mol\ CO_2}\right) = 0.660\ g\ CO_2$$

$$100.\ mL\left(\frac{1\ L}{1000\ mL}\right)\left(\frac{0.500\ mol\ HNO_3}{1\ L}\right)\left(\frac{1\ mol\ CO_2}{2\ mol\ HNO_3}\right)\left(\frac{44.01\ g\ CO_2}{1\ mol\ CO_2}\right) = 1.10\ g\ CO_2$$

Less CO_2 can be produced from 1.50 g of $CaCO_3$ than from 100. mL of 0.500 M HNO_3, so $CaCO_3$ is the limiting reactant, and 0.660 g of CO_2 can be formed.

Titrations

4.83. To calculate the volume of 0.812 M HCl needed, calculate the moles of NaOH in 1.45 g, then use the stoichiometry of the balanced equation:

$HCl(aq) + NaOH(aq) \rightarrow NaCl(aq) + H_2O(\ell)$

$$1.45\ g\ NaOH \cdot \frac{1\ mol\ NaOH}{40.00\ g\ NaOH} \cdot \frac{1\ mol\ HCl}{1\ mol\ NaOH} \cdot \frac{1\ L}{0.812\ mol\ HCl} \cdot \frac{1000\ mL}{1\ L} = 44.6\ mL$$

4.84. Volume of 0.955 M HCl to react with 2.152 g of Na_2CO_3:

$$2.152\ g\ Na_2CO_3 \cdot \frac{1\ mol\ Na_2CO_3}{105.99\ g\ Na_2CO_3} \cdot \frac{2\ mol\ HCl}{1\ mol\ Na_2CO_3} \cdot \frac{1\ L}{0.955\ mol\ HCl} \cdot \frac{1000\ mL}{1\ L} = 42.5\ mL$$

4.85. Calculate:

1. moles of Na_2CO_3 corresponding to 2.150 g Na_2CO_3
2. moles of HCl that react with that number of moles (using the balanced equation)
3. the molarity of HCl containing that number of moles of HCl in 38.55 mL.

The balanced equation is:

$$Na_2CO_3 \,(aq) + 2\,HCl(aq) \rightarrow 2\,NaCl(aq) + H_2O(\ell) + CO_2\,(g)$$

$$2.150\text{ g Na}_2\text{CO}_3 \cdot \frac{1\text{ mol Na}_2\text{CO}_3}{105.99\text{ g Na}_2\text{CO}_3} \cdot \frac{2\text{ mol HCl}}{1\text{ mol Na}_2\text{CO}_3} = 0.04057\text{ mol HCl}$$

$$\frac{0.04057\text{ mol HCl}}{38.55\text{ mL}\left(\dfrac{1\text{ L}}{1000\text{ mL}}\right)} = 1.052\text{ M HCl}$$

4.86. Molarity of NaOH solution:

$$0.902\text{ g KHC}_8\text{H}_4\text{O}_4 \cdot \frac{1\text{ mol KHC}_8\text{H}_4\text{O}_4}{204.22\text{ g KHC}_8\text{H}_4\text{O}_4} \cdot \frac{1\text{ mol NaOH}}{1\text{ mol KHC}_8\text{H}_4\text{O}_4} = 0.00442\text{ mol NaOH}$$

$$\frac{0.00442\text{ mol NaOH}}{26.45\text{ mL}\left(\dfrac{1\text{ L}}{1000\text{ mL}}\right)} = 0.167\text{ M NaOH}$$

4.87. Molar Mass of an acid if 36.04 mL of 0.509 M NaOH will titrate 0.954 g of an acid H_2A:
Note that the acid is a diprotic acid, and will require 2 mol of NaOH for each mol of the acid.

Calculate the # moles of NaOH: $\dfrac{0.509\text{ mol NaOH}}{1\text{ L}} \cdot 0.03604\text{ L} = 0.0183\text{ mol NaOH}$

The number of moles of acid will be **half** of the number of moles of NaOH:

$$0.0183\text{ mol NaOH} \cdot \frac{1\text{ mol H}_2\text{A}}{2\text{ mol NaOH}} = 0.00917\text{ mol H}_2\text{A}$$

Since you know the mass corresponding to this number of moles of acid, you can calculate the molar mass (# g/mol): $\dfrac{0.954\text{ g H}_2\text{A}}{0.00917\text{ mol H}_2\text{A}} = 104\text{ g/mol}$

4.88. Identify the unknown as citric or tartaric acid:

Moles of NaOH used: $29.1\text{ mL} \cdot \dfrac{1\text{ L}}{10^3\text{ mL}} \cdot \dfrac{0.513\text{ mol NaOH}}{1\text{ L}} = 0.0149\text{ mol NaOH}$

First, assume the acid is citric acid:
Mol of citric acid corresponding to 0.0149 mol NaOH:

$$0.0149\text{ mol NaOH} \cdot \frac{1\text{ mol citric acid}}{3\text{ mol NaOH}} = 0.00498\text{ mol citric acid}$$

If this is true, the molar mass of citric acid would be:

Molar Mass of citric acid predicted: $\dfrac{0.956\text{ g}}{0.00498\text{ mol}} = 192\text{ g/mol}$

Now assume the acid is tartaric acid:
Mol of tartaric acid corresponding to 0.0149 mol NaOH:

$$0.0149 \text{ mol NaOH} \cdot \frac{1 \text{ mol tartaric acid}}{2 \text{ mol NaOH}} = 0.00746 \text{ mol tartaric acid}$$

If this is true, the molar mass of tartaric acid would be:

Molar Mass of tartaric acid predicted: $\frac{0.956 \text{ g}}{0.00746 \text{ mol}} = 128 \text{ g/mol}$

The calculated molar mass matches that of citric acid (192 g/mol) but not that of tartaric acid (150. g/mol), so the unknown acid is citric acid.

4.89. Mass percent of iron in a 0.598-gram sample that requires 22.25 mL of 0.0123 M KMnO$_4$:

Note from the balanced, net ionic equation that 1 mol MnO$_4^-$ requires 5 mol of Fe^{2+}.

$$22.25 \text{ mL} \cdot \frac{1 \text{ L}}{1000 \text{ mL}} \cdot \frac{0.0123 \text{ mol MnO}_4^-}{1 \text{ L}} \cdot \frac{5 \text{ mol Fe}^{2+}}{1 \text{ mol MnO}_4^-} \cdot \frac{55.845 \text{ g Fe}}{1 \text{ mol Fe}^{2+}} = 0.0764 \text{ g Fe}$$

The mass percent of iron is then: $\frac{0.0764 \text{ g Fe}}{0.598 \text{ g sample}} \cdot 100\% = 12.8\% \text{ Fe}$

4.90. Mass of Vitamin C in tablet:

1) Calculate amount (moles) of Br$_2$ required

2) Use balanced equation to determine moles of vitamin C:

$$27.85 \text{ mL} \cdot \frac{1 \text{ L}}{1000 \text{ mL}} \cdot \frac{0.102 \text{ mol Br}_2}{1 \text{ L}} \cdot \frac{1 \text{ mol C}_6\text{H}_8\text{O}_6}{1 \text{ mol Br}_2} \cdot \frac{176.12 \text{ g C}_6\text{H}_8\text{O}_6}{1 \text{ mol C}_6\text{H}_8\text{O}_6} = 0.500 \text{ g C}_6\text{H}_8\text{O}_6$$

Spectrophotometry

4.91. The following data were collected for a dye:

Dye Concentration ($\times 10^6$)M	Absorbance (at 475 nm)
0.50	0.24
1.5	0.36
2.5	0.44
3.5	0.59
4.5	0.70

(a) The calibration plot, slope, and intercept:

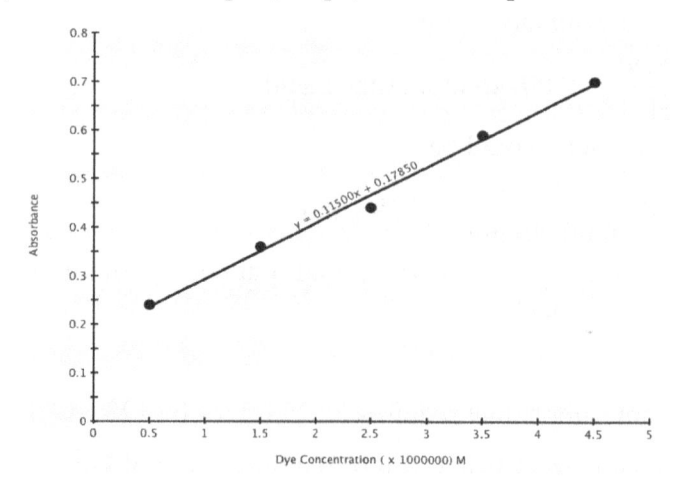

[Graph: Absorbance vs Dye Concentration (×1000000) M, with best-fit line y = 0.11500x + 0.17850]

The equation for the best-fit straight line is: $y = 0.115x + 0.1785$. Note that this graph has the dye concentrations multiplied by 10^6. Multiplying the slope (above) by this factor gives a slope of 1.2×10^5 M^{-1} with an intercept of 0.18 M.

(b) The dye concentration in a solution with A = 0.52:
One can calculate the value by substituting into the equation and rearranging:

$0.52 = 0.115x + 0.179$, so $x = (0.52 - 0.179)/0.115 = 3.0$ (2 sf).
Recall that all concentrations had been multiplied by the factor (10^6), so the actual concentration is 3.0 M/$10^6 = 3.0 \times 10^{-6}$ M

4.92. Concentration of NO$_2^-$ ion:

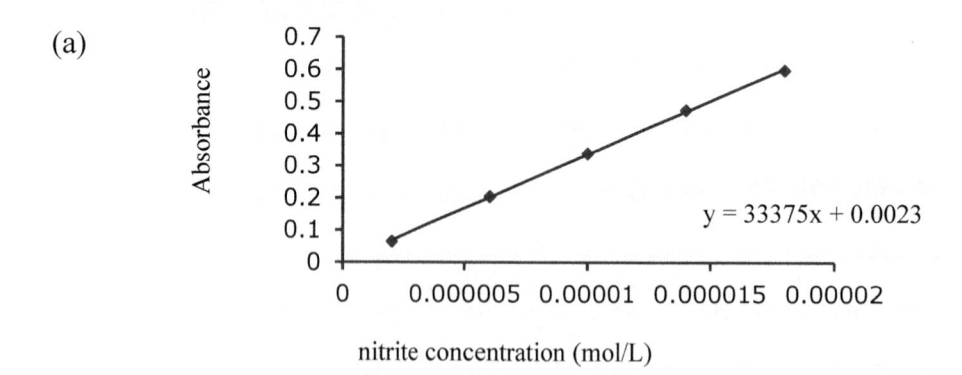

(a) [Graph: Absorbance vs nitrite concentration (mol/L), with y = 33375x + 0.0023]

(b) From the equation: $0.402 = 3.3375 \times 10^4 x + 0.0023$

Solving for the concentration: $x = 1.20 \times 10^{-5}$ M

General Questions

4.93. For the reaction of benzene with oxygen:

(a) the products of the reaction:

Combination of C_6H_6 with O_2 gives the oxide of C and the oxide of H: $CO_2(g) + H_2O(g)$

(b) the balanced equation for the reaction is:

2 C_6H_6 (ℓ) + 15 $O_2(g)$ → 12 $CO_2(g)$ + 6 $H_2O(g)$

(c) mass of oxygen, in grams, needed to completely consume the 16.04 g C_6H_6.

$$16.04 \text{ g } C_6H_6 \cdot \frac{1 \text{ mol } C_6H_6}{78.114 \text{ g } C_6H_6} \cdot \frac{15 \text{ mol } O_2}{2 \text{ mol } C_6H_6} \cdot \frac{31.998 \text{ g } O_2}{1 \text{ mol } O_2} = 49.28 \text{ g } O_2$$

(d) total mass of products expected:
This could be solved in several ways. Perhaps the simplest is to recognize that, according to the Law of Conservation of Matter, mass is conserved in a reaction. If 49.28 g of oxygen reacts with 16.04 g of benzene, the total products would also have a mass of (49.28 +16.04) g or 65.32 g.

4.94. Mass of acetone from 125 mg of acetoacetic acid:

$$125 \text{ mg acetoacetic acid} \cdot \frac{1 \text{ g}}{10^3 \text{ mg}} \cdot \frac{1 \text{ mol acetoacetic acid}}{102.1 \text{ g acetoacetic acid}} \cdot \frac{1 \text{ mol acetone}}{1 \text{ mol acetoacetic acid}}$$

$$\cdot \frac{58.08 \text{ g acetone}}{1 \text{ mol acetone}} = 0.0711 \text{ g acetone } (= 71.1 \text{ mg acetone})$$

4.95. To produce urea, what mass of arginine was used to produce 95 mg of urea?

$$95 \text{ mg urea} \cdot \frac{1 \text{ g}}{1000 \text{ mg}} \cdot \frac{1 \text{ mol urea}}{60.06 \text{ g urea}} \cdot \frac{1 \text{ mol arginine}}{1 \text{ mol urea}} \cdot \frac{174.2 \text{ g arginine}}{1 \text{ mol arginine}} = 0.28 \text{ g arginine}$$

What mass of ornithine was produced?

$$95 \text{ mg urea} \cdot \frac{1 \text{ g}}{1000 \text{ mg}} \cdot \frac{1 \text{ mol urea}}{60.06 \text{ g urea}} \cdot \frac{1 \text{ mol ornithine}}{1 \text{ mol urea}} \cdot \frac{132.2 \text{ g ornithine}}{1 \text{ mol ornithine}} = 0.21 \text{ g ornithine}$$

4.96. Iron reacting with chlorine to form iron(III) chloride:

(a) Balanced equation: 2 Fe(s) + 3 $Cl_2(g)$ → 2 $FeCl_3$(s)

(b) Mass of Cl_2 required: $10.0 \text{ g Fe} \cdot \frac{1 \text{ mol Fe}}{55.85 \text{ g Fe}} \cdot \frac{3 \text{ mol } Cl_2}{2 \text{ mol Fe}} \cdot \frac{70.90 \text{ g } Cl_2}{1 \text{ mol } Cl_2} = 19.0 \text{ g } Cl_2$

Mass of FeCl₃ produced:

$$10.0 \text{ g Fe} \cdot \frac{1 \text{ mol Fe}}{55.85 \text{ g Fe}} \cdot \frac{2 \text{ mol FeCl}_3}{2 \text{ mol Fe}} \cdot \frac{162.2 \text{ g FeCl}_3}{1 \text{ mol FeCl}_3} = 29.0 \text{ g FeCl}_3$$

(c) Percent yield: $\frac{18.5 \text{ g}}{29.0 \text{ g}} \cdot 100\% = 63.7\%$ yield

(d) Theoretical yield: 10.0 g Fe requires 19.0 g Cl₂ for complete reaction, so chlorine is the limiting reactant if 10.0 g of each reactant is combined.

$$10.0 \text{ g Cl}_2 \cdot \frac{1 \text{ mol Cl}_2}{70.90 \text{ g Cl}_2} \cdot \frac{2 \text{ mol FeCl}_3}{3 \text{ mol Cl}_2} \cdot \frac{162.2 \text{ g FeCl}_3}{1 \text{ mol FeCl}_3} = 15.3 \text{ g FeCl}_3$$

4.97. For the reaction of TiCl₄(ℓ) + 2 H₂O(ℓ) → TiO₂(s) + 4 HCl(g):

(a) Names of the compounds:

TiCl₄(ℓ) – titanium(IV) chloride—also called titanium tetrachloride

H₂O(ℓ) – water (non-systematic name)

TiO₂(s) – titanium(IV) oxide—also known as titanium dioxide
HCl(g) – hydrogen chloride

(b) Mass of water to react with 14.0 mL of TiCl₄ (d = 1.73 g/mL)

$$\frac{14.0 \text{ mL TiCl}_4}{1} \cdot \frac{1.73 \text{ g TiCl}_4}{1 \text{ mL TiCl}_4} \cdot \frac{1 \text{ mol TiCl}_4}{189.7 \text{ g TiCl}_4} \cdot \frac{2 \text{ mol H}_2\text{O}}{1 \text{ mol TiCl}_4} \cdot \frac{18.02 \text{ g H}_2\text{O}}{1 \text{ mol H}_2\text{O}} = 4.60 \text{ g H}_2\text{O}$$

(c) Mass of products expected:

$$\frac{14.0 \text{ mL}}{1} \cdot \frac{1.73 \text{ g TiCl}_4}{1 \text{ mL TiCl}_4} \cdot \frac{1 \text{ mol TiCl}_4}{189.7 \text{ g TiCl}_4} \cdot \frac{1 \text{ mol TiO}_2}{1 \text{ mol TiCl}_4} \cdot \frac{79.87 \text{ g TiO}_2}{1 \text{ mol TiO}_2} = 10.2 \text{ g TiO}_2$$

$$\frac{14.0 \text{ mL TiCl}_4}{1} \cdot \frac{1.73 \text{ g TiCl}_4}{1 \text{ mL TiCl}_4} \cdot \frac{1 \text{ mol TiCl}_4}{189.7 \text{ g TiCl}_4} \cdot \frac{4 \text{ mol HCl}}{1 \text{ mol TiCl}_4} \cdot \frac{36.46 \text{ g HCl}}{1 \text{ mol HCl}} = 18.6 \text{ g HCl}$$

4.98. Production of NO from NH₃ and O₂:

(a) According to the text (p. 197), O₂ is the limiting reactant.

$$750. \text{ g O}_2 \left(\frac{1 \text{ mol O}_2}{32.00 \text{ g O}_2} \right) \left(\frac{6 \text{ mol H}_2\text{O}}{5 \text{ mol O}_2} \right) \left(\frac{18.02 \text{ g H}_2\text{O}}{1 \text{ mol H}_2\text{O}} \right) = 507 \text{ g H}_2\text{O}$$

(b) Mass of O₂ required:

$$750. \text{ g NH}_3 \left(\frac{1 \text{ mol NH}_3}{17.03 \text{ g NH}_3} \right) \left(\frac{5 \text{ mol O}_2}{4 \text{ mol NH}_3} \right) \left(\frac{32.00 \text{ g O}_2}{1 \text{ mol O}_2} \right) = 1760 \text{ g O}_2 \text{ (3 sf)}$$

4.99. Production of sodium azide: NaNO₃ + 3 NaNH₂ → NaN₃ + 3 NaOH + NH₃

Mass of NaN₃ produced when 15.0 g of NaNO₃ reacts with 15.0 g of NaNH₂:

$$15.0 \text{ g NaNO}_3 \left(\frac{1 \text{ mol NaNO}_3}{84.99 \text{ g NaNO}_3}\right)\left(\frac{1 \text{ mol NaN}_3}{1 \text{ mol NaNO}_3}\right)\left(\frac{65.01 \text{ g NaN}_3}{1 \text{ mol NaN}_3}\right) = 11.5 \text{ g NaN}_3$$

$$15.0 \text{ g NaNH}_2 \left(\frac{1 \text{ mol NaNH}_2}{39.01 \text{ g NaNH}_2}\right)\left(\frac{1 \text{ mol NaN}_3}{3 \text{ mol NaNH}_2}\right)\left(\frac{65.01 \text{ g NaN}_3}{1 \text{ mol NaN}_3}\right) = 8.33 \text{ g NaN}_3$$

Less NaN₃ can be produced from 15.0 g NaNH₂ than from 15.0 g NaNO₃, so NaNH₂ is the limiting reactant, and 8.33 g of NaN₃ can be produced.

4.100. Regarding the formation of I₂:
(a) Reactant names: sodium iodate, sodium hydrogen sulfite (or sodium bisulfite)

(b) Masses of reactants required:

$$1.00 \text{ kg I}_2 \cdot \frac{10^3 \text{ g}}{1 \text{ kg}} \cdot \frac{1 \text{ mol I}_2}{253.8 \text{ g I}_2} \cdot \frac{2 \text{ mol NaIO}_3}{1 \text{ mol I}_2} \cdot \frac{197.9 \text{ g NaIO}_3}{1 \text{ mol NaIO}_3} = 1560 \text{ g NaIO}_3$$

$$1.00 \text{ kg I}_2 \cdot \frac{10^3 \text{ g}}{1 \text{ kg}} \cdot \frac{1 \text{ mol I}_2}{253.8 \text{ g I}_2} \cdot \frac{5 \text{ mol NaHSO}_3}{1 \text{ mol I}_2} \cdot \frac{104.1 \text{ g NaHSO}_3}{1 \text{ mol NaHSO}_3} = 2050 \text{ g NaHSO}_3$$

(c) Yield of I₂ from NaIO₃:

$$15.0 \text{ g NaIO}_3 \cdot \frac{1 \text{ mol NaIO}_3}{197.9 \text{ g NaIO}_3} \cdot \frac{1 \text{ mol I}_2}{2 \text{ mol NaIO}_3} \cdot \frac{253.8 \text{ g I}_2}{1 \text{ mol I}_2} = 9.62 \text{ g I}_2$$

Yield of I₂ from NaHSO₃:

$$125 \text{ mL} \cdot \frac{1 \text{ L}}{1000 \text{ mL}} \cdot \frac{0.853 \text{ mol NaHSO}_3}{1 \text{ L}} \cdot \frac{1 \text{ mol I}_2}{5 \text{ mol NaHSO}_3} \cdot \frac{253.8 \text{ g I}_2}{1 \text{ mol I}_2} = 5.41 \text{ g I}_2$$

Less I₂ can be prepared from 125 mL of 0.853 M NaHSO₃ than from 15.0 g NaIO₃, so NaHSO₃ is the limiting reatant. The maximum mass that can be produced is 5.41 g I₂.

4.101. Determine the mass percent of saccharin in the sample of sweetener:
1. Determine the S in BaSO₄:

$$0.2070 \text{ g BaSO}_4 \cdot \frac{32.06 \text{ g S}}{233.39 \text{ g BaSO}_4} = 0.02843 \text{ g S}$$

2. Determine the mass of saccharin that contains this mass of S:

$$0.02843 \text{ g S} \cdot \frac{183.18 \text{ g saccharin}}{32.06 \text{ g S}} = 0.1625 \text{ g saccharin}$$

3. The mass percent of saccharin is then:

$$\frac{0.1625 \text{ g saccharin}}{0.2140 \text{ g sweetener}} \cdot 100\% = 75.92\% \text{ saccharin}$$

4.102. Empirical formula of a boron hydride:

Mass of B in the compound:

$$0.422 \text{ g B}_2\text{O}_3 \cdot \frac{1 \text{ mol B}_2\text{O}_3}{69.62 \text{ g B}_2\text{O}_3} \cdot \frac{2 \text{ mol B}}{1 \text{ mol B}_2\text{O}_3} \cdot \frac{10.81 \text{ g B}}{1 \text{ mol B}} = 0.131 \text{ g B}$$

Mass of H in the compound: 0.148 g B$_x$H$_y$ – 0.131 g B = 0.017 g H

Mol:Mol ratio of B:H in the compound:

$$0.131 \text{ g B} \cdot \frac{1 \text{ mol B}}{10.81 \text{ g B}} = 0.0121 \text{ mol B} \qquad 0.017 \text{ g H} \cdot \frac{1 \text{ mol H}}{1.01 \text{ g H}} = 0.017 \text{ mol H}$$

$$\frac{0.017 \text{ mol H}}{0.0121 \text{ mol B}} = \frac{1.4 \text{ mol H}}{1 \text{ mol B}} = \frac{7 \text{ mol H}}{5 \text{ mol B}} \text{ so the empirical formula is B}_5\text{H}_7$$

4.103. To determine the empirical formula, you need the ratios of silicon and hydrogen atoms in the compound. Knowing that 6.22 g of the compound produced 11.64 g SiO₂ calculate the moles of silicon in 11.64 g.

$$11.64 \text{ g SiO}_2 \cdot \frac{1 \text{ mol SiO}_2}{60.083 \text{ g SiO}_2} \cdot \frac{1 \text{ mol Si}}{1 \text{ mol SiO}_2} = 0.1937 \text{ mol Si}$$

The mass of water formed can be used to calculate the amount of H in the compound.

$$6.980 \text{ g H}_2\text{O} \cdot \frac{1 \text{ mol H}_2\text{O}}{18.015 \text{ g H}_2\text{O}} \cdot \frac{2 \text{ mol H}}{1 \text{ mol H}_2\text{O}} = 0.7749 \text{ mol H}$$

Now determine the ratio of the moles of atoms of the elements present:

$$\frac{0.7749 \text{ mol H}}{0.1937 \text{ mol Si}} = \frac{4.000 \text{ mol H}}{1 \text{ mol Si}}$$

The ratio of Si:H is 1 mol Si:4 mol H, giving an empirical formula of SiH₄.

4.104. Empirical formula of menthol:

Calculate the mass of C and H in the original compound:

$$269 \text{ mg CO}_2 \cdot \frac{1 \text{ g}}{10^3 \text{ mg}} \cdot \frac{1 \text{ mol CO}_2}{44.01 \text{ g}} \cdot \frac{1 \text{ mol C}}{1 \text{ mol CO}_2} = 0.00611 \text{ mol C}$$

$$0.00611 \text{ mol C} \cdot \frac{12.01 \text{ g}}{1 \text{ mol C}} = 0.0734 \text{ g C}$$

$$111 \text{ mg H}_2\text{O} \cdot \frac{1 \text{ g}}{10^3 \text{ mg}} \cdot \frac{1 \text{ mol H}_2\text{O}}{18.02 \text{ g H}_2\text{O}} \cdot \frac{2 \text{ mol H}}{1 \text{ mol H}_2\text{O}} = 0.0123 \text{ mol H}$$

$$0.0123 \text{ mol H} \cdot \frac{1.008 \text{ g H}}{1 \text{ mol H}} = 0.0124 \text{ g H}$$

Mass O = sample mass – mass of C – mass of H

$$= (95.6 \text{ mg} \cdot \frac{1 \text{ g}}{10^3 \text{ mg}}) - 0.0734 \text{ g} - 0.0124 \text{ g}$$

$$= 0.0098 \text{ g O}$$

$$0.0098 \text{ g O} \cdot \frac{1 \text{ mol O}}{16.0 \text{ g O}} = 0.00061 \text{ mol O}$$

Mol:Mol ratios: $\frac{0.00611 \text{ mol C}}{0.00061 \text{ mol O}} = \frac{10 \text{ mol C}}{1 \text{ mol O}}$ and $\frac{0.0123 \text{ mol H}}{0.00061 \text{ mol O}} = \frac{20 \text{ mol H}}{1 \text{ mol O}}$

The empirical formula is $C_{10}H_{20}O$.

4.105. Empirical formula of Benzoquinone:

$$0.257 \text{ g CO}_2 \cdot \frac{1 \text{ mol CO}_2}{44.01 \text{ g CO}_2} \cdot \frac{1 \text{ mol C}}{1 \text{ mol CO}_2} = 0.00584 \text{ mol C}$$

$$0.00584 \text{ mol C} \cdot \frac{12.01 \text{ g C}}{1 \text{ mol C}} = 0.0701 \text{ g C}$$

$$0.0350 \text{ g H}_2\text{O} \cdot \frac{1 \text{ mol H}_2\text{O}}{18.02 \text{ g H}_2\text{O}} \cdot \frac{2 \text{ mol H}}{1 \text{ mol H}_2\text{O}} = 0.00388 \text{ mol H}$$

$$0.00388 \text{ mol H} \cdot \frac{1.008 \text{ g H}}{1 \text{ mol H}} = 0.00392 \text{ g H}$$

The mass of O present is: 0.105 g compound – (0.0701g C + 0.00392 g H) = 0.031 g O

$$0.031 \text{ g O} \cdot \frac{1 \text{ mol O}}{16.00 \text{ g O}} = 0.0019 \text{ mol O}$$

The mole ratios are

$$\frac{0.00583 \text{ mol C}}{0.0019 \text{ mol O}} = \frac{3.0 \text{ mol C}}{1 \text{ mol O}} \qquad \frac{0.00388 \text{ mol H}}{0.0019 \text{ mol O}} = \frac{2.0 \text{ mol H}}{1 \text{ mol O}}$$

The empirical formula is C_3H_2O.

4.106. For the reaction of iron(II) chloride and sodium sulfide:
(a) Balanced equation: FeCl$_2$(aq) + Na$_2$S(aq) → FeS(s) + 2 NaCl(aq)

(b) Limiting reactant: Determine mass of FeS with each reactant:

$$40.\text{ g Na}_2\text{S} \cdot \frac{1 \text{ mol Na}_2\text{S}}{78.0 \text{ g Na}_2\text{S}} \cdot \frac{1 \text{ mol FeS}}{1 \text{ mol Na}_2\text{S}} \cdot \frac{87.9 \text{ g FeS}}{1 \text{ mol FeS}} = 45 \text{ g FeS}$$

$$40.\text{ g FeCl}_2 \cdot \frac{1 \text{ mol FeCl}_2}{127 \text{ g FeCl}_2} \cdot \frac{1 \text{ mol FeS}}{1 \text{ mol FeCl}_2} \cdot \frac{87.9 \text{ g FeS}}{1 \text{ mol FeS}} = 28 \text{ g FeS}$$

FeCl$_2$ is the limiting reactant.

(c) The theoretical yield is the mass produced by the limiting reactant, so 28 g FeS is the theoretical yield.

(d) Mass of excess reactant:

$$40.\text{ g FeCl}_2 \cdot \frac{1 \text{ mol FeCl}_2}{127 \text{ g FeCl}_2} \cdot \frac{1 \text{ mol Na}_2\text{S}}{1 \text{ mol FeCl}_2} \cdot \frac{78.0 \text{ g Na}_2\text{S}}{1 \text{ mol Na}_2\text{S}} = 25 \text{ g Na}_2\text{S required}$$

40. g Na$_2$S available – 25 g Na$_2$S required = 15 g Na$_2$S remains

(e) Mass of FeCl$_2$ required:

$$40.\text{ g Na}_2\text{S} \cdot \frac{1 \text{ mol Na}_2\text{S}}{78.0 \text{ g Na}_2\text{S}} \cdot \frac{1 \text{ mol FeCl}_2}{1 \text{ mol Na}_2\text{S}} \cdot \frac{127 \text{ g FeCl}_2}{1 \text{ mol FeCl}_2} = 65 \text{ g FeCl}_2$$

4.107. Theoretical yield of sulfuric acid from 3.00 kg of Cu$_2$S:
Determine the # mol of cuprite, then the # mol of H$_2$SO$_4$ (since there is 1 H$_2$SO$_4$ molecule per molecule of cuprite needed). The mass of H$_2$SO$_4$ can be determined from the # mol of the acid.

$$3.00 \text{ kg Cu}_2\text{S} \cdot \frac{1000 \text{ g}}{1 \text{ kg}} \cdot \frac{1 \text{ mol Cu}_2\text{S}}{159.2 \text{ g Cu}_2\text{S}} \cdot \frac{1 \text{ mol H}_2\text{SO}_4}{1 \text{ mol Cu}_2\text{S}} \cdot \frac{98.07 \text{ g H}_2\text{SO}_4}{1 \text{ mol H}_2\text{SO}_4} \cdot \frac{1 \text{ kg}}{1000 \text{ g}}$$

$$= 1.85 \text{ kg H}_2\text{SO}_4$$

4.108. Identify metal:

Determine amount of MCO$_3$: $0.376 \text{ g CO}_2 \cdot \frac{1 \text{ mol CO}_2}{44.01 \text{ g}} \cdot \frac{1 \text{ mol MCO}_3}{1 \text{ mol CO}_2} = 0.00854 \text{ mol MCO}_3$

Molar Mass would be $\frac{1.056 \text{ g}}{0.00854 \text{ mol}} = 124 \text{ g/mol}$

124 g/mol (MCO$_3$) – 60 g/mol (CO$_3$) = 64 g/mol so the metal is (b) copper, Cu

4.109. A metal forms an oxide, MO$_2$.

The mass of oxide (0.452 g) – mass of metal (0.356 g) = 0.096 g of oxygen.

$$0.096 \text{ g O} \cdot \frac{1 \text{ mol O}}{16.00 \text{ g O}} \cdot \frac{1 \text{ mol M}}{2 \text{ mol O}} = 0.0030 \text{ mol M}$$

The third factor is arrived at by the formula (given in the problem).

This number of moles (0.0030 mol) corresponds to 0.356 g, so you can calculate the approximate atomic weight of the metal: $\dfrac{0.356 \text{ g M}}{0.0030 \text{ mol M}} = 1.2 \times 10^2$ g/mol.

The metal with atomic weight of approximately 120 is tin, Sn (118.71 g/mol)

4.110. Empirical formula of the titanium oxide:

Amount of Ti in oxide: $1.598 \text{ g TiO}_2 \cdot \dfrac{1 \text{ mol TiO}_2}{79.865 \text{ g TiO}_2} \cdot \dfrac{1 \text{ mol Ti}}{1 \text{ mol TiO}_2} = 0.02001 \text{ mol Ti}$

Mass of Ti: $0.02001 \text{ mol Ti} \cdot \dfrac{47.867 \text{ g}}{1 \text{ mol Ti}} = 0.9578 \text{ g Ti}$

Mass of O in unknown oxide: 1.438 g Ti$_x$O$_y$ – 0.9577 g Ti = 0.480 g O

Amount of O in unknown oxide: $0.480 \text{ g O} \cdot \dfrac{1 \text{ mol O}}{16.00 \text{ g O}} = 0.0300 \text{ mol O}$

Mole ratio: $\dfrac{0.0300 \text{ mol O}}{0.02001 \text{ mol Ti}} = \dfrac{1.50 \text{ mol O}}{1 \text{ mol Ti}} = \dfrac{3 \text{ mol O}}{2 \text{ mol Ti}}$ so the empirical formula is Ti$_2$O$_3$

4.111. The problem seems complex owing to the series of reactions that must occur to produce the desired product: KClO$_4$. Ask the question, "How do I produce KClO$_4$?" The answer is with the third equation given: 4 KClO$_3$ → 3 KClO$_4$ + KCl. Then you can trace back through the other equations until you get to Cl$_2$. To produce 3 mol KClO$_4$ requires 4 mol KClO$_3$, which can be produced with the 2nd equation given. To produce 1 mol KClO$_3$ requires 3 mol KClO, which can be produced with the 1st equation given. To produce 1 mol KClO requires 1 mol Cl$_2$. The full solution is:

$$234 \text{ kg KClO}_4 \cdot \frac{1000 \text{ g}}{1 \text{ kg}} \cdot \frac{1 \text{ mol KClO}_4}{138.54 \text{ g KClO}_4} \cdot \frac{4 \text{ mol KClO}_3}{3 \text{ mol KClO}_4} \cdot \frac{3 \text{ mol KClO}}{1 \text{ mol KClO}_3}$$

$$\cdot \frac{1 \text{ mol Cl}_2}{1 \text{ mol KClO}} \cdot \frac{70.90 \text{ g Cl}_2}{1 \text{ mol Cl}_2} \cdot \frac{1 \text{ kg}}{1000 \text{ g}} = 479 \text{ kg Cl}_2$$

4.112. Preparation of commercial sodium hydrosulfite:

(a) $125 \text{ kg Zn} \cdot \dfrac{1000 \text{ g}}{1 \text{ kg}} \cdot \dfrac{1 \text{ mol Zn}}{65.38 \text{ g Zn}} \cdot \dfrac{1 \text{ mol ZnS}_2\text{O}_4}{1 \text{ mol Zn}} \cdot \dfrac{1 \text{ mol Na}_2\text{S}_2\text{O}_4}{1 \text{ mol ZnS}_2\text{O}_4}$

$\cdot \dfrac{174.1 \text{ g Na}_2\text{S}_2\text{O}_4}{1 \text{ mol Na}_2\text{S}_2\text{O}_4} = 3.33 \times 10^5 \text{ g Na}_2\text{S}_2\text{O}_4$

$500. \text{ g SO}_2 \cdot \dfrac{1 \text{ mol SO}_2}{64.06 \text{ g SO}_2} \cdot \dfrac{1 \text{ mol ZnS}_2\text{O}_4}{2 \text{ mol SO}_2} \cdot \dfrac{1 \text{ mol Na}_2\text{S}_2\text{O}_4}{1 \text{ mol ZnS}_2\text{O}_4} \cdot \dfrac{174.1 \text{ g Na}_2\text{S}_2\text{O}_4}{1 \text{ mol Na}_2\text{S}_2\text{O}_4}$

$= 679 \text{ g Na}_2\text{S}_2\text{O}_4$

The maximum mass produced is 679 g Na$_2$S$_2$O$_4$, as SO$_2$ limits the amount of product.

(b) $679 \text{ g Na}_2\text{S}_2\text{O}_4 \cdot \dfrac{100 \text{ g commercial product}}{90.1 \text{ g Na}_2\text{S}_2\text{O}_4} = 754 \text{ g commercial product}$

4.113. Mass of lime obtainable from 125 kg of limestone:

The concentration of CaO in the limestone is 95.0% (95.0 g of CaO per 100 g of limestone).

Find the mass of CaCO$_3$ in 125 kg of limestone, and then the mass of CaO in CaCO$_3$

$125 \text{ kg limestone} \cdot \dfrac{1000 \text{ g}}{1 \text{ kg}} \cdot \dfrac{95.0 \text{ g CaCO}_3}{100 \text{ g limestone}} \cdot \dfrac{56.08 \text{ g CaO}}{100.1 \text{ g CaCO}_3} \cdot \dfrac{1 \text{ kg}}{1000 \text{ g}} = 66.5 \text{ kg CaO}$

4.114. Maximum mass of Ag$_2$MoS$_4$:

Determine mass of compound possible using each of the three reactants:

$8.63 \text{ g Ag} \cdot \dfrac{1 \text{ mol Ag}}{107.9 \text{ g Ag}} \cdot \dfrac{1 \text{ mol Ag}_2\text{MoS}_4}{2 \text{ mol Ag}} \cdot \dfrac{439.9 \text{ g Ag}_2\text{MoS}_4}{1 \text{ mol Ag}_2\text{MoS}_4} = 17.6 \text{ g Ag}_2\text{MoS}_4$

$3.36 \text{ g Mo} \cdot \dfrac{1 \text{ mol Mo}}{95.95 \text{ g Mo}} \cdot \dfrac{1 \text{ mol Ag}_2\text{MoS}_4}{1 \text{ mol Mo}} \cdot \dfrac{439.9 \text{ g Ag}_2\text{MoS}_4}{1 \text{ mol Ag}_2\text{MoS}_4} = 15.4 \text{ g Ag}_2\text{MoS}_4$

$4.81 \text{ g S} \cdot \dfrac{1 \text{ mol S}}{32.06 \text{ g S}} \cdot \dfrac{1 \text{ mol Ag}_2\text{MoS}_4}{4 \text{ mol S}} \cdot \dfrac{439.9 \text{ g Ag}_2\text{MoS}_4}{1 \text{ mol Ag}_2\text{MoS}_4} = 16.5 \text{ g Ag}_2\text{MoS}_4$

Less Ag$_2$MoS$_4$ can be obtained from 3.36 g Mo than from 8.63 g Ag or 4.81 g S, so Mo is the limiting reactant. The maximum mass that can be obtained is 15.4 g Ag$_2$MoS$_4$

4.115. Mass of ammonium nitrate produced:

$$6.00 \text{ g N}_2 \left(\frac{1 \text{ mol N}_2}{28.01 \text{ g N}_2}\right)\left(\frac{2 \text{ mol NH}_4\text{NO}_3}{2 \text{ mol N}_2}\right)\left(\frac{80.04 \text{ g NH}_4\text{NO}_3}{1 \text{ mol NH}_4\text{NO}_3}\right) = 17.1 \text{ g NH}_4\text{NO}_3$$

$$6.00 \text{ g H}_2\text{O} \left(\frac{1 \text{ mol H}_2\text{O}}{18.02 \text{ g H}_2\text{O}}\right)\left(\frac{2 \text{ mol NH}_4\text{NO}_3}{4 \text{ mol H}_2\text{O}}\right)\left(\frac{80.04 \text{ g NH}_4\text{NO}_3}{1 \text{ mol NH}_4\text{NO}_3}\right) = 13.3 \text{ g NH}_4\text{NO}_3$$

$$6.00 \text{ g O}_2 \left(\frac{1 \text{ mol O}_2}{32.00 \text{ g O}_2}\right)\left(\frac{2 \text{ mol NH}_4\text{NO}_3}{1 \text{ mol O}_2}\right)\left(\frac{80.04 \text{ g NH}_4\text{NO}_3}{1 \text{ mol NH}_4\text{NO}_3}\right) = 30.0 \text{ g NH}_4\text{NO}_3$$

Less NH₄NO₃ can be obtained from 6.00 g H₂O than from 6.00 g N₂ or 6.00 g O₂, so H₂O is the limiting reactant, and 13.3 g NH₄NO₃ can be obtained.

4.116. Mass of titanium(IV) chloride produced:

$$100. \text{ g FeTiO}_3 \left(\frac{1 \text{ mol FeTiO}_3}{151.71 \text{ g FeTiO}_3}\right)\left(\frac{2 \text{ mol TiCl}_4}{2 \text{ mol FeTiO}_3}\right)\left(\frac{189.67 \text{ g TiCl}_4}{1 \text{ mol TiCl}_4}\right) = 125 \text{ g TiCl}_4$$

$$175. \text{ g Cl}_2 \left(\frac{1 \text{ mol Cl}_2}{70.90 \text{ g Cl}_2}\right)\left(\frac{2 \text{ mol TiCl}_4}{7 \text{ mol Cl}_2}\right)\left(\frac{189.67 \text{ g TiCl}_4}{1 \text{ mol TiCl}_4}\right) = 134 \text{ g TiCl}_4$$

$$30.0 \text{ g C} \left(\frac{1 \text{ mol C}}{12.01 \text{ g C}}\right)\left(\frac{2 \text{ mol TiCl}_4}{6 \text{ mol C}}\right)\left(\frac{189.67 \text{ g TiCl}_4}{1 \text{ mol TiCl}_4}\right) = 158 \text{ g TiCl}_4$$

Less TiCl₄ can be obtained from 100. g FeTiO₃ than from 175 g Cl₂ or 30.0 g C, so FeTiO₃ is the limiting reactant, and 125 g of TiCl₄ can be obtained.

4.117. A mixture of butene, C₄H₈, and butane, C₄H₁₀, has a mass of 2.86 g.

8.800 g of CO₂ and 4.095 g of H₂O result upon combustion.
What is the weight percent of butene and butane in the mixture? The balanced equations for the combustions: Butene: C₄H₈(g) + 6 O₂(g) → 4 CO₂(g) + 4 H₂O(g)

Butane: 2 C₄H₁₀(g) + 13 O₂(g) → 8 CO₂(g) + 10 H₂O(g)

Establish 2 equations with 2 unknowns: Let x = g C₄H₈ and y = g C₄H₁₀

Then $x + y = 2.860$ g and

$$\left(x \cdot \frac{1 \text{ mol C}_4\text{H}_8}{56.108 \text{ g C}_4\text{H}_8} \cdot \frac{4 \text{ mol CO}_2}{1 \text{ mol C}_4\text{H}_8}\right) + \left(y \cdot \frac{1 \text{ mol C}_4\text{H}_{10}}{58.124 \text{ g C}_4\text{H}_{10}} \cdot \frac{8 \text{ mol CO}_2}{2 \text{ mol C}_4\text{H}_{10}}\right)$$

$$= 8.800 \text{ g CO}_2 \cdot \frac{1 \text{ mol CO}_2}{44.009 \text{ g CO}_2}$$

Note that the fractions $\dfrac{4 \text{ mol CO}_2}{1 \text{ mol C}_4\text{H}_8}$ and $\dfrac{8 \text{ mol CO}_2}{2 \text{ mol C}_4\text{H}_{10}}$ result from the stoichiometry of the two combustion equations.

Rearrange the first equation: $y = 2.860 \text{ g} - x$ and substitute into the 2nd equation:

$$\left(x \cdot \dfrac{1 \text{ mol C}_4\text{H}_8}{56.108 \text{ g C}_4\text{H}_8} \cdot \dfrac{4 \text{ mol CO}_2}{1 \text{ mol C}_4\text{H}_8} \right) + \left((2.860 \text{ g} - x) \cdot \dfrac{1 \text{ mol C}_4\text{H}_{10}}{58.124 \text{ g C}_4\text{H}_{10}} \cdot \dfrac{8 \text{ mol CO}_2}{2 \text{ mol C}_4\text{H}_{10}} \right)$$

$$= 8.800 \text{ g CO}_2 \cdot \dfrac{1 \text{ mol CO}_2}{44.009 \text{ g CO}_2}$$

Grouping terms and rearranging gives:

$$\dfrac{4x}{56.108} \dfrac{\text{mol CO}_2}{\text{g}} + \dfrac{8(2.860 \text{ g} - x)}{2(58.124)} \dfrac{\text{mol CO}_2}{\text{g}} = \dfrac{8.800}{44.009} \text{ mol CO}_2$$

$$0.071291x \dfrac{\text{mol CO}_2}{\text{g}} + 0.19682 \text{ mol CO}_2 - 0.068818x \dfrac{\text{mol CO}_2}{\text{g}} = 0.2000 \text{ mol CO}_2$$

$$0.002473x \dfrac{\text{mol CO}_2}{\text{g}} = 0.0031 \text{ mol CO}_2$$

solving gives
x = 1.3 g C₄H₈ and (2.860 g – x) or 1.6 g C₄H₁₀

4.118. Mass of (CH₃)₂SiCl₂ needed:

Thickness = 250 layers $\cdot \dfrac{0.60 \text{ nm}}{1 \text{ layer}} \cdot \dfrac{1 \text{ m}}{10^9 \text{ nm}} \cdot \dfrac{100 \text{ cm}}{1 \text{ m}} = 1.5 \times 10^{-5}$ cm thick

$3.00 \text{ m}^2 \cdot \left(\dfrac{100 \text{ cm}}{1 \text{ m}} \right)^2 = 3.00 \times 10^4 \text{ cm}^2$

Volume = Area × Thickness = $(3.00 \times 10^4 \text{ cm}^2)(1.5 \times 10^{-5} \text{ cm}) = 0.45 \text{ cm}^3$

Mass = $0.45 \text{ cm}^3 \cdot \dfrac{1.0 \text{ g}}{1 \text{ cm}^3} = 0.45$ g (CH₃)₂SiCl₂

4.119. The weight percent of CuS and Cu₂S in the ore:

You need to determine the mass of Cu originally contained in CuS (and similarly in Cu₂S). Knowing that the ore is 11.00% impure, write: 100.00 g ore – 11.00 g impurity = 89.00 g of the copper sulfides. The mass of **pure** copper (when reduced) is 89.50% of 75.40 g Cu. (or 67.48 g). Since there are two unknowns (the mass of CuS and the mass of Cu₂S) two equations are needed:

(1) Cu (from CuS) + Cu (from Cu₂S) = 67.48 g.

(2) If we let x = mass of CuS, then $89.00 - x$ = mass of Cu$_2$S.

Further we know that the % of Cu in each of the salts can be calculated from the formulas, and represented by (mass Cu)/(mass Cu$_x$S salt).

Rewriting equation (1): $\dfrac{63.546 \text{ g Cu}}{95.61 \text{ g CuS}} \cdot x + \dfrac{127.09 \text{ g Cu}}{159.15 \text{ g Cu}_2\text{S}} \cdot (89.00 - x) = 67.48$

[For those who like to keep track of units, note that x and the term $(89.00 - x)$ will have units of g Cu salt, so those units will "cancel" with the denominator of the respective fractions.] For simplicity's sake, reduce the two fractions to a decimal:

$0.6646x + 0.79855(89.00 - x) = 67.48$ and $0.6646x + 71.07 - 0.79855x = 67.48$

Combining and simplifying: $-0.1339x = -3.59$ and $x = 3.59/0.1339$ or 26.8 g CuS.

Knowing the mass of CuS, the mass of Cu$_2$S = $89.00 - 26.8 = 62.2$ g Cu$_2$S

The weight percent of CuS in the ORE: = 26.8% and of Cu$_2$S = 62.2%

4.120. Mass of NaHCO$_3$ required:

Equation: H$_3$C$_6$H$_5$O$_7$(aq) + 3 NaHCO$_3$(aq) → 3 H$_2$O(ℓ) + 3 CO$_2$(g) + Na$_3$C$_6$H$_5$O$_7$(aq)

100. mg H$_3$C$_6$H$_5$O$_7$ · $\dfrac{1 \text{ g}}{1000 \text{ mg}}$ · $\dfrac{1 \text{ mol H}_3\text{C}_6\text{H}_5\text{O}_7}{192.1 \text{ g H}_3\text{C}_6\text{H}_5\text{O}_7}$ · $\dfrac{3 \text{ mol NaHCO}_3}{1 \text{ mol H}_3\text{C}_6\text{H}_5\text{O}_7}$ · $\dfrac{84.01 \text{ g NaHCO}_3}{1 \text{ mol NaHCO}_3}$

= 0.131 g NaHCO$_3$

4.121. Limiting reagent between 125 mL of 0.15 M CH$_3$CO$_2$H and 15.0 g NaHCO$_3$:

15.0 g NaHCO$_3$ · $\dfrac{1 \text{ mol NaHCO}_3}{84.01 \text{ g NaHCO}_3}$ · $\dfrac{1 \text{ mol NaCH}_3\text{CO}_2}{1 \text{ mol NaHCO}_3}$ · $\dfrac{82.03 \text{ g NaCH}_3\text{CO}_2}{1 \text{ mol NaCH}_3\text{CO}_2}$

= 14.6 g NaCH$_3$CO$_2$

125 mL · $\dfrac{1 \text{ L}}{1000 \text{ mL}}$ · $\dfrac{0.15 \text{ mol CH}_3\text{CO}_2\text{H}}{1 \text{ L}}$ · $\dfrac{1 \text{ mol NaCH}_3\text{CO}_2}{1 \text{ mol CH}_3\text{CO}_2\text{H}}$ · $\dfrac{82.03 \text{ g NaCH}_3\text{CO}_2}{1 \text{ mol NaCH}_3\text{CO}_2}$

= 1.5 g NaCH$_3$CO$_2$

Less NaCH$_3$CO$_2$ can be prepared from 125 mL of 0.15 M CH$_3$CO$_2$H than from 15.0 g NaHCO$_3$, so CH$_3$CO$_2$H is the limiting reactant, and 1.5 g NaCH$_3$CO$_2$ can be formed.

4.122. Mass of citric acid per 100. mL:

33.51 mL · $\dfrac{1 \text{ L}}{1000 \text{ mL}}$ · $\dfrac{0.0102 \text{ mol NaOH}}{1 \text{ L}}$ · $\dfrac{1 \text{ mol H}_3\text{C}_6\text{H}_5\text{O}_7}{3 \text{ mol NaOH}}$ · $\dfrac{192.12 \text{ g H}_3\text{C}_6\text{H}_5\text{O}_7}{1 \text{ mol H}_3\text{C}_6\text{H}_5\text{O}_7}$

= 0.0219 g H$_3$C$_6$H$_5$O$_7$

4.123. Weight percent of Na₂S₂O₃ in 3.232 g sample of material:

$$40.21 \text{ mL} \cdot \frac{1 \text{ L}}{1000 \text{ mL}} \cdot \frac{0.246 \text{ mol I}_2}{1 \text{ L}} \cdot \frac{2 \text{ mol Na}_2\text{S}_2\text{O}_3}{1 \text{ mol I}_2} \cdot \frac{158.10 \text{ g Na}_2\text{S}_2\text{O}_3}{1 \text{ mol Na}_2\text{S}_2\text{O}_3}$$

$$= 3.13 \text{ g Na}_2\text{S}_2\text{O}_3$$

Now calculate the percent of the compound in the impure mixture:

$$\frac{3.13 \text{ g Na}_2\text{S}_2\text{O}_3}{3.232 \text{ g mixture}} \cdot 100\% = 96.8\% \text{ Na}_2\text{S}_2\text{O}_3$$

4.124. Mass percent of oxalic acid in mixture:

$$29.58 \text{ mL} \cdot \frac{1 \text{ L}}{1000 \text{ mL}} \cdot \frac{0.550 \text{ mol NaOH}}{1 \text{ L}} \cdot \frac{1 \text{ mol H}_2\text{C}_2\text{O}_4}{2 \text{ mol NaOH}} \cdot \frac{90.034 \text{ g H}_2\text{C}_2\text{O}_4}{1 \text{ mol H}_2\text{C}_2\text{O}_4}$$

$$= 0.732 \text{ g H}_2\text{C}_2\text{O}_4$$

Percent of oxalic acid: $\frac{0.732 \text{ g}}{4.554 \text{ g}} \cdot 100\% = 16.1\%$ H₂C₂O₄

4.125. For a solution of HCl:

(a) The pH of a 0.105 M HCl solution: Since HCl is considered a strong acid, the concentration of the hydronium ion will also be 0.105 M, and the pH= –log[0.105] or 0.979.

(b) What is the hydronium ion concentration of a solution with pH = 2.56?
Since the pH= 2.56, [H₃O⁺]= 10⁻²·⁵⁶ or 2.8 × 10⁻³ M. With a pH less than 7, this solution is acidic.

(c) Solution has a pH of 9.67:
Hydronium ion concentration = 10⁻⁹·⁶⁷ or 2.1 × 10⁻¹⁰ M
Is solution acidic or basic? With a pH greater than 7, this solution is considered **basic**.

(d) pH of solution formed by diluting 10.0 mL of 2.56 M HCl to 250. mL:
The HCl solution to be diluted has an amount of HCl that is: $c \times V$ = 2.56 mol/L × 0.0100 L or 0.0256 mol HCl
This amount will be contained in 250. mL, so the new concentration is 0.0256 mol/0.250 L or 0.102 M, and a pH of –log[0.102] or 0.990 (close to that of the solution in part (a)).

4.126. Mass of NaHCO₃ to react with HCl:

Net ionic equation: H₃O⁺(aq) + HCO₃⁻(aq) → 2 H₂O(ℓ) + CO₂(g)

[H₃O⁺] = 10⁻ᵖᴴ = 10⁻²·⁵⁶ = 0.0028 M H₃O⁺

$$0.125 \text{ L} \cdot \frac{0.0028 \text{ mol H}_3\text{O}^+}{1 \text{ L}} \cdot \frac{1 \text{ mol NaHCO}_3}{1 \text{ mol H}_3\text{O}^+} \cdot \frac{84.01 \text{ g NaHCO}_3}{1 \text{ mol NaHCO}_3} = 0.029 \text{ g NaHCO}_3$$

4.127. You need to keep track of the hydronium ion being contributed by the two solutions. Calculate the amount of H₃O⁺ in each of the solutions, add those amounts, and determine the concentration in 750. mL:

$$\frac{(0.500 \text{ L})(2.50 \text{ mol/L}) + (0.250 \text{ L})(3.75 \text{ mol/L})}{0.750 \text{ L}} = 2.92 \text{ M}$$

The pH is then: –log[2.92] = –0.465

4.128. pH of solution upon mixing HCl and NaOH:

HCl(aq) + NaOH(aq) → NaCl(aq) + H₂O(ℓ)

[H₃O⁺] = 10⁻¹·⁹² = 0.012 M H₃O⁺ = 0.012 M HCl.

The #mol of HCl: $0.250 \text{ L} \cdot \frac{0.012 \text{ mol}}{1 \text{ L}} = 0.0030 \text{ mol HCl}$

$0.250 \text{ L} \cdot \frac{0.0105 \text{ mol NaOH}}{1 \text{ L}} \cdot \frac{1 \text{ mol HCl}}{1 \text{ mol NaOH}} = 0.00263 \text{ mol HCl reacted}$

Remaining amount of HCl: 0.0030 mol – 0.00263 mol = 0.0004 mol HCl

With this number of mol in 500 mL, the pH = $-\log\left(\frac{0.0004 \text{ mol}}{0.500 \text{ L}}\right) = 3.1$

4.129. 2.56 g CaCO₃ in a beaker containing 250. mL 0.125 M HCl:

After reaction, does any CaCO₃ remain?

The amount of HCl present is: $\frac{0.125 \text{ mol HCl}}{1 \text{ L}} \cdot \frac{0.250 \text{ L}}{1} = 0.0313 \text{ mol HCl}$

The amount of CaCO₃ that reacts with this amount of HCl:

$0.0313 \text{ mol HCl} \cdot \frac{1 \text{ mol CaCO}_3}{2 \text{ mol HCl}} \cdot \frac{100.09 \text{ g CaCO}_3}{1 \text{ mol CaCO}_3} = 1.56 \text{ g CaCO}_3$

This is less than the mass of CaCO₃ present, so some CaCO₃ remains. This calculation shows that the mass of CaCO₃ remaining is then:

2.56 g – 1.56 g = 1.00 g CaCO₃.

Since HCl is the limiting reagent, the amount of CaCl$_2$ that can be produced is:

$$0.0313 \text{ mol HCl} \left(\frac{1 \text{ mol CaCl}_2}{2 \text{ mol HCl}}\right)\left(\frac{110.98 \text{ g CaCl}_2}{1 \text{ mol CaCl}_2}\right) = 1.74 \text{ g CaCl}_2$$

4.130. Regarding cisplatin:

(a) Balanced equation: (NH$_4$)$_2$PtCl$_4$(aq) + 2 NH$_3$(aq) → Pt(NH$_3$)$_2$Cl$_2$(aq) + 2 NH$_4$Cl(aq)

(b) Mass of (NH$_4$)$_2$PtCl$_4$ required:

$$12.50 \text{ g Pt(NH}_3)_2\text{Cl}_2 \cdot \frac{1 \text{ mol Pt(NH}_3)_2\text{Cl}_2}{300.04 \text{ g Pt(NH}_3)_2\text{Cl}_2} \cdot \frac{1 \text{ mol (NH}_4)_2\text{PtCl}_4}{1 \text{ mol Pt(NH}_3)_2\text{Cl}_2}$$

$$\cdot \frac{372.96 \text{ g (NH}_4)_2\text{PtCl}_4}{1 \text{ mol (NH}_4)_2\text{PtCl}_4} = 15.54 \text{ g (NH}_4)_2\text{PtCl}_4$$

Volume of 0.125 M NH$_3$ required:

$$12.50 \text{ g Pt(NH}_3)_2\text{Cl}_2 \cdot \frac{1 \text{ mol Pt(NH}_3)_2\text{Cl}_2}{300.04 \text{ g Pt(NH}_3)_2\text{Cl}_2} \cdot \frac{2 \text{ mol NH}_3}{1 \text{ mol Pt(NH}_3)_2\text{Cl}_2} \cdot \frac{1 \text{ L}}{0.125 \text{ mol NH}_3}$$

$$= 0.667 \text{ L}$$

(c) Unused pyridine is: $0.0370 \text{ L} \cdot \frac{0.475 \text{ mol HCl}}{1 \text{ L}} \cdot \frac{1 \text{ mol C}_5\text{H}_5\text{N}}{1 \text{ mol HCl}} = 0.0176 \text{ mol C}_5\text{H}_5\text{N}$

C$_5$H$_5$N originally added : $1.50 \text{ mL} \cdot \frac{0.979 \text{ g}}{1 \text{ mL}} \cdot \frac{1 \text{ mol C}_5\text{H}_5\text{N}}{79.10 \text{ g C}_5\text{H}_5\text{N}} = 0.0186 \text{ mol C}_5\text{H}_5\text{N}$

Amount of pyridine that reacted: 0.0186 mol – 0.0176 mol = 0.0010 mol C$_5$H$_5$N

$$0.150 \text{ g Pt(NH}_3)_2\text{Cl}_2 \cdot \frac{1 \text{ mol Pt(NH}_3)_2\text{Cl}_2}{300.04 \text{ g Pt(NH}_3)_2\text{Cl}_2} = 5.00 \times 10^{-4} \text{ mol Pt(NH}_3)_2\text{Cl}_2$$

Ratio of mol Pt(NH$_3$)$_2$Cl$_2$ to pyridine:

$$\frac{0.0010 \text{ mol C}_5\text{H}_5\text{N}}{5.00 \times 10^{-4} \text{ mol Pt(NH}_3)_2\text{Cl}_2} = \frac{2 \text{ mol C}_5\text{H}_5\text{N}}{1 \text{ mol Pt(NH}_3)_2\text{Cl}_2}$$

The compound formula is Pt(NH$_3$)$_2$Cl$_2$(C$_5$H$_5$N)$_2$

4.131. Determine the # mol of methylene blue:

$$1.0 \text{ g C}_{16}\text{H}_{18}\text{ClN}_3\text{S} \left(\frac{1 \text{ mol C}_{16}\text{H}_{18}\text{ClN}_3\text{S}}{319.85 \text{ g C}_{16}\text{H}_{18}\text{ClN}_3\text{S}}\right) = 0.0031 \text{ mol C}_{16}\text{H}_{18}\text{ClN}_3\text{S}$$

The measurement indicates that the molar concentration of methylene blue is 4.1×10^{-8} M. The concentration of methylene blue in the measured sample will be equal to that of methylene blue in the pool--assuming that the methylene blue has been uniformly distributed throughout the pool water before the sample was taken. Recalling that c = # mol/V, and knowing BOTH the molarity and the # mol, you can calculate the volume.

$$0.0031 \text{ mol } C_{16}H_{18}ClN_3S \left(\frac{1 \text{ L}}{4.1 \times 10^{-8} \text{ mol } C_{16}H_{18}ClN_3S} \right) = 76{,}000 \text{ L (2 sf)}$$

Note that we ignored the 50 mL of solution that originally contained the 1.0 g of methylene blue because this volume would appear in a digit that is insignificant in the large volume of the pool.

4.132. Weight percent of CaO and MgO in sample:

Amount of metal oxide: $0.125 \text{ L} \cdot \dfrac{2.55 \text{ mol HCl}}{1 \text{ L}} \cdot \dfrac{1 \text{ mol metal oxide}}{2 \text{ mol HCl}} = 0.159 \text{ mol metal oxide}$

Let x = mass of CaO and y = mass of MgO

Set up two equations with two unknowns:

$$0.159 \text{ mol metal oxide} = \left(x \text{ g CaO} \cdot \frac{1 \text{ mol CaO}}{56.08 \text{ g CaO}} \right) + \left(y \text{ g MgO} \cdot \frac{1 \text{ mol MgO}}{40.30 \text{ g MgO}} \right)$$

7.695 g metal oxide = x g CaO + y g MgO

Solve this 2nd equation for x g CaO (x = 7.695 – y) and substitute (7.695 – y) for x in the first equation.

$$0.159 \text{ mol metal oxide} = \left((7.695 - y) \text{ g CaO} \cdot \frac{1 \text{ mol CaO}}{56.08 \text{ g CaO}} \right) + \left(y \text{ g MgO} \cdot \frac{1 \text{ mol MgO}}{40.30 \text{ g MgO}} \right)$$

$0.159 = 0.1372 - 0.01783y + 0.02481y$

$0.022 = 0.00698y$

$y = 3.2$

y = 3.2 g MgO; Then x = 7.695 g metal oxide – 3.2 g MgO = 4.5 g CaO

% of oxides: $\dfrac{4.5 \text{ g CaO}}{7.695 \text{ g sample}} \cdot 100\% = 58\% \text{ CaO}$; $\dfrac{3.2 \text{ g MgO}}{7.695 \text{ g sample}} \cdot 100\% = 42\% \text{ MgO}$

4.133. For the reaction in which Au is dissolved by treatment with sodium cyanide:
(a) Oxidizing agent: elemental O_2 Reducing agent: elemental Au

Substance oxidized: elemental Au (oxidation state changes from 0 to +1)

Substance reduced: elemental O_2 (oxidation state changes from 0 to –2)

(b) Volume of 0.075 M NaCN to extract gold from 1000 kg of rock:

Mass of gold in the rock: $\dfrac{0.019 \text{ g gold}}{100 \text{ g rock}} \cdot \dfrac{10^3 \text{ g rock}}{1 \text{ kg rock}} \cdot 10^3 \text{ kg rock} = 190 \text{ g gold}$

Amount of NaCN needed (from the balanced equation):

$$190 \text{ g Au} \cdot \frac{1 \text{ mol Au}}{196.97 \text{ g Au}} \cdot \frac{8 \text{ mol NaCN}}{4 \text{ mol Au}} = 1.9 \text{ mol NaCN}$$

Volume of 0.075 M NaCN that contains that amount of NaCN:

$$1.9 \text{ mol NaCN} \cdot \frac{1 \text{ L}}{0.075 \text{ mol NaCN}} = 26 \text{ L}$$

4.134. Iron(III) chloride reacts with NaOH:

Balanced equation: FeCl₃(aq) + 3 NaOH(aq) → Fe(OH)₃(s) + 3 NaCl(aq)

(a) Maximum amount of Fe(OH)₃ using each reagent:

$$0.0250 \text{ L} \cdot \frac{0.234 \text{ mol FeCl}_3}{1 \text{ L}} \cdot \frac{1 \text{ mol Fe(OH)}_3}{1 \text{ mol FeCl}_3} \cdot \frac{106.9 \text{ g Fe(OH)}_3}{1 \text{ mol Fe(OH)}_3} = 0.625 \text{ g Fe(OH)}_3$$

$$0.0425 \text{ L} \cdot \frac{0.453 \text{ mol NaOH}}{1 \text{ L}} \cdot \frac{1 \text{ mol Fe(OH)}_3}{3 \text{ mol NaOH}} \cdot \frac{106.9 \text{ g Fe(OH)}_3}{1 \text{ mol Fe(OH)}_3} = 0.686 \text{ g Fe(OH)}_3$$

Less Fe(OH)₃ can be produced from 25.0 mL of 0.234 M FeCl₃ than from 42.5 mL of 0.453 M NaOH, so Fe(OH)₃ is the limiting reactant, and 0.625 g Fe(OH)₃ precipitates.

(b) The excess reactant is NaOH.

Amount NaOH required: $0.0250 \text{ L} \cdot \frac{0.234 \text{ mol FeCl}_3}{1 \text{ L}} \cdot \frac{3 \text{ mol NaOH}}{1 \text{ mol FeCl}_3} = 0.0176 \text{ mol NaOH}$

Excess NaOH: $(0.0425 \text{ L} \cdot \frac{0.453 \text{ mol NaOH}}{1 \text{ L}}) - 0.0176 \text{ mol NaOH} = 0.0017 \text{ mol NaOH}$

This is now dissolved in the total volume of the combined solutions, 250. mL + 425 mL = 675 mL.

Concentration of remaining NaOH: $\frac{0.0017 \text{ mol}}{0.0675 \text{ L}} = 0.025 \text{ M NaOH}$

4.135. The % atom economy for the desired product, CH₃CH₂CH₂CH₂Br:

You need the molar masses for all atoms used in the reaction:

C₄H₉OH: [(4 × C) + (10 × H) + (1 × O)] = 74.123

NaBr [(1 × Na) + (1 × Br)] = 102.894

H₂SO₄ [(2 × H) + (1 × S) + (4 × O)] = 98.07   Adding these gives 275.09

Calculate the molar mass of the desired product: [(4 × C) + (9 × H) + (1 × Br)] = 137.020

% atom economy = $\frac{137.020}{275.09} \cdot 100\% = 48.809\%$

4.136. % atom economy via two synthetic processes:

(a) "chlorohydrin route":
% atom economy = $M(C_2H_4O)/[M(C_2H_4) + M(Cl_2) + M(Ca(OH)_2)] \cdot 100\%$

$$= \frac{44.05 \text{ g/mol}}{(28.05 \text{ g/mol} + 70.90 \text{ g/mol} + 74.09 \text{ g/mol})} \cdot 100 = 25.46\%$$

"catalytic route": All atoms in reactants are in product. % atom economy is 100%.

The modern catalytic reaction is more efficient.

(b) Theoretical yield = $867 \text{ g C}_2\text{H}_4 \cdot \dfrac{1 \text{ mol C}_2\text{H}_4}{28.05 \text{ g C}_2\text{H}_4} \cdot \dfrac{1 \text{ mol C}_2\text{H}_4\text{O}}{1 \text{ mol C}_2\text{H}_4} \cdot \dfrac{44.05 \text{ g C}_2\text{H}_4\text{O}}{1 \text{ mol C}_2\text{H}_4\text{O}}$

$= 1.36 \times 10^3$ g C₂H₄O

Percent yield: $\dfrac{762 \text{ g}}{1.36 \times 10^3 \text{ g}} \cdot 100 = 56.0\%$

In The Laboratory

4.137. The concentration of the resulting solution can be determined by two identical processes.

First dilution: $\dfrac{0.110 \text{ mol Na}_2\text{CO}_3}{1 \text{ L}} \cdot \dfrac{25.0 \text{ mL}}{100.0 \text{ mL}} = 0.0275 \text{ M}$

Note that the ratio of volumes can both be expressed in units of milliliters, without a conversion to units of liters.

Second dilution: $\dfrac{0.0275 \text{ mol Na}_2\text{CO}_3}{1 \text{ L}} \cdot \dfrac{10.0 \text{ mL}}{250. \text{ mL}} = 1.10 \times 10^{-3} \text{ M}$

4.138. Weight percent of (NH₄)₂SO₄ in 0.475 g sample:

Total HCl present: $0.0500 \text{ L} \cdot \dfrac{0.100 \text{ mol HCl}}{1 \text{ L}} = 0.00500 \text{ mol HCl}$

Remaining HCl: $0.0111 \text{ L} \cdot \dfrac{0.121 \text{ mol NaOH}}{1 \text{ L}} \cdot \dfrac{1 \text{ mol HCl}}{1 \text{ mol NaOH}} = 0.00134 \text{ mol HCl}$

Amount HCl reacted: 0.00500 mol HCl − 0.00134 mol HCl = 0.00366 mol HCl

$0.00366 \text{ mol HCl} \cdot \dfrac{1 \text{ mol NH}_3}{1 \text{ mol HCl}} \cdot \dfrac{1 \text{ mol (NH}_4)_2\text{SO}_4}{2 \text{ mol NH}_3} \cdot \dfrac{132.1 \text{ g (NH}_4)_2\text{SO}_4}{1 \text{ mol (NH}_4)_2\text{SO}_4} = 0.241$ g (NH₄)₂SO₄

Weight percent of (NH₄)₂SO₄: $\dfrac{0.241 \text{ g (NH}_4)_2\text{SO}_4}{0.475 \text{ g sample}} \cdot 100\% = 50.8\%$

4.139. (a) Balanced net ionic equation for the reaction of silver nitrate with chloride ions:

$Ag^+(aq) + NO_3^-(aq) + Cl^-(aq) \rightarrow AgCl(s) + NO_3^-(aq)$. Noting that the nitrate ions **do not change**, we can delete them, leaving: $Ag^+(aq) + Cl^-(aq) \rightarrow AgCl(s)$

(b) Complete and net ionic equation for the reaction of $AgNO_3$ with K_2CrO_4:
$2\ AgNO_3(aq) + K_2CrO_4(aq) \rightarrow Ag_2CrO_4(s) + 2\ KNO_3(aq)$ (complete)
$2\ Ag^+(aq) + 2\ NO_3^-(aq) + 2\ K^+(aq) + CrO_4^{2-}(aq) \rightarrow Ag_2CrO_4(s) + 2\ K^+(aq) + 2\ NO_3^-(aq)$
Removing all species from **both** sides that **do not change** leaves:
$2\ Ag^+(aq) + CrO_4^{2-}(aq) \rightarrow Ag_2CrO_4(s)$

(c) Concentration of chloride ion in the sample?:
Amount of Ag^+:
$25.60\ \text{mL} \cdot \dfrac{1\ \text{L}}{1000\ \text{mL}} \cdot \dfrac{0.001036\ \text{mol AgNO}_3}{1\ \text{L}} \cdot \dfrac{1\ \text{mol Ag}^+}{1\ \text{mol AgNO}_3} = 2.652 \times 10^{-5}\ \text{mol Ag}^+$

The stoichiometry (part (a)) indicates that 1 mol of silver ion reacts with 1 mol of chloride ion: $2.652 \times 10^{-5}\ \text{mol Ag}^+ \cdot \dfrac{1\ \text{mol Cl}^-}{1\ \text{mol Ag}^+} = 2.652 \times 10^{-5}\ \text{mol Cl}^-$

Knowing the # of mol of chloride ion, and the volume in which they are contained, the concentration of chloride ion is: $\dfrac{2.652 \times 10^{-5}\ \text{mol Cl}^-}{50.0 \times 10^{-3}\ \text{L}} = 5.30 \times 10^{-4}\ \text{M Cl}^-$

We need to express this concentration in units of ppm:
$\dfrac{2.652 \times 10^{-5}\ \text{mol Cl}^-}{50.0 \times 10^{-3}\ \text{L}} \cdot \dfrac{35.45\ \text{g Cl}^-}{1\ \text{mol Cl}^-} \cdot \dfrac{1000\ \text{mg}}{1\ \text{g}} = 18.8\ \text{mg Cl}^-/\text{L}$, exceeding 8 mg/L. This level is sufficient to promote oyster-bed growth.

4.140. Regarding the compound $YBa_2Cu_3O_{7-x}$: Phases of substances are omitted for clarity:

To determine the stoichiometry ratios, examine the reactions:

(1) $YBa_2Cu_3O_{7-x} + 13\ H^+ \rightarrow Y^{3+} + 2\ Ba^{2+} + 3\ Cu^{2+} + \frac{1}{4}(1-2x)\ O_2 + 13/2\ H_2O$

(2) $2\ Cu^{2+} + 5\ I^- \rightarrow 2\ CuI + I_3^-$

(3) $I_3^- + 2\ S_2O_3^{2-} \rightarrow 3\ I^- + S_4O_6^{2-}$

From equation 3, you can see that 1 mol of I_3^- requires 2 mol of $S_2O_3^{2-}$.

From equation 2, you can see that 1 mol of I_3^- requires 2 mol of Cu^{2+}.

From equation 1, you can see that 3 mol of Cu^{2+} is produced when 1 mol of $YBa_2Cu_3O_{7-x}$ reacts.

Using these stoichiometric relationships, deduce the amount of $YBa_2Cu_3O_{7-x}$.

$1.542 \times 10^{-4}\ \text{mol S}_2O_3^{2-} \cdot \dfrac{1\ \text{mol I}_3^-}{2\ \text{mol S}_2O_3^{2-}} \cdot \dfrac{2\ \text{mol Cu}^{2+}}{1\ \text{mol I}_3^-} \cdot \dfrac{1\ \text{mol YBa}_2Cu_3O_{7-x}}{3\ \text{mol Cu}^{2+}}$
$= 5.140 \times 10^{-5}\ \text{mol YBa}_2Cu_3O_{7-x}$

If the compound is YBa₂Cu₃O₇, then 5.140×10^{-5} mol · (666.197 g/mol)

$$= 0.03424 \text{ g YBa}_2\text{Cu}_3\text{O}_7 = 34.24 \text{ mg YBa}_2\text{Cu}_3\text{O}_7$$

The initial mass of the compound was 34.02 mg, so 34.24 mg – 34.02 mg = 0.22 mg O deficient

What # mol of O does this mass of O represent?

$$0.22 \times 10^{-3} \text{ g O} \cdot \frac{1 \text{ mol O}}{15.999 \text{ g O}} = 1.4 \times 10^{-5} \text{ mol O}$$

How many mol O per mol of YBa₂Cu₃O₇₋ₓ does this represent?

$x = 1.4 \times 10^{-5}$ mol/5.140×10^{-5} mol = 0.27, so x is in the expected range of 0 to 0.50.

4.141. Weight percent of Cu in 0.251 g of a copper-containing alloy:

(a) Oxidizing and reducing agents in the two equations:

$$2 \text{ Cu}^{2+}(aq) + 5 \text{ I}^-(aq) \rightarrow 2 \text{ CuI}(s) + \text{I}_3^-(aq)$$

I⁻ reduces the Cu(II) ion to the Cu(I) ion— I⁻ acts as the reducing agent.

Cu²⁺ oxidizes I⁻ to I₃⁻ (I₂ and I⁻)–and acts as the oxidizing agent.

$$\text{I}_3^-(aq) + 2 \text{ S}_2\text{O}_3^{2-}(aq) \rightarrow \text{S}_4\text{O}_6^{2-}(aq) + 3 \text{ I}^-(aq)$$

S₂O₃²⁻ reduces the I₃⁻ ion (I₂ and I⁻) to the I⁻ ion—so S₂O₃²⁻ acts as the reducing agent.

I₃⁻ oxidizes S₂O₃²⁻ to S₄O₆²⁻ –and acts as the oxidizing agent.

In S₄O₆²⁻, the oxidation state of S can be thought of as (+2.5), and in S₂O₃²⁻ as (+2)

(b) Several steps in this problem:

(1) Note the relationship between Cu (Cu²⁺) and I₃⁻ formed (2 mol Cu²⁺ :1 mol I₃⁻)

(2) Excess I₃⁻ reacts with S₂O₃²⁻ in a 1 mol I₃⁻ to 2 mol S₂O₃²⁻ ratio

Begin by calculating the amount of thiosulfate in 26.32 mL of 0.101 M solution:

$$\frac{0.101 \text{ mol S}_2\text{O}_3^{2-}}{1 \text{ L}} \cdot 0.02632 \text{ L} = 0.00266 \text{ mol S}_2\text{O}_3^{2-} \text{ (3 sf)}$$

This amount can now be used to calculate I₃⁻ produced (step 2), and the amount of I₃⁻ related to the amount of Cu present (step 1):

$$0.00266 \text{ mol S}_2\text{O}_3^{2-} \cdot \frac{1 \text{ mol I}_3^-}{2 \text{ mol S}_2\text{O}_3^{2-}} \cdot \frac{2 \text{ mol Cu}^{2+}}{1 \text{ mol I}_3^-} \cdot \frac{1 \text{ mol Cu}}{1 \text{ mol Cu}^{2+}} \cdot \frac{63.546 \text{ g Cu}}{1 \text{ mol Cu}}$$

$$= 0.169 \text{ g Cu}$$

The weight percent is: $\frac{0.169 \text{ g Cu}}{0.251 \text{ g alloy}} \cdot 100\% = 67.3\% \text{ Cu}$

4.142. Determine correct formula—2 or 3 oxalate ions per formula:

$$0.03450 \text{ L} \cdot \frac{0.108 \text{ mol KMnO}_4}{1 \text{ L}} \cdot \frac{5 \text{ mol C}_2\text{O}_4^{2-}}{2 \text{ mol KMnO}_4} = 0.00932 \text{ mol C}_2\text{O}_4^{2-}$$

Now you can use mol $C_2O_4^{2-}$ to determine which formula is correct:

Establish stoichiometric ratios, assuming first 2 mol oxalate and then 3 mol oxalate to determine what mass of compound you would get (recall that the initial mass was 1.356 g):

$$0.00932 \text{ mol C}_2\text{O}_4^{2-} \cdot \frac{1 \text{ mol K[Fe(C}_2\text{O}_4)_2(\text{H}_2\text{O})_2]}{2 \text{ mol C}_2\text{O}_4^{2-}} \cdot \frac{307.0 \text{ g}}{1 \text{ mol K[Fe(C}_2\text{O}_4)_2(\text{H}_2\text{O})_2]}$$

$$= 1.43 \text{ g compound}$$

$$0.00932 \text{ mol C}_2\text{O}_4^{2-} \cdot \frac{1 \text{ mol K}_3[\text{Fe(C}_2\text{O}_4)_3]}{3 \text{ mol C}_2\text{O}_4^{2-}} \cdot \frac{437.2 \text{ g}}{1 \text{ mol K}_3[\text{Fe(C}_2\text{O}_4)_3]} = 1.36 \text{ g compound}$$

From these masses, you can deduce that the correct formula is $K_3[Fe(C_2O_4)_3]$.

4.143. The equation: $Cr(NH_3)_xCl_3(aq) + x$ HCl (aq) $\rightarrow x$ NH_4^+(aq) + Cr^{3+}(aq) + $(x + 3)Cl^-$(aq)

The amount of HCl: 24.26 mL of 1.500 M HCl (1.500 mol/L × 0.02426 L) = 0.03639 mol
You don't know the ratio between moles of salt and moles of HCl. What we do know is that for each mol of HCl that reacts, 1 mol of NH_3 is present in the salt (and 1 mol of NH_4^+ is produced in the reaction). So, knowing the amount of NH_3 calculate its mass.

$$0.03639 \text{ mol NH}_3 \cdot \frac{17.031 \text{ g NH}_3}{1 \text{ mol NH}_3} = 0.6198 \text{ g NH}_3$$

Given the mass of the salt (1.580 g), determine the mass of the remaining components (Cr and Cl). 1.580 g $CrCl_3$ – 0.6198 g NH_3 = 0.9602 g $CrCl_3$.

Now calculate the amount of the $CrCl_3$ salt: MM = 51.996 + 3(35.45) = 158.35 g/mol

$$0.9602 \text{ g CrCl}_3 \cdot \frac{1 \text{ mol CrCl}_3}{158.35 \text{ g CrCl}_3} = 6.064 \times 10^{-3} \text{ mol CrCl}_3$$

Now determine the ratio of ammonia to chromium(III) chloride:

$$\frac{3.639 \times 10^{-2} \text{ mol NH}_3}{6.064 \times 10^{-3} \text{ mol CrCl}_3} = 6.001, \text{ so the formula for the salt is: } Cr(NH_3)_6Cl_3.$$

4.144. Thioridazine content in a tablet:

$$0.310 \text{ g BaSO}_4 \cdot \frac{1 \text{ mol BaSO}_4}{233.4 \text{ g BaSO}_4} \cdot \frac{1 \text{ mol S}}{1 \text{ mol BaSO}_4} \cdot \frac{1 \text{ mol thioridazine}}{2 \text{ mol S}}$$

$$\cdot \frac{370.6 \text{ g thioridazine}}{1 \text{ mol thioridazine}} \cdot \frac{10^3 \text{ mg}}{1 \text{ g}} = 239 \text{ mg thioridazine}$$

The mass per tablet is: 239/12 tablets or 19.9 mg thioridazine per tablet.

4.145. A 1.236-g sample of the herbicide liberated the chlorine as Cl⁻, and the ion precipitated as AgCl, with a mass of 0.1840 g. The mass percent of 2,4-D in the sample:
Determine the amount (moles) of chlorine present. Noting that each molecule of 2,4-D has 2 atoms of Cl, allows you to calculate the number of moles of 2,4-D (each with a molar mass of 221.04 g/mol).

$$0.1840 \text{ g AgCl} \cdot \frac{1 \text{ mol AgCl}}{143.32 \text{ g AgCl}} \cdot \frac{1 \text{ mol Cl}}{1 \text{ mol AgCl}} \cdot \frac{1 \text{ mol 2,4-D}}{2 \text{ mol Cl}} \cdot \frac{221.03 \text{ g 2,4-D}}{1 \text{ mol 2,4-D}}$$

$$= 0.1419 \text{ g 2,4-D}$$

The mass percent of 2,4-D in the sample is: $\frac{0.1419 \text{ g 2,4-D}}{1.236 \text{ g sample}} \cdot 100\% = 11.48\%$ 2,4-D

4.146. Determine the concentration of H₂SO₄:

$$H_2SO_4(aq) + 2 \text{ NaOH}(aq) \rightarrow Na_2SO_4(aq) + 2 H_2O(\ell)$$

(a) Average volume (NaOH sample) = $\frac{(20.15 + 21.30 + 20.40 + 20.35) \text{ mL}}{4} = 20.55 \text{ mL}$

$$0.02055 \text{ L} \cdot \frac{0.1760 \text{ mol NaOH}}{1 \text{ L}} \cdot \frac{1 \text{ mol H}_2\text{SO}_4}{2 \text{ mol NaOH}} = 0.001808 \text{ mol H}_2\text{SO}_4$$

$$\frac{0.001808 \text{ mol H}_2\text{SO}_4}{0.01000 \text{ L}} = 0.1808 \text{ M H}_2\text{SO}_4 \text{ (diluted sample)}$$

So, the original concentration:

$$c_c = \frac{c_d V_d}{V_c} = \frac{(0.1808 \text{ M H}_2\text{SO}_4)(500. \text{ mL})}{5.00 \text{ mL}} = 18.1 \text{ M H}_2\text{SO}_4$$

(b) $\frac{18.1 \text{ mol H}_2\text{SO}_4}{1 \text{ L solution}} \cdot \frac{98.07 \text{ g H}_2\text{SO}_4}{1 \text{ mol H}_2\text{SO}_4} \cdot \frac{1.00 \text{ L solution}}{1.84 \text{ kg solution}} \cdot \frac{1 \text{ kg}}{10^3 \text{ g}} \cdot 100\% = 96.4\%$ H₂SO₄

4.147. The amount of water per mol of CaCl₂ can be determined by realizing that in the 150 g sample of the partially hydrated material there is **both** CaCl₂ and H₂O. Identify the amount of CaCl₂ remaining in the water:

Determine the solubility/80g (the data provides solubility/100 g).
$\frac{74.5 \text{ g CaCl}_2}{100 \text{ g H}_2\text{O}} = \frac{x \text{ g CaCl}_2}{80 \text{ g H}_2\text{O}}$ and $x = 59.6$ g CaCl₂

From the amount precipitated: $\frac{74.9 \text{ g CaCl}_2 \bullet 6 \text{ H}_2\text{O}}{1} \cdot \frac{110.98 \text{ g CaCl}_2}{219.07 \text{ g CaCl}_2 \bullet 6 \text{ H}_2\text{O}} = 37.9$ g CaCl₂

Total mass of CaCl₂: (59.6 + 37.9) = 97.5 g CaCl₂

The partially hydrated material had a mass of 150 g, so by difference we can calculate the amount of water present in that sample: 150 g – 97.5 = 52.5 g of water.

Calculate the moles of water and of CaCl₂

$$\frac{52.5 \text{ g H}_2\text{O}}{1} \cdot \frac{1 \text{ mol H}_2\text{O}}{18.02 \text{ g H}_2\text{O}} = 2.91 \text{ mol H}_2\text{O} \text{ and}$$

$$\frac{97.5 \text{ g CaCl}_2}{1} \cdot \frac{1 \text{ mol CaCl}_2}{110.98 \text{ g CaCl}_2} = 0.879 \text{ mol CaCl}_2$$

with a mol ratio of 2.91 mol H₂O: 0.879 mol CaCl₂ or 3.3 mol water/mol CaCl₂

4.148. Mass percent of Fe and Fe₂O₃ in the sample:

$$5 \text{ Fe}^{2+}(aq) + \text{MnO}_4^-(aq) + 8 \text{ H}_3\text{O}^+(aq) \rightarrow \text{Mn}^{2+}(aq) + 5 \text{ Fe}^{3+}(aq) + 12 \text{ H}_2\text{O}(\ell)$$

Amount of Fe in original mixture:

$$0.03750 \text{ L} \cdot \frac{0.04240 \text{ mol MnO}_4^-}{1 \text{ L}} \cdot \frac{5 \text{ mol Fe}^{2+}}{1 \text{ mol MnO}_4^-} \cdot \frac{1 \text{ mol Fe}}{1 \text{ mol Fe}^{2+}} = 0.007950 \text{ mol Fe}$$

Now we can set up two equations with two unknowns:

(1) $0.007950 \text{ mol Fe} = (x \text{ g Fe})\left(\frac{1 \text{ mol Fe}}{55.845 \text{ g Fe}}\right) + (y \text{ g Fe}_2\text{O}_3)\left(\frac{1 \text{ mol Fe}_2\text{O}_3}{159.69 \text{ g Fe}_2\text{O}_3}\right)\left(\frac{2 \text{ mol Fe}}{1 \text{ mol Fe}_2\text{O}_3}\right)$

(2) $0.5510 \text{ g mixture} = x \text{ g Fe} + y \text{ g Fe}_2\text{O}_3$

Solve for *x* in equation (2) and substitute: (0.5510 – *y*) for *x* into equation (1) and solve.

$$0.007950 = \frac{0.5510 - y}{55.845} + \frac{2y}{159.69}$$

0,007950 = 0.009867 – 0.017907*y* + 0.0012524*y*

–0.001917 = –0.005382*y*

y = 0.3561 g Fe₂O₃; mass percent = $\frac{0.3561 \text{ g Fe}_2\text{O}_3}{0.5510 \text{ g mixture}} \cdot 100\%$ = 64.62% Fe₂O₃

x = 0.1949 g Fe; mass percent = $\frac{0.1949 \text{ g Fe}}{0.5510 \text{ g mixture}} \cdot 100\%$ = 35.38% Fe

4.149 The data given are plotted below:

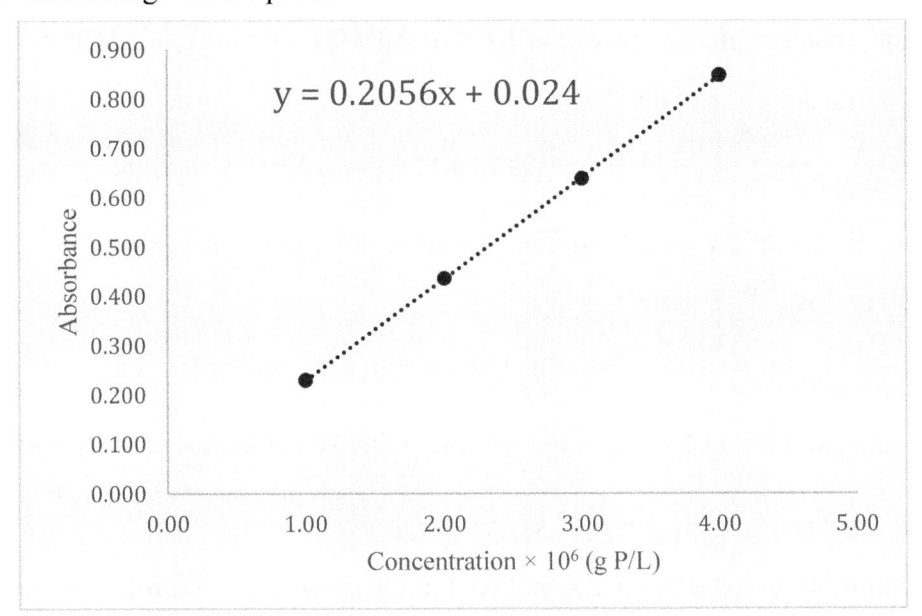

[Graph showing Absorbance vs Concentration × 10⁶ (g P/L), with linear fit y = 0.2056x + 0.024]

(a) The equation for the line is shown: $y = 0.2056x + 0.024$ in which y = Absorbance

Note that the data on the x-axis is multiplied by 10^6, so the slope is really $2.056 \times 10^5 = 2.06 \times 10^5$ (3 sf) and the y-intercept is 0.024.

(b) The mass of P/L of urine: The absorbance for the urine sample is 0.518. Using the equation for the line, you can calculate the concentration of phosphorus/L.

$0.518 = (2.056 \times 10^5)x + 0.024$

$x = 2.40 \times 10^{-6}$ g P/L

Knowing that this concentration was arrived at by diluting 1.00 mL of the 1122 mL sample to 50.00 mL, you can calculate the concentration in the original (1122 mL) sample:

$$\frac{2.40 \times 10^{-6} \text{ g P}}{1 \text{ L}} \cdot \frac{50.00 \text{ mL}}{1.00 \text{ mL}} = 1.20 \times 10^{-4} \text{ g P/L}$$

(c) Mass of Phosphate excreted per day:
The concentration of P calculated above x volume of urine in 24 hr (1122 mL) gives an overall mass of: 1.20×10^{-4} g P/L × 1.122 L = 1.35×10^{-4} g P,
Since the question asks for mass of *phosphate*, you need to calculate the mass of PO_4^{3-} that corresponds to this mass of phosphorus:

$$1.35 \times 10^{-4} \text{ g P} \cdot \frac{94.97 \text{ g PO}_4^{3-}}{30.97 \text{ g P}} \cdot \frac{1000 \text{ mg}}{1 \text{ g}} = 0.413 \text{ mg PO}_4^{3-}$$

4.150. Mass percent of KCl and KClO₄

Use a Mohr titration to determine the amount of KClO₄: $Ag^+(aq) + Cl^-(aq) \rightarrow AgCl(s)$

$$\frac{0.04100 \text{ L}\left(\frac{0.0750 \text{ mol Ag}^+}{1 \text{ L}}\right)\left(\frac{1 \text{ mol Cl}^-}{1 \text{ mol Ag}^+}\right)}{0.05000 \text{ L}} = 0.0615 \text{ M Cl}^- \text{ (as KCl) in mixture}$$

KClO₄ is reduced to KCl with the vanadium salt, and the resulting KCl is titrated:

$$\frac{0.03812 \text{ L}\left(\frac{0.0750 \text{ mol Ag}^+}{1 \text{ L}}\right)\left(\frac{1 \text{ mol Cl}^-}{1 \text{ mol Ag}^+}\right)}{0.02500 \text{ L}} = 0.114 \text{ Cl}^- \text{ (total) in mixture}$$

To determine the concentration of KClO₄ in the mixture, take the difference of these two Cl⁻ determinations:

0.114 M Cl⁻ (total) – 0.0615 M Cl⁻ (as KCl) = 0.053 M KClO₄

Convert the concentrations to mass of KClO₄ and KCl in the original 250.00 mL solution:

$$0.25000 \text{ L}\left(\frac{0.0615 \text{ mol KCl}}{1 \text{ L}}\right)\left(\frac{74.55 \text{ g KCl}}{1 \text{ mol KCl}}\right) = 1.15 \text{ g KCl}$$

$$0.25000 \text{ L}\left(\frac{0.053 \text{ mol KClO}_4}{1 \text{ L}}\right)\left(\frac{138.5 \text{ g KClO}_4}{1 \text{ mol KClO}_4}\right) = 1.8 \text{ g KClO}_4$$

The mass percents are:

$$\left(\frac{1.15 \text{ g KCl}}{4.000 \text{ g mixture}}\right) = 28.7\% \text{ KCl} \qquad \left(\frac{1.8 \text{ g KClO}_4}{4.000 \text{ g mixture}}\right)100\% = 46\% \text{ KClO}_4$$

Summary And Conceptual Questions

4.151. The total mass of the beakers and solutions after reaction will be equal to the mass of the beakers and solutions before the reaction: 167.170 g. No gases were produced in this reaction, and there is conservation of mass in chemical reactions.

4.152. For the reaction of Fe with Br₂:

(a) The graph indicates 10.8 g of product. Subtracting 2.0 g of Fe gives 8.8 g Br₂.

(b) Mol ratio of Fe:Br₂:

$$2.0 \text{ g Fe}\left(\frac{1 \text{ mol Fe}}{55.8 \text{ g Fe}}\right) = 0.036 \text{ mol Fe}$$

$$8.8 \text{ g Br}_2\left(\frac{1 \text{ mol Br}_2}{160. \text{ g Br}_2}\right)\left(\frac{2 \text{ mol Br}}{1 \text{ mol Br}_2}\right) = 0.11 \text{ mol Br}$$

$$\frac{0.11 \text{ mol Br}}{0.036 \text{ mol Fe}} = \frac{3 \text{ mol Br}}{1 \text{ mol Fe}}$$

(c) Empirical formula: Based upon part (b), the empirical formula would be FeBr$_3$.

(d) Balanced equation: 2 Fe(s) + 3 Br$_2$(ℓ) → 2 FeBr$_3$(s)

(e) Name of product: iron(III) bromide

(f) Best statement summarizing the experiments:

(i) When 1.00 g of Fe is added to the Br$_2$, Fe is the limiting reagent.

4.153. For the reaction of hydrogen and oxygen to form water:

(a)

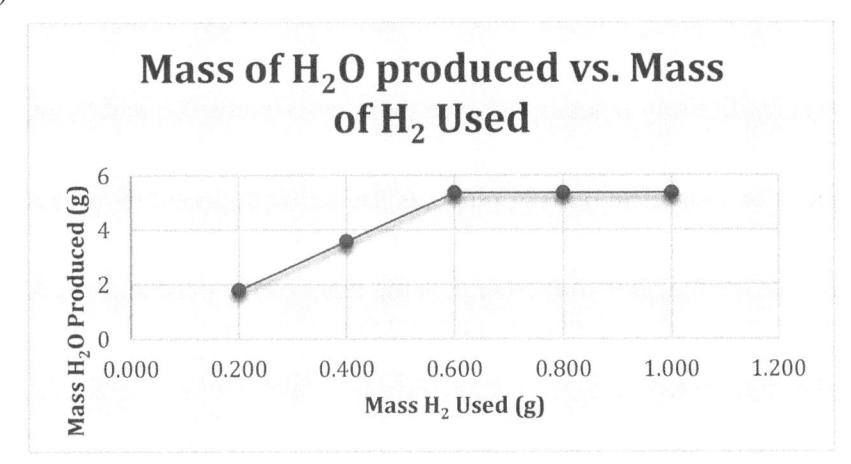

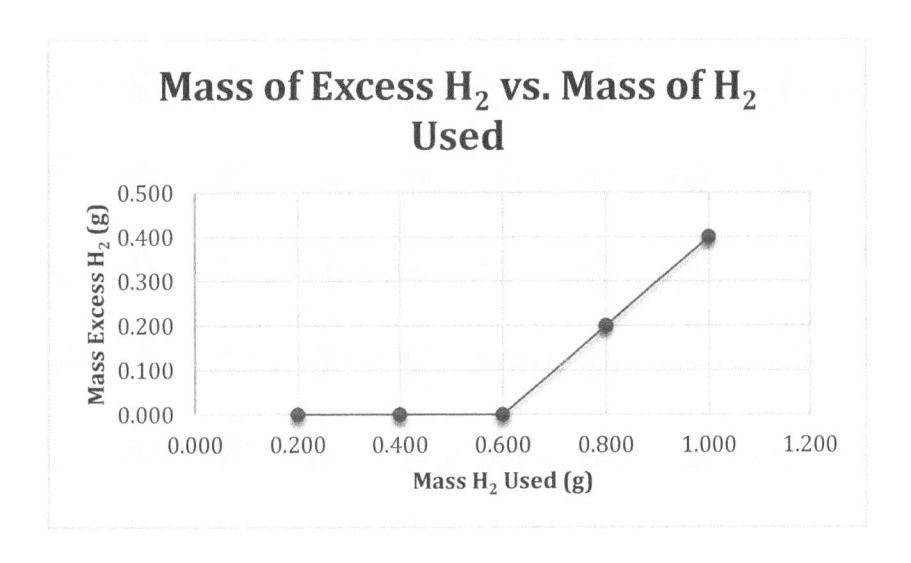

Mass of H$_2$O produced vs. Mass of H$_2$ Used

Mass of Excess H$_2$ vs. Mass of H$_2$ Used

Mass of Excess O₂ vs. Mass of H₂ Used

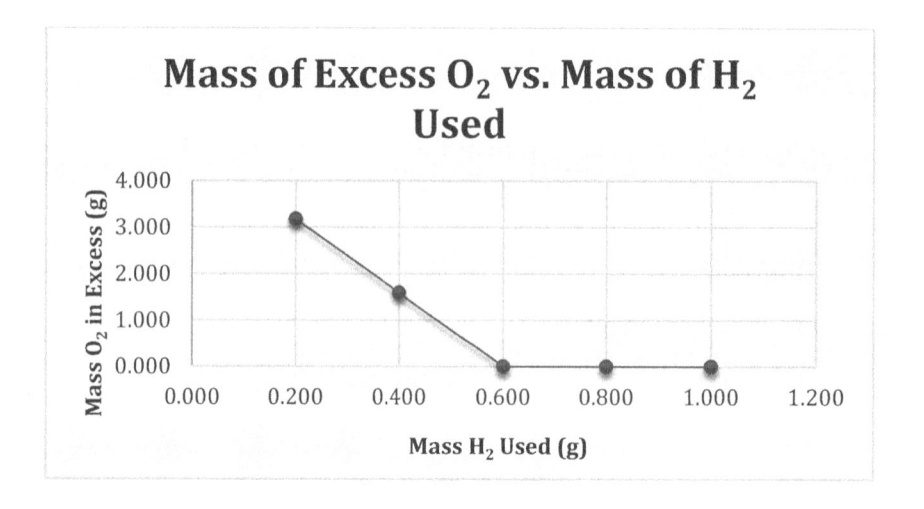

(Graph: Mass O₂ in Excess (g) on y-axis from 0.000 to 4.000; Mass H₂ Used (g) on x-axis from 0.000 to 1.200. Data points: approximately (0.200, 3.17), (0.400, 1.59), (0.600, 0.000), (0.800, 0.000), (1.000, 0.000).)

(b) In experiments 1 and 2, H₂ is the limiting reactant because there is no excess H₂ and there is excess O₂.

(c) In experiments 4 and 5, O₂ is the limiting reactant because there is no excess O₂ and there is excess H₂.

(d) Experiment 3 has the correct stoichiometric ratio of reagents. In that experiment there is no excess hydrogen or oxygen.

(e) In each experiment, the mass of the reactants mixed equals the mass of the product produced plus the mass of any excess reactant. Mass is conserved in each experiment.

Experiment	Mass Before Reaction	Mass After Reaction
1	0.200 g H₂ + 4.76 g O₂ = 4.96 g	1.79 g H₂O + 3.17 g O₂ = 4.96 g
2	0.400 g H₂ + 4.76 g O₂ = 5.16 g	3.57 g H₂O + 1.59 g O₂ = 5.16 g
3	0.600 g H₂ + 4.76 g O₂ = 5.36 g	5.36 g H₂O
4	0.800 g H₂ + 4.76 g O₂ = 5.56 g	5.36 g H₂O + 0.200 g H₂ = 5.56 g
5	1.00 g H₂ + 4.76 g O₂ = 5.76 g	5.36 g H₂O + 0.400 g H₂ = 5.76 g

4.154. Reaction of aluminum and bromine:

2 Al(s) + 3 Br₂(ℓ) → Al₂Br₆(s)

(a) In each experiment, the mass of Br₂ used was the same. This maximum mass of Al₂Br₆ that this can form is

$$35.5 \text{ g Br}_2 \left(\frac{1 \text{ mol Br}_2}{159.808 \text{ g Br}_2} \right) \left(\frac{1 \text{ mol Al}_2\text{Br}_6}{3 \text{ mol Br}_2} \right) \left(\frac{533.388 \text{ g Al}_2\text{Br}_6}{1 \text{ mol Al}_2\text{Br}_6} \right) = 39.5 \text{ g Al}_2\text{Br}_6$$

The mass of Al used in each experiment is different.

Experiment 1:

$$1.00 \text{ g Al} \left(\frac{1 \text{ mol Al}}{26.982 \text{ g Al}} \right) \left(\frac{1 \text{ mol Al}_2\text{Br}_6}{2 \text{ mol Al}} \right) \left(\frac{533.388 \text{ g Al}_2\text{Br}_6}{1 \text{ mol Al}_2\text{Br}_6} \right) = 9.88 \text{ g Al}_2\text{Br}_6$$

Less Al₂Br₆ can be formed from 1.00 g Al than from 35.5 g Br₂, so Al is the limiting reactant, and 9.88 g of Al₂Br₆ can be formed. All of the Al is used, so none remains.

The mass of Br₂ required is

$$1.00 \text{ g Al} \left(\frac{1 \text{ mol Al}}{26.982 \text{ g Al}} \right) \left(\frac{3 \text{ mol Br}_2}{2 \text{ mol Al}} \right) \left(\frac{159.808 \text{ g Br}_2}{1 \text{ mol Br}_2} \right) = 8.88 \text{ g Br}_2 \text{ required}$$

35.5 g Br₂ – 8.88 g Br₂ required = 26.5 g Br₂ remains

Experiment 2

$$2.00 \text{ g Al} \left(\frac{1 \text{ mol Al}}{26.982 \text{ g Al}} \right) \left(\frac{1 \text{ mol Al}_2\text{Br}_6}{2 \text{ mol Al}} \right) \left(\frac{533.388 \text{ g Al}_2\text{Br}_6}{1 \text{ mol Al}_2\text{Br}_6} \right) = 19.8 \text{ g Al}_2\text{Br}_6$$

Less Al₂Br₆ can be formed from 2.00 g Al than from 35.5 g Br₂, so Al is the limiting reactant, and 19.8 g of Al₂Br₆ can be formed. All of the Al is used, so none remains.

The mass of Br₂ required is

$$2.00 \text{ g Al} \left(\frac{1 \text{ mol Al}}{26.982 \text{ g Al}} \right) \left(\frac{3 \text{ mol Br}_2}{2 \text{ mol Al}} \right) \left(\frac{159.808 \text{ g Br}_2}{1 \text{ mol Br}_2} \right) = 17.8 \text{ g Br}_2 \text{ required}$$

35.5 g Br₂ – 17.8 g Br₂ required = 17.7 g Br₂ remains

Experiment 3:

$$4.00 \text{ g Al} \left(\frac{1 \text{ mol Al}}{26.982 \text{ g Al}} \right) \left(\frac{1 \text{ mol Al}_2\text{Br}_6}{2 \text{ mol Al}} \right) \left(\frac{533.388 \text{ g Al}_2\text{Br}_6}{1 \text{ mol Al}_2\text{Br}_6} \right) = 39.5 \text{ g Al}_2\text{Br}_6$$

In this case, the same mass of product is predicted from both reactants, so neither reactant is limiting; they were mixed in the proper stoichiometric ratio. The mass of Al₂Br₆ that can be formed is 39.5 g. All of each reactant is used, so no excess of either remains.

Experiment 4:

$$6.00 \text{ g Al} \left(\frac{1 \text{ mol Al}}{26.982 \text{ g Al}} \right) \left(\frac{1 \text{ mol Al}_2\text{Br}_6}{2 \text{ mol Al}} \right) \left(\frac{533.388 \text{ g Al}_2\text{Br}_6}{1 \text{ mol Al}_2\text{Br}_6} \right) = 59.3 \text{ g Al}_2\text{Br}_6$$

Now, less Al₂Br₆ can be formed from 35.5 g Br₂ than from 6.00 g Al, so Br₂ is the limiting reactant, and 39.5 g of Al₂Br₆ can be formed. All of the Br₂ is used, so none remains.

$$35.5 \text{ g Br}_2 \left(\frac{1 \text{ mol Br}_2}{159.808 \text{ g Br}_2} \right) \left(\frac{2 \text{ mol Al}}{3 \text{ mol Br}_2} \right) \left(\frac{26.982 \text{ g Al}}{1 \text{ mol Al}} \right) = 4.00 \text{ g Al}$$

6.00 g Al – 4.00 g Al required = 2.00 g Al remains.

These results are summarized in bold in the table.

Experiment	Mass Al Used	Mass Al₂Br₆ Produced	Mass Al in Excess	Mass Br₂ in Excess
1	1.00 g	**9.88 g**	**0.0 g**	**26.6 g**
2	2.00 g	**19.8 g**	**0.0 g**	**17.7 g**
3	4.00 g	**39.5 g**	**0.0 g**	**0.0 g**
4	6.00 g	**39.5 g**	**2.00 g**	**0.0 g**

(b) Aluminum is the limiting reactant in experiments 1 and 2. Note that 0.0 g of Al is in excess, but Br₂ is in excess.

(c) Bromine is the limiting reactant in experiment 4. No bromine is left in excess, but there is excess Al.

(d) The same mass of Br₂ was used in each experiment. The mass of aluminum bromide increases with the mass of aluminum if aluminum is limiting reactant. In experiment 4 aluminum is now in excess and bromine is the limiting reactant which halts the increase in aluminum bromide mass.

4.155. Each of the 3 flasks contains 0.100 mol of HCl. The mass of Zn differs. Calculate the # of mol Zn in each flask.

Flask 1: $7.00 \text{ g Zn} \cdot \dfrac{1 \text{ mol Zn}}{65.38 \text{ g Zn}} = 0.107 \text{ mol Zn}$

Flask 2: $3.27 \text{ g Zn} \cdot \dfrac{1 \text{ mol Zn}}{65.38 \text{ g Zn}} = 0.0500 \text{ mol Zn}$

Flask 3: $1.31 \text{ g Zn} \cdot \dfrac{1 \text{ mol Zn}}{65.38 \text{ g Zn}} = 0.0200 \text{ mol Zn}$

The balanced equation tells you that for each mol of Zn, we need 2 mol of HCl. In flask 1, 0.107 mol Zn exceeds the amount of HCl available. The reaction consumes all the HCl, and leaves unreacted Zn metal. In flask 2, the 0.0500 mol Zn reacts **exactly** with the 0.100 mol of HCl, leaving **no** unreacted Zn or HCl. In flask 3, the 0.0200 mol of Zn react with 0.0400 mol of HCl—completely consuming **all the Zn** and leaving unreacted HCl. The smaller amount of Zn present (0.0200 mol) produces a smaller amount of H₂ (0.0200 mol)—thus not totally inflating the balloon.

4.156. Regarding antacids:

(a) Compounds that form a gas when reacted with HCl:

$NaHCO_3 + HCl(aq) \rightarrow NaCl(aq) + H_2O(\ell) + CO_2(g)$

KHCO$_3$ + HCl(aq) → KCl(aq) + H$_2$O(ℓ) + CO$_2$(g)

CaCO$_3$ + 2 HCl(aq) → CaCl$_2$(aq) + H$_2$O(ℓ) + CO$_2$(g)

Mg(OH)$_2$ + 2 HCl(aq) → MgCl$_2$(aq) + 2 H$_2$O(ℓ)

Al(OH)$_3$ + 3 HCl(aq) → AlCl$_3$(aq) + 3 H$_2$O(ℓ)

States of the original reactant are unspecified, though most commonly they are (s). The compounds that form a gas when reacted with HCl are NaHCO$_3$, KHCO$_3$, and CaCO$_3$.

(b) Tums tablet with 500. mg of CaCO$_3$:

   (i) balanced equation—see (a)

   (ii) Volume of 0.500 M HCl to react with 1 Tums™ tablet:

   $$0.500 \text{ g CaCO}_3 \cdot \frac{1 \text{ mol CaCO}_3}{100.1 \text{ g CaCO}_3} \cdot \frac{2 \text{ mol HCl}}{1 \text{ mol CaCO}_3} \cdot \frac{1 \text{ L}}{0.500 \text{ mol HCl}} \cdot \frac{1000 \text{ mL}}{1 \text{ L}}$$

   $$= 20.0 \text{ mL}$$

(c) Rolaids® antacid containing 550 mg of CaCO$_3$:

   (i) balanced equation—see (a)

   (ii) Mass of Mg(OH)$_2$ in a tablet of Rolaids™:

   Total amount of HCl consumed: $\frac{0.500 \text{ mol HCl}}{1 \text{ L}} \cdot 0.02952 \text{ L} = 0.0148 \text{ mol HCl}$

   Amount of HCl consumed by CaCO$_3$:

   $$0.550 \text{ g CaCO}_3 \cdot \frac{1 \text{ mol CaCO}_3}{100.1 \text{ g CaCO}_3} \cdot \frac{2 \text{ mol HCl}}{1 \text{ mol CaCO}_3} = 0.0110 \text{ mol HCl}$$

   HCl consumed by Mg(OH)$_2$:

   0.0148 mol − 0.0110 mol = 0.0038 mol HCl

   Mass of Mg(OH)$_2$ in tablet:

   $$0.0038 \text{ mol HCl} \cdot \frac{1 \text{ mol Mg(OH)}_2}{2 \text{ mol HCl}} \cdot \frac{58.32 \text{ g Mg(OH)}_2}{1 \text{ mol Mg(OH)}_2} = 0.11 \text{ g Mg(OH)}_2$$

(d) Volume of 0.500 M HCl to react with 1 teaspoon of Maalox:

   HCl consumed by Al(OH)$_3$:

   $$0.200 \text{ g Al(OH)}_3 \cdot \frac{1 \text{ mol Al(OH)}_3}{78.00 \text{ g Al(OH)}_3} \cdot \frac{3 \text{ mol HCl}}{1 \text{ mol Al(OH)}_3} = 0.00769 \text{ mol HCl}$$

   HCl consumed by Mg(OH)$_2$:

   $$0.200 \text{ g Mg(OH)}_2 \cdot \frac{1 \text{ mol Mg(OH)}_2}{58.32 \text{ g Mg(OH)}_2} \cdot \frac{2 \text{ mol HCl}}{1 \text{ mol Mg(OH)}_2} = 0.00686 \text{ mol HCl}$$

Total moles HCl: 0.00769 + 0.00686 = 0.01455 mol HCl

Volume of 0.500M HCl consumed:

$$0.01455 \text{ mol HCl} \cdot \frac{1 \text{ L}}{0.500 \text{ mol HCl}} \cdot \frac{1000 \text{ mL}}{1 \text{ L}} = 29.1 \text{ mL}$$

(e) Greater amount of HCl consumed: Compare (b-ii), (c-ii), and (d):

According to the calculations, Rolaids™ consumes the greatest amount of HCl.

4.157. To decide the relative concentrations, calculate the dilutions. Assume that the HCl has a concentration of 0.100 M. Then calculate the diluted concentrations in each case.

Student 1: 20.0 mL of 0.100 M HCl is diluted to 40.0 mL total

The diluted molarity is: $\frac{0.100 \text{ mol HCl}}{1 \text{ L}} \cdot 0.0200 \text{ L} = 0.400 \text{ L} \cdot c_d$

$c_d = 0.050 \frac{\text{mol HCl}}{\text{L}}$

Student 2: 20.0 mL of 0.100 M HCl is diluted to 80.0 mL total

The diluted molarity is: $\frac{0.100 \text{ mol HCl}}{1 \text{ L}} \cdot 0.0200 \text{ L} = 0.0800 \text{ L} \cdot c_d$

$c_d = 0.025 \frac{\text{mol HCl}}{\text{L}}$

So, if the students compare the concentrations of the diluted solutions, the second student's prepared HCl solution is **(c) half the concentration of the first student**'s. **However**, if they compare the concentrations they calculated for the initial solution, they will calculate **the same concentration for the original HCl solution (e)** since the total number of moles of HCl in both solutions *is identical.*

4.158. Receive a ticket for exceeding BAL?

$$\frac{0.033 \text{ mol C}_2\text{H}_5\text{OH}}{1 \text{ L}} \cdot \frac{46.1 \text{ g}}{1 \text{ mol C}_2\text{H}_5\text{OH}} \cdot \frac{10^3 \text{ mg}}{1 \text{ g}} \cdot \frac{1 \text{ L}}{10 \text{ dL}} = 150 \text{ mg/dL (2 sf)}$$

The person is intoxicated and will be ticketed.

4.159. Formation of maleic anhydride from benzene:

(a) What is the % atom economy for the synthesis of maleic anhydride from benzene by this reaction?

Molar masses for all atoms used in the reaction:

Benzene [(6 × C) + (6 × H)] = 78.11
Oxygen [(9/2 × O$_2$) = 9 × O] = 143.99

Adding these gives 222.10.
Calculating the molar mass of the desired product:
$[(4 \times C) + (2 \times H) + (3 \times O)] = 98.06$.

So, the % atom economy = $\dfrac{98.06}{222.10} \cdot 100\% = 44.15\%$

(b) Percent yield of the anhydride:

Theoretical yield:

$1.00 \text{ kg } C_6H_6 \cdot \dfrac{1000 \text{ g}}{1 \text{ kg}} \cdot \dfrac{1 \text{ mol } C_6H_6}{78.11 \text{ g } C_6H_6} \cdot \dfrac{1 \text{ mol } C_4H_2O_3}{1 \text{ mol } C_6H_6} \cdot \dfrac{98.06 \text{ g } C_4H_2O_3}{1 \text{ mol } C_4H_2O_3}$

$= 1.26 \times 10^3 \text{ g } C_4H_2O_3$

Percent yield = $\dfrac{972 \text{ g } C_4H_2O_3}{1.26 \times 10^3 \text{ g } C_4H_2O_3} \cdot 100\% = 77.4\%$

What mass of the byproduct CO_2 is produced?

Theoretical yield:

$1.00 \text{ kg } C_6H_6 \cdot \dfrac{1000 \text{ g}}{1 \text{ kg}} \cdot \dfrac{1 \text{ mol } C_6H_6}{78.11 \text{ g } C_6H_6} \cdot \dfrac{2 \text{ mol } CO_2}{1 \text{ mol } C_6H_6} \cdot \dfrac{44.01 \text{ g } CO_2}{1 \text{ mol } CO_2}$

$= 1.13 \times 10^3 \text{ g } CO_2 \ (= 1.13 \text{ kg } CO_2)$

4.160. Regarding production of maleic anhydride from butane:

(a) % atom economy = $\dfrac{\text{Molar mass } (C_4H_2O_3)}{[\text{Molar mass } (C_4H_8) + 3 \cdot \text{Molar mass } (O_2)]} \cdot 100\%$

$= \dfrac{98.06 \text{ g/mol}}{(56.11 \text{ g/mol} + 3 \cdot 32.00 \text{ g/mol})} \cdot 100\% = 64.47\%$

(b) Theoretical yield =

$1.00 \times 10^3 \text{ g } C_4H_8 \cdot \dfrac{1 \text{ mol } C_4H_8}{56.11 \text{ g } C_4H_8} \cdot \dfrac{1 \text{ mol } C_4H_2O_3}{1 \text{ mol } C_4H_8} \cdot \dfrac{98.06 \text{ g } C_4H_2O_3}{1 \text{ mol } C_4H_2O_3}$

$= 1.75 \times 10^3 \text{ g } C_4H_2O_3 = 1.75 \text{ kg}$

Percent yield = $1.02 \text{ kg}/1.75 \text{ kg} \cdot 100\% = 58.4\%$

Mass of H_2O produced:

$1.00 \times 10^3 \text{ g } C_4H_8 \cdot \dfrac{1 \text{ mol } C_4H_8}{56.11 \text{ g } C_4H_8} \cdot \dfrac{3 \text{ mol } H_2O}{1 \text{ mol } C_4H_8} \cdot \dfrac{18.02 \text{ g } H_2O}{1 \text{ mol } H_2O}$

$= 963 \text{ g } H_2O = 0.963 \text{ kg } H_2O$

Solution and Answer Guide

Kotz Treichel Townsend Treichel, Chemistry and Chemical Reactivity 11e, 978-0-357-85140-1, Chapter 5: Principles of Chemical Reactivity: Energy and Chemical Reactions

TABLE OF CONTENTS

Applying Chemical Principles	184
Practicing Skills	186
General Questions	205
In the Laboratory	218
Summary and Conceptual Questions	220

Applying Chemical Principles

Gunpowder

5.1.1. Concerning black powder:

(a) The enthalpy change for: $2\ KNO_3(s) + 3\ C(s) + S(s) \rightarrow K_2S(s) + N_2(g) + 3\ CO_2(g)$

$\Delta_r H° = [(1\,mol)(–376.6\ kJ/mol) + (1\,mol)(0\ kJ/mol) + (3\ mol)(–393.509\ kJ/mol)]$
$\qquad – [(2\ mol)(–494.6\ kJ/mol) + (3\ mol)(0\ kJ/mol) + (1\ mol)(0\ kJ/mol)]$

$\Delta_r H° = [(–1557.127\ kJ) – (–989.2\ kJ)] = –567.9\ kJ$

(b) Using the assumption that 1 mol of black powder is: 2 mol KNO_3, 3 mol of C, and 1 mol of S, black powder would have a MM of $(2 × 101.102) + (3 × 12.011) + (1 × 32.06) = 270.30$ g/mol.

$$\frac{-567.9\ kJ}{1\ mol\ powder} \cdot \frac{1\ mol\ powder}{270.3\ g} = -2.10\ kJ/g$$

5.1.2. Energy of reaction per gram of guncotton:
$q = $ –(heat absorbed by water + heat absorbed by calorimeter):

$q = –[(4.184\ J/g·K)(1.200 × 10^3\ g)(1.32\ K) + (691\ J/K)(1.32\ K)] = –7.54 × 10^3\ J$
and to express this per gram of guncotton, divide by the mass of guncotton:
$–7.54 × 10^3\ J/0.725\ g = –1.04 × 10^4$ J/g guncotton

5.1.3. Regarding the decomposition of nitroglycerin:

(a) Balanced equation for the decomposition of nitroglycerin:

4 C$_3$H$_5$N$_3$O$_9$(ℓ) → 12 CO$_2$(g) + 10 H$_2$O(g) + 6 N$_2$(g) + O$_2$(g)

If you begin with a coefficient of 1 for "nitro", 3 C in "nitro" gives 3 CO$_2$, 2.5 H$_2$O, 1.5 N$_2$, and 0.25 O$_2$. So the question is, "What coefficient converts ALL these coefficients into integers"? Multiplying all coefficients by 4 solves things nicely!

(b) The decomposition releases 6.23 kJ/g nitro. Add the atomic weights of the involved atoms in nitroglycerine to get 227.09 g/mol. The energy released PER MOLE of nitro is: −6.23 kJ/g • 227.09 g/mol nitro = −1414.8 kJ/mol nitro or −1410 kJ/mol (3 sf)

Solve the equation for the Enthalpy change for the process:

4 C$_3$H$_5$N$_3$O$_9$(ℓ) → 12 CO$_2$(g) + 10 H$_2$O(g) + 6 N$_2$(g) + O$_2$(g)

Note that the $\Delta_r H°$ = 4 × −1414.8 kJ/mol nitroglycerine = −5659 kJ

$\Delta_r H°$ = [(12 mol)(−393.509 kJ/mol)]) + (10 mol)(−241.83 kJ/mol) + (6 mol)(0 kJ/mol) + (1 mol)(0 kJ/mol)]   − [(4 mol)($\Delta_f H°$ nitro)]

−5659 kJ = [(−7140.408 kJ) − (4 mol)($\Delta_f H°$ nitro)] =

−1481 kJ =   4 × $\Delta_f H°$ nitro, and dividing by 4 gives:
−370. kJ/mol nitro = $\Delta_f H°$ nitro

The Fuel Controversy—Alcohol and Gasoline

5.2.1.   Compare $\Delta_r H°$ for ethanol and octane on a per mole and per gram basis:

For octane: C$_8$H$_{18}$(ℓ) + 25/2 O$_2$(g) → 8 CO$_2$(g) + 9 H$_2$O(ℓ)

$\Delta_r H°$ = [(8 mol)(−393.509 $\frac{kJ}{mol}$) + (9 mol)(−285.83 $\frac{kJ}{mol}$)] − [(1 mol)(−250.1 $\frac{kJ}{mol}$) + 0 kJ]

$\Delta_r H°$ = −5470.4 kJ for 1 mol of octane

Expressed on a per gram basis: −5470.4 $\frac{kJ}{mol}$ • $\frac{1 \text{ mol C}_8\text{H}_{18}}{114.2 \text{ g C}_8\text{H}_{18}}$ = −47.90 kJ/g

For ethanol: C$_2$H$_5$OH(ℓ) + 3 O$_2$(g) → 2 CO$_2$(g) + 3 H$_2$O(ℓ)

$\Delta_r H°$ = [(2 mol)(−393.509 kJ/mol) + (3 mol)(−285.83 kJ/mol)] − [(1 mol)(−277.0 kJ/mol) + 0 kJ]

= [(−787.0 kJ) + (−857.49 kJ)] + 277.0 kJ

= −1367.5 kJ for the combustion of 1 mol of ethanol (coefficient in equation)

Calculate the energy change per gram of ethanol:

$\frac{-1367.5 \text{ kJ}}{1 \text{ mol}}$ • $\frac{1 \text{ mol}}{46.068 \text{ g}}$ = -29.68 kJ/g ethanol

So, octane provides more energy/mol AND per gram.

5.2.2. Compare energy produced per liter of fuel:

For octane: −47.90 kJ/g × 0.699 g/1 mL × 1000 mL/1 L = −33483 kJ/L
(or −3.35 × 10⁴ kJ/L to 3sf)

For ethanol: −29.68 kJ/g ethanol × 0.785 g/1 mL × 1000 mL/1 L = −23302 kJ/L
(or −2.33 × 10⁴ kJ/L to 3 sf)

5.2.3. Mass of CO_2 produced per liter of fuel:

For octane: 1 mol octane produces 8 mol CO_2:

$$\frac{1000 \text{ mL}}{1 \text{ L}} \cdot \frac{0.699 \text{ g C}_8\text{H}_{18}}{1.0 \text{ mL C}_8\text{H}_{18}} \cdot \frac{1 \text{ mol C}_8\text{H}_{18}}{114.2 \text{ g C}_8\text{H}_{18}} \cdot \frac{8 \text{ mol CO}_2}{1 \text{ mol C}_8\text{H}_{18}} \cdot \frac{44.0 \text{ g CO}_2}{1 \text{ mol CO}_2} = \frac{2150 \text{ g CO}_2}{1 \text{ L C}_8\text{H}_{18}}$$

For ethanol: 1 mol ethanol produces 2 mol CO_2:

$$\frac{1000 \text{ mL}}{1 \text{ L}} \cdot \frac{0.785 \text{ g C}_2\text{H}_5\text{OH}}{1.0 \text{ mL C}_2\text{H}_5\text{OH}} \cdot \frac{1 \text{ mol C}_2\text{H}_5\text{OH}}{46.068 \text{ g C}_2\text{H}_5\text{OH}} \cdot \frac{2 \text{ mol CO}_2}{1 \text{ mol C}_2\text{H}_5\text{OH}} \cdot \frac{44.0 \text{ g CO}_2}{1 \text{ mol CO}_2} = \frac{1.50 \times 10^3 \text{ g CO}_2}{1 \text{ L C}_2\text{H}_5\text{OH}}$$

5.2.4. Volume of ethanol with the same energy as 1.00 L of octane:

Begin with the energy produced by 1.00 L of octane:

$$3.34 \times 10^4 \text{ kJ} \cdot \frac{1.00 \text{ L C}_2\text{H}_5\text{OH}}{2.33 \times 10^4 \text{ kJ}} = 1.43 \text{ L C}_2\text{H}_5\text{OH}$$

Fuel producing more CO_2:

1.00 L of octane produces 2150 g CO_2 (See 5.2.3)

1.43 L of C_2H_5OH produces $\frac{1.43 \text{ L C}_2\text{H}_5\text{OH}}{1} \cdot \frac{1500 \text{ g CO}_2}{1 \text{ L C}_2\text{H}_5\text{OH}} = 2150 \text{ g CO}_2$

Octane produces more CO_2 for the same amount of energy.

Practicing Skills

Energy: Some Basic Principles

5.1. Define the terms system and surroundings.

SYSTEM: Is nothing more than the "thing" we are studying in a particular situation. If we are concerned with the energy change associated with heating water, then the SYSTEM is the water.
SURROUNDINGS: The surroundings are defined as all that matter which is in thermal contact with the system. So, if we were burning our stick of wood in a fireplace, the surroundings would include the air, and the fireplace materials in thermal contact with the stick of wood.

So, when we say that a system and its surroundings are in thermal equilibrium, we mean that the energy (heat) flow between the system and the surroundings is in a steady state—that the amount of energy "leaving" the system and energy "leaving" the surroundings is equal.

5.2. Energy as heat is transferred from an object at a higher temperature to one at a lower temperature.

5.3. Identify whether the following processes are exothermic or endothermic.
   (a) combustion of methane: Since we frequently burn hydrocarbons as a source of energy (think propane grill), the combustion of methane is **exothermic**; $q_{sys} = -$
   (b) melting of ice: Ice melts (solid becomes liquid) as it absorbs thermal energy from the surroundings—a process we call **endothermic**; $q_{sys} = +$
   (c) raising the temperature of water from 25 °C to 100 °C: We do this by **increasing** the energy of the water molecules—and this occurs in an **endothermic** process; $q_{sys} = +$
   (d) heating $CaCO_3(s)$ to form $CaO(s)$ and $CO_2(g)$: Just as in heating water (c above) we add thermal energy to $CaCO_3(s)$ in an **endothermic** process; $q_{sys} = +$

5.4. Identify processes as exothermic or endothermic:
   (a) reaction of Na and Cl_2: exothermic; $q_{sys} = -$
   (b) cooling and condensing liquid nitrogen: exothermic; $q_{sys} = -$
   (c) cooling a soft drink: exothermic; $q_{sys} = -$
   (d) heating HgO to form Hg and O_2: endothermic; $q_{sys} = +$

Heat Capacity and Specific Heat Capacity

5.5. Heat capacity of the wood block:

$\Delta T = 39.4\ °C - 27.2\ °C = 12.2\ °C = 12.2\ K$

$q = C \cdot \Delta T, \text{thus } C = \dfrac{q}{\Delta T}$

$C = \dfrac{168\ J}{12.2\ K} = 13.8\ J/K$

5.6. Energy required to heat metal:

$\Delta T = 82.9\ °C - 22.8\ °C = 60.1\ °C = 60.1\ K$

$q = C \cdot \Delta T$

$q = 23.1\ J/K(60.1\ K) = 1388\ J = 1390\ J\ (3\ sf)$

5.7. Specific heat capacity of copper:

$C = m \cdot c = 25.0 \text{ g Cu}(0.385 \text{ J/g} \cdot \text{K}) = 9.63 \text{ J/K}$

5.8. Specific heat capacity of iron:

$C = m \cdot c = 45.0 \text{ g Fe}(0.449 \text{ J/g} \cdot \text{K}) = 20.2 \text{ J/K}$

5.9. Convert molar heat capacity to specific heat capacity:

Molar heat capacity of Hg = $28.1 \text{ J/mol} \cdot \text{K}$

Molar mass of Hg = 200.59 g/mol

Specific heat capacity of Hg = $\dfrac{28.1 \text{ J}}{\text{mol} \cdot \text{K}} \left(\dfrac{1 \text{ mol}}{200.59 \text{ g}} \right) = 0.140 \text{ J/g} \cdot \text{K}$

5.10. Molar heat capacity of benzene:

$\dfrac{1.74 \text{ J}}{\text{g} \cdot \text{K}} \cdot \dfrac{78.11 \text{ g}}{1 \text{ mol C}_6\text{H}_6} = 136 \text{ J/mol} \cdot \text{K}$

5.11. Heat energy to warm 168 g copper from –12.2 °C to 25.6 °C:

Heat = mass × specific heat capacity × ΔT

For copper = $(168 \text{ g})(\dfrac{0.385 \text{ J}}{\text{g} \cdot \text{K}})[25.6 \text{ °C} - (-12.2) \text{ °C}] \cdot \dfrac{1 \text{ K}}{1 \text{ °C}} = 2.44 \times 10^3 \text{ J}$ or 2.44 kJ

5.12. Energy to raise temperature of 50.00 mL of water:

$q = (50.00 \text{ mL} \cdot \dfrac{0.997 \text{ g}}{1 \text{ mL}})(4.184 \dfrac{\text{J}}{\text{g} \cdot \text{K}})(301.90 \text{ K} - 298.67 \text{ K}) = 674 \text{ J}$

5.13. The final temperature of a 344 g sample of iron when 2.25 kJ of heat are added to a sample originally at 18.2 °C. The energy added is:

q_{Fe} = (mass)(specific heat capacity)(ΔT)

2.25×10^3 J = (344 g)(0.449 $\dfrac{J}{g \cdot K}$)(x) and solving for x we get:

14.57 K = x and since 1 K = 1 °C, ΔT = 14.57 °C.

The final temperature is (14.57 + 18.2) °C or 32.8 °C.

5.14. Initial temperature of Cu:

1.850×10^3 J = (500. g)(0.385 J/g·K)(310. K – T_i) and T_i = 300. K (27 °C)

5.15. Final T of copper-water mixture:

We must **assume** that **no energy** will be transferred to or from the beaker containing the water. Then the **magnitude** of energy lost by the hot copper and the energy gained by the cold water will be equal (but opposite in sign).

$q_{copper} = -q_{water}$

Using the heat capacities of H_2O and copper, and expressing the temperatures in Kelvin (K = °C + 273.15) we can write:

(62.8 g)(0.385 $\dfrac{J}{g \cdot K}$)(T_{final} – 371.45 K) = –(164 g)(4.184 $\dfrac{J}{g \cdot K}$)(T_{final} – 292.75K)

Simplifying each side gives:

24.178 $\dfrac{J}{K}$ · T_{final} – 8981 J = –686.176 $\dfrac{J}{K}$ · T_{final} + 200,878 J

710.354 $\dfrac{J}{K}$ · T_{final} = 209,859 J

T_{final} = 295.43 K or (295.43 – 273.15) or 22.3 °C

Don't forget: **Round numbers only at the end.**

5.16. Final water temperature:

$q_{cool\ water} + q_{warm\ water} = 0$
[(305 g)(4.184 J/g·K)(T_f – 296 K)] + [(174 g)(4.184 J/g·K)(T_f – 369 K)] = 0
T_f = 323 K (50. °C)

5.17. Heat capacity of coffee cup:

82.4 °C = 355.5 K   79.3 °C = 352.4 K   25.0 °C = 298.1 K

heat lost by coffee = 225 g(4.184 J/g·K)(352.4 K – 355.5 K) = –2918 J heat gained by cup = $x(352.4 \text{ K} - 298.1 \text{ K}) = x(54.3 \text{ K})$

heat gained by cup = −heat lost by coffee

$x(54.3 \text{ K}) = --(-2918 \text{ J})$

$x = C_{cup} = 54$ J/K

5.18 Heat capacity of drinking glass:

28.0 ° = 301.1 K   14.2 °C = 287.3 K   12.8 °C = 285.9 K

heat gained by water = 125 g(4.184 J/g·K)(287.3 K − 285.9 K) = 732 J

heat lost by glass = $x(287.3 \text{ K} - 301.1 \text{ K}) = x(-13.8 \text{ K})$

heat lost by glass = −heat gained by water

$x(-13.8 \text{ K}) = -732$ J

$x = C_{glass} = 53$ J/K

5.19. As in SQ5.15, $-q_{gold} = q_{water}$. Calculate the amount of heat gained by the water: (21.6 g)(4.184 J/K·mol)(304.05 K − 301.55 K) = q_{water} = 226 J, which corresponds to the heat lost by gold. (174 g)(0.128 J/K·mol)(304.05 K − T_i) = $-q_{gold}$

Let's represent the temperature change for gold (304.05 K − T_i) by ΔT.

(174 g)(0.128 J/K·mol)ΔT = $-q_{gold}$ = −226 J

Now solve for ΔT: ΔT = −226 J/[(174 g((0.128 J/K·mol)] = 10.1 K

Δ$T = T_f - T_i = -10.1$; $T_i = \Delta T + T_f = 10.1$ K + 304.05 = 314.2 K (314 K to 2 sf).

Expressing this temperature in the Celsius scale gives: 314.2 K − 273.15 = 41 °C (2 sf).

5.20. Initial temperature of second sample:
The final temperature is greater than 24.7 °C, so the 132 g sample of water must be warmer than the 225 g sample.

$q_{\text{cool water}} + q_{\text{warm water}} = 0$

$[(225 \text{ g})(4.184 \text{ J/g·K})(318.0 - 297.9 \text{ K})] + [(132 \text{ g})(4.184 \ \frac{\text{J}}{\text{g·K}})(318.0 \ \text{K} - T_i)] = 0$

T_i = 352.2 K (79.1 °C)

5.21. Here the warmer Zn is losing heat to the water: $q_{metal} = -q_{water}$

Remembering that Δ$T = T_f - T_i$, we can calculate the change in temperature for the water and the metal. Further, since we know the final and initial for both the metal and the water, we can calculate the temperature difference in units of Celsius degrees, since the **change** in temperature on the **Kelvin** scale would be numerically identical.

For the metal: $\Delta T = T_f - T_i = (27.1°C - 98.8 °C)$ or $-71.7 °C$ or -71.7 K.

For the water: $\Delta T = T_f - T_i = (27.1 °C - 25.0 °C)$ or $2.1 °C$ or 2.1 K (recalling that a Celsius degree and a Kelvin are the same "size").

(13.8 g)(c_{metal})(−71.7 K) = −(45.0 g)(4.184 J/g·K)(2.1 K)

−989.46 g · K(c_{metal}) = −395 J   and

c_{metal} = 0.40 J/g·K   (2 sf)

5.22. Specific heat capacity of Mo:
$q_{metal} + q_{water} = 0$
[(237 g)(c_{Mo})(288.5 K − 373.2 K)] + [(244 g)(4.184 J/g·K)(288.5 K − 283.2 K)] = 0
c_{Mo} = 0.27 J/g·K

Changes of State

5.23. Quantity of energy evolved when 1.0 L of water at 0 °C solidifies to ice:

The mass of water involved: If we assume a density of liquid water of 1.00 g/mL, 1.0 L of water (1000 mL) would have a mass of 1.0×10^3 g.

To freeze 1.0×10^3 g water: 1.0×10^3 g ice · 333 J/g ice = 333×10^3 J or 330 kJ (2sf)

5.24. Energy needed to melt a tray of ice cubes:

16 cubes · $\dfrac{62.0 \text{ g}}{1 \text{ cube}}$ · 333 J/g = 3.30×10^5 J

5.25. Heat required to vaporize (convert liquid to gas) 175 g C_6H_6:
The heat of vaporization of benzene is 30.8 kJ/mol.
Convert mass of benzene to moles of benzene: 175 g · 1 mol/78.11 g = 2.24 mol
Heat required: 2.24 mol C_6H_6 · 30.8 kJ/mol = 69.0 kJ
NOTE: No sign has been attached to the amount of heat, since we wanted to know the **quantity**. If we want to assign a **direction** of heat flow in this question, we add a (+) to 69.0 kJ to indicate that heat is being **added** to the liquid benzene.

5.26. Energy to convert 117 g of liquid to vapor:

117 g $\left(\dfrac{1 \text{ mol } CH_3Cl}{50.49 \text{ g}}\right)\left(\dfrac{21.40 \text{ kJ}}{1 \text{ mol } CH_3Cl}\right) = 49.6$ kJ

5.27. To calculate the quantity of heat for the process described, think of the problem in two steps:

1) cool liquid from 23.0 °C to liquid at –38.8 °C

2) freeze the liquid at its freezing point (–38.8 °C)

Note that the specific heat capacity is expressed in units of mass, so convert the volume of liquid mercury to **mass**. 1.00 mL · 13.6 g/mL = 13.6 g Hg (Recall: 1 cm^3 = 1 mL)
1) The energy to cool 13.6 g of Hg from 23.0 °C to liquid at –38.8 °C is:
ΔT = (234.35 K − 296.15 K) or −61.8 K

13.6 g Hg · 0.140 $\dfrac{J}{g \cdot K}$ · −61.8 K = −117.7 J

To convert liquid mercury to solid Hg at this temperature:
−11.4 J/g · 13.6 g = −155 J (The (−) sign indicates that heat is being removed from the Hg.) The total energy released by the Hg is: [−117.7 J + −155.0 J] = −273 J and since $q_{mercury} = -q_{surroundings}$, the quantity released to the surroundings is 273 J.

5.28. Quantity of energy to raise tin to its melting point and melt it:
q_{total} = energy to heat metal to melting point + energy to change phase from solid to liquid
$q_{heat\ metal}$ = (454 g)(0.227 J/g·K)(505.1 K − 298.2 K) = 2.13 × 10^4 J
$q_{phase\ change}$ = (454 g)(59.2 J/g) = 2.69 × 10^4 J
q_{total} = 2.13 × 10^4 J + 2.69 × 10^4 J = 4.82 × 10^4 J

5.29. To accomplish the process, one must:

1) heat the ethanol from 20.0 °C to 78.29 °C (ΔT = 58.29 °C)

2) boil the ethanol (convert from liquid to gas) at 78.29 °C

Using the specific heat for ethanol, the energy for the first step is:

(2.44 $\dfrac{J}{g \cdot K}$)(1.00 × 10^3 g)(58.29 K) = 142,227.6 J (142,000 J to 3 sf)

To boil the ethanol at 78.29 °C, we need: 855 J/g · 1.00 × 10^3 g = 855,000 J

The total heat energy needed (in J) is (142,000 + 855,000) = 997,000 or 9.97 × 10^5 J

5.30. Energy released as benzene is cooled and frozen:

25.0 mL · $\dfrac{0.80 \text{ g}}{1 \text{ mL}}$ = 20. g C$_6$H$_6$

q_{total} = energy to cool liquid to freezing point + energy to change phase from liquid to solid
$q_{cool\ liquid}$ = (20. g)(1.74 J/g·K)(278.7 K − 293.1 K) = −5.0 × 10^2 J
$q_{phase\ change}$ = (20. g)(−127 J/g) = −2500 J
q_{total} = −5.0 × 10^2 J + (−2500 J) = −3.0 × 10^3 J (3.0 × 10^3 J released to the surroundings)

5.31. Energy required to convert ice into water requires three steps:

1) Heat ice from −12.4 °C to 0.0 °C

ΔT = 273.2 K − 260.8 K = 12.4 K

115 g ice(2.06 J/g·K)(12.4 K) = 2938 J

2) Melt 115 g ice to liquid water

115 g(333 J/g) = 38295 J

3) Heat water from 0.0 °C to 32.4 °C

ΔT = 305.6 K − 273.2 K = 32.4 K

115 g water (4.184 J/g·K)(32.4 K) = 15590 J

Energy required is sum of the three amounts = 2938 J + 38295 J + 15590 J = 56823 J = 56.8 kJ (3 sf)

NOTE: No sign has been attached to the amount of heat since we wanted to know the **quantity**. Because heat is added to the system the sign would be positive (+).

5.32. Energy required to convert water into steam requires three steps:

1) Heat water from 22.7 °C to 100. °C

ΔT = 373.2 K − 295.9 K = 77.3 K

75.0 g water(4.184 J/g·K)(77.3 K) = 2.426 × 10⁴ J

2) Evaporate 75 g water to steam

75.0 g(2256 J/g) = 1.692 × 10⁵ J

3) Heat steam from 100. °C to 135.4 °C

ΔT = 408.6 K − 373.2 K = 35.4 K

75.0 g steam (1.86 J/g·K)(35.4 K) = 4938 J

Energy required is sum of the three amounts = 2.426 × 10⁴ J + 1.692 × 10⁵ J + 4938 J = 198 × 10⁵ J = 198 kJ (3 sf)

NOTE: No sign has been attached to the amount of heat since we wanted to know the **quantity**. Because heat is added to the system the sign would be positive (+).

Heat, Work, and Internal Energy

5.33. Calculate work: $w = -P\Delta V$
The pressure is 1.01 × 10⁵ Pa. Look at the change in volume:
Converting the volumes to m³, (1.25 L − 2.50 L) · 1 m³/1000 L = −1.25 × 10⁻³ m³.
Calculating: $w = -P\Delta V$ = −(1.01 × 10⁵ Pa)(−1.25 × 10⁻³ m³) NOTE that 1 Pa = 1 kg/m·s².

Substituting this value into the equation:

−(1.01 × 10⁵ kg/m·s²)−1.25 × 10⁻³ m³) = 126 kg·m²/s² = 126 J (since 1 J = 1 kg·m²/s²)

w = 126 J

5.34. Work done by balloon:
$w = -P\Delta V = -P(V_f - V_i) = -(1.01 \times 10^5 \text{ Pa})(1.20 \text{ L} - 0.75 \text{ L})(1 \text{ m}^3/1000 \text{ L}) -45$ kg m²/s² = −45 J

5.35. Balloon does 324 J of work. The pressure is constant at 7.33 × 10⁴ Pa. What is the change in volume?   This question is very similar to SQ#33, except here we have the work done, and want to know ΔV.
Since w = −PΔV, we calculate the volume change by dividing the P into the w.

$$-\left(7.33 \times 10^4 \frac{\text{kg}}{\text{m} \cdot \text{s}^2}\right)(\Delta V) = -324 \frac{\text{kg} \cdot \text{m}^2}{\text{s}^2} \text{ and}$$

$$\Delta V = \frac{-324 \frac{\text{kg} \cdot \text{m}^2}{\text{s}^2}}{-\left(7.33 \times 10^4 \frac{\text{kg}}{\text{mg} \cdot \text{s}^2}\right)} = 4.42 \times 10^{-3} \text{ m}^3 \text{ and}$$

since 1 m³ = 1000 L, 4.42 × 10⁻³ m³ · 1000 L/1 m³ = 4.42 L

5.36. Change in volume of gas:
w = −PΔV so ΔV = −w/P = −(+1.34 × 10³ kg m²/s²)/1.33 × 10⁵ kg/m s² = −1.01 × 10⁻² m³
and expressing this in units of L: (−1.01 × 10⁻² m³)(10 dm/m)³ = −10.1 L

5.37. Change in *U*:
The internal energy, Δ*U* = q + w;   Δ*U* = q + w = +825 J + (−417 J) = 408 J. What's critical about this solution is to understand that heat is being added to the gas (system), so q = +. With the gas expansion, it is doing work, so w = −.

5.38. Calculate work done by gas; does volume increase or decrease:

(a) Δ*U* = −1.40 × 10³ J and q = −1.72 × 10³ J as heat is transferred to surroundings

Δ*U* = q + w rearranging: w = Δ*U* − q = −1.40 × 10³ J − (−1.72 × 10³ J)

= 0.32 × 10³ J = 0.32 kJ

(b) work is done on system; thus, gas contracts, and volume decreases

5.39. Since the internal energy, $\Delta U = q_p + w_p$, and we know that $\Delta U = +1.11$ kJ, we can calculate w, if we know q. Energy is transferred as heat into the system, so $q = +125$ kJ, and we can calculate w:

$\Delta U = q_p + w_p$; $+1.11 \times 10^3$ J $= +1.25 \times 10^3$ J $+ w_p$; and $w_p = (1.11-1.25) \times 10^3$ J $= -0.14 \times 10^3$ J

Since $w = -P\Delta V$, our equation becomes -0.14×10^3 J $= -P\Delta V$. Multiplying both sides by -1 gives 0.14×10^3 J $= P\Delta V$, and solving for ΔV (dividing by P) gives:

$$\Delta V = \frac{w}{P} = \frac{0.14 \times 10^3 \text{ J}}{1.22 \times 10^5 \text{ Pa}} = \frac{1.4 \times 10^2 \frac{\text{kg} \cdot \text{m}^2}{\text{s}^2}}{1.22 \times 10^5 \frac{\text{kg}}{\text{m} \cdot \text{s}^2}} = 1.15 \times 10^{-3} \text{ m}^3$$

Volume change $= +1.15 \times 10^{-3}$ m$^3 \cdot$ 1000 L/1 m$^3 = 1.15$ L.   The original volume of Ar $= 1.50$ L, so the final volume $= 1.50 + 1.15 = 2.6$ L to 2 sf

5.40. Change in internal energy of nitrogen gas:

$\Delta U = q + w$; $w = -P\Delta V$ $\Delta U = q - P\Delta V = -695$ J $- (9.95 \times 10^4$ Pa$)(-1.88$ L$)(1$ m/10 dm$)^3$

$= -695$ J $+ 187$ J $= -508$ J

Enthalpy Changes

Note that in this chapter, I have left negative signs with the value for heat released.

(heat released = –; heat absorbed = +)

5.41. For a process in which the $\Delta H°$ is negative, that process is **exothermic**.

To calculate heat released when 1.50 g NO react, note that the energy shown (-114.1 kJ) is released when **2** moles of NO react, so we'll need to account for that:

$$1.50 \text{ g NO} \left(\frac{1 \text{ mol NO}}{30.01 \text{ g NO}}\right)\left(\frac{-114.1 \text{ kJ}}{2 \text{ mol NO}}\right) = -2.85 \text{ kJ}$$

5.42. Is reaction endo- or exothermic? What is the enthalpy change?

Endothermic, as $\Delta H° = +$

Enthalpy change: $15.0 \text{ g CaO} \cdot \frac{1 \text{ mol CaO}}{56.08 \text{ g}} \cdot \frac{1 \text{ mol-rxn}}{1 \text{ mol CaO}} \cdot \frac{464.8 \text{ kJ}}{1 \text{ mol-rxn}} = 124 \text{ Kj}$

5.43. The combustion of isooctane (IO) is **exothermic**. The molar mass of IO is: 114.2 g/mol. The heat evolved is:

$$1.00 \text{ L of IO} \cdot \frac{1 \times 10^3 \text{ mL}}{1 \text{ L}} \cdot \frac{0.69 \text{ g IO}}{1 \text{ mL}} \cdot \frac{1 \text{ mol IO}}{114.2 \text{ g IO}} \cdot \frac{-10922 \text{ kJ}}{2 \text{ mol IO}} = -3.3 \times 10^4 \text{ kJ}$$

5.44. Enthalpy change to produce 1.00 L of acetic acid:

$$1.00 \times 10^3 \text{ mL} \cdot \frac{1.044 \text{ g}}{1 \text{ mL}} \cdot \frac{1 \text{ mol CH}_3\text{CO}_2\text{H}}{60.05 \text{ g}} \cdot \frac{1 \text{ mol-rxn}}{1 \text{ mol CH}_3\text{CO}_2\text{H}} \cdot \frac{-134.6 \text{ kJ}}{1 \text{ mol-rxn}} = -2.34 \times 10^3 \text{ kJ}$$

Calorimetry

5.45. 100.0 mL of 0.200 M CsOH and 50.0 mL of 0.400 M HCl each supply 0.0200 moles of base and acid respectively. If we assume the specific heat capacities of the solutions are 4.2 J/g · K, the **heat evolved** for 0.200 moles of CsOH is:

q = (4.2 J/g · K)(150.0 g)(24.28 °C − 22.50 °C) [and since 1.78 °C = 1.78 K]

q = (4.2 J/g · K)(150.0 g)(1.78 K)

q = 120 J

The molar enthalpy of neutralization is: $\frac{-1120 \text{ J}}{0.0200 \text{ mol CsOH}} = -56000 \text{ J/mol (2 sf)}$

or −56 kJ/mol.

5.46. Enthalpy of neutralization per mole of CsOH:

$q_r + q_{solution} = 0$

q_r = −{[(125 mL + 50.0 mL)(1 g/mL)](4.2 J/g·K)(297.55 K − 294.65 K)} = −2130 J.

How many mol of CsOH? $0.125 \text{ L} \cdot \frac{0.250 \text{ mol CsOH}}{1 \text{ L}} = 0.0313 \text{ mol CsOH}$

Enthalpy change/mol: −2130 J/(0.0313 mol CsOH) = −68 kJ/mol

5.47. For this problem, we'll assume that the coffee-cup calorimeter absorbs **no** heat.

Since $q_{metal} = -q_{water}$

Remembering that $\Delta T = T_{final} - T_{initial}$, we can calculate the change in temperature for the water and the metal. Further, since we know the final and initial temperatures for both the metal and the water, we can calculate the temperature difference in units of Celsius degrees, since the **change** in temperature on the **Kelvin** scale would be numerically identical.

For the metal: $\Delta T = T_{final} - T_{initial} = (24.3 - 99.5)$ or $-75.2\,°C$ or $-75.2\,K$.

For the water: $\Delta T = T_{final} - T_{initial} = (24.3 - 21.7)$ or $2.6\,°C$ or $2.6\,K$
(recalling that a Celsius degree and a Kelvin are the same "size".

$(20.8\text{ g})(c_{metal})(-75.2\text{ K}) = -(75.0\text{ g})(4.184\,\dfrac{J}{g\cdot K})(2.6\text{ K})$

$-1564.16\text{ g}\cdot K(c_{metal}) = -816\text{ J}$

$$c_{metal} = 0.52\,\dfrac{J}{g\cdot K}\quad (2\text{ sf})$$

5.48. Calculate the specific heat capacity of Cr:

$q_{metal} + q_{water} = 0$
$[(24.26\text{ g})(c_{Cr})(298.8\text{ K} - 371.5\text{ K})] + [(82.3\text{ g})(4.184\text{ J/g}\cdot K)(298.8\text{ K} - 296.5\text{ K})] = 0$
$c_{Cr} = 0.45\,\dfrac{J}{g\cdot K}$

5.49. Enthalpy change when 5.44 g of NH_4NO_3 is dissolved in 150.0 g water at 18.6 °C.

Calculate the heat released by the solution: $\Delta T = (16.2\,°C - 18.6\,°C) = -2.4\,°C$ or $-2.4\,K$

$(155.4\text{ g})(4.2\,\dfrac{J}{g\cdot K})(-2.4\text{ K}) = -1566\text{ J}\quad \text{or} -1600\text{ J (2 sf)}$

Calculate the amount of NH_4NO_3: $5.44\text{ g }NH_4NO_3 \cdot \dfrac{1\text{ mol }NH_4NO_3}{80.04\text{ g }NH_4NO_3} = 0.0680\text{ mol}$

Recall that the energy that was released by the solution is **absorbed** by the ammonium nitrate, so we change the sign from (−) to (+).

The enthalpy change has been requested in units of kJ, so divide the energy (in J) by 1000:

Enthalpy of dissolving = $\dfrac{1.566\text{ kJ}}{0.0680\text{ mol}} = 23.0\text{ kJ/mol}$ or 23 kJ/mol (2 sf)

5.50. Calculate the enthalpy change when H_2SO_4 is dissolved in water:

Assume $c_{solution} = 4.2\text{ J/g}\cdot K$
$q_{dissolving} = -[(140.2\text{ g})(4.2\text{ J/g}\cdot K)(302.0\text{ K} - 293.4\text{ K})]$
$q_{dissolving} = -5064\text{ J}$

$5.2\text{ g }H_2SO_4 \cdot \dfrac{1\text{ mol }H_2SO_4}{98.1\text{ g}} = 0.0530\text{ mol }H_2SO_4$

$\Delta H = q_{dissolving} = \left(\dfrac{-5064\text{ J}}{0.0530\text{ mol }H_2SO_4}\right)\left(\dfrac{1\text{ kJ}}{1000\text{ J}}\right) = -96\text{ kJ/mol }H_2SO_4$

5.51. Calculate the heat evolved/mol SO_2 for the reaction of sulfur with oxygen to form SO_2. Keep in mind that $q_r + q_{water} + q_{bomb} = 0$, and so $q_r = -(q_{water} + q_{bomb})$

There are several steps:

1) Calculate the heat transferred to the water:
   815 g (4.184 J/g·K)(26.72 °C – 21.25 °C)(1K/1 °C) = 18,650 J

2) Calculate the heat transferred to the bomb calorimeter
   923 J/K · (26.72 – 21.25) °C · 1K/1 °C = 5,049.J

3) Amount of sulfur present: 2.56 g (1 mol S_8/256.48 g S_8) = 0.009979 mol S_8

   Note from the equation that 8 mol of SO_2 form from each mole of S_8

4) Calculating ΔU per mol of SO_2 yields:

$$\frac{-(18650 \text{ J} + 5049 \text{ J})}{0.07983 \text{ mol } SO_2} = -297,000 \text{ J/mol or } -297 \text{ kJ/mol}$$

5.52. Calculate ΔU per mol of carbon:

$q_{water} + q_{bomb} + q_r = 0$
$q_r = -([(775 \text{ g})(4.184 \text{ J/g·K})(300.53 \text{ K} - 298.15 \text{ K})] + [(893 \text{ J/K})(300.53 \text{ K} - 298.15 \text{ K})])$
$q_r = -9840 \text{ J}$

0.300 g C · $\frac{1 \text{ mol C}}{12.01 \text{ g}}$ = 0.0250 mol C

$\Delta U = \frac{-9840 \text{ J}}{0.0250 \text{ mol C}} \cdot \frac{1 \text{ kJ}}{10^3 \text{ J}} = -394 \text{ kJ/mol C}$

5.53. Quantity of heat evolved in the combustion of benzoic acid:
Let's approach this in several steps:
1) Calculate the heat transferred to the water:

   775 g (4.184 J/g·K)(31.69 °C – 22.50 °C)(1K/1 °C) = 29800 J

2) Calculate the heat transferred to the bomb calorimeter

   893 J/K · (31.69 – 22.50) °C · 1K/1 °C = 8210 J

3) Amount of benzoic acid (HBz):

   1.500 g HBz $\left(\frac{1 \text{ mol HBz}}{122.12 \text{ g HBz}}\right)$ = 0.01228 mol HBz

4) ΔU per mol of benzoic acid is:

$$\frac{-(29800 \text{ J} + 8210 \text{ J})}{0.01228 \text{ mol}} = -3.09 \times 10^6 \text{ J/mol or } 3.09 \times 10^3 \text{ kJ/mol}$$

5.54. Calculate ΔU per mol of glucose:

$q_{water} + q_{bomb} + q_r = 0$
$q_r = -([(575\ g)(4.184\ J/g\cdot K)(298.37\ K - 294.85\ K)] + [(650\ J/K)(298.37\ K - 294.85\ K)])$
$q_r = -1.08 \times 10^4\ J$

and the change/mol of glucose: $0.692\ g\ C_6H_{12}O_6 \cdot \dfrac{1\ mol\ C_6H_{12}O_6}{180.2\ g} = 0.00384\ mol\ C_6H_{12}O_6$

$\Delta U = \dfrac{-1.08 \times 10^4\ J}{0.00384\ mol\ C_6H_{12}O_6} \cdot \dfrac{1\ kJ}{1000\ J} = -2.80 \times 10^3\ \dfrac{kJ}{mol\ C_6H_{12}O_6}$

5.55. Heat absorbed by the ice: $\dfrac{333\ J}{1.00\ g\ ice} \cdot 3.54\ g\ ice = 1{,}180\ J$ (3 sf)

Since this energy (1180 J) is released by the metal, we can calculate the specific heat capacity of the metal: heat = specific heat capacity × mass × ΔT

$-1180\ J = c_{metal} \times 50.0\ g \times (273.2\ K - 373.0\ K)$ [Note that ΔT is negative!]

$0.236\ J/g\cdot K = c_{metal}$

Note that the heat released (left side of equation) has a negative sign to indicate the **directional flow** of the energy.

5.56. Calculate the specific heat capacity of Pt:
$q_{Pt} + q_{ice} = 0$
$[(9.36\ g)(c_{Pt})(273.2\ K - 371.8\ K)] + [(0.37\ g)(333\ J/g)] = 0$
$c_{Pt} = 0.13\ J/g\cdot K$

Hess's Law

5.57. (a) Hess's Law allows us to calculate the overall enthalpy change by the appropriate combination of several equations. In this case we add the two equations, reversing the second one (with the concomitant reversal of sign).

$CH_4(g) + 2\ O_2(g) \to CO_2(g) + 2\ H_2O(g)$	$\Delta H° = -802.4\ kJ/mol\text{-}rxn$
$CO_2(g) + 2\ H_2O(g) \to CH_3OH(g) + 3/2\ O_2(g)$	$\Delta H° = +676\ kJ/mol\text{-}rxn$
$CH_4(g) + ½\ O_2(g) \to CH_3OH(g)$	$\Delta H° = -126\ kJ/mol\text{-}rxn$

(b) A graphic description of the energy change:

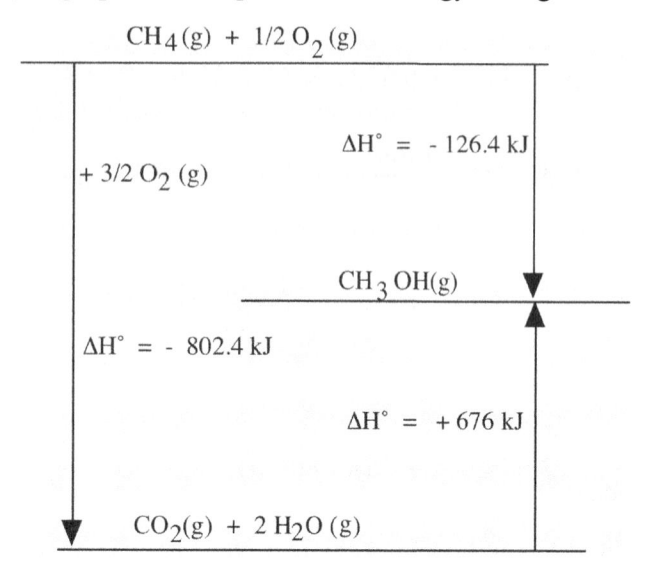

$$CH_4(g) + 1/2\ O_2(g)$$

$$+ 3/2\ O_2(g) \quad \Delta H° = -126.4\ kJ$$

$$CH_3OH(g)$$

$$\Delta H° = -802.4\ kJ \quad \Delta H° = +676\ kJ$$

$$CO_2(g) + 2\ H_2O(g)$$

5.58. Calculate enthalpy change for the conversion of ethylene to ethyl alcohol:

(a) $C_2H_4(g) + 3\ O_2(g) \rightarrow 2\ CO_2(g) + 2\ H_2O(\ell)$ $\Delta_rH° = -1411.1$ kJ/mol-rxn

     $2\ CO_2(g) + 3\ H_2O(\ell) \rightarrow C_2H_5OH(\ell) + 3\ O_2(g)$ $\Delta_rH° = -(-1367.5$ kJ/mol-rxn$)$

     $C_2H_4(g) + H_2O(\ell) \rightarrow C_2H_5OH(\ell)$ $\Delta_rH° = -43.6$ kJ/mol-rxn

(b) energy level diagram

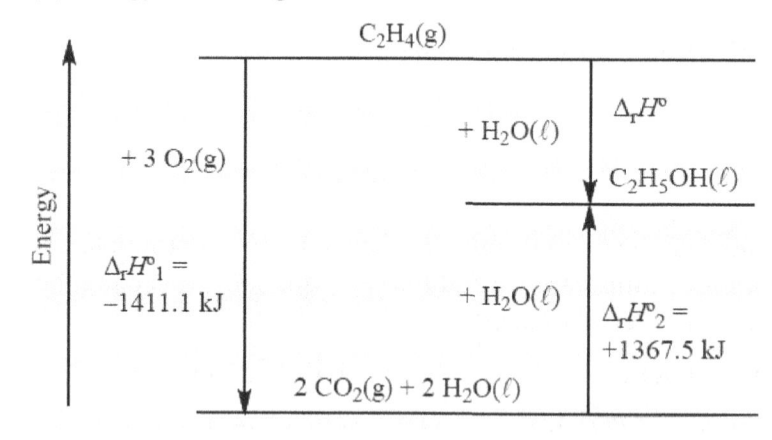

$C_2H_4(g)$
$+ 3\ O_2(g)$
$+ H_2O(\ell)$
$\Delta_rH°$
$C_2H_5OH(\ell)$
$\Delta_rH°_1 = -1411.1$ kJ
$+ H_2O(\ell)$
$\Delta_rH°_2 = +1367.5$ kJ
$2\ CO_2(g) + 2\ H_2O(\ell)$

5.59. The overall enthalpy change for ½ N_2(g) + ½ O_2(g) → NO(g)

For the overall equation, note that elemental nitrogen and oxygen are on the "left" side of the equation, and NO on the "right" side of the equation. Noting that equation 2 has 4 ammonia molecules consumed, let's multiply equation 1 by 2:

$2\ N_2(g) + 6\ H_2(g) \rightarrow 4\ NH_3(g)$ $\Delta_rH° = (2)(-91.8$ kJ/mol-rxn$)$

The second equation has NO on the right side:

4 NH₃(g) + 5 O₂(g) → 4 NO(g) + 6 H₂O(g) $\Delta_r H° = -906.2$ kJ/mol-rxn

The third equation has water as a product, and we need to "consume" the water formed in equation two, so let's reverse equation 3—changing the sign—AND multiply it by 6

6 H₂O(g) → 6 H₂(g) + 3 O₂(g) $\Delta_r H° = (+241.8)(6)$ kJ/mol-rxn

Adding these 3 equations gives

2 N₂(g) + 2 O₂(g) → 4 NO(g) $\Delta_r H° = 361$ kJ/mol-rxn

Dividing all the coefficients by 4 provides the desired equation with

ΔH = +361 kJ/mol-rxn · 0.25 or 90.3 kJ/mol-rxn

5.60. Enthalpy change upon formation of 1.00 mol of PCl₃ from phosphorus and chlorine:

P₄(s) + 10 Cl₂(g) → 4 PCl₅(s) $\Delta_r H° = -1774.0$ kJ/mol-rxn
4 PCl₅(s) → 4 PCl₃(ℓ) + 4 Cl₂(g) $\Delta_r H° = -(-123.8$ kJ/mol-rxn$) \times 4$
―――――――――――――――――――――――――――――
P₄(s) + 6 Cl₂(g) → 4 PCl₃(ℓ) $\Delta_r H° = -1278.8$ kJ/mol-rxn

$$1.00 \text{ mol PCl}_3 \cdot \frac{1 \text{ mol-rxn}}{4 \text{ mol PCl}_3} \cdot \frac{-1278.8 \text{ kJ}}{1 \text{ mol-rxn}} = -320. \text{ kJ}$$

Standard Enthalpies of Formation

5.61. The equation requested requires that we form **one** mol of product liquid CH₃OH from its elements—each in their standard state.

Begin by writing a balanced equation: 2 C(s, graphite) + O₂(g) + 4 H₂(g) → 2 CH₃OH(ℓ)

Now express the reaction so that you form one mole of CH₃OH—divide coefficients by 2.

C(s, graphite) + ½ O₂(g) + 2 H₂(g) → CH₃OH(ℓ)

From Appendix L, the $\Delta_f H°$ is reported as –238.4 kJ/mol-rxn.

5.62. Equation for formation of CaCO₃:

Ca(s) + C(s, graphite) + ³⁄₂ O₂(g) → CaCO₃(s) and from Appendix L: $\Delta_f H° = -1207.6$ kJ/mol-rxn

5.63. Balanced equation for formation of Cr₂O₃:

(a) The equation of the formation of Cr₂O₃ (s) from the elements:

2 Cr(s) + ³/₂ O₂(g) → Cr₂O₃(s)
from Appendix L $\Delta_fH°$ is reported as: −1134.7 kJ/mol for the oxide.

(b) The enthalpy change if 2.4 g of Cr is oxidized to Cr₂O₃ (g) is:

$$2.4 \text{ g Cr} \cdot \frac{1 \text{ mol Cr}}{52.0 \text{ g Cr}} \cdot \frac{-1134.7 \text{ kJ}}{2 \text{ mol Cr}} = -26 \text{ kJ (2 sf)}$$

5.64. Balanced equation for formation of MgO:

(a) Mg(s) + ½ O₂(g) → MgO(s) $\Delta_fH° = -601.24$ kJ/mol

(b) The enthalpy change if 2.5 mol of Mg is oxidized to MgO:

$$2.5 \text{ mol Mg} \cdot \frac{1 \text{ mol MgO}}{1 \text{ mol Mg}} \cdot \frac{-601.24 \text{ kJ}}{1 \text{ mol MgO}} = -1500 \text{ kJ}$$

5.65. Calculate $\Delta_rH°$ for the following processes:

(a) 1.0 g of white phosphorus burns:
P₄(s) + 5 O₂(g) → P₄O₁₀(s) from Appendix L: $\Delta_fH° = -2984.0$ kJ/mol

$$1.0 \text{ g P}_4 \cdot \frac{1.0 \text{ mol P}_4}{123.90 \text{ g P}_4} \cdot \frac{-2984.0 \text{ kJ}}{1 \text{ mol P}_4} = -24 \text{ kJ}$$

(b) 0.20 mol NO(g) decomposes to N₂(g) and O₂(g):
From Appendix L: $\Delta_fH°$ for NO = 90.29 kJ/mol

Since the reaction requested is the **reverse** of $\Delta_fH°$, we change the sign to –90.29 kJ/mol.

The enthalpy change is then $\frac{-90.29 \text{ kJ}}{1 \text{ mol}} \cdot 0.20 \text{ mol} = -18 \text{ kJ}$

(c) 2.40 g NaCl is formed from elemental Na and elemental Cl₂:

From Appendix L: $\Delta_fH°$ for NaCl(s) = −411.12 kJ/mol

The amount of NaCl is: $2.40 \text{ g NaCl} \cdot \frac{1 \text{ mol NaCl}}{58.44 \text{ g NaCl}} = 0.0411 \text{ mol}$

The overall energy change is: −411.12 kJ/mol · 0.0411 mol = −16.9 kJ

(d) 250 g of Fe oxidized to Fe₂O₃(s):

From Appendix L: $\Delta_fH°$ for Fe₂O₃(s) = −825.5 kJ/mol

The overall energy change is:

$$250 \text{ g Fe} \cdot \frac{1 \text{ mol Fe}}{55.845 \text{ g Fe}} \cdot \frac{1 \text{ mol Fe}_2\text{O}_3}{2 \text{ mol Fe}} \cdot \frac{-825.5 \text{ kJ}}{1 \text{ mol Fe}_2\text{O}_3} = -1.8 \times 10^3 \text{ kJ}$$

5.66. Calculate enthalpy changes:

(a) $0.054 \text{ g S} \cdot \dfrac{1 \text{ mol S}}{32.1 \text{ g}} \cdot \dfrac{1 \text{ mol SO}_2}{1 \text{ mol S}} \cdot \dfrac{-296.84 \text{ kJ}}{1 \text{ mol SO}_2} = -0.50 \text{ kJ}$

(b) $0.20 \text{ mol HgO} \cdot \dfrac{90.83 \text{ kJ}}{1 \text{ mol HgO}} = 18 \text{ kJ}$

(c) $2.40 \text{ g NH}_3 \cdot \dfrac{1 \text{ mol NH}_3}{17.03 \text{ g}} \cdot \dfrac{-45.90 \text{ kJ}}{1 \text{ mol NH}_3} = -6.47 \text{ kJ}$

(d) $1.05 \times 10^{-2} \text{ mol C} \cdot \dfrac{1 \text{ mol CO}_2}{1 \text{ mol C}} \cdot \dfrac{-393.509 \text{ kJ}}{1 \text{ mol CO}_2} = -4.13 \text{ kJ}$

5.67. Regarding the first step in the production of nitric acid from ammonia:

(a) The enthalpy change for the reaction:

$$4 \text{ NH}_3(g) + 5 \text{ O}_2(g) \rightarrow 4 \text{ NO}(g) + 6 \text{ H}_2\text{O}(g)$$

$\Delta_f H°$ (kJ/mol)   −45.90   0   90.29   −241.83

$\Delta_r H° = [(4 \text{ mol})(+90.29 \, \frac{\text{kJ}}{\text{mol}}) + (6 \text{ mol})(-241.83 \, \frac{\text{kJ}}{\text{mol}})] -$

$$[(4 \text{ mol})(-45.90 \, \tfrac{\text{kJ}}{\text{mol}}) + (5 \text{ mol})(0)]$$

$= (-1089.82 \text{ kJ}) - (-183.6 \text{ kJ})$

$= -906.2 \text{ kJ}.$ **The reaction is exothermic.**

(b) Heat **evolved** when 10.0 g NH₃ react:

The balanced equation shows that 4 mol NH₃ result in the release of 906.2 kJ.

$$10.0 \text{ g NH}_3 \cdot \frac{1 \text{ mol NH}_3}{17.03 \text{ g NH}_3} \cdot \frac{-906.2 \text{ kJ}}{4 \text{ mol NH}_3} = -133 \text{ kJ}$$

5.68. Regarding the formation of CaCO₃ from Ca(OH)₂:

(a) $\Delta_r H° = \Delta_f H°[\text{CaCO}_3(s)] + \Delta_f H°[\text{H}_2\text{O}(g)] - (\Delta_f H°[\text{Ca(OH)}_2(s)] + \Delta_f H°[\text{CO}_2(g)])$

$\Delta_r H° = 1 \text{ mol } (-1207.6 \text{ kJ/mol}) + 1 \text{ mol } (-241.83 \text{ kJ/mol})$
$\qquad\qquad - [1 \text{ mol } (-986.09 \text{ kJ/mol}) + 1 \text{ mol } (-393.509 \text{ kJ/mol})]$

$\Delta_r H° = -69.8 \text{ kJ/mol-rxn}$

(b) $1.00 \times 10^3 \text{ g Ca(OH)}_2 \cdot \dfrac{1 \text{ mol Ca(OH)}_2}{74.09 \text{ g}} \cdot \dfrac{1 \text{ mol-rxn}}{1 \text{ mol Ca(OH)}_2} \cdot \dfrac{-69.8 \text{ kJ}}{1 \text{ mol-rxn}} = -942 \text{ kJ}$

The negative sign indicates that heat is evolved.

5.69. Regarding the conversion of barium peroxide to barium oxide:

(a) The enthalpy change for the reaction:

2 BaO$_2$(s) → 2 BaO(s) + O$_2$(g)

Given $\Delta_f H°$ for BaO is: −553.5 kJ/mol and $\Delta_f H°$ for BaO$_2$ is: −634.3 kJ/mol

This equation can be seen as the summation of the two equations:

(1) 2 Ba(s) + O$_2$(g) → 2 BaO(s)

(2) 2 BaO$_2$(s) → 2 Ba(s) + 2 O$_2$(g)

Equation (1) corresponds to the formation of BaO × 2 while equation(2) corresponds to (2×) the **reverse** of the formation of Ba O$_2$

$\Delta_r H° = (2\, \Delta_f H°$ for BaO) + $-2\Delta_f H°$ for BaO$_2$) =

$\Delta_r H° = 161.6$ kJ and the reaction is **endothermic**.

(b) Energy level diagram for the equations in question:

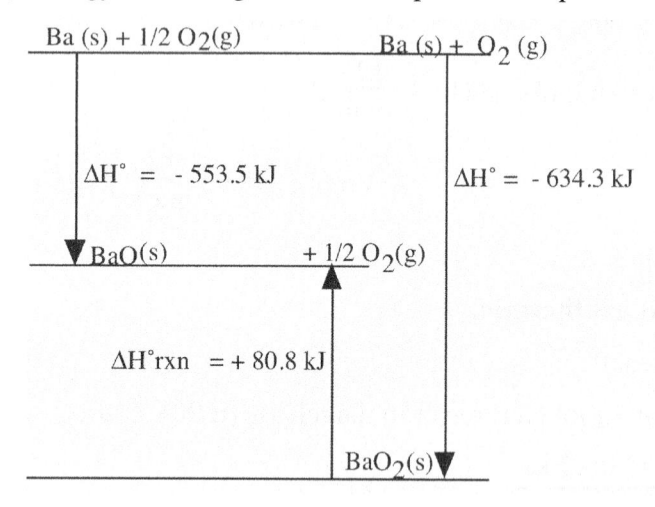

```
Ba (s) + 1/2 O2(g)              Ba (s) + O2 (g)
_____                  _____
       |                               |
       |                               |
       | ΔH° = -553.5 kJ               | ΔH° = -634.3 kJ
       |                               |
       ↓                               |
   BaO(s)      + 1/2 O2(g)             |
   _____↑_____          |
                     |                 |
                     |                 |
             ΔH°rxn = +80.8 kJ         |
                     |                 |
                     |                 |
                    BaO2(s) ↓
   _____
```

5.70. Regarding the oxidation of SO$_2$ to SO$_3$: SO$_2$(g) + 1/2 O$_2$(g) → SO$_3$(g)

(a) $\Delta_f H°[O_2(g)] = 0$ kJ/mol
$\Delta_r H° = \Delta_f H°[SO_3(g)] − \{\Delta_f H°[SO_2(g)] + 1/2\, \Delta_f H°[SO_2(g)]\}$
$\Delta_r H° = 1$ mol (−395.77 kJ/mol) − {1 mol (−296.84 kJ/mol) + 1/2 mol (0 kJ/mol)}
$\Delta_r H° = -98.93$ kJ/mol-rxn with the negative value of $\Delta_r H°$, the reaction is exothermic

(b) energy level diagram:

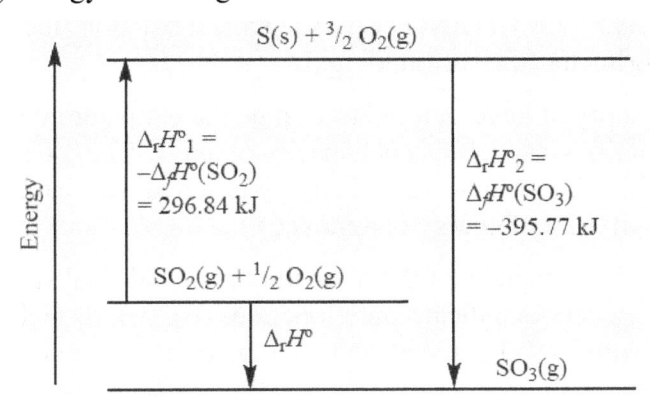

$$\begin{array}{c}
S(s) + {}^3/_2\, O_2(g) \\
\Delta_r H^\circ_1 = -\Delta_f H^\circ(SO_2) = 296.84 \text{ kJ} \\
\Delta_r H^\circ_2 = \Delta_f H^\circ(SO_3) = -395.77 \text{ kJ} \\
SO_2(g) + {}^1/_2\, O_2(g) \\
\Delta_r H^\circ \\
SO_3(g)
\end{array}$$

5.71. The molar enthalpy of formation of naphthalene can be calculated since we're given the enthalpic change for the reaction:

$$C_{10}H_8(s) + 12\, O_2(g) \rightarrow 10\, CO_2(g) + 4\, H_2O(\ell)$$

$\Delta_f H^\circ$ (kJ/mol)   ?   0   −393.509   −285.83

$\Delta_r H^\circ = \sum n \Delta_f H^\circ \text{ products} - \sum n \Delta_f H^\circ \text{ reactants}$

$-5156.1 \text{ kJ} = [(10 \text{ mol})(-393.509 \tfrac{\text{kJ}}{\text{mol}}) + (4 \text{ mol})(-285.83 \tfrac{\text{kJ}}{\text{mol}})] - [\Delta_f H^\circ\, C_{10}H_8]$

$-5156.1 \text{ kJ} = (-5078.41 \text{ kJ}) - \Delta_f H^\circ\, C_{10}H_8$

$-77.7 \text{ kJ} = -\Delta_f H^\circ\, C_{10}H_8$ and multiplying by −1: 77.7 kJ = $\Delta_f H^\circ\, C_{10}H_8$

5.72. Calculate the enthalpy of formation of styrene (kJ/mol):

$\Delta_f H^\circ [O_2(g)] = 0$ kJ/mol (so this term will be omitted from the ΔH° calculation)

$\Delta_r H^\circ = 8\, \Delta_f H^\circ[CO_2(g)] + 4\, \Delta_f H^\circ[H_2O(\ell)] - \Delta_f H^\circ[C_8H_8(\ell)]$

$-4395.0 \text{ kJ} = (8 \text{ mol})(-393.509 \text{ kJ/mol}) + (4 \text{ mol})(-285.83 \text{ kJ/mol}) - \Delta_f H^\circ[C_8H_8(\ell)]$

$\Delta_f H^\circ[C_8H_8(\ell)] = 103.6$ kJ/mol

General Questions

5.73. Define and give an example of:

(a) Exothermic and Endothermic—the suffix "thermic" talks about heat, and the prefixes "exo" and "endo" tell us whether heat is transferred from the system to the surroundings (exo) or transferred from the surroundings to the system (endo). Combustion reactions (e.g., gasoline burning in your automobile) are EXOthermic reactions, while ice melting is an ENDOthermic reaction.

(b) System and Surroundings—The "system" is the object or collection of objects being studied while the "surroundings" is EVERYTHING else. A chemical reaction (the system) taking place inside a calorimeter (the surroundings).

(c) Specific heat capacity—is the quantity of heat required to change the temperature of 1g of a substance by 1 °Celsius. Water has a specific heat capacity of about 4.2 J/g·K, meaning that 1 gram of water at 15 °C, to which 4.2 J of energy is added, will have a temperature of 16 °C. (or 14 °C—if 4.2 J of energy is removed).

(d) State function—Any parameter that is dependent ONLY on the initial and final states. Chemists typically use CAPITAL letters to indicate state functions (e.g., H, S,) while non-state functions are indicated with LOWER CASE letters (e.g., q, w). Your checking account balance is a state function!

(e) Standard state—Defined as the MOST STABLE (PHYSICAL) STATE for a substance at a pressure of 1 bar and at a specified temperature (typically 298 K). The standard state for elemental nitrogen at 25 °C (298 K) is **gas**.

(f) Enthalpy change—the heat transferred in a process that is carried out under constant pressure conditions is the enthalpy change, ΔH.

The enthalpy change upon the formation of 1 mol of water(ℓ) is –285.8 kJ, meaning that 285.8 kJ is released upon the formation of 1 mol of liquid water from 1 mol of hydrogen (g) and 1/2 mol oxygen (g).

(g) Standard Enthalpy of Formation—the enthalpy change for the formation of 1 mol of a compound directly from its component elements, each in their standard states. The standard enthalpy of formation of nitrogen gas (N$_2$) = 0 kJ/mol.

5.74. Tell if each of the following processes are exothermic or endothermic:

(a) H$_2$O(ℓ) → H$_2$O(s) 　　　　　Exothermic—moving to lower energy state

(b) 2 H$_2$(g) + O$_2$(g) → 2 H$_2$O(g) 　Exothermic—moving to lower energy state

(c) H$_2$O(ℓ, 25 °C) → H$_2$O(ℓ, 15 °C) Exothermic—moving to lower energy state

(d) H$_2$O(ℓ) → H$_2$O(g) 　　　　　Endothermic—moving to higher energy state

5.75. Define system and surroundings for each of the following, and give direction of heat transfer:

(a) Methane is burning in a gas furnace in your home:
(System) methane + oxygen (Surroundings) components of furnace and the air in your home. The heat flows from the methane + oxygen to the furnace and air.

(b) Water drops on your skin evaporate:
(System) water droplets (Surroundings) your skin and the surrounding air. The heat flows from your skin and the air to the water droplet.

(c) Liquid water at 25 °C is placed in freezer:

(System) water (Surroundings) freezer. The heat flows from the water to the freezer.

(d) Aluminum and Fe₂O₃ react in a flask on a lab bench:

(System) Al and Fe₂O₃ (Surroundings) flask, lab bench, and air around flask. The heat flows from the reaction of Al and Fe₂O₃ into the surroundings.

5.76. Define *standard state*. Identify standard states of the substances at 298 K:
Standard state is the most stable form of a substance in the physical state that exists at a pressure of 1 bar and at a specified temperature.
Standard states at 298 K: H₂O (liquid), NaCl (solid), Hg (liquid), CH₄ (gas)

5.77. Standard Enthalpies of Formation for O(g), O₂ (g), O₃ (g).

Substance	$\Delta_f H°$ (at 298 K) kJ/mol
O(g)	249.170
O₂(g)	0
O₃(g)	142.67

What is the standard state of oxygen? The standard state of oxygen is a gas, O₂(g).

Is the formation of O from O₂ exothermic?

$\Delta_r H° = \Sigma \Delta_f H°_{products} - \Sigma \Delta_f H°_{reactants}$ (O₂ → 2O)

$\Delta_r H° = $ (2 mol)(249.170 kJ/mol) – (1 mol)(0 kJ/mol) = 498.340 kJ (endothermic)

What is the $\Delta_r H°$ for 3/2 O₂(g) → O₃(g)

$\Delta_r H° = \Sigma \Delta_f H°_{products} - \Sigma \Delta_f H°_{reactants}$

$\Delta_r H° = $ (1 mol)(142.67 kJ/mol) – (3/2 mol)(0 kJ/mol) = 142.67 kJ

5.78. Effects on the internal energy of the system by the following changes:

(a) Temperature is raised from 80 °C to 90 °C: Internal energy will be increased.

(b) Vapor is condensed to liquid at 40 °C: Internal energy will be decreased.

5.79. Determine whether heat is evolved or required, and whether work was done on the system or whether the system does work on the surroundings, in the following processes at constant pressure:

(a) Liquid water at 100 °C is converted to steam at 100 °C. **Heat will be required** to convert liquid water to gaseous water. The gaseous water will occupy more volume than the liquid water, hence **work is done on the surroundings by the system**.

(b) Dry ice, $CO_2(s)$, sublimes to give $CO_2(g)$. **Heat will be required** to convert solid carbon dioxide into gaseous carbon dioxide. In the same way that the gasification of water (in (a.) above) results in a volume expansion, the gaseous carbon dioxide will occupy a greater volume than the solid, hence **work is done on the surroundings by the system.**

5.80. Determine whether heat is evolved or required, and whether work was done on the system or whether the system does work on the surroundings, in the following processes at constant pressure:

(a) Ozone decomposition is exothermic. Energy is transferred as heat from the system to the surroundings and work is done on the surroundings as the number of moles of gas of products is greater than the number of moles of gas of reactant.

(b) Methane combustion is exothermic. Energy is transferred as heat from the system to the surroundings, and work is done on the system by the surroundings as the number of moles of **gas** in the products (1 mol) is less than the number of moles of gas in the reactants (3 mol).

5.81. Enthalpy change that occurs when 1.00 g of $SnCl_4(\ell)$ reacts with excess $H_2O(\ell)$:

$SnCl_4(\ell) + 2\ H_2O(\ell) \rightarrow SnO_2(s) + 4\ HCl(aq)$

$\Delta_r H° = \Sigma\ n\Delta_f H°_{products} - \Sigma\ n\Delta_f H°_{reactants}$

$\Delta_r H° = [(1\ mol)(-577.63\ \frac{kJ}{mol}) + (4\ mol)(-167.159\ \frac{kJ}{mol})] -$
$[(1\ mol)(-511.3\ \frac{kJ}{mol}) + (2\ mol)(-285.83\ \frac{kJ}{mol})]$

$\Delta_r H° = -1246.266\ kJ - (-1082.96\ kJ) = -163.306\ kJ$ for 1 mol of $SnCl_4(\ell)$.

Convert this amount into an amount **per gram** of $SnCl_4(\ell)$, by dividing by the mass of $SnCl_4$.

$\frac{-163.306\ kJ}{1\ mol\ SnCl_4} \cdot \frac{1\ mol\ SnCl_4}{260.51\ g} = -0.627\ \frac{kJ}{g}$ (3 sf)

5.82. Compare energy release of 50.0 g water to 100. g ethanol:
$q_{water} = (50.0\ g)(4.184\ J/g \cdot K)(-40\ K) = -8400\ J$
$q_{ethanol} = (100.\ g)(2.46\ J/g \cdot K)(-40\ K) = -9800\ J$ Ethanol releases more heat.

Given the temperature change, the amounts should be expressed with 1 sf, or −8 kJ and −10 kJ respectively.

5.83. If 187 J raises the temperature of 93.45 g of Ag from 18.5 °C to 27.0 °C, what is the specific heat capacity of silver?

Recall that q = m · c_{Ag} · ΔT; so 187 J = 93.45 g · c_{Ag} · (27.0 – 18.5) °C, and

$$c_{Ag} = \frac{187 \text{ J}}{93.45 \text{ g} \cdot (27.0 - 18.5)°C} = 0.24 \text{ J/g} \cdot \text{K}$$

5.84. Calculate quantity of energy required:

q = energy to melt ice + energy to warm liquid + energy to vaporize liquid
$q_{melt\ ice}$ = (60.1 g)(333 J/g) = 2.00 × 10^4 J
$q_{warm\ liquid}$ = (60.1 g)(4.184 J/g·K)(373.2 K – 273.2 K) = 2.51 × 10^4 J
$q_{vaporize\ liquid}$ = (60.1 g)(2256 J/g) = 1.36 × 10^5 J
q_{total} = 2.00 × 10^4 J + 2.51 × 10^4 J + 1.36 × 10^5 J = 1.81 × 10^5 J

5.85. How much ice has melted upon addition of 100.0 g of water at 60 °C to 100.0 g of ice at 0.00 °C? The water cools to 0.00 °C.

As the ice absorbs heat from the water, two processes occur: (1) the ice melts and (2) the water cools. We can express this with the equation q_{water} = $-q_{ice}$

The melting of ice can be expressed with the heat of fusion of ice, 333 J/g, as q = m · 333 J/g. The cooling of the water may be expressed: q = m · c · ΔT = 100.0 g · (4.184 J/g ·K) · (0.00 – 60.0)K.

Setting these quantities equal gives: 100.0 g · (4.184 J/g ·K) · (0.00 – 60.0)K = $-x$ g · 333 J/g

[x = quantity of ice that melts. Note that since Celsius degrees and kelvin are the same "size", ΔT is –60.0 °C or –60.0 K]

100.0 g · (4.184 J/g ·K) · (–60.0 K) = $-x$ · 333 J/g

–25104 J = - x • 333 J/g or $\frac{-25104 \text{ J}}{- 333 \text{ J/g}} = x$ or 75.4 g of ice.

5.86. How much ice melts when three 45 g ice cubes at 0 °C are dropped in 5.00 × 10^2 mL tea?

Assume the density of the tea is 1.00 g/mL and its specific heat capacity is 4.184 J/g·K
$q_{tea} + q_{ice} = 0$
[(5.00 × 10^2 g) (4.184 J/g·K)(273.2 K – 293.2 K)] + [($m_{ice\ melted}$)(333 J/g)] = 0
$m_{ice\ melted}$ = 126 g and (3 × 45 g) – 126 g = 9 g ice remaining

q_{tea} describes the warming of the tea and q_{ice} describes the melting of the solid ice.

5.87. 90. g (two 45 g cubes) of ice cubes (at 0 °C) are dropped into 500. mL tea at 20.0 °C (Assume a density of 1.00 g/mL for tea). What is the final temperature of the mixture if all the ice melts?

$q_{water} = -q_{ice}$

NOTE however, that not only does all the ice melt, but the melted ice warms to a temperature above 0 °C, so q_{ice} has two parts: 1) the heat needed to melt the ice and 2) the heat needed to heat the liquid that forms.

$m_{tea} \cdot c \cdot \Delta T_{tea} = -[m_{ice} \cdot 333 \text{ J/g} + m_{ice} \cdot c \cdot \Delta T_{ice}]$

500. g · (4.184 J/g·K) · (T_{final} − 293.2 K) = − [(90. g · 333 J/g) +

(90. g · (4.184 J/g·K) · (T_{final} − 273.2 K))]

where T_{final} is the final temperature of the tea and melted ice.

2092 T_{final} J −613,374 J = −[29970 J + 377 T_{final} J − 102,876 J]

Simplifying:

2092 T_{final} J − 613,374 J + 29970 J + 377 T_{final} J − 102,876 J = 0

(2092 T_{final} J + 377 T_{final} J) + (− 613,374 J + 29970 J − 102,876 J) = 0

2469 T_{final} J + −686,280 = 0 or 2469 T_{final} J = 686,280 and T_{final} = (686,280/2469) = 278 K

and as 45 g of ice cube has 2 sf, we report a final temperature of 280 K or 5 °C.

5.88. Description of the system when one ice cube is added to a cold beverage:

Assume the density of the cola is 1.00 g/mL and its specific heat capacity is 4.184 J/g·K
Energy required to cool cola to 0 °C:

q_{cola} = (240 g) (4.184 J/g·K)(273.2 K − 283.7 K) = −1.1 × 10^4 J
Energy supplied by melting one ice cube:

q_{ice} = (45 g) (333 J/g) = 1.5 × 10^4 J
So the best description is: (a) The temperature is 0 °C and some ice remains.

How much ice remains?

$q_{cola} + q_{melt\ ice} = 0$

−1.1 × 10^4 J + ($m_{ice\ melted}$)(333 J/g) = 0

$m_{ice\ melted}$ = 32 g

45 g − 32 g = 13 g ice remaining

5.89. One can arrive at the desired answer if you recall the **definition** of $\Delta_f H°$. The definition is the enthalpy change associated with the formation of **one mole** of the substance (in this case B$_2$H$_6$) from its elements—each in their standard state (s for boron and g for hydrogen).

(a) Note that the 1ˢᵗ equation given uses **four** moles of B as a reactant — and we'll need only 2, so divide the first equation by 2 to give:

2 B(s) + 3/2 O₂(g) → B₂O₃(s) $\Delta_r H°$ = 1/2(−2543.8 kJ/mol-rxn) = −1271.9 kJ/mol-rxn

The formation of 1 mole of B₂H₆ will require the use of 6 moles of H (or 3 moles of H₂), so, multiply the second equation by 3 to give:

3 H₂(g) + 3/2 O₂(g) → 3 H₂O(g) $\Delta_r H°$ = 3(−241.8 kJ/mol-rxn) = −725.4 kJ/mol-rxn

Finally, the third equation given has B₂H₆ as a **reactant and not a product**. So reverse the third equation to give:
B₂O₃(s) + 3 H₂O(g) → B₂H₆(g) + 3 O₂(g) $\Delta_r H°$ = −(−2032.9 kJ/mol-rxn)
= +2032.9 kJ/mol-rxn

(b) Adding the three equations gives the equation:
2 B(s) + 3 H₂(g) → B₂H₆(g) with a $\Delta_r H°$ = (−1271.9 + −725.4 + 2032.9) kJ/mol-rxn
or a $\Delta_f H°$ for B₂H₆(g) of + 35.6 kJ/mol-rxn

(c) Energy level diagram for the reactions:

(d) Formation of B₂H₆ (g) is **endothermic**

5.90. Calculate the energy change for a reaction, and draw an energy level diagram:

(a) $\Delta_r H°$ = $\Delta_f H°$[CH₃Cl(g)] + $\Delta_f H°$[HCl(g)] − ($\Delta_f H°$[CH₄(g)] + 2 $\Delta_f H°$[Cl(g)])
$\Delta_r H°$ = 1 mol (−83.68 kJ/mol) + 1 mol (−92.31 kJ/mol)
            −[1 mol (−74.87 kJ/mol) + 2 mol (121.3 kJ/mol)]
$\Delta_r H°$ = −343.7 kJ/mol-rxn   The reaction is exothermic.

(b) the energy level diagram:

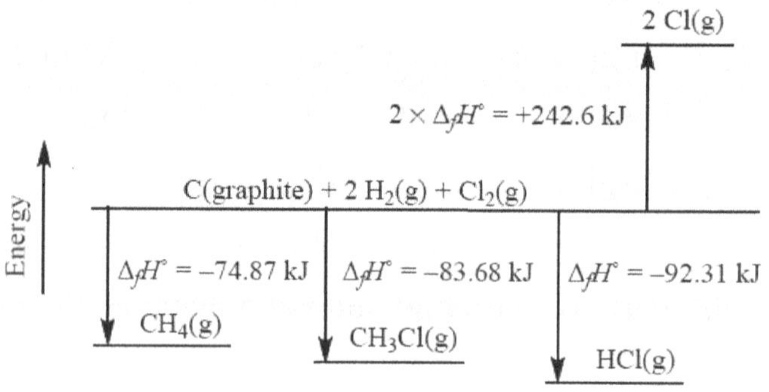

```
                                              2 Cl(g)
                                                ▲
                                                │
                              2 × Δ_fH° = +242.6 kJ
                                                │
              C(graphite) + 2 H₂(g) + Cl₂(g)
  ▲      ┌─────────┬──────────────┬──────────────┐
  │      │         │              │
Energy   │Δ_fH°=−74.87 kJ  Δ_fH°=−83.68 kJ  Δ_fH°=−92.31 kJ
  │      ▼         ▼              ▼
         CH₄(g)   CH₃Cl(g)
                                  HCl(g)
```

5.91. (a) Enthalpy change for: C(s) + H₂O(g) → CO(g) + H₂(g)

	C(s) +	H₂O(g) →	CO(g) +	H₂(g)
Δ_fH° (kJ/mol)	0	−241.83	−110.525	0

$\Delta_rH° = [(1 \text{ mol})(-110.525 \frac{\text{kJ}}{\text{mol}}) + 0 \text{ kJ}] - [0 \text{ kJ} + (1 \text{ mol})(-241.83 \frac{\text{kJ}}{\text{mol}})] =$ +131.31 kJ

(b) The process is **endothermic**.

(c) Heat involved when 1.0 metric ton (1000.0 kg) of C is converted to coal gas:

$1000.0 \text{ kg C} \cdot \frac{1000 \text{ g C}}{1 \text{ kg C}} \cdot \frac{1 \text{ mol C}}{12.011 \text{ g C}} \cdot \frac{+131.31 \text{ kJ}}{1 \text{ mol C}} = 1.0932 \times 10^7 \text{ kJ}$

5.92. Calculate enthalpy of combustion for propane, butane, gasoline, and ethanol:
This problem was solved assuming H₂O(ℓ) is a product in the combustion reactions.
Propane: C₃H₈(g) + 5 O₂(g) → 3 CO₂(g) + 4 H₂O(ℓ)
Δ_fH°[O₂(g)] = 0 kJ/mol (so this term will be omitted from the Δ_rH calculation)
Δ_rH° = 3 Δ_fH°[CO₂(g)] + 4 Δ_fH°[H₂O(ℓ)] − Δ_fH°[C₃H₈(g)]
Δ_rH° = 3 mol (−393.509 kJ/mol) + 4 mol (−285.83 kJ/mol) − 1 mol (−104.7 kJ/mol)
Δ_rH° = −2219.1 kJ/mol-rxn

$\frac{-2219.1 \text{ kJ}}{1 \text{ mol-rxn}} \cdot \frac{1 \text{ mol-rxn}}{1 \text{ mol C}_3\text{H}_8} \cdot \frac{1 \text{ mol C}_3\text{H}_8}{44.097 \text{ g}} = -50.324 \text{ kJ/g}$

Butane: C₄H₁₀(g) + ¹³/₂ O₂(g) → 4 CO₂(g) + 5 H₂O(ℓ)
Δ_fH°[O₂(g)] = 0 kJ/mol (so this term will be omitted from the Δ_rH° calculation)
Δ_rH° = 4 Δ_fH°[CO₂(g)] + 5 Δ_fH°[H₂O(ℓ)] − Δ_fH°[C₄H₁₀(g)]
Δ_rH° = 4 mol (−393.509 kJ/mol) + 5 mol (−285.83 kJ/mol) − 1 mol (−127.1 kJ/mol)
Δ_rH° = −2876.1.1 kJ/mol-rxn

$\frac{-2876.1 \text{ kJ}}{1 \text{ mol-rxn}} \cdot \frac{1 \text{ mol-rxn}}{1 \text{ mol C}_4\text{H}_{10}} \cdot \frac{1 \text{ mol C}_4\text{H}_{10}}{58.124 \text{ g}} = -49.482 \text{ kJ/g}$

Gasoline: $C_8H_{18}(\ell) + {}^{25}/_2\ O_2(g) \rightarrow 8\ CO_2(g) + 9\ H_2O(\ell)$

$\Delta_fH°[O_2(g)] = 0$ kJ/mol (so this term will be omitted from the ΔH calculation)

$\Delta_rH° = 8\ \Delta_fH°[CO_2(g)] + 9\ \Delta_fH°[H_2O(\ell)] - \Delta_fH°[C_8H_{18}(\ell)]$

$\Delta_rH° = 8$ mol $(-393.509$ kJ/mol$) + 9$ mol $(-285.83$ kJ/mol$) - 1$ mol $(-259.3$ kJ/mol$)$

$\Delta_rH° = -5461.2$ kJ/mol-rxn

$$\frac{-5461.2\ \text{kJ}}{1\ \text{mol-rxn}} \cdot \frac{1\ \text{mol-rxn}}{1\ \text{mol}\ C_8H_{18}} \cdot \frac{1\ \text{mol}\ C_8H_{18}}{114.232\ \text{g}} = -47.808\ \text{kJ/g}$$

Ethanol: $C_2H_5OH(\ell) + 3\ O_2(g) \rightarrow 2\ CO_2(g) + 3\ H_2O(\ell)$

$\Delta_fH°[O_2(g)] = 0$ kJ/mol (so this term will be omitted from the ΔH calculation)

$\Delta_rH° = 2\ \Delta_fH°[CO_2(g)] + 3\ \Delta_fH°[H_2O(\ell)] - \Delta_fH°[C_2H_5OH(\ell)]$

$\Delta_rH° = 2$ mol $(-393.509$ kJ/mol$) + 3$ mol $(-285.83$ kJ/mol$) - 1$ mol $(-277.0$ kJ/mol$)$

$\Delta_rH° = -1367.5$ kJ/mol-rxn

$$\frac{-1367.5\ \text{kJ}}{1\ \text{mol-rxn}} \cdot \frac{1\ \text{mol-rxn}}{1\ \text{mol}\ C_2H_5OH} \cdot \frac{1\ \text{mol}\ C_2H_5OH}{46.069\ \text{g}} = -29.683\ \text{kJ/g}$$

The hydrocarbons give off more heat per gram than the ethanol, which is already partially oxidized. Of the hydrocarbons, the smaller ones give off more heat per gram than the larger ones.

5.93. For the combustion of C_8H_{18}:

$C_8H_{18}(\ell) + {}^{25}/_2\ O_2(g) \rightarrow 8\ CO_2(g) + 9\ H_2O(\ell)$

$\Delta_rH° = [(8\ \text{mol})(-393.509\ \frac{\text{kJ}}{\text{mol}}) + (9\ \text{mol})(-285.83\ \frac{\text{kJ}}{\text{mol}})] - [(1\ \text{mol})(-259.3\ \frac{\text{kJ}}{\text{mol}}) + 0$ kJ$]$

$\Delta_rH° = -5461.2$ kJ

Expressed on a gram basis:

$$\frac{-5461.2\ \text{kJ}}{1\ \text{mol}\ C_8H_{18}} \left(\frac{1\ \text{mol}\ C_8H_{18}}{114.232\ \text{g}} \right) = -47.808\ \text{kJ/g}$$

For the combustion of CH_3OH:

$CH_3OH(\ell) + 3/2\ O_2(g) \rightarrow CO_2(g) + 2\ H_2O(\ell)$

$\Delta_rH° = [(1\ \text{mol})(-393.509$ kJ/mol$) + (2\ \text{mol})(-285.83$ kJ/mol$)]$
$\qquad\qquad - [(1\ \text{mol})(-238.4$ kJ/mol$) + 0$ kJ$]$

$= -726.8$ kJ

Express this on a per gram basis:

$$\frac{-726.8\ \text{kJ}}{1\ \text{mol}\ CH_3OH} \cdot \frac{1\ \text{mol}\ CH_3OH}{32.042\ \text{g}\ CH_3OH} = -22.68\ \text{kJ/g}$$

On a per gram basis, **octane liberates the greater amount** of heat energy.

5.94. Which fuel provides more energy per gram?

NOTE: $\Delta_fH°[O_2(g)] = \Delta_fH°[N_2(g)] = 0$ kJ/mol (so these terms will be omitted from the $\Delta_rH°$ calculation)

Hydrazine:

$\Delta_rH° = 2\ \Delta_fH°[H_2O(g)] - \Delta_fH°[N_2H_4(\ell)]$

$\Delta_rH° = 2$ mol $(-241.83$ kJ/mol$) - 1$ mol $(50.6$ kJ/mol$)$

$\Delta_rH° = -534.3$ kJ/mol-rxn

$\dfrac{-534.3 \text{ kJ}}{1 \text{ mol-rxn}} \cdot \dfrac{1 \text{ mol-rxn}}{1 \text{ mol N}_2\text{H}_4} \cdot \dfrac{1 \text{ mol N}_2\text{H}_4}{32.046 \text{ g}} = -16.67$ kJ/g N$_2$H$_4$

1,1-Dimethylhydrazine:

$\Delta_rH° = 2\ \Delta_fH°[CO_2(g)] + 4\ \Delta_fH°[H_2O(g)] - \Delta_fH°[N_2H_2(CH_3)_2(\ell)]$

$\Delta_rH° = 2$ mol $(-393.509$ kJ/mol$) + 4$ mol $(-241.83$ kJ/mol$) - 1$ mol $(48.9$ kJ/mol$)$

$\Delta_rH° = -1803.2$ kJ/mol-rxn

$\dfrac{-1803.2 \text{ kJ}}{1 \text{ mol-rxn}} \cdot \dfrac{1 \text{ mol-rxn}}{1 \text{ mol N}_2\text{H}_2(\text{CH}_3)_2} \cdot \dfrac{1 \text{ mol N}_2\text{H}_2(\text{CH}_3)_2}{60.098 \text{ g N}_2\text{H}_2(\text{CH}_3)_2} = -30.004$ kJ/g N$_2$H$_2$(CH$_3$)$_2$

1,1-Dimethylhydrazine gives more heat per gram when reacting with oxygen

5.95. (a) Enthalpy change for formation of 1.00 mol of SrCO$_3$

Sr(s) + C(graphite) + $^3/_2$ O$_2$(g) → SrCO$_3$(s) using the data:

Sr(s) + $^1/_2$ O$_2$(g) → SrO(s)	$\Delta_fH° = -592$ kJ/mol-rxn
SrO(s) + CO$_2$(g) → SrCO$_3$(s)	$\Delta_rH° = -234$ kJ/mol-rxn
C(graphite) + O$_2$(g) → CO$_2$(g)	$\Delta_fH° = -394$ kJ/mol-rxn

Let's add the equations to give our desired overall equation.

Sr(s) + $^1/_2$ O$_2$(g) → ~~SrO(s)~~	$\Delta_fH° = -592$ kJ/mol-rxn
~~SrO(s)~~ + ~~CO$_2$(g)~~ → SrCO$_3$(s)	$\Delta_rH° = -234$ kJ/mol-rxn
C (graphite) + O$_2$(g) → ~~CO$_2$ (g)~~	$\Delta_fH° = -394$ kJ/mol-rxn
Sr (s) + C (graphite) + $^3/_2$ O$_2$ (g) → SrCO$_3$(s)	$\Delta_rH° = -1220.$ kJ/mol-rxn

(b) Energy diagram relating the energy quantities:

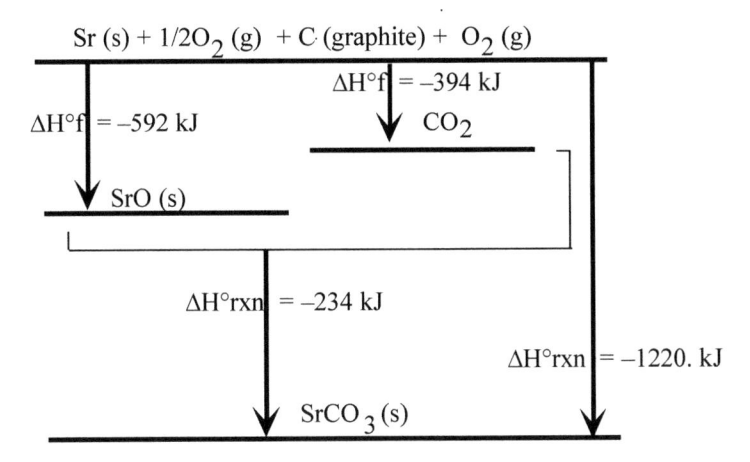

```
Sr (s) + 1/2 O₂ (g) + C (graphite) + O₂ (g)
   |              ΔH°f = –394 kJ
   |                   ↓   CO₂
ΔH°f = –592 kJ     _____
   ↓
   SrO (s)
   _____
        |
        ΔH°rxn = –234 kJ
                         ΔH°rxn = –1220. kJ
        ↓    SrCO₃ (s)
        _____
```

5.96. Regarding diet soda and your body energy:

(a) $q_{soda} = (355\ g)(4.184\ J/g \cdot K)(310.\ K - 278\ K) = 4.8 \times 10^4$ J

$q_{body} = -q_{soda} = -4.8 \times 10^4$ J (or 4.8×10^4 J expended by your body)

(b) Value above expressed in Cal: $-4.8 \times 10^4\ J \cdot \dfrac{1\ cal}{4.184\ J} \cdot \dfrac{1\ Cal}{10^3\ cal} = -11$ Cal

net energy change = 1 Cal + (–11 Cal) = –10. Cal (or 10 Cal expended by your body)

(c) net energy change = 140 Cal + (–11 Cal) = 129 Cal (or 130 Cal absorbed by your body)

5.97. The desired equation is: $CH_4(g) + 3\ Cl_2(g) \rightarrow 3\ HCl(g) + CHCl_3(g)$

Begin with equation 1 (the combustion of methane)

$CH_4(g) + 2\ O_2(g) \rightarrow 2\ H_2O(\ell) + CO_2(g)$ $\Delta_r H° = -890.4$ kJ = –890.4 kJ/mol-rxn

Noting that we form HCl as one of the products, using the second equation, we need to **reverse** it and (to adjust the coefficient of HCl to 3), multiply by 3/2 to give:

$^3/_2\ H_2(g) + ^3/_2\ Cl_2(g) \rightarrow 3\ HCl(g)$ $\Delta_r H° = -^3/_2\ (+184.6)$ kJ/mol-rxn = –276.9 kJ

Note that CO₂ formed in equation 1 doesn't appear in the overall equation so let's use the equation for the formation of CO₂ (reversed) to "consume" the CO₂:

$CO_2(g) \rightarrow C(graphite) + O_2(g)$ $\Delta_r H° = -1(-393.5)$ kJ = +393.5 kJ

Noting also that equation 1 produces 2 water molecules, let's "consume" them by using the equation for the formation of water (reversed) multiplied by 2:

$2\ H_2O(\ell) \rightarrow 2\ H_2(g) + O_2(g)$ $\Delta_r H° = -2(-285.8)$ kJ = +571.6 kJ

and finally we need to produce CHCl₃ which we can do with the equation that represents the $\Delta_f H°$ for CHCl₃:

$C(graphite) + ^1/_2\ H_2(g) + ^3/_2\ Cl_2(g) \rightarrow CHCl_3(g)$ $\Delta_f H° = -103.1$ kJ

The overall enthalpy change would then be:
$\Delta_r H° = -890.4$ kJ $- 276.9$ kJ $+ 393.5$ kJ $+ 571.6$ kJ -103.1 kJ $= -305.3$ kJ

5.98. Mass of C needed to provide energy to convert 1.00 kg of C to water gas:

Reaction of interest: $C(s) + H_2O(g) \rightarrow CO(g) + H_2(g)$
NOTE: $\Delta_fH°[C(s)] = \Delta_fH°[H_2(g)] = 0$ kJ/mol
$\Delta_rH° = \Delta_fH°[CO(g)] - \Delta_fH°[H_2O(g)]$
$\Delta_rH° = (1 \text{ mol})(-110.525 \text{ kJ/mol}) - (1 \text{ mol})(-241.83 \text{ kJ/mol})$
$\Delta_rH° = 131.31$ kJ/mol-rxn

$1.00 \text{ kg} \cdot \dfrac{10^3 \text{ g}}{1 \text{ kg}} \cdot \dfrac{1 \text{ mol C}}{12.011 \text{ g}} \cdot \dfrac{1 \text{ mol-rxn}}{1 \text{ mol C}} \cdot \dfrac{131.31 \text{ kJ}}{1 \text{ mol-rxn}} = 1.09 \times 10^4$ kJ

and for the reaction: $C(s) + O_2(g) \rightarrow CO_2(g)$ $\Delta_rH° = \Delta_fH°[CO_2(g)] = -393.509$ kJ/mol-rxn

$1.09 \times 10^4 \text{ kJ} \cdot \dfrac{1 \text{ mol-rxn}}{393.509 \text{ kJ}} \cdot \dfrac{1 \text{ mol C}}{1 \text{ mol-rxn}} \cdot \dfrac{12.011 \text{ g}}{1 \text{ mol C}} = 334$ g C

5.99. For the combustion of 100.0 g of ethanol the energy released:

Reaction of interest: $C_2H_5OH(\ell) + 3 \text{ } O_2(g) \rightarrow 2 \text{ } CO_2(g) + 3 \text{ } H_2O(g)$

$\Delta_rH° = [(2 \text{ mol})(-393.509 \text{ kJ/mol}) + (3 \text{ mol})(-241.83 \text{ kJ/mol})]$
$\hspace{3cm} - [(1 \text{ mol})(-277.0 \text{ kJ/mol}) + 0 \text{ kJ}]$

$= [(-787.0) + (-725.5)] \text{ } + 277.0 \text{ kJ}$

$= -1235.5$ kJ for the combustion of 1 mol of ethanol (coefficient in equation, right?)

Let's calculate the energy change per gram of ethanol:

$\dfrac{-1235.5 \text{ kJ}}{1 \text{ mol}} \cdot \dfrac{1 \text{ mol}}{46.069 \text{ g}} \cdot \dfrac{100.0 \text{ g}}{1} = -2682 \text{ kJ} = -2680$ kJ (3 sf)

5.100. Regarding the energy content of corn oil:

(a) Energy of 100. g: $(3766 \times 10^3 \text{ J})(1 \text{ cal}/4.184 \text{ J})(1 \text{ Calorie}/1000 \text{ cal}) = 900.1$ Calorie

(b) Tablespoons = 1500 Cal: (1500 Calorie)(100. g/900.1 Calorie)(1.0 Tbsp/14 g) = 11.9 Tbsp = 12 Tbsp (2 sf)

(c) Mass of water that can be heated:

(1.00 Tbsp)(1500 Calorie/12 Tbsp)(1000 cal/Calorie)(4.184 J/cal) = 5.23×10^5 J = 523 kJ

Energy to change temperature of water by 75.0 K: (75.0 K)(4.184 J/g K) = 314 J/g water

Amount of water that can be heated: $(5.23 \times 10^5 \text{ J})(1 \text{ g}/314 \text{ J}) = 1.67 \times 10^3$ g = 1.67 kg

5.101. Changes in enthalpy and internal energy for melting water:

(a) Enthalpy of fusion of ice:

$\left(\dfrac{333 \text{ J}}{\text{g H}_2\text{O}}\right)\left(\dfrac{1 \text{ kJ}}{1000 \text{ J}}\right)\left(\dfrac{18.02 \text{ g H}_2\text{O}}{1 \text{ mol H}_2\text{O}}\right) = 6.00$ kJ/mol H_2O

(b) Work done when ice melts:

1 mol H$_2$O = 18.02 g H$_2$O

Volume of ice = 18.02 g H$_2$O $\left(\dfrac{1 \text{ cm}^3}{0.9168 \text{ g}}\right) = 19.66 \text{ cm}^3 = 19.66$ mL or 0.01966 L

Volume of water = 18.02 g H$_2$O $\left(\dfrac{1 \text{ cm}^3}{0.9999 \text{ g}}\right) = 18.02 \text{ cm}^3 = 18.02$ mL or 0.01802 L

$\Delta V = V_{final} - V_{initial} = 0.01802$ L $- 0.01966$ L $= -0.00164$ L $\left(\dfrac{1 \text{ m}^3}{1000 \text{ L}}\right) = -1.64 \times 10^{-6}$ m^3

$w = -P\Delta V = -1.01 \times 10^5$ Pa $(-1.64 \times 10^{-6} \text{ m}^3) = +0.166$ J

(c) Change in internal energy:

$\Delta U = q + w = 6.00$ kJ $+ 0.000166$ kJ $= 6.000166$ kJ $= 6.00$ kJ (3 sf)

Note that the work done in melting is so small as to be negligible compared to the heat (enthalpy) involved in melting.

5.102. Changes in enthalpy and internal energy for vaporizing water:

(a) Enthalpy of vaporization of ice:

$\left(\dfrac{2256 \text{ J}}{\text{g H}_2\text{O}}\right)\left(\dfrac{1 \text{ kJ}}{1000 \text{ J}}\right)\left(\dfrac{18.02 \text{ g H}_2\text{O}}{1 \text{ mol H}_2\text{O}}\right) = 40.65$ kJ/mol H$_2$O

(b) Work done when water vaporizes:

1 mol H$_2$O = 18.02 g H$_2$O

Volume of water = 18.02 g H$_2$O $\left(\dfrac{1 \text{ cm}^3}{0.959 \text{ g}}\right) = 18.79 \text{ cm}^3 = 18.79$ mL or 0.01879 L

Volume of water = 18.02 g H$_2$O $\left(\dfrac{1 \text{ cm}^3}{0.000588 \text{ g}}\right) = 30{,}650 \text{ cm}^3 = 30{,}650$ mL or 30.65 L

$\Delta V = V_{final} - V_{initial} = 30.65$ L $- 0.01879$ L $= 30.63$ L $\left(\dfrac{1 \text{ m}^3}{1000 \text{ L}}\right) = 0.03063$ m^3

$w = -P\Delta V = -1.01 \times 10^5$ Pa $(0.03063 \text{ m}^3) = -3{,}094$ J $= -3.09$ kJ (3 sf)

(c) Change in internal energy:

$\Delta U = q + w = 40.65$ kJ $- 3.09$ kJ $= 37.56$ kJ

Note that the work done evaporating water is large enough to impact the internal energy of the process.

In the Laboratory

5.103. q_{metal} = specific heat capacity × mass × ΔT

$q_{metal} = c_{metal} \cdot 27.3 \text{ g} \cdot (299.47 \text{ K} - 372.05 \text{ K})$

Note that ΔT is negative, since T_{final} of the metal is LESS THAN $T_{initial}$

and q_{water} = 15.0 g · 4.184 J/gK · (299.47 K − 295.65K) = 239.7 J
Setting $q_{metal} = -q_{water}$

$c_{metal} \cdot 27.3 \text{ g} \cdot (-72.58 \text{ K}) = -239.7 \text{ J}$ and solving for c gives: $c_{metal} = 0.121 \dfrac{\text{J}}{\text{g} \cdot \text{K}}$

5.104. Final temperature of Cu and water:

$q_{Cu} + q_{water} = 0$

$[(192 \text{ g})(0.385 \text{ J/g·K})(T_f - 373.2 \text{ K})] + [(751 \text{ g})(4.184 \text{ J/g·K})(T_f - 277.2 \text{ K})] = 0$

T_f = 279 K (6 °C)

5.105. Calculate the enthalpy change for the precipitation of AgCl (in kJ/mol):

1) How much AgCl is being formed?
   250. mL of 0.16 M $AgNO_3$ will contain (0.250 L · 0.16 mol/L) 0.040 mol of $AgNO_3$
   125 mL of 0.32 M NaCl will contain (0.125L · 0.32 mol/L) 0.040 mol of NaCl.

   Given the stoichiometry, we anticipate the formation of 0.040 mol of AgCl.

2) How much energy is evolved?
   (250. mL + 125 mL) · 1.0 g/mL = 375 g of solution

   $375 \text{ g} \cdot 4.2 \dfrac{\text{J}}{\text{g} \cdot \text{K}} \cdot (296.05 \text{ K} - 294.30 \text{ K}) = 2{,}800 \text{ J (2 sf)}$

   The enthalpy change is −2800 J (since the reaction **releases** heat).
   The change in kJ/mol is $\dfrac{-2800 \text{ J}}{0.040 \text{ mol}} \cdot \dfrac{1 \text{ kJ}}{1000 \text{ J}} = -69 \text{ kJ/mol}$

5.106. Enthalpy change of $PbBr_2(s)$:

$q_r + q_{solution} = 0$
$q_r = -[(200. \text{ g} + 200. \text{ g})(4.2 \text{ J/g·K})(2.44 \text{ K})]$ and $q_r = -4.1 \times 10^3$ J
Reactants are present in the correct stoichiometric amounts.

$0.200 \text{ L} \cdot \dfrac{0.75 \text{ mol Pb(NO}_3)_2}{1 \text{ L}} \cdot \dfrac{1 \text{ mol PbBr}_2}{1 \text{ mol Pb(NO}_3)_2} = 0.15 \text{ mol PbBr}_2$

$$\Delta H = q_r = \frac{-4.1 \times 10^3 \text{ J}}{0.15 \text{ mol PbBr}_2} \cdot \frac{1 \text{ kJ}}{10^3 \text{ J}} = -27 \text{ kJ/mol PbBr}_2$$

5.107. Heat evolved when ammonium nitrate is decomposed:
$\Delta T = (20.72 \text{ °C} - 18.90 \text{ °C}) = 1.82 \text{ °C}$ (or 1.82 K).

Heat absorbed by the calorimeter: $155 \text{ J/K} \cdot 1.82 \text{ K} = 282 \text{ J}$

Heat absorbed by the water: $415 \text{ g} \cdot 4.184 \frac{\text{J}}{\text{g} \cdot \text{K}} \cdot 1.82 \text{ K} = 3160 \text{ J}$

$q_r = -(q_{water} + q_{bomb}) = -(3160 \text{ J} + 282 \text{ J}) = -3440 \text{ J}$ (3 sf)

7.647 g NH$_4$NO$_3$ = 0.09554 mol, so $\Delta U = \dfrac{-3440 \text{ J}}{0.09554 \text{ mol}} = -3.60 \times 10^4 \text{ J/mol} = -36.0 \text{ kJ/mol}$

5.108. Calculate ΔU for the combustion of ethanol:

$q_{water} + q_{bomb} + q_r = 0$
$q_r = -([(650 \text{ g})(4.184 \text{ J/g·K})(295.5 \text{ K} - 291.7 \text{ K})] + [(550 \text{ J/K})(295.5 \text{ K} - 291.7 \text{ K})])$
$q_r = -1.2 \times 10^4 \text{ J}$

$4.20 \text{ g C}_2\text{H}_5\text{OH} \cdot \dfrac{1 \text{ mol C}_2\text{H}_5\text{OH}}{46.07 \text{ g}} = 0.0912 \text{ mol C}_2\text{H}_5\text{OH}$

$\Delta_{com}U = \dfrac{-1.2 \times 10^4 \text{ J}}{0.0912 \text{ mol}} \cdot \dfrac{1 \text{ kJ}}{10^3 \text{ J}} = -140 \text{ kJ/mol C}_2\text{H}_5\text{OH}$

5.109. The enthalpy change for the reaction:

$$\text{Mg(s)} + 2 \text{ H}_2\text{O}(\ell) \rightarrow \text{Mg(OH)}_2\text{(s)} + \text{H}_2\text{(g)}$$

$\Delta_f H°$(kJ/mol)   0   −285.83   −924.54   0

$\Delta_r H° = (1 \text{ mol})(-924.54 \frac{\text{kJ}}{\text{mol}}) - (2 \text{ mol})(-285.83 \frac{\text{kJ}}{\text{mol}}) = -352.88 \text{ kJ}$ or -3.5288×10^5 J

Each mole of magnesium releases 352.88 kJ of heat energy.

Calculate the heat required to warm 25 mL of water from 25 °C to 85 °C.

Heat = specific heat capacity × mass × ΔT

$= (4.184 \text{ J/g·K})(25 \text{ mL})(\dfrac{1.00 \text{ g}}{1 \text{ mL}})(60. \text{ K})$

$= 6276$ or 6300 J or 6.3 kJ   (2 sf)

Magnesium required:

$$6.3 \text{ kJ} \cdot \frac{1 \text{ mol Mg}}{352.88 \text{ kJ}} \cdot \frac{24.3 \text{ g Mg}}{1 \text{ mol Mg}} = 0.43 \text{ g Mg}$$

5.110. Mass of Fe needed to supply energy:

NOTE: $\Delta_f H°$ Fe(s) = $\Delta_f H°$ O$_2$(g) = 0 kJ/mol (so these terms will be omitted from the ΔH calculation)
$\Delta_r H° = 2 \Delta_f H°$ Fe$_2$O$_3$(s) = 2 mol (–825.5 kJ/mol)
$\Delta_r H° = –1651.0$ kJ/mol-rxn
q_{water} = (15 mL)(1.00 g/mL)(4.184 J/g·K)(310. K – 296 K) = 880 J

$$880 \text{ J} \cdot \frac{1 \text{ kJ}}{10^3 \text{ J}} \cdot \frac{1 \text{ mol-rxn}}{1651.0 \text{ kJ}} \cdot \frac{4 \text{ mol Fe}}{1 \text{ mol-rxn}} \cdot \frac{55.85 \text{ g}}{1 \text{ mol Fe}} = 0.12 \text{ g Fe}$$

Summary and Conceptual Questions

5.111. Without doing calculations, decide whether each is exo- or endothermic:

(a) combustion of natural gas—oxidation reactions of hydrocarbons typically release heat-- this process is exothermic.

(b) Decomposition of glucose to form carbon and water- When you burn glucose to form carbon and water, heat is evolved--the reaction is exothermic.

5.112. Identify state functions:

Of these four options, (a), (c), and (d) are state functions.

5.113. Determine the value of $\Delta H°$ for the reaction:

Ca(s) + S(s) + 2 O$_2$(g) → CaSO$_4$(s)

Imagine this as the sum of several processes:

1) Ca(s) + ½ O$_2$(g) → CaO(s)

2) ⅛ S$_8$ (s) + 3/2 O$_2$(g) → SO$_3$(g)

3) CaO(s) + SO$_3$(g) → CaSO$_4$(s) $\Delta_r H° = –403.7$ kJ

Note that the SUM of the three processes is the DESIRED equation (the formation of CaSO$_4$(s)). The OVERALL $\Delta H°$ is the SUM of the $\Delta H°$ for process (1) and $\Delta H°$ for process (2). We know that the $\Delta_r H°$ for (3) = –403.7 kJ or

$\Delta_r H° = \Delta_f H°$CaSO$_4$(s) – [$\Delta_f H°$ CaO(s) + $\Delta_f H°$ SO$_3$(g)]. Since we know the $\Delta_r H°$ and BOTH the $\Delta_f H°$ for CaO(s) and SO$_3$(g), we can calculate the $\Delta_f H°$ CaSO$_4$(s).

From Appendix L we find,
$\Delta_f H°$ for CaO(s) = –635.09 kJ/mol and $\Delta_f H°$ for SO$_3$ (g) = −395.77 kJ/mol

$\Delta_r H° = \Delta_f H°$ CaSO$_4$(s) – [$\Delta_f H°$ CaO(s) + $\Delta_f H°$ SO$_3$(g)]

−403.7 kJ = $\Delta_f H°$ CaSO$_4$ (s) – [−635.09 kJ/mol + −395.77 kJ/mol]

−1,434.6 kJ = $\Delta_f H°$ CaSO$_4$ (s)

5.114. There is a rough non-linear correspondence, which shows that the specific heat increases with decreasing atomic weight. Using the relationship $c_{metal} \propto \dfrac{1}{\text{atomic weight}}$, or $c_{metal} \times$ atomic weight = constant ≈ 26, $c_{Pt} \approx 26/195 \approx 0.130$ J/g·K. This is in good agreement with the literature value (0.133 J/g·K).

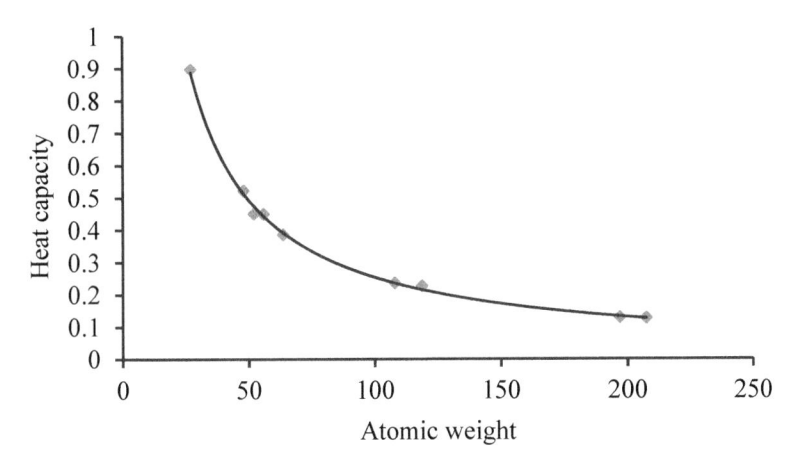

5.115. The molar heat capacities for Al, Fe, Cu, and Au are:

$0.897 \dfrac{J}{g \cdot K} \cdot \dfrac{26.98 \text{ g Al}}{1 \text{ mol Al}} = 24.2 \dfrac{J}{mol \cdot K}$

$0.449 \dfrac{J}{g \cdot K} \cdot \dfrac{55.85 \text{ g Fe}}{1 \text{ mol Fe}} = 25.1 \dfrac{J}{mol \cdot K}$

$0.385 \dfrac{J}{g \cdot K} \cdot \dfrac{63.55 \text{ g Cu}}{1 \text{ mol Cu}} = 24.5 \dfrac{J}{mol \cdot K}$

$0.129 \dfrac{J}{g \cdot K} \cdot \dfrac{197.0 \text{ g Au}}{1 \text{ mol Au}} = 25.4 \dfrac{J}{mol \cdot K}$

The graph shown is a plot of specific heat capacity versus atomic weight.

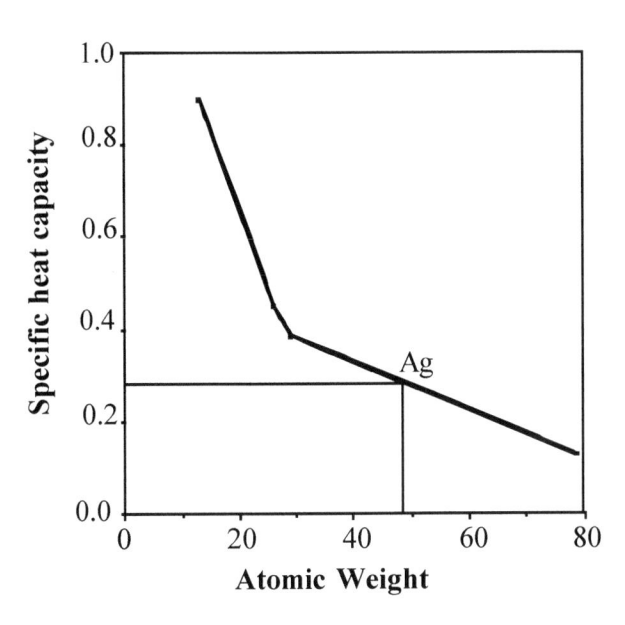

As you can see, no simple linear relationship exists for these metals. The plot of the specific heat of Cu (atomic weight 63.55) and Au (atomic weight 197) does show a **decreasing** value of specific heat capacity as the atomic weight of the element increases. If you estimate the atomic weight to be about 100 (exact value is about 108), one could **estimate** a value of approximately 0.28 as the specific heat (compared to the experimental value of 0.236). Alternatively, a quick examination of the values for the molar heat capacities of the four metals above indicates that they are **quite similar**, with an average of $24.8 \frac{J}{mol \cdot K}$. This translates into: $24.8 \frac{J}{mol \cdot K} \cdot \frac{1 \text{ mol Ag}}{107.9 \text{ g Ag}} = 0.230 \frac{J}{g \cdot K}$.

5.116. Why does the room warm?

To extract heat from the inside of the refrigerator, work has to be done. That work (by the condenser and motor) releases heat to the environment (your room). So, while the temporary relief of cool air from the inside of the refrigerator is pleasant, the motor has to do work—and heats your room.

5.117. Mass of methane needed to heat the air from 15.0 °C to 22.0 °C:

Calculate the volume of air, then with the density and average molar mass, the moles of air present:

$$275 \text{ m}^2 \cdot 2.50 \text{ m} \cdot \frac{1000 \text{ L}}{1 \text{ m}^3} \cdot \frac{1.22 \text{ g air}}{1 \text{ L air}} \cdot \frac{1 \text{ mol air}}{28.9 \text{ g air}} = 2.90 \times 10^4 \text{ mol air}$$

The energy needed to change the temperature of that amount of air by (22.0 – 15.0) °C:

$$2.90 \times 10^4 \text{ mol air} \cdot 29.1 \frac{J}{mol \cdot K} \cdot 7.0 \text{ K} = 5.9 \times 10^6 \text{ J}$$

What quantity of energy does the combustion of methane provide?

The reaction may be written: $CH_4(g) + 2\ O_2(g) \rightarrow 2\ H_2O(g) + CO_2(g)$

Using data from Appendix L:

$\Delta_r H° = [(2 \text{ mol})(-241.83 \text{ kJ/mol}) + (1 \text{mol})(-393.509 \text{ kJ/mol})]$
$\qquad\qquad - [(1 \text{ mol})(-74.87 \text{ kJ/mol}) + (2 \text{ mol})(0 \text{ kJ/mol})] = -802.30 \text{ kJ}$

The amount of methane necessary is:

$$5.9 \times 10^6 \text{ J} \cdot \frac{1 \text{ kJ}}{1000 \text{ J}} \cdot \frac{1 \text{ mol } CH_4}{802.3 \text{ kJ}} \cdot \frac{16.0 \text{ g } CH_4}{1 \text{ mol } CH_4} = 120 \text{ g } CH_4 \text{ (2 sf)}$$

5.118. Regarding the production of water:

(a) CaBr$_2$(s) + H$_2$O(g) → CaO(s) + 2 HBr(g)
Hg(ℓ) + 2 HBr(g) → HgBr$_2$(s) + H$_2$(g)
HgBr$_2$(s) + CaO(s) → HgO(s) + CaBr$_2$(s)
HgO(s) → Hg(ℓ) + ½ O$_2$(g)
_____
H$_2$O(g) → H$_2$(g) + ½ O$_2$(g)

(b) 1000. kg · $\dfrac{10^3 \text{ g}}{1 \text{ kg}}$ · $\dfrac{1 \text{ mol H}_2\text{O}}{18.015 \text{ g}}$ · $\dfrac{1 \text{ mol H}_2}{1 \text{ mol H}_2\text{O}}$ · $\dfrac{2.016 \text{ g}}{1 \text{ mol H}_2}$ = 1.119 × 10^5 g (111.9 kg)

(c) Step 1:

$\Delta_r H°$ = $\Delta_f H°$[CaO(s)] + 2 $\Delta_f H°$[HBr(g)] − ($\Delta_f H°$[CaBr$_2$(s)] + $\Delta_f H°$[H$_2$O(g)])
$\Delta_r H°$ = 1 mol (−635.09 kJ/mol) + 2 mol (−36.29 kJ/mol)
  − [1 mol (−683.2 kJ/mol) + 1 mol (−241.83 kJ/mol)]
$\Delta_r H°$ = 217.4 kJ/mol-rxn   endothermic

Step 2:
$\Delta_r H°$ = $\Delta_f H°$[HgBr$_2$(s)] − 2 $\Delta_f H°$[HBr(g)]
$\Delta_r H°$ = 1 mol (−169.5 kJ/mol) − 2 mol (−36.29 kJ/mol)
$\Delta_r H°$ = −96.9 kJ/mol-rxn   exothermic

Step 3:
$\Delta_r H°$ = $\Delta_f H°$[HgO(s)] + $\Delta_f H°$[CaBr$_2$(s)] − ($\Delta_f H°$[HgBr$_2$(s)] + $\Delta_f H°$[CaO(s)])
$\Delta_r H°$ = 1 mol (−90.83 kJ/mol) + 1 mol (−683.2 kJ/mol)
  − [1 mol (−169.5 kJ/mol) + 1 mol (−635.09 kJ/mol)]
$\Delta H°$ = 30.6 kJ/mol-rxn   endothermic

Step 4:
$\Delta_r H°$ = −$\Delta_f H°$[HgO(s)]
$\Delta_r H°$ = 90.83 kJ/mol-rxn   endothermic

(d) The commercial feasibility of this process is limited by the three endothermic steps.

5.119. Calculate the quantity of heat transferred to the surroundings from the water vapor condensation as rain falls.

Calculate the volume of water that falls, and then the mass of that water:

The problem states that an area of one square mile corresponds to 2.59 × 10^6 m^2. In square cm, this area is 2.59 × 10^6 m^2 (100 cm/1 m)^2 = 2.59 × 10^{10} cm^2.

1 in = 2.54 cm so the VOLUME of water is 2.59 × 10^{10} cm^2 × 2.54 cm = 6.6 × 10^{10} cm^3.

The mass of water is: 6.6 × 10^{10} cm^3 × 1.0 g/cm^3 or 6.6 × 10^{10} g of water.

The amount of heat: $\dfrac{6.6 \times 10^{10} \text{ g water}}{1}$ · $\dfrac{1 \text{ mol water}}{18.02 \text{ g water}}$ · $\dfrac{44.0 \text{ kJ}}{1 \text{ mol water}}$ = 1.6 × 10^{11} kJ

Note the much larger energy for this process than for the detonation of a ton of dynamite.

5.120. Number of peanuts to provide energy to boil a cup of water:

Begin by asking "How much energy is needed to boil a cup of water, beginning with water at 25 °C:

(a) Heat the water to boiling: $250. \text{ mL} \cdot \dfrac{1 \text{ g H}_2\text{O}}{1 \text{ mL H}_2\text{O}} \cdot \dfrac{4.184 \text{ J}}{\text{g} \cdot \text{K}} \cdot \dfrac{75\ ^0\text{C}}{1} \cdot \dfrac{1 \text{ kJ}}{1000 \text{ J}} = 78.5 \text{ kJ}$

(b) Vaporize at 100 °C (liquid → gas): $250. \text{ mL} \cdot \dfrac{1 \text{ g H}_2\text{O}}{1 \text{ mL H}_2\text{O}} \cdot \dfrac{2256 \text{ J}}{1 \text{ g H}_2\text{O}} = 564{,}000 \text{ J} = 564 \text{ kJ}$

Total energy to boil a cup of water: 78.5 kJ + 564 kJ = 642 kJ

"What amount of energy is released upon combustion of the peanut?"

Calculate: $\Delta_r H°$ for combustion of peanut oil and starch:

Peanut oil: $C_{16}H_{22}O_2 (s) + \dfrac{41}{2} O_2 (g) \rightarrow 16 \text{ CO}_2 (g) + 11 \text{ H}_2\text{O}(g)$

NOTE: $\Delta_f H°$ for O_2(g) = 0 kJ/mol

$\Delta_r H° = [(16 \text{ mol})(\Delta_f H° \text{ CO}_2) + (11 \text{ mol})(\Delta_f H° \text{ H}_2\text{O})] – [(1 \text{ mol})(\Delta_f H° \text{ C}_{16}\text{H}_{22}\text{O}_2)]$

$\Delta_r H° = [(16 \text{ mol})(–393.509 \text{ kJ/mol}) + (11 \text{ mol})(–241.83 \text{ kJ/mol})] – [(1 \text{ mol})(–848.4 \text{ kJ/mol})]$

$\Delta_r H° = –8107.874$ kJ for each mol of peanut oil

Starch: $C_6H_{10}O_5 (s) + 6 O_2 (g) \rightarrow 6 \text{ CO}_2 (g) + 5 \text{ H}_2\text{O}(g)$

$\Delta_r H° = [(6 \text{ mol})(\Delta_f H° \text{ CO}_2) + (5 \text{ mol})(\Delta_f H° \text{ H}_2\text{O})] – [(1 \text{ mol})(\Delta_f H° \text{ C}_6\text{H}_{10}\text{O}_5)]$

$\Delta_r H° = [(6 \text{ mol})(–393.509 \text{ kJ/mol}) + (5 \text{ mol})(–241.83 \text{ kJ/mol})] – [(1 \text{ mol})(–960 \text{ kJ/mol})]$

$\Delta_r H° = −2610.204$ kJ for each mol of starch

Each peanut has a mass of 0.73 g and is 49% peanut oil and 21% starch or

(0.73 g peanut)(0.49 g oil/g peanut) = 0.3577 g oil and

(0.73 g peanut)(0.21 g starch/g peanut) = 0.1533 g starch

So, each peanut releases $\dfrac{-8107.874 \text{ kJ}}{1 \text{ mol oil}} \cdot \dfrac{1 \text{ mol oil}}{246.4 \text{ g oil}} \cdot \dfrac{0.3577 \text{ g oil}}{1 \text{ peanut}} = -11.77 \text{ kJ/peanut}$

and $\dfrac{-2610.2 \text{ kJ}}{1 \text{ mol starch}} \cdot \dfrac{1 \text{ mol starch}}{162.1 \text{ g starch}} \cdot \dfrac{0.1533 \text{ g starch}}{1 \text{ peanut}} = -2.469 \text{ kJ/peanut}$

Each peanut would release a *total* of (–11.77 + –2.469 kJ) or 14.2 kJ/peanut

Since 642 kJ is the energy required, divide that energy by the energy per peanut.

$\dfrac{642 \text{ kJ}}{1 \text{ cup water}} \cdot \dfrac{1 \text{ peanut}}{14.2 \text{ kJ}} = 45 \text{ peanuts/cup water}$

5.121. Regarding the energy level content of three isomers:

(a) The diagram is:

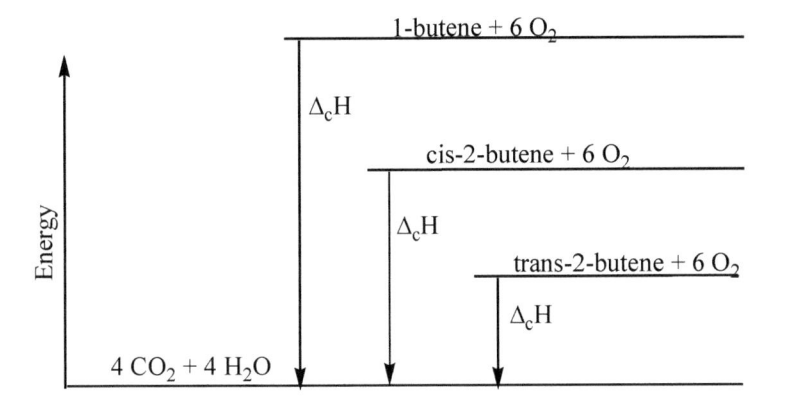

[Energy level diagram showing 1-butene + 6 O₂ at the highest level, cis-2-butene + 6 O₂ below it, trans-2-butene + 6 O₂ below that, and 4 CO₂ + 4 H₂O at the bottom, with Δ_cH arrows pointing down from each butene isomer to the products.]

(b) For the combustion reaction: $C_4H_8(g) + 6\ O_2(g) \rightarrow 4\ CO_2(g) + 4\ H_2O(\ell)$

$\Delta_cH° = 4 \cdot \Delta_fH°(CO_2) + 4 \cdot \Delta_fH°(H_2O) - [\Delta_fH°(C_4H_8) + 6 \cdot \Delta_fH°(O_2)]$ Note that the last term will be 0 in all cases. Substitute the thermodynamic data for each of the three isomers:

cis-2-butene:

$\Delta_cH° = 4 \cdot \Delta_fH°(CO_2) + 4 \cdot \Delta_fH°(H_2O) - \Delta_fH°(C_4H_8)$

1 mol · −2709.8 kJ/mol =

    4 mol · −393.509 kJ/mol + 4 mol · −285.83 kJ/mol − 1 mol · $\Delta_fH°(C_4H_8)$

and solving for $\Delta_fH°(C_4H_8)$ yields −7.6 kJ

trans-2-butene:

1 mol · −2706.6 kJ/mol =

    4 mol · −393.509 kJ/mol + 4 mol · −285.83 kJ/mol − 1 mol · $\Delta_fH°(C_4H_8)$

and solving for $\Delta_fH°(C_4H_8)$ yields −10.8 kJ

1-butene:

1 mol · −2716.8 kJ/mol =

    4 mol · −393.509 kJ/mol + 4 mol · −285.83 kJ/mol − 1 mol · $\Delta_fH°(C_4H_8)$

and solving for $\Delta_fH°(C_4H_8)$ yields −0.6 kJ

(c) Relation of Enthalpies of Isomers to the elements:

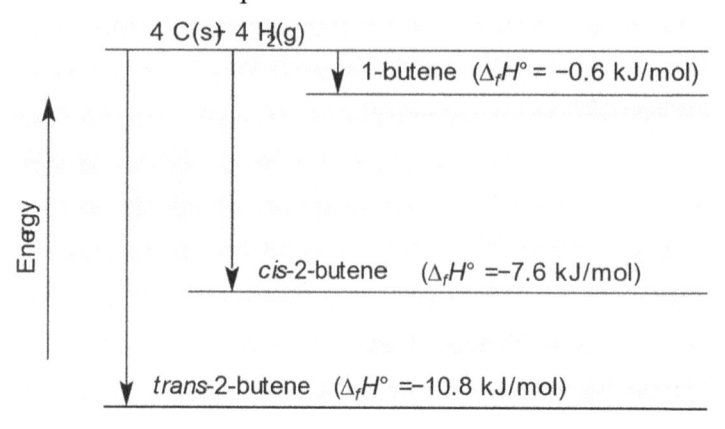

(d) Enthalpy change of cis-2-butene to trans-2-butene:

$\Delta_r H° = \Delta_f H°$(trans-2-butene) $- \Delta_f H°$(cis-2-butene) $= (-10.8$ kJ$) - (-7.6$ kJ$) = -3.2$ kJ

5.122. Regarding the enthalpy of vaporization of Br₂:

(a) $Br_2(\ell) \rightarrow Br_2(g)$

$\Delta_f H°[Br_2(\ell)] = 0$ kJ/mol

$\Delta_r H° = \Delta_f H°[Br_2(g)] = 30.9$ kJ/mol-rxn

(b) $Br_2(g) \rightarrow 2\, Br(g)$
$\Delta_r H° = 2\, \Delta_f H°[Br(g)] - \Delta_f H°[Br_2(g)] = 2$ mol $(111.9$ kJ/mol$) - 1$ mol $(30.9$ kJ/mol$)$
$\Delta_r H° = 192.9$ kJ/mol-rxn

5.123. Concerning the oxidation of Mg:

(a) A sample of 0.850 g Mg corresponds to 0.0350 mol Mg.
The amount of heat (evolved) is −25.4 kJ, corresponding to −25.4kJ/0.0350mol Mg
= −726 kJ/mol. Since $\Delta V = 0$, $\Delta U = -726$ kJ/mol.

(b) Final temperature of water and bomb calorimeter:

Heat evolved = −Heat absorbed

-25400 J $= -[(820.$ J/K$)\Delta T + (750.$g$)(4.184$ J/g·K$)\Delta T]$

-25400 J $= -[(820.$ J/K$)\Delta T + (3138$ J/K$)\Delta T]$ and -25400 J $= -3958$ J/K ΔT

-25400 J$/-3958$ J/K $= 6.41$ K (or 6.41 °C—since a K and a °C are the same "size")

The new temperature of water will be 18.6 °C + 6.41 °C = 25.0 °C

5.124. Will temperature change when Au or Cu are added to water?

$q_{Au} + q_{Cu} + q_{water} = 0$
$[(10.0\text{ g})(0.129\text{ J/g·K})(T_f - 373.2\text{ K})] + [(10.0\text{ g})(0.385\text{ J/g·K})(T_f - 273.2\text{ K})]$
$\quad\quad + [(150.\text{ g})(4.184\text{ J/g·K})(T_f - 293.2\text{ K})] = 0$
$T_f = 293.19\text{ K} = 20.04\text{ °C}$

The calculation indicates that the temperature increases slightly, but the calculated value for the kelvin temperature is only good to three significant figures (the ones digit), so the temperature is not significantly changed.

5.125. Regarding the energy level of methane and methanol:

(a) The energy diagram shown here indicates that methane liberates 890.2 kJ/mol while methanol liberates only 726.7 kJ/mol.

(b) Energy per gram:

For methane: $\dfrac{-890.2\text{ kJ}}{1\text{ mol}} \cdot \dfrac{1\text{ mol}}{16.043\text{ g}} = -55.49\text{ kJ/g}$

For methanol: $\dfrac{-726.7\text{ kJ}}{1\text{ mol}} \cdot \dfrac{1\text{ mol}}{32.041\text{ g}} = -22.68\text{ kJ/g}$

(c) Enthalpy conversion from methane to methanol: The diagram indicates that the difference in enthalpy for these two substances is the difference between the two "top boxes". Hence $\Delta H = -890.2\text{ kJ} - (-726.7\text{ kJ}) = -163.5\text{ kJ/mol}$

(d) The equation for conversion of methane to methanol: $CH_4(g) + 1/2\ O_2(g) \rightarrow CH_3OH(\ell)$

5.126. Regarding the formation of liquid ethanol:

The process can be viewed as the sum of three steps:

$2\ C(s) + 2\ O_2(g) \rightarrow 2\ CO_2(g)$	$\Delta_r H° = 2(-393.5\text{ kJ})$
$3\ H_2(g) + {}^3/_2\ O_2(g) \rightarrow 3\ H_2O(\ell)$	$\Delta_r H° = {}^3/_2(-571.6\text{ kJ})$
$2\ CO_2(g) + 3\ H_2O(\ell) \rightarrow C_2H_5OH(\ell) + 3\ O_2(g)$	$\Delta_r H° = -(-1367.5\text{ kJ})$
$2\ C(s) + 3\ H_2(g) + ½\ O_2(g) \rightarrow C_2H_5OH(\ell)$	$\Delta_r H° = -276.9\text{ kJ/mol-rxn}$

5.127. Piece of metal to heat and to cool to achieve a maximum T? Final temperature of water?

(a) To convey maximum heat per gram, one needs a metal with the greatest specific heat—so of these 3 metals, Al, is the best candidate, and the larger piece of metal (1) would convey MORE heat than the smaller piece (2) of Al. To minimize the heat absorbed by the "cooler" metal, one needs a metal with the lesser specific heat—and the smaller the better—so the smaller piece of Au (4) is a prime candidate. As to final T:
Heat loss (by warm metal) = Heat gain (by cool metal and water)

Note that the SIGNS of the two will be opposite, so let's (arbitrarily) place a (−) sign in front of the "heat loss" side.

−(100.0 g)(0.9002 J/g·K)(T_f − 373) =

$\qquad$ (50.0 g)(0.1289 J/g·K)(T_f − 263) + (300.g)(4.184 J/g·K)(T_f − 294)

−90.02T_f + 33577 = 6.445T_f − 1695.035 + 1255.2T_f − 369028.8

Collecting T_f terms:

−90.02T_f + −6.445T_f + −1255.2T_f = −369028.8 + −33577 + −1695.035 or

−1351.665 T_f = −404300.835 and T_f = 299 K or (299 − 273) = 26°C

(b) Process is similar to that in (a) but we want *minimal* T change:

Consider the following table of data and calculations:

Specific heat	Mass	Heat Capacity	Metal	ΔT for metal heated to 100 °C	ΔT for metal cooled to −10 °C	Heat lost upon cooling to 21 °C	Heat gained upon warming to 21 °C
0.9002	50.0	45.0	Al	79	31	3555.79	1395.31
0.386	50.0	19.3	Zn	79	31	1524.7	598.3

Note that the amount of heat lost by cooling 50.0 g Zn and the heat gained by warming 50.0 g of Al is approximately equal. You could also do these calculations for all the combinations of metals (both type and mass). Let's see how these two compute!

−Heat lost = Heat gained

−(50.0 g)(0.3860 J/g·K)(T_f − 373) =

$\qquad$ (50.0 g)(0.9002 J/g·K)(T_f − 263) + (300. g)(4.184 J/g·K)(T_f − 294)

−19.3T_f + 71989 = 45.01T_f − 11837.6 + 1255.2T_f − 369028.8

Collecting T_f terms:

−19.3T_f + −45.01T_f + −1255.2T_f = −369028.8 + −11837.6 − 71989 or

−1319.54T_f = −388065.3 and T_f = 294 K or (294 − 273) = 21°C

5.128. Predict how the following events will affect the value for $\Delta_rH°$:

(a) $\Delta_rH°$ is too high as less Ca(OH)$_2$ is added to the calorimeter.

(b) $\Delta_rH°$ is unaffected (assuming the actual volume of HCl is recorded) as the HCl is not the limiting reactant.

(c) $\Delta_rH°$ is too high as excess water results in less Ca(OH)₂ being added (see (a)).

(d) $\Delta_rH°$ is unaffected as weight of Ca(OH)₂ is accurate.

(e) $\Delta_rH°$ is too high as maximum temperature will be missed due to delay.

(f) $\Delta_rH°$ is too high as lost energy is not reflected in the temperature change.

(g) $\Delta_rH°$ is too high as the added heat (to the stirrer and thermometer) is neglected in your calculation.

5.129. The work done on the surroundings as carbon dioxide sublimes:

When the solid is converted to gas, the **change** in volume is essentially 0.36 L (since 1.0 g of carbon dioxide(s) will occupy only a small volume. First convert 0.36 L to m³.

$$0.36 \text{ L}\left(\frac{1 \text{ m}^3}{1000 \text{ L}}\right) = 3.6 \times 10^{-4} \text{ m}^3$$

The work is then:

$w = -P \times \Delta V$
$w = (-1.01 \times 10^5 \text{ Pa})(3.6 \times 10^{-4} \text{ m}^3) = -36 \text{ J (2 sf)}$

5.130. For the gas phase reaction of hydrogen and oxygen to form water vapor:

$$-22.4 \text{ L}\left(\frac{1 \text{ m}^3}{1000 \text{ L}}\right) = -2.24 \times 10^{-2} \text{ m}^3$$

(a) $w = -P\Delta V = (-1.01 \times 10^5 \text{ Pa})(-2.24 \times 10^{-2} \text{ L}) = 2.26 \times 10^3 \text{ J} = 2.26 \text{ kJ}$

(b) $\Delta_rU = \Delta_rH° - P\Delta V = -483.6 \text{ kJ} + 2.26 \text{ kJ} = -481.3 \text{ kJ}$

5.131. Kilauea's effect on lake:

(a) Volume of lake water:

$$V_{sphere} = \frac{4}{3}\pi r^3 = \frac{4}{3}(3.1416)(30.5 \text{ m})^3 = 1.188 \times 10^5 \text{ m}^3$$

$$V_{hemisphere} = V_{lake} = \frac{1}{2}(1.188 \times 10^5 \text{ m}^3) = 5.942 \times 10^4 \text{ m}^3 = 5.94 \times 10^4 \text{ m}^3 \text{ (3 sf)}$$

(b) Mass in g of lake water:

$$5.94 \times 10^4 \text{ m}^3 \left(\frac{1 \times 10^6 \text{ cm}^3}{1 \text{ m}^3}\right)\left(\frac{0.999 \text{ g}}{1 \text{ cm}^3}\right) = 5.936 \times 10^{10} \text{ g} = 5.94 \times 10^{10} \text{ g (3 sf)}$$

(c) Energy necessary to heat water to boiling point and evaporate water:

Energy to raise lake temperature from 18 °C to 100 °C:

$$5.94 \times 10^{10} \text{ g} \left(\frac{4.184 \text{ J}}{\text{g} \cdot \text{K}} \right) (373 \text{ K} - 291 \text{ K}) = 2.04 \times 10^{13} \text{ J}$$

Energy to evaporate water from lake:

$$5.94 \times 10^{10} \text{ g} \left(\frac{2256 \text{ J}}{\text{g}} \right) = 1.34 \times 10^{14} \text{ J}$$

Total energy required:

2.04×10^{13} J $+ 1.34 \times 10^{14}$ J $= 1.544 \times 10^{14}$ J $= 1.54 \times 10^{14}$ J (3 sf)

Solution and Answer Guide

Kotz Treichel Townsend Treichel, Chemistry and Chemical Reactivity 11e, 978-0-357-85140-1, Chapter 6: The Structure of Atoms

TABLE OF CONTENTS

Applying Chemical Principles 231
Practicing Skills..233
General Questions...244
In the Laboratory...251
Summary and Conceptual Questions..252

Applying Chemical Principles

Sunburn, Sunscreens, and Ultraviolet Radiation

6.1.1. Regarding wavelength, frequency, and energy per photon for visible or UV light:

A quick glance at Figure 6.2 reveals that **visible** light has the longer wavelength.

UV light has the higher frequency and therefore the higher energy per photon.

6.1.2. Energy per mole of photons (kJ/mol) for red light (wavelength = 700 nm):

We know that the energy for electromagnetic radiation can be expressed:

$E = h\nu = h \cdot \dfrac{c}{\lambda}$ Solving for E yields:

$\dfrac{6.626 \times 10^{-34} \text{ J·s/photon}}{1} \cdot \dfrac{2.998 \times 10^{8} \text{ m/s}}{700.00 \times 10^{-9} \text{ m}} = 2.838 \times 10^{-19} \dfrac{\text{J}}{\text{photon}}$

for a mol: $(2.838 \times 10^{-19} \text{ J/photon})(6.022 \times 10^{23} \text{ photons/mol}) = 1.709 \times 10^{5}$ J/mol

or 170.9 kJ/mol

Now the energy with radiation with wavelength = 300 nm:

$\dfrac{6.626 \times 10^{-34} \text{ J·s/photon}}{1} \cdot \dfrac{2.998 \times 10^{8} \text{ m/s}}{300.00 \times 10^{-9} \text{ m}} = 6.622 \times 10^{-19} \dfrac{\text{J}}{\text{photon}}$

for a mol: $(6.622 \times 10^{-19} \text{ J/photon})(6.022 \times 10^{23} \text{ photons/mol}) = 3.988 \times 10^{5}$ J/mol

Energy of 300 nm/700 nm = $\dfrac{3.988 \times 10^{5} \text{ J/photon}}{1.709 \times 10^{5} \text{ J/photon}} = 2.33$ times more energetic

What Makes the Colors in Fireworks?

6.2.1. Of the lines in the emission spectrum for sodium, which is responsible for the yellow color? 589 nm and 590 nm are the sodium D lines that are responsible for that color.

6.2.2. Compare main emission lines from Sr and Na salts:

The major lines for Na were cited in the problem above. Sr emits a red light, so the major lines for Sr are going to be at longer wavelengths than those for sodium.

6.2.3. The balanced equation for the reaction of Mg with KClO₄:

4 Mg(s) + KClO₄(s) → KCl(s) + 4 MgO(s)

Chemistry of the Sun

6.3.1. What is the frequency of light with wavelength = 587.6 nm?
We know that wavelength and frequency of light are related by the equation: $\lambda \cdot \nu = c$.

$$\nu = \frac{c}{\lambda} = \frac{2.998 \times 10^8 \text{ m/s}}{587.6 \times 10^{-9} \text{ m}} = 5.102 \times 10^{14} \text{ s}^{-1}$$

6.3.2. What is the wavelength of the light with frequency 5.688×10^{14} s⁻¹.

$$\lambda = \frac{c}{\nu} = \frac{2.998 \times 10^8 \text{ m/s}}{5.688 \times 10^{14} \text{ s}^{-1}} = 5.271 \times 10^{-7} \text{ m} = 527.1 \text{ nm}$$

6.3.3. What is the energy (in Joules) of photons at 589.00 nm and 589.59 nm?

We know that the energy is related by the equation: $E = h\nu = h \cdot \frac{c}{\lambda}$

$$\frac{6.626 \times 10^{-34} \text{ J} \cdot \text{s/photon}}{1} \cdot \frac{2.998 \times 10^8 \text{ m/s}}{589.00 \times 10^{-9} \text{ m}} = 3.3726 \times 10^{-19} \frac{\text{J}}{\text{photon}}$$

and $\frac{6.626 \times 10^{-34} \text{ J} \cdot \text{s/photon}}{1} \cdot \frac{2.998 \times 10^8 \text{ m/s}}{589.59 \times 10^{-9} \text{ m}} = 3.3692 \times 10^{-19} \frac{\text{J}}{\text{photon}}$

The difference in energy between the two photons:

$\Delta E = (3.3726 - 3.3692) \times 10^{-19}$ J/photon or 3.4×10^{-22} J/photon

6.3.4. The energy (in kJ/mol of photons) with wavelength of 434.1 nm:

$$E = \frac{hc}{\lambda} = \frac{(6.626 \times 10^{-34} \text{ J} \cdot \text{s/photon})(2.998 \times 10^8 \text{ m/s})}{434.1 \times 10^{-9} \text{ m}} = 4.576 \times 10^{-19} \frac{\text{J}}{\text{photon}}$$

For a mole: $\left(4.576 \times 10^{-19} \frac{\text{J}}{\text{photon}}\right)\left(6.022 \times 10^{23} \frac{\text{photons}}{\text{mol}}\right) = 2.756 \times 10^5$ J/mol

or 275.6 kJ/mol

6.3.5. What are the final and initial electronic states (n) for the hydrogen line (in the Balmer series) labeled F in the figure? The line labeled F is at approximately 485 nm (4861 Å). What is certain is that the FINAL n value for this line is n = 2 (which is true for all lines in the Balmer series). Figure 6.10 in your text indicates that the line at 4861 Å results from the transition of an electron from n = 4 to n = 2.

Practicing Skills

Electromagnetic Radiation

6.1. Using Figure 6.2:

(a) Microwave radiation is less energetic than X-ray radiation.

(b) Red light has a higher frequency than radar.

(c) Infrared radiation has a longer wavelength than ultraviolet radiation.

6.2. Concerning light:

(a) Color of light with less energy than green light: red, orange, yellow
(b) Color of light with photons of greater energy: blue
(c) Color of light with higher frequency: blue

6.3. (a) The higher frequency light is the green 500. nm light.
Recall that frequency and wavelength are inversely related.

(b) The frequency of amber light (595 nm) is:

$$\text{frequency} = \frac{\text{speed of light}}{\text{wavelength}} = \frac{2.9979 \times 10^8 \text{m/s}}{595 \text{ nm}} \cdot \frac{1.00 \times 10^9 \text{nm}}{1.00 \text{ m}} = 5.04 \times 10^{14} \text{ s}^{-1}$$

The frequency of green light (500 nm) is:

$$\text{frequency} = \frac{\text{speed of light}}{\text{wavelength}} = \frac{2.9979 \times 10^8 \text{ m/s}}{500. \text{ nm}} \cdot \frac{1.00 \times 10^9 \text{nm}}{1.00 \text{ m}} = 6.00 \times 10^{14} \text{ s}^{-1}$$

As predicted, the green light has a higher frequency.

6.4. Distance from the transmitter tower:

(a) $\lambda = \dfrac{c}{\nu} = \dfrac{2.998 \times 10^8 \text{ m}\cdot\text{s}^{-1}}{1150 \times 10^3 \text{ s}^{-1}} = 261$ m $\qquad 225 \text{ m} \cdot \dfrac{1 \text{ wavelength}}{261 \text{ m}} = 0.863$ wavelengths

(b) $\lambda = \dfrac{c}{\nu} = \dfrac{2.998 \times 10^8 \text{ m}\cdot\text{s}^{-1}}{98.1 \times 10^6 \text{ s}^{-1}} = 3.06$ m $\qquad 225 \text{ m} \cdot \dfrac{1 \text{ wavelength}}{3.06 \text{ m}} = 73.6$ wavelengths

Electromagnetic Radiation and Planck's Equation

6.5. To calculate the energy of one photon of light with 500 nm wavelength, we need to first calculate the frequency of the radiation:

$$\text{frequency} = \dfrac{\text{speed of light}}{\text{wavelength}} = \dfrac{2.9979 \times 10^8 \text{ m/s}}{5.0 \times 10^2 \text{ nm}} \cdot \dfrac{1.00 \times 10^9 \text{ nm}}{1.00 \text{ m}} = 6.0 \times 10^{14} \text{ s}^{-1}$$

the energy is $E = h\nu$ or $(6.626 \times 10^{-34} \text{ J} \cdot \text{s} \cdot \text{photons}^{-1})(6.0 \times 10^{14} \text{ s}^{-1}) = 4.0 \times 10^{-19}$ J photons^{-1}

Energy of 1.00 mol of photons = 4.0×10^{-19} J/photon $\left(\dfrac{6.022 \times 10^{23} \text{ photons}}{1.00 \text{ mol photons}} \right) = 2.4 \times 10^5$ J/mol photon

6.6. Regarding violet light:

$410 \text{ nm} \cdot \dfrac{10^{-9} \text{ m}}{1 \text{ nm}} = 4.1 \times 10^{-7}$ m $\qquad \nu = \dfrac{c}{\lambda} = \dfrac{2.998 \times 10^8 \text{ m}\cdot\text{s}^{-1}}{4.1 \times 10^{-7} \text{ m}} = 7.3 \times 10^{14} \text{ s}^{-1}$

$E = h\nu = (6.626 \times 10^{-34} \text{ J}\cdot\text{s})(7.3 \times 10^{14} \text{ s}^{-1}) = 4.8 \times 10^{-19}$ J/photon

Energy of 1.00 mol of photons = 4.8×10^{-19} J/photon $\left(\dfrac{6.022 \times 10^{23} \text{ photons}}{1.00 \text{ mol photons}} \right) = 2.9 \times 10^5$ J/mol photon

According to the text, red light has an energy of 1.75×10^5 J/mol photons

$\dfrac{2.9 \times 10^5 \text{ J}}{1.75 \times 10^5 \text{ J}} = 1.7$ so violet light is 1.7 times more energetic than red light.

6.7. The frequency of the line at 396.15 nm:

$$\text{frequency} = \dfrac{\text{speed of light}}{\text{wavelength}} = \dfrac{2.9979 \times 10^8 \text{ m/s}}{3.9615 \times 10^2 \text{ nm}} \cdot \dfrac{1.00 \times 10^9 \text{ nm}}{1.00 \text{ m}} = 7.5676 \times 10^{14} \text{ s}^{-1}$$

The energy of a photon of this light: $E = h\nu$

Planck's constant, h, has a value of 6.626×10^{-34} J $\cdot$ s $\cdot$ photons^{-1}

$E = (6.6261 \times 10^{-34}$ J $\cdot$ s $\cdot$ photon$^{-1})(7.5676 \times 10^{14}$ s$^{-1}) = 5.0144 \times 10^{-19}$ J $\cdot$ photon^{-1}

Energy of 1.00 mol of photons = 5.0144×10^{-19} J $\cdot$ photon^{-1} $\cdot$ $\dfrac{6.0221 \times 10^{23} \text{ photons}}{1.00 \text{ mol photons}}$

$= 3.0197 \times 10^5$ J/mol photon or 301.97 kJ/mol photon.

6.8. Region of the electromagnetic spectrum in which lines are found:

285.2 nm is in the ultraviolet region, 383.8 nm is just at the edge of the visible region, and 518.4 nm is in the visible region. The most energetic line has the shortest wavelength, 285.2 nm.

$$285.2 \text{ nm} \cdot \frac{10^{-9} \text{ m}}{1 \text{ nm}} = 2.852 \times 10^{-7} \text{ m}$$

$$E = \frac{hc}{\lambda} = \frac{(6.6261 \times 10^{-34} \text{ J} \cdot \text{s})(2.9979 \times 10^{8} \text{ m} \cdot \text{s}^{-1})}{2.852 \times 10^{-7} \text{ m}} \cdot \frac{6.0221 \times 10^{23} \text{ photons}}{1 \text{ mol}}$$

$$= 4.194 \times 10^{5} \text{ J/mol}$$

6.9. Since energy is proportional to frequency ($E = h\nu$), we can arrange the radiation in order of increasing energy per photon by listing the types of radiation in increasing frequency (or decreasing wavelength).

→	Energy increasing	→	
FM music	microwave	yellow light	x-rays

→ Frequency (ν) increasing →

← Wavelength (λ) increasing ←

6.10. Arrange in order of increasing energy per photon:

(b) radio waves (a) microwaves (d) red light (e) ultraviolet radiation (c) gamma-rays
—increasing energy per photon →

Photoelectric Effect

6.11. Energy = 2.0×10^{2} kJ/mol $\cdot \dfrac{1 \text{ mol}}{6.0221 \times 10^{23} \text{ photons}} \cdot \dfrac{1.00 \times 10^{3} \text{ J}}{1.00 \text{ kJ}} = 3.3 \times 10^{-19}$ J/photon

What wavelength of light would provide this energy?

$E = h\nu = \dfrac{hc}{\lambda}$ and rearranging:

$$\lambda = \frac{hc}{E} = \frac{(6.626 \times 10^{-34} \text{ J} \cdot \text{s} \cdot \text{photon}^{-1})(2.9979 \times 10^{8} \text{ m/s})}{3.3 \times 10^{-19} \text{ J} \cdot \text{photon}^{-1}} = 6.0 \times 10^{-7} \text{ m or}$$

6.0×10^{2} nm. Radiation of this wavelength (**visible** region) of the electromagnetic spectrum, would appear **orange**.

6.12. Does light with wavelength of 550 nm or greater possess energy to activate a switch?

Wavelength in m: $540 \text{ nm} \cdot \dfrac{10^{-9} \text{ m}}{1 \text{ nm}} = 5.4 \times 10^{-7} \text{ m}$

$E = \dfrac{hc}{\lambda} = \dfrac{(6.626 \times 10^{-34} \text{ J} \cdot \text{s})(2.998 \times 10^8 \text{ m} \cdot \text{s}^{-1})}{5.4 \times 10^{-7} \text{ m}} = 3.7 \times 10^{-19} \text{ J/photon}$

This radiation does not have enough energy to activate the switch. This is also true for radiation with wavelengths greater than 540 nm.

6.13. Minimum energy required to remove an electron from sodium:

$E = \dfrac{hc}{\lambda} = \dfrac{(6.626 \times 10^{-34} \text{ J} \cdot \text{s})(2.998 \times 10^8 \text{ m/s})}{544 \times 10^{-9} \text{ m}} = 3.65 \times 10^{-19} \text{ J}$

Minimum energy to remove a mole of photons from sodium:

$3.65 \times 10^{-19} \text{ J/photon} \left(\dfrac{6.022 \times 10^{23} \text{ photons}}{1 \text{ mol photons}} \right) = 2.20 \times 10^5 \text{ J/mol photons}$

6.14. Minimum energy required to remove an electron from selenium:

$E = \dfrac{hc}{\lambda} = \dfrac{(6.626 \times 10^{-34} \text{ J} \cdot \text{s})(2.998 \times 10^8 \text{ m/s})}{243 \times 10^{-9} \text{ m}} = 8.17 \times 10^{-19} \text{ J}$

Minimum energy to remove a mole of photons from selenium:

$8.17 \times 10^{-19} \text{ J/photon} \left(\dfrac{6.022 \times 10^{23} \text{ photons}}{1 \text{ mol photons}} \right) = 4.92 \times 10^5 \text{ J/mol photons}$

Atomic Spectra and the Bohr Atom

6.15. (a) The **most energetic light** would be represented by the light of **shortest wavelength (253.652 nm)**.

(b) The frequency of this light is: $\dfrac{2.9979 \times 10^8 \text{ m/s}}{253.652 \text{ nm}} \cdot \dfrac{1.00 \times 10^9 \text{ nm}}{1.00 \text{ m}} = 1.18190 \times 10^{15} \text{ s}^{-1}$

The energy of 1 photon with this wavelength is:

$E = h\nu = (6.62608 \times 10^{-34} \dfrac{\text{J} \cdot \text{s}}{\text{photon}})(1.18190 \times 10^{15} \text{ s}^{-1}) = 7.83139 \times 10^{-19} \dfrac{\text{J}}{\text{photon}}$

(c) The line emission spectrum of mercury shows the visible region between ≈ 400 and 750 nm. The lines at 404 and 436 nm are present while the lines at 253 nm, 365 nm and 10^{13} nm lie outside the visible region.

6.16. Regarding the lines in the neon spectrum:

(a) The infrared region

(b) None of the lines mentioned are in the spectrum shown in Figure 6.6.

None of the lines listed are in the visible region.

(c) The most energetic line has the shortest wavelength, 837.761 nm.

(d) Wavelength in m: $865.438 \text{ nm} \cdot \dfrac{10^{-9} \text{ m}}{1 \text{ nm}} = 8.65438 \times 10^{-7} \text{ m}$

$\nu = \dfrac{c}{\lambda} = \dfrac{2.997925 \times 10^8 \text{ m} \cdot \text{s}^{-1}}{8.65438 \times 10^{-7} \text{ m}} = 3.46406 \times 10^{14} \text{ s}^{-1}$

$E = h\nu = (6.626069 \times 10^{-34} \text{ J} \cdot \text{s})(3.46406 \times 10^{14} \text{ s}^{-1}) = 2.29531 \times 10^{-19}$ J/photon

6.17. The Balmer series of lines terminates with $n_f = 2$. According to Figure 6.10, the transition originates at $n_i = 6$. Light of wavelength 410.2 nm would be violet.

6.18. Wavelength and frequency of least energetic emission line in the Lyman series:

$\lambda = 121.6$ nm or 1.216×10^{-7} m

$\nu = \dfrac{c}{\lambda} = \dfrac{2.9979 \times 10^8 \text{ m} \cdot \text{s}^{-1}}{1.216 \times 10^{-7} \text{ m}} = 2.465 \times 10^{15} \text{ s}^{-1}$ and $n_{\text{initial}} = 2$ and $n_{\text{final}} = 1$

6.19. (a) <u>Transitions from</u>   <u>to</u>

$n = 5$	$n = 4, 3, 2,$ or 1	(4 transitions)
$n = 4$	$n = 3, 2,$ or 1	(3 transitions)
$n = 3$	$n = 2$ or 1	(2 transition)
$n = 2$	$n = 1$	(1 transition)

Ten transitions are possible from these five quantum levels, providing 10 emission lines.

(b) Photons of the highest **frequency** are emitted in a transition from level of $n = 5$ to a level with $n = 1$. Recalling that energy is directly proportional to frequency ($E = h\nu$), the highest frequency will correspond to the highest energy—a transition from the two levels that differ most in energy.

(c) Emission line having the longest wavelength corresponds to a transition from level of $n = 5$ to a level with $n = 4$. Levels 4 and 5 are closer in energy than other possibilities given here, so a transition between the two would be of longest wavelength (lowest E).

6.20. Considering transitions from $n = 1$ to $n = 4$:

(a) From $n = 4$ to $n = 3, 2,$ or 1 = 3 lines
    From $n = 3$ to $n = 2$ or 1 = 2 lines
    <u>From $n = 2$ to $n = 1$</u>   <u>= 1 line</u>
    Total   = 6 lines possible

(b) Lowest energy $n = 4$ to $n = 3$
(c) Shortest wavelength (highest energy) $n = 4$ to $n = 1$

6.21. Concerning energy during electronic transitions:

(a) Photons of the lowest energy will be emitted in a transition from the level with $n = 3$ to the level $n = 2$. This is easily seen with the aid of the equation

$$\Delta E = -Rhc \left(\frac{1}{n_f^2} - \frac{1}{n_i^2} \right)$$

Since R, h, and c are constant for any transition, the smaller change in energy results if $n_f = 3$ and $n_i = 2$. (The fractions in the equation above would correspond to $(\frac{1}{4} - \frac{1}{16})$ for the $4 \rightarrow 2$ transition or $(\frac{1}{4} - \frac{1}{9})$ for the $3 \rightarrow 2$ transition.

(b) Once again, using the equation above, the fraction in parenthesis (and hence ΔE) will be greater for the transition from $n_i = 4$ to $n_f = 1$ than for the transition from $5 \rightarrow 2$.

6.22. Transitions of longer wavelength than 102.6 nm: (a) $n = 2$ to $n = 4$ and (d) $n = 3$ to $n = 5$.

6.23. The wavelength of emitted light for the transition $n = 3$ to $n = 1$.

$\Delta E = -Rhc(\frac{1}{1^2} - \frac{1}{3^2})$ and the value of $Rhc = 1312$ kJ/mol, so

$\Delta E = -1312$ kJ/mol $(\frac{1}{1^2} - \frac{1}{3^2})$ or $= -1312$ kJ/mol $(\frac{8}{9}) = -1166$ kJ/mol

To calculate the frequency and wavelength, use $E = h\nu$. First express the energy **per photon** (as opposed to a mole of photons)

$\Delta E = \dfrac{-1166 \text{ kJ/mol photons}}{6.022 \times 10^{23} \text{ photons/1 mol photons}} \cdot \dfrac{10^3 \text{J}}{1 \text{ kJ}} = 1.936 \times 10^{-18}$ J/photon

and then solve for frequency, $\nu = \dfrac{1.936 \times 10^{-18} \text{J/photon}}{6.626 \times 10^{-34} \text{ J} \cdot \text{s/photon}} = 2.923 \times 10^{15}$ s^{-1}

Substituting into $\lambda\nu = c$: $\dfrac{2.998 \times 10^8 \text{ m}}{2.923 \times 10^{15} \text{ s}^{-1}} = 1.0257 \times 10^{-7}$m (102.6 nm—far UV)

6.24. Wavelength and frequency of light for the electronic transition of $n = 4$ to $n = 3$:

$$\Delta E = -Rhc\left(\frac{1}{n_{final}^2} - \frac{1}{n_{initial}^2}\right) = -1312 \text{ kJ/mol}\left(\frac{1}{3^2} - \frac{1}{4^2}\right) = -63.78 \text{ kJ/mol}$$

$$63.78 \text{ kJ/mol} \cdot \frac{1 \text{ mol}}{6.0221 \times 10^{23} \text{ photons}} \cdot \frac{10^3 \text{ J}}{1 \text{ kJ}} = 1.059 \times 10^{-19} \text{ J/photon}$$

$$\nu = \frac{E}{h} = \frac{1.059 \times 10^{-19} \text{ J}}{6.6261 \times 10^{-34} \text{ J} \cdot \text{s}} = 1.598 \times 10^{14} \text{ s}^{-1}$$

$$\lambda = \frac{c}{\nu} = \frac{2.9979 \times 10^8 \text{ m} \cdot \text{s}^{-1}}{1.598 \times 10^{14} \text{ s}^{-1}} = 1.876 \times 10^{-6} \text{ m (infrared region)}$$

De Broglie and Matter Waves

6.25. Mass of an electron: 9.11×10^{-31} kg

Planck's constant: 6.626×10^{-34} J · s · photon^{-1}

Velocity of the electron: 2.5×10^8 cm · s^{-1} or 2.5×10^6 m · s^{-1}

$$\lambda = \frac{h}{mv} = \frac{6.626 \times 10^{-34} \text{ J} \cdot \text{s}}{(9.11 \times 10^{-31} \text{ kg} \cdot 2.5 \times 10^6 \text{ m} \cdot \text{s}^{-1})}$$

$$= 2.9 \times 10^{-10} \text{ m} = 2.9 \text{ Angstroms} = 0.29 \text{ nm}$$

6.26. Wavelength of electrons with a speed of 1.3×10^8 m/s:

$$\lambda = \frac{h}{mv} = \frac{6.626 \times 10^{-34} \text{ J} \cdot \text{s}}{(9.11 \times 10^{-31} \text{ kg} \cdot 1.3 \times 10^8 \text{ m} \cdot \text{s}^{-1})} = 5.6 \times 10^{-12} \text{ m}$$

6.27. The wavelength can be determined exactly as in Question 23:

$$\lambda = \frac{h}{mv} = \frac{6.626 \times 10^{-34} \text{ J} \cdot \text{s}}{(0.046 \text{ kg})(30. \text{ m} \cdot \text{s}^{-1})} = 4.8 \times 10^{-34} \text{ m or } 4.8 \times 10^{-25} \text{ nm}$$

Velocity to have a wavelength of 5.6×10^{-3} nm:
Convert the wavelength to units of meters: 5.6×10^{-3} nm $\cdot \frac{1 \text{ m}}{1 \times 10^9 \text{ nm}} = 5.6 \times 10^{-12}$ m

Then rewriting the above equation to solve for v:

$$v = \frac{h}{m\lambda} = \frac{6.626 \times 10^{-34} \text{ J} \cdot \text{s}}{0.046 \text{ kg} \cdot 5.6 \times 10^{-12} \text{ m}} = 2.6 \times 10^{-21} \frac{\text{m}}{\text{s}}$$

6.28. Wavelength associated with a bullet:

$$\frac{7.00 \times 10^2 \text{ mile}}{1 \text{ hour}} \cdot \frac{1 \text{ km}}{0.6214 \text{ mile}} \cdot \frac{10^3 \text{ m}}{1 \text{ km}} \cdot \frac{1 \text{ hour}}{3600 \text{ s}} = 313 \text{ m·s}^{-1}$$

$$\lambda = \frac{h}{mv} = \frac{6.626 \times 10^{-34} \text{ J·s}}{(1.50 \times 10^{-3} \text{ kg})(313 \text{ m·s}^{-1})} = 1.41 \times 10^{-33} \text{ m}$$

6.29. Wavelength of protons in Large Hadron Collider (99.9999% the speed of light):

Mass of a proton: 1.673×10^{-27} kg

Planck's constant: 6.626×10^{-34} J · s · photon^{-1}

Velocity of the proton: 99.9999% × 2.998 × 10^8 m · s^{-1} = 2.998 × 10^8 m · s^{-1}

$$\lambda = \frac{h}{mv} = \frac{6.626 \times 10^{-34} \text{ J·s}}{(1.673 \times 10^{-27} \text{ kg})(2.998 \times 10^8 \text{ m/s})} = 1.321 \times 10^{-15} \text{ m or } 1.321 \times 10^{-6} \text{ nm}$$

6.30. Wavelength of protons in Large Hadron Collider (50.0% the speed of light):

Mass of a proton: 1.673×10^{-27} kg

Planck's constant: 6.626×10^{-34} J · s · photon^{-1}

Velocity of the proton: 50.0% × 2.998 × 10^8 m · s^{-1} = 1.50 × 10^8 m · s^{-1}

$$\lambda = \frac{h}{mv} = \frac{6.626 \times 10^{-34} \text{ J·s}}{(1.673 \times 10^{-27} \text{ kg})(1.50 \times 10^8 \text{ m/s})} = 2.64 \times 10^{-15} \text{ m}$$

Quantum Mechanics

6.31. (a) $n = 4$ possible ℓ values = 0, 1, 2, 3 ($\ell = 0, 1,... (n-1)$)

(b) $\ell = 2$ possible m_ℓ values = −2, −1, 0, +1, +2 ($-\ell$ ..., 0,....+ ℓ)

(c) orbital = 4s possible values for $n = 4$; $\ell = 0$; $m_\ell = 0$

(d) orbital = 4f possible values for $n = 4$; $\ell = 3$; m_ℓ = −3, −2, −1, 0, +1, +2, +3

6.32. Regarding orbitals:

(a) The orbital type is a 4d orbital. $m_\ell = -1$ could be any one of the five d orbitals.

(b) When $n = 5$, $\ell = 0, 1, 2, 3$, and 4

$\ell = 0$ 1 s orbital

$\ell = 1$ 3 p orbitals

$\ell = 2$ 5 d orbitals

$\ell = 3$ 7 f orbitals

$\ell = 4$ 9 g orbitals

There are a total of 25 orbitals in the $n = 5$ electron shell.

(c) In an f subshell there are 7 orbitals: $m_\ell = 0, \pm 1, \pm 2,$ and ± 3

6.33. An electron in a 4p orbital must have $n = 4$ and $\ell = 1$. The possible m_ℓ values give rise to the following sets of n, ℓ, and m_ℓ

n	ℓ	m_ℓ
4	1	−1
4	1	0
4	1	+1

Note that the **three values** of m_ℓ describe **three orbital orientations**.

6.34. Possible sets of quantum numbers for electron in 5d orbital:

n	ℓ	m_ℓ
5	2	−2
5	2	−1
5	2	0
5	2	1
5	2	2

6.35. Subshells in the electron shell with $n = 4$:

There are 4: s, p, d, and f sublevels corresponding to $\ell = 0, 1, 2,$ and 3 respectively.

Recall that values of ℓ from 0 to a maximum of $(n − 1)$ are possible.

6.36. Number of subshells within the shell $n = 5$:

When $n = 5$, there are five subshells: 5s, 5p, 5d, 5f, and 5g.

6.37. Explain why each of the following is not a possible set of quantum numbers for an electron in an atom.

(a) $n = 2$, $\ell = 2$, $m_\ell = 0$ For $n = 2$, the maximum value of ℓ is one (1).

(b) $n = 3$, $\ell = 0$, $m_\ell = -2$ For $\ell = 0$, the only possible value of m_ℓ is 0.

(c) $n = 6$, $\ell = 0$, $m_\ell = 1$ For $\ell = 0$, the only possible value of m_ℓ is 0.

6.38. Explain why each of the following is not a possible set of quantum numbers for an electron in an atom.

(a) incorrect; When $n = 3$, the maximum value of ℓ is 2.
(b) valid
(c) valid
(d) incorrect; When $\ell = 3$, m_ℓ can only have values of 0, ±1, ±2, or ±3.

6.39. Quantum number designation   Maximum number of orbitals

(a) $n = 3$; $\ell = 0$; $m_\ell = +1$ none; for $\ell = 0$, the only possible value of $m_\ell = 0$

(b) $n = 5$; $\ell = 1$ 3 (p orbitals)

(c) $n = 7$; $\ell = 5$ eleven; the number of orbitals is "2ℓ +1"

(d) $n = 4$; $\ell = 2$; $m_\ell = -2$ 1 (one of the three 4p orbitals)

6.40. Maximum number of orbitals identified:

(a) 7 orbitals
(b) 25 orbitals
(c) None; ℓ cannot have a value equal to n (maximum value is n−1)
(d) 1 orbital

6.41. Explain why the following sets of quantum numbers are not valid:

(a) $n = 4$, $\ell = 2$, $m_\ell = 0$, $m_S = 0$:

The possible values of m_S can only be +1/2 or −1/2. Change m_S to either +1/2 or −1/2.

(b) $n = 3$, $\ell = 1$, $m_\ell = -3$, $m_S = -1/2$:

The possible values for m_ℓ are −ℓ......0.....+ℓ. Changing m_ℓ to −1, 0, +1 would give a valid set of quantum numbers.

(c) $n = 3$, $\ell = 3$, $m_\ell = -1$, $m_s = +1/2$

The maximum value of ℓ is $(n - 1)$. Changing ℓ to 2 would provide a valid set of quantum numbers.

6.42. Explain why the following sets of quantum numbers are not valid:

(a) the maximum value of ℓ is $(n - 1)$; make $\ell = 1$
(b) m_s can only have values of $\pm 1/2$; change m_s to $+1/2$
(c) when $\ell = 1$, m_ℓ can only have values of $0, \pm 1$ make $m_\ell = 0$

6.43. Which of the following orbitals cannot exist and why:

2s exists	$n = 2$	permits ℓ values as large as 1 ($\ell = 0$ is an s sublevel)
2d cannot exist	$\ell = 2$	is not permitted for n < 3 ($\ell = 2$ is a d sublevel)
3p exists	$n = 3$	permits ℓ values as large as 2 ($\ell = 1$ is a p sublevel)
3f cannot exist	$\ell = 3$	is not permitted for n < 4 ($\ell = 3$ is an f sublevel)
4f exists	$\ell = 4$	permits ℓ values as large as 3 ($\ell = 3$ is an f sublevel)
5s exists	$n = 5$	permits ℓ values as large as 4 ($\ell = 0$ is an s sublevel)

6.44. Which of the following orbitals cannot exist and why:

3p can exist

4s can exist

2f does not follow the $\ell = 0, 1, 2, \ldots, n - 1$ rule

1p does not follow the $\ell = 0, 1, 2, \ldots, n - 1$ rule

6.45. The complete set of quantum numbers for:

	n	ℓ	m_ℓ	
(a) 2p	2	1	−1, 0, +1	(3 orbitals)
(b) 3d	3	2	−2, −1, 0, +1, +2	(5 orbitals)
(c) 4f	4	3	−3, −2, −1, 0, +1, +2, +3	(7 orbitals)

6.46. The complete set of quantum numbers for:

	n	ℓ	m_ℓ	
(a) 5f	5	3	−3, −2, −1, 0, +1, +2, +3	(7 orbitals)
(b) 4d	4	2	−2, −1, 0, +1, +2	(5 orbitals)
(c) 2s	2	1	0	(1 orbital)

6.47. With an $n = 4$, and $\ell = 2$, this orbital belongs in the 4th level ($n = 4$) and with an $\ell = 2$, this must be a *d* type orbital—so (d) 4d.

6.48. Which orbital could not have a magnetic quantum number of −1?

(d) *s* orbital

6.49. The number of nodal surfaces possessed by an orbital is equal to the value of the "ℓ" quantum number, so for each of the following:

orbital	number of planar nodes
(a) 2s ($\ell = 0$)	0
(b) 5d ($\ell = 2$)	2
(c) 5f ($\ell = 3$)	3

6.50. The number of nodal surfaces possessed by an orbital is equal to the value of the "ℓ" quantum number, so for each of the following:

orbital	number of planar nodes
(a) 4f: ($\ell = 3$)	3
(b) 2p: ($\ell = 1$)	1
(c) 6s: ($\ell = 0$)	0

General Questions

6.51. Concerning the photoelectric effect:

(a) Light is electromagnetic radiation—correct.

(b) Intensity of light beam related to frequency—incorrect.

The intensity is related to the number of photons of light.

(c) Light can be thought of as massless particles—correct.

6.52. Region of the Lyman series? Balmer series?

The Lyman series is in the ultraviolet region.

The Balmer series is in the visible region.

6.53. Number of nodal surfaces for the following orbital types:

Orbital type	Nodal surfaces
s	0 (because $\ell = 0$)
p	1 (because $\ell = 1$)
d	2 (because $\ell = 2$)
f	3 (because $\ell = 3$)

6.54. Maximum s, p, d, and f orbitals in a given shell:

Orbital	maximum number
s	1
p	3
d	5
f	7

6.55. Orbital types associated with values of "ℓ"

Orbital type	Values of "ℓ"
f	$\ell = 3$
s	$\ell = 0$
p	$\ell = 1$
d	$\ell = 2$

6.56. Boundary surfaces of s and p_x:

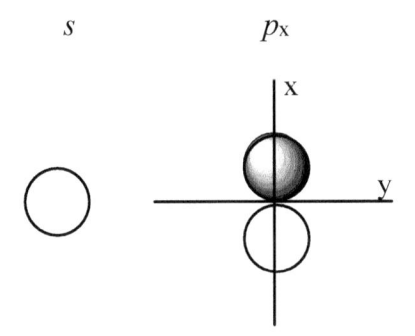

6.57.

Orbital Type	Number of orbitals in a given Subshell	Number of Nodal Surfaces
s	1	0
p	3	1
d	5	2
f	7	3

6.58. Wavelength and frequency of lowest energy line in the Pfund series:

$$\Delta E = -Rhc\left(\frac{1}{n_{final}^2} - \frac{1}{n_{initial}^2}\right) = -1312 \text{ kJ/mol}\left(\frac{1}{5^2} - \frac{1}{6^2}\right) = -16.04 \text{ kJ/mol}$$

$$16.04 \text{ kJ/mol} \cdot \frac{1 \text{ mol}}{6.0221 \times 10^{23} \text{ photons}} \cdot \frac{10^3 \text{ J}}{1 \text{ kJ}} = 2.663 \times 10^{-20} \text{ J/photon}$$

$$\nu = \frac{E}{h} = \frac{2.663 \times 10^{-20} \text{ J}}{6.6261 \times 10^{-34} \text{ J} \cdot \text{s}} = 4.019 \times 10^{13} \text{ s}^{-1}$$

$$\lambda = \frac{c}{\nu} = \frac{2.9979 \times 10^8 \text{ m} \cdot \text{s}^{-1}}{4.019 \times 10^{13} \text{ s}^{-1}} = 7.460 \times 10^{-6} \text{ m}$$

6.59. Concerning red and green light:

(a) Green light has the shorter wavelength and therefore higher energy photons.

(b) Green light has higher energy photons than red light, so the shorter wavelength (500 nm) is green and the 680 nm light is red.

(c) Since frequency and wavelength are inversely related, the shorter wavelength (green) must have the higher frequency.

6.60. Amount of energy from 1.00 mol photons with wavelength of 375 nm:

$$375 \text{ nm} \cdot \frac{10^{-9} \text{ m}}{1 \text{ nm}} = 3.75 \times 10^{-7} \text{ m}$$

$$E = \frac{hc}{\lambda} = \frac{(6.626 \times 10^{-34} \text{ J} \cdot \text{s})(2.998 \times 10^8 \text{ m} \cdot \text{s}^{-1})}{3.75 \times 10^{-7} \text{ m}} \cdot \frac{6.022 \times 10^{23} \text{ photons}}{1.00 \text{ mol}} \cdot \frac{1 \text{ kJ}}{10^3 \text{ J}}$$

$$= 319 \text{ kJ/mol}$$

6.61. For radiation of 850 MHz:

(a) Convert frequency to wavelength.

$$\lambda = \frac{c}{\nu} = \frac{2.998 \times 10^8 \text{ m/s}}{850 \times 10^6 \text{ s}^{-1}} = 0.353 \text{ m} = 0.35 \text{ m (2 s.f.)}$$

(b) The energy of 1.0 mol of photons with ν = 850 MHz:

$$E = N_A h\nu = (6.022 \times 10^{23} \text{ photons/mol})(6.626 \times 10^{-34} \text{ J·s·photon}^{-1})(850 \times 10^6 \text{ s}^{-1})$$

$$= 0.339 \text{ J/mol} = 0.34 \text{ J/mol (2 s.f.)}$$

(c) The energy of a mole of photons of 420 nm light:

$$E = N_A \frac{hc}{\lambda} = (6.022 \times 10^{23} \text{ photons/mol}) \frac{(6.626 \times 10^{-34} \text{ J·s·photon}^{-1})(2.998 \times 10^8 \text{ m/s})}{420 \times 10^{-9} \text{ m}}$$

$$= 2.85 \times 10^5 \text{ J/mol} = 2.8 \times 10^5 \text{ J/mol or 280 kJ/mol (2 s.f.)}$$

(d) The energy of a mole of photons of violet light is 840,000 times greater than that of the corresponding photons from a cell phone.

$$\frac{2.85 \times 10^5 \text{ J/mol}}{0.339 \text{ J/mol}} = 8.4 \times 10^5$$

6.62. Number of photons of blue light (λ = 470 nm) reaching your eyes:

$$E = \frac{hc}{\lambda} = \frac{(6.626 \times 10^{-34} \text{ J·s})(2.998 \times 10^8 \text{ m·s}^{-1})}{4.7 \times 10^{-7} \text{ m}} = 4.2 \times 10^{-19} \text{ J/photon}$$

$$\frac{2.50 \times 10^{-14} \text{ J}}{4.2 \times 10^{-19} \text{ J/photon}} = 5.9 \times 10^4 \text{ photons}$$

6.63. Calculate ionization energy for He$^+$ ion:

Example 6.3 in the text illustrates the calculation of the ionization energy for H's electron

$$E = -\frac{Z^2 Rhc}{n^2} = -2.179 \times 10^{-18} \text{ J/atom} \Rightarrow -1312 \text{ kJ/mol}$$

For He$^+$ the calculation yields

$$E = \frac{-(2)^2 (1.097 \times 10^7 \text{ m}^{-1})(6.626 \times 10^{-34} \text{ J·s})(2.998 \times 10^8 \text{ m·s}^{-1})}{(1)^2} = -8.717 \times 10^{-18} \text{ J/ion}$$

and expressing this energy for a mol of He$^+$ ions

$$= \frac{-8.717 \times 10^{-18} \text{ J}}{\text{ion}} \cdot \frac{6.0221 \times 10^{23} \text{ atoms}}{\text{mol}} \cdot \frac{1 \text{ kJ}}{1000 \text{ J}} = -5248 \frac{\text{kJ}}{\text{mol}}$$

The energy to remove the electron is 5248 kJ/mol of ions.

Note that this energy is four times that for H.

6.64. Concerning H atom transitions from $n = 7$ level:

Transition producing photon with:

(i) smallest energy: (b) $n = 7$ to $n = 6$

(ii) highest frequency: (a) $n = 7$ to $n = 1$

(iii) shortest wavelength: (a) $n = 7$ to $n = 1$

6.65. Orbitals in a H atom in order of increasing energy: $1s < 2s = 2p < 3s = 3p = 3d < 4s$

For the hydrogen atom, energy levels increase with increasing values of "n".

6.66. Number of orbitals corresponding to the designations:

(a) $3p$: 3 orbitals
(b) $4p$: 3 orbitals
(c) $4p_x$: 1 orbital
(d) $6d$: 5 orbitals
(e) $5d$: 5 orbitals
(f) $5f$: 7 orbitals
(g) $n = 5$: 25 orbitals
(h) $7s$: 1 orbital

6.67. First, convert energy from MeV to joules.

$$E(J) = 1.173 \times 10^6 \text{ eV} \left(\frac{1.6022 \times 10^{-19} \text{ J}}{1 \text{ eV}} \right) = 1.8794 \times 10^{-13} \text{ J}$$

Using the energy relationship: $E = h\nu$, solve for frequency:

$$\nu = \frac{E}{h} = \left(\frac{1.8794 \times 10^{-13} \text{ J}}{6.626 \times 10^{-34} \text{ J} \cdot \text{s}} \right) = 2.836 \times 10^{20} \text{ s}^{-1}$$

Substituting into the wavelength-frequency relationship, we have:

$$\lambda = \frac{c}{\nu} = \left(\frac{2.998 \times 10^8 \text{ m/s}}{2.836 \times 10^{20} \text{ s}^{-1}} \right) = 1.057 \times 10^{-12} \text{ m} = 1.057 \times 10^{-3} \text{ nm}$$

6.68. Photons with $\lambda = 12$ cm to raise temperature of eye by 3 °C :

To raise T of eye by 3 °C: $q_{eye} = (11 \text{ g})(4.0 \text{ J/g·K})(3.0 \text{ K}) = 130$ J

$$E = \frac{hc}{\lambda} = \frac{(6.626 \times 10^{-34} \text{ J} \cdot \text{s})(2.998 \times 10^8 \text{ m} \cdot \text{s}^{-1})}{0.12 \text{ m}} = 1.7 \times 10^{-24} \text{ J/photon}$$

$$\frac{130 \text{ J}}{1.7 \times 10^{-24} \text{ J/photon}} = 8.0 \times 10^{25} \text{ photons}$$

6.69. Time for Webb telescope's signal to travel to Earth:

Since light travels at 2.9979×10^8 m · s^{-1}, you can calculate the time:

$$(1.5 \times 10^6 \text{ km})\left(\frac{1 \times 10^3 \text{ m}}{1 \text{ km}}\right)\left(\frac{1 \text{ s}}{2.9979 \times 10^8 \text{ m}}\right) = 5.0 \text{ s}$$

6.70. Concerning lines in the chromium spectrum:

(a) The most energetic line has the shortest wavelength, 357.9 nm.
(b) The color of light with λ = 425.4 nm is blue-indigo.

6.71. Questions to summarize concepts in the chapter:

(a) The quantum number n describes the **size (and energy)** of an atomic orbital.

(b) The shape of an atomic orbital is given by the quantum number ℓ.

(c) A photon of green light has **more energy** than a photon of orange light.

(d) The maximum number of orbitals that may be associated with the quantum numbers n = **4**, ℓ = **3** is **seven.** (Corresponding to m_ℓ values of ± 3, ± 2, ± 1, and 0.)

(e) The maximum number of orbitals that may be associated with the quantum numbers n = **3,** ℓ = **2,** and m_ℓ = −2 is **one**.

(f) The orbital on the left is a **d orbital** and the one on the right is a **p orbital**, while the orbital in the middle is an **s orbital**.

(g) When n = 5, the possible values of ℓ **are 0, 1, 2, 3, and 4**. (Range is 0 ... (n − 1))

(h) The maximum number of orbitals that can be assigned to the n = 4 shell is **16**.

n = 4	ℓ = 0	m_ℓ = 0	1 orbital
n = 4	ℓ = 1	m_ℓ = −1, 0, +1	3 orbitals
n = 4	ℓ = 2	m_ℓ = −2, −1, 0, +1, +2	5 orbitals
n = 4	ℓ = 3	m_ℓ = −3, −2, −1, 0, +1, +2, +3	<u>7 orbitals</u>
			16 orbitals

6.72. Questions to summarize concepts in the chapter:

(a) Quantum number n describes **size and energy**; ℓ describes the **shape**

(b) Possible ℓ values when n = 3: 0, 1, 2

(c) Type of orbital corresponding to ℓ = 3: f

(d) Value of n for a 4d orbital = 4; ℓ value = 2; possible m_ℓ value = –2

(e)   letter p d
      ℓ value   1   2
      nodal planes   1   2

(f) Orbital with 3 planar nodes: f

(g) Orbitals that cannot exist according to quantum theory: 2d, 3f

(h) Sets with invalid quantum numbers:

$n = 2, \ell = 1, m_\ell = 2, m_s = +\frac{1}{2}$ is not valid

$n = 4, \ell = 3, m_\ell = 0, m_s = 0$ is not valid

(i) maximum number of orbitals associated with:

(i) $n = 2$ and $\ell = 1$: **3** (the 2p orbitals)

(ii) $n = 3$: **9** (the s, three p, and five d orbitals)

(iii) $n = 3$ and $\ell = 3$:   **none** (max value of ℓ is n−1)

(iv) $n = 2, \ell = 1$ and $m_\ell = 0$:   **1**

6.73. Energy of photon emitted in transition from n = 5 to n = 2:

$$\Delta E = -Rhc\left(\frac{1}{2^2} - \frac{1}{5^2}\right)$$ and the value of Rhc = 1312 kJ/mol, so

$$\Delta E = -1312 \text{ kJ/mol}\left(\frac{1}{4} - \frac{1}{25}\right) \text{ or } = -1312 \text{ kJ/mol } (0.21) = -275.5 \text{ kJ/mol}$$

To calculate the frequency and wavelength, we use $E = h\nu$. Recall that you must first express the energy **per photon** (as opposed to a mole of photons).

$$\Delta E = \frac{-275.5 \text{ kJ/mol photons}}{6.022 \times 10^{23} \text{ photons/1 mol photons}} \cdot \frac{10^3 \text{J}}{1 \text{ kJ}} = 4.576 \times 10^{-19} \text{J/photon}$$

Solve for frequency: $\nu = \dfrac{4.576 \times 10^{-19} \text{J/photon}}{6.626 \times 10^{-34} \text{ J} \cdot \text{s /photon}} = 6.906 \times 10^{14} \text{ s}^{-1}$

Substituting into $\lambda\nu = c$, you get $\dfrac{2.998 \times 10^8 \text{ m}}{6.906 \times 10^{14} \text{ s}^{-1}} = 4.341 \times 10^{-7}$ m (or 434.1 nm)

In the Laboratory

6.74. Frequency of light of λ = 540 nm; energy of one mol of photons with this wavelength:

$$540 \text{ nm} \cdot \frac{10^{-9} \text{ m}}{1 \text{ nm}} = 5.4 \times 10^{-7} \text{ m} \qquad \nu = \frac{c}{\lambda} = \frac{2.998 \times 10^8 \text{ m} \cdot \text{s}^{-1}}{5.4 \times 10^{-7} \text{ m}} = 5.6 \times 10^{14} \text{ s}^{-1}$$

$$E = h\nu = (6.626 \times 10^{-34} \text{ J} \cdot \text{s})(5.6 \times 10^{14} \text{ s}^{-1}) \frac{6.022 \times 10^{23} \text{ photons}}{1.00 \text{ mol photons}} = 2.2 \times 10^5 \text{ J/mol photons}$$

6.75. The pickle glows since the materials in the pickle are being "excited" by the addition of the energy (electric current). Since the pickle has been soaked in brine (NaCl), the electrons of sodium are excited and release energy as they "return" to lower energy states, providing "yellow" light. The same color of light is visible in many streetlamps.

6.76. Concerning the spectrum for aspirin:

Wavelength of light in m: $278 \text{ nm} \cdot \frac{10^{-9} \text{ m}}{1 \text{ nm}} = 2.78 \times 10^{-7} \text{ m}$

Frequency is: $\nu = \frac{c}{\lambda} = \frac{2.998 \times 10^8 \text{ m} \cdot \text{s}^{-1}}{2.78 \times 10^{-7} \text{ m}} = 1.08 \times 10^{15} \text{ s}^{-1}$

Energy: $E = h\nu = (6.626 \times 10^{-34} \text{ J} \cdot \text{s})(1.08 \times 10^{15} \text{ s}^{-1}) \frac{6.022 \times 10^{23} \text{ photons}}{1.00 \text{ mol photons}}$

$$= 4.30 \times 10^5 \text{ J/mol photons}$$

The region is: ultraviolet. Aspirin is colorless, as it does not absorb visible light.

6.77. Concerning the infrared spectrum for methanol:

(a) The wavelength of light at 2000 cm⁻¹? λ = 1/2000 cm⁻¹, or 0.0005 cm

(b) Which is the low energy and high energy end of the spectrum?

Since E is proportional to the reciprocal wavelength, the larger the value of the wavenumber, the greater the energy. The low energy is the **right side** while the higher energy end is the **left side** of the spectrum.

(c) Which interaction (3300-3400 cm⁻¹) or (2800-3000 cm⁻¹) requires more energy?
Since E is proportional to wavenumber, the 3300-3400 cm⁻¹ for the O—H absorption requires more energy.

Summary and Conceptual Questions

6.78. Criticize the Bohr model of the atom: Bohr's circular orbit model contradicts the laws of classical physics, and Bohr had to artificially introduce the concept of quantization to explain how these electron orbits could be stable.

6.79. The light visible from a sodium or mercury streetlight arises owing to (c) electrons are moving from a given energy level to one of lower n (and hence of lower energy).

6.80. Interpretation of the wavefunction squared? Units for $4\pi r^2\psi^2$?

The square of the wave function is the probability of finding the electron within a given region of space, also known as the electron density. This region of space where an electron of a given energy is most probably located is its orbital. The units for $4\pi r^2\psi^2$ are 1/distance.

6.81. "Wave-particle duality" refers to the different ways of describing the behavior of electrons: the particulate nature of the electron (e.g., as observed in the photoelectric effect) and the wave-like properties of the electron (e.g., the observation of diffraction patterns in the Davisson-Germer experiment). Both are important to our explanation of natural phenomena. The implications of this duality arise in our understanding of electrons in atoms as waves existing in regions of high probability.

6.82. Which of the following are observable?

(b) frequency of radiation emitted by H atoms

(e) diffraction patterns produced by electrons

(f) diffraction patterns produced by light

(g) energy required to remove an electron from an H atom

(h) an atom

(i) a molecule

(j) a water wave

6.83. The Heisenberg Uncertainty Principle answers this question for us. This principle says that it is impossible to **simultaneously determine** both the position and energy of an electron. Reducing the uncertainty about one of these parameters increases the uncertainty of the measurement of the other parameter. So we can determine (a) OR (b), but not (c).

6.84. Number of orbitals in the first three electron shells:

N = 1, L = 1, M = –1, 0, +1   3 orbitals
N = 2, L = 2, M = –1, 0, +1   3 orbitals
N = 3, L = 3, M = –1, 0, +1   3 orbitals
A total of 9 orbitals in the first three electron shells

6.85. A photon with a wavelength of 93.8 nm can excite the electron up to level $n = 6$ (Figure 6.10). Transitions from $n = 6$ to lower values of n are possible.

Transitions from level 6 → 5, 4, 3, 2, 1. **5** transitions;
from level 5 → 4, 3, 2, 1. **4** transitions;
from level 4 → 3, 2, 1. **3** transitions;
from level 3 → 2, 1. **2** transitions;
from level 2 → 1. **1** transition.
Total transitions: 5 + 4 + 3 + 2 + 1 = 15 transitions.

6.86. Possible to measure the wavelength of a golf ball in flight?

According to de Broglie's equation, $\lambda = \dfrac{h}{mv}$, any moving particle has an associated wavelength. However, a heavy particle such as a golf ball has an incredibly small wavelength that cannot be measured with any instrument now available.

6.87. Regarding Technetium:

(a) Technetium is in group 7B (also known as Group 7; IUPAC) of the fifth period.

(b) Quantum numbers for an electron in the 5s subshell: $n = 5$, $\ell = 0$, $m_\ell = 0$

(c) First, convert energy from MeV to joules.

$$E(J) = 0.141 \times 10^6 \text{ eV} \left(\dfrac{1.6022 \times 10^{-19} \text{ J}}{1 \text{ eV}} \right) = 2.259 \times 10^{-14} \text{ J}$$

Using $E = h\nu$, solve for frequency:

$$\nu = \dfrac{E}{h} = \dfrac{2.259 \times 10^{-14} \text{ J}}{6.636 \times 10^{-34} \text{ J} \cdot \text{s}} = 3.409 \times 10^{19} \text{ s}^{-1} = 3.41 \times 10^{19} \text{ s}^{-1} \text{ (2 s.f.)}$$

Substituting into the wavelength-frequency relationship, we have:

$$\lambda = \dfrac{c}{\nu} = \left(\dfrac{2.998 \times 10^8 \text{ m/s}}{3.409 \times 10^{19} \text{ s}^{-1}} \right) = 8.79 \times 10^{-12} \text{ m} = 8.79 \times 10^{-3} \text{ nm}$$

(d) In the preparation of NaTcO₄:

(i) The balanced equation: HTcO₄(aq) + NaOH(aq) → NaTcO₄(aq) + H₂O(ℓ)

(ii) Mass of NaTcO₄ from 4.5 mg of ^{99m}Tc:

$$4.5 \times 10^{-3} \text{ g } ^{99m}\text{Tc} \cdot \frac{1 \text{ mol } ^{99m}\text{Tc}}{99 \text{ g } ^{99m}\text{Tc}} \cdot \frac{1 \text{ mol Na}^{99m}\text{TcO}_4}{1 \text{ mol } ^{99m}\text{Tc}} \cdot \frac{186 \text{ g Na}^{99m}\text{TcO}_4}{1 \text{ mol Na}^{99m}\text{TcO}_4}$$

$$= 8.5 \times 10^{-3} \text{ g Na}^{99m}\text{TcO}_4$$

Mass of NaOH required:

$$4.5 \times 10^{-3} \text{ g } ^{99m}\text{Tc} \cdot \frac{1 \text{ mol } ^{99m}\text{Tc}}{99 \text{ g } ^{99m}\text{Tc}} \cdot \frac{1 \text{ mol H}^{99m}\text{TcO}_4}{1 \text{ mol } ^{99m}\text{Tc}} \cdot \frac{1 \text{ mol NaOH}}{1 \text{ mol H}^{99m}\text{TcO}_4} \cdot \frac{40.0 \text{ g NaOH}}{1 \text{ mol NaOH}}$$

$$= 1.8 \times 10^{-3} \text{ g NaOH}$$

(e) What mass of NaTcO₄ corresponds to 1.5 micromole?

$$1.5 \times 10^{-6} \text{ mol Na}^{99m}\text{TcO}_4 \cdot \frac{186 \text{ g Na}^{99m}\text{TcO}_4}{1 \text{ mol Na}^{99m}\text{TcO}_4} = 0.00028 \text{ g or } 0.28 \text{ mg Na}^{99m}\text{TcO}_4$$

The concentration would be: 1.5×10^{-6} mol NaTcO₄/0.010 L or 1.5×10^{-4} M.

6.88. Regarding the probability of finding a H 1s electron:

(a) Probability $= \dfrac{4r^2 \left(e^{-r/a_o}\right)^2 (d)}{a_o^3}$

For $r = 1\ a_0 = 52.9$ pm and $d = 1.0$ pm,

Probability $= \dfrac{4(1)^2 \left(e^{-1}\right)^2 (1.0 \text{ pm})}{52.9 \text{ pm}} = 0.010$

(b) For $r = 0.50\ a_0$ and $d = 1.0$ pm

Probability $= \dfrac{4(0.50)^2 \left(e^{-0.50}\right)^2 (1.0 \text{ pm})}{52.9 \text{ pm}} = 0.0070$

For $r = 4\ a_0$ and $d = 1.0$ pm

Probability $= \dfrac{4(4)^2 \left(e^{-4}\right)^2 (1.0 \text{ pm})}{52.9 \text{ pm}} = 0.00041$

The probabilities are in accord with the surface density plot shown in Figure 6.12b.

6.89. Photoelectric effect:

(a) Slope of the line

$$\text{slope} = \frac{\text{rise}}{\text{run}} = \frac{(2 \text{ eV} - 1 \text{ eV})\left(\frac{1.602 \times 10^{-19} \text{ J}}{1 \text{ eV}}\right)}{(9.2 - 7.0) \times 10^{14} \frac{1}{\text{s}}} \approx 7 \times 10^{-34} \text{ J} \cdot \text{s}$$

The slope approximates Planck's constant indicating that the emitted electrons are quantized.

(b) Note that below 683 nm there are no emitted electrons. Once the light energy exceeds the amount necessary to remove an electron from sodium photoelectrons are emitted.

Solution and Answer Guide

Kotz Treichel Townsend Treichel, Chemistry and Chemical Reactivity 11e, 978-0-357-85140-1, Chapter 7: The Structure of Atoms and Periodic Trends

TABLE OF CONTENTS

Applying Chemical Principles	256
Practicing Skills	258
Quantum Numbers and Electron Configurations	260
General Questions	268
In the Laboratory	275
Summary and Conceptual Questions	276

Applying Chemical Principles

The Not-So-Rare Earths

7.1.1. The most common oxidation state of a rare earth element is +3.

(a) What is the ground state electron configuration of Sm^{3+}?

The ground state configuration for elemental Sm is $[Xe]4f^6 6s^2$. [See Table 7.3] Losing three electrons would give an electron configuration of $[Xe]4f^5$, as the atom loses three higher energy electrons.

(b) Write a balanced chemical equation for the reaction of Sm(s) and O_2(g).
4 Sm(s) + 3 O_2(g) → 2 Sm_2O_3(s)

7.1.2. Y has a $[Kr]4d^1 5s^2$ configuration, while La has a $[Xe]5d^1 6s^2$ configuration and Lu has a $[Xe]4f^{14}5d^1 6s^2$ configuration. Based on electronic structure, both La and Lu are appropriately located under yttrium. All three atoms have electronic structures that finish with a filled outermost s-orbital and one electron in the outermost d-orbital.

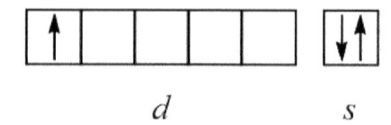

7.1.3. Gadolinium has eight unpaired electrons, the greatest number of any rare earth element.

(a) Orbital box diagram of the ground state electron configuration of Gd: The ground state configuration for elemental Gd is [Xe]$4f^7 5d^1 6s^2$. The orbital box diagram would look like:

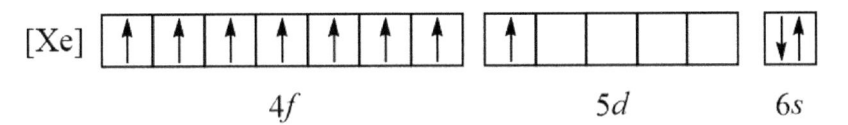

[Xe] [↑][↑][↑][↑][↑][↑][↑] [↑][ ][ ][ ][ ] [↑↓]
                  4f   5d   6s

(b) The most common oxidation state for Gd would be +3, with the electron configuration for Gd^{3+} being [Xe]$4f^7$.

7.1.4. (a) The electron configuration of the outermost shell of La and Lu is $6s^2$. Lutetium's 14 additional electrons are located in the 4f orbitals. These electrons are poorly shielding so that the outermost electrons feel a large portion of the charge from the additional 14 protons in the nucleus.

(b) The elements in the 5d block have similar radii to the atoms directly above them in the 4d block. Lanthanum and yttrium have similar radii, so lanthanum might be considered a more appropriate fit below yttrium on the periodic table.

7.1.5. What is the frequency and energy (in J/photon) of light with wavelength 612 nm?

frequency = $\dfrac{\text{speed of light}}{\text{wavelength}} = \dfrac{2.9979 \times 10^8 \text{ m/s}}{612 \text{ nm}} \cdot \dfrac{1.00 \times 10^9 \text{nm}}{1.00 \text{ m}} = 4.90 \times 10^{14} \text{ s}^{-1}$

The energy of a photon of this light may be determined using $E = h\nu$.

Planck's constant, h, has a value of 6.626×10^{-34} J · s · photons^{-1}

$E = (6.626 \times 10^{-34}$ J · s · photon$^{-1})(4.90 \times 10^{14}$ s$^{-1}) = 3.25 \times 10^{-19}$ J · photon^{-1}

7.1.6. Molar mass of Nd$_2$Fe$_{14}$B = 1081.12 g/mol

% Nd = (2 mol)(144.24 g/mol Nd)/(1081.12 g) × 100 % = 26.683 %

Metals in Biochemistry and Medicine

7.2.1. Electron configurations for Fe, Fe^{2+}, Fe^{3+}:

Fe: [Ar]$3d^6 4s^2$ Fe^{2+}: [Ar]$3d^6$ Fe^{3+}:[Ar]$3d^5$

Note that the formation of ions of transition metals begins with the removal of electrons with highest "n," the 4s electrons in this case. Losing 2 electrons (from 4s) yields the 2+ ion, and loss of a 3rd electron gives the 3+ ion.

7.2.2. As there are 6 electrons in a d sublevel (in Fe^{2+}), there will be 4 unpaired electrons, and, with the removal of the 3rd electron (to give Fe^{3+}), there are 5 unpaired electrons. Both species are paramagnetic.

7.2.3. Why are Cu atoms slightly larger than Fe atoms?

Cu has the configuration [Ar]$3d^{10}4s^1$, while Fe is [Ar]$3d^64s^2$: The increased number of electrons in the *d* sublevel (for Cu) will provide shielding for the 4*s* electrons to a greater extent than those of Fe. This leads to a slightly larger radius for Cu.

7.2.4. The reduction of Fe^{3+} to Fe^{2+} will add one more electron to the iron, causing an increase in the size of the iron, and a distortion of the planar configuration.

Practicing Skills

Writing Electron Configurations of Atoms

7.1. The orbital box and *spdf* notation for P and Cl:

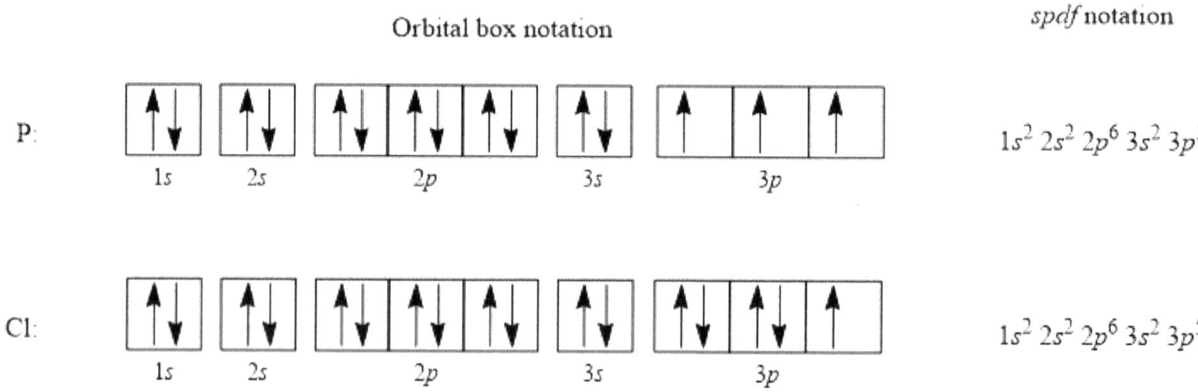

P: $1s^2\ 2s^2\ 2p^6\ 3s^2\ 3p^3$

Cl: $1s^2\ 2s^2\ 2p^6\ 3s^2\ 3p^5$

Note that Cl is in group 7A (17) indicating that there are SEVEN electrons in the outer shell, while P is in group 5A (15) indicating that there are FIVE electrons in the outer shell. Both Cl and P are on the "right side" of the periodic table—where elements have their "outermost" electrons in *p* subshells.

7.2. Mg: $1s^22s^22p^63s^2$

Magnesium is in Group 2A (2) and has two electrons in its outer shell.

Ar: $1s^22s^22p^63s^23p^6$

Argon is in Group 8A (18) and has eight electrons in its outer shell.

7.3. Electron configuration of chromium and iron:

(a) Cr: $1s^22s^22p^63s^23p^63d^54s^1$

(b) Fe: $1s^22s^22p^63s^23p^63d^64s^2$

See note (4) In "A Closer Look: Orbital Energies, Z*, and Electron Configurations." The unexpected configurations for Cr and Cu arise from a factor called "exchange energy."

7.4. Electron configuration of vanadium: $1s^22s^22p^63s^23p^63d^34s^2$

7.5. Use *spdf* and noble gas notation for arsenic and krypton:

(a) Arsenic's electron configuration: (33 electrons)

spdf notation: $1s^22s^22p^63s^23p^63d^{10}4s^24p^3$

spdf with noble gas notation: $[Ar]3d^{10}4s^24p^3$

(b) Krypton's electron configuration: (36 electrons)

spdf notation: $1s^22s^22p^63s^23p^63d^{10}4s^24p^6$

spdf with noble gas notation: [Kr]

7.6. Use *spdf* and noble gas notation for the requested elements:

(a) Sr: *spdf* notation: $1s^22s^22p^63s^23p^63d^{10}4s^24p^65s^2$

(b) Zr: *spdf* notation: $1s^22s^22p^63s^23p^63d^{10}4s^24p^64d^25s^2$

(c) Rh: *spdf* notation: $1s^22s^22p^63s^23p^63d^{10}4s^24p^64d^75s^2$

(actual configuration: $1s^22s^22p^63s^23p^63d^{10}4s^24p^64d^85s^1$)

(d) Sn: *spdf* notation: $1s^22s^22p^63s^23p^63d^{10}4s^24p^64d^{10}5s^25p^2$

7.7. Electron configurations for the requested elements:

(a) Tantalum's *spdf* with noble gas notation (73 electrons)

spdf with noble gas notation: $[Xe]4f^{14}5d^36s^2$

(b) Silver's *spdf* with noble gas notation (47 electrons)

spdf with noble gas notation: $[Kr]4d^{10}5s^1$

Tantalum's configuration is expected, with Ta in period 5 and in Group 5B (5). Silver, atomic number 47 is in the 5th period and is the first element in Group 1B (11). It is predicted to have a krypton core followed by 11 electrons distributed in the 5s and 4d subshells. You might have expected to have filled the 5s orbital and then added 9 electrons to the 4d orbitals. However, as happens in other cases, the atom is more stable with the 4d orbitals filled, leaving a single electron in 4s. Actual configuration (Table 7.3): $[Kr]4d^{10}5s^1$

7.8. Reasonable electron configurations for:
(a) Sm: [Xe]$4f^5 5d^1 6s^2$ (actual configuration in Table 7.3 [Xe]$4f^6 6s^2$)
(b) Yb: [Xe]$4f^{13} 5d^1 6s^2$ (actual configuration in Table 7.3 [Xe]$4f^{14} 6s^2$)

7.9. Americium's noble gas and *spdf* with noble gas notation (95 electrons)

spdf with noble gas notation: [Rn]$5f^6 6d^1 7s^2$ (actual configuration in Table 7.3 [Rn]$5f^7 7s^2$)

7.10. Predict electronic configurations for the following actinides:
(a) Pu: [Rn]$5f^5 6d^1 7s^2$ (actual configuration in Table 7.3 [Rn]$5f^6 7s^2$)
(b) Cm: [Rn]$5f^7 6d^1 7s^2$

Quantum Numbers and Electron Configurations

7.11. Maximum number of electrons associated with the following sets of quantum numbers:

	Maximum # of electrons	Explanation
(a) $n = 3$, $\ell = 3$, $m_\ell = 1$	NONE	ℓ cannot be equal to n, it runs from 0 to (n – 1)
(b) $n = 6$, $\ell = 1$, $m_\ell = -1$, $m_s = -1/2$	1	Correct set of quantum numbers for a 6p electron
(c) $n = 4$, $\ell = 3$, $m_\ell = -3$	2	Correct set of quantum numbers for 4f electrons

7.12. Maximum number of electrons associated with the following sets of quantum numbers:

	Maximum # of electrons	Explanation
(a) $n = 4$	32	Maximum number of electrons in a shell is $2n^2$ or $2(4^2) = 32$
(b) $n = 3$, $\ell = 1$	6	Correct set of quantum numbers for 3p electrons
(c) $n = 4$, $\ell = 0$, $m_\ell = -1$, $m_s = +1/2$	None.	If $\ell = 0$, the only possible value for m_ℓ is 0.
(d) $n = 5$, $\ell = 1$, $m_\ell = -1$, $m_s = +1/2$	1	Correct set of quantum numbers for one 5p electron

7.13. The electron configuration for Mg using the orbital box method:

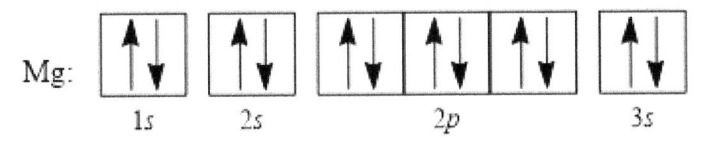

Mg: [↑↓] [↑↓] [↑↓][↑↓][↑↓] [↑↓]
     1s   2s   2p   3s

Electron number: 11 12

The noble gas notation: [Ne]$3s^2$

The quantum numbers for the two 3s electrons are

n	ℓ	m_ℓ	m_s
3	0	0	$+\frac{1}{2}$
3	0	0	$-\frac{1}{2}$

7.14. The electron configuration for P using the orbital box with noble gas method:

P: [Ne] [↑↓] [↑][↑][↑]
         3s   3p

For the electrons beyond those of neon, a possible set of quantum numbers:

$n = 3, \ell = 0, m_\ell = 0, m_s = +\frac{1}{2}$

$n = 3, \ell = 0, m_\ell = 0, m_s = -\frac{1}{2}$

$n = 3, \ell = 1, m_\ell = -1, m_s = +\frac{1}{2}$

$n = 3, \ell = 1, m_\ell = 0, m_s = +\frac{1}{2}$

$n = 3, \ell = 1, m_\ell = +1, m_s = +\frac{1}{2}$

For the 3p electrons, another set is possible that has the same values for n, ℓ, and m_ℓ as shown but that has $m_s = -\frac{1}{2}$ for all three electrons

7.15. The electron configuration for gallium using the orbital box with noble gas method:

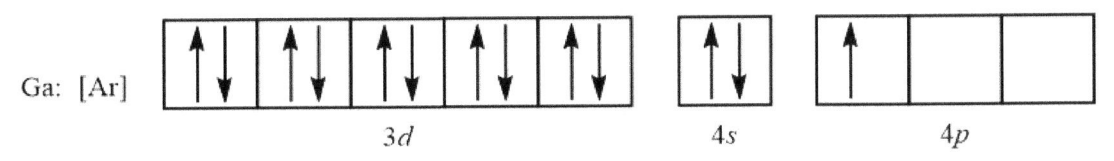

Ga: [Ar] [↑↓][↑↓][↑↓][↑↓][↑↓] [↑↓] [↑][ ][ ]
              3d   4s   4p

A possible set of quantum numbers for the highest energy electron:

$n = 4$, $\ell = 1$, $m_\ell = -1$, $m_s = +½$

Other possible sets have the same values of n and ℓ, but m_ℓ may have values of -1, 0, or $+1$, and m_s may be either $+½$ or $-½$.

7.16. The electron configuration for zirconium using the orbital box with noble gas method:

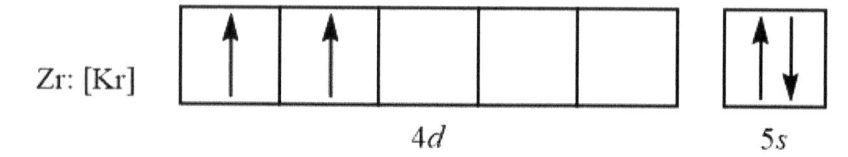

Zr: [Kr] ↑ ↑ _ _ _   ↑↓
         4d   5s

A possible set of quantum numbers for the electrons beyond noble gas:

$n = 4$, $\ell = 2$, $m_\ell = -2$, $m_s = +½$

$n = 4$, $\ell = 2$, $m_\ell = -1$, $m_s = +½$

$n = 5$, $\ell = 0$, $m_\ell = 0$, $m_s = +½$

$n = 5$, $\ell = 0$, $m_\ell = 0$, $m_s = -½$

For the 4d electrons, the value of m_ℓ may be any integer value between -2 and $+2$, so long as the two electrons have different values for m_ℓ, and the value of m_s may be either $+½$ or $-½$ so long as the two electrons have the same value for m_s.

Subshell Energies and Electron Assignments

7.17. Variation of effective nuclear charge among elements from Li to F:

Best described by (a) regular increase from Li to F

7.18. Describe effective nuclear charge for the electrons:

(a) the 2s electron at a large distance from a Li atom: (i) $Z^* = 1$

(b) the 2s electron at the most probable distance from the nucleus:

(ii) Z^* is between 1 and 3 (average is about 1.28)

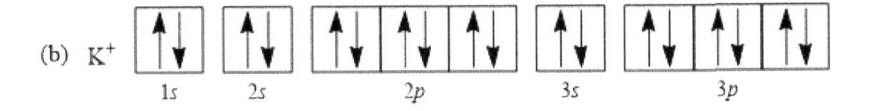

7.19. First five subshells in order of filling, according to the Aufbau principle:

1s, 2s, 2p, 3s, 3p (according to Figure 7.1)

7.20. First orbital to fill according to n and ℓ: (See Figure 7.1b)

In general, the lower $(n + \ell)$ fills first. Review those values for the three mentioned orbitals:

4f (4 + 3), 5d (5 + 2), 6s (6 + 0). Based upon these values, 6s fills first.

7.21. The orbital box representations for the following ions:

(a) Mg^{2+} [orbital diagram: 1s, 2s, 2p all filled]

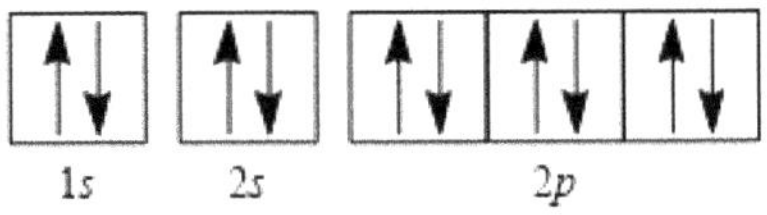

(b) K^+ [orbital diagram: 1s, 2s, 2p, 3s, 3p all filled]

(c) Cl^- [orbital diagram: 1s, 2s, 2p, 3s, 3p all filled]

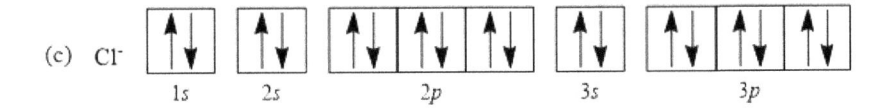

(d) O^{2-} [orbital diagram: 1s, 2s, 2p all filled]

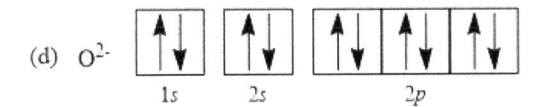

7.22. Use orbital box diagrams to show electron configurations for:

(a) Na⁺ [↑↓]₁ₛ [↑↓]₂ₛ [↑↓][↑↓][↑↓]₂ₚ

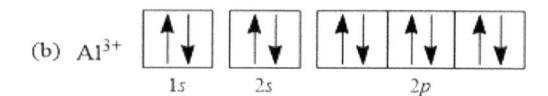

(b) Al³⁺ [↑↓]₁ₛ [↑↓]₂ₛ [↑↓][↑↓][↑↓]₂ₚ

(c) P³⁻ [↑↓]₁ₛ [↑↓]₂ₛ [↑↓][↑↓][↑↓]₂ₚ [↑↓]₃ₛ [↑↓][↑↓][↑↓]₃ₚ

(d) F⁻ [↑↓]₁ₛ [↑↓]₂ₛ [↑↓][↑↓][↑↓]₂ₚ

7.23. Electron configurations of:

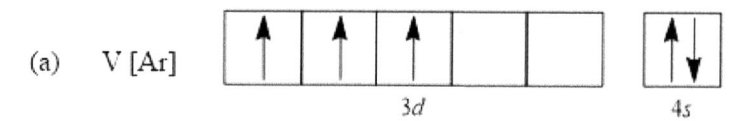

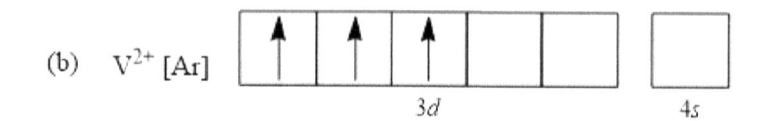

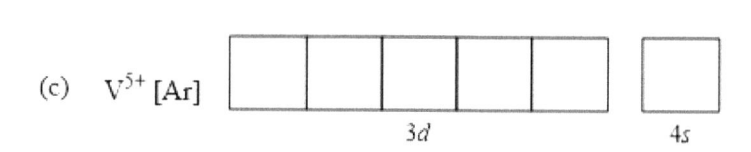

(a) V [Ar] [↑][↑][↑][ ][ ] 3d   [↑↓] 4s

(b) V²⁺ [Ar] [↑][↑][↑][ ][ ] 3d   [ ] 4s

(c) V⁵⁺ [Ar] [ ][ ][ ][ ][ ] 3d   [ ] 4s

Note that V and the V²⁺ ion contain unpaired electrons and are therefore paramagnetic.

7.24. Use orbital box diagrams to show electron configurations for: Ti, Ti²⁺, Ti⁴⁺

(a) Ti   [Ar] [↑][↑][ ][ ][ ] 3d   [↑↓] 4s

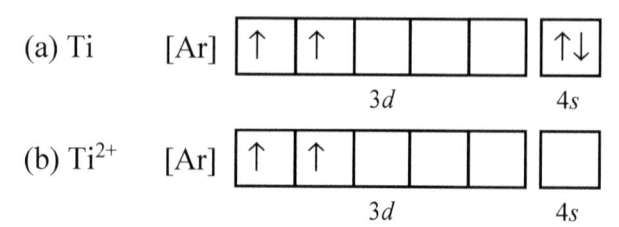

(b) Ti²⁺ [Ar] [↑][↑][ ][ ][ ] 3d   [ ] 4s

(c) Ti^{4+} [Ar] ▯▯▯▯▯ ▯
         3d   4s

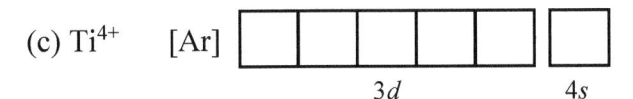

Not that Ti and the Ti^{2+} ion are paramagnetic with two unpaired electrons.

7.25. Manganese's orbital box and noble gas diagrams:

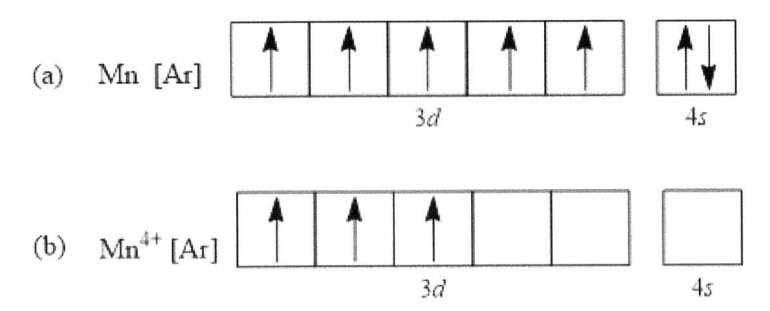

(a) Mn [Ar] ↑ ↑ ↑ ↑ ↑ | ↑↓
              3d   4s

(b) Mn^{4+} [Ar] ↑ ↑ ↑ ▯ ▯ | ▯
                    3d   4s

(c) The ion is paramagnetic.
(d) As seen in part b, the ion contains 3 unpaired electrons.

7.26. Electron configurations for Ni^{2+}, Ni^{3+}. Are either of the ions paramagnetic?

Ni^{2+} [Ar] ↑↓ ↑↓ ↑↓ ↑ ↑ | ▯   Paramagnetic
                3d   4s

Ni^{3+} [Ar] ↑↓ ↑↓ ↑ ↑ ↑ | ▯   Paramagnetic
                3d   4s

Periodic Properties

7.27. Elements arranged in order of increasing size: C < B < Al < Na < K

Radii from Figure 7.5 (in pm)   77 < 83 < 143 < 186 < 227

Since K is in period 4, we anticipate it being larger than Na, its analog in Group 1A (1) (period 3).

Al is to the right of Na, so we expect it to be smaller than Na.

B and C are in period 2, with B to the left of C, and therefore larger than C.

7.28. Elements arranged in order of increasing size: Ca, Rb, P, Ge, and Sr
Radii from Figure 7.5 gives the order: P < Ge < Ca < Sr < Rb

7.29. The species in each pair with the larger radius:

(a) Cl⁻ is larger than Cl: The ion has more electrons/proton than the atom.
(b) Al is larger than O: Al is in period 3, while O is in period 2.
(c) In is larger than I: Atomic radii decrease, in general, across a period.

7.30. The species in each pair with the larger radius:

(a) Cs is larger than Rb: Cs is in period 6, Rb is in period 5. Both are in Group 1A (1)

(b) O²⁻ is larger than O: Anions are always larger than the neutral atom

(c) As is larger than Br: Elements tend to decrease in radius as one moves to the right in a period.

7.31. The group of elements with correctly ordered increasing ionization energy (IE):
(c) Li < Si < C < Ne.

Neon would have the greatest IE. Silicon, being slightly larger in atomic radius than carbon, has a smaller IE. Lithium, the largest atom of this group, would have the smallest IE.

7.32. Arrange the elements in increasing ionization energy (IE):

K < Li < C < N. K being the largest of the four elements has the smallest IE. The remaining elements are arranged across Period 2, with N being the smallest atom and Li being the largest atom. Therefore, the IE increases in the order given.

7.33. For the elements Na, Mg, O, and P:

(a) The smallest atomic radius: O

The smaller the period number, the smaller the atom. Radius also decreases to the right in a given period.

(b) The most negative electron attachment enthalpy: O

Electron attachment enthalpy is greatest for elements in Groups 6A (16) and 7A (17).

(c) Increasing ionization energy: Na < Mg < P < O

The ionization energy varies inversely with the atomic radius. The smaller the atom, the greater the IE.

7.34. Concerning the elements B, Al, C, Si:

(a) Most metallic character: Al. Metallic character tends to increase down a group and to the left in a period.

(b) Smallest atomic radius: C. Atomic radius tends to increase down a group, and to the left in a period.

(c) Most negative electron attachment energy: C (based on periodic trends) but according to experimental data (Figure 7.10) Si has the most negative value.

(d) Al < B < C

7.35. Concerning the periodic trends:

(a) Increasing ionization energy: S < O < F

Ionization energy is inversely proportional to atomic size.

(b) Largest ionization energy of O, S, or Se: O

Oxygen is the smallest of these Group 6A (16) elements, and hence has the largest IE.

(c) Most negative electron attachment enthalpy of Se, Cl, or Br: Cl

Chlorine is the smallest of these three elements. Electron attachment enthalpy tends to become more negative on a diagonal from the lower left of the periodic table to the upper right.

(d) Oxide ion, O^{2-}, has the largest radius of the three species O^{2-}, F^-, F.

If one considers the attraction of the nucleus for the extranuclear electrons, the greater the proton-to-electron ratio the smaller the ion owing to a larger proton-electron attraction. The oxide ion (with 8 protons and 10 electrons) has more electrons than protons (of these three species) and so is the largest.

7.36. Explain the following trends:

(a) F < O < S The trend is for atomic radius to increase to the left and down the periodic table.

(b) Based on a knowledge of first-order periodic trends, you might predict that S should have the largest IE of the elements P, Si, S, and Se. However, recall that the O atom IE is smaller than that of N, so it is not surprising that the same effect carries over into the third period. That is, the order of ionization energies should be Si < S < P for the 3rd period elements. (See Figure 7.10 and Table 7.5) and so P has the largest IE of these four elements.

(c) F^- < O^{2-} < N^{3-} These ions are isoelectronic and size increases as the number of electrons relative to nuclear protons increases.

(d) Cs < Ba < Sr The trend for ionization energy is to decrease down the periodic table and to the left of a period. Cs has a first ionization energy of 357.7 kJ/mol, whereas Ba (502.9 kJ/mol) and Sr (549.5 kJ/mol) are much less.

7.37. Identify the elements corresponding to the data given:

(a) With 3 electrons, the element is Li (atomic number 3)

(b) With 19 electrons, (2 + 2 + 6 + 2 + 6 + 1), the element is K (atomic number 19)

(c) With 21 electrons, (2 + 2 + 6 + 2 + 6 + 1 + 2), the element is Sc (atomic number 21)

7.38. Identify element that is consistent with the PES data:

(a) Na: element has 11 electrons

(b) Ca: element has 20 electrons

(c) V: element has 23 electrons

7.39. The hydrogen and helium spectra show one peak, since these atoms have electrons in only the 1s subshell, while Li has an electron in the 2s subshell. Since the intensity of the peaks corresponds to the number of electrons in a specific subshell, the He peak would have **twice** the intensity of the H peak, corresponding to 2 electrons (for He) versus 1 electron (for H). Likewise for Li, the two peaks would have a 2:1 ratio (corresponding to the 1s and 2s electrons).

7.40. Expected peaks for atoms of nitrogen:

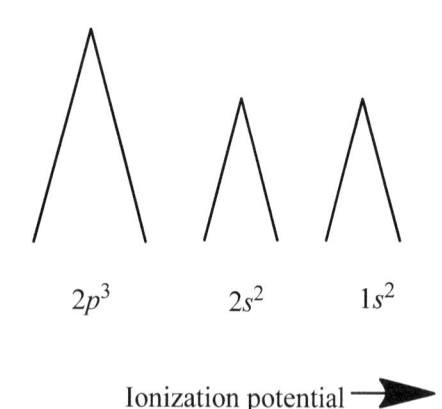

The three peaks correspond to the 7 electrons in atomic N. The size of the peaks indicate the relative number of electrons in each of the energy levels.

$2p^3$ $2s^2$ $1s^2$

Ionization potential →

General Questions

7.41. Concerning the electron configuration of Cr and Cr^{3+}:

(a) For Cr: [Ar]$3d^5 4s^1$ Note that Cr is an *exception* to the Aufbau principle in filling orbitals, with 5 electrons in the 3d subshell and 1 in the 4s subshell.
For Cr^{2+}: [Ar]$3d^4$
For Cr^{3+}: [Ar]$3d^3$
Recall that *electrons are always removed first from the electron shell of highest n.*

(b) The **3+** ion has unpaired electrons (all of the same spin), so it is **paramagnetic**. The **2+** ion is **also paramagnetic**, since the four electrons in the 3d sublevel would exist in a (1,1,1,1,0) configuration (here numbers to represent electrons in the 5 orbitals).

(c) The Al^{3+} ion has a radius of 57 pm, while that for Cr^{3+} is 64 pm. The numbers confirm our "scientific guess" that an ion in the 4th period would probably be larger than one from the 3rd period—though the difference in radii is not large. This small difference is important because it means that Cr^{3+} ions can substitute for Al^{3+} ions in a crystal, and it is this that is the origin of the red color of rubies.

7.42. Electron configuration for Fe^{2+} and Ti^{4+}:

Fe^{2+}, $[Ar]3d^6$; 26 electrons – 2 electrons give this Fe^{2+} ion.

Ti^{4+}, $[Ar]$; 22 electrons – 4 electrons gives an ion that is isoelectronic with argon.

7.43. The electron configuration for Np and Np^{3+}:

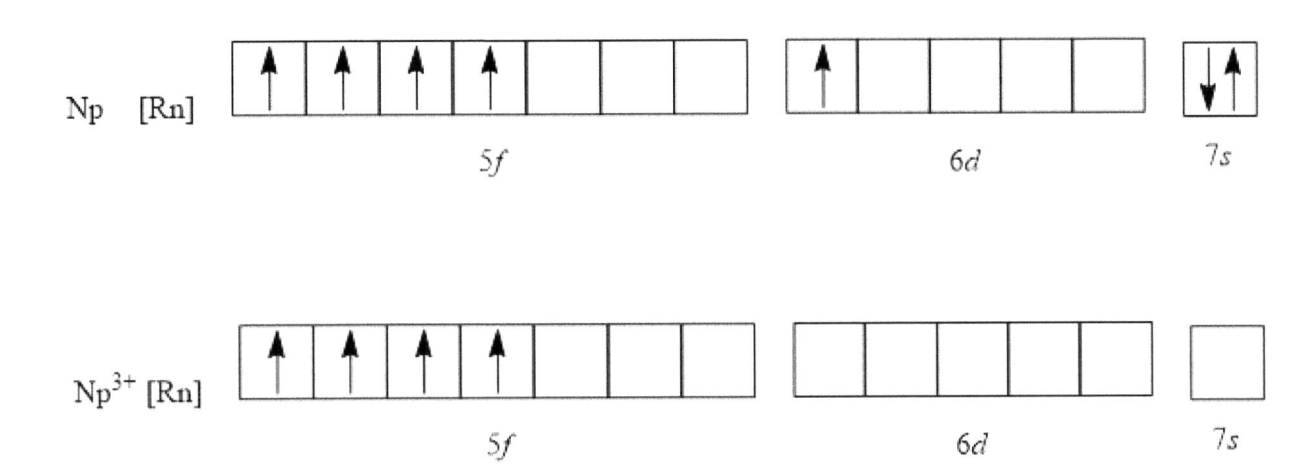

Np [Rn]   5f: ↑ ↑ ↑ ↑ _ _ _   6d: ↑ _ _ _ _   7s: ↑↓

Np^{3+} [Rn]   5f: ↑ ↑ ↑ ↑ _ _ _   6d: _ _ _ _ _   7s: _

Both species have unpaired electrons and are therefore **paramagnetic**.

7.44. Provide the orbital box notation for Ce, Ce^{3+}, Ho, and Ho^{3+}:

(a) Ce: [Xe]   4f: ↑ _ _ _ _ _ _   5d: ↑ _ _ _ _   6s: ↑↓

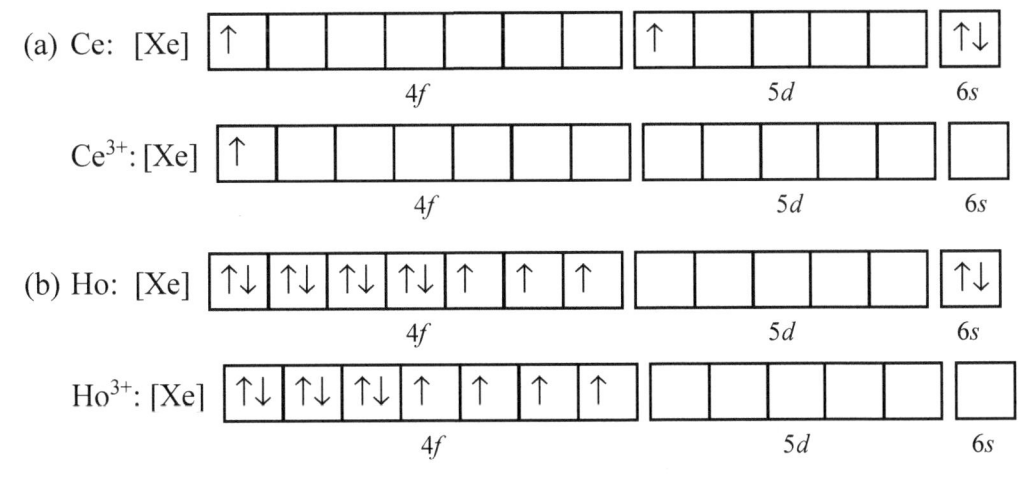

Ce^{3+}: [Xe]   4f: ↑ _ _ _ _ _ _   5d: _ _ _ _ _   6s: _

(b) Ho: [Xe]   4f: ↑↓ ↑↓ ↑↓ ↑↓ ↑ ↑ ↑   5d: _ _ _ _ _   6s: ↑↓

Ho^{3+}: [Xe]   4f: ↑↓ ↑↓ ↑↓ ↑ ↑ ↑ ↑   5d: _ _ _ _ _   6s: _

Note that in each case, the ion is formed by loss of electrons in the highest *n* level first.

7.45. Using the *spdf* notation the atom described would have an electron configuration: $1s^22s^22p^63s^23p^64s^2$

(a) Adding the electrons gives a sum of 20. Since this is a **neutral** atom, the number of protons and electrons would be equal. Twenty protons give this element an atomic number = 20.

(b) There are 2 *s* electrons in each of 4 shells: 8 *s* electrons total

(c) There are 6 *p* electrons in each of 2 shells: 12 *p* electrons total

(d) There are 0 *d* electrons

(e) With its outer electrons in an "*s*" sublevel (specifically the 4*s* sublevel), the element is a **metal,** specifically elemental **calcium**.

7.46. Electron configuration of Mt using *spdf* and noble gas notation:

Mt:[Rn]$5f^{14}6d^77s^2$ Other elements in this group: Co, Rh, Ir

7.47. Consider the sets of quantum numbers below:

	n	ℓ	m_ℓ	m_s	Possible elements (if allowable)
(a)	2	0	0	−½	Li or Be
(b)	1	1	0	+½	Not allowable, since max. ℓ is $n-1$ (and $n=1$)
(c)	2	1	−1	−½	B, C, N, O, F or Ne
(d)	4	2	+2	−½	Y, Zr, Nb, Mo, Tc, Ru, Rh, Pd, Ag or Cd

7.48. Possible sets of quantum numbers for a 4*p* electron:

$n = 4, \ell = 1, m_\ell = -1, m_s = +½$

$n = 4, \ell = 1, m_\ell = -1, m_s = -½$

$n = 4, \ell = 1, m_\ell = 0, m_s = +½$

$n = 4, \ell = 1, m_\ell = 0, m_s = -½$

$n = 4, \ell = 1, m_\ell = +1, m_s = +½$

$n = 4, \ell = 1, m_\ell = +1, m_s = -½$

7.49. (a) Electron configuration for Nd, Fe, and B

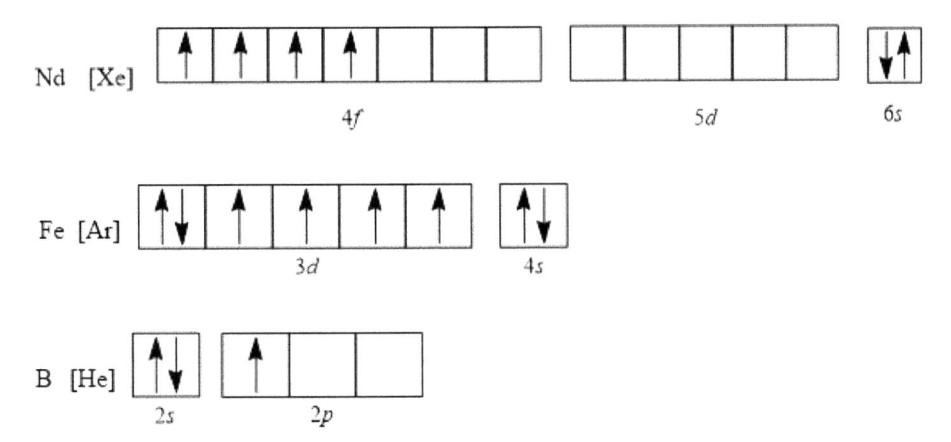

Nd [Xe] 4f: ↑ ↑ ↑ ↑ _ _ _   5d: _ _ _ _ _   6s: ↑↓

Fe [Ar] 3d: ↑↓ ↑ ↑ ↑ ↑   4s: ↑↓

B [He] 2s: ↑↓   2p: ↑ _ _

(b) Paramagnetic or diamagnetic?

All three species have unpaired electrons, so all are paramagnetic.

(c) For the ions Nd^{3+} and Fe^{3+} their electron configurations are:

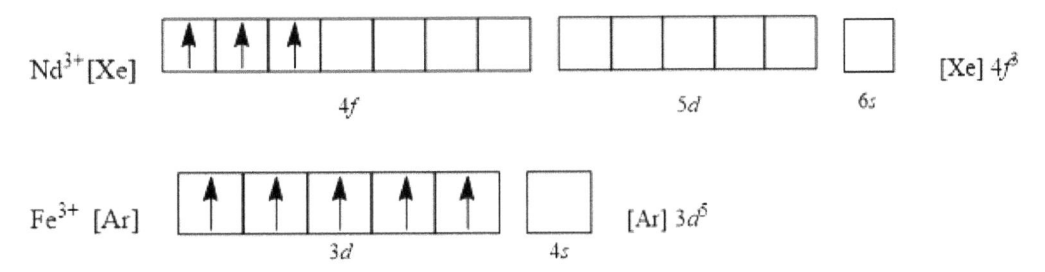

Nd^{3+} [Xe] 4f: ↑ ↑ ↑ _ _ _ _   5d: _ _ _ _ _   6s: _   [Xe] $4f^3$

Fe^{3+} [Ar] 3d: ↑ ↑ ↑ ↑ ↑   4s: _   [Ar] $3d^5$

Both ions have unpaired electrons and are paramagnetic.

7.50. Name the element with the following characteristics:

(a) element with configuration $1s^2 2s^2 2p^6 3s^2 3p^3$: P, phosphorus

(b) alkaline earth element with smallest radius: Be, beryllium

(c) element with largest IE in Group 5A (15): N, nitrogen

(d) element with 2+ ion having the configuration $[Kr]4d^5$: Tc, technetium

(e) element with most negative electron attachment enthalpy in Group 7A (17): Cl, chlorine

(f) element with configuration $[Ar]3d^{10}4s^2$: Zn, zinc

7.51. Arranged in order of increasing ionization energy for K, Ca, Si, and P.

As we move across Period 4, we find K and then Ca. In Period 3, we find Si and then P. With IE increasing across a period (which means that K < Ca and Si < P), and decreasing down a group (so elements in Period 4 generally have lower IE than analogues in Period 3), we expect the ionization energy to increase in the order: K < Ca < Si < P.

7.52. Rank in order of increasing IE for Cl, Ca^{2+}, Cl^-: $Cl^- < Cl < Ca^{2+}$

Ca^{2+} and Cl^- are isoelectronic. Ca^{2+} has a larger IE than Cl^- because the calcium ion has a 2+ charge and removing another electron from that ion would take more energy. Removing an electron from the chloride ion would require less energy than the removal of an electron from the Cl atom due to electron repulsion forces.

7.53. For element A = $[Kr]5s^1$ and B = $[Ar]3d^{10}4s^24p^4$

(a) Element A (Rb) is a metal (1 electron in the *s* sublevel).

(b) Element B (Se) has the greater IE. B's atomic radius would be smaller than that of A.

(c) Element A has the less negative $\Delta_{ea}H$. In general $\Delta_{ea}H$ for nonmetals is more positive than that of the metals in the same period.

(d) A would have the larger atomic radius. Its outermost electrons are in the 5th shell.

(e) As A has but one electron in the outer level, it would tend to form a 1+ ion. Similarly, B has 6 electrons in the outer level, and would tend to form a 2– ion. The formula for the compound between these two ions would be A₂B. (Rb₂Se)

7.54. Answer questions regarding elements A and B:

(a) Element A is an alkaline earth metal, as it has 2 electrons in an *s* subshell.

(b) Element B is a nonmetal (halogen), as it has 7 electrons in the outer shell.

(c) Element with larger IE: B, as it has the smaller atomic radius.

(d) Element with smaller atomic radius: B

7.55. Ions not likely to be found include: In^{4+}, Fe^{6+}, and Sn^{5+}

Indium [Group 3A (13)] is expected to form a cation by losing **three** electrons, forming the In^{3+} cation.

Iron can lose either 2 or 3 electrons, forming Fe^{2+} and Fe^{3+} ions. Loss of more electrons would be difficult as the proton-to-electron ratio increases and with it there is increased attraction between the nucleus and the remaining electrons.

Tin can lose up to 4 electrons (2 *s* and 2 *p*) to form the 4+ cation. Loss of more electrons would require removal of electrons from a filled *d* subshell.

7.56. Arrange ions in order of decreasing size: $S^{2-} > Cl^- > K^+ > Ca^{2+}$

These ions all have 18 electrons, with S^{2-} having 16 protons, Cl having 17 protons, K^+ having 19 protons, and Ca^{2+} having 20 protons. The increasing proton-to-electron ratio results in smaller ions.

7.57. Periodic trends among the listed elements:

(a) Element with the largest atomic radius: Se has the largest radius as it is in period 4 [Group 6A (16)]. S would be smaller, and Cl would be smaller than S. Recall that atomic radius decreases across a period.

(b) Br⁻ is larger than Br. The anion has more electrons per proton than the neutral atom.

(c) Na would have the largest **difference** between the 1st and 2nd IE. Removing a second electron requires removal from a completed electron shell much lower in energy.

(d) Element with the largest ionization energy: N (IE is inversely proportional to atomic radius).

(e) Largest radius: N³⁻ would be the largest, since the number of electrons/proton is greatest for the N³⁻ ion and smallest for the F⁻ ion.

7.58. Rank the isoelectronic species in increasing size, IE, electron attachment enthalpy. In general, the proton-to-electron ratio determines these trends.

(a) Size: $Ca^{2+} < K^+ < Cl^-$ Ca^{2+} has 20 protons for 18 electrons, whereas Cl⁻ has 17 protons for 18 electrons.

(b) Ionization energy: $Cl^- < K^+ < Ca^{2+}$. Chloride ion has the smallest IE as it has the fewest protons for the number of electrons.

(c) Electron attachment enthalpy: $Cl^- < K^+ < Ca^{2+}$. This follows the same trend as the IE.

7.59. For the elements Na, B, Al, and C:

(a) Largest atomic radius: Na and Al have electrons in $n = 3$, while B and C have valence electrons in $n = 2$. So, Na and Al are larger than B and C. With 3 electrons in the outer level, Al is smaller than Na (with only 1 electron), so **Na** is the largest atom.

(b) Largest negative electron attachment enthalpy ($\Delta_{ea}H$): Since, in general, elements with higher ionization energy have a more negative $\Delta_{ea}H$, you can anticipate that B and C should have more negative values of $\Delta_{ea}H$ than Na or Al. C is the smaller of the two species (B, C), so it is expected to have the higher IE and the more negative $\Delta_{ea}H$.

(c) Increasing order of ionization energy: B and C are in period 2, and Na and Al are in period 3, it is expected that B and C will have greater IEs than that of Na and Al. With Na in Group 1A (1), and Al in Group 3A (13), Al is expected to have a greater IE than Na. Boron is in Group 3A (13) and carbon is in Group 4A (14). Thus, boron is expected to have a larger ionization energy than aluminum, and carbon is greater than boron. The order of increasing IE is Na < Al < B < C.

7.60. Elements in the second transition series (Period 5) with four unpaired electrons in their 3+ ions:

Tc and Rh. Examine the electronic configuration for the atoms and remove 3 electrons.

7.61. According to the provided electron configuration:

(a) The element contains 27 electrons and is therefore **cobalt.**

(b) The sample contains three unpaired electrons—so it is paramagnetic

(c) The 3+ ion would be formed by the loss of the two 4s electrons and 1 of the (paired) d electrons— leaving four unpaired electrons.

7.62. Given the electron configuration, answer the questions about the element:

(a) With 23 electrons, the element is V, vanadium.

(b) Vanadium is in Group 5B (5), Period 4.

(c) With valence electrons in the 3d electrons, vanadium is a transition metal.

(d) Vanadium is paramagnetic with three unpaired electrons.

(e) Quantum numbers for the valence electrons:

$n = 3, \ell = 2, m_\ell = -2, m_s = +\frac{1}{2}$

$n = 3, \ell = 2, m_\ell = -1, m_s = +\frac{1}{2}$

$n = 3, \ell = 2, m_\ell = 0, m_s = +\frac{1}{2}$

$n = 4, \ell = 0, m_\ell = 0, m_s = +\frac{1}{2}$

$n = 4, \ell = 0, m_\ell = 0, m_s = -\frac{1}{2}$

In addition to the possibilities shown above for the 3d electrons, other possibilities exist. The value of m_ℓ may be any integer values between −2 and +2, so long as each electron has a different value, and the values of m_s may be either $+\frac{1}{2}$ or $-\frac{1}{2}$ so long as they are the same value for each electron.

(f) Electron configuration for the 2+ ion: V^{2+} [Ar]$3d^3$
The two 4s electrons are removed and the resulting ion is paramagnetic.

7.63. Regarding elements A and B:

(a) Element A is in the 5th period and has 2 valence electrons in the 5s orbital. It belongs in Group 2A (2) of the periodic table and is a **metal**.

(b) Element B is in the 5th period but in Group 7A (17). As this is also a 5th period element, and as ionization energies increase on moving across a period, B should have a **greater ionization energy** than A.

(c) Element A will have the larger atomic radius. Atomic radii tend to decline from the left to right in a given period.

(d) Since the electron attachment enthalpy generally becomes more negative from the left to the right side of a given period, Element B will have the more negative electron attachment enthalpy.

(e) Element A is metallic, while Element B is nonmetallic. Element A will form a dipositive (2+) cation.

(f) From part (e) you expect that Element A will form a 2+ ion. Element B has 7 electrons in the outermost shell, so you should expect B will form a 1– anion. A neutral compound formed between these two ions would have the formula AB$_2$.

7.64. Concerning ground state electron configurations:

(a) Element with the configuration [Ar]$3d^6 4s^2$: Fe

(b) Element whose 2+ ion has configuration [Ar]$3d^5$: Mn, paramagnetic

(c) Number of unpaired electrons in a Ni^{2+} ion: 2

(d) With 33 electrons, the element is As, arsenic; the element has three unpaired electrons and so is paramagnetic. The 3– ion has zero unpaired electrons.

(e) With 42 electrons, the element would be predicted to be Mo, molybdenum. However, Mo actually has the configuration [Kr]$4d^5 5s^1$ (similar to Cr a period above). Thus, no element has this configuration in its ground state.

In the Laboratory

7.65. Concerning nickel(II) formate:

(a) The theoretical yield of nickel(II) formate from 0.500g of nickel(II) acetate and excess formic acid:

$$\frac{0.500 \text{ g Ni(CH}_3\text{CO}_2)_2}{1} \cdot \frac{1 \text{ mol Ni(CH}_3\text{CO}_2)_2}{176.78 \text{ g Ni(CH}_3\text{CO}_2)_2} \cdot \frac{1 \text{ mol Ni(HCO}_2)_2}{1 \text{ mol Ni(CH}_3\text{CO}_2)_2} \cdot \frac{148.73 \text{ g Ni(HCO}_2)_2}{1 \text{ mol Ni(HCO}_2)_2}$$
$$= 0.421 \text{ g Ni(HCO}_2)_2$$

(b) Nickel(II) formate is paramagnetic. The Ni(II) ion has 8 d electrons (2 of which are unpaired).

(c) Mass of nickel derived by heating 253 mg of nickel(II) formate:

$$\frac{0.253 \text{ g Ni(HCO}_2)_2}{1} \cdot \frac{1 \text{ mol Ni(HCO}_2)_2}{148.73 \text{ g Ni(HCO}_2)_2} \cdot \frac{58.69 \text{ g Ni}}{1 \text{ mol Ni(HCO}_2)_2} = 0.0998 \text{ g Ni}$$

or 99.8 mg Ni

Elemental nickel has two unpaired electrons and is paramagnetic.

7.66. Concerning the spinels:

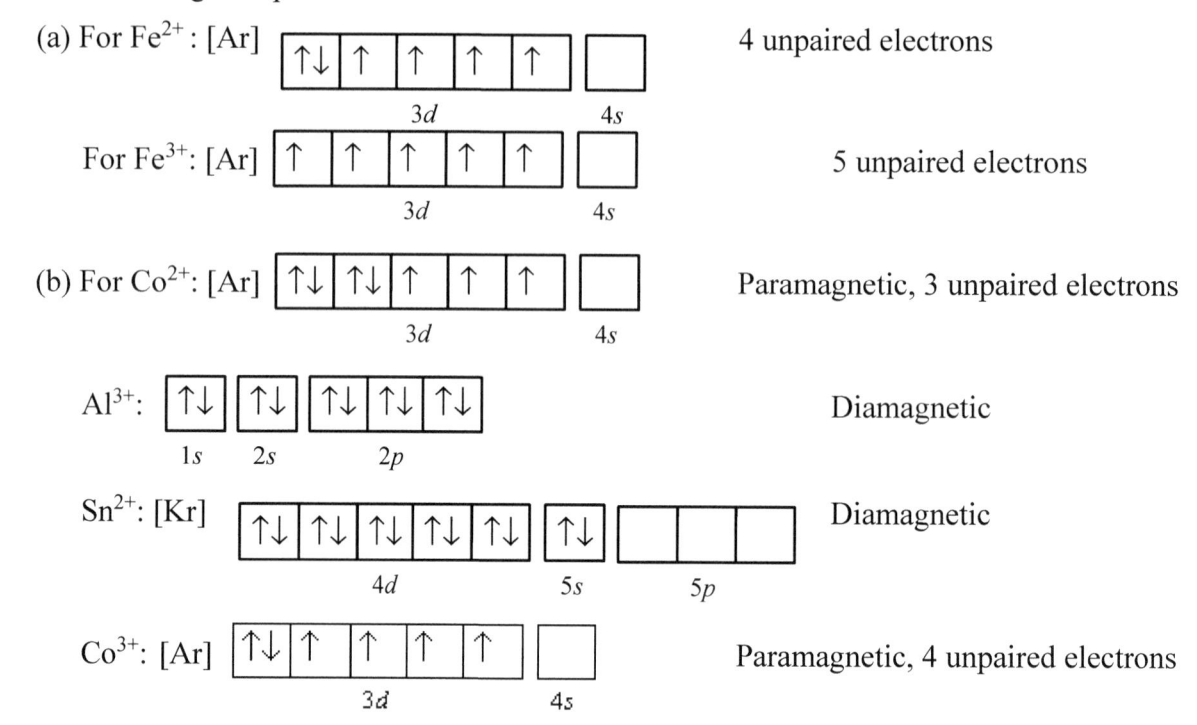

(a) For Fe²⁺: [Ar] [↑↓][↑][↑][↑][↑] [ ]   4 unpaired electrons
                 3d   4s

For Fe³⁺: [Ar] [↑][↑][↑][↑][↑] [ ]   5 unpaired electrons
                 3d   4s

(b) For Co²⁺: [Ar] [↑↓][↑↓][↑][↑][↑] [ ]   Paramagnetic, 3 unpaired electrons
                 3d   4s

Al³⁺: [↑↓][↑↓][↑↓][↑↓][↑↓]   Diamagnetic
     1s   2s   2p

Sn²⁺: [Kr] [↑↓][↑↓][↑↓][↑↓][↑↓] [↑↓] [ ][ ][ ]   Diamagnetic
                4d   5s   5p

Co³⁺: [Ar] [↑↓][↑][↑][↑][↑] [ ]   Paramagnetic, 4 unpaired electrons
                3d   4s

Summary and Conceptual Questions

7.67. The radius of Li⁺ is smaller than that of Li because the lithium ion has two electrons in the 1s energy level, whereas atomic Li has three electrons in the 1s and 2s energy levels. Also note that Li⁺ has 3 protons and only 2 electrons, whereas atomic Li has 3 protons and 3 electrons. The Li⁺ ion has the greater number of protons per electron, and so it the smaller of the two species. Using similar logic, the radius of F⁻ is anticipated to be larger than that of F, since F⁻ has 9 protons and 10 electrons, whereas atomic F has 9 protons and 9 electrons. There are fewer protons per electron in the F⁻ ion and so it has the greater radius.

7.68. Unlikely ions found in chemical compounds:

K²⁺ Potassium has only one outer shell electron, so it is unlikely to form a 2+ ion

Al⁴⁺ Aluminum has three outer shell electrons, so it is unlikely to form a 4+ ion

F²⁻ Fluorine has seven outer shell electrons, so adding one electron (to a 2p orbital) forms an anion with the same number of electrons as the nearest noble gas (Ne). Adding another electron to the next higher energy level (n = 3) would require a significant amount of energy.

7.69. Regarding first IE:

(a) The single 3p electron of Al is at a higher energy level than the 3s electrons of Mg. Thus, the 3p electron of Al experiences a somewhat lower effective nuclear charge than the 3s electrons of Mg (Figure 7.3). Because of this the 3p electron of Al is easier to remove than the 3s electrons of Mg.

(b) P has the configuration [Ne]3s²3p³. There is a single electron in each 3p orbital. In contrast, S has the configuration [Ne]3s²3p⁴. Here one p orbital has a pair of electrons. Although the effective nuclear charge for S is slightly larger than for P (Figure 7.3), electron-electron repulsions in the 3p orbital mean that the ionization energy of S is slightly lower than expected.

7.70. Regarding IE for He: Substitute into the Bohr equation to yield:

Assumption 1: IE = –1312 kJ/mol $\left(\frac{2^2}{1^2}\right)$ = –5248 kJ/mol

Assumption 2: IE = –1312 kJ/mol $\left(\frac{1^2}{1^2}\right)$ = –1312 kJ/mol

The actual IE value is between the two extreme limits. This suggests that in He one electron probably screens the other from the full nuclear charge to some extent, but not completely.

7.71. Hund's Rule applies to this question. The rule states that the most stable (lowest energy) configuration is obtained when electrons occupy orbitals singly with similar spins as shown in (d). The highest energy (least stable) would be (a) in which the electrons both have the same spin in one orbital before all three orbitals have at least one electron.

7.72. Estimate bond distances:

Sn—Cl: Calculated bond distance = 141 pm + ½(200 pm) = 241 pm
    Actual bond distance = 233 pm

Sn—Br: Calculated bond distance = 141 pm + ½(228 pm) = 255 pm
    Actual bond distance = 250 pm

Sn—I: Calculated bond distance = 141 pm + ½(266 pm) = 274 pm
    Actual bond distance = 270 pm

7.73. Electron configurations for the two ionizations of the K atom:
Electron configuration for K: $1s^22s^22p^63s^23p^64s^1$. The loss of one electron (1st ionization) gives a configuration of: $1s^22s^22p^63s^23p^6$. The loss of the second electron (2nd ionization) gives a configuration of: $1s^22s^22p^63s^23p^5$. The 1st ionization results in a cation that is isoelectronic with a noble gas (Ar) and is not anticipated to require a large amount of energy. However, the 2nd electron must be removed from the much lower energy, filled 3p orbitals, so a much greater amount of energy is required for this step than for the first ionization.

7.74. Trend in IE proceeding down a group:
For *s*- and *p*-block elements, first IE generally decreases down a group because the electron removed is increasingly farther from the nucleus, thus reducing the nucleus-electron attractive force.

7.75. Explain the trends in size:

(a) The decrease in atomic size across a period is due to the increasing nuclear charge with an increasing number of protons (that is, with an increase in effective nuclear charge). Given that the electrons are in the same outer energy level, the nuclear attraction for electrons results in a diminishing atomic radius.

(b) The size of a transition metal atom is measured by the size of the *ns* orbitals. For Fe, for example, with [Ar]$3d^6 4s^2$, its size depends on the extent of the 4*s* orbital. As $(n-1)d$ electrons are added across a transition metal series, the repulsive effect of adding *d* electrons is balanced by increasing nuclear charge. The extent of the 4*s* orbital, and so the size of the metal atom is affected only slightly. See "A Closer Look: Orbital Energies, Z*, and Electron Configuration."

7.76. Greatest difference between 1st IE and 2nd IE: Li, lithium
Of the four elements shown, only Li is in Group 1A (1). The loss of the first electron results in an ion with a filled outer shell. Removal of a second electron (from the 1*s* electron shell) would require a much larger amount of energy. In each of the other cases, the second electron is removed from the same subshell as the first electron.

7.77. The existence of Mg^{2+} and O^{2-} rather than their monopositive and mononegative counterparts can be argued from the experimental evidence that members of Group 2A (2) (Mg) typically lose 2 electrons while members of group 6A (16) (O) typically gain 2 electrons—both with the aim of becoming isoelectronic with a noble gas. One form of experimental evidence is in measuring the relative melting points of MgO versus a +1 salt (e.g. NaCl). MgO melts at about 2850 °C, while NaCl (a +1/–1) salt melts at about 800 °C.

The greater charges of the +2/–2 ions are one factor for the much greater melting point of MgO over that of the +1/–1 NaCl salt.

7.78. Explain the 1st and 2nd IE for Ca and K:
Ca is smaller than K, so we would expect the first IE of Ca to be greater than that of K. Once K has lost one electron, it has a noble gas (Ar) configuration. Removal of a second electron (from a filled electron shell) requires much additional energy. Ca, on the other hand, can lose a "second" electron to obtain a noble gas configuration with a much smaller amount of energy (smaller IE).

7.79. Regarding orbital energies:

(a) Orbital energies decrease across period 2, owing to the increased effective nuclear charge. With the electrons in "*p*" sublevels—and hence about the same average distance from the nucleus—the increasing nuclear charge (with increasing numbers of protons in the nucleus) results in a greater attraction for the valence electrons and a lower orbital energy.

(b) As the orbital energies decrease (become more negative), the amount of energy needed to remove an electron (the ionization energy) increases. With this decreasing orbital energy and increase in effective nuclear charge, an atom has a greater affinity to **gain** an electron.

(c) The data show:

Li	–520.0 kJ/mol	(2*s*)
Be	–899.2 kJ/mol	(2*s*)
B	–800.8 kJ/mol	(2*p*)
C	–1029 kJ/mol	(2*p*)

Li has the lowest effective nuclear charge and so the lowest ionization energy (IE). Removing a 2*s* electron results in a noble gas configuration for the Li$^+$ ion. Be has a higher IE because the effective nuclear charge is greater. The slightly lower IE for B (compared to Be) is because the 2*p* orbital energy is slightly less negative than for the 2*s* orbital, thus requiring slightly less energy to remove the 2*p* electron than the 2*s* electron. The general trend in increasing ionization energy resumes when moving from B to C.

7.80. Rationalize trends in IE for the elements: Si, P, S, Cl:

Generally, the increasing effective nuclear charge across a period causes ionization energy to increase. For sulfur, however, two of its four 2*p* electrons are paired in the same orbital. The electron-electron repulsion experienced by these electrons makes it easier to remove one of them, and the ionization energy of sulfur is lower than expected.

7.81. A plot of the atomic radii of the elements K—V shows a decrease in radius. With the mass of these elements increasing from K through V (as more protons, neutrons, and electrons are added), the density is expected to increase.

7.82. The fifth and sixth period transition metals have similar atomic radii. The sixth period transition metals have a higher mass, resulting in higher density for the sixth period transition metals.

7.83. Regarding elements 113 and 115:

(a) Electron configurations for elements 113 and 115

Element 113 is located in period 7, Group 3A (13) while 115 is in Group 5A (15).

The anticipated electron configurations are:

113: [Rn]$5f^{14}6d^{10}7s^27p^1$

115: [Rn]$5f^{14}6d^{10}7s^27p^3$

(b) Name an element in the same periodic group as 113 and 115.

Thallium is a group member of 3A (13) (like 113), while bismuth is a group member of 5A (15) (like 115).

(c) What atom could be used as a projectile to bombard Am to produce the element 113?

With 95 protons in an americium nucleus, 18 protons need to be added to form the 113 nucleus. Hence, an element with 18 protons would be called for (argon).

7.84. To form CaF$_3$, calcium would have to form a 3+ cation. Since calcium [in Group 2A (2)] normally forms 2+ cations (with a noble gas configuration), the formation of the 3+ cation would require considerable energy and is highly unlikely.

7.85. Regarding the formation of thionyl chloride:

(a) Orbital box notation for sulfur:

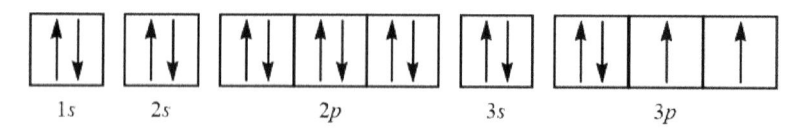

(b) Quantum numbers for the "highest energy electron":

$n = 3$, $\ell = 1$, $m_\ell = +1$ (or 0 or –1), $m_s = -½$

(c) Element with the smallest ionization energy: S

Element with the smallest radius: O

(d) S: Negative ions are always larger than the element from which they are derived.

(e) Mass of SCl$_2$ to prepare 675 g of SOCl$_2$:

$$675 \text{ g SOCl}_2 \cdot \frac{1 \text{ mol SOCl}_2}{119.0 \text{ g SOCl}_2} \cdot \frac{1 \text{ mol SCl}_2}{1 \text{ mol SOCl}_2} \cdot \frac{102.96 \text{ g SCl}_2}{1 \text{ mol SCl}_2} = 584 \text{ g SCl}_2$$

(f) Theoretical yield of SOCl$_2$ if 10.0 g of SO$_3$ and 10.0 g of SCl$_2$ are used:

Calculate the mass of SOCl$_2$ that could be obtained from each reactant if it were the limiting reactant:

$$10.0 \text{ g SO}_3 \cdot \frac{1 \text{ mol SO}_3}{80.06 \text{ g SO}_3} \cdot \frac{1 \text{ mol SOCl}_2}{1 \text{ mol SO}_3} \cdot \frac{119.0 \text{ g SOCl}_2}{1 \text{ mol SOCl}_2} = 14.9 \text{ g SOCl}_2$$

$$10.0 \text{ g SCl}_2 \cdot \frac{1 \text{ mol SCl}_2}{102.96 \text{ g SCl}_2} \cdot \frac{1 \text{ mol SOCl}_2}{1 \text{ mol SCl}_2} \cdot \frac{119.0 \text{ g SOCl}_2}{1 \text{ mol SOCl}_2} = 11.6 \text{ g SOCl}_2$$

Less SOCl₂ can be obtained from 10.0 g of SCl₂, so **SCl₂ is the limiting reactant**, and the theoretical yield of SOCl₂ is 11.6 g.

(g) For the reaction:

$SO_3(g) + SCl_2(g) \rightarrow SOCl_2(g) + SO_2(g)$ $\Delta_rH° = -96.0$ kJ/mol SOCl₂

$\Delta_rH°$ = [(1 mol SOCl₂/mol-rxn) · $\Delta_fH°$ SOCl₂ (g) + (1 mol SO₂/mol-rxn) · $\Delta_fH°$ SO₂(g)] – [(1 mol SO₃/mol-rxn) · $\Delta_fH°$ SO₃(g) + (1 mol SCl₂/mol-rxn) · $\Delta_fH°$ SCl₂ (g)]

–96.0 kJ/mol-rxn = [(1 mol SOCl₂/mol-rxn) · (–212.5 kJ/mol SOCl₂) + (1 mol SO₂/mol-rxn) · (–296.84 kJ/mol SO₂)] – [(1 mol SO₃/mol-rxn). · (–395.77 kJ/mol SO₃) + (1 mol SCl₂/mol-rxn) · ($\Delta_fH°$ SCl₂(g))]

–96.0 kJ/mol-rxn = [–509.34 kJ/mol-rxn] – [–395.77 kJ/mol-rxn +

(1 mol SCl₂/mol-rxn) · $\Delta_fH°$ SCl₂(g)]

–96.0 kJ/mol-rxn = –113.57 kJ/mol-rxn – (1 mol SCl₂/mol-rxn) · $\Delta_fH°$ SCl₂(g)

17.6 kJ/mol-rxn = –(1 mol SCl₂/mol-rxn) · $\Delta_fH°$ SCl₂(g)

$\Delta_fH°$ SCl₂(g) = –17.6 kJ/mol

7.86. For the reaction of sodium with chlorine to form sodium chloride:

(a) The reducing agent is Na. The low ionization energy of sodium plays a major role in making it a good reducing agent.

(b) The oxidizing agent is Cl₂. Among other properties, the element has a high electron affinity.

(c) Na₂Cl would have a Cl²⁻ ion. Adding a second electron to Cl⁻ means placing an electron in a higher energy electron shell. Conversely, NaCl₂ would have a Na²⁺ ion. Here one would have to remove the second electron from the atom's core.

7.87. Examine the effective nuclear charge for:

Using the rules (1)-(4):

(a) Calculate Z* for F, Ne: Relate the Z* to atomic radii and ionization energy

Z* (for F): 9 – [(2 · 0.85) + (6 · 0.35)] = 5.20

Z* (for Ne): 10 – [(2 · 0.85) + (7 · 0.35)] = 5.85

In general, there is a decrease in atomic radius and an increase in the ionization energy as the Z* increases.

(b) Calculate Z* for Mn (3d) electron and for a (4s) electron.

For a 3d electron in Mn, we calculate:

Mn: Rule 4 states that electrons in the same d or f group contribute 0.35, while all others (to the left) contribute 1.00. Those to the right (e.g. (4s²), do not shield. So we have 18 electrons which contribute 1.00 [$(1s^2)(2s^2\,2p^6)(3s^2\,3p^6)$] and 4 electrons [$(3d^5) - 1$] that contribute 0.35 each:

$Z^* = 25 - [(18 \cdot 1.00) + (4 \cdot 0.35)] = 5.60$

For a 4s electron:

$Z^* = 25 - [(10 \cdot 1.00) + (13 \cdot 0.85) + (1 \cdot 0.35)] = 3.60$

Examine the electron groupings for Mn: [$(1s^2)(2s^2\,2p^6)(3s^2\,3p^6)(3d^5)(4s^2)$]

There is 1 **other** 4s electron, 13(n – 1) electrons [$(3s^2\,3p^6)(3d^5)$], 10(n – 2) or lower electrons [$(1s^2)(2s^2\,2p^6)$].

In a manner parallel to O, F, and Ne in part (e), the lower Z^* for the 4s electrons indicates that Mn loses those electrons more easily than the 3d electrons—owing to the greater effective nuclear charge on the 3d electrons.

Solution and Answer Guide

Kotz Treichel Townsend Treichel, Chemistry and Chemical Reactivity 11e, 978-0-357-85140-1, Chapter 8: Bonding and Molecular Structure

TABLE OF CONTENTS

Applying Chemical Principles 283
Practicing Skills..285
General Questions..308
In the Laboratory...321
Summary and Conceptual Questions: 323

Applying Chemical Principles

Ibuprofen, A Study in Green Chemistry

8.1.1. Net change in the reaction:

The change in the bonding for the reaction shows a breaking of a C–O bond and a C≡O bond (from the CO) and forming C–C, C=O, and C–O bonds.

Energy input: 1 mol C≡O: 1 mol C≡O/mol-rxn · 1046 kJ/mol C≡O = 1046 kJ/mol-rxn
　　　　　　　1 mol C–O: 1 mol C–O/mol-rxn · 358 kJ/mol-rxn   = 358 kJ/mol-rxn
　　　　　　　　　　　　　　　　　　　　　　　　　　　Total input   = 1404 kJ/mol-rxn

Energy release: 1 mol C=O: 1 mol C=O/mol-rxn · 745 kJ/mol C–O = 745 kJ/mol-rxn
　　　　　　　　1 mol C–O: 1 mol C–O/mol-rxn · 358 kJ/mol C–O   = 358 kJ/mol-rxn
　　　　　　　　1 mol C–C: 1 mol C–C/mol-rxn · 346 kJ/mol C–C   = 346 kJ/mol-rxn
　　　　　　　　　　　　　　　　　　　　　　　　　Total released = 1449 kJ/mol-rxn

Energy change: = 1404 kJ/mol-rxn – 1449 kJ/mol-rxn = – 45 kJ/mol-rxn, so the reaction is exothermic.

8.1.2. All atoms in the molecule have a zero formal charge.

8.1.3. Most polar bond in molecule:

	H–C	C–C	C=O	O–H
$\Delta\chi$	0.3	0	1.0	1.3

Because the carboxylic acid loses the H⁺ in reaction, and from the calculations above, the O–H bond is the most polar in the molecule.

8.1.4. The polar "end" of the molecule (the carboxylic acid moiety) causes the molecule to be polar.

8.1.5. Shortest bond in the molecule: From the tables, we see:

Bond	**O–H**	C–H	C=O	C=C	C–C
Length(pm)	**94**	110	122	134	154

8.1.6. Highest bond order bonds: The C=O bond of the carboxylic acid moiety has bond order of two. The C=C bonds in the ring have bond order of 1.5 owing to the resonance of electrons in the benzene ring.

8.1.7. There are no 180° angles in the molecule, but there are several 120° angles: the C atoms in the ring (with C=C bonds), and the C=O in the carboxylic acid group all have 120° angles.

8.1.8. Volume of 0.0259 M NaOH solution:

$$200 \times 10^{-3} \text{ g IB} \left(\frac{1 \text{ mol IB}}{206.3 \text{ g IB}} \right) \left(\frac{1 \text{ mol NaOH}}{1 \text{ mol IB}} \right) \left(\frac{1 \text{ L}}{0.0259 \text{ mol IB}} \right) \left(\frac{1 \times 10^3 \text{ mL}}{1 \text{ L}} \right) = 37.4 \text{ mL}$$

van Arkel Triangles and Bonding

8.2.1. Using the electronegativity values, we construct the table below:

	GaAs	SBr₂	Mg₃N₂	BP	C₃N₄	CuZn	SrBr₂
Difference	0.4	0.4	1.7	0.2	0.5	0.3	2.0
Average	2.0	2.8	2.15	2.1	2.75	1.75	2.0
Region	Semi	Covalent	Ionic	Semi	Covalent	Metallic	Ionic

For the compounds, plot the difference and average on the van Arkel plot.
(a) Metallic compounds: CuZn

(b) Semiconductors: GaAs, BP, (B and As are metalloids)

(c) Ionic: Mg₃N₂, SrBr₂: Compounds are composed of metal and nonmetal.

(d) Bonding predicted for C₃N₄ is covalent.

(e) Covalent: SBr₂ and C₃N₄: In both compounds the elements are nonmetals.

8.2.2. Where does BeCl₂ lie on the van Arkel diagram?

For BeCl₂, the difference in electronegativities (y) = 1.6, and the average (x) = 2.4, which puts it at the same place on the van Arkel diagram as AlCl₃.—between ionic and covalent.

Linus Pauling and the Concept of Electronegativity

8.3.1. $(\chi_{Cl} - \chi_H) = ?$

Using the formula: $(\chi_{Cl} - \chi_H) = 0.102\sqrt{D(Cl-H) - \dfrac{D(Cl-Cl) + D(H-H)}{2}}$

$= 0.102\sqrt{432 - \dfrac{242 + 436}{2}} = 0.98$

Compare this to the value obtained from Figure 8.2: $\Delta\chi(Cl - H) = 3.2 - 2.2 = 1.0$

8.3.2. Predict bond dissociation energy for N–I bond:

Using the equation (found in 8.3.1) the NI bond energy(X) may be calculated as:

$3.0 - 2.7 = 0.102[x - (163 kJ/mol + 151 kJ/mol)/2]^{1/2}$ or $x = 2 \times 10^2$ kJ/mol

8.3.3. Calculate electronegativity for S:

$\chi_S = 1.97 \times 10^{-3}$ (IE $- \Delta_{ea}$H) + 0.19 = 1.97×10^{-3} (1000 $-$ ($-$200.41)) + 0.19

$\chi_S = 1.97 \times 10^{-3}$ (1200.41) + 0.19 = 2.55. This compares favorably with the value of 2.6 listed in Figure 8.2.

Practicing Skills

Valence Electrons and the Octet Rule

8.1.
Element	Group Number	Number of Valence Electrons
(a) O	6A (16)	6
(b) B	3A (13)	3
(c) Na	1A (1)	1
(d) Mg	2A (2)	2
(e) F	7A (17)	7
(f) S	6A (16)	6

8.2.
Element	Group Number	Number of Valence Electrons
(a) C	4A (14)	4
(b) Cl	7A (17)	7
(c) Ne	8A (18)	8

(d) Si	4A (14)	4
(e) Se	6A (16)	6
(f) Al	3A (13)	3

8.3. Lewis dot symbols

(a) ·Ċ·   (b) :Ö·   (c) ·Ṅ·   (d) :F̈·

8.4. Lewis dot symbols

(a) ·Ġa·   (b) ·Ġe·   (c) ·Äs·   (d) :B̈r·

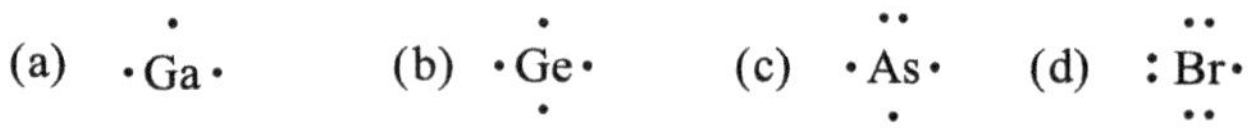

Electronegativity and Bond Polarity

8.5. Indicate the more polar bond (Arrow points toward the more negative atom in the dipole).

(a) C–O → > C–N →   (b) P–Cl → > P–Br →
(c) B–O → > B–S →   (d) B–F → > B–I →

8.6. For the bonds below, which atom is more negatively charged:

Bond   Atom more negatively charged
(a) C—N   N
(b) C—H   C
(c) C—Br   Br
(d) S—O   O

8.7 Increasing ionic bond character

H–H < C–H < O–H < O–Ca

Differences in electronegativity are greatest for O-Ca and least for H-H.

8.8 Increasing ionic bond character

O–F < S–O < Co–Cl < Na–F

Differences in electronegativity are greatest for Na–F and least for O–F.

8.9. For the bonds in acrolein the differences in electronegativity are:

	H–C	C–C	C=O	
Δχ	0.3	0	1.0	(Note that χ represents electronegativity)

(a) The C–C bonds are nonpolar, C–H bonds are slightly polar, and the C=O bond is polar.

(b) The most polar bond in the molecule is the C=O bond, with the oxygen atom being the negative end of the dipole.

8.10. Concerning the bonds in urea:

(a) All of the bonds in urea are polar.

(b) The C=O bond is the most polar bond; the O atom is the negative end of the dipole.

Lewis Electron Dot Structures

8.11.
Group Number	Number of Bonds
4A (14)	4
5A (15)	3
6A (16)	2
7A (17)	1

8.12 Number of bonds for each element

Element	Number of Bonds
O	2
N	3
C	4
F	1

8.13. Elements capable of forming hypervalent compounds:
(b) P, (e) Cl, (g) Se, and (h) Sn can accommodate more than four valence electron pairs. This behavior is possible in periods 3 and beyond.

8.14. Elements capable of forming hypervalent compounds:
(b) S and (c) Br can accommodate more than four valence electron pairs. This behavior is possible in periods 3 and beyond.

8.15. Lewis structures for molecules and ions
(a) NF$_3$: [1(5) + 3(7)] = 26 valence electrons

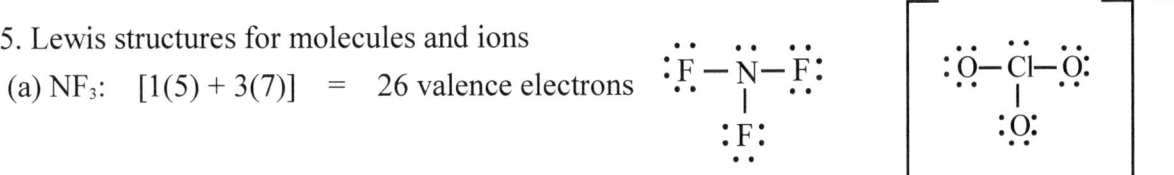

(b) ClO₃⁻: [1(7) + 3(6) + 1] = 26 valence electrons
            ↑
   add electron due to ion charge

(c) HOBr: [1(1) + 1(6) + 1(7)] = 14 valence electrons

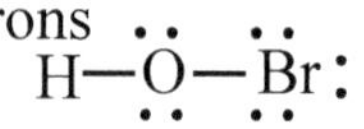

H—Ö—B̈r:

(d) SO₃²⁻: [1(6) + 3(6) + 2] = 26 valence electrons
            ↑
          ion charge

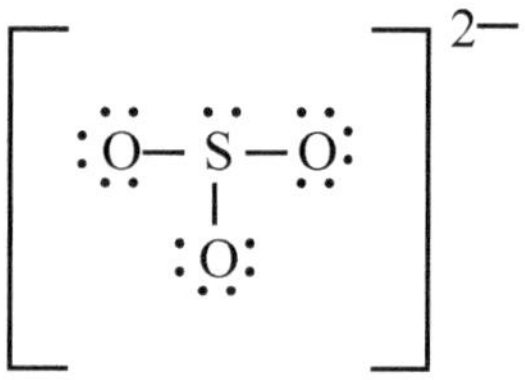

$$\left[\begin{array}{c}:\ddot{O}-\ddot{S}-\ddot{O}:\\|\\:\ddot{O}:\end{array}\right]^{2-}$$

8.16. Lewis structure for the following:

(a) CS₂: [1(4) + 2(6)] = 16 valence electrons

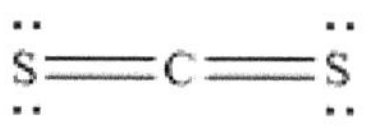

S̈=C=S̈

(b) BF₄⁻: [1(3) + 4(7) + 1] = 32 valence electrons

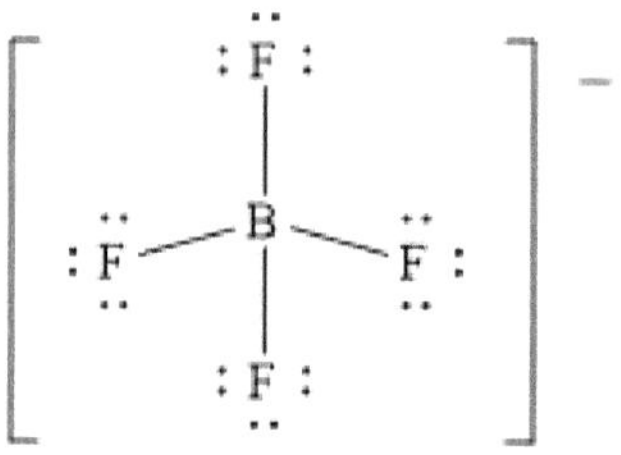

$$\left[\begin{array}{c}:\ddot{F}:\\|\\:\ddot{F}-B-\ddot{F}:\\|\\:\ddot{F}:\end{array}\right]^{-}$$

(c) HNO₂: [1(1) + 1(5) + 2(6)] = 18 valence electrons

H—Ö—N=Ö

(d) OSCl₂: [1(6) + 1(6) + 2(7)] = 26 valence electrons

:Ö:
|
:C̈l—S̈—C̈l:

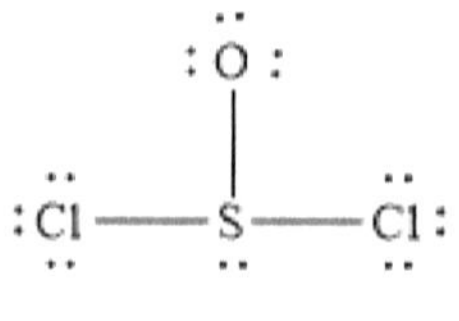

8.17. Lewis structure for the following:

(a) CHClF₂ : [1(4) +1(1) + 1(7) + 2(7)] = 26 valence electrons

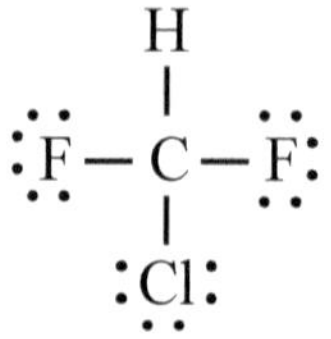

```
       H
       |
  :F̈—C—F̈:
       |
      :C̈l:
```

(b) C₂H₅CO₂H: [5(1) + 3(4) + 2(6) + 1(1)] = 30 valence electrons

```
      H   H   :O:
      |   |   ||
  H — C — C — C — Ö — H
      |   |       ¨
      H   H
```

(c) H₃CCN: [3(1) + 2(4) + 1(5)] = 16 valence electrons

```
       H
       |
   H — C — C ≡ N:
       |
       H
```

(d) H₂CCCH₂: [4(1) + 3(4)] = 16 valence electrons

```
       H       H
       |       |
   H — C = C = C — H
```

8.18. Lewis structures for the following:
(a) CH₃OH: [1(4) + 4(1) + 1(6)] = 14 valence electrons

```
       H
       |     ¨
   H — C — Ö — H
       |     ¨
       H
```

(b) C₂H₃Cl: [2(4) + 3(1) + 1(7)] = 18 valence electrons

```
    H       H
     \     /
      C = C
     /     \
    H      :Cl:
            ¨
```

(c) CH₂CHCN: [3(4) + 3(1) + 1(5)] = 20 valence electrons

```
    H       H
     \     /
      C = C
     /     \
    H       C
            ‖
            N:
            ¨
```

8.19. Show resonance structures for:

(a) SO₂: :Ö—S̈=O: ⟷ :O=S̈—Ö:

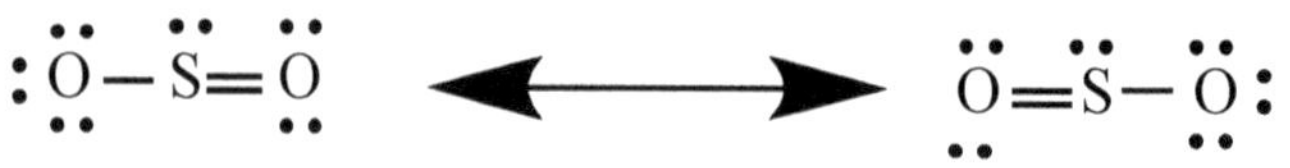

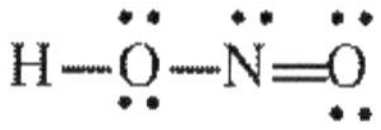

(b) HNO₂: H—Ö—N̈=O:

(c) HSCN: :N̈—C≡S—H ⟷ :N̈=C=S̈—H ⟷ :N≡C—S̈—H

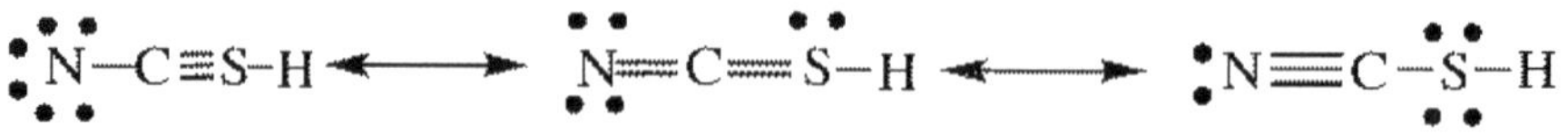

8.20. Show resonance structures for:

(a) nitrate ion:

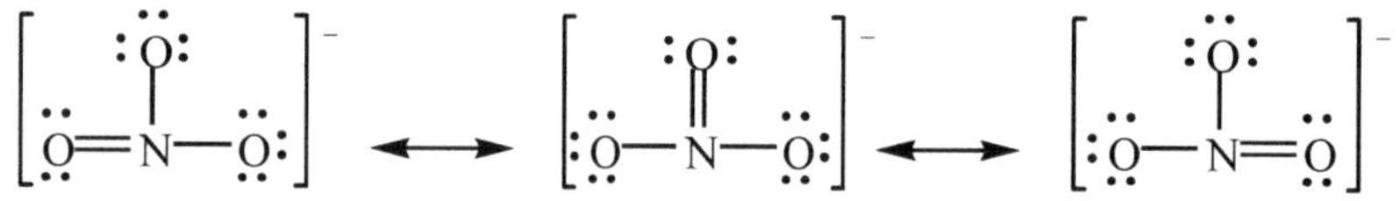

[:Ö—N(=Ö:)—Ö:]⁻ ⟷ [:Ö—N(=Ö:)—Ö:]⁻ ⟷ [:Ö—N(=Ö:)—Ö:]⁻

(b) nitric acid:

:Ö=N(—Ö:)—Ö—H ⟷ :Ö—N(=Ö:)—Ö—H

(c) dinitrogen monoxide:

:N≡N—Ö: ⟷ :N̈=N=Ö: ⟷ :N̈—N≡O:

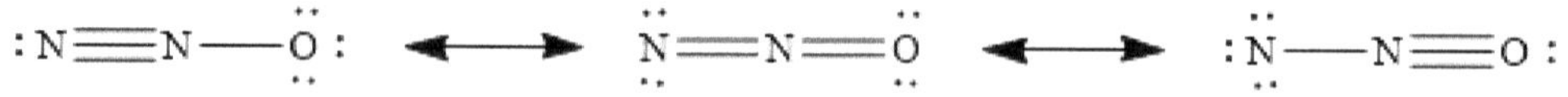

8.21. (a) BrF₃ :   [1(7) + 3(7)] = 28 valence electrons

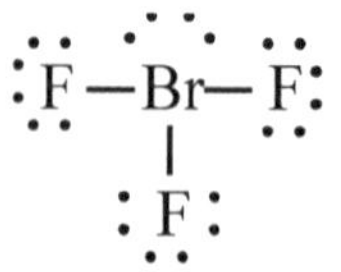

:F̈—Br—F̈:
    |
   :F̈:

(b) I₃⁻ :   [3(7) + 1] = 22 valence electrons

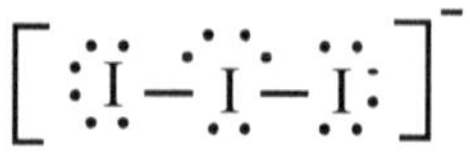

[:Ï—Ï—Ï:]⁻

(c) XeO₂F₂ :   [1(8) + 2(6) + 2(7)] = 34 valence electrons

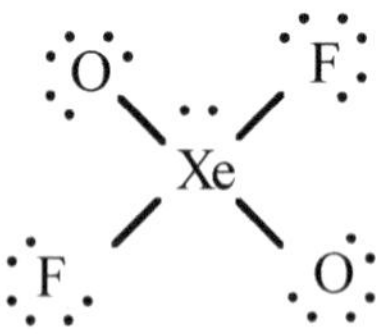

:Ö   F̈:
  \\ /
   Xe
  / \\
:F̈   Ö:

290

(d) XeF$_3^+$: [1(8) + 3(7) – 1] = 28 valence electrons

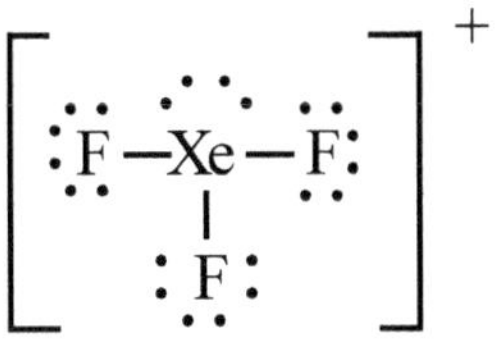

$$\left[\begin{array}{c} ..\\ :\ddot{F}-Xe-\ddot{F}:\\ |\\ :\ddot{F}:\\ .. \end{array} \right]^+$$

8.22. Lewis structure for the following:

(a) BrF$_5$: [1(7) + 5(7)] = 42 valence electrons

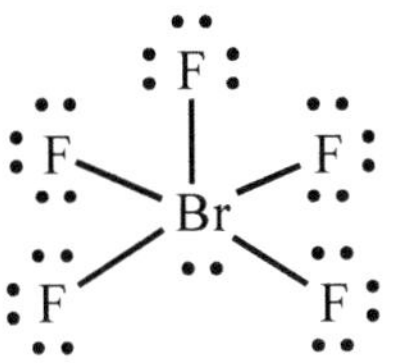

(b) IF$_3$: [1(7) + 3(7)] = 28 valence electrons

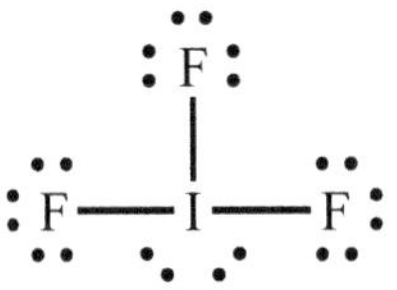

(c) IBr$_2^-$: [1(7) + 2(7) + 1] = 22 valence electrons

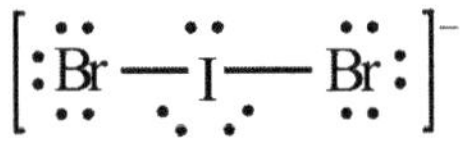

$$\left[:\ddot{Br}-\ddot{I}-\ddot{Br}: \right]^-$$

(d) BrF$_2^+$: [1(7) + 2(7) – 1] = 20 valence electrons

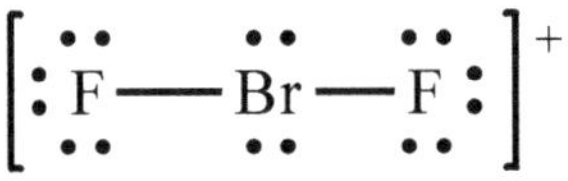

$$\left[:\ddot{F}-\ddot{Br}-\ddot{F}: \right]^+$$

Formal Charge

8.23. Formal charge on each atom in the following:

(a) N$_2$H$_4$

Atom	Formal Charge
H	1 – 1/2(2) = 0
N	5 – [2 + 1/2(6)] = 0

H–N̈–N̈–H with H H on top

(b) PO₄³⁻   Atom   Formal Charge
             P $5 - 1/2(8) = +1$
             O $6 - [6 +- 1/2(2)] = -1$
                     Sum = –3 (charge on ion)

(c) BH₄⁻   Atom   Formal Charge
             B $3 - 1/2(8) = -1$
             H $1 - 1/2(2) = 0$
                     Sum = –1 (charge on ion)

(d) NH₂OH   Atom   Formal Charge
             N $5 - [2 + 1/2(6)] = 0$
             H $1 - 1/2(2) = 0$
             O $6 - [4 + 1/2(4)] = 0$

8.24. Formal charge on each atom in the following:

(a) SCO   Atom   Formal Charge
             S $6 - [4 + {}^1/_2(4)] = 0$
             C $4 - [0 + {}^1/_2(8)] = 0$
             O $6 - [4 + {}^1/_2(4)] = 0$

(b) CHO₂⁻   Atom   Formal Charge
             H $1 - [0 + {}^1/_2(2)] = 0$
             C $4 - [0 + {}^1/_2(8)] = 0$
             O1 $6 - [4 + {}^1/_2(4)] = 0$
             O2 $6 - [6 + {}^1/_2(2)] = -1$

In the resonance hybrid, the average formal charge on each O is –1/2.

(c) CO₃²⁻   Atom   Formal Charge
             O1 $6 - [6 + {}^1/_2(2)] = -1$
             C $4 - [0 + {}^1/_2(8)] = 0$
             O2 $6 - [4 + {}^1/_2(4)] = 0$
             O3 $6 - [6 + {}^1/_2(2)] = -1$

In the resonance hybrid, the average formal charge on each O is –2/3.

(d) HCO₂H   Atom   Formal Charge
             H of C–H $1 - [0 + {}^1/_2(2)] = 0$
             C $4 - [0 + {}^1/_2(8)] = 0$
             O of C=O $6 - [4 + {}^1/_2(4)] = 0$

O of O–H $6 - [4 + \frac{1}{2}(4)] = 0$

H of O–H $1 - [0 + \frac{1}{2}(2)] = 0$

8.25. Formal charge on each atom in the following:

(a) NO_2^+

Atom	Formal Charge
O	$6 - [4 + 1/2(4)] = 0$
N	$5 - [0 + 1/2(8)] = +1$

$[\ddot{O}=N=\ddot{O}]^+$

(b) NO_2^-

Atom	Formal Charge
O1	$6 - [4 + 1/2(4)] = 0$
O2	$6 - [6 + 1/2(2)] = -1$
N	$5 - [2 + 1/2(6)] = 0$

$[\ddot{O}=N-\ddot{O}\!:]^-$
       1   2

There is a second equivalent resonance structure in which O1 has a formal charge of –1 and O2 has a formal charge of 0. Thus, the average formal charge on each oxygen is –1/2.

(c) NF_3

Atom	Formal Charge
F	$7 - [6 + 1/2(2)] = 0$
N	$5 - [2 - 1/2(6)] = 0$

$:\!\ddot{F}-\ddot{N}-\ddot{F}\!:$
            $|$
           $:\!\ddot{F}\!:$

(d) HNO_3

Atom	Formal Charge
O1	$6 - [4 + 1/2(4)] = 0$
O2	$6 - [6 + 1/2(2)] = -1$
O3	$6 - [4 + 1/2(4)] = 0$
N	$5 - [0 - 1/2(8)] = +1$
H	$1 - 1/2(2) = 0$

$H:\ddot{O}_1-N=\ddot{O}_3$
           $\ddot{O}_2$

There is a second equivalent resonance structure in which O2 has a formal charge of 0 and O3 has a formal charge of –1. Thus, the average formal charge on each of these oxygens is approximately –1/2.

8.26. Formal charge on each atom in the following:

(a) SO_2

Atom	Formal Charge
O of S–O	$6 - [6 + \frac{1}{2}(2)] = -1$
S	$6 - [2 + \frac{1}{2}(6)] = +1$
O of S=O	$6 - [4 + \frac{1}{2}(4)] = 0$

$:\!\ddot{O}-\ddot{S}=\ddot{O}$

In the resonance hybrid, the charge of each O is approximately –1/2.

(b) $OSCl_2$

Atom	Formal Charge
Cl	$7 - [6 + \frac{1}{2}(2)] = 0$
S	$6 - [2 + \frac{1}{2}(6)] = +1$
O	$6 - [6 + \frac{1}{2}(2)] = -1$

      $:\!\ddot{O}\!:$
        $|$
$:\!\ddot{Cl}-S-\ddot{Cl}\!:$

(c) O$_2$SCl$_2$

Atom	Formal Charge
O	$6 - [6 + \frac{1}{2}(2)] = -1$
S	$6 - [0 + \frac{1}{2}(8)] = +2$
Cl	$7 - [6 + \frac{1}{2}(2)] = 0$

(d) FSO$_3^-$

Atom	Formal Charge
O	$6 - [6 + \frac{1}{2}(2)] = -1$
S	$6 - [0 + \frac{1}{2}(8)] = +2$
F	$7 - [6 + \frac{1}{2}(2)] = 0$

8.27. Atom(s) on which the negative charge resides in:

(a) OH$^-$ Formal charges: O = –1 ; H = 0

Oxygen is much more electronegative than H so the negative charge should reside on oxygen. Thus, both formal charge and bond polarity considerations predict that the negative charge will reside on the O.

(b) BH$_4^-$

Formal charges: B = –1: H = 0

Even though formal charge considerations would place the negative charge on the B, hydrogen is slightly more electronegative than B (2.2 compared to 2.0), so the negative charge would reside on the H atoms (although the B–H bonds are **not very polar**).

(c) CH$_3$CO$_2^-$

Formal charges: H = 0; C = 0; O1 = 0; O2 = –1

There is a second equivalent resonance structure in which the formal charge on O1 = –1 and on O2 = 0, so each O is predicted to have a formal charge of –1/2.

Oxygen is more electronegative than C, so the charge would reside on the oxygens as opposed to the C.

Both formal charge and bond polarity considerations predict that the negative charge will be on the oxygen atoms.

8.28. Atom(s) on which the positive charge resides in:

(a) Even though the formal charge on O is +1 and on H is 0, H is less electronegative. The three H atoms therefore likely bear the +1 charge of the ion. The O—H bonds are polar with the H atom the positive end.

(b) Even though the formal charge on N is +1 and on H is 0, H is less electronegative. The four H atoms therefore likely bear the +1 charge of the ion. The N—H bonds are polar with the H atom the positive end.

(c) The formal charge on N is +1 and on O it is 0. This conforms with the relative electronegativities. The bonds are polar with N the positive end.

(d) The formal charge on N is +1 and on F it is 0. This conforms with the relative electronegativities. The bonds are polar with N the positive end.

8.29. Nitrosyl chloride resonance structures and formal charges:

(a) Structure 1   Structure 2

(b) Structure 1

Atom	Formal Charge
O	$6 - [4 + \frac{1}{2}(4)] = 0$
N	$5 - [2 + \frac{1}{2}(6)] = 0$
Cl	$7 - [6 + \frac{1}{2}(2)] = 0$

Structure 2

Atom	Formal Charge
O	$6 - [6 + \frac{1}{2}(2)] = -1$
N	$5 - [2 + \frac{1}{2}(6)] = 0$
Cl	$7 - [4 + \frac{1}{2}(4)] = +1$

(c) Structure 1 is dominant. The formal charges for structure 1 are all 0.

8.30. Resonance structures for thiocyanate ion:

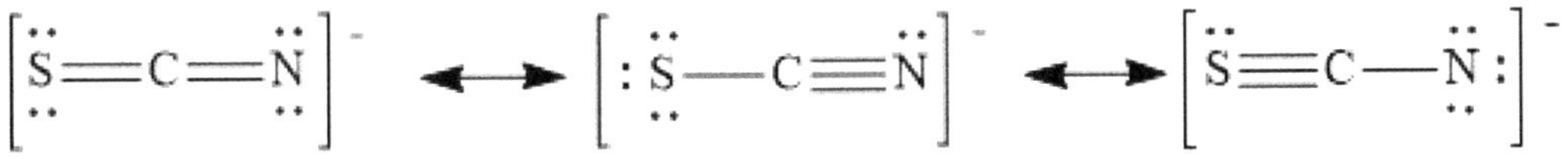

(a) Structure 1   Structure 2   Structure 3

(b) Structure 1

Atom	Formal Charge
S	$6 - [4 + \frac{1}{2}(4)] = 0$
C	$4 - [0 + \frac{1}{2}(8)] = 0$
N	$5 - [4 + \frac{1}{2}(4)] = -1$

Structure 2

Atom	Formal Charge
S	$6 - [6 + \frac{1}{2}(2)] = -1$
C	$4 - [0 + \frac{1}{2}(8)] = 0$
N	$5 - [2 + \frac{1}{2}(6)] = 0$

Structure 3 | Atom | Formal Charge |
|---|---|
| S | $6 - [2 + \frac{1}{2}(6)] = +1$ |
| C | $4 - [0 + \frac{1}{2}(8)] = 0$ |
| N | $5 - [6 + \frac{1}{2}(2)] = -2$ |

(c) Most reasonable representation of bonding: structure 1 with –1 formal charge on N

(d) Compare bonding and structure of SCN⁻ with OCN⁻: The central carbon has a double bond to both the sulfur and nitrogen in the dominant structure of SCN⁻ while the carbon has a triple bond to nitrogen and a single bond to oxygen in the dominant structure of OCN⁻.

8.31. Compare hydrogen carbonate ion and nitric acid electron dot structures:

(a) Structures isoelectronic?

HCO₃⁻ has 1 + 4 + 3(6) + 1 valence electrons = 24 valence electrons

HNO₃ has 1 + 5 + 3(6) valence electrons = 24 valence electrons

so **the structures are isoelectronic.**

(b) Resonance structures for each substance:

[Lewis structures shown: H–O₁–C(=O₃)–O₂ with bracket and –1 charge for HCO₃⁻; H–O₁–N(=O₃)–O₂ for HNO₃]

Note that the pictures above are **one** of **two** resonance structures that each substance possesses. The second structure for each would have two electron pairs between the C (or N) and the O labeled 2 in the diagrams and one electron pair between the C (or N) and the O labeled 3 in the diagrams.

(c) Formal charges of atoms in the structures: Let's calculate formal charges for the two structures shown above.

HNO₃ Atom	Formal Charge [Calculated as Valence e – (LP e + ½Bond e)]
O1	$6 - [4 + 1/2(4)] = 0$
O2	$6 - [6 + 1/2(2)] = -1$ Owing to resonance of the double bond,
O3	$6 - [4 + 1/2(4)] = 0$ O2 and O3 share a bonding pair of e's
N	$5 - [0 + 1/2(8)] = +1$ giving rise to –1/2 formal charge for each.
H	$1 - 1/2(2) = 0$

HCO₃⁻ Atom	Formal Charge
O1	$6 - [4 + 1/2(4)] = 0$
O2	$6 - [6 + 1/2(2)] = -1$ Owing to resonance of the double
O3	$6 - [4 + 1/2(4)] = 0$ bond, O2 and O3 share a bonding
C	$4 - [0 + 1/2(8)] = 0$ pair of e's giving rise to –1/2 formal
H	$1 - 1/2(2) = 0$ charge for each.

(d) Nitric acid is a strong acid, and loses the hydrogen readily, forming a –1 ion. Hydrogen carbonate bears a –1 charge, and the loss of the hydrogen causes the formation of a –2 ion (i.e. one of increased negative charge—and hence less "desirable"). So, hydrogen carbonate is a weak acid! Conversely, HCO_3^- is much more basic than HNO_3; the negative charge of HCO_3^- will attract a positively charged H^+.

8.32. Compare electron dot structures for carbonate and borate ions:

(a) Isoelectronic? Yes, both ions contain 24 valence electrons.

(b) CO_3^{2-} has three reasonable resonance structures, and BO_3^{2-} has four

(c)

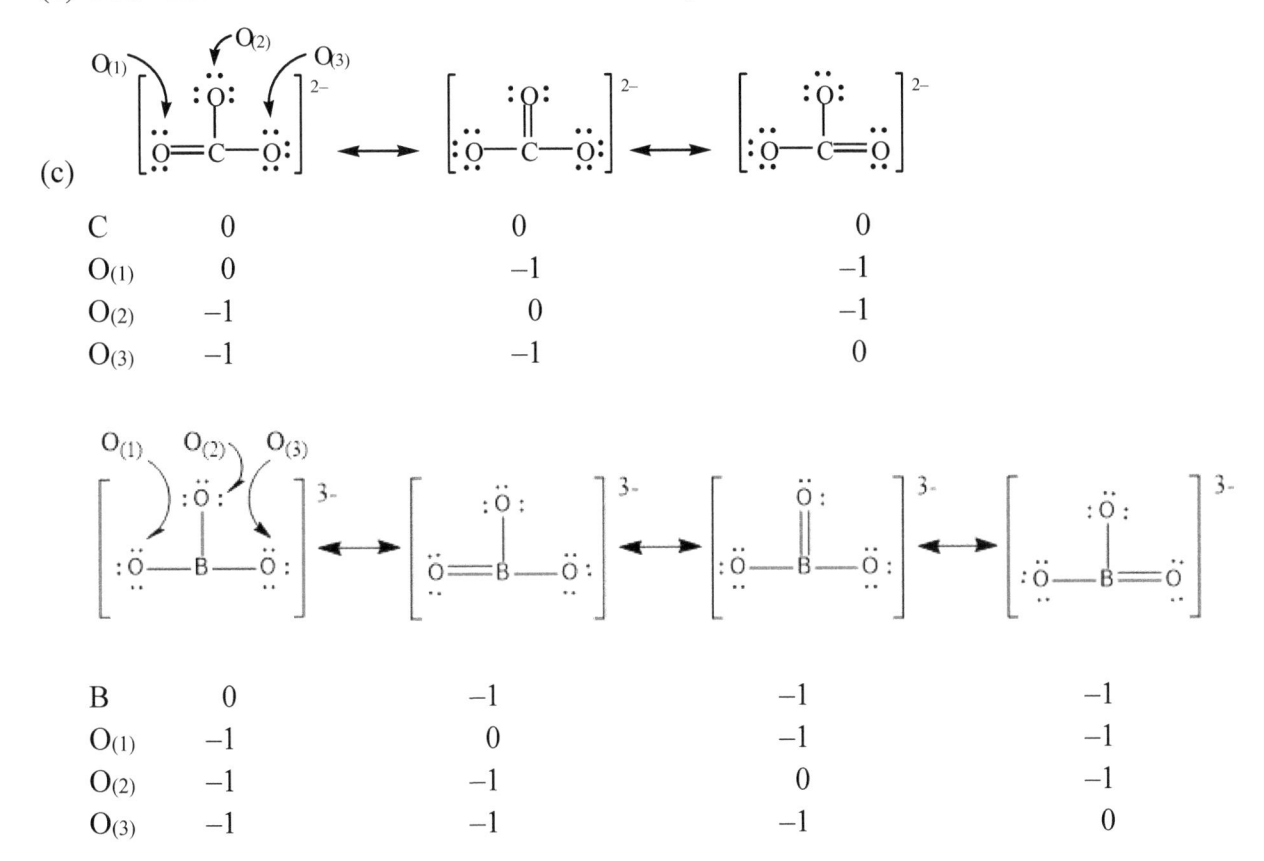

C	0	0	0	
O(1)	0	–1	–1	
O(2)	–1	0	–1	
O(3)	–1	–1	0	

B	0	–1	–1	–1
O(1)	–1	0	–1	–1
O(2)	–1	–1	0	–1
O(3)	–1	–1	–1	0

(d) The H^+ ion would attach to the oxygen atom.

8.33. For the nitrite ion and HNO_2:

(a)

Formal Charges:
O₁ 6 − [4 + 1/2(4)] = 0   6 − [6 + 1/2(2)] = −1
O₂ 6 − [6 + 1/2(2)] = −1   6 − [4 + 1/2(4)] = 0
N   5 − [2 + 1/2(6)] = 0   5 − [2 + 1/2(6)] = 0

(b) The preferred bonding of the H⁺ ion for O is predictable owing to the negative formal charges on the O atoms as opposed to the 0 charge on the N atom.

(c) Resonance structures for HNO₂:

$$\left[\ddot{\underset{..}{O}}=\ddot{N}-\underset{..}{\ddot{O}}-H\right] \longleftrightarrow \left[:\underset{..}{\ddot{O}}-\ddot{N}=\underset{..}{\ddot{O}}-H\right]$$
   1   2   1   2

Formal Charges:
O₁ 6 − [4 + 1/2(4)] = 0   6 − [6 + 1/2(2)] = −1
O₂ 6 − [4 + 1/2(4)] = 0   6 − [2 + 1/2(6)] = 1
N   5 − [2 + 1/2(6)] = 0   5 − [2 + 1/2(6)] = 0
H   1 − 1/2(2) = 0   1 − 1/2(2) = 0

The left structure is preferred, owing to the 0 formal charge on all atoms.

8.34. Regarding resonance structures for the formate ion:

$$\left[H-\underset{2}{C}\overset{\overset{\displaystyle :\ddot{O}:_1}{\|}}{-}\underset{..}{\ddot{O}:}\right]^- \longleftrightarrow \left[H-\underset{2}{C}\overset{\overset{\displaystyle :\ddot{O}:_1}{|}}{=}\underset{..}{\ddot{O}}\right]^-$$

H = 1 − [0 + ½(2)] = 0   H = 1 − [0 + ½(2)] = 0
C = 4 − [0 + ½(8)] = 0   C = 4 − [0 + ½(8)] = 0
O₁ = 6 − [4 + ½(4)] = 0   O₁ = 6 − [6 + ½(2)] = −1
O₂ = 6 − [6 + ½(2)] = −1   O₂ = 6 − [4 + ½(4)] = 0

The H⁺ ion would attach to the oxygen.

Molecular Geometry

8.35. Using the Lewis structure describe the electron-pair and molecular geometry:

(a). H−N̈−C̈l̈:   Electron-pair: tetrahedral
         | Molecular: trigonal pyramidal
         H

(b) :C̈l−Ö−C̈l: Electron-pair: tetrahedral
                         Molecular: bent or angular

(c) $[\ddot{\text{N}}=\text{C}=\ddot{\text{S}}]^-$ Electron-pair: linear
Molecular: linear

(d) H—Ö—F̈: Electron-pair: tetrahedral
Molecular: bent or angular

8.36. Using the Lewis structure describe the electron-pair and molecular geometry:

(a) $[\ddot{\text{F}}-\text{Cl}-\ddot{\text{F}}]^+$ Electron-pair: tetrahedral
Molecular: bent

(b) [Cl—Sn—Cl with Cl above]⁻ Electron-pair: tetrahedral
Molecular: trigonal pyramidal

(c) [PO₄]³⁻ Electron-pair: tetrahedral
Molecular: tetrahedral

(d) S=C=S Electron-pair: linear
Molecular: linear

8.37. Using the Lewis structure describe the electron-pair and molecular geometry:

(a) O=C=O Electron-pair: linear
Molecular: linear

(b) [Ö=N—Ö:]⁻

Electron-pair: trigonal planar
Molecular: bent or angular

(c) :Ö—Ö=Ö
Electron-pair: trigonal planar
Molecular: bent or angular

(d) [:Ö—Cl—Ö:]⁻ Electron-pair: tetrahedral
Molecular: bent or angular

For the species shown above, having at least one lone pair on the central atom **changes the molecular geometry from that of the electron-pair geometry.**

8.38. Using the Lewis structure describe the electron-pair and molecular geometry:

(a) [Lewis structure of CO₃²⁻]   Electron-pair: trigonal planar
Molecular: trigonal planar

(b) [Lewis structure of NO₃⁻]   Electron-pair: tetrahedral
Molecular: trigonal pyramidal

(c) [Lewis structure of SO₃²⁻]   Electron-pair: trigonal planar
Molecular: trigonal planar

(d) [Lewis structure of ClO₃⁻]   Electron-pair : tetrahedral
Molecular : trigonal pyramidal

8.39. Using the Lewis structure describe the electron-pair and molecular geometry. [Lone pairs on F have been omitted for clarity.]

(a) [ClF₂⁻]   Electron-pair: trigonal bipyramidal
Molecular: linear

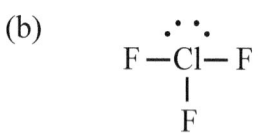

(b) [ClF₃]   Electron-pair: trigonal bipyramidal
Molecular: T-shaped

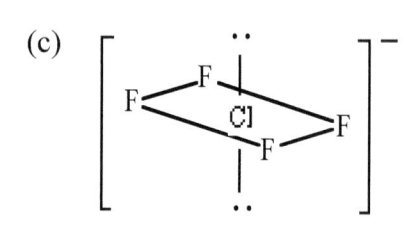

(c) [ClF₄⁻]   Electron-pair: octahedral
Molecular: square planar

(d)

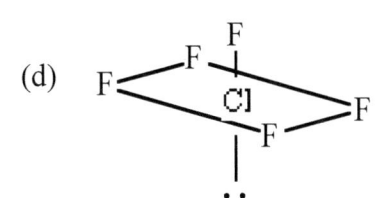

Electron-pair: octahedral
Molecular: square pyramidal

8.40. Using the Lewis structure describe the electron-pair and molecular geometry.

(a) 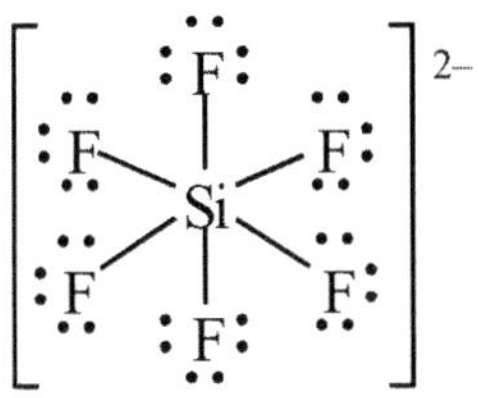 [SiF₆]²⁻

Electron-pair: octahedral
Molecular: octahedral

(b) 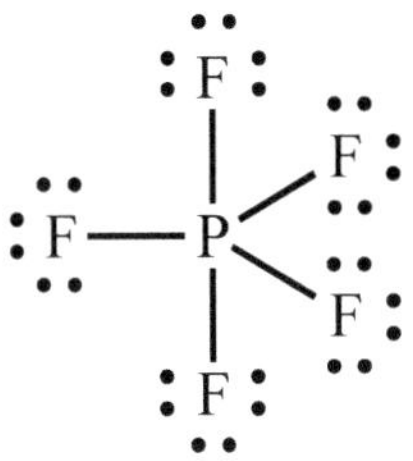 PF₅

Electron-pair: trigonal bipyramidal
Molecular: trigonal bipyramidal

(c) 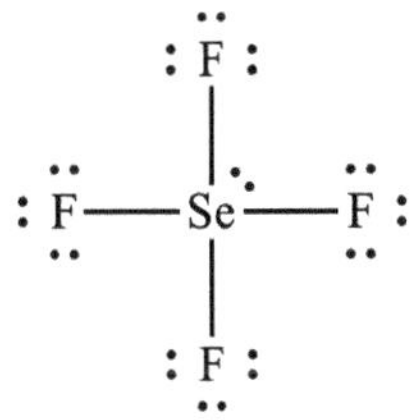 SeF₄

Electron-pair: trigonal bipyramidal
Molecular: seesaw

(d) 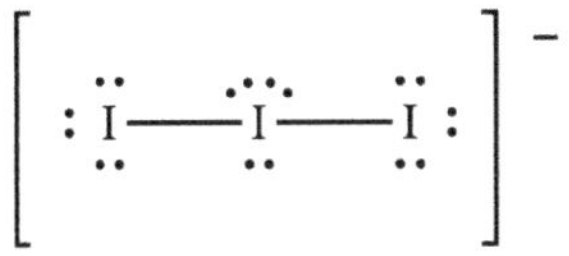 [I–I–I]⁻

Electron-pair: trigonal bipyramidal
Molecular: linear

8.41. Supply approximate values for the indicated bond angles:

(a) O–S–O angle in SO₂ : Slightly less than 120°; The lone pair of S should reduce the predicted 120° angle slightly.

(b) F–B–F angle in BF₃ : 120°

(c) Cl–C–Cl in Cl₂CO Around 120°

(d) (1) H–C–H angle in CH₃CN: 109°

(2) C–C≡N angle in CH₃CN: 180°

8.42. Supply approximate values for the indicated bond angles:

(a) Cl—S—Cl = 109°

(b) N—N—O = 180°

(c) Vinyl alcohol: angle 1 = 120°; angle 2 = 120°; angle 3 = 109°

8.43. Estimate the values of the angles indicated in the model of phenylalanine below:

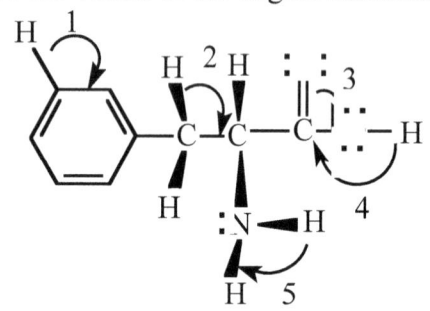

Angle 1: H–C–C   120°   three groups around the C atom
Angle 2: H–C–C   109°   four groups around the C atom
Angle 3: O–C–O   120°   three groups around the C atom
Angle 4: C–O–H   109°   four groups around the O atom
Angle 5: H–N–H   109°   four groups around the N atom

The CH$_2$–CH(NH$_2$)–CO$_2$H cannot be a straight line, since the first two carbons will have bond angles of 109° (with their connecting atoms) and the third C (the C of the CO$_2$H group) has a 120° angle with the C and O on either side.

8.44. Estimate the values of the angles indicated in the model of acetylacetone:

Angle 1 = 109°; Angle 2 = 120°; Angle 3 = 120°

Molecular Polarity

8.45. For the molecules:

	H$_2$O	NH$_3$	CO$_2$	ClF	CCl$_4$

(i) Using the electronegativities to determine bond polarity:
Δχ   1.3   0.8   1.0   0.8   0.7

The H–O bonds in water are the most polar of these bonds.

(ii) The nonpolar compounds are:

CO$_2$ The O–C–O bond angle is 180°, thereby canceling the C–O dipoles.

CCl$_4$ The Cl–C–Cl bond angles are approximately 109°, with the Cl atoms directed at the corners of a tetrahedron. Such an arrangement results in a net dipole moment of zero.

(iii) The F atom in ClF is more negatively charged. (Electronegativity of F = 4.0, Cl = 3.2)

8.46. Consider the bonding in the following molecules:

Molecule	$\Delta\chi$ for bond
CH_4	C—H = 2.5 – 2.2 = 0.3
NH_2Cl	N—Cl = 3.2 – 3.0 = 0.2
	N—H = 3.0 – 2.2 = 0.8
BF_3	F—B = 4.0 – 2.0 = 2.0
CS_2	C—S = 2.6 – 2.5 = 0.1

(i) The B—F bonds in BF_3 are the most polar.

(ii) CH_4, BF_3, and CS_2 are nonpolar molecules.

(iii) Positive, as electronegativity of N is greater than that of H.

8.47. Molecular polarity of the following: (a) $BeCl_2$, (b) HBF_2, (c) CH_3Cl, (d) SO_3

$BeCl_2$ and SO_3 are nonpolar—since the linear geometry (of $BeCl_2$) and the trigonal planar geometry (of SO_3) would give a net dipole moment of zero. For HBF_2, the hydrogen and fluorine atoms are arranged at the corners of a triangle. The "negative end" of the molecule lies on the plane between the fluorine atoms, and the H atom is the "positive end."
For CH_3Cl, with the H and Cl atoms arranged at the corners of a tetrahedron, the chlorine atom is the negative end and the H atoms form the positive end.

8.48. BCl_3, CF_4, and GeH_4 are nonpolar molecules, as they are symmetric, and any bond dipoles would cancel. For PCl_3, χ for chlorine is greater than χ for phosphorus. The phosphorus possesses the partial positive charge, while the chlorines possess partial negative charges. The center of the negative charges is centered between the three chlorine atoms.

Bond Order and Bond Length

8.49.

	Species	Bond Order : Bonded Atoms
(a)	H_2CO	1: C–H 2: C=O
(b)	SO_3^{2-}	1: S–O
(c)	NO_2^+	2: N=O
(d)	$CH_3C\equiv CH$	1: C–H 1: C–C 3: C≡C

8.50.

	Species	Bond Order : Bonded Atoms		
(a)	CN^-	3: C≡N		
(b)	CH_3CN	1: C–H	1: C–C	3: C≡N
(c)	SO_3	1: S–O	2: S=O	

with resonance structures, overall bond order = 1.33

(d) $CH_3CH=CH_2$ 1: C–H   1: C–C   2: C=C

8.51. In each case the shorter bond length should be between the atoms with smaller radii--if we assume that the bond orders are equal.

(a) B–Cl   B is smaller than Ga   (b) C–O   C is smaller than Sn
(c) P–O   O is smaller than S   (d) C=O   O is smaller than N

8.52. In each case the shorter bond length should be between the atoms with smaller radii--if we assume that the bond orders are equal.
(a) Si—O   O is smaller than N   (c) C—F   F is smaller than Br
(b) C—O   C is smaller than Si   (d) C≡N   Multiple bonds are shorter than single bonds.

8.53. Which has the longest bond length:

In order, the bond lengths are C–C > C–O > C=C > C=O

Multiple bonds are shorter than single bonds of similar elements. There is also a periodic trend involved. Oxygen atoms are smaller than carbon atoms.

8.54. Which has the shortest bond length:

In order, the bond lengths are N≡N < C≡C < N−N < C−C

Multiple bonds are shorter than single bonds of similar elements. There is also a periodic trend involved. Nitrogen atoms are smaller than carbon atoms.

8.55. The bond order for NO_2^+ is 2, for NO_2^- is 3/2 while the bond order for NO_3^- is 4/3. The Lewis dot structure for the NO_2^+ ion indicates that both NO bonds are double, while in NO_3^-, any resonance structure (there are three) shows one double bond and two single bonds. The nitrite ion has—in either resonance structure (there are two)—one double and one single bond. Hence the **NO bonds in the nitrate ion will be longest** while those in the **NO_2^+ ion will be shortest**.

8.56. Compare HCO_2^-, CO_3^{2-} ions, and CH_3OH:

In each of two resonance structures for HCO_2^- there is 1 CO single bond and one CO double bond, with a CO bond order = $^3/_2$.

In CH_3OH there is 1 CO single bond, with a CO bond order = 1.

In each of three resonance structures for CO_3^{2-}, there are 2 CO single bonds, and 1 CO double bond, with a CO bond order = $^4/_3$.

So CH_3OH has the longest CO bond (lowest bond order), and HCO_2^- has the shortest CO bonds (highest bond order).

Bond Strength and Bond Dissociation Enthalpy

8.57. The CO triple bond in carbon monoxide is shorter than the CO double bond in formaldehyde. The CO bond in carbon monoxide is a **triple bond**, thus it requires more energy to break than the CO double bond in H_2CO.

8.58. For hydrazine and N_2O, compare the N-N and N-O bonds for length and strength:

The nitrogen–nitrogen bond in hydrazine is a single bond with a bond order of 1. The nitrogen–nitrogen bond in N_2O has a bond order greater than one, so it has the shorter and stronger nitrogen–nitrogen bond.

8.59. Estimate the enthalpy change for the reaction: $H_2C=CH_2(g) + H_2O(g) \rightarrow CH_3CH_2OH(g)$

Begin by determining which bonds are broken, and which are formed.

Those are tabulated below using dissociation enthalpies from Table 8.8

Bond	Enthalpy	# bonds	Total
C–H	413	4	1652
C=C	610	1	610
O–H	463	2	926
Energy of bonds broken		total	3188
C–H	413	5	2065
C–C	346	1	346
C–O	358	1	358
O–H	463	1	463
Energy of bonds formed		total	3232

The overall enthalpy change is:

$\Delta_r H° = \Sigma D$(bonds broken) $- \Sigma D$(bonds formed) $= 3188$ kJ/mol-rxn $- 3232$ kJ/mol-rxn $= -44$ kJ/mol-rxn

How does this value compare to the value calculated from enthalpies of formation?

For the reaction: H$_2$C=CH$_2$(g) + H$_2$O(g) → CH$_3$CH$_2$OH(g)

$\Delta_r H° = (\Delta_f H°$ for CH$_3$CH$_2$OH(g)) $- [(\Delta_f H°$ for C$_2$H$_4$(g)) $+ (\Delta_f H°$ for H$_2$O(g))]$

$\Delta_r H° = (-235.3$ kJ/mol$)(1$ mol/mol-rxn$) - [(52.47$ kJ/mol$)(1$ mol/mol-rxn$) + (-241.83$ kJ/mol$)(1$ mol/mol-rxn$)]$

$\Delta_r H° = -45.9$ kJ/mol-rxn

8.60. Enthalpy change for the oxidation of methane to methanol:

Begin by determining which bonds are broken, and which are formed. Those are tabulated below using dissociation enthalpies from Table 8.8

Bond	Enthalpy	# bonds	Total
C–H	413	8	3304
O=O	498	1	498
Energy of bonds broken		*total*	*3802*
C–H	413	6	2478
C–O	358	2	716
O–H	463	2	926
Energy of bonds formed		*total*	*4120*

The overall enthalpy change is:

$\Delta_r H° = \Sigma D$(bonds broken) $- \Sigma D$(bonds formed) $= 3802$ kJ/mol-rxn $- 4120$ kJ/mol-rxn $= -318$ kJ/mol-rxn

How does this value compare to the value calculated from enthalpies of formation?

For the reaction: 2 CH$_4$(g) + O$_2$(g) → 2 CH$_3$OH(g)

$\Delta_r H° = 2$ mol/mol-rxn $\cdot \Delta_f H°$ (CH$_3$OH(g)) $- (2$ mol/mol-rxn $\cdot \Delta_f H°$ (CH$_4$(g)) $+ 1$ mol/mol-rxn $\cdot \Delta_f H°$ (O$_2$(g)))

$\Delta_r H° = (2$ mol/mol-rxn$)(-201.0$ kJ/mol$) - [(2$ mol/mol-rxn$)(-74.87$ kJ/mol$) + (1$ mol/mol-rxn$)(0$ kJ/mol$)]$

$\Delta_r H° = -402.0$ kJ/mol-rxn $- (-149.74$ kJ/mol-rxn $+ 0$ kJ/mol-rxn$) = -252.3$ kJ/mol-rxn

In this case, the estimate from bond-dissociation enthalpies and the value calculated using enthalpies of formation are significantly different.

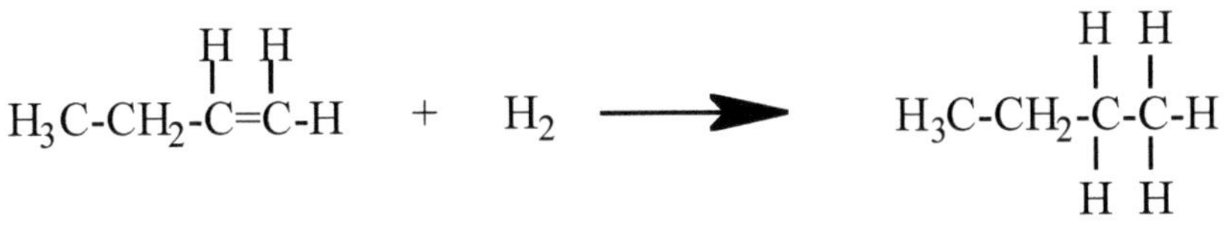

$$\text{H}_3\text{C-CH}_2\text{-C=C-H} \overset{\text{H H}}{\underset{}{|\ |}} + \text{H}_2 \longrightarrow \text{H}_3\text{C-CH}_2\text{-C-C-H} \overset{\text{H H}}{\underset{\text{H H}}{|\ |}}$$

8.61.
    Energy input: 1 mol C=C: (1 mol C=C/mol-rxn) · 610 kJ/mol C=C = 610 kJ/mol-rxn
                 1 mol H–H: (1 mol H–H/mol-rxn) · 436 kJ/mol H–H = 436 kJ/mol-rxn
                                                 Total input   = 1046 kJ/mol-rxn
    Energy released: 1 mol C–C: (1 mol C–C/mol-rxn) · 346 kJ/mol C–C = 346 kJ/mol-rxn
                    2 mol C–H: (2 mol C–H/mol-rxn) · 413 kJ/mol C–H = 826 kJ/mol-rxn
                                               Total released   = 1172 kJ/mol-rxn
    Energy change: 1046 kJ/mol-rxn – 1172 kJ/mol-rxn = – 126 kJ/mol-rxn

8.62. Using bond dissociation energies, estimate the enthalpy change for the formation of phosgene:

    Energy input: 1 mol C≡O: (1 mol C≡O/mol-rxn) · 1046 kJ/mol C≡O = 1046 kJ/mol-rxn
                          1 mol Cl–Cl: (1 mol Cl–Cl/mol-rxn) · 242 kJ/mol Cl–Cl = 242 kJ/mol-rxn
                                             Total input   = 1288 kJ/mol-rxn
    Energy released: 1 mol C=O: (1 mol C=O/mol-rxn) · 745 kJ/mol C=O = 745 kJ/mol-rxn
                      2 mol C–Cl: (2 mol C–Cl/mol-rxn) · 339 kJ/mol C–Cl = 678 kJ/mol-rxn
                                            Total released = 1423 kJ/mol-rxn
    Energy change: 1288 kJ/mol-rxn – 1423 kJ/mol-rxn = –135 kJ/mol-rxn

8.63.   OF$_2$ (g) + H$_2$O (g) → O$_2$ (g) + 2 HF (g)   Δ$_r$H° = –318 kJ/mol-rxn
    Energy input : 2 mol O–F: (2 mol O–F/mol-rxn) x (where x = O–F bond energy)
                     2 mol O–H: (2 mol O–H/mol-rxn) · 463 kJ/mol O–H = 926 kJ/mol-rxn
                                      Total input   = (926 + 2x) kJ/mol-rxn
    Energy released: 1 mol O=O: (1 mol O–O/mol-rxn) · 498 kJ/mol O=O = 498 kJ/mol-rxn
                    2 mol H–F: (2 mol H–F/mol-rxn) · 565 kJ/mol H–F = 1130 kJ/mol-rxn
                                       Total released   = 1628 kJ/mol-rxn
    –318 kJ kJ/mol-rxn = (926 kJ + 2x) kJ/mol-rxn – 1628 kJ/mol-rxn
    384 kJ/mol-rxn = 2x
    x = 192 kJ/mol = O–F bond energy

8.64. Using bond dissociation energies, estimate the enthalpy change for each carbon-carbon bond in benzene:

Energy input :

6 mol benzene carbon–carbon bonds: 6*x* (where *x* = benzene carbon–carbon bond energies)

6 mol C–H bonds: 6 mol C–H/mol-rxn · 413 kJ/mol C–H = 2478 kJ/mol-rxn

3 mol H–H bonds: 3 mol H–H/mol-rxn · 436 kJ/mol H–H = 1308 kJ/mol-rxn

Total input   = (3786 + 6*x*) kJ/mol-rxn

Energy released:

6 mol C–C bonds: 6 mol C–C/mol-rxn · 346 kJ/mol C–C = 2076 kJ/mol-rxn

12 mol C–H bonds: 12 mol C–H/mol-rxn · 413 kJ/mol C–H = 4956 kJ/mol-rxn

Total released = 7032 kJ/mol-rxn

Energy change: (3786 kJ + 6*x*) kJ/mol-rxn – 7032 kJ/mol-rxn

Set this quantity equal to $\Delta_r H° = -206$ kJ/mol/mol-rxn and solve for *x*.

3786 kJ/mol-rxn + 6*x* – 7032 kJ/mol-rxn = –206 kJ/mol-rxn

6*x* = 3040 kJ/mol-rxn

x = 506.7 kJ/mol

C-C single bond energy is 346 kJ/mol, C=C double bond energy is 610 kJ/mol. Bond order for carbon–carbon bonds in benzene is 1.5. Note that the energy calculated in this problem is between that of a single and double bond but closer to a double bond. This is due to the resonance stabilization of the carbon–carbon bonds in benzene.

General Questions

8.65. Number of valence electrons for Li, Ti, Zn, Si, and Cl:

Species	Li	Ti	Zn	Si	Cl
# valence electrons	1	4	2	4	7

The number of valence electrons for any main group (representative) element is determined by viewing the GROUP NUMBER in which that element resides. Ti—which is a transition element, has two *d* electrons and two *s* electrons, accounting for the 4—as the number of valence electrons. Zn has a *filled d* sublevel and two *s* electrons—giving rise to 2 valence electrons.

8.66. Show B-Cl electron distribution in BCl₃ and the formation of a bond to increase number of B bonds.

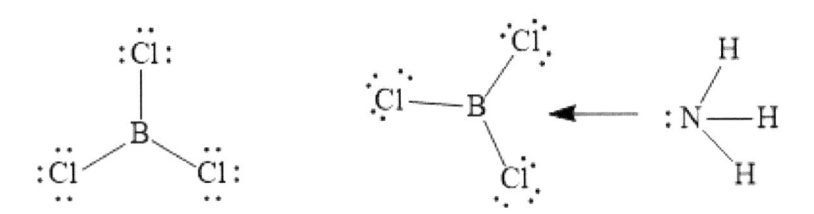

8.67. Which of the following do not have an octet surrounding the central atom: BF₄⁻, SiF₄, SeF₄, BrF₄⁻, XeF₄

BF₄⁻: 32 valence electrons   SiF₄: 32 valence electrons   SeF₄: 34 valence electrons

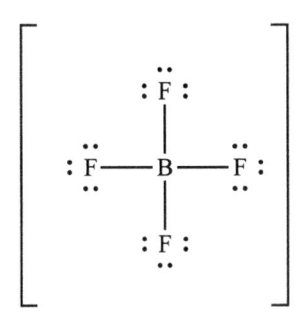

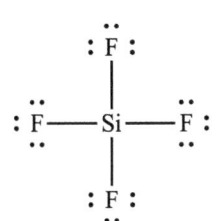

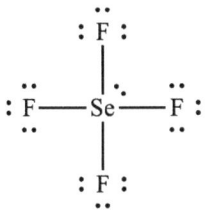

Octet   Octet   Exceeds octet

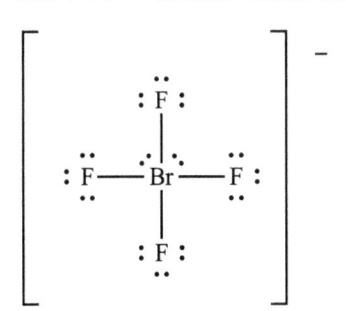

BrF₄⁻: 36 valence electrons   XeF₄: 36 valence electrons

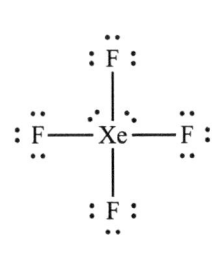

Exceeds octet   Exceeds octet

8.68. Central species obeying octet rule for the species: NO₂, SF₄, NH₃, SO₃, ClO₂ and ClO₂⁻

NO₂: 17 valence electrons

:Ö=N—Ö: ⟷ :Ö—N=Ö:

Odd-electron species; no octet on central N

SF₄: 34 valence electrons

(structure with S central, four F atoms)

Exceeds octet

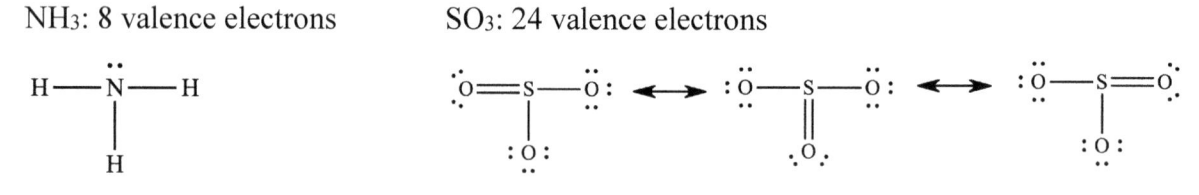

NH₃: 8 valence electrons

H—N(H)—H with lone pair on N

Octet

SO₃: 24 valence electrons

:Ö=S—Ö: ⟷ :Ö—S—Ö: ⟷ :Ö—S=Ö:
   | || |
  :Ö: :Ö: :Ö:

Octet

ClO₂: 19 valence electrons

:Ö—Cl—Ö:

Odd-electron species; no octet on central Cl

ClO₂⁻: 20 valence electrons

[:Ö—Cl—Ö:]⁻

Octet

NH₃, SO₃, and ClO₂⁻ follow the octet rule

8.69. For the formate ion, resonance structures and C–O bond order:

[Ö=C(H)—Ö:]⁻ ⟷ [:Ö—C(H)=Ö]⁻
   1   2   1   2

Bond order is calculated: (number of bonding electron pairs shared between two atoms). For the left structure: C–O(1) bond order = 2 (a double bond); C–O(2) bond order = 1 (a single bond). For the right structure, the opposite pattern is present. The average bond order is then 3/2 order. Alternatively, you can calculate this by the fact that there are three electron pairs shared between two carbon–oxygen linkages: 3/2.

8.70. Order the bonds in increasing bond length:

As these are all **single** bonds, the bond length is proportional to the atomic radii of the elements, hence: C—F < C—O < C—N < C—C < C—B

8.71. What bond dissociation enthalpies are needed for the reaction:

$$O_2(g) + 2 H_2(g) \rightarrow 2 H_2O(g)$$

One will need the H–H, the O–H and the O=O bond-dissociation enthalpies.

Calculation for the enthalpy change for the reaction:

Energy input : (1 mol O=O bonds/mol-rxn)(498 kJ/mol O=O bonds) = 498 kJ/mol-rxn
(2 mol H–H bonds/mol-rxn)(436 kJ/mol H–H bonds) = 872 kJ/mol-rxn
1370 kJ/mol-rxn

Energy released: (4 mol H–O bonds/mol-rxn)(463 kJ/mol H–O bonds) = 1852 kJ/mol-rxn

Energy change: 1370 kJ/mol-rxn – 1852 kJ/mol-rxn = –482 kJ/mol-rxn

The $\Delta_rH°$ for the reaction is ΣD(bonds broken) – ΣD(bonds formed). Note that by subtracting the bond energies of those bonds formed, the **effect** will be to *change the sign* of the bond energies of those bonds formed (since bond formation is always exothermic).

8.72. Principle of electroneutrality:

The electroneutrality principle states that the electrons in a molecule are distributed in such a way that the charges on the atoms are as close to zero as possible and that when a negative charge occurs it should be placed on the most electronegative atom. Similarly, a positive charge should be on the least electronegative atom.

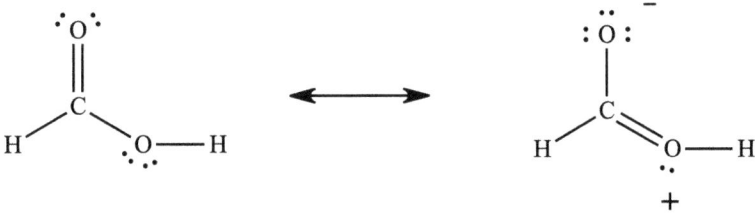

The right resonance structure places a positive formal charge on a very electronegative element, oxygen, and can be excluded.

8.73. Lewis structure(s) for the following: What are similarities and differences?
(a) CO₂   CH₄

formal charges   −1 0 +1   0 0 0   +1 0 −1

:Ö—C≡O: ⟷ :Ö=C=Ö: ⟷ :O≡C—Ö:

(b) N₃⁻

formal charges   −2 +1 0   −1 +1 −1   0 +1 −2

[:N̈—N≡N:]⁻ ⟷ [:N̈=N=N̈:]⁻ ⟷ [:N≡N—N̈:]⁻

(c) OCN⁻

formal charges   −1 0 0   0 0 −1   +1 0 −2

[:Ö—C≡N:]⁻ ⟷ [:Ö=C=N̈:]⁻ ⟷ [:O≡C—N̈:]⁻

Each of these species has 16 valence electrons, and each is linear. Carbon dioxide is neutral while the azide ion and isocyanate ion are charged. The isocyanate ion is polar, while carbon dioxide and azide are not. For CO₂ and N₃⁻, the structures with two double bonds are dominant, whereas for OCN⁻, the structure with an O–C single bond and a C≡N triple bond is dominant.

8.74. Regarding the SO₂ molecule: Ö=S—Ö: ⟷ :Ö—S=Ö
                                   δ− δ+ δ−   δ− δ+ δ−

Formal charges: 0 +1 −1   −1 +1 0

SO₂ has polar bonds, and is a polar molecule with a net dipole moment. The prediction is confirmed by the electrostatic potential surface (negative region around oxygen atoms and positive region around sulfur).

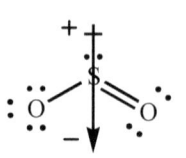

8.75. Bond orders in NO₂⁻ and NO₂⁺:

The Lewis dot structure for the NO₂⁺ ion (SQ8-25a) indicates that both NO bonds are double bonds (bond order = 2). The nitrite ion (NO₂⁻) (SQ8-25b) has—in either resonance structure —one double and one single bond (bond order = 1.5). The greater the bond order, the shorter the bond length, so 110 pm is the N–O bond distance in NO_2^+ and 124 pm is the N–O bond distance in NO_2^-.

8.76. Greater O–N–O bond angle:
NO_2^- has a smaller bond angle (about 120°) than NO_2^+ (180°). The former has a trigonal-planar electron pair geometry, whereas the latter is linear.

8.77. Compare the F–Cl–F angles in ClF_2^+ and ClF_2^-

$$\left[F - \ddot{\underset{..}{Cl}} - F \right]^+ \text{ and } \left[F - \overset{..}{\underset{..}{Cl}} - F \right]^-$$

(lone pairs around the F atoms are not shown)
The cation has 4 groups around the Cl atom—at a F–Cl–-F bond angle of 109°. The anion has 5 groups around the Cl atom, with a F–Cl–F bond angle of 180°.

8.78. Using the electron dot formulation for CN^-, should a proton attach to the C or the N:

$[:C{\equiv}N:]^-$ The negative (formal) charge resides on C, so the H^+ should attach to that atom and form HCN.

8.79. Draw the electron dot for SO_3^{2-}. Does the H^+ ion attach to the S or an O atom?

The Lewis dot structure for sulfite ion shows single bonds between each O and S atom, and one lone pair of electrons on the S atom. The formal charges show: for the S atom $(6 – [2 – 1/2(6)] = +1)$.

[group # – [lone pair electrons + 1/2(bonding electrons)]].

For the O atoms, the formal charges (all are identical) are $6 – [6 + 1/2(2)] = –1$.
So, one would predict that the H^+ would attach itself to the more negative O atoms.

8.80. Combustion of propane:

(a) $C_3H_8(g) + 5\ O_2(g) \longrightarrow 3\ CO_2(g) + 4\ H_2O(g)$

(b)

$$\underset{\underset{H}{|}}{\overset{\overset{H}{|}}{H:C}}:\underset{\underset{H}{|}}{\overset{\overset{H}{|}}{C}}:\underset{\underset{H}{|}}{\overset{\overset{H}{|}}{C}}:H \;+\; 5\ \ddot{O}::\ddot{O} \;\longrightarrow\; 3\ \ddot{O}::C::\ddot{O} \;+\; 4\ H:\ddot{O}:H$$

(c) Energy input:

8 mol C–H bonds	8 mol C–H/mol-rxn × 413 kJ/mol C–H =	3304 kJ/mol-rxn
2 mol C–C bonds	2 mol C–H/mol-rxn × 346 kJ/mol C–H =	692 kJ/mol-rxn
5 mol O=O bonds	5 mol O=O/mol-rxn × 498 kJ/mol O=O =	2490 kJ/mol-rxn
	Total energy input:	6486 kJ/mol-rxn

Energy released:

6 mol C=O bonds	6 mol C=O/mol-rxn × 803 kJ/mol C=O =	4818 kJ/mol-rxn
8 mol O–H bonds	8 mol O–H/mol-rxn × 463 kJ/mol O–H =	3704 kJ/mol-rxn
	Total energy release:	8522 kJ/mol-rxn

Energy change: 6486 kJ/mol-rxn – 8522 kJ/mol-rxn = –2036 kJ/mol-rxn

(d) $\Delta_rH° =$ [(3 mol CO$_2$/mol-rxn)(–393.51 kJ/mol CO$_2$) + (4 mol H$_2$O/mol-rxn)(–241.83 kJ/mol H$_2$O)] – [1 mol C$_3$H$_8$/mol-rxn)(–104.7 kJ/mol C$_3$H$_8$) + (5 mol O$_2$/mol-rxn)(0 kJ/mol O$_2$)

= –2043.15 kJ/mol-rxn

8.81. (a) Estimate the enthalpy change for the reaction.

2 CH$_3$OH(g) + 3 O$_2$(g) → 2 CO$_2$(g) + 4 H$_2$O(g)

Energy input: 6 mol C–H: (6 mol C–H/mol-rxn) · 413 kJ/mol C–H = 2478 kJ/mol-rxn

2 mol C–O: (2 mol C–O/mol-rxn) · 358 kJ/mol C–O = 716 kJ/mol-rxn

2 mol O–H: (2 mol O–H/mol-rxn) · 463 kJ/mol O–H = 926 kJ/mol-rxn

3 mol O=O: (3 mol O=O/mol-rxn) · 498 kJ/mol O=O = 1494 kJ/mol-rxn

Total input = 5614 kJ/mol-rxn

Energy release: 4 mol C=O: (4 mol C=O/mol-rxn) · 803 kJ/mol C=O = 3212 kJ/mol-rxn

8 mol H–O: (8 mol H–O/mol-rxn) · 463 kJ/mol H–O = 3704 kJ/mol-rxn

Total released = 6916 kJ/mol-rxn

Energy change: = 5614 kJ/mol-rxn – 6916 kJ/mol-rxn = –1302 kJ/mol-rxn

(b) The calculation using $\Delta_f H°$ values:

$\Delta_r H° = [2 \Delta_f H° (CO_2(g)) + 4 \Delta_f H° (H_2O(g))] - [2 \Delta_f H° (CH_3OH(g)) + 3 \Delta_f H° (O_2(g))]$

$\Delta_r H° = [(2 \text{ mol } CO_2/\text{mol-rxn}) \cdot -393.509 \text{ kJ/mol } CO_2 + (4 \text{ mol } H_2O/\text{mol-rxn}) \cdot -241.83 \text{ kJ/mol } H_2O] - [(2 \text{ mol } CH_3OH/\text{mol-rxn}) \cdot -201.0 \text{ kJ/mol} + (3 \text{ mol } O_2/\text{mol-rxn})) \cdot 0 \text{ kJ/mol } O_2]$

$\Delta_r H° = [-787.018 \text{ kJ/mol-rxn} + -967.32 \text{ kJ/mol-rxn}] - [-402.0 \text{ kJ/mol-rxn}]$

$\Delta_r H° = -1352.3 \text{ kJ/mol-rxn}$

8.82. Concerning acrylonitrile:

(a) Angle 1 = 120°; Angle 2 = 180°; Angle 3 = 120°

(b) The C=C bond is shorter than the C—C bond

(c) The C=C bond is stronger than the C—C bond

(d) Positive regions around H atoms, negative region around N

(e) The C≡N bond is the most polar; $\Delta\chi = 3.0 - 2.5 = 0.5$

(f) The molecule is polar, owing to its asymmetry.

8.83. Concerning the cyanate ion:

(a) Resonance structures for CNO⁻ with formal charges.

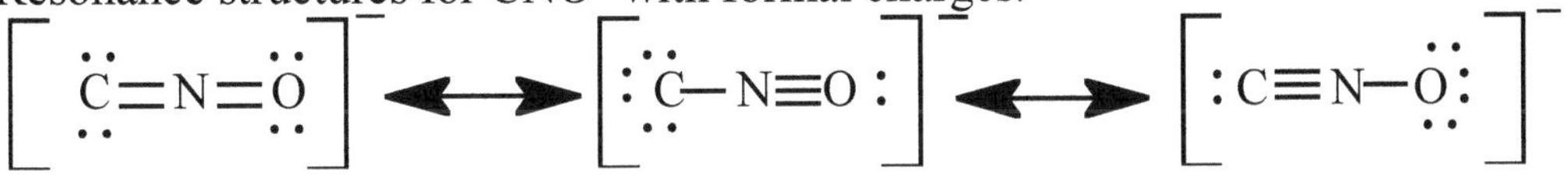

[Lewis structures: [C̈=N=Ö]⁻ ↔ [:C̈—N≡O:]⁻ ↔ [:C≡N—Ö:]⁻]

Formal Charges:
−2   +1   0   −3   +1   +1   −1   +1   −1

(b) The most reasonable structure is the one at the far right because it is the one in which the formal charges on the atoms are at a minimum, and oxygen has the negative formal charge.

(c) The instability of the ion could be attributed to the fact that the least electronegative atom in the ion bears a negative charge in all resonance structures.

8.84. Concerning vanillin:

(a) Angle 1 = 120°; Angle 2 = 109°; Angle 3 = 120°

(b) The C=O bond is the shortest carbon–oxygen bond in the molecule.

(c) The O—H bond is the most polar; $\Delta\chi = 3.5 - 2.2 = 1.3$

8.85. Explain the following geometries:

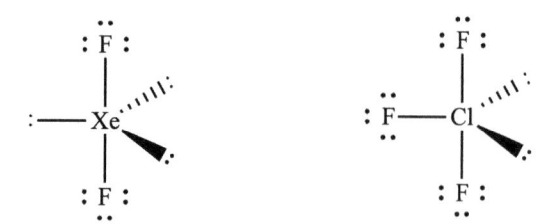

In the trigonal-bipyramidal electron-pair geometry, lone pairs occupy equatorial positions to reduce the number of unfavorable 90° interactions between the lone pairs and bonding pairs.

(a) In XeF$_2$ the bonding pairs occupy the axial positions (of a trigonal bipyramid) with the lone pairs located in the equatorial plane.

(b) In ClF$_3$ two of the three equatorial positions (of a trigonal bipyramid) are occupied by the lone pairs of electrons on the Cl atom.

8.86. Concerning nitryl chloride, ClNO$_2$:

(a) Lewis structures:

(b) NO bond order = (3 pairs linking NO)/(2 NO linkages) = $^3/_2$

(c) The electron-pair geometry is trigonal planar, the molecular geometry is trigonal planar, and all bond angles are approximately 120°.

(d) The NO bonds are the most polar; $\Delta\chi = 3.5 - 3.0 = 0.5$

The molecule is polar.

(e) Atom A = Cl; Atom B = O; Atom C = N in accordance with predictions.

8.87. Hydroxyproline has the structure:

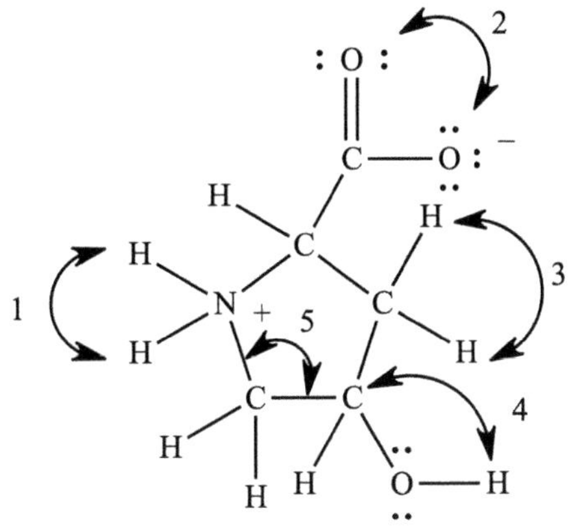

(a) Values for the selected angles:

Angle 1: 109° since the N has four groups of electrons around the atom.

Angle 2: 120° since the C atom has three groups of electrons around it.

Angle 3: 109° since the C atom has four groups of electrons around it.

Angle 4: 109° since the O atom has four groups of electrons around it

(2 bonding and 2 non-bonding pairs)

Angle 5: 109° since the C atom has four groups of electrons around it.

(b) The most polar bond in the molecule is the O–H bond. The electronegativity of O is 3.5, while that of H is 2.2 ($\Delta\chi = 1.3$). This difference represents the **greatest** difference in electronegativity in the hydroxyproline molecule.

8.88. For the amide whose skeletal structure is shown:

(a) a second resonance structure:

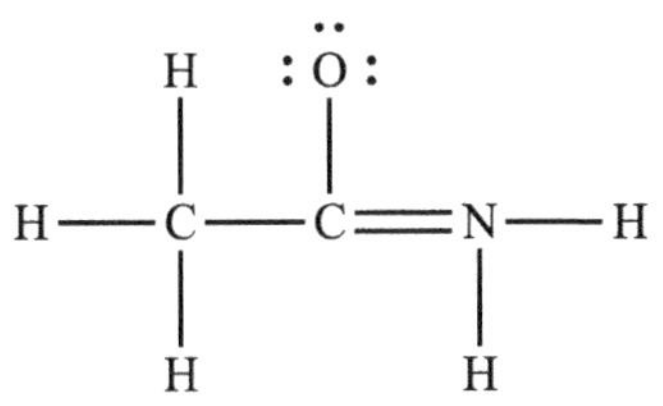

In this resonance structure there is a +1 formal charge on N and a –1 formal charge on O, whereas the other resonance structure has 0 formal charge on both O and N. Thus, the normally drawn resonance structure should be the major contributor as formal charges are minimized.

(b) The 120° bond angle suggests that this resonance structure may contribute to the resonance hybrid.

8.89. Enthalpy change for decomposition of urea:

$$H-\ddot{N}(H)-\overset{:\ddot{O}:}{\underset{||}{C}}-\ddot{N}(H)-H \longrightarrow H-\underset{\ddot{}}{N}(H)-\underset{\ddot{}}{N}(H)-H + :C\equiv O:$$

Break N–C bonds (2): (2 mol N–C/mol-rxn) · 305 kJ/mol N–C
 C=O bond (1):   1 mol C=O/mol-rxn · 745 kJ/mol C=O
 bonds broken: 610 kJ/mol-rxn + 745 kJ/mol-rxn = 1355 kJ/mol-rxn

Make N–N bond (1):   (1 mol N–N/mol-rxn) · 163 kJ/mol N–N
 C≡O bond (1):   (1 mol C≡O/mol-rxn) · 1046 kJ/mol C≡O
 bonds made: 163 kJ/mol-rxn + 1046 kJ/mol-rxn = 1209 kJ/mol-rxn

Change in energy: 1355 kJ kJ/mol-rxn – 1209 kJ/mol-rxn = 146 kJ/mol-rxn

Since there was no change in the number of N-H bonds (in urea) or hydrazine, those bond energies were omitted in the calculation.

8.90. Regarding 2-furylmethanethiol:

(a) Formal charges: S: $6 - [4 + \frac{1}{2}(4)] = 0$ O: $6 - [4 + \frac{1}{2}(4)] = 0$

(b) Angle 1 = 109°; Angle 2 = 109°; Angle 3 = 120°

(c) The C=C bonds are shorter than the C—C bonds, as multiple bonds are shorter than single bonds.

(d) The C–O bond is most polar, owing to differences in electronegativity.

(e) The molecule is polar.

(f) The four C atoms are planar and trigonal, so the ring as a whole is planar.

8.91. (a) Bond energies for the conversion of acetone to dihydroxyacetone:

acetone + O₂ ⟶ dihydroxyacetone

There are two mol of C–H bonds and one mol of O=O bond broken, and two mol O–C and two mol H–O bonds formed.

Energy input:

(2 mol C–H/mol-rxn) · 413 kJ/mol C–H = 826 kJ/ mol-rxn

(1 mol O=O/mol-rxn) · 498 kJ/mol O=O = 498 kJ/mol-rxn

Total input = 1324 kJ/mol-rxn

Energy release:
(2 mol O–C/mol-rxn) · 358 kJ/mol O–C = 716 kJ/mol-rxn
(2 mol H–O/mol-rxn) · 463 kJ/mol H–O = 926 kJ/mol-rxn
$$\text{Total released} = 1642 \text{ kJ/mol-rxn}$$

Energy change = 1324 kJ/mol-rxn – 1642 kJ/mol-rxn = –318 kJ for the reaction, so the reaction is **exothermic**.

(b) The central carbon in each molecule is trigonal planar. The three groups surrounding this carbon are not identical, so the molecules are polar.

(c) The O–H hydrogen atoms are the most positive in dihydroxyacetone because O and H have the greatest electronegativity difference and O is more electronegative than H.

8.92. Assess resonance structures for HNO_3:

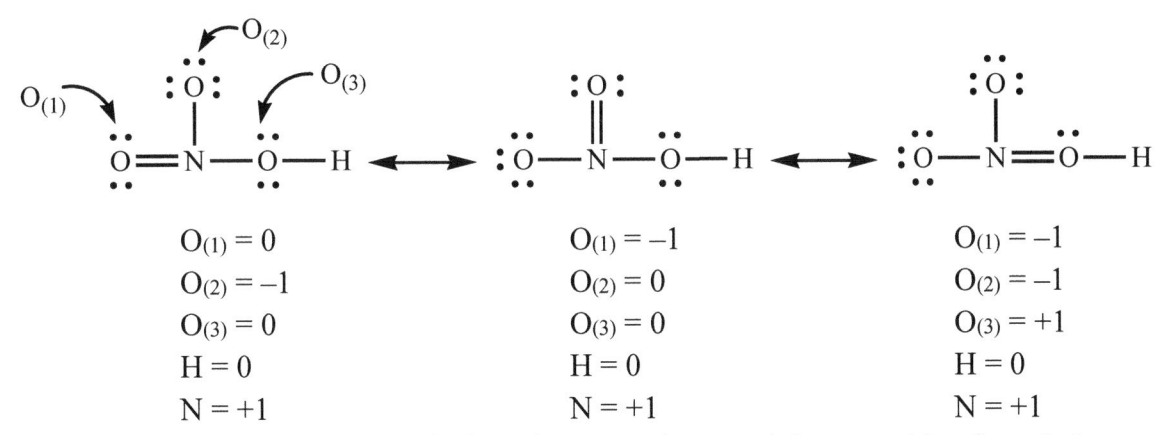

$O_{(1)} = 0$	$O_{(1)} = -1$	$O_{(1)} = -1$
$O_{(2)} = -1$	$O_{(2)} = 0$	$O_{(2)} = -1$
$O_{(3)} = 0$	$O_{(3)} = 0$	$O_{(3)} = +1$
$H = 0$	$H = 0$	$H = 0$
$N = +1$	$N = +1$	$N = +1$

The third resonance structure is the least important because it has a positive formal charge on one of the oxygen atoms.

8.93. For the synthesis of acrolein from ethylene:

(a) The C=C bond in acrolein is stronger than the C–C bond. Double bonds are stronger than single bonds between the same two atoms.

(b) The C–C bond is longer than a C=C bond. The stronger the bond, the shorter the bond length.

(c) Polarity of ethylene and acrolein:
For the bonds in acrolein the differences in electronegativity are as follows:

	H–C	C–C	C=O
$\Delta\chi$	0.3	0	1.0

The C–C and C=C bonds are nonpolar, the C–H bonds are very slightly polar, and the C=O bond is polar. The structure of ethylene (coupled with the relative lack of polarity of the C–H and C=C bonds) results in a molecule that is **nonpolar** (net dipole moment of 0). Acrolein however substitutes the polar C=O group (essentially inserted into a C–H bond) and gives a **polar** molecule.

(d) Is the reaction endothermic or exothermic? :

Examination of the structures reveals that to form acrolein from ethylene, we must break a C–H bond (which reforms in the product), and a C≡O bond (which forms the C=O bond in the product). In addition, a C–C bond forms. Using bond energies, the energy change is:

Break

C≡O bond (1): (1 mol C≡O/mol-rxn) · 1046 kJ/mol C≡O   = 1046 kJ/mol-rxn

C–H bond (1): (1 mol C–H/mol-rxn) · 413 kJ/mol C–H   = 413 kJ/mol-rxn

Energy input: 1459 kJ/mol-rxn

Make

C–C bond (1): (1 mol C–C/mol-rxn) · 346 kJ/mol C–C   = 346 kJ/mol-rxn

C–H bond (1): (1 mol C–H/mol-rxn) · 413 kJ/mol C–H   = 413 kJ/mol-rxn

C=O bond (1): (1 mol C=O/mol-rxn) · 745 kJ/mol C=O   = 745 kJ/mol-rxn

Energy released: 1504 kJ/mol-rxn

Change in energy: 1459 kJ/mol-rxn – 1504 kJ/mol-rxn = –45 kJ/mol-rxn , so the reaction is **exothermic**.

8.94. Regarding glycolaldehyde:

(a) Regarding molecular polarity: The molecule is polar.

(b) The positive partial charges are near the H atoms;

the negative partial charges are near the O atoms.

(c)   H—C≡C—C≡N:

8.95. For the reaction of acetylene with chlorine:

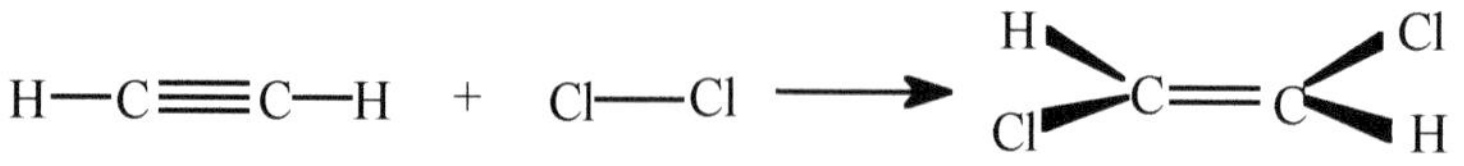

H—C≡C—H + Cl—Cl ⟶ (H)(Cl)C=C(Cl)(H)

Energy input:

1 mol C≡C: (1 mol C≡C/mol-rxn) · 835 kJ/mol C≡H =   835 kJ/mol-rxn

1 mol Cl–Cl: (1 mol Cl–Cl/mol-rxn) · 242 kJ/mol Cl–Cl = <u>242 kJ/mol-rxn</u>

Total input 1077 kJ/mol-rxn

Energy release:

1 mol C=C: (1 mol C=C/mol-rxn) · 610 kJ/mol C=C = 610 kJ/mol-rxn

2 mol C–Cl: (2 mol C–Cl/mol-rxn) · 339 kJ/mol C–Cl = <u>678 kJ/mol-rxn</u>

Total released   1288 kJ/mol-rxn

Energy change: = 1077 kJ/mol-rxn – 1288 kJ/mol-rxn = –211 kJ/mol-rxn

8.96. For the molecule epinephrine:

(a) Angle 1 = 109°; Angle 2 = 120°; Angle 3 = 120°; Angle 4 = 109°; Angle 5 = 109°

(b) The O—H bonds are most polar.

In the Laboratory

8.97. Methanol is the more polar solvent, as it contains the polar O–H bond, and an angular (or bent) geometry around the C–O–H bonds. The O also contains 2 lone pairs of electrons. Toluene on the other hand contains only C–C and C–H bonds, neither of which are particularly polar. No lone pairs of electrons are present in the molecule.

8.98. Regarding methylacetamide:

(a) Regarding molecular polarity: The molecule is polar.

(b) The partial negative charge is near the O atom and the most positive partial charge is near the N—H hydrogen atom, as confirmed by the electrostatic potential surface.

8.99. Regarding dimethylsulfide (DMS):

(a) Lewis structure of DMS and bond angles:

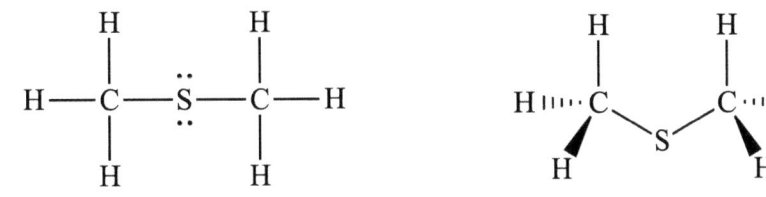

All bond angles are approximately 109°

(b) Location of positive and negative charges in the molecule.

The bonds are not especially polar, owing to the similar electronegativities of C, H, and S. There will be a *slightly* negative charge on the S. The C atoms would be only *slightly* positive. Due to the bent geometry around the sulfur, the molecule will be polar.

(c) Molecules of DMS present in 1.0 m³ of seawater with a concentration of 2.7nM DMS:

$$\frac{2.7 \times 10^{-9} \text{ mol DMS}}{1 \text{ L}} \left(\frac{1 \text{ L}}{1.00 \times 10^{-3} \text{ m}^3} \right) \left(\frac{6.02 \times 10^{23} \text{ molecules DMS}}{1 \text{ mol DMS}} \right) = 1.6 \times 10^{18} \text{ molecules DMS}$$

8.100. Regarding uracil:

(a) The O—C—N angle = 120° (with 3 groups around the C).

The C—N—H angle is predicted to be approximately 109° (with 4 groups around the N), but in reality it is close to 120°.

(b) As drawn, the C=C bond would be expected to be shorter than the C—C bond (as is typical of multiple bonds compared to single bonds between the same two atoms), but in reality they are approximately equal in length.

(c) The proton will attach at oxygen, the most negative area on the potential surface.

8.101. Regarding guanine:

(a) The most polar bond in the molecule is the C=O bond (as determined from differences in electronegativity).

(b) The N–C=N angle in the 6-member ring will be approximately 120° (with 3 groups around the C).

(c) The N–C=N angle in the 5-member ring will be approximately 120° (with 3 groups around the C).

(d) The bond angle around the N atom of the NH₂ group is approximately 109° (with 4 groups around the N).

8.102. Regarding deoxyribose:

(a) Rank bonds in terms of increasing polarity:

C–C < C–O < O–H bonds as is determined from electronegativity of the atoms.

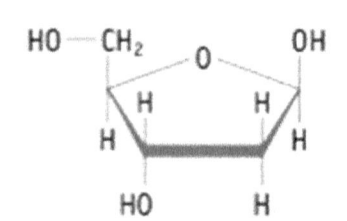

(b) Expected bond angles in the sugar ring:

All bonds in the sugar ring have either C or O connected to four groups, giving rise to bond angles of approximately 109°.

Summary and Conceptual Questions:

8.103. (a) To form PCl₃ from P₄ and Cl₂: P₄ + 6 Cl₂ → 4 PCl₃

(b) Δ_rH° = (4 · Δ_fH° for PCl₃(g)) − [(Δ_fH° for P₄ (g)) + (6 · Δ_fH° for Cl₂ (g))]

Δ_rH° = (−287.0 kJ/mol PCl₃)(4 mol PCl₃/mol-rxn) − [(+58.9 kJ/mol P₄)(1mol P₄/mol-rxn) + (0 kJ/mol Cl₂)(6 mol Cl₂/mol-rxn)]

Δ_rH° = (−1148.0 kJ/mol-rxn) − (+58.9 kJ/mol-rxn) = −1206.9 kJ/mol-rxn

(c) Estimate P–Cl bond energy from Enthalpy of reaction:

Δ_rH° = Σ D(bonds broken) − Σ D(bonds formed)

Δ_rH° = [(6 mol P–P bonds/mol-rxn)(201 kJ/mol P–P bonds) + (6 mol Cl–Cl bonds/mol-rxn)(242 kJ/mol/mol Cl–Cl bonds)] − (12 mol P–Cl bonds/mol-rxn)(X)

Note that each mol of PCl₃ contains 3 P–Cl bonds!

Also note that P₄ is a tetrahedron, with a **total** of 6 P–P bonds!

Using the enthalpy change from part A and calculating the total known bond energies:

−1206.9 kJ/mol-rxn = (2658 kJ/mol-rxn) − 12 mol P–Cl bonds/mol-rxn · X

12 mol P–Cl bonds/mol-rxn · X = (2658 kJ/mol-rxn) + (1206.9 kJ/mol-rxn)

12 mol P–Cl bonds/mol-rxn · X = 3865 kJ/mol-rxn

and X = (3865 kJ/mol-rxn)/(12 mol P–Cl bonds/mol-rxn) = 322.1 kJ/mol P–Cl bonds

The P–Cl bond energy from Table 8.8 is 326 kJ/mol P–Cl bond.

8.104. Concerning the formation of PCl₅:

Calculate enthalpy change for PCl₃(g) + Cl₂(g) → PCl₅(g)

Δ_rH° = (1 mol PCl₅/mol-rxn · Δ_fH°(PCl₅(g)) − [(1 mol PCl₃/mol-rxn · Δ_fH°(PCl₃(g)) + (1 mol Cl₂/mol-rxn · Δ_fH°(Cl₂(g))]

Δ_rH° = −374.9 kJ/mol-rxn − (−287.0 kJ/mol-rxn + 0 kJ/mol-rxn) = −87.9 kJ/mol-rxn

Using the Δ_rH° value, estimate the energy of formation of a P–Cl bond:
Net: Bonds broken: Cl–Cl; Bonds formed: 2 P–Cl; let x = D(P–Cl)

Δ_rH° = −87.9 kJ = (1 mol Cl–Cl/mol-rxn · D(Cl–Cl)) − (2 mol P–Cl/mol-rxn · x = 242 kJ/mol-rxn − 2 mol P–Cl/mol-rxn · x

x = 165 kJ/mol P–Cl

Thus, the bond-dissociation enthalpy for a P–Cl bond is 165 kJ/mol. Forming this bond releases this same amount of energy, so the enthalpy change for forming a P–Cl bond is –165 kJ/mol.

8.105. (a) Odd-electron molecules:

	HBr	BrO	HOBr
Valence electrons	1 + 7 = 8	7 + 6 = 13	1 + 6 + 7 = 14

BrO is an odd-electron molecule.

(b) Estimate energy reactions of the three reactions:

$Br_2(g) \rightarrow 2\ Br(g)$

1 · Br–Br bond energy = (1 mol Br–Br/mol-rxn) · 193 kJ/mol Br–Br = 193 kJ/mol-rxn

$2\ Br(g) + O_2(g) \rightarrow 2\ BrO(g)$

Energy input: 1 mol O=O: (1 mol O=O/mol-rxn) · 498 kJ/mol O=O = 498 kJ/mol-rxn
Energy released: 2 mol Br–O: (2 mol Br–O/mol-rxn) · 201 kJ/mol Br–O = 402 kJ/mol-rxn
  Energy change: 498 kJ/mol-rxn – 402 kJ/mol-rxn = 96 kJ/mol-rxn

$BrO(g) + H_2O(g) \rightarrow HOBr(g) + OH(g)$

Energy input:
1 mol Br–O: (1 mol Br–O/mol-rxn) · 201 kJ/mol Br–O = 201 kJ/mol-rxn
2 mol H–O: (2 mol H–O/mol-rxn) · 463 kJ/mol H–O = 926 kJ/mol-rxn
                                        Total energy input: 1127 kJ/mol-rxn

Energy released:
1 mol Br–O: (1 mol Br–O/mol-rxn) · 201 kJ/mol Br–O = 201 kJ/mol-rxn
2 mol H–O: (2 mol H–O/mol-rxn) · 463 kJ/mol H–O =   926 kJ/mol-rxn
                                        Total energy released: 1127 kJ/mol-rxn
Energy change: 1127 kJ/mol-rxn –1127 kJ/mol-rxn   = 0 kJ mol-rxn

(c) Molar enthalpy of formation of HOBr:

Consider the formation of two moles of HOBr from the elements; this corresponds to twice the reaction for the enthalpy of formation of HOBr.

$Br_2(g) + O_2(g) + H_2(g) \rightarrow 2\ HOBr(g)$

Energy input:
1 mol Br–Br: (1 mol Br–Br/mol-rxn) · 193 kJ/mol Br–Br = 193 kJ/mol-rxn
1 mol O=O: (1 mol O=O/mol-rxn) · 498 kJ/mol O=O =   498 kJ/mol-rxn
1 mol H–H: (1mol H–H/mol-rxn) · 436 kJ/mol H–H =   436 kJ/mol-rxn
                                        Total energy input: 1127 kJ/mol-rxn

Energy released:

2 mol Br–O: (2 mol Br–O/mol-rxn) · 201 kJ/mol Br–O = 402 kJ/mol-rxn

2 mol H–O: (2 mol H–O/mol-rxn) · 463 kJ/mol H–O = 926 kJ/mol-rxn

Total energy released: 1328 kJ/mol-rxn

Energy change: 1127 kJ/mol-rxn – 1328 kJ/mol-rxn = –201 kJ/mol-rxn

As we form 2 moles of HOBr in this process, divide to obtain –101 kJ/mol HOBr.

(d) The first two reactions in (b) are **endothermic**, while the 3rd is neither endo- nor exothermic. In (c), the process releases energy, so it is **exothermic**.

8.106. Regarding acrylamide:

(a) The bond angles around N are approximately 109°. All other angles are 120°

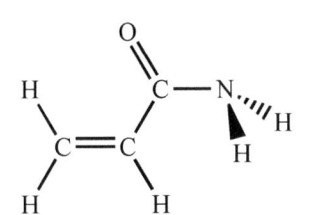

(b) The C=C bond is stronger than the C—C bond.

(c) The molecule is polar.

(d) $28 \text{ g chips} \cdot \dfrac{1.7 \text{ mg acrylamide}}{1000 \text{ g chips}} \cdot \dfrac{1 \text{ g}}{1000 \text{ mg}} \cdot \dfrac{1 \text{ mol acrylamide}}{71.1 \text{ g}} = 6.7 \times 10^{-7}$ mol acrylamide

8.107. (a) Three resonance structures for N$_2$O:

Structure 1   Structure 2   Structure 3

(b) Formal charge calculations:

Structure 1   First N formal charge = 5 – [4 + ½ (4)] = – 1

Middle N formal charge = 5 – [0 + ½ (8)] = +1

Oxygen formal charge = 6 – [4 + ½ (4)] = 0

Structure 2   First N formal charge = 5 – [2 + ½ (6)] = 0

Middle N formal charge = 5 – [0 + ½ (8)] = +1

Oxygen formal charge = 6 – [6 + ½ (2)] = – 1

Structure 3   First N formal charge = 5 – [6 + ½ (2)] = – 2

Middle N formal charge = 5 – [0 + ½ (8)] = +1

Oxygen formal charge = 6 – [2 + ½ (6)] = + 1

(c) Oxygen is more electronegative than N. Structures 1 and 3 have lower formal charges than structure 3. Structure 2 has the negative formal charge on the oxygen atom as well as the positive formal charge on the central N atom. It is the dominant resonance structure.

(d) For the reaction 2 N₂O(g) → 2 N₂(g) + O₂(g)

Energy input:

2 mol N≡N: (2 mol N≡N/mol-rxn) · 945 kJ/mol N≡N = 1890 kJ/mol-rxn

2 mol N–O: (2 mol N–O/mol-rxn) · 201 kJ/mol N–O =   402 kJ/mol-rxn
Total energy input: 2292 kJ/mol-rxn

Energy released:

2 mol N≡N: (2 mol N≡N/mol-rxn) · 945 kJ/mol N≡N = 1890 kJ/mol-rxn

1 mol O=O: (1 mol O=O/mol-rxn) · 498 kJ/mol O=O =   498 kJ/mol-rxn
Total energy released: 2388 kJ/mol-rxn

Energy change = 2292 kJ/mol-rxn – 2388 kJ/mol-rxn = –96 kJ/mol-rxn

This estimate is likely to be less accurate than most because the other resonance structures are also contributors to the molecular dynamics.

8.108. For the reaction O₃(g) + O(g) → 2 O₂(g):

The bond order for ozone, O₃, is 1.5. There are two oxygen–oxygen bonds in ozone. Let x = D(oxygen–oxygen bond in ozone)

(a) Energy input: (2 mol ozone bonds/mol-rxn) · x

Total energy input: (2 mol ozone bonds/mol-rxn) x
Energy released: (2 mol O=O/mol-rxn) · 498 kJ/mol O=O = 996 kJ/mol-rxn
Total energy released: 996 kJ
Energy change = (2 mol ozone bonds/mol-rxn) x – 996 kJ/mol-rxn
$\Delta_rH°$ = –392 kJ/mol = (2 mol ozone bonds/mol-rxn) x – 996 kJ/ mol-rxn
x = 302 kJ/mol ozone bond which is the estimate for the bond-dissociation enthalpy of the oxygen–oxygen bonds in ozone

(b) Bond dissociation energy for O–O is 146 kJ/mol

Bond dissociation energy for O=O is 498 kJ/mol

This calculation indicates that bond dissociation energy is in-between that of single and double bonds as expected.

(c) For the reaction O₂(g) → 2 O(g):

Energy input: (1 mol O=O/mol-rxn) · 498 kJ/mol O=O =   498 kJ/mol-rxn
Total energy input: 498 kJ/mol-rxn

Energy released: There are no bonds in O(g) so there is no energy released.

$$\text{Total energy released: } 0 \text{ kJ/mol-rxn}$$

Energy change = 498 kJ/mol-rxn – 0 kJ/mol-rxn = 498 kJ/mol-rxn

Determine the energy involved in a single reaction:

$$\frac{498 \text{ kJ}}{1 \text{ mol reactions}} \left(\frac{1 \text{ mol reactions}}{6.022 \times 10^{23} \text{ reactions}} \right) = 8.27 \times 10^{-22} \text{ kJ}/1 \text{ reaction}$$

Calculate the wavelength corresponding to this energy:

$$E = \frac{hc}{\lambda} \text{ thus } \lambda = \frac{hc}{E}$$

$$\lambda = \frac{(6.626 \times 10^{-34} \text{ J} \cdot \text{s})(2.998 \times 10^8 \text{ m/s})}{(8.27 \times 10^{-22} \text{ kJ})\left(\frac{1000 \text{ J}}{1 \text{ kJ}}\right)} = 2.40 \times 10^{-7} \text{ m} = 240. \text{ nm}$$

(d) For the reaction $O_3(g) \rightarrow O_2(g) + O(g)$:

Energy input: (2 mol ozone bonds/mol-rxn) · 302 kJ/mol ozone bonds = 604 kJ/mol-rxn
Total energy input: 604 kJ/mol-rxn
Energy released: (1 mol O=O/mol-rxn) · 498 kJ/mol O=O = 498 kJ/mol-rxn
There are no bonds in O(g) so there is no energy released: 0 kJ/mol-rxn
Total energy released: 498 kJ/mol-rxn
Energy change = 604 kJ/mol-rxn – 498 kJ/mol-rxn = 106 kJ/mol-rxn

Determine the energy involved in a single reaction:

$$\frac{106 \text{ kJ}}{1 \text{ mol reactions}} \left(\frac{1 \text{ mol reactions}}{6.022 \times 10^{23} \text{ reactions}} \right) = 1.76 \times 10^{-22} \text{ kJ}/1 \text{ reaction}$$

Calculate the wavelength corresponding to this energy:

$$E = \frac{hc}{\lambda} \text{ thus } \lambda = \frac{hc}{E}$$

$$\lambda = \frac{(6.626 \times 10^{-34} \text{ J} \cdot \text{s})(2.998 \times 10^8 \text{ m/s})}{(1.76 \times 10^{-22} \text{ kJ})\left(\frac{1000 \text{ J}}{1 \text{ kJ}}\right)} = 1.13 \times 10^{-6} \text{ m} = 1130 \text{ nm}$$

(e) The reaction $O_2(g) \rightarrow 2\ O(g)$ occurs in UV-C range and protects us from most of the radiation in the UV-C region entering the Earth's atmosphere.

The reaction $O_3(g) \rightarrow O_2(g) + O(g)$ theoretically should protect us from all three types of UV radiation. But the ozone UV absorption spectrum indicates that ozone does not absorb much UV radiation at wavelengths greater than 320 nm. So, ozone protects us from UV-C and much of the UV-B region but not much from UV-A.

Solution and Answer Guide

Kotz Treichel Townsend Treichel, Chemistry and Chemical Reactivity 11e, 978-0-357-85140-1, Chapter 9: Bonding and Molecular Structure: Orbital Hybridization and Molecular Orbitals

TABLE OF CONTENTS

Applying Chemical Principles 328
Practicing Skills 329
General Questions 338
In the Laboratory 346
Summary and Conceptual Questions 350

Applying Chemical Principles

Probing Molecules with Photoelectron Spectroscopy

9.1.1. In the photoelectric effect, electrons are ejected when light strikes the surface of a metal.

9.1.2. $E = h\nu = hc/\lambda = (6.626 \times 10^{-34}$ J · s$)(2.998 \times 10^8$ m/s$)/(58.4 \times 10^{-9}$ m$)$
$= 3.40 \times 10^{-18}$ J/photon
$(3.40 \times 10^{-18}$ J/photon$)(6.022 \times 10^{23}$ photons/mol$) = 2.05 \times 10^6$ J/mol $= 2.05 \times 10^3$ kJ/mol

9.1.3. Using the figure provided, the σ_{2p} has an ionization energy of 15.6 eV.

9.1.4. From problem 9.1.2, the energy is: $h\nu = 3.40 \times 10^{-18}$ J.
IE $= h\nu -$ KE $= 3.40 \times 10^{-18}$ J $- 4.23 \times 10^{-19}$ J $= 2.98 \times 10^{-18}$ J/electron
On a per mole basis:
$(2.98 \times 10^{-18}$ J/electron$)(6.022 \times 10^{23}$ electron/mol$)(1$ kJ/1000 J$) = 1.79 \times 10^3$ kJ/mol
The energy in eV: $(2.98 \times 10^{-18}$ J$)(1$ eV/1.60218×10^{-19} J$) = 18.6$ eV

9.1.5. The ionization energy corresponding to 18.6 eV is from an antibonding orbital. The ionization energies of 15.6 eV and 16.7 eV correspond to electrons from bonding orbitals. Loss of these bonding electrons results in a longer (and weaker) bond.

Green Chemistry, Safe Dyes, and Molecular Orbitals

9.2.1. Empirical formula of Tyrian purple:

The molecular formula is $C_{16}H_8Br_2N_2O_2$. The empirical formula would be C_8H_4BrNO.

9.2.2. Molecule absorbing higher energy light:

The energy of light is proportional to its frequency and inversely proportional to the wavelength. Hence, the nitrated form of butter yellow absorbs light at the shorter wavelength of 408 nm, that is, higher energy light.

9.2.3. Number of alternating double bonds in Tyrian purple? In nitrated butter yellow?

The alternating double bonds are principally those in the two 6-member rings (3 in each ring), with a C=C and two C=O bonds for a total of 9. Nitrated butter yellow contains a N=O bond and an N=N bond that would contribute to the overall resonance structure (in addition to the rings), so one could say that there are 8 alternating double bonds in nitrated butter yellow.

Practicing Skills

Valence Bond Theory

9.1. The Lewis electron dot structure of $CHCl_3$:

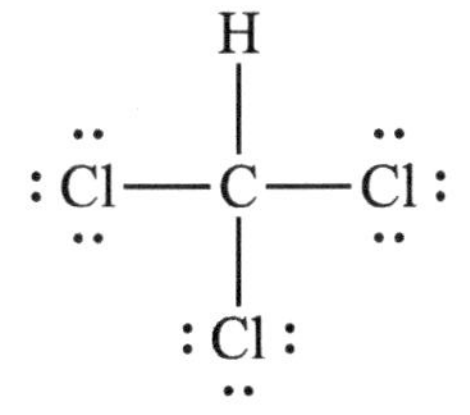

The electron pair geometry and molecular geometry are both tetrahedral. The H–C bond is a result of the overlap of a hydrogen 1s orbital with an sp^3 hybrid orbital on carbon. The Cl–C bonds are each formed by the overlap of an sp^3 hybrid orbital on carbon with a 3p orbital on chlorine.

9.2. Regarding NF_3:

The electron-pair geometry is tetrahedral and the molecular geometry is trigonal pyramidal. The N atom is sp^3 hybridized. Three of these hybrid orbitals each overlap a fluorine 2p orbital to form three N—F sigma bonds.

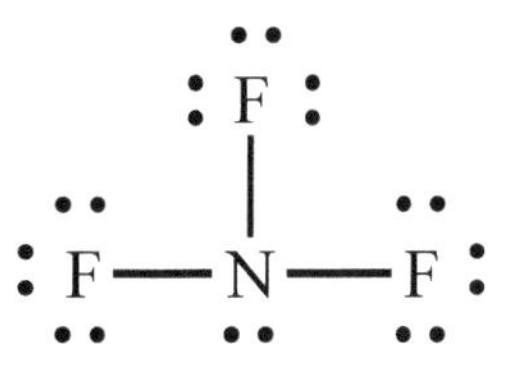

9.3. Lewis structure for hydroxylamine:

Nitrogen and oxygen will both use sp^3 hybridization in the molecule, as they both require 4 orbitals. N has bonds to 3 atoms and 1 LP of electrons. O has bonds to 2 atoms and 2 lone pairs of electrons.

One of the sp^3 hybrid orbitals on N will overlap with one of the sp^3 hybrid orbitals on O to form the N–O bond.

9.4. Concerning 1,1-dimethylhydrazine:

The nitrogen atoms are sp^3 hybridized. The bond between the nitrogen atoms is formed by overlap of an sp^3 hybrid orbital from each N atom.

9.5. Lewis structure for carbonyl fluoride, COF$_2$.

Electron-pair geometry on C is trigonal planar.

The molecular geometry on C is trigonal planar. The C utilizes sp^2 hybridization to form 3 equivalent orbitals. The O atom also utilizes sp^2 hybridization to form 3 equivalent orbitals. An sp^2 hybrid orbital for both C and O overlap to form the σ bond while the unhybridized $2p$ orbital on the C atom and on the O atom overlap to form the π bond.

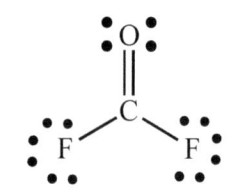

9.6. Concerning acetamide:

The electron-pair geometry and molecular geometry are trigonal planar at the "carbonyl" carbon, which is sp^2 hybridized. The σ bond is formed between a carbon sp^2 orbital and an oxygen sp^2 orbital. The π bond is formed between $2p$ orbitals on the carbon and oxygen atoms. The "methyl" C is sp^3 hybridized, with both electron-pair and molecular geometries being tetrahedral. C–C σ bond is formed by overlap of an sp^3 orbital on the "methyl" C with an sp^2 orbital on the "carbonyl" C, while the C–H σ bonds are formed by overlap of sp^3 orbitals on C with the $1s$ orbitals on H.

9.7. Electron-Pair and Molecular Geometry—Orbital sets used by the underlined atoms:

	Atom Molecule	Electron-Pair Geometry	Molecular Geometry	Hybrid Orbitals	Groups to be attached to the atom
(a)	$\underline{B}Br_3$	trigonal planar	trigonal planar	sp^2	3
(b)	$\underline{C}O_2$	linear	linear	sp	2
(c)	$\underline{C}H_2Cl_2$	tetrahedral	tetrahedral	sp^3	4
(d)	$\underline{C}O_3^{2-}$	trigonal planar	trigonal planar	sp^2	3

9.8. Electron-Pair and Molecular Geometry—Orbital sets used by the underlined atoms:

	Atom Molecule	Electron-Pair Geometry	Molecular Geometry	Hybrid Orbitals	Groups to be attached to the atom
(a)	$\underline{C}Se_2$	linear	linear	sp	2
(b)	$\underline{S}O_2$	trigonal planar	bent	sp^2	3
(c)	$\underline{C}H_2O$	trigonal planar	trigonal planar	sp^2	3
(d)	$\underline{N}H_4^+$	tetrahedral	tetrahedral	sp^3	4

9.9. Hybrid orbital sets used by the underlined atoms:

(a) the C atoms and the O atom in dimethyl ether: $\underline{C}$: sp^3; $\underline{O}$: sp^3

In the case of either the carbon or oxygen atoms in dimethyl ether, each atom is bound to *four other groups*. This would require **four orbitals**.

(b) The carbon atoms in propene: $\underline{C}H_3$: sp^3; $\underline{C}H$ and $\underline{C}H_2$: sp^2

The methyl carbon (CH_3) is attached to four groups (3 H and 1 C), so it needs four orbitals. The methylene (CH_2) and methine (CH) carbon atoms have bonds to three groups (2 H and 1 C) in the case of CH_2 and three groups (1 H and 2 C) in the case of CH.

(c) The C atoms and the N atom in glycine: $\underline{N}$: sp^3; $\underline{C}H_2$: sp^3; $\underline{C}$=O: sp^2

The N atom is attached to 4 groups (3 atoms and 1 lone pair), the CH_2 carbon has 4 groups attached (2 H atoms and 1 N atom and 1 C atom). The carbonyl carbon is attached to only 3 groups (1 C and 2 O atoms).

9.10. Describe the hybrid orbital set used by each of the following:

	Underlined atom	Hybrid orbitals
(a)	N	both N atoms are sp^3 hybridized
	C	sp^2
(b)	C of CH₃	sp^3
	C of C=C and C=O	both C atoms are sp^2 hybridized
(c)	C of C=C	sp^2
	C of C≡N	sp

9.11. Lewis structures for HSO₃F and SO₃F⁻:

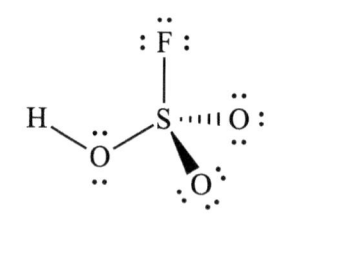

Molecular geometry	tetrahedral (around S)	tetrahedral (around S)
Hybridization of S	sp^3	sp^3

9.12. For the acid HPO₂F₂ and the anion PO₂F₂⁻:

Structure

Molecular geometry	tetrahedral (around P)	tetrahedral (around P)
Hybridization of P	sp^3	sp^3

9.13. For the molecule COCl₂: Hybridization of C = sp^2

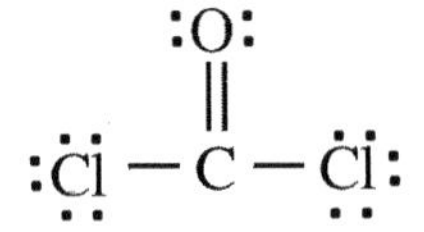

$$\begin{array}{c} :\!\ddot{O}\!: \\ \| \\ :\!\ddot{C}\!l\!-\!C\!-\!\ddot{C}\!l\!: \end{array}$$

1 σ bond between each chlorine and carbon (sp^2 hybrid orbital on carbon atom and $3p$ orbital on chlorine atom)

1 σ bond between carbon and oxygen (sp^2 hybrid orbitals on each)

1 π bond between carbon and oxygen ($2p$ orbital on each)

9.14. Hybridization of C atoms in benzene? What about σ and π bonding?

The C atoms are sp^2 hybridized. Each carbon uses two sp^2 orbitals to form C–C σ bonds with an sp^2 orbital of each adjacent carbon atom and an sp^2 orbital to form a C–H σ bond with a $1s$ orbital of a hydrogen atom. Each carbon uses a $2p$ orbital to form a π bond with another carbon atom's $2p$ orbital. (Each $2p$ orbital overlaps equally well with the $2p$ orbitals of both adjacent carbons, and the π interaction is unbroken around the six-member ring.)

9.15. Describe the S in SO₂Cl₂:

The electron-pair geometry and molecular geometry are tetrahedral. The hybridization of the sulfur is sp^3.

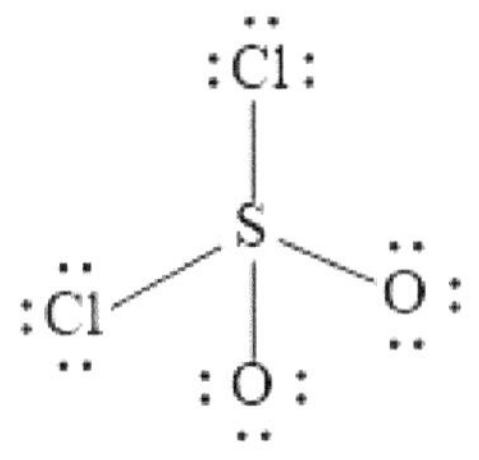

9.16. Electron pair and molecular geometry around the S atom in thionyl chloride:

Note that the S atom possesses a lone pair of electrons. With 4 groups around the S (3 atoms and 1 LP), the S has tetrahedral electron-pair and trigonal-pyramidal molecular geometry. The hybridization of sulfur is sp^3.

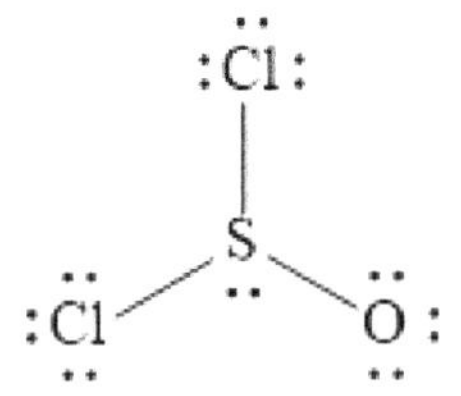

9.17. For the following compounds, the other isomer is:

(a)	H₃C, H C=C H, CH₃	H₃C, CH₃ C=C H, H
(b)	Cl, CH₃ C=C H, H	Cl, H C=C H, CH₃

9.18. Decide if an isomer is possible:

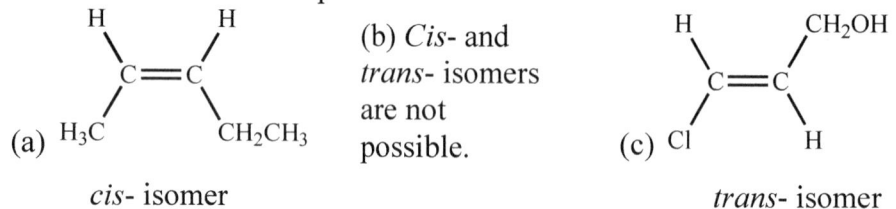

(a) H, H C=C H₃C, CH₂CH₃ — *cis*- isomer

(b) *Cis*- and *trans*- isomers are not possible.

(c) H, CH₂OH C=C Cl, H — *trans*- isomer

Molecular Orbital Theory

9.19. Configuration for H₂⁺: $(\sigma_{1s})^1$

Bond order for H₂⁺: 1/2 (# bonding e⁻ – # antibonding e⁻) = 1/2(1 – 0) = 1/2

The bond order for molecular hydrogen (H₂) is <u>one</u> (1), and so the H–H bond is <u>stronger</u> in the H₂ molecule than in the H₂⁺ ion.

9.20. Electron configurations for the ions in molecular orbital terms:

Li₂⁺: $(\sigma_{1s})^2(\sigma^*_{1s})^2(\sigma_{2s})^1$ Bond order = ½(3 – 2) = ½
Li₂⁻: $(\sigma_{1s})^2(\sigma^*_{1s})^2(\sigma_{2s})^2(\sigma^*_{2s})^1$ Bond order = ½(4 – 3) = ½
Li₂: $(\sigma_{1s})^2(\sigma^*_{1s})^2(\sigma_{2s})^2$ Bond order = ½(4 – 2) = 1

The bond order of Li₂ is greater than that of either of its ions.

9.21. The molecular orbital diagram for C_2^{2-}, the acetylide ion: There are 10 valence electrons in C_2^{2-}: four from each carbon atom and two due to the 2– charge.

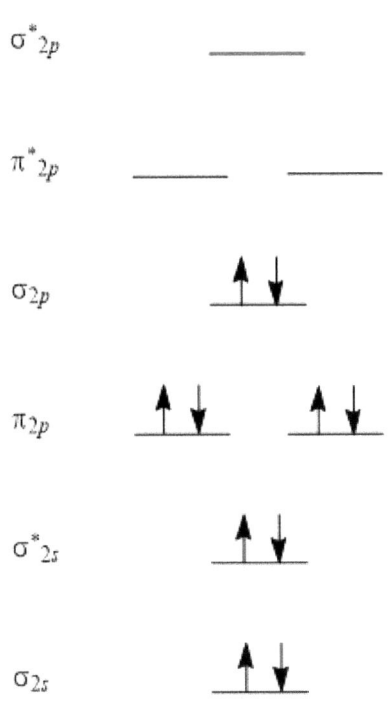

σ^*2p	____
π^*2p	____ ____
$\sigma 2p$	↑↓
$\pi 2p$	↑↓ ↑↓
σ^*2s	↑↓
$\sigma 2s$	↑↓

(a) There are 2 net pi bonds and 1 net sigma bond in the ion.

(b) The bond order is 3.

(c) Both the valence bond and molecular orbital models predict 2 net pi bonds and one net sigma bond.

(d) On adding two electrons to C_2 (added to $\sigma 2p$) to obtain C_2^{2-}, the bond order increased by one.

(e) The ion is *not paramagnetic*. All the electrons are paired in the molecular orbital diagram.

9.22. Molecular orbital diagram for O_2^+:

There are 11 valence electrons in this species (6 from each O – 1 because of the 1+ charge).

MO diagram (left to right, bottom to top): σ_{2s} (↑↓), σ^*_{2s} (↑↓), σ_{2p} (↑↓), π_{2p} (↑↓)(↑↓), π^*_{2p} (↑)(—), σ^*_{2p} (—).

Net σ and π bonds: 1 σ and 1 ½ π bonds

O–O bond order is 2½.

Bond order increases by ½ upon removing an electron.

The presence of the unpaired electron in π^*2p makes this ion paramagnetic.

9.23. Write the electron configuration for peroxide ion in molecular orbital terms, and then compare it with the electron configuration of the O_2 molecule.

For oxygen, the electron configuration O_2 is: $(\sigma_{2s})^2(\sigma^*_{2s})^2(\sigma_{2p})^2(\pi_{2p})^4(\pi^*_{2p})^2$

The peroxide ion, O_2^{2-}, has 2 more electrons: $(\sigma_{2s})^2(\sigma^*_{2s})^2(\sigma_{2p})^2(\pi_{2p})^4(\pi^*_{2p})^4$

(a) Compare the molecule with the peroxide ion:

Magnetic character: molecular oxygen is paramagnetic (2 unpaired electrons) whereas peroxide has no unpaired electrons and is therefore diamagnetic.

Net number of σ and π bonds: 1 σ in peroxide versus 1 σ and 1π in oxygen.

Bond order for O_2 = 2 [which is 1/2 (bonding electrons – antibonding electrons)]. Bond order for O_2^{2-} = 1

Oxygen–oxygen bond length: The O_2^{2-} bond will be longer than the O_2 bond owing to the decreased bond order in the peroxide ion.

(b) Compare the valence bond and molecular orbital descriptions:

The Lewis structure of the peroxide ion is

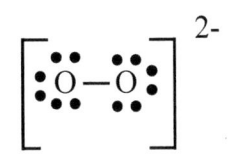

$$\left[:\ddot{\underset{..}{O}} - \ddot{\underset{..}{O}}: \right]^{2-}$$

Both the valence bond and molecular orbital models predict one net σ bond and one net π bond with an overall bond order of 2 for O_2 and one net σ bond and a bond order of 1 for O_2^{2-}. Both models predict that O_2^{2-} should be diamagnetic, but molecular orbital theory predicts that O_2 will be paramagnetic whereas valence bond theory does not.

9.24. Superoxide ion: $(\sigma_{1s})^2(\sigma^*_{1s})^2(\sigma_{2s})^2(\sigma^*_{2s})^2(\sigma_{2p})^2(\pi_{2p})^4(\pi^*_{2p})^3$

(a) O_2^- and O_2 are both paramagnetic, but there is 1 unpaired e⁻ in O_2^- vs. 2 in O_2

(b) 1 σ bond and ½ π bond in O_2^- vs. 1 σ bond and 1 π bond in O_2

(c) 1 1/2 in O_2^- vs. 2 in O_2

(d) longer bond in O_2^- than in O_2

9.25. Which has the shortest and longest bond: Li_2, B_2, C_2, N_2, O_2?

Molecule	MO occupation	Bond order
Li_2	$(\sigma_{2s})^2$	1
B_2	$(\sigma_{2s})^2(\sigma^*_{2s})^2(\pi_{2p})^2$	1
C_2	$(\sigma_{2s})^2(\sigma^*_{2s})^2(\pi_{2p})^4$	2
N_2	$(\sigma_{2s})^2(\sigma^*_{2s})^2(\pi_{2p})^4(\sigma_{2p})^2$	3
O_2	$(\sigma_{2s})^2(\sigma^*_{2s})^2(\pi_{2p})^4(\sigma_{2p})^2(\pi^*_{2p})^2$	2

In general bond length and bond order are inversely related. The greater the bond order, the shorter the bond. Of the molecules above, N_2 has the shortest bond. Table 9.1 confirms this logic. In general, with MO theory *as our only metric*, the Li_2 and B_2 molecules would have equally long bonds. (Also considering the sizes of the atoms, Li_2 would be expected to have the longest bond we lithium atoms are larger than boron atoms.)

9.26. Regarding the list of molecules and ions:

(a) Species with bond order of 3: CN^-, CO, NO^+, C_2^{2-}

(b) Paramagnetic species: O_2^-, O_2, NO

(c) Species with a fractional bond order: O_2^-, NO

9.27. (a) The electron configuration showing the 13 valence electrons for ClO:

σ^*_p ──────

π^*_p ↓↑ ↑

σ_p ↑↓

π_p ↑↓ ↑↓

σ^*_s ↑↓

σ_s ↑↓

(b) The HOMO is the π^*_p. The LUMO is σ^*_p.

(c) There is one unpaired electron, so ClO is *paramagnetic*.

(d) There is one net sigma bond, and 1/2 net pi bonds for an overall bond order of 1½.

9.28. Consider the nitrosyl ion:

(a) There are 10 valence electrons in NO⁺. The molecular orbital electron configuration is [core electrons]$(\sigma_{2s})^2(\sigma^*_{2s})^2(\pi_{2p})^4(\sigma_{2p})^2$. All of the electrons are paired in this configuration, so NO⁺ is diamagnetic.

(b) The HOMO is σ_{2p}. The LUMO is π^*_{2p}.

(c) Bond order for NO⁺ = ½(8 − 2) = 3

(d) The bond order of NO is 2.5, whereas that of NO⁺ is 3. Therefore, NO⁺ has a stronger bond than NO.

General Questions

9.29. Lewis structure for AlF₄⁻:

Electron-pair geometry: tetrahedral (4 groups)

Molecular geometry: tetrahedral

Orbitals on Al and F that overlap: on Al: sp^3 and on F: $2p$ orbitals

Formal charges on the atoms?

Formal charge on Al: $3 - [0 + 1/2(8)] = -1$ No lone pairs (0) and four bonds (8 electrons)

Formal charge on F (all are alike): $7 - [6 + 1/2(2)] = 0$. Three lone pairs (6) and one bond (2 electrons)

Is this a reasonable charge distribution?

No. Fluorine has a much greater electronegativity than does Al, so a reasonable charge distribution would have a parital *negative* charge on the F atoms, and a partial *positive* charge on the Al atom.

9.30. What are the O–S–O bond angles and hybrid orbitals used in the following molecules and ions?

(a) SO_2 120° sp^2
(b) SO_3 120° sp^2
(c) SO_3^{2-} 109° sp^3
(d) SO_4^{2-} 109° sp^3

SO_2 and SO_3 have the same bond angle and the S atom in each uses the same hybrid orbitals. SO_3^{2-} and SO_4^{2-} have the same bond angle and the S atom in each uses the same hybrid orbitals.

9.31. Resonance structures for the nitrite ion:

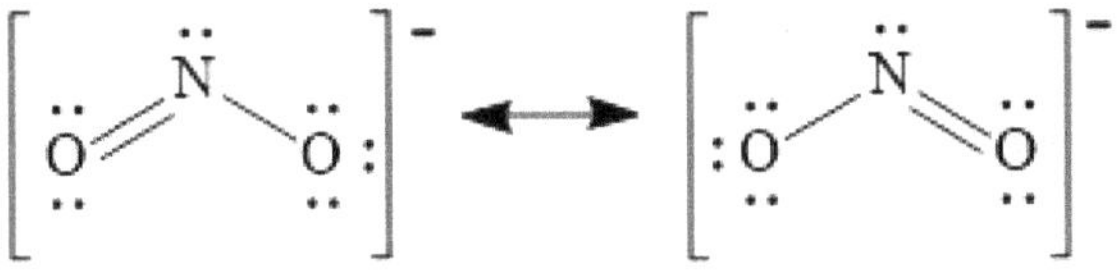

The electron-pair geometry of the ion is *trigonal planar*, and the molecular geometry is *bent* (or angular). The O–N–O bond angle is about 120°. The average bond order is 3/2 (three bonds connecting the two O atoms). The hybridization of the N atom is sp^2.

9.32. Resonance structures for the nitrate ion:

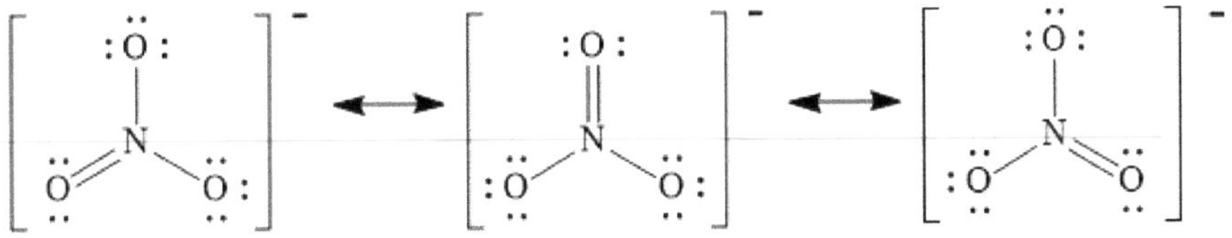

The hybridization of the N atom is predicted to be the same in each resonance structure. The nitrogen sp^2 hybrid orbitals are used to make three N–O sigma bonds. The p orbital, not used in the N atom hybridization, is used to form the N–O pi bond.

9.33. Resonance structures for N₂O:

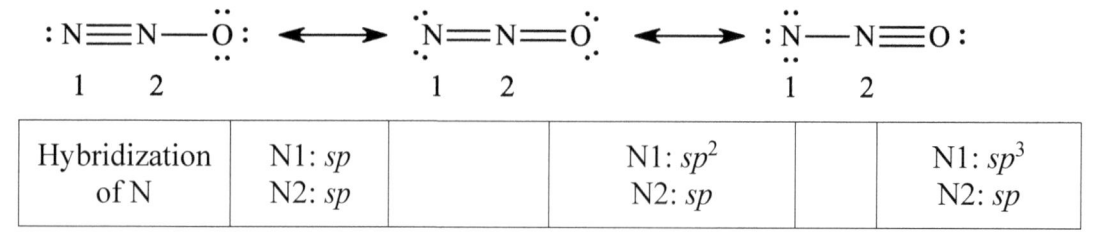

:N≡N—Ö: ⟷ :N̈=N=Ö: ⟷ :N̈—N≡O:
 1   2   1   2   1   2

Hybridization of N	N1: *sp* N2: *sp*	N1: *sp²* N2: *sp*	N1: *sp³* N2: *sp*

The central N atom is predicted to be *sp* hybridized in all three structures. The two *sp* hybrid orbitals on the central N atom are used to form N–N and N–O σ bonds. The two *p* orbitals not used in the N atom hybridization are used to form the required π bonds.

9.34. Compare carbon dioxide and the carbonate ion:

	O—C—O bond angle	CO bond order	C atom hybridization
CO₂	180° (2 groups attached)	2	*sp*
CO₃²⁻	120° (3 groups attached)	4/3	*sp²*

9.35. Properties of ethylene oxide, acetaldehyde, and vinyl alcohol:

	ethylene oxide	acetaldehyde	vinyl alcohol
(a) Formula:	C₂H₄O	C₂H₄O	C₂H₄O—these **are** isomers of one another
(b) Hybridization	C1 and C2 *sp³*	C1 *sp³* C2 *sp²*	C1 and C2 *sp²*
(c) H–C–H Bond angles	109°	109°	120°
(d) Polarity	Polar	Polar	Polar
(e) Strongest bond		C–O bond (double bond)	C–C bond (double bond)

9.36. Regarding acrolein:

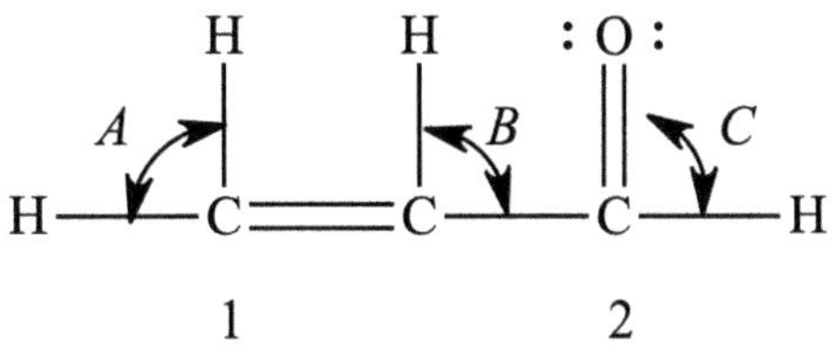

(a) Hybridization of C atoms 1 and 2: With 3 groups attached, both C(1) and C(2) have 120° bond angles, with accompanying sp^2 hybridization.

(b) Angles A, B, and C are all equal to 120°.

(c) With C(1) having two hydrogen atoms, *cis-trans* isomerism is not possible.

9.37. For the oxime shown:

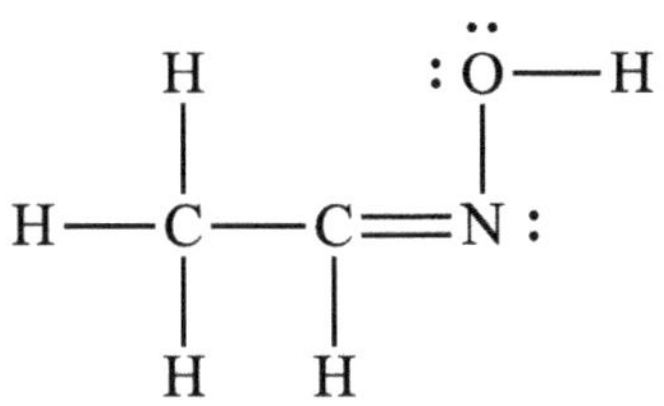

(a) Hybridization of the C atoms and the N atom: The leftmost C in the diagram has four groups around it (3 H atoms and a C atom), so it will have tetrahedral electron-pair geometry. This C atom is sp^3 hybridized.

The rightmost C atom has three groups around it (C, H, and N atoms), so it will have trigonal-planar electron-pair geometry. This C atom is sp^2 hybridized.

The N atom has three groups around it (1 lone pair, a C atom, and an O atom), so it will have trigonal planar electron-pair geometry. This N atom is sp^2 hybridized.

(b) Approximate C–N–O angle: With an N atom that is sp^2 hybridized, we would anticipate a bond angle of approximately 120°.

9.38. Using the diagram provided for acetylsalicylic acid:

(a) Angle A = 120°; Angle B = 109°; Angle C = 109°; Angle D = 120°

(b) Carbon 1: sp^2; Carbon 2: sp^2; Carbon 3: sp^3

9.39. For phosphoserine:

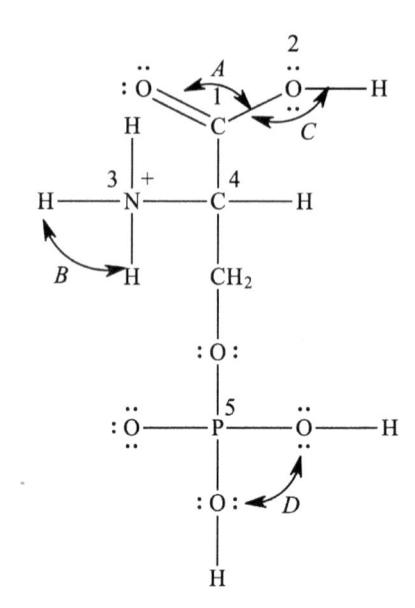

(a) Hybridizations of atoms 1-5:

Atom 1: 3 groups attached (3 atoms): sp^2

Atom 2: 4 groups attached (2 atoms; 2 LP): sp^3

Atom 3: 4 groups attached (4 atoms): sp^3

Atom 4: 4 groups attached (4 atoms): sp^3

Atom 5: 4 groups attached (4 atoms): sp^3

(b) Approximate bond angles A-D:

Angle A: For an sp^2 hybridized C atom: 120°

Angle B: for an sp^3 hybridized N atom: 109°

Angle C: for an sp^3 hybridized O atom: 109°

Angle D: for an sp^3 hybridized P atom: 109°

(c) Most polar bonds in the molecule: Owing to differing electronegativities, we expect the P–O bonds and the O–H bonds to be the most polar.

9.40. Concerning lactic acid:

(a) Bonding: 1 π bond (the C=O bond) and 11 σ bonds.

(b) Hybridization of atoms: C(1) = sp^3, C(2) = sp^2, O(3) = sp^3

(c) Shortest and strongest bond: The C=O bond is the shortest and strongest carbon–oxygen bond, owing to the bond order.

(d) Bond angles: Angle A = 109°, Angle B = 109°, Angle C = 120°

9.41. For the molecule cinnamaldehyde:
(a) The C=O is the most polar bond (electronegativity differences).
(b) 2 π bonds (outside the ring) and 3 π bonds (in the aromatic ring); There are 18 σ bonds.
(c) *Cis-trans* isomerism is possible. The two isomers are shown below.

cis-isomer   *trans*-isomer

(d) Since each C is attached to 3 atoms and has no lone pairs, all the C atoms are sp^2 hybridized.
(e) Since the angles indicated have an sp^2 hybridized C at the center, all the angles are 120°.

9.42. Concerning the ion Si_2^-:
(a) Assume this ion is similar to C_2^- (see Study Question 9.21 for C_2^{2-}) except that the 3s and 3p orbitals are used to form molecular orbitals, and there is only one electron in σ_{3p}.
The electron configuration is [core electrons]$(\sigma_{3s})^2(\sigma^*_{3s})^2(\pi_{3p})^4(\sigma_{3p})^1$
The bond order is 1/2(7 – 2) = 2.5.
(b) paramagnetic (ion has an odd number of electrons)
(c) σ_{3p}

9.43. (a) A Lewis dot structure of the peroxide ion indicates that the bond order is 1.

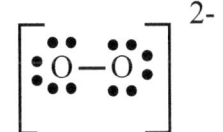

(b) The molecular orbital electron configuration of the peroxide ion is:

[core electrons]$(\sigma_{2s})^2(\sigma^*_{2s})^2(\sigma_{2p})^2(\pi_{2p})^4(\pi^*_{2p})^4$.
The MO configuration shows that there are 8 bonding valence electrons and 6 antibonding valence electrons. The bond order = 1/2(8 –6) = 1.

(c) Both theories show *no unpaired electrons*, and a bond order of 1. The magnetic character of both is the same: diamagnetic.

9.44. Compare N_2, N_2^+ and N_2^-:

N_2 [core electrons]$(\sigma_{2s})^2(\sigma^*_{2s})^2(\pi_{2p})^4(\sigma_{2p})^2$

N_2^+ [core electrons]$(\sigma_{2s})^2(\sigma^*_{2s})^2(\pi_{2p})^4(\sigma_{2p})^1$

N_2^- [core electrons]$(\sigma_{2s})^2(\sigma^*_{2s})^2(\pi_{2p})^4(\sigma_{2p})^2(\pi^*_{2p})^1$

	N₂	N₂⁺	N₂⁻
(a) magnetic character	diamagnetic	paramagnetic	paramagnetic
(b) σ and π bonds net σ bonds net π bonds	1 σ bond 2 π bonds	½ σ bond 2 π bonds	1 σ bond 1½ π bonds
(c) bond order	3	2½	2½
(d) bond length	\multicolumn{3}{c}{N₂ < N₂⁺ ≈ N₂⁻}
(e) bond strength	\multicolumn{3}{c}{N₂⁺ ≈ N₂⁻ < N₂}
HOMO	σ$_{2p}$	σ$_{2p}$	π*$_{2p}$

9.45. Consider the diatomic molecules of Li₂ through Ne₂:

molecule	valence electron configuration	magnetic property	bond order
Li₂	$(\sigma_{2s})^2$	diamagnetic	1
Be₂	$(\sigma_{2s})^2(\sigma^*_{2s})^2$	diamagnetic	0
B₂	$(\sigma_{2s})^2(\sigma^*_{2s})^2(\pi_{2p})^2$	paramagnetic	1
C₂	$(\sigma_{2s})^2(\sigma^*_{2s})^2(\pi_{2p})^4$	diamagnetic	2
N₂	$(\sigma_{2s})^2(\sigma^*_{2s})^2(\pi_{2p})^4(\sigma_{2p})^2$	diamagnetic	3
O₂	$(\sigma_{2s})^2(\sigma^*_{2s})^2(\sigma_{2p})^2(\pi_{2p})^4(\pi^*_{2p})^2$	paramagnetic	2
F₂	$(\sigma_{2s})^2(\sigma^*_{2s})^2(\sigma_{2p})^2(\pi_{2p})^4(\pi^*_{2p})^4$	diamagnetic	1
Ne₂	$(\sigma_{2s})^2(\sigma^*_{2s})^2(\sigma_{2p})^2(\pi_{2p})^4(\pi^*_{2p})^4(\sigma^*_{2p})^2$	diamagnetic	0

B₂ and O₂ are paramagnetic. Li₂, B₂, and F₂ have a bond order of 1. C₂ and O₂ have a bond order of 2. N₂ has the highest bond order.

9.46. Describe magnetic character and identify HOMO for the following:

Molecule or ion	Magnetic behavior	HOMO
(a) NO	paramagnetic	π^*_{2p}
(b) OF⁻	diamagnetic	π^*_{2p}
(c) O₂²⁻	diamagnetic	π^*_{2p}
(d) Ne₂⁺	paramagnetic	σ^*_{2p}

9.47. The CN molecule has 9 valence electrons (4 electrons from C and 5 from N). The electron configuration is predicted to be:

[core electrons]$(\sigma_{2s})^2(\sigma^*_{2s})^2(\pi_{2p})^4(\sigma_{2p})^1$

(a) The highest energy MO to which an electron is assigned is the σ_{2p}. The lowest energy MO that is unoccupied is π^*_{2p}.

(b) Bond order = 1/2 (# bonding electrons – # non-bonding electrons)
= 1/2 (7 – 2) or 5/2 or 2.5.

(c) There is a **net of 1/2** sigma bond —$(\sigma_{2p})^1$, and 2 net π bonds—$(\pi_{2p})^4$.

(d) The molecule has an unpaired electron, so it is *paramagnetic.*

9.48. Amphetamine:

(a) C₆ ring carbon atoms: *sp²*; side chain carbon atoms: *sp³*; N atom: *sp³*
(b) Angle A = 120°; Angle B = 109°; Angle C = 109°
(c) 23 σ bonds and 3 π bonds
(d) The molecule is polar.
(e) The H⁺ ion attaches to the most electronegative atom in the molecule, N, and this is confirmed by the electrostatic potential map (most negative region near N).

9.49. Menthol:

(a) Every carbon is surrounded by 4 electron pairs, so the hybridization for each carbon is *sp³*.

(b) The O atom has 4 electron pairs (two lone pairs and two bond pairs), so the C–O–H bond angle is predicted to be the tetrahedral angle of 109°.

(c) With the OH group attached to the ring, the molecule will be slightly polar.

(d) The ring is *not planar*. Each of the carbons is bonded to two other carbons with a 109° bond angle, and *unlike the benzene ring—which is planar*, this ring will be puckered.

9.50. Regarding structures for B₂F₄, C₂H₄, N₂H₄, O₂H₂:

B is *sp²* hybridized, F—B—B = 120°		C is *sp²* hybridized, H—C—C = 120°
[Lewis structure of B₂F₄ with two F atoms on each B, B—B bond]		[Lewis structure of C₂H₄ with C=C double bond and two H on each C]

N is *sp³* hybridized, H—N—N = 109°		O is *sp³* hybridized, H—O—O = 109°
[Lewis structure of N₂H₄ with two H on each N, N—N bond]		[Lewis structure of H₂O₂ with H—O—O—H]

In the Laboratory

9.51. For BF₃ and NH₃BF₃:

(a) With 3 groups attached to the B in BF₃, the geometry is trigonal planar. With 4 groups attached to the B in NH₃BF₃, the geometry is tetrahedral.

(b) In BF₃ the B is *sp²* hybridized; in NH₃BF₃ the B is *sp³* hybridized.

(c) The B atom is partially positive (because it is bonded to 3 *very electronegative* F atoms). The N atom in NH₃ is partially negative (with a lone pair of electrons residing on it). So, it is anticipated that the partially negative N atom will donate electrons to the positive B atom.

(d) Given the partially positive charge on the B atom, the B atom will attract a lone pair of electrons on the O atom of water:

[Reaction scheme: :F—B(—F:)—F: + H₂O → H₂O:→B(F)(F)(F) adduct]

9.52. Concerning ethylene oxide:

(a) Even though the atoms are sp^3 hybridized, the bond angles in the three-member ring must be 60°. Because the angles are significantly less than the expected value of 109°, the ring structure is strained and relatively easy to break. (The molecule is reactive.)

(b) The molecule is polar. The negative partial charge is near oxygen and the hydrogen atoms carry a partial positive charge.

9.53. The sulfamate ion can be pictured as being formed by the reaction of the amide ion and sulfur trioxide:

(a) Geometries of amide ion and of sulfur trioxide: With 4 groups (2 atoms and 2 lone pairs) around the nitrogen, the amide ion has a molecular geometry that is angular (or bent). The hybridization of the N would be sp^3. For sulfur trioxide, there are three groups around the S, so the molecular geometry would be trigonal planar and the hybridization of the S would be sp^2.

(b) The bond angles around the N and S in the sulfamate ion would be approximately 109°, since both N and S would have 4 groups around them in the ion.

(c) While the N atom has sp^3 hybridization in both the amide and sulfamate ions, the hybridization of the S changes from sp^2 in the molecule to sp^3 in the sulfamate ion.

(d) The SO₃ molecule is the acceptor of the electron pair during the reaction, as expected. The negative area of the electron pair is attracted to the positive area of the sulfur trioxide molecule.

9.54. Regarding acetylacetone:

(a) The *keto* and *enol* forms are not resonance structures because *both* electron pairs and atoms have been rearranged.

(b) The terminal —CH₃ carbon atoms are sp^3 hybridized and the three central carbon atoms are sp^2 hybridized in the *enol* form. In the *keto* form, the terminal —CH₃ carbon atoms and the central C atom are sp^3 hybridized and the two C=O carbon atoms are sp^2 hybridized.

(c) *Enol* form:
- —CH₃ carbon atoms have tetrahedral electron-pair and molecular geometries.
- central three carbon atoms have trigonal planar electron-pair and molecular geometries.
- C=O carbon atom has trigonal planar electron-pair and molecular geometries.

Keto form:
- —CH₃ carbon atoms have tetrahedral electron-pair and molecular geometries.
- central —CH₂— carbon atom has tetrahedral electron-pair and molecular geometries.
- C=O carbon atoms have trigonal planar electron-pair and molecular geometries. Only the center carbon atom changes geometry, from trigonal planar to tetrahedral.

(d)

[structures showing three resonance forms of the acetylacetonate anion]

(e) *Cis-trans* isomerism is possible for the *enol* form.

(f) *Enol* form is polar. The negative partial charges are on the oxygen atoms. The positive partial charges are on the hydrogen atoms, especially the O—H hydrogen.

9.55. Regarding the acetate anion resonance structures:

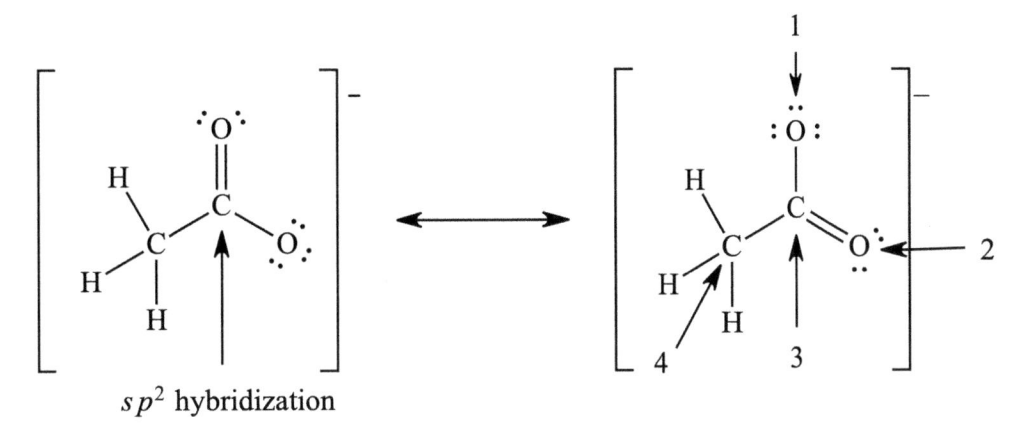

[Structure showing acetate anion resonance: left structure with CH₃ group bonded to C with double bond to O (top) and single bond to O⁻ (right), labeled sp² hybridization. Right structure showing the resonance form with labels 1 (top O), 2 (right O), 3 (central C), 4 (methyl C).]

sp² hybridization

Using the right-most structure:

C(3) uses an *sp²* hybrid to overlap with an *sp²* hybrid from O(2) to form the σ bond.

The π bond between C(3) and O(2) results from the overlap of the unhybridized *p* orbitals on the respective C and O atoms.

The σ bond between C(4) and C(3) results from overlapping an *sp²* hybrid orbital from C(3) with an *sp³* hybrid orbital from C(4).

O(1) has four orbitals occupied (1 for the bond to the C atom and 3 for the lone pairs of electrons) so it would appear to use *sp³* hybridization in this resonance structure and overlap one of those *sp³* hybrid orbitals with an *sp²* hybrid orbital on C(3) to form the σ bond. However, in the resonance hybrid, this O must also have *sp²* hybridization.

9.56. Regarding carbon dioxide, dinitrogen monoxide, azide ion, and cyanate ion:

(a) The central atoms are *sp* hybridized in all four molecules/ions. The most stable resonance structures are shown below (i.e., those that minimize formal charges and place negative charges on the most electronegative atom).

[Lewis structures shown: :Ö=C=Ö: [:Ö—C≡N:]⁻   :N≡N—Ö: [:N̈=N=N:]⁻]

(b) For each of the species, three resonance structures can be drawn: (i) one with two double bonds, (ii) one with a triple bond on the left and a single bond on the right, and (iii) one with a triple bond on the right and a single bond on the left. The structure that minimizes formal charges and places a negative formal charge on the most electronegative atom or a positive formal charge on the least electronegative atom is the resonance structure that should contribute the most to the actual structure.

For CO_2, the formal charge on each atom in the structure in part (a) is zero; the bond lengths are expected to be equivalent and to be typical of a carbon–oxygen double bond.

For N₃⁻, the formal charge on each outer atom is –1 and on the central atom is +1 in the structure in part (a); the bond lengths are expected to be equivalent and to be typical of a nitrogen–nitrogen double bond.

For OCN⁻, the formal charges are oxygen, –1; carbon, 0; and nitrogen, 0. The carbon–oxygen bond should be close to single bond length, and the carbon–nitrogen bond should be close to triple bond length.

For N₂O, the central nitrogen has a formal charge of +1, the outer nitrogen has a formal charge of 0, and the oxygen has a formal charge of –1. The nitrogen–nitrogen bond should approach triple bond length, while the N–O bond should approach a single bond length.

9.57. Resonance structures for SO₂:

In total the lone pairs and bonding pairs in the σ framework of SO₂ account for 7 of the 9 valence electron pairs. The 3 unhybridized *p* orbitals (one from S and one from each O) form 3 π MOs. One of the remaining pairs of electrons occupies a bonding π$_p$ MO, the other pair occupies a nonbonding π$_p$ MO. With one pair of electrons spread over two bonds, the bond order for those electrons is 0.5. The electrons in the σ framework account for a bond order of 1, making the bond order for the bonds holding the O's to the S = 1.5. See Figure 9.17 for an analogous description of O₃.

9.58. Concerning the molecule diimide:

The electron-pair geometry at the nitrogen atoms is trigonal planar, and the molecular geometry is bent. The nitrogen atoms are *sp²* hybridized. The N=N double bond is composed of a σ bond (made from the overlap of *sp²* hybrid orbitals on the nitrogen atoms) and a π bond (made from overlap of unhybridized 2*p* orbitals on the nitrogen atoms).

Summary and Conceptual Questions

9.59. C has four valence orbitals (one *s* and three *p* orbitals). Hence hybridizing the **minimum** number would provide 2 orbitals (*sp*), while the **maximum** number of hybrid orbitals would involve all the orbitals and result in **4 hybrid** orbitals (*sp³*).

9.60. Concerning BF₄⁻, SiF₄, and SF₄:

(a) CF₄ is isoelectronic with BF₄⁻ (32 valence electrons).
(b) SiF₄ (32 valence electrons) and SF₄ (34 valence electrons) are not isoelectronic.
(c) Hybridization of the central atom: BF₄⁻: *sp*³   SiF₄: *sp*³

9.61. The essential part of the Lewis structure for the peptide linkage is shown here:

(a) The carbonyl carbon (C=O) is attached to 3 groups and so has a hybridization of *sp*². The N atom is connected to 4 groups (3 atoms, 1 lone pair) and so is predicted to have a hybridization of *sp*³.

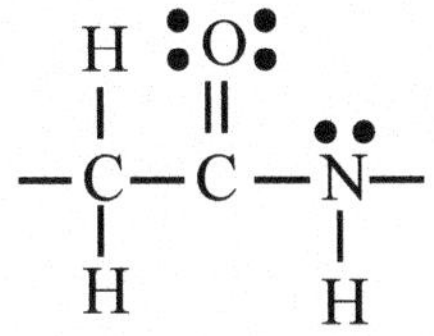

(b) Another structure is feasible:

This structure would have a formal charge on the carbonyl carbon of 0, on the oxygen of –1, and on the N a formal charge of +1. Since the "preferred" structure has 0 formal charges on **all three atoms**, the first structure is the more favorable.

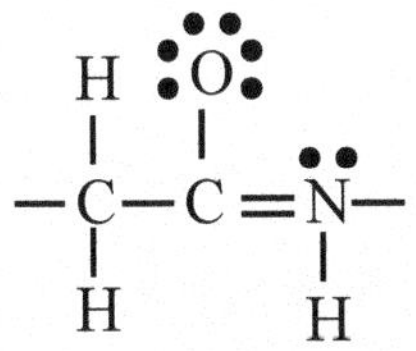

(c) If the "less preferred" resonance structure is viewed as a contributor, both the carbonyl carbon and the N are *sp*² hybridized. This leaves one "*p*" orbital on O, C, and N unhybridized, and capable of side-to-side overlap (also known as π type overlap) between the C and the O and between the C and the N, forming a planar region in the molecule. As illustrated by the electrostatic potential diagram for this dipeptide, the regions of negative charge are the O atoms and the nitrogen of the "free" –NH₂ group.

9.62. The connection between bond order, bond length, and bond energy:

The higher the bond order, the shorter the bond length and the greater the bond energy. Acetylene has a carbon–carbon triple bond (bond order = 3), so it has the shortest and strongest carbon–carbon bond. Ethane has a carbon–carbon single bond (bond order = 1), so it has the longest and weakest carbon–carbon bond.

9.63. Valence bond theory rests on an assumption that bonds are formed by pairing of electrons in orbitals on neighboring atoms. This theory has difficulties when one considers odd-electron molecules. VB theory also fails in explaining the magnetic properties of some molecules, for example, O₂. MO theory provides a better explanation of the effect of adding energy to molecules. A molecule can absorb energy and an electron can be promoted to a higher level. Using MO theory makes this transition more understandable. MO theory is also a better model to predict molecular paramagnetism.

9.64. Rationalize the O–O bond order of 1.5 in ozone. See Figure 9.17.

Valence bond theory uses resonance to explain the bond order of 1.5 in O_3. Three electron pairs are shared between two O–O linkages. Molecular orbital theory uses three π molecular orbitals (bonding, nonbonding, and antibonding) combined with sigma bonding and antibonding molecular orbitals to explain the bond order. There is one net σ bond. The π electrons are located in the bonding and nonbonding orbitals (two in each) giving rise to one net π bond.

9.65. Place the three molecular orbitals for cyclobutadiene in order of increasing energy.

The energy increases with the number of nodes. Orbital C has one node, orbital B has two nodes, and orbital A has three nodes. Thus, the energy increases in the order:
Orbital C < Orbital B < Orbital A

9.66. Regarding hybridization:

(a) The number of hybrid orbitals is always equal to the number of atomic orbitals used.

(b) Hybrid orbitals always involve an *s* orbital.

(c) The energy of the hybrid orbital set is the weighted average of the energy of the combining atomic orbitals.

9.67. The anion from borax has the structure pictured in the text:

B atom with	3 electron pairs	4 electron pairs
Electron-pair geometry	trigonal planar	tetrahedral
Molecular geometry	trigonal planar	tetrahedral
Hybridization	sp^2	sp^3
Formal charge	3 − [0 + ½(6)] = 0	3 − [0 + ½(8)] = −1

Geometry: With 3 pairs of electrons around a B atom, VSEPR theory indicates trigonal planar geometry, utilizing 3 equivalent orbitals—hence sp^2 hybridization. For 4 pairs of electrons, tetrahedral geometry is needed with 4 equivalent orbitals—and sp^3 hybridization. Recall that formal charge is calculated by subtracting from the group number (3 for B) all the lone pair electrons and half the bonding electrons. There are no lone pairs on these B atoms.

9.68. Concerning the molecule allene:

(a) Two parallel *p* orbitals on C(1) and C(2) overlap to form a pi bond. C(3) must be rotated so that the *p* orbitals for it and C(2) align to form another pi bond.

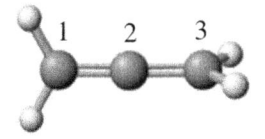

(b) Hybridization: C(1) *sp²*; C(2) *sp*; C(3) *sp²*

(c) The overlaps of C(1) and C(3) atoms' *sp²* orbitals with the C(2) atom's *sp* orbitals form the σ bonds. The overlap of the carbon atoms' 2*p* orbitals forms the π bonds.

9.69. Refer to "A Closer Look" in the section "Resonance and MO Theory" for a pictorial representation of the HF_2^- ion. The bond order for the ion is 0.5, since two bonding electrons are spread over 2 bonds. The HF molecule has a bond order of 1. So the energy required to break an H–F bond in the HF_2^- ion should be **less than** the energy required to break the H–F bond in the molecule.

9.70. Regarding melamine:

(a) A resonance structure can be drawn with double bonds between the alternate carbons and nitrogens of the ring; the average of the two resonance structures would result in identical bonds with a bond order of 1.5 between each C and N of the ring.

(b) Enthalpy change for the formation of melamine:

$\Delta_r H° =$ [(1 mol/mol-rxn)($\Delta_f H°$(melamine(s))) + (6 mol/mol-rxn)($\Delta_f H°$(NH₃(g))) + (3 mol/mol-rxn)($\Delta_f H°$(CO₂(g)))] – (6 mol/mol-rxn)($\Delta_f H°$(urea(s)))

= [(1 mol/mol-rxn)(–66.1 kJ/mol) + (6 mol/mol-rxn)(–45.90 kJ/mol) + (3 mol/mol-rxn)(–393.509 kJ/mol)] – (6 mol/mol-rxn)(–333.1 kJ/mol)

= +476.6 kJ/mol-rxn

The reaction is endothermic.

9.71. Concerning the oxides of bromine:

(a) A 100-g sample of the oxide contains 90.90 g Br and 9.10 g O. The amount of each atom would be: $\dfrac{90.90 \text{ g Br}}{1} \cdot \dfrac{1 \text{ mol Br}}{79.904 \text{ g Br}} = 1.1376$ mol Br;

$\dfrac{9.10 \text{ g O}}{1} \cdot \dfrac{1 \text{ mol O}}{15.999 \text{ g O}} = 0.5688$ mol O

The ratio of Br:O is: $\dfrac{1.1376 \text{ mol Br}}{0.5688 \text{ mol O}} = 2.000$ mol Br/mol O. This leads to an empirical formula of Br₂O.

A proposed Lewis structure for Br₂O: :Br—O—Br:

The hybridization of the O atom here would be *sp*³ (with two atoms and two lone pairs of electrons attached).

(b) Using Table 9.1 as a template, 13 valence electrons (7 from Br, and 6 from O) need to be accounted for in BrO. Placing the electrons in the MO would give: $(\sigma_s)^2(\sigma^*_s)^2(\sigma_p)^2(\pi_p)^4(\pi^*_p)^3$. The HOMO would be the π^*_p orbital.

9.72. In the reaction of urea with malonic acid to form barbituric acid:

(a) The structures are shown below, with the bonds to be broken indicated by dashed lines.

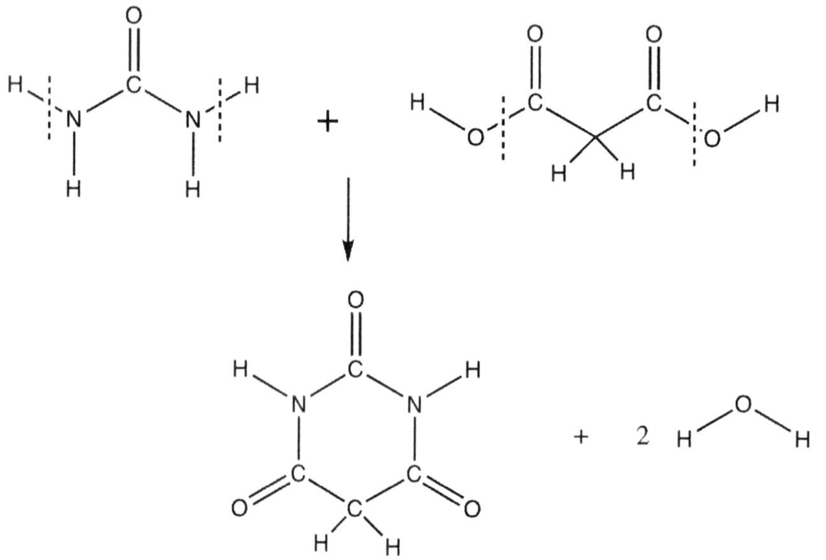

Total bonds broken (2 C–O) and (2 N–H). Bonds formed (2 C–N) and (2 O–H).

Using bond enthalpies:

Bonds broken: 2 C–O + 2 N–H:
(2 mol C–O/mol-rxn)(+358 kJ/mol C–O) + (2 mol N–H/mol-rxn)(+391 kJ/mol N–H)
= 1498 kJ/mol-rxn

Bonds formed: 2 C–N + 2 O–H:
(2 mol C–N/mol-rxn)(+305 kJ/mol C–N) + (2 mol O–H/mol-rxn)(+463 kJ/mol O–H)
= 1536 kJ/mol-rxn

The bond enthalpies for bonds **formed** indicate the energy released (–), and the enthalpies for bonds **broken** indicate the energy absorbed (+). Therefore, the **net energy change** will be +1498 kJ/mol-rxn –1536 kJ/mol-rxn = –38 kJ/mol-rxn. The reaction is **exothermic**.

(b) H₂NCONH₂ + CH₂(CO₂H)₂ → C₄H₄N₂O₃ + 2 H₂O

(c) For barbituric acid, the angles are shown:

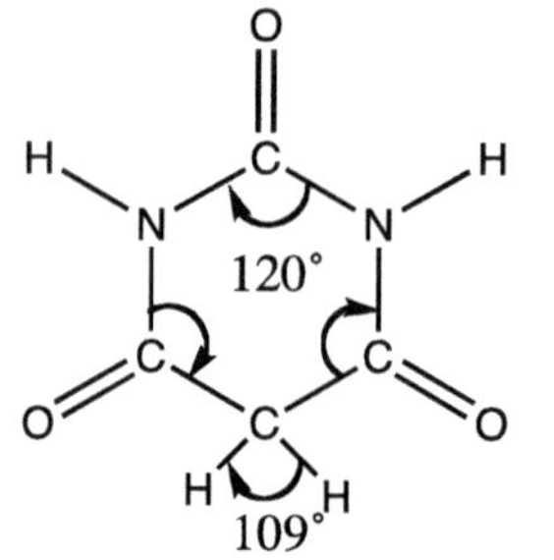

(d) For the C=O atoms in the ring, the hybridizations are *sp*² and the CH₂ carbon is *sp*³.

(e) With an electronegativity difference of 1.0 (between C and O), the C=O bonds are the most polar in the molecule.

(f) The molecule is not totally symmetric and so has a net dipole moment. It is polar.

9.73. (a) Electron dot structure for SO₃.

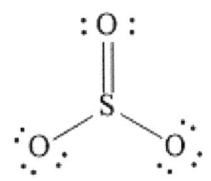

(b) Formal charges of the atoms and bond order of the SO bonds in SO₃

Formal charge on S atom = 6 − [0 + ½(8)] = +2

Formal charge of O with double bond = 6 − [4 + ½(4) = 0

Formal charge of O with single bond = 6 − [6 + ½(2)] = −1

There are 3 sigma bonds and 1 pi bond in the molecule. There are thus four electrons pairs spread out between three linkages, giving a bond order of 4/3 =1.33

(c) Electron-pair geometry and molecular geometry are trigonal planar

(d) S atom is *sp²* hybridized.

(e) S₃O₉ plausible electron dot structure

9.74. Acetylene should have a total of 4 π molecular orbitals (two that are occupied and two that are unoccupied). This LUMO is an antibonding π orbital. The *p* orbitals produce an antibonding molecular orbital in this case because they were combined with lobes of opposite sign aligned (as indicated by the colors of the overlapping orbitals). See Figures 9.12 and, especially, 9.15.

Solution and Answer Guide

Kotz Treichel Townsend Treichel, Chemistry and Chemical Reactivity 11e, 978-0-357-85140-1, Chapter 10: Gases & Their Properties

TABLE OF CONTENTS

Applying Chemical Principles 356
Practicing Skills..357
General Questions..374
In the Laboratory...388
Summary and Conceptual Questions..394

Applying Chemical Principles

The Atmosphere and Altitude Sickness

10.1.1. Atmospheric pressure at summit of Mt. Everest:

(a) 0.29(760 mm Hg) = 220.4 mm Hg = 220 mm Hg (2 sf)

(b) Air contains 21% oxygen.

0.21(220.4 mm Hg) = 46 mm Hg

10.1.2. Variation of atmospheric pressure vs altitude: Assuming that atmosphere pressure is 760 torr at 0 m altitude, the graph at right results. The plot shows a non-linear relationship between atmospheric pressure and altitude.

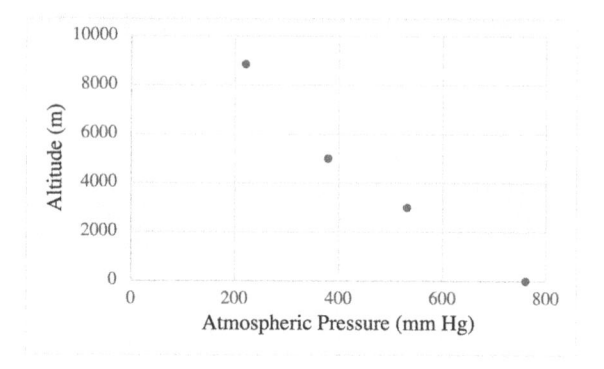

10.1.3. The partial pressure of oxygen is 90 mm at approximately 4400 m.
The percentage of oxygen is 21%. The question becomes: If the percent of O_2 is 0.21, and the partial pressure of O_2 is 90 mm, what is the total pressure? Once that is known, you can read the altitude from the graph, approximately 4400 m. (or 4.4×10^3 m to 2sf)

$\left(\dfrac{21}{100}\right)x = 90$ mm Hg and $x = \dfrac{(90)(100)}{21} = 428$ mm Hg or 400 mm Hg to 1sf

The Chemistry of Airbags

10.2.1. Calculate the mol of N₂ needed to fill the 75 L bag.

$$n = \frac{(3.0 \text{ atm})(75 \text{ L})}{(0.082057 \text{ L} \cdot \text{atm/K} \cdot \text{mol})(298 \text{ K})} = 9.2 \text{ mol N}_2$$

This is the **total** amount of N₂ to fill the bag. Note that the Na produced during decomposition of the azide ALSO produces N₂ in a 10 mol Na: 1 mol N₂ ratio.

(a) $(9.2 \text{ mol N}_2)\left(\dfrac{2 \text{ mol NaN}_3}{3.2 \text{ mol N}_2}\right)\left(\dfrac{65.01 \text{ g NaN}_3}{1 \text{ mol NaN}_3}\right) = 373.8 \text{ g NaN}_3$ or 370 g (2 sf)

Note that the decomposition of azide produces 3 mol N₂/2 mol NaN₃.

2 mol of azide also produces 2 mol of Na. The reaction with KNO₃ produces 1 mol N₂ when 10 mol of Na react. In essence, 10 mol of Na produce 1 mol of N₂. So 2 mol of Na would produce 0.2 mol of N₂. What this means is that 2 mol of NaN₃ produce (3 + 0.2) mol N₂—hence the ratio in the equation above.

(b) Mass of KNO₃ needed to consume the Na produced:

$$373.8 \text{ g NaN}_3 \left(\dfrac{1 \text{ mol NaN}_3}{65.01 \text{ g NaN}_3}\right)\left(\dfrac{1 \text{ mol Na}}{1 \text{ mol NaN}_3}\right)\left(\dfrac{2 \text{ mol KNO}_3}{10 \text{ mol Na}}\right)\left(\dfrac{101.10 \text{ g KNO}_3}{1 \text{ mol KNO}_3}\right) = 116.3 \text{ g KNO}_3$$

10.2.2. Fill the airbag using ammonium nitrate:

(a) Mass of ammonium nitrate:

From 10.2.1, you know that 9.2 mol of N₂ are needed to fill the airbag. Obviously 9.2 mol of any ideal gas will also fill the airbag. The decomposition of NH₄NO₃ proceeds as follows: NH₄NO₃(s) →1 N₂O(g) + 2 H₂O(g). This ratio of 3 mol gas: 1 mol NH₄NO₃ permits you to calculate the amount of NH₄NO₃ needed:

$$9.2 \text{ mol gas}\left(\dfrac{1 \text{ mol NH}_4\text{NO}_3}{3 \text{ mol gas}}\right)\left(\dfrac{80.05 \text{ g NH}_4\text{NO}_3}{1 \text{ mol NH}_4\text{NO}_3}\right) = 245.5 \text{ g NH}_4\text{NO}_3 (250 \text{ to 2sf})$$

(b) Partial pressures of N₂O and H₂O in the airbag:

As the total pressure is 3.0 atm, and the two gases are produced in a 1:2 mol ratio,

The partial pressure of N₂O is 1.0 atm, and that of H₂O is 2.0 atm.

Practicing Skills

Pressure

10.1. Express pressure in different units (using appropriate conversion factors):

(a) $0.633 \text{ atm}\left(\dfrac{760 \text{ mm Hg}}{1 \text{ atm}}\right) = 481.1 \text{ mm Hg} = 481 \text{ mm Hg } (3 \text{ sf})$

(b) $0.633 \text{ atm}\left(\dfrac{1 \text{ bar}}{0.98692 \text{ atm}}\right) = 0.6414 \text{ bar} = 0.641 \text{ bar } (3 \text{ sf})$

(c) $0.633 \text{ atm}\left(\dfrac{101.325 \text{ kPa}}{1 \text{ atm}}\right) = 64.14 \text{ kPa} = 64.1 \text{ kPa } (3 \text{ sf})$

10.2. Express pressure in different units:

(a) $278 \text{ mm Hg}\left(\dfrac{1 \text{ atm}}{760 \text{ mm Hg}}\right) = 0.3658 \text{ atm} = 0.366 \text{ atm } (3 \text{ sf})$

(b) $278 \text{ mm Hg}\left(\dfrac{1 \text{ atm}}{760 \text{ mm Hg}}\right)\left(\dfrac{1 \text{ bar}}{0.98692 \text{ atm}}\right) = 0.3709 \text{ bar} = 0.371 \text{ bar } (3 \text{ sf})$

(c) $278 \text{ mm Hg}\left(\dfrac{1 \text{ atm}}{760 \text{ mm Hg}}\right)\left(\dfrac{101.325 \text{ kPa}}{1 \text{ atm}}\right) = 37.06 \text{ kPa} = 37.1 \text{ kPa } (3 \text{ sf})$

10.3. The higher pressure in each of the pairs (using appropriate conversion factors):

(a) 534 mm Hg or 0.754 bar

$534 \text{ mm Hg}\left(\dfrac{1 \text{ atm}}{760 \text{ mm Hg}}\right)\left(\dfrac{1 \text{ bar}}{0.98692 \text{ atm}}\right) = 0.712 \text{ bar}$ so 0.754 bar is higher

(b) 534 mm Hg or 650 kPa

$534 \text{ mm Hg}\left(\dfrac{101.325 \text{ kPa}}{760 \text{ mm Hg}}\right) = 71.2 \text{ kPa}$ so 650 kPa is higher

(c) 1.34 bar or 934 kPa

$1.34 \text{ bar}\left(\dfrac{101.325 \text{ kPa}}{1.01325 \text{ bar}}\right) = 134 \text{ kPa}$ so 934 kPa is higher

10.4. Arrange in order of increasing pressure:

$363 \text{ mm Hg}\left(\dfrac{1 \text{ atm}}{760 \text{ mm Hg}}\right) = 0.478 \text{ atm}$ $363 \text{ kPa}\left(\dfrac{1 \text{ atm}}{101.325 \text{ kPa}}\right) = 3.58 \text{ atm}$

$0.523 \text{ bar}\left(\dfrac{0.98692 \text{ atm}}{1 \text{ bar}}\right) = 0.516 \text{ atm}$

0.256 atm < 363 mm Hg < 0.523 bar < 363 kPa

Boyle's Law and Charles's Law

10.5. **Boyle's law** states that the pressure a gas exerts is inversely proportional to the volume it occupies, or for a given amount of gas, PV = constant. You can write this as: $P_1V_1 = P_2V_2$

So (67.5 mm Hg)(500. mL) = (P_2)(125 mL) and $\dfrac{(67.5 \text{ mmHg})(500.\text{ mL})}{125 \text{ mL}} = 270.$ mm Hg

Note that **volume decreased** by a factor of **4,** and the **pressure increased** by a factor of **4.**

10.6. Transfer sample of hydrogen gas to new container:

$$P_2 = \dfrac{P_1V_1}{V_2} = \dfrac{(0.43 \text{ atm})(125 \text{ mL})}{375 \text{ mL}} = 0.14 \text{ atm}$$

10.7. Transfer sample of methane gas to new container:

$$V_2 = \dfrac{P_1V_1}{P_2} = \dfrac{(3.88 \times 10^4 \text{ Pa})(50.0 \text{ mL})}{3.15 \times 10^4 \text{ Pa}} = 61.6 \text{ mL}$$

10.8. Transfer sample of CO_2 gas to new container:

$$V_2 = \dfrac{P_1V_1}{P_2} = \dfrac{(56.5 \text{ mm Hg})(125 \text{ mL})}{62.3 \text{ mm Hg}} = 113 \text{ mL}$$

10.9. **Charles' law** states that $V \propto T$ (in Kelvin):

Note that with the **increase in T** there has been an **increase in volume**.

$$\dfrac{V_2}{V_1} = \dfrac{T_2}{T_1}, \text{ so } V_2 = \dfrac{T_2 V_1}{T_1}$$

$$V_2 = \dfrac{(3.5 \text{ L})(310 \text{ K})}{295 \text{ K}} = 3.7 \text{ L}$$

10.10. Volume change of CO_2 upon cooling in ice bath:

$$\dfrac{V_2}{V_1} = \dfrac{T_2}{T_1}, \text{ so } V_2 = \dfrac{T_2 V_1}{T_1}$$

$$V_2 = \dfrac{(5.0 \text{ mL})(273 \text{ K})}{295 \text{ K}} = 4.6 \text{ mL}$$

10.11. Temperature change of N_2 upon volume change:

32 °C = 305 K

$\dfrac{V_2}{V_1} = \dfrac{T_2}{T_1}$, so $T_2 = \dfrac{V_2 T_1}{V_1}$

$T_2 = \dfrac{(18.5 \text{ mL})(305 \text{ K})}{25.0 \text{ mL}} = 226 \text{ K} = -47 \text{ °C}$

10.12. Temperature change of CO_2 upon volume change:

27.5 °C = 300.6 K

$\dfrac{V_2}{V_1} = \dfrac{T_2}{T_1}$, so $T_2 = \dfrac{V_2 T_1}{V_1}$

$T_2 = \dfrac{(525 \text{ mL})(300.6 \text{ K})}{625.0 \text{ mL}} = 252.5 \text{ K} = -21 \text{ °C}$

The General Gas Law

10.13. Pressure of 3.6 L of H_2 at 380 mm Hg and 25 °C, if transferred to a 5.0 L flask at 0.0 °C:

The following relationship will allow you to calculate P_2.

$\dfrac{P_1 V_1}{T_1} = \dfrac{P_2 V_2}{T_2}$

Rearranging the equation:

$P_2 = P_1 \times \dfrac{T_2 V_1}{T_1 V_2} = 380 \text{ mm Hg} \times \dfrac{(273 \text{ K})(3.6 \text{ L})}{(298 \text{ K})(5.0 \text{ L})} = 250 \text{ mm Hg}$

10.14. Volume of flask B:

The following relationship will allow you to calculate V_2.

$\dfrac{P_1 V_1}{T_1} = \dfrac{P_2 V_2}{T_2}$

Rearranging the equation:

$V_2 = V_1 \times \dfrac{T_2 P_1}{T_1 P_2} = 25.0 \text{ mL} \times \dfrac{(297.7 \text{ K})(436.5 \text{ mm Hg})}{(293.7 \text{ K})(94.3 \text{ mm Hg})} = 117 \text{ mL}$

10.15. Temperature and pressure are directly proportional:

$$\frac{P_2}{P_1} = \frac{T_2}{T_1}$$

so,

$$P_2 = \frac{P_1 T_2}{T_1} = \frac{(360 \text{ mm Hg})(268.2 \text{ K})}{(298.7 \text{ K})} = 320 \text{ mm Hg}$$

10.16. Pressure of gas in 252-mL flask at 0.0 °C:

$$P_2 = \frac{P_1 V_1 T_2}{V_2 T_1} = \frac{(388 \text{ mm Hg})(165 \text{ mL})(273.2 \text{ K})}{(252 \text{ mL})(305.6 \text{ K})} = 227 \text{ mm Hg}$$

10.17. Using the general gas law, you can write:

$$\frac{P_1 V_1}{T_1} = \frac{P_2 V_2}{T_2}$$

The volume is changing from 400. cm³ to 50.0 cm³, and the temperature from 15 °C to 77 °C. Rearrange the equation to solve for the new pressure:

$$P_2 = \frac{P_1 V_1 T_2}{V_2 T_1} = \frac{(1.00 \text{ atm})(400. \text{ cm}^3)(350. \text{ K})}{(50.0 \text{ cm}^3)(288 \text{ K})} = 9.27 \text{ atm}$$

10.18. Volume of He-filled balloon at 2-mile altitude:

$$V_2 = \frac{V_1 P_1 T_2}{P_2 T_1} = \frac{(1.2 \times 10^7 \text{ L})(748 \text{ mm Hg})(265.4 \text{ K})}{(600. \text{ mm Hg})(298.0 \text{ K})} = 1.3 \times 10^7 \text{ L}$$

Avogadro's Hypothesis

10.19. Reaction of nitrogen monoxide with oxygen:

(a) The balanced equation indicates that 1 O_2 is needed for 2 NO. At the same conditions of T and P, the amount of O_2 is 1/2 the amount of NO, so 75 mL of O_2 is required.

(b) The amount of NO_2 produced will have the same volume as the amount of NO (since their coefficients are equal in the balanced equation)—150 mL of NO_2.

10.20. Combustion of ethane:

$$8.4 \text{ L C}_2\text{H}_6 \cdot \frac{7 \text{ L O}_2}{2 \text{ L C}_2\text{H}_6} = 29 \text{ L O}_2 \quad \text{and} \quad 8.4 \text{ L C}_2\text{H}_6 \cdot \frac{6 \text{ L H}_2\text{O}}{2 \text{ L C}_2\text{H}_6} = 25 \text{ L H}_2\text{O}$$

Ideal Gas Law

10.21. The pressure of 1.25 g of gaseous carbon dioxide can be calculated using the ideal gas law:

$$1.25 \text{ g CO}_2 \cdot \frac{1 \text{ mol CO}_2}{44.01 \text{ g CO}_2} = 0.0284 \text{ mol CO}_2. \quad \text{Rearranging } PV = nRT \text{ to solve for } P:$$

$$P = \frac{nRT}{V} = \frac{(0.0284 \text{ mol})\left(0.082057 \frac{\text{L} \cdot \text{atm}}{\text{K} \cdot \text{mol}}\right)(295.7 \text{ K})}{0.750 \text{ L}} = 0.919 \text{ atm}$$

10.22. Volume of He in balloon:

$$2.50 \text{ kg He} \left(\frac{1000 \text{ g}}{1 \text{ kg}}\right)\left(\frac{1 \text{ mol He}}{4.0026 \text{ g}}\right) = 624.6 \text{ mol He}$$

$$V = \frac{nRT}{P} = \frac{(624.6 \text{ mol He})(0.082057 \text{ L} \cdot \text{atm/mol} \cdot \text{K})(300. \text{ K})}{0.985 \text{ atm}} = 1.56 \times 10^4 \text{ L}$$

10.23. The volume of the flask may be calculated by realizing that the gas will expand to fill the flask:

$$2.2 \text{ g CO}_2 \left(\frac{1 \text{ mol CO}_2}{44.0 \text{ g CO}_2}\right) = 0.050 \text{ mol CO}_2; \ P = 318 \text{ mm Hg} \left(\frac{1 \text{ atm}}{760 \text{ mm Hg}}\right) = 0.418 \text{ atm}$$

$$V = \frac{(0.050 \text{ mol})\left(0.082057 \frac{\text{L} \cdot \text{atm}}{\text{K} \cdot \text{mol}}\right)(295 \text{ K})}{0.418 \text{ L}} = 2.9 \text{ L}$$

10.24. Pressure of ethanol vapor in steel cylinder:

$$1.50 \text{ g} \cdot \frac{1 \text{ mol C}_2\text{H}_5\text{OH}}{46.07 \text{ g}} = 0.0326 \text{ mol C}_2\text{H}_5\text{OH}$$

$$251 \text{ cm}^3 \cdot \frac{1 \text{ mL}}{1 \text{ cm}^3} \cdot \frac{1 \text{ L}}{10^3 \text{ mL}} = 0.251 \text{ L}$$

$$P = \frac{nRT}{V} = \frac{(0.0326 \text{ mol C}_2\text{H}_5\text{OH})(0.082057 \text{ L} \cdot \text{atm/K} \cdot \text{mol})(523 \text{ K})}{0.251 \text{ L}} = 5.57 \text{ atm}$$

10.25. Mass of He in long-distance flying balloon:

Rearranging $PV = nRT$ to solve for n, you obtain $n = \dfrac{PV}{RT}$

Converting 737 mm Hg to atmospheres, you obtain

737 mm Hg $\cdot \dfrac{1 \text{ atm}}{760 \text{ mm Hg}} = 0.970$ atm

$n = \dfrac{(0.970 \text{ atm})(1.2 \times 10^7 \text{ L})}{(0.082057 \text{ L} \cdot \text{atm/K} \cdot \text{mol})(298.2 \text{ K})} = 4.8 \times 10^5$ moles of He

and since each mole of He has a mass of 4.0026 g,

4.8×10^5 mol He $\cdot \dfrac{4.0026 \text{ g He}}{1 \text{ mol He}} = 1.9 \times 10^6$ g He

10.26. Mass of He in 3.0 L balloon:

$n = \dfrac{PV}{RT} = \dfrac{(1.1 \text{ atm})(3.0 \text{ L})}{(0.082057 \text{ L} \cdot \text{atm/mol} \cdot \text{K})(295 \text{ K})} = 0.136$ mol He

0.136 mol He $\left(\dfrac{4.0026 \text{ g He}}{1 \text{ mol He}} \right) = 0.55$ g He

Gas Density and Molar Mass

10.27. Air density forty miles above Earth's surface:

Write the ideal gas law as: Molar Mass $= \dfrac{dRT}{P}$ where d = density in grams per liter.

Solving for d: $\dfrac{(\text{Molar Mass}) \cdot P}{R \cdot T} = d$

The average molar mass for air is approximately 28.96 g/mol.

$\dfrac{(28.96 \text{ g/mol})(0.20 \text{ mm Hg})\left(\dfrac{1 \text{ atm}}{760 \text{ mm Hg}} \right)}{(0.082057 \dfrac{\text{L} \cdot \text{atm}}{\text{K} \cdot \text{mol}})(250 \text{ K})} = 3.7 \times 10^{-4}$ g/L

10.28. Density of diethyl ether vapor:

233 mm Hg $\left(\dfrac{1 \text{ atm}}{760 \text{ mm Hg}} \right) = 0.3066$ atm

$$d = \frac{PM}{RT} = \frac{(0.3066 \text{ atm})(74.12 \text{ g/mol})}{(0.082057 \text{ L·atm/K·mol})(298 \text{ K})} = 0.929 \text{ g/L}$$

10.29. Molar mass of organofluorine compound:

$$\text{Molar mass} = \frac{(0.355 \text{ g/L})(0.082057 \frac{\text{L·atm}}{\text{K·mol}})(290.\text{K})}{(189 \text{ mm Hg})\left(\frac{1 \text{atm}}{760 \text{ mm Hg}}\right)} = 34.0 \text{ g/mol}$$

10.30. Molar mass of chloroform:

$$195 \text{ mm Hg} \left(\frac{1 \text{ atm}}{760 \text{ mm Hg}}\right) = 0.257 \text{ atm}$$

$$M = \frac{dRT}{P} = \frac{(1.25 \text{ g/L})(0.082057 \text{ L·atm/K·mol})(298.2 \text{ K})}{0.257 \text{ atm}} = 119 \text{ g/mol}$$

10.31. Molar mass of unknown gas:

Rearranging $PV = nRT$ to solve for n, you obtain:

$$n = \frac{PV}{RT}$$

Converting 715 mm Hg to atmospheres: $715 \text{mm Hg} \cdot \frac{1 \text{ atm}}{760 \text{ mm Hg}} = 0.941 \text{ atm}$

$$n = \frac{(0.941 \text{ atm})(0.452 \text{ L})}{(0.082057 \frac{\text{L·atm}}{\text{K·mol}})(296.2 \text{ K})} = 0.0175 \text{ moles of unknown gas.}$$

Since this number of moles of the gas has a mass of 1.007 g, you can calculate the molar mass. $\frac{1.007 \text{ g of unknown gas}}{0.0175 \text{ moles of unknown gas}} = 57.5 \text{ g/mol}$

10.32. Molecular formula of compound whose empirical formula is C₄H₅:

$$d = \frac{0.0130 \text{ g}}{0.165 \text{ L}} = 0.0788 \text{ g/L}$$

and converting pressure to units of atm: $13.7 \text{ mm Hg} \cdot \frac{1 \text{ atm}}{760 \text{ mm Hg}} = 0.0180 \text{ atm}$

$$M = \frac{dRT}{P} = \frac{(0.0788 \text{ g/L})(0.082057 \text{ L·atm/K·mol})(295.7 \text{ K})}{0.0180 \text{ atm}} = 106 \text{ g/mol}$$

$$\frac{106 \text{ g/mol}}{53 \text{ g/mol}} = 2 \quad \text{The molecular formula is (C}_4\text{H}_5)_2 \text{ or C}_8\text{H}_{10}$$

10.33. To calculate the molar mass:

$$M = \frac{dRT}{P} = \frac{(0.0125 \text{ g}/0.125 \text{ L})(0.082057 \text{ L} \cdot \text{atm/K} \cdot \text{mol})(298.2 \text{ K})}{[24.8 \text{ mm Hg}(1 \text{ atm}/760 \text{ mm Hg})]} = 75.0 \text{ g/mol}$$

B₆H₁₀ has a molar mass of 74.9 grams.

10.34. Determine molecular formula of chlorofluorocarbon:

$$173 \text{ mm Hg} \left(\frac{1 \text{ atm}}{760 \text{ mm Hg}} \right) = 0.2276 \text{ atm}$$

$$n = \frac{PV}{RT} = \frac{(0.2276 \text{ atm})(0.375 \text{ L})}{(0.082057 \text{ L} \cdot \text{atm/K} \cdot \text{mol})(398.2 \text{ K})} = 0.00261 \text{ mol}$$

Molar mass = 0.533 g/0.00261 mol = 204 g/mol

CCl₂F has an empirical mass of 101.9 g/unit

$$\frac{204 \text{ g/mol}}{101.9 \text{ g/unit}} = 2 \text{ empirical units/mol}$$

Molecular formula is $C_2Cl_4F_2$

Gas Laws and Chemical Reactions

10.35. Determine the amount of H₂ generated when 2.2 g Fe reacts:

$$2.2 \text{ g Fe} \left(\frac{1 \text{ mol Fe}}{55.845 \text{ g Fe}} \right) \left(\frac{1 \text{ mol H}_2}{1 \text{ mol Fe}} \right) = 0.0394 \text{ mol H}_2 \text{ (0.039 to 2 sf)}$$

Note that the latter factor is achieved by examining the balanced equation!

The pressure of this amount of H₂ is:

$$P = \frac{nRT}{V} = \frac{(0.0394 \text{ mol})(0.082057 \text{ L} \cdot \text{atm/K} \cdot \text{mol})(298 \text{ K})}{10.0 \text{ L}} = 0.096 \text{ atm (or 73 mm Hg)}$$

10.36. Reaction of silane and oxygen:

$$356 \text{ mm Hg} \cdot \frac{1 \text{ atm}}{760 \text{ mm Hg}} = 0.468 \text{ atm} \quad \text{and} \quad 425 \text{ mm Hg} \cdot \frac{1 \text{ atm}}{760 \text{ mm Hg}} = 0.559 \text{ atm}$$

$$n = \frac{PV}{RT} = \frac{(0.468 \text{ atm})(5.20 \text{ L})}{(0.082057 \text{ L} \cdot \text{atm/K} \cdot \text{mol})(298 \text{ K})} = 0.0996 \text{ mol SiH}_4$$

$$0.0996 \text{ mol SiH}_4 \cdot \frac{2 \text{ mol O}_2}{1 \text{ mol SiH}_4} = 0.199 \text{ mol O}_2$$

$$V = \frac{nRT}{P} = \frac{(0.199 \text{ mol O}_2)(0.082057 \text{ L} \cdot \text{atm/K} \cdot \text{mol})(298 \text{ K})}{0.559 \text{ atm}} = 8.71 \text{ L}$$

10.37. Mass of sodium azide required to fill airbag:

Calculate the moles of N₂ needed:

$$n = \frac{PV}{RT} = \frac{(1.3 \text{ atm})(75.0 \text{ L})}{(0.082057 \frac{\text{L} \cdot \text{atm}}{\text{K} \cdot \text{mol}})(298 \text{ K})} = 3.99 \text{ mol N}_2$$

The mass of NaN₃ needed is obtained from the stoichiometry of the equation:

$$3.99 \text{ mol N}_2 \left(\frac{2 \text{ mol NaN}_3}{3 \text{ mol N}_2}\right)\left(\frac{65.01 \text{ g NaN}_3}{1 \text{ mol NaN}_3}\right) = 173 \text{ g NaN}_3 = 170 \text{ g NaN}_3 \text{ (2 sf)}$$

10.38. Pressure of O₂ and water vapor after heptane combustion:

$$0.0325 \text{ g} \left(\frac{1 \text{ mol C}_7\text{H}_{16}}{100.2 \text{ g}}\right) = 3.24 \times 10^{-4} \text{ mol C}_7\text{H}_{16}$$

$$3.24 \times 10^{-4} \text{ mol C}_7\text{H}_{16} \left(\frac{8 \text{ mol H}_2\text{O}}{1 \text{ mol C}_7\text{H}_{16}}\right) = 0.00259 \text{ mol H}_2\text{O}$$

$$P_{\text{H}_2\text{O}} = \frac{n_{\text{H}_2\text{O}}RT}{V} = \frac{(0.00259 \text{ mol H}_2\text{O})(0.082057 \text{ L} \cdot \text{atm/K} \cdot \text{mol})(303.2 \text{ K})}{5.00 \text{ L}} = 0.0129 \text{ atm}$$

$$3.24 \times 10^{-4} \text{ mol C}_7\text{H}_{16} \left(\frac{11 \text{ mol O}_2}{1 \text{ mol C}_7\text{H}_{16}}\right) = 0.00356 \text{ mol O}_2$$

$$P_{\text{O}_2} = \frac{n_{\text{O}_2}RT}{V} = \frac{(0.00356 \text{ mol O}_2)(0.082057 \text{ L} \cdot \text{atm/K} \cdot \text{mol})(318.2 \text{ K})}{5.00 \text{ L}} = 0.0186 \text{ atm}$$

10.39. Oxygen pressure to consume 1.00 kg of hydrazine:

N₂H₄(g) + O₂(g) → N₂(g) + 2 H₂O(ℓ)

$$1.00 \text{ kg N}_2\text{H}_4 \left(\frac{1 \times 10^3 \text{ g}}{1 \text{ kg}}\right)\left(\frac{1 \text{ mol N}_2\text{H}_4}{32.05 \text{ g N}_2\text{H}_4}\right)\left(\frac{1 \text{ mol O}_2}{1 \text{ mol N}_2\text{H}_4}\right) = 3.12 \times 10^1 \text{ mole O}_2$$

$$P_{O_2} = \frac{n_{O_2} \cdot R \cdot T}{V} = \frac{(3.12 \times 10^1 \text{ mol})(0.082057 \frac{\text{L} \cdot \text{atm}}{\text{K} \cdot \text{mol}})(296 \text{ K})}{(450 \text{ L})}$$

$$P_{O_2} = 1.67 \text{ atm or } 1.7 \text{ atm } (2 \text{ sf})$$

10.40. Mass of potassium superoxide to consume carbon dioxide:

$$767 \text{ mm Hg} \left(\frac{1 \text{ atm}}{760 \text{ mm Hg}}\right) = 1.009 \text{ atm}$$

$$n = \frac{PV}{RT} = \frac{(1.009 \text{ atm})(8.90)}{(0.082057 \text{ L} \cdot \text{atm/K} \cdot \text{mol})(295.2 \text{ K})} = 0.3708 \text{ mol CO}_2$$

$$0.3708 \text{ mol CO}_2 \left(\frac{4 \text{ mol KO}_2}{2 \text{ mol CO}_2}\right)\left(\frac{71.10 \text{ g}}{1 \text{ mol KO}_2}\right) = 52.7 \text{ g KO}_2$$

Gas Mixtures and Dalton's Law

10.41. You know that the total pressure will be equal to the sum of the pressure of each gas (also called the *partial pressure* of each gas).

$$1.0 \text{ g H}_2 \left(\frac{1 \text{ mol H}_2}{2.016 \text{ g H}_2}\right) = 0.496 \text{ mol H}_2$$

$$8.0 \text{ g Ar} \left(\frac{1 \text{ mol Ar}}{39.95 \text{ g Ar}}\right) = 0.200 \text{ mol Ar}$$

You can calculate the **total pressure** using the total moles (0.696 moles) and the ideal gas law:

$$P = \frac{n_{\text{total}} RT}{V} = \frac{(0.696 \text{ mol})(0.082057 \text{ L} \cdot \text{atm/K} \cdot \text{mol})(300. \text{ K})}{3.0 \text{ L}} = 5.71 \text{ atm} = 5.7 \text{ atm } (2 \text{ sf})$$

The pressure of **each gas** can be calculated by multiplying the total pressure (5.7 atm) by the mole fraction of the gas.

$$P(\text{H}_2) = \frac{0.496 \text{ mol H}_2}{0.496 \text{ mol H}_2 + 0.200 \text{ mol Ar}}(5.71 \text{ atm}) = 4.1 \text{ atm}$$

$$P(\text{Ar}) = \frac{0.200 \text{ mol Ar}}{0.496 \text{ mol H}_2 + 0.200 \text{ mol Ar}}(5.71 \text{ atm}) = 1.6 \text{ atm}$$

Note that the total pressure is indeed (4.1 + 1.6) or 5.7 atm

10.42. Partial pressures of each gas in cylinder:

%N_2 = 100.0 − (6.5% H_2S + 1.5% CO_2) = 92.0% N_2

The partial pressure of each gas is proportional to its percentage:

P_{N_2} = (37 atm)(0.920) = 34 atm

P_{H_2S} = (37 atm)(0.065) = 2.4 atm

P_{CO_2} = (37 atm)(0.015) = 0.55 atm

10.43. Halothane – oxygen mixture problem:

$P_{total} = P_{halothane} + P_{oxygen}$ = 170 mm Hg + 570 mm Hg = 740 mm Hg

(a) Since you know that the pressure a gas exerts is **proportional** to the moles of gas present you can calculate the ratio of moles by using their partial pressures:

$$\frac{\text{moles of halothane}}{\text{moles of oxygen}} = \frac{170 \text{ mm Hg}}{570 \text{ mm Hg}} = 0.30$$

(b)

$$160 \text{ g oxygen} \left(\frac{1.0 \text{ mol oxygen}}{32.0 \text{ g oxygen}} \right) \left(\frac{0.30 \text{ mol halothane}}{1.0 \text{ mol oxygen}} \right) \left(\frac{197.38 \text{ g halothane}}{1.0 \text{ mol halothane}} \right) = 3.0 \times 10^2 \text{g halothane (2sf)}$$

10.44. Balloon filled with helium and oxygen:

(a) $n = \dfrac{PV}{RT} = \dfrac{(1.00 \text{ atm})(12.5 \text{ L})}{(0.082057 \text{ L} \cdot \text{atm/K} \cdot \text{mol})(294.7 \text{ K})} = 0.517$ mol He

$0.517 \text{ mol He} \left(\dfrac{4.003 \text{ g}}{1 \text{ mol He}} \right) = 2.07$ g He

(b) $P = \dfrac{nRT}{V} = \dfrac{(0.517 \text{ mol He})(0.082057 \text{ L} \cdot \text{atm/K} \cdot \text{mol})(294.7 \text{ K})}{26 \text{ L}} = 0.48$ atm

(c) $P_{O_2} = P_{total} - P_{He}$ = 1.00 atm − 0.48 atm = 0.52 atm

(d) $X_{He} = \dfrac{0.48 \text{ atm}}{1.00 \text{ atm}} = 0.48$ $X_{O_2} = \dfrac{0.52 \text{ atm}}{1.00 \text{ atm}} = 0.52$

Kinetic-Molecular Theory

10.45. Comparison of two gases:

(a) The average kinetic energy depends on the temperature so the average kinetic *energies of the molecules of CO_2 will be greater than those of H_2*.

(b) The average molecular velocity, v, is related to the kinetic energy by the expression: KE = 1/2 mv^2 and KE is related to the temperature by the expression KE = (3/2)*RT*.

Setting the two quantities that are equal to KE gives $1/2\, mv^2 = (3/2)RT$ Multiplying both sides of the equation by 2 gives $mv^2 = 3RT$, and solving for v gives $v = \sqrt{\dfrac{3RT}{m}}$

If you think in terms of moles of the gas, you can replace m with M (the molar mass).

For CO$_2$ this would be $v_{CO_2} = \sqrt{\dfrac{3R \cdot 298K}{44.0\, g/mol}}$ and for H$_2$ the expression would be:

$v_{H_2} = \sqrt{\dfrac{3R \cdot 273K}{2.0\, g/mol}}$. Note that $3R$ is equal in the two expressions, so you treat $3R$ as a constant (which won't affect the relative velocities).

The two velocities are then proportional to the term $\sqrt{\dfrac{298}{44}}$ or 2.6 for CO$_2$.

For H$_2$ $\sqrt{\dfrac{273}{2}}$ or 11.7 so the hydrogen molecules have an average velocity greater than the CO$_2$ molecules.

(c) Since the volumes are equal for these two gas samples, the pressure is proportional to the amount of gas present.

$V_A = \dfrac{n_A RT_A}{P_A}$ and $V_B = \dfrac{n_B RT_B}{P_B}$ and $V_A = V_B$ so

$\dfrac{n_A RT_A}{P_A} = \dfrac{n_B RT_B}{P_B}$ and rearranging to solve for the ratio of molecules present.

$\dfrac{P_B T_A}{T_B P_A} = \dfrac{n_B}{n_A}$ substituting yields: $\dfrac{2\,atm \cdot 273\,K}{298\,K \cdot 1\,atm} = \dfrac{n_B}{n_A} = 1.8$

There are 1.8 times as many moles (and molecules) of gas in Flask B (CO$_2$) as there are in Flask A (H$_2$).

(d) Since Flask B contains 1.8 times as many moles of CO$_2$ as Flask A contains of H$_2$, the *ratio* of masses of gas present are:

$\dfrac{\text{Mass (Flask B)}}{\text{Mass (Flask A)}} = \dfrac{(1.8\, mole\, CO_2)(44\, g\, CO_2/mol\, CO_2)}{(1\, mol\, H_2)(2\, g\, H_2/mol\, H_2)} = \dfrac{40}{1}$

Note that any number of moles of CO$_2$ and H$_2$ (in the ratio of 1.8:1) would provide the same answer.

10.46. Comparison of N_2 and Ar:

The molar mass of Ar (40 g/mol) is greater than the molar mass of N_2 (28 g/mol). Therefore, for samples with equal mass there are more moles of N_2 present than moles of Ar.
(a) True. There are more moles of N_2 present, so there are more molecules of N_2 present.
(b) False. Pressure is directly related to the number of moles of gas present. The pressure in the nitrogen flask is greater because there are more moles of N_2 present.
(c) False. The gas with the smaller molar mass (N_2) will have a greater velocity than the gas with the greater molar mass (Ar).
(d) True. The nitrogen molecules have a greater velocity than the argon molecules and there are more molecules of nitrogen present, so they will collide more frequently with the walls of the flask.

10.47. RMS speed of oxygen molecules:

Since two gases at the same temperature have the same kinetic energy:

$$KE_{O_2} = KE_{CO_2} \text{ and since the average } KE = 1/2\, M\overline{U}^2$$

where $\overline{U}$ is the average speed of a molecule, you can write.

$$\frac{1}{2}M_{O_2}\overline{U^2_{O_2}} = \frac{1}{2}M_{CO_2}\overline{U^2_{CO_2}} \text{ or } M_{O_2}\overline{U^2_{O_2}} = M_{CO_2}\overline{U^2_{CO_2}} \text{ and } \frac{M_{O_2}}{M_{CO_2}} = \frac{\overline{U^2_{CO_2}}}{\overline{U^2_{O_2}}}$$

and solving for the average velocity of CO_2:

$$\frac{\overline{U^2_{CO_2}}}{1} = \frac{\overline{U^2_{O_2}}M_{O_2}}{M_{CO_2}}. \text{ Taking the square root of both sides}$$

$$\frac{\overline{U_{CO_2}}}{1} = \sqrt{\frac{M_{O_2}}{M_{CO_2}}} \cdot \overline{U}_{O_2} = \sqrt{\frac{32.0 \text{ g } O_2 / \text{mol } O_2}{44.0 \text{ g } CO_2 / \text{mol } CO_2}} \cdot 4.28 \times 10^4 \text{ cm/s} = 3.65 \times 10^4 \text{ cm/s}$$

10.48. Calculate the rms for NO molecules at 15 °C (288 K):

$$\sqrt{\overline{u^2}} = \sqrt{\frac{3RT}{M}} = \sqrt{\frac{3(8.3145 \text{ J/K} \cdot \text{mol})(288 \text{ K})}{0.03001 \text{ kg/mol}}} = 489 \text{ m/s}$$

$$\frac{\text{rms speed NO}}{\text{rms speed Ar}} = \sqrt{\frac{39.95}{30.01}} = 1.154$$

10.49. The species will have average molecular speeds that are inversely proportional to their molar masses.   Slowest   Fastest

	CH_2F_2	<	Ar	<	N_2	<	CH_4
Molar masses	54		40		28		16 (integral values)

10.50. Increasing average molecular speed: $OSCl_2$ < Cl_2O < Cl_2 < SO_2
molar mass (g/mol):   119   87   71   64

Diffusion and Effusion

10.51. Relative rates of effusion for the following pairs of gases:
(a) CO_2 or F_2: *Fluorine* effuses faster (MM F_2 = 38 g/mol; CO_2 = 44 g/mol)
(b) O_2 or N_2: *Nitrogen* effuses faster. (MM N_2 = 28 g/mol; O_2 = 32 g/mol)
(c) C_2H_4 or C_2H_6: *Ethylene* effuses faster. (MM C_2H_4 = 28 g/mol; C_2H_6 = 30 g/mol)
(d) $CFCl_3$ or $C_2Cl_2F_4$: $CFCl_3$ effuses faster. (MM $CFCl_3$ = 137 g/mol; $C_2Cl_2F_4$ = 171 g/mol)

10.52. Argon and helium comparison:

He will effuse faster $\dfrac{\text{Rate of He effusion}}{\text{Rate of Ar effusion}} = \sqrt{\dfrac{39.9 \text{ g/mol}}{4.00 \text{ g/mol}}} = 3.16 \text{ times faster.}$

10.53. Determine the molar mass of a gas which effuses at a rate 1/3 that of He:

$$\dfrac{\text{He effusion rate}}{\text{Unknown effusion rate}} = \dfrac{\text{He effusion rate}}{1/3 \text{ He effusion rate}} = 3$$

$$3 = \sqrt{\dfrac{\text{MM of unknown}}{\text{MM of He}}} = \sqrt{\dfrac{\text{MM of unknown}}{4.0 \text{ g/mol}}}$$

$$9 = \dfrac{\text{MM of unknown}}{4.0 \text{ g/mol}}$$

MM of unknown = $9(4.0 \text{ g/mol}) = 36$ g/mol

10.54. Molar mass and molecular formula of unknown gas:

$$\dfrac{CH_4 \text{ effusion rate}}{\text{Unknown effusion rate}} = \dfrac{CH_4 \text{ effusion rate}}{1/2 \ CH_4 \text{ effusion rate}} = 2$$

$$2 = \sqrt{\dfrac{\text{MM of unknown}}{M \text{ of } CH_4}} = \sqrt{\dfrac{\text{MM of unknown}}{16.0 \text{ g/mol}}}$$

$$4 = \dfrac{\text{MM of unknown}}{16.0 \text{ g/mol}}$$

MM of unknown = $4(16.0 \text{ g/mol}) = 64$ g/mol

CHF has an empirical mass of 32 g/empirical unit.

$\dfrac{64 \text{ g/mol}}{32 \text{ g/unit}} = 2$ empirical units/mol

Molecular formula is $C_2H_2F_2$

10.55. Molecular weight of unknown compound:

$$\frac{\text{Rate of CO}_2}{\text{Rate of unknown}} = \sqrt{\frac{MM_{unknown}}{MM_{CO_2}}}$$

$$\frac{\left(\dfrac{0.521 \text{ g CO}_2}{1 \text{ day}} \times \dfrac{1 \text{ mol CO}_2}{44.01 \text{ g}}\right)}{\left(\dfrac{0.413 \text{ g}}{1 \text{ day}} \times \dfrac{1}{MM_{unknown}}\right)} = \sqrt{\frac{MM_{unknown}}{44.01 \text{ g/mol}}}$$

$$(0.0287)^2 (MM_{unknown})^2 = \frac{MM_{unknown}}{44.01 \text{ g/mol}}$$

$$MM_{unknown} = 27.6 \text{ g/mol}$$

10.56. What is the molecular weight of uranium fluoride:

$$\frac{\text{Rate of I}_2}{\text{Rate of uranium fluoride}} = \sqrt{\frac{MM_{uranium\ fluoride}}{MM_{I_2}}}$$

$$\frac{\left(\dfrac{0.0150 \text{ g}}{1 \text{ hr}} \times \dfrac{1 \text{ mol I}_2}{253.8 \text{ g}}\right)}{\left(\dfrac{0.0177 \text{ g}}{1 \text{ hr}} \times \dfrac{1}{MM_{uranium\ fluoride}}\right)} = \sqrt{\frac{MM_{uranium\ fluoride}}{253.8 \text{ g/mol}}}$$

$$(0.00334)^2 (MM_{uranium\ fluoride})^2 = \frac{MM_{uranium\ fluoride}}{253.8 \text{ g/mol}}$$

$$MM_{uranium\ fluoride} = 353 \text{ g/mol}$$

Nonideal Gases

10.57. Conditions in which CO_2 will deviate most from ideal gas behavior:

(c) 10 atm, 0 °C--- gases deviate at higher pressures (when they will occupy smaller volumes—and the volumes of the gas particles become a more significant part of the overall volume); gases also deviate at lower temperatures (when the velocities are lower, and interparticle forces can have a greater effect).

10.58. Conditions under which Cl_2 will deviate least from ideal gas behavior:
(b) 0.1 atm, 100 °C (See 10.51 for rationale.)

10.59. According to the Ideal Gas Law, the pressure would be:

$$P = \frac{n \cdot R \cdot T}{V} = \frac{(4.00 \text{ mol})(0.082057 \frac{\text{L atm}}{\text{K mol}})(373 \text{ K})}{4.00 \text{ L}} = 30.6 \text{ atm}$$

The van der Waal's equation is: $\left[P + a\left(\frac{n}{V}\right)^2\right][V - bn] = nRT$. Substituting, you get:

$$\left[P + 6.49 \frac{\text{atm} \cdot \text{L}^2}{\text{mol}^2}\left(\frac{4.00 \text{ mol}}{4.00 \text{ L}}\right)^2\right]\left[4.00 \text{ L} - 0.0562 \frac{\text{L}}{\text{mol}}(4.00 \text{ mol})\right] = 4.00 \text{ mol} \cdot 0.082057 \frac{\text{atm} \cdot \text{L}}{\text{K} \cdot \text{mol}} \cdot 373 \text{K}$$

Simplifying: $[P + 6.49 \text{ atm}][4.00 \text{ L} - 0.22 \text{ L}] = 122.43 \text{ atm} \cdot \text{L}$ and

$$P = \frac{122.43 \text{ atm} \cdot \text{L}}{3.78 \text{ L}} - 6.49 \text{ atm} = 25.9 \text{ atm}$$

10.60. Storage of CO_2 in tank at room temperature:

$$165 \text{ g}\left(\frac{1 \text{ mol CO}_2}{44.01 \text{ g}}\right) = 3.75 \text{ mol CO}_2$$

(a) $P = \frac{nRT}{V} = \frac{(3.75 \text{ mol})(0.082057 \text{ L} \cdot \text{atm/K} \cdot \text{mol})(298 \text{ K})}{12.5 \text{ L}} = 7.33 \text{ atm}$

(b) $\left(P + a\left[\frac{n}{V}\right]^2\right)(V - bn) = nRT$

$$\left(P + (3.59 \text{ atm} \cdot \text{L}^2/\text{mol}^2)\left[\frac{3.75 \text{ mol}}{12.5 \text{ L}}\right]^2\right)(12.5 \text{ L} - [0.0427 \text{ L/mol}][3.75 \text{ mol}])$$
$$= (3.75 \text{ mol})(0.08206 \text{ L} \cdot \text{atm/K} \cdot \text{mol})(298 \text{ K})$$

$P = 7.11 \text{ atm}$

10.61. Conditions: 5.00 L; 325 g H_2O; 275 °C (or 548 K)

(a) pressure using the ideal gas law:

$$P = \frac{n \cdot R \cdot T}{V} = \frac{(\frac{325 \text{ g H}_2\text{O}}{18.02 \text{ g H}_2\text{O/mol}}\text{H}_2\text{O})(0.082057 \frac{\text{L} \cdot \text{atm}}{\text{K} \cdot \text{mol}})(548 \text{ K})}{5.00 \text{ L}} = 162 \text{ atm (3sf)}$$

Note that 325 g of water corresponds to 18.0 moles of water.

pressure using the van der Waals equation:

Substituting into $\left[P + a\left(\dfrac{n}{V}\right)^2\right][V - bn] = nRT$

$\left[P + 5.46\dfrac{atm \cdot L^2}{mol^2}\left(\dfrac{18.0 \text{ mol}}{5.00 \text{ L}}\right)^2\right]\left[5.00 \text{ L} - 0.0305\dfrac{L}{mol} \cdot 18.0 \text{ mol}\right] = 18.0 \text{ mol} \cdot 0.082057 \dfrac{atm \cdot L}{K \cdot mol} \cdot 548 \text{ K}$

Simplifying: [P + 70.76 atm][5.00 L – 0.549 L] = 809 atm · L

[P + 70.76 atm][4.451 L] = 809 atm · L

and $P = \dfrac{809 \text{ atm} \cdot L}{4.451 \text{ L}}$ - 70.76 atm = 111 atm (3 sf)

(b) Since the term $a(n/V)^2$ affects the pressure, it has the greater influence.

10.62. For a 5.00-L tank with 375 g Ar at 25°C:

(a) 25 °C = (298 K) (375 g)(1 mol Ar/39.948 g) = 9.39 mol

$P = nRT/V$ = (9.39 mol)(0.0821 L · atm/mol · K)(298 K)/(5.00 L) = 45.9 atm

$\dfrac{(9.39 \text{ mol})(0.082057 \text{ L} \cdot atm/K \cdot mol)(298 \text{ K})}{5.00 \text{ L} - (0.0322 \text{ L/mol})(9.39 \text{ mol})}$ - 1.34 atm · L^2 / mol$^2 \left(\dfrac{9.39 \text{ mol}}{5.00 \text{ L}}\right)^2$

= 48.9 atm - 4.73 atm = 44.2 atm

(b) $a(n/V)^2$

General Questions

10.63.

	atm	mm Hg	kPa	bar
Standard atmosphere:	1	1 atm · $\dfrac{760. \text{ mm Hg}}{1 \text{ atm}}$ = 760. mm Hg	1 atm · $\dfrac{101.325 \text{ kPa}}{1 \text{ atm}}$ = 101.325 kPa	1 atm · $\dfrac{1.013 \text{ bar}}{1 \text{ atm}}$ = 1.013 bar
Partial pressure of N$_2$ in the atmosphere	593 mm Hg · $\dfrac{1 \text{ atm}}{760. \text{ mm Hg}}$ = 0.780 atm	**593 mm Hg**	0.780 atm · $\dfrac{101.3 \text{ kPa}}{1 \text{ atm}}$ = 79.1 kPa	0.780 atm · $\dfrac{1.013 \text{ bar}}{1 \text{ atm}}$ = 0.791 bar
Tank of compressed H$_2$	133 bar · $\dfrac{1 \text{ atm}}{1.013 \text{ bar}}$ = 131 atm	131 atm · $\dfrac{760. \text{ mm Hg}}{1 \text{ atm}}$ = 9.98 ×10^4 mm Hg	131 atm · $\dfrac{101.3 \text{ kPa}}{1 \text{ atm}}$ = 1.33 ×10^4 kPa	**133 bar**

| Atmospheric pressure at top of Mt. Everest | 33.7 kPa · $\dfrac{1\ \text{atm}}{101.3\ \text{kPa}}$ = 0.333 atm | 0.333 atm · $\dfrac{760.\ \text{mm Hg}}{1\ \text{atm}}$ = 253 mm Hg | **33.7 kPa** | 0.333 atm · $\dfrac{1.013\ \text{bar}}{1\ \text{atm}}$ = 0.337 bar |

10.64. For the compound, $C_xH_yN_z$, combustion gives 2.0 L CO₂, 3.5 L H₂O, and 0.5 L N₂. At STP, 1 mol of an ideal gas occupies 22.414 L:

$$2.0\ \text{L CO}_2 \left(\dfrac{1\ \text{mol CO}_2}{22.414\ \text{L}}\right)\left(\dfrac{1\ \text{mol C}}{1\ \text{mol CO}_2}\right) = 0.089\ \text{mol C}$$

$$3.5\ \text{L H}_2\text{O} \left(\dfrac{1\ \text{mol H}_2\text{O}}{22.414\ \text{L}}\right)\left(\dfrac{2\ \text{mol H}}{1\ \text{mol H}_2\text{O}}\right) = 0.31\ \text{mol H}$$

$$0.50\ \text{L N}_2 \left(\dfrac{1\ \text{mol N}_2}{22.414\ \text{L}}\right)\left(\dfrac{2\ \text{mol N}}{1\ \text{mol N}_2}\right) = 0.045\ \text{mol N}$$

$\dfrac{0.31\ \text{mol H}}{0.045\ \text{mol N}} = \dfrac{7\ \text{mol H}}{1\ \text{mol N}}$ $\dfrac{0.089\ \text{mol C}}{0.045\ \text{mol N}} = \dfrac{2\ \text{mol C}}{1\ \text{mol N}}$ The empirical formula is C_2H_7N.

10.65. To increase the average speed of helium atoms by 10.0%, you must know the average speed initially. $\sqrt{u^2} = \sqrt{\dfrac{3RT}{M}}$ and substituting for He:

$$\sqrt{u^2} = \sqrt{\dfrac{3(8.314\ \text{J/K}\cdot\text{mol})(240\ \text{K})}{4.00 \times 10^{-3}\ \text{kg/mol}}} = 1220\ \text{m/s}$$

NOTE: A **Joule** is a kg·m/s, so it's necessary to express the molar mass of helium in kg/mol.

The new average speed = 110% (1220 m/s) = 1350 m/s.

Substituting this value as u:

$$\dfrac{Mu^2}{3R} = T;\ \dfrac{4.00 \times 10^{-3}\ \text{kg/mol}(1350\ \text{m/s})^2}{3(8.314\ \text{J/K}\cdot\text{mol})} = 290.\ \text{K}\quad \text{or } (290. - 273)\ °\text{C or } 17\ °\text{C}.$$

10.66. Mass of oxygen to inflate a balloon at 5 °C:

$$n_2 = n_1\left(\dfrac{T_1}{T_2}\right)\ \text{and}\ n = \dfrac{\text{mass (g)}}{32.00\ \text{g/mol}}$$

$$\text{mass}_2 = \text{mass}_1\left(\dfrac{T_1}{T_2}\right) = (16.0\ \text{g})\left(\dfrac{305.\ \text{K}}{277.2\ \text{K}}\right) = 17.6\ \text{g}$$

10.67. The balanced equation for the combustion of C₄H₉SH is:

2 C₄H₉SH(g) + 15 O₂(g) → 2 SO₂(g) + 8 CO₂(g) + 10 H₂O(g).

Note that this stoichiometry indicates that 2 mol of C₄H₉SH produces a total of (2 + 8 + 10) or 20 mol of product gases. The amount of C₄H₉SH present in 95.0 mg is:

$$95.0 \text{ mg C}_4\text{H}_9\text{SH} \left(\frac{1 \text{ g C}_4\text{H}_9\text{SH}}{1000 \text{ mg C}_4\text{H}_9\text{SH}}\right)\left(\frac{1 \text{ mol C}_4\text{H}_9\text{SH}}{90.184 \text{ g C}_4\text{H}_9\text{SH}}\right)\left(\frac{10 \text{ mol gas}}{1 \text{ mol C}_4\text{H}_9\text{SH}}\right) = 0.01053 \text{ mol}$$

gas (0.01053 mol to 3 sf)

$$P_{\text{total}} = \frac{nRT}{V} = \frac{(0.01053 \text{ mol})(0.082057 \text{ L} \cdot \text{atm/K} \cdot \text{mol})(298)}{5.25 \text{ L}} = 0.04906 \text{ atm} = 37.29 \text{ mm Hg}$$

P_{total} = 37.3 mm Hg (3 sf)

As water is 10/20 (or 0.50) mol of the product gases, $P(\text{H}_2\text{O})$ = 18.6 mm Hg

SO₂ is responsible for 2/20 (or 0.10) mol of the product gases, $P(\text{SO}_2)$ = 3.73 mm Hg

CO₂ is responsible for 8/20 (or 0.40) mol of the product gases, $P(\text{CO}_2)$ = 14.9 mm Hg

Note alternatively that the pressure of carbon dioxide may be calculated:

37.3 mm Hg = 18.6 mm Hg + 3.73 mm Hg + $P(\text{CO}_2)$ or

37.3 mm Hg – (18.6 mm Hg + 3.73 mm Hg) = 14.9 mm Hg.

10.68. Temperature needed to blowout a bicycle tire:

$$T = \frac{PV}{nR} = \frac{(7.25 \text{ atm})(1.52 \text{ L})}{(0.406 \text{ mol})(0.082057 \text{ L} \cdot \text{atm/K} \cdot \text{mol})} = 331 \text{ K} = 58 \text{ °C}$$

10.69. Using the Ideal Gas Law,

$$n = \frac{P \cdot V}{R \cdot T} = \frac{\left(8 \text{ mm Hg} \cdot \frac{1 \text{ atm}}{760 \text{ mm Hg}}\right)\left(10. \text{ m}^3 \cdot \frac{1000 \text{ L}}{1 \text{ m}^3}\right)}{0.082057 \frac{\text{L} \cdot \text{atm}}{\text{K} \cdot \text{mol}} \cdot 300. \text{ K}} = 4 \text{ mol (1 sf)}$$

10.70. Regarding dimethyldichlorosilane:

$$1.90 \text{ g Si}\left(\frac{1 \text{ mol Si}}{28.085 \text{ g Si}}\right) = 0.06765 \text{ mol Si}$$

$$n_{\text{CH}_3\text{Cl}} = \frac{\left(554 \text{ mm Hg}\left(\frac{1 \text{ atm}}{760 \text{ mm Hg}}\right)\right)(7.25 \text{ L})}{(0.082057 \text{ L} \cdot \text{atm/K} \cdot \text{mol})(298 \text{ K})} = 0.216 \text{ mol CH}_3\text{Cl}$$

$$\frac{0.216 \text{ mol CH}_3\text{Cl}}{0.0677 \text{ mol Si}} = \frac{3.19 \text{ mol CH}_3\text{Cl}}{1 \text{ mol Si}} > \frac{2 \text{ mol CH}_3\text{Cl}}{1 \text{ mol Si}} \quad \text{Si is the limiting reactant}$$

$$0.06765 \text{ mol Si} \left(\frac{1 \text{ mol (CH}_3)_2\text{SiCl}_2}{1 \text{ mol Si}} \right) \left(\frac{129.1 \text{ g}}{1 \text{ mol (CH}_3)_2\text{SiCl}_2} \right) = 8.73 \text{ g (CH}_3)_2\text{SiCl}_2$$

$$P_{(CH_3)_2SiCl_2} = \frac{(0.06765 \text{ mol})(0.082057 \text{ L} \cdot \text{atm/K} \cdot \text{mol})(348)}{7.25 \text{ L}} = 0.266 \text{ atm}$$

10.71. Volume of oxygen (at STP) needed to oxidize 0.400 mol of P_4:

The balanced equation indicates that 1 mol of P_4 requires 5 mol of O_2.

So, 0.400 mol of P_4 would require (5 × 0.400 mol) or 2.00 mol of O_2.

Now the molar gas volume says that (at STP) 1 mol of a gas occupies 22.4 L, so the volume occupied by 2.00 mol of O_2 would be 22.4 × 2.00 or 44.8 L.

10.72. Regarding the decomposition of nitroglycerin:

(a) P_T = 4.2 atm

4 $C_3H_5(NO_3)_3(\ell) \rightarrow$ 6 $N_2(g)$ + $O_2(g)$ + 12 $CO_2(g)$ + 10 $H_2O(g)$

n_T = 6 mol $N_2(g)$ + 1 mol $O_2(g)$ + 12 mol $CO_2(g)$ + 10 mol $H_2O(g)$ = 29 mol

$$P_{N_2} = 4.2 \text{ atm} \cdot \left(\frac{6 \text{ mol N}_2}{29 \text{ mol}} \right) = 0.87 \text{ atm}$$

(b) Rearranging the Ideal gas Law: $n = PV/RT$ =

(0.87 atm N_2)(1.5 L)/(0.082057 L·atm/mol·K)(723.2 K) = 0.0217 mol N_2 = 0.022 mol N_2

0.0217 mol N_2 × (4 mol nitroglycerin)/(6 mol N_2) × (227.09 g/mol nitroglycerin) = 3.3 g

10.73. This problem has two parts: 1) How many moles of Ni are present?

2) How many moles of CO are present?

1) # moles of Ni present:

$$0.450 \text{ g Ni} \cdot \frac{1 \text{ mol Ni}}{58.693 \text{ g Ni}} = 7.67 \times 10^{-3} \text{ mol Ni and since 1 mol Ni(CO)}_4$$

is formed for **each mol of Ni**, one can form 7.67×10^{-3} mol Ni(CO)$_4$

2) moles of CO present:

Using the Ideal Gas Law:

$$n = \frac{\left(\frac{418}{760}\text{ atm}\right)(1.50\text{ L})}{0.082057\frac{\text{L} \cdot \text{atm}}{\text{K} \cdot \text{mol}} \cdot 298\text{ K}} = 0.0337\text{ mol CO}$$ which would be capable of forming

$$0.0337\text{ mol CO} \cdot \frac{1\text{ mol Ni(CO)}_4}{4\text{ mol CO}} = 8.43 \times 10^{-3}\text{ mol Ni(CO)}_4$$

Since the amount of *nickel limits the maximum amount of Ni(CO)₄ that can be formed*, the maximum mass of Ni(CO)₄ is then:

$$7.67 \times 10^{-3}\text{ mol Ni(CO)}_4 \cdot \frac{170.7\text{ g Ni(CO)}_4}{1\text{ mol Ni(CO)}_4} = 1.31\text{ g Ni(CO)}_4$$

10.74. Concerning the four gases involved in the reaction:
(a) molar mass (g/mol): H_2O, 18; CO_2, 44; O_2, 32; C_2H_6, 30

Increasing RMS speed: $CO_2 < O_2 < C_2H_6 < H_2O$

(b) The C_2H_6 and the O_2 gases are present in the flask at the same temperature. The moles of each gas is proportional to their pressures. If 7 mol O_2 react with 2 mol C_2H_6, then the pressure of the O_2 must be 7/2(256 mm Hg) or 896 mm Hg (1.18 atm).

10.75. For the four samples given:
(1) 1.0 L of H_2 at STP
(2) 1.0 L of Ar at STP
(3) 1.0 L of H_2 at 27 °C and 760 mm Hg
(4) 1.0 L of He at 0 °C and 900 mm Hg

For samples (1) and (2), the calculation is identical for number of particles:

Rearranging the Ideal Gas Law: $n = \frac{PV}{RT} = \frac{(1\text{ atm})(1.0\text{ L})}{(0.082057\frac{\text{L} \cdot \text{atm}}{\text{K} \cdot \text{mol}})(273\text{ K})} = 4.4 \times 10^{-2}\text{ mol}$

Note that you can alternatively use the factor: 22.4 L = 1 mol (since samples 1 and 2 are at STP). For sample (3) $n = \frac{(1\text{ atm})(1.0\text{ L})}{(0.082057\frac{\text{L} \cdot \text{atm}}{\text{K} \cdot \text{mol}})(300\text{ K})} = 4.1 \times 10^{-2}\text{ mol of H}_2$

For sample (4) $n = \frac{(\frac{900}{760}\text{ atm})(1.0\text{ L})}{(0.082057\frac{\text{L} \cdot \text{atm}}{\text{K} \cdot \text{mol}})(273\text{ K})} = 5.3 \times 10^{-2}\text{ mol of He}$

(a) Which sample has the greatest number of gas particles? Sample 4

(b) Which sample has the fewest number of gas particles? Sample 3

(c) Which sample represents the largest mass?

Given the number of moles of each gas, you can calculate the masses:

Sample (1): 4.4×10^{-2} mol H$_2 \cdot$ 2.0 g H$_2$/mol H$_2$ = 8.8×10^{-2} g H$_2$

Sample (2): 4.4×10^{-2} mol Ar $\cdot$ 39.95 g Ar/mol Ar = 1.7 g Ar

Sample (3): 4.1×10^{-2} mol H$_2 \cdot$ 2.0 g H$_2$/mol H$_2$ = 8.2×10^{-2} g H$_2$

Sample (4): 5.3×10^{-2} mol of He $\cdot$ 4.00 g He/mol He = 0.21 g He

Sample 2 has the greatest mass.

10.76. Partial pressures of propane and oxygen:

$$P_{C_3H_8} = \frac{1 \text{ mol } C_3H_8}{6 \text{ mol gas}} \cdot 288 \text{ mm Hg} = 48.0 \text{ mm Hg}$$

$$P_{O_2} = \frac{5 \text{ mol } O_2}{6 \text{ mol gas}} \cdot 288 \text{ mm Hg} = 240. \text{ mm Hg}$$

Assuming the reaction consumes all the oxygen, and knowing that the moles of gases are proportional to their pressures:

$$P_{H_2O} = 240. \text{ mm Hg} \cdot \frac{4 \text{ mol } H_2O}{5 \text{ mol } O_2} = 192 \text{ mm Hg}$$

10.77. To determine the theoretical yield of Fe(CO)$_5$, you need to know the # of moles of both Fe and of CO. Use the Ideal Gas Law to calculate moles of CO.

$$n = \frac{(\frac{732}{760} \text{ atm})(5.50 \text{ L})}{(0.082057 \frac{L \cdot atm}{K \cdot mol})(296 \text{ K})} = 0.218 \text{ mol CO}$$

For Fe: $\frac{3.52 \text{ g Fe}}{1} \cdot \frac{1 \text{ mol Fe}}{55.85 \text{ g Fe}} = 0.0630 \text{ mol Fe}$

The equation is: Fe + 5 CO → Fe(CO)$_5$, indicating that you need 5 mol of CO per mol of Fe.

Moles available = $\frac{0.218 \text{ mol CO}}{0.0630 \text{ mol Fe}} = 3.46$, indicating that CO is the Limiting Reagent.

The theoretical yield of Fe(CO)$_5$ will be:

$$0.218 \text{ mol CO} \left(\frac{1 \text{ mol Fe(CO)}_5}{5 \text{ mol CO}} \right) \left(\frac{195.90 \text{ g Fe(CO)}_5}{1 \text{ mol Fe(CO)}_5} \right) = 8.54 \text{ g Fe(CO)}_5$$

10.78. Molecular formula of chlorofluorocarbon:

100.00% − (14.05% C + 41.48% Cl) = 44.47% F
Assume 100.00 g compound

14.05 g $\left(\dfrac{1 \text{ mol C}}{12.011 \text{ g}}\right)$ = 1.170 mol C

41.48 g $\left(\dfrac{1 \text{ mol Cl}}{35.45 \text{ g}}\right)$ = 1.170 mol Cl

44.47 g $\left(\dfrac{1 \text{ mol F}}{18.998 \text{ g}}\right)$ = 2.341 mol F

$\dfrac{1.170 \text{ mol C}}{1.170 \text{ mol Cl}} = \dfrac{1 \text{ mol C}}{1 \text{ mol Cl}}$ $\dfrac{2.341 \text{ mol F}}{1.170 \text{ mol C}} = \dfrac{2 \text{ mol F}}{1 \text{ mol C}}$

The empirical formula is CClF$_2$ with a mass of 85.5 g/unit.

$M = \dfrac{dRT}{P} = \dfrac{\left(\dfrac{0.277 \text{ g}}{0.195 \text{ L}}\right)(0.082057 \text{ L·atm}/\text{K·mol})(298 \text{ K})}{154 \text{ mm Hg}\left(\dfrac{1 \text{ atm}}{760 \text{ mm Hg}}\right)} = 171 \text{ g/mol}$

$\dfrac{171 \text{ g/mol}}{85.5 \text{ g/mol}} = 2$ The molecular formula is (CCl$_2$F)$_2$ or C$_2$Cl$_4$F$_2$.

10.79. Determine the empirical formula of the S$_x$F$_y$ compound:
Given the data, calculate the # of mol of the compound:

Note that you convert 89 mL to the volume in liters (0.089 L). You also need to express 83.8 mm Hg in units of atmospheres. With those conversions, you can calculate the # mol of the compound.

$n = \dfrac{(\dfrac{83.8}{760} \text{ atm})(0.089 \text{ L})}{(0.082057 \dfrac{\text{L·atm}}{\text{K·mol}})(318 \text{ K})} = 3.8 \times 10^{-4}$ mol compound (2 sf)

The molecular weight is then: 0.0955 g / 3.8 × 10^{-4} mol = 253.9 g/mol or 250 (2 sf).

The formula is 25.23% S (and therefore 74.77 % F). So you can anticipate that of the 250 g, 25.23% is S (0.2523 · 250) = 64.1 g S.   The amount of F is (0.7477 · 250) = 190 g F.

Now you can calculate the # of moles of each element present:

$\dfrac{64.1 \text{ g S}}{1} \cdot \dfrac{1 \text{ mol S}}{32.07 \text{ g S}} = 2.0$ mol (2 sf)

and for F: $\dfrac{190 \text{ g F}}{1} \cdot \dfrac{1 \text{ mol F}}{19.00 \text{ g}} = 10.$ mol (2 sf). So the formula is S$_2$F$_{10}$.

10.80. Regarding the pressures of gases resulting from the decomposition of $(NH_4)_2Cr_2O_7$:

$$0.95 \text{ g} \cdot \frac{1 \text{ mol } (NH_4)_2Cr_2O_7}{252 \text{ g}} \cdot \frac{5 \text{ mol gas}}{1 \text{ mol } (NH_4)_2Cr_2O_7} = 0.019 \text{ mol gas produced}$$

$$P_{\text{total}} = \frac{n_{\text{total}}RT}{V} = \frac{(0.019 \text{ mol})(0.082057 \text{ L} \cdot \text{atm/K} \cdot \text{mol})(296 \text{ K})}{15.0 \text{ L}} = 0.031 \text{ atm}$$

$$P_{N_2} = 0.031 \text{ atm} \cdot \frac{1 \text{ mol } N_2}{5 \text{ mol gas}} = 0.0061 \text{ atm}$$

$$P_{H_2O} = 0.031 \text{ atm} \cdot \frac{4 \text{ mol } H_2O}{5 \text{ mol gas}} = 0.024 \text{ atm}$$

10.81. Regarding the atmosphere:

(a) Average molar mass of air at 20 km above the earth's surface:

Converting volume to L, you note the conversion factor, 1 L = 1 × 10⁻³ m³ so 1 m³ is 1000 L.

Also noting that –63 °C will be 210 K, you can substitute:

$$n = \frac{(\frac{42}{760} \text{ atm})(1000 \text{ L})}{(0.082057 \frac{\text{L} \cdot \text{atm}}{\text{K} \cdot \text{mol}})(210 \text{ K})} = 3.2 \text{ mol air.}$$ Given the density as 92 g/m³, you can calculate the molar mass: 92 g/3.2 mol = 29 g/mol.   (2 sf)

Calculations for part (b) will be made using 2 sf. Using the un-rounded molar mass in part (a), the answers in part (b) will be different from those shown here.

(b) If the atmosphere is only O_2 and N_2, what is the mole fraction of each gas?

You know that the $X_{O_2} + X_{N_2} = 1$. You know that the mixture of oxygen and nitrogen has a molar mass of 29 g/mol. You can then calculate the % (or mole fraction) of this weighted average.

$(X_{O_2})(32.0 \text{ g/mol}) + (X_{N_2})(28.0 \text{ g/mol}) = 28.7 \text{ g/mol}$.

Using the mf relationship from above, you get:

$(X_{O_2})(32.0 \text{ g/mol}) + (1 - X_{O_2})(28.0 \text{ g/mol}) = 28.7 \text{ g/mol}$.

For simplicity use x to represent mf O_2.

$x \cdot 32.0 \text{ g/mol} + (1 - x) \cdot 28.0 \text{ g/mol} = 28.7 \text{ g/mol}$.

$32.0x + 28.0 - 28.0x = 28.7$ and

$4.0 x = (28.7 - 28.0)$ and $x = (28.7 - 28.0)/4.0$ or 0.17

The $X_{O_2} = 0.17$ and the $X_{N_2} = 0.83$

10.82. Partial pressures of He and Hg in the flasks:

He: $P_2 = \dfrac{P_1 V_1}{V_2} = \dfrac{(145 \text{ mm Hg})(3.0 \text{ L})}{5.0 \text{ L}} = 87 \text{ mm Hg}$

Hg: $P_2 = \dfrac{P_1 V_1}{V_2} = \dfrac{(355 \text{ mm Hg})(2.0 \text{ L})}{5.0 \text{ L}} = 140 \text{ mm Hg}$

P_{total} = 87 mm Hg + 140 mm Hg = 230 mm Hg

10.83. Calculate the moles of gas present in the flask:

$n = \dfrac{(\dfrac{17.2}{760} \text{ atm})(1.850 \text{ L})}{(0.082057 \dfrac{\text{L} \cdot \text{atm}}{\text{K} \cdot \text{mol}})(294 \text{ K})} = 1.74 \times 10^{-3} \text{ mol Cl}_x\text{O}_y\text{F}_z$

This amount of the gas has a mass of 0.150 g, so the molar mass is 0.150 g/1.74 × 10⁻³ mol or 86.4 g/mol. If the gas contains Cl, O, and F, the "molar mass" of ClOF = (35.5 + 16 + 19) = 70.5 g/mol. With our calculated value of 86.4, you hypothesize the formula to be ClO₂F.

10.84. Determine the empirical formula of xenon fluoride:

$P_{F_2 \text{(consumed)}} = P_{\text{total}} - P_{Xe} - P_{F_2 \text{(unreacted)}}$ = 0.72 atm − 0.12 atm − 0.36 atm = 0.24 atm

$n_F = \dfrac{(0.24 \text{ atm})(0.25 \text{ L})}{(0.082057 \text{ L} \cdot \text{atm/K} \cdot \text{mol})(273.2 \text{ K})} \cdot \dfrac{2 \text{ mol F}}{1 \text{ mol F}_2} = 0.0054 \text{ mol F}$

$n_{Xe} = \dfrac{(0.12 \text{ atm})(0.25 \text{ L})}{(0.082057 \text{ L} \cdot \text{atm/K} \cdot \text{mol})(273.2 \text{ K})} = 0.0013 \text{ mol Xe}$

$\dfrac{0.0054 \text{ mol F}}{0.0013 \text{ mol Xe}} = \dfrac{4 \text{ mol F}}{1 \text{ mol Xe}}$ The empirical formula is XeF₄.

10.85. So, 0.37 g of gas at 21°C is released into a flask of 732 mL. The pressure in the flask is 209 mm Hg. You want to know the identity of the gas. Start by calculating the moles of gas.

$n = \dfrac{\left(209 \text{ mm Hg}\left(\dfrac{1 \text{ atm}}{760 \text{ mm Hg}}\right)\right)(0.732 \text{ L})}{(0.082057 \text{ L} \cdot \text{atm/K} \cdot \text{mol})(294 \text{ K})} = 8.34 \times 10^{-3} \text{ mol}$

You know that this amount of gas had a mass of 0.37 g, so you calculate the molar mass by dividing the mass by the # of moles of gas:

$\dfrac{0.37 \text{ g gas}}{8.34 \times 10^{-3} \text{ mol gas}}$ = 44 g/mol, so, the unknown gas is might be CO₂.

10.86. Of the gases He, SO$_2$, CO$_2$, and Cl$_2$:

(a) Gas with the largest density: Cl$_2$—gas with the largest molar mass

(b) Gas that effuses fastest: He—gas with the smallest molar mass

10.87. Which of the following is not correct?

(a) Diffusion of gases occurs more rapidly at higher temperatures.
Correct. Since the kinetic energy of the gases will be greater at higher temperatures, the gases will diffuse more rapidly.

(b) Effusion of H$_2$ is faster than effusion of He (assume similar conditions and a rate expressed in units of mol/h).
Correct. Graham's Law tells us that effusion occurs at a rate that is inversely proportional to the square root of the molar masses of gases involved. H$_2$ is less massive than He (by a factor of 2), so should effuse faster than He.

(c) Diffusion will occur faster at low pressure than at high pressure.
Correct. At lower pressures, the concentration of gas is lower than at higher pressures. Collisions will be less frequent, so diffusion should occur faster.

(d) The rate of effusion of a gas (mol/h) is directly proportional to molar mass.
Incorrect. See part (b).

10.88. Use the Ideal Gas Law and the van der Waals equation to calculate pressure:

(a) (12.0 g CO$_2$)(1mol/44.01 g) = 0.273 mol

$$P = \frac{nRT}{V} = \frac{(0.273 \text{ mol})(0.082057 \text{ L} \cdot \text{atm/K} \cdot \text{mol})(298 \text{ K})}{0.500 \text{ L}} = 13.4 \text{ atm}$$

The Van der Waals' equation: $P = \frac{nRT}{(V-bn)} - a\left(\frac{n}{V}\right)^2$

$$P = \frac{(0.273 \text{ mol})(0.082057 \text{ L} \cdot \text{atm/K} \cdot \text{mol})(298 \text{ K})}{(0.500 \text{ L} - (0.0427 \text{ L/mol})(0.273 \text{ mol}))}$$

$$- (3.59 \text{ atm} \cdot \text{L}^2/\text{mol}^2)\left(\frac{0.273 \text{ mol}}{0.500 \text{ L}}\right)^2$$

$P = 12.6$ atm

(13.4 atm – 12.6 atm)/12.6 atm x 100 % = 6 %

(b) $P = \dfrac{nRT}{V} = \dfrac{(0.273 \text{ mol})(0.082057 \text{ L} \cdot \text{atm/K} \cdot \text{mol})(203 \text{ K})}{0.500 \text{ L}} = 9.10$ atm

Using the van der Waal's equation:

$$P = \frac{(0.273 \text{ mol})(0.082057 \text{ L·atm/K·mol})(203 \text{ K})}{(0.500 \text{ L} - (0.0427 \text{ L/mol})(0.273 \text{ mol}))} - (3.59 \text{ atm·L}^2/\text{mol}^2)\left(\frac{0.273 \text{ mol}}{0.500 \text{ L}}\right)^2$$

$P = 8.25$ atm

(9.10 atm – 8.25 atm)/8.25 atm × 100 % = 10 %

10.89. Carbon dioxide, CO_2, was shown to effuse through a porous plate at the rate of 0.033 mol/min. The same quantity of an unknown gas effuses through the same porous barrier in 104 seconds. Calculate the molar mass of the unknown gas.

$$\frac{\text{Rate of effusion of } CO_2}{\text{Rate of effusion of unknown}} = \sqrt{\frac{\text{MM of unknown}}{\text{MM of } CO_2}}$$

Note that the rates should be expressed in the same units of time, so convert 104 seconds into minutes (1.733 min). The rate of effusion of the unknown is then 0.033mol/1.733 minutes or 0.0190 mol/min.

Substituting:

$$\frac{0.033 \text{ mol/min}}{0.0190 \text{ mol/min}} = \sqrt{\frac{\text{Mm of unknown}}{44.01 \text{ g/mol}}}$$ and squaring both sides of the equation yields:

$$3.0166 = \frac{\text{Mm of unknown}}{44.01 \text{ g/mol}}$$ and (3.0166) (44.01 g/mol) = 130 g/mol (2 sf)

10.90. Compare the rates of effusion of CH_4 and CF_4:

$$\frac{0.48 \text{ mol}}{5.4 \text{ min}} = 0.089 \text{ mol}/\text{min}$$

$$\frac{\text{rate}_{CH_4}}{\text{rate}_{CF_4}} = \sqrt{\frac{M_{CF_4}}{M_{CH_4}}} \quad \text{substituting:} \quad \frac{\text{rate}_{CH_4}}{0.089 \text{ mol/min}} = \sqrt{\frac{88.0 \text{ g/mol}}{16.0 \text{ g/mol}}} = 0.21 \text{ mol/min}$$

and for the time: 0.48 mol (1/0.21 mol/min) = 2.3 min

10.91. The number of moles of He in the balloon can be calculated with the Ideal Gas Law. First calculate the *P* of He in the balloon:

gauge = total - barometric and rearranging:

gauge + barometric = total pressure = 22 mm Hg + 755 mm Hg = 777 mm Hg = 1.02 atm

$$n = \frac{PV}{RT} = \frac{\left(777 \text{ mm Hg}\left(\frac{1 \text{ atm}}{760 \text{ mm Hg}}\right)\right)(0.305 \text{ L})}{(0.082057 \text{ L} \cdot \text{atm/K} \cdot \text{mol})(298 \text{ K})} = 0.0128 \text{ mol He}$$

10.92. Molecules of water in the vapor phase:

$$\frac{n}{V} = \frac{P}{RT} = \frac{23.8 \text{ mm Hg} \cdot \frac{1 \text{ atm}}{760 \text{ mm Hg}}}{(0.082057 \text{ L} \cdot \text{atm/K} \cdot \text{mol})(298 \text{ K})} \cdot \frac{6.022 \times 10^{23} \text{ molecules}}{1 \text{ mol}} \cdot \frac{1 \text{ L}}{10^3 \text{ cm}^3}$$

$$= 7.71 \times 10^{17} \text{ molecules/cm}^3$$

10.93. Calculate moles of O_2 to determine the amount of $KClO_3$ originally present.

$$n = \frac{(\frac{735}{760} \text{ atm})(0.327 \text{ L})}{(0.082057 \frac{\text{L} \cdot \text{atm}}{\text{K} \cdot \text{mol}})(292 \text{ K})} = 1.32 \times 10^{-2} \text{ mol O}_2$$

The mole ratio from the equation tells you that for each mol of O_2, you had 2/3 mole of $KClO_3$. The number of moles of $KClO_3$ will be:

$$1.32 \times 10^{-2} \text{ mol O}_2 \left(\frac{2 \text{ mol KClO}_3}{3 \text{ mol O}_2}\right) = 8.80 \times 10^{-3} \text{ mol of KClO}_3.$$

The mass to which this corresponds is:

$$8.80 \times 10^{-3} \text{ mol KClO}_3 \left(\frac{122.55 \text{ g KClO}_3}{1 \text{ mol KClO}_3}\right) = 1.08 \text{ g}$$

The percent perchlorate in the original mixture: 1.08 g $KClO_3$/1.56 g mixture · 100 = 69.1 %

10.94. Partial pressure of N_2 in a mixture of gases:

$P_{\text{total}} = P_{H2O} + P_{O2} + P_{CO2} + P_{N2}$

P_{N2} = 253 mm Hg – 47.1 mm Hg – 35 mm Hg – 7.5 mm Hg

P_{N2} = 163 mm Hg

10.95. Reaction of nitrogen monoxide and oxygen:

(a) NO, O_2, and NO_2, in increasing velocity at 298 K. You know that the velocity of gases is inversely related to the molar masses. So the NO molecules would be moving fastest, and the NO_2 molecules slowest, with O_2, molecules intermediate in velocity.

(b) Partial pressure of O₂, when mixed in the appropriate ratio with NO: The equation shows that 2 molecules of NO are needed per molecule of O₂. With the partial pressure of NO = 150 mm Hg, the partial pressure of O₂, would be 150/2 or 75-mm Hg.

(c) The partial pressure of NO₂ after reaction should be equal to that of NO before reaction (150 mm Hg). Since the partial pressures are proportional to the number of moles, 2 moles of NO and 1 mole of O₂, would form 2 moles of NO₂.

10.96. Regarding the synthesis of NH₃ from N₂ and H₂

(a) $562 \text{ g} \cdot \left(\dfrac{1 \text{ mol NH}_3}{17.03 \text{ g}}\right)\left(\dfrac{3 \text{ mol H}_2}{2 \text{ mol NH}_3}\right) = 49.5 \text{ mol H}_2$

$V_{H_2} = \dfrac{nRT}{P} = \dfrac{(49.5 \text{ mol})(0.082057 \text{ L} \cdot \text{atm/K} \cdot \text{mol})(329 \text{ K})}{745 \text{ mm Hg} \times \dfrac{1 \text{ atm}}{760 \text{ mm Hg}}} = 1360 \text{ L}$

(b) $562 \text{ g} \cdot \left(\dfrac{1 \text{ mol NH}_3}{17.03 \text{ g}}\right)\left(\dfrac{1 \text{ mol N}_2}{2 \text{ mol NH}_3}\right)\left(\dfrac{100.0 \text{ mol air}}{78.1 \text{ mol N}_2}\right) = 21.1 \text{ mol air}$

$V_{air} = \dfrac{nRT}{P} = \dfrac{(21.1 \text{ mol})(0.082057 \text{ L} \cdot \text{atm/K} \cdot \text{mol})(302 \text{ K})}{745 \text{ mm Hg} \times \dfrac{1 \text{ atm}}{760 \text{ mm Hg}}} = 534 \text{ L}$

10.97. For the process: 4 NH₃(g) + 3 F₂(g) → 3 NH₄F(s) + NF₃(g), the correct mix of NH₃ with F₂ is a 4:3 mole ratio. This will provide (4 + 3) or 7 total moles of gas with a total pressure of 120 mm Hg, the gases will exert a pressure in proportion to the mole fraction of each gas.

$P_{F_2} = \dfrac{3 \text{ mol F}_2}{7 \text{ mol gas}}(120 \text{ mm Hg}) = 51 \text{ mm Hg}$. While one can perform a similar calculation for NH₃, one can also note that the two gases provide a total 120 mm Hg pressure, so NH₃ exerts a pressure of (120-51) or 69 mm Hg.
One mole of gas will be produced for each 3 mol of F₂, so you anticipate a pressure that is 1/3 of that exerted by the F₂, or 1/3(51) = 17 mm Hg. (Assuming *T* is constant.)

10.98. In the reaction of ClF₃ with NiO:

(a) $n_{ClF_3} = \dfrac{PV}{RT} = \dfrac{\left(250 \text{ mm Hg} \cdot \dfrac{1 \text{ atm}}{760 \text{ mm Hg}}\right)(2.5 \text{ L})}{(0.082057 \text{ L} \cdot \text{atm/K} \cdot \text{mol})(293 \text{ K})} = 0.034 \text{ mol ClF}_3$

$0.034 \text{ mol ClF}_3 \cdot \dfrac{6 \text{ mol NiO}}{4 \text{ mol ClF}_3} \cdot \dfrac{74.7 \text{ g}}{1 \text{ mol NiO}} = 3.8 \text{ g NiO}$

(b) Partial pressures:

$$0.034 \text{ mol ClF}_3 \left(\frac{3 \text{ mol O}_2}{4 \text{ mol ClF}_3} \right) = 0.026 \text{ mol O}_2$$

$$0.034 \text{ mol ClF}_3 \left(\frac{2 \text{ mol Cl}_2}{4 \text{ mol ClF}_3} \right) = 0.017 \text{ mol Cl}_2$$

$$P_{O_2} = \frac{nRT}{V} = \frac{(0.026 \text{ mol})(0.082057 \text{ L} \cdot \text{atm/K} \cdot \text{mol})(293 \text{ K})}{2.5 \text{ L}} = 0.25 \text{ atm} = 190 \text{ mm Hg}$$

$$P_{Cl_2} = \frac{nRT}{V} = \frac{(0.017 \text{ mol})(0.082057 \text{ L} \cdot \text{atm/K} \cdot \text{mol})(293 \text{ K})}{2.5 \text{ L}} = 0.16 \text{ atm} = 120 \text{ mm Hg}$$

$$P_{total} = 190 \text{ mm Hg} + 120 \text{ mm Hg} = 310 \text{ mm Hg}$$

10.99. Mass of water per liter of air under the conditions:

(a) At 20 °C and 45% relative humidity. Recalling the definition of Relative Humidity (RH):

$$RH = \frac{P(H_2O) \text{ in air}}{\text{Vapor Pressure(VP) of water}}$$

You can calculate the P(H₂O) in air by multiplying the VP of water by the RH.

At 20 °C, the VP of water is 17.5 torr (Data from Appendix at back of textbook)

P(H₂O) = RH · VP = 0.45 · 17.5 torr

Recalling that water vapor is a gas, you can use the Ideal Gas Law to calculate the # of moles—and the mass of water present in 1 L of air. Substituting the P(H₂O) that you calculated above into "P", rearranging the Ideal Gas Law: $\frac{P \cdot V}{R \cdot T} = n$, and recalling that n = mass(g)/MW you can further rearrange to produce: $\frac{P \cdot V \cdot MW}{R \cdot T} = \text{mass}$.

Given that VP is given in torr, use the value of $R = 62.4 \frac{L \cdot \text{torr}}{K \cdot \text{mol}}$.

[Alternatively convert 17.5 torr to atm, by dividing by 760, and using the value of $R = 0.082057 \frac{L \cdot \text{atm}}{K \cdot \text{mol}}$]

$$\frac{(0.45 \cdot 17.5 \text{ torr}) \cdot 1 \text{ L} \cdot 18.02 \text{ g/mol}}{62.4 \frac{L \cdot \text{torr}}{K \cdot \text{mol}} \cdot 293 \text{ K}} = \text{mass} = 7.8 \times 10^{-3} \text{ g per liter. (2 sf)}$$

(b) At 0 °C and 95% relative humidity: (VP of water at 0 °C = 4.6 torr)

$$\frac{(0.95 \cdot 4.6 \text{ torr}) \cdot 1 \text{ L} \cdot 18.02 \text{ g/mol}}{62.4 \frac{L \cdot \text{torr}}{K \cdot \text{mol}} \cdot 273 \text{ K}} = \text{mass} = 4.6 \times 10^{-3} \text{ g (2 sf)}$$

10.100. Mass of water vapor present in a dormitory room:

$$P_{H_2O} = \text{(relative humidity)}(H_2O \text{ vapor pressure}) = \left(\frac{55 \text{ mm Hg}}{100 \text{ mm Hg}}\right)(21.1 \text{ mm Hg}) = 12 \text{ mm Hg}$$

$$n = \frac{PV}{RT} = \frac{\left(12 \text{ mm Hg}\left(\frac{1 \text{ atm}}{760 \text{ mm Hg}}\right)\right)\left((4.5 \text{ m}^2)(3.5 \text{ m})\left(\frac{1 \text{ L}}{10^{-3} \text{ m}^3}\right)\right)}{(0.082057 \text{ L} \cdot \text{atm/K} \cdot \text{mol})(296 \text{ K})} = 10. \text{ mol } H_2O$$

$$10. \text{ mol } H_2O \cdot \frac{18.0 \text{ g } H_2O}{1 \text{ mol } H_2O} = 1.8 \times 10^2 \text{ g } H_2O$$

In the Laboratory

10.101. Calculate the pressure of O_2 in the tank:

Mass of O_2 = 0.0870 g or (0.0870 g O_2 · 1 mol O_2/32.00 g O_2) = 0.00272 mol O_2

$$P = \frac{(2.72 \times 10^{-3} \text{ mol})\left(0.082057 \frac{L \cdot atm}{K \cdot mol}\right)(297 \text{ K})}{0.550 \text{ L}} = 0.12 \text{ atm (2 sf)}$$

With the total pressure (of O_2, CO, and CO_2) = 1.56 atm, the pressure of CO and CO_2 will be 1.56 atm = $P(O_2) + P(CO) + P(CO_2)$ = 0.12 atm + $P(CO) + P(CO_2)$.

Then 1.44 atm = $P(CO) + P(CO_2)$.

You know that $P(CO) + P(O_2)$ = 1.34 atm. So, $P(CO)$ + 0.12 = 1.34 atm and $P(CO)$ = 1.22 atm.

Since 1.44 atm = $P(CO) + P(CO_2)$, and $P(CO)$ = 1.22 atm, then $P(CO_2)$ = 0.22 atm.

The masses of CO and CO_2 can be found by substitution into the Ideal Gas Law:

$$n = \frac{(1.22 \text{ atm})(0.550 \text{ L})}{(0.082057 \frac{L \cdot atm}{K \cdot mol})(297 \text{ K})} = 2.75 \times 10^{-2} \text{ mol CO} \times 28 \text{ g CO/mol} = 0.77 \text{ g CO}$$

$$n = \frac{(0.22 \text{ atm})(0.550 \text{ L})}{(0.082057 \frac{L \cdot atm}{K \cdot mol})(297 \text{ K})} = 4.96 \times 10^{-3} \text{ mol } CO_2 \times 44 \text{ g/mol} = 0.22 \text{ g of } CO_2$$

10.102. For the combustion of methane: $CH_4(g) + 2 O_2(g) \rightarrow CO_2(g) + 2 H_2O(g)$

Assume a one-minute time-period:

$$n_{CH_4} = \frac{\left(773 \text{ mm Hg} \cdot \frac{1 \text{ atm}}{760 \text{ mm Hg}}\right)(5.0 \text{ L})}{(0.082057 \text{ L} \cdot \text{atm/K} \cdot \text{mol})(301 \text{ K})} = 0.21 \text{ mol } CH_4$$

$$0.21 \text{ mol CH}_4 \cdot \frac{2 \text{ mol O}_2}{1 \text{ mol CH}_4} = 0.41 \text{ mol O}_2$$

$$V_{O_2} = \frac{(0.41 \text{ mol O}_2)(0.082057 \text{ L}\cdot\text{atm/K}\cdot\text{mol})(299 \text{ K})}{742 \text{ mm Hg} \cdot \frac{1 \text{ atm}}{760 \text{ mm Hg}}} = 10. \text{ L O}_2$$

The oxygen must be supplied to the burner at a rate of 10 L/min.

10.103. The CO₂ evolved is:

$$n = \frac{(\frac{44.9 \text{ atm}}{760})(1.50 \text{ L})}{(0.082057 \frac{\text{L}\cdot\text{atm}}{\text{K}\cdot\text{mol}})(298 \text{ K})} = 3.62 \times 10^{-3} \text{ mol CO}_2$$

Note that the CO in the compound is oxidized to CO₂, so you know the original number of moles of CO in the compound (3.62×10^{-3} mol).

The mass of this amount of CO would be: (3.62×10^{-3} mol · 28.0 g CO/mol) = 0.101 g CO.

From 0.142 g sample of the compound, the mass of Fe is (0.142 − 0.101) or 4.0×10^{-2} grams Fe corresponding to (4.0×10^{-2} g Fe · 1 mol Fe/55.85 g Fe) = 7.26×10^{-4} mol Fe. The ratio of Fe: CO would be: 7.26×10^{-4} mol Fe: 3.62×10^{-3} mol CO or 1:5.

The empirical formula would be Fe(CO)₅

10.104. Regarding heating a metal carbonate to evolve carbon dioxide:

$$n_{CO_2} = n_{MCO_3} = \frac{\left(69.8 \text{ mm Hg} \cdot \frac{1 \text{ atm}}{760 \text{ mm Hg}}\right)(0.285 \text{ L})}{(0.082057 \text{ L}\cdot\text{atm/K}\cdot\text{mol})(298 \text{ K})} = 0.00107 \text{ mol MCO}_3$$

$$\frac{0.158 \text{ g MCO}_3}{0.00107 \text{ mol MCO}_3} = 148 \text{ g/mol}$$

148 g/mol = M_metal + M_CO3 = M_metal + 60.0 g/mol

M_metal = 88 g/mol

The metal is probably Sr (87.6 g/mol)

10.105. (a) Pressure of B₂H₆ formed:

$$\frac{0.136 \text{ g NaBH}_4}{1} \cdot \frac{1 \text{ mol NaBH}_4}{37.83 \text{ g NaBH}_4} = 3.60 \times 10^{-3} \text{ mol NaBH}_4.$$

The equation shows that you get 1 mol of B₂H₆ for each 2 mol of NaBH₄.

Using the mol of NaBH₄, you can calculate the pressure of the B₂H₆.

$$P = \frac{(1.80 \times 10^{-3} \text{ mol})\left(0.082057 \frac{\text{L}\cdot\text{atm}}{\text{K}\cdot\text{mol}}\right)(298\ K)}{2.75\ \text{L}} = 0.0160 \text{ atm (3 sf)}$$

(b) The total pressure is that of B₂H₆ and H₂: The balanced equation shows 2 mol of H₂ gas for each 1 mol of B₂H₆. If the pressure of B₂H₆ = 0.0160 atm, the pressure of H₂ will be twice that pressure (0.0320 atm); total pressure = 0.0160 + 0.0320 = 0.0480 atm.

10.106. Amount of NaNO₂ present in a mixture:

$$n_{N_2} = \frac{PV}{RT} = \frac{\left(713\ \text{mm Hg}\left(\frac{1\ \text{atm}}{760\ \text{mm Hg}}\right)\right)(0.295\ \text{L})}{(0.082057\ \text{L}\cdot\text{atm/K}\cdot\text{mol})(294.2\ K)} = 0.0115 \text{ mol } N_2$$

$$0.0115 \text{ mol } N_2 \left(\frac{1 \text{ mol NaNO}_2}{1 \text{ mol } N_2}\right)\left(\frac{69.00\ \text{g}}{1 \text{ mol NaNO}_2}\right) = 0.791 \text{ g NaNO}_2$$

$$\frac{0.791 \text{ g NaNO}_2}{1.232 \text{ g sample}} \cdot 100\% = 64.2\%$$

10.107. The amount of HCl used is: (1.50 mol HCl/L · 0.0120 L) = 0.0180 mol HCl.

Use x = mass of Na₂CO₃, and (1.249 – x) = mass of NaHCO₃.

You know that each mol of Na₂CO₃ consumes 2 mol HCl and each mol of Na₂CO₃ consumes 1 mol of HCl---and that the total number of mol of HCl consumed is 0.0180 mol.

mol HCl (consumed by Na₂CO₃) + mol HCl (consumed by Na₂CO₃) = 0.0180 mol [1]

Express the mol HCl consumed by each substance:

$$x\left(\frac{1 \text{ mol Na}_2\text{CO}_3}{106 \text{ g Na}_2\text{CO}_3}\right)\left(\frac{2 \text{ mol HCl}}{1 \text{ mol Na}_2\text{CO}_3}\right) = \text{mol HCl consumed by Na}_2\text{CO}_3 \text{ and}$$

$$(1.249 - x)\left(\frac{1 \text{ mol NaHCO}_3}{84.0 \text{ g NaHCO}_3}\right)\left(\frac{1 \text{ mol HCl}}{1 \text{ mol NaHCO}_3}\right) = \text{mol HCl consumed by Na}_2\text{CO}_3$$

Substituting into equation [1]:

$$x\left(\frac{1 \text{ mol Na}_2\text{CO}_3}{106 \text{ g Na}_2\text{CO}_3}\right)\left(\frac{2 \text{ mol HCl}}{1 \text{ mol Na}_2\text{CO}_3}\right) + 1.249 - x\left(\frac{1 \text{ mol NaHCO}_3}{84.0 \text{ g NaHCO}_3}\right)\left(\frac{1 \text{ mol HCl}}{1 \text{ mol NaHCO}_3}\right)$$

$$= 0.0180 \text{ mol}$$

Simplifying: $\frac{2x}{106} + \left[\frac{1.249}{84} - \frac{x}{84}\right] = 0.0180$

0.01887 x + 0.01487 − 0.01190 x = 0.0180 and

0.006965 x = 0.00313 and x = 0.449 g (Na₂CO₃), indicating that mass of Na₂CO₃ = (1.249 – 0.449) = 0.800 g Na₂CO₃.

Converting to moles of each substance:

$$0.449 \text{ g Na}_2\text{CO}_3 \left(\frac{1 \text{ mol Na}_2\text{CO}_3}{106 \text{ g Na}_2\text{CO}_3} \right) = 0.00424 \text{ mol Na}_2\text{CO}_3.$$

$$0.800 \text{ g NaHCO}_3 \left(\frac{1 \text{ mol NaHCO}_3}{84.0 \text{ g NaHCO}_3} \right) = 0.00951 \text{ mol NaHCO}_3$$

The two equations tell you that 1 mol of either substance produces 1 mol of CO₂.

The amount of CO₂ produced is then: 0.00424 mol Na₂CO₃ + 0.00951 mol NaHCO₃ or 0.0138 mol CO₂.

This gas would occupy a volume of:

$$V = \frac{(0.0138 \text{ mol})(0.082057 \frac{\text{L} \cdot \text{atm}}{\text{K} \cdot \text{mol}})(298 \text{ K})}{\left(\frac{745}{760} \text{ atm}\right)} = 0.343 \text{ L or } 343 \text{ mL}$$

10.108. Weight percent of Na₂CO₃ and NaHCO₃ in a mixture:

$$n_{\text{CO}_2} = \frac{PV}{RT} = \frac{\left(735 \text{ mm Hg}\left(\frac{1 \text{ atm}}{760 \text{ mm Hg}}\right)\right)(0.665 \text{ L})}{(0.082057 \text{ L} \cdot \text{atm/K} \cdot \text{mol})(298 \text{ K})} = 0.0263 \text{ mol CO}_2$$

Set up two equations with two unknowns.

$$0.0263 \text{ mol CO}_2 = x \text{ mol Na}_2\text{CO}_3 \cdot \frac{1 \text{ mol CO}_2}{1 \text{ mol Na}_2\text{CO}_3} + y \text{ mol Na}_2\text{CO}_3 \cdot \frac{1 \text{ mol CO}_2}{1 \text{ mol NaHCO}_3}$$

$$2.50 \text{ g} = x \text{ mol NaHCO}_3 \cdot \frac{84.01 \text{ g}}{1 \text{ mol NaHCO}_3} + y \text{ mol Na}_2\text{CO}_3 \cdot \frac{106.0 \text{ g}}{1 \text{ mol Na}_2\text{CO}_3}$$

Substitute (0.0263 – y) for x and solve.
y = 0.0132 mol Na₂CO₃; x = 0.0131 mol NaHCO₃

$$\frac{0.0131 \text{ mol}\left(\frac{84.01 \text{ g}}{1 \text{ mol NaHCO}_3}\right)}{2.50 \text{ g mixture}} \cdot 100\% = 44.0\% \text{ NaHCO}_3$$

$$\frac{0.0132 \text{ mol}\left(\frac{106.0 \text{ g}}{1 \text{ mol Na}_2\text{CO}_3}\right)}{2.50 \text{ g mixture}} \cdot 100\% = 56.0\% \text{ Na}_2\text{CO}_3$$

10.109. For the decomposition of copper(II) nitrate one may write the equation:

2 Cu(NO$_3$)$_2$(s) → 2 CuO(s) + 4 NO$_2$(g) + O$_2$(g).

The ideal gas law provides the information:

$$n = \frac{(\frac{725 \text{ atm}}{760})(0.125 \text{ L})}{(0.082057 \frac{\text{L} \cdot \text{atm}}{\text{K} \cdot \text{mol}})(308 \text{ K})} = 4.72 \times 10^{-3} \text{ mol of gas}$$

The average molar mass is: 0.195 g gas/4.72 × 10^{-3} mol of gas = 41.3 g/mol

To solve for the mole fractions of the two gases, you need to solve a system of 2 equations:

(1) # mol NO$_2$ + # mol O$_2$ = 0.00472 mol

(2) mass NO$_2$ + mass O$_2$ = 0.195 g   Represent the mass of NO$_2$ as x.

Substitute equation (2) into equation (1), expressing the mass of O$_2$ in terms of the mass of NO$_2$:

$$\frac{x \text{ g}}{46.00 \text{ g/mol}} + \frac{(0.195 - x) \text{ g}}{32.00 \text{ g/mol}} = 0.00472$$

$$\frac{(32.00 \text{ g/mol } x) + (46.01 \text{ g/mol})(0.195 - x)}{(46.01 \text{ g/mol})(32.00 \text{ g/mol})} = 0.00472 \text{ mol}$$

And (omitting units for clarity of the mathematics):

$32.00x - 46.01x + (0.195)(46.01) = 0.00472(46.01)(32.00)$

Solving for x gives $x = 0.144$ g NO$_2$ and $(0.195 - 0.144)$ g O$_2$ or 0.050 g O$_2$

Calculating the amounts (moles) of each substance:

$$\frac{0.144 \text{ g NO}_2}{1} \cdot \frac{1 \text{ mol NO}_2}{46.01 \text{ g NO}_2} = 3.13 \times 10^{-3} \text{ mol NO}_2$$

$$\frac{0.050 \text{ g O}_2}{1} \cdot \frac{1 \text{ mol O}_2}{32.00 \text{ g O}_2} = 1.57 \times 10^{-3} \text{ mol O}_2$$

Calculate the mol fractions of each substance:

$$\frac{0.00313}{(0.00313 + 0.00157)} = 0.666 \text{ mf NO}_2 \text{ and mf O}_2 = 1.000 - 0.666 = 0.334$$

Noting that the balanced equation makes one anticipate a 4mol:1mol ratio of NO$_2$:O$_2$ the amount of NO$_2$ is **not** as expected. Clearly the dimerization of NO$_2$ plays a role in this reaction.

10.110. Empirical and molecular formulas:

$$0.09912 \text{ g H}_2\text{O} \left(\frac{1 \text{ mol H}_2\text{O}}{18.015 \text{ g}}\right)\left(\frac{2 \text{ mol H}}{1 \text{ mol H}_2\text{O}}\right) = 0.01100 \text{ mol H}$$

$$0.01100 \text{ mol H} \left(\frac{1.0079 \text{ g}}{1 \text{ mol H}}\right) = 0.01109 \text{ g H}$$

H₂CO₃(aq) + 2 NaOH(aq) → 2 H₂O(ℓ) + Na₂CO₃(aq)

$$0.02881 \text{ L} \cdot \frac{0.3283 \text{ mol NaOH}}{1 \text{ L}} \cdot \frac{1 \text{ mol H}_2\text{CO}_3}{2 \text{ mol NaOH}} \cdot \frac{1 \text{ mol C}}{1 \text{ mol H}_2\text{CO}_3} = 0.004729 \text{ mol C}$$

$$0.004729 \text{ mol C} \left(\frac{12.011 \text{ g}}{1 \text{ mol C}}\right) = 0.05680 \text{ g C}$$

$$n_{N_2} = \frac{PV}{RT} = \frac{\left(65.12 \text{ mm Hg} \cdot \frac{1 \text{ atm}}{760 \text{ mm Hg}}\right)(0.2250 \text{ L})}{(0.082057 \text{ L} \cdot \text{atm/K} \cdot \text{mol})(298 \text{ K})} = 7.88 \times 10^{-4} \text{ mol}$$

$$7.88 \times 10^{-4} \text{ mol N}_2 \left(\frac{2 \text{ mol N}}{1 \text{ mol N}_2}\right) = 0.00158 \text{ mol N}; \quad 0.00158 \text{ mol N}\left(\frac{14.007 \text{ g}}{1 \text{ mol N}}\right) = 0.0221 \text{ g N}$$

0.1152 g compound − (0.01109 g H) − (0.05680 g C) − (0.0221 g N) = 0.0252 g O

$$0.0252 \text{ g O} \left(\frac{1 \text{ mol O}}{16.00 \text{ g}}\right) = 0.00158 \text{ mol O}$$

$$\frac{0.01100 \text{ mol H}}{0.00158 \text{ mol N}} = \frac{7 \text{ mol H}}{1 \text{ mol N}} \quad \frac{0.004729 \text{ mol C}}{0.00158 \text{ mol N}} = \frac{3 \text{ mol C}}{1 \text{ mol N}}$$

and for O/N: $\dfrac{0.00158 \text{ mol O}}{0.00158 \text{ mol N}} = \dfrac{1 \text{ mol O}}{1 \text{ mol N}}$ The empirical formula is C_3H_7NO (73.1 g/mol).

$$\frac{150 \text{ g/mol}}{73.1 \text{ g/mol}} = 2.1. \text{ The molecular formula is } C_6H_{14}N_2O_2.$$

10.111. What is the molar mass of the phosphorus–fluorine compound?

(a) What is the molar mass of the gas whose density is 5.60 g/L at a pressure of 0.971 atm and a T = 18.2 °C.

$$\text{Molar mass} = \frac{(5.60 \text{ g/L})(0.082057 \frac{\text{L} \cdot \text{atm}}{\text{K} \cdot \text{mol}})([18.2+273.15]\text{K})}{(0.971 \text{ atm})} = 138 \text{ g/mol}$$

(b) If the unknown phosphorus fluoride effuses at a rate of 0.028 mol/min gas and CO₂ effuses at a rate of 0.050 mol/min, what is the molar mass of the unknown gas?

$$\frac{\text{CO}_2 \text{ effusion rate}}{\text{Unknown effusion rate}} = \sqrt{\frac{\text{Mm unknown}}{\text{Mm CO}_2}}$$

$$\frac{0.050 \text{ mol/min}}{0.028 \text{ mol/min}} = \sqrt{\frac{\text{Mm of unknown}}{44.01 \text{ g/mol}}} \text{ and squaring both sides gives:}$$

$$\left(\frac{0.050}{0.028}\right)^2 (44.01 \text{ g/mol}) = \text{Mm of unknown} = 140 \text{ g/mol (2 sf)}$$

The molar masses of the gases given are: PF$_3$ (87.9), PF$_5$ (125.9) and P$_2$F$_4$ (137.9), so the unknown gas is most likely P$_2$F$_4$.

10.112. Concerning the reaction of propane with oxygen:

(a) C$_3$H$_8$(g) + 5 O$_2$(g) → 3 CO$_2$(g) + 4 H$_2$O(ℓ)

(b) $n = \dfrac{PV}{RT} = \dfrac{(0.10 \text{ atm})(1.50 \text{ L})}{(0.08206 \text{ L·atm/mol·K})(293 \text{ K})} = 0.0062 \text{ mol}$

(c) $X = \dfrac{P_{C_3H_8}}{P_T} = \dfrac{0.10 \text{ atm}}{(0.10 \text{ atm} + 5.0 \text{ atm})} = 0.020$

(d) $n = \dfrac{PV}{RT} = \dfrac{(5.0 \text{ atm})(1.50 \text{ L})}{(0.08206 \text{ L·atm/mol·K})(293 \text{ K})} = 0.31 \text{ mol}$

moles O$_2$ remaining = 0.31 mol – 0.019 mol = 0.29 mol O$_2$

(e) (3 mol CO$_2$/1 mol C$_3$H$_8$)(0.0062 mol C$_3$H$_8$) = 0.019 mol CO$_2$

$$P_{CO_2} = \frac{(0.019 \text{ mol})(0.08206)(296.4 \text{ K})}{(1.50 \text{ L})} = 0.31 \text{ atm}$$

(f) $P_{O_2} = \dfrac{(0.29 \text{ mol})(0.08206)(296.4 \text{ K})}{(1.50 \text{ L})} = 4.7 \text{ atm}$

Summary and Conceptual Questions

10.113. In a 1.0-L flask containing 10.0 g each of O$_2$ and CO$_2$ at 25 °C.

(a) The gas with the greater partial pressure:

Partial pressure is a relative measure of the number of moles of gas present. The molar mass of oxygen is approximately 32 g/mol while that of CO$_2$ is approximately 44 g/mol. There will be a greater number of moles of oxygen in the flask—hence the **partial pressure of O$_2$ will be greater**.

(b) The gas with the greater average speed:

The kinetic energy of each gas is given as: KE = 1/2 mu^2, where u is the average speed of the molecules. Since the average KE of both gases are the same (They are at the same *T*), The lighter of the gases **(O$_2$) will have a greater average speed**.

(c) The gas with the greater average kinetic energy:

The KE depends upon temperature, and since both gases are at the same T, the **average KE of the two gases is the same**.

10.114. Identify the true statements, and correct the false statement:

(a) True—N₂ has a small molar mass, so there would be a greater number of mol of N₂ than mol of O₂.

(b) False--Nitrogen (N₂) has a smaller molar mass than O₂, so an equal mass of N₂ contains more moles and more molecules than the O₂ sample.

10.115. Two containers with 1.0 kg of CO and 1.0 kg C₂H₂

(a) Cylinder with the greater pressure:
Since P is proportional to the amount of substance present, calculate the moles of each gas.

Molar Mass of CO = 28 g/mol   Molar Mass of C₂H₂ = 26 g/mol

While you could calculate the # of moles, note that since acetylene has a small molar mass, 1.0 kg of acetylene would have more moles of gas (and *a greater pressure*) than 1.0 kg of CO.

(b) Cylinder with the greater number of molecules:

Since you know that the cylinder with acetylene has more moles of gas, the cylinder *with the acetylene* will have the greater number of molecules--since to convert between moles of gas and molecules of gas, you multiply by the constant, Avogadro's number.

10.116. At a constant pressure, the number of moles of a gas is inversely proportional to the temperature of the gas. Therefore, flask B (at a lower temperature) contains more moles (and more molecules) of oxygen.

10.117. Which of the following samples is a gas?

(a) Material expands 1% when a sample of gas originally at 100 atm is suddenly allowed to exist at one atmosphere pressure: This sample is **not a gas**, since a gas would expand in volume by 100-fold (to exist at 1 atm).

(b) A 1.0-ml sample of material weighs 8.2 g: This sample is **not a gas** since the density of the sample (at 8.2 g/mL) is too great.

(c) Material is transparent and pale green in color: **Insufficient information** to tell. Liquids could also be pale green and transparent.

(d) One cubic meter of material contains as many molecules as an equal volume of air at the same temperature and pressure: This material **is a gas**, since one cubic meter of a liquid or solid would contain a greater number of molecules than a cubic meter of air.

10.118. Concerning four flasks filled with different gases:

(a) All four flasks have the same pressure, temperature, and volume. They have the same number of gas molecules.

(b) 160. g/16.0 g = 10 times heavier

(c) All the molecules have the same kinetic energy (they have the same temperature). Helium is the lightest gas of the four, so its molecules have the greatest average speed.

10.119. Concerning two balloons filled with hydrogen and helium:

(a) To determine the balloon containing the greater number of molecules you need only to calculate the ratio of the amounts of gas present in the balloons.
The Ideal Gas Law allows you to solve for the number of moles of each gas:

Pick a volume of gas—say 5 L for He. Since the hydrogen balloon is twice the size of the He balloon, you will use 10 L for the volume of H. You can then establish a ratio of this expression for the two gases.
$$\frac{n_{He}}{n_{H_2}} = \frac{\frac{P_{He} \cdot V_{He}}{R \cdot T_{He}}}{\frac{P_{H_2} \cdot V_{H_2}}{R \cdot T_{H_2}}} = \frac{\frac{2 \text{ atm} \cdot 5 \text{ L}}{296 \text{ K}}}{\frac{1 \text{ atm} \cdot 10 \text{ L}}{268 \text{ K}}} = \frac{268 \text{ K}}{296 \text{ K}} = \frac{0.9 \text{ mol He}}{1 \text{ mol H}_2}$$

Note that **R** cancels in the numerator and denominator, simplifying the calculation.

(b) To calculate which balloon contains the greater mass of gas, you use the ratio found:
0.9 mol He · 4.0 g/mol He = 3.6 g He; 1 mol H₂ · 2.0 g/mol H₂ = 2 g H₂

The balloon containing the *HELIUM* has the greater mass. Note that the *ratio* of moles of helium and hydrogen are **independent** of the sizes of the balloons.

10.120. Regarding sodium azide in automobile airbags:

(a) $65.0 \text{ g} \cdot \frac{1 \text{ mol Na}}{23.00 \text{ g}} = 2.83 \text{ mol Na}$

$n_{N_2O} = \frac{(2.12 \text{ atm})(35.0 \text{ L})}{(0.082057 \text{ L} \cdot \text{atm/K} \cdot \text{mol})(296 \text{ K})} = 3.05 \text{ mol N}_2\text{O}$

$\frac{3.05 \text{ mol N}_2\text{O}}{2.83 \text{ mol Na}} = \frac{1.08 \text{ mol N}_2\text{O}}{1 \text{ mol Na}} > \frac{0.75 \text{ mol N}_2\text{O}}{1 \text{ mol Na}}$ so Na is the limiting reactant.

$2.83 \text{ mol Na} \left(\frac{1 \text{ mol NaN}_3}{4 \text{ mol Na}} \right) \left(\frac{65.02 \text{ g}}{1 \text{ mol NaN}_3} \right) = 45.9 \text{ g NaN}_3$

(b) [:N≡N—N̈:]⁻ ↔ [N̈=N=N̈]⁻ ↔ [:N̈—N≡N:]⁻
    0 +1 −2   −1 +1 −1   −2 +1 0

The formal charges (shown below the resonance structures) indicate that the center resonance structure is most likely.

(c) The azide ion is linear.

10.121. The change of average speed of a gaseous molecule when T doubles:

Since the speed of a gaseous molecule is related to the square root of the *absolute T*, a change by a factor of 2 in the absolute *T* will lead to a change of $\sqrt{2}$ in the speed.

10.122. Regarding chlorine dioxide and ozone as disinfectants:

(a) Valence electrons in ClO$_2$: 7 + 2(6) = 19 valence electrons

(b) Electron dot structure for chlorite ion: $\left[:\ddot{\underset{..}{O}}-\ddot{Cl}-\ddot{\underset{..}{O}}: \right]^-$

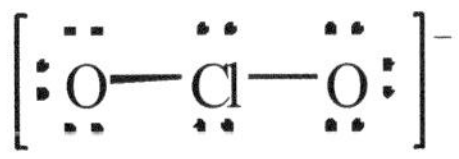

(c) Hybridization of Cl in ClO$_2^-$: sp^3 the ion is bent.

(d) Ozone has a larger bond angle. The central atom is sp^2 hybridized. $\ddot{O}=\ddot{O}-\ddot{\underset{..}{O}}:$

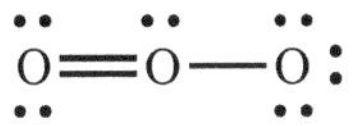

(e) $15.6 \text{ g} \left(\dfrac{1 \text{ mol NaClO}_2}{90.44 \text{ g}} \right) = 0.172 \text{ mol NaClO}_2$

$$n_{Cl_2} = \dfrac{PV}{RT} = \dfrac{\left(1050 \text{ mm Hg} \cdot \dfrac{1 \text{ atm}}{760 \text{ mm Hg}}\right)(1.45 \text{ L})}{(0.082057 \text{ L} \cdot \text{atm/K} \cdot \text{mol})(295 \text{ K})} = 0.0828 \text{ mol Cl}_2$$

$\dfrac{0.172 \text{ mol NaClO}_2}{0.0828 \text{ mol Cl}_2} = \dfrac{2.08 \text{ mol NaClO}_2}{1 \text{ mol Cl}_2} > \dfrac{2 \text{ mol NaClO}_2}{1 \text{ mol Cl}_2}$ so Cl$_2$ is the limiting reactant

Mass of ClO$_2$ that can be produced:

$0.0828 \text{ mol Cl}_2 \left(\dfrac{2 \text{ mol ClO}_2}{1 \text{ mol Cl}_2} \right) \left(\dfrac{67.45 \text{ g}}{1 \text{ mol ClO}_2} \right) = 11.2 \text{ g ClO}_2$

Solution and Answer Guide

Kotz Treichel Townsend Treichel, Chemistry and Chemical Reactivity 11e, 978-0-357-85140-1, Chapter 11: Intermolecular Forces and Liquids

TABLE OF CONTENTS

Applying Chemical Principles 398
Practicing Skills..399
General Questions...405
In the Laboratory...408
Summary and Conceptual Questions..410

Applying Chemical Principles

Chromatography

11.1.1. Assume that a mixture of the three molecules below are separated on a C-18 column using a methanol/water mixture as the mobile phase.

(a) The *most polar* of the 3 molecules, 1,5-pentanediol, will be most attracted to the polar mobile phase. The diol has two –OH groups, so hydrogen bonding will be a major intermolecular force. Since the molecule is polar, dipole-dipole forces will be present, as are the omnipresent London dispersion forces.

(b) The molecule most attracted to the stationary phase will be the *least polar* of the three molecules— ethyl propyl ether. The ethyl propyl ether is most attracted to the stationary phase by London dispersion forces, but there are also some dipole-induced dipole interactions.

(c) In what order will the three molecules exit (or "elute from") the column?

The only question remaining is how 1-pentanol fits into the elution order. This alcohol possesses that important –OH functionality (but only one, unlike the pentanediol). You should anticipate that the pentanol will be attracted to the mobile phase *to a greater extent* than the ether, so the elution order would be: 1,5-pentanediol first, followed by the 1-pentanol, with ethyl propyl ether eluting last.

11.1.2. Predict the order of elution: (first to last) pentane, hexane, heptane, octane.

This is expected as greater London dispersion forces occur for larger molecules.

A Pet Food Catastrophe

11.2.1. Weight percent of nitrogen in melamine and cyanuric acid:

Melamine	# atoms	Mass	Cyanuric acid	# atoms	Mass
C	3	36.03	C	3	36.03
H	6	6.05	H	3	3.02
N	6	84.04	N	3	42.02
O	0		O	3	48.00
	MM	126.12		MM	129.07
	%N	66.64%		%N	32.56%

Both substances have a larger percentage of N than protein.

11.2.2. Mass of melamine in one pound of infant formula with 0.14 ppm melamine:

$$454 \text{ g formula} \left(\frac{0.14 \text{ g melamine}}{1,000,000 \text{ g formula}} \right) = 6.4 \times 10^{-5} \text{ g melamine}$$

Practicing Skills

Intermolecular Forces

11.1. Intermolecular forces that must be overcome to:

change	intermolecular force(s)
(a) melt ice	hydrogen bonds, dipole-dipole, and induced dipole-induced dipole (molecule with OH bonds)
(b) sublime solid I_2	induced dipole-induced dipole (nonpolar molecule)
(c) convert $NH_3(\ell)$ to $NH_3(g)$	hydrogen bonds, dipole-dipole, and induced dipole-induced dipole (molecule with NH bonds)

11.2. Intermolecular forces:

Forces to overcome in solid iodine: induced dipole-induced dipole
Forces to overcome in CH₃OH: hydrogen bonding, dipole-dipole, and induced dipole-induced dipole
Forces between I₂ and CH₃OH in solution: dipole-induced dipole

11.3. To convert species from a liquid to a gas the intermolecular forces that must be overcome:

species	intermolecular force
(a) liquid N₂	induced dipole-induced dipole (nonpolar molecule)
(b) CHCl₃	dipole-dipole and induced dipole-induced dipole (polar molecule)
(c) CH₄	induced dipole-induced dipole (nonpolar molecule)
(d) HCO₂H	hydrogen bonding, dipole-dipole, and induced dipole-induced dipole (polar molecule with OH bonds)

11.4. To convert species from a liquid to a gas the intermolecular forces that must be overcome:

species	intermolecular force
(a) CH₃CO₂H	hydrogen bonding, dipole-dipole, and induced dipole-induced dipole (polar molecule with OH bonds)
(b) N₂H₄	hydrogen bonding, dipole-dipole, and induced dipole-induced dipole (polar molecule with NH bonds)
(c) SF₆	induced dipole-induced dipole (nonpolar molecule)
(d) Cl₂	induced dipole-induced dipole (nonpolar molecule)

11.5. Considering intermolecular forces, which exists as a gas at 25 °C and 1 atm:

(a) Ne   (b) CH₄   (c) CO   (d) CCl₄

Neon and methane are nonpolar species and possess only induced dipole-induced dipole interactions. Carbon monoxide is a polar molecule, and molecules of CO would be attracted to each other by dipole-dipole interactions, but the CO molecule does not possess a very large dipole moment. The CCl₄ molecule is a non-polar molecule, but very massive (compared to the other three). Hence the greater London forces in CCl₄ result in the strongest attractions of this set of substances—and the only substance expected **not** to be a gas at 25 °C and 1 atmosphere.

11.6. Considering intermolecular forces, which exists as a gas at 25 °C and 1 atm:

(a) (butane) $CH_3CH_2CH_2CH_3$ (b) (methanol) CH_3OH (c) Ar

Ar and $CH_3CH_2CH_2CH_3$ are gases at 25 °C and 1 atm as both are nonpolar. CH_3OH is polar, utilizes hydrogen bonding, and is expected to be a liquid under these conditions.

11.7. Compounds which can form hydrogen bonds are those containing polar O-H, N-H, or F-H bonds and lone pairs of electrons on N, O, or F.

(a) CH_3-O-CH_3 no; no "polar H's" and the C-O bond is not very polar

(b) CH_4 no

(c) HF   yes: lone pairs of electrons on F and a "polar hydrogen".

(d) CH_3COOH yes: lone pairs of electrons on O atoms, and a "polar hydrogen" attached to one of the oxygen atoms

(e) Br_2 no: nonpolar molecules

(f) CH_3OH yes: "polar H" and lone pairs of electrons on O

11.8. Compounds which can form hydrogen bonds:

(a) H_2Se no; no "polar H's"

(b) HCO_2H yes: lone pairs of electrons on O atoms, and a "polar hydrogen"

(c) HI   no: no "polar hydrogen".

(d) $(CH_3)_2CO$ no: lone pairs of electrons on O atom, but no "polar hydrogen"

11.9. For each pair, which has the more negative enthalpy of hydration?

(a) $MgCl_2$; since Mg^{2+} is a smaller cation than Ba^{2+}.

(b) $Ca(NO_3)_2$; two effects here. Ca^{2+} is a smaller cation than K^+, and secondly calcium has a 2+ charge while K has only a 1+ charge.

(c) ZnI_2 — same effects here as in (b) above. Zn^{2+} is a smaller cation than Cs^+, and is doubly charged.

11.10. Ca^{2+} is most strongly hydrated because of its small size and large positive charge. Rb^+ is least strongly hydrated because of its large size and smaller positive charge.

Liquids

11.11. Heat required: $125 \text{ mL} \left(\dfrac{0.7849 \text{ g}}{1 \text{ mL}} \right) \left(\dfrac{1 \text{ mol}}{46.07 \text{ g}} \right) \left(\dfrac{42.32 \text{ kJ}}{1 \text{ mol}} \right) = 90.1 \text{ kJ}$

11.12. Heat required:

$$0.500 \text{ mL} \left(\frac{13.6 \text{ g}}{1 \text{ mL}}\right)\left(\frac{1 \text{ mol}}{200.59 \text{ g}}\right)\left(\frac{59.11 \text{ kJ}}{1 \text{ mol}}\right) = 2.00 \text{ kJ}$$

11.13. Using Figure 11.12:

(a) The equilibrium vapor pressure of water at 60 °C is approximately 150 mm Hg. (Appendix G lists this value as 149.4 mm Hg.)

(b) Water has a vapor pressure of 600 mm Hg at 93 °C.

(c) At 70 °C the vapor pressure of water is approximately 225 mm Hg while that of ethanol is approximately 520 mm Hg.

11.14. Using Figure 11.12:

(a) The equilibrium vapor pressure of diethyl ether at 20 °C is approximately 400 mm Hg.
(b) Increasing intermolecular forces: diethyl ether < ethanol < water

(c) At 400 mm Hg, and 40 °C, diethyl ether is a gas, and ethanol and water are in the liquid phase.

11.15. The vapor pressure of $(C_2H_5)_2O$ at 30 °C is **590 mm Hg**.
Calculate the amount of $(C_2H_5)_2O$ to furnish this vapor pressure at 30 °C (303 K).

$$n = \frac{PV}{RT} = \frac{590 \text{ mm Hg}\left(\frac{1 \text{ atm}}{760 \text{ mm Hg}}\right)(0.100 \text{ L})}{\left(0.082057 \frac{\text{L} \cdot \text{atm}}{\text{K} \cdot \text{mol}}\right)(303 \text{ K})} = 3.1 \times 10^{-3} \text{ mol}$$

Mass of ether: $3.1 \times 10^{-3} \text{ mol} \cdot \frac{74.1 \text{ g}}{1 \text{ mol}} = 0.23 \text{ g}$

The mass of $(C_2H_5)_2O$ [FW = 74.1 g] needed to create this pressure is about 0.23 g. Since there is adequate ether to provide this pressure, the pressure in the flask is expected to be approximately 590 mm Hg. As the flask is cooled from 30. °C to 0 °C, **some of the gaseous ether will condense** to form liquid ether.

11.16. Using Figure 11.12:

(a) As the water cools, the vapor pressure of the water decreases, resulting in a decrease in the total pressure inside the bottle. The bottle will collapse because the pressure inside is less than the external pressure.

(b) At 30 °C and 400 mm Hg there will be only vapor diethyl ether. At 30 °C and 800 mm Hg there will be only liquid diethyl ether.

11.17. Member of each pair with the higher boiling point:

(a) O_2 would have a higher boiling point than N_2. Both are nonpolar and only have London dispersion forces, but O_2 has more electrons and is more polarizable.

(b) SO_2 would boil higher since SO_2 is a polar molecule while CO_2 is non-polar.

(c) HF would boil higher since strong hydrogen bonds exist in HF but not in HI.

(d) GeH_4 would boil higher. While both molecules are non-polar, germane is the more massive molecule with a greater number of electrons, and therefore stronger London dispersion forces.

11.18. Organize in order of increasing boiling point: $Ne < C_2H_6 < C_5H_{12} < CCl_4$

—increasing boiling point→

Ne, being a nonpolar atom, has the lowest bp. Ethane and pentane are nonpolar molecules with a greater number of electrons (and, thus, larger London dispersion forces), so higher boiling points than Ne. Pentane is a larger molecule, so a higher bp than ethane. Finally, CCl_4 is also nonpolar, and the most massive molecule of the set.

11.19. Using the vapor pressure curves shown:

(a) The vapor pressure of CS_2 is approximately 620 mmHg and for nitromethane is approximately 80 mm Hg.

(b) The intermolecular force for CS_2 (a nonpolar molecule) is **induced dipole-induced dipole**; for nitromethane (polar molecule) the forces are **dipole-dipole** and **induced dipole-induced dipole**.

(c) The normal boiling point from the figure for CS_2 is 46 °C and for CH_3NO_2, 100 °C.

(d) The temperature at which the vapor pressure of CS_2 is 600 mm Hg is about 39 °C.

(e) The vapor pressure of CH_3NO_2 is 600 mm Hg at approximately 95 °C.

11.20. Organize the substances from least to greatest intermolecular forces: A < B < C

C, having the highest bp has the strongest intermolecular forces. A, with a higher vapor pressure than B has weaker intermolecular forces than B.

11.21. Regarding benzene:

(a) Using the data provided, note that the vapor pressure of benzene is 760 mm Hg at 80.1 °C (the definition of the normal boiling point).

(b) The graph of the data:

The temperature at which the liquid has a vapor pressure of 250 mm Hg is about 48 °C. The temperature at which the vapor pressure is 650 mm Hg is about 75 °C.

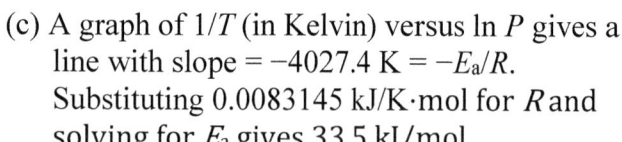

(c) A graph of $1/T$ (in Kelvin) versus $\ln P$ gives a line with slope = −4027.4 K = −E_a/R. Substituting 0.0083145 kJ/K·mol for R and solving for E_a gives 33.5 kJ/mol.

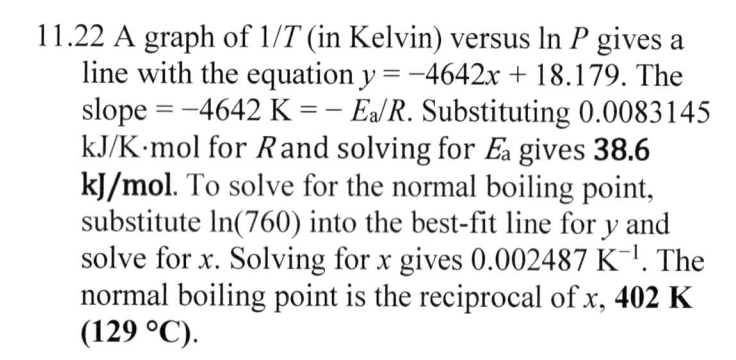

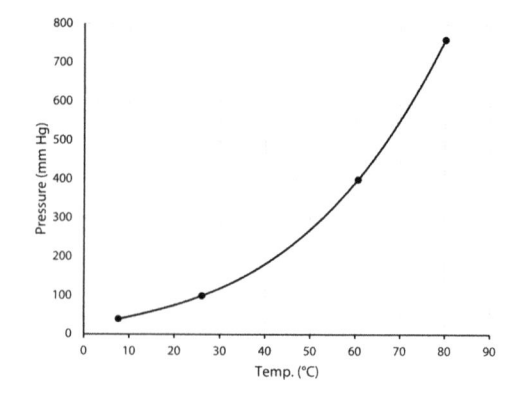

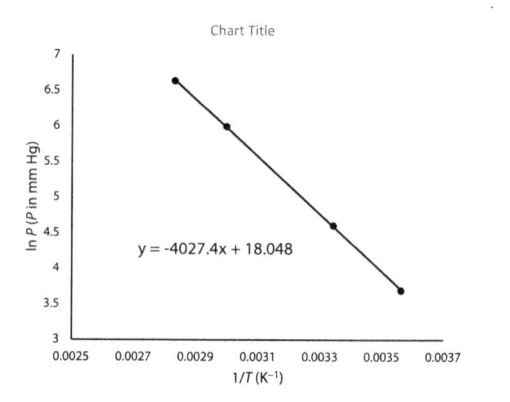

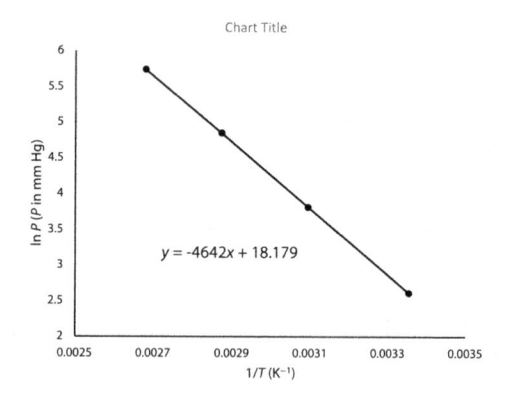

11.22 A graph of $1/T$ (in Kelvin) versus $\ln P$ gives a line with the equation $y = -4642x + 18.179$. The slope = −4642 K = − E_a/R. Substituting 0.0083145 kJ/K·mol for R and solving for E_a gives **38.6 kJ/mol**. To solve for the normal boiling point, substitute ln(760) into the best-fit line for y and solve for x. Solving for x gives 0.002487 K^{-1}. The normal boiling point is the reciprocal of x, **402 K (129 °C)**.

11.23. A gas can be liquefied at or below its critical temperature. The critical temperature for CO is 132.9 K (or −140 °C) so CO **cannot** be liquefied at room temperature (25 °C or 298 K).

11.24. The critical temperature for propane is well above room temperature, so propane can be liquefied at room temperature.

11.25. Surface tension is the energy needed to break through the surface of a liquid. Surface tension is manifest in many areas—the spherical form of a soap bubble is one. The ability of the "water strider" insect to "walk" across the surface of a pond is another. Surface tension is the consequence of intermolecular forces, since—at the surface, the molecules (e.g., of water) are being tightly held together to neighboring molecules on all sides, and to molecules *below* them, but **not** to molecules above them. This "unbalanced" force results in a taut surface.

11.26. Factors affecting viscosity:

Strength of intermolecular forces—the stronger the intermolecular forces, the greater the viscosity. The shape of the molecules--long molecules that can become entangled similar to spaghetti have greater viscosities.

Regarding viscosity—in decreasing order: glycerol > ethylene glycol > ethanol > water; length of molecules from left to right allows for more intermolecular interaction.

Increasing temperature provides more energy for intermolecular forces to be overcome, so temperature does affect viscosity.

11.27. The climbing of water upwards into the paper is an example of capillary action, brought about by the attraction of the water molecules in the container to the -OH bonds on the cellulose.

11.28. Regarding the shape of the meniscus:

Demonstration of adhesion vs. cohesion. Hg…Hg interactions are stronger then Hg…SiO$_2$ interactions. Interactions between glass surface and Hg are minimized to maximize Hg…Hg interactions. For water, H$_2$O…SiO$_2$ interactions are stronger than H$_2$O…H$_2$O interactions.

For ethylene glycol, a concave meniscus is predicted.

General Questions

11.29. These three molecules differ solely in the attachment of the hydroxy, -OH, group relative to the carboxyl, -COOH, group. In 2-hydroxybenzoic acid the hydroxy group is adjacent to the carboxyl group. The relative closeness of the two groups interferes with the formation of strong, multiple hydrogen bonds from one 2-hydroxybenzoic acid molecule to other molecules in the solid state. The effect is to lower the melting point relative to the other two molecules. Notice that 4-hydroxybenzoic acid, where the hydroxy and carboxyl groups are as distant as possible on the benzene ring, has the highest melting point because there is the least interference of the two functional groups to form hydrogen bonds with adjacent molecules.

11.30. Lewis structures for C_2H_6O.

Isomer that forms hydrogen bonds and representation of hydrogen bonding.

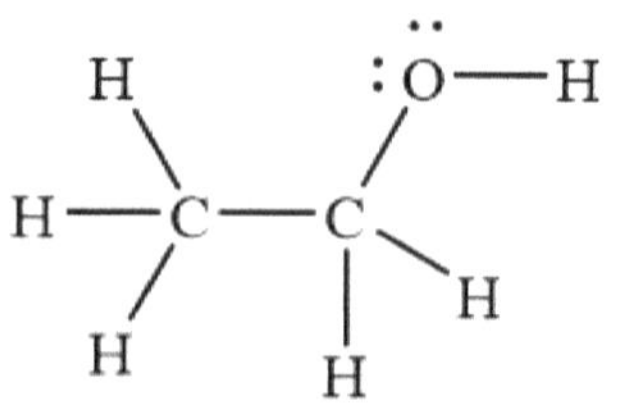

ethanol

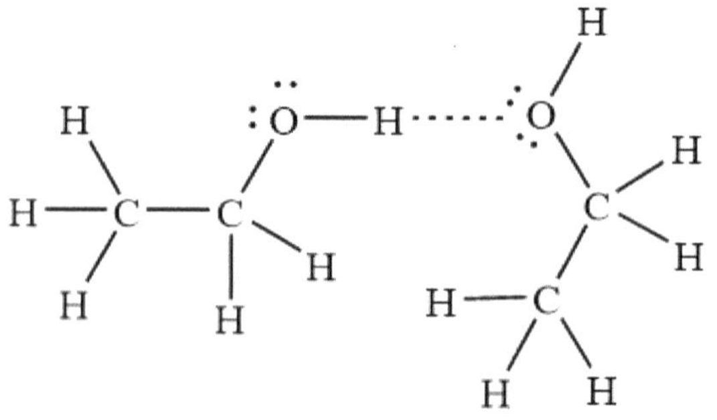

Isomer that does not form hydrogen bonds. Dipole-dipole forces are the strongest intermolecular attraction for this molecule.

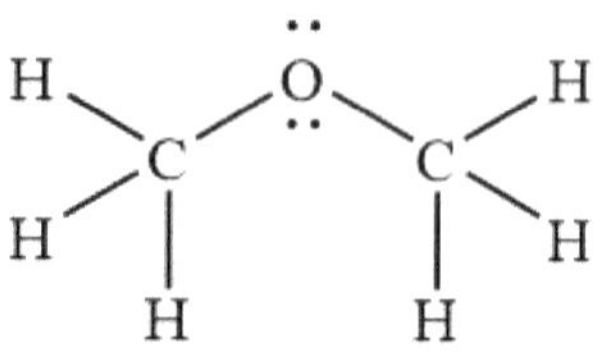

dimethyl ether

11.31. Types of intermolecular forces important in the liquid phases of:

(a) H_2O_2: polar molecule, so **induced dipole-induced dipole**, **dipole-dipole**, and **hydrogen bonding**

(b) CH_2Cl_2: polar molecule, so **induced dipole-induced dipole** and **dipole-dipole**

(c) H_2CO: polar molecule, so **induced dipole-induced dipole** and **dipole-dipole**

11.32. Intermolecular forces acting:

(a) liquid phase of C_2H_6: induced dipole-induced dipole forces—nonpolar molecule

(b) liquid phase of isopropyl alcohol: hydrogen bonding, dipole-dipole, and induced dipole-induced dipole—polar molecule with O-H bonds

11.33. Which salt has the *more exothermic* enthalpy of hydration? Between the two salts, the lithium cation—being the smaller of the two—will have greater interaction with the water—and a resulting greater negative (exothermic) enthalpy of hydration.

11.34. Substance with the higher boiling point:

(a) ICl (polar molecule)

(b) krypton (greater molar mass)

(c) CH₃CH₂OH (hydrogen bonding)

11.35. Using the vapor pressure curves:

(a) The vapor pressure of ethanol at 60 °C is: 350 mm Hg (to the limits of this reader's ability to read the graph)

(b) The stronger intermolecular forces in the liquid state: Ethanol has a lower vapor pressure than carbon disulfide at every temperature—and hence stronger intermolecular forces. This is expected since ethanol has hydrogen bonding as an intermolecular force while CS₂ has **only** induced dipole forces.

(c) The temperature at which heptane has a vapor pressure of 500 mm Hg is: 84 °C

(d) The approximate normal boiling points of the three substances are:

Carbon disulfide	bp = 46 °C (literature value 46.5 °C)
Ethanol	bp = 78 °C (literature value 78.5 °C)
Heptane	bp = 99 °C (literature value 98.4 °C)

(e) At a pressure of 400 mm Hg and 70 °C, the state of the three substances is:

Carbon disulfide	state = gas
Ethanol	state = gas
Heptane	state = liquid

11.36. Ionic compound with the most negative enthalpy of hydration:

Compounds (a), (b), and (d) contain small, highly charged metal ions. These salts will most likely experience strong ion-dipole forces between the metal cation and water molecules and exist as hydrated solids. (d) Al(NO₃)₃ should have the most negative enthalpy of hydration because the Al³⁺ cation is the smallest of the three highly charged ions..

11.37. Increasing molar enthalpy of vaporization:

Owing to the H-bonding of CH₃OH, this substance is likely to have the greatest molar enthalpy of vaporization. Ethane, on the other extreme, being a nonpolar compound, and of the three substances with the lowest molecular mass, will have the lowest. HCl—with a intermediate molecular mass and being polar—will have an intermediate molar enthalpy of vaporization.   C₂H₆ < HCl < CH₃OH

11.38. Rank in order of increasing boiling point:

$$CH_3F < CH_3Cl < CH_3CH_2CH_2CH_2CH_3$$
—increasing boiling point→

11.39. Calculate the number of atoms represented by the vapor:

The pressure of Hg is $0.00169 \text{ mm Hg} \left(\dfrac{1 \text{ atm}}{760 \text{ mm Hg}} \right) = 2.22 \times 10^{-6}$ atm

The number of moles/L is $\dfrac{P}{RT} = \dfrac{n}{V} = \dfrac{2.22 \times 10^{-6} \text{ atm}}{\left(0.082057 \dfrac{\text{L atm}}{\text{K mol}} \right)(297 \text{K})} = 9.12 \times 10^{-8}$ mol/L

Converting this to atoms/m³:

$\dfrac{9.12 \times 10^{-8} \text{ mol}}{\text{L}} \left(\dfrac{1000 \text{ L}}{1 \text{ m}^3} \right) \left(\dfrac{6.022 \times 10^{23} \text{ atoms}}{1 \text{ mol}} \right) = 5.49 \times 10^{19} \dfrac{\text{atoms}}{\text{m}^3}$

Note that the information that the air was saturated with mercury vapor obviates the need to calculate the volume of the room.

11.40. For the equilibrium vapor pressure of limonene:

(a) Plot data of ln(vapor pressure) vs $1/T$:

(b) Using the equation for the straight line, $T = 135$ °C when $P = 250$ mm Hg and $T = 168$ °C when $P = 650$ mm Hg

(c) $T = 173$ °C when $P = 760$ mm Hg

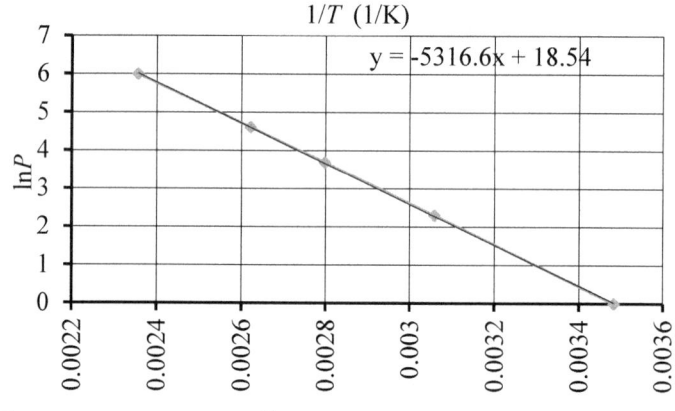

ln(vapor pressure) vs. 1/temperature
$1/T$ (1/K)
$y = -5316.6x + 18.54$

(d) $\ln \dfrac{400. \text{ mm Hg}}{100. \text{ mm Hg}} = \dfrac{-\Delta H}{0.0083145 \dfrac{\text{kJ}}{\text{K} \cdot \text{mol}}} \left[\dfrac{1}{424.6 \text{ K}} - \dfrac{1}{381.5 \text{ K}} \right]$ and $\Delta H = 43.3$ kJ/mol

In the Laboratory

11.41. Regarding dichlorodimethylsilane:

(a) The normal boiling point is easily determined since it is *defined* as the temperature at which the vapor pressure of a substance is equal to 760 mm Hg. The normal boiling point is 70.3 °C.

(b) Plotting the data:

Using the equation for the straight-line, ln P = 17.95 – 3885(1/T), you can calculate T at P values of 250 and 650 mm Hg. Substituting P = 250 mm Hg, ln P = 5.52 and T = 312.6 K (39.5 °C).

Similarly, for P = 650 mm Hg, ln P = 6.48 and T = 338.7 K (65.5 °C).

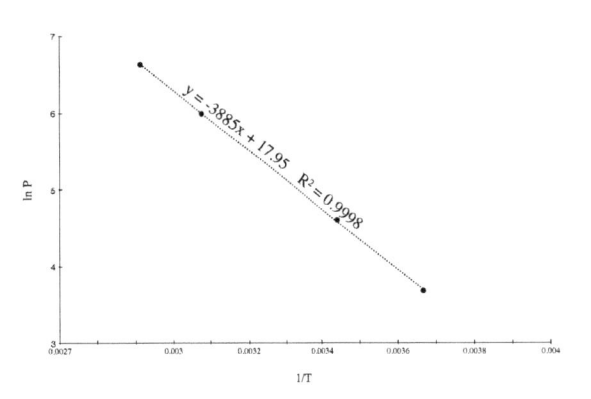

(c) The molar enthalpy of vaporization for dichlorodimethylsilane from the Clausius-Clapeyron equation:

$$\ln\left(\frac{P_2}{P_1}\right) = \frac{\Delta H_{vap}}{R}\left[\frac{1}{T_1} - \frac{1}{T_2}\right]$$ Using the T,P data from part (b):

$$\ln\left(\frac{650. \text{ mm Hg}}{250. \text{ mm Hg}}\right) = \frac{\Delta H}{8.3145 \times 10^{-3} \text{kJ/K} \cdot \text{mol}}\left[\frac{1}{312.6 K} - \frac{1}{338.7 K}\right]$$

$$\ln 2.60 \cdot 8.3145 \times 10^{-3} \text{kJ/K} \cdot \text{mol} = \Delta H\left[\frac{1}{312.6 \text{ K}} - \frac{1}{338.7 \text{ K}}\right]$$

$$\ln 2.60 \cdot 8.3145 \times 10^{-3} \frac{\text{kJ}}{\text{K} \cdot \text{mol}} = \Delta H (2.46 \times 10^{-4} \text{ K}^{-1})$$

$$\Delta H = \frac{7.9445 \times 10^{-3} \text{ kJ/K} \cdot \text{mol}}{2.46 \times 10^{-4} \text{ K}^{-1}} = 32.3 \text{ kJ/mol}$$

11.42. Regarding the hand boiler:

(a) The volatile liquid has a low boiling point. Warming the lower compartment in your hands increases the number of molecules in the vapor phase, raising the vapor pressure of the liquid and forcing the liquid to the upper chamber.

(b) Of the liquids suggested: CCl_3F since it boils just below body temperature

11.43. Regarding the collapsing can:

(a) The can collapses because of the condensation of the gas in the can—which has filled the heated can—to a liquid. The distances between the particles of liquid are **much less** than the distances between the particles of gas. The resulting decrease in pressure inside the can causes the greater pressure outside the can to crush it.

(b) Before the can is heated, most of the water exists in the liquid form, while some exists as the vapor (above the liquid surface). After heating, the liquid has been converted into the gaseous form, filling the volume of the can.

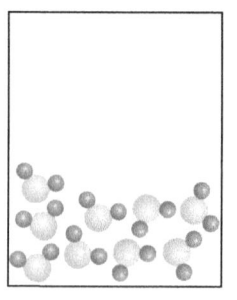

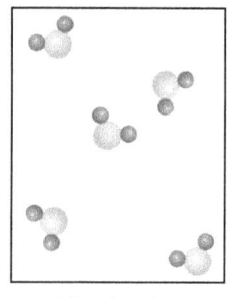

Before heating   After heating

11.44. Regarding ethanol in a small lab room:

Amount of ethanol: $1.0 \times 10^3 \text{ mL} \left(\dfrac{1 \text{ cm}^3}{1 \text{ mL}}\right)\left(\dfrac{0.785 \text{ g}}{1 \text{ cm}^3}\right)\left(\dfrac{1 \text{ mol C}_2\text{H}_5\text{OH}}{46.07 \text{ g}}\right) = 17 \text{ mol C}_2\text{H}_5\text{OH}$

Volume of room: $V_{room} = 3.0 \text{ m} \times 2.5 \text{ m} \times 2.5 \text{ m}\left(\dfrac{100 \text{ cm}}{1 \text{ m}}\right)^3\left(\dfrac{1 \text{ mL}}{1 \text{ cm}^3}\right)\left(\dfrac{1 \text{ L}}{1000 \text{ mL}}\right) = 1.9 \times 10^4 \text{ L}$

Amount of ethanol needed to fill this volume at this P, T:

$n = \dfrac{PV}{RT} = \dfrac{\left(59 \text{ mm Hg} \cdot \dfrac{1 \text{ atm}}{760 \text{ mm Hg}}\right)(1.9 \times 10^4 \text{ L})}{(0.082057 \text{ L} \cdot \text{atm/K} \cdot \text{mol})(298 \text{ K})} = 60. \text{ mol ethanol}$

Less ethanol is available (17 mol) than would be required to completely fill the room with vapor (60. mol), so all the ethanol will evaporate.

Summary and Conceptual Questions

11.45. Acetone readily absorbs water due to **hydrogen bonding** between the C = O oxygen atom and the O—H bonds of water.

11.46. Regarding the polarity of motor oil:

Since motor oil does not mix with water, the two substances have different types of intermolecular forces. Water is a polar molecule, and strong hydrogen bonding occurs between water molecules. The molecules in motor oil are nonpolar.

11.47. The **viscosity of ethylene glycol would be predicted to be greater** than that of ethanol since the glycol possesses two O-H groups per molecule while ethanol possesses one. Two OH groups/molecule would provide more hydrogen bonding and increase the viscosity.

11.48. The meniscus is concave since there are adhesive forces between the methanol and the silicate of the glass. See SQ11.28.

11.49. Explain the fact that:

(a) Ethanol has a lower boiling point than water. Both molecules are polar, and both can form hydrogen bonds. Water has **two** polar H atoms, and two lone pairs of electrons on the O atom. Ethanol has only **one** polar H atom, with the accompanying two lone pairs on the O atom. Given the increased ability of water to form hydrogen bonds with other water molecules, one would expect that water would have a higher boiling point.

(b) Mixing 50 mL of water with 50 mL of ethanol results in less than 100 mL of solution. The H-bonding that occurs not only between (water molecules and other water molecules) and (ethanol molecules and other ethanol molecules) also occurs between ethanol and water molecules. The attraction results in the molecules occupying less space than one would anticipate—and a non-additive volume.

11.50. Compare the boiling points of 1-propanol versus the bp of methyl ethyl ether:

1-propanol has stronger intermolecular forces (hydrogen bonding) than methyl ethyl ether (dipole-dipole)—hence a higher boiling point.

11.51. Evidence that water molecules in the liquid state exert attractive forces on one another:

(a) Water has a relatively large specific heat capacity (4.18 J/g ·K). This large value reflects the strong forces that hold water molecules together requiring a large amount of energy to overcome.

(b) Water is a liquid at room temperature—even though it has a molar mass of approximately 18 g/mol. With this molar mass, one would anticipate a boiling point well below 0 °C (and a *gaseous* physical state).

11.52. Why use a non-frost-free refrigerator:

The frost–free refrigerator's freezer is warmed periodically to near the freezing point of water. This allows the vapor pressure of the ice to increase so that sublimation readily occurs, keeping the frost from accumulating in the freezer. It would therefore make the hailstones disappear over time. No such mechanism exists in the older non–frost–free refrigerator.

11.53. Referring to Figure 11.8:

(a) The hydrogen halide with the **largest total intermolecular force** is HI.

(b) The dispersion forces are greater for HI than for HCl owing to the fact that HI (specifically I) has a larger volume than HCl (or Cl).

(c) Dipole-dipole forces are greater for HCl than for HI owing to the more polar H-Cl bond.

The electronegativities (H = 2.2; Cl = 3.2; I = 2.7) indicate this difference in polarity (3.2-2.2) versus (2.7-2.2).

(d) The HI molecule has the largest dispersion forces of the molecules shown in the Figure. This is quite reasonable given the large volume of the iodine atom (see part b).

11.54. Regarding H_2S:

(a) H_2O has stronger hydrogen bonding than H_2S.

(b) Partial pressure of H_2S:

(725 mm Hg)(1 atm/760 mm Hg) = 0.954 atm

$$P_{H_2S} = \frac{80 \text{ mol}}{1,000,000 \text{ mol}}(0.954 \text{ atm}) = 7.63 \times 10^{-5} \text{ or } 8 \times 10^{-5} \text{ atm (1 sf)}$$

(c) Using Avogadro's hypothesis, $n_1V_1 = n_2V_2$,

5.2 L(2 mol O_2/1 mol H_2S) = 10.4 L

11.55. The critical temperature for CF_4 (−45.7 °C) is below room temperature (25 °C). Since room temperature is greater than T_c for CF_4, the substance **cannot be** liquefied at room temperature.

11.56. Regarding a cylinder of dichlorodifluoromethane:

(a) Normal boiling point is −27 °C, as the vapor pressure at this T is 1 atm.

(b) Pressure of the vapor is approximately 6.5 atm, as determined from the graph at this T.

(c) The flow is rapid at first because the liquid has vaporized to form the maximum amount of vapor possible at that temperature, with a pressure of approximately 6.5 atm. As the gas leaves the cylinder, additional liquid vaporizes in an attempt to reach liquid-vapor equilibrium. The flow becomes much slower because the pressure in the cylinder has dropped and cannot increase unless the cylinder is closed or enough heat is supplied to vaporize the remaining liquid. This vaporization requires energy, so the temperature of the cylinder drops and ice forms on the outside of the cylinder.

(d) Procedure to rapidly empty the container: (2) Cool the cylinder to −78 °C and open the valve. Cooling the cylinder will condense the vapor, allowing it to be easily removed as a liquid.

11.57. The electrostatic potential surface is shown below:

Hydrogen bonding occurs between H-atoms and the **lone pairs** of electrons on very electronegative atoms (F, O, N). The electropotential surface indicates two oxygen atoms (red spheres in the original diagram—and indicated by arrows below), each of which will contain two lone pairs of electrons at which hydrogen bonding is likely to occur. Also the N atom of the amide group will also exhibit H-bonding.

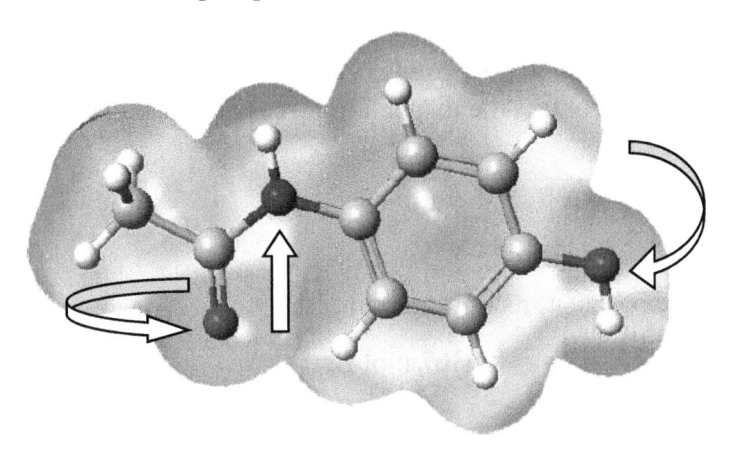

11.58. Regarding cytosine and guanine:

The nitrogen and oxygen atoms and the hydrogen atoms attached to nitrogen can be involved in hydrogen bonding.

11.59. Four properties of liquids that are directly determined by intermolecular forces:

- vapor pressure—the pressure exerted by a liquid in equilibrium with its gaseous phase.

- boiling point—the temperature at which the vapor pressure of the liquid equals the ambient atmospheric pressure.

- surface tension—energy required to break through the surface of a liquid.

- viscosity—the resistance of liquids to flow

11.60. Ions in order of hydration energies:
$$K^+ < Na^+ < Ca^{2+} < Mg^{2+}$$
—increasing enthalpy of hydration→

The smallest ion with the greatest charge (Mg^{2+}) has the most negative enthalpy of hydration. The largest ion with the lowest charge (K^+) has the least negative enthalpy of hydration.

11.61. The branched chain isomers have a smaller exposed surface than their unbranched analogues. With decreased molecular contact the dispersion forces are weakened, with the resulting lower boiling points.

11.62. Describe the contents of the 1.00 L flask at 58.8 °C:

The amount of bromine that will vaporize at its boiling point is
$$n = \frac{PV}{RT} = \frac{(1.00 \text{ atm})(1.00 \text{ L})}{\left(0.082057 \frac{L \cdot atm}{K \cdot mol}\right)(332.0 \text{ K})} = 0.0367 \text{ mol Br}_2$$

The amount of Br_2 available is $8.82 \text{ g} \cdot \frac{1 \text{ mol Br}_2}{159.8 \text{ g}} = 0.0552 \text{ mol Br}_2$

More Br_2 is available than can vaporize at the boiling point, so the flask will contain both bromine liquid and bromine vapor in equilibrium.

11.63. Rank the halogens (F_2, Cl_2, Br_2, I_2) and the noble gases (He, Ne, Ar, Kr, Xe) in order of polarizability (from least polarizable to most polarizable):

London dispersion forces (and polarizability) increase *with increasing atomic or molecular volume*, so use the position of the elements in the Periodic Table (and hence their respective volumes) to rank them:

Least polarizable → Most polarizable

Halogens: $F_2 < Cl_2 < Br_2 < I_2$

Noble gases: He < Ne < Ar < Kr < Xe

11.64. Molecules in which hydrogen bonding is expected:

(d) Benzoic acid, and (e) acetamide are expected to exhibit hydrogen bonding, owing to polar H with lone pairs of electrons.

11.65. Use the Clausius-Clapeyron equation to calculate the temperature at which water boils in the pressure cooker:

$$\ln\frac{P_2}{P_1} = -\frac{\Delta H^0_{vap}}{R}\left[\frac{1}{T_2}-\frac{1}{T_1}\right] \text{ and substituting}$$

$$\ln\frac{(14.70 + 15)\text{psi}}{14.70 \text{ psi}} = -\frac{40.7\,\frac{\text{kJ}}{\text{mol}}}{0.0083145\,\frac{\text{kJ}}{\text{K}\cdot\text{mol}}}\left[\frac{1}{T_2}-\frac{1}{373\text{ K}}\right]$$

$$0.7033 = -4895\left[\frac{1}{T_2}-\frac{1}{373\text{ K}}\right] \text{ and } \frac{0.7033}{-4895} = \left[\frac{1}{T_2}-\frac{1}{373\text{ K}}\right]$$

$$-0.000143675 = \left[\frac{1}{T_2}-\frac{1}{373\text{ K}}\right] \text{ and } -0.000143675 + \frac{1}{373\text{ K}} = \frac{1}{T_2};$$

solving for T_2 gives 394 K or 121 °C.

11.66. Use the data to calculate the enthalpy of vaporization for ammonia:

Plot $\ln P$ vs $1/T$:

ln(vapor pressure) vs. 1/temperature

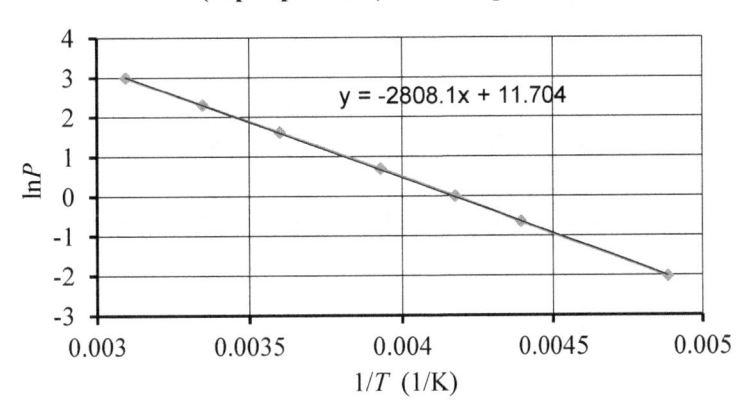

y = −2808.1x + 11.704

Using the equation for the best line: slope = -2808.1 K^{-1} = $\frac{-\Delta H}{R} = \frac{-\Delta H}{0.0083145\,\frac{\text{kJ}}{\text{K}\cdot\text{mol}}}$

and ΔH = 23.3 kJ/mol

11.67. Using the data in Question 66, the plot shown in 11.66 resulted:

The best straight line for the data is: $y = -2808.1x + 11.704$

Substituting 20 °C (293.15K) into the equation gives a "y" value of: 2.1249
So $\ln P$ = 2.1249 and P is: 8.4 atm (2 sf)

11.68. From the given plot:

(a) The best straight line has the equation: $y = -5141.8\,x + 21.283$

(b) The slope of the line is equal to $\dfrac{-\Delta H}{R}$. Set up an equation and solve for ΔH.

$\dfrac{-\Delta H}{R} = -5141.8\ 1/K$ thus $\Delta H = R \times 5141.8\ 1/K$

(c) $\Delta H = \left(0.0083145\ \dfrac{kJ}{K\,mol}\right) 5141.8\ 1/K$

$\Delta H = 42.75\ \dfrac{kJ}{mol}$

(d) Using the equation for the straight line, $P = 11.6$ mm Hg when $T = 0.00\ °C$ and $P = 1804$ mm Hg when $T = 100\ °C$.

11.69. Water (10.0 g) is placed in a thick-walled glass tube whose internal volume is 50.0 cm³. Then all the air is removed, the tube is sealed, and then the tube and contents are heated to 100 °C.

(a) What is the appearance of the system at 100 °C.

There will be liquid water in the tube, and some of the liquid will be converted to the gaseous state (water vapor).

(b) What is the pressure inside the tube?

Since the water is at 100 °C, the vapor pressure of water is 760 mm Hg or 1.00 atm (the temperature at which water boils at 1 atm is 100 °C).

(c) What volume of liquid water is in the tube at this temperature?
[$d_{liquid\ water}$ = 0.958 g/cm³] If all the water is in the liquid state, the volume of water would be: $10.0\ g \cdot \dfrac{1\ cm^3}{0.958\ g} = 10.4\ cm^3$ (3 sf)

(d) Some of the water is in the vapor state. Determine the mass of water in the gaseous state. The volume of the tube is 50.0 cm³. The water (in the liquid state) would occupy 10.4 cm³. So the gas would have a volume of (50.0-10.4) or 39.6 cm³.

Substituting into the Ideal Gas Law:

$n = \dfrac{PV}{RT} = \dfrac{(1.00\,atm)(0.0396\,L)}{\left(0.082057\,\dfrac{L\cdot atm}{K\cdot mol}\right)(373.15\,K)} = 0.001293\ mol$

which corresponds to 18.02 g/mol × 0.001293 mol = 0.0233 g H₂O.

11.70. Comparing allyl alcohol to acetone:

(a) Vapor pressure of allyl alcohol would be lower. Intermolecular forces are greater for allyl alcohol because it can hydrogen bond.

(b) Calculate the enthalpy of vaporization for acetone using the data. Plot ln P vs $1/T$

ln P (P in mm Hg)	$1/T$ (1/K)
3.69	0.003791
4.605	0.003560
5.991	0.003198
6.633	0.003033

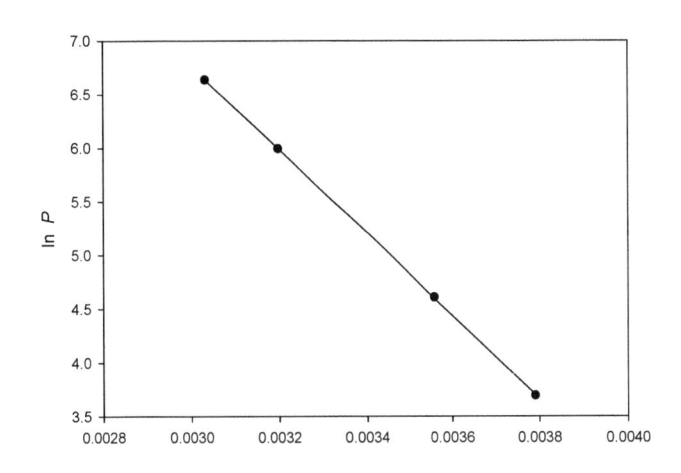

Equation for the best straight line is $y = -3873 x + 18.38$ so slope $= -\Delta H°/R = -3873$ K

$\Delta H° = -\text{slope} \times R = -(-3873 \text{ K}) \times 0.0083145 \dfrac{\text{kJ}}{\text{K} \cdot \text{mol}} = 32.2$ kJ/mol

(c) Identify the fluorinated compound using the data:

(722 mm Hg)(1 atm/760 mm Hg) = 0.950 atm; 22 °C = 295 K

Substitute into the Ideal Gas Law: n $= \dfrac{PV}{RT} = \dfrac{(0.950 \text{ atm})(0.264 \text{ L})}{\left(0.0821 \dfrac{\text{L} \cdot \text{atm}}{\text{mol} \cdot \text{K}}\right)(295 \text{ K})} = 0.0104$ mol

So, 1.53 g/0.0104 mol = 148 g/mol. The gas is C_3HF_5O (148 g/mol).

Solution and Answer Guide

Kotz Treichel Townsend Treichel, Chemistry and Chemical Reactivity 11e, 978-0-357-85140-1, Chapter 12: The Solid State

TABLE OF CONTENTS

Applying Chemical Principles 418
Practicing Skills..421
General Questions 432
In the Laboratory...441
Summary and Conceptual Questions 443

Applying Chemical Principles

Lithium and "Green Cars"

12.1.1. Mass of Li from 73 million metric tons of lithium carbonate:

The molar mass of Li_2CO_3 is 73.89 g/mol. Of this, (2×6.94) g/mol = 13.88 g/mol is attributable to Li. Thus, in 73.89 g of Li_2CO_3, there is 13.88 g of Li and in 73.89 metric tons of Li_2CO_3, there is likewise 13.88 metric tons of Li.

$$73 \times 10^6 \text{ metric tons } Li_2CO_3 \left(\frac{13.88 \text{ metric tons Li}}{73.89 \text{ metric tons } Li_2CO_3} \right) = 1.4 \times 10^7 \text{ metric tons Li}$$

(= 14 million metric tons)

12.1.2. Unit cell of lithium: The figure indicates that Li has a body-centered cubic lattice.

12.1.3. Calculate the density of lithium, given that the unit cell has an edge length of 351 pm: [See Study Question 12.3 for another example of this problem involving potassium.]

Calculate the volume of the unit cell:

The length in cm: $351 \text{ pm} \left(\frac{1 \text{ m}}{1 \times 10^{12} \text{ pm}} \right) \left(\frac{100 \text{ cm}}{1 \text{ m}} \right) = 3.51 \times 10^{-8} \text{ cm} = \text{length}$

Volume = $(3.51 \times 10^{-8} \text{ cm})^3 = 4.32 \times 10^{-23} \text{ cm}^3$

Li crystallizes as a body-centered cube, so there are 2 Li atoms/unit cell.

$$\text{Mass} = 2 \text{ atoms Li} \left(\frac{1 \text{ mol Li}}{6.022 \times 10^{23} \text{ atoms Li}} \right) \left(\frac{6.94 \text{ g Li}}{1 \text{ mol Li}} \right) = 2.30 \times 10^{-23} \text{ g Li}$$

$$\text{Density} = \frac{\text{mass}}{\text{volume}} = \frac{2.30 \times 10^{-23} \text{ g Li}}{4.32 \times 10^{-23} \text{ cm}^3 \text{ Li}} = 0.533 \text{ g/cm}^3$$

12.1.4. To produce LiCl from Li$_2$CO$_3$, one can react the carbonate with aqueous HCl:
Li$_2$CO$_3$(aq) + 2 HCl(aq) → 2 LiCl(aq) + H$_2$O(ℓ) + CO$_2$(g)

12.1.5. Calculate the density of MnO$_2$ based on unit cell dimensions:
From the unit cell image, you can see there are 8 Mn atoms on the corners and 1 Mn atom in the cell's body. There are also 4 O atoms in the faces and 2 O atoms inside the unit cell.

Number of Mn atoms per unit cell:

Corners: $8 \times \dfrac{1}{8} = 1$ Mn atom

Body: $1 \times 1 = 1$ Mn atom

Total: 2 Mn atoms per unit cell

Number of O atoms per unit cell:

Faces: $4 \times \dfrac{1}{2} = 2$ O atoms

Body: $2 \times 1 = 2$ O atoms

Total: 4 O atoms per unit cell

Next determine the mass of the atoms in the unit cell:

$$2 \text{ Mn atoms}\left(\dfrac{1 \text{ mol Mn}}{6.02214 \times 10^{23} \text{ Mn atoms}}\right)\left(\dfrac{54.938 \text{ g Mn}}{1 \text{ mol Mn}}\right) = 1.8245 \times 10^{-22} \text{ g Mn}$$

$$4 \text{ O atoms}\left(\dfrac{1 \text{ mol O}}{6.02214 \times 10^{23} \text{ O atoms}}\right)\left(\dfrac{15.999 \text{ g O}}{1 \text{ mol O}}\right) = 1.0627 \times 10^{-22} \text{ g O}$$

Total mass per unit cell = 1.8245×10^{-22} g + 1.0627×10^{-22} g = 2.8872×10^{-22} g/unit cell

You need the volume of one unit cell in units of cm^3 for appropriate density units. So, convert nm to cm for the cell dimensions and then calculate the volume of this tetragonal unit cell.

For sides *a* and *b*: $0.44008 \text{ nm}\left(\dfrac{1 \text{ m}}{1 \times 10^9 \text{ nm}}\right)\left(\dfrac{100 \text{ cm}}{1 \text{ m}}\right) = 4.4008 \times 10^{-8}$ cm

For side *c*: $0.28745 \text{ nm}\left(\dfrac{1 \text{ m}}{1 \times 10^9 \text{ nm}}\right)\left(\dfrac{100 \text{ cm}}{1 \text{ m}}\right) = 2.8745 \times 10^{-8}$ cm

$V_{\text{unit cell}} = (4.4008 \times 10^{-8} \text{ cm})(4.4008 \times 10^{-8} \text{ cm})(2.8745 \times 10^{-8} \text{ cm}) = 5.5671 \times 10^{-23}$ cm^3/unit cell

Finally, calculate the density of MnO$_2$ from the mass and volume:

density = $\dfrac{2.8872 \times 10^{-22} \text{ g}}{5.5671 \times 10^{-23} \text{ cm}^3} = 5.1862$ g/cm^3

To three significant figures, this is 5.19 g/cm^3.

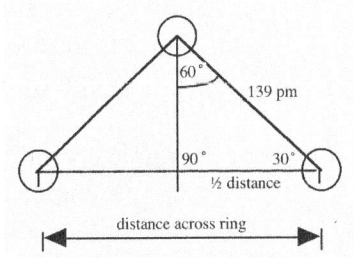

Nanotubes and Graphene

12.2.1. Based on the C–C length of 139 pm, what is the distance across a planar C-6 ring?

If you consider three of the C atoms in the C-6 ring, you can construct the figure at the right. The hypotenuse of each right triangle is 139 pm. As this is part of a regular hexagon, the angle at the top is 120 degrees.

Bisecting that angle produces a 30/60/90 triangle in which 1/2 the distance across the ring is related to the cos(30°). [Check out your trigonometry book here: the cosine of an angle is equal to the ratio of the lengths of the side adjacent to the angle and the hypotenuse.] That distance corresponds to cos(30°)(139 pm) = 120.4 pm. Thus, the distance across the ring is 2 × 120.4 pm = 240.8 pm = 241 pm (3 sf).

12.2.2. Assume that the ring is approximately 241 pm across (based upon calculation in 12.2.1).

$$\left(\frac{1 \text{ ring}}{240.8 \text{ pm}}\right)\left(\frac{1 \text{ pm}}{1\times10^{-12} \text{ m}}\right)\left(\frac{1\times10^{-6} \text{ m}}{1 \text{ micrometer}}\right)(1.0 \text{ micrometer}) = 4153 \text{ rings or } 4200 \text{ rings}$$

With each ring approximately 241 pm across, 4200 rings will fit side-by-side in a 1.0 micrometer sheet.

12.2.3. Estimate the thickness of a sheet of graphene:
If the sheet is planar, then the thickness of a sheet will be equal to the diameter of a C atom. The covalent radius of a C atom is 77 pm. The diameter is 2 × 77 pm = 154 pm = 150 pm (2 sf)

12.2.4. Describe the bonding in graphene:

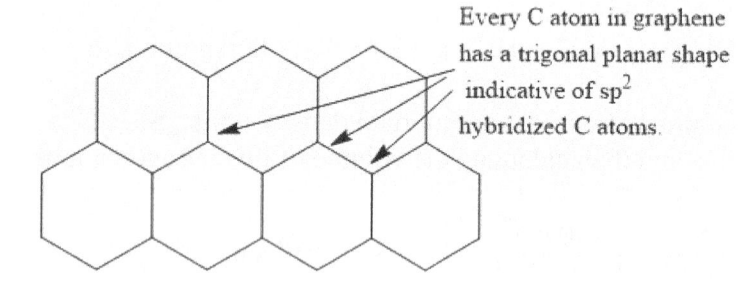

Every C atom in graphene has a trigonal planar shape indicative of sp^2 hybridized C atoms.

Tin Disease

12.3.1. How many tin atoms are contained in the tetragonal crystal lattice unit cell of β-tin? How many tin atoms are contained in the cubic crystal lattice unit cell of α-tin?

The tetragonal crystal lattice unit cell for β-tin (Figure 2) shows:
8 corner atoms (shared by 8 other cells) × 1/8 = 1
4 face atoms (each shared by 2 other cells) × ½ = 2
1 body atom (not shared) × 1 = 1
Total tin atoms = 4

For the cubic crystal lattice of α-tin:
8 corner atoms (shared by 8 other cells) × 1/8 = 1
6 face atoms (each shared by 2 other cells) × ½ = 3
4 body atoms (not shared) × 1 = 4
Total tin atoms = 8

12.3.2. Calculate the density of **white** or tetragonal β-tin:

The tetragonal unit cell (Figure 2) is 583 pm × 583 pm × 318 pm.

The volume = 583 pm · 583 pm · 318 pm = 1.08×10^8 pm³

$$\left(\frac{4 \text{ atoms Sn}}{1.08 \times 10^8 \text{pm}^3}\right)\left(\frac{1 \text{ mol}}{6.022 \times 10^{23} \text{atoms}}\right)\left(\frac{118.7 \text{ g}}{1 \text{ mol}}\right)\left(\frac{1 \times 10^{10} \text{pm}}{1 \text{ cm}}\right)^3 = 7.29 \text{ g/cm}^3$$

12.3.3. Dimensions of the unit cell of **gray** or cubic α-tin:

Density for gray tin = 5.769 g/cm³. As noted in 12.3.1, there are 8 tin atoms in the unit cell of the cubic crystal lattice of grey tin.

The mass/unit cell is: $\frac{8 \text{ Sn atoms}}{1 \text{ unit cell}}\left(\frac{118.71 \text{ g Sn}}{6.0221 \times 10^{23} \text{ atoms}}\right) = 1.57699 \times 10^{-21}$ g/unit cell

The volume of the unit cell of gray tin is:

$$\frac{1.57699 \times 10^{-21} \text{ g}}{1 \text{ unit cell}}\left(\frac{1 \text{ cm}^3}{5.769 \text{ g}}\right) = 2.734 \times 10^{-22} \text{ cm}^3$$

Converting this volume to "atomic dimensions":

$$2.734 \times 10^{-22} \text{ cm}^3 \left(\frac{1 \text{ m}}{100 \text{ cm}}\right)^3 \left(\frac{1 \times 10^{12} \text{ pm}}{1 \text{ m}}\right)^3 = 2.734 \times 10^8 \text{ pm}^3$$

Since the cell is a cube, you can find the edge length by taking the cube root of this volume: Edge of gray tin = $[2.734 \times 10^8 \text{ pm}^3]^{1/3}$ = 649.0 pm

12.3.4. To find the space occupied by tin atoms in tetragonal and cubic crystal lattices use the volumes calculated in 12.3.2 and 12.3.3:

The unit cell of white or β-tin (tetragonal lattice) contains 4 atoms. The volume of space occupied by the atom is: $V_{\text{Sn atoms}} = \frac{4}{3}\pi r^3 \times (4 \text{ atoms}) = \frac{4}{3}\pi(141 \text{ pm})^3 \times (4 \text{ atoms})$

$$= 4.70 \times 10^7 \text{ pm}^3$$

$$\frac{V_{\text{Sn atoms}}}{V_{\beta\text{-tin}}} \times 100\% = \frac{4.70 \times 10^7 \text{ pm}^3}{1.08 \times 10^8 \text{ pm}^3} \times 100\% = 43.5\%$$

The unit cell of α-tin (cubic lattice) contains 8 atoms. The volume of space occupied by the atom is: $V_{\text{Sn atoms}} = \frac{4}{3}\pi r^3 \times (8 \text{ atoms}) = \frac{4}{3}\pi(141 \text{ pm})^3 \times (8 \text{ atoms}) = 9.39 \times 10^7 \text{ pm}^3$

$$\frac{V_{\text{Sn atoms}}}{V_{\alpha\text{-tin}}} \times 100\% = \frac{9.39 \times 10^7 \text{ pm}^3}{2.734 \times 10^8 \text{ pm}^3} \times 100\% = 34.4\%$$

Practicing Skills

Metallic and Ionic Solids

12.1. This compound would have the formula AB₈ since each black square (A) has eight corresponding white squares (B).

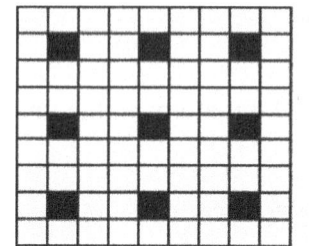

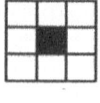

The unit shown above is the smallest "cookie cutter" that would reproduce the larger pattern shown at left. The formula for such a compound would be AB₈.

12.2. The area inside the box is a unit cell. Each unit cell contains 2 A squares and 2 B squares, so the simplest formula is AB.

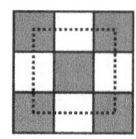

12.3. Regarding the unit cell for potassium:
(a) According to the diagram, this unit cell is body-centered cubic. (See Figures 12.4 and 12.5.)

(b) Calculate the density of metallic potassium if one edge of the unit cell is 533 pm: Figure 12.4 confirms there are 2 net atoms per unit cell for substances in the body-centered cubic system (1 in the center and 8 atoms in each corner [each contributing 1/8]).
Calculate the volume of the unit cell:

$$533 \text{ pm} \left(\frac{1 \text{ m}}{1 \times 10^{12} \text{ pm}}\right)\left(\frac{100 \text{ cm}}{1 \text{ m}}\right) = 5.33 \times 10^{-8} \text{ cm} = \text{length}$$

Volume = $(5.33 \times 10^{-8} \text{ cm})^3 = 1.514 \times 10^{-22} \text{ cm}^3$

Mass = $\frac{39.098 \text{ g K}}{1 \text{ mol K}}\left(\frac{1 \text{ mol K}}{6.02214 \times 10^{23} \text{ atoms K}}\right)\left(\frac{2 \text{ atoms K}}{\text{unit cell}}\right)$

$= 1.2985 \times 10^{-22}$ g K/unit cell

Density = mass/volume = 1.2985×10^{-22} g K/1.514×10^{-22} cm³ = 0.858 g/cm³ (3sf)

12.4. Calculate the density of SiC:
(a) According to the diagram, the C atoms in the unit cell occupy a face-centered cubic arrangement.

(b) There are 8(1/8) + 6(1/2) = 4 face-centered cubic C atoms, and there are 4 Si atoms inside the unit cell. Thus, the unit cell has an atom ratio of 4 Si atoms per 4 C atoms or a formula of Si₄C₄, which is equivalent to SiC.

(c) The edge of the unit cell is 436.0 pm, what is the calculated density?
Calculate the volume of the unit cell:

$$436.0 \text{ pm} \left(\frac{1 \text{ m}}{1 \times 10^{12} \text{ pm}}\right)\left(\frac{100 \text{ cm}}{1 \text{ m}}\right) = 4.360 \times 10^{-8} \text{ cm} = \text{edge length}$$

Volume = $(4.360 \times 10^{-8} \text{ cm})^3 = 8.288 \times 10^{-23} \text{ cm}^3$.

You have to consider the mass of BOTH the Si and C atoms:

$$\text{Mass Si} = 4 \text{ atoms Si}\left(\frac{1 \text{ mol Si}}{6.02214 \times 10^{23} \text{ atoms Si}}\right)\left(\frac{28.085 \text{ g Si}}{1 \text{ mol Si}}\right) = 1.8654 \times 10^{-22} \text{ g Si}$$

$$\text{Mass C} = 4 \text{ atoms C}\left(\frac{1 \text{ mol C}}{6.02214 \times 10^{23} \text{ atoms C}}\right)\left(\frac{12.011 \text{ g C}}{1 \text{ mol C}}\right) = 7.9779 \times 10^{-23} \text{ g C}$$

$$\text{Density} = \text{mass/volume} = \frac{1.8654 \times 10^{-22} \text{ g Si} + 7.9779 \times 10^{-23} \text{ g C}}{8.288 \times 10^{-23} \text{ cm}^3} = 3.213 \text{ g/cm}^3$$

12.5. To determine the perovskite formula, determine the number of each atom belonging uniquely to the unit cell shown. The Ca atom is wholly contained within the unit cell. There are Ti atoms at each of the eight corners. Since each of these atoms belong to eight unit cells, the portion of each Ti atom belonging to the pictured unit cell is 1/8; 8 Ti atoms × 1/8 = 1 Ti atom. The O atoms on an edge belong to 4 unit cells, so the fraction contained within the pictured cell is 1/4. There are twelve such O atoms, leading to 12 × 1/4 = 3 O atoms. Thus, the formula is CaTiO₃.

12.6. Formula units of TiO₂ in the rutile unit cell:

8 corner Ti × 1/8	= 1 Ti	4 face O × 1/2	= 2 O
1 internal Ti	= 1 Ti	2 internal O	= 2 O
	= 2 Ti total		= 4 O total

There are two TiO₂ units per unit cell.

12.7. For cuprite:
(a) Formula for cuprite: There are 8 oxygen atoms at the corners of the cell (each 1/8 within the cell) and 1 oxygen atom wholly within the cell.
The number of oxygen atoms within the cell is then [(8 × 1/8) + (1)] or 2.
There are 4 Cu atoms wholly within the cell, so the ratio is Cu₂O.
(b) With a formula of Cu₂O, and the oxidation state of O = –2, the oxidation state of copper must be +1.

12.8. Fluorite unit cell and formula:

(a) Type of unit cell: Ca^{2+} ions are in a face-centered cubic environment.
(b) F^- ions located in tetrahedral holes
(c) Formula of fluorite:
(8 corner Ca^{2+} × 1/8) + (6 face Ca^{2+} × 1/2) = 4 Ca^{2+}
8 internal F^- = 8 F^-.
The formula = CaF_2.

12.9. Calculate the radius of a calcium atom, given that the density of solid Ca is 1.54 g/cm³ and that Ca crystallizes in face-centered cubic (fcc) cell:

- Because calcium crystallizes in an fcc unit cell, you know there are 4 calcium atoms in the unit cell: (6 faces × ½ in a face) + (8 corners × 1/8 at each corner) = 4 Ca atoms.
- Using the density, next calculate the volume of the unit cell:

$$\frac{4 \text{ atoms Ca}}{1 \text{ unit cell}} \left(\frac{1 \text{ mol Ca}}{6.022 \times 10^{23} \text{ atoms Ca}}\right)\left(\frac{40.078 \text{ g Ca}}{1 \text{ mol Ca}}\right)\left(\frac{1 \text{ cm}^3}{1.54 \text{ g Ca}}\right) = 1.73 \times 10^{-22} \text{ cm}^3$$

- The volume of the unit cell is (edge length)³, so the length of the edge is $(1.73 \times 10^{-22} \text{ cm}^3)^{1/3}$ or 5.57×10^{-8} cm
- The face diagonal of the unit cell is 4 × atomic radius or $\sqrt{2}$ × edge of the unit cell. (Refer to the diagram found in Study Questions 12.40 and 12.41 in this manual.)

$4 \times \text{radius} = \sqrt{2} \times 5.57 \times 10^{-8}$ cm

$\text{radius} = \dfrac{\sqrt{2} \times 5.57 \times 10^{-8} \text{ cm}}{4} = 1.97 \times 10^{-8}$ cm or 197 pm

12.10. Identify the unit cell for copper metal given its density and the radius of the Cu atom:

Determine mass of one Cu atom = $\dfrac{63.546 \text{ g}}{1 \text{ mol Cu}}\left(\dfrac{1 \text{ mol Cu}}{6.02214 \times 10^{23} \text{ atoms}}\right) = 1.0552 \times 10^{-22}$ g/atom

Radius of atom in cm = $127.8 \text{ pm}\left(\dfrac{1 \text{ m}}{1 \times 10^{12} \text{ pm}}\right)\left(\dfrac{100 \text{ cm}}{1 \text{ m}}\right) = 1.278 \times 10^{-8}$ cm

<u>Determine possible cell edge length for primitive, body-centered, and face-centered cubic unit cells:</u>

Primitive cubic unit cell: edge length = 2 × radius of atom = $2(1.278 \times 10^{-8} \text{ cm}) = 2.556 \times 10^{-8}$ cm

Body-centered cubic unit cell: edge length = $\dfrac{4 \times \text{radius of atom}}{\sqrt{3}} = \dfrac{4(1.278 \times 10^{-8} \text{ cm})}{\sqrt{3}} = 2.951 \times 10^{-8}$ cm

Face-centered cubit unit cell: edge length = $\dfrac{4 \times \text{radius of atom}}{\sqrt{2}} = \dfrac{4(1.278 \times 10^{-8} \text{ cm})}{\sqrt{2}} = 3.615 \times 10^{-8}$ cm

Determine possible volumes for primitive, body-centered, and face-centered cubic unit cells:

Primitive cubic unit cell volume: $V = (2.256 \times 10^{-8} \text{ cm})^3 = 1.670 \times 10^{-23} \text{ cm}^3$

Body-centered cubic unit cell volume: $V = (2.951 \times 10^{-8} \text{ cm})^3 = 2.571 \times 10^{-23} \text{ cm}^3$

Face-centered cubic unit cell volume: $V = (3.615 \times 10^{-8} \text{ cm})^3 = 4.723 \times 10^{-23} \text{ cm}^3$

Determine possible masses of primitive, body-centered, and face-centered cubit unit cells:

Primitive cubic unit cell mass: mass = 1 atom/unit cell$(1.0552 \times 10^{-22} \text{ g/atom}) = 1.0552 \times 10^{-22}$ g

Body-centered cubic unit cell mass: mass = 2 atoms/unit cell$(1.0552 \times 10^{-22} \text{ g/atom}) = 2.1104 \times 10^{-22}$ g

Face-centered cubic unit cell mass: mass = 4 atoms/unit cell$(1.0552 \times 10^{-22} \text{ g/atom}) = 4.2208 \times 10^{-22}$ g

Determine possible densities of primitive, body-centered, and face-centered cubit unit cells:

Primitive cubic unit cell density: density = $\dfrac{1.0552 \times 10^{-22} \text{ g}}{1.670 \times 10^{-23} \text{ cm}^3} = 6.319 \text{ g/cm}^3$

Body-centered cubic unit cell density: density = $\dfrac{2.1104 \times 10^{-22} \text{ g}}{2.571 \times 10^{-23} \text{ cm}^3} = 8.209 \text{ g/cm}^3$

Face-centered cubic unit cell density: density = $\dfrac{4.2208 \times 10^{-22} \text{ g}}{4.723 \times 10^{-23} \text{ cm}^3} = 8.937 \text{ g/cm}^3$

Because the face-centered unit cell density is closest to the density of Cu metal, the unit cell is most likely face-centered cubic.

12.11. Calculate the length of the side of a unit cell of KI given its density (3.12 g/cm^3):

KI has a face-centered cubic lattice of I$^-$ ions with K$^+$ ions in the octahedral holes. The unit cell will thus have a net of 4 I$^-$ ions and 4 K$^+$ ions. (See the NaCl lattice in Figure 12.8.)

Now calculate the volume of a unit cell of KI:

$\left(\dfrac{1 \text{ cm}^3}{3.12 \text{ g KI}}\right)\left(\dfrac{166.0 \text{ g KI}}{1 \text{ mol KI}}\right)\left(\dfrac{1 \text{ mol KI}}{6.02214 \times 10^{23} \text{ KI pairs}}\right)\left(\dfrac{4 \text{ KI pairs}}{1 \text{ unit cell}}\right) = 3.53 \times 10^{-22} \text{ cm}^3$

The unit cell length is the cube root of the volume. (The volume of a cube = length^3):
Edge length = $\sqrt[3]{3.53 \times 10^{-22} \text{ cm}^3} = 7.07 \times 10^{-8}$ cm

Edge of the KI unit cell (in pm) = $7.07 \times 10^{-8} \text{ cm}\left(\dfrac{1 \text{ m}}{100 \text{ cm}}\right)\left(\dfrac{1 \text{ pm}}{1 \times 10^{-12} \text{ m}}\right) = 707$ pm

12.12. Calculate the radius of the Cs$^+$ ion:
You know the CsCl lattice has 8 corner Cl$^-$ ions with 1/8 of each within the cell = 1 net Cl$^-$ ion. You also know the lattice has one Cs$^+$ ion within the cell.

- Calculate the unit cell volume:

$$\frac{1 \text{ CsCl}}{\text{unit cell}} \left(\frac{1 \text{ mol CsCl}}{6.02214 \times 10^{23} \text{ formula units}}\right) \left(\frac{168.36 \text{ g CsCl}}{1 \text{ mol CsCl}}\right) \left(\frac{1 \text{ cm}^3}{3.99 \text{ g CsCl}}\right) = 7.01 \times 10^{-23} \text{ cm}^3$$

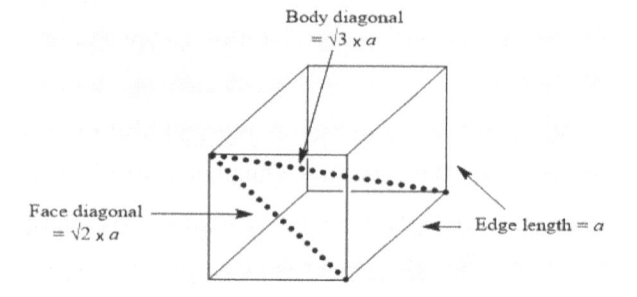

[Diagram of a cube showing: Body diagonal = $\sqrt{3} \times a$, Face diagonal = $\sqrt{2} \times a$, Edge length = a]

- Calculate the unit cell edge length:

(cell edge length)^3 = cell volume = 7.01×10^{-23} cm^3

edge length = $\sqrt[3]{7.01 \times 10^{-23} \text{ cm}^3}$ = 4.12×10^{-8} cm

4.12×10^{-8} cm $\left(\frac{1 \text{ m}}{100 \text{ cm}}\right)\left(\frac{1 \times 10^{12} \text{ pm}}{1 \text{ m}}\right)$ = 412 pm

- Calculate the face diagonal distance and from it the body diagonal distance.

(face diagonal)^2 = (edge length)^2 + (edge length)^2

face diagonal = $\sqrt{2}$ × edge length

(cube diagonal)^2 = (edge length)^2 + (face diagonal)^2

(cube diagonal)^2 = (edge length)^2 + ($\sqrt{2}$ · edge length)^2

cube diagonal = $\sqrt{3}$ · (edge length)

cube diagonal = $\sqrt{3}$ × 412 pm = 714 pm

- From the body diagonal distance calculate the Cs$^+$ radius.

cube diagonal = 714 pm = (2 × Cl$^-$ radius) + (2 × Cs$^+$ radius)

714 pm = (2 × 181 pm) + (2 × Cs$^+$ radius)

Therefore, the Cs$^+$ radius is **176 pm.**

Ionic Bonding and Lattice Energy

12.13. Arrange lattice energies from least negative to most negative:
Since CaO involves 2+ and 2– ions, the lattice energy for this compound is more negative than the other compounds. Given that lattice energy is **inversely** related to the distance between the ions, the lattice energy for LiI is more negative than that for RbI (because the Rb$^+$ ion is larger than the Li$^+$ ion). The small diameter of the fluoride ion (compared to

iodide) indicates that the lattice energy for LiF would be more negative than for either LiI or RbI. The lattice energies are then:

least ----- RbI ----- LiI ----- LiF ----- CaO **most**

negative   **negative**

12.14. Lattice energy depends directly on ion charges and inversely on the distance between ions. The sizes of the Cl$^-$, Br$^-$, and I$^-$ ions fall in a relatively narrow range (181, 196, and 220 pm, respectively), and the ion sizes change by only 15–24 pm from one ion to the next. Therefore, their lattice energies are expected to decrease in a narrow range. The F$^-$ ion (133 pm), however, is only 73% as large as the Cl$^-$ ion, so the lattice energy of NaF is much more negative.

12.15. Since melting a solid involves disassembling the crystal lattice of cations and anions, the smaller the distance between cations and anions, the greater the attraction between the cation and anion, and the harder it becomes to disassemble the lattice. Hence, the smaller the distance between the nuclei of cations and anions the **higher the melting point**.

12.16. Compound in each of the pairs with the higher melting point:

(a) NaCl   Na$^+$ has a smaller radius than the Rb$^+$ ion
(b) MgO   Mg^{2+} has a smaller radius than the Ba^{2+} ion
(c) MgS   Mg^{2+} and S^{2-} are divalent whereas NaCl is composed of monovalent ions. The larger the ion charge the greater the attraction.

12.17. Calculate the molar enthalpy of formation of solid lithium fluoride knowing its lattice enthalpy and other thermodynamic information:

1/2 F$_2$(g)	→	F(g)	$\Delta_fH°$ =	+78.99 kJ/mol
F(g) + e$^-$	→	F$^-$(g)	$\Delta_{ea}H$ =	–328.0 kJ/mol
Li(s)	→	Li(g)	$\Delta_fH°$ =	+159.37 kJ/mol
Li(g)	→	Li$^+$(g) + e$^-$	IE =	+520. kJ/mol
Li$^+$(g) + F$^-$(g)	→	LiF(s)	$\Delta_{lattice}H$ =	–1037 kJ/mol
Li(s) + 1/2 F$_2$(g)	→	LiF(s)	$\Delta_fH°$ =	–607 kJ/mol

12.18. Calculate the lattice enthalpy of RbCl knowing the enthalpy of formation of RbCl and other thermodynamic information.

1/2 Cl$_2$(g)	→	Cl(g)	$\Delta_f H°$	=	+121.3 kJ/mol
Cl(g) + e$^-$	→	Cl$^-$(g)	$\Delta_{ea} H$	=	–349.0 kJ/mol
Rb(s)	→	Rb(g)	$\Delta_f H°$	=	+80.9 kJ/mol
Rb(g)	→	Rb$^+$(g) + e$^-$	IE	=	+403 kJ/mol
Rb$^+$(g) + Cl$^-$(g)	→	RbCl(s)	$\Delta_{lattice} H$	=	x
Rb(s) + 1/2 Cl$_2$(g)	→	RbCl(s)	$\Delta_f H°$	=	–435.4 kJ/mol

$\Delta_f H°$(RbCl) = $\Delta_f H°$(Cl) + $\Delta_{ea} H$(Cl) + $\Delta_f H°$(Rb(g)) + IE(Rb) + $\Delta_{lattice} H$
Solving for $\Delta_{lattice} H$:
$\Delta_{lattice} H$ = x = $\Delta_f H°$(RbCl) – $\Delta_f H°$(Cl) – $\Delta_{ea} H$(Cl) – $\Delta_f H°$(Rb(g)) – IE(Rb)
$\Delta_{lattice} H$ = –435.4 kJ/mol – 121.3 kJ/mol – (–349.0 kJ/mol) – 80.9 kJ/mol – 403 kJ/mol
= – 692 kJ/mol

Metals and Semiconductors

12.19. Using the 2s atomic orbitals of 1000 Li atoms, 1000 molecular orbitals can be formed. (See Figure 12.13.) In the lowest energy state, half of these orbitals will be populated by pairs of electrons and half will be empty.

12.20. Molecular orbitals:
(4 orbitals/Mg atom)(6.022 × 10^{23} Mg atoms/mol) = 2.409 × 10^{24} orbitals/mol.
Each Mg atom has a total of two electrons occupying its four valence orbitals. The molecular orbitals will similarly be ¼ full.

12.21. How does band theory for metallic bonding explain conductivity?
Metals have very closely spaced molecular orbitals. Small amounts of energy result in promoting electrons into higher (unoccupied) levels, creating a space (called a hole) that the electron had occupied. Electrical conductivity arises from the presence of the holes and "promoted" electrons in singly occupied states in the presence of an applied electric field. Current results as the "promoted" electrons move toward the "+" side of the applied field, and the holes move toward the "–" side of the applied electric field.

12.22. Explanation of shininess of metals:
A metal can absorb energy of nearly any wavelength, causing an electron to move to a higher energy state. The now-excited system can immediately emit a photon of the same energy as the electron returns to the original energy level. This rapid and efficient absorption and reemission of light makes metals appear shiny.

12.23. Diamond is an insulator and silicon is a semi-conductor:
The band gap for silicon is small enough to permit electrons to be promoted from the valence band to the conduction band, while the band gap for carbon is too great to permit this to occur.

12.24. The band gap gets smaller with increased metallic character, so the band gap order is C > Si > Ge > Sn.

12.25. Intrinsic semiconductors are those in which the semiconducting phenomenon is naturally occurring (e.g., silicon), while extrinsic semiconductors are those in which dopants are added. For example, extrinsic conductors are made by the addition of a Group 3A (13) or Group 5A (15) atom to a "base" material from Group 4A (14).

12.26. Classify aluminum-doped silicon as an *n*- or *p*-type semiconductor:
Aluminum has one less valence electron than silicon. The Si–Al bonds form a discrete but empty energy band called the acceptor band at a level higher in energy than the valence band but lower than the conduction band. Electrons can be promoted from the valence band to the acceptor band with concurrent generation of positive holes in the valence band. This is a *p*-type semiconductor.

Other Types of Solids

12.27. Allotropes of carbon that are **not** network solids: (c) buckyballs (C_{60}). With a molecular structure with a finite number of atoms, buckyballs are not network solids, in contrast to the other substances listed.

12.28. With a low melting point, this solid is **not** a network or ionic solid. Given that it has a wide melting point range, one could classify this substance as (c) an amorphous solid.

12.29. For the unit cell of diamond:

(a) The unit cell has 8 corner atoms (1/8 in the cell), 6 face atoms (1/2 in the cell), and 4 atoms wholly within the cell, for a total of **8 carbon atoms**.

(b) Diamond uses a fcc unit cell (The structure shown also has 4 atoms occupying holes in the lattice). The holes are **tetrahedral**.

12.30. Describe graphite:

(a) Intermolecular forces between layers: induced dipole–induced dipole
(b) Explain the slippery nature of graphite: Since there are only weak intermolecular forces between the layers in graphite, the layers slide over each other easily, giving graphite a slippery feel. Pushing a pencil tip against paper causes some of the carbon layers to rub off, leaving a black mark.

12.31. Classify the following substances:

	Classification	Particles	Attractive Forces	Physical Property
(a) gallium arsenide	Network	Atoms	Covalent bonds with some ionic character	semiconductor
(b) polystyrene	Amorphous	Atoms	Covalent bonds	Wide melting range
(c) silicon carbide	Network	Atoms	Covalent bonds	Very hard substance
(d) perovskite	Ionic	Cations and Anions	Ion–Ion attractions	Brittle; high-melting

12.32. Classify each as one of the types of categories in Table 12.2:

	Classification	Particles	Attractive Forces	Physical Property
(a) Si doped with P	Network	Atoms	Covalent bonds	semiconductor
(b) graphite	Network	Atoms	Covalent bonds	Brittle; high-melting
(c) benzoic acid	Molecular	Molecules	Covalent bonds	Poor conductor
(d) Na$_2$SO$_4$	Ionic	Cations and Anions	Ion–Ion attractions	Brittle; high-melting

Phase Changes and Phase Diagrams

12.33. The energy evolved as heat when 15.5 g of benzene freezes at 5.5 °C:

The enthalpy of fusion is 9.95 kJ/mol. This is the enthalpy change for melting one mole of benzene. Freezing is the opposite process as melting, so the enthalpy change for that process, the enthalpy of crystallization, is –9.95 kJ/mol.

$$15.5 \text{ g benzene} \left(\frac{1 \text{ mol benzene}}{78.11 \text{ g benzene}}\right)\left(\frac{-9.95 \text{ kJ}}{1 \text{ mol benzene}}\right) = -1.97 \text{ kJ}$$

Note that the negative sign indicates that energy is evolved as heat.

The energy needed to remelt this 15.5 g sample of benzene would be +1.97 kJ.

12.34. Quantity of energy to change 5.00 g of Ag(s) at 25 °C to a liquid at 962 °C:
The total energy required is the sum of the energy required to (1) heat the solid from 25 °C to its melting point and (2) liquefy the solid at its melting point.

q_{solid} = (5.00 g Ag)(0.235 J/g · K)(1235 K − 298 K) = 1.10×10^3 J

$q_{melting} = 5.00 \text{ g Ag} \cdot \dfrac{1 \text{ mol Ag}}{107.9 \text{ g Ag}} \cdot \dfrac{11.3 \text{ kJ}}{1 \text{ mol Ag}} \cdot \dfrac{10^3 \text{ J}}{1 \text{ kJ}} = 524 \text{ J}$

$q_{total} = (1.10 \times 10^3 \text{ J}) + 524 \text{ J} = 1.62 \times 10^3$ J or 1.62 kJ

12.35. Regarding the phase diagram for carbon dioxide:
(a) The positive slope of the solid/liquid equilibrium line means the liquid CO_2 is **less dense** than solid CO_2.
(b) At 5 atm and 0 °C, CO_2 is in the **gaseous phase**.
(c) The phase diagram for CO_2 shows the critical pressure for CO_2 to be 73 atm, and the critical temperature to be +31 °C, so CO_2 cannot be liquefied at 45 °C.

12.36. Using the phase diagram, describe the state of this substance under different conditions of temperature and pressure:

(a) The substance is a gas at room temperature and 1.0 atm pressure.
(b) The substance is a liquid at 0.75 atm pressure and −114 °C.
(c) When the pressure on a sample of liquid substance is 380 mm Hg (0.5 atm), the temperature is between −117 and −119 °C.
(d) At −122 °C, the vapor pressure of solid substance is 0.25 atm.
(e) The solid phase is denser than the liquid phase because the solid–liquid equilibrium line has a positive slope.

12.37. The heat required is a summation of three steps:
(a) heat the liquid (at −50.0 °C) to its boiling point (−33.3 °C),
(b) boil the liquid, converting it to a gas at its boiling point, and
(c) warm the gas from −33.3 °C to 0.0 °C.

1. To heat the liquid (at −50.0 °C) to its boiling point (−33.3 °C):

$q_{liquid} = (1.2 \times 10^4 \text{ g})(4.7 \dfrac{\text{J}}{\text{g} \cdot \text{K}})(239.9 \text{ K} - 223.2 \text{ K}) = 9.4 \times 10^5$ J

2. To boil the liquid:

$q_{boil} = (23.33 \times 10^3 \dfrac{\text{J}}{\text{mol}})(\dfrac{1 \text{ mol NH}_3}{17.03 \text{ g NH}_3})(1.2 \times 10^4 \text{ g}) = 1.6 \times 10^7$ J

3. To heat the gas from −33.3 °C to 0.0 °C:

$q_{gas} = (1.2 \times 10^4 \text{ g})(2.2 \dfrac{\text{J}}{\text{g} \cdot \text{K}})(273.2 \text{ K} - 239.9 \text{ K}) = 8.8 \times 10^5$ J

The total heat required is:
$9.4 \times 10^5 \text{ J} + 1.6 \times 10^7 \text{ J} + 8.8 \times 10^5 \text{ J} = 1.8 \times 10^7$ J or 1.8×10^4 kJ

12.38. The heat evolved is a summation of three steps:
(a) cool the gas from 40.0 °C to its boiling point,
(b) condense the gas at its boiling point, and
(c) cool the liquid to –40.0 °C.

1. To cool the gas (at 40.0 °C) to its boiling point (–29.8 °C):

$$q_{gas} = 20.0 \text{ g} \left(\frac{1 \text{ mol CCl}_2\text{F}_2}{120.9 \text{ g CCl}_2\text{F}_2} \right) \left(\frac{117.2 \text{ J}}{\text{mol} \cdot \text{K}} \right) (243.4 \text{ K} - 313.2 \text{ K}) = -1350 \text{ J}$$

2. To condense the gas to a liquid:

$$q_{condensation} = 20.0 \text{ g} \left(\frac{1 \text{ mol CCl}_2\text{F}_2}{120.9 \text{ g CCl}_2\text{F}_2} \right) \left(\frac{-20.11 \text{ kJ}}{\text{mol}} \right) \left(\frac{10^3 \text{ J}}{1 \text{ kJ}} \right) = -3330 \text{ J}$$

3. Cool the liquid to –40.0 °C.

$$q_{liquid} = 20.0 \text{ g} \left(\frac{1 \text{ mol CCl}_2\text{F}_2}{120.9 \text{ g CCl}_2\text{F}_2} \right) \left(\frac{72.3 \text{ J}}{\text{mol} \cdot \text{K}} \right) (233.2 \text{ K} - 243.4 \text{ K}) = -122 \text{ J}$$

The total energy released as heat is:
$-1350 \text{ J} + (-3330 \text{ J}) + (-122 \text{ J}) = -4.80 \times 10^3 \text{ J}$

General Questions

12.39. The estimated vapor pressure at 77 K

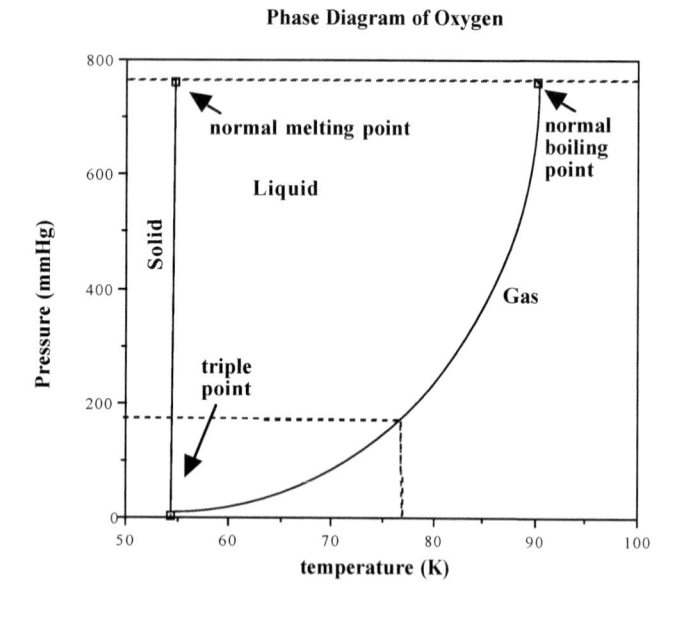

Phase Diagram of Oxygen

(–196 °C) is between 150–200 mm Hg.

The very slight positive slope of the solid/liquid equilibrium line indicates the **solid is denser than the liquid.**

12.40. Concerning tungsten:

(a) unit cell: body-centered cubic
(b) atoms/unit cell: (8 corner W × 1/8) + (1 internal W) = 2 W atoms/unit cell
(c) radius of W atom:
(cube diagonal)^2 = (edge length)^2 + (face diagonal)^2
(cube diagonal)^2 = (edge length)^2 + ($\sqrt{2}$ × edge length)^2
cube diagonal = $\sqrt{3}$ × (edge length)
cube diagonal = $\sqrt{3}$ × 316.5 pm
diagonal = 548 pm

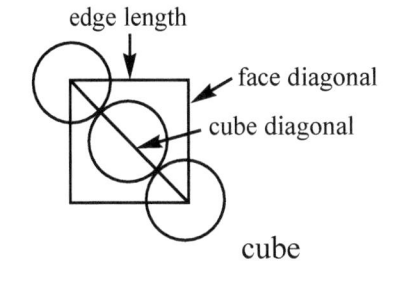

radius = $\dfrac{548 \text{ pm}}{4}$ = 137 pm

The reported radius of the tungsten atom is 137 pm.

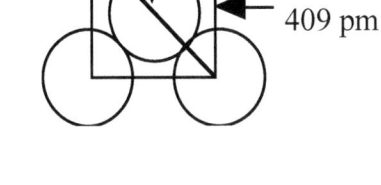

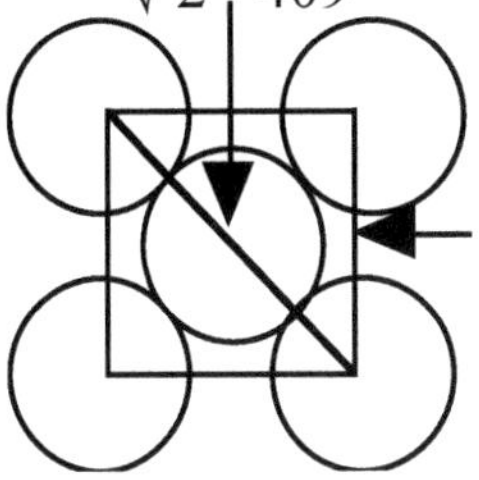

12.41. Silver crystallizes in a face-centered cubic cell with a side of 409 pm. The radius of a silver atom can be found by examining the geometry of a face of the cell.
The diagram shows such a face with an edge of 409 pm. The diagonal distance across the face is the hypotenuse of a right triangle, two sides of which are 409 pm.
hypotenuse^2 = (409 pm)^2 + (409 pm)^2
Let d be the hypotenuse, d^2 = 2 × (409 pm)^2. The distance d equals 1 diameter (of the center atom) and two-halves of two other atoms (or a total of 4 silver radii).
d = $\sqrt{2}$ × 409 pm or (1.414 × 409 pm) = 578 pm (3sf)
So 4 radii = 578 pm and 1 radius = 145 pm (3sf). This radius compares favorably with the reported radius of Ag.

12.42. Use the unit cell of calcium carbide to decide on the number of Ca atoms and the number of C atoms in the cell.

8 corner Ca × 1/8 = 1 Ca   8 edge C × 1/4 = 2 C
1 internal Ca   = 1 Ca   2 internal C   = 2 C
                  = 2 Ca total   = 4 C total

With a ratio of two calcium ions to four carbon ions, the formula is CaC$_2$.

12.43. Determine density of austenite:
Determine the length of the cell diagonal:
cell diagonal length = 2×radius of Ni + 2×radius of Ti

cell diagonal length = $2(125 \text{ pm}) + 2(145 \text{ pm}) = 540 \text{ pm} = 5.40 \times 10^{-8}$ cm

Determine unit cell length from edge diagonal length
This unit cell is a modified body-centered cubic unit cell

cell edge length = $\dfrac{\text{cell diagonal length}}{\sqrt{3}} = \dfrac{5.40 \times 10^{-8} \text{ cm}}{\sqrt{3}} = 3.12 \times 10^{-8}$ cm

Determine unit cell volume
$V = (3.12 \times 10^{-8} \text{ cm})^3 = 3.03 \times 10^{-23} \text{ cm}^3$

Determine mass of unit cell
This unit cell contains 1 Ni atom plus 8 × 1/8 = 1 Ti atom.

mass of 1 Ni atom = $\dfrac{58.693 \text{ g Ni}}{1 \text{ mol Ni}} \left(\dfrac{1 \text{ mol Ni}}{6.02214 \times 10^{23} \text{ Ni atoms}} \right) = 9.7462 \times 10^{-23}$ g/Ni atom

mass of 1 Ti atom = $\dfrac{47.867 \text{ g Ti}}{1 \text{ mol Ti}} \left(\dfrac{1 \text{ mol Ti}}{6.02214 \times 10^{23} \text{ Ti atoms}} \right) = 7.9485 \times 10^{-23}$ g/Ni atom

mass of unit cell = 9.7462×10^{-23} g + 7.9485×10^{-23} g = 1.76947×10^{-22} g

Determine density of austenite
density = $\dfrac{\text{mass}}{\text{volume}} = \dfrac{1.76947 \times 10^{-22} \text{ g}}{3.03 \times 10^{-23} \text{ cm}^3} = 5.84 \text{ g/cm}^3$

12.44. Questions based upon nitrogen properties:
(a) Since the critical point for N_2 is 33.5 atm and 126.0 K, at temperatures and pressures above that N_2 is a **supercritical fluid**.
(b) N_2 does not exist as a liquid at pressures below the triple point pressure of **0.127 atm**.
(c) At 18.5 atm and 55.9 K N_2 is a **solid**. Temperature is below the melting and triple points.
(d) At 0.127 atm and 84.6 K N_2 is a **gas**. Temperature is above the boiling and triple points.
(e) At 1.00 atm and 71.4 K N_2 is a **liquid**. Temperature is below the boiling point.
(f) Solid N_2 is **more** dense than liquid N_2. Positive slope for solid-liquid equilibrium indicates that solid is denser than the liquid.

12.45. Density for iridium = 22.56 g/cm³. What is the radius of an iridium atom?
Ir crystallizes in the fcc lattice so there are 4 atoms in the unit cell.
(8 × 1/8 corner atoms + 6 × 1/2 face atoms).

The mass/unit cell is: $\dfrac{4 \text{ Ir atoms}}{1 \text{ unit cell}} \left(\dfrac{192.22 \text{ g Ir}}{6.0221 \times 10^{23} \text{ atoms}} \right) = 1.2768 \times 10^{-21}$ g/unit cell

The volume of the unit cell is:

$$\frac{1.2768 \times 10^{-21} \text{ g}}{1 \text{ unit cell}} \left(\frac{1 \text{ cm}^3}{22.56 \text{ g}}\right) = 5.659 \times 10^{-23} \text{ cm}^3$$

Since the cell is a cube, determine an edge dimension by taking the cube root of this volume:

$$\text{Edge} = \sqrt[3]{5.659 \times 10^{-23} \text{ cm}^3} = 3.839 \times 10^{-8} \text{ cm}$$

The face diagonal of the unit cell (= 4 × Ir radius) = $\sqrt{2}\,(3.839 \times 10^{-8} \text{ cm})$

$$\text{radius} = \frac{\sqrt{2}\,(3.839 \times 10^{-8} \text{ cm})}{4} = 1.357 \times 10^{-8} \text{ cm} = 135.7 \text{ pm}.$$

This value compares favorably with the literature value of 136 pm.

12.46. Determine if vanadium has a primitive, body-centered, or face-centered cubic unit cell.

$$\text{Determine mass of one V atom} = \frac{50.942 \text{ g}}{1 \text{ mol V}}\left(\frac{1 \text{ mol V}}{6.02214 \times 10^{23} \text{ atoms}}\right) = 8.4591 \times 10^{-23} \text{ g/V atom}$$

$$132 \text{ pm}\left(\frac{1 \text{ m}}{1 \times 10^{12} \text{ pm}}\right)\left(\frac{100 \text{ cm}}{1 \text{ m}}\right) = 1.32 \times 10^{-8} \text{ cm}$$

Determine possible cell edge length for primitive, body-centered, and face-centered cubic unit cells

Primitive unit cells: edge length = 2 × radius of atom = $2(1.32 \times 10^{-8} \text{ cm}) = 2.64 \times 10^{-8}$ cm

Body-centered unit cells : edge length = $\dfrac{4 \times \text{radius of atom}}{\sqrt{3}} = \dfrac{4(1.32 \times 10^{-8} \text{ cm})}{\sqrt{3}} = 3.05 \times 10^{-8}$ cm

Face-centered unit cells : edge length = $\dfrac{4 \times \text{radius of atom}}{\sqrt{2}} = \dfrac{4(1.32 \times 10^{-8} \text{ cm})}{\sqrt{2}} = 3.73 \times 10^{-8}$ cm

Determine possible volumes for primitive, body-centered, and face-centered cubic unit cells

Primitive cubic unit cell volume: $V = (2.64 \times 10^{-8} \text{ cm})^3 = 1.84 \times 10^{-23} \text{ cm}^3$

Body-centered cubic unit cell volume: $V = (3.05 \times 10^{-8} \text{ cm})^3 = 2.83 \times 10^{-23} \text{ cm}^3$

Face-centered cubic unit cell volume: $V = (3.73 \times 10^{-8} \text{ cm})^3 = 5.20 \times 10^{-23} \text{ cm}^3$

Determine possile masses of primitive, body-centered, and face-centered cubic unit cells

Primitive cubic unit cell mass: mass = 1 atom/unit cell$(8.4591 \times 10^{-23} \text{ g/atom}) = 8.4591 \times 10^{-23}$ g

Body-centered cubic unit cell mass: mass = 2 atoms/unit cell$(8.4591 \times 10^{-23} \text{ g/atom}) = 1.6918 \times 10^{-22}$ g

Face-centered cubic unit cell mass: mass = 4 atoms/unit cell$(8.4591 \times 10^{-23} \text{ g/atom}) = 3.3836 \times 10^{-22}$ g

Determine possible densities of primitive, body-centered, and face-centered cubic unit cells

Primitive cubic unit cell density: density = $\dfrac{8.4591 \times 10^{-23} \text{ g}}{1.84 \times 10^{-23} \text{ cm}^3}$ = 4.60 g/cm^3

Body-centered cubic unit cell density: density = $\dfrac{1.6918 \times 10^{-22} \text{ g}}{2.83 \times 10^{-23} \text{ cm}^3}$ = 5.97 g/cm^3

Face-centered cubic unit cell density: density = $\dfrac{3.3836 \times 10^{-22} \text{ g}}{5.20 \times 10^{-23} \text{ cm}^3}$ = 6.50 g/cm^3

The chemical literature reports that the unit cell is body-centered cubic with a cell edge of 305 pm and a densty of 6.11 g/cm^3.

12.47. Given that the edge of a CaF$_2$ unit cell is 5.46295 × 10^{-8} cm, and the density of the solid is 3.1805 g/cm^3, calculate the value of Avogadro's number.

The unit cell contains 4 calcium ions and 8 fluoride ions (or 4 CaF$_2$ ion pairs). With the density and the length of the unit cell, you can calculate the mass of these 4 ion pairs.

$(5.46295 \times 10^{-8} \text{ cm})^3 \left(\dfrac{3.1805 \text{ g}}{1 \text{ cm}^3}\right) = 5.1853 \times 10^{-22}$ g

The mass of one ion pair would be 1/4 of that mass or 1.2963 × 10^{-22} g. If you add the atomic masses of Ca and 2 F you get a molar mass of 78.074 g. Since this is the mass corresponding to Avogadro's number of formula units, you can calculate the number of formula units of CaF$_2$ per mole:

$\dfrac{78.074 \text{ g CaF}_2}{1 \text{ mol CaF}_2}\left(\dfrac{1 \text{ CaF}_2 \text{ formula unit}}{1.2963 \times 10^{-22} \text{ g CaF}_2}\right) = 6.0227 \times 10^{23}$ formula units/mol CaF$_2$

12.48. Calculate the mass of one iron atom and compare this to the mass of one mole of iron.

$V_{\text{unit cell}} = \left[286.65 \text{ pm}\left(\dfrac{1 \text{ m}}{10^{12} \text{ pm}}\right)\left(\dfrac{100 \text{ cm}}{1 \text{ m}}\right)\right]^3 = 2.3554 \times 10^{-23}$ cm^3

$2.3554 \times 10^{-23} \text{ cm}^3 \left(\dfrac{7.874 \text{ g}}{1 \text{ cm}^3}\right)\left(\dfrac{1 \text{ unit cell}}{2 \text{ Fe atoms}}\right) = 9.2730 \times 10^{-23} \dfrac{\text{g}}{\text{Fe atom}}$

$\dfrac{55.845 \dfrac{\text{g}}{\text{mol Fe}}}{9.2730 \times 10^{-23} \dfrac{\text{g}}{\text{Fe atom}}} = 6.0223 \times 10^{23}$ Fe atoms/mol

12.49. For the two unit cells, determine the amount of filled space in each.

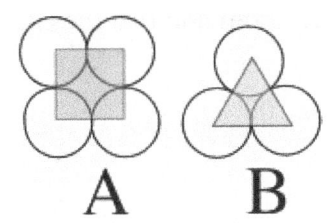

Assign a diameter of each of the atoms = d.

For unit cell A:
The length of the square inscribed is equal to 2 radii or 1 diameter (d).
The area associated with the square is then $d \times d$ or d^2.
Each circle has **one-fourth** of its area covered by the inscribed square.
The area of a circle is πr^2. So the r (the radius) = $d/2$.
Only one-fourth of each circle is covered by the portion of the inscribed square, so the area covered by the inscribed square is $1/4 \times (\pi d^2)/4$.
Noting that there are 4 circles, each of which has the area we've calculated, the **total** area covered by the inscribed square is $(\pi d^2)/4$.
The area **not covered** is the *difference* between the area of the square (d^2) and the area of the circles inside the square ($\pi d^2/4$): $d^2 - \pi d^2/4$.
Arbitrarily assign a length to d (say 2 inches). Then the area not covered is $(2)^2 - 3.14(2)^2/4$ or $4 - 3.14 = 0.86$ sq. in.
The amount of coverage is 3.14/4.00 or 78.5%.

For unit cell B:

Once again, assign a diameter of d to each circle. The equilateral triangle shown here and in B above covers a portion of each of the three neighboring atoms in the cell. Each interior angle (at the corners of the triangle) is 60 degrees.

Since there are 360 degrees in a circle, this corresponds to 1/6 of each of the circles covered by the triangle. With three circles involved, $3 \times 1/6 = 1/2$ of the area of the 3 circles.
The area of a circle (in terms of the diameter) is $(\pi d^2)/4$, so the total area of the 3 circles in B covered by the inscribed triangle is $1/2 \cdot (\pi d^2)/4$ or $(\pi d^2)/8$.
The area of a triangle is $(1/2\ b \times h)$, where b = base and h = height.
The base of the triangle is equal to $2r$, as shown in the diagram above.
The height can be calculated using the Pythagorean Theorem, since you know the length of the hypotenuse (d) and the base of the triangle, r.
$d^2 = r^2 + h^2$ and rearranging, $d^2 - r^2 = h^2$.
Noting that the radius, r, is $(d/2)$, $d^2 - (d/2)^2 = h^2$ or $3/4(d)^2 = h^2$.
Substituting the arbitrary value of 2 for d, gives $3/4(2)^2 = h^2$ or $3 = h^2$ and $1.732 = h$.
The area of the triangle is: $1/2 \times d \times h$, and A = $1/2 \times 2 \times 1.732 = 1.732$ sq. inches
The area of the circles (covered by the triangle) is $(\pi d^2)/8$, and substituting the arbitrary value 2 for d gives: $(\pi \times 4)/8$ or $3.14/2 = 1.57$ sq in.
The percent occupied is then $1.57/1.732 \cdot 100 = 90.7\%$.

12.50. Calculate the percentage of occupied space in the three cubic unit cells.
  (a) Recall that a primitive cubic unit cell contains one net atom and that the edge length of the cube = 2r.

  $$\% \text{ occupied space} = \frac{\text{volume occupied by spheres}}{\text{total volume of unit cell}} \times 100\%$$

  $$= \frac{1 \text{ sphere}\left(\frac{4}{3}\pi r^3\right)}{(2r)^3} \times 100\%$$

  $$= 52\%$$

  (b) Recall that a body-centered cubic unit cell contains two net atoms. In addition, the diagonal across the cube = $4r$ = edge × $\sqrt{3}$, so edge = $\frac{4r}{\sqrt{3}}$.

  $$\% \text{ occupied space} = \frac{\text{volume occupied by spheres}}{\text{total volume of unit cell}} \times 100\%$$

  $$= \frac{2 \text{ spheres}\left(\frac{4}{3}\pi r^3\right)}{\left(\frac{4r}{\sqrt{3}}\right)^3} \times 100\%$$

  $$= 68\%$$

  Recall that a face-centered cubic unit cell contains four net atoms. In addition, the diagonal across a face = $4r$ = edge × $\sqrt{2}$, so edge = $\frac{4r}{\sqrt{2}}$.

  $$\% \text{ occupied space} = \frac{\text{volume occupied by spheres}}{\text{total volume of unit cell}} \times 100\%$$

  $$= \frac{4 \text{ spheres}\left(\frac{4}{3}\pi r^3\right)}{\left(\frac{4r}{\sqrt{2}}\right)^3} \times 100\%$$

  $$= 74\%$$

  Both will be denser than a primitive cell for a given atom or molecule. The face-centered cubic unit cell is the densest of the three, and the primitive cubic unit cell is the least dense.

12.51. The structure of silicon is shown here:
  (a) Note that the unit cell has Si atoms at each corner and atoms in each face. The cell is **face-centered cubic (fcc).**

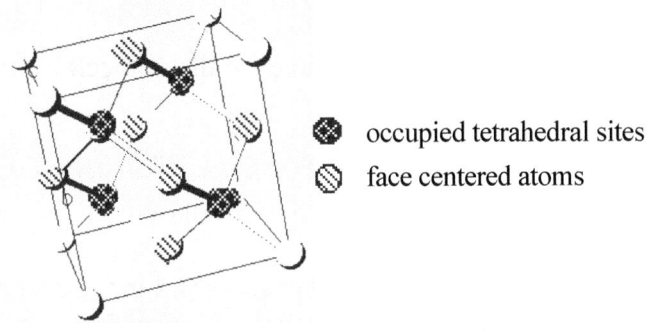

(b) Four of the eight possible tetrahedral holes are occupied.
(c) With 8 Si atoms at the corner (each with a contribution of 1/8), and 6 atoms in the faces (each with a contribution of 1/2), and 4 Si atoms (in the occupied tetrahedral sites), there is a total of 8 Si atoms in a unit cell.
(d) The density can be calculated by first calculating the mass and volume of a unit cell. The volume is: (unit cell length)^3.

$$543.1 \text{ pm} \left(\frac{1 \times 10^{-12} \text{ m}}{1 \text{ pm}} \right) \left(\frac{100 \text{ cm}}{1 \text{ m}} \right) = 5.431 \times 10^{-8} \text{ cm}$$

$V = (5.431 \times 10^{-8} \text{ cm})^3 = 1.602 \times 10^{-22} \text{ cm}^3$

The mass of the unit cell:

$$\frac{28.085 \text{ g}}{1 \text{ mol Si}} \left(\frac{1 \text{ mol Si}}{6.02214 \times 10^{23} \text{ atoms Si}} \right) \left(\frac{8 \text{ atoms Si}}{1 \text{ unit cell}} \right) = 3.7309 \times 10^{-22} \text{ g/unit cell}$$

The density is: 3.7309×10^{-22} g / 1.602×10^{-22} cm^3 = 2.329 g/cm^3

(e) Estimate the radius of the Si atom.
Note that a Si atom in a tetrahedral hole is bonded to a Si atom at a corner and in a face.

The diagonal distance across a cell face is equal to $\sqrt{2}(543.1 \text{ pm}) = 768.1$ pm. Therefore, the distance from a Si atom at a corner and one in a face is 384.0 pm.

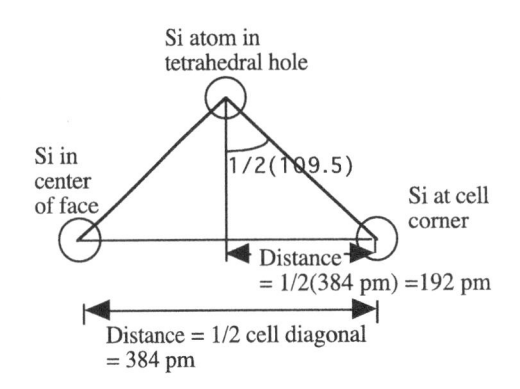

The Si atoms in the face do not touch one another. You need the distance between an atom in a hole and one at a corner or in the face at an angle of 109.5°. To calculate this distance, recognize that the sine of the angle of 54.75 (half of 109.5) is equal to the ratio of 192.0 pm/ Si-Si distance (where 192.0 pm is ½ the Si separation on the face).

$$\sin(54.75) = \frac{192 \text{ pm}}{\text{Distance}} \text{ (where } \sin(54.75) = 0.8166\text{)}$$

Therefore, the Si–Si bond distance = 192.0 pm/0.8166 = or 235.1 pm.
Note that this distance corresponds to two Si radii, so radius of one Si atom is 235.1 pm/2 or 117.6 pm. (The literature value for the Si radius is 117 pm.)

12.52. Solid state structure of silicon carbide:

(a) Atoms of each type in the unit cell: 4 Si are in tetrahedral holes in the unit cell.
8 C atoms are on the corners of the unit cell: 8 × 1/8 = 1 C
6 C atoms are on the faces of the unit cell: 6 × 1/2 = 3 C
Net C atoms in the unit cell = 4
In total, there is a net of 4 Si atoms and 4 C atoms, giving the formula SiC.

(b) Use the geometric relationship between the C atoms at cell corners and in the center of the cell faces and the Si atoms in tetrahedral holes along with the Si–C distance to calculate the length of the face diagonal. (See the solution to 12.51 above.)

$$\sin\left(\frac{109.5°}{2}\right) = \frac{1/4(\text{face diagonal})}{\text{Si–C distance}} = \frac{1/4(\text{face diagonal})}{188.8 \text{ pm}}$$

face diagonal = 616.7 pm

$$\text{edge length} = \frac{\text{face diagonal}}{\sqrt{2}} = \frac{616.7 \text{ pm}}{\sqrt{2}} = 436.1 \text{ pm}$$

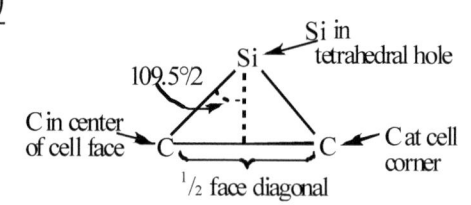

Volume = $(4.361 \times 10^{-8} \text{ cm})^3 = 8.293 \times 10^{-23} \text{ cm}^3$

$$\frac{4 \text{ molecules SiC}}{1 \text{ unit cell}} \left(\frac{40.096 \text{ g}}{1 \text{ mol SiC}}\right)\left(\frac{1 \text{ mol SiC}}{6.0221 \times 10^{23} \text{ molecules SiC}}\right) = 2.663 \times 10^{-22} \frac{\text{g}}{\text{unit cell}}$$

$$\text{Density} = \frac{2.663 \times 10^{-22} \text{ g}}{8.293 \times 10^{-23} \text{ cm}^3} = 3.211 \text{ g/cm}^3$$

12.53. Normal spinel crystal structure:
(a) Mg^{2+} and Al^{3+} ions are present in $MgAl_2O_4$. Mg^{2+} ions occupy the 1/8 of the tetrahedral sites and Al^{3+} ions are in 1/2 of the octahedral sites.

(b) For chromite, Fe^{2+} and Cr^{3+} ions are present. Fe^{2+} occupies 1/8 of the tetrahedral sites and Cr^{3+} occupies ½ of the octahedral sites.

12.54. Second electron affinity [$\Delta_{ea}H(2)$] of oxygen:

1/2 O_2(g)	→	O(g)	Δ_fH	=	+249.1 kJ/mol
O(g) + e⁻	→	O⁻(g)	$\Delta_{ea}H(1)$	=	−141.0 kJ/mol
O⁻(g) + e⁻	→	O^{2-}(g)	$\Delta_{ea}H(2)$	=	x
2 Na(s)	→	2 Na(g)	2 $\Delta_{atom}H$	=	2(+107.3) kJ/mol
2 Na(g)	→	Na⁺(g) + 2 e⁻	2 IE	=	2(+495.9) kJ/mol
2 Na⁺(g) + O^{2-}(g)	→	Na_2O(s)	$\Delta_{lattice}H$	=	−2481 kJ/mol
2 Na(s) + 1/2 O_2(g)	→	Na_2O(s)	Δ_fH	=	−418.0 kJ/mol

–418.0 kJ/mol = 249.1 kJ/mol + (–141.0 kJ/mol) + x + 2(107.3 kJ/mol) + 2 (+495.9 kJ/mol) + –2481 kJ/mol

x = –418.0 kJ/mol – 249.1 kJ/mol – (–141.0 kJ/mol) – 2(107.3 kJ/mol)
 – 2(495.9 kJ/mol) – (–2481 kJ/mol)

$x = \Delta_{ea}H(2)(O)$ = 749 kJ/mol

12.55. The band gap in gallium arsenide is 140 kJ/mol. What is the maximum wavelength of light needed to excite an electron to move from the valence band to the conduction band?

Calculate λ given the energy: $E = h\dfrac{c}{\lambda}$

First convert 140 kJ/mol to J/photon.

$\dfrac{140 \times 10^3 \text{ J}}{1 \text{ mol}} \cdot \dfrac{1 \text{ mol}}{6.02 \times 10^{23} \text{ photons}} = (6.626 \times 10^{-34} \text{ J} \cdot \text{s})\dfrac{2.9979 \times 10^8 \text{ m/s}}{\lambda}$

$\dfrac{1}{\lambda} = \dfrac{2.3 \times 10^{-19} \text{ J/photon}}{(6.626 \times 10^{-34} \text{ J} \cdot \text{s})(2.9979 \times 10^8 \text{ m/s})} = 1.2 \times 10^6 \text{ m}^{-1}$

Therefore, λ = 8.5×10^{-7} m
Or, expressed in nanometers, the wavelength is:
(8.5×10^{-7} m)(1×10^9 nm/m) = 8.5×10^2 nm

12.56. Increasing T leads to an increase in the energy available for electrons to jump the band gap.

12.57. Germanium has a greater metallic tendency, so it will show a higher conductivity at 298 K.

12.58. Identify as either *n*-type or *p*-type semiconductors:

(a) germanium doped with As: *n*-type
(b) silicon doped with P: *n*-type
(c) germanium doped with In: *p*-type
(d) germanium doped with Sb: *n*-type

12.59. Figure 12.17 helps with this question. Boron is a Group 3A (13) element, and has 3 valence electrons, while carbon in Group 4A (14) has 4. Doping carbon with boron leads to a *p*-type semiconductor, since B is electron-deficient and leads to an acceptor level between the valence and conduction bands.

12.60. Molecular solids are made of molecules held together by intermolecular forces. Network solids are made of atoms held together by covalent bonds. Amorphous solids are networks of atoms held together by covalent bonds but lacking long range order.

In the Laboratory

12.61. Galena (PbS) does not have the same solid structure as ZnS. While the sulfide anions in PbS occupy an fcc structure like the sulfide ions in ZnS, the Pb ions occupy both tetrahedral holes and edges in galena unlike Zn^{2+} ions in ZnS.

The face-centered cubic unit cell for galena *shown here* has 4 lead ions (larger spheres) in the unit cell [(12 atoms with 1/4 in the cell) = 3 ions and 1 ion in the center] and 4 sulfide ions (smaller yellow spheres) [(8 atoms in the corners with 1/8 in the cell) = 1 ion and (6 atoms with 1/2 in the cell) = 3 ions], giving the formula PbS.

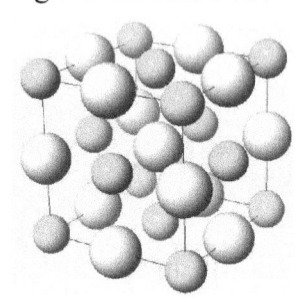

12.62. Regarding the unit cell for $CaTiO_3$:

(a) Volume of the unit cell =
$$\frac{1 \ CaTiO_3}{unit \ cell}\left(\frac{135.94 \ g \ CaTiO_3}{1 \ mol \ CaTiO_3}\right)\left(\frac{1 \ mol \ CaTiO_3}{6.0221 \times 10^{23} \ CaTiO_3}\right)\left(\frac{1 \ cm^3}{4.10 \ g}\right) = 5.51 \times 10^{-23} \ cm^3$$

Edge length = $\sqrt[3]{5.51 \times 10^{-23} \ cm^3} = 3.80 \times 10^{-8} \ cm = 380. \ pm$

(b) Assume that the O^{2-} and Ti^{4+} ions touch along center of cell.
edge length = $2(O^{2-}$ radius$) + 2(Ti^{4+}$ radius$)$
The radius of an O^{2-} ion is 140. pm (Figure 7.12).
380. pm = 2(140. pm) + 2(Ti^{4+} radius)
Ti^{4+} radius = 50. pm. The literature value of the Ti^{4+} radius = 69 pm.

12.63. Calculate the density of KBr given that it has the same lattice structure as NaCl.

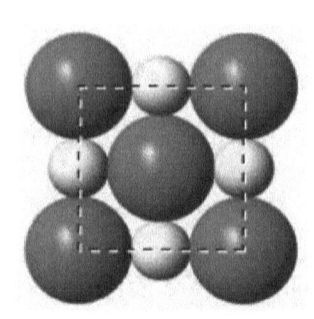

An examination of the structure of the face-centered cubic lattice associated with NaCl (and KBr) shows that an edge of the unit cell consists of 2 radii each of the cation and anion (dotted line).

This distance is: $2 \times K^+ + 2 \times Br^- = (2 \times 133 \ pm) + (2 \times 196 \ pm) = 658 \ pm$. Convert this distance to centimeters:

$$658 \ pm \left(\frac{1 \times 10^{-10} \ cm}{1 \ pm}\right) = 6.58 \times 10^{-8} \ cm.$$

This distance cubed gives a volume for the KBr unit cell: $2.85 \times 10^{-22} \ cm^3$.

The mass of the unit cell is found by calculating the mass of 4 KBr ion pairs. [The text in Section 12.2 explains the fact that there are 4 KBr ion pairs in a unit cell for the face-centered cubic lattice system.]

Mass of 1 mol of potassium and bromide ions: (39.098 g + 79.904 g) = 119.002 g
Divide by Avogadro's number to obtain the mass of **each** ion pair, and then multiply by 4.

$$4 \text{ KBr ion pairs} \left(\frac{119.002 \text{ g KBr}}{6.022 \times 10^{23} \text{ KBr ion pairs}} \right) = 7.904 \times 10^{-22} \text{g}.$$

The density is 7.904×10^{-22} g / 2.85×10^{-22} cm³ = 2.77 g/cm³.

12.64. Calculate the lattice enthalpy of $CaCl_2$:

$Cl_2(g)$	→	$2\ Cl(g)$	$2\ \Delta_f H°$	=	$2(+121.3$ kJ/mol$)$
$2\ Cl(g) + e^-$	→	$2\ Cl^-(g)$	$2\ \Delta_{ea}H$	=	$2(-349.0$ kJ/mol$)$
$Ca(s)$	→	$Ca(g)$	$\Delta_f H°$	=	$+178.2$ kJ/mol
$Ca(g)$	→	$Ca^+(g) + e^-$	IE_1	=	$+599$ kJ/mol
$Ca^+(g)$	→	$Ca^{2+}(g) + e^-$	IE_2	=	$+1145$ kJ/mol
$Ca^{2+}(g) + 2\ Cl^-(g)$	→	$CaCl_2(s)$	$\Delta_{lattice}H$	=	x
$Ca(s) + Cl_2(g)$	→	$CaCl_2(s)$	$\Delta_f H°$	=	-795.8 kJ/mol

-795.8 kJ/mol = $2(121.3$ kJ/mol$) + 2(-349.0$ kJ/mol$) + 178.2$ kJ/mol + 599 kJ/mol
$\qquad\qquad\qquad\qquad\qquad\qquad\qquad\qquad\qquad\qquad\qquad\qquad\quad + 1145$ kJ/mol + x

$x = \Delta_{lattice}H = -795.8$ kJ/mol $- (242.6$ kJ/mol + -698.0 kJ/mol + 178.2 kJ/mol + 599 kJ/mol
$\qquad\qquad\qquad\qquad\qquad\qquad\qquad\qquad\qquad\qquad\qquad\qquad\qquad\quad + 1145$ kJ/mol$)$

$\qquad = -2263$ kJ/mol

Summary and Conceptual Questions

12.65. Regarding boron phosphide:

(a) The balanced equation for the synthesis of BP:
$BBr_3(g) + PBr_3(g) + 3\ H_2(g) \rightarrow BP(s) + 6\ HBr(g)$

(b) How many tetrahedral holes are filled with P in each cell? BP uses the zinc blende structure in which B atoms are in face-centered cubic pattern, and P atoms occupy half the tetrahedral sites. Note that in the fcc lattice, there are 4 B atoms [(8 × 1/8) + (6 × 1/2)]. With the formula BP, with 4B atoms, there must be 4 P atoms.

(c) With a length of 478 pm, what is the density (in g/cm³)?

First, determine the volume of the unit cell (where 478 pm = 4.78×10^{-8} cm):

$V = (4.78 \times 10^{-8}$ cm$)^3 = 1.09 \times 10^{-22}$ cm³

Knowing that the molar mass of BP = 41.78 g/mol, and that there are 4 BP pairs in the cell, calculate the mass of the boron phosphide unit cell:

$$\frac{41.78 \text{ g BP}}{1 \text{ mol BP}} \left(\frac{1 \text{ mol BP}}{6.02214 \times 10^{23} \text{ BP pairs}} \right) \left(\frac{4 \text{ BP pairs}}{1 \text{ unit cell}} \right) = 2.775 \times 10^{-22} \text{ g BP/unit cell}$$

$$\text{Density} = \frac{2.775 \times 10^{-22} \text{ g BP}}{1.09 \times 10^{-22} \text{ cm}^3} = 2.54 \text{ g/cm}^3$$

(d) Calculate the closest distance between a B and a P atom in the unit cell:

With a unit cell length of 478 pm, the face diagonal will be $\sqrt{2}(478 \text{ pm}) = 676$ pm

The B–P distance corresponds to the hypotenuse of a right triangle. Recognizing that the sine of the angle of 54.75° (half of 109.5°) is equal to the ratio of 1/4(676 pm)/BP distance,

$$\sin(54.75) = \frac{169 \text{ pm}}{\text{Distance}}$$

Distance = 169 pm/0.8166 = or 207 pm.

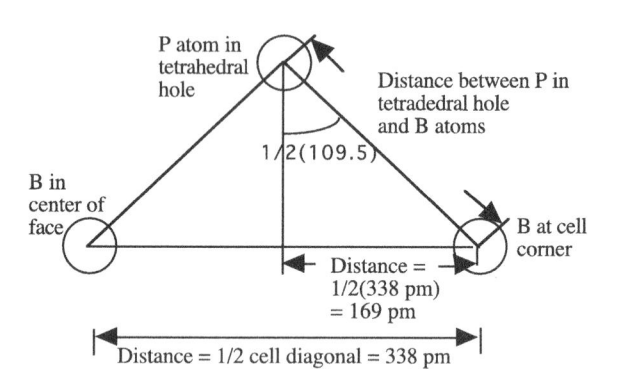

12.66. Why can M₃X not exist in a face-centered cubic lattice?

In a face−centered cubic lattice of anions (X), there is a total of four anions per unit cell.

Type of holes occupied by cation (M)	Number of cations per unit cell	Formula of salt
all tetrahedral holes	8	M₂X
half of the tetrahedral holes	4	MX
all octahedral holes	4	MX

It is not possible to have a cation:anion ratio of 3:1.

12.67. Nickel aluminide structure:
Sketch of solid-state structure.

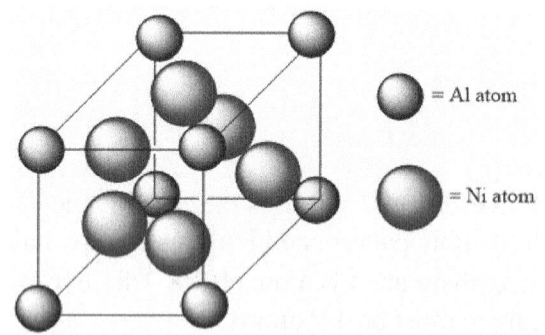

(a) There are six Ni atoms in the cube faces. Each contributes ½ an atom to the unit cell.
Number of Ni atoms per unit cell = 6 × ½ = 3 Ni atoms = x in Ni$_x$Al$_y$.
There are eight Al atoms at the cube corners. Each contributes 1/8 an atom to the unit cell.
Number of Al atoms = 8 × 1/8 = 1 Al atoms = y in Ni$_x$Al$_y$.
The formula is Ni₃Al
(b) Density = 7.71 g/cm³. What is unit cell edge length?

$$\text{mass of 1 Ni atom} = \frac{58.693 \text{ g Ni}}{1 \text{ mol Ni}} \left(\frac{1 \text{ mol Ni}}{6.02214 \times 10^{23} \text{ Ni atoms}} \right) = 9.7462 \times 10^{-23} \text{ g/Ni atom}$$

$$\text{mass of 1 Al atom} = \frac{26.982 \text{ g Al}}{1 \text{ mol Al}} \left(\frac{1 \text{ mol Al}}{6.02214 \times 10^{23} \text{ Al atoms}} \right) = 4.4805 \times 10^{-23} \text{ g/Al atom}$$

$$\text{mass of unit cell} = 3 \left(9.7462 \times 10^{-23} \text{ g} \right) + 4.4805 \times 10^{-23} \text{ g} = 3.3719 \times 10^{-22} \text{ g}$$

$$V_{\text{unit cell}} = 3.3719 \times 10^{-22} \text{ g} \left(\frac{1 \text{ cm}^3}{7.71 \text{ g}} \right) = 4.37 \times 10^{-23} \text{ cm}^3$$

$$\text{Edge length} = \sqrt[3]{4.37 \times 10^{-23} \text{ cm}^3} = 3.52 \times 10^{-8} \text{ cm}$$

12.68. Consider the spinels: $CoAl_2O_4$ and $SnCo_2O_4$:

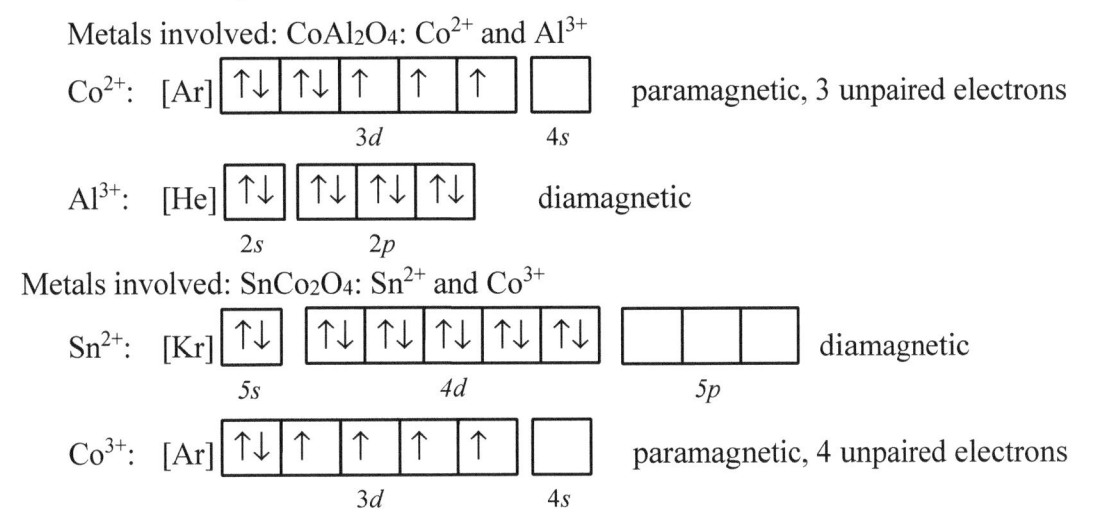

Metals involved: $CoAl_2O_4$: Co^{2+} and Al^{3+}

Co^{2+}: [Ar] 3d, 4s — paramagnetic, 3 unpaired electrons

Al^{3+}: [He] 2s, 2p — diamagnetic

Metals involved: $SnCo_2O_4$: Sn^{2+} and Co^{3+}

Sn^{2+}: [Kr] 5s, 4d, 5p — diamagnetic

Co^{3+}: [Ar] 3d, 4s — paramagnetic, 4 unpaired electrons

12.69. A procedure to calculate the percent of space occupied by the atoms in a fcc arrangement:

- First, calculate the volume of the cell by calculating the length of one side, and then cube that dimension to get the cell volume.
- For the volume occupied by the atoms, recall that there are 4 atoms within the unit cell of a face-centered cubic lattice (Section 12.2 of the text). Calculate the volume of those 4 atoms, remembering that the volume of a sphere = $4/3\pi r^3$.
- Divide the volume occupied by the atoms [$4 \times 4/3\pi r^3$] by the cell volume [the length of one side cubed], and multiply by 100.

12.70. Regarding the compound $Ca_2(Mg,Fe)_5(Si_4O_{11})_2(OH)_2$:

(a) Charge on the $(Si_4O_{11})^{n-}$ ion: 2 Ca^{2+} provide 4+; 5 Fe^{2+} (or Mg^{2+} or both) provide 10+; 2 OH^- provide 2− for a total of 12+. So, 2 $(Si_4O_{11})^{n-}$ have to offset 12+, so $n = 6-$.

(b) Oxidation state of Si:
Let x equal the oxidation number of each Si. The oxidation number of each O is –2. The sum of the oxidation numbers must equal the charge of the ion:
$$-6 = 4(x) + 11(-2)$$
$$x = +4$$
The oxidation number of Si is +4 in this ion.

(c) Percent of Fe: Molar mass is 914.872 g/mol

$$\frac{(5 \times 0.65 \text{ mol Fe})\left(55.845 \frac{\text{g Fe}}{\text{mol Fe}}\right)}{914.86 \frac{\text{g compound}}{\text{mol compound}}} \times 100\% = 20.\ \% \text{ Fe (2 sf)}$$

(d) Fe^{2+}: [Ar] [↑↓][↑][↑][↑][↑] [ ]   paramagnetic, 4 unpaired electrons
        3d   4s

12.71. Using the diagram provided with this question:
(a) Number of triple points and the phases in equilibrium: 2
The "upper" point has diamond/liquid C/graphite in equilibrium, while the "lower" triple point has graphite/liquid C/vapor C in equilibrium.
(b) There is **no** point at which all four phases "meet," that is, are in equilibrium.
(c) Viewing the phase diagram, you can see that as the pressure increases (higher vertical points), the diamond form—the denser phase—prevails.
(d) The stable phase of C at room temperature and 1 atm pressure: graphite

12.72. Graph lattice energy *vs.* sum of ionic radii:

$\Delta_{lattice}U$	sum of radii(pm)	$\Delta_{lattice}U$	sum of radii(pm)	$\Delta_{lattice}U$	sum of radii(pm)
LiF −1037	78 + 133 = 211	NaF −926	98 + 133 = 231	KF −821	133 + 133 = 266
LiCl −852	78 + 181 = 259	NaCl −786	98 + 181 = 279	KCl −717	133 + 181 = 314
LiBr −815	78 + 196 = 274	NaBr −752	98 + 196 = 294	KBr −689	133 + 196 = 329
LiI −761	78 + 220 = 298	NaI −702	98 + 220 = 318	KI −649	133 + 220 = 353

[Scatter plot: Sum of Ionic Radii (pm) vs Lattice Energy (kJ/mol)]

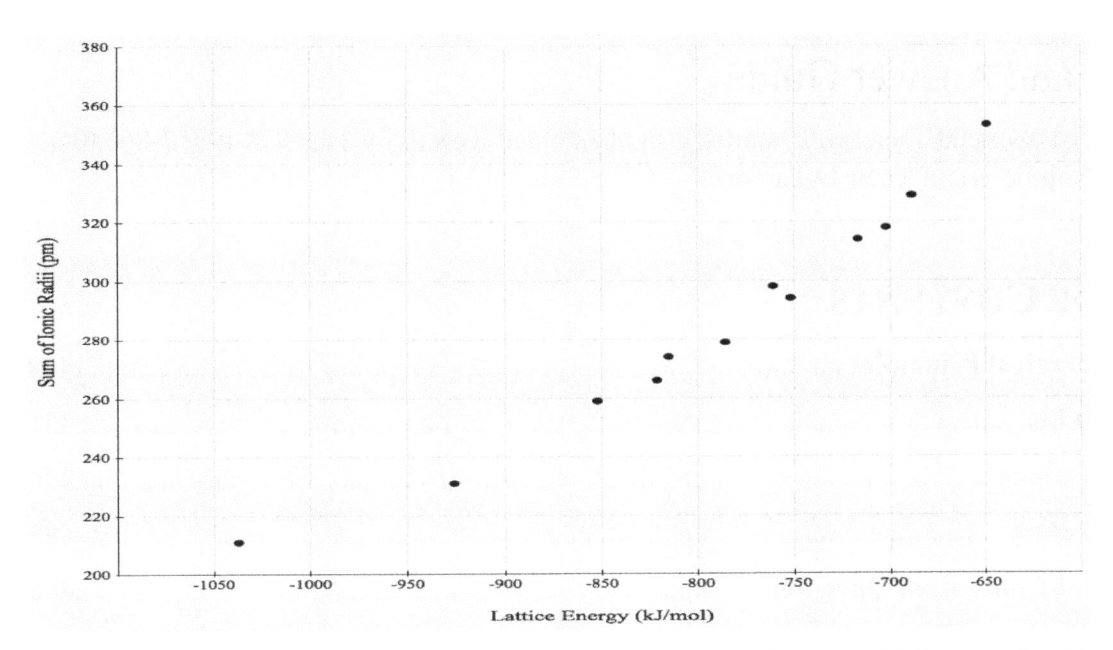

Smaller sum of radii results in greater (more exothermic) $\Delta_{lattice}U$, in accord with a smaller distance between ions leading to greater force of attraction in Coulomb's Law.

12.73. See Study Question 12.50 for help with this problem. If one assumes that the packing pattern used in the two pools is identical, the percentage of space occupied would not be a function of the size of the spheres, so the two pools will have identical levels of water when the ice melts.

Solution and Answer Guide

Kotz Treichel Townsend Treichel, Chemistry and Chemical Reactivity 11e, 978-0-357-85140-1, Chapter 13: Solutions and Their Behavior

TABLE OF CONTENTS

Applying Chemical Principles 448
Practicing Skills 451
General Questions 470
In the Laboratory 483
Summary and Conceptual Questions 486

Applying Chemical Principles

Distillation

13.1.1. What is the approximate mole fraction of hexane in the vapor phase after two evaporation-condensation cycles?

A copy of Figure B is attached: Letter a indicates the first evaporation-condensation cycle, and letter b the second evaporation-condensation cycle. So, the mole fraction of hexane in the 2nd evaporation-condensation cycle is approximately 0.6—perhaps around 0.59.

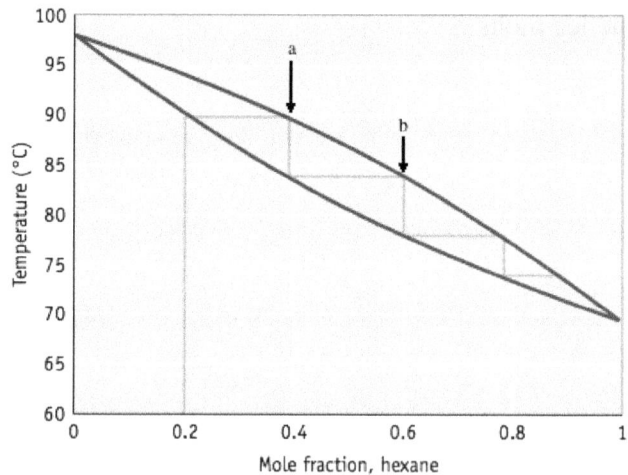

13.1.2. Number of theoretical plates to produce a solution with a mole fraction of hexane greater than 0.90: 4 plates. From the Figure above, one plate produces a solution with approximately 0.40 mole fraction hexane; two plates produce a solution with 0.60 mole fraction; etc.

13.1.3. What is the mass percent of hexane in a mixture with heptane if the mole fraction of hexane is 0.20?
We need the masses of both hexane and heptane! We know BOTH mole fractions: since mole fraction (hexane) = 0.20, and this is a binary mixture, the mole fraction(heptane) is 1.00 − 0.20 = 0.80.

© 2024 Cengage Learning, Inc. All Rights Reserved. May not be scanned, copied or duplicated, or posted to a publicly accessible website, in whole or in part.

Calculate the mass of one mole of each compound:

Hexane: C_6H_{14} has a molar mass of 86.18 g/mol; Heptane: C_7H_{16} has a molar mass of 100.2 g/mol

The mass of hexane present would be: 0.20 × 86.18 g/mol = 17.2 g and the mass of heptane present would be: 0.80 × 100.2 g/mol = 80.2 g. The total mass is: 17.2 g + 80.2 g = 97.4 g, and the mass percent of hexane is (17.2 g/97.4 g) × 100 = 18% (2 sf).

13.1.4. Enthalpy of vaporization for heptane:

$$\ln\left(\frac{P_2}{P_1}\right) = -\left(\frac{\Delta_{vap}H}{R}\right)\left(\frac{1}{T_2} - \frac{1}{T_1}\right)$$

$$\ln\left(\frac{361.5 \text{ mm Hg}}{760.0 \text{ mm Hg}}\right) = -\left(\frac{\Delta_{vap}H}{0.0083145 \frac{\text{kJ}}{\text{K} \cdot \text{mol}}}\right)\left(\frac{1}{348.2 \text{ K}} - \frac{1}{371.6 \text{ K}}\right)$$

and solving: $\Delta_{vap}H$ = 34.2 kJ/mol

Henry's Law and Exploding Lakes

13.2.1. Amount of CO_2 contained in 25 mL of headspace at 25°C and 4.0 atm:

This is equivalent to solving an Ideal Gas Law problem:

$PV = nRT$ so n = PV/RT

$$n = \frac{PV}{RT} = \frac{(4.0 \text{ atm})(0.025 \text{ L})}{\left(0.082057 \frac{\text{L} \cdot \text{atm}}{\text{K} \cdot \text{mol}}\right)(298 \text{ K})} = 4.1 \times 10^{-3} \text{ mol } CO_2$$

13.2.2. What volume will this amount of CO_2 occupy if the pressure is reduced to 3.7×10^{-4} atm:
Note the equivalence of a gas law problem involving a change of P at constant T:
So $P_1V_1 = P_2V_2$, and (4.0 atm)(25 × 10⁻³ L) = (3.7 × 10⁻⁴ atm)(V_2) and
V_2 = (4.0 atm)(25 × 10⁻³ L)/ (3.7 × 10⁻⁴ atm) = 270 L

Factor by which gas expanded: = 270 L/0.025 L = 11,000 (2 sf)

13.2.3. The solubility of CO_2 in water at 25°C when $P_{CO_2} = 3.7 \times 10^{-4}$ bar:

Solubility = $k_H \times P_{CO_2}$ = (0.034 mol/kg·bar)(3.7 × 10⁻⁴ bar) = 1.3 × 10⁻⁵ mol CO_2/kg water.

13.2.4. Mass of CO₂ released to the atmosphere:

Calculate the solubility of CO₂ at 4.0 atm (bar):

Solubility = $k_H \times P_{CO_2}$ = (0.034 mol/kg·bar)(4.0 bar) = 0.136 mol/kg (or 0.14 to 2 sf)

We know from 13.2.3 above the solubility of CO₂ when the $P_{CO_2} = 3.7 \times 10^{-4}$ atm is 1.3×10^{-5} mol CO₂/kg. The difference is 0.14 mol CO₂/kg – 1.3×10^{-5} mol CO₂/kg = 0.14 mol CO₂/kg

If we assume that the density is 1.0 g/cm³, then 1.0 kg = 1.0 L, and the mass of CO₂ released is: 0.14 mol × 44.0 g/mol = 6.0 g CO₂.

Narcosis and the Bends

13.3.1. External pressure experienced by a diver at 10.0 m depth:

Partial pressures of O₂ and N₂:

Calculate the pressure of the water: density × acceleration due to gravity × depth =

$(1.00 \times 10^3 \text{ kg/m}^3)\left(9.81 \dfrac{m}{s^2}\right)(10.0 \text{ m}) = 9.81 \times 10^4 \text{Pa} = 98.1 \text{ kPa}$

[The density of water (1 g/cm³) converted into kg/m³ = 1.00×10^3 kg/m³]

Express this pressure in atm: 98.1 kPa × 1 atm/101.325 kPa = 0.968 atm

Since pressure on your body is 1.0 atm + 0.968 atm the total pressure is 1.968 atm.

$P_{N_2} = X_{N_2} \times P_{total}$ = (0.7808)(1.968 atm) = 1.537 atm

$P_{O_2} = X_{O_2} \times P_{total}$ = (0.2095)(1.968 atm) = 0.4123 atm

The mole fractions of nitrogen and oxygen are taken from the table showing the composition of earth's atmosphere.

13.3.2. Solubility of O₂ in water at 3.0 atm pressure:

At 3.0 atm the $P_{O_2} = X_{O_2} \times P_{total}$ = (0.2095)(3.0 atm) = 0.629 atm

Solubility = $k_H \times P_{O_2} = (1.3 \times 10^{-3} \text{ mol/kg} \cdot \text{bar})(0.629 \text{ atm})(1 \text{ atm}/0.98692 \text{ atm})$

Solubility = 8.28×10^{-4} mol O₂/kg (or 8.3×10^{-4} to 2 sf)

Mass of O₂ that will dissolve in 1.0 L at this pressure:

$\left(\dfrac{8.28 \times 10^{-4} \text{ mol O}_2}{\text{kg}}\right)\left(\dfrac{32.00 \text{ g O}_2}{1 \text{ mol O}_2}\right)\left(\dfrac{1 \text{ kg}}{1 \text{ L}}\right) = 0.026 \text{ g O}_2$

13.3.3. Total amount of N₂ and O₂ in 1.0 L of water shaken in air (at 25 °C).

From 13.3.1 we know that the amount of N₂ is 78.08 % (0.7808) of the atmospheric pressure and that O₂ is 20.95% (0.2095).

So, the pressures exerted by each will be the percentage (expressed as a decimal) × 1.0 atm

$P_{N_2} = (0.7808)(1.0 \text{ atm}) = 0.7808 \text{ atm}$

$P_{O_2} = (0.2095)(1.0 \text{ atm}) = 0.2095 \text{ atm}$

Now the solubilities of each:

The amount of gas in 1.0 L (or 1.0 kg) of water is (4.75 × 10⁻⁴ mol + 2.76 × 10⁻⁴ mol) or 7.51 × 10⁻⁴ mol gas.

$S_{N_2} = k_H \times P_{N_2} = (6.0 \times 10^{-4} \text{ mol/kg} \cdot \text{bar})(0.7808 \text{ atm})(1 \text{ bar}/0.98692 \text{ atm}) = 4.75 \times 10^{-4} \text{ mol N}_2/\text{kg}$

$S_{O_2} = k_H \times P_{O_2} = (1.3 \times 10^{-3} \text{ mol/kg} \cdot \text{bar})(0.2095 \text{ atm})(1 \text{ bar}/0.98692 \text{ atm}) = 2.76 \times 10^{-4} \text{ mol O}_2/\text{kg}$

If the gases are expelled, what is the mole fraction of O₂?

$X_{O_2} = \dfrac{2.76 \times 10^{-4} \text{ mol O}_2}{7.51 \times 10^{-4} \text{ mol gas}} = 0.37 \text{ (2 sf)}$

Practicing Skills

Concentration

13.1. For 1.50 g of succinic acid in 500. mL of water:

$1.50 \text{ g C}_2\text{H}_4(\text{CO}_2\text{H})_2 \left(\dfrac{1 \text{ mol C}_2\text{H}_4(\text{CO}_2\text{H})_2}{118.09 \text{ g C}_2\text{H}_4(\text{CO}_2\text{H})_2} \right) = 0.0127 \text{ mol C}_2\text{H}_4(\text{CO}_2\text{H})_2$

The molality of the solution: Molality = #mole solute/kg solvent:

With a density of water of 1.00 g/cm³, 500. mL = 0.500 kg

$\text{Molality} = \dfrac{0.0127 \text{ mol C}_2\text{H}_4(\text{CO}_2\text{H})_2}{0.500 \text{ kg}} = 0.0254 \text{ molal}$

The mole fraction of succinic acid in the solution:

For mole fraction we need *both* the # moles of solute *and* # moles of solvent.

$\text{Moles of water} = 500. \text{ g H}_2\text{O} \left(\dfrac{1 \text{ mol H}_2\text{O}}{18.015 \text{ g H}_2\text{O}} \right) = 27.8 \text{ mol H}_2\text{O}$

$\text{The mole fraction of acid} = \dfrac{0.0127 \text{ mol C}_2\text{H}_4(\text{CO}_2\text{H})_2}{0.0127 \text{ mol acid} + 27.8 \text{ mol water}} = 4.57 \times 10^{-4}$

The weight percentage of succinic acid in the solution:

The fraction of *total* mass of solute + solvent which is solute:

$$\text{Weight percentage} = \frac{1.50 \text{ g } C_2H_4(CO_2H)_2}{501.50 \text{ g acid + water}} \times 100\% = 0.299\%$$

13.2. Molality of camphor solution:

$$20.0 \text{ g } C_{10}H_{16}O \left(\frac{1 \text{ mol } C_{10}H_{16}O}{152.2 \text{ g}} \right) = 0.131 \text{ mol } C_{10}H_{16}O$$

$$250. \text{ mL } C_2H_5OH \left(\frac{0.785 \text{ g}}{1 \text{ mL}} \right) \left(\frac{1 \text{ mol } C_2H_5OH}{46.07 \text{ g } C_2H_5OH} \right) = 4.26 \text{ mol } C_2H_5OH$$

Molality of solution

$$m = \frac{\text{amount of solute}}{\text{kg of solvent}} = \frac{0.131 \text{ mol } C_{10}H_{16}O}{250. \text{ mL } C_2H_5OH} \left(\frac{1 \text{ mL}}{0.785 \text{ g}} \right) \left(\frac{1000 \text{ g}}{1 \text{ kg}} \right) = 0.670 \, m$$

Mole fraction of camphor:

$$X_{C_{10}H_{16}O} = \left(\frac{0.131 \text{ mol } C_{10}H_{16}O}{0.131 \text{ mol } C_{10}H_{16}O + 4.26 \text{ mol } C_2H_5OH} \right) = 0.0299$$

Weight percent:

$$250. \text{ mL } C_2H_5OH \left(\frac{0.785 \text{ g}}{1 \text{ mL}} \right) = 196 \text{ g } C_2H_5OH$$

$$\text{Weight \% } C_{10}H_{16}O = \left(\frac{20.0 \text{ g } C_{10}H_{16}O}{20.0 \text{ g } C_{10}H_{16}O + 196 \text{ g } C_2H_5OH} \right) = 9.25\% \, C_{10}H_{16}O$$

13.3. Complete the following transformations for

NaI:

Weight percent:

$$\left(\frac{0.15 \text{ mol NaI}}{1 \text{ kg solvent}} \right) \left(\frac{149.9 \text{ g NaI}}{1 \text{ mol NaI}} \right) = \frac{22.5 \text{ g NaI}}{1 \text{ kg solvent}}$$

$$\frac{22.5 \text{ g NaI}}{22.5 \text{ g NaI} + 1000 \text{ g solvent}} \times 100\% = 2.2\% \text{ (2sf)}$$

Mole fraction:

$$1000 \text{ g } H_2O \left(\frac{1 \text{ mol } H_2O}{18.02 \text{ g } H_2O} \right) = 55.49 \text{ mol } H_2O$$

$$X_{NaI} = \frac{0.15 \text{ mol NaI}}{0.15 \text{ mol NaI} + 55.49 \text{ mol } H_2O} = 2.7 \times 10^{-3}$$

C₂H₅OH:

Molality:

$$\left(\frac{5.0 \text{ g C}_2\text{H}_5\text{OH}}{100 \text{ g solution}}\right)\left(\frac{1 \text{ mol C}_2\text{H}_5\text{OH}}{46.07 \text{ g C}_2\text{H}_5\text{OH}}\right)\left(\frac{100 \text{ g solution}}{95 \text{ g solvent}}\right)\left(\frac{1000 \text{ g solvent}}{1 \text{ kg solvent}}\right) = 1.1 \, m$$

Mole fraction:

$$5.0 \text{ g C}_2\text{H}_5\text{OH}\left(\frac{1 \text{ mol C}_2\text{H}_5\text{OH}}{46.07 \text{ g C}_2\text{H}_5\text{OH}}\right) = 0.11 \text{ mol C}_2\text{H}_5\text{OH}$$

$$95.0 \text{ g H}_2\text{O}\left(\frac{1 \text{ mol H}_2\text{O}}{18.02 \text{ g H}_2\text{O}}\right) = 5.27 \text{ mol H}_2\text{O}$$

$$X_{\text{C}_2\text{H}_5\text{OH}} = \frac{0.11 \text{ mol C}_2\text{H}_5\text{OH}}{0.11 \text{ mol C}_2\text{H}_5\text{OH} + 5.27 \text{ mol H}_2\text{O}} = 0.020$$

C₁₂H₂₂O₁₁:

Weight percent:

$$\left(\frac{0.15 \text{ mol C}_{12}\text{H}_{22}\text{O}_{11}}{1 \text{ kg solvent}}\right)\left(\frac{342.3 \text{ g C}_{12}\text{H}_{22}\text{O}_{11}}{1 \text{ mol C}_{12}\text{H}_{22}\text{O}_{11}}\right) = \frac{51.3 \text{ g C}_{12}\text{H}_{22}\text{O}_{11}}{1 \text{ kg solvent}}$$

$$\frac{51.3 \text{ g C}_{12}\text{H}_{22}\text{O}_{11}}{51.3 \text{ g C}_{12}\text{H}_{22}\text{O}_{11} + 1000 \text{ g solvent}} \times 100\% = 4.9\%$$

Mole fraction:

$$1000 \text{ g H}_2\text{O}\left(\frac{1 \text{ mol H}_2\text{O}}{18.02 \text{ g H}_2\text{O}}\right) = 55.49 \text{ mol H}_2\text{O}$$

$$X_{\text{C}_{12}\text{H}_{22}\text{O}_{11}} = \frac{0.15 \text{ mol C}_{12}\text{H}_{22}\text{O}_{11}}{0.15 \text{ mol C}_{12}\text{H}_{22}\text{O}_{11} + 55.49 \text{ mol H}_2\text{O}} = 2.7 \times 10^{-3}$$

13.4. Fill in the blanks in the table: (given info is emboldened)

Compound	Molality	Weight percent	Mole fraction
KNO₃	1.10	**10.0**	0.0194
CH₃CO₂H	**0.183**	1.09	3.29 × 10⁻³
HOC₂H₄OH	2.85	**15.0**	0.0488

KNO₃: 10.0 g KNO₃ dissolved in 90.0 g H₂O

$$10.0 \text{ g KNO}_3\left(\frac{1 \text{ mol KNO}_3}{101.1 \text{ g KNO}_3}\right) = 0.0989 \text{ mol KNO}_3$$

$$90.0 \text{ g H}_2\text{O} \left(\frac{1 \text{ mol H}_2\text{O}}{18.02 \text{ g H}_2\text{O}} \right) = 4.99 \text{ mol H}_2\text{O}$$

$$m = \frac{0.0989 \text{ mol KNO}_3}{0.0900 \text{ kg H}_2\text{O}} = 1.10 \, m \text{ KNO}_3$$

$$X_{\text{KNO}_3} = \frac{0.0989 \text{ mol KNO}_3}{0.0989 \text{ mol KNO}_3 + 4.99 \text{ mol H}_2\text{O}} = 0.0194$$

CH$_3$CO$_2$H: 0.183 mol CH$_3$CO$_2$H dissolved in 1.00 kg H$_2$O

$$0.183 \text{ mol CH}_3\text{CO}_2\text{H} \left(\frac{60.05 \text{ g CH}_3\text{CO}_2\text{H}}{1 \text{ mol CH}_3\text{CO}_2\text{H}} \right) = 11.0 \text{ g CH}_3\text{CO}_2\text{H}$$

$$1000. \text{ g H}_2\text{O} \left(\frac{1 \text{ mol H}_2\text{O}}{18.02 \text{ g H}_2\text{O}} \right) = 55.49 \text{ mol H}_2\text{O}$$

$$\text{Weight \% CH}_3\text{CO}_2\text{H} = \frac{11.0 \text{ g CH}_3\text{CO}_2\text{H}}{11.0 \text{ g CH}_3\text{CO}_2\text{H} + 1000. \text{ g H}_2\text{O}} \times 100\% = 1.09\% \text{ CH}_3\text{CO}_2\text{H}$$

$$X_{\text{CH}_3\text{CO}_2\text{H}} = \frac{0.183 \text{ mol CH}_3\text{CO}_2\text{H}}{0.183 \text{ mol CH}_3\text{CO}_2\text{H} + 55.49 \text{ mol H}_2\text{O}} = 0.00329$$

HOCH$_2$CH$_2$OH: 15.0 g HOCH$_2$CH$_2$OH dissolved in 85.0 g H$_2$O

$$15.0 \text{ g HOCH}_2\text{CH}_2\text{OH} \left(\frac{1 \text{ mol HOCH}_2\text{CH}_2\text{OH}}{62.07 \text{ g HOCH}_2\text{CH}_2\text{OH}} \right) = 0.242 \text{ mol HOCH}_2\text{CH}_2\text{OH}$$

$$85.0 \text{ g H}_2\text{O} \left(\frac{1 \text{ mol H}_2\text{O}}{18.02 \text{ g H}_2\text{O}} \right) = 4.72 \text{ mol H}_2\text{O}$$

$$m = \frac{0.242 \text{ mol HOCH}_2\text{CH}_2\text{OH}}{0.0850 \text{ kg H}_2\text{O}} = 2.84 \, m \text{ HOCH}_2\text{CH}_2\text{OH}$$

$$X_{\text{HOCH}_2\text{CH}_2\text{OH}} = \frac{0.242 \text{ mol HOCH}_2\text{CH}_2\text{OH}}{0.242 \text{ mol HOCH}_2\text{CH}_2\text{OH} + 4.72 \text{ mol H}_2\text{O}} = 0.0487$$

13.5. To prepare a solution that is 0.200 m Na$_2$CO$_3$:

$$0.125 \text{ kg H}_2\text{O} \left(\frac{0.200 \text{ mol Na}_2\text{CO}_3}{1 \text{ kg H}_2\text{O}} \right) \left(\frac{105.99 \text{ g Na}_2\text{CO}_3}{1 \text{ mol Na}_2\text{CO}_3} \right) = 2.65 \text{ g Na}_2\text{CO}_3$$

The mole fraction of Na$_2$CO$_3$ in the resulting solution:

$$\text{mol Na}_2\text{CO}_3 = 0.125 \text{ kg H}_2\text{O} \left(\frac{0.200 \text{ mol Na}_2\text{CO}_3}{1 \text{ kg H}_2\text{O}} \right) = 0.0250 \text{ mol Na}_2\text{CO}_3$$

$$125 \text{ g H}_2\text{O} \left(\frac{1 \text{ mol H}_2\text{O}}{18.02 \text{ g H}_2\text{O}} \right) = 6.94 \text{ mol H}_2\text{O}$$

$$X_{Na_2CO_3} = \frac{0.0250 \text{ mol Na}_2\text{CO}_3}{0.0250 \text{ mol Na}_2\text{CO}_3 + 6.94 \text{ mol H}_2\text{O}} = 0.00359$$

13.6. Mass of KNO₃ needed:

$$500. \text{ g H}_2\text{O} \left(\frac{1 \text{ kg}}{1000 \text{ g}}\right)\left(\frac{0.0512 \text{ mol KNO}_3}{1 \text{ kg H}_2\text{O}}\right)\left(\frac{101.10 \text{ g KNO}_3}{1 \text{ mol KNO}_3}\right) = 2.59 \text{ g KNO}_3$$

$$1000. \text{ g H}_2\text{O}\left(\frac{1 \text{ mol H}_2\text{O}}{18.02 \text{ g H}_2\text{O}}\right) = 55.5 \text{ mol H}_2\text{O}$$

$$X_{KNO_3} = \frac{0.0512 \text{ mol KNO}_3}{0.0512 \text{ mol KNO}_3 + 55.5 \text{ mol H}_2\text{O}} = 9.22 \times 10^{-4}$$

13.7. To calculate the number of mol of C₃H₅(OH)₃:

$$\text{mol H}_2\text{O} = 425 \text{ g H}_2\text{O}\left(\frac{1 \text{ mol H}_2\text{O}}{18.02 \text{ g H}_2\text{O}}\right) = 23.58 \text{ mol H}_2\text{O}$$

$$X_{C_3H_5(OH)_3} = \frac{x \text{ mol C}_3\text{H}_5(\text{OH})_3}{x \text{ mol C}_3\text{H}_5(\text{OH})_3 + 23.58 \text{ mol H}_2\text{O}} = 0.090$$

$$x \text{ mol C}_3\text{H}_5(\text{OH})_3 = (x \text{ mol C}_3\text{H}_5(\text{OH})_3 + 23.58 \text{ mol H}_2\text{O})\,0.090$$

$$x = 2.3 \text{ mol C}_3\text{H}_5(\text{OH})_3$$

Grams of glycerol needed: $2.3 \text{ mol C}_3\text{H}_5(\text{OH})_3 \times \frac{92.1 \text{ g}}{1 \text{ mol}} = 210 \text{ g C}_3\text{H}_5(\text{OH})_3$ (2 sf)

The molality of the solution is (2.3 mol C₃H₅(OH)₃/0.425 kg H₂O) = 5.5 *m*

13.8. Mass of ethylene glycol required:

$$\text{mol H}_2\text{O} = 955 \text{ g H}_2\text{O}\left(\frac{1 \text{ mol H}_2\text{O}}{18.02 \text{ g H}_2\text{O}}\right) = 53.0 \text{ mol H}_2\text{O}$$

$$X_{C_3H_5(OH)_3} = \frac{x \text{ mol HOCH}_2\text{CH}_2\text{OH}}{x \text{ mol HOCH}_2\text{CH}_2\text{OH} + 53.0 \text{ mol H}_2\text{O}} = 0.125$$

$$x \text{ mol HOCH}_2\text{CH}_2\text{OH} = (x \text{ mol HOCH}_2\text{CH}_2\text{OH} + 53.0 \text{ mol H}_2\text{O})\,0.125$$

$$x = 7.57 \text{ mol HOCH}_2\text{CH}_2\text{OH}$$

$$m = \frac{7.57 \text{ mol HOCH}_2\text{CH}_2\text{OH}}{0.955 \text{ kg H}_2\text{O}} = 7.93 \text{ } m$$

13.9. Concentrated HCl is 12.0 M and has a density of 1.18 g/cm³.

(a) The molality of the solution:

Molality is defined as moles HCl/kg solvent, so begin by deciding the mass of 1 L, and the mass of water in that 1 L. Since the density = 1.18 g/mL, then 1 L (1000 mL) will have a mass of 1180 g.

The mass of HCl present in 12.0 mol HCl =

$$12.0 \text{ mol HCl}\left(\frac{36.46 \text{ g HCl}}{1 \text{ mol HCl}}\right) = 437.52 \text{ g HCl}$$

Since 1 L has a mass of 1180 g and 437.52 g is HCl, the difference (1180 – 437.52) is solvent. So, 1 L of solution has 742.48 g or 0.74248 kg of water.

$$\frac{12.0 \text{ mol HCl}}{0.74248 \text{ kg H}_2\text{O}} = 16.2 \, m \text{ HCl}$$

(b) Weight percentage of HCl:

12.0 mol HCl has a mass of 437.52 g, and the 1 L of solution has a mass of 1180 g.

$$\text{Weight \% HCl} = \left(\frac{437.52 \text{ g HCl}}{1180 \text{ g solution}}\right) \times 100\% = 37.1\%$$

(c) mole fraction HCl:

$$\text{mol H}_2\text{O} = 742.48 \text{ g H}_2\text{O}\left(\frac{1 \text{ mol H}_2\text{O}}{18.02 \text{ g H}_2\text{O}}\right) = 41.2 \text{ mol H}_2\text{O}$$

$$X_{\text{HCl}} = \frac{12.0 \text{ mol HCl}}{12.0 \text{ mol HCl} + 41.2 \text{ mol H}_2\text{O}} = 0.226$$

13.10. Molality and Molarity of concentrated sulfuric acid:

Molality:

$$\text{mol H}_2\text{SO}_4 = 95.0 \text{ g H}_2\text{SO}_4\left(\frac{1 \text{ mol H}_2\text{SO}_4}{98.07 \text{ g H}_2\text{SO}_4}\right) = 0.969 \text{ mol H}_2\text{SO}_4$$

mass H2O = 100.0 g solution –95.0 g H2SO4 = 5.0 g H2O = 0.0050 kg H2O

$$\text{Molality} = \frac{0.969 \text{ mol H}_2\text{SO}_4}{0.0050 \text{ kg H}_2\text{O}} = 190 \, m \text{ (2 sf)}$$

Molarity:

$$100.0 \text{ g solution}\left(\frac{1 \text{ cm}^3 \text{ solution}}{1.84 \text{ g solution}}\right)\left(\frac{1 \text{ mL}}{1 \text{ cm}^3}\right)\left(\frac{1 \text{ L}}{1000 \text{ mL}}\right) = 0.0543 \text{ L solution}$$

$$\text{Molarity} = \frac{0.969 \text{ mol H}_2\text{SO}_4}{0.0543 \text{ L solution}} = 17.8 \text{ M H}_2\text{SO}_4$$

13.11. The concentration of ppm expressed in grams is:

[Note that this answer assumes that 1 kg of seawater (the solution) is equivalent to 1 kg of solvent (H₂O). That is, mass solvent = 1.0×10^6 g solution – 0.18 g solute = 1.0×10^6 g solvent for this dilute solution.]

$$0.18 \text{ ppm} = \frac{0.18 \text{ g solute}}{1.0 \times 10^6 \text{ g solvent}} = \frac{0.18 \text{ g solute}}{1.0 \times 10^3 \text{ kg solvent}} \text{ or } \frac{0.00018 \text{ g solute}}{1 \text{ kg water}}$$

$$\left(\frac{0.00018 \text{ g Li}^+}{1 \text{ kg water}}\right)\left(\frac{1 \text{ mol Li}^+}{6.94 \text{ g Li}^+}\right) = 2.6 \times 10^{-5} \; m \text{ Li}^+$$

13.12. Silver ion is present at an average level of 28 ppb in U.S. water supplies.

[Note that this answer assumes that 1 kg of water in U.S. water supplies (the solution) is equivalent to 1 kg of solvent (H₂O). That is, mass solvent = 1.0×10^9 g solution – 28 g solute = 1.0×10^9 g solvent for this dilute solution.]

(a) What is 28 ppb expressed in units of molality?

$$\left(\frac{28 \text{ g Ag}}{1 \times 10^9 \text{ g H}_2\text{O}}\right)\left(\frac{1 \text{ mol Ag}}{107.87 \text{ g Ag}}\right)\left(\frac{1000 \text{ g}}{1 \text{ kg}}\right) = 2.6 \times 10^{-7} \; m$$

(b) Necessary volume of water to treat:

$$1.0 \times 10^2 \text{ g Ag}\left(\frac{1 \times 10^9 \text{ g H}_2\text{O}}{28 \text{ g Ag}}\right)\left(\frac{1 \text{ cm}^3}{1.0 \text{ g}}\right)\left(\frac{1 \text{ L}}{1000 \text{ cm}^3}\right) = 3.6 \times 10^6 \text{ L H}_2\text{O}$$

The Solution Process

13.13. Pairs of liquids that will be miscible:

(a) H₂O/CH₃CH₂CH₂CH₂CH₃
Will **not** be miscible. Water is a polar substance, while pentane is nonpolar.

(b) C₆H₆/CCl₄
Will **be** miscible. Both liquids are nonpolar and are expected to be miscible.

(c) H₂O/CH₃CO₂H

Will **be** miscible. Both substances can hydrogen bond, and we know that they mix—since a 5% aqueous solution of acetic acid is sold as "vinegar".

13.14. Why is acetone soluble in water?
Acetone is a polar molecule, so the strong dipole–dipole interactions between acetone and water molecules lead to a high solubility of acetone in water.

13.15. The enthalpy of solution for NaF:

The process can be represented as NaF(s) → NaF(aq)

The $\Delta_{soln}H° = \Sigma \Delta_fH°$ (product) - $\Sigma \Delta_fH°$ (reactant)
= (−572.8 kJ/mol)(1 mol) − (−573.6 kJ/mol)(1 mol) = +0.8 kJ

The similar calculation for NH₄NO₃ is +25.7 kJ. The enthalpy of solution data is from Table 13.1.

13.16. Enthalpy of solution for NaClO₄:

The process can be represented as: NaClO₄(s) → NaClO₄(aq)
$\Delta_{soln}H° = \Delta_fH°[NaClO_4(aq)] - \Delta_fH°[NaClO_4(s)]$
= −369.5 kJ/mol − (−382.9 kJ/mol) = +13.4 kJ/mol

13.17. Raising the temperature of the solution will increase the solubility of NaCl in water. To increase the amount of dissolved NaCl in solution one must **(c) raise the temperature of the solution and add some NaCl**.

13.18. Observations from saturated solutions of LiCl and Li₂SO₄:
As temperature increases the solubility of Li₂SO₄ decreases; additional solid should appear in the beaker. As temperature increases the solubility of LiCl increases; additional solid should dissolve.

Henry's Law

13.19. Mass of O₂ dissolved.

$$40 \text{ mm Hg} \left(\frac{1 \text{ atm}}{760 \text{ mm Hg}} \right) \left(\frac{1 \text{ bar}}{0.98692 \text{ atm}} \right) = 0.053 \text{ bar}$$

Solubility of O₂ = $k_H \cdot P_{O_2}$

$$\text{Solubility of } O_2 = \left(1.3 \times 10^{-3} \frac{\text{mol } O_2}{\text{kg} \cdot \text{bar}} \right)(0.053 \text{ bar}) = 6.9 \times 10^{-5} \frac{\text{mol } O_2}{\text{kg}}$$

Assume that 1.0 kg of an aqueous solution has a volume of 1.0 L, so the solubility is 6.9 × 10⁻⁵ mol O₂/L and $\left(\dfrac{6.9 \times 10^{-5} \text{ mol } O_2}{1 \text{ L solution}}\right)\left(\dfrac{32.0 \text{ g } O_2}{1 \text{ mol } O_2}\right) = 2 \times 10^{-3} \dfrac{\text{g } O_2}{\text{L solution}}$ (1 sf)

13.20. Henry's Law constant for O_2 at 50 °C:
Since the solubility of a gas generally decreases with increasing temperature, (a) is the only reasonable choice because it is less than the value of the constant at 25 °C.

13.21. Solubility = $k_H \cdot P_{CO_2}$; $0.0506\,m = (0.034 \dfrac{\text{moles}}{\text{kg} \cdot \text{bar}})(P_{CO_2})$

P_{CO_2} = 1.49 bar = 1.5 bar (2 sf) or expressed in units of atmospheres:

1.49 bar × $\dfrac{1 \text{ atm}}{1.01325 \text{ bar}}$ = 1.47 atm or, since 1 atm = 760 mm Hg, 1100 mm Hg (2 sf)

13.22. Concentration of H_2 in the water in g/mL:

$P_{H_2} = P_{total} - P_{H_2O}$ = 1.00 bar − (23.8 mm Hg)(1.013 bar/760 mm Hg) = 0.97 bar

Solubility = $k_H \times P_{H_2} = \dfrac{7.8 \times 10^{-4} \text{ mol}}{\text{kg} \cdot \text{bar}} \times 0.97 \text{ bar} = 7.6 \times 10^{-4} \dfrac{\text{mol } H_2}{\text{kg}}$

$\dfrac{7.6 \times 10^{-4} \text{ mol } H_2}{1 \text{ kg}}\left(\dfrac{2.016 \text{ g } H_2}{1 \text{ mol } H_2}\right)\left(\dfrac{1 \text{ kg}}{1 \text{ L}}\right)\left(\dfrac{1 \text{ L}}{1000 \text{ mL}}\right) = 1.5 \times 10^{-6}$ g/mL

Le Chatelier's Principle

13.23. Partial pressure of O_2 gas = 1.5 atm in a sealed flask at 25°C:

(a) Concentration of O_2:

According to Henry's Law, $S = k_H P_g$; $S = (1.3 \times 10^{-3} \dfrac{\text{moles}}{\text{kg} \cdot \text{bar}})(1.5 \text{ atm})(\dfrac{1.01325 \text{ bar}}{1 \text{ atm}})$

= 1.98×10^{-3} mol/kg or 2.0×10^{-3} mol/kg (2 sf)

(b) If the pressure of O_2 gas is raised to 1.7 atm, the amount of dissolved oxygen will increase! Similarly, as the pressure of O_2 gas is lowered, the amount of dissolved oxygen will decrease. (Aquatic life is susceptible to such decreases!)

13.24. Regarding the solubility of butane in water:

(a) $S = k_H P_g$; S = (0.0011 mol/ kg · bar)(0.21 atm)(1 bar/0.9869 atm) = 2.3×10^{-4} mol/kg

(b) When the pressure of butane is increased to 1.0 atm, the concentration of butane increases.

Raoult's Law

13.25. Since $P_{water} = X_{water} P°_{water}$, to determine the vapor pressure of the solution (P_{water}), we need the mole fraction of water.

$$35.0 \text{ g glycol} \left(\frac{1 \text{ mol glycol}}{62.07 \text{ g glycol}}\right) = 0.564 \text{ mol glycol}$$

$$500.0 \text{ g water} \left(\frac{1 \text{ mol water}}{18.015 \text{ g water}}\right) = 27.75 \text{ mol water}$$

The mole fraction of water is then:

$$\frac{27.75 \text{ mol H}_2\text{O}}{(27.75 \text{ mol} + 0.564 \text{ mol})} = 0.9801$$

and $P_{water} = X_{water} P°_{water} = 0.9801(35.7 \text{ mm Hg}) = 35.0 \text{ mm Hg}$

13.26. Vapor pressure of solution at 24 °C:

$$9.00 \text{ g urea} \left(\frac{1 \text{ mol urea}}{60.06 \text{ g urea}}\right) = 0.150 \text{ mol urea}$$

$$10.0 \text{ mL water} \left(\frac{1.00 \text{ g}}{1 \text{ mL}}\right)\left(\frac{1 \text{ mol water}}{18.02 \text{ g water}}\right) = 0.555 \text{ mol water}$$

$$X_{water} = \frac{0.555 \text{ mol water}}{0.555 \text{ mol water} + 0.150 \text{ mol urea}} = 0.787$$

$$P_{H_2O} = X_{H_2O} P°_{H_2O} = (0.787)(22.4 \text{ mm Hg}) = 17.6 \text{ mm Hg}$$

13.27. Using Raoult's Law, we know that the vapor pressure of pure water ($P°$) multiplied by the mole fraction (X) of the solvent gives the vapor pressure of the solvent above the solution (P).

$P_{water} = X_{water} P°_{water}$ The vapor pressure of pure water at 90 °C is 525.8 mmHg (from Appendix G). Since the P_{water} is given as 457 mmHg, the mole fraction of the water is:

$$X_{water} = \frac{457 \text{ mmHg}}{525.8 \text{ mmHg}} = 0.869$$

The 2.00 kg of water corresponds to a mole fraction of 0.869. This mass of water corresponds to:

$$2.00 \times 10^3 \text{ g H}_2\text{O} \left(\frac{1 \text{ mol H}_2\text{O}}{18.02 \text{ g H}_2\text{O}}\right) = 111 \text{ mol water.}$$

Representing moles of ethylene glycol as x we can write:

$$X_{water} = \frac{\text{mol water}}{\text{mol water + mol ethylene glycol}} = \frac{111 \text{ mol}}{111 \text{ mol} + x} = 0.869$$

$0.869(111 \text{ mol} + x) = 111 \text{ mol}$

$x = 16.7$ mol ethylene glycol

$16.7 \text{ mol ethylene glycol} \left(\frac{62.07 \text{ g ethylene glycol}}{1 \text{ mol ethylene glycol}} \right) = 1.04 \times 10^3$ g ethylene glycol

13.28. Vapor pressure of CCl_4—I_2 solution at 65 °C:

Moles of each substance:

$105 \text{ g } I_2 \left(\frac{1 \text{ mol } I_2}{253.8 \text{ g } I_2} \right) = 0.414 \text{ mol } I_2$ $325 \text{ g } CCl_4 \left(\frac{1 \text{ mol } CCl_4}{153.8 \text{ g } CCl_4} \right) = 2.11 \text{ mol } CCl_4$

$X_{CCl_4} = \frac{2.11 \text{ mol}}{2.11 \text{ mol} + 0.414 \text{ mol}} = 0.836$ and using Raoult's Law:

$P_{CCl_4} = X_{CCl_4} P°_{CCl_4} = (0.836)(531 \text{ mm Hg}) = 444$ mm Hg

Boiling Point Elevation

13.29. Benzene normally boils at a temperature of 80.10 °C.

Calculate the ΔT, using the equation $\Delta T = K_{bp} \times m_{solute}$:

The molality of the solution is $\frac{0.200 \text{ mol}}{0.125 \text{ kg solvent}}$ or 1.60 m; K_{bp} for benzene is +2.53 °C/m

So $\Delta T = K_{bp} \times m_{solute} = +2.53$ °C/m × 1.60 m = 4.05 °C.

The new boiling point is: 80.10 °C + 4.05 °C = 84.15 °C

13.30. Boiling point of solution with 15.0 g of urea in 0.500 kg water:

$15.0 \text{ g urea} \left(\frac{1 \text{ mol urea}}{60.06 \text{ g urea}} \right) = 0.250$ mol urea

$m_{urea} = \frac{0.250 \text{ mol urea}}{0.500 \text{ kg water}} = 0.500 \ m$

$\Delta T_{bp} = (0.5121 \text{ °C}/m)(0.500 \ m) = 0.256$ °C and the new boiling point:

$T_{bp} = 100.00$ °C + 0.256 °C = 100.26 °C

13.31. Calculate the molality of acenaphthene, $C_{12}H_{10}$, in the solution.

$$0.515 \text{ g } C_{12}H_{10} \times \frac{1 \text{ mol } C_{12}H_{10}}{154.2 \text{ g } C_{12}H_{10}} = 3.34 \times 10^{-3} \text{ mol } C_{12}H_{10}$$

and the molality is: $\dfrac{3.34 \times 10^{-3} \text{ mol acenaphthene}}{0.0150 \text{ kg CHCl}_3} = 0.223$ molal

the boiling point *elevation* is: $\Delta T = m \cdot K_{bp} = (0.223 \, m)(+3.63 \, °C/m) = 0.808 \, °C$

and the boiling point will be $61.70 + 0.808 = 62.51 \, °C$

13.32. Mass of glycerol in solution:

$\Delta T_{bp} = 104.4 \, °C - 100.0 \, °C = 4.4 \, °C$

$$m_{solute} = \frac{\Delta T_{bp}}{K_{bp}} = \frac{4.4 \, °C}{0.5121 \, °C/m} = 8.6 \, m$$

$$0.735 \text{ kg water} \left(\frac{8.6 \text{ mol glycerol}}{1 \text{ kg water}} \right) = 6.3 \text{ mol glycerol}$$

$$6.3 \text{ mol glycerol} \left(\frac{92.1 \text{ g glycerol}}{1 \text{ mol glycerol}} \right) = 580 \text{ g glycerol}$$

$$735 \text{ g water} \left(\frac{1 \text{ mol water}}{18.02 \text{ g water}} \right) = 40.8 \text{ mol water}$$

$$X_{glycerol} = \frac{6.3 \text{ mol glycerol}}{6.3 \text{ mol glycerol} + 40.8 \text{ mol water}} = 0.13$$

Freezing Point Depression

13.33. The solution freezes 16.0 °C lower than pure water.

(a) We can calculate the molality of the ethanol: $\Delta T_{fp} = mK_{fp}$

$-16.0 \, °C = m \, (-1.86 \, °C/\text{molal})$ and $\dfrac{-16.0 \, °C}{-1.86 \, \frac{°C}{\text{molal}}} = 8.60 = $ molality of the alcohol

(b) If the molality is 8.60 then there are 8.60 moles of C_2H_5OH (8.60 mol × 46.07 g/mol = 396 g) in 1000 g of H_2O.

The weight percent of alcohol is $\dfrac{396 \text{ g}}{1396 \text{ g}} \times 100 = 28.4 \%$ ethanol

13.34. Mass of ethylene glycol added:

$$m_{solution} = \frac{\Delta T_{fp}}{K_{fp}} = \frac{-15.0 \,°C}{-1.86 \,°C/m} = 8.06 \, m$$

$$\left(\frac{8.06 \text{ mol ethylene glycol}}{1 \text{ kg water}}\right)\left(\frac{62.07 \text{ g ethylene glycol}}{1 \text{ mol ethylene glycol}}\right)(5.0 \text{ kg water}) = 2500 \text{ g ethylene glycol}$$

13.35. Freezing point of a solution containing 15.0 g sucrose in 225 g water:

(1) Calculate the molality of sucrose in the solution:

$$15.0 \text{ g sucrose}\left(\frac{1 \text{ mol sucrose}}{342.30 \text{ g sucrose}}\right) = 0.0438 \text{ mol sucrose}$$

$$\frac{0.0438 \text{ mol } C_{12}H_{22}O_{11}}{0.225 \text{ kg } H_2O} = 0.195 \text{ molal}$$

(2) Use the ΔT_{fp} equation to calculate the freezing point change:

$\Delta T_{fp} = mK_{fp} = (0.195 \text{ molal})(-1.86 \,°C/\text{molal}) = -0.362 \,°C$

The solution is expected to begin freezing at $-0.362 \,°C$.

13.36. Will the wine freeze at $-20 \,°C$:

In an 11% ethanol solution, there are 11 g ethanol and 100 g – 11 g = 89 g of water for each 100 g of solution.

Mole of ethanol: $11 \text{ g} \times \frac{1 \text{ mol } C_2H_5OH}{46.1 \text{ g}} = 0.24 \text{ mol } C_2H_5OH$

Molality of the solution: $m_{solution} = \frac{0.24 \text{ mol ethanol}}{0.089 \text{ kg water}} = 2.7 \, m$

$\Delta T_{fp} = (-1.86 \,°C/m)(2.7 \, m) = -5.0 \,°C$
$T_{fp} = -5.0 \,°C$, so the wine will freeze if it is chilled to $-20 \,°C$.

Osmosis

13.37. Net water movement in different sucrose solutions:

Water will flow through the semipermeable membrane from the 0.01 M solution to the 0.1 M solution to dilute the more concentrated sucrose solution.

13.38. Direction of water movement between pure water and aqueous solution:

Water will flow through the semipermeable membrane from the pure water to the aqueous solution to dilute the aqueous solution.

13.39. Osmotic pressure of glycerol solution:

25 °C + 273.15 K = 298 K

$$5.00 \text{ g glycerol} \left(\frac{1 \text{ mol glycerol}}{92.09 \text{ g glycerol}} \right) = 0.0543 \text{ mol glycerol}$$

$$\text{Molarity} = \frac{0.0543 \text{ mol glycerol}}{0.250 \text{ L}} = 0.217 \text{ M glycerol}$$

$$\Pi = cRT = (0.217 \text{ M})\left(0.082057 \frac{\text{L} \cdot \text{atm}}{\text{K} \cdot \text{mol}} \right)(298 \text{ K}) = 5.31 \text{ atm}$$

13.40. Osmotic pressure of sucrose solution:

$$\Pi = cRT = (0.0010 \text{ M})(0.082057 \text{ L} \cdot \text{atm/K} \cdot \text{mol})(298 \text{ K}) = 0.0245 \text{ atm}$$

$$0.0245 \text{ atm} \left(\frac{760 \text{ mm Hg}}{1 \text{ atm}} \right) = 18.6 \text{ mm Hg}$$

13.41. Assume we have 100 g of this solution, the number of moles of phenylalanine is

$$3.00 \text{ g phenylalanine} \times \frac{1 \text{ mol phenylalanine}}{165.2 \text{ g phenylalanine}} = 0.0182 \text{ mol phenylalanine}$$

The molality of the solution is: $\frac{0.0182 \text{ mol phenylalanine}}{0.09700 \text{ kg water}} = 0.187 \text{ molal}$

(a) The freezing point:

$\Delta T_{fp} = mK_{fp} = (0.187 \ m)(-1.86 \ °C/m) = -0.348 \ °C$

The new freezing point is 0.0 °C – 0.348 °C = –0.348 °C.

(b) The boiling point of the solution

$\Delta T_{bp} = mK_{bp} = (0.187 \ m)(0.5121 \ °C/m) = +0.0959 \ °C$

The new boiling point is 100.000 + 0.0959 = +100.0959 °C

(c) The osmotic pressure of the solution:

If we assume that the **Molarity** of the solution is equal to the **molality**, then the osmotic pressure should be:

$$\Pi = (0.187 \text{ mol/L})(0.0821 \frac{\text{L} \cdot \text{atm}}{\text{K} \cdot \text{mol}})(298 \text{ K}) = 4.58 \text{ atm}$$

The osmotic pressure will be most easily measured, since the magnitudes of osmotic pressures (large values) result in decreased experimental error.

13.42. Mass of sucrose dissolved in solution:

$$28.7 \text{ mm Hg}\left(\frac{1 \text{ atm}}{760 \text{ mm Hg}}\right) = 0.0378 \text{ atm}$$

$$\Pi = cRT \text{ thus } c = \frac{\Pi}{RT}$$

$$c = \frac{0.0378 \text{ atm}}{\left(0.082057 \text{ L·atm}/\text{K·mol}\right)(298 \text{ K})} = 0.00154 \text{ M}$$

$$0.00154 \text{ M}(0.500 \text{ L}) = 0.00772 \text{ moles sucrose}$$

$$0.00772 \text{ moles sucrose}\left(\frac{342.3 \text{ g sucrose}}{1 \text{ mole sucrose}}\right) = 0.264 \text{ g sucrose}$$

Colligative Properties and Molar Mass Determination

13.43. The change in the temperature of the boiling point is (80.26 – 80.10) °C or 0.16 °C.

Using the equation $\Delta T = m_{solute}K_{bp}$; 0.16 °C = m_{solute} (+2.53 °C/m), and the molality is:

$$\frac{0.16 \text{ °C}}{+2.53 \text{ °C/m}} = 0.063 \text{ molal}$$

The solution contains 11.12 g of solvent (or 0.01112 kg solvent). We can calculate the # of moles of the orange compound, since we know the molality:

$$0.063 \text{ molal} = \frac{x \text{ mol compound}}{0.01112 \text{ kg solvent}}$$

$$x = (0.063 \text{ molal})(0.01112 \text{ kg solvent}) = 7.0 \times 10^{-4} \text{ mol compound}$$

This number of moles of compound has a mass of 0.255 g, so 1 mol of compound is:

$$\frac{0.255 \text{ g compound}}{7.0 \times 10^{-4} \text{ mol}} = 360 \text{ g/mol}$$

The empirical formula, $C_{10}H_8Fe$, has a molar mass of 184 g, so the # of "empirical formula units" in one molecular formula is:

$$\frac{360 \text{ g/mol molecular formula}}{184 \text{ g/mol empirical formula}} = 2 \text{ mol empirical formula/molecular formula or a molecular formula of } C_{20}H_{16}Fe_2.$$

13.44. Molar mass of BHA:

ΔT_{bp} = 62.22 °C – 61.70 °C = 0.52 °C so molality is: $m_{BHA} = \dfrac{\Delta T_{bp}}{K_{bp}} = \dfrac{0.52 \text{ °C}}{3.63 \text{ °C/m}} = 0.14 \text{ } m$

From the definition of molality: $0.0250 \text{ kg CHCl}_3 \left(\dfrac{0.14 \text{ mol BHA}}{1 \text{ kg CHCl}_3} \right) = 0.0036 \text{ mol BHA}$

and $\dfrac{0.640 \text{ g BHA}}{0.0036 \text{ mol BHA}} = 178.7 \text{ g/mol} = 180 \text{ g/mol}$ (2sf)

13.45. The change in the temperature of the boiling point is (61.82 – 61.70) °C or 0.12 °C.
Using the equation $\Delta T = mK_{bp}$; $0.12 \text{ °C} = m_{solute}(+3.63 \text{ °C}/m)$, and the molality is:
$\dfrac{0.12 \text{ °C}}{3.63 \text{ °C}/m}$ = 0.033 molal. The solution contains 25.0 g of solvent (or 0.0250 kg solvent).

We can calculate the # of moles of benzyl acetate:

$0.033 \text{ molal} = \dfrac{x \text{ mol compound}}{0.0250 \text{ kg solvent}}$ or 8.3×10^{-4} mol compound.

This number of moles of benzyl acetate has a mass of 0.125 g, so 1 mol of benzyl acetate is:
$\dfrac{0.125 \text{ g compound}}{8.3 \times 10^{-4} \text{ mol}} = 150$ g/mol. (2 sf)

13.46. Molecular formula for anthracene: $\Delta T_{bp} = 80.34 \text{ °C} - 80.10 \text{ °C} = 0.24 \text{ °C}$

$m_{anthracene} = \dfrac{\Delta T_{bp}}{K_{bp}} = \dfrac{0.24 \text{ °C}}{2.53 \text{ °C}/m} = 0.095 \ m$

$\dfrac{0.095 \text{ mol anthracene}}{1 \text{ kg benzene}} \cdot 0.0300 \text{ kg benzene} = 0.0028 \text{ mol anthracene}$

$\dfrac{0.500 \text{ g anthracene}}{0.0028 \text{ mol anthracene}} = 180 \text{ g/mol}$

The molar mass of the empirical formula (C_7H_5) is 89 g/mol.

$\dfrac{180 \text{ g/mol}}{89 \text{ g/mol}} = 2$ so, the molecular formula is 2 × (C_7H_5) or $C_{14}H_{10}$

13.47. To determine the molar mass, first determine the molality of the solution.

$-0.040 \text{ °C} = m_{solute}(-1.86 \text{ °C}/m)$

Molality = 0.0215 m (or 0.022 to 2sf). Now we can solve for MM:

$0.022 \text{ molal} = \dfrac{\dfrac{0.180 \text{ g solute}}{MM}}{0.0500 \text{ kg water}}$ and MM = 167 or 170 (2sf)

13.48. Molar mass of aluminon:

$$m_{solute} = \frac{\Delta T_{fp}}{K_{fp}} = \frac{-0.197\ °C}{-1.86\ °C/m} = 0.106\ m$$

With the molality, we can solve for mol of aluminon:

$$0.0500\ \text{kg water}\left(\frac{0.106\ \text{mol aluminon}}{1\ \text{kg water}}\right) = 0.00530\ \text{mol aluminon}$$

$$\frac{2.50\ \text{g aluminon}}{0.00530\ \text{mol aluminon}} = 472\ \text{g/mol}$$

13.49. The molar mass of bovine insulin with a solution having an osmotic pressure of 3.1 mm Hg:

$$3.1\ \text{mm Hg} \times \frac{1\ \text{atm}}{760\ \text{mm Hg}} = (c)(0.082057\ \frac{L \cdot atm}{K \cdot mol})(298\ K)$$

$$c = 1.67 \times 10^{-4}\ M\ \text{or}\ 1.7 \times 10^{-4}\ M\ (2\ sf)$$

The definition of molarity is #mol/L. Substituting into the definition we obtain:

$$\frac{1.7 \times 10^{-4}\ \text{mol bovine insulin}}{L} = \frac{\frac{1.00\ \text{g bovine insulin}}{MM}}{1\ L};\ \text{solving for MM} = 6.0 \times 10^3\ \text{g/mol}$$

13.50. Molar mass of maltose:

$$136\ \text{mm Hg}\left(\frac{1\ \text{atm}}{760\ \text{mm Hg}}\right) = 0.179\ \text{atm}$$

$$\Pi = cRT,\ \text{thus}\ c = \frac{\Pi}{RT}$$

$$c = \frac{0.179\ \text{atm}}{(0.082057\ L \cdot atm/K \cdot mol)(298\ K)} = 0.00732\ M$$

$$0.00732\ \frac{\text{mol maltose}}{L}(0.500\ L) = 0.00366\ \text{mol maltose}$$

$$\frac{1.25\ \text{g maltose}}{0.00366\ \text{mol maltose}} = 342\ \text{g/mol}$$

Colligative Properties of Ionic Compounds

13.51. The number of moles of LiF is: $52.5 \text{ g LiF} \cdot \dfrac{1 \text{ mol LiF}}{25.94 \text{ g LiF}} = 2.02 \text{ mol LiF}$

So $\Delta T_{fp} = \dfrac{2.02 \text{ mol LiF}}{0.306 \text{ kg H}_2\text{O}} (-1.86 \text{ °C/molal})(2) = -24.6 \text{ °C}$ (LiF has an i = 2)

The anticipated freezing point is then 24.6 °C lower than pure water (0.0 °C) or –24.6 °C

13.52. Mass of NaCl to add to 3.0 kg of water:

$\Delta T_{fp} = K_{fp} \cdot m_{NaCl} \cdot i$

$m_{NaCl} = \dfrac{\Delta T_{fp}}{K_{fp} \cdot i} = \dfrac{-10. \text{ °C}}{(-1.86 \text{ °C}/m)(1.83)} = 2.9 \, m$

$3.0 \text{ kg water} \left(\dfrac{58.44 \text{ g NaCl}}{1 \text{ mol NaCl}} \right) \left(\dfrac{2.9 \text{ mol NaCl}}{1 \text{ kg water}} \right) = 5.2 \times 10^2 \text{ g NaCl}$

13.53. Solutions given in order of increasing melting point (lowest melting point listed first):

The solution with the **greatest concentration** of particles will have the lowest melting point. The effective molality (= molality × i) of solutions is:

	Solution	Particles / formula unit	Identity of particles	Total molality
(a)	0.1 m sugar	1	covalently bonded molecules	0.1 m × 1 = 0.1 m
(b)	0.1 m NaCl	2	Na$^+$, Cl$^-$	0.1 m × 2 = 0.2 m
(c)	0.08 m CaCl$_2$	3	Ca^{2+}, 2 Cl$^-$	0.08 m × 3 = 0.24 m
(d)	0.04 m Na$_2$SO$_4$	3	2 Na$^+$, SO$_4^{2-}$	0.04 m × 3 = 0.12 m

The melting points would increase in the order:

0.08 m CaCl$_2$ < 0.1 m NaCl < 0.04 m Na$_2$SO$_4$ < 0.1 m sugar

13.54. Arrange solutions in order of decreasing freezing point:

	Solution	Particles / formula unit	Total molality
(a)	0.20 *m* Ethylene glycol	1	0.20 *m* × 1 = 0.20 *m*
(b)	0.12 *m* K$_2$SO$_4$	3	0.12 *m* × 3 = 0.36 *m*
(c)	0.10 *m* MgCl$_2$	3	0.10 *m* × 3 = 0.30 *m*
(d)	0.12 *m* KBr	2	0.12 *m* × 2 = 0.24 *m*

Freezing point decreases as the particle concentration increases:
0.20 *m* ethylene glycol > 0.12 *m* KBr > 0.10 *m* MgCl$_2$ > 0.12 *m* K$_2$SO$_4$

13.55. Estimate the osmotic pressure of human blood:

Concentration of ions in solution = (0.154 M)(1.85) = 0.285 M
$\Pi = cRT$ = (0.285 mol/L)(0.082057 L·atm/K·mol)(310. K) = 7.25 atm

13.56. Osmotic pressure of NaCl solution in water at 0°C:

Concentration of ions in solution = (0.0120 M)(1.94) = 0.0233 M
$\Pi = cRT$ = (0.0233 mol/L)(0.082057 $\frac{\text{L·atm}}{\text{K·mol}}$)(273 K) = 0.522 atm

Colloids

13.57. AgCl colloidal dispersion:

AgCl precipitating from solution is an example of a solid dispersing in a liquid which is a sol.

13.58. Mayonnaise colloidal dispersion:

Mayonnaise is an example of a liquid, vinegar, dissolved in a liquid, vegetable oil, which is an emulsion. These normally insoluble liquids are able to form a stable mixture due to the addition of an emulsifying agent found in egg yolks.

13.59. Upon mixing of solutions of barium chloride and sodium sulfate:

(a) Equation: BaCl$_2$(aq) + Na$_2$SO$_4$(aq) → BaSO$_4$(s) + 2 NaCl(aq)

(b) The BaSO$_4$ initially formed is of a colloidal size — not large enough to precipitate fully.

(c) The particles of BaSO$_4$ grow with time, owing to a gradual loss of charge, and become large enough to have gravity affect them—and settle to the bottom.

13.60. Regarding the spheres of a dispersed phase:

With a diameter = 1.0×10^2 nm and radius = 50. nm

(a) Volume of spheres: $V = 4/3 \, \pi \, r^3 = 4/3 \, \pi \, (50. \text{ nm})^3 = 5.2 \times 10^5$ nm^3
Area of spheres: $A = 4 \, \pi \, r^2 = 4 \, \pi \, (50. \text{ nm})^2 = 3.1 \times 10^4$ nm^2

(b) Spheres to provide the total volume:

$$1.0 \text{ cm}^3 \left(\frac{1 \times 10^9 \text{ nm}}{100 \text{ cm}}\right)^3 \left(\frac{1 \text{ sphere}}{5.2 \times 10^2 \text{ nm}^3}\right) = 1.9 \times 10^{15} \text{ spheres}$$

Total surface area in square meters:

$$1.9 \times 10^{15} \text{ spheres} \left(\frac{3.1 \times 10^4 \text{ nm}^2}{1 \text{ sphere}}\right) \left(\frac{1 \text{ m}}{1 \times 10^9 \text{ nm}}\right)^2 = 60. \text{ m}^2$$

General Questions

13.61. A solution of 0.52 g of phenylcarbinol in 25.0 g of water melts at – 0.36 °C.

$\Delta T_{fp} = m_{solute} K_{fp})$

–0.36 °C = m_{solute}(–1.86 °C/molal), and solving for molality:

$$m_{solute} = \frac{-0.36 \, °C}{-1.86 \, °C/m} = 0.19 \, m$$

$0.19 \, \dfrac{\text{mol phenylcarbinol}}{\text{kg water}} \times 0.0250 \text{ kg water} = 4.8 \times 10^{-3}$ mol phenylcarbinol

The molar mass of phenylcarbinol is $\dfrac{0.52 \text{ g phenylcarbinol}}{4.8 \times 10^{-3} \text{ mol phenylcarbinol}} = 110$ g/mol (2 sf)

13.62. Regarding the aqueous solutions:

(a) The higher boiling point: 0.10 m Na$_2$SO$_4$ solution has a higher particle concentration (0.30 m) so it should have the higher boiling point.

(b) The higher vapor pressure: The 0.15 m Na$_2$SO$_4$ solution has a lower particle concentration (0.45 m) so it should have the higher water vapor pressure.

13.63. Arranged the solutions in order of (i) increasing vapor pressure of water and (ii) increasing boiling points:

Both of these colligative properties depend on the concentration of particles dissolved in solution. For each solution, calculate the effective molality of particles ($m_{solute} \times i$):

$\quad$ ethylene glycol = 0.35 m × 1 = 0.35 m
$\quad$ sugar $\quad\quad\quad\quad$ = 0.50 m × 1 = 0.50 m
$\quad$ KBr $\quad\quad\quad\quad\;$ = 0.20 m × 2 = 0.40 m
$\quad$ Na$_2$SO$_4$ $\quad\quad\;\;$ = 0.20 m × 3 = 0.60 m

(i) The solution with the highest water vapor pressure would have the **lowest particle concentration**, since according to Raoult's Law, the vapor pressure of the water in the solution is directly proportional to the mole fraction of the water. The lower the number of particles, the greater the mole fraction of water, and the greater the vapor pressure. Hence the order of *increasing* vapor pressure is:

0.20 *m* Na₂SO₄ < 0.50 *m* sugar < 0.20 *m* KBr < 0.35 *m* HOCH₂CH₂OH (ethylene glycol)

(ii) The solution with the highest boiling point will have the highest particle concentration since ΔT_{bp} is proportional to the concentration of particles. Arranged in *increasing* boiling points:

0.35 *m* ethylene glycol < 0.20 *m* KBr < 0.50 *m* sugar < 0.20 *m* Na₂SO₄

13.64. Weight percent, mole fraction, and molality of NaCl in water:

(a) $\dfrac{1130 \text{ g NaCl}}{1130 \text{ g NaCl} + 7250 \text{ g H}_2\text{O}} \times 100\% = 13.5\%$

(b)

$1130 \text{ g NaCl} \left(\dfrac{1 \text{ mol NaCl}}{58.44 \text{ g NaCl}} \right) = 19.3 \text{ mol NaCl}$

$7250 \text{ g H}_2\text{O} \left(\dfrac{1 \text{ mol H}_2\text{O}}{18.02 \text{ g H}_2\text{O}} \right) = 402 \text{ mol H}_2\text{O}$

$X_{\text{NaCl}} = \dfrac{19.3 \text{ mol NaCl}}{19.3 \text{ mol NaCl} + 402 \text{ mol H}_2\text{O}} = 0.0459$

(c) $m_{\text{NaCl}} = \dfrac{19.3 \text{ mol NaCl}}{7.25 \text{ kg H}_2\text{O}} = 2.67 \, m$

13.65. For DMG, (CH₃CNOH)₂, the MM is 116.1 g/mol

So, 53.0 g is: $53.0 \text{ g} \times \dfrac{1 \text{ mol DMG}}{116.12 \text{ g DMG}} = 0.456 \text{ mol DMG}$

525 g of C₂H₅OH is: $525 \text{ g} \times \dfrac{1 \text{ mol C}_2\text{H}_5\text{OH}}{46.07 \text{ g C}_2\text{H}_5\text{OH}} = 11.4 \text{ mol C}_2\text{H}_5\text{OH}$

(a) mole fraction of DMG: $\dfrac{0.456 \text{ mol}}{(11.4 + 0.456) \text{ mol}} = 0.0385$ mole fraction DMG

(b) molality of the solution: $\dfrac{0.456 \text{ mol DMG}}{0.525 \text{ kg}} = 0.869$ molal DMG

(c) $P_{\text{alcohol}} = P°_{\text{alcohol}} \times X_{\text{alcohol}} = (760.0 \text{ mm Hg})(1 - 0.0385) = 730.7 \text{ mm Hg}$

(d) boiling point of the solution:

$\Delta T_{bp} = m_{solute} \times K_{bp} = (0.869)(+1.22 \,°C/m) = 1.06 \,°C$

The new boiling point is 78.4 °C + 1.06 °C = 79.46 °C or 79.5 °C

13.66. Begin with a 10.7 m solution of NaOH, density of 1.33 g/cm³:

$$10.7 \text{ mol NaOH}\left(\frac{40.00 \text{ g NaOH}}{1 \text{ mol NaOH}}\right) = 428 \text{ g NaOH}$$

$$1\times10^3 \text{ g H}_2\text{O}\left(\frac{1 \text{ mol H}_2\text{O}}{18.02 \text{ g H}_2\text{O}}\right) = 55.5 \text{ mol H}_2\text{O}$$

$$(428 \text{ g} + 1000 \text{ g})\left(\frac{1 \text{ cm}^3}{1.33 \text{ g}}\right)\left(\frac{1 \text{ mL}}{1 \text{ cm}^3}\right)\left(\frac{1 \text{ L}}{1000 \text{ mL}}\right) = 1.07 \text{ L solution}$$

(a) $X_{NaOH} = \dfrac{10.7 \text{ mol NaOH}}{10.7 \text{ mol NaOH} + 55.5 \text{ mol H}_2\text{O}} = 0.162$

(b) Weight % NaOH $= \left(\dfrac{428 \text{ g NaOH}}{428 \text{ g NaOH} + 1000 \text{ g H}_2\text{O}}\right)100\% = 30.0\%$ NaOH

(c) Molarity $= \dfrac{10.7 \text{ mol NaOH}}{1.07 \text{ L}} = 9.97$ M NaOH

13.67. Concentrated NH_3 is 14.8 M and has a density of 0.90 g/cm³.

(1) The molality of the solution:

Molality is defined as moles NH_3/kg solvent, so begin by deciding the mass of NH_3 in 1 L of solution, and the mass of water in that 1 L. Since the density = 0.90 g/mL, then 1 L (1000 mL) has a mass of 900 g.

The mass of NH_3 present in 14.8 mol NH_3 = 14.8 mol $NH_3\left(\dfrac{17.03 \text{ g NH}_3}{1 \text{ mol NH}_3}\right) = 252$ g NH_3

Since 1 L has a mass of 900 g (2 sf) and 252 g is NH_3, the difference (900 g – 252 g) is solvent.
So, 1 L of solution has 648 g water.

$$\left(\frac{14.8 \text{ mol NH}_3}{1 \text{ L}}\right)\left(\frac{1 \text{ L}}{648 \text{ g H}_2\text{O}}\right)\left(\frac{1000 \text{ g H}_2\text{O}}{1 \text{ kg H}_2\text{O}}\right) = 22.8 \,m \text{ or } 23 \,m \text{ (2 sf)}$$

(2) The mole fraction of ammonia is:

Calculate the # of moles of water present:

$$648 \text{ g H}_2\text{O} \cdot \frac{1 \text{ mol H}_2\text{O}}{18.02 \text{ g H}_2\text{O}} = 35.96 \text{ mol H}_2\text{O} \text{ (retaining some extra digits for now)}$$

The mole fraction NH₃ is: $X_{\text{NH}_3} = \dfrac{14.8 \text{ mol NH}_3}{14.8 \text{ mol NH}_3 + 35.96 \text{ mol H}_2\text{O}} = 0.29$

(3) Weight percentage of NH₃:

14.8 mol NH₃ has a mass of 252 g, and 1 L of the solution has a mass of 900 g.

$$\text{Weight \% NH}_3 = \left(\frac{252 \text{ g NH}_3}{900 \text{ g solution}}\right) 100\% = 28\% \text{ (2 sf)}$$

13.68. Molality of Ca(NO₃)₂; Total ion molality:

$$2.00 \text{ g Ca(NO}_3)_2 \left(\frac{1 \text{ mol Ca(NO}_3)_2}{164.1 \text{ g Ca(NO}_3)_2}\right) = 0.0122 \text{ mol Ca(NO}_3)_2$$

$$m_{\text{Ca(NO}_3)_2} = \frac{0.0122 \text{ mol Ca(NO}_3)_2}{0.75 \text{ kg H}_2\text{O}} = 0.016 \, m$$

$$m_{\text{ions}} = 0.016 \, m \text{ Ca(NO}_3)_2 \left(\frac{3 \text{ mol ions}}{1 \text{ mol Ca(NO}_3)_2}\right) = 0.049 \, m$$

13.69. To make a 0.100 m solution of ions, we need a ratio of #moles of ions/kg solvent that is 0.100.

$$0.100 \, m = \frac{x}{0.125 \text{ kg solvent}}$$

x = 0.0125 mol ions

The salt will dissociate into 3 ions per formula unit (2 Na⁺ and 1 SO₄²⁻).

The amount of Na₂SO₄ is:

$$0.0125 \text{ mol ions} \left(\frac{1 \text{ mol Na}_2\text{SO}_4}{3 \text{ mol ions}}\right)\left(\frac{142.04 \text{ g Na}_2\text{SO}_4}{1 \text{ mol Na}_2\text{SO}_4}\right) = 0.592 \text{ g Na}_2\text{SO}_4$$

13.70. Solution with highest boiling point; lowest freezing point; highest water vapor pressure:

Substance	Concentration	Concentration × i	Particle concentration
(i) HOCH$_2$CH$_2$OH	0.20 m	0.20 × 1	= 0.20 m
(ii) CaCl$_2$	0.10 m	0.10 × 3	= 0.30 m
(iii) KBr	0.12 m	0.12 × 2	= 0.24 m
(iv) Na$_2$SO$_4$	0.12 m	0.12 × 3	= 0.36 m

(a) 0.12 m Na$_2$SO$_4$ Boiling point increases as the particle concentration increases.

(b) 0.12 m Na$_2$SO$_4$ Freezing point decreases as the particle concentration increases.

(c) 0.20 m HOCH$_2$CH$_2$OH Vapor pressure increases as particle concentration decreases.

13.71. Solution properties:

(a) The solution with the higher boiling point:

Recall that $\Delta T_{bp} = m_{solute} \times K_{bp} \times i$. The difference in these solutions will be in the product ($m_{solute} \cdot i$). The products for these solutions are:

sugar: 0.30 m × 1 = 0.30 m (sugar molecules remain as one unit)

KBr: 0.20 m × 2 = 0.40 m (KBr dissociates into K$^+$ and Br$^-$ ions)

The KBr solution will provide the larger ΔT_{bp}.

(b) The solution with the lower freezing point:

Using the same logic as in part (a), NH$_4$NO$_3$ provides 2 ions/formula unit while Na$_2$CO$_3$ provides 3. The product, ($m_{solute} \times i$), is larger for Na$_2$CO$_3$, so the **Na$_2$CO$_3$** solution gives the lower freezing point.

13.72. Boiling point of the NaCl solution: Calculate the molality:

$$39 \text{ g NaCl} \left(\frac{1 \text{ mol NaCl}}{58.44 \text{ g NaCl}} \right) = 0.67 \text{ mol NaCl}$$

$$m_{NaCl} = \frac{0.67 \text{ mol NaCl}}{0.100 \text{ kg H}_2\text{O}} = 6.7 \ m$$

$\Delta T_{bp} = m_{NaCl} \times K_{bp} \times i = (6.7 \ m)(0.5121 \ °C/m)(1.8) = 6.2 \ °C$

$T_{bp} = 100.0 \ °C + 6.2 \ °C = 106.2 \ °C$

13.73. The change in temperature of the freezing point is: $\Delta T_{fp} = m_{solute} \times K_{fp} \times i$

Calculate the molality:

$$35.0 \text{ g CaCl}_2 \left(\frac{1 \text{ mol CaCl}_2}{111.0 \text{ g CaCl}_2} \right) = 0.315 \text{ mol CaCl}_2$$

$$\text{Molality} = \frac{0.315 \text{ mol CaCl}_2}{0.150 \text{ kg H}_2\text{O}} = 2.10 \, m$$

$\Delta T_{fp} = m_{solute} \times K_{fp} \times i = (2.10 \, m)(-1.86 \, °C/molal)(2.7) = -11$ (2 sf)

The freezing point of the solution is $0.0\,°C - 11\,°C = -11\,°C$

13.74. Molecular formula of the solute:

$\Delta T_{bp} = 61.82\,°C - 61.70\,°C = 0.12\,°C$.
Calculate the molality:

$$m_{solute} = \frac{\Delta T_{bp}}{K_{bp}} = \frac{0.12\,°C}{3.63\,°C/m} = 0.033 \, m$$

$$0.0250 \text{ kg CHCl}_3 \left(\frac{0.033 \, m \text{ solute}}{1 \text{ kg CHCl}_3} \right) = 8.3 \times 10^{-4} \text{ mol solute}$$

$$\frac{0.135 \text{ g solute}}{8.3 \times 10^{-4} \text{ mol solute}} = 160 \text{ g/mol (2 sf)}$$

$$\frac{160 \text{ g/mol molecular formula}}{82 \text{ g/mol empirical formula}} = 2.0 \text{ empirical formulas/molecular formula}$$

The molecular formula is $2 \times (C_5H_6O)$ or $C_{10}H_{12}O_2$.

13.75. The molar mass of hexachlorophene if 0.640 g of the compound in 25.0 g of $CHCl_3$ boils at 61.93 °C: (*i* for hexachlorophene = 1)

$\Delta T_{bp} = m_{solute} \times K_{bp} = m_{solute}(+3.63\,°C/molal) = (61.93 - 61.70)\,°C$

Solving for m_{solute}: $\dfrac{0.23\,°C}{3.63\,°C/m} = 0.063 \, m$

Substitute into the definition for molality: molality = #mol/kg solvent

$$0.063 \text{ molal} = \frac{\dfrac{0.640 \text{ g hexachloraphene}}{MM}}{0.0250 \text{ kg}} \text{ and solving; } 4.0 \times 10^2 \text{ g/mol} = MM$$

13.76. Mass of ammonium formate that precipitates:

At 80 °C 1092 g NH4CHO2 will dissolve in 200. g of water (546 g/100 g).
At 0 °C only 204 g NH4CHO2 will dissolve in 200. g of water (102 g/100 g).

So, 1092 g – 204 g = 888 g NH4CHO2 precipitates at 0 °C.

13.77. Amount of N2 that can dissolve in water at 25°C:

Solubility of N2 = $k_H \times P_{N_2}$

= $(6.0 \times 10^{-4} \dfrac{mol}{kg \cdot bar}) \times 585 \text{ mmHg} \times \dfrac{1.01325 \text{ bar}}{760 \text{ mmHg}}$ = 4.7×10^{-4} mol N2/kg

13.78. Percent by mass of glycerol to lower the vapor pressure of water:

$X_{H_2O} = \dfrac{P_{H_2O}}{P°_{H_2O}} = 0.55$ and $X_{C_3H_5(OH)_3} = 1 - 0.55 = 0.45$

0.55 mol H$_2$O $\left(\dfrac{18.02 \text{ g H}_2\text{O}}{1 \text{ mol H}_2\text{O}} \right) = 9.9$ g H$_2$O

0.45 mol C$_3$H$_5$(OH)$_3$ $\left(\dfrac{92.09 \text{ g C}_3\text{H}_5\text{(OH)}_3}{1 \text{ mol C}_3\text{H}_5\text{(OH)}_3} \right) = 41$ g C$_3$H$_5$(OH)$_3$

Weight % = $\left(\dfrac{41 \text{ g C}_3\text{H}_5\text{(OH)}_3}{41 \text{ g C}_3\text{H}_5\text{(OH)}_3 + 9.9 \text{ g H}_2\text{O}} \right)$ 100 % = 81 %

13.79. Concerning the aqueous solution of 10.0 g of starch/liter:

(a) *Average* MM of starch if 10.0 g starch/L has an osmotic pressure = 3.8 mm Hg at 25 °C.

3.8 mm Hg × $\dfrac{1 \text{ atm}}{760 \text{ mm Hg}}$ = $(c)(0.08206 \dfrac{L \cdot atm}{K \cdot mol})(298 \text{ K})$

$c = 2.0 \times 10^{-4}$ (2 sf)

The definition of molarity is #mol/L.

Substituting into the definition we obtain:

$2.0 \times 10^{-4} \dfrac{\text{mol starch}}{1 \text{ L}} = \dfrac{\dfrac{10.0 \text{ g starch}}{MM}}{1 \text{ L}}$; solving for MM = 4.9×10^4 g/mol

(b) Freezing point of the solution:

$\Delta T_{fp} = m_{solute} \times K_{fp} \times i$ (assume that *i*=1 and that Molarity = molality for a dilute solution)

$\Delta T_{fp} = 2.0 \times 10^{-4}\,m \times (-1.86\,°C/m) = -3.8 \times 10^{-4}\,°C$. In essence the starch solution will freeze at the temperature of pure water. From this data we can assume that *it will NOT be easy* to measure the molecular weight of starch using this technique.

13.80. Vinegar is a 5% solution of acetic acid:

Mole fraction of acetic acid:

$$5\text{ g CH}_3\text{CO}_2\text{H}\left(\frac{1\text{ mol CH}_3\text{CO}_2\text{H}}{60.05\text{ g CH}_3\text{CO}_2\text{H}}\right) = 0.08\text{ mol CH}_3\text{CO}_2\text{H}$$

$$95\text{ g H}_2\text{O}\left(\frac{1\text{ mol H}_2\text{O}}{18.02\text{ g H}_2\text{O}}\right) = 5.3\text{ mol H}_2\text{O}$$

$$X_{\text{CH}_3\text{CO}_2\text{H}} = \frac{0.08\text{ mol CH}_3\text{CO}_2\text{H}}{0.08\text{ mol CH}_3\text{CO}_2\text{H} + 5.3\text{ mol H}_2\text{O}} = 0.02$$

Molality of acetic acid:

$$m_{\text{CH}_3\text{CO}_2\text{H}} = \frac{0.08\text{ mol CH}_3\text{CO}_2\text{H}}{0.0950\text{ kg H}_2\text{O}} = 0.9\,m$$

$$\frac{5\text{ g CH}_3\text{CO}_2\text{H}}{100\text{ g solution}} \times 1{,}000{,}000 = 5 \times 10^4\text{ ppm}$$

Calculating molarity requires knowing the total volume of the solution. Without knowing the density of the acetic acid solution, it is impossible to calculate the molarity of the solution.

13.81. The enthalpies of solution for Li₂SO₄ and K₂SO₄:

The process is MX(s) → MX(aq)

Using the data for **Li₂SO₄**:

$\Delta_{solution}H° = \Delta_fH°(aq) - \Delta_fH°(s) = (-1464.4\text{ kJ/mol}) - (-1436.4\text{ kJ/mol}) = -28.0\text{ kJ/mol}$

Using the data for **K₂SO₄**:

$\Delta_{solution}H° = \Delta_fH°(aq) - \Delta_fH°(s) = (-1413.0\text{ kJ/mol}) - (-1437.7\text{ kJ/mol}) = 24.7\text{ kJ/mol}$

Note that for Li₂SO₄ *the process is* **exothermic** *while for* K₂SO₄ *the process is endothermic.*

Similar data for LiCl and KCl:

For **LiCl**: $\Delta_fH°(aq) - \Delta_fH°(s) = (-445.6\text{ kJ/mol}) - (-408.7\text{ kJ/mol}) = -36.9\text{ kJ/mol}$

For **KCl**: $\Delta_fH°(aq) - \Delta_fH°(s) = (-419.5\text{ kJ/mol}) - (-436.7\text{ kJ/mol}) = 17.2\text{ kJ/mol}$

Note the similarities of the chloride salts, with the lithium salt being **exothermic** *while the potassium salt is* **endothermic**.

13.82. The molality and molarity of pure water at 25 °C:

The density of water is 0.997 g/cm³, so 1000. mL of water will have a mass of 997 g.

$$\left(\frac{997 \text{ g H}_2\text{O}}{1 \text{ L}}\right)\left(\frac{1 \text{ mol H}_2\text{O}}{18.02 \text{ g H}_2\text{O}}\right) = 55.3 \text{ M}$$

Now the molality: $\dfrac{55.3 \text{ mol H}_2\text{O}}{0.997 \text{ kg}} = 55.5 \, m$

13.83. Dalton's law says that the pressure of a mixture of gases (benzene and toluene) is the sum of the partial pressures. Using Raoult's Law:

$P_{benzene} = X_{benzene} \times P°_{benzene}$ and similarly for toluene, $P_{toluene} = X_{toluene} \times P°_{toluene}$.

The total pressure is: $P_{total} = P_{benzene} + P_{toluene}$

$$P_{total} = \left(\frac{2.0 \text{ mol benzene}}{3.0 \text{ mol}} \times 75 \text{ mm Hg}\right) + \left(\frac{1.0 \text{ mol toluene}}{3.0 \text{ mol}} \times 22 \text{ mm Hg}\right) = 57 \text{mm Hg}$$

What is the mole fraction of each component in the liquid and in the vapor?

The **mole fraction of the components in the liquid** are benzene: 2/3 = 0.67 and toluene: 1/3 = 0.33

The **mole fraction of the components in the vapor** are proportional to their pressures in the vapor state.

The mole fraction of benzene is: $\dfrac{50 \text{ mm Hg}}{57 \text{ mm Hg}} = 0.87$

The mole fraction of toluene would be (1 – 0.87) or 0.13.

13.84. Total vapor pressure over the solution:

Calculate moles of each component:

$$50.0 \text{ mL C}_2\text{H}_5\text{OH}\left(\frac{0.789 \text{ g C}_2\text{H}_5\text{OH}}{1 \text{ mL C}_2\text{H}_5\text{OH}}\right)\left(\frac{1 \text{ mol C}_2\text{H}_5\text{OH}}{46.07 \text{ g C}_2\text{H}_5\text{OH}}\right) = 0.856 \text{ mol C}_2\text{H}_5\text{OH}$$

$$50.0 \text{ mL H}_2\text{O}\left(\frac{0.998 \text{ g H}_2\text{O}}{1 \text{ mL H}_2\text{O}}\right)\left(\frac{1 \text{ mol H}_2\text{O}}{18.02 \text{ g H}_2\text{O}}\right) = 2.77 \text{ mol H}_2\text{O}$$

The mole fraction ethanol: $X_{C_2H_5OH} = \dfrac{0.856 \text{ mol C}_2\text{H}_5\text{OH}}{0.856 \text{ mol C}_2\text{H}_5\text{OH} + 2.77 \text{ mol H}_2\text{O}} = 0.236$

$P_{total} = P_{C_2H_5OH} + P_{H_2O} = X_{C_2H_5OH}P°_{C_2H_5OH} + (1 - X_{C_2H_5OH})P°_{H_2O}$

$P_{total} = (0.236)(43.6 \text{ mm Hg}) + (1 - 0.236)(17.5 \text{ mm Hg}) = 23.7 \text{ mm Hg}$

13.85. A 2.0 % aqueous solution of novocainium chloride (NC) is also 98.0 % in water. Assume that we begin with 100 g of solution. The molality of the solution is:

$$2.0 \text{ g NC} \left(\frac{1 \text{ mol NC}}{272.8 \text{ g NC}} \right) = 0.0073 \text{ mol NC}$$

$$\text{Molality} = \frac{0.0073 \text{ mol NC}}{0.0980 \text{ kg H}_2\text{O}} = 0.075 \text{ m}$$

Using the delta T equation: $\Delta T_{fp} = m_{solute} \times K_{fp} \times i$,

solve for i: $i = \dfrac{\Delta T_{fp}}{m_{solute} \times K_{fp}} = \dfrac{-0.237 \text{ °C}}{(0.075 \text{ m})(-1.86 \text{ °C}/m)} = 1.7$

So approximately **2 moles of ions are present per mole of compound**.

13.86. A solution is 4.00 % maltose and 96.00 % water:

(a) Molar mass of maltose:

$$m_{maltose} = \frac{\Delta T_{fp}}{K_{fp}} = \frac{-0.229 \text{ °C}}{-1.86 \text{ °C}/m} = 0.123 \text{ m}$$

Using the molality, calculate the MM of maltose:

$$0.09600 \text{ kg H}_2\text{O} \left(\frac{0.123 \text{ mol maltose}}{1 \text{ kg H}_2\text{O}} \right) = 0.0118 \text{ mol maltose}$$

$$\frac{4.00 \text{ g maltose}}{0.0118 \text{ mol maltose}} = 338 \text{ g/mol}$$

(b) Osmotic pressure of the solution:

$$100.00 \text{ g solution} \left(\frac{1 \text{ mL solution}}{1.014 \text{ g solution}} \right) = 98.62 \text{ mL solution}$$

$$c_{maltose} = \frac{0.0118 \text{ mol maltose}}{0.09682 \text{ L}} = 0.120 \frac{\text{mol}}{\text{L}}$$

$$\Pi = cRT = \left(0.120 \frac{\text{mol}}{\text{L}} \right) \left(0.082057 \frac{\text{L} \cdot \text{atm}}{\text{K} \cdot \text{mol}} \right) (298 \text{ K}) = 2.93 \text{ atm}$$

13.87. Concerning seawater:

(a) We can calculate the freezing point of sea water if we calculate the molality of the solution. Let's imagine that we have 1,000,000 (or 10^6) g of sea water. The amounts of the ions are then equal to the concentration (in ppm).

Cl^-: $1.95 \times 10^4 \text{ g Cl}^- \times \dfrac{1 \text{ mol Cl}^-}{35.45 \text{ g Cl}^-} = 550. \text{ mol Cl}^-$

Na⁺: $1.08 \times 10^4 \text{ g Na}^+ \times \dfrac{1 \text{ mol Na}^+}{22.99 \text{ g Na}^+} = 470. \text{ mol Na}^+$

Mg²⁺: $1.29 \times 10^3 \text{ g Mg}^{2+} \times \dfrac{1 \text{ mol Mg}^{2+}}{24.31 \text{ g Mg}^{2+}} = 53.1 \text{ mol Mg}^{2+}$

SO₄²⁻: $9.05 \times 10^2 \text{ g SO}_4^{2-} \times \dfrac{1 \text{ mol SO}_4^{2-}}{96.06 \text{ g SO}_4^{2-}} = 9.42 \text{ mol SO}_4^{2-}$

Ca²⁺: $4.12 \times 10^2 \text{ g Ca}^{2+} \times \dfrac{1 \text{ mol Ca}^{2+}}{40.08 \text{ g Ca}^{2+}} = 10.3 \text{ mol Ca}^{2+}$

K⁺: $3.80 \times 10^2 \text{ g K}^+ \times \dfrac{1 \text{ mol K}^+}{39.10 \text{ g K}^+} = 9.72 \text{ mol K}^+$

Br⁻: $67 \text{ g Br}^- \times \dfrac{1 \text{ mol Br}^-}{79.90 \text{ g Br}^-} = 0.84 \text{ mol Br}^-$

For a total of: 1103 mol ions

The concentration per gram is: $\dfrac{1103 \text{ mol ions}}{10^6 \text{ g H}_2\text{O}}$

The *change* in the freezing point of the sea water is:

$\Delta T_{fp} = m_{\text{solute}} \times K_{fp} = \left(\dfrac{1103 \text{ mol ions}}{10^6 \text{ g H}_2\text{O}} \times \dfrac{1000 \text{ g H}_2\text{O}}{1 \text{ kg H}_2\text{O}}\right)(-1.86\ °C/m) = -2.05\ °C$

So, we expect sea water to begin freezing at –2.05 °C.

(b) The osmotic pressure (in atmospheres) can be calculated if *we assume the density of sea water is 1.00 g/mL.*

$\Pi = cRT = \dfrac{1.103 \text{ mol}}{1 \text{ L}} \times 0.082057 \dfrac{\text{L} \cdot \text{atm}}{\text{K} \cdot \text{mol}} \times 298 \text{ K} = 27.0 \text{ atm}$

The pressure needed to purify sea water by reverse osmosis would then be a pressure greater than 27.0 atm.

13.88. Concerning a 10.0 m tree:

(a) The 10 m tree is equal to a column of water 10^4 mm tall. The equivalent column of mercury would be

$10.0 \text{ m H}_2\text{O}\left(\dfrac{1000 \text{ mm}}{1 \text{ m}}\right)\left(\dfrac{1 \text{ mm Hg}}{13.6 \text{ mm H}_2\text{O}}\right) = 735 \text{ mm Hg}$

$735 \text{ mm Hg}\left(\dfrac{1 \text{ atm}}{760 \text{ mm Hg}}\right) = 0.967 \text{ atm}$

$\Pi = cRT$

$c = \dfrac{\Pi}{RT} = \dfrac{0.967 \text{ atm}}{\left(0.082057 \dfrac{\text{L} \cdot \text{atm}}{\text{K} \cdot \text{mol}}\right)(293 \text{ K})} = 0.0402 \text{ mol/L}$

(b) In 1 L of sap, there is 0.0402 mol sucrose and assuming the density of sap is 1.0 g/mL, the mass of 1 L of sap (the solution) is 1.0×10^3 g

$\text{Mass \%} = \dfrac{(0.0402 \text{ mol sucrose})(342.3 \text{ g sucrose/mol sucrose})}{1.0 \times 10^3 \text{ g solution}} = 1.4\%$

13.89. A 2.00 % aqueous solution of sulfuric acid is also 98.00 % in water. Assume that we begin with 100 g of solution.

(a) We calculate the van't Hoff factor by first calculating the molality of the solution:

$2.00 \text{ g H}_2\text{SO}_4 \times \dfrac{1 \text{ mol H}_2\text{SO}_4}{98.07 \text{ g H}_2\text{SO}_4} = 0.0204 \text{ mol H}_2\text{SO}_4$

$\text{Molality} = \dfrac{0.0204 \text{ mol H}_2\text{SO}_4}{0.09800 \text{ kg H}_2\text{O}} = 0.208 \, m$

Using the delta T equation:

$\Delta T_{fp} = m_{solute} \times K_{fp} \times i$, we can solve for i:

$i = \dfrac{\Delta T_{fp}}{m_{solute} \times K_{fp}} = \dfrac{-0.796 \text{ °C}}{(0.208 \, m)(-1.86 \text{ °C}/m)} = 2.06$

(b) Given the van't Hoff factor of 2.06 in part a, the best representation of a dilute solution of sulfuric acid in water must be: $H_3O^+ + HSO_4^-$.

13.90. Identify the halide ion in KX:

$\Delta T_{fp} = m_{solute} \times K_{fp} \times i$

$m_{solute} = \dfrac{\Delta T_{fp}}{i \times K_{fp}} = \dfrac{-1.28 \text{ °C}}{(2)(-1.86 \text{ °C}/m)} = 0.344 \, m$

$0.100 \text{ kg H}_2\text{O} \left(\dfrac{0.344 \text{ mol KX}}{1 \text{ kg H}_2\text{O}}\right) = 0.0344 \text{ mol KX}$

$\dfrac{4.00 \text{ g KX}}{0.0344 \text{ mol KX}} = 116 \text{ g/mol}$

116 g/mol − 39 g/mol = 77 g/mol ; The halide ion is probably Br⁻.

13.91. The Henry's law constant for N_2O is 2.4×10^{-2} mol/kg · bar. What mass of N_2O will dissolve in 500. mL of water, under an N_2O pressure of 1.00 bar?

Solubility of $N_2O = k_H \cdot P_{N_2O}$

$= (2.4 \times 10^{-2}$ mol/kg · bar$) \times 1.00$ bar $= 2.4 \times 10^{-2}$ mol N_2O/kg

and $\dfrac{2.4 \times 10^{-2} \text{ mol } N_2O}{1 \text{ kg}} \times \dfrac{44.01 \text{ g } N_2O}{1 \text{ mol } N_2O} \times 0.500 \text{ kg} = 0.53 \text{ g } N_2O$

What is the concentration of N_2O in this solution, expressed in ppm ($d_{H_2O} = 1.00$ g/mL)?

Since the density of water is 1.00 g/mL, 500. mL of water will have a mass of 500. grams.

Concentration of N_2O is: $\dfrac{0.53 \text{ g } N_2O}{500. \text{ g } H_2O} = \dfrac{x}{1,000,000 \text{ g } H_2O} = 1.1 \times 10^3$ (2 sf)

Note that the mass of N_2O is negligible and therefore ignored in the mass of the solution!

13.92. Concentration of CO_2 in the beverage:

Solubility $= k_H P_{CO_2}$; $S = (0.034 \dfrac{\text{mol}}{\text{kg} \cdot \text{bar}})(1.5 \text{ bar}) = 0.051$ mol/kg

From the ACP (Henry's Law and Exploding Lakes) the partial pressure of CO_2 in the atmosphere is $P_{CO_2} = 3.75 \times 10^{-4}$ atm

$S_{CO_2} = k_H P_{CO_2} = \left(0.034 \dfrac{\text{mol}}{\text{kg} \cdot \text{bar}}\right)(3.75 \times 10^{-4} \text{ atm})\left(\dfrac{1.01325 \text{ bar}}{1 \text{ atm}}\right) = 1.3 \times 10^{-5}$ mol/kg

Fraction of gas that escapes:
(0.051 mol/kg – 1.3×10^{-5} mol/kg)/ 0.051 mol/kg × 100 % = 1.0×10^2%
~100% lost

13.93. Tests to determine if the contents of a flask are a solution or a colloid:

One very simple test is to shine a beam of light through the liquid. A colloid will show the "path of the light" as the light is reflected off the particles of the colloid—the Tyndall effect.

In a colloid, the dispersed material will usually either not crystallize or crystallized only with difficulty. Yet another test is to attempt to pass the colloid through a membrane. The colloidal particles will diffuse through the membrane either slowly or not at all, owing to the sizes of the suspended medium.

13.94. What happens to a needle floating on the surface of water when soap is added:
The needle sinks. Soap breaks up the surface layer of water molecules, reducing the surface tension; thus, the surface can no longer support the denser needle.

In the Laboratory

13.95. Using the freezing point depression and boiling point elevation equations, calculate the term (m_{solute}).

At the freezing point: $\Delta T_{fp} = m_{solute} \times K_{fp}$; (3.1 °C – 5.50 °C) = m_{solute} × (–5.12 °C/molal)

and $\dfrac{-2.4\ °C}{-5.12\ °C/molal} = m_{solute}$ so 0.47 molal = m_{solute}

At the boiling point: $\Delta T_{bp} = m_{solute} \times K_{bp}$; (82.6 °C – 80.10 °C) = m_{solute} × (+2.53 °C/molal)

and $\dfrac{+2.5\ °C}{+2.53\ °C/molal} = m_{solute}$ so 0.99 molal = m_{solute}

A higher molality at the higher temperature indicates more molecules are dissolved. Therefore, assuming benzoic acid forms dimers, dimer formation is more prevalent at the lower temperature. In this process, two molecules become one entity, lowering the number of separate species in solution and lowering the effective molality.

13.96. Masses of I_2 in the water and CCl_4 layers:

The solubility of I_2 in the two solvents can be expressed:

$$\dfrac{I_2\ \text{solubility in}\ CCl_4}{I_2\ \text{solubility in}\ H_2O} = \dfrac{85}{1}$$

Let x = mass of I_2 that remains dissolved in H_2O, then we can write:

$$\dfrac{(5.0\ mg - x)/10.0\ mL}{x/25\ mL} = \dfrac{85}{1}$$

$$\dfrac{25(5.0\ mg - x)}{10.0(x)} = 85$$

$$\dfrac{125\ mg - 25x}{10.0x} = 85$$

125 mg – 25x = 850x

125 mg = 875x

Solving for x = 0.14 mg I_2 dissolved in H_2O, and 5.0 – x = 4.9 mg I_2 dissolved in CCl_4.

13.97. The apparent molecular weight of acetic acid in benzene, determined by the depression of benzene's freezing point:

$\Delta T_{fp} = m_{solute} \times K_{fp}$; (3.37 °C – 5.50 °C) = m_{solute} × (–5.12 °C/molal)

$$m_{solute} = \dfrac{-2.13\ °C}{-5.12\ °C/m} = 0.416\ m$$

and the apparent molecular weight is:

$$0.416\ m = \frac{\frac{5.00\ \text{g acetic acid}}{\text{MM}}}{0.100\ \text{kg}}$$ and solving for MM; 120. g/mol = MM

The apparent molecular weight of acetic acid in water:

$\Delta T_{fp} = m_{solute} \times K_{fp}$; $(-1.49\ °C - 0.00\ °C) = m_{solute} \times (-1.86\ °C/molal)$

and $m_{solute} = \dfrac{-1.49\ °C}{-1.86\ °C/m} = 0.801\ m$ and the apparent molecular weight is:

$$0.801\ \text{molal} = \frac{\frac{5.00\ \text{g acetic acid}}{\text{MM}}}{0.100\ \text{kg}}$$ and solving for MM; 62.4 g/mol = MM

The accepted value for acetic acid's molar mass is approximately 60.1 g/mol. Hence the value for *i* isn't much larger than 1, indicating that the degree of dissociation of acetic acid molecules in water is not great, a finding consistent with the designation of acetic acid as a weak acid. The apparently doubled molar mass of acetic acid in benzene indicates that the acid must exist primarily as a dimer in benzene.

13.98. Composition of the mixture:

Use the osmotic pressure to calculate the molarity of the solution:

$539\ \text{mm Hg}\left(\dfrac{1\ \text{atm}}{760\ \text{mm Hg}}\right) = 0.709\ \text{atm}$

$\Pi = cRT$

$c = \dfrac{\Pi}{RT} = \dfrac{0.709\ \text{atm}}{\left(0.082057\ \dfrac{\text{L} \cdot \text{atm}}{\text{K} \cdot \text{mol}}\right)(298\ \text{K})} = 0.0290\ \dfrac{\text{mol}}{\text{L}}$

With this molarity, the # mol involved is: (0.0290 M)(0.1000 L) = 0.00290 mol
Set up two equations with two unknowns.
Let *x* = mass of $C_{12}H_{23}O_5N$ in the 1.00-g sample and *y* = mass of $C_{12}H_{22}O_{11}$ in the 1.00-g sample.

$x + y = 1.00\ \text{g}$

$0.00290\ \text{mol} = \left(x \times \dfrac{1\ \text{mol}\ C_{21}H_{23}O_5N}{369.4\ \text{g}\ C_{21}H_{23}O_5N}\right) + \left(y \times \dfrac{1\ \text{mol}\ C_{12}H_{22}O_{11}}{342.3\ \text{g}\ C_{12}H_{22}O_{11}}\right)$

Substitute (1.00 g – *x*) for *y* and solve.

$$0.00290 \text{ mol} = \left(x \times \frac{1 \text{ mol C}_{21}\text{H}_{23}\text{O}_5\text{N}}{369.4 \text{ g C}_{21}\text{H}_{23}\text{O}_5\text{N}} \right) + \left((1.00 \text{ g} - x) \times \frac{1 \text{ mol C}_{12}\text{H}_{22}\text{O}_{11}}{342.3 \text{ g C}_{12}\text{H}_{22}\text{O}_{11}} \right)$$

$$0.00290 \text{ mol} = \frac{x \text{ mol}}{369.4 \text{ g}} + \frac{(1.00 \text{ g} - x) \text{ mol}}{342.3 \text{ g}}$$

0.00290 mol $= 0.002707x$ mol/g $+ 0.00292$ mol $- 0.002921x$ mol/g

-0.00002 mol $= -0.000214x$ mol/g

$x = 0.1$ g

$x = 0.1$ g $C_{21}H_{23}O_5N$ (1 sf) and from the first equation: $y = 0.9$ g $C_{12}H_{22}O_{11}$ (1 sf)

Thus, the mixture is (0.1 g heroin/1.00 g sample) × 100% = 10% heroin and 90% lactose (1 sf for each answer).

13.99. The vapor pressure data should permit us to calculate the molar mass of the compound.

$P_{\text{benzene}} = X_{\text{benzene}} \cdot P°_{\text{benzene}}$

94.16 mm Hg $= X_{\text{benzene}} \times 95.26$ mm Hg, and rearranging: $X_{\text{benzene}} = \dfrac{94.16 \text{ mm Hg}}{95.26 \text{ mm Hg}}$

$X_{\text{benzene}} = 0.98845$

Now we need to know the # of moles of the compound, so use the mole fraction of benzene to find that: 10.0 g benzene $\times \dfrac{1 \text{ mol benzene}}{78.11 \text{ g benzene}} = 0.128$ mol benzene

$0.98845 = \dfrac{0.128 \text{ mol benzene}}{0.128 \text{ mol benzene} + x \text{ mol C}_x\text{H}_y\text{N}_z}$

$0.98845(0.128 + x) = 0.128$ and solving for $x = 0.001496$ mol $C_xH_yN_z$

Knowing that this # of moles of compound have a mass of 0.177 g, we can calculate the molar mass:

$\dfrac{0.177 \text{ g}}{0.001496 \text{ mol}} = 118$ g/mol

Now we can calculate the empirical formula, since we know that the compound is 71.17 % C, 5.12 % H and (100.00 – 71.17 – 5.12) or 23.71 % N.

In 100 g of the compound there are 71.17 g C $\times \dfrac{1 \text{ mol C}}{12.011 \text{ g C}} = 5.925$ mol C and

23.71 g N $\times \dfrac{1 \text{ mol N}}{14.007 \text{ g N}} = 1.693$ mol N and 5.12 g H $\times \dfrac{1 \text{ mol H}}{1.008 \text{ g H}} = 5.079$ mol H

Determine the ratio of moles of these three elements, by dividing the three # of mol by the smallest (1.693):

$\dfrac{5.925 \text{ mol C}}{1.693 \text{ mol N}} = 3.50$ mol C/mol N; $\dfrac{5.079 \text{ mol H}}{1.693 \text{ mol N}} = 3.00$ mol H/mol N

If we double each mol:mol ratio we obtain C₇H₆N₂ as an empirical formula.

Dividing the molar mass by the mass of the empirical formula, yields:

$$\frac{118 \text{ g/mol}}{118 \text{ g/empirical formula}} = 1 \text{ empirical formula/mol}$$, so the molecular formula is C₇H₆N₂.

13.100. Determine the empirical and molecular formula of the compound:

The compound contains 73.94% C, 8.27% H, and 17.79% Cr, so determine the # moles of each element:

$$73.94 \text{ g C}\left(\frac{1 \text{ mol C}}{12.011 \text{ g C}}\right) = 6.156 \text{ mol C} \quad 8.27 \text{ g H}\left(\frac{1 \text{ mol H}}{1.008 \text{ g H}}\right) = 8.20 \text{ mol H}$$

$$17.79 \text{ g Cr}\left(\frac{1 \text{ mol Cr}}{51.996 \text{ g Cr}}\right) = 0.3421 \text{ mol Cr}$$

The mole: mole ratio of these elements is:

$$\frac{6.156 \text{ mol C}}{0.3421 \text{ mol Cr}} = \frac{18 \text{ mol C}}{1 \text{ mol Cr}} \qquad \frac{8.20 \text{ mol H}}{0.3421 \text{ mol Cr}} = \frac{24 \text{ mol H}}{1 \text{ mol Cr}}$$

The empirical formula is C₁₈H₂₄Cr. Use the osmotic pressure to determine the number of moles of the compound:

$$3.17 \text{ mm Hg}\left(\frac{1 \text{ atm}}{760 \text{ mm Hg}}\right) = 0.00417 \text{ atm}$$

$$\Pi = cRT$$

$$\Pi = \left(\frac{n}{V}\right)RT$$

$$n = \frac{\Pi V}{RT} = \frac{(0.00417 \text{ atm})(0.100 \text{ L})}{\left(0.082057 \frac{\text{L} \cdot \text{atm}}{\text{K} \cdot \text{mol}}\right)(298 \text{ K})} = 1.71 \times 10^{-5} \text{ mol}$$

$$\frac{5.00 \times 10^{-3} \text{ g}}{1.71 \times 10^{-5} \text{ mol}} = 293 \text{ g/mol}$$

The empirical formula weight is 292.4 g/mol, so the molecular formula is also C₁₈H₂₄Cr.

Summary and Conceptual Questions

13.101. Greater hydration of Be²⁺, Mg²⁺, and Ca²⁺: In general, the smaller the ion, the greater the degree of hydration, so Be²⁺ will be most strongly hydrated, while Ca²⁺ will be least strongly hydrated.

13.102. A cucumber shrivels when placed in a solution of salt:

The solution inside the cucumber has a higher solvent concentration than the concentrated salt solution. The solvent molecules flow out of the cucumber, and the cucumber shrivels.

13.103. Equimolar amounts of CaCl₂ and NaCl lower freezing points differently. The formulas tell us that CaCl₂ provides 3 particles per formula unit while NaCl provides only two. Hence, we expect van't Hoff factors of 3 and 2 respectively, so CaCl₂ should have a freezing point depression that is about 50% greater than that of NaCl.

13.104. When NaCl is placed in water, describe the equilibrium that exists:

At the particulate level, Na^+ and Cl^- ions are leaving the solid state and entering solution. Concurrently, solid NaCl is forming from Na^+ and Cl^- ions in solution. These reverse processes are occurring at equal rates.

13.105. Solutes likely to dissolve in water; and solutes likely to dissolve in benzene:

Substances likely to dissolve in water are polar (and many ionic compounds) and moreover polar substances capable of hydrogen bonding. Substances likely to dissolve in benzene are nonpolar substances.

Water is a polar solvent. The ionic compounds, NaNO₃ and NH₄Cl, are both soluble in water (NaNO₃ solubility (at 25 °C) = 91 g per 100 g water; NH₄Cl solubility (at 25 °C) = 40 g per 100 g water). The slightly polar compound diethyl ether is soluble to a small extent in water (about 6 g per 100 g water at 25 °C). Nonpolar naphthalene is not soluble in water (solubility at 25 °C = 0.003 g per 100 g water).

Benzene is a nonpolar solvent. Thus, ionic substances such as NaNO₃ and NH₄Cl will certainly not dissolve. However, naphthalene is also nonpolar and resembles benzene in its structure; it should dissolve very well. (A chemical handbook gives a solubility of 33 g naphthalene per 100 g benzene.) Diethyl ether is weakly polar but is miscible to some extent with benzene due to significant regions being nonpolar with C–C and C–H bonds.

13.106. Account for the differing solubilities of alcohols in water:

All alcohols contain a polar —OH group that can interact with polar water molecules. The smaller alcohols are miscible with water because of this polar group. However, with an increase in the size of the hydrocarbon group, the organic group (the nonpolar part of the molecule) has become a larger fraction of the molecule, and properties associated with nonpolarity begin to dominate.

13.107. Since hydrophilic colloids are those that "love water," we would expect starch to form a hydrophilic colloid since it contains the OH bonds that can hydrogen bond to water. Hydrocarbons, on the other hand, have nonpolar bonds that should have little-to-no attraction to water molecules, and form a hydrophobic colloid.

13.108. Should sucrose or ethylene glycol affect the vapor pressure of water more:

Equal masses of sucrose and ethylene glycol have different amounts (moles) because their molar masses are different. 10.0 g of ethylene glycol contains a greater amount (moles) of solute because ethylene glycol has a smaller molar mass, so it has a greater influence on the vapor pressure.

13.109. Semipermeable membrane dividing container into two parts; one side containing 5.85 g NaCl in 100. mL solution, and the other side containing 8.88 g KNO3 in 100. mL solution. Both solutes produce 2 mol of ions per mol of compound (the value of *i* is approximately the same), so their molarities can be compared directly.

Molarity for NaCl: $\dfrac{5.85 \text{ g NaCl}}{0.100 \text{ L}}\left(\dfrac{1 \text{ mol NaCl}}{58.44 \text{ g NaCl}}\right) = 1.00 \text{ M}$

Molarity for KNO3: $\dfrac{8.88 \text{ g KNO}_3}{0.100 \text{ L}}\left(\dfrac{1 \text{ mol KNO}_3}{101.1 \text{ g KNO}_3}\right) = 0.878 \text{ M}$

The KNO3 solution has a higher solvent concentration (lower solute concentration), so solvent will flow from the KNO3 solution into the NaCl solution.

13.110. Will a protozoan shrivel or burst when placed in fresh water:
Solvent (water) concentration is higher in fresh water than in ocean water, solvent will flow into the protozoan causing it to burst.

13.111. For the mixture of ethanol and water:

(a) mole fraction of ethanol and water:

A 12% solution of ethanol contains 12 g ethanol (and 88 g water) in 100 g solution.

$12. \text{ g ethanol}\left(\dfrac{1 \text{ mol ethanol}}{46.07 \text{ g ethanol}}\right) = 0.260 \text{ mol ethanol}$

and $88. \text{ g water}\left(\dfrac{1 \text{ mol water}}{18.02 \text{ g water}}\right) = 4.883 \text{ mol water}$

$X_{C_2H_5OH} = \dfrac{0.260 \text{ mol C}_2\text{H}_5\text{OH}}{0.260 \text{ mol C}_2\text{H}_5\text{OH} + 4.883 \text{ mol H}_2\text{O}} = 0.0506 = 0.051 \text{ (2 sf)}$

and 1.000 – 0.051 or 0.949 mole fraction water.

(b) What are the equilibrium vapor pressures of ethanol and water at this temperature:

Given that normal boiling point is defined as the temperature at which the vapor pressure of a substance is equal to atmospheric pressure, we have:

(0.0506)(760 mm Hg) = 38 mm Hg (2 sf)

Appendix G lists 327.3 mm Hg as the vapor pressure of water at 78 °C and 341.0 mm Hg as the vapor pressure of water at 79 °C. A linear interpolation between these two values gives a value for the vapor pressure at 78.5 °C.

$$\frac{341.0 \text{ mm Hg} - 327.3 \text{ mm Hg}}{79 \text{ °C} - 78 \text{ °C}} = \frac{x - 327.3 \text{ mm Hg}}{78.5 \text{ °C} - 78 \text{ °C}}$$

$x = 334.2$ mm Hg

Since the mixture has a mole fraction of water that is 0.949, then the vapor pressure of water is 0.949 × 334.2 mm Hg or 317 mm Hg.

(c) The mole fractions can be found by using the equilibrium vapor pressures calculated above: mole fraction ethanol = (38 mm Hg)/(38 mm Hg + 317 mm Hg) = 0.108 = 0.11 (2 sf) and because the two mole fractions muist equal 1, the mole fraction water = 1 – 0.108 = 0.892 = 0.89 (2 sf).

(d) The mole fraction of ethanol originally was 0.051. The mole fraction of ethanol in the vapor (and hence in the condensate) is 0.11. The mole fraction of ethanol was increased by more than 2! The mass percent of ethanol can be calculated by asking "How much ethanol and how much water are present?"

$$0.108 \text{ mol } C_2H_5OH \left(\frac{46.07 \text{ g } C_2H_5OH}{1 \text{ mol } C_2H_5OH} \right) = 4.98 \text{ g } C_2H_5OH$$

and

$$0.892 \text{ mol } H_2O \left(\frac{18.02 \text{ g } H_2O}{1 \text{ mol } H_2O} \right) = 16.1 \text{ g } H_2O$$

so the mass % is 4.98 g/(4.98 g + 16.1 g) × 100% = 24 % of ethanol (2 sf).

13.112. Compare the freezing points of solutions of NaCl and $CaCl_2$;

Freezing point for the NaCl solution:

$$\frac{200. \text{ g NaCl}}{1.00 \text{ kg water}} \left(\frac{1 \text{ mol NaCl}}{58.44 \text{ g NaCl}} \right) = 3.42 \ m \text{ and } i = 2 \text{ for NaCl}$$

For such a solution, $\Delta T_{fp} = (-1.86 \text{ °C}/m)(3.42 \ m)(2) = -12.7$ °C

Freezing point for the $CaCl_2$ solution:

$$\frac{200. \text{ g } CaCl_2}{1.00 \text{ kg water}} \left(\frac{1 \text{ mol } CaCl_2}{110.98 \text{ g } CaCl_2} \right) = 1.80 \ m \text{ and } i = 3 \text{ for } CaCl_2$$

For this solution, $\Delta T_{fp} = (-1.86 \text{ °C}/m)(1.80 \ m)(3) = -10.1$ °C

On the sole basis of lowering of the freezing point, NaCl wins based on using equal masses of the two compounds.

13.113. For NaCl, 5.00% NaCl contains 5.00 g NaCl and 95.00 g water (by definition). The amount (moles) of NaCl is then: $5.00 \text{ g NaCl} \left(\dfrac{1 \text{ mol NaCl}}{58.44 \text{ g NaCl}} \right) = 0.0856 \text{ mol NaCl}$ and the molality is: $\dfrac{0.0856 \text{ mol NaCl}}{0.0950 \text{ kg water}} = 0.901\ m$ and calculating the ΔT_{fp},

$\Delta T_{fp} = (-1.86\ °C/m)(0.901 m) = -1.68°C$. Given that the measured $\Delta T_{fp} = -3.05°C$, the van't Hoff factor is $(-3.05°C/-1.68°C) = 1.82$.

For Na₂SO₄: 5.00% Na₂SO₄ contains 5.00 g Na₂SO₄ and 95.00 g water (by definition).

The molality of the solution is: $\dfrac{5.00 \text{ g Na}_2\text{SO}_4}{0.0950 \text{ kg water}} \left(\dfrac{1 \text{ mol Na}_2\text{SO}_4}{142.04 \text{ g Na}_2\text{SO}_4} \right) = 0.371\ m$

and calculating the ΔT,
$\Delta T_{fp} = (-1.86\ °C/m)(0.371 m) = -0.689°C$.

Given that the measured $\Delta T = -1.36\ °C$, the van't Hoff factor is $(-1.36/-0.689) = 1.97$.

Looking at the values of the factor in Table 13.4, these values of *i* are consistent with the values expected—being less than for the 2% solutions of NaCl and Na₂SO₄ noted in Table 13.4.

13.114. Calculate molality, freezing points, and van't Hoff factors for solutions described:

(a) Emboldened entries indicate provided data.

Acid, 1.00 mass% (MM)	Molality	T_measured °C	T_calculated °C	i
HNO₃ (63.01)	0.160	**−0.56**	−0.298	1.9
CH₃CO₂H (60.05)	0.168	**−0.32**	−0.313	1.0
H₂SO₄ (98.07)	0.103	**−0.42**	−0.192	2.2
H₂C₂O₄ (90.03)	0.112	**−0.30**	−0.209	1.4
HCO₂H (46.03)	0.219	**−0.42**	−0.408	1.0
CCl₃CO₂H (163.38)	0.0618	**−0.21**	−0.115	1.8

The generic solution for this problem consists of the following:

Definition of a 1.00 mass% solution = $\dfrac{1.00 \text{ g compound}}{1.00 \text{ g compound} + 99.00 \text{ g H}_2\text{O}} \cdot 100\%$

Mass of solvent: 99.00 g H₂O = 0.09900 kg H₂O

Molality: m_{solute} = (1.00 g)(1 mol compound/MM of compound)/(0.09900 kg)

Molality for each solution was calculated and entered in the 2nd column of the table above. $T_{calculated}$ is the lowering of the fp (i.e. ΔT_{fp}) for each solution calculated by ΔT_{fp} = (−1.86 °C /m)(m_{solute}) where m_{solute} represents the calculated molality for each acid. Finally the "*i*" values are calculated by $T_{measured}/T_{calculated}$.

(b) The "*i*" values indicate that HNO₃, H₂SO₄, and CCl₃CO₂H are "strong acids" as noted by their ability to produce more than 1 particle per formula unit (that particle corresponding to the hydrated H⁺ ion). While H₂C₂O₄ is a weak acid, it has a moderate strength with *i* = 1.4. Finally, CH₃CO₂H and HCO₂H are both decidedly weak with *i* values very close to 1.

13.115. This question really asks "Is −16.46 °C equivalent to 0°F?" So do the conversion!

$\left(\dfrac{9\text{ °F}}{5\text{ °C}}\right)(-16.46\text{ °C}) + 32\text{ °F} = 2.372\text{ °F}$

This is still above a temperature of 0 °F. Not a very strong endorsement of the story!

13.116. What is the concentration of ions in seawater to provide an osmotic pressure of 27 atm?

Π = *cRT*. Substitute the data, and solve for *c*: 27 atm = $c(0.082057\ \dfrac{\text{L} \cdot \text{atm}}{\text{K} \cdot \text{mol}})(298\text{ K})$

c = 1.1 M; A total concentration of about 1.1 M would provide the requested osmotic pressure of 27 atm.

Solution and Answer Guide

Kotz Treichel Townsend Treichel, Chemistry and Chemical Reactivity 11e, 978-0-357-85140-1, Chapter 14: Chemical Kinetics The Rates of Chemical Reactions

TABLE OF CONTENTS

Applying Chemical Principles 492
Kinetics and Mechanisms: a 70-Year-Old Mystery Solved 493
Practicing Skills 494
General Questions 513
In the Laboratory 526
Summary and Conceptual Questions 529

Applying Chemical Principles

Enzymes-Nature's Catalysts

14.1.1. Time to decompose H_2O_2 catalytically:

The uncatalyzed time for decomposition of H_2O_2 is one year (365 days). If the catalyzed rate is 10^7 times faster, then the time for decomposition will be $1/10^7$. In units of days:

$$\frac{365 \text{ days}}{10,000,000} = 3.65 \times 10^{-5} \text{ or } 4 \times 10^{-5} \text{ days (1sf) or 3 s (1sf).}$$

14.1.2 Moles of carbonic anhydrase (CA) present:

\# mol CA in 100 mL of a solution with 2 mg CA/mL:

$$\frac{2 \text{ mg CA}}{1 \text{ mL}}\left(\frac{100 \text{ mL}}{1}\right)\left(\frac{1 \text{ g CA}}{1000 \text{ mg CA}}\right)\left(\frac{1 \text{ mol CA}}{29,000 \text{ g CA}}\right) = 6.90 \times 10^{-6} \text{ or } 7 \times 10^{-6} \text{ mol CA (1sf)}$$

\# mol CO_2 converted to HCO_3^{1-} in 1 s:

$$1 \text{ s}\left(\frac{1 \times 10^6 \text{ mol CO}_2}{1 \text{ mol CA} \cdot \text{s}}\right)\left(\frac{6.90 \times 10^{-6} \text{ mol CA}}{1}\right) = 6.9 \text{ or } 7 \text{ mol CO}_2 \text{ (1sf)}$$

Converted to units of mass: 6.9 mol × 44.0 g CO_2/mol = 3 × 10^2 g CO_2

Kinetics and Mechanisms: a 70-Year-Old Mystery Solved

14.2.1. Sullivan used 578 nm light to dissociate I_2 molecules to I atoms.

a) What is the energy (in kJ/mol) of 578 nm light?

$E = hc/\lambda = (6.626 \times 10^{-34}\text{ J·s})(2.998 \times 10^8\text{ m/s})/578 \times 10^{-9}\text{ m} = 3.44 \times 10^{-19}$ J/photon

Now we need this energy in kJ/mol of photons, so multiply by Avogadro's number, and divide by 1000 (to convert J to kJ):

$(3.44 \times 10^{-19}$ J/photon$)(6.022 \times 10^{23}$ photon/mol$)(1$ kJ/1000 J$) = 207$ kJ/mol

b) Breaking an I_2 bond requires 151 kJ/mol of energy. What is the longest wavelength of light that has enough energy to dissociate I_2?

In part (a) we solved for the energy of light with a specific wavelength. In part (b) we reverse the procedure, and ask what wavelength will have at least 151 kJ/mol. If we divide the required energy (151 kJ/mol) by the 2nd and 3rd terms in the expression above, we'll have the energy in J/photon

$\left(\dfrac{151\text{ kJ/mol}}{6.022 \times 10^{23}\text{ photons/mol}}\right)\left(\dfrac{1000\text{ J}}{1\text{ kJ}}\right) = 2.51 \times 10^{-19}$ J/photon

Now we can use the energy equation ($E = hc/\lambda$) to solve for λ.

Note that we'll divide hc/E to give the wavelength:

$\lambda = \dfrac{(6.626 \times 10^{-34}\text{ J·s})(2.998 \times 10^8\text{ m/s})}{2.51 \times 10^{-19}\text{ J/photon}} = 792 \times 10^{-9}$ m or 792 nm

14.2.2. Verify Sullivan's two-step mechanism yields second-order rate law:
For the fast equilibrium, the rate of the forward reaction equals the rate of the reverse reaction; thus, $k_1[I_2] = k_{-1}[I]^2$ where k_1 and k_{-1} are the forward and reverse rate constants, respectively. The rate limiting step determines the rate law, Rate = $k_2[H_2][I]^2$

Substituting for $[I]^2$ gives Rate = $k_1 k_2 k_{-1}^{-1}[H_2][I_2] = k[H_2][I_2]$

14.2.3. Why is a termolecular elementary step likely to be the slowest in a mechanism?
The probability of a 3-body collision with appropriate geometry is low!

14.2.4. Determine Activation Energy for reaction of H₂ and I₂ to produce HI:
A graph of ln(k) versus 1/T, (where T is in K) is below:

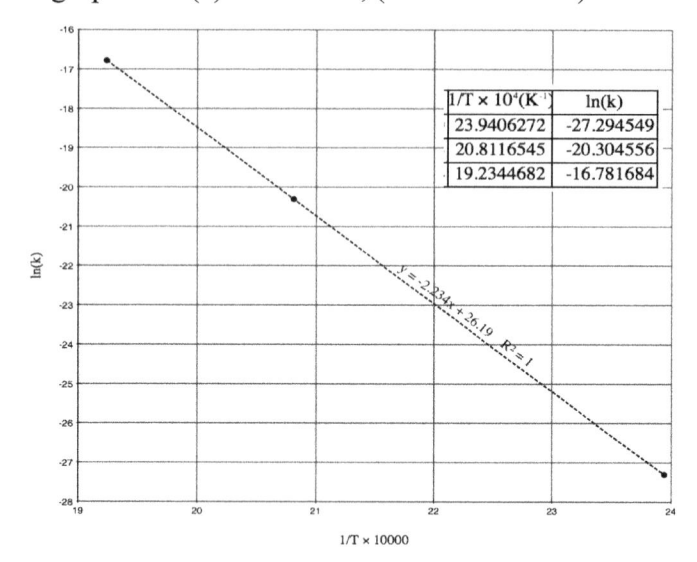

1/T × 10⁴(K⁻¹)	ln(k)
23.9406272	-27.294549
20.8116545	-20.304556
19.2344682	-16.781684

This plot yields a straight line with a slope of -2.234×10^4. Note that we had the T values × 10^4, so we express the slope in that order of magnitude as well. Since the slope = E_a/R, we can substitute: $-2.234 \times 10^4 = -E_a/R$ where R = 8.314510×10^{-3} kJ/K · mol and E_a = 186 kJ/mol.

Practicing Skills

Reaction Rates

14.1. Relative Rates of Disappearance and Formation:

(a) $2 O_3(g) \rightarrow 3 O_2(g)$ $\text{Rate} = -\frac{1}{2}\frac{\Delta[O_3]}{\Delta t} = +\frac{1}{3}\frac{\Delta[O_2]}{\Delta t}$

(b) $2 HOF(g) \rightarrow 2 HF(g) + O_2(g)$ $\text{Rate} = -\frac{1}{2}\frac{\Delta[HOF]}{\Delta t} = +\frac{1}{2}\frac{\Delta[HF]}{\Delta t} = +\frac{\Delta[O_2]}{\Delta t}$

14.2. Relative Rates of Disappearance and Formation:

(a) $NO(g) + \frac{1}{2} O_2(g) \rightarrow NO_2(g)$ $\text{Rate} = -\left(\frac{\Delta[NO]}{\Delta t}\right) = -\frac{1}{2}\frac{\Delta[O_2]}{\Delta t} = \left(\frac{\Delta[NO_2]}{\Delta t}\right)$

(b) $2 NH_3(g) \rightarrow N_2(g) + 3 H_2(g)$ $\text{Rate} = -\frac{1}{2}\left(\frac{\Delta[NH_3]}{\Delta t}\right) = \frac{\Delta[N_2]}{\Delta t} = \frac{1}{3}\left(\frac{\Delta[H_2]}{\Delta t}\right)$

14.3. For the reaction, $2 O_3 (g) \rightarrow 3 O_2 (g)$, the rate of formation of O_2 is 1.5×10^{-3} mol/L·s.

SQ14.1(a) offers a clear assist, indicating that O_2 forms at a rate 1.5 times the rate that ozone decomposes. (2 ozones produce 3 oxygens.) Hence the rate of decomposition of O_3 is -1.0×10^{-3} mol/L·s.

14.4. What is $\Delta[NH_3]/\Delta t$:

$$\frac{\Delta[NH_3]}{\Delta t} = -\frac{2}{3}\frac{\Delta[H_2]}{\Delta t} = -\frac{2}{3} \times (6.4 \times 10^{-4} \text{ mol/L·min}) = -4.3 \times 10^{-4} \text{ mol/L·min}$$

14.5. Plot the data for the hypothetical reaction $A \rightarrow 2B$

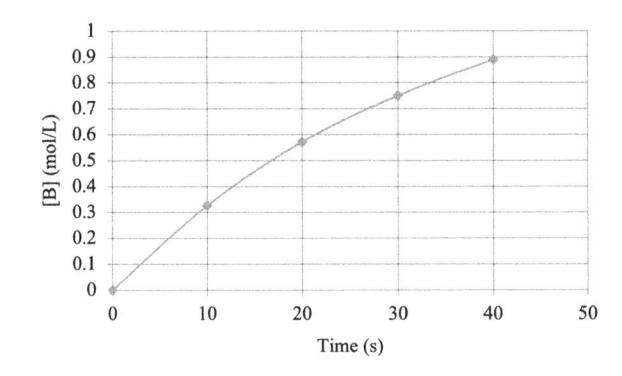

(a) Rate = $\frac{\Delta[B]}{\Delta t} = \frac{(0.326 - 0.000)}{10.0 - 0.00} = +\frac{0.326}{10.0} = +0.0326 \frac{\text{mol}}{\text{L·s}}$

$= \frac{(0.572 - 0.326)}{20.0 - 10.00} = +\frac{0.246}{10.0} = +0.0246 \frac{\text{mol}}{\text{L·s}}$

$= \frac{(0.750 - 0.572)}{30.0 - 20.00} = +\frac{0.178}{10.0} = +0.0178 \frac{\text{mol}}{\text{L·s}}$

$= \frac{(0.890 - 0.750)}{40.0 - 30.00} = +\frac{0.140}{10.0} = +0.0140 \frac{\text{mol}}{\text{L·s}}$

The rate of change decreases from one time interval to the next *due to a continuing decrease* in the amount of reacting material (A).

(b) Since each A molecule forms 2 molecules of B, the concentration of A will decrease at a rate that is **half** of the rate at which B appears. The negative signs here indicate a **decrease in [A], not a negative concentration of A.**

t	[B]	[A] = 1/2([B]₀ − [B])
10.0 s	0.326	−1/2(0.326) = −0.163
20.0 s	0.572	−1/2(0.572) = −0.286

Rate at which A changes $= \dfrac{\Delta[A]}{\Delta t} = \dfrac{(-0.286 - -0.163)}{20.0 - 10.00} = \dfrac{-0.123}{10.0}$ or $-0.0123 \dfrac{mol}{L \cdot s}$

Note that the **negative sign** indicates a <u>reduction in the concentration of A</u> as the reaction proceeds. Compare this change with the change in [B] for the same interval above (+ 0.0246 $\dfrac{mol}{L \cdot s}$). The disappearance of A is half that of the appearance of B.

14.6. For the reaction of phenyl acetate with water:

(a) The curve is a decreasing exponential curve.

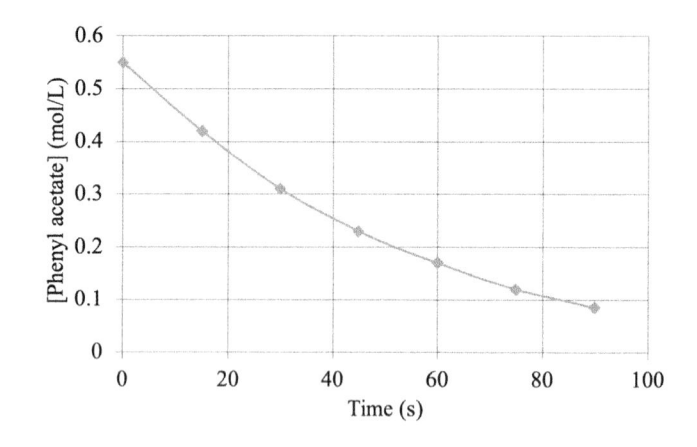

(b) $-\dfrac{\Delta[\text{phenyl acetate}]}{\Delta t} = -\dfrac{(0.31 - 0.42) \text{ mol/L}}{(30.0 - 15.0) \text{ s}} = 0.0073 \text{ mol/L} \cdot \text{s}$

$-\dfrac{\Delta[\text{phenyl acetate}]}{\Delta t} = -\dfrac{(0.085 - 0.12) \text{ mol/L}}{(90.0 - 75.0) \text{ s}} = 0.0023 \text{ mol/L} \cdot \text{s}$

The second rate is slower than the first because the rate depends on the concentration of phenyl acetate, which is smaller during the second time period.

Concentration and Rate Equations

14.7. For the rate equation: Rate = $k[A]^2[B]$, the reaction is 2nd order in A (superscript 2 with A), 1st order in B (implied superscript of 1 with B), and (2+1) or 3rd order overall.

14.8. For the rate equation: Rate = $k[A]^2[B]^2$, the reaction is 2nd order in A (superscript 2 with A), 2nd order in B (superscript of 2 with B), and (2+2) or 4th order overall.

14.9. For the rate equation: Rate = $k[A][B]^2$, if [A] is doubled the rate doubles. If the [B] doubles, the rate quadruples. If both [A] and [B] are doubled, the rate increases by 2 × 4 = 8 times faster.

14.10. The rate will increase by a factor of 9 if the concentration of A is tripled—owing to the 2nd order dependence on [A]. The rate will decrease by a factor of ¼ if the concentration of A is halved—using the same logic.

14.11. For the reaction between ozone and nitrogen dioxide:

(a) The rate equation: Rate = $k[NO_2][O_3]$

(b) Since k is constant, if [O_3] is held constant, the rate would be tripled if the concentration of NO_2 is tripled. Let **C** represent the concentration of NO_2. Substituting into the rate equation:

Rate$_1$ = $k[\mathbf{C}][O_3]$

Rate$_2$ = k[3**C**][O_3] = 3 · $k[\mathbf{C}][O_3]$ or 3 · Rate$_1$

(c) Halving the concentration of O_3 —assuming [NO_2] is constant, would halve the rate.

Rate$_1$ = $k[NO_2][\mathbf{C}]$

Rate$_2$ = $k[NO_2][1/2\ \mathbf{C}]$ = 1/2[NO_2][**C**] or 1/2 · Rate$_1$

14.12. For the reaction forming nitrosyl bromide, NOBr, from NO and Br_2:
(a) Rate = $k[NO]^2[Br_2]$
(b) If the concentration of Br_2 is tripled, the rate will triple.
(c) If the concentration of NO decreases by ½, the rate will decrease by a factor of ¼.

14.13. (a) If we designate the three experiments (data sets in the table as i, ii, and iii respectively),

Experiment	[NO]	[O_2]	$-\dfrac{\Delta[NO]}{\Delta t}\left(\dfrac{mol}{L \cdot s}\right)$
i	0.010	0.010	2.5×10^{-5}
ii	0.020	0.010	1.0×10^{-4}
iii	0.010	0.020	5.0×10^{-5}

Note that experiment ii proceeds at a rate four times that of experiment i.

$$\frac{\text{experiment ii rate}}{\text{experiment i rate}} = \frac{1.0 \times 10^{-4} \frac{\text{mol}}{\text{L} \cdot \text{s}}}{2.5 \times 10^{-5} \frac{\text{mol}}{\text{L} \cdot \text{s}}} = 4$$

This rate change was the result of doubling the concentration of NO. The *order of dependence of NO must be second-order*. Comparing experiments i and iii, we see that changing the concentration of O$_2$ by a factor of two, also affects the rate by a factor of two. The *order of dependence of O$_2$ must be first-order*.

(b) Using the results above we can write the rate equation: Rate = k[NO]^2[O$_2$]^1

(c) To calculate the rate constant we have to have a rate. Note the data provided gives the rate of disappearance of NO. The relation of this concentration to the rate is:

$$\text{Rate} = -\frac{\Delta[\text{NO}]}{\Delta t}; \text{ Using experiment ii, the disappearance of NO is at the rate}$$

$1.0 \times 10^{-4} \frac{\text{mol}}{\text{L} \cdot \text{s}}$ which we can substitute into the rate law:

$$1.0 \times 10^{-4} \frac{\text{mol}}{\text{L} \cdot \text{s}} = k[0.020 \frac{\text{mol}}{\text{L}}]^2 [0.010 \frac{\text{mol}}{\text{L}}] \text{ gives } 25 \frac{\text{L}^2}{\text{mol}^2 \cdot \text{s}} = k$$

(d) Rate when [NO] = 0.015 M and [O$_2$] = 0.0050 M

$$\text{Rate} = k[\text{NO}]^2[\text{O}_2]$$

$$= 25 \frac{\text{L}^2}{\text{mol}^2 \cdot \text{s}} \left(0.015 \frac{\text{mol}}{\text{L}}\right)^2 \left(0.0050 \frac{\text{mol}}{\text{L}}\right)$$

$$= 2.8 \times 10^{-5} \frac{\text{mol}}{\text{L} \cdot \text{s}}$$

(e) The relation between reaction rate and concentration changes:

$$\text{Rate} = -\frac{\Delta[\text{NO}]}{\Delta t} = -\frac{2}{1} \cdot \frac{\Delta[\text{O}_2]}{\Delta t} = +\frac{\Delta[\text{NO}_2]}{\Delta t}$$

So, when NO is reacting at $1.0 \times 10^{-4} \frac{\text{mol}}{\text{L} \cdot \text{s}}$ then O$_2$ will be reacting at $5.0 \times 10^{-5} \frac{\text{mol}}{\text{L} \cdot \text{s}}$ and NO$_2$ will be forming at $1.0 \times 10^{-4} \frac{\text{mol}}{\text{L} \cdot \text{s}}$

14.14. For the reaction of NO with H$_2$:

(a) In the first two sets of data, [H$_2$] is constant while [NO] is halved. The rate decreases by a factor of four on going from the first set of data to the second, so the reaction is second-order in NO. Looking at the second and third data sets, [NO] is constant and [H$_2$]

doubles. The rate also doubles from the second data set to the third, so the reaction is first-order in H₂.

(b) Rate = $k[NO]^2[H_2]$.

(c) Rate = $\frac{\Delta[N_2]}{\Delta t}$ = 0.136 mol/L·s

$k = \frac{\text{Rate}}{[NO]^2[H_2]} = \frac{0.136 \text{ mol/L·s}}{(0.420 \text{ mol/L})^2(0.122 \text{ mol/L})} = 6.32 \text{ L}^2/\text{mol}^2 \cdot \text{s}$

(d) Rate = (6.32 L²/mol²·s)(0.350 mol/L)²(0.205 mol/L) = 0.159 mol/L·s

14.15. For the reaction: NO(g) + ½ O₂(g) → NO₂(g):

(a) The rate law can be determined by examining the effect on the rate by changing the concentration of *either* NO *or* O₂

In Data sets 1 and 2, the [O₂] doubles, and the rate doubles—a first-order dependence.

In Data sets 2 and 3, the [NO] is halved, and the rate is quartered —a second-order dependence.

The rate law will be: Rate = $k[O_2][NO]^2$

(b) To calculate the rate constant we have to have a rate. Note the data provided gives the rate of disappearance of NO. We need to adjust the rate by a factor of 1/2.
The relation of this concentration to the rate is:
Rate = $-\frac{\Delta[NO]}{\Delta t}$;

Rate: $3.4 \times 10^{-8} = k[5.2 \times 10^{-3} \frac{\text{mol}}{\text{L}}][3.6 \times 10^{-4} \frac{\text{mol}}{\text{L}}]^2$

Solving for *k*: *k* = 50. L²/mol² · h.

Note that I selected the data from Experiment 1. Any of the data sets, (1, 2, or 3) would have provided the same value of *k*.

(c) The initial rate for Experiment 4 is determined by substitution into the rate law (with the value of *k* determined in (b)):

Rate = 50. L²/mol² · h[$5.2 \times 10^{-3} \frac{\text{mol}}{\text{L}}$][$1.8 \times 10^{-4} \frac{\text{mol}}{\text{L}}$]² = 8.4 × 10⁻⁹ mol/L·h.

14.16. For the reaction of CO(g) + NO₂(g):

(a) In the first two data sets, [CO] is constant while [NO₂] is halved. The rate also halves from experiment 1 to experiment 2, so the reaction is first-order in NO₂. Comparing experiments 1 and 3, [NO₂] is constant while [CO] doubles. The rate doubles from experiment 1 to experiment 3, so the reaction is first-order in NO,
or Rate = $k[CO][NO_2]$

(b) $k = \dfrac{\text{Rate}}{[\text{CO}][\text{NO}_2]} = \dfrac{3.4 \times 10^{-8} \text{ mol/L} \cdot \text{h}}{(5.0 \times 10^{-4} \text{ mol/L})(0.36 \times 10^{-4} \text{ mol/L})} = 1.9 \text{ L/mol} \cdot \text{h}$

(c) Rate = $(1.9 \text{ L/mol} \cdot \text{h})(1.5 \times 10^{-3} \text{ mol/L})(0.72 \times 10^{-4} \text{ mol/L}) = 2.1 \times 10^{-7} \text{ mol/L} \cdot \text{h}$

Concentration-Time Relationships

14.17. Note that the reaction is first-order. We write the rate expression:

$\ln\left(\dfrac{[C_{12}H_{22}O_{11}]}{[C_{12}H_{22}O_{11}]_0}\right) = -kt$

Substitute the concentrations of sucrose at $t = 0$ and $t = 27$ minutes into the equation:

$\ln\left(\dfrac{[0.0132 \text{ mol/L}]}{[0.0146 \text{ mol/L}]_0}\right) = -k(27 \text{ min})$ and solve for k to obtain: $k = 3.7 \times 10^{-3}$ min^{-1}

14.18. The value of the rate constant for the decomposition of N$_2$O$_5$:

For a first-order reaction: $\ln\dfrac{[N_2O_5]}{[N_2O_5]_0} = -kt$

and substituting: $\ln\left(\dfrac{2.50 \text{ mg}}{2.56 \text{ mg}}\right) = -k(4.26 \text{ min})$ and $k = 0.00557$ min^{-1}

14.19. Since the reaction is first-order, we can write:

$\ln\left(\dfrac{[SO_2Cl_2]}{[SO_2Cl_2]_0}\right) = -kt$. Given the rate constant, 2.8×10^{-3} min^{-1}, we can calculate the time required for the concentration to fall from 1.24×10^{-3} M to 0.31×10^{-3} M

$\ln\left(\dfrac{0.31 \times 10^{-3} \text{ M}}{1.24 \times 10^{-3} \text{ M}}\right) = -(2.8 \times 10^{-3} \text{ min}^{-1}) t$ and $\dfrac{\ln(0.25)}{(-2.8 \times 10^{-3} \text{ min}^{-1})} = 495$ min

or 5.0×10^2 min (2 sf)

14.20. Time for [cyclopropane] to decrease by a factor of eight:

For a first-order process: $\ln\dfrac{[C_3H_6]}{[C_3H_6]_0} = -kt$ and substituting the concentrations and rate constant: $\ln\left(\dfrac{0.010 \text{ M}}{0.080 \text{ M}}\right) = -(2.42 \times 10^{-2} \text{ h}^{-1})t$ and solving for t: $t = 86$ h

14.21. For the decomposition of H_2O_2:

(a) Since the reaction is first-order, we can write:

$\ln\left(\dfrac{[H_2O_2]}{[H_2O_2]_0}\right) = -kt$. Given the rate constant, 1.06×10^{-3} min^{-1}, we can calculate the time required for the concentration to fall from the original concentration to 85% of that value. Note that the concentrations *per se* are not that critical.

Let's assume the initial concentration is 1.00 M and after the passage of t time the concentration is 0.85 M (that's 15% decomposed).

$\ln\left(\dfrac{0.85}{1.00}\right) = -1.06 \times 10^{-3} \text{ min}(t)$

$t = 153$ min $= 150$ min (2 sf).

(b) For 85% of the sample to decompose, we repeat the process, substituting 0.15 for the $[H_2O_2]$ remaining:

$\ln\left(\dfrac{0.15}{1.00}\right) = -1.06 \times 10^{-3} \text{ min}(t)$

$t = 1790$ min (3 sf).

14.22. For the decomposition of N_2O:

(a) Since the reaction is first-order, we can write:

$\ln\dfrac{[N_2O]}{[N_2O]_0} = -kt$. Since the rate constant is 0.0131 min^{-1}, we can calculate the time required for the concentration to fall from the original concentration to 75% of that value.

$\ln\left(\dfrac{0.75}{1.00}\right) = -0.0131 \text{ min}(t)$

$t = 21.96$ min $= 22$ min (2 sf)

(b) For 75% of the sample to decompose, we repeat the process, substituting 0.25 for the $[N_2O]$ remaining:

$\ln\left(\dfrac{0.25}{1.00}\right) = -0.0131 \text{ min}(t)$

$t = 105.8$ min $= 106$ min (3 sf)

14.23. Dimerization of C_2F_4:

Integrated rate equation for second order reactions:

$$\frac{1}{[R]_t} - \frac{1}{[R]_0} = kt$$

$$\frac{1}{\left[0.22 \text{ mol}/L\right]_t} - \frac{1}{\left[0.44 \text{ mol}/L\right]_0} = \left(0.052 \text{ L}/\text{mol} \cdot \text{s}\right)t$$

$t = 43.7$ s or 44 s (2sf)

14.24. The decomposition of NO_2 is 2nd order in NO_2:

For a 2nd order process: $\dfrac{1}{[NO_2]} - \dfrac{1}{[NO_2]_0} = kt$ and substituting the concentrations:

$$\frac{1}{1.50 \text{ mol/L}} - \frac{1}{2.00 \text{ mol/L}} = (3.40 \text{ L/mol} \cdot \text{min})t \text{ and solving for } t: t = 0.0490 \text{ min}$$

14.25. $[NO_2] = 1.9 \times 10^{-2}$ mol/L initially; Decomposition is 2nd order in NO_2; $k = 1.1$ L/mol·s

The integrated rate equation for a 2nd order process is: $\dfrac{1}{[NO_2]_t} - \dfrac{1}{[NO_2]_0} = kt$

We want to know the time, t, needed for 75% of the original NO_2 to decompose. That reduced concentration would be $(0.25)(1.9 \times 10^{-2}) = 4.75 \times 10^{-3}$.

Substituting into the integrated rate equation gives:

$$\frac{1}{\left[4.75 \times 10^{-3}\right]_t} - \frac{1}{\left[1.9 \times 10^{-2}\right]_0} = (1.1 \text{ L}/\text{mol} \cdot \text{s})t$$

$210.5 - 52.6 = 1.1$ L/mol·s $\times t$ and solving for t gives: 143.5 s (or 140 s to 2 sf)

14.26. Calculate the rate constant for the dimerization of butadiene:

21% consumed means 79% remaining; that concentration is: (0.79)(0.0087 M) = 0.0069 M

As second-order, $\dfrac{1}{[\text{butadiene}]_t} - \dfrac{1}{[\text{butadiene}]_0} = kt$

Substituting: $\dfrac{1}{\left[0.0069 \text{ M}\right]_t} - \dfrac{1}{\left[0.0087 \text{ M}\right]_0} = k \, (600. \text{ s})$ and $k = 0.05$ $M^{-1} \cdot s^{-1}$ (1 sf)

14.27. Decomposition of NH₃ on a metal surface:

Integrated rate equation for zero-order reactions:

$[R]_t - [R]_0 = -kt$

$0.16 \text{ g NH}_3 \left(\dfrac{1 \text{ mol NH}_3}{17.03 \text{ g}} \right) = 0.00939 \text{ mol or } 0.0094 \text{ mol (2 sf)}$

Concentration in a 1.0 L container = $\dfrac{0.0094 \text{ mol}}{1.0 \text{ L}} = 0.0094 \text{ M}$

$0.00 \text{ M} - 0.0094 \text{ M} = -\left(1.5 \times 10^{-3} \text{ mol}/\text{L} \cdot \text{s}\right) t$

$t = 6.26 \text{ s or } 6.3 \text{ s (2sf)}$

14.28. Decomposition of HI on a metal surface:

Integrated rate equation for zero-order reactions:

$[R]_t - [R]_0 = -kt$

$0.050 \text{ mol}/\text{L} - 0.280 \text{ mol}/\text{L} = -\left(3.8 \times 10^{-3} \text{ mol}/\text{L} \cdot \text{s}\right) t$

$t = 60.5 \text{ s or } 61 \text{ s (2sf)}$

14.29. Rearrangement of ammonium cyanate to yield urea:

$2.0 \text{ days} \left(\dfrac{24 \text{ hours}}{1 \text{ day}} \right) \left(\dfrac{60 \text{ min}}{1 \text{ hour}} \right) = 2{,}880 \text{ min}$

Integrated rate equation for second order reactions:

$\dfrac{1}{[R]_t} - \dfrac{1}{[R]_0} = kt$

$\dfrac{1}{\left[1.5 \times 10^{-3} \text{ mol}/\text{L} \right]_t} - \dfrac{1}{\left[R \text{ mol}/\text{L} \right]_0} = \left(0.012 \text{ L}/\text{mol} \cdot \text{min} \right) 2{,}880 \text{ min}$

$666.7 \text{ L}/\text{mol} - \dfrac{1}{\left[R \text{ mol}/\text{L} \right]_0} = 34.56 \text{ L}/\text{mol}$

$666.7 \text{ L}/\text{mol} - 34.56 \text{ L}/\text{mol} = \dfrac{1}{\left[R \text{ mol}/\text{L} \right]_0}$

$632.1 \text{ L}/\text{mol} = \dfrac{1}{\left[R \text{ mol}/\text{L} \right]_0}$ and $R = 0.00158 \text{ mol}/\text{L}$

14.30. For the decomposition of HI:

Reaction is 2nd order; As second-order, $\dfrac{1}{[HI]_t} - \dfrac{1}{[HI]_0} = kt$

Substituting: $\dfrac{1}{[HI]_t} - \dfrac{1}{[0.025]_0} = (30.\ M^{-1}\cdot min^{-1})(15.\ min)$

and $[HI]_t = 0.0020\ M = 2.0 \times 10^{-3}\ M$

Half-Life

14.31. The decomposition of 75% of the reactant requires 2 half-lives. The half-life for the decomposition must be half of 270 s or 135 s. Rounded to two significant figures, the half-life is 140 s.

14.32. First-order decomposition reaction:
If the concentration decreases from 0.480 M to 0.060 M after 12.6 days we can solve for k using those values and the rate law:

$\ln \dfrac{[0.060\ mol/L]_t}{[0.480\ mol/L]_0} = -k(12.6\ days)$ and $k = 1.065 \times 10^{-1}\ d^{-1}$ or $1.1 \times 10^{-1}\ d^{-1}$ (2sf)

Then use the value of k to determine the half-life:

$t_{1/2} = \dfrac{0.693}{k} = \dfrac{0.693}{1.1 \times 10^{-1}\ d^{-1}} = 4.2\ days$

You might also notice that the decrease from 0.480 M to 0.060 M is 3 half-lives. Thus, 12.6 days/3 = 4.2 days.

14.33. Given that the reaction is first-order we can use the integrated form of the rate law:

$\ln(\dfrac{[N_2O_5]}{[N_2O_5]_0}) = -kt$

(a) Since the **definition of half-life** is "the time required for half of a substance to react", the fraction on the left side = 1/2, and $\ln(0.50) = -0.693$

Given the rate constant $6.7 \times 10^{-5}\ s^{-1}$ we can solve for t:
$-0.693 = -(6.7 \times 10^{-5}\ s^{-1})t$ and $t = 1.0 \times 10^4$ seconds

(b) Time required for the concentration to drop to 1/10 of the original value:
Substitute the ratio 1/10 for the concentration of N_2O_5:
$\ln(0.10) = -(6.7 \times 10^{-5}\ s^{-1})t$ and $t = 3.4 \times 10^4$ seconds

14.34. Mass of azomethane remaining after 0.0500 hr for the first-order decomposition:

$$\ln\left(\frac{x}{2.00 \text{ g}}\right) = -(0.0216 \text{ min}^{-1})(0.0500 \text{ h})\left(\frac{60 \text{ min}}{1 \text{ h}}\right)$$

$$\ln(x) - \ln(2.00 \text{ g}) = -(0.0216 \text{ min}^{-1})(3.00 \text{ min}) \text{ and } \ln(x) = 0.628$$

$$x = e^{0.628} = 1.87 \text{ g azomethane remains}$$

Mass of N_2 forming in this time:

$$(2.00 \text{ g} - 1.87 \text{ g}) \cdot \frac{1 \text{ mol CH}_3\text{NNCH}_3}{58.08 \text{ g}} \cdot \frac{1 \text{ mol N}_2}{1 \text{ mol CH}_3\text{NNCH}_3} \cdot \frac{28.01 \text{ g}}{1 \text{ mol N}_2} = 0.063 \text{ g N}_2$$

14.35. Since the decomposition is first-order: $\ln\frac{[SO_2Cl_2]_t}{[SO_2Cl_2]_0} = -kt$

We know the half-life (245 min), so we can calculate the rate constant k:

$$k = \frac{0.693}{245 \text{ min}} = 2.83 \times 10^{-3} \text{ min}^{-1}.$$

Now we can substitute into the integrated rate expression and solve for t.

$$\ln\frac{[2.00 \times 10^{-4} \text{M}]_t}{[3.6 \times 10^{-3} \text{M}]_0} = -(2.83 \times 10^{-3} \text{ min}^{-1})t \,;\, \ln(0.0555) = -(2.83 \times 10^{-3} \text{ min}^{-1})t$$

$$\frac{-2.890}{-2.83 \times 10^{-3} \text{ min}^{-1}} = t \text{ and } t = 1022 \text{ min or } 1.0 \times 10^3 \text{ min (2sf)}$$

14.36. Time for decomposition of $Xe(CH_3)_2$:

Solve for the rate constant:

$$k = \frac{0.693}{t_{1/2}} = \frac{0.693}{30. \text{ min}} = 0.023 \text{ min}^{-1}$$

Knowing the rate constant value, we can use the integrated rate expression:

$$\ln\left(\frac{0.25 \text{ mg}}{7.50 \text{ mg}}\right) = -(0.023 \text{ min}^{-1})t \text{ and solving, } t = 150 \text{ min}$$

14.37. Since this is a first-order process, $\ln \dfrac{[Cu^{2+}]}{[Cu^{2+}]_0} = -kt$ and $k = -\dfrac{0.693}{12.70 \text{ hr}}$

What fraction of the copper remains after time = 64 hr?

Radioactive decay is a first-order process, so we use the equation and substitute 64 hr for t:

$\ln \dfrac{[Cu^{2+}]}{[Cu^{2+}]_0} = -\dfrac{0.693}{12.70 \text{ hr}}$ (64 hr)

$\ln \dfrac{[Cu^{2+}]}{[Cu^{2+}]_0} = -3.49$ and $\dfrac{[Cu^{2+}]}{[Cu^{2+}]_0} = e^{-3.49}$ or 0.03 which means 3% remains.

14.38. Mass of gold-198 remaining after 1.0 day:

Given the half-life, calculate the rate constant $k = \dfrac{0.693}{t_{1/2}} = \dfrac{0.693}{2.7 \text{ days}} = 0.26 \text{ days}^{-1}$

$\ln\left(\dfrac{x}{5.6 \text{ mg}}\right) = -(0.26 \text{ days}^{-1})(1.0 \text{ day})$ and $\ln(x) - \ln(5.6 \text{ mg}) = -(0.26 \text{ days}^{-1})(1.0 \text{ day})$

$\ln(x) = 1.46$ and $x = e^{1.46} = 4.3 \text{ mg}$

Graphical Analysis: Rate Equations and *k*

14.39. For the decomposition of N₂O:

Since **ln[N₂O] vs *t*** gives a **straight line**, we know that the reaction is first-order with respect to N₂O, and the line has a slope = −*k*. The equation for the line (calculated using Excel's trendline function is $y = -0.0128x - 2.304$. The rate constant, *k*, is 0.0128 min⁻¹.

The rate equation is: Rate = *k* [N₂O]. The rate of decomposition when [N₂O] = 0.035 mol/L:

Rate = $(0.0128 \text{ min}^{-1})(0.035 \dfrac{\text{mol}}{\text{L}}) = 4.5 \times 10^{-4}$ $\dfrac{\text{mol}}{\text{L} \cdot \text{s}}$

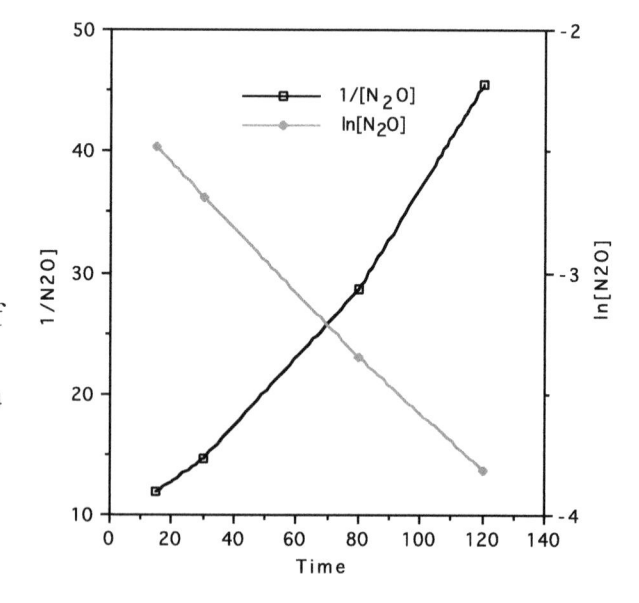

14.40. Plotting [NH₃] decomposition in two ways:

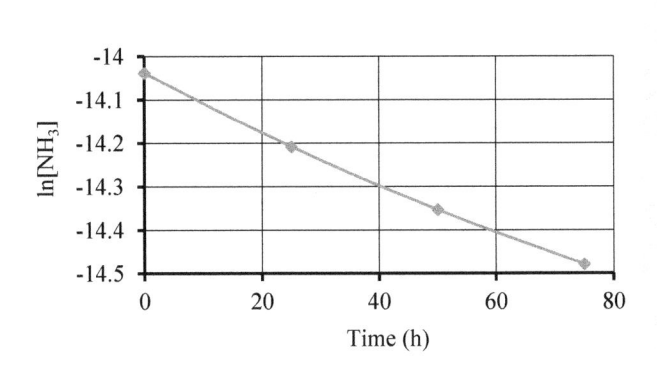

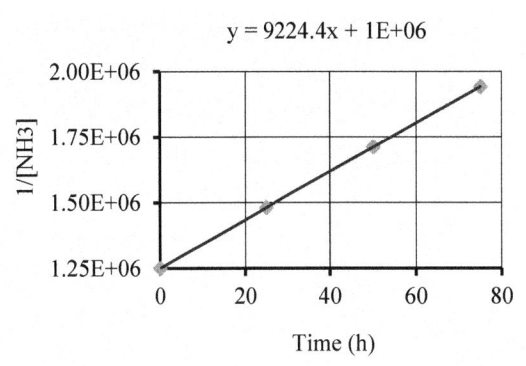

The plot of 1/[NH₃] versus time is linear, indicating that the reaction is second−order in [NH₃]. The slope of the line = k = 9220 L/mol·h

14.41. Since the graph of reciprocal concentration gives a straight line, we know the reaction is **second-order** with respect to N₂O. The **slope** of the line is equal to the rate constant, so k = 1.1 L/mol · s. The rate law is Rate = $k[N_2O]^2$

14.42. Determine the rate law and rate constant for the decomposition of HOF:

The plot of ln[HOF] versus time is linear, indicating that the reaction is first-order in HOF. So the rate law is: Rate = k[HOF], and k = –slope = –(–0.025 min⁻¹) = 0.025 min⁻¹

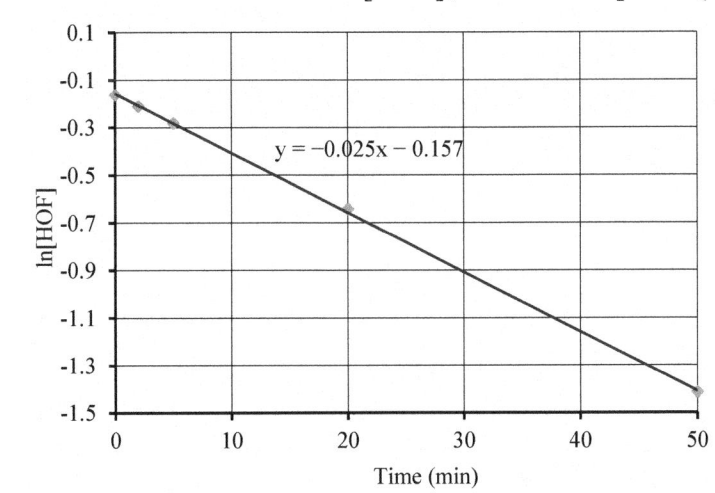

14.43. The straight line obtained when the reciprocal concentration of C₂F₄ is plotted vs t indicates that the reaction is second-order in C₂F₄.

The rate law is: Rate = $\dfrac{-\Delta[C_2F_4]}{\Delta t} = 0.04 \dfrac{L}{mol \cdot s}[C_2F_4]^2$

14.44. For the dimerization of butadiene:

(a) The plot of 1/[C_4H_6] versus time is linear, indicating that the reaction is second-order in C_4H_6.

(b) Given that the slope is 0.0583, then k = slope = 0.0583 L/mol·s.

y = 0.0583x + 100.93

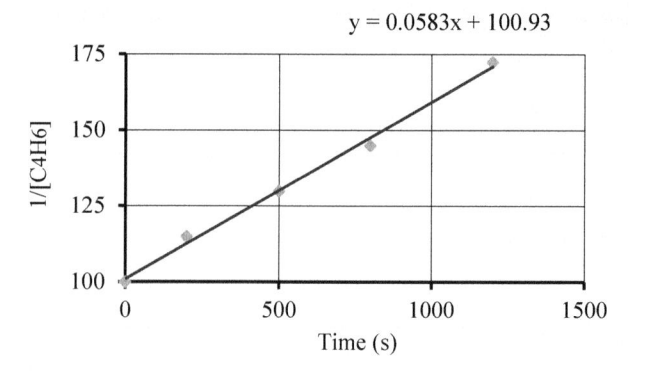

Kinetics and Energy

14.45. The E_a for the reaction 2 N_2O_5 (g) → 4 NO_2 (g) + O_2 (g)

Given k at 25 °C = 3.46 × 10^{-5} s^{-1} and k at 55 °C = 1.5 × 10^{-3} s^{-1}

The Arrhenius equation is helpful here.

$$\ln\frac{k_2}{k_1} = -\frac{E_a}{R}\left(\frac{1}{T_2} - \frac{1}{T_1}\right) ; \ln\frac{1.5 \times 10^{-3} \text{ s}^{-1}}{3.46 \times 10^{-5} \text{ s}^{-1}} = -\frac{E_a}{8.31 \times 10^{-3} \text{ kJ/mol}\cdot\text{K}}\left(\frac{1}{328 \text{ K}} - \frac{1}{298 \text{ K}}\right)$$

and solving for E_a yields a value of 102 kJ/mol for E_a or 1.0 × 10^2 kJ/mol (2 sf).

14.46. The activation energy for the reaction:

$$\ln\frac{k_2}{k_1} = -\frac{E_a}{R}\left(\frac{1}{T_2} - \frac{1}{T_1}\right)$$

$$\ln\left(\frac{3k_1}{k_1}\right) = -\frac{E_a}{8.3145 \times 10^{-3} \text{ kJ/K}\cdot\text{mol}}\left(\frac{1}{330.\text{ K}} - \frac{1}{315.\text{ K}}\right)$$ and solving for E_a,

E_a = 63 kJ/mol

14.47. Using the Arrhenius equation: $\ln\frac{k_2}{k_1} = -\frac{E_a}{R}\left(\frac{1}{T_2} - \frac{1}{T_1}\right)$, T_1 = 800 K, and T_2 = 850 K

Given E_a = 260 kJ/mol and k_1 = 0.0315 s^{-1}, we can calculate k_2.

$$\ln\frac{k_2}{0.0315 \text{ s}^{-1}} = -\frac{260 \text{ kJ/mol}}{8.3145 \times 10^{-3} \text{ kJ/mol}\cdot\text{K}}\left(\frac{1}{850 \text{ K}} - \frac{1}{800 \text{ K}}\right)$$

$\ln\frac{k_2}{0.0315 \text{ s}^{-1}} = 2.30 = \ln k_2 - \ln(0.0315 \text{ s}^{-1})$

2.30 + ln(0.0315 s^{-1}) = ln k_2 = 2.30 − 3.458 = −1.16

$k_2 = e^{-1.16}$ = 0.3 s^{-1} (1 sf owing to a temperature (800K) with 1 sf)

14.48. Determine E_a for conversion of cyclopropane to propene:

Using the equation: $\ln\dfrac{k_2}{k_1} = -\dfrac{E_a}{R}\left(\dfrac{1}{T_2} - \dfrac{1}{T_1}\right)$, solve for E_a

$\ln\left(\dfrac{1.02 \times 10^{-3}}{1.10 \times 10^{-4}}\right) = -\dfrac{E_a}{8.3145 \times 10^{-3}\ \text{kJ/mol·K}}\left(\dfrac{1}{783\ \text{K}} - \dfrac{1}{743\ \text{K}}\right)$ and $E_a = 270$ kJ/mol

14.49. Energy progress diagram:

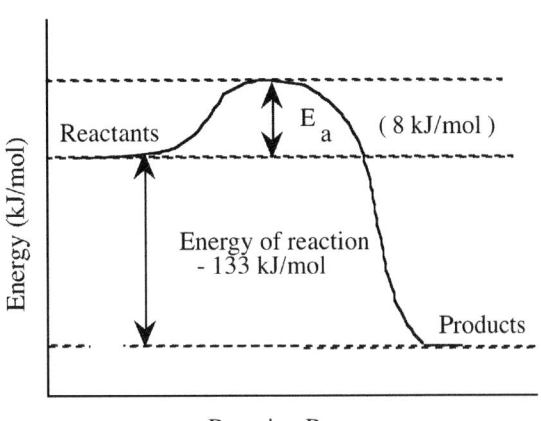

14.50. Based on the graphic:

(a) Is the reaction endo- or exo-thermic: The reaction is endothermic—that is, the Energy of products is greater than energy of reactants.

(b) Does the reaction occur in one or more-than-one step: The reaction occurs in two steps, as evidenced by the "two humps", each with their own activation energy.

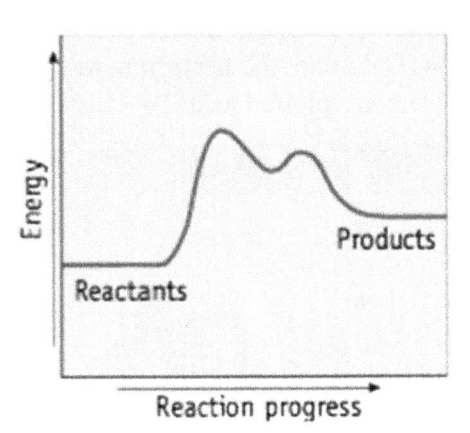

Enzymes

14.51. Compare lock-and-key and induced fit models for substrate binding to an enzyme:

The lock-and-key model pictures the active site of the enzyme as complementary to the structure of the substrate (fits like a "key in a lock"). The induced-fit model pictures the enzyme as being flexible, so that it becomes complementary to the substrate by changing its shape.

14.52. Species to which an enzyme should bind best: the *transition state.*

14.53. Plot 1/[S] vs 1/Rate and determine Rate_max:

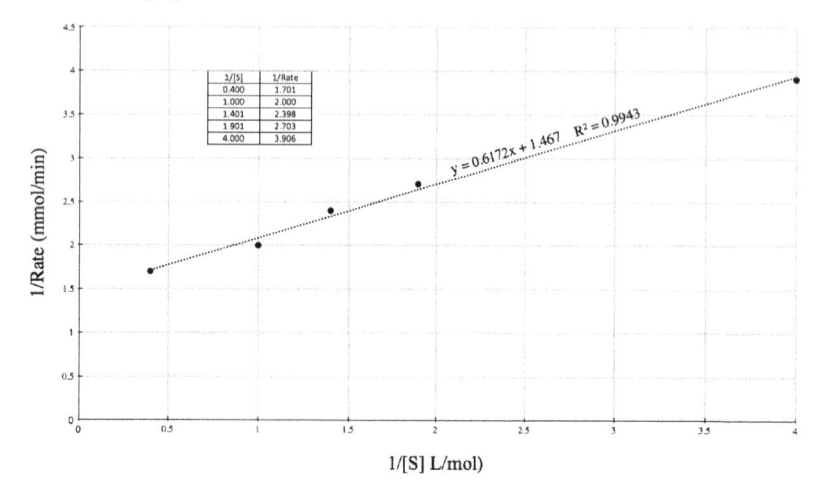

1/[S]	1/Rate
0.400	1.701
1.000	2.000
1.401	2.398
1.901	2.703
4.000	3.906

V_{max} occurs when $1/[S] = 0$. Recall that $x = 1/[S]$ and $y = 1/\text{Rate}$.

The equation for the best-fit straight line is $y = 0.6172x + 1.467$.

So, in the above equation, substitute 0 for x, and the resulting $y = 1.47$.

So 1/RATE = 1.47 and RATE (V_{max}) = 1/1.47 or 0.68 mmol/min.

14.54. Calculate the maximum rate of transformation of CO_2 into HCO_3^- ions:
Data are plotted as in 14.45 to obtain the graph below:

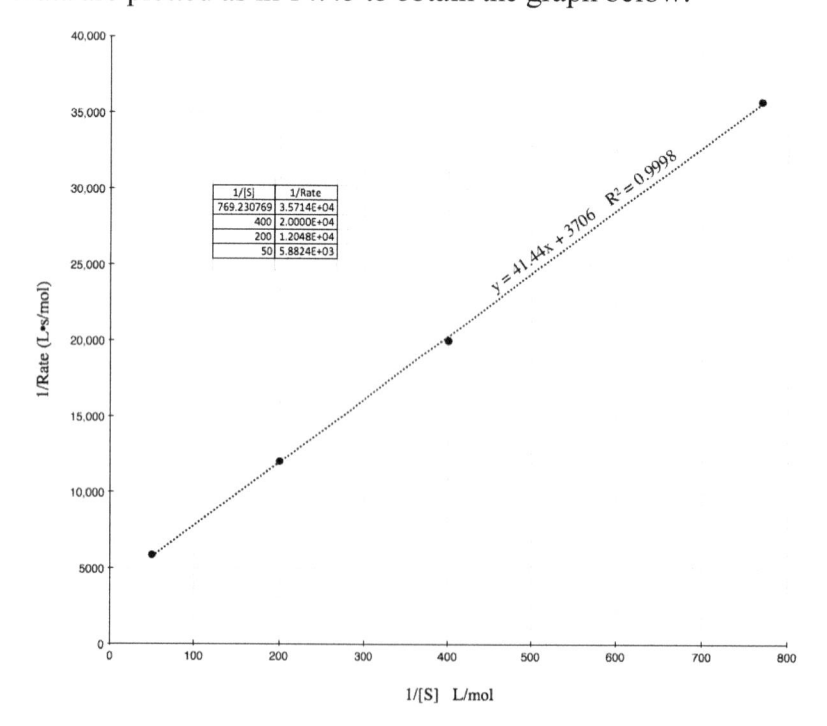

1/[S]	1/Rate
769.230769	3.5714E+04
400	2.0000E+04
200	1.2048E+04
50	5.8824E+03

$y = 41.44x + 3706$ $R^2 = 0.9998$

V_{max} occurs when $1/[S] = 0$. Recall that $x = 1/[S]$ and $y = 1/$Rate.

The equation for the best-fit straight line is $y = 41.44x + 3706$.

So, in the above equation, substitute 0 for x, and the resulting $y = 3706$

So, 1/RATE = 3706 and RATE (V_{max}) = 1/3706 or $2.70 \times 10^{-4} \, \dfrac{\text{mol}}{\text{L} \cdot \text{s}}$.

Reaction Mechanisms

14.55. Elementary Step   Rate law

(a) $NO(g) + NO_3(g) \rightarrow 2\, NO_2(g)$ Rate = $k[NO][NO_3]$
    Reaction is bimolecular

(b) $Cl(g) + H_2(g) \rightarrow HCl(g) + H(g)$ Rate = $k[Cl][H_2]$
    Reaction is bimolecular

(c) $(CH_3)_3CBr(aq) \rightarrow (CH_3)_3C^+(aq) + Br^-(aq)$ Rate = $k(CH_3)_3CBr]$
    Reaction is unimolecular

14.56. Rate law for each of the following elementary reactions:

Elementary Step   Rate Law

(a) $NO_2(g) + CO(g) \rightarrow NO(g) + CO_2(g)$ Rate = $k[NO_2][CO]$

(b) $N_2O_4(g) \rightarrow 2\, NO_2(g)$ Rate = $k[N_2O_4]$

(c) $2\, NO_2(g) \rightarrow NO(g) + NO_3(g)$ Rate = $k[NO_2]^2$

14.57. For the reaction reflecting the decomposition of ozone:

(a) The *second step* is the slow step, and therefore rate-determining.

(b) The rate equation involves *only* these substances that affect the rate, (since they participate in the rate determining step), Rate = $k[O_3][O]$.

14.58. For the reaction of NO_2 and CO:

(a) elementary steps add to overall equation:

Step 1 Slow $\cancel{NO_2(g)} + NO_2(g) \rightarrow NO(g) + \cancel{NO_3(g)}$

Step 2 Fast $\cancel{NO_3(g)} + CO(g) \rightarrow \cancel{NO_2(g)} + CO_2(g)$

Overall Reaction $NO_2(g) + CO(g) \rightarrow NO(g) + CO_2(g)$

(b) molecularity of each step:

Both steps 1 and 2 are bimolecular (involve only two molecules as reactants)

(c) experimental rate equation:

Rate = $k[NO_2]^2$ as Step 1 is the slow step and involves 2 molecules of NO_2

(d) intermediates:

NO_3 is an intermediate—being formed in step 1 and consumed in step 2—not appearing in the net reaction.

14.59. For the reaction of NO_2 and CO:

(a) Classify the species:

$NO_2(g)$	Reactant (step 1); Product (step 2)
$CO(g)$	Reactant (step 2)
$NO_3(g)$	Intermediate (produced & consumed subsequently)
$CO_2(g)$	Product
$NO(g)$	Product

(b) A reaction coordinate diagram

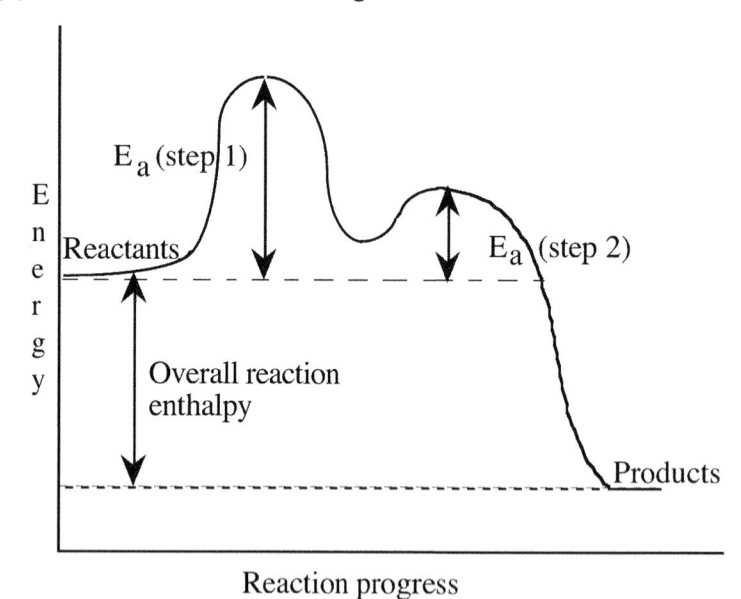

14.60. For the reaction of CH₃OH and HBr:

(a) Overall equation: $CH_3OH + H^+ + Br^- \rightarrow CH_3Br + H_2O$

(b) Reaction coordinate diagram:

(c) Show rate law:

Using the rate–determining step, the rate law is $-\dfrac{\Delta[CH_3OH]}{\Delta t} = k_2[CH_3OH_2^+][Br^-]$.

However, this rate law contains an intermediate, $CH_3OH_2^+$ Rate of production of $CH_3OH_2^+ = k_1[CH_3OH][H^+]$

$-\dfrac{\Delta[CH_3OH]}{\Delta t} = k_2\{k_1[CH_3OH][H^+]\}[Br^-] = k[CH_3OH][H^+][Br^-]$ (where $k = k_1 \cdot k_2$)

General Questions

14.61. What happens to the reaction rate for a reaction with the rate equation: Rate = $k[A]^2[B]$

Rate₁ = $k[A]^2[B]$. Double the concentration of A—call it 2A, and halve the concentration of B—call it B/2.

Rate₂ = $k[2A]^2[B/2]$. Reducing the concentrations gives Rate₂ = $(2)^2 \cdot (1/2) k[A]^2[B]$

Rate₂ = $2 \cdot k [A]^2[B]$. So Rate₂ is **two times** that of Rate₁.

14.62. Fraction of reactants remaining after 6 half-lives:

$(1/2)^6 = 1/64$ remains

14.63. To determine second-order dependence, after acquisition of the pH vs time data, plot 1/[OH⁻] versus time. The reaction is second-order in OH⁻ if a straight line is obtained.

14.64. Comparing experiment 1 to experiment 2, [Br₂] remains constant, [NO] increases by a factor of four, and the rate increases by a factor of sixteen. The reaction is second-order in NO. Comparing experiment 1 to experiment 3, [NO] is constant, [Br₂] increases by a factor of 2.5, and the rate increases by a factor of 2.5. The reaction is first-order in Br₂. The reaction is third-order overall.

To determine the rate constant, k, use the data from any one of the three experiments and the overall rate law as determined above, Rate = $k[NO]^2[Br_2]$.

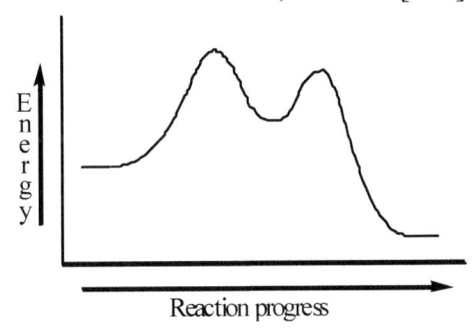

Rate = $k[NO]^2[Br_2]$

From experiment 1

Rate = 6.0×10^{-2} mol/L·s [NO] = 2.5×10^{-2} M [Br$_2$] = 2.0×10^{-2} M

6.0×10^{-2} mol/L·s = $k[2.5 \times 10^{-2} \text{ M}]^2 [2.0 \times 10^{-2} \text{ M}]$

6.0×10^{-2} mol/L·s = $k[6.25 \times 10^{-4} \text{ M}^2][2.0 \times 10^{-2} \text{ M}]$

6.0×10^{-2} mol/L·s = $k[1.25 \times 10^{-5} \text{ M}^3]$

$k = \dfrac{6.0 \times 10^{-2} \text{ mol/L·s}}{1.25 \times 10^{-5} \text{ M}^3} = 4.8 \times 10^3 \ 1/\text{M}^2\text{s}$

14.65. For first-order kinetics, we know that $\ln\left(\dfrac{[HCO_2H]}{[HCO_2H]_0}\right) = -kt$.

Substituting into the equation, solve for **k**.

$\ln\left(\dfrac{[25]}{[100]_0}\right) = -k(72 \text{ s})$ Rearranging to solve for k gives $\dfrac{-1.386}{-72 \text{ s}} = k = 0.01925 \text{ s}^{-1}$,

and since we're determining the $t_{1/2}$, $t_{1/2} = 0.693/k$ so $t_{1/2} = 0.693/0.01925 \text{ s}^{-1} = 36$ s.

A **much simpler** route to this answer is to recognize that 1 half-life would consume 50% of the original sample, and the 2nd half-life would consume half of the remaining amount (25%). So 2 half-lives would result in the consumption of 75% of the original sample—or 1/2(72 s).

14.66. For the isomerization of CH$_3$NC:

(a) A plot of ln[CH$_3$NC] versus time gives a straight line with a negative slope. The reaction is first-order in CH$_3$NC. Rate = k[CH$_3$NC]
(b) ln[CH$_3$NC]$_t$ = $-kt$ + ln[CH$_3$NC]$_0$
(c) Estimating slope from the graph (see text), k = –slope = –(–2 × 10^{-4} s^{-1}) = 2 × 10^{-4} s^{-1}
(d) Time for half the sample to isomerize: $t_{1/2} = \dfrac{0.693}{k} = \dfrac{0.693}{2 \times 10^{-4} \text{ s}^{-1}} = 3 \times 10^3$ s
(e) Concentration of CH$_3$NC after 10,000 s:

From the graph, we have the initial concentration: [CH$_3$NC]$_0$ = $e^{-4.1}$ = 0.0166 mol/L

$\ln\dfrac{[CH_3NC]}{0.0166 \text{ mol/L}} = -(2 \times 10^{-4} \text{ s}^{-1})(10,000 \text{ s})$

ln[CH$_3$NC] – ln(0.0166 mol/L) = –(2 × 10^{-4} s^{-1})(10,000 s) and solving for [CH$_3$NC]:

[CH$_3$NC] = 0.002 mol/L

14.67. For the dimerization to form octafluorocyclobutane, the following plot is obtained:

(a) The plot of reciprocal concentration vs time gives a straight line. Such behavior is indicative of **a second-order process.**
The rate law is: Rate = $k[C_2F_4]^2$

(b) The rate constant is equal to the slope of the line: Using a graphical package (such as Excel) the equation for the line is:
$y = 0.04482x + 9.990$.
So $k = 0.045$ L/mol · s

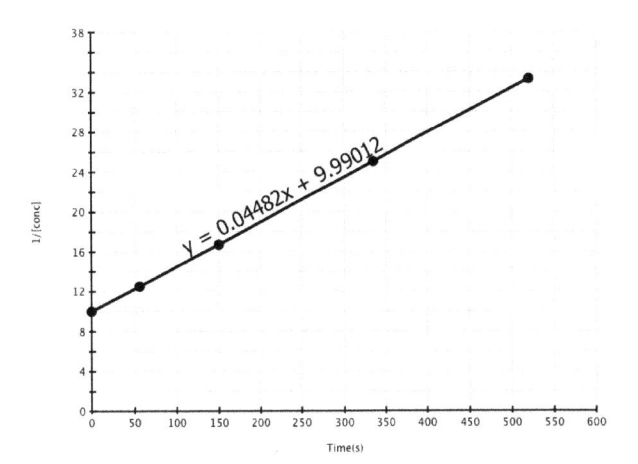

(c) The concentration after 600 s is found by using the integrated rate equation for second-order processes:

$$\frac{1}{[C_2F_4]} = 0.045 \text{ L/mol·s}(600 \text{ s}) + 9.990$$

$[C_2F_4]_{600} = 0.027$ (or 0.03 M to 1 sf)

(d) Time required for 90% completion: Using the integrated equation for a second-order decay reaction, and substituting 10% of the initial concentration as our "concentration at time t", solve for t.

$$\frac{1}{[C_2F_4]_t} - \frac{1}{[C_2F_4]_0} = 0.045 \text{ L/mol·s}(t)$$

$$\frac{1}{[0.010]} - \frac{1}{[0.100]} = 0.045 \text{ L/mol·s}(t)$$

$$100 - 10 = 0.045 \text{ L/mol·s}(t)$$

and $t = 2000$ s

14.68. For the reaction of NO_2 and CO:

(a) Comparing experiment 1 to experiment 2, [CO] remains constant, $[NO_2]$ doubles, and the rate doubles. The reaction is first-order in NO_2. Comparing experiment 1 to experiment 4, $[NO_2]$ is constant, [CO] doubles, and the rate doubles. The reaction is first-order in CO.

(b) Using the orders derived in the logic of part (a), the rate equation is:
Rate = $k[CO][NO_2]$

(c) $k = \dfrac{\text{Rate}}{[CO][NO_2]} = \dfrac{3.4 \times 10^{-8} \text{ mol/L·h}}{(5.1 \times 10^{-4} \text{ mol/L})(0.35 \times 10^{-4} \text{ mol/L})} = 1.9 \text{ L/mol·h}$

14.69. For the formation of urea from ammonium cyanate:

(a) Plot the data to determine the order.

The upper graph shown here is a plot of the 1/[NH₄NCO] vs time, and the lower graph ln[NH₄NCO] versus time. Note that the plot of reciprocal concentration gives a straight line, indicating the reaction is *second-order in ammonium cyanate*.

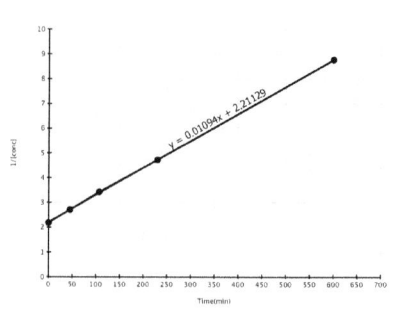

(b) k is the slope of the line, which is 0.0109 L/mol·min

(c) The half-life can be calculated using the integrated rate equation:

$$\frac{1}{[0.229]_t} - \frac{1}{[0.458]_0} = 0.0109 \frac{L}{mol \cdot min}(t)$$

The concentration at time t is 1/2 that of the original concentration of NH₄NCO (the definition of half-life) and solving for t = 200. minutes.

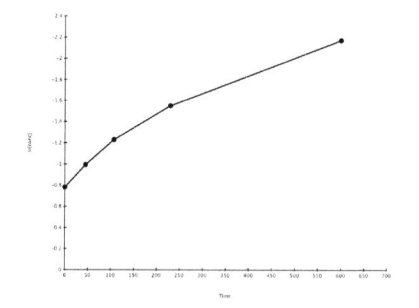

(d) The concentration of ammonium cyanate after 12.0 hours (720. min) is found by using the integrated rate equation. Since we know k and t, we can solve for concentration at time t = (12.0 hours).

$$\frac{1}{[NH_4NCO]_t} - \frac{1}{[0.458]_0} = 0.0109 \frac{L}{mol \cdot min}(t)$$

$$\frac{1}{[NH_4NCO]_t} = 0.0109 \frac{L}{mol \cdot min}(720.\ min) + \frac{1}{[0.458]_0}$$

Solving for [NH₄NCO] we obtain [NH₄NCO] = 0.0997 M

14.70. Regarding the decomposition of NOₓ:

(a) Determine the value of k:

$$k = \frac{0.693}{t_{1/2}} = \frac{0.693}{3.9\ h} = 0.178\ h^{-1}$$

For a first-order process:

$$\ln\left(\frac{x}{1.50\ mg}\right) = -0.178\ h^{-1}(5.25\ h)$$

Rewriting to solve for ln(x): ln(x) − ln(1.50) = −0.178 h⁻¹(5.25 h)

And solving for x: x = 0.59 mg (2 sf).

(b) Hours of daylight needed to reduce the concentration to the stated values:

$$\ln\left(\frac{2.50 \times 10^{-6} \text{ mg}}{1.50 \text{ mg}}\right) = -0.178 \text{ h}^{-1}(t)$$

and solving for t: t = 75 h (2 sf)

14.71. The reaction between carbon monoxide and nitrogen dioxide has a rate equation that is second-order in NO₂. This means that the *slowest step* in the mechanism involves **2** molecules of nitrogen dioxide. Mechanism 2 has a SLOW step that fulfills this requirement. Note that Mechanisms 1 and 3 are only first-order in nitrogen dioxide.

14.72. For the formation of nitryl fluoride:

(a) The rate equation can be written, once we determine the order for each reactant.
In experiments 1 and 2, [F₂] and [NO₂F] are constant while the concentration of NO₂ doubles. The rate also doubles, so the reaction is first-order in NO₂.
In experiments 3 and 4, [F₂] doubles while [NO₂F] and [NO₂] remain constant. The rate also doubles, so the reaction is first-order in F₂.
Finally, in experiments 5 and 6 [F₂] and [NO₂] remain constant while [NO₂F] doubles. The rate remains constant, so the reaction is zero-order in NO₂F. Rate = k[F₂][NO₂]

(b) The reaction is first-order in F₂, first-order in NO₂, and zero-order in NO₂F.

(c) $k = \dfrac{2.0 \times 10^{-4} \text{ mol/L} \cdot \text{s}}{(0.001 \text{ mol/L})(0.005 \text{ mol/L})} = 40 \text{ L/mol} \cdot \text{s}$

14.73. The decomposition of dinitrogen pentoxide has a first-order rate equation. Determine the rate constant and the half-life by substitution into the integrated rate equation for first-order reactions: $\ln\left(\dfrac{[N_2O_5]_t}{[N_2O_5]_0}\right) = -kt$

The decomposition is 20.5% complete in 13.0 hours at 298K.

The amount of N₂O₅ remaining after 13.0 h is 79.5% of the original concentration. The left hand term is then 79.5/100.

$$\ln\left(\frac{79.5}{100}\right) = -k \cdot 13.0 \text{ h and } k = 0.0176 \text{ h}^{-1}$$

Calculation of the half-life is accomplished by noting that after a half-life, the left-hand side of the integrated rate equation will have a value of: $\ln(0.50) = -0.693$. Substituting the value of k from above:

$$\frac{-0.693}{-0.0176 \text{ hr}^{-1}} = 39.3 \text{ h}$$

14.74. For the decomposition of N₂O₅: Plot ln k versus 1/T

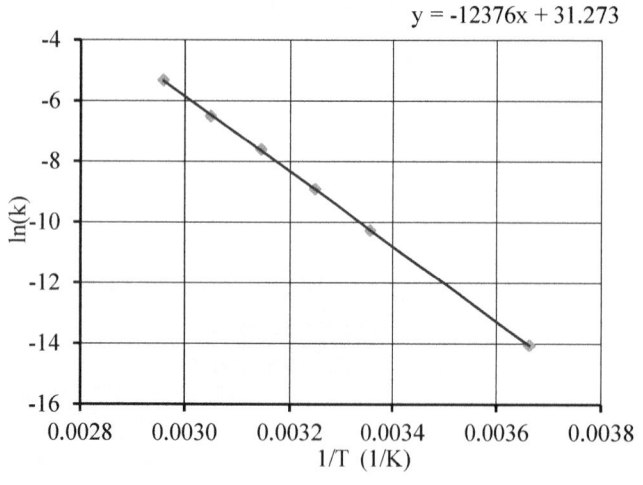

y = -12376x + 31.273

Slope = $-12376 \text{ K}^{-1} = -\dfrac{E_a}{R} = -\dfrac{E_a}{8.3145 \times 10^{-3} \text{ kJ/K} \cdot \text{mol}}$ and solving for E_a

E_a = 103 kJ/mol

14.75. For the decomposition of dimethyl ether:

(a) The mass of dimethyl ether remaining after 125 min and after 145 min:

The half-life is 25.0 min. A period of 125 minutes is 5 half-lives. The fraction remaining after n half-lives is $\left(\dfrac{1}{2}\right)^n$ and with n = 5, the fraction remaining is 0.03125.

(1/32 of the original amount). The mass remaining is (0.03125)(8.00 g) = 0.250 g dimethyl ether. Note that 145 minutes is *almost* 6 half-lives. So you should be able to "guess" at a value for the amount remaining. The mass should be slightly greater than 1/64 of the original amount. The exact amount can be found by substitution into the first-order rate equation to solve for the rate constant:

$k = \dfrac{0.693}{25.0 \text{ min}} = 0.02772 \text{ min}^{-1}$

Note that the "ln term" is simplified by remembering that a "half-life" is a time in which 50% decomposes. Then ln(0.50) = −0.693   {A HANDY THING TO REMEMBER}

Substituting the value of k into the first-order integrated rate equation, and solving for the mass remaining after 145 min:

$\ln\left(\dfrac{x \text{ g}}{8.00 \text{ g}}\right) = -0.02772 \text{ min}^{-1}(145 \text{ min})$

Solving for x: x = 0.0144 g = 0.14 g (2 sf).

(b) The time required for 7.60 ng of ether to be reduced to 2.25 ng:
Substitute into the first-order equation (as we've done above). Now we know the value for the "left side" and we know k; we can solve for t

$$\frac{\ln\left(\frac{[2.25 \text{ ng}]_t}{[7.60 \text{ ng}]_0}\right)}{-0.0277 \text{ min}^{-1}} = t = \frac{\ln(0.296)}{-0.0277 \text{ min}^{-1}} = \frac{-1.217}{-0.0277 \text{ min}^{-1}} = 43.9 \text{ min}$$

(c) The fraction remaining after 150 minutes is easily calculated by noting that 150 minutes is *exactly* 6 half-lives. The fraction remaining is 1/64 or 0.016 (2 sf).

14.76. For the thermal decomposition of phosphine:

(a) When ¾ of the PH₃ has decomposed, ¼ remains and two half−lives have passed.
$2(t_{1/2}) = 2(37.9 \text{ s}) = 75.8 \text{ s}$

(b) $k = \dfrac{0.693}{t_{1/2}} = \dfrac{0.693}{37.9 \text{ s}} = 0.0183 \text{ s}^{-1}$

$\ln \dfrac{[PH_3]}{[PH_3]_0} = -(0.0183 \text{ s}^{-1})(2 \text{ min})\left(\dfrac{60 \text{ s}}{1 \text{ min}}\right)$

fraction remaining $= \dfrac{[PH_3]}{[PH_3]_0} = 0.11$

14.77. The data are tabulated below, with the [C₄H₂] being converted into both ln[C₄H₂] and 1/[C₄H₂] columns. The *x* axes are time (in ms).

Time(ms)	ln[C₄H₂,mol/L]	1/[C₄H₂, mol/L]x 1000
0	−9.190537745	9.8
50	−9.893537222	19.8
100	−10.56126759	38.6
150	−10.81479074	49.8
200	−11.14828235	69.4
250	−11.2505612	76.9

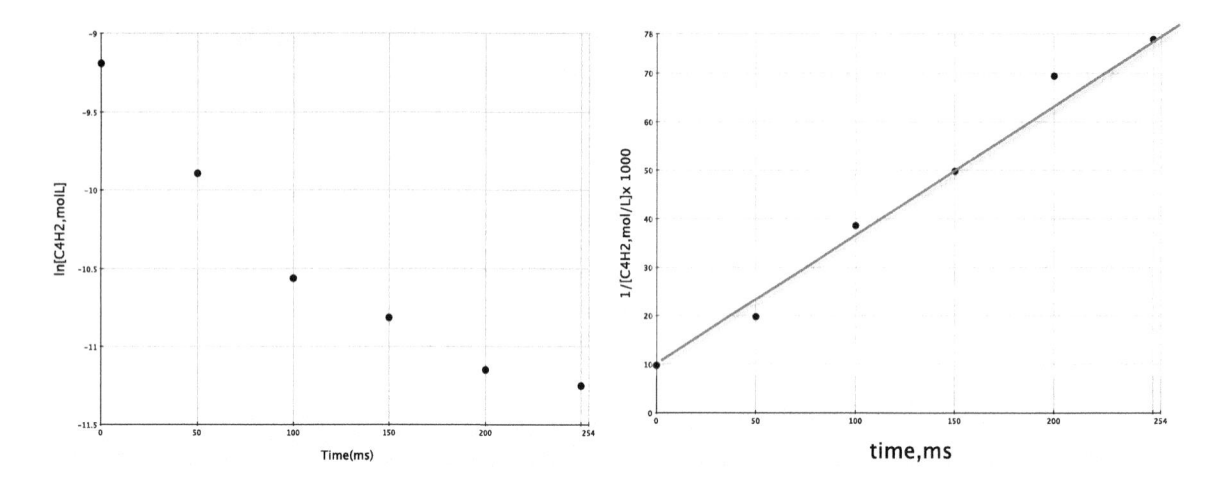

The plot of 1/[C₄H₂] vs time gives a straight line, while the plot of ln[C₄H₂] vs time is not, indicating that the reaction is 2nd order.

14.78. For the thermal decomposition of diacetylene, calculate E_a:

A plot of ln k vs. 1/T generated a straight line with a slope of -1.99×10^4 K.

Slope = $-E_a/R$ Substituting: -1.99×10^4 K = $-E_a/(8.31 \times 10^{-3}$ kJ/K)

and E_a = 165 kJ

14.79. Show consistency of mechanism with the rate law: Rate = $k[O_3]^2/[O_2]$:

The rate law for the **slow** step is Rate = $k[O_3][O]$, but the concentration of O is affected by the preceding equilibrium step. Solve for the [O] in that equilibrium step:
The equilibrium constant expression for the fast step is:

$$K = \frac{[O_2][O]}{[O_3]}; \text{ solving for } [O] = \frac{K \cdot [O_3]}{[O_2]}$$

Substitute this concentration into the Rate law for the slow step above:

$$\text{Rate} = k[O_3][O] \text{ ; Rate} = \frac{K \cdot k[O_3][O_3]}{[O_2]}$$

Combining K · k as k', Rate = $\dfrac{k'[O_3]^2}{[O_2]}$

14.80. Concerning the ozone interactions described:

The slowest reaction has the smallest k (d) Cl + CH₂FCl → HCl + CHFCl
The fastest reaction has the largest k (c) Cl + C₃H₈ → HCl + C₃H₇

14.81. Calculate E_a from the provided data:

We can calculate E_a if we know values of rate constants at two temperatures:

$$\ln\frac{k_2}{k_1} = -\frac{E_a}{R}\left(\frac{1}{T_2} - \frac{1}{T_1}\right)$$

The slope was determined to be –6370.

Since slope = $-E_a/R$,

$$-6370 = -\frac{E_a}{8.314 \times 10^{-3}\,\frac{kJ}{K\cdot mol}} = 53.0\,\frac{kJ}{mol}$$

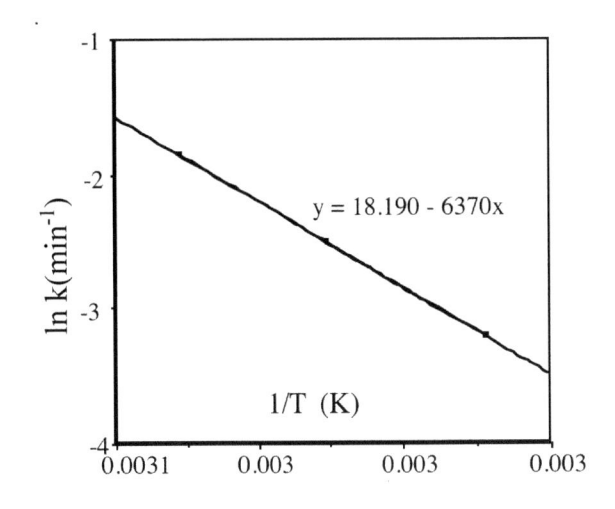

14.82. Find the rate constant at 313.0 K, for the decomposition of N_2O_5:

We can use this equation to find k at a 2nd temperature (313.0 K) if we know it at one temperature: $\ln\frac{k_2}{k_1} = -\frac{E_a}{R}\left(\frac{1}{T_2} - \frac{1}{T_1}\right)$. Substitute the known k value, and the two temperatures to obtain: $\ln\left(\frac{k_2}{0.0900\,\text{min}^{-1}}\right) = -\frac{103\,\text{kJ/mol}}{8.3145 \times 10^{-3}\,\text{kJ/K}\cdot\text{mol}}\left(\frac{1}{313.0\,\text{K}} - \frac{1}{328.0\,\text{K}}\right)$

Solving for k_2 gives a value of 0.0147 min^{-1}.

14.83. For the reaction:

(i) Which is the rate determining step?
Since Step 2 is the **slow** step, it is rate determining.

(ii) Is there an intermediate in the reaction?
N_2O_2 is formed in Step 1, and destroyed in Step 2, so N_2O_2 is an intermediate.

(iii) What is the experimentally determined rate law?

The rate is determined by the rate law: Rate = $k[N_2O_2][O_2]$. The concentration of N_2O_2 however is determined by the equilibrium in the first step, which we can write:

$K = \frac{[N_2O_2]}{[NO]^2}$, and solving for $[N_2O_2]$, $K\cdot[NO]^2 = [N_2O_2]$. Now we substitute into the rate law from the first line: Rate $= -\frac{1}{2}\cdot\frac{\Delta[NO]}{\Delta t} = kK[NO]^2[O_2]$, and since both k and K are constants, we combine them into one constant k, making Rate = $k[NO]^2[O_2]$

14.84. Concerning the decomposition of SO_2Cl_2:

(a) As the decomposition is first-order in SO_2Cl_2, we can write:
Rate = (0.17/hr)(0.010 M) = 1.7×10^{-3} M/hr

(b) The half-life of the reaction: $t_{1/2}$ = 0.693/k = 0.693/0.17 hr^{-1} = 4.1 hr

(c) After one half-life, the pressure of SO_2Cl_2 drops by ½ to 0.025 atm. From stoichiometry, the pressure of both products, SO_2 and Cl_2, must be 0.025 atm. Thus, total pressure = 0.025 atm + 0.025 atm + 0.025 atm = 0.075 atm.

14.85. The decomposition of NO_2 is 2nd order in NO_2:

(a) Determine the rate constant for the reaction:

The rate expression: $\dfrac{1}{[NO_2]_t} - \dfrac{1}{[NO_2]_0} = kt$, and substituting:

$\dfrac{1}{[0.100 \text{ M}]_t} - \dfrac{1}{[0.250 \text{ M}]_0} = k(1.76 \text{ min})$

(10.0 – 4.00) 1/M = k (1.76 min) and solving for k:

k = 3.4 L/mol·min

(b) The rate equation for the original equation is: Rate $= -\dfrac{\Delta[NO_2]}{\Delta t} = k[NO_2]^2$

For the rewritten equation, the rate equation is equal to: $-\dfrac{1}{2} \cdot \dfrac{\Delta[NO_2]}{\Delta t} = k'[NO_2]^2$ or

$-\dfrac{\Delta[NO_2]}{\Delta t} = 2k'[NO_2]^2$ therefore $k = 2k'$ or $k' = \dfrac{1}{2}\left(3.41 \dfrac{L}{mol \cdot min}\right) = 1.70 \dfrac{L}{mol \cdot min}$

14.86. For the decomposition of H_2O_2:

(a) As doubling [H_2O_2] doubles the rate, the reaction is first-order in the reactant, H_2O_2
Rate = $k[H_2O_2] = \Delta[O_2]/\Delta t$

(b) k = Rate/[H_2O_2]; Using data from the first experiment, k = 5.30×10^{-5} M/min/0.0500 M
$k = 1.06 \times 10^{-3}$ min^{-1}

(c) Value for the "new" rate constant: k' = 2k = 2(1.06 × 10^{-3} min^{-1}) = 2.12 × 10^{-3} min^{-1}

14.87. Since the time required to prepare the egg will be inversely proportional to the rate constant, we can calculate the ratio of the time, by calculating the ratio of the rate constants.

Using the Arrhenius equation, $\ln \dfrac{k_2}{k_1} = -\dfrac{E_a}{R}\left(\dfrac{1}{T_2} - \dfrac{1}{T_1}\right)$

Recalling the reciprocal relationship between k and t, we write:

$$\ln\left(\frac{t_{90}}{t_{100}}\right) = -\frac{E_a}{R}\left[\frac{1}{T_2} - \frac{1}{T_1}\right] \text{ or } \ln\left(\frac{t_{90}}{t_{100}}\right) = -\frac{52.0 \text{ kJ/mol}}{8.314\times 10^{-3} \text{ kJ/K·mol}}\left(\frac{1}{373 \text{ K}} - \frac{1}{363 \text{ K}}\right)$$

$$\ln\left(\frac{t_{90}}{t_{100}}\right) = -6254 \text{ K}\left(-7.385\times 10^{-5} \text{ K}^{-1}\right) = 0.4619$$

$\frac{t_{90}}{t_{100}} = 1.59$ or $t_{90} = 1.59(t_{100})$ so $t_{90} = 1.59(3 \text{ min}) = 4.76$ min or 5 min (1sf)

14.88. Determine the order of the reaction and the rate constant for the reaction of butadiene:

Calculate $P_{C_4H_6 \text{(unreacted)}}$ at each time (min) and plot the data to determine the reaction order:

(1) $P_{C_4H_6 \text{(unreacted)}} = P_{\text{total}(t=0)} - P_{C_4H_6 \text{(reacted)}} = P_{\text{total}(t=0)} - (2\times P_{C_8H_{12}})$

(2) $P_{\text{total}(t=\text{time})} = P_{C_4H_6 \text{(unreacted)}} + P_{C_8H_{12}}$; Solving for $P_{C_4H_6 \text{(unreacted)}}$:

Start with eq. (1): $P_{C_4H_6 \text{(unreacted)}} = P_{\text{total}(t=0)} - (2\times P_{C_8H_{12}})$

Rearrange eq. (2) to solve for $P_{C_4H_6 \text{(unreacted)}}$: $P_{C_4H_6 \text{(unreacted)}} = P_{\text{total}(t=\text{time})} - P_{C_8H_{12}}$ multiply by 2

$2\times P_{C_4H_6 \text{(unreacted)}} = 2\times P_{\text{total}(t=\text{time})} - 2\times P_{C_8H_{12}}$ now subtract eq.1 from this equation.

$2\times P_{C_4H_6 \text{(unreacted)}} = 2\times P_{\text{total}(t=\text{time})} - 2\times P_{C_8H_{12}}$
$- P_{C_4H_6 \text{(unreacted)}} = -P_{\text{total}(t=0)} + (2\times P_{C_8H_{12}})$
$\overline{P_{C_4H_6 \text{(unreacted)}} = 2\times P_{\text{total}(t=\text{time})} - P_{\text{total}(t=0)}}$

Substituting the pressures from the table a results:
The reaction is second-order;
k = slope = 2.36×10^{-5} atm^{-1}·min^{-1}

y = 2.36E-05x + 2.30E-03

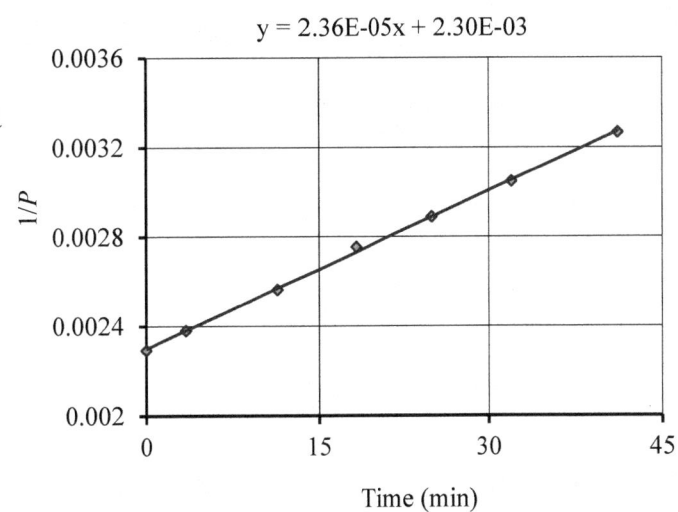

(plot of 1/P vs Time (min))

14.89. The decomposition of HOF proceeds in a first-order reaction with a half-life of 30 minutes.

The equation is: HOF(g) → HF(g) + 1/2 O₂ (g)

If the initial pressure of HOF is 1.00×10^2 mm Hg at 25°C, what is the total pressure in the flask and the partial pressure of HOF after 30 minutes?

The rate constant is calculated: $k = \dfrac{0.693}{t_{1/2}} = \dfrac{0.693}{30 \text{ min}} = 0.0231 \text{ min}^{-1}$

To calculate the concentration (pressure in the case of gases) at **any** time, the integrated first-order rate law can be used: $\ln \dfrac{[\text{HOF}]_t}{[\text{HOF}]_0} = -kt$

For the 30 minute time-frame, the process is simplified. The concentration fraction is 1/2 (since 30 minutes is equal to a half-life). So P_{HOF} = 1/2(100. mm Hg) or 50.0 mm Hg.

The stoichiometry indicates that we get 1 HF and 1/2 O_2 for each HOF. Then the P_{HF} = 50.0 mm Hg, and P_{O_2} = 25.0 mm Hg. The **total** pressure is then (50.0 + 50.0 + 25.0) or 125.0 mm Hg.

The corresponding values after 45 minutes:

$\ln \dfrac{[\text{HOF}]_t}{[\text{HOF}]_0} = -kt$ becomes $\ln \dfrac{[\text{HOF}]_t}{[100. \text{ mm Hg}]_0} = -0.0231 \text{ min}^{-1} (45 \text{ min})$

$\dfrac{[\text{HOF}]_t}{100. \text{ mm Hg}} = e^{-(0.0231 \text{ min}^{-1} \times 45 \text{ min})}$ and $[\text{HOF}]_t = 0.3536 \cdot 100.$ mm Hg = 35.4 mm Hg

P_{HOF} = 35.4 mm Hg; P_{HF} = (100.0 - 35.4) mm Hg or 65 mm Hg; P_{O_2} = 32 mm Hg

The **total** pressure is then (35.4 mm Hg + 65 mm Hg + 32 mm Hg) or 132 mm Hg.

14.90. For the decomposition of SO_2Cl_2 at 600 K:

After 245 minutes (one half-life): Half of the 25 mm Hg decomposes with the resulting P:
$P_{SO_2Cl_2}$ = 13 mm Hg P_{SO_2} = 13 mm Hg P_{Cl_2} = 13 mm Hg

The total pressure will be (3 × 12.5) = 37.5 mm or P_{total} = 38 mm Hg.

After 12 hours (720 min):

$k = \dfrac{0.693}{t_{1/2}} = \dfrac{0.693}{245 \text{ min}} = 0.00283 \text{ min}^{-1}$

$\ln\left(\dfrac{x}{25 \text{ mm Hg}}\right) = -(0.00283 \text{ min}^{-1})(720 \text{ minutes})$

$\ln(x) - \ln(25 \text{ mm Hg}) = -(0.00283 \text{ min}^{-1})(720 \text{ minutes})$

And solving for x yields: 3.3 mm Hg
$P_{SO_2Cl_2}$ = 3.3 mm Hg; P_{SO_2} = $P_{SO_2Cl_2}$ consumed = 25 – 3.3 = 22 mm Hg

P_{Cl_2} = P_{SO_2} = 22 mm Hg and P_{total} = 47 mm Hg

14.91. Regarding the decomposition of nitramide:

(a) Given that the experimental rate law is Rate = $\dfrac{k[NO_2NH_2]}{[H_3O^+]}$, the apparent order will be first order in NO_2NH_2 and –1 in hydronium ion. Since the hydronium ion concentration will be constant in a buffered solution, the apparent order will be 1.

(b) Note that the rate expression shows that H_3O^+ is a factor in the rate determining step. Examine **Mechanism 3**. Note that the rate determining step in it has the apparent rate law: Rate = $k_5[NO_2NH^-]$

As this specie (NO_2NH^-) is an intermediate, we can—using a steady state approximation—express the concentration as:

$[NO_2NH^-] = \dfrac{k_4[NO_2NH_2]}{k_4'[H_3O^+]}$ and substituting into the apparent rate law:

Rate = $\dfrac{k_5 k_4 [NO_2NH_2]}{k_4'[H_3O^+]}$ and since k_4, k_4', and k_5 are constants, we can combine them into

a k' or Rate = $\dfrac{k'[NO_2NH_2]}{[H_3O^+]}$ as expressed by the experimental rate law.

(c) Note that $\dfrac{k_5 k_4}{k_4'}$ is another constant k'.

(d) Note that in the mechanism, in the "very fast reaction" step, hydronium ion reacts with OH^-, speeding up the production of water. The consumption of hydronium ion (raise the pH) would accelerate the overall rate (since the hydronium ion concentration is in the denominator in the rate law) and increase the $[NO_2NH]^-$.

14.92. Rate law resulting from the three steps:

Overall reaction: HA + X → A^- + products
Step 3 is rate−determining: Rate = $k[XH^+]$ but XH^+ is an intermediate

Step 1: $K_1 = \dfrac{[H^+][A^-]}{[HA]}$, so $[H^+] = \dfrac{K_1[HA]}{[A^-]}$

Step 2: $K_2 = \dfrac{[XH^+]}{[X][H^+]}$, so $[XH^+] = K_2[X][H^+]$

Substitute into rate equation: Rate = $k[XH^+]$ = $k(K_2[X])\left(\dfrac{K_1[HA]}{[A^-]}\right) = kK_1K_2 \dfrac{[X][HA]}{[A^-]}$

The reaction is first-order with respect to HA
Doubling the HA concentration would double the rate.

In the Laboratory

14.93. The data are plotted as follows:

Time (s)	[Phenolphthalein] x 1000 (mol/L)
0.0	5.0
10.5	4.5
22.3	4.0
35.7	3.5
51.1	3.0
69.3	2.5
91.6	2.0
120.4	1.5
160.9	1.0
230.3	0.5
299.6	0.3

(a) For the average rate for the first few seconds of the reaction, we have approximately:

$$\frac{0.0050 - 0.0045\,M}{0 - 10.5\,s} = 4.7 \times 10^{-5}\,M/s$$

For the time from 100 to 125 s (we have data at 91.6 and 120.4s)

$$\frac{0.0020 - 0.0015\,M}{91.6 - 120.4\,s} = 1.6 \times 10^{-5}\,M/s$$

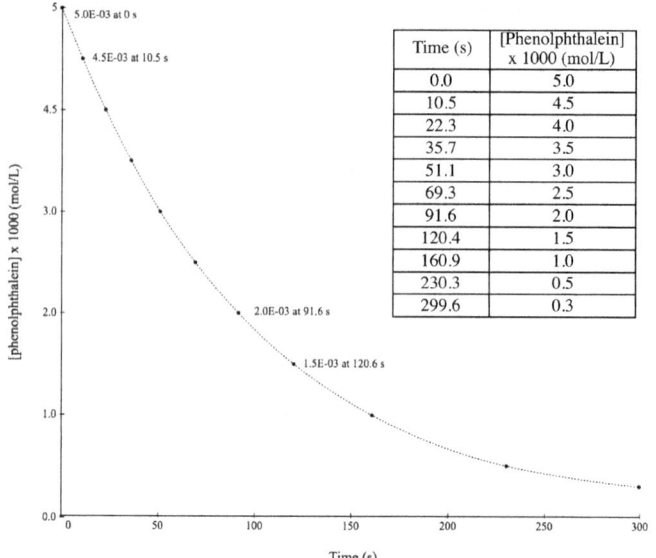

Note that the rate *decreases* with time as expected, since we anticipate rate to be proportional to the concentration of the phenolphthalein.

(b) To determine the order graphically, we plot [phenolphthalein] vs time (graph above), 1/[phenolphthalein] vs time, and ln[phenolphthalein] vs time (graphs below).

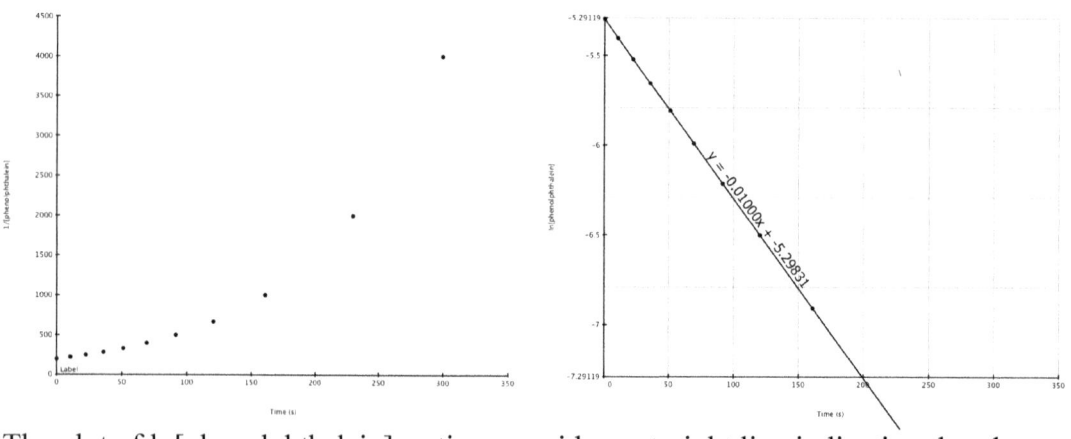

The plot of ln[phenolphthalein] vs time provides a straight line indicating that the reaction is **1st order.**

The rate law is Rate = k[phenolphthalein].
The slope of the line is −0.0100, so $k = 1.0 \times 10^{-2}$ M/s

(c) The half-life, $t_{1/2} = 0.693/k$ or $0.693/0.010 = 69.3$ s

14.94. Concerning the hydrolysis of the cobalt complex:
(a) Rate = k[*trans*−Co(en)$_2$Cl$_2$]$^+$

(b) The solution will appear gray when the concentrations of the green and the red complexes are equal. This occurs when exactly half of the green starting material has reacted, and thus the time to do so represents the reaction half-life. The half-life of a 1st order process is independent of initial concentration, whereas that for zero-order or 2nd-order reactions depend on initial concentration.

(c) The rate constant is directly related to 1/time so the reaction is first-order.

A plot of ln(1/time) vs. $1/T(K)$ is below:

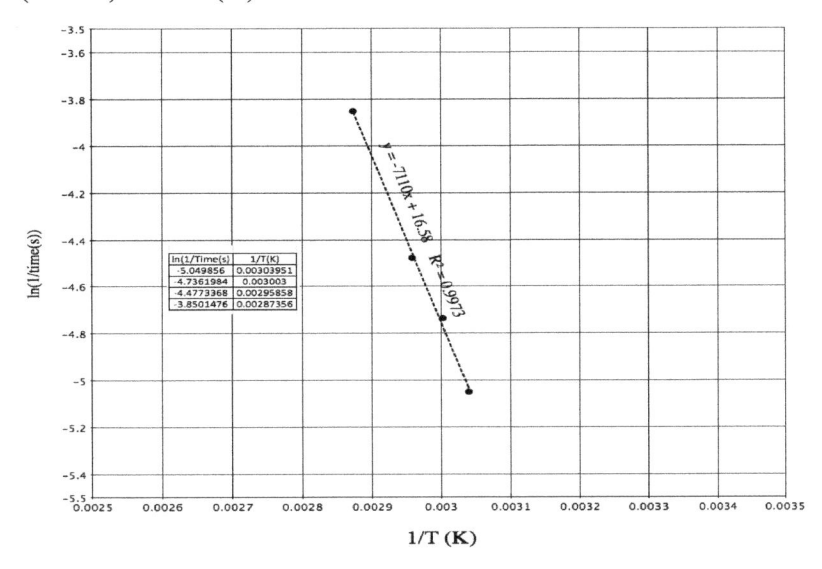

The plot is a straight line with an equation of $y = -7110 x - 16.58$.

slope = $-7110 \text{ K}^{-1} = \dfrac{-E_a}{8.3145 \times 10^{-3} \text{ kJ/K} \cdot \text{mol}}$ and $E_a = 59$ kJ/mol

14.95. Calculate the maximum rate of the reaction, V_{max}, using the data provided:

Michaelis-Menten indicates that a plot of 1/[concentration] vs 1/rate has an intercept equal to $1/V_{max}$.

The intercept for this graph is 7.6×10^4 min/M giving $V_{max} = 1.3 \times 10^{-5}$ M/min.

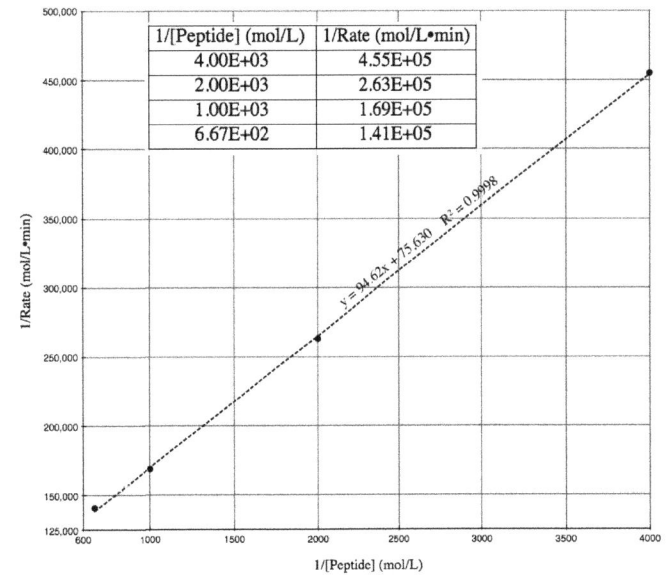

14.96. For the substitution reaction of $Ni(CO)_4$ with L:

(a) Molecularity of each step: The slow step is unimolecular and the fast step is bimolecular.

(b) Rate = $k[Ni(CO)_4]$ Yes, this rate law matches the stoichiometry of the slow step in the mechanism—one molecule reacting.

(c) $\ln\dfrac{[Ni(CO)_4]}{[Ni(CO)_4]_0} = -kt$

$\ln\dfrac{[Ni(CO)_4]}{0.025 \text{ mol/L}} = -(9.3 \times 10^{-3} \text{ s}^{-1})(5.0 \text{ min})\left(\dfrac{60 \text{ s}}{1 \text{ min}}\right)$

$\ln[Ni(CO)_4] - \ln(0.025 \text{ mol/L}) = -(9.3 \times 10^{-3} \text{ s}^{-1})(5.0 \text{ min})\left(\dfrac{60 \text{ s}}{1 \text{ min}}\right)$

$[Ni(CO)_4] = 0.0015$ mol/L

$[Ni(CO)_3L] = [Ni(CO)_4]$ consumed $= 0.025$ mol/L $- 0.0015$ mol/L $= 0.024$ mol/L

14.97. For the reaction of the iodide ion with the hypochlorite ion:

(a) Determine the rate law for the reaction:

Experiment	Initial Concentrations (mol/L)			Initial Rate
	[ClO⁻]	[I⁻]	[OH⁻]	(mol IO⁻/L · s)
1	4.0×10^{-3}	2.0×10^{-3}	1.0	4.8×10^{-4}
2	2.0×10^{-3}	4.0×10^{-3}	1.0	5.0×10^{-4}
3	2.0×10^{-3}	2.0×10^{-3}	1.0	2.4×10^{-4}
4	2.0×10^{-3}	2.0×10^{-3}	0.50	4.6×10^{-4}

Compare experiments 3 and 4. Note that the initial rate **doubles** when the [OH⁻] is halved, indicating a reciprocal relationship between [OH⁻] and the rate. Compare experiments 1 and 3. Halving the [ClO⁻] results in an initial rate that is **halved**, indicating a 1ˢᵗ order relationship. Comparing experiments 2 and 3, in which we halve the [I⁻], we **halve** the initial rate, indicating a 1ˢᵗ order relationship.

These data, taken together, indicate a rate law of the following: Rate $= k\dfrac{[ClO^-][I^-]}{[OH^-]}$

(b) The slow step, Step 2 would have the rate law: Rate $= k[I^-][HOCl]$. Since HOCl is generated in Step 1 in a fast equilibrium, we can write:

$K = \dfrac{[HOCl][OH^-]}{[OCl^-]}$. Solving for [HOCl] we obtain: $[HOCl] = \dfrac{K[OCl^-]}{[OH^-]}$.

Substitute into the rate law for Step 2 (from above):

Rate = $\dfrac{k[\text{I}^-]K[\text{OCl}^-]}{[\text{OH}^-]}$ and Rate = $\dfrac{kK[\text{I}^-][\text{OCl}^-]}{[\text{OH}^-]}$, making this mechanism match the experimentally-determined rate law in (a).

14.98. For the acid-catalyzed iodination of acetone:

Approximately doubling [H$^+$] between experiments 1 and 2 results in approximately doubling the rate of reaction; thus, the reaction is first-order in [H$^+$]. Approximately halving [CH$_3$COCH$_3$] between experiments 2 and 3 and between experiments 3 and 4 results in the rate decreasing by approximately one−half; thus, the reaction is first-order in [CH$_3$COCH$_3$]. Changing [I$_2$] between experiments 4 and 5 results in no change within error for the rate; thus, the reaction is zero-order in [I$_2$].
The resulting rate equation: Rate = k[CH$_3$COCH$_3$][H$^+$]

Summary and Conceptual Questions

14.99. Finely divided rhodium has a larger surface area than a block of the metal with the same mass. Since hydrogenation reactions depend on adsorption of H$_2$ on the catalyst surface, the greater the surface area, the greater the locations for such adsorption to occur.

14.100. Consider 1000 blocks 1.0 cm on a side:

Fraction of the cubes with 1 surface on the outside surface:

Accounting for the edge and corner cubes, there are 488 cubes with at least one surface on the outside surface of the 10. cm by 10. cm cube (48.8%).

Form eight cubes 5.0 cm on a side:

Splitting the cubes into eight piles of 125 blocks results in 784 blocks with at least one surface on the outside surface of the smaller cubes (78.4%).

This model demonstrates that the surface area for a finely divided substance is much greater than that for a large lump of substance having the same mass.

14.101. For the reaction: H$_2$(g) + I$_2$(g) → 2 HI(g), the rate law is Rate = k[H$_2$][I$_2$], which of the following statements is false—and why it is incorrect.

(a) The reaction **must** occur in a single step. **False**. Only an elaboration of the mechanism can clarify this fact. While it might occur in a single step, it *need not occur in a single step*.

(b) This is a 2nd order reaction overall. **True**. Addition of the order of H$_2$ and I$_2$ (1 +1) gives an overall order of 2.

(c) Raising the temperature will cause the value of k to decrease. **False**. Increases in T lead to increases in the value of k.

(d) Raising the temperature lowers the activation energy for this reaction. **False**. Temperature does not affect activation energy.

(e) If the concentrations of both reactants are doubled the rate will double. **False**. Doubling the concentration of *either reactant* will double the rate. Doubling the concentration of both reactants will increase the rate by a factor of 4.

(f) Adding a catalyst in the reaction will cause the initial rate to increase. **True**. A catalyst should increase the rate at which the reaction occurs.

14.102. For the ozone reactions that occur in the stratosphere:

$$\cancel{Cl} + O_3 \rightarrow \cancel{ClO} + O_2$$
$$\cancel{ClO} + \cancel{O} \rightarrow \cancel{Cl} + O_2$$
$$\underline{O_3 \rightarrow \cancel{O} + O_2}$$
Net: $2\,O_3 \rightarrow 3\,O_2$

The overall process consumes two moles of ozone, and since the Cl atoms are not consumed in the process (Cl is a catalyst), it can be repeated many times. ClO is an intermediate.

14.103. Describe the following statements as true or false:

(a) True—the rate-determining step in a mechanism **is the slowest step**.

(b) True—the rate constant is a proportionality constant that relates the concentration and rate *at a given temperature* and varies with temperature.

(c) False—As a reaction proceeds, the concentration of reactants diminishes and the rate of the reaction slows as fewer collisions occur.

(d) False—If the slow (single) step involves a termolecular process, then only one step is necessary. However termolecular collisions are **not highly likely**.

14.104. Describe the following statements as true or false:

(a) Incorrect. Reactions are faster at a higher temperature because the fraction of molecules with higher energies increases.
(b) Correct
(c) Correct
(d) Incorrect. The function of a catalyst is to provide a different pathway with a lower activation energy for the reaction.

14.105. For the reaction: cyclopropane → propene, describe the **change** on the following quantities.

(a) [cyclopropane] – Concentrations of reactants *decrease* as reactions proceed.

(b) [propene] – Concentrations of products *increase* as reactions proceed.

(c) [catalyst]—As catalysts are not consumed during a process, the concentration of the catalyst *will not change.*

(d) rate constant—The rate constant *will not change* as the reaction proceeds.

(e) The order of the reaction—*will not change* as the reaction proceeds.

(f) the half-life of cyclopropane—Since the half-life of cyclopropane is a function of the rate constant (which doesn't change—see (d) above), the half-life *does not change.*

14.106. Use the Oxygen-18 isotope to identify source of O in H₂O:

Assume that you begin with labeled oxygen in methanol, $CH_3{}^{18}OH$. If we represent the labeled oxygen with a *, you can see from the equation below that labeled oxygen (in the water) results if the O originated from the methanol.

14.107. For the reaction coordinate diagram shown:

(a) Number of steps: **3**- Note that each "peak" indicates a transition between a "reactant" and a "product".

(b) Is the reaction exo- or endothermic? Noting that the energy of the products is *lower* than that of the reactants, the reaction is **exothermic**.

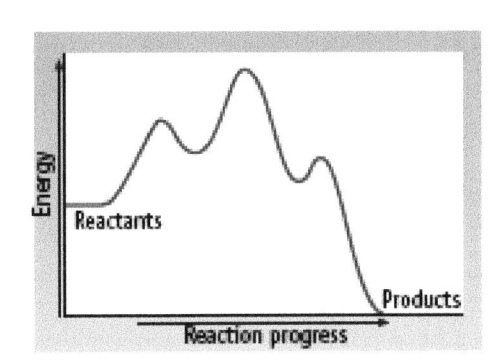

14.108. A reaction coordinate diagram for an exothermic single-step reaction:

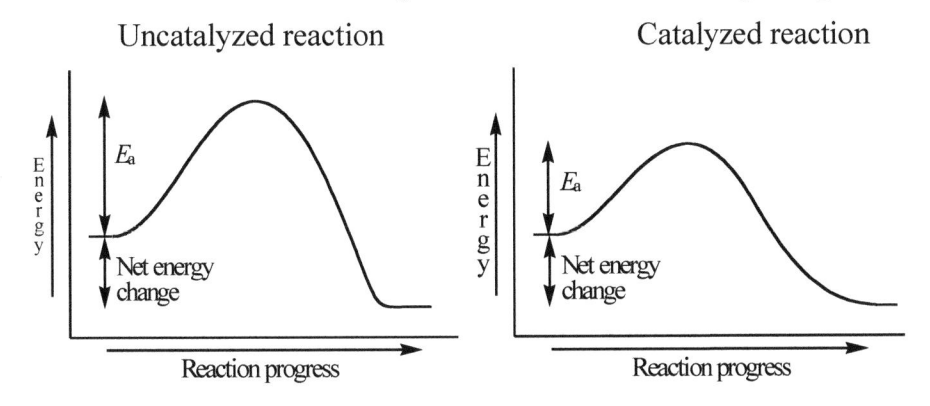

Uncatalyzed reaction   Catalyzed reaction

The activation energy in the catalyzed reaction is less than the activation energy in the uncatalyzed reaction. The net energy change is the same in both reactions.

14.109. Orientation for the best possibility of effective collision between O_3 and NO:

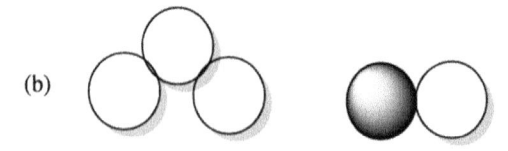

(b)

With the dark circle representing a N atom, this orientation would provide the best chance for a collision resulting in the formation of an O-N bond (to form NO_2), and the cleavage of the O-O bond in O_3 to form O_2.

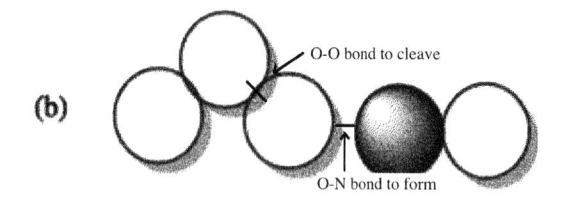

(b)

O-O bond to cleave
O-N bond to form

Solution and Answer Guide

Kotz Treichel Townsend Treichel, Chemistry and Chemical Reactivity 11e, 978-0-357-85140-1, Chapter 15: Principles of Chemical Reactivity: Equilibria

TABLE OF CONTENTS

Applying Chemical Principles 533
Practicing Skills 535
General Questions 546
In the Laboratory 565
Summary and Conceptual Questions 567

Applying Chemical Principles

Applying Equilibrium Concepts-The Haber-Bosch Ammonia Process

15.1.1. Regarding ammonia:

(a) React the HNO_3 with NH_3:

$NH_3 + HNO_3 \rightarrow NH_4NO_3$

(b) Would high pressure favor the production of urea? Would high T favor?

High pressure **favors** the production of urea because 3 moles of gas form 1 mole of gas. To answer the temperature question, we must calculate the $\Delta_r H$ for the reaction:

$\Delta_r H° = [\Delta_f H°((NH_2)_2CO) + \Delta_f H°(H_2O)] - [2\Delta_f H°(NH_3) + \Delta_f H°(CO_2)]$

$\Delta_r H = [-333.1 \text{ kJ} + -241.8 \text{ kJ}] - [2 \times -45.90 \text{ kJ} + -393.5 \text{ kJ}]$

$\Delta_r H = [-574.9 \text{ kJ}] - [-485.3 \text{ kJ}] = -89.6 \text{ KJ}$

Since this reaction is exothermic, it would be favored at **low temperatures**.

15.1.2. For the steam reforming process:

(a) Are the reactions endo- or exothermic? As in 5.1.1.b above, we must calculate the $\Delta_r H$ for the reactions:

For the formation of CO from methane and water:

$\Delta_r H° = [\Delta_f H°(CO) + 3\Delta_f H°(H_2)] - [\Delta_f H°(CH_4) + \Delta_f H°(H_2O)]$

$\Delta_r H° = [-110.5 \text{ kJ} + 3 \times 0] - [-74.87 \text{ kJ} + -241.8 \text{ kJ}] = 206.2 \text{ kJ}$

For the formation of CO_2 from carbon monoxide and water:

$\Delta_rH° = [\Delta_fH°(CO_2) + \Delta_fH°(H_2)] - [\Delta_fH°(CO) + \Delta_fH°(H_2O)]$

$\Delta_rH° = [-393.5 \text{ kJ} + 0] - [-110.5 \text{ kJ} + -241.8 \text{ kJ}] = -41.2 \text{ kJ}$

The first reaction is **endothermic**, while the second is **exothermic.**

(b) Mass of CH_4 consumed and mass of CO_2 produced to make 15 billion kg of NH_3:

$15 \times 10^{12} \text{g NH}_3 \left(\dfrac{1 \text{ mol NH}_3}{17.03 \text{ g NH}_3}\right)\left(\dfrac{3 \text{ mol H}_2}{2 \text{ mol NH}_3}\right)\left(\dfrac{1 \text{ mol CH}_4}{4 \text{ mol H}_2}\right)\left(\dfrac{16.04 \text{ g CH}_4}{1 \text{ mol CH}_4}\right) = 5.3 \times 10^{12} \text{ g CH}_4$

The mass of CO_2 is less straightforward. The net formation of CO_2 is found by adding the two equations in 15.1.2a above: $CH_4 + 2 H_2O \rightarrow CO_2 + 4 H_2$. This provides the stoichiometric relation between CO_2 + 4 H_2.

$15 \times 10^{12} \text{g NH}_3 \left(\dfrac{1 \text{ mol NH}_3}{17.03 \text{ g NH}_3}\right)\left(\dfrac{3 \text{ mol H}_2}{2 \text{ mol NH}_3}\right)\left(\dfrac{1 \text{ mol CO}_2}{4 \text{ mol H}_2}\right)\left(\dfrac{44.01 \text{ g CO}_2}{1 \text{ mol CO}_2}\right) = 1.5 \times 10^{13} \text{ g CO}_2$

Trivalent Carbon

15.2.1. Freezing point depression is one means of determining the molar mass of a compound. The freezing point depression constant of benzene is –5.12 °C/m.

(a) When a 0.503 g sample of the white crystalline dimer is dissolved in 10.0 g benzene, the freezing point of benzene is decreased by 0.542 °C. Verify that the molar mass of the dimer is 475 g/mol when determined by freezing point depression. Assume no dissociation of the dimer occurs.
If we write the freezing point expression as $\Delta T = i \times m \times K_{fp}$

$0.542 \text{ °C} = i \times \dfrac{\dfrac{0.503 \text{ g}}{475 \text{ g/mol}}}{0.010 \text{ kg}} \times 5.12 \text{ °C}/m$ and solving, $i = 1$. So the molar mass is verified as being 475 g/mol and no dissociation is assumed ($i = 1$).

(b) The correct molar mass of the dimer is 487 g/mol.

Explain why the dissociation equilibrium causes the freezing point depression calculation to yield a lower molar mass for the dimer. K for this equilibrium is **much** smaller than 1, so there is a greater abundance of the monomer (with molar mass much less than 487)—giving rise to the lower calculated molar mass.

15.2.2. Monomer concentration that exists at equilibrium with 0.015 mol/L dimer at 20 °C?

$K_{eq} = \dfrac{[2x]^2}{(0.015 - x)} = 4.1 \times 10^{-4}$ Solve the quadratic equation to give $x = 0.0012$ M.

The concentration of the monomer is $2x$ or 0.0024 M.

15.2.3. A 0.64 g sample of the white crystalline dimer (4) is dissolved in 25.0 mL of benzene at 20 °C. Use the equilibrium constant to calculate the concentrations of monomer (2) and dimer (4) in this solution.

$K = \dfrac{[\text{monomer}]^2}{[\text{dimer}]} = 4.1 \times 10^{-4}$ Assuming that the correct molar mass of the dimer is 487 (as in b above), you can calculate the [dimer] as $\dfrac{\dfrac{0.64 \text{ g dimer}}{487 \text{ g/mol dimer}}}{0.025 \text{ L}} = 0.053 \text{ M (2 sf)}$

Knowing the [dimer], you can substitute into the K expression:

$\dfrac{[2x]^2}{(0.053 - x)} = 4.1 \times 10^{-4}$ and multiplying both sides by the denominator:

$4x^2 = 4.1 \times 10^{-4}(0.053 - x)$; $4x^2 = 2.16 \times 10^{-5} - 4.1 \times 10^{-4}x$

Solving via the quadratic equation gives $x = 2.27 \times 10^{-3}$ M

The [monomer] = $2 \times 2.27 \times 10^{-3}$ M = 4.5×10^{-3} M (2 sf).
The concentration of the dimer at equilibrium would be $(0.053 - 0.00227) = 0.050$ M (2 sf).

15.2.4. Is the dissociation of the dimer exo- or endothermic?

The monomer is yellow in color. Heating produces more monomer (i.e., product), so the reaction is endothermic.

15.2.5. Which of the organic species mentioned in this story is paramagnetic?
(a) triphenylmethyl chloride
(b) triphenylmethyl radical
(c) the triphenylmethyl dimer

A radical, **by definition,** has an unpaired electron, so (b) is paramagnetic.

Practicing Skills

Writing Equilibrium Constant Expressions

15.1. Equilibrium constant expressions:

(a) $K = \dfrac{[H_2O]^2[O_2]}{[H_2O_2]^2}$ (b) $K = \dfrac{[CO_2]}{[CO][O_2]^{1/2}}$ (c) $K = \dfrac{[CO]^2}{[CO_2]}$ (d) $K = \dfrac{[CO_2]}{[CO]}$

Note that **solids** in the equations are *omitted* in the equilibrium constant expressions.

15.2. Equilibrium constant expressions:

(a) $K = \dfrac{[O_3]^2}{[O_2]^3}$ $K_p = \dfrac{P_{O_3}^2}{P_{O_2}^3}$

(b) $K = \dfrac{[Fe(CO)_5]}{[CO]^5}$ $K_p = \dfrac{P_{Fe(CO)_5}}{P_{CO}^5}$

(c) $K = [NH_3]^2[CO_2][H_2O]$ $K_p = P_{NH_3}^2 P_{CO_2} P_{H_2O}$

(d) $K = [Ag^+]^2[SO_4^{2-}]$

15.3. Balanced chemical equations from equilibrium constants:

(a) 2 SO₂(g) + O₂(g) ⇌ 2 SO₃(g)

(b) 2 CO(g) + O₂(g) ⇌ 2 CO₂(g)

(c) PbCl₂(s) ⇌ Pb²⁺(aq) + 2 Cl⁻(aq)   (Remember solids aren't included in K_{eq}.)

(d) HF(aq) + H₂O(ℓ) ⇌ H₃O⁺(aq) + F⁻(aq)   (Remember pure liquids and solvents aren't included in K_{eq}.)

15.4. Balanced chemical equations from equilibrium constants:

(a) 2 CO₂(g) ⇌ 2 CO(g) + O₂(g)

(b) 2 CH₂Cl₂(g) ⇌ CH₄(g) + CCl₄(g)

(c) Mg(OH)₂(s) ⇌ Mg²⁺(aq) + 2 OH⁻(aq)   (Remember solids aren't included in K_{eq}.)

(d) F⁻(aq) + H₂O(ℓ) ⇌ HF(aq) + OH⁻(aq) (Remember liquids aren't included in K_{eq}.)

The Equilibrium Constant and Reaction Quotient

15.5. The equilibrium expression for the reaction: I₂(g) ⇌ 2 I(g) has a $K = 5.6 \times 10^{-12}$

Substituting the molar concentrations into the equilibrium expression:

$$Q = \dfrac{[I]^2}{[I_2]} = \dfrac{(2.00 \times 10^{-8})^2}{(2.00 \times 10^{-2})} = 2.0 \times 10^{-14}$$

Since Q < K, the system **is not at equilibrium** and will move to **the right to make more product (and reach equilibrium)**.

15.6. For the dimerization of NO₂:

$$[NO_2] = \dfrac{2.0 \times 10^{-3} \text{ mol}}{10.\text{ L}} = 2.0 \times 10^{-4} \text{ mol/L}$$

$$[N_2O_4] = \dfrac{1.5 \times 10^{-3} \text{ mol}}{10.\text{ L}} = 1.5 \times 10^{-4} \text{ mol/L}$$

$$Q = \frac{[N_2O_4]}{[NO_2]^2} = \frac{1.5 \times 10^{-4}}{(2.0 \times 10^{-4})^2} = 3800 \text{ so, } Q > K \text{ The reaction is not at equilibrium.}$$

The concentration of NO_2 will increase as the system proceeds to equilibrium.

15.7. Is the system at equilibrium?

[SO₂]	5.0×10^{-3} M
[O₂]	1.9×10^{-3} M
[SO₃]	6.9×10^{-3} M

Substituting into the equilibrium expression:

$$\frac{[SO_3]^2}{[SO_2]^2[O_2]} = \frac{[6.9 \times 10^{-3}]^2}{[5.0 \times 10^{-3}]^2[1.9 \times 10^{-3}]} = 1000 \text{ or } 1.0 \times 10^3 \text{ (2sf)}$$

Since $Q > K$, the system is **not** at equilibrium, and will move to the left, to make more reactants to reach equilibrium.

15.8. Is the system at equilibrium?

$$Q = \frac{[NO]^2[Cl_2]}{[NOCl]^2} = \frac{(2.5 \times 10^{-3})^2(2.0 \times 10^{-3})}{(5.0 \times 10^{-3})^2} = 5.0 \times 10^{-4}$$

$Q < K$ The reaction is not at equilibrium.

The reaction will proceed to the right, to form more products.

Calculating an Equilibrium Constant

15.9. For the equilibrium: $PCl_5(g) \rightleftarrows PCl_3(g) + Cl_2(g)$, calculate K

The equilibrium concentrations are: $[PCl_5] = 4.2 \times 10^{-5}$ M
$[PCl_3] = 1.3 \times 10^{-2}$ M
$[Cl_2] = 3.9 \times 10^{-3}$ M

The equilibrium expression is: $\frac{[PCl_3][Cl_2]}{[PCl_5]} = \frac{[1.3 \times 10^{-2}][3.9 \times 10^{-3}]}{[4.2 \times 10^{-5}]} = 1.2$

15.10. Calculate K_c: $K = \frac{[SO_3]^2}{[SO_2]^2[O_2]} = \frac{(4.13 \times 10^{-3})^2}{(3.77 \times 10^{-3})^2(4.30 \times 10^{-3})} = 279$

15.11. The equilibrium expression is $K = \dfrac{[CO]^2}{[CO_2]}$

The quantities given are **moles**, so first calculate the **concentrations at equilibrium**.

$[CO] = \dfrac{1.0 \text{ mol}}{200.0 \text{ L}} = 0.0050 \text{ M}$ $[CO_2] = \dfrac{0.20 \text{ mol}}{200.0 \text{ L}} = 0.0010 \text{ M}$

(a) $K = \dfrac{[CO]^2}{[CO_2]} = \dfrac{[0.0050 \text{ M}]^2}{[0.0010 \text{ M}]} = 0.025$ or 2.5×10^{-2}

(b & c) The only change here is in the amount of carbon. Since C does not appear in the equilibrium expression, K would not change.

15.12. Calculate K_c at 986 °C for $H_2 + CO_2 \rightleftharpoons H_2O + CO$:

(a) $K = \dfrac{[H_2O][CO]}{[H_2][CO_2]} = \dfrac{(0.11 \text{ mol}/50.0 \text{ L})(0.11 \text{ mol}/50.0 \text{ L})}{(0.087 \text{ mol}/50.0 \text{ L})(0.087 \text{ mol}/50.0 \text{ L})} = 1.6$

(b)

	H_2	CO_2	H_2O	CO
Initial (M)	8.0×10^{-5} M	8.0×10^{-5} M	0	0
Change (M)	$-x$	$-x$	$+x$	$+x$
Equilibrium (M)	$8.0 \times 10^{-5} - x$	$8.0 \times 10^{-5} - x$	x	x

Note the amounts of H_2 and CO_2 have been converted to concentrations in the table above.

$K = 1.6 = \dfrac{(x)(x)}{(8.0 \times 10^{-5} - x)(8.0 \times 10^{-5} - x)} = \dfrac{x^2}{(8.0 \times 10^{-5} - x)^2}$

$\sqrt{1.6} = \sqrt{\dfrac{x^2}{(8.0 \times 10^{-5} - x)^2}} = \dfrac{x}{8.0 \times 10^{-5} - x}$ and $x = 4.5 \times 10^{-5}$

$300.0 \text{ L} \left(\dfrac{4.5 \times 10^{-5} \text{ mol}}{1 \text{ L}} \right) = 0.014 \text{ mol} = 0.014 \text{ mol } H_2O = 0.014 \text{ mol } CO$

15.13. The reaction is $CO(g) + Cl_2(g) \rightleftharpoons COCl_2(g)$

	CO	Cl_2	$COCl_2$
Initial	0.0102 M	0.00609 M	0 M
Change	−0.00308 M	−0.00308 M	+0.00308 M
Equilibrium	0.0071 M	0.00301 M	0.00308 M

(a) Equilibrium concentrations of CO and COCl₂ are found by noting that 0.00308 M Cl₂ is consumed (to leave 0.00301M at equilibrium), an equimolar amount of CO is consumed and an equal amount of COCl₂ is produced.
So [COCl₂] = 0.00308 M and [CO] = 0.0071 M.

(b) $K = \dfrac{[+0.00308]}{[0.0071][0.00301]} = 140$ (2 sf)

15.14. Calculate K_c for the reaction: 2 SO₃ ⇌ 2 SO₂ + O₂ at 1150 K:

	SO₃	SO₂	O₂
Initial	0.00375 M	0 M	0 M
Change	−2(0.00073) M	+2(0.00073) M	+0.00073 M
Equilibrium	0.00229 M	0.0015 M	+0.00073 M

Note that the concentrations in the table above are calculated below:

$[SO_3] = \dfrac{0.0300 \text{ mol}}{8.00 \text{ L}} = 0.00375$ M $[O_2] = \dfrac{0.0058 \text{ mol}}{8.00 \text{ L}} = 0.00073$ M

Now substitute into the K_c expression: $K_c = \dfrac{(0.0015)^2(0.00073)}{(0.00229)^2} = 0.00031$

Using Equilibrium Constants

15.15. For the system butane ⇌ isobutene, $K = 2.5$

Equilibrium concentrations may be found using a table. First note that the amount of butane must be converted to molar concentrations. So, 0.017 mol butane/0.50 L = 0.034 M.

	butane	isobutane
Initial	0.034	0
Change	−x	+x
Equilibrium	0.034 − x	x

Substituting these equilibrium concentrations into the equilibrium expression:

$K = \dfrac{[\text{isobutene}]}{[\text{butane}]}$; $K = \dfrac{x}{0.034 - x} = 2.5$ and $x = 2.5(0.034 - x)$;

multiplying $x = 0.085 - 2.5x$ and solving for x gives $x = 2.4 \times 10^{-2}$
and [isobutane] = 2.4×10^{-2} M, [butane] = $0.034 - x = 1.0 \times 10^{-2}$ M

15.16. Find [cyclohexane] and [methylcyclopentane] at equilibrium:

First, determine the initial concentration of cyclohexane: 0.058 mol/2.8 L = 0.021 M

	cyclohexane	methylcyclopentane
Initial	0.0207 M	0
Change	$-x$	$+x$
Equilibrium	$0.0207 - x$	x

Substitute the values into the K expression:

$$K = 0.12 = \frac{[C_3H_9CH_3]}{[C_6H_{12}]} = \frac{x}{0.207 - x}$$

$[C_3H_9CH_3] = x = 0.00222$ M $= 0.0022$ M (2 sf)

$[C_6H_{12}] = 0.0207 - 0.00222 = 0.0184$ M $= 0.018$ M (2 sf)

15.17. For the equilibrium, the equilibrium constant is $\frac{[I]^2}{[I_2]} = 3.76 \times 10^{-3}$

The initial concentration of I_2 is 0.105 mol/12.3 L = 8.54×10^{-3} M. The equation indicates that 2 mol of I form for each mol of I_2 that reacts. If some amount, say x M, of I_2 reacts, then the amount of I that forms is $2x$, and the amount of I_2 remaining at equilibrium is $(8.54 \times 10^{-3} - x)$. The equilibrium concentrations can be substituted into the equilibrium expression.

$$\frac{(2x)^2}{8.54 \times 10^{-3} - x} = 3.76 \times 10^{-3} \text{ or } 4x^2 = 3.76 \times 10^{-3}(8.54 \times 10^{-3} - x).$$

Rearranging we get: $4x^2 + 3.76 \times 10^{-3}x - 3.21 \times 10^{-5} = 0$.

Solve this using the quadratic equation. Using the positive solution to that equation we find that $x = 2.40 \times 10^{-3}$.

So, at equilibrium:

$[I_2] = 8.54 \times 10^{-3} - 2.40 \times 10^{-3} = 6.14 \times 10^{-3}$ M

$[I] = 2(2.40 \times 10^{-3}) = 4.80 \times 10^{-3}$ M

15.18. For the equilibrium between N₂O₄ and NO₂:

First calculate the initial concentration N₂O₄ of:

$$[N_2O_4] = \frac{14.8 \text{ g}}{5.0 \text{ L}}\left(\frac{1 \text{ mol N}_2\text{O}_4}{92.01 \text{ g}}\right) = 0.0322 \text{ M}$$

Insert the variables into a table for clarity:

	N₂O₄	NO₂
Initial	0.0322 M	0
Change	−x	+2x
Equilibrium	0.0322 − x	2x

Substitute into the K expression:

$$K = \frac{[NO_2]^2}{[N_2O_4]} = 5.9\times 10^{-3} = \frac{(2x)^2}{0.0322-x} \text{ and multiplying by the denominator}$$

$$0 = 4x^2 + 5.9\times 10^{-3}x - 0.000190$$

Solve using the quadratic equation: $x = 0.00619$ M; [NO₂] = $2x$ = 0.0124 M
[N₂O₄] = 0.0322 − x = 0.0260 M

(a) Amount of NO₂ present at equilibrium = $2x$ = 0.0124 M = 0.012 M (2 sf)
= 0.0124 M(5.0 L) = 0.062 moles NO₂

(b) % N₂O₄ dissociated = $\dfrac{0.00619 \text{ mol}}{0.0322 \text{ mol}} \times 100\% = 19\%$ (2 sf)

15.19. For I₂ distributed between aqueous and CCl₄ layers, how much is in the aqueous layer:

Determine the initial concentration of I₂ in the aqueous layer:

$$[I_2(aq)] = \frac{0.0340 \text{ g}}{0.1000 \text{ L}}\left(\frac{1 \text{ mol I}_2}{253.8 \text{ g}}\right) = 0.00134 \text{ M}$$

	I₂(aq)	I₂(CCl₄)
Initial	0.00134 M	0
Change	−x	+x
Equilibrium	0.00134 − x	x

Substitute into the K expression:

$K = 85.0 = \dfrac{x}{0.00134 - x}$ and $x = 0.00132$

[I$_2$(aq)] = 0.00134 – x = 1.6 × 10^{-5} M = 2 × 10^{-5} M (1 sf) and expressed as mass:

Amount of I$_2$ remaining in water = (1.6 × 10^{-5} mol/L)(0.1000 L)(253.8 g/mol) = 4 × 10^{-4} g I$_2$

15.19. Given at equilibrium: $K = \dfrac{[CO][Br_2]}{[COBr_2]} = 0.190$ at 73°C

First, calculate concentrations: [COBr$_2$] = $\dfrac{0.0500 \text{ mol}}{2.00 \text{ L}}$ = 0.0250 M

Note that the stoichiometry of the equation tells us that for **each CO, we obtain 1** Br$_2$.

	COBr$_2$	CO	Br$_2$
Initial	0.0250 M	0 M	0 M
Change	–x	+x	+x
Equilibrium	0.0250 – x	+x	+x

You can rewrite the expression to read: $\dfrac{[CO][Br_2]}{0.0250-x} = \dfrac{x^2}{0.0250-x} = 0.190$

Solve this using the quadratic equation. $x^2 + 0.190x - 0.00475 = 0$

Using the positive solution to that equation, x = 0.0224 M = [CO] = [Br$_2$] and [COBr$_2$] = (0.0250 - 0.0224) = 0.0026 M.

The percentage of COBr$_2$ that has decomposed is: $\dfrac{0.0224 \text{ M}}{0.0250 \text{ M}}(100\%) = 89.6\%$

Manipulating Equilibrium Constant Expressions

15.21. To compare the two equilibrium constants, write the equilibrium expressions for the two.

$\dfrac{[C]^2}{[A][B]} = K_1$ and $\dfrac{[C]^4}{[A]^2[B]^2} = K_2$

Note that the first expression *squared* is equal to the second expression, so **(b) $K_2 = K_1^2$**

15.22. To compare the two equilibrium constants, write the equilibrium expressions for the two.

$$\frac{[C]^2}{[A]^2[B]} = K_1 \text{ and } \frac{[C]}{[A][B]^{1/2}} = K_2$$

Note that the square root of the first expression is equal to the second expression, so

(c) $K_2 = K_1^{1/2}$

15.23. To compare the two equilibrium constants, write the equilibrium expressions for the two.

$$\frac{[C]}{[A][B]^2} = K_1 \text{ and } \frac{[A]^2[B]^4}{[C]^2} = K_2$$

Note that the inverse and square of the first expression is equal to the second expression, so

(d) $K_2 = 1/K_1^2$

15.24. The expression that correctly shows the relationship between the K's for the two equations:

The second equation has been reversed and multiplied by 1/2 so (a) $K_2 = \sqrt{1/(K_1)}$

15.25. Comparing the two equilibria:
(1) $SO_2(g) + 1/2\ O_2(g) \rightleftarrows SO_3(g)$ K_1
(2) $2\ SO_3(g) \rightleftarrows 2\ SO_2(g) + O_2(g)$ K_2

The expression that relates K_1 and K_2 is (e): $K_2 = \dfrac{1}{K_1^2}$

Reversing equation 1 gives: $SO_3(g) \rightleftarrows SO_2(g) + 1/2\ O_2(g)$ with the eq. constant $\dfrac{1}{K_1}$

Multiplying the (reversed equation 1) × 2 gives: $2\ SO_3(g) \rightleftarrows 2\ SO_2(g) + O_2(g)$ and the modified equilibrium constant $\dfrac{1}{K_1^2}$.

15.26. Calculate K for the equilibrium $2\ CO(g) + O_2(g) \rightleftarrows 2\ CO_2(g)$

The second equation has been reversed and multiplied by 2.

$K_{new} = \dfrac{1}{K_{orig}^2} = 1/(6.66 \times 10^{-12})^2 = 2.25 \times 10^{22}$

15.27. Calculate *K* for the reaction: $SnO_2(s) + 2\ CO(g) \rightleftharpoons Sn(s) + 2\ CO_2(g)$ given:

(1) $SnO_2(s) + 2\ H_2(g) \rightleftharpoons Sn(s) + 2\ H_2O(g)$ *K* = 8.12

(2) $H_2(g) + CO_2(g) \rightleftharpoons H_2O(g) + CO(g)$ *K* = 0.771

Take equation 2 and reverse it, then multiply by 2 to give:

$2\ H_2O(g) + 2\ CO(g) \rightleftharpoons 2\ H_2(g) + 2\ CO_2(g)$ $K = \dfrac{1}{K_{orig}^2} = \dfrac{1}{(0.771)^2} = 1.68$

add equation 1: $SnO_2(s) + 2\ H_2(g) \rightleftharpoons Sn(s) + 2\ H_2O(g)$ *K* = 8.12

$SnO_2(s) + 2\ CO(g) \rightleftharpoons Sn(s) + 2\ CO_2(g)$ K_{net} = 8.12 × 1.68 = 13.7

15.28. Calculate *K*, given the *K* values for two other equations:

$H_2O(g) + CO(g) \rightleftharpoons H_2(g) + CO_2(g)$ *K* = 1.6

Reverse the above reaction and determine its *K* value. Then add the two equations.

$FeO(s) + CO(g) \rightleftharpoons Fe(s) + CO_2(g)$ *K* = 0.67

$H_2(g) + CO_2(g) \rightleftharpoons H_2O(g) + CO(g)$ *K* = 1/1.6

$FeO(s) + H_2(g) \rightleftharpoons Fe(s) + H_2O(g)$ K_{net} = (0.67)/(1.6) = 0.42

15.29. Relationship between K_p and K_c

(a) K_p = 0.16 at 25°C. Calculate K_c:

Recall that the two *K*'s are related by the equation: $K_p = K_c (RT)^{\Delta n}$

The equation: $2\ NOBr(g) \rightleftharpoons 2\ NO(g) + Br_2(g)$ has a $\Delta n = (3 - 2) = 1$

Note that Δn = (Σ gaseous products - Σ gaseous reactants). Applying the relationship:

$0.16 = K_c (0.08205 \cdot 298)^1$ [units are omitted for clarity here]

$\dfrac{0.16}{(0.08205 \cdot 298)^1} = K_c = 6.5 \times 10^{-3}$

(b) For the equation: $2\ CH_2Cl_2(g) \rightleftharpoons CH_4(g) + CCl_4(g)$, $\Delta n = (2 - 2) = 0$.
So the two *K*'s are equal.

15.30. Relationship between K_p and K_c

(a) $\Delta n = (2 - 1) = +1$; $K_p = K_c(RT)^{\Delta n}$
K_p = (170)(0.0821 L · atm/mol · K)(298 K) = 4.16×10^3

(b) $\Delta n = (2 - 0) = +2$; $K_p = K_c(RT)^{\Delta n}$
$K_p = (1.8 \times 10^{-4})[(0.0821 \text{ L} \cdot \text{atm/mol} \cdot \text{K})(298 \text{ K})]^2 = 0.11$

Disturbing a Chemical Equilibrium

15.31. The equilibrium may be represented: $N_2O_3(g) + \text{heat} \rightleftarrows NO_2(g) + NO(g)$

Since the process is **endothermic** ($\Delta H = +$), heat is absorbed in the "left to right" reaction.

The effect of:

(a) Adding more $N_2O_3(g)$: An increase in the pressure of N_2O_3 (adding more N_2O_3) will shift the equilibrium to the **right**, producing more NO_2 and NO.

(b) Adding more $NO_2(g)$: An increase in the pressure of NO_2 (adding more NO_2) will shift the equilibrium to the **left**, producing more N_2O_3.

(c) Increasing the volume of the reaction flask: If the volume of the flask is increased, the pressure will drop. (Remember that P for gases is inversely related to volume.) A drop in pressure will favor that side of the equilibrium with the "larger total number of moles of gas"—so this equilibrium will shift to the **right**, producing more NO_2 and NO.

(d) Lowering the temperature: You should note that a change in T (up or down) will result in a **change in the equilibrium constant**. (None of the three changes mentioned above change K.) However, the same principle applies. The removal of heat (a decrease in T) favors the exothermic process (shifts the equilibrium to the **left**) producing more N_2O_3.

15.32. Predict changes of the following on the equilibrium: $2 \text{ NOBr}(g) \rightleftarrows 2 \text{ NO}(g) + Br_2(g)$

(a) Removing some $Br_2(g)$ will shift the equilibrium to the right.

(b) Adding more $NOBr(g)$ will shift the equilibrium to the right.

(c) Increasing the temperature will shift the equilibrium to the right.

(d) Decreasing the container volume will shift the equilibrium to the left.

15.33. K for butane $\rightleftarrows$ isobutane is 2.5.

(a) Equilibrium concentrations if 0.50 mol/L of isobutane is added:

	butane	isobutane
Original concentration	1.0	2.5
Change immediately after addition	1.0	2.5 + 0.50
Change (going to equilibrium)	$+x$	$-x$
Equilibrium concentration	$1.0 + x$	$3.0 - x$

Substituting into the equilibrium expression: $K = \dfrac{[\text{isobutane}]}{[\text{butane}]} = \dfrac{3.0 - x}{1.0 + x} = 2.5$

$3.0 - x = 2.5(1.0 + x)$ and $0.14 = x$

The equilibrium concentrations are:
[butane] = 1.0 + x = 1.1 M and [isobutane] = 3.0 − x = 2.9 M

(b) Equilibrium concentrations if 0.50 mol/L of butane is added:

	butane	isobutane
Original concentration	1.0	2.5
Change immediately after addition	1.0 + 0.50	2.5
Change (going to equilibrium)	−x	+x
Equilibrium concentration	1.5 − x	2.5 + x

$K = \dfrac{[\text{isobutane}]}{[\text{butane}]} = \dfrac{2.5 + x}{1.5 - x} = 2.5$ and solving: $2.5 + x = 2.5(1.5 - x)$ and $x = 0.36$

The equilibrium concentrations are: [butane] = 1.5 − 0.36 = 1.1 M and [isobutane] = 2.5 + 0.36 = 2.9 M

15.34. Predict changes of the following on the equilibrium: $NH_4HS(s) \rightleftarrows NH_3(g) + H_2S(g)$

When the temperature is raised, the reaction adjusts to the added heat by consuming reactants and forming more products. The equilibrium shifts to the right.

Adding NH_4HS, a solid, will have no effect on the equilibrium.

Adding $NH_3(g)$, a product, will shift the equilibrium to the left.

Removing H_2S, a product, will shift the reaction to the right, increasing the NH_3 pressure.

General Questions

15.35. For the equilibrium: $Br_2(g) \rightleftarrows 2Br(g)$, first calculate the concentration of Br_2.

$[Br_2] = \dfrac{0.086 \text{ mol}}{1.26 \text{ L}} = 0.0683 \text{ M}$

Now complete an equilibrium table:

	Br₂	Br
Initial	0.0683 M	0
Change	−(0.037)(0.0683) M	+2(0.037)(0.0683) M
Equilibrium	0.0657 M	+0.00505 M

The changes reflect the dissociation of Br₂ into 2 Br. Note the effect of the stoichiometry on the change in the [Br].

Substituting into the K expression:

$$K = \frac{[Br]^2}{[Br_2]} = \frac{(0.00505)^2}{0.0657} = 3.9 \times 10^{-4}$$

15.36. Changes in K when the equation is rewritten:

(a) The equation has been multiplied by 1/2 so $K_{new} = K^{1/2} = (1.7 \times 10^{-3})^{1/2} = 0.041$

(b) The equation has been reversed and doubled so $K_{new} = 1/K^2 = 1/(1.7 \times 10^{-3})^2 = 3.5 \times 10^5$

15.37. K_p is 6.5×10^{11} at 25°C for CO(g) + Cl₂(g) ⇌ COCl₂(g)

What is the value of K_p for COCl₂(g) ⇌ CO(g) + Cl₂(g)?

Writing the equilibrium expressions for these two processes will show that one is the inverse of the other. So K_p for the two will be mathematically related by the reciprocal relationship. K_p for the second reaction is $1/K_p$ for the first reaction or $1/6.5 \times 10^{11}$ or 1.5×10^{-12}.

15.38. For the equilibrium: 2 CH₂Cl₂(g) ⇌ CH₄(g) + CCl₄(g), determine [CCl₄] at eq:

$$K = \frac{[CH_4][CCl_4]}{[CH_2Cl_2]^2}$$ and substituting the values: $1.05 = \frac{(0.0163)[CCl_4]}{(0.0206)^2}$

Solving for $[CCl_4]$: $[CCl_4] = 0.0273$ M

15.39. Calculate K for the system CS₂(g) + 3 Cl₂(g) ⇌ S₂Cl₂(g) + CCl₄(g)

Calculate the initial concentrations of CS₂ and Cl₂:

$$[CS_2] = \frac{0.12 \text{ mol}}{10.0 \text{ L}} = 0.012 \text{ M} \;;\; [Cl_2] = \frac{0.36 \text{ mol}}{10.0 \text{ L}} = 0.036 \text{ M};$$

At equilibrium, the concentrations of CCl₄ (and S₂Cl₂) are 0.090 mol/10.0 L = 0.0090 M

	CS₂	Cl₂	S₂Cl₂	CCl₄
Initial	0.012 M	0.036 M	0	0
Change	−0.0090	−3 · 0.0090	+0.0090	+0.0090
Equilibrium	0.0030	0.0090	0.0090	0.0090

Equilibrium concentrations are determined by noting the stoichiometry of the reaction. If 0.0090 mol of CCl₄ are formed (to reach equilibrium), an equal amount of S₂Cl₂ is formed and an equal amount of CS₂ reacted. The amount of Cl₂ that reacts is found by noting the 1:3 ratio of CS₂ to Cl₂. If 0.090 mol of CS₂ reacts, then 3 × 0.090 mol of Cl₂ reacts. REMEMBER: You need *molar concentrations* in our *K* expression, so you divide the number of moles by 10.0 L.

Substituting into the equilibrium expression:

$$\frac{[S_2Cl_2][CCl_2]}{[CS_2][Cl_2]^3} = \frac{(0.0090)(0.0090)}{(0.0030)(0.0090)^3} = 3.7 \times 10^4 = 4 \times 10^4 \text{ (1 sf)}$$

15.40. Calculate K_c for the reaction: H₂(g) + I₂(g) ⇌ 2 HI(g)

	H₂	I₂	HI
Initial	0.0088 M	0.0088 M	0
Change	−(0.786)(0.0088)	−(0.786)(0.0088)	+2(0.786)(0.0088)
Equilibrium	0.00188 M	0.00188 M	0.0138 M

Very much like SQ15.35, you need to pay heed to the 2:1 stoichiometry of HI: H₂ or I₂. Substituting these values into the *K* expression:

$$K = \frac{[HI]^2}{[H_2][I_2]} = \frac{(0.0138)^2}{(0.00188)(0.00188)} = 54 \text{ (2 sf)}$$

15.41. *K* for butane ⇌ isobutane is 2.5. Is the system at equilibrium if 1.75 mol of butane and 1.25 mol of isobutane are mixed?

Substituting into the equilibrium expression: $K = \frac{[\text{isobutane}]}{[\text{butane}]} = \frac{1.25 \text{ mol}}{1.75 \text{ mol}} = 0.714$.

The system is not at equilibrium (since Q is not equal to K). Since Q is less than K, the system will need to shift to the right (producing more isobutane) to reach equilibrium.

	butane	isobutane
Original concentration	1.75	1.25
Change (going to equilibrium)	$-x$	$+x$
Equilibrium concentration	$1.75 - x$	$1.25 + x$

$$K = \frac{[\text{isobutane}]}{[\text{butane}]} = \frac{1.25 + x}{1.75 - x} = 2.5 \text{ and solving for } x: 2.5(1.75 - x) = (1.25 + x)$$

Expanding: $4.375 - 2.5x = 1.25 + x$; $(4.375 - 1.25) = 3.5x$; $3.125 = 3.5x$; $x = 0.89$

The equilibrium concentrations are: [butane] = $1.75 - x$ = 0.86 M and
[isobutane] = $1.25 + x$ = 2.14 M

15.42. For the equilibrium: $N_2(g) + O_2(g) \rightleftarrows 2\,NO(g)$

(a) Solve for Q:

$$Q = \frac{[NO]^2}{[N_2][O_2]} = \frac{(0.056)^2}{(0.25)(0.25)} = 0.050$$

$Q > K$, so the system is not at equilibrium.

(b) $Q > K$; the reaction proceeds to the left.

(c) Substitute concentrations into a table:

	N₂	O₂	NO
Initial	0.25 M	0.25 M	0.056 M
Change	$+x$	$+x$	$-2x$
Equilibrium	$0.25 + x$	$0.25 + x$	$0.056 - 2x$

$$K = 1.7 \times 10^{-3} = \frac{[NO]^2}{[N_2][O_2]} = \frac{(0.056 - 2x)^2}{(0.25 + 2x)^2}$$

Take the square root of both sides of the equation and solve for x. $x = 0.0219$
[N₂] = [O₂] = $0.25 + x$ = 0.27 M
[NO] = $0.056 - 2x$ = 0.012 M

15.43. For the two equilibria shown, which correctly relates the two equilibrium constants?

K_1 NOCl(g) ⇌ NO(g) + 1/2 Cl$_2$(g)

K_2 2 NO(g) + Cl$_2$(g) ⇌ 2 NOCl(g)

One way to answer the question is to write the equilibrium expressions for the two equilibria:

$$K_1 = \frac{[NO][Cl_2]^{1/2}}{[NOCl]} \text{ and } K_2 = \frac{[NOCl]^2}{[NO]^2[Cl_2]}.$$

Note two differences between the K expressions: (1) they are reciprocals of each other—in one case with NOCl in the denominator, and in the other with NOCl in the numerator. Mentally invert K_2. Note the 2nd difference. K_2 is $(K_1)^2$. So, the relationship between the two is: (c) $K_2 = \dfrac{1}{K_1^2}$

15.44. For the equilibrium: COBr$_2$(g) ⇌ CO(g) + Br$_2$(g):

(a) Equilibrium concentrations of all species:

Calculate the initial concentration of COBr$_2$: 0.50 mol/9.50 L = 0.053 M COBr$_2$

Setup a table:

	COBr$_2$	CO	Br$_2$
Initial	0.053 M	0	0
Change	$-x$	$+x$	$+x$
Equilibrium	0.053 – x	x	x

Substitute into the equilibrium constant expression:

$$K = \frac{x^2}{(0.053 - x)} = 0.190 \text{ and simplifying: } 0.190(0.053 - x) = x^2$$

Rewrite as a quadratic expression: $x^2 + 0.190x - 0.0101 = 0$, and solve using the quadratic equation:

x = 0.0429; [Br$_2$] = [CO] = 0.043 M

and [COBr$_2$] = 0.053 M – x = 0.010 M

(b) If volume is decreased to 4.5 L, what are the new equilibrium concentrations:

Calculate the initial concentration of COBr$_2$: 0.50 mol/4.50 L = 0.11 M COBr$_2$

Repeat the steps from (a):

	COBr₂	CO	Br₂
Initial	0.11 M	0	0
Change	−x	+x	+x
Equilibrium	0.11 − x	x	x

$K = \dfrac{x^2}{(0.11 - x)} = 0.190$ and simplifying: $0.190(0.11 - x) = x^2$

Solving the quadratic equation yields: $x = 0.0786$ M.

[Br₂] = [CO] = 0.079 M and [COBr₂] = 0.11 M − x = 0.03 M

(c) A decrease in volume (increase in pressure) favors the side of reaction with fewer moles of gas; equilibrium shifts to left toward more moles of reactants relative to moles of products.

15.45. For the equilibrium: $BaCO_3(s) \rightleftarrows CO_2(g) + BaO(s)$, the effect of:

(a) adding BaCO₃(s): (i) no effect. Since the solid does not appear in the equilibrium expression, the addition of more solid has **no effect** on the position of equilibrium.

(b) adding CO₂: Increasing the concentration of CO₂ would shift the equilbrium (ii) to the **left**—consuming CO₂ until the system returned to equilibrium.

(c) Adding BaO: As in (a) above, since the solid, BaO, does not appear in the equilibrium expression, adding BaO would have (i) **no effect**.

(d) Raising the temperature: Raising T would have the effect of accelerating the decomposition of the carbonate, producing more CO₂—shifting the equilibrium (iii) to the **right**.

(e) Increasing the volume of the flask: This decreases the concentration of CO₂, and (iii) shifts the equilibrium to the **right**—just as in (b) increasing the CO₂ concentration shifted the equilibrium to the left.

15.46. For the equilibrium: $COBr_2(g) \rightleftarrows CO(g) + Br_2(g)$

(a) The effect of adding CO: The addition of CO will shift the equilibrium to the left.

(b) After equilibrium is reestablished, the concentrations:

This problem can be solved by ignoring the equilibrium that was established prior to the addition more CO. Thus, the initial conditions are 0.500 mol COBr₂ in 2.00 L (0.250 M) and 2.00 mol CO in 2.00 L (1.00 M).

	COBr₂	CO	Br₂
Initial (M)	0.250 M	1.00 M	0 M
Change	−x	+x	+x
Equilibrium	0.250 − x	1.00 + x	x

$$K = 0.190 = \frac{(1.00+x)(x)}{0.250-x}$$

Rearranging the expression: $0 = x^2 + 1.190x - 0.04750$

Solve using the quadratic equation: $x = 0.03866$.

[COBr₂] = 0.250 − 0.03866 = 0.211 M
[CO] = 1.00 + 0.03866 = 1.039 M and [Br₂] = 0.0387 M

(c) The addition of CO(g) caused the reaction to shift to the left. Thus, adding CO(g) has decreased the amount of COBr₂ decomposed.

15.47. For the equilibrium: PCl₅(g) ⇌ PCl₃(g) + Cl₂(g), calculate K

Calculate the concentration of added Cl₂: (1.418g/70.90 g/mol)/10.0 L = 0.002000 M

	PCl₅	PCl₃	Cl₂
Equilibrium (expressed as mol)	3.120g/208.22 g/mol = 0.01498 mol	3.845g/137.32 g/mol = 0.02800 mol	1.787g/70.90 g/mol = 0.02520 mol
Concentration	0.001498 M	0.002800 M	0.002520 M
Addition			+0.002000 M
Concentration immediately after addition	0.001498 M	0.002800 M	0.002520 M + 0.002000 M = 0.004520 M
Change (going to equilibrium)	+x	−x	−x
New equilibrium	0.001498 + x	0.002800 − x	0.004520 − x

(a) The addition of the chlorine will shift the equilibrium to the **left** as the system re-establishes equilibrium.

(b) The new equilibrium concentrations will be found using the values from the table, but you must first know the value of K, which we can get by using the equilibrium concentrations given in the problem.

$$K = \frac{[PCl_3][Cl_2]}{[PCl_5]} = \frac{(0.002800)(0.002520)}{(0.001498)} = 4.710 \times 10^{-3}$$

Substituting the values from our table above:

$$\frac{(0.002800 - x)(0.004520 - x)}{(0.001498 + x)} = 4.710 \times 10^{-3}$$

The resulting quadratic equation can be solved for x. The "sensible" root for $x = 0.0004851$ M. Substituting into our "New equilibrium concentrations" we get the resulting equilibrium concentrations:

[PCl₅] = 0.001498 + x = 0.001498 + 0.000481 M = 0.001983 M

[PCl₃] = 0.002800 – x = 0.002800 – 0.000481 M = 0.002315 M

[Cl₂] = 0.004520 – x = 0.004520 – 0.000481 M = 0.004035 M

15.48. Total pressure in the flask for the decomposition of NH₄HS(s):

$K_p = P_{NH_3} P_{H_2S} = 0.11$

$P_{NH_3} = P_{H_2S}$, so $K_p = 0.11 = P^2_{NH_3} = P^2_{H_2S}$ and taking the square root:

$P_{NH_3} = P_{H_2S} = \sqrt{0.11} = 0.33$ atm, and $P_{total} = 0.33$ atm + 0.33 atm = 0.66 atm

15.49. For the system: NH₄I(s) ⇌ NH₃(g) + HI(g), what is K_p?

Given that the total pressure is attributable **only** to NH₃ + HI, **and** that the stoichiometry of the equation tells us that equal amounts of the two substances are formed, we can easily determine the pressure of **each** of the gases. Since we desire K_p in atmospheres, we must convert 705 mm Hg to units of atmospheres: 705 mm Hg $\left(\dfrac{1 \text{ atm}}{760 \text{ mmHg}}\right)$ = 0.928 atm

$P_{total} = P(NH_3) + P(HI) = 0.928$ atm so the pressure of each gas is (1/2)(0.928) or 0.464 atm.

$K_p = P(NH_3) \times P(HI) = (0.464 \text{ atm})^2 = 0.215$

15.50. For the decomposition of ammonium carbamate:

The decomposition forms 2 mol of NH₃(g) and 1 mol of CO₂(g) for each mol of carbamate, so we expect the pressure of CO₂ to be twice that of the NH₃.

$P_{total} = 0.116$ atm $= P_{NH_3} + P_{CO_2}$, so $P_{CO_2} = (1/3)P_{total} = 0.0387$ atm and $P_{NH_3} = 0.0774$ atm

$K_p = P^2_{NH_3} \times P_{CO_2} = (0.0774)^2(0.0387) = 2.31 \times 10^{-4}$

15.51. Given K_p for $N_2O_4(g) \rightleftharpoons 2\ NO_2(g)$ is 0.148 at 25 °C.

(a) If total P is 1.50 atm, what fraction of N_2O_4 has dissociated?

$P_{total} = P_{NO_2} + P_{N_2O_4} = 1.50$ atm; rearranging $P_{N_2O_4} = 1.50 - P_{NO_2}$

$K_p = \dfrac{P^2_{NO_2}}{P_{N_2O_4}} = 0.148$ and $K_p = \dfrac{P^2_{NO_2}}{1.50 - P_{NO_2}} = 0.148$ or $0.148(1.50 - P_{NO_2}) = P^2_{NO_2}$

and rearranging, $0.222 - 0.148\ P_{NO_2} = P^2_{NO_2}$

Solve for P_{NO_2} with the quadratic equation: $P_{NO_2} = 0.403$ atm and $P_{N_2O_4} = 1.10$ atm

To determine the fraction of N_2O_4 that has dissociated, you need to know the amount of N_2O_4 that was originally present. The stoichiometry tells you that you get 2 NO_2 for each N_2O_4 that dissociates. The 0.403 atm of NO_2 that exist indicate that (0.403/2) atm of N_2O_4 dissociated. The **original pressure** of N_2O_4 will then be 1.10 atm + 0.20 atm or 1.30 atm. The fraction that has dissociated is then: (1.30 − 1.10)/1.30 or 0.15.

(b) The fraction dissociated if the total equilibrium pressure falls to 1.00 atm:

$P_{total} = P_{NO_2} + P_{N_2O_4} = 1.00$ atm ; $P_{N_2O_4} = 1.00 - P_{NO_2}$

$K_p = \dfrac{P^2_{NO_2}}{P_{N_2O_4}} = 0.148 = \dfrac{P^2_{NO_2}}{1.00 - P_{NO_2}}$; $0.148(1.00 - P_{NO_2}) = P^2_{NO_2}$

Solving via the quadratic equation: $P_{NO_2} = 0.318$ atm and $P_{N_2O_4} = 0.682$ atm

The equilibrium pressure of 0.682 atm for N_2O_4 tells us that the **original pressure of N_2O_4 was 0.682 + 1/2(0.318) or 0.841 atm**.
[Recall that 0.318 atm of NO_2 represents 0.159 atm of N_2O_4 decomposing].

The fraction dissociated is: 0.159 atm/0.841 atm = 0.189 or approximately 19%.

15.52. For the equilibrium between monomers and the dimer of acetic acid:

	CH₃CO₂H	(CH₃CO₂H)₂
Initial	5.4×10^{-4}	0
Change	$-2x$	$+x$
Equilibrium	$5.4 \times 10^{-4} - 2x$	x

(a) Percentage of the monomer that is converted to the dimer:

Substituting into the K expression:

$K = 3.2 \times 10^4 = \dfrac{x}{(5.4 \times 10^{-4} - 2x)^2}$

$0 = (1.3 \times 10^5)x^2 - 70x + 9.3 \times 10^{-3}$

Solve using the quadratic equation: $x = 3.2 \times 10^{-4}$ and 2.3×10^{-4}

$\dfrac{2(2.3 \times 10^{-4} \text{ M})}{5.4 \times 10^{-4} \text{ M}} \cdot 100\% = 84\%$

(b) Increasing the temperature will shift the equilibrium to the left.

15.53. For the decomposition of ammonia at 723 K, $K_c = 6.3$.

What are equilibrium concentrations of NH$_3$, N$_2$, and H$_2$ and what is the **total** pressure in the flask?

Initial concentration of NH$_3$: 3.60 mol NH$_3$ in a 2.00 L vessel is 1.80 M NH$_3$.

Establish a reaction table:

	[NH$_3$]	[N$_2$]	[H$_2$]
Initial concentrations (mol/L)	1.80	0	0
Change (going to equilibrium)	$-x$	$+x/2$	$+3x/2$
New equilibrium	$1.80 - x$	$x/2$	$3x/2$

Substituting into the K expression:

$6.3 = \dfrac{\left[\dfrac{x}{2}\right]\left[\dfrac{3x}{2}\right]^3}{[1.80 - x]^2}$ and simplifying gives: $6.3 = \dfrac{\dfrac{27x^4}{16}}{[1.80 - x]^2}$

Simplifying further: $6.3 = \dfrac{1.6875 x^4}{[1.80 - x]^2}$, taking the square root of both sides will simplify the

math: $2.51 = \dfrac{1.299 x^2}{[1.80 - x]}$ and $2.51[1.80 - x] = 1.299 x^2$.

Expanding gives: $4.4519 - 2.51x = 1.299x^2$ and using the quadratic equation to solve gives $x = 1.13$. Using the table above, we get:

[NH$_3$] = 1.80 – x = 0.67 M; [N$_2$] = x/2 = 0.57 M; [H$_2$]= 3x/2 = 1.7 M

Total pressure in flask: $P = nRT/V$ and since M = n/V, we can use $P = MRT$

$$P = (0.67 + 0.57 + 1.7)(0.082057 \frac{L \cdot atm}{K \cdot mol})(723 \text{ K}) = 180 \text{ atm (2 sf)}$$

15.54. Calculate partial pressure of NO$_2$ and N$_2$O$_4$ in the mixture:

$P_{Total} = P_{NO_2} + P_{N_2O_4}$ and rearranging: $P_{N_2O_4} = 0.36 - P_{NO_2}$ Substituting into the K expression:

$K_p = 7.1 = \dfrac{0.36 - P_{NO_2}}{P_{NO_2}^2}$ and rearranging into the form of a quadratic equation:

$0 = 7.1 P_{NO_2}^2 + P_{NO_2} - 0.36$

Solve using the quadratic equation.

$P_{NO_2} = 0.17$ and $P_{N_2O_4} = 0.36 - P_{NO_2} = 0.19$ atm

15.55. K_c for the decomposition of NH$_4$HS into ammonia and hydrogen sulfide gases =1.8 × 10^{-4}.

(a) When pure salt decomposes, the equilibrium concentrations of [NH$_3$] and [H$_2$S]:

K_p = [NH$_3$][H$_2$S] = 1.8 × 10^{-4}. Noting the stoichiometry of the reaction, the two concentrations will be equal. [NH$_3$]^2 = [H$_2$S]^2 = 1.8 × 10^{-4} and

[NH$_3$] = [H$_2$S] = 1.3 × 10^{-2} M.

(b) If NH$_4$HS is placed into a flask containing [NH$_3$] = 0.020 M, when system achieves equilibrium, what are equilibrium concentrations?

The equilibrium table:

	[NH$_3$]	[H$_2$S]
Initial concentration	0.020	0
	[NH$_3$]	[H$_2$S]
Change (going to equilibrium).	+x	+x
New equilibrium	0.020 + x	x

K_p = [NH$_3$][H$_2$S] = 1.8 × 10^{-4} and [0.020 + x][x] = 1.8 × 10^{-4}

Simplifying gives: $0.020x + x^2 = 1.8 \times 10^{-4}$. This will require the quadratic equation to resolve the equation: $x^2 + 0.020x - 1.8 \times 10^{-4} = 0$. Solving gives $x = 6.7 \times 10^{-3}$ M.

The equilibrium concentrations are then $[NH_3] = 0.020 + 6.7 \times 10^{-3}$ or 0.0267M (or 0.027 M to 2 sf) and for $[H_2S] = 6.7 \times 10^{-3}$ M.

15.56. For the equilibrium: $COCl_2(g) \rightleftarrows CO(g) + Cl_2(g)$:

Initial concentration of $COCl_2$: 0.050 mol $COCl_2$/12.5 L = 0.0040 M $COCl_2$

	$[COCl_2]$	$[CO]$	$[Cl_2]$
Initial concentrations (mol/L)	0.0040	0	0
Change (going to equilibrium)	$-x$	$+x$	$+x$
Equilibrium	$0.0040 - x$	x	x

Substitute into the K expression: $K_c = 0.0071 = \dfrac{[CO][Cl_2]}{[COCl_2]} = \dfrac{(x)(x)}{(0.0040-x)}$

Rearranging, the equation forms the quadratic equation: $x^2 + 0.0071x - 0.0000284 = 0$

Solving gives: $x = 0.0029$ M $= [CO] = [Cl_2]$

$[COCl_2] = 0.0040$ M $- x = 0.0011$ M

Now we need to calculate pressure, so convert these concentrations into moles:

(0.0029 mol/L)(12.5 L) = 0.036 mol CO = 0.036 mol Cl_2

(0.0011 mol/L)(12.5 L) = 0.014 mol $COCl_2$

Total mol = 0.014 mol + 0.036 mol + 0.036 mol = 0.086 mol

The ideal gas law provides the pressure: $PV = nRT$

$P = \dfrac{nRT}{V} = \dfrac{(0.086 \text{ mol})(0.082057 \frac{L \cdot atm}{K \cdot mol})(600K)}{12.5 \text{ L}} = 0.3$ atm (1 sf)

15.57. What is P_{Total} for a mixture of NO_2 and N_2O_4 (total mass = 6.44 g) in a 15 L flask at 300 K?

Note that K_p for 2 $NO_2(g) \rightleftarrows N_2O_4(g)$ is 7.1 at 300 K so $K_p = \dfrac{P_{N_2O_4}}{P^2_{NO_2}} = 7.1$ (eq 1)

Mass of NO_2 + Mass of N_2O_4 = 6.44 g (eq 2) and

$P_{Total} = P_{NO_2} + P_{N_2O_4}$ (eq 3)

Begin by substituting into equation 2:

Mass of NO₂ + Mass of N₂O₄ = 6.44 g. Calculate the mass of each by noting that the product of # mol · MM = mass (in grams).
mol · MM (for NO₂) + # mol · MM (for N₂O₄) = 6.44

[# mol NO₂ · 46.00 g/mol NO₂] + [# mol N₂O₄ · 92.01 g/mol N₂O₄] = 6.44

Recall that $P = \dfrac{nRT}{V} = \dfrac{\# mol \left(0.082 \dfrac{L \cdot atm}{K \cdot mol}\right)(300 \text{ K})}{15 \text{ L}}$

Using equation 1: $P_{N2O4} = 7.1 \cdot P_{NO2}^2$

$$\dfrac{\# \text{mol N}_2\text{O}_4 \left(0.082 \dfrac{L \cdot atm}{K \cdot mol}\right)(300 \text{ K})}{15 \text{ L}} = 7.1 \left(\dfrac{\# \text{mol NO}_2 \cdot 0.082 \dfrac{L \cdot atm}{K \cdot mol} \cdot 300 \text{ K}}{15 \text{ L}}\right)^2$$

Cancelling similar terms on both sides gives:

$$\# \text{mol N}_2\text{O}_4 = \dfrac{7.1 (\# \text{mol NO}_2)^2 \left(0.082 \dfrac{L \cdot atm}{K \cdot mol}\right)(300 \text{ K})}{15 \text{ L}}$$

Multiplying all the constants on the right-hand side gives:

mol N₂O₄ = 11.644 (# mol NO₂)² and substituting into the equation below:

[# mol NO₂ (46.00 g/mol NO₂)] + [# mol N₂O₄ (92.01 g/mol N₂O₄)] = 6.44 g gives

[# mol NO₂ (46.00 g/mol NO₂)] + [11.644 (# mol NO₂)² (92.01)] = 6.44 g and leaving off all units (for clarity) and multiplying (11.644 × 92.01) gives
46.01n + 1071.36444n^2 = 6.44, which can be solved by the quadratic equation to give n = 0.0590 mol NO₂.

Since each mol of NO₂ has a mass of 46.00 g, the mass of NO₂ = 0.0590 · 46.00 = 2.714 g. From equation 2: 6.44 g − 2.714 g = 3.73 g of N₂O₄.

The pressure of NO₂ is:

$P = \dfrac{nRT}{V} = \dfrac{(0.0590 \text{ mol NO}_2)\left(0.082 \dfrac{L \cdot atm}{K \cdot mol}\right)(300 \text{ K})}{15 \text{ L}} = 0.096$ atm

Substituting into equation 1, we can solve for the pressure of N₂O₄.

$P_{N2O4} = 7.1 \cdot P_{NO2}^2 = 7.1(0.096)^2 = 0.066$ atm (2sf)

The total pressure is: 0.096 + 0.066 = 0.16 atm

15.58. For the decomposition of lanthanum oxalate:

(a) $P_{Total} = 0.200$ atm $= P_{CO} + P_{CO_2}$

$P_{CO} = P_{CO_2} = \frac{1}{2}(0.200$ atm$) = 0.100$ atm

$K_p = (0.100$ atm$)^3(0.100$ atm$)^3 = 1.00 \times 10^{-6}$

(b) $n_{CO} = \dfrac{PV}{RT} = \dfrac{(0.100 \text{ atm})(10.0 \text{ L})}{(0.082057 \text{ L} \cdot \text{atm/K} \cdot \text{mol})(373 \text{ K})} = 0.0327$ mol CO

$0.0327 \text{ mol CO}\left(\dfrac{1 \text{ mol La}_2(C_2O_4)_3}{3 \text{ mol CO}}\right) = 0.0109$ mol $La_2(C_2O_4)_3$

$(0.100$ mol $- 0.0109$ mol$) = 0.089$ mol $La_2(C_2O_4)_3$ remains unreacted

15.59. For the equilibrium of H_2, I_2, and HI:

(a) For the reaction of hydrogen and iodine to give HI, if $K_c = 56$ at 435 °C, what is K_p?

The relation between the equilibrium constants is: $K_p = K_c(RT)^{\Delta n}$. The equation is:

$$H_2(g) + I_2(g) \rightleftharpoons 2 HI(g)$$

Note that the **total** number of moles of gas is 2 on both sides of the equilibrium, so that $\Delta n = 0$, therefore $K_p = 56$.

(b) Mix 0.045 mol of each gas in a 10.0 L flask at 435 °C, what is the total pressure of the mixture before and after equilibrium?

Before equilibrium: $P = nRT/V$ or

$\dfrac{0.090 \text{ mol}\left(0.082057 \dfrac{\text{L} \cdot \text{atm}}{\text{K} \cdot \text{mol}}\right)(708 \text{ K})}{10.0 \text{ L}} = 0.52$ atm

Note that we can simply add the amount of each gas since the P of either gas is 0.26 atm.

At equilibrium:

Substituting into the equilibrium expression gives:

$K_p = \dfrac{P_{HI}^2}{P_{H_2} \cdot P_{I_2}} = \dfrac{(2x)^2}{(0.26 - x)^2} = 56$ Simplify the expression by taking the square root of both sides to give: $\dfrac{(2x)}{(0.26 - x)} = 7.48$. Solving for x gives $x = 0.205$ atm, so

$P_{HI} = 2 \cdot 0.205$ or 0.42 atm (2 sf), $P_{H_2} = P_{I_2} = (0.26 - 0.205) = 0.05$ atm

The total pressure is: 0.42 atm $+ 0.05$ atm $+ 0.05$ atm $= 0.52$ atm

Note that the pressure **does not change** in reaching equilibrium—not a surprise since the total number of moles of gas does not change!

(b) The partial pressures, as noted above, are: $P_{HI} = 0.42$ atm, $P_{H_2} = P_{I_2} = 0.05$ atm.

15.60. For the equilibrium: $SO_2Cl_2(g) \rightleftharpoons SO_2(g) + Cl_2(g)$, $K_c = 0.045$ at 375 °C:

(a) concentration of species at equilibrium when 6.70 g of SO_2Cl_2 is placed in a 12.0 L flask:

$$[SO_2Cl_2] = \frac{7.80 \text{ g } SO_2Cl_2}{12.0 \text{ L}} \left(\frac{1 \text{ mol } SO_2Cl_2}{134.9 \text{ g}} \right) = 0.00482 \text{ M}$$

Substituting into the K expression: $K = 0.045 = \dfrac{x^2}{0.00482 - x}$

$0 = x^2 + 0.0045x - 0.000217$

Solving using the quadratic equation yields a value for $x = 0.0044$
$[SO_2Cl_2] = 0.00482 - x = 0.0004$ M
$[SO_2] = [Cl_2] = x = 0.0044$ M

fraction dissociated $= \dfrac{0.0044}{0.00482} = 0.91$

(b) concentration of species at equilibrium when 6.70 g of SO_2Cl_2 and 0.10 atm of Cl_2 is placed in a 12.0 L flask:

Determine the concentration of chlorine gas:

$$[Cl_2] = \frac{n}{V} = \frac{P}{RT} = \frac{0.10 \text{ atm}}{(0.082057 \text{ L} \cdot \text{atm/K} \cdot \text{mol})(648 \text{ K})} = 0.00188 \text{ M}$$

A table may help to get the data straight:

	$[SO_2Cl_2]$	$[SO_2]$	$[Cl_2]$
Initial concentrations (mol/L)	0.00482 M	0	0.00188 M
Change (going to equilibrium)	$-x$	$+x$	$+x$
Equilibrium	$0.00482 - x$	x	$0.00188 + x$

Substitute these data into the K expression: $K = 0.045 = \dfrac{x(0.00188 + x)}{0.00482 - x}$

Simplifying, we get the quadratic equation: $0 = x^2 + 0.04688x - 0.000217$

From the quadratic equation, we obtain the value $x = 0.0042$. Using our equilibrium concentrations from the table:

$[SO_2Cl_2] = 0.00482 - x = 0.0006$ M

$[SO_2] = x = 0.0042$ M

$[Cl_2] = 0.00188 + x = 0.0061$ M

The fraction dissociated = $\dfrac{0.0042}{0.00482} = 0.87$

(c) Le Chatelier's principle predicts that the addition of $Cl_2(g)$ would shift the equilibrium to the left, a prediction that is confirmed by these calculations.

15.61. For the reaction of hemoglobin (Hb) with CO we can write: $K = \dfrac{[HbCO][O_2]}{[HbO_2][CO]} = 2.0 \times 10^2$

If $\dfrac{[HbCO]}{[HbO_2]} = 1$, substitution into the K expression shows that $\dfrac{[O_2]}{[CO]} = 2.0 \times 10^2$

If $[O_2] = 0.20$ atm, then $\dfrac{0.20 \text{ atm}}{[CO]} = 2.0 \times 10^2$ and solving for $[CO] = 1.0 \times 10^{-3}$ atm.

So, a partial pressure of $[CO] = 1.0 \times 10^{-3}$ atm would likely be fatal.

15.62. Quantity of $CaCO_3$ that decomposes to reach equilibrium:

The equilibrium expression is: $K_p = P_{CO_2} = 3.87$ atm

What amount of CO_2 would provide the pressure of 3.87 atm at 1000°C?

$n_{CO_2} = \dfrac{PV}{RT} = \dfrac{(3.87 \text{ atm})(5.00 \text{ L})}{(0.082057 \text{ L·atm/K·mol})(1273 \text{ K})} = 0.185$ mol CO_2

The mass of $CaCO_3$ that corresponds to this number of mol of CO_2.

$0.185 \text{ mol} \left(\dfrac{1 \text{ mol } CaCO_3}{1 \text{ mol } CO_2} \right) \left(\dfrac{100.09 \text{ g}}{1 \text{ mol } CaCO_3} \right) = 18.5$ g $CaCO_3$

15.63. How many O atoms are present if 0.050 mol of O_2 is placed in a 10. L vessel at 1800 K?

Given K_p for $O_2(g) \rightleftarrows 2 O(g) = 1.2 \times 10^{-10}$

Since we need to express the # of O atoms in a 10. L vessel, convert K_p into K_c:

Since $K_p = K_c(RT)^{\Delta n}$ then $K_c = K_p/(RT)^{\Delta n} = 1.2 \times 10^{-10}/(0.082057 \dfrac{L \cdot atm}{K \cdot mol}(1800 \text{ K}))^1$

and solving, $K_c = 8.12 \times 10^{-13}$. Substituting into the K_c expression, we get:

$K = \dfrac{[O]^2}{[O_2]} = 8.12 \times 10^{-13}$ If we let a mol/L of O_2 dissociate into atoms, we get:

$K = \dfrac{[2a]^2}{[0.005 - a]} = 8.12 \times 10^{-13}$

So $4a^2 = 8.12 \times 10^{-13} (0.005 - a)$ and solving for a: $a = 3.2 \times 10^{-8}$. Note here that with the small value of K, the denominator could be simplified to 0.005, and an approximate value of a calculated. The [O] is then $2a$ or 6.4×10^{-8} M. To calculate the # of O atoms:

10. L $(6.4 \times 10^{-8} \text{mol/L})(6.02 \times 10^{23} \text{atoms/mol}) = 3.9 \times 10^{17}$ O atoms.

15.64. What is the value of K_p for NOBr dissociation at 25°C?

As values of K_p are derived from pressures in atmospheres, convert the pressure to units of atm: $190 \text{ mm Hg} \left(\dfrac{1 \text{ atm}}{760 \text{ mm Hg}} \right) = 0.25$ atm

The equilibrium is: NOBr ⇌ NO + ½ Br$_2$

As we don't know the amount of NOBr in the flask, let's arbitrarily assign a $P_{NOBr} = x$ atm

Setup a chart to detail pressures:

	NOBr	NO	Br$_2$
Initial	x	0	0
Change	$-0.34x$	$+0.34x$	$1/2(+0.34x)$
Equilibrium	$0.66x$	$0.34x$	$0.17x$

We also know that the $P_{total} = 0.25$ atm $= P_{NOBr} + P_{NO} + P_{Br2} = 0.66x + 0.34x + 0.17x$

$P_{total} = 1.17x = 0.25$ atm, so $x = 0.25/1.17 = 0.21$

$P_{NOBr} = (0.66)(0.21) = 0.14$ atm

$P_{NO} = (0.34)(0.21) = 0.073$ atm

$P_{Br2} = (0.17)(0.21) = 0.036$ atm

Now we can substitute these values into the K expression:

$$K_p = \dfrac{P_{NO} \cdot P^{1/2}_{Br_2}}{P_{NOBr}} = \dfrac{(0.073) \cdot (0.036)^{1/2}}{(0.14)} = 0.098$$

15.65. For the equilibrium with boric acid (BA) + glycerin(gly), $K = 0.90$.

B(OH)$_3$ (aq) + glycerin(aq) ⇌ B(OH)$_3$·glycerin(aq)

	BA	gly	BA•gly
Initial	0.10 M	?	0
Change	−0.60 · 0.10 M		+ 0.60 · 0.10M
Equilibrium	0.040M	x	0.060 M

Substituting into the equilibrium expression yields:

$$\frac{[BA \cdot gly]}{[BA][gly]} = 0.90 = \frac{[0.060]}{[0.040][gly]}$$

Rearranging: $\frac{[0.060]}{[0.040][0.90]} = [gly] = 1.67$ M (or 1.7 to 2 sf)

Solving for [gly] gives the *equilibrium* amount of glycerin.

Recall, however, that the equilibrium amount of glycerin represents the amount of glycerin remaining uncomplexed from the original amount. Recall that *some* glycerin is consumed in making the complex (in this case 0.060 M). So, the initial amount of glycerin present would be 1.67 + 0.060 or 1.73 M—which is 1.7 M (2 sf).

15.66. $K_p = 1.16$ at 800°C for the dissociation of calcium carbonate:

(a) What is K_c?

$K_p = K_c(RT)^{\Delta n}$; Solving for K_c:

$$K_c = \frac{K_p}{(RT)^{\Delta n}} = \frac{1.16}{[(0.082057 \text{ L} \cdot \text{atm/K} \cdot \text{mol})(1100 \text{ K})]^1} = 0.013$$

(b) Pressure of CO$_2$ in the container:

$P_{CO_2} = K_p = 1.16$ atm since CO$_2$ is the only gaseous product.

(c) Percentage of original sample undissociated:

[CO$_2$] = K_c = 0.013 mol/L

$$\frac{0.013 \text{ mol CO}_2}{1 \text{ L}} \cdot 9.56 \text{ L} \cdot \frac{1 \text{ mol CaCO}_3}{1 \text{ mol CO}_2} \cdot \frac{100.1 \text{ g}}{1 \text{ mol CaCO}_3} = 12 \text{ g CaCO}_3 \text{ decomposed}$$

$\left(\frac{22.5 \text{ g} - 12 \text{ g}}{22.5 \text{ g}}\right) 100\% = 45\%$ undecomposed

15.67. For the reaction of N_2O_4 decomposing into NO_2, a sample of N_2O_4 at a pressure of 1.00 atm reaches equilibrium with 20.0 % of the N_2O_4 having been converted into NO_2.

(a) What is K_p?

	N_2O_4	NO_2
Initial	1.00 atm	0 atm
Change	−0.200(1.00 atm)	+ 2(0.200)(1.00 atm)
Equilibrium	0.80 atm	0.400 atm

Substituting into the equilibrium expression gives:

$$K_p = \frac{P^2(NO_2)}{P(N_2O_4)} = \frac{(0.400)^2}{(0.80)} = 0.20$$

(b) If the initial pressure is 0.10 atm, the percent dissociation:

Now we know that $K_p = 0.20$. Substituting into the equilibrium expression:

$$K_p = \frac{P^2(NO_2)}{P(N_2O_4)} = \frac{(2x)^2}{(0.10 - x)} = 0.20 \text{ and rearranging: } 4x^2 = (0.10 - x)(0.20)$$

or $4x^2 = 0.020 - 0.20x$, and solving for the quadratic equation, $x = 0.050$ atm.

This represents half (or 50%) of the original N_2O_4.

Does this agree with LeChatelier's Principle? YES. We began with a lower concentration (pressure), so we expect that a larger percentage of the N_2O_4 will dissociate (50% compared to 20%), and it does.

15.68. For the reaction of O_3 with NO:

(a) Is the system at equilibrium?

$$Q = \frac{[O_2][NO_2]}{[O_3][NO]} = \frac{(8.2 \times 10^{-3})(2.5 \times 10^{-4})}{(1.0 \times 10^{-6})(1.0 \times 10^{-5})} = 2.1 \times 10^5$$

$Q < K$ so the reaction will proceed to the right to reach equilibrium

(b) Will concentration of products increase or decrease?

$\Delta H° = \Delta_f H°[NO_2(g)] - (\Delta_f H°[O_3(g)] + \Delta_f H°[NO(g)])$
$\Delta H° = 1$ mol (33.1 kJ/mol) – [1 mol (142.67 kJ/mol) + 1 mol (90.29 kJ/mol)]
$\Delta H° = -199.9$ kJ/mol

As the process is exothermic, increasing the temperature shifts the equilibrium to the left. The product concentrations will decrease.

In the Laboratory

15.69. For the equilibrium: $(NH_3)[B(CH_3)_3] \rightleftharpoons B(CH_3)_3 + NH_3 \quad K_p = 4.62$

Substituting $(CH_3)_3P$ for NH_3 gives an equilibrium with $K_p = 0.128$

while substituting $(CH_3)_3N$ gives an equilibrium with $K_p = 0.472$.

(a) To determine which of the three systems would provide the largest concentration of $B(CH_3)_3$, one needs to ask, "What does K tell me?"

One answer to that question is "the extent of reaction—that is, the larger the value of K, the more products are formed. Hence **the system with the largest K_p would give the largest partial pressure of $B(CH_3)_3$ at equilibrium.**

(b) Given $(NH_3)[B(CH_3)_3] \rightleftharpoons B(CH_3)_3 + NH_3 \quad K_p = 4.62$

Since we have a K_p, we need to express "concentrations" as pressures:

$$P = \frac{0.010 \text{ mol} \left(0.082057 \frac{L \cdot atm}{K \cdot mol}\right)(373 \text{ K})}{1.25 \text{ L}} = 0.24 \text{ atm}$$

Organize the information:

	$(NH_3)[B(CH_3)_3]$	$B(CH_3)_3$	NH_3
Initial:	0.24 atm	0 M	0 M
Change	$-x$	$+x$	$+x$
Equilibrium	$0.24 - x$	x	x

Substituting into the equilibrium expression: $K_p = \frac{(x)(x)}{0.24 - x} = 4.62$

and simplifying gives: $x^2 = 4.62(0.24 - x)$ and $x^2 + 4.62x - 1.13 = 0$
Solving via the quadratic formula gives $x = 0.232$ atm (or 0.23 to 2 sf).

So, the **concentrations at equilibrium** are:

$[B(CH_3)_3] = [NH_3] = 0.23$ atm and $[(NH_3)[B(CH_3)_3]] = 0.012$ atm

What is the total pressure?

$P_{total} = P_{B(CH_3)_3} + P_{NH_3} + P_{(NH_3)[B(CH_3)_3]} = 0.23 \text{ atm} + 0.23 \text{ atm} + 0.012 \text{ atm} = 0.48 \text{ atm}$

The percent dissociation of $(NH_3)[B(CH_3)_3]$ is:

$\frac{\text{amount changed}}{\text{original amount}} = \frac{0.23 \text{ atm}}{0.24 \text{ atm}} \times 100 = 95\%$ dissociated

15.70. For the equilibrium between chromate and dichromate ions:

(a) When H_3O^+ is added, Le Chatelier's principle predicts the equilibrium shifts right. The excess acid is consumed by the yellow chromate ion, and orange dichromate ion is produced.

(b) When NaOH is added, it reacts with H_3O^+, decreasing the concentration of this reactant. The equilibrium shifts left, producing additional H_3O^+ and yellow chromate ion. The solution becomes more yellow (less orange).

15.71. For the equilibrium between iron(II) and thiocyanate ions:

(a) The addition of KSCN results in the color becomes even more red. Why? This is to be expected if the solution had not yet reached equilibrium prior to the addition of the KSCN. As more Fe^{2+} is added, more of the $(Fe(H_2O)_5SCN)^+$ complex ion would form—making the solution more red (in accordance with Le Chatelier's principle).

$$[Fe(H_2O)_6]^{2+} + SCN^- \rightarrow [(Fe(H_2O)_5SCN)]^+$$

(b) Once again, Le Chatelier's principle would answer the question. The addition of silver ion would initiate formation of the white AgSCN solid, removing SCN^- ions from solution—and reducing the concentration of the red $(Fe(H_2O)_5SCN)^+$ complex ion.

$Ag^+ + SCN^- \rightarrow AgSCN(s)$ resulting in $[Fe(H_2O)_6]^{2+} + SCN^- \leftarrow [(Fe(H_2O)_5SCN)]^+$

15.72. For the reaction of Ni(II) with NH_3 and ethylenediamine:

For simplicity, ethylenediamine ($NH_2CH_2CH_2NH_2$) will be represented as en

(a) Consider the two reactions:

$[Ni(H_2O)_6]^{2+}(aq) + 6\ NH_3(aq) \rightleftarrows [Ni(NH_3)_6]^{2+}(aq) + 6\ H_2O(\ell)$ $\qquad K_1$

$[Ni(NH_3)_6]^{2+}(aq) + 3\ en(aq) \rightleftarrows [Ni(en)_3]^{2+}(aq) + 6\ NH_3\ (aq)$ $\qquad K_2$

Adding these two equations gives:

$[Ni(H_2O)_6]^{2+}(aq) + 3\ en(aq) \rightleftarrows [Ni(en)_3]^{2+}(aq)$

The equilibrium expression is:

$$K = \frac{\left[Ni(en)_3^{2+}\right]}{\left[Ni(H_2O)_6^{2+}\right]\left[en\right]^3} = K_1 \times K_2$$

(b) $[Ni(NH_2CH_2CH_2NH_2)_3]^{2+}$(aq) is the most stable, as evidenced by the color of the solution.

Summary and Conceptual Questions

15.73. Decide upon the truth of each of the statements:

(a) Adding two chemical equations results in an equilibrium constant that is the product of the equilibrium constants of the summed equations. **true**

(b) The equilibrium constant for a reaction has the same magnitude but opposite sign as K for the reverse reaction. **false** K for the reverse reaction is the inverse of the original K.

(c) Only the concentration of CO_2 appears in the equilibrium constant for the decomposition of calcium carbonate reaction. **true**

(d) For the equilibrium involving the decomposition of $CaCO_3$, the value of K is independent of the expression of the amount of CO_2, **false;** $\boldsymbol{K_p = K_c(RT)^{\Delta n}}$, with Δn equal to a non-zero number, K_p and K_c will differ by the factor $(RT)^{\Delta n}$.

(e) If the standard enthalpy of reaction is positive, does increasing the temperature increase the value of K? **true** Because reaction is endothermic heating increases the product concentration and decreases the reactant concentration. The overall effect is to increase the value of K.

15.74. Pb^{2+} concentration in the two beakers:

The dissolution of $PbCl_2$ has a larger K than the dissolution of PbF_2, so solutions of $PbCl_2$ have a greater concentration of Pb^{2+}.

15.75. Characterize each of the following as product- or reactant-favored:

(a) with $K_p = 1.2 \times 10^{45}$; A large K indicates the equilibrium is **product-favored**.

(b) with $K_p = 9.1 \times 10^{-41}$; A small K indicates the equilibrium would "lie to the left"—that is **reactant-favored**.

(c) with $K_p = 6.5 \times 10^{11}$; equilibrium would "lie to the right"—that is **product-favored.**

15.76. Color changes in the $NO_2 \rightleftarrows N_2O_4$ equilibrium upon changing volume:

$N_2O_4(g) \rightleftarrows 2\ NO_2(g)$

colorless $\rightleftarrows$ brown

(a) Assume volume change occurs faster than reaction occurs. Molecules occupy less space and color is more intense.

(b) Reduced volume corresponds to increased pressure. Apply stress to product side of reaction, that with most moles of gas. The intensity of the brown color diminishes.

15.77. There are several ways to prove the dynamic nature of this equilibrium. Begin the experiment with a mixture of (a) deuterated H₂ (represented as D₂), (b) elemental N₂, and (c) NH₃. Let the mixture reach equilibrium. Separate the three different gases and measure the amount of D₂. If the system is dynamic, the amount of elemental D₂ will be reduced (as D is incorporated into the NH₃ molecules), AND the NH₃ present will consist of varying amounts of $NH_{(3-x)}D_x$ as a result of the incorporation of D atoms into the NH₃ molecular species.

15.78. For the equilibrium involving Co(II) and HCl:
(a) Conversion of red cation to blue anion: Endothermic as applying heat shifts equilibrium toward the products (blue anion).
(b) HCl is a reactant. Adding HCl shifts the equilibrium toward the products (blue anion). Water is a product in the reaction. Adding water shifts the equilibrium toward the reactants (red cation).
(c) The position of the equilibrium can be shifted in either direction by the addition of reactants or products.

15.79. What is K_p for the system: $2\ H_2S(g) \rightleftharpoons 2\ H_2(g) + S_2(g)$ if a tank, initially containing 10.00 atm of H₂S at 800 K, contains 0.020 atm of S₂ upon reaching equilibrium.

	H₂S	H₂	S₂
Initial	10.00 atm	0	0
Change	−2(0.020 atm)	+2(0.020 atm)	+0.020 atm
Equilibrium	9.96 atm	0.040 atm	0.020 atm

Substitute these values into the equilibrium constant expression:

$$K_p = \frac{P^2_{H_2} \cdot P_{S_2}}{P^2_{H_2S}} = \frac{(0.040)^2 (0.020)}{(9.96)^2} = 3.2 \times 10^{-7}$$

15.80. Calculate K_p for the decomposition of PCl₅:

	PCl₅	PCl₃	Cl₂
Initial (atm)	2.000	0	0
Change (atm)	−x	+x	+x
Equilibrium (atm)	2.000 − x	x	x

At equilibrium, the $P_{Cl_2} = 0.814$ atm, so P_{PCl_3} also must be 0.814 atm

$P_{PCl_5} = 2.000$ atm − 0.814 atm = 1.186 atm

Substitute into the K expression:

$$K_p = \frac{P_{Cl_2} P_{PCl_3}}{P_{PCl_5}} = \frac{(0.814)(0.814)}{1.186} = 0.559$$

Solution and Answer Guide

Kotz Treichel Townsend Treichel, Chemistry and Chemical Reactivity 11e, 978-0-357-85140-1, Chapter 16: Principles of Chemical Reactivity: The Chemistry of Acids and Bases

TABLE OF CONTENTS

Applying Chemical Principles 569
Practicing Skills..571
General Questions...594
In the Laboratory..603
Summary and Conceptual Questions..606

Applying Chemical Principles

Would You Like Some Belladonna Juice in Your Drink?

16.1.1. Moles of atropine in 100. mg compound:

$$100 \text{ mg } C_{17}H_{23}NO_3 \left(\frac{1 \text{ g}}{1000 \text{ mg}}\right)\left(\frac{1 \text{ mol } C_{17}H_{23}NO_3}{289.4 \text{ g}}\right) = 3.46 \times 10^{-4} \text{ mol}$$

16.1.2. The proton will attach to the only N atom in atropine. That N is the most basic portion of the molecule.

16.1.3. Compare the pK_a of the conjugate acid of atropine with those of ammonia, methylamine, and aniline. The pK_a is similar to that of the anilinium ion, and significantly lower than that of the ammonium ion and the methyammonium ion.

pK_a	ion
4.35	atropinium
9.26	ammonium
10.70	methylammonium
4.60	anilinium

The Leveling Effect, Nonaqueous Solvents, and Superacids

16.2.1. Convert the pK values to K values for the dissociations of HCl, HClO$_4$, and H$_2$SO$_4$ in glacial acetic acid. Rank these acids in order from strongest to weakest.

pK for HCl = 8.8 so $K = 10^{-8.8}$ and $K = 2 \times 10^{-9}$

pK for HClO$_4$ = 5.3 so $K = 10^{-5.3}$ and $K = 5 \times 10^{-6}$

pK for H$_2$SO$_4$ = 6.8 so $K = 10^{-6.8}$ and $K = 2 \times 10^{-7}$

Ranking: HClO$_4$ > H$_2$SO$_4$ > HCl

16.2.2. For glacial acetic acid:

(a) Autoionization of glacial acetic acid: 2 CH$_3$CO$_2$H $\rightleftarrows$ CH$_3$CO$_2^-$ + CH$_3$CO$_2$H$_2^+$

(b) Given the K for glacial acetic acid, the concentration of [CH$_3$CO$_2$H$_2^+$] = $\sqrt{3.2 \times 10^{-15}}$ = 5.7×10^{-8} M.

16.2.3. Write an equation for the reaction of the amide ion (a stronger base than OH$^-$) and water. Does equilibrium favor products or reactants?

NH$_2^-$(aq) + H$_2$O(ℓ) $\rightleftarrows$ NH$_3$(aq) + OH$^-$(aq); the stronger base extracts H$^+$ from H$_2$O.

As the amide ion is a much stronger base than the hydroxide ion, the equilibrium favors the products.

16.2.4. HClO$_4$ in glacial acetic acid is a weak conductor of electricity as it only partially ionizes. (See K in 16.2.2.b.)

16.2.5. To measure the relative strengths of bases stronger than OH$^-$, it is necessary to choose a solvent that is a weaker acid than water. One such solvent is liquid ammonia.

(a) Write a chemical equation for the autoionization of ammonia.
NH$_3$(aq) + NH$_3$(aq) $\rightleftarrows$ NH$_2^-$(aq) + NH$_4^+$(aq)

(b) What is the strongest acid and base that can exist in liquid ammonia?
The strongest acid that can exist will be NH$_4^+$; the strongest base will be NH$_2^-$.

(c) Will a solution of HCl in liquid ammonia be a strong electrical conductor, a weak conductor, or a nonconductor?

The HCl will react with the NH$_4^+$ ions to form the soluble salt, NH$_4$Cl, and make a **strongly conducting** solution.

(d) Oxide ion (O^{2-}) is a stronger base than the amide ion (NH$_2^-$).
Write an equation for the reaction of O^{2-} with NH$_3$ in liquid ammonia.
O^{2-}(aq) + NH$_3$(aq) $\rightleftarrows$ NH$_2^-$(aq) + OH$^-$(aq)
Since we have a stronger base (O^{2-}) forming a weaker base (NH$_2^-$), the equilibrium will favor the products.

Practicing Skills

The Brönsted Concept

16.1. | Conjugate base of: | Formula | Name |
|---|---|---|
| (a) HCN | CN⁻ | cyanide ion |
| (b) HSO₄⁻ | SO₄²⁻ | sulfate ion |
| (c) HF | F⁻ | fluoride ion |

16.2. | Conjugate base of: | Formula | Name |
|---|---|---|
| (a) HCO₂H | HCO₂⁻ | formate ion |
| (b) H₃PO₄ | H₂PO₄⁻ | dihydrogen phosphate ion |
| (c) HCO₃⁻ | CO₃²⁻ | carbonate ion |

16.3. | Conjugate acid of: | Formula | Name |
|---|---|---|
| (a) CO₃²⁻ | HCO₃⁻ | hydrogen carbonate ion |
| (b) ClO₂⁻ | HClO₂ | chlorous acid |
| (c) H₂PO₄⁻ | H₃PO₄ | phosphoric acid |

16.4. | Conjugate acid of: | Formula | Name |
|---|---|---|
| (a) NH₃ | NH₄⁺ | ammonium ion |
| (b) HCO₃⁻ | H₂CO₃ | carbonic acid |
| (c) Br⁻ | HBr | hydrobromic acid |

16.5. Products of acid-base reactions:

(a) HNO₃(aq)   +   H₂O(ℓ)   →   H₃O⁺(aq)   +   NO₃⁻(aq)

   acid   base   conjugate acid   conjugate base

(b) HSO₄⁻(aq) +   H₂O(ℓ)   →   H₃O⁺(aq)   +   SO₄²⁻(aq)

   acid   base   conjugate acid   conjugate base

(c) H₃O⁺(aq)   +   F⁻(aq)   →   HF(aq)   +   H₂O(ℓ)

   acid   base   conjugate acid   conjugate base

16.6. Products of the acid base reactions:

(a) $HClO_4$ + H_2O → H_3O^+ + ClO_4^-
    acid   base   conjugate acid   conjugate base

(b) NH_4^+ + H_2O → NH_3 + H_3O^+
    acid   base   conjugate base   conjugate acid

(c) HCO_3^- + OH^- → CO_3^{2-} + H_2O
    acid   base   conjugate base   conjugate acid

16.7. Hydrogen oxalate acting as Brönsted acid and Brönsted base:

Brönsted acid: $HC_2O_4^-(aq) + H_2O(\ell) \rightleftarrows C_2O_4^{2-}(aq) + H_3O^+(aq)$

Brönsted base: $HC_2O_4^-(aq) + H_2O(\ell) \rightleftarrows H_2C_2O_4(aq) + OH^-(aq)$

Hydrogen oxalate ion is an amphoteric (amphiprotic) substance. Note the characteristic of many such substances:

(a) a negative charge—making it attractive to positively charged hydronium ions, and

(b) the presence of an acidic H, making it capable of donating a proton—and acting as a Brönsted acid.

16.8. Equations showing HPO_4^{2-} acting as a Brönsted acid and as a Brönsted base:

Brönsted acid: $HPO_4^{2-}(aq) + H_2O(\ell) \rightleftarrows H_3O^+(aq) + PO_4^{3-}(aq)$

Brönsted base: $HPO_4^{2-}(aq) + H_2O(\ell) \rightleftarrows OH^-(aq) + H_2PO_4^-(aq)$

16.9. Identify Brönsted acid and base (on left) and conjugate partners (on right):

(a) $HCO_2H(aq) + H_2O(\ell) \rightleftarrows HCO_2^-(aq) + H_3O^+(aq)$

    acid   base   conjugate   conjugate
                                 of HCO_2H of H_2O

(b) $NH_3(aq) + H_2S(aq) \rightleftarrows NH_4^+(aq) + HS^-(aq)$

    base   acid   conjugate   conjugate
                                 of NH_3 of H_2S

(c) $HSO_4^-(aq) + OH^-(aq) \rightleftarrows SO_4^{2-}(aq) + H_2O(\ell)$

    acid   base   conjugate   conjugate
                                 of HSO_4^- of OH^-

16.10.
Brönsted acid	Brönsted base	conjugate base	conjugate acid
(a) HNO_2	C_5H_5N	NO_2^-	$C_5H_5NH^+$
(b) HCO_3^-	N_2H_4	CO_3^{2-}	$N_2H_5^+$
(c) $[Al(H_2O)_6]^{3+}$	OH^-	$[Al(H_2O)_5(OH)]^{2+}$	H_2O

pH Calculations

16.11. Since pH = 3.75, $[H_3O^+] = 10^{-pH}$ or $10^{-3.75}$ or 1.8×10^{-4} M. To solve for $[OH^-]$: pOH = 14.00 − pH = 14.00 − 3.75 = 10.25. $[OH^-] = 10^{-pOH} = 10^{-10.25} = 5.6 \times 10^{-11}$.

Since the $[H_3O^+]$ is greater than 1×10^{-7} (pH< 7), the solution is acidic.

16.12. Saturated solution of $Mg(OH)_2$:

hydronium ion: $[H_3O^+] = 10^{-pH} = 10^{-10.52} = 3.0 \times 10^{-11}$ M

hydroxide ion: $[OH^-] = \dfrac{K_w}{[H_3O^+]} = \dfrac{1.0 \times 10^{-14}}{3.0 \times 10^{-11}} = 3.3 \times 10^{-4}$ M

The solution is basic (pH > 7).

16.13. pH of a solution of 0.0075 M HCl:

Since HCl is considered a strong acid, a solution of 0.0075 M HCl has

$[H_3O^+]$ = 0.0075 or 7.5×10^{-3}; pH = $-\log[H_3O^+]$ = $-\log[7.5 \times 10^{-3}]$ = 2.12

The hydroxide ion concentration is readily determined since $[H_3O^+][OH^-] = 1.0 \times 10^{-14}$.

$[OH^-] = 1.0 \times 10^{-14}/7.5 \times 10^{-3} = 1.3 \times 10^{-12}$ M

16.14. pH of a solution of 4.5×10^{-3} M KOH:

KOH is a strong base so $[OH^-] = [KOH] = 4.5 \times 10^{-3}$

$[H_3O^+] = \dfrac{K_w}{[OH^-]} = \dfrac{1.0 \times 10^{-14}}{4.5 \times 10^{-3}} = 2.2 \times 10^{-12}$ M

pH = $-\log[H_3O^+]$ = $-\log(2.2 \times 10^{-12})$ = 11.65

16.15. pH of a solution of 0.0015 M $Ba(OH)_2$:

Soluble metal hydroxides are strong bases. To the extent that it dissolves [$Ba(OH)_2$ is not that soluble], $Ba(OH)_2$ gives two OH^- for each formula unit of $Ba(OH)_2$.

0.0015 M $Ba(OH)_2$ would provide 0.0030 M OH^-. Since $[H_3O^+] \times [OH^-] = 1.0 \times 10^{-14}$

[H₃O⁺] would then be:

$$[H_3O^+] = \frac{1.0 \times 10^{-14}}{3.0 \times 10^{-3}} = 3.3 \times 10^{-12} \text{ M and pH} = -\log[3.3 \times 10^{-12}] = 11.48.$$

16.16. Concentration of hydroxide ion in Ba(OH)₂ solution:

pH is 10.66, so pOH = 14.00 – pH = 3.34
[OH⁻] = 10⁻ᵖᴼᴴ = 10⁻³·³⁴ = 4.6 × 10⁻⁴ M
Mass of Ba(OH)₂ dissolved in 125 mL:

$$0.125 \text{ L} \left(\frac{4.6 \times 10^{-4} \text{ mol OH}^-}{1 \text{ L}}\right)\left(\frac{1 \text{ mol Ba(OH)}_2}{2 \text{ mol OH}^-}\right)\left(\frac{171.3 \text{ g Ba(OH)}_2}{1 \text{ mol Ba(OH)}_2}\right) = 4.9 \times 10^{-3} \text{ g}$$

16.17. KOH solution has pOH = 2.55:

What is pH? pH = 14.00 – pOH = 14.00 – 2.55 = 11.45

What is [H₃O⁺]? [H₃O⁺] = 10⁻ᵖᴴ = 10⁻¹¹·⁴⁵ = 3.5 × 10⁻¹² M

16.18 NH₃ solution has pOH = 5.81:

What is pH? pH = 14 – pOH = 14 – 5.81 = 8.19

What is [H₃O⁺]? [H₃O⁺] = 10⁻ᵖᴴ = 10⁻⁸·¹⁹ = 6.5 × 10⁻⁹ M

Equilibrium Constants for Acids and Bases

16.19. Equilibrium constant expression for the reaction of ammonium ion with water:
NH₄⁺ + H₂O ⇌ NH₃ + H₃O⁺

$$K = \frac{[NH_3][H_3O^+]}{[NH_4^+]}$$

16.20. Equilibrium constant expression for the reaction of methylamine with water:
CH₃NH₂ + H₂O ⇌ CH₃NH₃⁺ + OH⁻

$$K = \frac{[CH_3NH_3^+][OH^-]}{[CH_3NH_2]}$$

16.21. Concerning the following acids:

Phenol		Formic acid		Hydrogen oxalate ion	
C_6H_5OH	1.3×10^{-10}	HCO_2H	1.8×10^{-4}	$HC_2O_4^-$	6.4×10^{-5}

(a) The strongest acid is formic. The weakest acid is phenol. Acid strength is proportional to the magnitude of K_a.

(b) Since K_w is a constant, the greater the magnitude of K_a, the smaller the K_b value of the conjugate base. The strongest acid (HCO_2H) has the weakest conjugate base.

(c) The weakest acid (phenol) has the strongest conjugate base.

16.22. Regarding the acids listed:

(a) The strongest acid is HF (largest K_a) and the weakest acid is HPO_4^{2-} (smallest K_a).

(b) The conjugate base of the acid HF is F^-.

(c) The strongest acid (HF) has the weakest conjugate base (F^-).

(d) The weakest acid (HPO_4^{2-}) has the strongest conjugate base (PO_4^{3-}).

16.23. The substance with the smallest value for K_a will have the strongest conjugate base.

One can prove this quantitatively with the relationship: $K_a \times K_b = K_w$. An examination of Appendix H shows that, of these three substances, HClO has the smallest K_a, thus ClO^- is the strongest conjugate base.

16.24. Species in the list with the weakest conjugate base:
The strongest acid is H_2S (largest K_a), so it has the weakest conjugate base (HS^-).

16.25. Substance with the weakest conjugate acid: Write the species and their conjugate acids:

Specie	Conjugate Acid	K_a of conjugate acid
HCO_3^-	H_2CO_3	4.2×10^{-7}
F^-	HF	7.2×10^{-4}
NO_2^-	HNO_2	4.5×10^{-4}

The weakest conjugate acid has the smallest K_a value. Hence the hydrogen carbonate ion has the weakest conjugate acid.

16.26. Species in the list with the strongest conjugate acid:

Substance with the weakest conjugate acid: Write the species and their conjugate acids:

Specie	Conjugate Acid	K_a of conjugate acid
CN^-	HCN	4.0×10^{-10}
NH_3	NH_4^+	5.6×10^{-10}
ClO^-	HClO	3.5×10^{-8}

The strongest conjugate acid has the largest K_a value. Hence the hypochlorite ion has the strongest conjugate acid.

16.27. The equation for potassium carbonate dissolving in water:
$K_2CO_3(aq) \rightarrow 2\ K^+(aq) + CO_3^{2-}(aq)$

Soluble salts, like K_2CO_3, dissociate in water. The carbonate ion formed in this process is a base, and reacts with water:

$CO_3^{2-}(aq) + H_2O(\ell) \rightleftarrows HCO_3^-(aq) + OH^-(aq)$

The production of the hydroxide ion, a strong base, in this second step is responsible for the basic nature of solutions of water-soluble carbonates.

16.28. Equation for reaction of sodium hydrogen phosphate with water to produce a basic solution:

$Na_2HPO_4(aq) \rightarrow 2\ Na^+(aq) + HPO_4^{2-}(aq)$

The hydrogen phosphate ion is basic reacting with water to produce hydroxide ions, the ultimate source of the basic solution:

$HPO_4^{2-}(aq) + H_2O(\ell) \rightleftarrows H_2PO_4^-(aq) + OH^-(aq)$

16.29. Equation for reaction of hexaaquairon(III) ion with water to produce an acidic solution:

$[Fe(H_2O)_6]^{3+}(aq) + H_2O(\ell) \rightleftarrows [Fe(H_2O)_5(OH)]^{2+}(aq) + H_3O^+(aq)$

The small Fe^{3+} ion is acidic reacting with water to produce acidic hydronium ions in solution.

16.30. Balanced equation to produce an acidic solution when NH_4Br dissolves in water:
Ammonium bromide is soluble in water, producing $NH_4^+(aq)$ and $Br^-(aq)$.
$NH_4^+(aq) + H_2O(\ell) \rightleftarrows H_3O^+(aq) + NH_3(aq)$

16.31. Most of the salts shown are sodium salts. Since Na⁺ does not hydrolyze, we can estimate the acidity (or basicity) of such solutions by looking at the extent of reaction of the anions with water (hydrolysis).

The Al^{3+} ion is acidic, as is the $H_2PO_4^-$ ion. From Table 16.2 we see that the K_a for the hydrated aluminum ion is greater than that for the $H_2PO_4^-$ ion, making the Al^{3+} solution more acidic providing the lowest pH of these solutions. All the other salts will produce basic solutions and since the S^{2-} ion has the largest K_b, we anticipate that the Na_2S solution will be most basic having the highest pH.

16.32. Substance providing a basic solution:

(a) $NaNO_3$ —Neutral. Neither ion affects the pH of the solution.

(b) $NaC_7H_5O_2$ —Basic. Na⁺ has no effect on pH, but $C_7H_5O_2^-$, the conjugate base of the weak acid $HC_7H_5O_2$, makes the solution basic.

(c) Na_2HPO_4 —Basic. Na⁺ has no effect on pH. However, HPO_4^{2-} is amphiprotic; it can behave as an acid or a base. The value of K_b (1.6 × 10⁻⁷) is larger than K_a (3.6 × 10⁻¹³), so the solution will be basic.

pK_a: A Logarithmic Scale of Acid Strength

16.33. The pK_a for an acid with a K_a of 8.3 × 10⁻⁹. pK_a = –log(8.3 × 10⁻⁹) = 8.08

16.34. What is the pK_a?

pK_a = –log(K_a) = –log(6.4 × 10⁻⁷) = 6.19

16.35. K_a for epinephrine hydrochloride, whose pK_a = 9.53. K_a = 10⁻^{Ka} so K_a = 10⁻⁹·⁵³

or 3.0 × 10⁻¹⁰

From Table 16.2, we see that epinephrine belongs between:

Hexaaquairon(II) ion $Fe(H_2O)_6^{2+}$ 3.2 × 10⁻¹⁰

Hydrogen carbonate ion HCO_3^- 4.8 × 10⁻¹¹

16.36. What is the K_a?

K_a = 10⁻^{pKa} = 10⁻⁸·⁹⁵ = 1.1 × 10⁻⁹

The acid is weaker than [$Co(H_2O)_6^{2+}$] and stronger than $B(OH)_3H_2O$.

16.37. *2-chlorobenzoic acid* has a smaller pK_a than benzoic acid, so it is the stronger acid.

16.38. Identify the stronger of the two acids:

Acetic acid $pK_a = -\log(1.8 \times 10^{-5}) = 4.74$

The acid with the smaller pK_a has the larger K_a. (b) Chloroacetic acid is the stronger acid.

Ionization Constants for Weak Acids and Their Conjugate Bases

16.39. The K_b for the chloroacetate ion:

Recall the relationship between acids and their conjugate bases: $K_a \times K_b = K_w$

K_b for the chloroacetate ion will be $\dfrac{1.00 \times 10^{-14}}{1.41 \times 10^{-3}} = 7.1 \times 10^{-12}$

16.40. Calculate the K_a for the conjugate acid of the base with $K_b = 5.7 \times 10^{-8}$:

K_a for the conjugate acid: $K_a = \dfrac{K_w}{K_b} = \dfrac{1.0 \times 10^{-14}}{5.7 \times 10^{-8}} = 1.8 \times 10^{-7}$

16.41. The K_a for $(CH_3)_3NH^+$ is 10^{-pK_a} or $10^{-9.80} = 1.6 \times 10^{-10}$ so $K_b = \dfrac{1.0 \times 10^{-14}}{1.6 \times 10^{-10}} = 6.3 \times 10^{-5}$

16.42. What is K_b for the conjugate ion?

With $K_a = 10^{-pK_a} = 10^{-3.95} = 1.1 \times 10^{-4}$ then $K_b = \dfrac{K_w}{K_a} = \dfrac{1.0 \times 10^{-14}}{1.1 \times 10^{-4}} = 8.9 \times 10^{-11}$

Predicting the Direction of Acid-Base Reactions

16.43. $HNO_2(aq) + HCO_3^-(aq) \rightleftarrows NO_2^-(aq) + H_2CO_3(aq)$

Since nitrous acid is a stronger acid than carbonic acid (Table 16.2), the equilibrium lies predominantly to the right.

16.44. Balanced equation for the possible reaction between NH_4Cl and NaH_2PO_4:
$NH_4^+(aq) + H_2PO_4^-(aq) \rightleftarrows NH_3(aq) + H_3PO_4(aq)$
Since H_3PO_4 is a stronger acid than NH_4^+, the equilibrium will lie predominantly to the left.

16.45. Predict whether the equilibrium lies predominantly to the left or to the right:

(a) $CH_3CO_2H(aq) + Br^-(aq) \rightleftarrows CH_3CO_2^-(aq) + HBr(aq)$

HBr is a stronger acid than CH_3CO_2H; equilibrium lies to left.

(b) $H_3PO_4(aq) + F^-(aq) \rightleftarrows H_2PO_4^-(aq) + HF(aq)$

H_3PO_4 is a stronger acid than HF; equilibrium lies to right.

(c) $[Ni(H_2O)_6]^{2+}(aq) + HS^-(aq) \rightleftarrows [Ni(H_2O)_5(OH)]^+(aq) + H_2S(aq)$

H_2S is a stronger acid than $[Ni(H_2O)_6]^{2+}$; equilibrium lies to left.

16.46. Predict whether the equilibrium lies predominantly to the left or to the right:

(a) $HS^-(aq) + HCO_3^-(aq) \rightleftarrows H_2S(aq) + CO_3^{2-}(aq)$

H_2S is a stronger acid than HCO_3^-, so the equilibrium lies predominantly to the left.

(b) $CN^-(aq) + HSO_4^-(aq) \rightleftarrows HCN(aq) + SO_4^{2-}(aq)$

HSO_4^- is a stronger acid than HCN, so the equilibrium lies predominantly to the right.

(c) $HSO_4^-(aq) + CH_3CO_2^-(aq) \rightleftarrows SO_4^{2-}(aq) + CH_3CO_2H(aq)$

HSO_4^- is a stronger acid than CH_3CO_2H, so the equilibrium lies predominantly to the right.

16.47. For the reaction of sodium hydroxide with sodium hydrogen phosphate:

(a) The net ionic equation for the reaction:

$OH^-(aq) + HPO_4^{2-}(aq) \rightleftarrows H_2O(\ell) + PO_4^{3-}(aq)$

(b) The equilibrium lies to the right (since HPO_4^{2-} is a stronger acid than H_2O and OH^- is a stronger base than PO_4^{3-}), but phosphate and hydroxide ions aren't too different in basic strength, so the position of equilibrium does not lie very far to the right.

16.48. For the reaction of NaOCl(aq) and HCl(aq):

(a) Balanced net ionic equation: $H_3O^+(aq) + OCl^-(aq) \rightleftarrows H_2O(\ell) + HOCl(aq)$

(b) The equilibrium lies to the right (since H_3O^+ is a stronger acid than HClO and OCl^- is a stronger base than H_2O).

16.49. For the reaction of CH_3CO_2H with Na_2HPO_4:

(a) The net ionic equation: $CH_3CO_2H(aq) + HPO_4^{2-}(aq) \rightleftarrows CH_3CO_2^-(aq) + H_2PO_4^-(aq)$

(b) The equilibrium lies to the right (since CH_3CO_2H is a stronger acid than $H_2PO_4^-$ and HPO_4^{2-} is a stronger base than $CH_3CO_2^-$).

16.50. For the reaction of NH_3 (aq) with NaH_2PO_4:

(a) The net ionic equation: $NH_3(aq) + H_2PO_4^-(aq) \rightleftarrows NH_4^+(aq) + HPO_4^{2-}(aq)$

(b) The equilibrium will lie predominantly to the right as $H_2PO_4^-$ is a stronger acid than NH_4^+ and NH_3 is a stronger base than HPO_4^{2-}.

Using pH to Calculate Ionization Constants

16.51. For a 0.015 M HOCN, pH = 2.67:

(a) Since pH = 2.67, $[H_3O^+] = 10^{-pH}$ or $10^{-2.67} = 2.14 \times 10^{-3} = 2.1 \times 10^{-3}$ M (2 sf).

(b) The K_a for the acid is in general: $K_a = \dfrac{[H_3O^+][A^-]}{[HA]}$

The acid is monoprotic, meaning that for each H^+ ion, one also gets a OCN^- ion.

The two numerator terms are then equal. The concentration of the molecular acid is the (original concentration – concentration that dissociates).

$$K_a = \frac{[2.14 \times 10^{-3}][2.14 \times 10^{-3}]}{[0.015 - 2.14 \times 10^{-3}]} = 3.6 \times 10^{-4}$$

16.52. Calculate K_a for chloroacetic acid:

Calculate $[H_3O^+]$: $[H_3O^+] = [ClCH_2CO_2^-] = 10^{-pH} = 10^{-1.95} = 0.0112$ M

	ClCH$_2$CO$_2$H	ClCH$_2$CO$_2^-$	H$_3$O$^+$
Initial	0.10	0	0
Change	–0.0112	+0.0112	+0.0112
Equilibrium	0.0888	0.0112	0.0112

For the equilibrium: $ClCH_2CO_2H + H_2O \rightleftarrows ClCH_2CO_2^- + H_3O^+$ we insert the equilibrium concentrations into the K_a expression:

$$K_a = \frac{[0.0112][0.0112]}{[0.10 - 0.0112]} = 1.4 \times 10^{-3} = 1 \times 10^{-3}$$

The solution has only one significant figure because the denominator is only known to 1 significant figure.

16.53. With a pH = 9.11, the solution has a pOH of 4.89 and $[OH^-] = 10^{-4.89} = 1.29 \times 10^{-5}$ M.

The equation for the base in water can be written:

$H_2NOH(aq) + H_2O(\ell) \rightleftarrows H_3NOH^+(aq) + OH^-(aq)$

At equilibrium, $[H_2NOH] = [H_2NOH] - [OH^-] = (0.025 - 1.3 \times 10^{-5}) \approx 0.025$ M.

$$K_b = \frac{[H_3NOH^+][OH^-]}{[H_2NOH]} = \frac{(1.29 \times 10^{-3})^2}{0.025} = 6.6 \times 10^{-9}$$

16.54. The value of K_b for methylamine:

With a pH of 11.70, pOH = 14.00 − pH = 2.30

As methylamine reacts exactly as the hydroxylamine (SQ16.47),

$[CH_3NH_3^+] = [OH^-] = 10^{-pOH} = 10^{-2.30} = 0.0050$ M. Substituting into the K_b expression:

$$K_b = \frac{[CH_3NH_3^+][OH^-]}{[CH_3NH_2]} = \frac{(0.0050)(0.0050)}{(0.065 - 0.0050)} = 4.2 \times 10^{-4}$$

16.55. For the unknown acid:

(a) With a pH = 3.80 the solution has a $[H_3O^+] = 10^{-3.80}$ or 1.6×10^{-4} M

(b) Writing the equation for the unknown acid, HA, in water we obtain:
HA(aq) + H₂O(ℓ) ⇌ H₃O⁺(aq) + A⁻(aq)
$[H_3O^+] = 1.6 \times 10^{-4}$ implying that [A⁻] is also 1.6×10^{-4}. Therefore, the equilibrium concentration of acid, HA, is $(2.5 \times 10^{-3} - 1.6 \times 10^{-4})$ or ≈ 2.3×10^{-3}.

$$K_a = \frac{[H_3O^+][A^-]}{[HA]} = \frac{(1.6 \times 10^{-4})^2}{2.3 \times 10^{-3}} = 1.1 \times 10^{-5}$$

This acid is moderately weak.

16.56. For the base, calculate [H₃O⁺] and [OH⁻]:

(a) $[H_3O^+] = 10^{-pH} = 10^{-9.94} = 1.1 \times 10^{-10}$ M and $[OH^-] = \frac{K_w}{[H_3O^+]} = \frac{1.0 \times 10^{-14}}{1.1 \times 10^{-10}} = 9.1 \times 10^{-5}$

(b) Is the base strong, moderately weak, or weak: $K_b = \frac{(9.1 \times 10^{-5})(9.1 \times 10^{-5})}{(0.015 - 9.1 \times 10^{-5})} = 5.5 \times 10^{-7}$

The base is moderately weak.

Using Ionization Constants

16.57. For the equilibrium system: CH_3CO_2H (aq) + $H_2O(\ell) \rightleftarrows CH_3CO_2^- + H_3O^+$(aq)

The table of species:

	CH_3CO_2H	$CH_3CO_2^-$	H_3O^+
Initial	0.20 M		
Change	$-x$	$+x$	$+x$
Equilibrium	$0.20 - x$	x	x

Substituting into the K_a expression:

$$K_a = \frac{[H_3O^+][CH_3CO_2^-]}{[CH_3CO_2H]} = \frac{x^2}{0.20-x} = 1.8 \times 10^{-5}$$ and solving for $x = 1.9 \times 10^{-3}$

So $[CH_3CO_2H]$ = 0.20 M; $[CH_3CO_2^-]$ = $[H_3O^+]$ = 1.9×10^{-3} M

16.58. Equilibrium concentrations of all species in a 0.060 M solution of the acid HA:

	[HA]	[H$^+$]	[A$^-$]
Initial	0.040	0	0
Change	$-x$	$+x$	$+x$
Equilibrium	$0.040 - x$	x	x

The equation for the weak acid: HA + $H_2O \rightleftarrows A^- + H_3O^+$.

Substituting the equilibrium concentrations into the K_a expression for the weak acid:

$$K_a = \frac{[H_3O^+][A^-]}{[HA]} = \frac{x^2}{0.060 - x} = 8.0 \times 10^{-10}$$ and if we assume that x is much smaller than

0.060, the expression simplifies to $\frac{x^2}{0.060} = 8.0 \times 10^{-10}$ and solving for x:

$x = [A^-] = [H_3O^+] = 6.9 \times 10^{-6}$ M and [HA] = 0.060 M

16.59. HCN(aq) + H$_2$O(ℓ) $\rightleftarrows$ H$_3$O$^+$(aq) + CN$^-$(aq)

	HCN	H$_3$O$^+$	CN$^-$
Initial	0.025M		
Change	$-x$	$+x$	$+x$
Equilibrium	$0.025 - x$	x	x

Substituting these values into the K_a expression for HCN:

$$K_a = \frac{[H_3O^+][CN^-]}{[HCN]} = \frac{x^2}{0.025 - x} = 4.0 \times 10^{-10}$$

Assuming that the denominator may be approximated as 0.025 M, we obtain:

$$\frac{x^2}{0.025} = 4.0 \times 10^{-10} \text{ and } x = 3.2 \times 10^{-6}.$$

The equilibrium concentrations of [H$_3$O$^+$] = [CN$^-$] = 3.2 × 10^{-6} M.
The equilibrium concentration of [HCN] = (0.025 − 3.2 × 10^{-6}) or 0.025 M.
Since x represents [H$_3$O$^+$], the pH = −log(3.2 × 10^{-6}) = 5.50.

16.60. Phenol acts as a weak monoprotic acid (K_a = 1.3 × 10^{-10}):

The concentration of phenol is: $\left(\dfrac{0.195 \text{ g phenol}}{0.125 \text{ L}}\right)\left(\dfrac{1 \text{ mol phenol}}{94.11 \text{ g phenol}}\right) = 0.0166$ M

The equation for phenol in water: C$_6$H$_5$OH + H$_2$O $\rightleftarrows$ C$_6$H$_5$O$^-$ + H$_3$O$^+$

Our table:

	C$_6$H$_5$OH	C$_6$H$_5$O$^-$	H$_3$O$^+$
Initial	0.0166	0	0
Change	$-x$	$+x$	$+x$
Equilibrium	$0.0166 - x$	x	x

Substituting into the K_a expression: $K_a = 1.3 \times 10^{-10} = \dfrac{[C_6H_5O^-][H_3O^+]}{[C_6H_5OH]} = \dfrac{x^2}{0.0166 - x}$

Assuming x is much smaller than 0.0166, we simplify: $\dfrac{x^2}{0.0166} = 1.3 \times 10^{-10}$

and x = [H$_3$O$^+$] = 1.5 × 10^{-6} M, with a pH = −log[H$_3$O$^+$] = 5.83

16.61. The equilibrium of ammonia in water can be written: NH$_3$ + H$_2$O ⇌ NH$_4^+$ + OH$^-$

	NH$_3$	NH$_4^+$	OH$^-$
Initial	0.15 M	0	0
Change	$-x$	$+x$	$+x$
Equilibrium	$0.15 - x$	x	x

Substituting into the K_b expression: $K_b = \dfrac{[\text{NH}_4^+][\text{OH}^-]}{[\text{NH}_3]} = \dfrac{x^2}{(0.15-x)} = 1.8\times 10^{-5}$

With the value of K_b, the extent to which ammonia reacts with water is slight.
We can approximate the denominator (0.15 - x) as 0.15 M, and solve the equation:

$\dfrac{x^2}{0.15} = 1.8\times 10^{-5}$ and $x = 1.6\times 10^{-3}$

So [NH$_4^+$] = [OH$^-$] = 1.6 × 10^{-3} M and [NH$_3$] = (0.15 − 1.6 × 10^{-3}) ≈ 0.15 M
The pH of the solution: pOH = −log(1.6 × 10^{-3}) = 2.78
and pH = (14.00 − 2.78) = 11.22.

16.62. Equilibrium concentrations of all species in a 0.15 M solution of the base:

The equation can be written: B(aq) + H$_2$O(ℓ) ⇌ BH$^+$(aq) + OH$^-$(aq)

Substituting into the K_b expression: $K_b = \dfrac{[\text{BH}^+][\text{OH}^-]}{[\text{B}]} = \dfrac{x^2}{(0.25-x)} = 7.0\times 10^{-4}$

The assumption $x \ll$ [B]$_{initial}$ is not valid. Solve using the quadratic equation.
x = [BH$^+$] = [OH$^-$] = 1.3 × 10^{-2} M   [B] = 0.15 − x = 0.14 M

16.63. For CH$_3$NH$_2$ (aq) + H$_2$O(ℓ) ⇌ CH$_3$NH$_3^+$(aq) + OH$^-$(aq) K_b = 4.2 × 10^{-4}

Using the approach of SQ16.62, the equilibrium expression can be written:

$K_b = \dfrac{[\text{CH}_3\text{NH}_3^+][\text{OH}^-]}{[\text{CH}_3\text{NH}_2]} = \dfrac{x^2}{(0.25-x)} = 4.2\times 10^{-4}$

Solve using the quadratic equation gives x = 1.0 × 10^{-2}; [OH$^-$] = 1.0 × 10^{-2} M
pOH would then be 2.00 and pH 12.00.

16.64. For the base aniline:

$$K_b = \frac{[C_6H_5NH_3^+][OH^-]}{[C_6H_5NH_2]} = \frac{x^2}{(0.18-x)} \approx \frac{x^2}{0.18} = 4.0 \times 10^{-10} \text{ and } x = 8.5 \times 10^{-6}$$

With a K_b this small, we can simplify the denominator (0.18 – x).

$x = [OH^-] = 8.5 \times 10^{-6}$ M and pOH = –log[OH⁻] = 5.07; pH = 14.00 – pOH = 8.93

16.65. pH of 1.0×10^{-3} M HF; $K_a = 7.2 \times 10^{-4}$

Using the same method as in SQ16.61, we can write the expression:

$$K_a = \frac{[H_3O^+][F^-]}{[HF]} = \frac{x^2}{0.0010-x} = 7.2 \times 10^{-4}$$

The concentration of the HF and the magnitude of K_a preclude the use of our usual approximation: ($1.0 \times 10^{-3} - x \approx 1.0 \times 10^{-3}$)

So multiply both sides of the equation by the denominator to get:

$x^2 = 7.2 \times 10^{-4}(1.0 \times 10^{-3} - x)$ and $x^2 = 7.2 \times 10^{-7} - 7.2 \times 10^{-4}$ x

Using the quadratic equation, solve for x:

and $x = 5.6 \times 10^{-4}$ M = [F⁻] = [H₃O⁺] and pH = 3.25

16.66. Regarding a solution of hydrofluoric acid:

The equation for the reaction: HF(aq) + H₂O(ℓ) ⇌ H₃O⁺(aq) + F⁻(aq)

The equation shows that [F⁻] = [H₃O⁺] = 10^(–pH) = 10^(–2.30) = 0.00501 M = 0.0050 M (2 sf)

Substituting into the K_a expression:

$$K_a = \frac{[H_3O^+][F^-]}{[HF]} = \frac{(0.00501)(0.00501)}{[HF]} = 7.2 \times 10^{-4}$$

Solving for the equilibrium concentration of HF, [HF] = 0.0349 M = 0.035 M (2 sf). Now you can solve for the initial concentration of HF:
[HF]_{initial} = 0.0349 M + 0.00501 M = 0.040 M

Acid-Base Properties of Salts

16.67. Hydrolysis of the NH₄⁺ produces H₃O⁺ according to the equilibrium:
NH₄⁺(aq) + H₂O(ℓ) ⇌ H₃O⁺(aq) + NH₃ (aq)

With the ammonium ion acting as an acid, to donate a proton, we can write the K_a expression: $K_a = \dfrac{[H_3O^+][NH_3]}{[NH_4^+]} = \dfrac{K_w}{K_b} = \dfrac{1.0\times10^{-14}}{1.8\times10^{-5}} = 5.6\times10^{-10}$

The concentrations of both terms in the numerator are equal, and the concentration of ammonium ion is 0.20 M. (Note the approximation for the **equilibrium** concentration of NH_4^+ to be equal to the **initial** concentration.) Substituting and rearranging we get

$[H_3O^+] = \sqrt{0.20(5.6\times10^{-10})} = 1.1\times10^{-5}$ and the pH = $-\log(1.1\times10^{-5}) = 4.98$.

16.68. For a solution of 0.015 M sodium formate:

The equation for the reaction: $HCO_2^-(aq) + H_2O(\ell) \rightleftarrows HCO_2H(aq) + OH^-(aq)$

Solving for the K_b value (as we know the K_a): $K_b = \dfrac{K_w}{K_a} = \dfrac{1.0\times10^{-14}}{1.8\times10^{-4}} = 5.6\times10^{-11}$

Substituting into the K_b expression:

$K_b = \dfrac{[HCO_2H][OH^-]}{[HCO_2^-]} = \dfrac{x^2}{(0.015-x)} \approx \dfrac{x^2}{0.015} = 5.6\times10^{-11}$ and $x = 9.2\times10^{-7}$

$x = [OH^-] = 9.2\times10^{-7}$ M   pOH = $-\log(9.2\times10^{-7}) = 6.04$
pH = 14.00 − pOH = 14.00 − 6.04 = 7.96

16.69. The hydrolysis of CN^- produces OH^- according to the equilibrium:

The equation for the reaction: $CN^-(aq) + H_2O(\ell) \rightleftarrows HCN(aq) + OH^-(aq)$

Calculating the initial concentrations of Na^+ and CN^-:

$[Na^+] = [CN^-] = \dfrac{10.8 \text{ g NaCN}}{0.500 \text{ L}} \left(\dfrac{1 \text{ mol NaCN}}{49.01 \text{ g NaCN}}\right) = 0.441$ M

Then $K_b = \dfrac{K_w}{K_a} = \dfrac{1.0\times10^{-14}}{4.0\times10^{-10}} = 2.5\times10^{-5} = \dfrac{[HCN][OH^-]}{[CN-]}$

Substituting the $[CN^-]$ concentration into the K_b expression and noting that:

At equilibrium: $[OH^-] = [HCN]$ we rearrange the equation to solve for $[OH^-]$:

$[OH^-] = \sqrt{(2.5\times10^{-5})(4.41\times10^{-1})} = 3.3\times10^{-3}$ M and

$[H_3O^+] = \dfrac{1.0\times10^{-14}}{3.3\times10^{-3}} = 3.0\times10^{-12}$ M

16.70. Calculate ion concentrations in a 0.10 M solution of sodium propanoate:

The equation for the reaction: $CH_3CH_2CO_2^-(aq) + H_2O(\ell) \rightleftharpoons CH_3CH_2CO_2H(aq) + OH^-(aq)$

$$K_b = \frac{[CH_3CH_2CO_2H][OH^-]}{[CH_3CH_2CO_2^-]} = \frac{1.0 \times 10^{-14}}{1.3 \times 10^{-5}} = 7.7 \times 10^{-10}$$

$$\frac{[CH_3CH_2CO_2H][OH^-]}{[CH_3CH_2CO_2^-]} = \frac{x^2}{0.10-x} \approx \frac{x^2}{0.10} = 7.7 \times 10^{-10} \text{ and } x = 8.8 \times 10^{-6}$$

$x = [CH_3CH_2CO_2H] = [OH^-] = 8.8 \times 10^{-6}$ M so
pOH = $-\log[OH^-]$ = 5.06 and pH = 14.00 − pOH = 8.94
finally $[H_3O^+] = 10^{-pH} = 1.1 \times 10^{-9}$ M

pH after an Acid-Base Reaction

16.71. The net reaction is: $CH_3CO_2H(aq) + NaOH(aq) \rightleftharpoons CH_3CO_2^-(aq) + Na^+(aq) + H_2O(\ell)$

Adding 22.0 mL of 0.15 M NaOH (3.3 mmol NaOH) to 22.0 mL of 0.15 M CH_3CO_2H (3.3 mmol CH_3CO_2H) produces water and the soluble salt, sodium acetate (3.3 mmol $CH_3CO_2^-Na^+$). The acetate ion is the anion of a weak acid and reacts with water according to the equation: $CH_3CO_2^-(aq) + H_2O(\ell) \rightleftharpoons CH_3CO_2H(aq) + OH^-(aq)$

The equilibrium constant expression is: $K_b = \dfrac{[CH_3CH_2CO_2H][OH^-]}{[CH_3CH_2CO_2^-]} = 5.6 \times 10^{-10}$

The concentration of acetate ion is: $\dfrac{3.3 \text{ mmol}}{(22.0+22.0)\text{mL}} = 0.075$ M

	$CH_3CO_2^-$	CH_3CO_2H	OH^-
Initial concentration	0.075		
Change	−x	+x	+x
Equilibrium	0.075 − x	x	x

Substituting into the K_b expression:

$$\frac{[CH_3CH_2CO_2H][OH^-]}{[CH_3CH_2CO_2^-]} = \frac{x^2}{0.075-x} \approx \frac{x^2}{0.075} = 5.6 \times 10^{-10} \text{ and } x = 6.5 \times 10^{-6}$$

$[OH^-] = 6.5 \times 10^{-6}$. The hydronium ion concentration is related to the hydroxyl ion concentration by the equation:

$K_w = [H_3O^+][OH^-] = 1.0 \times 10^{-14}$

$[H_3O^+] = \dfrac{1.0 \times 10^{-14}}{[OH^-]} = \dfrac{1.0 \times 10^{-14}}{6.5 \times 10^{-6}} = 1.5 \times 10^{-9}$ and pH = 8.81

The pH is greater than 7, as expected for a salt of a strong base and weak acid.

16.72. The equation for the reaction: $NH_3(aq) + H_3O^+(aq) \rightarrow NH_4^+(aq) + H_2O(\ell)$
Initial amounts of acid and base:
(0.0500 L NH_3)(0.40 mol/L) = 0.020 mol NH_3
(0.0500 L HCl)(0.40 mol/L) = 0.020 mol HCl
Amount of NH_4Cl formed: 0.020 mol $NH_3 \times \dfrac{1 \text{ mol } NH_4^+}{1 \text{ mol } NH_3} = 0.020$ mol NH_4^+

$[NH_4^+] = \dfrac{0.020 \text{ mol}}{(0.0500 + 0.0500) \text{ L}} = 0.20$ M

The ammonium ion reacts with water: $NH_4^+(aq) + H_2O(\ell) \rightleftarrows NH_3(aq) + H_3O^+(aq)$

Substituting into the K_a expression: $K_a = \dfrac{[H_3O^+][NH_3]}{[NH_4^+]} = \dfrac{x^2}{0.20 - x} = 5.6 \times 10^{-10}$

Solving for $x = [H_3O^+] = 1.1 \times 10^{-5}$ M and pH = $-\log[H_3O^+]$ = 4.98

16.73. Equal numbers of moles of acid and base are added in each case, leaving only the salt of the acid and base. The reaction (if any) of that salt with water (hydrolysis) will affect the pH.

pH of solution	Reacting Species	Reaction controlling pH
(a) >7	CH_3CO_2H/KOH	Hydrolysis of $CH_3CO_2^-$
(b) <7	HCl/NH_3	Hydrolysis of NH_4^+
(c) = 7	HNO_3/NaOH	No hydrolysis

16.74.
pH of solution	Reacting Species	Reaction controlling pH
(a) pH > 7	H_2SO_4/NaOH	Hydrolysis of SO_4^{2-}
(b) pH > 7	HCOOH/NaOH	Hydrolysis of HCO_2^-
(c) pH > 7	$H_2C_2O_4$/NaOH	Hydrolysis of $C_2O_4^{2-}$

Polyprotic Acids and Bases

16.75. Chemical equilibrium expressions for the two ionizations of oxalic acid:
(1st ionization) $H_2C_2O_4(aq) + H_2O(\ell) \rightleftarrows HC_2O_4^-(aq) + H_3O^+(aq)$

$$K_{a1} = \frac{[HC_2O_4^-][H_3O^+]}{[H_2C_2O_4]}$$

(2nd ionization) $HC_2O_4^-(aq) + H_2O(\ell) \rightleftarrows C_2O_4^{2-}(aq) + H_3O^+(aq)$

$$K_{a2} = \frac{[C_2O_4^{2-}][H_3O^+]}{[HC_2O_4^-]}$$

16.76. Chemical equilibrium expressions for the two base reactions of Na_2CO_3 with water:

$CO_3^{2-}(aq) + H_2O(\ell) \rightleftarrows HCO_3^-(aq) + OH^-(aq)$

The equilibrium constant expression is:

$$K_{b1} = \frac{[HCO_3^-][OH^-]}{[CO_3^{2-}]}$$

The "second" reaction: $HCO_3^-(aq) + H_2O(\ell) \rightleftarrows H_2CO_3(aq) + OH^-(aq)$

The equilibrium constant expression is: $K_{b2} = \dfrac{[H_2CO_3][OH^-]}{[HCO_3^-]}$

16.77. Prove that $K_{a1} \times K_{b2} = K_w$ for $H_2C_2O_4$:

For the 1st ionization step in SQ16.75, the K_{a1} expression is: $K_{a1} = \dfrac{[HC_2O_4^-][H_3O^+]}{[H_2C_2O_4]}$

The equation for the **second step** of the oxalate ion reacting with water is:

$HC_2O_4^-(aq) + H_2O(\ell) \rightleftarrows OH^-(aq) + H_2C_2O_4(aq)$

the K_{b2} expression is: $K_{b2} = \dfrac{[H_2C_2O_4][OH^-]}{[HC_2O_4^-]}$.

Multiply the two K expressions:

$$K_{a1} \times K_{b2} = \frac{[HC_2O_4^-][H_3O^+]}{[H_2C_2O_4]} \cdot \frac{[H_2C_2O_4][OH^-]}{[HC_2O_4^-]} = [H_3O^+][OH^-] = K_w$$

16.78. Prove that $K_{a3} \times K_{b1} = K_w$ for H_3PO_4:

$HPO_4^{2-}(aq) + H_2O(\ell) \rightleftarrows PO_4^{3-}(aq) + H_3O^+(aq)$	$K_{a3} = 3.6 \times 10^{-13}$
$PO_4^{3-}(aq) + H_2O(\ell) \rightleftarrows HPO_4^{2-}(aq) + OH^-(aq)$	$K_{b1} = 2.8 \times 10^{-2}$
$2\ H_2O(\ell) \rightleftarrows H_3O^+(aq) + OH^-(aq)$	$K_w = K_{a3} \times K_{b1} = 1.0 \times 10^{-14}$

16.79. Regarding H_2SO_3:

(a) pH of 0.45 M H_2SO_3: The equilibria for the diprotic acid are:

$$K_{a1} = \frac{[HSO_3^-][H_3O^+]}{[H_2SO_3]} = 1.2 \times 10^{-2} \text{ and } K_{a2} = \frac{[SO_3^-][H_3O^+]}{[HSO_3^-]} = 6.2 \times 10^{-8}$$

The pH of a fully protonated polyprotic acid is almost entrely controlled by the first ionization (K_{a1}). For the first step of dissociation:

	H_2SO_3	HSO_3^-	H_3O^+
Initial concentration	0.45 M		
Change	$-x$	$+x$	$+x$
Equilibrium	$0.45 - x$	x	x

Substituting into the K_{a1} expression: $K_{a1} = \dfrac{(x)(x)}{0.45 - x} = 1.2 \times 10^{-2}$

You must solve this expression with the quadratic equation since $(0.45 < 100 \cdot K_{a1})$.

The equilibrium concentrations for HSO_3^- and H_3O^+ ions are found to be 0.0677 M.

The further dissociation is indicated by K_{a2}. It can be shown that the further ionization of the acid has almost no affect on the pH of the solution. Using the equilibrium concentrations from the first step, substitute into the K_{a2} expression.

	HSO_3^-	SO_3^{2-}	H_3O^+
Initial concentration	0.0677	0	0.0677
Change	$-x$	$+x$	$+x$
Equilibrium	$0.0677 - x$	x	$0.0677 + x$

$$K_{a2} = \frac{[SO_3^{2-}][H_3O^+]}{[HSO_3^-]} = \frac{(x)(0.0677 + x)}{(0.0677 - x)} = 6.2 \times 10^{-8}$$

Note that x will be small in comparison to 0.0677, and simplify the expression:

$$K_{a2} = \frac{(x)(0.0677)}{(0.0677)} = 6.2 \times 10^{-8}$$ In summary, the concentrations of HSO_3^- and H_3O^+ ions have been virtually unaffected by the second dissociation. So $[H_3O^+] = 0.0677$ M and pH = 1.17

(b) The equilibrium concentration of SO_3^{2-}:

From the K_{a2} expression above: $[SO_3^{2-}] = 6.2 \times 10^{-8}$ M

16.80. Regarding ascorbic acid:

The pH of the solution is determined by the first ionization of the acid.
$C_6H_8O_6(aq) + H_2O(\ell) \rightleftarrows C_6H_7O_6^-(aq) + H_3O^+(aq)$

The concentration of ascorbic acid: $\left(\dfrac{0.0050 \text{ g}}{0.0010 \text{ L}}\right)\left(\dfrac{1 \text{ mol } C_6H_8O_6}{176.1 \text{ g}}\right) = 0.028$ M

Substituting into the K_{a1} expression:

$$K_{a1} = \frac{[C_6H_7O_6^-][H_3O^+]}{[C_6H_8O_6]} = \frac{(x)^2}{(0.028-x)} \approx \frac{(x)^2}{(0.028)} = 6.8 \times 10^{-5}$$

solving for x yields: $x = [H_3O^+] = 1.4 \times 10^{-3}$ M and pH = $-\log[H_3O^+]$ = 2.86

16.81. For the reaction of hydrazine with water:

(a) Concentrations of OH^-, $N_2H_5^+$, and $N_2H_6^{2+}$ in 0.010 M N_2H_4:

The K_{b1} equilibrium allows us to calculate $N_2H_5^+$ and OH^- formed by the reaction of N_2H_4 with H_2O. $K_{b1} = \dfrac{[N_2H_5^+][OH^-]}{[N_2H_4]} = 8.5 \times 10^{-7}$

	N_2H_4	$N_2H_5^+$	OH^-
Initial concentration	0.010	0	0
Change	$-x$	$+x$	$+x$
Equilibrium	$0.010 - x$	x	x

Substituting into the K_{b1} expression: $K_{b1} = \dfrac{(x)(x)}{(0.010-x)} = 8.5 \times 10^{-7}$

You can simplify the denominator (0.010 > 100 x K_{b1}).

$$\frac{(x)(x)}{(0.010)} = 8.5 \times 10^{-7} \text{ and } x = 9.2 \times 10^{-5} \text{ M} = [\text{N}_2\text{H}_5^+] = [\text{OH}^-]$$

The second equilibrium (K_{b2}) indicates further reaction of the N_2H_5^+ ion with water. The step should consume some N_2H_5^+ and produce more OH^-.

The magnitude of K_{b2} indicates that the equilibrium "lies to the left" and we anticipate that not much $\text{N}_2\text{H}_6^{2+}$ (or additional OH^-) will be formed by this interaction.

	N_2H_5^+	$\text{N}_2\text{H}_6^{2+}$	OH^-
Initial concentration	9.2×10^{-5}	0	9.2×10^{-5}
Change	$-x$	$+x$	$+x$
Equilibrium	$9.2 \times 10^{-5} - x$	x	$9.2 \times 10^{-5} + x$

$$K_{b2} = \frac{[\text{N}_2\text{H}_6^{2+}][\text{OH}^-]}{[\text{N}_2\text{H}_5^+]} = 8.9 \times 10^{-16} = \frac{x(9.2 \times 10^{-5} + x)}{(9.2 \times 10^{-5} - x)}$$

Simplifying yields $\dfrac{x(9.2 \times 10^{-5})}{(9.2 \times 10^{-5})} = 8.9 \times 10^{-16}$ and $x = 8.9 \times 10^{-16}$

In summary, the second stage produces a negligible amount of OH^- and consumes very little N_2H_5^+ ion. The equilibrium concentrations are:

$[\text{N}_2\text{H}_5^+] = 9.2 \times 10^{-5}$ M; $[\text{N}_2\text{H}_6^{2+}] = 8.9 \times 10^{-16}$ M; $[\text{OH}^-] = 9.2 \times 10^{-5}$ M

(b) The pH of the 0.010 M solution: $[\text{OH}^-] = 9.2 \times 10^{-5}$ M so pOH = 4.04 and pH = 14.0 − 4.04 = 9.96

16.82. The reaction of ethylenediamine with water:

The pH of the solution is determined by the first ionization.

$\text{H}_2\text{NCH}_2\text{CH}_2\text{NH}_2(aq) + \text{H}_2\text{O}(\ell) \rightleftharpoons \text{H}_2\text{NCH}_2\text{CH}_2\text{NH}_3^+(aq) + \text{OH}^-(aq)$

$$K_{b1} = \frac{[\text{H}_2\text{NCH}_2\text{CH}_2\text{NH}_3^+][\text{OH}^-]}{[\text{H}_2\text{NCH}_2\text{CH}_2\text{NH}_2]} = \frac{(x)(x)}{(0.15 - x)} \approx \frac{x^2}{0.15} = 8.5 \times 10^{-5}$$

Solving for x: $x = [\text{OH}^-] = 3.6 \times 10^{-3}$ M the pOH = 2.45 and pH = 11.55.

As observed in SQ16.81, the concentration of the product of the second ionization, $[\text{H}_3\text{NCH}_2\text{CH}_2\text{NH}_3^{2+}]$, is equal to K_{b2} or 2.7×10^{-8} M.

Molecular Structure, Bonding, and Acid-Base Behavior

16.83. HOCN will be the stronger acid. In HOCN the proton is attached to the very electronegative O atom. This great electronegativity will provide a very polar bond, weakening the O–H bond, and making the H more acidic than in HCN.

16.84. The ion with the more highly charged metal ion, $[V(H_2O)_6]^{3+}$, should be the stronger acid.

16.85. Benzenesulfonic acid is a Brönsted acid owing to the inductive effect of three oxygen atoms attached to the S (and through the S to the benzene ring). These very electronegative O atoms will—through the inductive effect—remove electron density between the O and the H—weakening the OH bond making the H acidic.

16.86. Ethylenediamine can act as a proton acceptor (Brönsted base) **and** an electron pair donor (Lewis base).

Lewis Acid and Bases

16.87. Classify each of the following as a Lewis acid or Lewis base:

(a) H₂NOH	electron rich (accepts H⁺)	Lewis base
(b) Fe²⁺	electron poor	Lewis acid
(c) CH₃NH₂	electron rich (accepts H⁺)	Lewis base

16.88. Classify each of the following as a Lewis acid or Lewis base:

(a) BCl₃	electron poor	Lewis acid
(b) H₂NNH₂	electron rich (accepts H⁺)	Lewis base
(c) Ag⁺	electron poor	Lewis acid
NH₃	electron rich (accepts H⁺)	Lewis base

16.89. CO is a Lewis base (donates electron pairs) in complexes with nickel and iron.

16.90. BH₃ is a Lewis acid.

General Questions

16.91. For the equilibrium: $HC_9H_7O_4(aq) + H_2O(\ell) \rightleftarrows C_9H_7O_4^-(aq) + H_3O^+(aq)$

Write the K_a expression: $K_a = \dfrac{[C_9H_7O_4^-][H_3O^+]}{[HC_9H_7O_4]} = 3.27 \times 10^{-4}$

The initial concentration of aspirin is:

$\dfrac{2 \text{ tablets}}{0.225 \text{ L}} \left(\dfrac{0.325 \text{ g}}{1 \text{ tablet}} \right) \left(\dfrac{1 \text{ mol } HC_9H_7O_4}{180.16 \text{ g } HC_9H_7O_4} \right) = 1.60 \times 10^{-2} \text{ M}$

Substituting into the K_a expression:

$K_a = \dfrac{x^2}{1.60 \times 10^{-2} - x} = 3.27 \times 10^{-4}$

Using the quadratic equation, $[H_3O^+] = 2.13 \times 10^{-3}$ M and pH = 2.671.

16.92. Consider aqueous solutions of the ions:
   (a) NH_4^+ ions make an acidic solution, CO_3^{2-} and PO_4^{3-} ions make basic solutions.
   (b) Cl^- and NO_3^- have no effect on the pH of a solution.
   (c) PO_4^{3-} ions are the strongest base.
   (d) $CO_3^{2-}(aq) + H_2O(\ell) \rightleftarrows HCO_3^-(aq) + OH^-(aq)$
   $PO_4^{3-}(aq) + H_2O(\ell) \rightleftarrows HPO_4^{2-}(aq) + OH^-(aq)$

16.93. Calculate the molar mass of each base: $Ba(OH)_2$ = 171.3 g/mol and $Sr(OH)_2$ = 121.6 g/mol. The pH gives you information about the $[OH^-]$ since pH + pOH = 14.00. With pH = 12.61, pOH = 14.00 − 12.61 = 1.39 and $[OH^-] = 10^{-1.39}$ or 0.0407 M (0.041 to 2 sf). Note that two bases, as they totally dissolve, provide two moles of hydroxide ion per mol of the base. So, you can calculate the # of moles of base by dividing the hydroxide ion concentration by 2.

[Base] = ½(0.0407 M) = 0.0204 or 0.021 M (2 sf).

Recall that 2.50 g of the solid sample provided this concentration of base, so we can calculate the molar mass of the base: $\dfrac{2.50 \text{ g base}}{0.0204 \text{ mol base}} = 120$ g/mol (2 sf). Comparing the molar masses of the two possible bases, this solid sample is most likely $Sr(OH)_2$.

16.94. There are many ways to solve this problem. One option is to set up two equations with two unknowns.

$0.11 = \dfrac{[H_3O^+]}{[CH_3CO_2H]_{initial}}$ and $K_a = 1.8 \times 10^{-5} = \dfrac{[H_3O^+]^2}{[CH_3CO_2H]_{initial} - [H_3O^+]}$

Solve for [CH₃CO₂H]ᵢₙᵢₜᵢₐₗ : ([H₃O⁺]/0.11) = [CH₃CO₂H]ᵢₙᵢₜᵢₐₗ and substitute into the second equation. $K_a = 1.8 \times 10^{-5} = \dfrac{[H_3O^+]^2}{\dfrac{[H_3O^+]}{0.11} - [H_3O^+]}$

Solving for [H₃O⁺]: [H₃O⁺] = 1.5 × 10⁻⁴ M and the pH = −log[H₃O⁺] = 3.84
using the [H₃O⁺] value obtained: [CH₃CO₂H]ᵢₙᵢₜᵢₐₗ = [H₃O⁺]/0.11 = 1.3 × 10⁻³ M

$1.00 \text{ L} \left(\dfrac{1.3 \times 10^{-3} \text{ mol CH}_3\text{CO}_2\text{H}}{1.00 \text{ L}} \right) \left(\dfrac{60.1 \text{ g CH}_3\text{CO}_2\text{H}}{1 \text{ mol CH}_3\text{CO}_2\text{H}} \right) = 0.078 \text{ g CH}_3\text{CO}_2\text{H}$

16.95. The reaction between H₂S and NaCH₃CO₂:

H₂S(aq) + CH₃CO₂⁻(aq) ⇌ HS⁻(aq) + CH₃CO₂H(aq)

An examination of Table 16.2 reveals that CH₃CO₂H ($K_a = 1.8 \times 10^{-5}$) is a stronger acid than H₂S ($K_a = 1 \times 10^{-7}$), so the equilibrium will lie to the left (reactants).

16.96. Predict the position of equilibrium in the equations:

(a) HCO₃⁻ is a weaker acid than HSO₄⁻, so the equilibrium lies predominantly to the left.

(b) HSO₄⁻ is a stronger acid than CH₃CO₂H, so the equilibrium lies predominantly to the right.

(c) [Co(H₂O)₆²⁺] is a weaker acid than CH₃CO₂H, so the equilibrium lies predominantly to the left.

16.97. Monoprotic acid has $K_a = 1.3 \times 10^{-3}$. Equilibrium concentrations of HX, H₃O⁺ and pH for 0.010 M solution of HX:

	HX	H⁺	X⁻
Initial concentration	0.010	0	0
Change	−x	+x	+x
Equilibrium	0.010 − x	x	x

$K_a = \dfrac{[H_3O^+][A]}{[HA]} = \dfrac{x^2}{0.010 - x} = 1.3 \times 10^{-3}$

Since K_a and the concentration are of the same order of magnitude, the quadratic equation will help: $x^2 + 1.3 \times 10^{-3}x - 1.3 \times 10^{-5} = 0$.
The "reasonable solution" is $x = 3.0 \times 10^{-3}$

Concentrations are then: [HX] = 0.010 − x = 0.010 − 0.0030 = 7.0 × 10⁻³ M

[H⁺] = +x = 3.0 × 10⁻³ M   [X⁻] = +x = 3.0 × 10⁻³ M

pH = −log(3.0 × 10⁻³) = 2.52

16.98. Arrange the following solutions in order of increasing pH:

$$HCl < NH_4Cl < NaCl < NaCH_3CO_2 < KOH$$
—increasing pH→

The outliers are easy: HCl is a strong acid, and KOH is a strong base.

NaCl will produce a neutral solution. This leaves NH₄Cl and NaCH₃CO₂.

NH₄Cl will produce an acidic solution (with pH higher than that of HCl), owing to the reaction of the ammonium ion with water. NaCH₃CO₂ will produce a basic solution owing to the reaction of the acetate ion with water.

16.99. The pK_a of *m*-nitrophenol:

The K_a expression for this weak acid is: $K_a = \dfrac{[H^+][A^-]}{[HA]}$

The pH of a 0.010 M solution of nitrophenol is 3.44, so we know: $[H^+] = 3.63 \times 10^{-4}$

Since one A⁻ in is created for each H⁺ ion, the concentrations of the two are equal.

The [HA] = (0.010 − 3.63 × 10⁻⁴).

Substituting those values into the K_a expression yields:

$$K_a = \dfrac{[H^+][A^-]}{[HA]} = \dfrac{[3.63 \times 10^{-4}][3.63 \times 10^{-4}]}{[0.010 - 3.63 \times 10^{-4}]} = 1.4 \times 10^{-5} \text{ (2sf)}$$

pK_a = −log(1.4 × 10⁻⁵) = 4.86

16.100. Concerning the butylammonium ion:

(a) $K_b = \dfrac{K_w}{K_a} = \dfrac{1.0 \times 10^{-14}}{2.3 \times 10^{-11}} = 4.3 \times 10^{-4}$

(b) Using Table 16.2, you see that the acid is placed directly below [Ni(H₂O)₆]²⁺. HPO₄²⁻ is a weaker acid than C₄H₉NH₃⁺. PO₄³⁻ is a stronger base than C₄H₉NH₂.

(c) pH of 0.15 M solution of C₄H₉NH₃Cl:

Setup a K_a expression.:

$$K_a = 2.3 \times 10^{-11} = \dfrac{[C_4H_9NH_2][H_3O^+]}{[C_4H_9NH_3^+]} = \dfrac{x^2}{0.015 - x} \text{ and solve for } x$$

Note the assumption that $0.15 - x \approx 0.15$
$x = [H_3O^+] = 5.9 \times 10^{-7}$ M and pH = −log[H₃O⁺] = 6.23

16.101. For Novocain, the pK_a = 8.85. The pH of a 0.0015 M solution is:

First calculate K_a: $K_a = 10^{-8.85}$ so $K_a = 1.41 \times 10^{-9}$

Now treat Novocain as any weak acid, with the appropriate K_a expression:

$$K_a = \frac{[H^+][A^-]}{[HA]} = \frac{x^2}{0.0015 - x} = 1.4 \times 10^{-9}$$

Since K_a is so small, we can assume that the denominator is approximated by 0.0015 M.

$$\frac{x^2}{0.0015} = 1.4 \times 10^{-9} \text{ and } x^2 = 1.45 \times 10^{-6}$$

Since $[H^+] = 1.45 \times 10^{-6}$; pH = $-\log(1.45 \times 10^{-6}) = 5.84$

16.102. pH of 0.025 M solution of pyridinium hydrochloride:

The equation for the reaction: $C_5H_5NH^+(aq) + H_2O(\ell) \rightleftharpoons H_3O^+(aq) + C_5H_5N(aq)$

As $K_b(C_5H_5N) = 1.5 \times 10^{-9}$, the K_a for the conjugate acid: $K_a(C_5H_5NH^+) = \dfrac{K_w}{K_b} = 6.7 \times 10^{-6}$

Setting up the K_a expression: $K_a = \dfrac{[H_3O^+][C_5H_5N]}{[C_5H_5NH^+]} = \dfrac{x^2}{0.025 - x} \approx \dfrac{x^2}{0.025} = 6.7 \times 10^{-6}$

Given the magnitude of K and the concentration (0.025), we simplify (0.025 – x) to (0.025). Solving gives $x = [H_3O^+] = 4.1 \times 10^{-4}$ M and the pH = $-\log[H_3O^+] = 3.39$

16.103. Regarding ethylamine and ethanolamine:

(a) Since ethylamine has the larger K_b, ethylamine is the stronger base.

(b) The pH of 0.10 M solution of ethylamine:

The equation for the reaction: $C_2H_5NH_2 + H_2O \rightleftharpoons C_2H_5NH_3^+ + OH^-$

	$C_2H_5NH_2$	$C_2H_5NH_3^+$	OH^-
Initial concentration	0.10	0	0
Change	$-x$	$+x$	$+x$
Equilibrium	$0.10 - x$	x	x

Substituting into the K_b expression:

$$K_b = \frac{[OH^-][C_2H_5NH_3^+]}{[C_2H_5NH_2]} = \frac{x^2}{0.010 - x} = 4.3 \times 10^{-4}$$

$[OH^-] = 6.56 \times 10^{-3}$ so pOH = $-\log(6.56 \times 10^{-3}) = 2.18$ and pH = $14.00 - 2.18 = 11.82$

16.104. pH of a solution of chloroacetic acid:

The equation: $ClCH_2CO_2H(aq) + H_2O(\ell) \rightleftarrows ClCH_2CO_2^-(aq) + H_3O^+(aq)$

Initial concentration of acid:

$$[ClCH_2CO_2H] = \frac{0.0945 \text{ g}}{0.125 \text{ L}}\left(\frac{1 \text{ mol ClCH}_2\text{CO}_2\text{H}}{94.49 \text{ g ClCH}_2\text{CO}_2\text{H}}\right) = 8.00 \times 10^{-3} \text{ M}$$

$$K_a = \frac{[H_3O^+][ClCH_2CO_2^-]}{[ClCH_2CO_2H]} = \frac{x^2}{0.0080 - x} = 1.40 \times 10^{-3}$$

The magnitude of K_a and the solute concentration renders the approximation ($x \ll 0.00800$) invalid. So, we must solve using the quadratic equation.
$x = [H_3O^+] = 2.72 \times 10^{-3}$ M and the pH = $-\log[H_3O^+]$ = 2.566

16.105. With a pK_a = 2.32, the K_a for saccharin is $10^{-2.32}$ = 4.8 × 10^{-3}

The equation for the reaction is: $C_7H_4NO_3S^- + H_2O \rightleftarrows HC_7H_4NO_3S + OH^-$

As we are dealing with the conjugate base of saccharin, $C_7H_4NO_3S^-$, we use the K_b of the conjugate base.

Since $K_a \times K_b = K_w$, $K_b = \dfrac{K_w}{K_a} = \dfrac{1.0 \times 10^{-14}}{4.8 \times 10^{-3}} = 2.1 \times 10^{-12}$.

The equilibrium constant expression would be:

$$\frac{[HC_7H_4NO_3S][OH^-]}{[C_7H_4NO_3S^-]} = 2.1 \times 10^{-12}$$

If we let both the [OH$^-$] and [$C_7H_4NO_3S^-$] = x, the equilibrium concentration of the molecular acid would be 0.10 – x. Substituting into the expression above:

$\dfrac{x^2}{0.10} \approx \dfrac{x^2}{0.10} = 2.1 \times 10^{-12}$ and $x = 4.6 \times 10^{-7} = [OH^-]$

pOH = $-\log[4.6 \times 10^{-7}]$ or 6.34 and pH = 14.00 − 6.34 = 7.66.

16.106. Classify the solutions as acidic or basic:

(a) 0.1 M NH$_4$NO$_3$ (e) 0.1 M N$_2$H$_4$

(b) 0.1 M Na$_2$CO$_3$ (f) 0.1 M KCl

(c) 0.1 M NaF   (g) 0.1 M HCO$_2$H

(d) 0.1 M Na$_3$PO$_4$

(i) Acidic solutions: (d) 0.1 M NH$_4$NO$_3$ and (g) 0.1 M HCO$_2$H

(ii) Basic solutions: (b) 0.1 M Na$_2$CO$_3$, (c) 0.1 M NaF, (d) 0.1 M Na$_3$PO$_4$, and (e) 0.1 M N$_2$H$_4$

(iii) Most acidic solution: (g) 0.1 M HCO$_2$H, as formic acid is a stronger acid than NH$_4^+$

16.107. pH of aqueous solutions of

	reaction	pH
(a) NaHSO$_4$	hydrolysis of HSO$_4^-$ produces H$_3$O$^+$	< 7
(b) NH$_4$Br	hydrolysis of NH$_4^+$ produces H$_3$O$^+$	< 7
(c) KClO$_4$	no hydrolysis occurs	= 7
(d) Na$_2$CO$_3$	hydrolysis of CO$_3^{2-}$ produces OH$^-$	> 7
(e) (NH$_4$)$_2$S	hydrolysis of S^{2-} produces OH$^-$	> 7
	(NH$_4^+$ is a weak acid, but S^{2-} is a stronger base)	
(f) NaNO$_3$	no hydrolysis occurs	= 7
(g) Na$_2$HPO$_4$	hydrolysis of HPO$_4^{2-}$ produces OH$^-$	> 7
	(HPO$_4^{2-}$ is amphiprotic, but K_{b2} is greater than K_{a2})	
(h) LiBr	no hydrolysis occurs	= 7
(i) FeCl$_3$	hydrolysis of Fe^{3+} produces H$_3$O$^+$	< 7

Highest pH = (NH$_4$)$_2$S; Lowest pH = NaHSO$_4$

16.108. Approximate pH of a 0.020 M solution of nicotine:

As has been observed in many pH problems with diprotic weak acids, or dibasic weak bases, the pH of the solution is determined by the first ionization of the base. To simplify the equation, we'll use Nic to represent the base:
Using K_{b1}, we write the equilibrium expression:

$$K_{b1} = \frac{[OH^-][NicH^+]}{[Nic]} = \frac{x^2}{0.020 - x} = 7.0 \times 10^{-7}$$

Given the magnitude of K compared to the solute concentrations, we can assume that $(0.020 - x) \approx (0.020)$, and solve for x: $x = [OH^-] = 1.2 \times 10^{-4}$ M and pOH = $-\log[OH^-] = 3.93$
pH = 14.00 − pOH = 10.07

16.109. For oxalic acid $K_{a1} = 5.9 \times 10^{-2}$ and $K_{a2} = 6.4 \times 10^{-5}$

1st step: H$_2$C$_2$O$_4$ + H$_2$O ⇌ HC$_2$O$_4^-$ + H$_3$O$^+$	$K_{a1} = 5.9 \times 10^{-2}$
2nd step: HC$_2$O$_4^-$ + H$_2$O ⇌ C$_2$O$_4^{2-}$ + H$_3$O$^+$	$K_{a2} = 6.4 \times 10^{-5}$
Sum: H$_2$C$_2$O$_4$ + 2 H$_2$O ⇌ C$_2$O$_4^{2-}$ + 2 H$_3$O$^+$	$K_{net} = (5.9 \times 10^{-2})(6.4 \times 10^{-5}) = 3.8 \times 10^{-6}$

16.110. Confirm the equilibrium constant for the reaction of HCl with NH$_3$:

NH$_3$(aq) + H$_2$O(ℓ) ⇌ NH$_4^+$(aq) + OH$^-$(aq) K_b = 1.8 × 10^{-5}

OH$^-$(aq) + H$_3$O$^+$(aq) ⇌ 2 H$_2$O(ℓ) K = 1/K_w

Sum: NH$_3$(aq) + H$_3$O$^+$(aq) ⇌ H$_2$O(ℓ) + NH$_4^+$(aq) $K = \dfrac{1.8 \times 10^{-5}}{1.0 \times 10^{-14}} = 1.8 \times 10^9$

16.111. Confirm that 1.8 × 10^{10} is the value of the equilibrium constant for the reaction of formic acid and sodium hydroxide:

To do this, you need two equations that have a **net** equation that corresponds to:

HCO$_2$H(aq) + OH$^-$(aq) ⇌ HCO$_2^-$(aq) + H$_2$O (ℓ)

Begin with equilibria associated with the weak acid in water, and the K_w for water:

HCO$_2$H(aq) + H$_2$O (ℓ) ⇌ HCO$_2^-$(aq) + H$_3$O$^+$(aq) K_a = 1.8 × 10^{-4}

H$_3$O$^+$(aq) + OH$^-$(aq) ⇌ 2 H$_2$O(ℓ)   1/K_w = 1.0 × 10^{14}

sum: HCO$_2$H(aq) + OH$^-$(aq) ⇌ HCO$_2^-$(aq) + H$_2$O(ℓ) and
the net K is: $K_a \cdot 1/K_w$ = (1.8 × 10^{-4})(1.0 × 10^{14}) = 1.8 × 10^{10}

16.112. pH of solution when formic acid and sodium hydroxide are mixed:

The equation for the reaction: HCO$_2$H(aq) + OH$^-$(aq) → HCO$_2^-$(aq) + H$_2$O(ℓ)
Amounts of reacting species:

(0.0250 L HCO$_2$H)(0.14 mol/L) = 0.0035 mol HCO$_2$H
(0.0500 L NaOH)(0.070 mol/L) = 0.0035 mol NaOH
As we have equal numbers of moles of acid and base, we are at the equivalence point, so the major species responsible for pH is the salt, in this case the formate ion:

The concentration of the formate ion: [HCO$_2^-$] = $\dfrac{0.0035 \text{ mol}}{(0.0250 + 0.0500) \text{ L}}$ = 0.047 M

The reaction of formate ion with water: HCO$_2^-$(aq) + H$_2$O(ℓ) ⇌ HCO$_2$H(aq) + OH$^-$(aq)

Using the K_a of formic acid from Table 16.2, calculate K_b:

$K_b = \dfrac{K_w}{K_a} = \dfrac{1.0 \times 10^{-14}}{1.8 \times 10^{-4}} = 5.6 \times 10^{-11}$ Now substitute concentrations into the K expression:

$\dfrac{[\text{HCO}_2\text{H}][\text{OH}^-]}{[\text{HCO}_2^-]} = \dfrac{x^2}{0.047 - x} = 5.6 \times 10^{-11}$ and using the approximation that

$(0.047 - x) \approx (0.047)$, $\dfrac{x^2}{0.047} = 5.6 \times 10^{-11}$ and $x = 1.6 \times 10^{-6} = [\text{OH}^-]$

So $[H_3O^+] = \dfrac{K_w}{[OH^-]} = 6.2 \times 10^{-9}$ M and pH = $-\log[H_3O^+]$ = 8.21

16.113. Volume to which 1.00×10^2 mL of a 0.20 M solution of a weak acid, HA, should be diluted to result in a doubling of the percent ionization:

Begin by writing the K_a expression: $K = \dfrac{[H^+][A^-]}{[HA]}$ (1)

The fraction of dissociation (which you want to double) is $\dfrac{[A^-]}{[HA]}$ or equivalently:

$\dfrac{[H^+]}{[HA]}$ (2)

Knowing that for a monoprotic acid, $[H^+] = [A^-]$, you can rearrange equation (1) to yield:

$[H^+]^2 = K_a[HA - H^+]$, where $[HA - H^+]$ represents the concentration of molecular acid.

Expanding gives: $[H^+]^2 = K_a[HA] - K_a[H^+]$, and setting this up as a quadratic equation gives: $[H^+]^2 + K_a[H^+] - K_a[HA] = 0$.

For the sake of argument, pick a value for K_a, say 1.0×10^{-5}.

Given the original concentration of acid is 0.2 M, substitute the values into the equation.

$1[H^+]^2 + (1.0 \times 10^{-5})[H^+] - (1.0 \times 10^{-5})[0.2] = 0$, and solve the equation, call it (3).

This gives $[H^+] = 1.41 \times 10^{-3}$ M. The fraction of dissociation is (from equation (2) above)

$\dfrac{[H^+]}{[HA]}$ or 0.00705. Now you want this **fraction to double**. So, substitute differing concentrations of acid (increasingly dilute) into equation (3), solve the quadratic equation that results, and calculate the fraction dissociated.

The table following shows such a series of calculations:

Acid concentration	fraction	Relative dissociation
0.20	0.00704611	1.0
0.15	0.0081317	1.2
0.10	0.00995012	1.4
0.05	0.01404249	2.0

Note that when the acid concentration falls to 1/4 of the original value, the relative dissociation has doubled. So, 100 mL of the acid would need to be diluted to 400 mL to double the percentage of dissociation.

16.114. pH of a 0.050 M solution of potassium hydrogen phthalate:

$[H_3O^+] = \sqrt{(1.12 \times 10^{-3})(3.91 \times 10^{-6})} = 6.62 \times 10^{-5}$ M so, pH = $-\log[H_3O^+]$ = 4.179

16.115. Oxalic acid, $H_2C_2O_4$, is diprotic, with K values: $K_{a1} = 5.9 \times 10^{-2}$ and $K_{a2} = 6.4 \times 10^{-5}$.

The K_{a1} equilibrium allows us to calculate $HC_2O_4^-$ and H_3O^+ formed by the reaction of $H_2C_2O_4$ with H_2O. $K_{a1} = \dfrac{[H_3O^+][HC_2O_4^-]}{[H_2C_2O_4]} = 5.9 \times 10^{-2}$

	$H_2C_2O_4$	$HC_2O_4^-$	H_3O^+
Initial concentration	0.10	0	
Change	$-x$	$+x$	$+x$
Equilibrium	$0.10 - x$	x	x

Substituting into the K_{a1} expression we obtain: $\dfrac{x^2}{0.10 - x} = 5.9 \times 10^{-2}$

We cannot simplify the denominator, since (0.10 < 100 × K_{a1}).

Solving via the quadratic equation, we obtain $x = 5.3 \times 10^{-2}$.

$[H_2C_2O_4] = 0.10 - x = 0.05$ M; $[HC_2O_4^-] = 5.3 \times 10^{-2}$ M and $[H_3O^+] = 5.3 \times 10^{-2}$ M

Very little H_3O^+ and $C_2O_4^{2-}$ are formed from the second ionization, but the reaction does slightly decrease the $HC_2O_4^-$ concentration and slightly increase the H_3O^+ concentration. So $[H_3O^+]$ is slightly greater than $[HC_2O_4^-]$ and $[C_2O_4^{2-}] = K_{a2} = 6.4 \times 10^{-5}$. The major species present, in decreasing concentrations are:

$H_2O > H_2C_2O_4 > H_3O^+ > HC_2O_4^- > C_2O_4^{2-} > OH^-$

16.116. Molecules and ions present in solution in a 0.10 M sodium oxalate solution in order of decreasing concentration.

$H_2O > Na^+ > C_2O_4^{2-} > OH^- > HC_2O_4^- > H_2C_2O_4 > H_3O^+$

As in SQ16.115, the concentrations of OH^- and $HC_2O_4^-$ are nearly equal. But, the slight ionization of the hydrogen oxalate ion creates additional hydroxide ion.

16.117. Molecules and ions in solution when equimolar amounts of HCl and NaF are mixed in order of decreasing concentration.

$H_2O > Na^+ = Cl^- > HF > F^- = OH^-$

16.118. Molecules and ions that exist in a solution when NaOH and acetic acid are mixed:

As in SQ16.115, an equal number of moles of NaOH and CH_3CO_2H are mixed.

So, the principal species are those of the salt, $NaCH_3CO_2$.

As a result, the ions/molecules are: $H_2O > Na^+ > CH_3CO_2^- > CH_3CO_2H = OH^- > H_3O^+$

—decreasing concentration→

In the Laboratory

16.119. To determine the relative basic strengths of the substances, one can measure the pH of aqueous solutions of an equal concentration (e.g., 0.50 M) of each solute. Remember the equilibrium associated with bases in water:

$$B + H_2O \rightleftarrows BH^+ + OH^-$$

The strongest base will have an equilibrium lying farther "to the right" than the weaker bases, resulting in a solution with the *greatest* concentration of hydroxide ions (and the greatest pH). The weakest base will form a solution with the *lowest* concentration of hydroxide ions (and the lowest pH).

16.120. Consider the relative strengths of the acids given:

(a) The increasing acidity as Br atoms replace H atoms is due to the inductive effect of the Br atoms.

(b) The strongest acid (Br_3CCO_2H) will have the lowest pH, and the weakest acid (CH_3CO_2H) will have the highest pH.

16.121. This puzzle is a great way to test your chemical knowledge. Organize what you know.

Cations	Anions
Na^+	Cl^-
NH_4^+	OH^-
H^+	

Experimentally we observe:

B + Y → acidic solution   B + Z → basic solution   A + Z → neutral solution

If A + Z give a neutral solution, then A + Z must be acid and base (in some order).

Since B + Z gives a basic solution, B + Z cannot be acid and base (as A + Z), otherwise B + Z would be neutral (as A + Z). Z must be basic (OH^-), meaning Y must be neutral (Cl⁻). Since B + Y gives an acid solution, then B must contain NH_4^+. If B contains ammonium, then A must contain H⁺, and C must contain Na⁺.

In summary then: A = H⁺; B = NH_4^+ ; C = Na⁺ ; Y = Cl⁻; Z = OH⁻.

Pairing these cations with chloride and potassium ions, we have:

A = HCl; B = NH₄Cl; C = NaCl; Y = KCl; Z = KOH.

16.122. Consider the substituted pyridines:

(a) The $CH_3C_5H_4NH^+$ solution would have the highest pH (smallest K_a value, weakest conjugate acid). The $NO_2C_5H_4NH^+$ solution would have the lowest pH (largest K_a value, strongest conjugate acid).

(b) The strongest conjugate acid ($NO_2C_5H_4NH^+$) has the weakest Brønsted base ($NO_2C_5H_4N$). The weakest conjugate acid ($CH_3C_5H_4NH^+$) has the strongest Brønsted base ($CH_3C_5H_4N$).

16.123. The K_a expression for nicotinic acid in water is:

$C_6H_5NO_2 + H_2O \rightleftharpoons C_6H_4NO_2^- + H_3O^+$

The pH allows us to determine the [H₃O⁺]; pH = $10^{-2.70}$ = 2.0 × 10⁻³ M. As this is a monoprotic acid, we also know that, at equilibrium the [$C_6H_4NO_2^-$] = 2.0 × 10⁻³ M. The concentration of acid that has ionized is: 2.0 × 10⁻³ M.

Calculating the initial concentration of nicotinic acid:

For $C_6H_5NO_2$ the molar mass = 123.1 g, so

$$[C_6H_5NO_2] = \frac{1.00 \text{ g}}{0.060 \text{ L}} \left(\frac{1 \text{ mol } C_6H_5NO_2}{123.1 \text{ g } C_6H_5NO_2} \right) = 0.14 \text{ M (2sf)}$$

Substituting into the K_a expression, we obtain:

$$\frac{[C_6H_4NO_2^-][H_3O^+]}{[C_6H_5NO_2]} = \frac{(2.0 \times 10^{-3})^2}{(0.14 - 2.0 \times 10^{-3})} = 3.0 \times 10^{-5}$$

16.124. Consider the dimethyl ether complex of BF_3:

(a) Describe the products: BF_3 is a Lewis acid, $(CH_3)_2O$ is a Lewis base.

(b) Pressure when equilibrium is established:

$$1.00 \text{ g } (CH_3)_2 \text{O-BF}_3 \left(\frac{1 \text{ mol } (CH_3)_2 \text{O-BF}_3}{113.9 \text{ g}(CH_3)_2 \text{O-BF}_3} \right) = 0.00878 \text{ mol } (CH_3)_2 \text{O-BF}_3$$

$$P_{(CH_3)_2 \text{O-BF}_3} = \frac{nRT}{V} = \frac{(0.00878 \text{ mol})(0.082057 \text{ L} \cdot \text{atm}/\text{K} \cdot \text{mol})(398 \text{ K})}{0.565 \text{ L}} = 0.508 \text{ atm}$$

Use this pressure to solve for pressures of BF$_3$ and (CH$_3$)$_2$O:

$$K_p = \frac{P_{BF_3} P_{(CH_3)_2O}}{P_{(CH_3)_2O\text{-}BF_3}} = \frac{x^2}{0.508-x} = 0.17 \text{ ; Solve using the quadratic equation.}$$

$x = P_{BF_3} = P_{(CH_3)_2O} = 0.22$ atm and $P_{(CH_3)_2OBF_3} = 0.508 - x = 0.29$ atm

P$_{total}$ = 0.22 atm + 0.22 atm + 0.29 atm = 0.73 atm

16.125. Regarding aniline:

(a) Aniline can be **either** a Brönsted base **or** a Lewis base:

(b) The pH of a solution of 1.25 g sodium sulfanilate in 125 mL: Sulfanilic acid is a weak acid, so its conjugate base should provide a slightly basic solution.
The pK_a = 3.23 so K_a = 10$^{-3.23}$ = 5.9 × 10^{-4}.

The formula for sulfanilic acid is complex (H$_2$NC$_6$H$_4$SO$_3$H), so use HSA to represent the acid, and SA to represent the anion.

The concentration of the sodium salt (represented as NaSA) will be needed:

$$\frac{1.25 \text{ g NaSA}}{0.125 \text{ L}} \left(\frac{1 \text{ mol NaSA}}{195.17 \text{ gNaSA}} \right) = 0.0512 \text{ mol NaSA}$$

In water, the NaSA salt will exist predominantly as sodium cations and SA anions. We can ignore the sodium cations. The hydrolysis reaction that occurs with the conjugate base and water is: SA + H$_2$O ⇌ HSA + OH$^-$

(Note the production of hydroxyl ion – the reason we expect the solution to be basic.)

As the anion is acting as a base, we'll need to calculate an appropriate K for the anion.

$$K_{b(\text{conjugate})} = \frac{1.0 \times 10^{-14}}{5.9 \times 10^{-4}} = 1.7 \times 10^{-11}$$

The K_b expression is: $K_{b(conjugate)} = \dfrac{[OH^-][HSA]}{[SA^-]} = \dfrac{x^2}{0.0512-x} = 1.7 \times 10^{-11}$

The size of K ($100 \times K < 0.0512$) tells us that we can safely approximate the denominator of the fraction as 0.0512.

The resulting equation is: $x^2 = (0.0512)(1.7 \times 10^{-11})$ and
$x = 9.32 \times 10^{-7}$ Since x represents the concentration of OH^- ion,
the pOH $= -\log(9.32 \times 10^{-7}) = 6.03$, and the pH $= 7.97$.

16.126. pH of 0.20 M solution of alanine hydrochloride: $K_a = 10^{-2.4} = 4 \times 10^{-3}$

$$K_a = \dfrac{[H_3O^+][NH_3CHCH_3CO_2^-]}{[NH_3CHCH_3CO_2H]} = \dfrac{x^2}{0.20-x} = 4 \times 10^{-3}$$

Given the magnitude of K and the concentration of solute, the approximation $x \ll 0.20$ is not valid, so we solve using the quadratic equation.
$x = [H_3O^+] = 0.03$ M and the pH $= -\log[H_3O^+] = 1.6$

Summary and Conceptual Questions

16.127. Water can be both a Brönsted base and a Lewis base. The Brönsted system requires that a base be able to accept a H^+ ion. Water does that in the formation of the H_3O^+ ion. The Lewis system defines a base as an electron pair donor. The two lone pairs of electrons on the O atom provide a source for those electrons. Examine the reaction of H_2O with H^+. The hydrogen ion accepts the electron pairs from the O atom of the water molecule—fulfilling water's role as a Lewis base and the H^+'s role as a Lewis acid.

A Brönsted acid furnishes H^+ to another specie. The autoionization of water provides one example of water acting as a Brönsted acid. Water however cannot function as a Lewis acid, as it has no capacity to accept an electron pair.

16.128. Explain the acidity of a Ni(II) solution:
The polar Ni—OH_2 bonds result in polarized O—H bonds, and the H^+ ion can be removed by H_2O in the reaction of $[Ni(H_2O)_6^{2+}]$ with H_2O.
$[Ni(H_2O)_6]^{2+}(aq) + H_2O(\ell) \rightleftarrows [Ni(H_2O)_5(OH)]^+(aq) + H_3O^+(aq)$

16.129. Concerning the three weak acids, HOX:
(a) The strongest acid of the three oxyacids is HOCl, with the pK_a = 7.46. Recall that the magnitude of the K_a indicates the degree of ionization for the acid. The greater the ionization, the greater the K_a, and the smaller the pK_a.

(b) The change in acid strength is understandable if you recall that in changing from I to Br to Cl, the electronegativity of the halogen increases, strengthening the O–X bond (where X represents the halogen) and reducing the electron density (and weakening) the H–O bond, making the acid stronger.

16.130. Trends in acidity for oxoacids:

(a) General trends: The acidity increases as the number of single oxygen atoms (not –OH groups) bonded to the central atom increases.

(b) Greater effect on acidity: The number of oxygen atoms directly bonded to the central atom has a greater effect on acidity than the number of –OH groups.

(c) Correlation of formal charge with acidity: As the formal charge on the central atom increases (becomes more positive) the acidity increases.

(d) Predicted K_a values: P(OH)$_3$ would have a pK_a around 7–8. The hydrogen atoms directly attached to oxygen atoms are lost as H$^+$ ions.

16.131. For perchloric acid:

(a) The reaction of perchloric acid with sulfuric acid:

HClO$_4$ + H$_2$SO$_4$ ⇌ ClO$_4^-$ + H$_3$SO$_4^+$

(b) Lewis dot structure for sulfuric acid:

Sulfuric acid has two oxygen atoms capable of "donating" an electron pair, to a proton, H$^+$, thereby acting as a base.

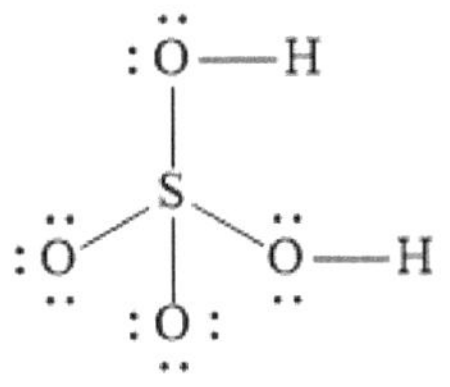

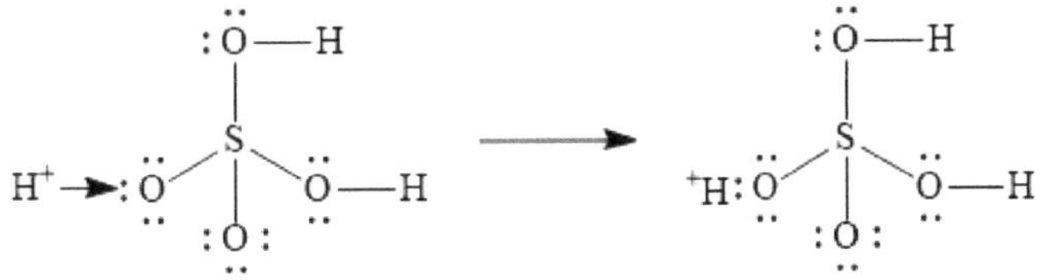

16.132. Why might a bottle of water be slightly acidic?

Dissolved CO$_2$(g) or dissolved metal ions can cause bottled water to be slightly acidic.

16.133. For the reaction of I$_2$ in an I$^-$ solution:

(a) Electron dot structure for I$_3^-$:

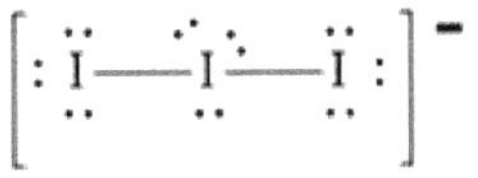

(b) I$^-$ + I$_2$ → I$_3^-$ in which I$^-$ donates the electron pair (Lewis acid), and I$_2$ accepts the electron pair (Lewis base).

16.134. Sites of H-bonding and sites of Lewis basicity:

Hydrogen bonding is possible at the carbonyl (C=O) oxygen atoms and the amine (N–H) nitrogen and hydrogen atoms. The amine nitrogens and carbonyl oxygens can act as Lewis bases. The carbon atom marked with * has a partial negative charge.

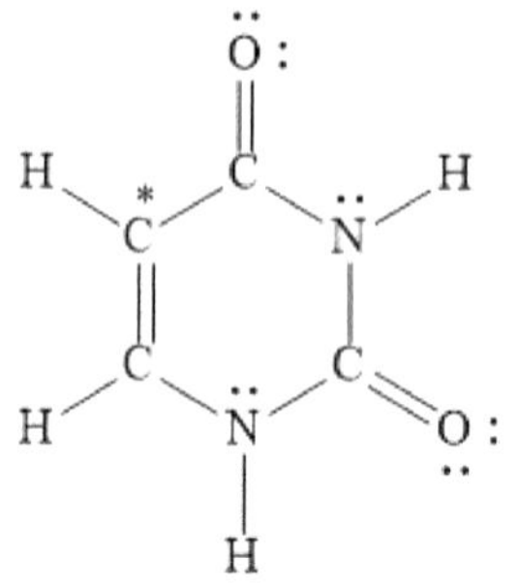

16.135. Regarding the degree of ionization:

(a) If the *degree of ionization* (α) is viewed as the part of the molecular specie (e.g., acid) that dissociates (or ionizes), then $(1 - \alpha)$ is the part of the molecular specie remaining at equilibrium. At equilibrium, for the hypothetical acid, HA, then concentrations are:
$[H^+] = \alpha C_o$ $[A^-] = \alpha C_o$ $[HA] = (1 - \alpha)C_o$
Substituting into an equilibrium constant expression:

$$K = \frac{[H^+][A^-]}{[HA]} = \frac{[\alpha C_o][\alpha C_o]}{[(1-\alpha)C_o]} \text{ and cancelling a } C_o \text{ term } K = \frac{[\alpha^2 C_o]}{[(1-\alpha)]}$$

(b) The degree of ionization for ammonium ion in 0.10 M NH₄Cl:

The K_a expression for the ammonium ion is: $K_{a(NH_4^+)} = \frac{[H^+][NH_3]}{[NH_4^+]} = 5.6 \times 10^{-10}$

Using our equation: $K_{a(NH_4^+)} = \frac{[\alpha^2(0.10)]}{[(1-\alpha)]} = 5.6 \times 10^{-10}$

Dividing both sides by 0.10, the expression becomes: $\frac{\alpha^2}{(1-\alpha)} = 5.6 \times 10^{-11}$

and $\alpha^2 = 5.6 \times 10^{-11}(1 - \alpha)$. Solving for α using the quadratic equation, $\alpha = 7.5 \times 10^{-6}$.

16.136. Concerning the degree of ionization from SQ16.135:

(a) Using the procedure in SQ16.135, calculate the α for the concentrations specified:

[HCO₂H]	α
0.0100	0.13
0.0200	0.090
0.0400	0.065
0.100	0.042
0.200	0.030
0.400	0.021
1.00	0.013
2.00	0.0094
4.00	0.0067

(b) A plot of α versus [HCO₂H] is not linear.

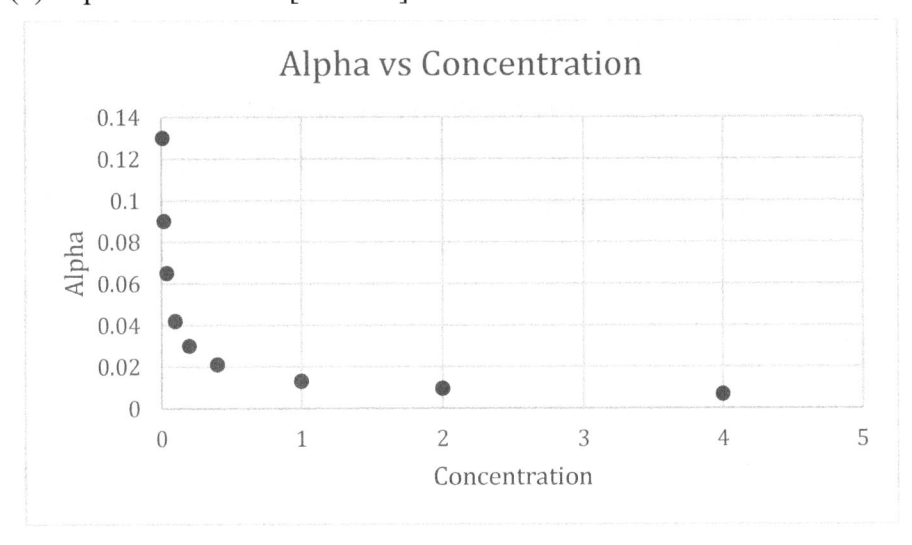

Alpha vs Concentration (scatter plot of Alpha versus Concentration)

A plot of log(C₀) vs α:

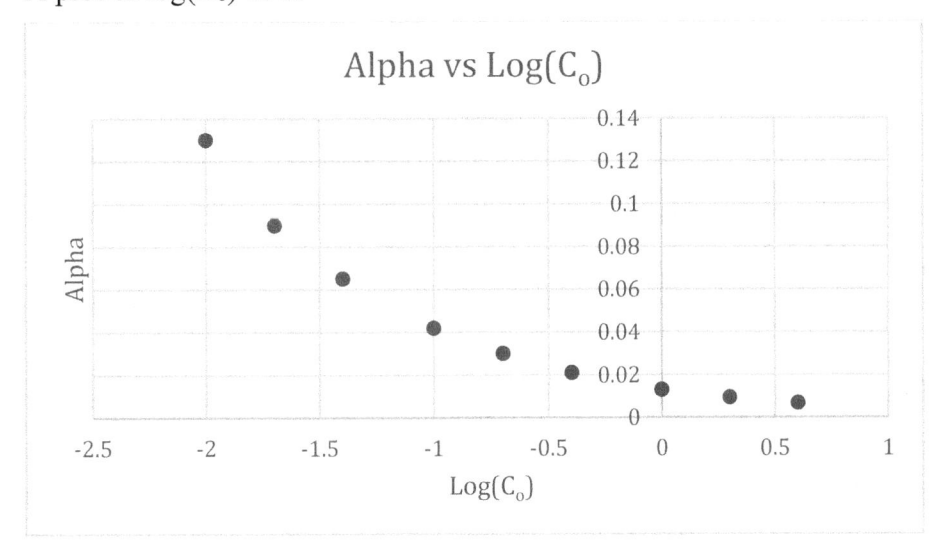

Alpha vs Log(C₀) (scatter plot of Alpha versus Log(C₀))

There is a more linear relationship between alpha and log(C₀).

(c) The degree of ionization is very high at low concentrations and decreases with increasing initial acid or base concentration.

16.137. Regarding the salt of a weak base and a weak acid:

(a) Determine the form for the equilibrium constant for the reaction:

$$NH_4^+(aq) + CN^-(aq) \rightleftarrows NH_3(aq) + HCN(aq)$$

What we need is a series of reactions that provide the above equation as the net.

Begin with K_a: K_a for NH_4^+: $NH_4^+(aq) + H_2O(\ell) \rightleftarrows NH_3(aq) + H_3O^+(aq)$ $K_1 = K_a$

Similarly, we can write the K_b expression for CN^-:

K_b for CN^-: $CN^-(aq) + H_2O(\ell) \rightleftarrows HCN(aq) + OH^-(aq)$ $K_2 = K_b$

Finally note that the components of water, $H_2O(\ell)$ and $OH^-(aq)$ and $H_3O^+(aq)$, are absent from our desired equation, but formed in each of the two equations above.

We can cancel the parts of water by recalling the equilibrium associated with the autoionization of water, K_w. To "cancel" the $H_3O^+ + OH^-$ as products, write the K_w equilibrium in the "reverse" sense:

$$H_3O^+(aq) + OH^-(aq) \rightleftarrows 2\,H_2O(\ell) \qquad K_3 = 1/K_w.$$

Summarizing:

$NH_4^+(aq) + H_2O(\ell) \rightleftarrows NH_3(aq) + H_3O^+(aq)$ $K_1 = K_a$

$CN^-(aq) + H_2O(\ell) \rightleftarrows HCN(aq) + OH^-(aq).$ $K_2 = K_b$

$\underline{H_3O^+(aq) + OH^-(aq) \rightleftarrows 2\,H_2O(\ell) \qquad\qquad K_3 = 1/K_w}$

$NH_4^+(aq) + CN^-(aq) \rightleftarrows NH_3(aq) + HCN\,(aq)$ $K_{net} = K_1 \times K_2 \times K_3 = \dfrac{K_a \times K_b}{K_w}$

(b) Calculate K_{net} values for each of the following: NH_4CN, $NH_4CH_3CO_2$, and NH_4F. Which salt has the largest value of K_{net} and why?

$$K_{net} = K_1 \times K_2 \times K_3 = \frac{K_a \times K_b}{K_w}$$

For NH_4CN: $K_{net} = \dfrac{(5.6 \times 10^{-10})(2.5 \times 10^{-5})}{1.0 \times 10^{-14}} = 1.4$

For $NH_4CH_3CO_2$: $K_{net} = \dfrac{(5.6 \times 10^{-10})(5.6 \times 10^{-10})}{1.0 \times 10^{-14}} = 3.1 \times 10^{-5}$

For NH₄F: $K_{net} = \dfrac{(5.6\times10^{-10})(1.4\times10^{-11})}{1.0\times10^{-14}} = 7.8\times10^{-7}$

The base cyanide is the strongest of the three bases, and therefore the base most capable of extracting the hydrogen ion, producing the greatest amount of product—and the largest K_{net}.

(c) No calculation is really needed. If the K's of the respective conjugate acid and base are equal, each reacts to an equivalent degree, producing a neutral solution—as is the case for NH₄CH₃CO₂. If the K_b for the base is greater than the K_a for the acid—as is the case for NH₄CN, the solution will be basic, while an acidic solution will result if the $K_a > K_b$, as in the case of NH₄F.

Solution and Answer Guide

Kotz Treichel Townsend Treichel, Chemistry and Chemical Reactivity 11e, 978-0-357-85140-1, Chapter 17: Principles of Chemical Reactivity: Other Aspects of Aqueous Equilibria

TABLE OF CONTENTS

Applying Chemical Principles 612
Practicing Skills 614
General Questions 647
In the Laboratory 656
Summary and Conceptual Questions 666

Applying Chemical Principles

Everything that Glitters

17.1.1. What volume (in mL) of 0.035 mass percent NaCN solution contains 0.10 g of sodium cyanide? Assume the density of the solution is 1.0 g/mL.

$$0.10 \text{ g NaCN} \left(\frac{100 \text{ g solution}}{0.035 \text{ g NaCN}}\right)\left(\frac{1.0 \text{ mL solution}}{1.0 \text{ g solution}}\right) = 290 \text{ mL} \quad (2 \text{ sf})$$

17.1.2. Minimum volume of 0.0071 M NaCN to dissolve the gold from 1.0 metric ton of ore if the ore is 0.012% gold:

$$1.0 \times 10^3 \text{kg ore}\left(\frac{1000 \text{ g}}{1 \text{ kg}}\right)\left(\frac{0.00012 \text{ g Au}}{1 \text{ g ore}}\right)\left(\frac{1 \text{ mol Au}}{196.97 \text{ g Au}}\right)\left(\frac{8 \text{ mol NaCN}}{4 \text{ mol Au}}\right)\left(\frac{1 \text{ L}}{0.0071 \text{ mol NaCN}}\right) = 1.7 \times 10^2 \text{ L}$$

17.1.3. What is the equilibrium concentration of Au⁺(aq) in a solution that is 0.0071 M CN⁻ and 1.1 × 10⁻⁴ M [Au(CN)₂]⁻. From Appendix K, K = 2.0 × 10³⁸.

The equilibrium expression is: $\dfrac{\left[\text{Au(CN)}_2^-\right]}{\left[\text{Au}^{+1}\right]\left[\text{CN}^{-1}\right]^2} = 2.0 \times 10^{38}$

Rearranging the equation to solve for [Au⁺¹]: $\dfrac{\left[\text{Au(CN)}_2^-\right]}{2.0 \times 10^{38}\left[\text{CN}^{-1}\right]^2} = \left[\text{Au}^{+1}\right]$

Substituting the given concentrations: $\dfrac{\left[1.1 \times 10^{-4}\right]}{2.0 \times 10^{38}\left[7.1 \times 10^{-3}\right]^2} = \left[\text{Au}^{+1}\right] = 1.1 \times 10^{-38} \text{ M}$

© 2024 Cengage Learning, Inc. All Rights Reserved. May not be scanned, copied or duplicated, or posted to a publicly accessible website, in whole or in part.

Examining the magnitude of the gold ion concentration above, it is reasonable to conclude that 100% of the gold in solution is present as the [Au(CN)₂]⁻ complex ion.

17.1.4. Concerning the reaction of Ag⁺ with CN⁻:

(a) Determine K for step 1: Step 1 represents the formation of the AgCN complex from the respective ions, so the K for that step is $K_{sp}^{-1} = \dfrac{1}{6.0 \times 10^{-17}} = 1.7 \times 10^{16}$

(b) Note what happens as you add the equilibria:

(1) Ag⁺(aq) + CN⁻(aq) ⇌ AgCN(s) $K_1 = 1.7 \times 10^{16}$
(2) AgCN(s) + CN⁻(aq) ⇌ [Ag(CN)₂]⁻(aq) $K_2 = ?$

(3) Ag⁺(aq) + 2 CN⁻(aq) ⇌ [Ag(CN)₂]⁻(aq) $K_f = 1.3 \times 10^{21}$

So $K_f = K_1 \times K_2$ and $K_2 = K_f/K_1 = 1.3 \times 10^{21}/1.7 \times 10^{16} = 7.8 \times 10^4$

(c) AgCN(s) equilibrates with 1.0 L of 0.0071 M CN⁻. What is [CN⁻] and [Ag(CN)₂⁻]?

The magnitude of K_2 indicates that the reaction will go to completion, with the only concentration of CN⁻ forming as a result of the dissociation of the [Ag(CN)₂]⁻ complex. Let's compose an equilibrium table for the reaction:

[Ag(CN)₂]⁻(aq) ⇌ AgCN(s) + CN⁻(aq) $K = 1/K_2$

	[Ag(CN)₂]⁻	CN⁻
Initial	0.0071	0
Change	−x	+x
Equilibrium	0.0071 − x	x

Substituting into the K expression:

$\dfrac{1}{K_2} = \dfrac{[CN^-]}{[Ag(CN)_2^-]}$; $\dfrac{1}{7.8 \times 10^4} = \dfrac{x}{0.0071 - x}$ and $x = 9.10 \times 10^{-8}$ M = [CN⁻]

and [Ag(CN)₂]⁻ = 0.0071 M

17.1.5. Write a balanced chemical equation for the reaction of NaAu(CN)₂(aq) and Zn(s):
Zn(s) + 2 Na⁺(aq) + 2 [Au(CN)₂]⁻(aq) ⇌ 2 Na⁺(aq) + 2 Au(s) + [Zn(CN)₄]²⁻(aq)

Take A Deep Breath

17.2.1. Ratio of HPO_4^{2-} to $H_2PO_4^-$ to control pH at 7.40:

$$pH = pK_a + \log\frac{[\text{conjugate base}]}{[\text{acid}]} \text{ or specifically } pH = pK_a + \log\frac{[HPO_4^{2-}]}{[H_2PO_4^-]}$$

$$7.40 = 7.20 + \log\frac{[HPO_4^{2-}]}{[H_2PO_4^-]} \text{ rearranging: } 0.20 = \log\frac{[HPO_4^{2-}]}{[H_2PO_4^-]}$$

and $10^{0.20} = 1.6$, so a ratio of 1.6:1 would provide a pH of 7.40.

17.2.2. A typical total phosphate concentration in a cell, $[HPO_4^{2-}] + [H_2PO_4^-]$, is 2.0×10^{-2} M. What are the concentrations of HPO_4^{2-} and $H_2PO_4^-$ at pH 7.40?

After solving 17.2.1, this problem is easier. Since $[HPO_4^{2-}] + [H_2PO_4^-]$ is 2.0×10^{-2} M, assign one of them, e.g. $[HPO_4^{2-}] = x$. Then $[H_2PO_4^-]$ is $(2.0 \times 10^{-2} \text{ M} - x)$. Substitute this into the expression: $\frac{[HPO_4^{2-}]}{[H_2PO_4^-]} = 1.6$.

$\frac{[x]}{[0.020 - x]} = 1.6$; solving for x yields $x = 0.012$ M and $[H_2PO_4^-] = (0.020 - 0.012) = 0.008$ M.

Practicing Skills

The Common Ion Effect and Buffer Solutions

17.1. To determine how pH is expected to change, examine the equilibria in each case:

(a) $NH_3(aq) + H_2O(\ell) \rightleftarrows NH_4^+(aq) + OH^-(aq)$

As the added NH_4Cl dissolves, ammonium ions are liberated—increasing the ammonium ion concentration and shifting the position of equilibrium to the left—reducing OH^- and *decreasing* the pH.

(b) $CH_3CO_2H(aq) + H_2O(\ell) \rightleftarrows CH_3CO_2^-(aq) + H_3O^+(aq)$

As sodium acetate dissolves, the additional acetate ion will shift the position of equilibrium to the left—reducing H_3O^+ and *increasing* the pH.

(c) $NaOH(aq) \rightarrow Na^+(aq) + OH^-(aq)$

NaOH is a strong base, and as such is totally dissociated. Since the added NaCl does not hydrolyze to any appreciable extent—*no change* in pH occurs.

17.2. Describe pH change when you:

(a) add solid sodium oxalate: pH increases ($C_2O_4^{2-}$ is a weak base)

(b) add solid NH₄Cl: pH decreases slightly (NH_4^+, a weak acid, is being added to a solution containing a strong acid)

(c) add NaCl: no change (NaCl is a neutral salt)

17.3. The pH of the buffer solution is $K_b = \dfrac{[NH_4^+][OH^-]}{[NH_3]} = \dfrac{(0.20)[OH^-]}{(0.20)} = 1.8 \times 10^{-5}$

solving for hydroxide ion yields: $[OH^-] = 1.8 \times 10^{-5}$; pOH = 4.75 and pH = 9.25

17.4. The pH of the buffer solution is $K_a = \dfrac{[HCO_2^-][H_3O^+]}{[HCO_2H]} = \dfrac{(0.25)[H_3O^+]}{(0.25)} = 1.8 \times 10^{-4}$

$[H_3O^+] = 1.8 \times 10^{-4}$; pH = $-\log(1.8 \times 10^{-4})$ = 3.74

17.5. pH of acetic acid – sodium acetate buffer solution:

$[CH_3CO_2^-] = \dfrac{16.4 \text{ g NaCH}_3CO_2}{1.0 \text{ L}}\left(\dfrac{1 \text{ mol NaCH}_3CO_2}{82.03 \text{ g NaCH}_3CO_2}\right) = 0.20 \text{ M}$

$K_a = \dfrac{[CH_3CO_2^-][H_3O^+]}{[CH_3CO_2H]} = \dfrac{(0.20)[H_3O^+]}{(0.20)} = 1.8 \times 10^{-5}$

$[H_3O^+] = 1.8 \times 10^{-5}$; pH = $-\log(1.8 \times 10^{-5})$ = 4.74

17.6. pH of ammonia – ammonium chloride buffer solution:

$[NH_4^+] = \dfrac{16.0 \text{ g NH}_4Cl}{1.0 \text{ L}}\left(\dfrac{1 \text{ mol NH}_4Cl}{53.48 \text{ g NH}_4Cl}\right) = 0.30 \text{ M}$

$K_b = \dfrac{[NH_4^+][OH^-]}{[NH_4Cl]} = \dfrac{(0.30)[OH^-]}{(0.30)} = 1.8 \times 10^{-5}$

$[OH^-] = 1.8 \times 10^{-5}$; pOH = $-\log(1.8 \times 10^{-5})$ = 4.74

pH = 14.00 - pOH = 9.26

17.7. pH of benzoic acid – potassium benzoate buffer solution:

$$30.0 \text{ mL KOH(aq)} \left(\frac{0.015 \text{ mmol KOH}}{1 \text{ mL KOH}} \right) = 0.45 \text{ mmol KOH}$$

$$50.0 \text{ mL benzoic acid} \left(\frac{0.015 \text{ mmol benzoic acid}}{1 \text{ mL benzoic acid}} \right) = 0.75 \text{ mmol benzoic acid}$$

benzoic acid is monoprotic so the reaction ratio is 1 mol benzoic acid:1 mol KOH

0.75 mmol benzoic acid reacts with 0.45 mmol KOH to produce 0.45 mmol benzoate ions and 0.30 mmol benzoic acid in excess

$V_{\text{Total}} = 50.0 \text{ mL} + 30.0 \text{ mL} = 80.0 \text{ mL}$

$$[\text{benzoate ion}] = \frac{0.45 \text{ mmol benzoate ions}}{80.0 \text{ mL}} = 0.0056 \text{ M}$$

$$[\text{benzoic acid}] = \frac{0.30 \text{ mmol benzoate ions}}{80.0 \text{ mL}} = 0.0038 \text{ M}$$

$$K_a = \frac{[\text{benzoate ion}][H_3O^+]}{[\text{benzoic acid}]} = \frac{(0.0056)[H_3O^+]}{(0.0038)} = 6.3 \times 10^{-5}$$

$$[H_3O^+] = \frac{(6.3 \times 10^{-5})(0.0038)}{(0.0056)} = 4.2 \times 10^{-5}$$

$\text{pH} = -\log(4.2 \times 10^{-5}) = 4.38$

17.8. pH when HCl and NH₃ are mixed:

The reaction: $NH_3(aq) + H_3O^+(aq) \rightarrow NH_4^+(aq) + H_2O(\ell)$

Amount of HCl present: (0.0250 L HCl)(0.12 mol/L) = 0.0030 mol HCl

Amount of NH₃ present: (0.0250 L NH₃)(0.43 mol/L) = 0.011 mol NH₃

Amount of NH₄⁺ produced: $0.0030 \text{ mol HCl} \left(\frac{1 \text{ mol NH}_4^+}{1 \text{ mol HCl}} \right) = 0.0030 \text{ mol NH}_4^+ \text{ produced}$

Amount of NH₃ consumed: $0.0030 \text{ mol HCl} \left(\frac{1 \text{ mol NH}_3}{1 \text{ mol HCl}} \right) = 0.0030 \text{ mol NH}_3 \text{ consumed}$

Amount of NH₃ remaining: (0.011 mol – 0.0030 mol) = 0.008 mol NH₃

$[NH_3] = \dfrac{0.008 \text{ mol}}{0.0500 \text{ L}} = 0.2 \text{ M}$ $[NH_4^+] = \dfrac{0.0030 \text{ mol}}{0.0500 \text{ L}} = 0.060 \text{ M}$

Insert these into the equilibrium: $NH_4^+(aq) + H_2O(\ell) \rightleftarrows H_3O^+(aq) + NH_3(aq)$

$$K_a = \frac{[H_3O^+][NH_3]}{[NH_4^+]} = \frac{(x)(0.2 + x)}{(0.060 - x)} \approx \frac{(x)(0.2)}{0.060} = 5.6 \times 10^{-10}$$

Solving for $x = [H_3O^+] = 2 \times 10^{-10}$ M and pH = $-\log[H_3O^+]$ = 9.7

17.9. The original pH of the 0.12 M NH$_3$ solution is:

$$\frac{[NH_4^+][OH^-]}{[NH_3]} = 1.8 \times 10^{-5} \text{ and } \frac{x^2}{0.12 - x} = 1.8 \times 10^{-5}$$

Assuming $(0.12 - x \approx 0.12)$, $x = 1.5 \times 10^{-3}$

$[OH^-] = 1.5 \times 10^{-3}$ and pOH = 2.83 with pH = 11.17

Adding 2.2 g of NH$_4$Cl (0.0411 mol) to 250 mL will produce an immediate increase of 0.165 M NH$_4^+$ (0.041 mol/0.250 L).

Substituting into the equilibrium expression as you did earlier, you get:

$$\frac{(x + 0.16)(x)}{0.12 - x} = 1.8 \times 10^{-5}$$, making the simplifying assumptions: $(x + 0.16 \approx 0.16)$ and $(0.12 - x \approx 0.12)$ reduces the fraction to $0.16x = (0.12)(1.8 \times 10^{-5})$ and $x = (0.12/0.165)(1.8 \times 10^{-5}) = 1.31 \times 10^{-5}$ M $= [OH^-]$

Note the hundred-fold decrease in [OH$^-$] over the initial ammonia solution, as predicted by Le Chatelier's principle. So, pOH = 4.88 and pH = 9.12 (lower than original).

17.10. Regarding lactic acid:

(a) The equation: CH$_3$CHOHCO$_2$H(aq) + H$_2$O(ℓ) $\rightleftarrows$ H$_3$O$^+$(aq) + CH$_3$CHOHCO$_2^-$(aq)

$$\frac{8.40 \text{ g NaCH}_3\text{CHOHCO}_2}{0.500 \text{ L}} \left(\frac{1 \text{ mol NaCH}_3\text{CHOHCO}_2}{112.06 \text{ g}} \right) \left(\frac{1 \text{ mol CH}_3\text{CHOHCO}_2^-}{1 \text{ mol NaCH}_3\text{CHOHCO}_2} \right) = 0.150 \text{ M}$$

$$K_a = \frac{[H_3O^+][CH_3CHOHCO_2^-]}{[CH_3CHOHCO_2H]} = \frac{(x)(0.150 + x)}{(0.100 - x)} \approx \frac{(x)(0.150)}{0.100} = 1.4 \times 10^{-4}$$

$x = [H_3O^+] = 9.3 \times 10^{-5}$ M and pH = $-\log[H_3O^+]$ = 4.03

(b) The buffer solution has a higher pH than the original lactic acid solution (pH = 2.43) because a weak base (CH$_3$CHOHCO$_2^-$) was added to the lactic acid solution.

17.11. Mass of sodium acetate needed to change 1.00 L solution of 0.10 M CH$_3$CO$_2$H to pH = 4.50: The equilibrium affected is that of acetic acid in water:

CH$_3$CO$_2$H + H$_2$O $\rightleftarrows$ CH$_3$CO$_2^-$ + H$_3$O$^+$ $\qquad$ K$_a$ = 1.8 × 10^{-5}

The equilibrium expression is $K_a = \dfrac{[CH_3CO_2^-][H_3O^+]}{[CH_3CO_2H]} = 1.8 \times 10^{-5}$

You know the concentration of acetic acid (0.10M), and you know the desired [H₃O⁺]:

pH = 4.50 so [H₃O⁺] = $10^{-4.50}$ = 3.2 × 10⁻⁵

Substituting these values into the equilibrium expression gives:

$$\dfrac{[CH_3CO_2^-][H_3O^+]}{[CH_3CO_2H]} = \dfrac{[CH_3CO_2^-][3.2 \times 10^{-5}]}{[0.10]} = 1.8 \times 10^{-5}$$

You can solve for [CH₃CO₂⁻] = 0.057 M

What mass of NaCH₃CO₂ would give this concentration of CH₃CO₂⁻?

$$1 \text{ L solution} \left(\dfrac{0.057 \text{ mol } C_2H_3O_2^-}{1 \text{ L}} \right) \left(\dfrac{1 \text{ mol NaCH}_3CO_2}{1 \text{ mol CH}_3CO_2^-} \right) \left(\dfrac{82.03 \text{ g NaCH}_3CO_2}{1 \text{ mol NaCH}_3CO_2} \right) = 4.7 \text{ g NaCH}_3CO_2$$

17.12. Mass of NH₄Cl to give a pH = 9.50:

[H₃O⁺] = 10^{-pH} = $10^{-9.50}$ = 3.16 × 10⁻¹⁰ ; Now that you know the [H₃O⁺], you can calculate the ammonium ion concentration to provide that:

$$K_a = \dfrac{[H_3O^+][NH_3]}{[NH_4^+]} = \dfrac{(3.16 \times 10^{-9})(0.10)}{[NH_4^+]} = 5.6 \times 10^{-10}$$

Solving for [NH₄⁺] = 0.0565 M

$$0.500 \text{ L solution} \left(\dfrac{0.0565 \text{ mol NH}_4^+}{1 \text{ L}} \right) \left(\dfrac{1 \text{ mol NH}_4Cl}{1 \text{ mol NH}_4^+} \right) \left(\dfrac{53.49 \text{ g}}{1 \text{ mol NH}_4Cl} \right) = 1.5 \text{ g NH}_4Cl$$

Using the Henderson-Hasselbalch Equation

17.13. The pH of a solution with 0.050 M acetic acid and 0.075 M sodium acetate:

$$pH = pK_a + \log \dfrac{[\text{conjugate base}]}{[\text{acid}]}$$

The pKₐ for acetic acid is −log(Kₐ) = −log(1.8 × 10⁻⁵) or 4.74

$$pH = 4.74 + \log \dfrac{[0.075]}{[0.050]} = 4.74 + 0.176 = 4.92$$

17.14. Calculate the pH of a solution of NH_3/NH_4Cl:

For the NH_4^+ ion, $pK_a = -\log(K_a) = -\log(5.6 \times 10^{-10}) = 9.25$

Substitute into the Henderson Hasselbalch equation: $pH = pK_a + \log\dfrac{[\text{conjugate base}]}{[\text{acid}]}$

$pH = pK_a + \log\dfrac{[NH_3]}{[NH_4^+]} = 9.25 + \log\dfrac{[0.045]}{[0.050]} = 9.20$

17.15. Ratio of acetic acid/acetate ion to have pH = 4.50:

$pH = pK_a + \log\dfrac{[\text{acetate ion}]}{[\text{acetic acid}]}$

You know that the pK_a for acetic acid is 4.74 (SQ17.13 above).

$4.50 = 4.74 + \log\dfrac{[\text{acetate ion}]}{[\text{acetic acid}]}$ and $-0.24 = \log\dfrac{[\text{acetate ion}]}{[\text{acetic acid}]}$

$10^{-0.24} = \dfrac{[\text{acetate ion}]}{[\text{acetic acid}]} = 0.58$, and the $\dfrac{[\text{acetic acid}]}{[\text{acetate ion}]} = \dfrac{1}{0.58} = 1.74 = 1.7$ (2 sf)

17.16. Ratio of conjugate pairs to provide a buffer with pH = 7.50:

$pH = pK_a + \log\dfrac{[\text{conjugate base}]}{[\text{acid}]}$ $K_a = 6.2 \times 10^{-8}$ and $pK_a = 7.21$

$7.50 = 7.21 + \log\dfrac{[HPO_4^{2-}]}{[H_2PO_4^-]}$ and $\log\dfrac{[HPO_4^{2-}]}{[H_2PO_4^-]} = 0.29$

$\dfrac{[HPO_4^{2-}]}{[H_2PO_4^-]} = 1.95$ giving $\dfrac{[H_2PO_4^-]}{[HPO_4^{2-}]} = \dfrac{1}{1.95} = 0.513 = 0.51$ (2 sf)

17.17. For the buffer of formic acid and sodium formate:

(a) The pK_a for formic acid is $= -\log(K_a) = -\log(1.8 \times 10^{-4}) = 3.74$

$pH = pK_a + \log\dfrac{[\text{conjugate base}]}{[\text{acid}]}$

$pH = 3.74 + \log\dfrac{0.035}{0.050} = 3.74 - 0.15$ or 3.59

(b) Ratio of conjugate pairs to increase pH by 0.5 (to pH= 4.09)

Substituting into the Henderson-Hasselbalch equation

$$4.09 = 3.74 + \log\frac{[\text{conjugate base}]}{[\text{acid}]} \quad \text{or} \quad 0.345 = \log\frac{[\text{conjugate base}]}{[\text{acid}]}$$

$$\text{so } -0.345 = \log\frac{[\text{acid}]}{[\text{conjugate base}]} \quad \text{and } 0.45 = \frac{[\text{acid}]}{[\text{conjugate base}]} \quad (2\text{ sf})$$

17.18. For the buffer of KH_2PO_4 and Na_2HPO_4:

(a) $pK_a = -\log(6.2 \times 10^{-8}) = 7.21$

$$5.677 \text{ g Na}_2\text{HPO}_4 \left(\frac{1 \text{ mol Na}_2\text{HPO}_4}{141.96 \text{ g}}\right) = 0.03999 \text{ mol Na}_2\text{HPO}_4$$

$$1.360 \text{ g KH}_2\text{PO}_4 \left(\frac{1 \text{ mol KH}_2\text{PO}_4}{136.08 \text{ g}}\right) = 0.009994 \text{ mol KH}_2\text{PO}_4$$

$$pH = 7.21 + \log\frac{[0.03999]}{[0.009994]} = 7.81$$

(b) Decreasing pH from 7.81 to 7.31:

$$7.31 = 7.21 + \log\frac{[HPO_4^{2-}]}{[H_2PO_4^-]} \quad \text{and}$$

$$0.10 = \log\frac{[HPO_4^{2-}]}{[H_2PO_4^-]} \quad \text{thus} \quad \frac{[HPO_4^{2-}]}{[H_2PO_4^-]} = 10^{0.10} = 1.3$$

$$\frac{[0.03999 \text{ mol}]}{[x]} = 1.3 \text{ and } x = 0.031 \text{ mol } H_2PO_4^-$$

$$0.031 \text{ mol KH}_2\text{PO}_4 \left(\frac{1 \text{ mol KH}_2\text{PO}_4}{1 \text{ mol H}_2\text{PO}_4^-}\right)\left(\frac{136.1 \text{ g KH}_2\text{PO}_4}{1 \text{ mol KH}_2\text{PO}_4}\right) = 4.2 \text{ g KH}_2\text{PO}_4$$

Mass of KH_2PO_4 to add = 4.2 g total − 1.360 g already in buffer = 2.8 g

Preparing a Buffer Solution

17.19. The best combination to provide a buffer solution of pH 5 is (c) the $CH_3CO_2H/NaCH_3CO_2$ system. Note that pK_a (for CH_3CO_2H) is 4.74. Buffer systems are good when the desired pH is ±1 unit from pK_a. The HCl and NaCl don't form a buffer, and the pK_a for the NH_4^+ is 9.26.

17.20. Best combination to buffer pH at about 7:

(a) H_3PO_4/NaH_2PO_4 $pK_a(H_3PO_4) = 2.12$

(b) NaH_2PO_4/Na_2HPO_4 $pK_a(H_2PO_4^-) = 7.21$

(c) Na_2HPO_4/Na_3PO_4 $pK_a(HPO_4^{2-}) = 12.44$

The best choice is (b), the NaH_2PO_4/Na_2HPO_4 buffer, as the pK_a is ≈ 7.

17.21. Preparation of a buffer of NaH₂PO₄ and Na₂HPO₄ for a pH = 7.5:

The principal species present (other than Na⁺ ions) are H₂PO₄⁻ and HPO₄²⁻

Use the K_{a2} expression for H₃PO₄.

$$\frac{[\text{HPO}_4^{2-}][\text{H}_3\text{O}^+]}{[\text{H}_2\text{PO}_4^-]} = 6.2 \times 10^{-8}$$

For a pH = 7.5, [H₃O⁺] = 3.2 × 10⁻⁸ M. So, the ratio of the conjugate pairs is:

$$\frac{[\text{HPO}_4^{2-}]}{[\text{H}_2\text{PO}_4^-]} = \frac{6.2 \times 10^{-8}}{3.2 \times 10^{-8}} = 1.94 \text{ (or 2 to 1 sf). The reciprocal of this value is } 1/2 = 0.5.$$

So, mixing 0.5 moles of NaH₂PO₄ with 1 mol Na₂HPO₄ would provide the desired pH.

17.22. Prepare a buffer of CH₃CO₂H and NaCH₃CO₂ with pH = 4.5:

pOH = 14.00 − 4.5 = 9.5 [OH⁻] = 10⁻ᵖᴼᴴ = 10⁻⁹·⁵ = 3 × 10⁻¹⁰ M

$$K_b = \frac{[\text{OH}^-][\text{CH}_3\text{CO}_2\text{H}]}{[\text{CH}_3\text{CO}_2^-]} = \frac{(3 \times 10^{-10})[\text{CH}_3\text{CO}_2\text{H}]}{[\text{CH}_3\text{CO}_2^-]} = 5.6 \times 10^{-10} \text{ and } \frac{[\text{CH}_3\text{CO}_2\text{H}]}{[\text{CH}_3\text{CO}_2^-]} = 1.8$$

So a solution in which the CH₃CO₂H concentration is twice that of NaCH₃CO₂ will have a pH of 4.5. For example, adding 2 mol CH₃CO₂H and 1 mol NaCH₃CO₂ to enough water to dissolve both compounds produces a buffer with pH = 4.5.

17.23. Volume (in mL) of 1.00 M NaOH added to 250 mL of 0.50 M CH₃CO₂H to produce a buffer of pH = 4.50:

So, you need a buffer of acetic acid and sodium acetate, like that in SQ17.22.

$$\text{pH} = pK_a + \log\frac{[\text{acetate ion}]}{[\text{acetic acid}]} \text{ and } 4.50 = 4.74 + \log\frac{[\text{acetate ion}]}{[\text{acetic acid}]}$$

Solving for the log fraction to obtain: $-0.24 = \log\frac{[\text{acetate ion}]}{[\text{acetic acid}]}$ and $0.575 = \frac{[\text{acetate ion}]}{[\text{acetic acid}]}$

Stoichiometry dictates that for each mole of NaOH that is added, one mole of acetic acid is consumed and one mole of sodium acetate ion is formed. Moles of acetic acid present initially: 0.250 L(0.50 mol acetic acid/L) = 0.125 mol acetic acid.

Treat the **ratio** of **concentrations** of the acetate ion and acetic acid as the **ratio** of **moles** of the two. If x moles of the acid is consumed, x moles of the acetate ion will be produced. Substituting into the equation above:

$0.575 = \dfrac{\text{mol acetate ion}}{\text{mol acetic acid}} = \dfrac{x}{0.125 - x}$ Rearrange: $0.575(0.125 - x) = x$ and solving for x = 0.0456 mol. Adding 0.0456 mol of NaOH will **consume** 0.0456 mol of CH₃CO₂H and **produce** 0.0456 mol of the acetate ion. Volume of NaOH needed: 0.0456 mol = V · 1.00 mol NaOH/L and V = 45.6 mL or expressed to 2 sf: 46 mL.

17.24. Volume of 1.00 M HCl to add:

$$pH = pK_a + \log\frac{[\text{conjugate base}]}{[\text{acid}]}$$ and with $pK_a = 7.21$,

$$7.00 = 7.21 + \log\frac{[HPO_4^{2-}]}{[H_2PO_4^-]} \text{ or } -0.21 = \log\frac{[HPO_4^{2-}]}{[H_2PO_4^-]} \text{ or } 0.617 = \frac{[HPO_4^{2-}]}{[H_2PO_4^-]}$$

Initial amount of Na₂HPO₄: (0.750 L)(0.50 mol/L) = 0.375 mol

Treat the **ratio** of **concentrations** of the hydrogen phosphate ion and dihydrogen phosphate ion as the **ratio** of **moles** of the two. If x moles of the HPO_4^{2-} is reacts with an equal number of moles of HCl, x moles of the $H_2PO_4^-$ ion will be produced. Substituting into the equation above:

$$0.617 = \frac{\text{mol } HPO_4^{2-}}{\text{mol } H_2PO_4^-} = \frac{0.375 - x}{x}$$

Solve for x: $x = 0.232$; the amount of $H_2PO_4^- = 0.232$ mol and the amount of $HPO_4^{2-} = 0.375$ mol − 0.232 = 0.143 mol.

Adding 0.232 mol of HCl will **consume** 0.232 mol of HPO_4^{2-} and **produce** 0.232 mol of the $H_2PO_4^-$. Volume of HCl needed: 0.232 mol = V • 1.00 mol HCl/L and V = 232 mL or expressed to 2 sf: 230 mL.

Adding an Acid or Base to a Buffer Solution

17.25. For the buffer solution of sodium acetate and acetic acid:

(a) Initial pH: Need to know the concentrations of the conjugate pairs:

The equilibrium expression shows the *ratio of the conjugate pairs*, you can calculate **moles** of the conjugate pairs, and know that the ratio of the # of moles of the species will have the same value as the ratio of their concentrations.

CH₃CO₂H = 0.250 L(0.150 M) = 0.0375 mol

NaCH₃CO₂ = 4.95 g NaCH₃CO₂ $\left(\dfrac{1 \text{ mol NaCH}_3CO_2}{82.03 \text{ g}}\right)$ = 0.0603 mol

Substituting into the K_a expression:

$$\frac{[CH_3CO_2^-][H_3O^+]}{[CH_3CO_2H]} = 1.8 \times 10^{-5} = \frac{[0.0603][H_3O^+]}{[0.0375]}$$

and solving for $[H_3O^+] = 1.1 \times 10^{-5}$ M; pH = 4.95

(b) pH after 82. mg NaOH is added to 100. mL of the buffer. The amount of the conjugate pair in 100/250 of the buffer is (100/250)(0.0375 mol) = 0.0150 mol CH₃CO₂H and (100/250)(0.0603 mol) = 0.0241 mol CH₃CO₂⁻

$$82 \text{ mg NaOH}(\frac{1 \text{ mmol NaOH}}{40.0 \text{ mg NaOH}}) = 2.05 \text{ mmol NaOH} = 0.00205 \text{ mol NaOH or to 2sf 2.1 mmol NaOH}$$

This base would consume an equivalent amount of CH_3CO_2H and produce an equivalent amount of $CH_3CO_2^-$.

After that process: (0.0150 − 0.0021) or 0.0129 mol CH_3CO_2H and (0.0241 + 0.0021) or 0.0262 mol $CH_3CO_2^-$ are present. Substituting into the K_a expression as in part (a)

$$\frac{[0.0262][H_3O^+]}{[0.0129]} = 1.8 \times 10^{-5} \text{ so } [H_3O^+] = 8.9 \times 10^{-6} \text{ and pH} = 5.05$$

17.26. Change in pH upon adding NaOH to a buffer:

$[H_2PO_4^-] = [HPO_4^{2-}]$ so $[H_3O^+] = K_a = 6.2 \times 10^{-8}$

Before adding NaOH: pH = −log[H_3O^+] = 7.21

$$[\text{NaOH}] = \left(\frac{0.425 \text{ g}}{2.00 \text{ L}}\right)\left(\frac{1 \text{ mol NaOH}}{40.00 \text{ g}}\right) = 0.00531 \text{ M}$$

The reaction is: $OH^-(aq) + H_2PO_4^-(aq) \rightarrow H_2O(\ell) + HPO_4^{2-}(aq)$

The volume will increase upon the addition of NaOH. However, buffers are relatively insensitive to volume changes, so this problem can be solved assuming no volume change.

	$H_2PO_4^-$	HPO_4^{2-}
Initial	0.132	0.132
Change	−0.00531	+0.00531
Equilibrium	0.127	0.137

$$[H_3O^+] = \frac{[H_2PO_4^-]}{[HPO_4^{2-}]} K_a \approx \left(\frac{0.127}{0.137}\right) 6.2 \times 10^{-8} = 5.7 \times 10^{-8}$$

After adding NaOH: pH = −log[H_3O^+] = 7.24

17.27. pH of buffer of NH_3 and NH_4Cl:

(a) The pH of the buffer solution is:

$$K_b = \frac{[NH_4^+][OH^-]}{[NH_3]} = \frac{[0.250][OH^-]}{[0.500]} = 1.8 \times 10^{-5}$$

Note: Here the data presented are given as moles (in the case of ammonium chloride) and molar concentration (in the case of ammonia). In SQ17.25 you substituted the # moles of the conjugate pairs into the *K* expression. Here you must first *decide* whether to substitute # moles or molar c*oncentrations* into the K_b expression. Either would work.

What is critical to remember is that you must have both species expression in one **or** the other form—not a mix of the two. Here I chose to convert moles of NH₄Cl into molar concentrations, and substitute.

Solving for hydroxide ion in the K_b expression above yields: [OH⁻] = 3.6 × 10⁻⁵ M and pOH = 4.45 so pH = 9.55

(b) pH after addition of 0.0100 mol HCl:

The basic component of the buffer (NH₃) will react with the HCl, producing more ammonium ion.

The composition of the solution is:	NH₃	NH₄Cl
Moles present (before HCl added)	0.250	0.125
Change (reaction)	−0.0100	+0.0100
Following reaction	0.240	0.135

The amounts of NH₃ and NH₄Cl following the reaction with HCl are only slightly different from the amounts prior to reaction. Converting these numbers into molar concentrations (Volume is 500. mL) and substituting the concentrations into the K_b expression yields:

$$K_b = \frac{[NH_4^+][OH^-]}{[NH_3]} = \frac{(0.270)[OH^-]}{(0.480)} = 1.8 \times 10^{-5}$$

[OH⁻] = 3.2 × 10⁻⁵; pOH = 4.50, and the new pH = 9.50.

17.28. pH change upon addition of NaOH to buffer:

Original pH of buffer: $pH = pK_a + \log\frac{[NH_3]}{[NH_4^+]} = 9.25 + \log\frac{[0.169]}{[0.183]} = 9.22$

Amount of NaOH added: (0.0200 L NaOH)(0.100 mol/L) = 0.00200 mol NaOH
Initial amount of NH₃ in buffer: (0.0800 L NH₃)(0.169 mol/L) = 0.0135 mol NH₃
Initial amount of NH₄⁺ in buffer: (0.0800 L NH₄⁺)(0.183 mol/L) = 0.0146 mol NH₄⁺
Total volume = 0.0200 L + 0.0800 L = 0.100 L
OH⁻(aq) + NH₄⁺(aq) → H₂O(ℓ) + NH₃(aq)

	NH₄⁺	NH₃
Initial (mol)	0.0146	0.0135
Change (mol)	−0.00200	+0.00200
Equilibrium (mol)	0.0126	0.0155

$$\text{pH} = pK_a + \log\frac{[\text{NH}_3]}{[\text{NH}_4^+]} = 9.25 + \log\frac{[0.0155]}{[0.0126]} = 9.34$$

The change in pH is 9.34 – 9.22 = 0.12.

17.29. Buffer made of NaH₂PO₄ and Na₂HPO₄:

(a) Initial pH:

Acidic buffers are made of a weak acid and its conjugate base. In this case, H₂PO₄⁻ is the weak acid and HPO₄²⁻ is the conjugate base. K_{a2} for phosphoric acid is the K_a of H₂PO₄⁻ = 6.2 × 10⁻⁸. $pK_a = -\log(6.3 \times 10^{-8}) = 7.21$

$$[\text{NaH}_2\text{PO}_4] = [\text{Na}_2\text{HPO}_4] = \frac{0.50 \text{ mol}}{1.0 \text{ L}} = 0.50 \text{ M}$$

$$\text{pH} = pK_a + \log\frac{[\text{conjugate base}]}{[\text{acid}]} = 7.21 + \log\frac{[0.50]}{[0.50]} = 7.21$$

(b) Volume 0.15 M NaOH required to raise pH by 1.0:

$$\text{pH} = pK_a + \log\frac{[\text{conjugate base}]}{[\text{acid}]}$$

$$8.21 = 7.21 + \log\frac{[\text{conjugate base}]}{[\text{acid}]}$$

$$1.0 = \log\frac{[\text{conjugate base}]}{[\text{acid}]} \quad \text{thus} \quad \frac{[\text{conjugate base}]}{[\text{acid}]} = 10^{1.0} = 10.0$$

To raise pH by 1.0 you must increase the [conjugate base] by a factor of 10 relative to the [acid]. Because they are in the same solution, this will have the same effect on the amounts (moles) of the species. The added amount of OH⁻ (x) will decrease the amount of H₂PO₄⁻ and increase the amount of HPO₄²⁻ as shown in the following amounts table.

Equation	OH⁻(aq)	+	H₂PO₄⁻(aq)	→	HPO₄²⁻(aq)	+	H₂O(ℓ)
Initial (mol)	x		0.50		0.50		
Change (mol)	–x		–x		+x		
Final (mol)	0		0.50 – x		0.50 + x		

$$10 = \frac{0.50 \text{ mol} + x}{0.50 \text{ mol} - x}$$

$$10(0.50 \text{ mol} - x) = 0.50 \text{ mol} + x$$

$$5.0 \text{ mol} - 10x = 0.50 \text{ mol} + x$$

$$4.5 = 11x$$

$$x = 0.41 \text{ mol}$$

$$0.41 \text{ mol NaOH} \left(\frac{1 \text{ L solution}}{0.15 \text{ mol NaOH}} \right) = 2.7 \text{ L of } 0.15 \text{ M NaOH solution}$$

17.30. Buffer made of CH₃CO₂H and NaCH₃CO₂:

(a) Initial pH:

$$[CH_3CO_2H] = [NaCH_3CO_2] = \frac{0.36 \text{ mol}}{1.0 \text{ L}} = 0.36 \text{ M}$$

$$pH = pK_a + \log \frac{[\text{conjugate base}]}{[\text{acid}]} = 4.74 + \log \frac{[0.36]}{[0.36]} = 4.74$$

(b) Volume 0.15 M NaOH required to raise pH by 1.0:

$$pH = pK_a + \log \frac{[\text{conjugate base}]}{[\text{acid}]}$$

$$3.74 = 4.74 + \log \frac{[\text{conjugate base}]}{[\text{acid}]}$$

$$-1.0 = \log \frac{[\text{conjugate base}]}{[\text{acid}]} \quad \text{thus} \quad \frac{[\text{conjugate base}]}{[\text{acid}]} = 10^{-1.0} = 0.01$$

To lower pH by 1.0 you must decrease the [conjugate base] by a factor of 10 relative to the [acid]. Because they are in the same solution, this will have the same effect on the amounts (moles) of the species. The added amount of HCl will decrease the amount of CH₃CO₂⁻ and increase the amount of CH₃CO₂H as shown in the following amounts table.

Equation	H₃O⁺(aq)	+	CH₃CO₂⁻(aq)	→	CH₃CO₂H(aq)	+	H₂O(ℓ)
Initial (mol)	x		0.36		0.36		
Change (mol)	–x		–x		+x		
Final (mol)	0		0.36 – x		0.36 + x		

$$0.10 = \frac{0.36 \text{ mol} - x}{0.36 \text{ mol} + x}$$

$0.10(0.36 \text{ mol} + x) = 0.36 \text{ mol} - x$

$0.036 \text{ mol} + 0.10 \, x = 0.36 \text{ mol} - x$

$0.324 \text{ mol} = 1.10 \, x$

$x = 0.29 \text{ mol}$

$0.29 \text{ mol HCl} \left(\dfrac{1 \text{ L solution}}{0.10 \text{ mol HCl}} \right) = 2.9 \text{ L of } 0.10 \text{ M HCl solution}$

More About Acid-Base Reactions: Titrations

17.31. Calculate the amount of phenol present (converting mass to moles):

$0.515 \text{ g } C_6H_5OH \left(\dfrac{1 \text{ mol } C_6H_5OH}{94.11 \text{ g } C_6H_5OH} \right) = 5.472 \times 10^{-3} \text{ mol } C_6H_5OH$

(a) The pH of the solution containing 5.472×10^{-3} mol C_6H_5OH in 125 mL water:

With $K_a = 1.3 \times 10^{-10}$, the K_a expression is: $K_a = \dfrac{[C_6H_5O^-][H_3O^+]}{[C_6H_5OH]} = 1.3 \times 10^{-10}$

The stoichiometry of the compound indicates that it ionizes to create one phenoxide ion for each hydronium ion.

The initial concentration is 5.472×10^{-3} mol/ 0.125 L = 0.04378 M. The K_a expression becomes:

$\dfrac{[x][x]}{[0.04378 - x]} = 1.3 \times 10^{-10}$

Given the magnitude of the K_a, you can safely approximate the denominator as 0.04378 M.

$x^2 = 1.3 \times 10^{-10} \cdot 0.04378$ and $x = 2.4 \times 10^{-6}$ M and pH = $-\log(2.4 \times 10^{-6}) = 5.62$.

(b) At the equivalence point 5.472×10^{-3} mol of NaOH will have been added. Phenol is a monoprotic acid, so one mol of phenol reacts with one mol of sodium hydroxide. The volume of 0.123 M NaOH needed to provide this amount of base is:
moles = M × V

5.472×10^{-3} mol NaOH = $\dfrac{0.123 \text{ mol NaOH}}{\text{L}}$ × V or 44.5 mL of the NaOH solution.

The total volume would be (125 + 44.5) or 169.5 mL (or 0.1695 L) solution.

Sodium phenoxide is a soluble salt hence the initial concentration of both sodium and phenoxide ions will be equal to:

$$\frac{5.472 \times 10^{-3} \text{ mol}}{0.1695 \text{ L}} = 3.229 \times 10^{-2} \text{ M}$$

The phenoxide ion however is the conjugate ion of a weak acid and undergoes hydrolysis. $C_6H_5O^-$ (aq) + $H_2O(\ell)$ ⇌ C_6H_5OH(aq) + OH^-(aq)

The concentrations at equilibrium:

	$C_6H_5O^-$	C_6H_5OH	OH^-
Initial (M)	3.239×10^{-2}		
Change (M)	$-x$	$+x$	$+x$
Equilibrium (M)	3.229×10^{-2} M $- x$	x	x

$$K_b = \frac{[C_6H_5OH][OH^-]}{[C_6H_5O^-]} = \frac{x^2}{3.229 \times 10^{-2} - x} = \frac{1.0 \times 10^{-14}}{1.3 \times 10^{-10}} = 7.69 \times 10^{-5}$$

Since $7.69 \times 10^{-5}(100) < 3.23 \times 10^{-2}$ you can simplify:

$$\frac{x^2}{3.229 \times 10^{-2}} = 7.69 \times 10^{-5}$$

and $x = 1.58 \times 10^{-3}$ M $= 1.6 \times 10^{-3}$ M (2 sf).

While the phenoxide ion reacts with water (hydrolyzes) to some extent, the Na^+ is a "spectator ion" and its concentration remains unchanged at 3.23×10^{-2} M.
The concentration of phenoxide is reduced (albeit slightly) and at equilibrium is $(3.229 \times 10^{-2} - x)$ or 3.08×10^{-2} M.

(c) At the equivalence point: $[OH^-] = 1.58 \times 10^{-3}$ M

pOH $= -\log(1.58 \times 10^{-3}) = 2.80$ and pH $= 11.20$

17.32. For the titration of benzoic acid with NaOH:

(a) Concentration of acid: $[C_6H_5CO_2H] = \frac{0.235 \text{ g}}{0.100 \text{ L}}\left(\frac{1 \text{ mol } C_6H_5CO_2H}{122.1 \text{ g}}\right) = 0.0192$ M

$$K_a = \frac{[C_6H_5CO_2^-][H_3O^+]}{[C_6H_5CO_2H]} = \frac{(x)^2}{(0.0192 - x)} \approx \frac{(x)^2}{0.0192} = 6.3 \times 10^{-5}$$

$x = [H_3O^+] = 0.0011$ M and the pH $= -\log[H_3O^+] = 2.96$

(b) For the reaction: $C_6H_5CO_2H + OH^- \rightleftharpoons C_6H_5CO_2^- + H_2O$

	$C_6H_5CO_2H$	OH^-	$C_6H_5CO_2^-$
Initial(mol)	0.00192	0.00192	0
Change(mol)	−0.00192	−0.00192	+0.00192
After reaction(mol)	0	0	0.00192

Total volume = 0.100 L + (0.00192 mol NaOH)(1 L/0.108 mol) = 0.118 L
Once the reaction has occurred, the equilibrium in control is:

$C_6H_5CO_2^-(aq) + H_2O(\ell) \rightleftharpoons C_6H_5CO_2H(aq) + OH^-(aq)$

Concentration of anion: $[C_6H_5CO_2^-] = \dfrac{0.00192 \text{ mol}}{0.118 \text{ L}} = 0.0163$ M

$K_b = \dfrac{K_w}{K_a} = \dfrac{[C_6H_5CO_2H][OH^-]}{[C_6H_5CO_2^-]} = \dfrac{(x)^2}{(0.0163 - x)} \approx \dfrac{(x)^2}{0.0163} = 1.6 \times 10^{-10}$

and solving for $x = 1.6 \times 10^{-6}$ M = $[OH^-]$

$[H_3O^+] = \dfrac{K_w}{[OH^-]} = 6.2 \times 10^{-9}$ M

$[Na^+] = [C_6H_5CO_2^-] = 0.0163$ M

(c) pH = $-\log[H_3O^+]$ = 8.21

17.33. For the titration of aqueous ammonia with aqueous HCl:

(a) At the equivalence point the moles of acid = moles of base.

(0.03678 L) (0.0105 M HCl) = 3.862×10^{-4} mol HCl

If this amount of base were contained in 25.0 mL of solution, the concentration of NH_3 in the original solution was 0.01545 M. Answer: 0.0154 M (2 sf).

(b) At the equivalence point NH_4Cl will hydrolyze according to the equation:

$NH_4^+(aq) + H_2O(\ell) \rightleftharpoons NH_3(aq) + H_3O^+(aq)$

$K_a = \dfrac{[H_3O^+][NH_3]}{[NH_4^+]} = \dfrac{(1.0 \times 10^{-14})}{(1.8 \times 10^{-5})} = 5.6 \times 10^{-10}$

The salt (3.862×10^{-4} mol) is contained in (25.0 + 36.78) or 61.78 mL. Its concentration will be (3.862×10^{-4} mol)/(0.06178 L) = 6.251×10^{-3} M.

Substituting into the K_a expression: $\dfrac{[H_3O^+]^2}{6.25 \times 10^{-3}} = 5.6 \times 10^{-10}$ and $[H_3O^+] = 1.9 \times 10^{-6}$

Since $[H_3O^+][OH^-] = 1.0 \times 10^{-14}$ then $[OH^-] = \dfrac{1.0 \times 10^{-14}}{1.9 \times 10^{-6}} = 5.3 \times 10^{-9}$ M

and $[NH_4^+] = 6.25 \times 10^{-3}$ M (i.e., the salt concentration)

(c) With $[H_3O^+] = 1.9 \times 10^{-6}$ the pH = 5.73.

17.34. For the titration of aniline with HCl:

(a) $[C_6H_5NH_2] = 0.02567 \text{ L} \left(\dfrac{0.175 \text{ mol HCl}}{1.00 \text{ L}}\right)\left(\dfrac{1 \text{ mol } C_6H_5NH_2}{1 \text{ mol HCl}}\right)\left(\dfrac{1}{0.0250 \text{ L}}\right) = 0.180$ M

(b) The equilibrium with aniline in water:
$C_6H_5NH_3^+(aq) + H_2O(\ell) \rightleftarrows H_3O^+(aq) + C_6H_5NH_2(aq)$

Total volume = 0.02567 L + 0.0250 L = 0.0507 L

$[C_6H_5NH_3^+] = 0.02567 \text{ L} \left(\dfrac{0.175 \text{ mol HCl}}{1.00 \text{ L}}\right)\left(\dfrac{1 \text{ mol } C_6H_5NH_3^+}{1 \text{ mol HCl}}\right)\left(\dfrac{1}{0.0507 \text{ L}}\right) = 0.0887$ M

$K_a = \dfrac{K_w}{K_b} = \dfrac{[C_6H_5NH_2][H_3O^+]}{[C_6H_5NH_3^+]} = \dfrac{(x)^2}{(0.0887 - x)} \approx \dfrac{(x)^2}{0.0887} = 2.5 \times 10^{-5}$

Solving for $x = 0.0015$ M = $[H_3O^+]$ and $[OH^-] = \dfrac{K_w}{[H_3O^+]} = 6.7 \times 10^{-12}$ M

and $[C_6H_5NH_3^+] \approx 0.0887$ M

(c) pH = $-\log[H_3O^+] = 2.83$

Titration Curves and Indicators

17.35. The titration of 0.10 M NaOH with 0.10 M HCl (a strong base vs a strong acid)

The initial pH of a 0.10 M NaOH would be pOH = $-\log[0.10]$ so pOH = 1.00 and pH = 13.00.

When 15.0 mL of 0.10 M HCl have been added, one-half of the NaOH initially present will be consumed, leaving 0.5 (0.030 L · 0.10 mol/L) or 1.50×10^{-3} mol NaOH in 45.0 mL—

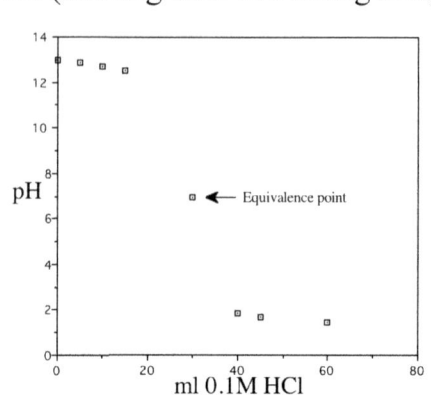

therefore a concentration of 0.0333 M NaOH.
pOH = 1.48 and pH = 12.52.

At the equivalence point (30.0 mL of the 0.10 M acid are added) there is only NaCl present. Since this salt does not hydrolyze, the pH at that point is exactly 7.0. The total volume present at this point is 60.0 mL.

Once a total of 60.0 mL of acid are added, there is an excess of 3.0×10^{-3} mol of HCl. Contained in a total volume of 90.0 mL of solution, the [HCl] = 0.0333 M and pH =1.5.

17.36. The titration of 0.050 M pyridine with 0.10 M HCl (a weak base vs a strong acid):

Initial pH ≈ 9, pH at equivalence point ≈ 3.

Total volume at equivalence point
= 50 mL + 25 mL = 75 mL.

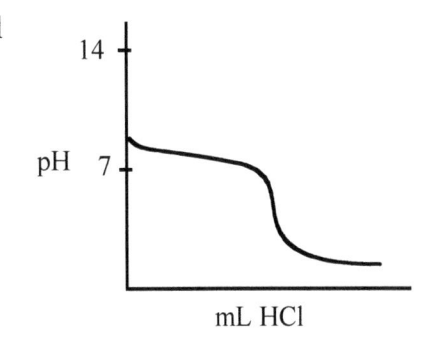

17.37. Titrate 0.10 M NH$_3$ with 0.10 M HCl:

(a) pH of 25.0 mL of 0.10 M NH$_3$:

For the weak base, NH$_3$, the equilibrium in water is represented as:

NH$_3$(aq) + H$_2$O(ℓ) ⇌ NH$_4^+$(aq) + OH$^-$(aq)

The slight dissociation of NH$_3$ would form equimolar amounts of NH$_4^+$ and OH$^-$ ions. K_b
$= \dfrac{[NH_4^+][OH^-]}{[NH_3]} = \dfrac{x^2}{[0.10 - x]} = 1.8 \times 10^{-5}$

Simplifying, you get: $\dfrac{x^2}{0.10} = 1.8 \times 10^{-5}$; $x = 1.4 \times 10^{-3}$ M = [OH$^-$]

pOH = 2.87 and pH = 11.13

(b) Addition of HCl will consume NH$_3$ and produce NH$_4^+$ (the conjugate) according to the net equation: NH$_3$(aq) + H$^+$(aq) ⇌ NH$_4^+$(aq)

The strong acid will drive this equilibrium to the right so you will assume this reaction to be complete. Calculate the moles of NH$_3$ initially present:

$0.0250 \text{ L} \left(\dfrac{0.10 \text{ mol NH}_3}{1.00 \text{ L}} \right) = 0.00250 \text{ mol NH}_3$

Reaction with the HCl will produce the conjugate acid, NH$_4^+$. The task is two-fold. First calculate the amounts of the conjugate pair present. Second substitute the concentrations into the K_b expression. [One time-saving hint: The ratio of concentrations and the ratio of the amounts (moles) will have the same numerical value. One can substitute the amounts of the conjugate pair into the K_b expression.]

$$K_b = \frac{[NH_4^+][OH^-]}{[NH_3]} = 1.8 \times 10^{-5}$$

When 25.0 mL of the 0.10 M HCl has been added (total solution volume = 50.0 mL), the reaction is at the equivalence point. All the NH$_3$ will be consumed, leaving the salt, NH$_4$Cl. The NH$_4$Cl (2.50 millimol) has a concentration of 5.0×10^{-2} M.

This salt, being formed from a weak base and strong acid, undergoes hydrolysis.

NH$_4^+$(aq) + H$_2$O(ℓ) ⇌ NH$_3$(aq) + H$_3$O$^+$ (aq)

	NH$_4^+$	NH$_3$	H$_3$O$^+$
Initial concentration	5.0×10^{-2}	0	0
Change	$-x$	$+x$	$+x$
Equilibrium	$5.0 \times 10^{-2} - x$	x	x

$$K_a = \frac{[H_3O^+][NH_3]}{[NH_4^+]} = \frac{x^2}{(5.0 \times 10^{-2} - x)} \approx \frac{x^2}{(5.0 \times 10^{-2})} = 5.6 \times 10^{-10}$$

and $x = 5.3 \times 10^{-6}$ = [H$_3$O$^+$] and pH = 5.28 (equivalence point)

(c) The halfway point of the titration occurs when 12.50 mL of the acid have been added. At that point the amount of base and salt present are equal. An examination of the K_b expression will show that under these conditions the [OH$^-$] = K_b.
So, pOH = 4.75 and pH = 9.25.

(d) From the table of indicators in your text, one indicator to use is methyl red. This indicator would be yellow prior to the equivalence point and red past that point. Bromocresol green would also be suitable, being blue prior to the equivalence point and yellow-green after the equivalence point.

(e)

mL of 0.10 M HCl added	millimol HCl added	millimol NH$_3$ after reaction	millimol NH$_4^+$ after reaction	[OH$^-$] after reaction	pH
5.00	0.50	2.0	0.50	7.2×10^{-5}	9.85
15.0	1.5	1.0	1.5	1.2×10^{-5}	9.08
20.0	2.0	0.50	2.0	4.5×10^{-6}	8.65
22.0	2.2	0.30	2.2	2.5×10^{-6}	8.39

For the pH after 30.0 mL have been added: Addition of acid in excess of 25.00 mL will result in a solution which is essentially a strong acid. After the addition of 30.0 mL, substances present are: millimol HCl added: 3.00;

millimol NH$_3$ present: 2.50; excess HCl present: (3.00 − 2.50) or 0.50 millimol

This HCl is present in a total volume of 55.0 mL of solution, hence the calculation for a strong acid proceeds as follows:
[H$_3$O$^+$] = 0.50 mmol HCl/55.0 mL = 9.1 × 10^{-3} M and pH = 2.04.

A summary of volume vs pH follows, and is graphed below:

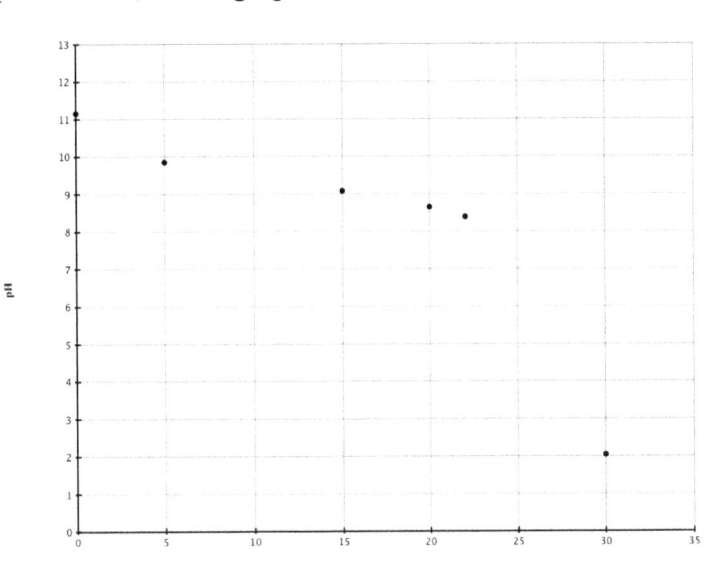

mL acid	pH
0.00	11.15
5.00	9.85
15.0	9.08
20.0	8.65
22.0	8.39
30.0	2.04

17.38. Titration of HCN with NaOH:

Rough plot of pH versus volume of base:

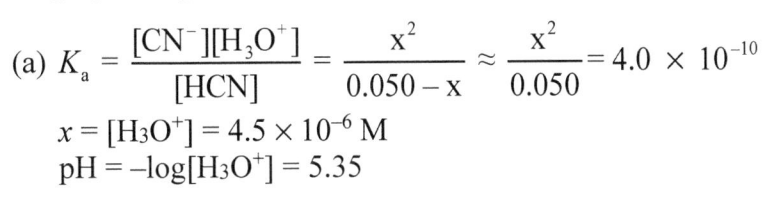

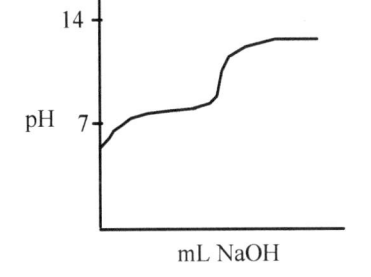

(a) $K_a = \dfrac{[CN^-][H_3O^+]}{[HCN]} = \dfrac{x^2}{0.050 - x} \approx \dfrac{x^2}{0.050} = 4.0 \times 10^{-10}$

$x = [H_3O^+] = 4.5 \times 10^{-6}$ M
pH = –log[H$_3$O$^+$] = 5.35

(b) At the half-neutralization point, [HCN] = [CN$^-$], and pH = pK_a = 9.40

(c) When 95% of NaOH has been added
mol CN$^-$ = 0.95(mol HCN)$_{initial}$ = 0.95(0.0250 L)(0.050 mol/L) = 0.0012 mol CN$^-$
mol HCN = 0.05(mol HCN)$_{initial}$ = 0.05(0.0250 L)(0.050 mol/L) = 6 × 10^{-5} mol HCN

$\text{pH} = \text{p}K_a + \log\dfrac{[CN^-]}{[HCN]} = -\log(4.0 \times 10^{-10}) + \log\left(\dfrac{1.2 \times 10^{-3}}{6 \times 10^{-5}}\right) = 10.7$

(d) Volume to reach equivalence point:
$\dfrac{0.050 \text{ mol HCN}}{1.0 \text{ L}} \cdot \dfrac{0.0250 \text{ L}}{1} \cdot \dfrac{1 \text{ mol NaOH}}{1 \text{ mol HCN}} \cdot \dfrac{1 \text{ L}}{0.075 \text{ mol NaOH}} = 0.017 \text{ L} = 17 \text{ mL}$

(e) pH at equivalence point:

Major species present is CN$^-$: CN$^-$(aq) + H$_2$O(ℓ) ⇌ HCN(aq) + OH$^-$(aq)

$[CN^-] = \dfrac{0.050 \text{ mol HCN}}{1.0 \text{ L}} \cdot 0.025 \text{ L} \cdot \dfrac{1 \text{ mol CN}^-}{1 \text{ mol HCN}} \cdot \dfrac{1}{0.042 \text{ L}} = 0.030 \text{ M}$

$$K_b = \frac{K_w}{K_a} = 2.5 \times 10^{-5} = \frac{[HCN][OH^-]}{[CN^-]} = \frac{x^2}{0.030-x} \approx \frac{x^2}{0.030}$$

$x = [OH^-] = 8.7 \times 10^{-4}$ M and pOH = $-\log[OH^-] = 3.06$ so pH = 10.94

(f) Alizarin yellow GG would be a reasonable choice for an indicator.

(g) When 105% of NaOH has been added, pH depends only on the excess OH⁻
excess OH⁻ = 0.05(0.017 L NaOH)(0.075 mol/L) = 6×10^{-5} mol OH⁻

$$[OH^-] = \frac{6 \times 10^{-5} \text{ mol}}{0.042 \text{ L}} = 0.0015 \text{ M} \text{ so pOH} = -\log[OH^-] = 2.83 \text{ and pH} = 11.17$$

17.39. Suitable indicators for titrations:

(a) HCl with pyridine: A solution of pyridinium chloride would have a pH of approximately 3. A suitable indicator would be thymol blue or bromophenol blue.

(b) NaOH with formic acid: The salt formed at the equivalence point is sodium formate. Hydrolysis of the formate ion would give rise to a basic solution (pH ≈ 8.5). Phenolphthalein would be a suitable indicator.

(c) Ethylenediamine and HCl: The base will have two endpoints—as it has two K's ($K_{b1} = 8.5 \times 10^{-5}$ and $K_{b2} = 2.7 \times 10^{-8}$). The first endpoint would contain, as the predominant specie, $H_2N(CH_2)_2NH_3^+$. This ion would hydrolyze with water. The K_a of this conjugate acid would be $1.0 \times 10^{-14}/ 8.5 \times 10^{-5}$ or 1.2×10^{-10}. Assume that you have a 0.1 M soln of the anion. The pH can be determined by the expression:

$$K_a = \frac{[H_3O^+]^2}{0.1} = 1.2 \times 10^{-10}$$ [Recall that the concentration of the anion and the hydronium ion would be identical.]
$[H_3O^+] = 3.5 \times 10^{-6}$ and pH = $-\log(3.5 \times 10^{-6})$ or 5.5. Since the indicator would need to change colors +/- 1 on either side of the endpoint, methyl red would be a suitable indicator for this endpoint.

The 2nd endpoint would be reached when the monoprotonated base (let's call it BH⁺) has accepted a 2nd proton (to form BH₂²⁺).

The material formed would hydrolyze: $BH_2^{2+} + H_2O \rightleftharpoons BH^+ + H_3O^+$, with a $K_a = 1.0 \times 10^{-14}/ 2.7 \times 10^{-8}$ or 3.7×10^{-7}. Substituting into the K_a expression:

$$K_a = \frac{[H_3O^+]^2}{0.1} = 3.7 \times 10^{-7}. \text{ So } [H_3O^+] = 1.9 \times 10^{-4} \text{ and pH} = 3.7.$$

A suitable indicator would be thymol blue or methyl red.

17.40. Suggest a suitable indicator:

Titration	pH at Equiv. Point	Possible Indicator
(a) HCO_3^- titrated with NaOH	> 7 (about 12-13)	alizarin yellow
(b) HClO with NaOH	> 7 (about 10-11)	thymolphthalein
(c) $(CH_3)_3N$ with HCl	< 7 (about 5-6)	methyl red

Solubility Guidelines

17.41. Two insoluble salts of

    (a) Cl^- Silver chloride, AgCl; lead(II) chloride, $PbCl_2$

    (b) Zn^{2+} Zinc carbonate, $ZnCO_3$; zinc sulfide, ZnS

    (c) Fe^{2+} Iron(II) carbonate, $FeCO_3$; iron(II) oxalate, FeC_2O_4

17.42. Two insoluble salts of:

    (a) CO_3^{2-} Iron(II) carbonate, $FeCO_3$; lead(II) carbonate, $PbCO_3$

    (b) Ba^{2+} Barium sulfate, $BaSO_4$; barium phosphate, $Ba_3(PO_4)_2$

    (c) I^- lead(II) iodide, PbI_2; silver iodide, AgI

17.43. Using the table of solubility guidelines, predict water solubility for the following:

    (a) $(NH_4)_2CO_3$ Ammonium salts are **soluble**.
    (b) $ZnSO_4$ Sulfates are generally **soluble**.
    (c) NiS   Sulfides are generally **insoluble**.
    (d) $BaSO_4$ Sr^{2+}, Ba^{2+}, and Pb^{2+} form **insoluble** sulfates.

17.44. Predict whether the salts are soluble or insoluble:

    (a) $Ni(OH)_2$ insoluble (most hydroxide salts are insoluble)
    (b) $Zn(NO_3)_2$ soluble (most nitrate salts are soluble)
    (c) $MgCl_2$ soluble (most chloride salts are soluble)
    (d) K_2S soluble (most potassium salts are soluble)

Writing Solubility Product Constant Expressions

17.45. <u>Salt dissolving</u> K_{sp} <u>expression</u>

    (a) $AgCN(s) \rightleftarrows Ag^+(aq) + CN^-(aq)$ $K_{sp} = [Ag^+][CN^-]$

    (b) $NiCO_3(s) \rightleftarrows Ni^{2+}(aq) + CO_3^{2-}(aq)$ $K_{sp} = [Ni^{2+}][CO_3^{2-}]$

    (c) $AuBr_3(s) \rightleftarrows Au^{3+}(aq) + 3\,Br^-(aq)$ $K_{sp} = [Au^{3+}][Br^-]^3$

17.46. Salt dissolving K_{sp} expression

(a) $CaCO_3(s) \rightleftarrows Ca^{2+}(aq) + CO_3^{2-}(aq)$ $K_{sp} = [Ca^{2+}][CO_3^{2-}]$

(b) $PbBr_2(s) \rightleftarrows Pb^{2+}(aq) + 2\,Br^-(aq)$ $K_{sp} = [Pb^{2+}][Br^-]^2$

(c) $Ag_2S(s) \rightleftarrows 2\,Ag^+(aq) + S^{2-}(aq)$ $K_{sp} = [Ag^+]^2[S^{2-}]$

Calculating K_{sp}

17.47. Here you need only to substitute the equilibrium concentrations into the K_{sp} expression:

$K_{sp} = [Tl^+][Br^-] = (1.9 \times 10^{-3})(1.9 \times 10^{-3}) = 3.6 \times 10^{-6}$

17.48. Calculate K_{sp} for silver acetate:

$[Ag^+] = [CH_3CO_2^-] = \dfrac{1.0 \text{ g}}{0.1000 \text{ L}} \cdot \dfrac{1 \text{ mol AgCH}_3CO_2}{167 \text{ g}} \cdot \dfrac{1 \text{ mol Ag}^+}{1 \text{ mol AgCH}_3CO_2} = 0.060 \text{ M}$

$K_{sp} = [Ag^+][CH_3CO_2^-] = (0.060)(0.060) = 0.0036$

17.49. What is K_{sp} for SrF_2?

The temptation is to calculate the number of moles of the solid that are added to water, but one must recall (a) not all the solid dissolves and (b) the concentration of the solid does not appear in the K_{sp} expression.

Note that the equilibrium concentration of $[Sr^{2+}] = 1.03 \times 10^{-3}$ M. The stoichiometry of the solid dissolving indicates two fluoride ions accompany the formation of one strontium ion, so $[F^-] = 2.06 \times 10^{-3}$ M. Substituting into the K_{sp} expression:

$K_{sp} = [Sr^{2+}][F^-]^2 = (1.03 \times 10^{-3})(2.06 \times 10^{-3})^2 = 4.37 \times 10^{-9}$.

17.50. What is K_{sp} for $Ca(OH)_2$?

Based on the amount of $Ca(OH)_2$ that dissolves:

$[Ca^{2+}] = \dfrac{0.824 \text{ g Ca(OH)}_2}{1 \text{ L}} \left(\dfrac{1 \text{ mol Ca(OH)}_2}{74.09 \text{ g}} \right) \left(\dfrac{1 \text{ mol Ca}^{2+}}{1 \text{ mol Ca(OH)}_2} \right) = 0.01112 \text{ M}$

$K_{sp} = [Ca^{2+}][OH^-]^2 = (0.01112)(2 \times 0.01112)^2 = 5.50 \times 10^{-6}$

17.51. For lead(II) hydroxide, the K_{sp} expression is $K_{sp} = [Pb^{+2}][OH^-]^2$.

Since you know the pH, you can calculate the $[OH^-]$.

pH = 9.15 and pOH = 14.00 − 9.15 = 4.85 so $[OH^-] = 1.4 \times 10^{-5}$.

For each mole of $Pb(OH)_2$ that dissolves, you get one mol of Pb^{+2} and two mol of OH^-. Since $[OH^-]$ at equilibrium = 1.41×10^{-5} M, then $[Pb^{+2}] = 1/2 \cdot 1.41 \times 10^{-5}$ M

$K_{sp} = [Pb^{+2}][OH^-]^2 = [7.0 \times 10^{-6}][1.4 \times 10^{-5}]^2 = 1.4 \times 10^{-15}$

17.52. Estimate the value of K_{sp} for Ca(OH)$_2$:

pOH = 14.00 − pH = 14.00 − 12.35 = 1.65 so [OH⁻] = 10^{-pOH} = 0.0224 M

[Ca^{2+}] = ½ × [OH⁻] = 0.0112

K_{sp} = [Ca^{2+}][OH⁻]^2 = (0.0112)(0.0224)^2 = 5.6 × 10^{-6}

Estimating Salt Solubility from K_{sp}

17.53. The solubility of AgI in water at 25°C in (a) mol/L and (b) g/L

(a) The K_{sp} for AgI (from Appendix J) is 8.5 × 10^{-17}. The K_{sp} expression is:

K_{sp} = [Ag$^+$][I⁻] = 8.5 × 10^{-17}

Since the solid dissolves to give one silver ion/one iodide ion, the concentrations of Ag$^+$ and I⁻ will be equal. You can write the K_{sp} as: [Ag$^+$]^2 = [I⁻]^2 = 8.5 × 10^{-17}.

Taking the square root of both sides you obtain: [Ag$^+$] = [I⁻] = 9.22 × 10^{-9} and recognizing that each mole of AgI that dissolves per liter gives one mol of silver ion, the solubility of AgI will be 9.2 × 10^{-9} moles per liter.

(b) The solubility in g/L: $\dfrac{9.2 \times 10^{-9} \text{ mol AgI}}{L} \left(\dfrac{234.77 \text{ g AgI}}{1 \text{ mol AgI}} \right)$ = 2.2 × 10^{-6} g AgI/L

17.54. Concentration of Au$^+$ in saturated AuCl:

The equation for the equilibrium in question: AuCl(s) ⇌ Au$^+$(aq) + Cl⁻(aq)

	Au$^+$	Cl⁻
Initial (M)	0	0
Change (M)	+x	+x
Equilibrium (M)	x	x

K_{sp} = [Au$^+$][Cl⁻] = (x)(x) = x^2 = 2.0 × 10^{-13}

x = [Au$^+$] = $\sqrt{K_{sp}}$ = $\sqrt{2.0 \times 10^{-13}}$ = 4.5 × 10^{-7} mol/L

17.55. Solubility of CaF$_2$ in g/L and mol/L:

The K_{sp} for CaF$_2$ = 5.3 × 10^{-11}

K_{sp} = [Ca^{2+}][F⁻]^2 = 5.3 × 10^{-11}

if a mol/L of CaF$_2$ dissolve, [Ca^{2+}] = x and [F⁻] = 2x

K_{sp} = (x)(2x)^2 = 4x^3 = 5.3 × 10^{-11} and x = 2.37 × 10^{-4}

(a) The molar solubility is 2.4 × 10⁻⁴ moles per liter. (2 sf)

(b) Solubility in g/L:

$$\frac{2.37 \times 10^{-4} \text{ mol CaF}_2}{1 \text{ L}} \left(\frac{74.09 \text{ g CaF}_2}{1 \text{ mol CaF}_2} \right) = 0.018 \text{ g CaF}_2/\text{L}$$

17.56. Solubility of PbI$_2$ in g/L and mol/L:

The equation for the equilibrium: PbI$_2$(s) ⇌ Pb^{2+}(aq) + 2 I⁻(aq)

	Pb^{2+}	I⁻
Initial (M)	0	0
Change (M)	+x	+2x
Equilibrium (M)	x	2x

(a) $K_{sp} = 9.8 \times 10^{-9} = [\text{Pb}^{2+}][\text{I}^-]^2 = (x)(2x)^2 = 4x^3$; $x = 1.35 \times 10^{-3}$.
One mole of Pb^{2+} is created for each mole of PbI$_2$ dissolved. So, the solubility is 1.4×10^{-3} mol/L. (2 sf)

(b) Mass of PbI$_2$ per liter:

$$\frac{1.35 \times 10^{-3} \text{ mol PbI}_2}{1 \text{ L}} \left(\frac{461.0 \text{ g PbI}_2}{1 \text{ mol PbI}_2} \right) = 0.62 \text{ g PbI}_2$$

17.57. $K_{sp} = [\text{Ra}^{2+}][\text{SO}_4^{2-}] = 4.2 \times 10^{-11}$ so $[\text{Ra}^{2+}] = [\text{SO}_4^{2-}] = 6.5 \times 10^{-6}$ M
RaSO$_4$ will dissolve to the extent of 6.5×10^{-6} mol/L
Express this as grams in 100. mL (or 0.1 L)

$$\frac{6.5 \times 10^{-6} \text{ mol RaSO}_4}{1 \text{ L}} \cdot \frac{0.1 \text{ L}}{1} \cdot \frac{322 \text{ g RaSO}_4}{1 \text{ mol RaSO}_4} = 2.1 \times 10^{-4} \text{ g}$$

Expressed as milligrams: 0.21 mg (2 sf) of RaSO$_4$ will dissolve. The 25 mg of RaSO$_4$ will not completely dissolve.

17.58. How much barium fluoride will dissolve?

The equation for the equilibrium in question: BaF$_2$(s) ⇌ Ba^{2+}(aq) + 2 F⁻(aq)

	Ba^{2+}	F⁻
Initial	0	0
Change	+x	+2x
Equilibrium	x	2x

$K_{sp} = [\text{Ba}^{2+}][\text{F}^-]^2 = (x)(2x)^2 = 4x^3$

$x = \sqrt[3]{\dfrac{K_{sp}}{4}} = \sqrt[3]{\dfrac{1.8 \times 10^{-7}}{4}} = 3.6 \times 10^{-3}$ mol/L

$25 \text{ mg BaF}_2 = 0.025 \text{ g BaF}_2 \left(\dfrac{1 \text{ mol BaF}_2}{175.33 \text{ g BaF}_2} \right) = 1.4 \times 10^{-4} \text{ mol BaF}_2$

$\dfrac{1.4 \times 10^{-4} \text{ mol BaF}_2}{0.250 \text{ L}} = 5.7 \times 10^{-4}$ mol/L

Because the solution concentration is less than the K_{sp} value all the BaF$_2$ will dissolve.

17.59. Which solute in each of the following pairs is more soluble.

	Compound	K_{sp}
(a)	**PbCl$_2$**	1.7×10^{-5}
	PbBr$_2$	6.6×10^{-6}
(b)	HgS (red)	4×10^{-54}
	FeS	6×10^{-19}
(c)	**Fe(OH)$_2$**	4.9×10^{-17}
	Zn(OH)$_2$	3×10^{-17}

To compare relative solubilities of two compounds of the same general formula, one can examine the K_{sp} value. The larger the value of K_{sp}, the more soluble the compound.

17.60. The more soluble compound of the pair:
(a) AgCl ($K_{sp} = 1.8 \times 10^{-10}$) is more soluble than AgSCN ($K_{sp} = 1.0 \times 10^{-12}$)
(b) PbSO$_4$ ($K_{sp} = 2.5 \times 10^{-8}$) is more soluble than PbCO$_3$ ($K_{sp} = 7.4 \times 10^{-14}$)
(c) MgF$_2$ ($K_{sp} = 5.2 \times 10^{-11}$) is more soluble than AgCl ($K_{sp} = 1.8 \times 10^{-10}$)
(d) PbF$_2$ ($K_{sp} = 3.3 \times 10^{-8}$) is more soluble than SrF$_2$ ($K_{sp} = 4.3 \times 10^{-9}$)

17.61. The more soluble compound of the pair:
(a) PbBr$_2$ (After solubility calculation, [TlBr] = 1.9×10^{-3} M and [PbBr$_2$] = 5.5×10^{-2} M.)
(b) Hg$_2$Cl$_2$ (After solubility calculation, [AuCl] = 4.5×10^{-7} M and [Hg$_2$Cl$_2$] = 7.0×10^{-7} M.)
(c) TlI (After solubility calculation, [TlI] = 2.3×10^{-4} M and [Hg$_2$I$_2$] = 1.9×10^{-10} M.)

17.62. The more soluble compound of the pair:

(a) Ag$_2$CO$_3$ (After solubility calculation, [BaCO$_3$] = 5.1x10^{-5} M and [Ag$_2$CO$_3$] = 1.3x10^{-4} M.)

(b) PbI$_2$ (After solubility calculation, [TlI] = 2.4x10^{-4} M and [PbI$_2$] = 1.4x10^{-3} M.)

(c) Hg$_2$Br$_2$ (After solubility calculation, [AgBr] = 7.3x10^{-7} M and [Hg$_2$Br$_2$] = 1.4x10^{-6} M.)

The Common Ion Effect and Salt Solubility

17.63. The equilibrium for AgSCN dissolving is: AgSCN(s) ⇌ Ag$^+$(aq) + SCN$^-$(aq).

As x mol/L of AgSCN dissolve in pure water, x mol/L of Ag$^+$ and x mol/L of SCN$^-$ are produced. The expression would be: K_{sp} = [Ag$^+$][SCN$^-$]

Substituting x for the concentrations of the ions: x^2 = 1.0 × 10^{-12} and x = 1.0 × 10^{-6} M

So, 1.0 × 10^{-6} mol AgSCN/L dissolve in pure water.

The equilibrium for AgSCN dissolving in NaSCN (0.010 M) is like that above. Equimolar amounts of Ag$^+$ and SCN$^-$ ions are produced as the solid dissolves. However, the [SCN$^-$] is augmented by the soluble NaSCN.

K_{sp} = [Ag$^+$][SCN$^-$] = $(x)(x + 0.010)$ = 1.0 × 10^{-12}

You can simplify the expression by assuming that $x + 0.010 \approx 0.010$. Note that the value of x above (1.0 × 10^{-6}) lends credibility to this assumption.

$(x)(0.010)$ = 1.0 × 10^{-12} and x = 1.0 × 10^{-10} M

The solubility of AgSCN in 0.010 M NaSCN is 1.0 × 10^{-10} M -- reduced by four orders of magnitude from its solubility in pure water.

17.64. Calculate the solubility of silver bromide:

K_{sp} = 5.4 × 10^{-13} = [Ag$^+$][Br$^-$] = $(x)(x) = x^2$
x = solubility of AgBr in pure water = 7.3 × 10^{-7} mol/L

$$[Br^-] = \left(\frac{0.15 \text{ g NaBr}}{0.225 \text{ L}}\right)\left(\frac{1 \text{ mol NaBr}}{103 \text{ g}}\right)\left(\frac{1 \text{ mol Br}^-}{1 \text{ mol NaBr}}\right) = 0.0065 \text{ M}$$

In water containing 0.0065 M Br$^-$: K_{sp} = 5.4 × 10^{-13} = $(x)(0.0065 + x) \approx x(0.0065)$
x = solubility of AgBr in water containing 0.0065 M Br$^-$ = 8.3 × 10^{-11} mol/L

17.65. Solubility in mg/mL of AgI in (a) pure water and (b) and in 0.020 M in Ag$^+$ (aq). K_{sp} = [Ag$^+$][I$^-$] = 8.5 × 10^{-17}

(a) Since the solid dissolves to give one silver ion/one iodide ion, the concentrations of Ag$^+$ and I$^-$ will be equal. You can write the K_{sp} as: [Ag$^+$]^2 = [I$^-$]^2 = 8.5 × 10^{-17}.

Taking the square root of both sides you obtain: $[Ag^+] = [I^-] = 9.22 \times 10^{-9}$ and recognizing that each mole of AgI that dissolves per liter gives one mol of silver ion, the solubility of AgI will be 9.2×10^{-9} moles per liter.

The solubility in mass per liter:

$$\frac{9.2 \times 10^{-9} \text{ mol AgI}}{L} \left(\frac{234.77 \text{ g AgI}}{1 \text{ mol AgI}} \right) = 2.2 \times 10^{-6} \text{ g AgI/L}$$

So, 2.2×10^{-6} mg will dissolve in water per mL of solution.

(b) When AgI dissolves in water, you know that the concentrations of Ag^+ and I^- are equal. With the addition of the 0.020 M silver solution, this is no longer a valid assumption. If you let x mol/L of the AgI dissolve, the concentrations of Ag^+ and I^- (from the salt dissolving) will be x mol/L. The $[Ag^+]$ will be amended by 0.020 M, so you write: $K_{sp} = [x + 0.020][x] = 8.5 \times 10^{-17}$. This could be solved with the quadratic equation, but a bit of thought will simplify the process. In the (a) part you discovered that the $[Ag^+]$ was approximately 10^{-9} M. This value is small compared to 0.020. Use that approximation to convert the K_{sp} expression to: $[0.020][x] = 8.5 \times 10^{-17}$ and $x = 4.25 \times 10^{-15}$ M.

This molar solubility translates into:

$$\frac{4.25 \times 10^{-15} \text{ mol AgI}}{L} \cdot \frac{234.77 \text{ g AgI}}{1 \text{ mol AgI}} = 9.98 \times 10^{-13} \text{ g AgI/L}$$

or 1.0×10^{-12} mg/mL (2 sf).

17.66. Calculate the solubility of barium fluoride:

(a) $K_{sp} = 1.8 \times 10^{-7} = [Ba^{2+}][F^-]^2 = (x)(2x)^2 = 4x^3$ $x = \sqrt[3]{\frac{K_{sp}}{4}} = \sqrt[3]{\frac{1.8 \times 10^{-7}}{4}} = 0.0036$ mol/L

$$\frac{0.0036 \text{ mol Ba}^{2+}}{1 \text{ L}} \left(\frac{1 \text{ mol BaF}_2}{1 \text{ mol Ba}^{2+}} \right) \left(\frac{175.33 \text{ g}}{1 \text{ mol BaF}_2} \right) \left(\frac{10^3 \text{ mg}}{1 \text{ g}} \right) \left(\frac{1 \text{ L}}{10^3 \text{ mL}} \right)$$

= 0.63 mg/mL in pure water

(b) $[F^-] = \left(\frac{0.0050 \text{ g KF}}{0.001 \text{ L}} \right) \left(\frac{1 \text{ mol}}{58.10 \text{ g}} \right) = 0.086$ M

$K_{sp} = 1.8 \times 10^{-7} = [Ba^{2+}][F^-]^2 = (x)(0.086 + 2x)^2 \approx (x)(0.086)^2$
$x = 2.4 \times 10^{-5}$ mol/L

$$\frac{2.4 \times 10^{-5} \text{ mol Ba}^{2+}}{1 \text{ L}} \left(\frac{1 \text{ mol BaF}_2}{1 \text{ mol Ba}^{2+}} \right) \left(\frac{175.10 \text{ g}}{1 \text{ mol BaF}_2} \right) \left(\frac{10^3 \text{ mg}}{1 \text{ g}} \right) \left(\frac{1 \text{ L}}{10^3 \text{ mL}} \right)$$

= 4.2×10^{-3} mg/mL in 5.0 mg/mL KF

17.67. Calculate the solubility (in mol/L) of Fe(OH)₂ in a solution buffered to pH = 7.00. The solubility expression for the iron(II) hydroxide is:

$K_{sp} = [Fe^{2+}][OH^-]^2 = 4.9 \times 10^{-17}$. Now you know the [OH⁻], since the pH = 7.00, the pOH is 7.00, and [OH⁻] = 1.0×10^{-7}. Substitute this concentration into the K_{sp} expression above: $[Fe^{2+}][1.0 \times 10^{-7}]^2 = 4.9 \times 10^{-17}$. Solving for [Fe²⁺] gives:

$$[Fe^{2+}] = \frac{4.9 \times 10^{-17}}{[1.0 \times 10^{-7}]^2} = 4.9 \times 10^{-3}.$$

So 4.9×10^{-3} mol/L of the Fe(OH)₂ will dissolve.

17.68. Solubility of Mg(OH)₂ in a solution that is buffered at pH = 9.50:

x = solubility of Mg(OH)₂ = [Mg²⁺]

The equilibrium: Mg(OH)₂(s) ⇌ Mg²⁺(aq) + 2 OH⁻(aq)

At pH = 9.50, pOH = 4.50, so $[OH^-] = 3.2 \times 10^{-5}$ M

The $K_{sp} = [Mg^{2+}][OH^-]^2 = 5.5 \times 10^{-12}$.

$x = [Mg^{2+}] = 5.5 \times 10^{-12}/[OH^-]^2 = 5.5 \times 10^{-12}/(3.2 \times 10^{-5})^2 = 5.5 \times 10^{-3}$ M

The Effect of Basic Anions on Salt Solubility

17.69. The salt that should be more soluble in nitric acid than in pure water from the pairs:

(a) PbCl₂ or PbS: PbS will be more soluble, since the S²⁻ ion will react with the nitric acid to produce H₂S, reducing the [S²⁻], and increasing the amount of PbS that dissolves.

(b) Ag₂CO₃ or AgI: Ag₂CO₃ will be more soluble. The CO₃²⁻ ion will react with the nitric acid and produce HCO₃⁻ and H₂CO₃ that will decompose to CO₂ and H₂O. The removal of carbonate shifts the equilibrium to the right, increasing the amount that dissolves.

(c) Al(OH)₃ or AgCl: Al(OH)₃ will be more soluble. As in (b) above, the OH⁻ will react with the H⁺ to form water. The reduction in [OH⁻] will increase the amount of the salt that dissolves.

17.70. The salt that should be more soluble in nitric acid than in pure water from the pairs:

(a) AgCN or AgBr: AgCN will be more soluble, since the CN⁻ ion will react with the nitric acid, reducing the [CN⁻], and increasing the amount of AgCN that dissolves.

(b) PbF₂ or PbI₂: PbF₂ will be more soluble. The F⁻ ion will react with the nitric acid, reducing the [F⁻], and increasing the amount of PbF₂ that dissolves.

(c) CuI or Cu(OH)₂: Cu(OH)₂ will be more soluble. The OH⁻ will react with the H⁺ to form water. The reduction in [OH⁻] will increase the amount of the salt that dissolves.

17.71. The salt that should be more soluble in pure water than predicted:

(a) AuCl or AuCN: AuCN will be more soluble, since the CN⁻ ion will react slightly with water to form HCN, reducing the [CN⁻], and increasing the amount of AgCN that dissolves.

(b) Ag₂C₂O₄ or AgBr: Ag₂C₂O₄ will be more soluble. The C₂O₄²⁻ ion will react slightly with water, producing HC₂O₄⁻, reducing the [C₂O₄²⁻], and increasing the amount of Ag₂C₂O₄ that dissolves.

(c) Hg₂I₂ or Hg₂CO₃: Hg₂CO₃ is more soluble as the CO₃²⁻ ion will react with the water and produce HCO₃⁻. The removal of carbonate shifts the equilibrium to the right, increasing the amount that dissolves.

17.72. More soluble in water than predicted from the K_{sp}:

(a) AgI or Ag₂CO₃: Ag₂CO₃ is more soluble as the CO₃²⁻ ion will react with the water and produce HCO₃⁻. The removal of carbonate shifts the equilibrium to the right, increasing the amount that dissolves.

(b) PbCO₃ or PbCl₂: PbCO₃ is more soluble—for the same reasons as noted in (a).

(c) AgCl or AgCN: AgCN is more soluble as interaction of the CN⁻ ion with water will produce HCN, reducing the concentration of CN⁻ and increasing the solubility of AgCN.

Precipitation Reactions

17.73. Given the equation for PbCl₂ dissolving in water: PbCl₂(s) ⇌ Pb⁺² (aq) + 2 Cl⁻(aq) you can write the K_{sp} expression: K_{sp} = [Pb⁺²][Cl⁻]² = 1.7 × 10⁻⁵

Substituting the ion concentrations into the K_{sp} expression you get:
Q = [Pb⁺²][Cl⁻]² = (0.0012)(0.010)² = 1.2 × 10⁻⁷

Since Q is less than K_{sp}, no PbCl₂ precipitates.

17.74. Will precipitation of NiCO₃ occur: [K_{sp} for NiCO₃ = 1.4 × 10⁻⁷]

(a) When [CO₃²⁻] = 1.0 × 10⁻⁶ M: Q = [Ni²⁺][CO₃²⁻] = (0.0024)(1.0 × 10⁻⁶) = 2.4 × 10⁻⁹

$Q < K_{sp}$, so NiCO₃ will not precipitate

(b) When [CO₃²⁻] = 1.0 × 10⁻⁴ M: Q = [Ni²⁺][CO₃²⁻] = (0.0024)(1.0 × 10⁻⁴) = 2.4 × 10⁻⁷
$Q > K_{sp}$, so NiCO₃ will precipitate

17.75. If Zn(OH)₂ is to precipitate, the reaction quotient (Q) must exceed the K_{sp} for the salt. 4.0 mg of NaOH in 10. mL corresponds to a concentration of:

$$[OH^-] = \left(\frac{4.0 \times 10^{-3} \text{ g NaOH}}{0.0100 \text{ L}}\right)\left(\frac{1 \text{ mol NaOH}}{40.00 \text{ g NaOH}}\right) = 0.010 \text{ M}$$

The value of Q is: $[Zn^{2+}][OH^-]^2 = (1.6 \times 10^{-4})(1.0 \times 10^{-2})^2 = 1.6 \times 10^{-8}$

The value of $Q > K_{sp}$ for the salt (3×10^{-17}), so $Zn(OH)_2$ precipitates.

17.76. Will $PbCl_2$ precipitate when NaCl is added: [K_{sp} for $PbCl_2 = 1.7 \times 10^{-5}$]

$$[Cl^-] = \frac{1.20 \text{ g}}{0.095 \text{ L}} \cdot \frac{1 \text{ mol NaCl}}{58.44 \text{ g}} \cdot \frac{1 \text{ mol Cl}^-}{1 \text{ mol NaCl}} = 0.216 \text{ M}$$

$Q = [Pb^{2+}][Cl^-]^2 = (0.0012)(0.216)^2 = 5.6 \times 10^{-5}$

$Q > K_{sp}$ so $PbCl_2$ will precipitate

17.77. The molar concentration of Mg^{2+} is:

$$\frac{1350 \text{ mg Mg}^{2+}}{1 \text{ L}} \left(\frac{1 \text{ g Mg}^{2+}}{1000 \text{ mg Mg}^{2+}}\right)\left(\frac{1 \text{ mol Mg}^{2+}}{24.305 \text{ g Mg}^{2+}}\right) = 5.55 \times 10^{-2} \text{M}$$

For $Mg(OH)_2$ to precipitate, Q must be greater than K_{sp} for the salt (5.6×10^{-12}).

$Q = [Mg^{2+}][OH^-]^2 = (5.55 \times 10^{-2})(OH^-)^2 = 5.6 \times 10^{-12}$ so

$(OH^-)^2 = \frac{5.6 \times 10^{-12}}{5.55 \times 10^{-2}}$ or 1.01×10^{-10}, and $[OH^-] = 1.0 \times 10^{-5}$ M.

17.78. Will $Ca(OH)_2$ precipitate when NaOH is added to a $CaCl_2$ solution:

$$[OH^-] = \frac{(0.025 \text{ L})(0.010 \text{ mol/L})}{0.150 \text{ L}} = 1.7 \times 10^{-3} \text{ M}$$

$$[Ca^{2+}] = \frac{(0.125 \text{ L})(0.10 \text{ mol/L})}{0.150 \text{ L}} = 8.3 \times 10^{-2} \text{ M}$$

$Q = [Ca^{2+}][OH^-]^2 = (0.083)(0.0017)^2 = 2.3 \times 10^{-7}$

The value of $Q < K_{sp}$ (5.5×10^{-6}) so $Ca(OH)_2$ will not precipitate

Equilibria Involving Complex Ions

17.79. An equilibrium to demonstrate that sufficient OH⁻ will dissolve $Zn(OH)_2$:

The equilibrium for the dissolution of $Zn(OH)_2$ is:

$Zn(OH)_2(s) \rightleftarrows Zn^{2+}(aq) + 2 OH^-(aq)$ $K = 3 \times 10^{-17}$

$Zn^{2+}(aq) + 4 OH^-(aq) \rightleftarrows [Zn(OH)_4]^{2-}(aq)$ $K = 4.6 \times 10^{17}$

Net $Zn(OH)_2(s) + 2 OH^-(aq) \rightleftarrows [Zn(OH)_4]^{2-}(aq)$

$K_{net} = (3 \times 10^{-17})(4.6 \times 10^{17}) = 10$ (1 sf)

Note that the magnitude of the net K indicates that sufficient OH⁻ will dissolve Zn(OH)₂ as the K indicates the equilibrium is product-favored.

17.80. Calculate K_{net} for AgI dissolving in a cyanide solution:

AgI(s) ⇌ Ag⁺(aq) + I⁻(aq)	$K_{sp} = 8.5 \times 10^{-17}$
Ag⁺(aq) + 2 CN⁻(aq) ⇌ [Ag(CN)₂]⁻(aq)	$K_{form} = 1.3 \times 10^{21}$
AgI(s) + 2 CN⁻(aq) ⇌ [Ag(CN)₂]⁻(aq) + I⁻(aq)	$K_{net} = K_{sp} \times K_{form} = 1.1 \times 10^{5}$

17.81. Moles of ammonia to dissolve 0.050 mol of AgCl suspended in 1.0 L of water:

The equilibria involved are:

AgCl(s) ⇌ Ag⁺(aq) + Cl⁻(aq)	$K = 1.8 \times 10^{-10}$
Ag⁺(aq) + 2 NH₃(aq) ⇌ [Ag(NH₃)₂]⁺(aq)	$K = 1.1 \times 10^{7}$
AgCl(s) + 2 NH₃(aq) ⇌ [Ag(NH₃)₂]⁺(aq) + Cl⁻(aq)	K_{net} where

$$K_{net} = (1.8 \times 10^{-10})(1.1 \times 10^{7}) = 1.98 \times 10^{-3}$$

The equilibrium expression for the net equation is: $\dfrac{[Ag(NH_3)_2]^+[Cl^-]}{[NH_3]^2} = 1.98 \times 10^{-3}$

Note that if you dissolve 0.050 mol AgCl (in 1.0 L), both the concentration of the complex ion AND the chloride ion will be equal to 0.050 M. Substituting these values into the expression above, and solving for the concentration of ammonia yields:

$$\frac{(0.050)(0.050)}{1.98 \times 10^{-3}} = [NH_3]^2 \text{ and } [NH_3] = 1.12 \text{ M (or 1.1 M to 2 sf)}$$

The equilibrium concentration of ammonia must be 1.1 M. Note that the stoichiometry of forming the complex ion requires (2 × 0.050 mol) or 0.10 mol NH₃.

The total amount of ammonia needed is: 0.10 mol + 1.1 mol or 1.2 mol NH₃.

17.82. Can you dissolve 15.0 mg of AuCl in 100.0 mL of water by adding NaCN?

The equation in question is: AuCl(s) + 2 CN⁻(aq) ⇌ [Au(CN)₂]⁻(aq) + Cl⁻(aq) for which the $K_{net} = K_{sp} \times K_{form} = (2.0 \times 10^{-13}) \times (2.0 \times 10^{38}) = 4.0 \times 10^{25}$

[[Au(CN)₂]⁻] = [Cl⁻] =

$$\frac{1.50 \times 10^{-2} \text{ g}}{0.1150 \text{ L}} \left(\frac{1 \text{ mol AuCl}}{242.42 \text{ g}} \right) \left(\frac{1 \text{ mol }[Au(CN)_2]^-}{1 \text{ mol AuCl}} \right) = 5.381 \times 10^{-4} \text{ M}$$

$$15 \times 10^{-3} \text{ g} \left(\frac{1 \text{ mol AuCl}}{232.40 \text{ g}} \right) \left(\frac{1 \text{ mol [Au(CN)}_2\text{]}^-}{1 \text{ mol AuCl}} \right) \left(\frac{1}{0.1000 \text{ L}} \right) = 6.45 \times 10^{-4} M$$

Set up an equilibrium expression and solve for the CN⁻ ion concentration:

$$4.0 \times 10^{25} = \frac{[\text{Au(CN)}_2^-][\text{Cl}^-]}{[\text{CN}^-]^2} = \frac{(5.381 \times 10^{-4})^2}{[\text{CN}^-]^2}$$

[CN⁻] = 8.5 × 10⁻¹⁷ M

Amount of CN⁻ needed: (1.0 × 10⁻¹⁶ M)(0.1000 L) = 1.0 × 10⁻¹⁷ mol CN⁻ needed to dissolve 15.0 mg AuCl in 100.0 mL water and an additional 2×(5.381 × 10⁻⁴ mol/L)(0.1150 L) = 1.238 × 10⁻⁴ moles to complex with the Au⁺ ion. A total amount of 1.2 × 10⁻⁴ moles of CN⁻ is needed.

Moles of CN⁻ available: (15.0 × 10⁻³ L)(6.00 M) = 0.0900 mol NaCN

The solid will dissolve.

17.83. Solubility of AgCl in water and in 1.0 M NH₃:

(a) Solubility of AgCl in water:

AgCl(s) ⇌ Ag⁺(aq) + Cl⁻(aq) K_{sp} = 1.8 × 10⁻¹⁰ = [Ag⁺][Cl⁻]

Noting that the concentration of the two ions will be identical: [Ag⁺]² = 1.8 × 10⁻¹⁰ and [Ag⁺] = 1.3 × 10⁻⁵ M so the solubility of AgCl is 1.3 × 10⁻⁵ M.

(b) Solubility of AgCl in 1.0 M NH₃:
The pertinent equilibrium is:
AgCl(s) + 2 NH₃(aq) ⇌ [Ag(NH₃)₂]⁺(aq) + Cl⁻(aq) $K = K_{sp} \times K_f$ = 1.98 × 10⁻³

The equilibrium concentration of ammonia is 1.0 M, with the small amount of NH₃ used in the reaction: 1.0 − 2x
Substitute into the K expression:

$$\frac{[\text{Ag(NH}_3)_2^+][\text{Cl}^-]}{[\text{NH}_3]^2} = \frac{[x][x]}{[1.0 - 2x]^2} = 2.0 \times 10^{-3}$$

The concentration of both the chloride and complex ion will be equal (x), so the terms in the numerator can be represented as x². Taking the square root of both sides yields:

$\frac{[x]}{[1.0 - 2x]} = 4.47 \times 10^{-2}$. Solving for x gives: x = 0.041. The equilibrium concentration of NH₃ is 1.0 − 2(0.041)= 0.92 or 0.9 M to 1sf.

The solubility of AgCl in 1.0 M NH₃ is 4.1 × 10⁻² mol/L.

17.84. The chemistry of AgCN:

(a) Solubility of AgCN in water: $K_{sp} = [Ag^+][CN^-] = \text{solubility}^2 = 6.0 \times 10^{-17}$ and solubility $= 7.8 \times 10^{-9}$ M

(b) Calculate K for: AgCN(s) + CN$^-$(aq) $\rightleftarrows$ [Ag(CN)$_2$]$^-$(aq)

$$K = \frac{\left[[Ag(CN)_2]^-\right]}{[CN^-]} = K_f \times K_{sp} = (1.3 \times 10^{21})(6.0 \times 10^{-17}) = 7.8 \times 10^4 \gg 1 \text{ thus,}$$

AgCN will dissolve.

(c) Calculate K for: AgCN(s) + 2 S$_2$O$_3^{2-}$(aq) $\rightleftarrows$ [Ag(S$_2$O$_3$)$_2$]$^{3-}$(aq) + CN$^-$(aq)

To form the complex: $K_f = \dfrac{\left[[Ag(S_2O_3)_2]^{3-}\right]}{[Ag^+][S_2O_3^{2-}]^2} = 2.9 \times 10^{13}$

$$K_f \times K_{sp} = \frac{\left[[Ag(S_2O_3)_2]^{3-}\right]}{[Ag^+][S_2O_3^{2-}]^2} \times \frac{[Ag^+][CN^-]}{1} = \frac{\left[[Ag(S_2O_3)_2]^{3-}\right][CN^-]}{[S_2O_3^{2-}]^2} =$$

Substituting values into the K_{net} expression:

$$K_f \times K_{sp} = (2.9 \times 10^{13})(6.0 \times 10^{-17}) = 1.7 \times 10^{-3}$$

Note that the [Ag(S$_2$O$_3$)$_2^{3-}$] is numerically equal to [Ag$^+$], so the numerator of the fraction above may be written as S^2 (where S is defined in part (a)) and the equation for the complex may be simplified to

$$\frac{\text{solubility}^2}{[S_2O_3^{2-}]^2} = \frac{\text{solubility}^2}{(0.10)^2} = 1.7 \times 10^{-3} \text{ and solubility}^2 = (1.7 \times 10^{-3})(0.10)^2 \text{ and}$$

solubility $= 4.1 \times 10^{-3}$

The solubility is increased.

General Questions

17.85. Solution producing precipitate:

The balanced equations are:

(a) AgNO$_3$(aq) + NaBr(aq) $\rightarrow$ AgBr(s) + NaNO$_3$(aq)   precipitate forms

(b) Pb(NO$_3$)$_2$(aq) + 2 KCl(aq) $\rightarrow$ PbCl$_2$(s) + 2 KNO$_3$(aq)   precipitate forms

One can make these decisions in one of two ways: (1) Recalling the solubility tables—probably learned earlier or (2) Reviewing the data from the K_{sp} tables (which contains compounds that are normally classified as insoluble—and precipitate from a solution containing the appropriate pairs of cations and anions (Ag$^+$ and Br$^-$ for example).

17.86. Solution producing precipitate:

(a) No precipitate forms as Na^+, Mg^{2+}, NO_3^-, and SO_4^{2-} ions do not form insoluble salts with each other.

(b) $K_3PO_4(aq) + FeCl_3(aq) \rightarrow 3\ KCl(aq) + FePO_4(s)$ precipitate forms

17.87. Will BaSO$_4$ precipitate?

Calculate the concentrations of barium and sulfate ions (after the solutions are mixed).

$$\frac{48\ mL}{72\ mL}\left(0.0012\ M\ Ba^{2+}\right) = 0.00080\ M\ Ba^{2+}\ \text{and}$$

$$\frac{24\ mL}{72\ mL}\left(1.0 \times 10^{-6}\ M\ SO_4^{2-}\right) = 3.3 \times 10^{-7}\ M\ SO_4^{2-}$$

Substituting in the K_{sp} expression: $Q = [Ba^{+2}][SO_4^{2-}] = (8.00 \times 10^{-4})(3.3 \times 10^{-7})$

$Q = 2.7 \times 10^{-10}$; $Q > K_{sp}$ for the solid (1.1×10^{-10}) so BaSO$_4$ precipitates.

17.88. Calculate pH and [H$_3$O$^+$] when CH$_3$CO$_2$H and NaOH are mixed:

The reaction that occurs: $CH_3CO_2H(aq) + OH^-(aq) \rightarrow CH_3CO_2^-(aq) + H_2O(\ell)$

Amount of acetic acid: (0.0200 L CH$_3$CO$_2$H)(0.15 mol/L) = 0.0030 mol CH$_3$CO$_2$H
Amount of NaOH: (0.0050 L NaOH)(0.17 mol/L) = 0.00085 mol NaOH
The reaction indicates that the # mol CH$_3$CO$_2$H consumed will be 0.00085 mol NaOH and that 0.00085 mol CH$_3$CO$_2^-$ are produced.
CH$_3$CO$_2$H remaining: 0.0030 mol CH$_3$CO$_2$H – 0.00085 mol = 0.0022 mol CH$_3$CO$_2$H
As this is a buffer, you can use the Henderson-Hasselbalch equation to solve for pH:

$$pH = pK_a + \log\frac{[CH_3CO_2^-]}{[CH_3CO_2H]} = -\log(1.8 \times 10^{-5}) + \log\left(\frac{0.00085}{0.0022}\right) = 4.33$$

and [H$_3$O$^+$] = 10^{-pH} = 4.7×10^{-5} M

17.89. The pH and [H$_3$O$^+$] when 50.0 mL of 0.40 M NH$_3$ is mixed with 25.0 mL of 0.20 M HCl:

The number of moles of each reactant:

(0.20 mol HCl/L)(0.025 L) = 0.0050 mol HCl and (0.40 mol NH$_3$/L)(0.050 L) = 0.020 mol NH$_3$. So 0.005 mol of NH$_3$ are consumed—and 0.005 mol NH$_4^+$ are produced.

	NH$_3$	NH$_4^+$
Initial	0.020 mol	0
Change	−0.0050 (consumed by HCl)	+ 0.0050 mol
Equilibrium	0.015 mol	0.0050 mol

You have a conjugate pair buffer (NH_3 and NH_4^+) present.

The K_b expression for NH_3 can be written: $\dfrac{[NH_4^+][OH^-]}{[NH_3]} = 1.8 \times 10^{-5}$

Substituting the # of mol into the K_b expression yields:

$\dfrac{[0.005 \text{ mol}][OH^-]}{[0.015 \text{ mol}]} = 1.8 \times 10^{-5}$ or $[OH^-] = 5.4 \times 10^{-5}$ M.

pOH = 4.27 and pH = 9.73, and a concomitant $[H_3O^+] = 10^{-9.73}$ or 1.9×10^{-10} M.

17.90. Decide whether the pH is less than, equal to, or greater than 7:

(a) Equal volumes of 0.20 M NH_3 and 0.20 M HCl: pH < 7
The solution will contain the conjugate acid of the weak base.

(b) Equal volumes of 0.10 M CH_3CO_2H and 0.10 M KOH: pH > 7
The solution will contain the conjugate base of the weak acid.

(c) Mixture of 25 mL of 0.015 M NH_3 and 12 mL of 0.015 M HCl: pH > 7

Upon mixing a reaction ensues that creates the NH_3/NH_4Cl buffer. The pK_a for NH_4^+ is 9.25, so the solution will be basic.

(d) Mixture of 150 mL of 0.20 M HNO_3 and 75 mL of 0.40 M NaOH: pH = 7
The solution contains equimolar amounts of ions that do not react with water to an appreciable extent.

(e) Mixture of 25 mL of 0.45 M H_2SO_4 and 25 mL of 0.90 M NaOH: pH > 7

The NaOH neutralized the diprotic H_2SO_4, producing SO_4^{2-} ion. The SO_4^{2-} ion is a weak base, so the pH will be above 7.

17.91. Compounds in order of increasing solubility in H_2O:

Compound	K_{sp}
$BaCO_3$	2.6×10^{-9}
Ag_2CO_3	8.5×10^{-12}
Na_2CO_3	not listed:

Na_2CO_3 is very soluble in water and is the most soluble salt of the three.

To determine the relative solubilities, find the molar solubilities.

For $BaCO_3$: $K_{sp} = [Ba^{2+}][CO_3^{2-}] = (x)(x) = 2.6 \times 10^{-9}$ and $x = 5.1 \times 10^{-5}$

The molar solubility of $BaCO_3$ is 5.1×10^{-5} M.

and for Ag_2CO_3: $K_{sp} = [Ag^+]^2[CO_3^{2-}] = (2x)^2(x) = 8.5 \times 10^{-12}$
$4x^3 = 8.5 \times 10^{-12}$ and $x = 1.3 \times 10^{-4}$

The molar solubility of Ag_2CO_3 is 1.3×10^{-4} M.

In order of increasing solubility: $BaCO_3 < Ag_2CO_3 < Na_2CO_3$

17.92. Maximum [F⁻] present without precipitating CaF₂:

For CaF₂ to precipitate, Q must exceed K_{sp} (5.3×10^{-11})

$K_{sp} = [Ca^{2+}][F^-]^2 = 5.3 \times 10^{-11} = (2.0 \times 10^{-3})[F^-]^2$. Solving for [F⁻]:

$$\frac{5.3 \times 10^{-11}}{2.0 \times 10^{-3}} = [F^-]^2, \text{ and } [F^-] = \sqrt{2.65 \times 10^{-8}}; [F^-] = 1.6 \times 10^{-4}$$

So [F⁻] must be greater than 1.6×10^{-4} M to initiate precipitation.

17.93. pH of a solution that contains 5.15 g NH₄NO₃ and 0.10 L of 0.15 M NH₃:

The concentration of the ammonium ion (from the nitrate salt) is:

$$5.15 \text{ g NH}_4\text{NO}_3 \left(\frac{1 \text{ mol NH}_4\text{NO}_3}{80.04 \text{ g NH}_4\text{NO}_3}\right)\left(\frac{1}{0.10 \text{ L}}\right) = 0.64 \text{ M}$$

You can calculate the pH using the K_b expression for ammonia:

$$K_b = \frac{[NH_4^+][OH^-]}{[NH_3]} = \frac{(0.64)(OH^-)}{(0.15)} = 1.8 \times 10^{-5} \text{ and solving for [OH⁻],}$$

$$x = \frac{(1.8 \times 10^{-5})(0.15)}{(0.64)} = 4.2 \times 10^{-6} \text{ M and pOH} = 5.38; \text{ so pH} = 8.62$$

Diluting the solution does not change the pH of the solution, since the dilution affects the concentration of both members of the conjugate pair. Since the pH of the buffer is a function of the *ratio* of the conjugate pair, the pH does not change.

17.94. Will 5.0 mg of SrSO₄ dissolve in 1.0 L of water?

$K_{sp} = [Sr^{2+}][SO_4^{2-}] = (x)(x) = x^2 = 3.4 \times 10^{-7}$

$x = \sqrt{K_{sp}} = \sqrt{3.4 \times 10^{-7}} = 5.83 \times 10^{-4}$ mol/L

$$1.0 \text{ L} \left(\frac{5.83 \times 10^{-4} \text{ mol Sr}^{2+}}{1 \text{ L}}\right)\left(\frac{1 \text{ mol SrSO}_4}{1 \text{ mol Sr}^{2+}}\right)\left(\frac{183.7 \text{ g}}{1 \text{ mol SrSO}_4}\right)\left(\frac{10^3 \text{ mg}}{1 \text{ g}}\right) = 107 \text{ mg SrSO}_4 \text{ can dissolve}$$

All the solid will dissolve.

17.95. The effect on pH of:

(a) Adding CH₃CO₂⁻Na⁺ to 0.100 M CH₃CO₂H:

The equilibrium affected is: CH₃CO₂H + H₂O ⇌ CH₃CO₂⁻ + H₃O⁺

Addition of sodium acetate increases the concentration of acetate ion. The equilibrium will shift to the left—reducing hydronium ion, and increasing the pH.

(b) Adding NaNO₃ to 0.100 M HNO₃: No effect on pH. Nitric acid is a strong acid; the equilibrium lies very far to the right. Addition of NO₃⁻ will not shift the equilibrium.

The effects differ owing to the nature of the two acids. Nitric acid (a strong acid) exists as hydronium and nitrate ions. Acetic acid exists as the molecular acid and acetate ion. The addition of conjugate base of both acids (CH₃CO₂⁻ and NO₃⁻) will affect the acetic acid equilibrium but not the nitric acid system.

17.96. Amount of NaOH added to a solution to change pH to 4.70:

$$pH = 4.70 = -\log(6.4 \times 10^{-5}) + \log\frac{[C_2O_4^{2-}]}{[HC_2O_4^-]}$$ giving the necessary ratio:

$$\frac{[C_2O_4^{2-}]}{[HC_2O_4^-]} = \frac{\text{mol } C_2O_4^{2-}}{\text{mol } HC_2O_4^-} = 3.2$$

So in 100. mL of solution, (0.100 L HC₂O₄⁻)(0.100 mol/L) = 0.0100 mol HC₂O₄⁻

One mole of C₂O₄²⁻ is formed for each mole of HC₂O₄⁻ consumed by reaction with NaOH. The necessary ratio of these is:

$$\frac{x}{0.0100 - x} = 3.2$$ solving for x = mol C₂O₄²⁻ = mol OH⁻ = 0.0076 mol

(0.0076 mol OH⁻)(1 L/0.120 mol) = 0.064 L will be needed.

17.97. Several approaches are valid. Perhaps the simplest is the use of the Henderson-Hasselbalch equation:

(a) pH of buffer solution: $pH = pK_a + \log\frac{[C_6H_5CO_2^-]}{[C_6H_5CO_2H]}$

Amount of sodium benzoate:

$$1.50 \text{ g } C_6H_5CO_2^-\left(\frac{1 \text{ mol } C_6H_5CO_2^-Na^+}{144.1 \text{ g } C_6H_5CO_2^-Na^+}\right) = 0.0104 \text{ mol } C_6H_5CO_2^-Na^+$$

Similarly, the amount of benzoic acid:

$$1.50 \text{ g } C_6H_5CO_2H\left(\frac{1 \text{ mol } C_6H_5CO_2H}{122.1 \text{ g } C_6H_5CO_2H}\right) = 0.0123 \text{ mol } C_6H_5CO_2H$$

Note that these are NOT the concentrations since these quantities are dissolved in 150.0 mL. Since you are using a **ratio** of the conjugate pairs, you can use the ratio of # mol exactly as you would have used the ratio of the molar concentrations:

$$pH = pK_a + \log\frac{[C_6H_5CO_2^-]}{[C_6H_5CO_2H]} = -\log(6.3 \times 10^{-5}) + \log\frac{[0.0104]}{[0.0123]} = 4.20 + -0.07 = 4.13$$

(b) To reduce the pH to 4.00, you will need to add the "acidic" component of the buffer. Substituting into the equation, as before, solving for the amount of benzoic acid:

$$\text{pH} = pK_a + \log\frac{[C_6H_5CO_2^-]}{[C_6H_5CO_2H]} \qquad 4.00 = 4.20 + \log\frac{[0.0104]}{[C_6H_5CO_2H]}$$

$$4.00 - 4.20 = \log\frac{[0.0104]}{[C_6H_5CO_2H]} \quad \text{and} \quad 10^{-0.20} = \frac{[0.0104]}{[C_6H_5CO_2H]}$$

Remember that you are using moles in the equation, mol $C_6H_5CO_2H$ = 0.0165 mol.

The amount of $C_6H_5CO_2H$ needed is $(0.0165 - 0.0123)$ mol × 122.1 g/mol = 0.51 g (0.5 g to 1 sf—subtraction.)

(c) Quantity of 2.0 M NaOH or HCl to change buffer pH to 4.00:

To reduce the pH of the buffer, you need to add 2.0 M HCl. The question is "how much" of the 2.0 M HCl:
you need to add enough HCl to reduce the concentration of the benzoate ion AND increase the concentration of benzoic acid appropriately:

$$4.00 - 4.20 = \log\frac{[C_6H_5CO_2^-Na^+]}{[C_6H_5CO_2H]} \quad \text{so} \quad 10^{-0.20} = \frac{[C_6H_5CO_2^-Na^+]}{[C_6H_5CO_2H]} = \frac{0.0104 - x}{0.0123 + x}$$

$$0.63 = \frac{0.0104 - x}{0.0123 + x} \quad \text{and } 0.63(0.0123 + x) = 0.0104 - x \text{ and } x = 0.00163 \text{ mol}$$

The amount of 2.0 M HCl = $0.00163 \text{ mol}\left(\dfrac{1 \text{ L}}{2.0 \text{ mol HCl}}\right) = 8.2 \times 10^{-4}$ L

17.98. Volume of 0.200 M HCl to change pH to 9.00:

$$\text{pH} = 9.00 = -\log(5.6 \times 10^{-10}) + \log\frac{[NH_3]}{[NH_4^+]} \quad \text{and solving for the ratio of the conjugate pairs:}$$

$$\frac{[NH_3]}{[NH_4^+]} = 0.56$$

Amount of NH_3 present: $(0.5000 \text{ L } NH_3)(0.250 \text{ mol/L}) = 0.125$ mol NH_3

Now substituting into the ratio: $\dfrac{[NH_3]}{[NH_4^+]} = \dfrac{0.125 - x}{x} = 0.56$.

Solving for x gives: 0.080 mol H_3O^+ to be added.

Amount of the 0.200 M HCl solution to be added:
$(0.080 \text{ mol } H_3O^+)(1 \text{ L}/0.200 \text{ mol}) = 0.40$ L of 0.200 M HCl should be added

17.99. The equation AgCl(s) + I⁻(aq) ⇌ AgI(s) + Cl⁻(aq) can be obtained by adding two equations:

1. AgCl(s) ⇌ Ag⁺(aq) + Cl⁻(aq) $K_{sp1} = 1.8 \times 10^{-10}$

2. Ag⁺(aq) + I⁻(aq) ⇌ AgI(s) $\dfrac{1}{K_{sp2}} = 1.2 \times 10^{16}$

The net equation: AgCl(s) + I⁻(aq) ⇌ AgI(s) + Cl⁻ (aq) $K_{net} = \dfrac{K_{sp1}}{K_{sp2}}$

$K_{net} = \dfrac{K_{sp1}}{K_{sp2}} = 2.12 \times 10^6$ or 2.1×10^6 (2 sf)

The equilibrium lies to the right. This indicates that AgI will form if I⁻ is added to a saturated solution of AgCl.

17.100. Calculate the *K* for the indicated equation:

Zn(OH)₂(s) ⇌ Zn²⁺(aq) + 2 OH⁻(aq) $K_{sp} = 3 \times 10^{-17}$

Zn²⁺(aq) + 2 CN⁻(aq) ⇌ Zn(CN)₂(s) $K = 1/K_{sp} = 1.3 \times 10^{11}$

Zn(OH)₂(s) + 2 CN⁻(aq) ⇌ Zn(CN)₂(s) + 2 OH⁻(aq)

$K_{net} = (3 \times 10^{-17})(1.3 \times 10^{11}) = 4 \times 10^{-6}$

The equilibrium lies predominantly to the left. The transformation of zinc hydroxide into zinc cyanide will not occur to a large extent because of the unfavorable equilibrium constant.

17.101. Mass of oxalic acid in 28 g of rhubarb that is 1.2% oxalic acid by weight:

28 g rhubarb $\left(\dfrac{1.2 \text{ g oxalic acid}}{100 \text{ g rhubarb}}\right) = 0.336$ g oxalic acid

(a) Volume of 0.25 M NaOH needed to titrate the oxalic acid:

0.336 g $H_2C_2O_4 \left(\dfrac{1 \text{ g mol } H_2C_2O_4}{90.03 \text{ g } H_2C_2O_4}\right) = 0.00373$ mol $H_2C_2O_4$

The diprotic oxalic acid will require **two** moles of NaOH for each **one** mol of $H_2C_2O_4$, so you need 0.00746 mol of NaOH.

0.00373 mol NaOH $\left(\dfrac{1 \text{ L}}{0.25 \text{ mol NaOH}}\right) = 0.0299$ L or 29.9 mL (30. mL to 2 sf)

(b) What mass of CaC₂O₄ could be formed from the oxalic acid?

$$0.00373 \text{ mol } H_2C_2O_4 \left(\frac{1 \text{ mol } CaC_2O_4}{1 \text{ mol } H_2C_2O_4}\right)\left(\frac{128.1 \text{ g } CaC_2O_4}{1 \text{ mol } CaC_2O_4}\right) = 0.48 \text{ g (2 sf)}$$

17.102. Solubility of calcium oxalate in grams/L:

CaC₂O₄(s) ⇌ Ca²⁺(aq) + C₂O₄²⁻(aq)

x = solubility of CaC₂O₄ = [Ca²⁺] = [C₂O₄²⁻]

$K_{sp} = x^2 = 4 \times 10^{-9}$ and $x = 6.3 \times 10^{-5}$ M

Expressed as g/L: (6.3 × 10⁻⁵ mol/L)(128.1 g/mol) = 0.008 g/L

17.103. Regarding the solubility of barium and calcium fluorides:

(a) [F⁻] that will precipitate maximum amount of calcium ion:

BaF₂ begins to precipitate when the ion product just exceeds the K_{sp}.

K_{sp} = [Ba²⁺][F⁻]² = 1.8 × 10⁻⁷, and the concentration of barium ion = 0.10 M

[F⁻] = (1.8 × 10⁻⁷/0.10)^(1/2) = 1.34 × 10⁻³ M

This would be the maximum fluoride concentration permissible.

(b) [Ca²⁺] remaining when BaF₂ just begins to precipitate:

The [Ca²⁺] ion remaining will be calculable from the K_{sp} expression for CaF₂.

K_{sp} = [Ca²⁺][F⁻]² = 5.3 × 10⁻¹¹ and you know that [F⁻] is 1.34 × 10⁻³ M.

So, K_{sp} = 5.3 × 10⁻¹¹ = [Ca²⁺][1.34 × 10⁻³]², [Ca²⁺] = 2.9 × 10⁻⁵ M

17.104. Describe the precipitation in a solution containing iodide and carbonate ions:

(a) K_{sp} = 9.8 × 10⁻⁹ = [Pb²⁺][I⁻]² = [Pb²⁺](0.10)²

[Pb²⁺] = 9.8 × 10⁻⁷ mol/L

K_{sp} = 7.4 × 10⁻¹⁴ = [Pb²⁺][CO₃²⁻] = [Pb²⁺](0.10) and [Pb²⁺] = 7.4 × 10⁻¹³ mol/L

PbCO₃ will precipitate first.

(b) You know from (a) that the [Pb²⁺] = 9.8 × 10⁻⁷ mol/L when PbI₂ begins to precipitate, so you can use the K_{sp} for PbCO₃ to determine the carbonate ion concentration:

K_{sp} = 7.4 × 10⁻¹⁴ = [Pb²⁺][CO₃²⁻] = (9.8 × 10⁻⁷)[CO₃²⁻]

[CO₃²⁻] = 7.6 × 10⁻⁸ mol/L

17.105. Regarding the precipitation of CaSO₄ and PbSO₄ from a solution 0.010 M in each metal:

(a) Examine the K_{sp} of the two salts:

K_{sp} for CaSO₄ = 4.9 × 10⁻⁵ and for PbSO₄ = 2.5 × 10⁻⁸

These data indicate that PbSO₄ will begin to precipitate first—as it is the less soluble of the two salts. This can be found by examining the general K_{sp} expression for 1:1 salts: K_{sp} = [M²⁺][SO₄²⁻].
With both metal concentrations at 0.010 M, the sulfate ion for the less soluble sulfate will be exceeded first.

(b) When the calcium salt just begins to precipitate, the [Pb²⁺] = ?

You know that the metal ion concentration is 0.010 M. So the sulfate ion concentration when the more soluble calcium sulfate begins to precipitate is:

K_{sp} = [0.010][SO₄²⁻] = 4.9 × 10⁻⁵ and [SO₄²⁻] = (4.9 × 10⁻⁵/0.010) = 4.9 × 10⁻³ M

At that point the [Pb²⁺] would be: K_{sp} = [Pb²⁺][SO₄²⁻] = 2.5 × 10⁻⁸

Then [Pb²⁺][4.9 × 10⁻³] = 2.5 × 10⁻⁸ and [Pb²⁺] = (2.5 × 10⁻⁸/4.9 × 10⁻³) or 5.1 × 10⁻⁶ M.

17.106. Calculate the buffer capacity of the solution:

The initial pH is 4.74.

$5.74 = pK_a + \log\dfrac{[CH_3CO_2^-]}{[CH_3CO_2H]} = 4.74 + \log\dfrac{[CH_3CO_2^-]}{[CH_3CO_2H]}$ and solving for $\dfrac{[CH_3CO_2^-]}{[CH_3CO_2H]} = 10$

The ratio must change from 1:1 to 10:1 for the pH to change by one unit.

$10 = \dfrac{0.10 + x}{0.10 - x}$, solving for x = 0.082 mol/L. The buffer capacity is therefore 0.082 mol.

17.107. Concerning the equilibrium of AgBr and [Ag(S₂O₃)₂]³⁻:

(a) AgBr(s) + 2 S₂O₃²⁻(aq) ⇌ [Ag(S₂O₃)₂]³⁻(aq) + Br⁻(aq)

$K_{net} = K_{sp} \cdot K_{form}$ = 15.7 = 16 (2 sf)

Note this expression is arrived at by **adding** the equations for the formation of the complex ion **to** the equation for AgBr dissolution, hence you multiply the K values.

(b) Mass of Na₂S₂O₃ to dissolve the AgBr:

$[[Ag(S_2O_3)_2]^{3-}] = [Br^-] = 1.00 \text{ g} \left(\dfrac{1 \text{ mol AgBr}}{187.8 \text{ g}}\right)\left(\dfrac{1 \text{ mol }[Ag(S_2O_3)_2]^{3-}}{1 \text{ mol AgBr}}\right)\left(\dfrac{1}{1.00 \text{ L}}\right)$

= 5.325 × 10⁻³ M

Now you can calculate the [S₂O₃²⁻] that is in equilibrium with the Br⁻ ion and the silver complex, by using the K_{net} from part (a).

$$\frac{[[\text{Ag}(\text{S}_2\text{O}_3)]^{3-}][\text{Br}^-]}{[\text{S}_2\text{O}_3^{2-}]^2} = \frac{(5.325 \times 10^{-3})^2}{[\text{S}_2\text{O}_3^{2-}]^2} = 15.7 \text{ and } [\text{S}_2\text{O}_3^{2-}] = 1.34 \times 10^{-3} \text{ M}$$

Now you can calculate the mass of Na₂S₂O₃ needed to make this concentration:

$$1.0 \text{ L} \left(\frac{1.34 \times 10^{-3} \text{ mol S}_2\text{O}_3^{2-}}{1 \text{ L}} \right) \left(\frac{1 \text{ mol Na}_2\text{S}_2\text{O}_3}{1 \text{ mol S}_2\text{O}_3^{2-}} \right) \left(\frac{158.1 \text{ g}}{1 \text{ mol Na}_2\text{S}_2\text{O}_3} \right) = 0.212 \text{ g Na}_2\text{S}_2\text{O}_3$$

Additional Na₂S₂O₃ is needed to complex with the Ag⁺ ion. The solution originally contained 5.325 × 10⁻³ mol Ag⁺ ion. Two moles of S₂O₃²⁻ ion complex for each mole of Ag⁺ ion, so an additional 2(5.325 × 10⁻³ mol Na₂S₂O₃)(158.1 g/mol) = 1.684 g of Na₂S₂O₃ are required.

The mass of Na₂S₂O₃ added is 1.684 g + 0.212 g = 1.896 g or 1.90 g Na₂S₂O₃.

17.108. Percentage of calcium ions removed: [Ca²⁺]initial = 0.010 M

At equilibrium, you can write: [Ca²⁺][CO₃²⁻] = 3.4 × 10⁻⁹

If the carbonate ion concentration is 0.050 M, then [Ca²⁺][0.050] = 3.4 × 10⁻⁹ and

$$[\text{Ca}^{2+}] = \frac{3.4 \times 10^{-9}}{[0.050]} = 6.8 \times 10^{-8} \text{ M}.$$ This represents the calcium ion concentration at equilibrium. The amount of calcium ion removed per liter is: (0.010 – 6.8 × 10⁻⁸) or 9.99 × 10⁻³ M. In essence, ALL the calcium ion (> 99.9%) has been removed.

In the Laboratory

17.109. Separate the following pairs of ions:

(a) Ba²⁺ and Na⁺: Most sodium salts are soluble. Find a barium salt which is not soluble-e.g., a sulfate salt. Addition of dilute sulfuric acid should provide a source of SO₄²⁻ ions in sufficient quantity to precipitate the barium ions, but not the sodium ions.

K_{sp}(BaSO₄) = 1.1 × 10⁻¹⁰
K_{sp}(Na₂SO₄) = not listed, owing to the large solubility of sodium sulfate.

(b) Ni²⁺ and Pb²⁺: Add HCl or another source of chloride ion. PbCl₂ will precipitate, but NiCl₂ is water-soluble.

17.110. Separate the following pairs of ions:

(a) Since AgCl is insoluble, and CuCl₂ is soluble, add HCl to precipitate the Ag⁺ as AgCl and leave Cu²⁺(aq) in solution.

(b) Add excess NaOH. The Fe³⁺ ions form Fe(OH)₃, an insoluble hydroxide salt. The Al³⁺ ions initially precipitate as Al(OH)₃ solid, but then dissolve as the soluble complex ion [Al(OH)₄]⁻. (See Chapter 16.10)

17.111. Describe the precipitation in a solution containing Ba^{2+} and Sr^{2+} ions:

(a) These metal sulfates are 1:1 salts. The K_{sp} expression has the general form:

$K_{sp} = [M^{2+}][SO_4^{2-}]$

To determine the $[SO_4^{2-}]$ necessary to begin precipitation, you divide the equation by the metal ion concentration to obtain: $\dfrac{K_{sp}}{[M^{2+}]} = [SO_4^{2-}]$

The concentration of the metal ions under consideration are each 0.10 M. Substitution of the appropriate K_{sp} for the sulfates and 0.10 M for the metal ion concentration yields the sulfate ion concentrations in the table below. As the soluble sulfate is added to the metal ion solution, the sulfate ion concentration increases from zero molarity. The lowest sulfate ion concentration is reached first, with higher concentrations reached later. The order of precipitation is listed in the last column of the table below.

Compound	K_{sp}	Maximum $[SO_4^{2-}]$	Order of Precipitation
BaSO₄	1.1×10^{-10}	1.1×10^{-9}	1
SrSO₄	3.4×10^{-7}	3.4×10^{-6}	2

(b) Concentration of Ba^{2+} when SrSO₄ begins to precipitate:

From (a) you note that $[SO_4^{2-}]$ will be 3.4×10^{-6} when SrSO₄ begins to precipitate.

So you calculate the concentration of Ba^{2+} by noting since the barium salt has been precipitating, it is a **saturated solution of BaSO₄**.

$K_{sp} = [Ba^{2+}][SO_4^{2-}] = 1.1 \times 10^{-10}$.
So $[Ba^{2+}] = 1.1 \times 10^{-10}/3.4 \times 10^{-6} = 3.2 \times 10^{-5}$.

17.112. Order of precipitation of the metal hydroxides:

$K_{sp} = 4.9 \times 10^{-17} = [Fe^{2+}][OH^-]^2 = (0.1)[OH^-]^2$
So $[OH^-]$ required to precipitate Fe(OH)₂ = 2.2×10^{-8} M
$K_{sp} = 1.4 \times 10^{-15} = [Pb^{2+}][OH^-]^2 = (0.1)[OH^-]^2$
So the $[OH^-]$ required to precipitate Pb(OH)₂ = 1.2×10^{-7} M
$K_{sp} = 1.3 \times 10^{-33} = [Al^{3+}][OH^-]^3 = (0.1)[OH^-]^3$
So the $[OH^-]$ required to precipitate Al(OH)₃ = 2.4×10^{-11} M

You observe that Al(OH)₃ will precipitate first, followed by Fe(OH)₂ and then Pb(OH)₂.

17.113. For the titration of aniline hydrochloride with NaOH:

(a) initial pH: Recall that the hydrochloride (represented in this example as HA) is a weak acid: ($K_a = 2.4 \times 10^{-5}$) Use a K_a expression to solve for H_3O^+:

$$K_a = \frac{[A^-][H_3O^+]}{[HA]} = 2.4 \times 10^{-5} \cdot 0.10 = x^2 = 2.4 \times 10^{-6} \text{ and solving for } x:$$

$x = 1.5 \times 10^{-3}$ M = $[H_3O^+]$ and pH = 2.81

(b) pH at the equivalence point:

At the equivalence point, you know that the number of moles of acid = # moles base (by definition). Since you have 0.00500 mol of HA (M·V), you must ask the question, "How much 0.185 M NaOH contains 0.00500 mol NaOH?" That volume is: 0.00500 mol OH⁻ · 1000 mL/ 0.185 mol NaOH = 27.0 mL of NaOH solution.

The total volume is 50.0 mL + 27.0 mL = 0.0770 L.

The [HA] = 0.00500 mol/0.0770 L = 0.0649 M

You can calculate the pH of the solution using the K_b expression for A:

$$K_b = \frac{[HA][OH^-]}{[A]} = \frac{1.0 \times 10^{-14}}{2.4 \times 10^{-5}} \text{ and knowing that the } [HA] = [OH^-]$$

$x^2 = 4.2 \times 10^{-12} \cdot 0.0649$ and $x = 5.2 \times 10^{-6}$ M; pOH = 5.28 and pH = 8.72

(c) pH at the midpoint:

This portion is can be solved if you examine the K_a expression or the Henderson-Hasselbalch expression for this system:

From the K_b expression above, note that the conjugate pairs are found in the numerator and denominator of the right side of the K_b term (the same applies to K_a expressions).

At the midpoint you have reacted half the acid, (using 13.5 mL—see part (b)) forming its conjugate base. The result is that the concentrations of the conjugate pairs are equal and the pOH = pK_b and pH = pK_a. So pH = −log(2.4 × 10⁻⁵) = 4.62.

(d) With a pH = 8.72 at the equivalence point, o-cresolphthalein or phenolphthalein would serve as an adequate indicator.

(e) The pH after the addition of 10.0, 20.0 and 30.0 mL base:

The volumes of base correspond to

10 mL (0.185 mol/L·0.0100 L)	0.00185 mol NaOH
20.0 mL (0.185 mol/L · 0.0200 L)	0.00370 mol NaOH
30.0 mL(0.185 mol/L · 0.0300L)	0.00555 mol NaOH

Each mol of NaOH consumes a mol of aniline hydrochloride(HA) and produces an equal number of mol of aniline(A).

Recall that you began with 0.00500 mol of the acid (which is represented as HA)

Substitution into the K_a expression yields:

$$\frac{[A^-][H_3O^+]}{[HA]} = 2.4 \times 10^{-5}$$ and rearranging them $$[H_3O^+] = 2.4 \times 10^{-5}\left(\frac{[HA]}{[A]}\right)$$

After 10 mL of base are added, 0.00185 mol HA are consumed, and 0.00185 mol A produced. The acid remaining is (000500 − 0.00185) mol.

Substituting into the rearranged equation:

$$[H_3O^+] = 2.4 \times 10^{-5}\left(\frac{[0.00315 \text{ mol}]}{[0.00185 \text{ mol}]}\right) = 4.1 \times 10^{-5}; \text{pH} = -\log(4.1 \times 10^{-5}) = 4.39$$

After 20 mL of base are added, 0.00370 mol HA are consumed, and 0.00370 mol A produced. The acid remaining is (000500 − 0.00370)mol.

Substituting into the rearranged equation:

$$[H_3O^+] = 2.4 \times 10^{-5}\left(\frac{[0.00130 \text{ mol}]}{[0.00370 \text{ mol}]}\right) = 8.4 \times 10^{-6}; \text{pH} = -\log(8.4 \times 10^{-6}) = 5.07$$

After 30 mL of base are added, all the acid is consumed, and excess strong base is present (0.00555 mol − 0.00500) mol. You can treat this solution as one of a strong base. The volume of the solution is (50.0 + 30.0 or 80.0 mL).

The concentration of the NaOH is:

0.00055 mol/ 0.080 L = 0.006875 M so pOH = −log(0.006875) and pOH= 2.16, the pH= 11.84.

Vol base	pH	Vol base	pH
0.0	2.81	20.0	5.07
10.0	4.39	27.0	8.72
13.5	4.62	30.0	11.84

(f) The approximate titration curve:

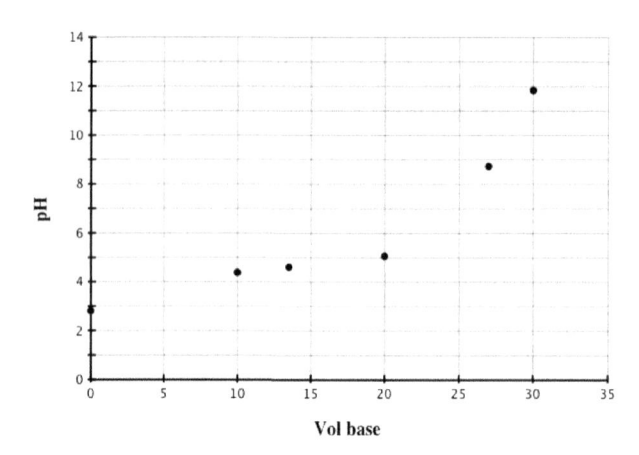

17.114. For the titration of ethanolamine with HCl:

(a) Initial pH is that of the base, so you use a K_b expression:
$$K_b = 3.2 \times 10^{-5} = \frac{[OH^-][HOCH_2CH_2NH_3^+]}{[HOCH_2CH_2NH_2]} = \frac{x^2}{0.010 - x} \approx \frac{x^2}{0.010}$$
$x = [OH^-] = 5.7 \times 10^{-4}$ M
pOH = –log[OH⁻] = 3.25
pH = 14.00 – pOH = 10.75

(b) pH at equivalence point:

You need the concentration of the ethanolammonium ion:

Amount: (0.0250 L ethanolamine)(0.010 mol/L) = 2.5×10^{-4} mol ethanolamine
Total volume = 0.025 L + (2.5×10^{-4} mol)(1 L/0.0095 mol) = 0.0513 L
Concentration: $[HOCH_2CH_2NH_3^+] = \frac{2.5 \times 10^{-4} \text{ mol}}{0.0513 \text{ L}} = 0.0049$ M

Use the K_a for the ion:
$$K_a = \frac{K_w}{K_b} = 3.1 \times 10^{-10} = \frac{[H_3O^+][HOCH_2CH_2NH_2]}{[HOCH_2CH_2NH_3^+]} = \frac{x^2}{0.0049 - x} \approx \frac{x^2}{0.0049}$$
Solving for $x = [H_3O^+] = 1.2 \times 10^{-6}$ M and pH = –log[H₃O⁺] = 5.91

(c) At titration midpoint [HOCH₂CH₂NH₂] = [HOCH₂CH₂NH₃⁺] and
pH = pK_a = –log(3.1 × 10⁻¹⁰) = 9.51

(d) Methyl red would detect the equivalence point.

(e) pH after addition of aliquots of HCl:

mL HCl added	mol H$_3$O$^+$ added	mol conjugate acid produced	mol base remaining	pH
5.00	4.8×10^{-5}	4.8×10^{-5}	2.0×10^{-4}	10.13
10.00	9.5×10^{-5}	9.5×10^{-5}	1.6×10^{-4}	9.72
20.00	1.9×10^{-4}	1.9×10^{-4}	6×10^{-5}	9.00
30.00	2.9×10^{-4}			3.20

When 30.00 mL HCl is added, pH depends only on the excess H$_3$O$^+$.
[H$_3$O$^+$] = (2.9×10^{-4} mol – 2.5×10^{-4} mol)/0.0550 L = 6.4×10^{-4} M.

(f)

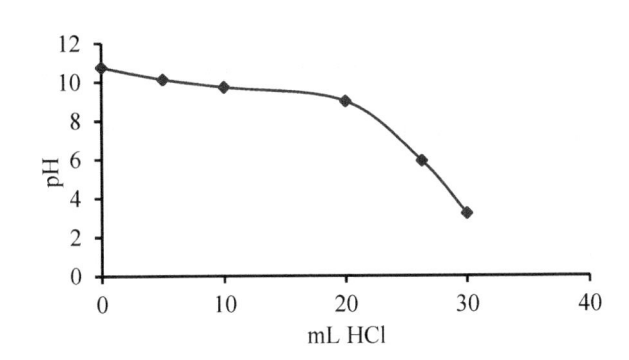

17.115. For the titration of 0.150 M ethylamine ($K_b = 4.27 \times 10^{-4}$) with 0.100 M HCl:

(a) pH of 50.0 mL of 0.150 M CH$_3$CH$_2$NH$_2$:

For the weak base, CH$_3$CH$_2$NH$_2$, the equilibrium in water is represented as:

CH$_3$CH$_2$NH$_2$(aq) + H$_2$O(ℓ) $\rightleftarrows$ CH$_3$CH$_2$NH$_3^+$ (aq) + OH$^-$(aq)

The slight dissociation of CH$_3$CH$_2$NH$_2$ would form equimolar amounts of CH$_3$CH$_2$NH$_3^+$ and OH$^-$ ions.

$$K_b = \frac{[CH_3CH_2NH_3^+][OH^-]}{[CH_3CH_2NH_2]} = \frac{(x)(x)}{0.150 - x} = 4.27 \times 10^{-4}$$

The quadratic equation can be used to find an exact solution.

$x^2 = 6.405 - 4.27 \times 10^{-4} x$

Rearranging: $x^2 + 4.27 \times 10^{-4} x - 6.405 \times 10^{-5} = 0$

and solving for $x = 7.79 \times 10^{-3}$; pOH = $-\log(7.79 \times 10^{-3}) = 2.11$ and pH = 11.89

(b) pH at the halfway point of the titration:

The volume of acid added isn't that important since the amounts of base and conjugate acid will be equal. In part (d), you find that 75.0 mL of acid are required to reach the equivalence point, so the volume of acid at this point is (0.5 · 75.0 mL)

Substituting that fact into the equilibrium expression you obtain:

$$\frac{[CH_3CH_2NH_3^+][OH^-]}{[CH_3CH_2NH_2]} = \frac{(x)[OH^-]}{x} = 4.27 \times 10^{-4}$$ so you can see that the

$[OH^-] = 4.27 \times 10^{-4}$ and pOH = 3.37 and pH = (14.00 − 3.37) = 10.63

(c) pH when 75% of the required acid has been added:

Amount of base initially present is (0.050 L · 0.150 M) or 0.0075 moles base.

So 75% of this amount is 0.00563 mol, requiring 0.00563 mol of HCl.

The volume of 0.100 M HCl containing that # mol is

0.00563 mol HCl/0.100 M = 0.0563 L or 56.3 mL, so 0.001875 mol base remain.

Substituting into the K_b expression:

$$\frac{[CH_3CH_2NH_3^+][OH^-]}{[CH_3CH_2NH_2]} = \frac{0.00563 \text{ mol}[OH^-]}{0.001875 \text{ mol}} = 4.27 \times 10^{-4}$$

Solving for hydroxyl ion concentration yields: 1.42×10^{-4} M and pOH = 3.85 and pH = 10.15.

(d) pH at the equivalence point:
At the equivalence point, there are equal # of moles of acid and base. The number of moles of ethylamine = (0.050 L · 0.150 M) or 0.00750 moles ethylamine.
That amount of HCl would be:

$$7.50 \times 10^{-3} \text{ mol HCl} \left(\frac{1 \text{ L}}{0.100 \text{ mol HCl}}\right) = 0.0750 \text{ L (or 75.0 mL)}$$

This total amount of solution would be: 75.0 + 50.0 = 125.0 mL
Since you have added equal amounts of acid and base, the reaction between the two will result in the existence of only the salt.

and the concentration of salt would be: $\dfrac{7.50 \times 10^{-3} \text{ mol}}{0.1250 \text{ L}} = 6.00 \times 10^{-2}$ M

Recall that the salt will act as a weak acid, and you can calculate the K_a: $CH_3CH_2NH_3^+$ (aq) + H_2O (ℓ) ⇌ $CH_3CH_2NH_2$ (aq) + H_3O^+ (aq)

The equilibrium constant $K_a = \dfrac{K_w}{K_b} = \dfrac{1.0 \times 10^{-14}}{4.27 \times 10^{-4}} = 2.3 \times 10^{-11}$

The equilibrium expression would be: $\dfrac{[CH_3CH_2NH_2][H_3O^+]}{[CH_3CH_2NH_3^+]} = 2.3 \times 10^{-11}$

Given that the salt would hydrolyze to form equal amounts of ethylamine and hydronium ion (as shown by the equation above). If you represent the concentrations of those species as x, then you can write (using our usual approximation):

$$\frac{x^2}{6.00 \times 10^{-2}} = 2.3 \times 10^{-11} \text{ and solving for } x = 1.19 \times 10^{-6}$$

Since x represents [H$_3$O$^+$], then pH = $-\log(1.19 \times 10^{-6})$ = 5.93

(e) pH after addition of 10.0 mL of HCl more than required:

Past the equivalence point, the excess strong acid controls the pH. To determine the pH, you need only calculate the concentration of HCl. In part (d) you found that the total volume at the equivalence point was 125.0 mL. The addition of 10.0 mL will bring that total volume to 135.0 mL of solution. The # of moles of excess HCl is:

(0.100 mol HCl/L)(0.0100 L) = 0.00100 mol HCl contained within 135.0 mL, for a concentration of 7.41×10^{-3} M HCl. Since HCl is a strong acid, the pH = $-\log(7.41 \times 10^{-3})$ or 2.13.

(f) The titration curve:

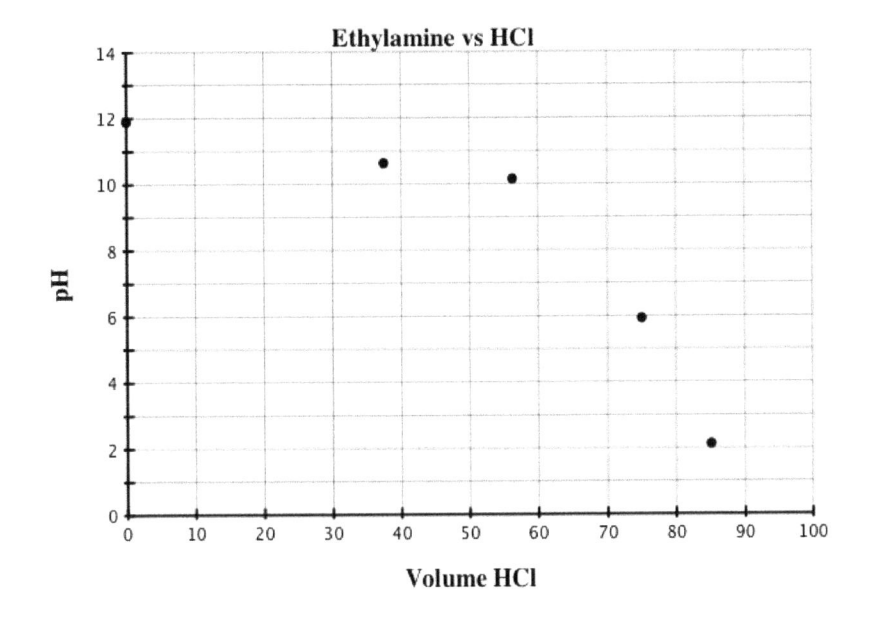

Volume HCl	pH
0.0	11.89
37.5	10.63
56.3	10.15
75.0	5.93
85.0	2.13

(g) Suitable indicator for endpoint: Alizarin or Bromocresol purple would be suitable indicators. (See Figure 17.11 for indicators.)

17.116. Consider a buffer composed of Na$_2$HPO$_4$ and Na$_3$PO$_4$:

(a) Use Henderson-Hasselbalch to determine larger component at pH = 12.00.

$$\text{pH} = 12.00 = -\log(3.6 \times 10^{-13}) + \log\frac{[\text{PO}_4^{3-}]}{[\text{HPO}_4^{2-}]}$$

$\dfrac{[PO_4^{3-}]}{[HPO_4^{2-}]} = 0.36$ so HPO_4^{2-} is present in a larger amount

(b) Mass of Na_2HPO_4: With $[PO_4^{3-}] = 0.400$, solve for $[HPO_4^{2-}]$

$\dfrac{0.400}{[HPO_4^{2-}]} = 0.36$ and $[HPO_4^{2-}] = 1.1$ mol/L

Mass needed for this concentration:

$0.2000 \text{ L} \left(\dfrac{1.1 \text{ mol } HPO_4^{2-}}{1 \text{ L}} \right) \left(\dfrac{1 \text{ mol } Na_2HPO_4}{1 \text{ mol } HPO_4^{2-}} \right) \left(\dfrac{141.95 \text{ g}}{1 \text{ mol } Na_2HPO_4} \right) = 31.2$ g Na_2HPO_4

(c) Additional base (PO_4^{3-}) must be added to raise the pH to 12.25. Use Henderson-Hasselbalch equation to calculate ratio needed:

$12.25 = -\log(3.6 \times 10^{-13}) + \log \dfrac{x}{1.1}$

$x = [PO_4^{3-}] = 0.70$ mol/L

(0.70 mol/L)(0.2000 L) − (0.400 mol/L)(0.2000 L) = 0.061 mol Na_3PO_4

$0.061 \text{ mol } Na_3PO_4 \cdot \dfrac{163.9 \text{ g}}{1 \text{ mol } Na_3PO_4} = 10.$ g Na_3PO_4 should be added.

17.117. Make a buffer of pH = 2.50 from 100. mL of 0.230 M H_3PO_4 and 0.150 M NaOH.

For the first ionization of the acid, $K_a = 7.5 \times 10^{-3}$

The pK_a of the acid = 2.12. The Henderson-Hasselbalch equation will help to determine the ratio of the conjugate pair to make the pH 2.50.

Assume that (since the second K_a is approximately 10^{-8}, you are dealing only with the first ionization. This *is an approximation*.

$pH = pK_a + \log \dfrac{[\text{conjugate base}]}{[\text{acid}]}$ so $2.50 = 2.12 + \log \dfrac{[\text{conjugate base}]}{[\text{acid}]}$

Solving: $0.38 = \log \dfrac{[A]}{[HA]}$ and the ratio of $\dfrac{[A]}{[HA]} = 2.37$

The initial amount of HA = (0.100 L · 0.230M HA) = 0.0230 mol HA.

Rearranging the ratio: [A] = 2.37[HA]. Knowing that the acid present is *either* the molecular acid, HA, or the conjugate base of the acid, A, you can write:

A + HA = 0.0230 mol and with A = 2.37 HA

2.37 HA + HA = 0.0230 mol; 3.37 HA = 0.0230 mol; and HA = 0.0068 mol.

You need to consume (0.0230 mol HA − 0.0068 mol HA) or 0.0162 mol HA.

This requires 0.0162 mol NaOH, and from 0.150 M NaOH you need:

0.0162 mol NaOH = 0.150 mol/L · V and V = 0.108 L or 110 mL of NaOH (2 sf).

17.118. Mass of Na₃PO₄ needed to obtain buffer with pH = 7.75:

Use the Henderson-Hasselbalch equation to determine the necessary ratio:

$$pH = 7.75 = -\log(6.2 \times 10^{-8}) + \log\frac{[HPO_4^{2-}]}{[H_2PO_4^-]}$$ and the ratio is:

$$\frac{[HPO_4^{2-}]}{[H_2PO_4^-]} = \frac{\text{mol } HPO_4^{2-}}{\text{mol } H_2PO_4^{2-}} = 3.49$$

Rearranging: (mol HPO₄²⁻) = 3.49 (mol H₂PO₄⁻)

Amount of HCl being added: (0.0800 L HCl)(0.200 mol/L) = 0.0160 mol H₃O⁺

The total H₃O⁺ can be used to produce HPO₄⁻ (1 H₃O⁺/PO₄³⁻) and H₂PO₄⁻ (2 H₃O⁺/PO₄³⁻)

0.0160 mol H₃O⁺ = (mol HPO₄²⁻) + 2(mol H₂PO₄⁻);

substitute the relationship from above for mol HPO₄²⁻:
0.0160 mol H₃O⁺ = 3.49(mol H₂PO₄⁻) + 2(mol H₂PO₄⁻)

So 0.0150 mol H₃O⁺ = 5.49(mol H₂PO₄⁻) or 1 mol H₃O⁺ = 0.00291 mol H₂PO₄⁻

The ratio of the two phosphates is 3.49:1, so mol of HPO₄²⁻ = 3.49 × mol H₂PO₄⁻

or # mol of HPO₄²⁻ = 3.49 × 0.0029 mol H₂PO₄⁻ = 0.0102 mol HPO₄²⁻

Since Total mol PO₄³⁻ needed = mol HPO₄²⁻ + mol H₂PO₄⁻

= 0.00291 mol + 0.0102 mol = 0.0131 mol PO₄³⁻

$$0.0131 \text{ mol Na}_3PO_4 \left(\frac{163.9 \text{ g}}{1 \text{ mol Na}_3PO}\right) = 2.14 \text{ g} = 2.1 \text{ g Na}_3PO4 \text{ (2 sf)}$$

17.119. Separate solutions containing Ag⁺, Cu²⁺, and Pb²⁺ into three separate test tubes:

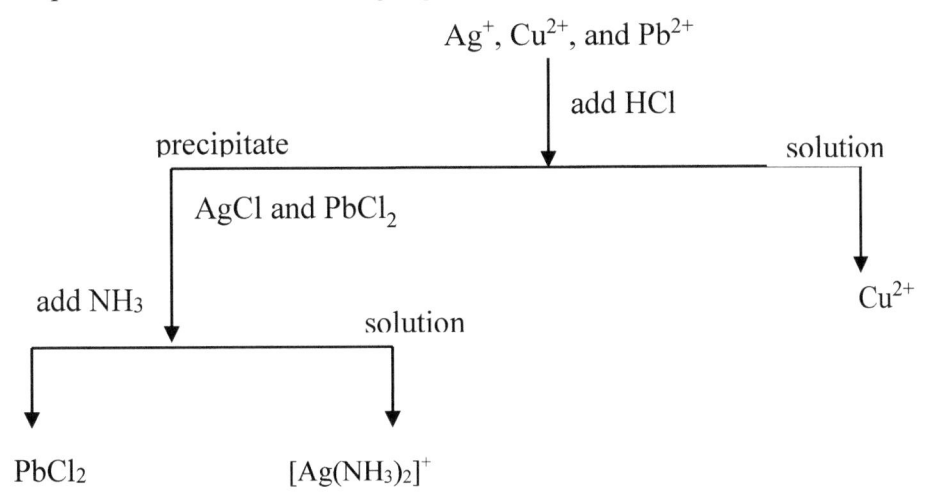

Ag⁺, Cu²⁺, and Pb²⁺
↓ add HCl
precipitate → AgCl and PbCl₂   solution → Cu²⁺
add NH₃ ↓
PbCl₂   [Ag(NH₃)₂]⁺

The addition of HCl precipitates both Ag⁺ and Pb²⁺ ions. Separating the filtrate from the precipitate removes the copper(II) ions.

Pb^{2+} and Ag^+ ions can be separated in one of two ways:

(a) Lead(II) chloride is soluble in very hot water, so the addition of hot water to the precipitate will dissolve the lead(II) chloride. Filtering the hot solution will remove the dissolved lead salt from the silver chloride.

(b) Ag^+ ions form a complex ion with ammonia, while Pb^{2+} ions do not. Adding NH_3 to the solution containing both Pb^{2+} and Ag^+ ions will dissolve the AgCl solid—as the silver complex ion forms. Filtering the remaining solid lead(II) chloride will separate the lead salt from the silver complex ion.

17.120. Confirmatory tests for Cu^{2+}, Pb^{2+}, and Ag^+:

(a) $PbCl_2(s) + CrO_4^{2-}(aq) \rightleftarrows PbCrO_4(s) + 2\ Cl^-(aq)$

$$K_{net} = K_{sp}(PbCl_2) \times \frac{1}{K_{sp}(PbCrO_4)} = 6.1 \times 10^7$$

You see from the magnitude of K_{net} that the chloride would be converted to the chromate salt.

(b) First confirm the presence of Cu^{2+} by adding dilute NaOH to form a precipitate of $Cu(OH)_2$. Then add excess ammonia to dissolve the precipitate as the dark blue $Cu(NH_3)_4^{2+}$ ion. Silver chloride will also dissolve in excess ammonia to form the $Ag(NH_3)_2^+$ complex ion.

Summary and Conceptual Questions

17.121. To separate CuS and $Cu(OH)_2$:

The K_{sp} for CuS is 6×10^{-37} and that for $Cu(OH)_2$ is 2.2×10^{-20}. The small size of the K_{sp} for CuS indicates that it is not very soluble. Addition of acid will cause the more soluble hydroxide to dissolve, leaving the CuS in the solid state. Another method would be to add NH_3. Again, the $Cu(OH)_2$ will dissolve, forming $Cu(NH_3)_4^+$, before the CuS.

17.122. Which salts should dissolve in HCl: $Ba(OH)_2$, $BaSO_4$, or $BaCO_3$:
$Ba(OH)_2$ and $BaCO_3$ should dissolve. The hydroxide would react with the HCl, with the lower concentration of $[OH^-]$ resulting in the dissolution of $Ba(OH)_2$. Likewise, $BaCO_3$ would react with HCl, much in the manner of all carbonate salts with HCl, to form H_2O and CO_2—dissolving the $BaCO_3$.

17.123. Silver phosphate can be more soluble in water than calculated from K_{sp} data owing to competing reactions. The phosphate anion is also the conjugate base of a weak acid and will undergo a reaction with water (hydrolysis), which lowers the $[PO_4^{3-}]$, and causes the solubility equilibrium to shift to the right, increasing the solubility of the phosphate salt. This is an example of Le Chatelier's Principle.

17.124. Which is the stronger acid: HA or HB:

Consider the equivalence point. The major species present (other than H₂O) is the salt (A⁻ or B⁻). If the solution of the salt of HA has a lower pOH (4.5 compared to 5.5), the [OH⁻] is greater, which means that the equilibrium: A⁻ + H₂O ⇌ HA + OH⁻ lies farther to the right—indicating that A⁻ is a stronger base than B⁻--so you have answered the (b) part of this question. Recalling the reciprocity between acids and their conjugate bases, if A⁻ is a stronger base than B⁻, then HB is a stronger acid than HA.

(a) HB is a stronger acid than HA.
(b) A⁻ is a stronger base than B⁻.

17.125. For the equilibrium between acetic acid and its conjugate base (acetate ion).

(a) One can understand the fraction of acetic acid present by viewing the K_a expression for the acid:

$$K_a = \frac{[CH_3CO_2^-][H_3O^+]}{[CH_3CO_2H]}$$ As pH increases, the concentration of H₃O⁺ decreases. Since K_a remains constant, the "mathematical" result is that the "numerator term" ([CH₃CO₂⁻]) must increase at the expense of the "denominator term" ([CH₃CO₂H]). Chemically, this means that the fraction of CH₃CO₂H decreases.

(a) Predominant species at pH = 4.0

This is answered by rearranging the K_a expression for the acid.

$$\frac{K_a}{[H_3O^+]} = \frac{[CH_3CO_2^-]}{[CH_3CO_2H]}$$. Substituting values for K_a and the [H₃O⁺]

$$\frac{1.8 \times 10^{-5}}{1.0 \times 10^{-4}} = \frac{[CH_3CO_2^-]}{[CH_3CO_2H]} = 0.18.$$ This means that the molecular acid is the predominant specie at pH = 4.

Predominant species at pH = 6.0?

Once again, examine the K_a expression, substituting 1 × 10⁻⁶ for the [H₃O⁺].

$$\frac{1.8 \times 10^{-5}}{1.0 \times 10^{-6}} = \frac{[CH_3CO_2^-]}{[CH_3CO_2H]} = 18.$$ Now the conjugate base is present at a level almost 20 times that of the molecular acid.

(c) The pH at the equivalence point (when the fraction of acid is equal to the fraction of conjugate base). Let's calculate the [H₃O⁺] at this point.

$$1.8 \times 10^{-5} = \frac{[0.5][H_3O^+]}{[0.5]}.$$ Note that I arbitrarily chose 0.5 as the concentration for both species—any number here will do. So [H₃O⁺] = 1.8 × 10⁻⁵ and pH = −log(1.8 × 10⁻⁵) or 4.74.

17.126. For the alpha plot of carbonic acid:

(a) As the pH increases (H₃O⁺ concentration decreases), the first ionization equilibrium H₂CO₃(aq) + H₂O(ℓ) ⇌ HCO₃⁻(aq) + H₃O⁺(aq) is shifted to the right and more HCO₃⁻ is produced. As the pH rises further, the second ionization occurs, HCO₃⁻(aq) + H₂O(ℓ) ⇌ CO₃²⁻(aq) + H₃O⁺(aq) and is shifted to the right with increasing pH, decreasing the amount of HCO₃⁻ present in the solution.

(b) From the graph, at pH = 6.0, the solution is 72% H₂CO₃ and 28% HCO₃⁻.
At pH = 10.0, the solution is 66% HCO₃⁻ and 34% CO₃²⁻.

(c) A solution buffered at pH = 11.0 should have a HCO₃⁻ to CO₃²⁻ ratio of 0.2 to 1.

17.127. For salicylic acid:

(a) Approximate values for the bond angles:

Bond	Angle
i	120
ii	120
iii	109
iv	120

(b) The hybridization of C in the ring is sp^2. With three atoms attached to any one carbon, the three groups separate by an angle of approximately 120°. This is also true of the C in the carboxylate group.

(c) What is K_a of acid?
Using SA to represent the molecular acid, the monoprotic acid would have a K_a expression of: $K_a = \dfrac{[H^+][A^-]}{[SA]}$.

You calculate the concentration of SA:

$$\dfrac{1.00 \text{ g SA}}{0.460 \text{ L}} \left(\dfrac{1 \text{ mol SA}}{138.1 \text{ g SA}} \right) = 1.57 \times 10^{-2} \text{ M SA}$$

Noting that pH = 2.4, lets us know that [H⁺] = $10^{-2.4}$ or 3.98×10^{-3} M.

Since SA is a monoprotic acid, the concentration of A⁻ is also 3.98×10^{-3} M.

$$K_a = \dfrac{(3.98 \times 10^{-3})^2}{(1.57 \times 10^{-2} - 3.98 \times 10^{-3})} = 1.34 \times 10^{-3} \text{ or } 1 \times 10^{-3} \text{ (1 sf)}$$

(d) If pH = 2.0, the % of salicylate ion present:

$$K_a = \frac{[H^+][A^-]}{[SA]} = 1 \times 10^{-3}$$; rearranging to isolate the salicylate ion: $\frac{[A^-]}{[SA]} = \frac{1 \times 10^{-3}}{1 \times 10^{-2}}$

At pH = 2.0, the salicylate ion is 10% of the acid form, with the molecular acid, SA, being 90% of the acid present.

(e) In a titration of 25.0 mL of a 0.014 M solution of SA titrated with 0.010 M NaOH:

(i) What is pH at the halfway point?

At the halfway point, the $[A^-]$ = [SA] so $[H^+] = 1 \times 10^{-3}$, so pH = 3.0

(ii) What is pH at the equivalence point?

At the equivalence point, only salt (A^-) remains. Begin by asking "What's the concentration of the salt?" The amount of SA present is
(0.0250 L)(0.014 mol SA/L) = 0.00035 mol.

The amount of 0.010 M NaOH that contains this number of moles of NaOH is (0.00035 mol)/(0.010 mol NaOH/L) = 0.035 L or 35 mL. At the equivalence point, the total volume is then (35 + 25.0) or 60. mL.

The concentration is: 3.5×10^{-4} mol/0.060 L = 5.83×10^{-3} M.
Substituting into the K_b expression for the salicylate anion reacting:

$A^-(aq) + H_2O(\ell) \rightleftarrows HA(aq) + OH^-(aq)$

$$K_b = \frac{K_w}{K_a} = \frac{1.0 \times 10^{-14}}{1.34 \times 10^{-3}} = 7.46 \times 10^{-12}$$

Since [HA] = [OH⁻], you can substitute into the K_b expression to get:

$$\frac{\left[OH^-\right]^2}{\left[5.83 \times 10^{-3}\right]} = 7.46 \times 10^{-12} \text{ and } \left[OH^-\right]^2 = 4.35 \times 10^{-14}; \left[OH^-\right] = 2.1 \times 10^{-7} \text{ M}$$

so pOH = 6.7, and pH = 7.3.

17.128. For Al(OH)₃:

(a) Al(OH)₃(aq) + H₃PO₄(aq) ⇄ AlPO₄(s) + 3 H₂O(ℓ)

(b) Theoretical yield of AlPO₄:

$152 \text{ g Al(OH)}_3 \left(\frac{1 \text{ mol Al(OH)}_3}{78.0 \text{ g}}\right)\left(\frac{1 \text{ mol AlPO}_4}{1 \text{ mol Al(OH)}_3}\right)\left(\frac{122.0 \text{ g}}{1 \text{ mol AlPO}_4}\right) = 237.7 \text{ g AlPO}_4$

$3.00 \text{ L} \left(\frac{0.750 \text{ mol H}_3\text{PO}_4}{1 \text{ L}}\right)\left(\frac{1 \text{ mol AlPO}_4}{1 \text{ mol H}_3\text{PO}_4}\right)\left(\frac{122.0 \text{ g}}{1 \text{ mol AlPO}_4}\right) = 275 \text{ g AlPO}_4$

The theoretical yield of AlPO₄ is 238 g.

(c) $K_{sp} = 1.3 \times 10^{-20} = [Al^{3+}][PO_4^{3-}] = (x)(x) = x^2$
$x = [Al^{3+}] = [PO_4^{3-}] = 1.1 \times 10^{-10}$ M

The amount of AlPO₄ added to the solution, $\dfrac{25.0 \text{ g AlPO}_4}{1.00 \text{ L}} \cdot \dfrac{1 \text{ mol AlPO}_4}{122.0 \text{ g}} = 0.205$ M, exceeds the solubility of the salt, so $[Al^{3+}] = [PO_4^{3-}] = 1.1 \times 10^{-10}$ M

(d) $[Al(H_2O)_6^{3+}]$ is a weak acid and PO_4^{3-} is a weak base, so adding H^+ will have complex results. However, PO_4^{3-} is a stronger base than the conjugate base of $[Al(H_2O)_6^{3+}]$, so adding a strong acid will increase the solubility of AlPO₄.

Solution and Answer Guide
Kotz Treichel Townsend Treichel, Chemistry and Chemical Reactivity 11e, 978-0-357-85140-1, Chapter 18: Thermodynamics-Entropy and Free Energy

TABLE OF CONTENTS

Applying Chemical Principles 671
Practicing Skills...673
General Questions...689
In the Laboratory...700
Summary and Conceptual Questions...702

Applying Chemical Principles

Thermodynamics and Living Things

18.1.1. Which is the favored process?

Consider the reactions:

(1) Creatine phosphate + H₂O → Creatine + HPᵢ	$\Delta_r G^{\circ\prime} = -43.3$ kJ/mol
(2) Adenosine + HPᵢ → Adenosine-5-monophosphate + H₂O	$\Delta_r G^{\circ\prime} = +9.2$ kJ/mol
(3) Creatine phosphate + Adenosine → Creatine + Adenosine-5-monophosphate	$\Delta_r G^{\circ\prime} = -34.1$ kJ/mol

The negative net ΔG (reaction 3) indicates that the transfer of phosphate from creatine to adenosine is product-favored.

18.1.2. Relationship between $\Delta_r G^{\circ\prime}$ and $\Delta_r G^\circ$:

$\Delta_r G^{\circ\prime} = \Delta_r G^\circ + RT \ln K$ and substituting the K expression: $K = \dfrac{[C][H_3O^+]}{[A][B]}$

$$\Delta_r G^{0\prime} = \Delta_r G^0 + 8.31 \times 10^{-3}(298\,\text{K}) \ln \frac{[C][H_3O^+]}{[A][B]}$$

$$\Delta_r G^{0\prime} = \Delta_r G^0 + 8.31 \times 10^{-3}(298\,\text{K}) \ln \frac{[1][1 \times 10^{-7}]}{[1][1]}$$

$$\Delta_r G^{0\prime} = \Delta_r G^0 + 8.31 \times 10^{-3}(298\,\text{K})(-16.12) = -39.9\,\text{kJ/mol}$$

$$\Delta_r G^{0\prime} = \Delta_r G^0 - 39.9\,\text{kJ/mol}$$

© 2024 Cengage Learning, Inc. All Rights Reserved. May not be scanned, copied or duplicated, or posted to a publicly accessible website, in whole or in part.

Are Diamonds Forever?

18.2.1. (a) Use $\Delta_fG°$ values from Appendix L to calculate $\Delta_rG°$ and K_{eq} for the reaction under standard conditions and 298.15 K.

$\Delta_rG° = \Delta_fG°(\text{graphite}) - \Delta_fG°(\text{diamond})$

$\Delta_rG° = 0.0 \text{ kJ/mol}(1\text{mol}) - 2.900 \text{ kJ/mol}(1\text{mol}) = -2.9 \text{ kJ}$ (2 sf)

$\Delta_rG° = -RT\ln K_{eq}$ and $-2.9 \times 10^3 \text{ J} = -(8.3145 \text{ J/K·mol})(298.15 \text{ K})\ln K_{eq}$

Solving for $\ln K_{eq}$: $\ln K_{eq} = \dfrac{-2.9 \times 10^3 \text{ J}}{-(8.3145 \text{ J/K·mol})(298.15 \text{ K})} = 1.1698$ and $K = 3.22$

(b) Use $\Delta_fH°$ and $S°$ values from Appendix L to calculate $\Delta_rG°$ and K_{eq} for the reaction at 1000 K. Assume that enthalpy and entropy values are valid at these temperatures. Does heating shift the equilibrium toward the formation of diamond or graphite?

$\Delta_rH° = \Delta_fH°(\text{graphite}) - \Delta_fH°(\text{diamond})$

$\Delta_rH° = 0.0 \text{ kJ/mol}(1\text{mol}) - 1.8 \text{ kJ/mol}(1\text{mol}) = -1.8 \text{ kJ}$

$\Delta_rS° = S°(\text{graphite}) - S°(\text{diamond})$

$\Delta_rS° = 5.6 \text{ J/K·mol } (1\text{mol}) - 2.377 \text{ J/K·mol } (1\text{mol}) = 3.2 \text{ J/K}$

$\Delta_rG° = \Delta_fH° - T\Delta_rS° = -1.8 \text{ kJ } (1000 \text{ J}/1 \text{ kJ}) - (1000 \text{ K})(3.2 \text{ J/K}) = -5000 \text{ J}$ or -5.0 kJ
$\Delta_rG° = -RT\ln K_{eq}$
$-5000 \text{ J} = -(8.3145 \text{ J/K·mol})(1000 \text{ K})\ln K_{eq}$ and

$\ln K_{eq} = \dfrac{-5000 \text{ J}}{-(8.3145 \text{ J/K·mol})(1000 \text{ K})} = 0.6013$ and $K_{eq} = 1.8$

Heating shifts the equilibrium toward the formation of diamond.

(c) Why is the formation of diamond favored at high pressures?

Examine the phase diagram and you'll note that as P increases, diamond is favored for any given T.

(d) Why is the conversion done at much higher T and P?
Higher temperatures will increase the rate for the process.

18.2.2. Regarding the conversion of buckminsterfullerene into diamond:

(a) Calculate $\Delta_rH°$ for conversion of $C_{60}(s)$ to C(diamond) at standard state at 298.2 K

$\Delta_rH° = 60 \text{ mol} \cdot \Delta H_f° \text{ C(diamond)(s)} - 1 \text{ mol} \cdot \Delta H_f° \text{ C}_{60}(s)$

$\Delta_rH° = (60 \text{ mol} \cdot 1.8 \text{ kJ/mol}) - (1 \text{ mol} \cdot 2320 \text{ kJ/mol}) = -2212 \text{ kJ}$

(b) Is conversion of buckminsterfullerene to diamond product-favored at room T?

$\Delta_rG° = \Delta_rH° - T\Delta_rS°$

You are told to assume that $\Delta_rS° = 0$

$\Delta_rG° = -2210 \text{ kJ} - (298.2 \text{ K} \cdot 0) = -2210 \text{ kJ}$, so conversion is product-favored

Units for thermodynamic processes are typically expressed for the balanced equation given. Hence the equation for the formation of HCl: $H_2 + Cl_2 \rightarrow 2$ HCl has a $\Delta_r G°$, $\Delta_r H°$, and $\Delta_r S°$ that represent the formation of 2 mol of HCl. You can express this as "energy change"/mol-rxn, where "energy change" typically has units of kJ (for G, H) and J/K (for S).

In this chapter, we shall frequently omit the "mol-rxn" notation in the interest of brevity. *Unless otherwise noted*, answers should be read: $\Delta_r G°$ and $\Delta_r H°$, units will be **kJ/mol-rxn**, and for $\Delta_r S°$, units will be **J/K· mol-rxn**.

Practicing Skills

Entropy and the 2nd and 3rd Laws of Thermodynamics

18.1. Solid NH_4NO_3 dissolves in water with a drop in temperature:

(a) The temperature of water drops, so dissolving is endothermic.

(b) The process was spontaneous. The solid dissolved.

(c) Entropy increases as the number of particles increase. Ions were dissociated.

(d) For all spontaneous processes, the entropy of the universe increases.

18.2. Acetic acid is added to water:

(a) The temperature of the water doesn't change, so the process is neither endothermic nor exothermic.

(b) The process was spontaneous.

(c) Entropy increases as the number of particles increase. Ions were formed.

(d) For all spontaneous processes, the entropy of the universe increases.

18.3. Identify as spontaneous or not spontaneous:

(a) Ice melting is spontaneous.

(b) Decomposition of HI was spontaneous.

(c) Ethanol and water mixing to form a solution is spontaneous.

(d) Spontaneous process for $PbCl_2$ dissolving.

18.4. Identify as spontaneous or not spontaneous:

(a) Freezing water is spontaneous.

(b) Compressing gas is not spontaneous.

(c) Reaction of Na with water is spontaneous.

(d) Formation of saturated solution is spontaneous.

18.5. Does the process result in increase in entropy of system?

(a) Condensation of gas to liquid decreases entropy.

(b) Formation of Al$_2$Br$_6$ from Al and Br$_2$ decreases entropy—number of particles decrease.

(c) Decomposition of CaCO$_3$(s) results in formation of greater number of particles—increase.

(d) Decomposition of AgCl(s) results in formation of greater number of particles—increase.

18.6. Does the process result in an increase in entropy of system?

(a) Release of ammonia gas into atmosphere increases entropy.

(b) Decrease in number of moles of gas as reaction proceeds decreases entropy.

(c) Overall increase in number of gaseous substances increases entropy. However, the increase will be small because the overall number of particles (all gaseous) remains the same.

(d) Solution formation results in greater number of particles increasing entropy.

18.7. Which of the following are reversible?

(a) Nitrogen expands into vacuum—not reversible. The gas cannot be returned to the compressed state along the same pathway as the expansion.

(b) Sublimation of dry ice—not reversible—reverse would require very low T and high P.

(c) Energy added to ice and water—reversible—removal of energy would cause some liquid water to freeze.

(d) Mixing of methanol and ethanol —not reversible. Separating the mixture cannot be accomplished by the same pathway as mixing.

18.8. Which of the following are reversible?

(a) Homogeneous mixture forms upon mixing of gases—not reversible.

(b) Sublimation of ice—not reversible at the given temperature and pressure.

(c) Energy transferred to system from mixture—reversible.

(d) Dissolution of CO$_2$ and diffusion into atmosphere—not reversible.

18.9. Calculate ΔS when 0.50 mol of ice melts at 0 °C:

Melting is an equilibrium process, so the following calculation applies.

$$0.50 \text{ mol H}_2\text{O} \left(\frac{18.02 \text{ g H}_2\text{O}}{1 \text{ mol H}_2\text{O}} \right) = 9.01 \text{ g H}_2\text{O}$$

$$9.01 \text{ g H}_2\text{O}\left(\frac{333 \text{ J}}{\text{g}}\right) = 3.00\times10^3 \text{ J}$$

$$0 \text{ °C} = 273 \text{ K}$$

$$\Delta S = \frac{q_{rev}}{T} = \frac{3.00\times10^3 \text{ J}}{273 \text{ K}} = 10.99 \text{ J/K} = 11 \text{ J/K (2 sf)}$$

18.10. Calculate ΔS when 1.00 mol of water (or 18.02 g H₂O) condenses at 100 °C (or 273 K):

Condensation is an equilibrium process, so the following calculation applies.

$$18.02 \text{ g H}_2\text{O}\left(\frac{-40.7 \text{ kJ}}{1 \text{ g}}\right) = -733.4 \text{ kJ} = -7.334\times10^5 \text{ J}$$

$$\Delta S_{condensation} = \frac{q_{rev}}{T} = \frac{-7.334\times10^5 \text{ J}}{373 \text{ K}} = -1.97\times10^3 \text{ J/K}$$

18.11. Entropy change going from 5 to 30 accessible microstates:

$$\Delta S = k(\ln W_{final} - \ln W_{initial})$$
$$\Delta S = k(\ln(30) - \ln(5))$$
$$\Delta S = 1.381\times10^{-23} \text{ J/K}(3.40-1.61)$$
$$\Delta S = 2.47\times10^{-23} \text{ J/K}$$

18.12. Entropy change when volume decreases from 3.0 L to 2.0 L:

$$\Delta S = nR\ln(V_{final}/V_{initial}) = (1 \text{ mol})(8.314 \text{ J/K·mol})\ln(3.0 \text{ L}/2.0 \text{ L}) = 3.37 \text{ J/K}$$

18.13. To calculate $S°$ for water, you need the $\Delta H°_{fusion}$ for water, and the energy needed to heat water, specific heat capacity, from 0 K to 273 K.

18.14. Trends in $S°$ values:

(a) For the halogens:

	F₂ (g)	Cl₂(g)	Br₂ (ℓ)	I₂(s)
S°(J/K·mol)	202.8	223.08	152.2	116.135

The increase in S from F₂ to Cl₂ is predictable because of the larger Cl₂ molecular size. The reduction in S for Br₂ and I₂ is because of the liquid and solid states, respectively.

(b) For the hydrocarbons:

	CH₄ (g)	C₂H₆(g)	C₃H₈ (g)
$S°$(J/K·mol)	186.26	229.2	270.3

An anticipated, the increase in $S°$ with larger molecules reflects the larger number of energy states available as well as the increased C–C bond motion.

Entropy

18.15. Substance with the higher entropy:

Correct response in bold.

(a) CO_2(s) at –78 °C vs **CO_2(g) at 0 °C**: Entropy increases with temperature and with changes in state from solid to gas.

(b) $H_2O(\ell)$ at 25 °C vs **$H_2O(\ell)$ at 50 °C**: Entropy increases with temperature.

(c) Al_2O_3(s) (pure) vs **Al_2O_3(s) (ruby)**: Entropy of a solution (even a solid one) is greater than that of a pure substance.

(d) **1 mol N_2(g) at 1 bar** vs 1 mol N_2(g) at 10 bar: With the increased P, molecules have greater order.

18.16. Substance with higher entropy:

(a) The sample of silicon containing trace impurities has a higher entropy.
(b) The sample of O_2(g) at the higher temperature (0 °C) has a higher entropy.
(c) The He(g) sample in a 2-L container has higher entropy than in a 1-L container.
(d) The sample of O_2(g) at the lower pressure (0.01 bar) has a higher entropy.

18.17. Entropy changes:

(a) KOH(s) → KOH(aq)

$\Delta_r S°$ = 91.6 J/K·mol(1 mol) – 78.9 J/K·mol(1 mol) = +12.7 J/K·mol-rxn

The increase in entropy reflects the greater disorder of the solution state.

(b) Na(g) → Na(s)

$\Delta_r S°$ = 51.21 J/K·mol(1 mol) – 153.765 J/K·mol(1 mol) = –102.55 J/K·mol-rxn

The lower entropy of the solid state is evidenced by the negative sign.

(c) $Br_2(\ell)$ → Br_2(g)

$\Delta_r S°$ = 245.42 J/K·mol(1 mol) – 152.2 J/K·mol(1 mol) = +93.2 J/K·mol-rxn

The increase in entropy is expected with the transition to the disordered state of a gas.

(d) $HCl(g) \rightarrow HCl(aq)$

$\Delta_r S° = 56.5$ J/K·mol(1 mol) – 186.2 J/K·mol(1 mol) = –129.7 J/K·mol-rxn

The lowered entropy reflects the greater order of the solution state over the gaseous state.

18.18. Calculate standard entropy change for each of the following:

(a) $\Delta_r S° = S°[NH_4Cl(aq)] - S°[NH_4Cl(s)]$
$\Delta_r S° = 1$ mol (169.9 J/K·mol) – 1 mol (94.85 J/K·mol) = 75.1 J/K·mol-rxn
A positive $\Delta_r S°$ indicates an increase in entropy.

(b) $\Delta_r S° = S°[CH_3OH(g)] - S°[CH_3OH(\ell)]$
$\Delta_r S° = 1$ mol (282.70 J/K·mol) – 1 mol (160.7 J/K·mol) = 122.0 J/K·mol-rxn
A positive $\Delta_r S°$ indicates an increase in entropy.

(c) $\Delta_r S° = S°[CCl_4(\ell)] - S°[CCl_4(g)]$
$\Delta_r S° = 1$ mol (214.39 J/K·mol) – 1 mol (309.65 J/K·mol) = –95.26 J/K·mol-rxn
A negative $\Delta_r S°$ indicates a decrease in entropy.

(d) $\Delta_r S° = S°[NaCl(g)] - S°[NaCl(s)]$
$\Delta_r S° = 1$ mol (229.79 J/K·mol) – 1 mol (72.11 J/K·mol) = 157.68 J/K·mol-rxn
A positive $\Delta_r S°$ indicates an increase in entropy.

18.19. Standard Entropy change for compound formation from elements:

(a) $HCl(g)$: $Cl_2(g) + H_2(g) \rightarrow 2\ HCl(g)$

$\Delta_r S° = 2 \cdot S°\ HCl(g) - [1 \cdot S°\ Cl_2(g) + 1 \cdot S°\ H_2(g)]$

$= (2\ mol)(186.2\ J/K·mol) - [(1\ mol)(223.08\ J/K·mol) + (1\ mol)(130.7\ J/K·mol)]$

$= +18.6$ J/K and +9.3 J/K·mol-rxn

(b) $Ca(OH)_2(s)$: $Ca(s) + O_2(g) + H_2(g) \rightarrow Ca(OH)_2(s)$

$\Delta_r S° = 1 \cdot S°\ Ca(OH)_2(s) - [1 \cdot S°\ Ca(s) + 1 \cdot S°\ O_2(g) + 1 \cdot S°\ H_2(g)]$

$= (1\ mol)(83.39\ J/K·mol) - [(1\ mol)(41.59\ J/K·mol) + (1\ mol)(205.07\ J/K·mol) +$

$(1\ mol)(130.7\ J/K·mol)]$

$= -293.97$ J/K·mol-rxn

18.20. Calculate standard entropy change for each of the following:

(a) $H_2(g) + C(graphite) \rightarrow CO_2(g)$
$\Delta_r S° = S°[CO_2(g)] - \{S°\ [H_2(g)] + S°[C(s)]\}$
$\Delta_r S° = 1$ mol (213.74 J/K·mol) – {1 mol (130.7 J/K·mol) + 1 mol (5.6 J/K·mol)}
$\Delta_r S° = 77.4$ J/K·mol-rxn

Entropy increases primarily due to conversion of C(s) to $CO_2(g)$.

(b) K(s) + ½ Cl$_2$(g) + $^3/_2$ O$_2$(g) → KClO$_3$(s)

$\Delta_r S°$ = $S°$[KClO$_3$(s)] – {$S°$[K(s)] + ½ $S°$[Cl$_2$(g)] + $^3/_2$ $S°$ [O$_2$(g)]}

$\Delta_r S°$ = 1 mol (143.1 J/K·mol) – {1 mol (64.63 J/K·mol) + ½ mol (223.08 J/K·mol) + $^3/_2$ mol (205.07 J/K·mol)}

$\Delta_r S°$ = –340.7 J/K·mol-rxn

Entropy decreases primarily due to conversion of O$_2$(g) and Cl$_2$(g) to KClO$_3$(s).

18.21. Standard molar entropy changes for:

(a) 2 Al(s) + 3 Cl$_2$(g) → 2 AlCl$_3$(s)

$\Delta_r S°$ = 2 · $S°$ AlCl$_3$(s) – [2 · $S°$ Al(s) + 3 · $S°$ Cl$_2$(g)]

= (2 mol)(109.29 J/K·mol) – [(2 mol)(28.3 J/K·mol) + (3 mol)(223.08 J/K·mol)]

= –507.3 J/K·mol-rxn

Entropy decreases as a gaseous reactant is incorporated into a solid compound

(b) 2 CH$_3$OH(ℓ) + 3 O$_2$(g) → 2 CO$_2$(g) + 4 H$_2$O(g)

$\Delta_r S°$ = [2 · $S°$ CO$_2$(g) + 4 · $S°$ H$_2$O(g)] – [2 · $S°$ CH$_3$OH(ℓ) + 3 · $S°$ O$_2$(g)]

= [(2 mol)(213.74 J/K·mol) + (4 mol)(188.84 J/K·mol)] –

[(2 mol)(127.19 J/K·mol) + (3 mol)(205.07 J/K·mol)]

= + 313.25 J/K·mol-rxn

Entropy increases as five molecules (three of them in the gas phase) form six molecules of products (all gases).

18.22. Standard entropy changes for:

(a) $\Delta_r S°$ = 2 $S°$[NaOH(aq)] + $S°$[H$_2$(g)] – {2 $S°$[Na(s)] + 2 $S°$[H$_2$O(ℓ)]}

$\Delta_r S°$ = 2 mol (48.1 J/K·mol) + 1 mol (130.7 J/K·mol)

– [2 mol (51.21 J/K·mol) + 2 mol (69.96 J/K·mol)]

$\Delta_r S°$ = –15.4 J/K·mol-rxn

A negative $\Delta_r S°$ indicates a small decrease in entropy (which is not expected due to the production of a gas (H$_2$) and a dissolved ionic compound (NaOH).

(b) $\Delta_r S°$ = 2 $S°$[NaCl(aq)] + $S°$[H$_2$O(ℓ)] + $S°$[CO$_2$(g)] – {$S°$[Na$_2$CO$_3$(s)] + 2 $S°$[HCl(aq)]}

$\Delta_r S°$ = 2 mol (115.5 J/K·mol) + 1 mol (69.95 J/K·mol) + 1 mol (213.74 J/K·mol)

– [1 mol (134.79 J/K·mol) + 2 mol (56.5 J/K·mol)]

$\Delta_r S°$ = 266.9 J/K·mol-rxn;

A positive $\Delta_r S°$ indicates an increase in entropy.

$\Delta_rS°$(universe) and Spontaneity

18.23. Is the reaction: Si(s) + 2 Cl$_2$(g) → SiCl$_4$(g) spontaneous?

$\Delta_rS°$ (system) = 1 · $S°$ SiCl$_4$(g) − {1 · $S°$ Si(s) + 2 · $S°$ Cl$_2$(g)}

= (1 mol)(330.86 J/K·mol) − [(1 mol)(18.82 J/K·mol) + (2 mol)(223.08 J/K·mol)]

= −134.12 J/K·mol-rxn

To calculate $\Delta_rS°$(surroundings), you calculate $\Delta_rH°$(system):

$\Delta_rH°$ = 1 · $\Delta H°$ SiCl$_4$(g) − [1 · $\Delta H°$ Si(s) + 2 · $\Delta H°$ Cl$_2$(g)]

= (1 mol)(− 662.75 kJ/mol) − [(1 mol)(0 kJ/mol)+ (2 mol)(0 kJ/mol)]

= −662.75 kJ/mol-rxn

$\Delta S°$(surroundings) = −$\Delta_rH°$/T = (662.75 × 10^3 J/mol)/298.15 K = 2222.9 J/K·mol

$\Delta S°$(universe) = $\Delta_rS°$(system) + $\Delta S°$(surroundings) =
(−134.12 + 2222.9) = 2088.8 J/K·mol-rxn

18.24. Is the reaction: C(graphite) + 2 H$_2$(g) → CH$_4$(g) spontaneous?

$\Delta_rS°$(system) = 1 · $S°$[CH$_4$(g)] − {1 · $S°$[C(graphite)] + 2 · $S°$[H$_2$(g)]}

= 1 mol (186.26 J/K·mol) − [1 mol (5.6 J/K·mol) + 2 mol (130.7 J/K·mol)]

= −80.7 J/K·mol-rxn

$\Delta S°$(surroundings)= −$\Delta H°$(system)/T = −$\Delta_fH°$[CH$_4$(g)]/(298.15 K)

= −[1 mol (−74.87 kJ/mol)/(298.15 K)] = +0.2511 kJ/K·mol-rxn

= 251.1 J/K

$\Delta S°$(universe) = $\Delta_rS°$(system) + $\Delta S°$(surroundings)

= −80.7 J/K·mol-rxn + (251.1 J/K·mol-rxn) = +170.4 J/K·mol-rxn

The reaction is spontaneous.

18.25. Is the reaction: 2 H$_2$O(ℓ) → 2 H$_2$(g) + O$_2$(g) product-favored?

$\Delta_rS°$(system) = [2 · $S°$ H$_2$(g) + 1 · $S°$ O$_2$(g)] − 2 · $S°$ H$_2$O(ℓ)

= [(2 mol)(130.7 J/K·mol) + (1 mol)(205.07 J/K·mol)] − (2 mol)(69.95 J/K·mol)

= +326.57 J/K and for decomposition of 1 mol of water = 163.3 J/K·mol-rxn

To calculate $\Delta S°$(surroundings), you calculate $\Delta_rH°$ system):

$\Delta_rH°$(system) = [2 · $\Delta H°$ H$_2$(g) + 1 · $\Delta H°$ O$_2$(g)] − 2 · $\Delta H°$ H$_2$O(ℓ)

= [(2 mol)(0 kJ/mol) + (1 mol)(0 kJ/mol)] − (2 mol)(−285.83 kJ/mol)

= +571.66 kJ/mol and for decomposition of 1 mol of water = 285.83 kJ/mol-rxn

$\Delta S°$(surroundings) $= -\Delta_rH°/T = -$ (285.83 × 10^3 J/mol)/298.15 K $= -958.68$ J/K·mol-rxn

$\Delta S°$(universe) $= \Delta_rS°$(system) $+ \Delta S°$(surroundings) $= 163.3 + (-958.68)$

$\Delta S°$ (universe) $= -795.4$ J/K·mol-rxn.

Since this value is less than zero, the process is not spontaneous (not product-favored at equilibrium). This is a reassuring result as life on Earth does not need water decomposing spontaneously.

18.26. Calculate $\Delta S°$(universe) for the formation of HCl(g) from hydrogen and chlorine:

½ H$_2$(g) + ½ Cl$_2$(g) → HCl(g)
$\Delta H°$(system) $= \Delta_fH°$[HCl(g)] $= 1$ mol (-92.31 kJ/mol) $= -92.31$ kJ/mol-rxn
$\Delta S°$(system) $= S°$[HCl(g)] $- \{$½ $S°$[H$_2$(g)] + ½ $S°$[Cl$_2$(g)]$\}$
$= 1$ mol (186.2 J/K·mol) $- \{$½ mol (130.7 J/K·mol) + ½ mol (223.08 J/K·mol)$\}$
$= 9.3$ J/K·mol-rxn
$\Delta S°$(universe) $= \Delta S°$(system) $+ \Delta S°$(surroundings)
$= (9.3$ J/K·mol-rxn$) + \{-(-92.31$ kJ/mol-rxn$)(10^3$ J/1 kJ$)/298.15$ K$)\} = 318.9$ J/K·mol-rxn
The reaction is spontaneous, as $\Delta S°$(universe) is positive.

18.27. Using Table 18.1 classify each of the reactions:
   (a) Fe$_2$O$_3$(s) + 2 Al(s) → 2Fe(s) + Al$_2$O$_3$(s)

   $\Delta H°$system $= -$; $\Delta S°$system $= -$; Product-favored at equilibrium at lower T

   (b) N$_2$(g) + 2 O$_2$(g) → 2NO$_2$(g)

   $\Delta H°$system $= +$; $\Delta S°$system $= -$; Not product-favored at equilibrium under any conditions

18.28. Using Table 18.1 classify each of the reactions:

   (a) C$_6$H$_{12}$O$_6$(s) + 6 O$_2$(g) → 6 CO$_2$(g) + 6 H$_2$O(ℓ)

   $\Delta_rH° < 0$, $\Delta_rS° > 0$; Product-favored at equilibrium under all conditions

   (b) MgO(s) + C(graphite) → Mg(s) + CO(g)

   $\Delta_rH° > 0$, $\Delta_rS° > 0$; depends on T and relative magnitudes of $\Delta_rH°$ and $\Delta_rS°$, more favorable at higher T.

Gibbs Free Energy

18.29. Calculate $\Delta_rG°$ for:

   (a) 2 Pb(s) + O$_2$(g) → 2 PbO(s)

   $\Delta_rH° = (2$ mol$)(-219$ kJ/mol$) - [0 + 0] = -438$ kJ

$\Delta_r S° = (2 \text{ mol})(66.5 \text{ J/K} \cdot \text{mol}) - [(2 \text{ mol})(64.81 \text{ J/K} \cdot \text{mol}) + (1 \text{ mol})(205.07 \text{ J/K} \cdot \text{mol})]$

$\Delta_r S° = -201.7 \text{ J/K}$

$\Delta_r G° = \Delta_r H° - T\Delta_r S° = -438 \text{ kJ} - (298.15 \text{ K})(-201.7 \text{ J/K})(1.000 \text{ kJ}/1000 \text{ J}) = -378 \text{ kJ}$

Reaction is product-favored at equilibrium since $\Delta G° < 0$. With the very large negative $\Delta H°$, the process is enthalpy driven.

(b) $NH_3(g) + HNO_3(aq) \rightarrow NH_4NO_3(aq)$

$\Delta_r H° = (1 \text{ mol})(-339.87 \text{ kJ/mol}) - [(1 \text{ mol})(-45.90 \text{ kJ/mol}) + (1 \text{ mol})(-207.36 \text{ kJ/mol}]$

$\Delta_r H° = -86.61 \text{ kJ}$

$\Delta_r S° = (1 \text{ mol})(259.8 \text{ J/K} \cdot \text{mol}) - [(1 \text{ mol})(192.77 \text{ J/K} \cdot \text{mol}) + (1 \text{ mol})(146.4 \text{ J/K} \cdot \text{mol})]$

$\Delta_r S° = -79.4 \text{ J/K} \cdot \text{mol-rxn}$

$\Delta_r G° = \Delta_r H° - T\Delta_r S° = -86.61 \text{ kJ} - (298.15 \text{ K})(-79.4 \text{ J/K})(1.000 \text{ kJ}/1000 \text{ J})$

$\Delta_r G° = -62.9 \text{ kJ/mol-rxn}$

Reaction is product-favored at equilibrium since $\Delta G° < 0$. With the very large negative $\Delta H°$, the process is enthalpy driven.

18.30. Calculate $\Delta G°$ for each of the following:

(a) $\Delta_r H° = 2\ \Delta_f H°[\text{NaOH(aq)}] - 2\ \Delta_f H°[\text{H}_2\text{O}(\ell)]$

$\Delta_r H° = 2 \text{ mol} (-469.15 \text{ kJ/mol}) - 2 \text{ mol} (-285.83 \text{ kJ/mol}) = -366.64 \text{ kJ/mol-rxn}$

$\Delta_r S° = 2\ S°[\text{NaOH(aq)}] + S°[\text{H}_2(g)] - \{2\ S°[\text{Na(s)}] + 2\ S°[\text{H}_2\text{O}(\ell)]\}$

$\Delta_r S° = 2 \text{ mol} (48.1 \text{ J/K} \cdot \text{mol}) + 1 \text{ mol} (130.7 \text{ J/K} \cdot \text{mol}) -$

$\{2 \text{ mol} (51.21 \text{ J/K} \cdot \text{mol}) + 2 \text{ mol} (69.95 \text{ J/K} \cdot \text{mol})\}$

$\Delta_r S° = -15.4 \text{ J/K} \cdot \text{mol-rxn}$
$\Delta_r G° = \Delta_r H° - T\Delta_r S° = -366.64 \text{ kJ/mol-rxn} - (298 \text{ K})(-15.4 \text{ J/K} \cdot \text{mol-rxn})(1 \text{ kJ}/10^3 \text{ J})$
$\Delta_r G° = -362.1 \text{ kJ/mol-rxn}$

The reaction is product-favored at equilibrium and enthalpy-driven.

(b) $\Delta H° = \Delta_f H°[C_6H_6(\ell)] = 1 \text{ mol} (48.95 \text{ kJ/mol}) = 48.95 \text{ kJ/mol-rxn}$

$\Delta S° = S°[C_6H_6(\ell)] - \{6\ S°[\text{C(graphite)}] + 3\ S°[\text{H}_2(g)]\}$

$\Delta S° = 1 \text{ mol} (173.26 \text{ J/K} \cdot \text{mol}) - \{6 \text{ mol} (5.6 \text{ J/K} \cdot \text{mol}) + 3 \text{ mol} (130.7 \text{ J/K} \cdot \text{mol})\}$

$= -252.4 \text{ J/K} \cdot \text{mol-rxn}$
$\Delta G° = \Delta H° - T\Delta S° = 48.95 \text{ kJ/mol-rxn} - (298 \text{ K})(-252.4 \text{ J/K} \cdot \text{mol-rxn})(1 \text{ kJ}/10^3 \text{ J})$

$= 124.2 \text{ kJ/mol-rxn}$

The reaction is reactant-favored at equilibrium and enthalpy-driven.

18.31. Calculate the molar free energies of formation for:

(a) $CS_2(g)$; The reaction is: $C(graphite) + 2\ S(s,rhombic) \rightarrow CS_2(g)$

$\Delta_rH° = (1\ mol)(116.7\ kJ/mol) – [0 + 0] = +116.7\ kJ$

$\Delta_rS° = (1\ mol)(237.8\ J/K·mol) – [(1\ mol)(5.6\ J/K·mol) + (2\ mol)(32.1\ J/K·mol)]$

$= +168.0\ J/K$

$\Delta_fG° = \Delta_fH° – T\Delta_rS° = (116.7\ kJ) – (298.15\ K)(168.0\ J/K)(1.000\ kJ/1000\ J)$

$= +66.6\ kJ$ Appendix value: 66.61 kJ/mol

(b) NaOH(s); The reaction is: $Na(s) + {}^1/_2\ O_2(g) + {}^1/_2\ H_2(g) \rightarrow NaOH(s)$

$\Delta_rH° = (1\ mol)(–425.93\ kJ/mol) – [0 + 0 + 0] = –425.93\ kJ/mol\text{-rxn}$

$\Delta_rS° = (1\ mol)(64.46\ J/K·mol) –$

$[(1\ mol)(51.21\ J/K·mol) + ({}^1/_2\ mol)(205.07\ J/K·mol) + ({}^1/_2\ mol)(130.7\ J/K·mol)]$

$= –154.6\ J/K·mol\text{-rxn}$

$\Delta_r G° = \Delta_rH° – T\Delta_rS° = (–425.93\ kJ) – (298.15\ K)(–154.6\ J/K)(1.000\ kJ/1000\ J)$

$= –379.82\ kJ$ Appendix value: –379.75 kJ/mol

(c) ICl(g); The reaction is: ${}^1/_2\ I_2(g) + {}^1/_2\ Cl_2(g) \rightarrow ICl(g)$

$\Delta_rH° = (1\ mol)(+17.51\ kJ/mol) – [0 + 0] = +17.51\ kJ/mol\text{-rxn}$

$\Delta_rS° = (1\ mol)(247.56\ J/K·mol) –$

$[({}^1/_2\ mol)(116.135\ J/K·mol) + ({}^1/_2\ mol)(223.08\ J/K·mol)]$

$= +77.95\ J/K·mol\text{-rxn}$

$\Delta_rG° = \Delta_rH° – T\Delta_rS° = (+17.51\ kJ) – (298.15\ K)(+77.95\ J/K)(1.000\ kJ/1000\ J)$

$= –5.73\ kJ$ Appendix value: –5.73 kJ/mol

Parts (b) and (c) are product-favored at equilibrium.

18.32. Calculate the molar free energies of formation for:

(a) $Ca(s) + O_2(g) + H_2(g) \rightarrow Ca(OH)_2(s)$

$\Delta_rH° = \Delta_fH°[Ca(OH)_2(s)] = 1\ mol\ (–986.09\ kJ/mol) = –986.09\ kJ/mol\text{-rxn}$
$\Delta_rS° = S°[Ca(OH)_2(s)] – \{S°[Ca(s)] + S°[O_2(g)] + S°[H_2(g)]\}$

$\Delta_rS° = 1\ mol\ (83.39\ J/K·mol) –$

$\{1\ mol\ (41.59\ J/K·mol) + 1\ mol\ (205.07\ J/K·mol) + 1\ mol\ (130.7\ J/K·mol)\}$

$\Delta_rS° = –294.0\ J/K·mol\text{-rxn}$

$\Delta_fG° = \Delta_rH° – T\Delta_rS° = –986.09\ kJ/mol\text{-rxn} – (298\ K)(–294.0\ J/K·mol\text{-rxn})(1\ kJ/10^3\ J)$
$\Delta_fG° = –898.5\ kJ/mol\text{-rxn}$ Appendix L value –898.43 kJ

(b) $\frac{1}{2}$ Cl$_2$(g) → Cl(g)

$\Delta_rH° = \Delta_fH°$ [Cl(g)] = 1 mol (121.3 kJ/mol) = 121.3 kJ/mol-rxn

$\Delta_rS° = S°$[Cl(g)] – $\frac{1}{2}$ $S°$[Cl$_2$(g)]

$\Delta_rS°$ = 1 mol (165.19 J/K·mol) – $\frac{1}{2}$ mol (223.08 J/K·mol) = 53.65 J/K·mol-rxn

$\Delta_fG° = \Delta_rH° – T\Delta_rS°$ = 121.3 kJ/mol-rxn – (298 K)(53.65 J/K·mol-rxn)(1 kJ/10^3 J)

   = 105.3 kJ/mol-rxn   Appendix L value 105.3 kJ

(c) 2 Na(s) + C(graphite) + $\frac{3}{2}$ O$_2$(g) → Na$_2$CO$_3$(s)

$\Delta_rH° = \Delta_fH°$[Na$_2$CO$_3$(s)] = 1 mol (–1130.77 kJ/mol) = –1130.77 kJ/mol-rxn

$\Delta_rS° = S°$[Na$_2$CO$_3$(s)] – {2 $S°$[Na(s)] + $S°$[C(graphite)]+ $\frac{3}{2}$ $S°$[O$_2$(g)]}

$\Delta_rS°$ = 1 mol (134.79 J/K·mol) – {[2 mol (51.21 J/K·mol) + 1 mol (5.6 J/K·mol)
    + $\frac{3}{2}$ mol (205.07 J/K·mol)}

$\Delta_rS°$ = –280.8 J/K·mol-rxn

$\Delta_fG° = \Delta_rH° – T\Delta_rS°$ = –1130.77 kJ/mol-rxn – (298 K)(–280.8 J/K·mol-rxn)(1 kJ/10^3 J)

$\Delta_fG°$ = –1047.1 kJ/mol-rxn   Appendix L value –1048.08 kJ

Reactions (a) and (c) are product-favored at equilibrium.

Free Energy of Formation

18.33. Calculate $\Delta_rG°$ for the following equations. Are they product-favored at equilibrium?

(a) 2 K(s) + Cl$_2$(g) → 2 KCl(s)

$\Delta_rG°$ = [2 · $\Delta_fG°$ KCl(s)] – [1 · $\Delta_fG°$ Cl$_2$(g) + 2 · $\Delta_fG°$ K(s)]

   = [(2 mol)(– 408.77 kJ/mol] – [(1 mol)(0 kJ/mol) +(2 mol)(0 kJ/mol)]

$\Delta_rG°$ = –817.54 kJ/mol-rxn

With a $\Delta_rG°$ < 0, the reaction is product-favored at equilibrium.

(b) 2 CuO(s) → 2 Cu(s) + O$_2$(g)

$\Delta_rG°$ = [2 · $\Delta_fG°$ Cu(s) + $\Delta_fG°$ O$_2$(g)] – [2 · $\Delta_fG°$ CuO(s)]

$\Delta_rG°$ = [(2 mol)(0 kJ/mol) + (1 mol)(0 kJ/mol)] – [(2mol)(–128.3 kJ/mol) = +256.6 kJ

With a $\Delta G°$ > 0, the reaction is not product-favored at equilibrium.

(c) 4 NH$_3$(g) + 7 O$_2$(g) → 4 NO$_2$(g) + 6 H$_2$O(g)

$\Delta_rG°$ = [4 · $\Delta_fG°$ NO$_2$(g) + 6 · $\Delta_fG°$ H$_2$O(g)] – [4 · $\Delta_fG°$ NH$_3$(g) + 7 · $\Delta_fG°$ O$_2$(g)]

$\Delta_rG°$ = [(4 mol)(+51.23 kJ/mol) + (6 mol)(–228.59 kJ/mol] –

   [(4 mol)(–16.37 kJ/mol) + (7 mol)(0 kJ/mol)] = –1101.14 kJ/mol-rxn

With a $\Delta_rG°$ < 0, the reaction is product-favored at equilibrium.

18.34. Calculate $\Delta_rG°$ for the following equations. Are they product-favored?

(a) $\Delta_rG° = \Delta_fG°[SO_2(g)] – \Delta_fG°[HgS(s)]$

$\Delta_rG° =$ 1 mol (–300.13 kJ/mol) – 1 mol (–50.6 kJ/mol) = –249.5 kJ/mol-rxn
The reaction is product-favored at equilibrium.

(b) $\Delta_rG° = 2\Delta_fG°[H_2O(g)] + 2\Delta_fG°[SO_2(g)] – 2\Delta_fG°[H_2S(g)]$
$\Delta_rG° =$ 2 mol (–228.59 kJ/mol) + 2 mol (–300.13 kJ/mol) – 2 mol (–33.56 kJ/mol)
$\Delta_rG° =$ –990.32 kJ/mol-rxn
The reaction is product-favored at equilibrium.

(c) $\Delta_rG° = 2\Delta_fG°[MgCl_2(s)] – \Delta_fG°[SiCl_4(g)]$
$\Delta_rG° =$ 2 mol (–592.09 kJ/mol) – 1 mol (–622.76 kJ/mol) = –561.42 kJ/mol-rxn
The reaction is product-favored at equilibrium.

18.35. Value for $\Delta_fG°$ of $BaCO_3(s)$:

$\Delta_rG° = [\Delta_fG° BaO(s) + \Delta_fG° CO_2(g)] – [\Delta_fG° BaCO_3(s)]$

+219.7 kJ = [(1 mol)(–520.38 kJ/mol) + (1 mol)(–394.359 kJ/mol)] – $\Delta_fG° BaCO_3(s)$

+219.7 kJ = –914.74 kJ – $\Delta_fG° BaCO_3(s)$

–1134.4 kJ/mol = $\Delta_fG° BaCO_3(s)$

18.36. Value for $\Delta_fG°$ of $TiCl_2(s)$:

$\Delta_rG° = \Delta_fG°[TiCl_4(\ell)] – \Delta_fG°[TiCl_2(s)]$

–272.8 kJ = 1 mol (–737.2 kJ/mol) – 1 mol $\Delta_fG°[TiCl_2(s)]$

$\Delta_fG°[TiCl_2(s)]$ = –464.4 kJ/mol

Effect of Temperature on ΔG

18.37. Entropy-favored or entropy-disfavored reactions?

(a) $N_2(g) + 2 O_2(g) \rightarrow 2 NO_2(g)$

$\Delta_rH° =$ (2 mol)(+33.1 kJ/mol) – [0 + 0] = +66.2 kJ/mol-rxn

$\Delta_rS° =$ (2 mol)(+240.04 J/K·mol)

– [(1 mol)(191.56 J/K·mol) + (2 mol)(+205.07 J/K·mol)]

= –121.62 J/K·mol-rxn

$\Delta_rG° = \Delta_rH° – T\Delta_rS° =$ +66.2 kJ/mol-rxn – (298.15 K)(–0.12162 kJ/K·mol-rxn)

= +102.5 kJ/mol-rxn

The reaction is entropy-disfavored *and* enthalpy-disfavored.
*There is **no** T at which $\Delta G° < 0$.*

(b) 2 C(s) + O$_2$(g) → 2 CO(g)

$\Delta_r H°$ = (2 mol)(–110.525 kJ/mol) – [0 + 0] = – 221.05 kJ/mol-rxn

$\Delta_r S°$ = (2 mol)(+197.674 J/K·mol) –

[(2 mol)(+ 5.6 J/K·mol) + (1mol)(+205.07 J/K·mol)] = +179.1 J/K·mol-rxn

$\Delta_r G°$ = $\Delta_r H°$ – $T\Delta_r S°$ = –221.05 kJ/mol-rxn –(298.15 K)(+0.1791 kJ/K·mol-rxn)

= –274.45 kJ/mol-rxn

This reaction is **both** entropy- and enthalpy-favored *at all temperatures.*

(c) CaO(s) + CO$_2$(g) → CaCO$_3$(s)

$\Delta_r H°$ = (1 mol)(–1207.6 kJ/mol) – [(1mol)(–635.0 kJ/mol) + (1mol)(–393.509 kJ/mol)]

= –179.0 kJ/mol-rxn

$\Delta_r S°$ = (1 mol)(+91.7 J/K·mol) – [(1mol)(38.2 J/K·mol) + (1mol)(+213.74 J/K·mol)]

= –160.2 J/K·mol-rxn

$\Delta_r G°$ = $\Delta_r H°$ – $T\Delta_r S°$ = –179.0 kJ/mol-rxn – (298.15 K)(–0.1602 kJ/K·mol-rxn)

= –131.23 kJ/mol-rxn

This reaction is entropy-disfavored but enthalpy-favored and will be *product-favored at low temperatures*.

(d) 2 NaCl(s) → 2 Na(s) + Cl$_2$(g)

$\Delta_r H°$ = [(2 mol)(0 kJ/mol) + (1mol)(0 kJ/mol)] – (2 mol)(–411.12 kJ/mol)]

= +822.24 kJ/mol-rxn

$\Delta_r S°$ = [(2 mol)(+51.21 J/K·mol) + (1 mol)(+223.08 J/K·mol)] –

(2mol)(+72.11 J/K·mol)] = +181.28 J/K·mol-rxn

$\Delta_r G°$ = $\Delta_r H°$ – $T\Delta_r S°$ = +822.24 kJ/mol-rxn – (298.15 K)(+0.18128 kJ/K·mol-rxn)

= +768.19 kJ/mol-rxn

This reaction is *entropy-favored* and will be *product-favored* at high temperatures.

18.38. Entropy-favored or entropy-disfavored reactions?

(a) $\Delta_r S°$ = 2 $S°$[I(g)] – $S°$[I$_2$(g)]
$\Delta_r S°$ = 2 mol (180.791 J/K·mol) – 1 mol (260.69 J/K·mol) = 100.89 J/K·mol-rxn
Entropy-favored; Increasing the temperature will make the reaction more product-favored at equilibrium.

(b) $\Delta_r S°$ = 2 $S°$[SO$_3$(g)] – {2 $S°$[SO$_2$(g)] + $S°$[O$_2$(g)]}
$\Delta_r S°$ = 2 mol (256.77 J/K·mol) – [2 mol (248.21 J/K·mol) + 1 mol (205.07 J/K·mol)]
$\Delta_r S°$ = –187.95 J/K·mol-rxn
Entropy-disfavored; Increasing the temperature will make the reaction more reactant-favored at equilibrium.

(c) $\Delta_rS° = S°[SiO_2(s)] + 4\, S°[HCl(g)] - \{S°[SiCl_4(g)] + 2\, S°[H_2O(\ell)]\}$

$\Delta_rS° = 1$ mol (41.46 J/K·mol) + 4 mol (186.2 J/K·mol) – [1 mol (330.86 J/K·mol) + 2 mol (69.95 J/K·mol)]

$\Delta_rS° = 315.5$ J/K·mol-rxn

Entropy-favored;. Increasing the temperature will make the reaction more product-favored at equilibrium.

(d) $\Delta_rS° = 4\, S°[PH_3(g)] - \{S°[P_4(s, \text{white})] + 6\, S°\,[H_2(g)]\}$

$\Delta_rS° = 4$ mol (210.24 J/K·mol) – [1 mol (41.1 J/K·mol) + 6 mol (130.7 J/K·mol)]

$\Delta_rS° = 15.7$ J/K·mol-rxn

Entropy-favored; Increasing the temperature will make the reaction more product-favored at equilibrium.

18.39. For the decomposition of $MgCO_3(s) \rightarrow MgO(s) + CO_2(g)$:

(a) $\Delta_rS°(\text{system}) = [1 \cdot S°\, MgO(s) + 1 \cdot S°\, CO_2(g)] - 1 \cdot S°\, MgCO_3(s)$

$= [(1\,\text{mol})(26.85\,\text{J/K·mol}) + (1\,\text{mol})(213.74\,\text{J/K·mol})] - (1\,\text{mol})(65.84\,\text{J/K·mol})$

$= +174.75$ J/K·mol-rxn

Calculate $\Delta H°(\text{system})$:

$\Delta_rH°(\text{system}) = [1 \cdot \Delta_fH°\, MgO(s) + 1 \cdot \Delta_fH°\, CO_2(g)] - 1 \cdot \Delta_fH°\, MgCO_3(s)$

$= [(1\,\text{mol})(-601.24\,\text{kJ/mol}) + (1\,\text{mol})(-393.509\,\text{kJ/mol})]$

$- (1\,\text{mol})(-1111.69\,\text{kJ/mol}) = +116.94$ kJ/mol-rxn

Using the $\Delta_rH°(\text{system})$ and $\Delta_rS°(\text{system})$, you can calculate $\Delta_rG°$.

$\Delta_rG° = \Delta H° - T\Delta S° = +116.94$ kJ/mol-rxn – (298.15 K)(+174.75 J/K·mol-rxn)(1 kJ/10³ J)

$= +64.84$ kJ/mol-rxn.

(b) The sign of $\Delta_rG°$ is +, so the process is not product-favored at equilibrium at 298 K.

(c) From Table 18.1 you observe that this type of reaction ($\Delta_rH° = +$ and $\Delta_rS° = +$) is product-favored at equilibrium at higher T.

18.40. Regarding the reaction of tin(IV) oxide with carbon:

$\Delta_rH° = \Delta_fH°[CO_2(g)] - \Delta_fH°[SnO_2(s)]$

$\Delta_rH° = 1$ mol (–393.509 kJ/mol) – 1 mol (–577.63 kJ/mol) = 184.12 kJ/mol-rxn

$\Delta_rS° = S°[Sn(s, \text{white})] + S°[CO_2(g)] - \{S°[SnO_2(s)] + S°[C(\text{graphite})]\}$

$\Delta_rS° = 1$ mol (51.08 J/K·mol) + 1 mol (213.74 J/K·mol)

$- [1$ mol (49.04 J/K·mol) + 1 mol (5.6 J/K·mol)]

$\Delta_rS° = 210.2$ J/K·mol-rxn

(a) $\Delta_rG° = \Delta_rH° - T\Delta_rS° = 184.12$ kJ/mol-rxn $- (298.15$ K$)(210.2$ J/K·mol-rxn$)(1$ kJ/10^3 J$)$

= 121.5 kJ/mol-rxn

The reaction is predicted to not be product-favored at equilibrium at 298 K ($\Delta_rG° > 0$).

(b) The reaction is predicted to be product-favored at equilibrium at higher temperatures.

Free Energy and Equilibrium Constants

18.41. Given the K_a for acetic acid, calculate $\Delta_rG°$:

$\Delta_rG° = -RT\ln K$ so $-(8.3145$ J/K · mol$)(298.15$ K$)\ln(1.8 \times 10^{-5})$

$\Delta_rG° = -(8.3145$ J/K · mol$)(298.15$ K$)(-10.925) = 27083$ J/mol or 27.1 kJ/mol, so the reaction is reactant-favored at equilibrium.

18.42. Given the K_f for formation of diamminesilver(I), calculate $\Delta_rG°$:

$\Delta_rG° = -RT\ln K$ so $-(8.3145$ J/K · mol$)(298.15$ K$)\ln(1.1 \times 10^7)$

$\Delta_rG° = -(8.3145$ J/K · mol$)(298.15$ K$)(16.213) = -40193$ J/mol or -40.2 kJ/mol, so the reaction is product-favored at equilibrium.

18.43. Calculate K_p for the reaction:

½ N_2(g) + ½ O_2(g) ⇌ NO(g) $\Delta_fG° = +86.58$ kJ/mol NO

$\Delta_fG° = -RT\ln K_p$ so 86.58×10^3 J/mol $= -(8.3145$ J/K · mol$)(298.15$ K$)$ ln K_p

$-34.926 = \ln K_p$ and $6.8 \times 10^{-16} = K_p$

Note that the + value of $\Delta_fG°$ results in a value of K_p which is small--reactants are favored at equilibrium. A negative value would result in a large K_p -- a process in which the products were favored at equilibrium.

18.44. Calculate K_p for the equilibrium: 3 O_2(g) ⇌ 2 O_3(g)

$\Delta_rG° = -RT\ln K_p$
163.2 kJ/mol $= -(8.3145 \times 10^{-3}$ kJ/K·mol$)(298$ K$)$ ln K_p
ln $K_p = -65.87$ so $K_p = 2.5 \times 10^{-29}$
The large, positive $\Delta_rG°$ value results in a K value much less than 1.

18.45. From the $\Delta_rG°$ and K_p, determine if the hydrogenation of ethylene is product-favored:

Using $\Delta_fG°$ data from the Appendix: C_2H_4(g) + H_2(g) ⇌ C_2H_6(g)

$\Delta_rG° = (1$ mol$)(-31.89$ kJ/mol$) - [(1$mol$)(68.35$ kJ/mol$) + (1$mol$)(0$ kJ/mol$)] = -100.24$ kJ

and since $\Delta_rG° = -RT\ln K_p$

-100.24×10^3 J/mol $= -(8.3145$ J/K · mol$)(298.15$ K$)$ ln K_p

$40.436 = \ln K_p$ and $K_p = 3.64 \times 10^{17} = 4 \times 10^{17}$ (1 sf)

The negative value of $\Delta_rG°$ means the reaction is product-favored at equilibrium.
The large value of $\Delta_rG°$ means that K_p is very large.

18.46. Calculate $\Delta_rG°$ for the formation of $C_2H_5OH(g)$:

$\Delta_rG° = \Delta_fG°[C_2H_5OH(g)] - \{\Delta_fG°[C_2H_4(g)] + \Delta_fG°[H_2O(g)]\}$
$\Delta_rG° = 1$ mol (–168.49 kJ/mol) – [1 mol (68.35 kJ/mol) + 1 mol (–228.59 kJ/mol)

$= -8.25$ kJ/mol-rxn

$\Delta_rG° = -RT\ln K_p$ and -8.25 kJ/mol $= -(8.3145 \times 10^{-3}$ kJ/K·mol)(298 K) ln K_p
ln $K_p = 3.33$ and $K_p = 28$.
Both the negative $\Delta_rG°$ value and the large K value indicate a product-favored reaction.

18.47. Calculate $\Delta_rG°$ for the decomposition of 1 mol of $CaCO_3(s)$ at 455 K:

$CaCO_3(s) \rightleftarrows CaO(s) + CO_2(g)$

$\Delta_rH° = \{\Delta_fH°[CaO(s)] + \Delta_fG°[CO_2(g)]\} - \{\Delta_fH°[CaCO_3(s)]\}$
$\Delta_rH° = \{1$ mol/1 mol-rxn (–635.09 kJ/mol) + 1mol/1 mol-rxn (–393.509 kJ/mol)} – {[1 mol/1 mol-rxn (–1207.6 kJ/mol)}

$= 179.00$ kJ/mol-rxn

$\Delta_rS° = \{S°[CaO(s)] + S°[CO_2(g)]\} - \{S°[CaCO_3(s)]\}$
$\Delta_rS° = \{1$ mol/1 mol-rxn (0.0382 kJ/K·mol) + 1mol/1 mol-rxn (0.21374 kJ/ K·mol)} – {[1 mol/1 mol-rxn (0.0917 kJ/ K·mol)}

$= 0.16024$ kJ/K·mol

$\Delta_rG° = \Delta_rH° - T\Delta_rS° = 179.00$ kJ/ mol-rxn + (455 K) 0.16024 kJ/K·mol-rxn = 106.09 kJ/mol-rxn = 106. 1 kJ/mol $CaCO_3(s)$ decomposed.

$\Delta_rG° = -RT\ln K_p$ and 106.09 kJ/mol $= -(8.3145 \times 10^{-3}$ kJ/K·mol)(455 K) ln K_p
ln $K_p = -28.04$ and $K_p = 6.6 \times 10^{-13} = 7 \times 10^{-13}$ (1 sf)
Both the positive $\Delta_rG°$ value and the very small K value indicate a reactant-favored reaction at equilibrium.

18.48. Calculate $\Delta_rG°$ for the formation of 1 mol of $NH_3(s)$ at 675 K:

$N_2(g) + 3 H_2(g) \rightleftarrows 2 NH_3(g)$

$\Delta_rG° = \{2$ mol $\Delta_fG°[NH_3(g)]\} - \{\Delta_fG°[N_2(g)] + 3$ mol $\Delta_fG°[H_2(g)]\}$
$\Delta_rG° = \{2$ mol (–16.37 kJ/mol)} – {[1 mol (0.0 kJ/mol) + 3 mol (0.0 kJ/mol)}

$= -32.74$ kJ/mol-rxn

$\Delta_rG° = -RT\ln K_p$ and -32.74 kJ/mol $= -(8.3145 \times 10^{-3}$ kJ/K·mol)(675 K) ln K_p
ln $K_p = 5.83$ and $K_p = 340$
Both the negative $\Delta_rG°$ value and the large K value indicate a product-favored reaction at equilibrium.

Free Energy and Reaction Conditions

18.49. For the synthesis of NH$_3$ from its elements:

(a) For the reaction N$_2$(g) + 3 H$_2$(g) → 2 NH$_3$(g), calculate $\Delta_rG°$ from $\Delta_fG°$:

Using $\Delta_fG°$ data from Appendix L:

$\Delta_rG° $ = (2 mol)(–16.37 kJ/mol) – [(1mol)(0 kJ/mol) + (3 mol)(0 kJ/mol)] = –32.74 kJ

With a $\Delta_rG° < 0$, the reaction is product-favored at equilibrium.

(b) What is Δ_rG when reactants and products are each present at 0.10 atm pressure?

$\Delta_rG = \Delta_rG° + RT \ln Q$ and $Q = \dfrac{P_{NH_3}^2}{P_{N_2} \cdot P_{H_2}^3} = \dfrac{(0.10)^2}{(0.10)(0.10)^3} = 100$

Δ_rG = –32.74 kJ + (8.31447 J/K·mol)(298.15 K)ln(100)

Δ_rG = –32.74 kJ + 11416.0 J/mol [Note the need to change J to kJ in the 2nd term]

Δ_rG = –32.74 kJ + 11.4160 kJ/mol = –21.33 kJ   (The process is spontaneous.)

18.50. For the decomposition of calcium carbonate at 25 °C:

(a) $\Delta_rG°$ = (1 mol/mol-rxn)($\Delta_fG°$[CaO(s)]) + (1 mol/mol−rxn)($\Delta_fG°$[CO$_2$(g)]) –

(1 mol/mol-rxn)(Δ_fG[CaCO$_3$(s)])

$\Delta_rG°$ = (1 mol/mol-rxn)(−603.42 kJ/mol) + (1 mol/mol-rxn)(−394.359 kJ/mol)

– (1 mol/mol-rxn)(−1129.16 kJ/mol)

= 131.38 kJ/mol-rxn, so it is reactant-favored at equilibrium.

(b) $\Delta G = \Delta G° + RT \ln Q = \Delta G° + RT \ln P_{CO_2}$

= 131.38 kJ/mol−rxn + (8.3145 J/K·mol−rxn)(1 kJ/1000 J)(298 K)ln(0.10)

= 125.7 kJ/mol−rxn; No, the reaction is not spontaneous.

General Questions

18.51. Compound with higher standard entropy:

Correct answer in **bold**.

(a) HF(g) vs HCl(g) vs **HBr(g)**: Entropy increases with molecular size (mass).

(b) NH$_4$Cl(s) vs **NH$_4$Cl(aq)**: Entropy of solutions is greater than that of the solid.

(c) **C$_2$H$_4$(g)** vs N$_2$(g): Entropy increases with molecular complexity.

(d) NaCl(s) vs **NaCl(g)**: Entropy of the gaseous state is very high. The solid state has lower entropy.

18.52. Calculate $\Delta_r S°$ for: $1/2\ N_2(g) + 3/2\ H_2(g) \rightleftarrows NH_3(g)$

$\Delta_r S° = S°[NH_3(g)] - \{1/2\ S°[N_2(g)] + 3/2\ S°[H_2(g)]\}$

$\Delta_r S° = 1\ mol\ (192.77\ J/K·mol) - [1/2\ mol\ (191.56\ J/K·mol) + 3/2\ mol\ (130.7\ J/K·mol)]$

$\Delta_r S° = -99.1\ J/K·mol\text{-rxn}$

18.53. For the reaction $C_6H_6(\ell) + 3\ H_2(g) \rightarrow C_6H_{12}(\ell)$, $\Delta_r H° = -206.7\ kJ$

and $\Delta_r S° = -361.5\ J/K$

$\Delta_r G° = \Delta_r H° - T\Delta_r S°$

$= -206.7\ kJ - (298.15\ K)(-361.5\ J/K)(1.000\ kJ/1000\ J)$

$= -206.7\ kJ - (-107.8\ kJ) = -98.9\ kJ/mol\text{-rxn}$

The negative value for $\Delta_r G°$ tells you that the reaction would be product-favored at equilibrium under standard conditions. The negative value for $\Delta_r H°$ tells you that the reaction is enthalpy driven.

18.54. Calculate $\Delta_r H°, \Delta_r S°$ and $\Delta_r G°$ for the hydrogenation of octene:

$\Delta_r H° = \Delta_f H°[C_8H_{18}(g)] - \Delta_f H°[C_8H_{16}(g)]$

$= 1\ mol\ (-208.45\ kJ/mol) - 1\ mol\ (-82.93\ kJ/mol) = -125.52\ kJ/mol\text{-rxn}$

$\Delta_r S° = S°[C_8H_{18}(g)] - \{S°[C_8H_{16}(g)] + S°[H_2(g)]\}$

$\Delta_r S° = 1\ mol\ (463.639\ J/K·mol) - [1\ mol\ (462.8\ J/K·mol) + 1\ mol\ (130.7\ J/K·mol)]$

$= -129.9\ J/K·mol\text{-rxn}$

$\Delta_r G° = \Delta_r H° - T\Delta_r S° = -125.52\ kJ/mol\text{-rxn} - (298\ K)(-129.9\ J/K·mol\text{-rxn})(1\ kJ/10^3\ J)$

$= -86.81\ kJ/mol\text{-rxn}$

The reaction is product-favored at equilibrium.

18.55. Calculate $\Delta_r H°$ and $\Delta_r S°$ for the combustion of ethane:

	$C_2H_6(g)$	$O_2(g)$	$CO_2(g)$	$H_2O(\ell)$
$\Delta_f H°$ (kJ/mol)	–83.85	0	–393.509	–285.83
$S°$ (J/K·mol)	+229.2	+205.07	+213.74	+69.95

$\Delta_r H° = [2 · \Delta_f H°\ CO_2(g) + 3 · \Delta_f H°\ H_2O(\ell)] - [1 · \Delta_f H°\ C_2H_6(g) + 7/2 · \Delta_f H°\ O_2(g)]$

$= [(2\ mol/mol\text{-rxn})(-393.509\ kJ/mol) + (3\ mol/mol\text{-rxn})(-285.83\ kJ/mol)] - [(1\ mol\text{-rxn})(-83.85\ kJ/mol) + 0]$

$= -1560.658\ kJ/mol\text{-rxn}$

$\Delta_r S° = [2 \cdot S°\ CO_2(g) + 3 \cdot S°\ H_2O(\ell)] - [1 \cdot S°\ C_2H_6(g) + 7/2 \cdot S°\ O_2(g)]$

$= [(2\ mol/mol\text{-}rxn)(213.74\ J/K\cdot mol) + (3\ mol/mol\text{-}rxn)(69.95\ J/K\cdot mol)] -$

$[(1\ mol/mol\text{-}rxn)(229.2\ J/K\cdot mol) + (7/2\ mol/mol\text{-}rxn)(205.07\ J/K\cdot mol)] = -309.62\ J/K\cdot mol\text{-}rxn$

$$\Delta S°(\text{surroundings}) = \frac{-\Delta_r H°}{T}$$

$= (1.560658 \times 10^6\ J/mol\text{-}rxn)/298\ K = 5237\ J/K\cdot mol\text{-}rxn$

$\Delta S°(\text{universe}) = \Delta_r S°(\text{system}) + \Delta S°(\text{surroundings})$

$\Delta S°(\text{universe}) = -309.62\ J/K\cdot mol\text{-}rxn + 5237\ J/K\cdot mol\text{-}rxn$

$= 4930\ J/K\cdot mol\text{-}rxn$

Since $\Delta S°$(universe) is positive, the process is product-favored at equilibrium. This calculation is consistent with our expectations. You know that hydrocarbons burn completely (in the presence of sufficient oxygen) to produce carbon dioxide and water.

18.56. Equation for formation of Fe_3O_4 and $\Delta_f G°$ for 1 lb Fe_3O_4:

$3\ Fe(s) + 2\ O_2(g) \rightarrow Fe_3O_4(s)$

from Appendix L, $\Delta_f G°[Fe_2O_3(s)] = -1015.4\ kJ/mol$, so for 1 lb Fe_2O_3:

$454\ g\ Fe_3O_4 \left(\dfrac{1\ mol\ Fe_3O_4}{231.53\ g\ Fe_3O_4}\right)\left(\dfrac{-1015.4\ kJ}{1\ mol\ Fe_3O_4}\right) = -1.99 \times 10^3\ kJ$

18.57. For the reaction of HCl and NH_3:

(a) Calculate $\Delta_r G°$ for $NH_3(g) + HCl(g) \rightarrow NH_4Cl(s)$

	$NH_3(g)$	$HCl(g)$	$NH_4Cl(s)$
$S°$ (J/K·mol)	192.77	186.2	94.85
$\Delta_f H°$ (kJ/mol)	−45.90	−92.31	−314.55

$\Delta_r S° = 1 \cdot S°\ NH_4Cl(s) - [1 \cdot S°\ NH_3(g) + 1 \cdot S°\ HCl(g)]$

$= (1\ mol)(94.85\ J/K\cdot mol) - [(1\ mol)(192.77\ J/K\cdot mol) + (1\ mol)(186.2\ J/K\cdot mol)]$

$= -284.1\ J/K$

$\Delta_r H° = 1 \cdot \Delta_f H°\ NH_4Cl(s) - [1 \cdot \Delta_f H°\ NH_3(g) + 1 \cdot \Delta_f H°\ HCl(g)]$

$= (1\ mol)(-314.55\ kJ/mol) - [(1\ mol)(-45.90\ kJ/mol) + (1\ mol)(-92.31\ kJ/mol)]$

$= -176.34\ kJ$

$\Delta_r G° = \Delta_r H° - T \Delta_r S°$

$= -176.34 \text{ kJ} - (298.15 \text{ K})(-284.1 \text{ J/K})(1.000 \text{ kJ}/1000 \text{ J})$

$= -176.34 + 84.67 = -91.64 \text{ kJ/mol-rxn}$

$\Delta S°(\text{surroundings}) = \dfrac{-\Delta_r H°}{T} = \dfrac{176.34 \text{ kJ}}{298.15 \text{ K}} \cdot \dfrac{1000 \text{ J}}{1 \text{ kJ}} = 591.45 \text{ J/K}$

$\Delta S°(\text{universe}) = \Delta_r S°(\text{system}) + \Delta S°(\text{surroundings}) = -284.1 \text{ J/K} + 591.45 \text{ J/K}$

$= +307.3 \text{ J/K·mol-rxn}$

The value for $\Delta_r G°$ for the equation is negative, indicating that is product-favored at equilibrium. The reaction is enthalpy driven ($\Delta_r H° < 0$).

(b) Calculate K_p for the reaction:

$\Delta_r G° = -RT \ln K_p$ so $-91.64 \times 10^3 \text{ J/mol} = -(8.3145 \text{ J/K} \cdot \text{mol})(298.15 \text{ K}) \ln K_p$

$36.97 = \ln K_p$ and $1.1 \times 10^{16} = K_p$

18.58. Calculate $\Delta S°(\text{system})$, $\Delta S°(\text{surroundings})$, and $\Delta S°(\text{universe})$:

(a) $\Delta S°(\text{system}) = S°[HNO_3(aq)] - S°[HNO_3(g)]$
$\Delta S°(\text{system}) = 1 \text{ mol } (146.4 \text{ J/K·mol}) - 1 \text{ mol } (266.38 \text{ J/K·mol}) = -120.0 \text{ J/K·mol-rxn}$
$\Delta_r H° = \Delta_f H°[HNO_3(aq)] - \Delta_f H°[HNO_3(g)]$
$\Delta_r H° = 1 \text{ mol } (-207.36 \text{ kJ/mol}) - 1 \text{ mol } (-135.06 \text{ kJ/mol}) = -72.30 \text{ kJ/mol-rxn}$

$\Delta S°(\text{surroundings}) = \dfrac{-\Delta_r H°}{T} = -[(-72.30 \text{ kJ/mol-rxn})(10^3 \text{ J}/1 \text{ kJ})/298 \text{ K}]$

$= 242.6 \text{ J/K·mol-rxn}$

$\Delta S°(\text{universe}) = \Delta S°(\text{system}) + \Delta S°(\text{surroundings})$

$= -120.0 \text{ J/K·mol-rxn} + 242.6 \text{ J/K·mol-rxn} = 122.6 \text{ J/K·mol-rxn}$

(b) $\Delta S°(\text{system}) = S°[NaOH(aq)] - S°[NaOH(s)]$
$\Delta S°(\text{system}) = 1 \text{ mol } (48.1 \text{ J/K·mol}) - 1 \text{ mol } (64.46 \text{ J/K·mol}) = -16.4 \text{ J/K·mol-rxn}$
$\Delta_r H° = \Delta_f H°[NaOH(aq)] - \Delta_f H°[NaOH(s)]$
$\Delta_r H° = 1 \text{ mol } (-469.15 \text{ kJ/mol}) - 1 \text{ mol } (-425.93 \text{ kJ/mol}) = -43.22 \text{ kJ/mol-rxn}$

$\Delta S°(\text{surroundings}) = \dfrac{-\Delta_r H°}{T} = -[(-43.22 \text{ kJ/mol-rxn})(10^3 \text{ J}/1 \text{ kJ})/298 \text{ K}]$

$= 145.0 \text{ J/K·mol-rxn}$

$\Delta S°(\text{universe}) = \Delta S°(\text{system}) + \Delta S°(\text{surroundings})$

$= -16.4 \text{ J/K·mol-rxn} + 145.0 \text{ J/K·mol-rxn} = 128.6 \text{ J/K·mol-rxn}$

Both systems are product-favored at equilibrium and enthalpy-driven.

18.59. Calculate K_p for the formation of methanol from its elements at 298K:

Begin by calculating a $\Delta_rG°$.

$\Delta_rG° = \Delta_fG$ (CH$_3$OH(ℓ)) – [Δ_fG C(graphite) +1/2 · Δ_fG O$_2$(g) + 2 · Δ_f GH$_2$(g)]

$\Delta_rG° = –166.14$ kJ – (0 kJ + 0 kJ + 0 kJ) = –166.14 kJ

$\Delta_rG° = –RT\ln K_p$

$–166.1 \times 10^3$ J/mol = – (8.3145 J/K·mol)(298.15 K) ln K_p

67.00 = ln K_p and $1.3 \times 10^{29} = K_p$

The large value of K_p indicates that this process is product-favored at 298 K. Judging by the relative numbers of gaseous particles, (without doing a calculation), one can see that $\Delta_rS°$ for the reaction is < 0, so higher temperatures would reduce the value of K_p.
Regarding the connection between $\Delta G°$ and K, the more negative the value of $\Delta G°$, the larger the value of K.

18.60. Calculate $\Delta_rS°$ for condensation of (C$_2$H$_5$)$_2$O(g) at 35.0 °C

(C$_2$H$_5$)$_2$O(g) $\rightleftarrows$ (C$_2$H$_5$)$_2$O(ℓ)   At equilibrium, $\Delta_rG° = 0$ and $\Delta_rS° = \dfrac{-\Delta_fH°}{T}$

$\Delta_rS° = \dfrac{-\Delta_{vap}H°}{T} = \dfrac{-26.0 \times 10^3 \text{ J/mol}}{308.2 \text{ K}} = -84.4$ J/K · mol-rxn

18.61. Calculate the $\Delta_rS°$ for the vaporization of ethanol at 78.0 °C.

$\Delta_rS° = \dfrac{-\Delta_{vap}H°}{T} = \dfrac{39.3 \times 10^3 \text{ J/mol}}{351 \text{ K}} = 112$ J/K · mol-rxn

18.62. Estimate the normal boiling point of ethanol: C$_2$H$_5$OH(ℓ) $\rightleftarrows$ C$_2$H$_5$OH(g)

$\Delta_rS° = S°$[C$_2$H$_5$OH(g)] – $S°$[C$_2$H$_5$OH(ℓ)] = 1 mol (282.70 J/K·mol) – 1 mol (160.7 J/K·mol)

$\Delta_rS° = 122.0$ J/K·mol-rxn

$\Delta_rH° = \Delta_fH°$[C$_2$H$_5$OH(g)] – $\Delta_fH°$[C$_2$H$_5$OH(ℓ)]

= 1 mol (–235.3 kJ/mol) – 1 mol (–277.0 J/K·mol) = 41.7 kJ/mol-rxn

At equilibrium, $\Delta_rG° = 0$ and $\Delta_rS° = \dfrac{\Delta_rH°}{T}$

$T = \dfrac{-\Delta_rH°}{-\Delta_rS°} = \dfrac{41.7 \times 10^3 \text{ J/mol-rxn}}{122.0 \times 10^3 \text{ J/K·mol-rxn}} = 342$ K (69 °C)

The calculated value is somewhat lower than the actual value (78 °C).

18.63. For the decomposition of phosgene:

$\Delta_r H° = [\Delta_f H° \, CO(g) + \Delta_f H° \, Cl_2(g)] - [\Delta_f H° \, COCl_2(g)]$

$\Delta_r H° = [(1 \text{ mol})(-110.525 \text{ kJ/mol}) + (1 \text{ mol})(0 \text{ kJ/mol})] - [(1 \text{ mol})(-218.8 \text{ kJ/mol})]$

$= 108.275 \text{ kJ/mol}$

$\Delta_r S° = [S° \, CO(g) + S° \, Cl_2(g)] - [S° \, COCl_2(g)]$

$\Delta_r S° = [(1 \text{ mol})(197.674 \text{ J/K·mol}) + (1 \text{ mol})(223.07 \text{ J/K·mol})] -$

$[(1 \text{ mol})(283.53 \text{ J/K·mol})] = 137.2 \text{ J/K·mol-rxn}$

From the $\Delta_r S°$ data, you can see that raising the temperature will favor the endothermic decomposition of this substance.

18.64. Estimate value of $\Delta_r S°$ for the reaction at 897 °C:

At 897 °C the system is at equilibrium.

$\Delta_r G° = -RT \ln K_p = -RT \ln (1.00) = 0$

Assuming $\Delta_r H°$ values are relatively constant as the temperature changes,

$\Delta_r S° = \dfrac{\Delta_r H°}{T} = \dfrac{1.790 \times 10^5 \text{ J}}{1170. \text{ K}} = 153.0 \text{ J/K·mol-rxn}$

18.65. For the reaction of sodium with water: $Na(s) + H_2O(\ell) \rightarrow NaOH(aq) + ½ \, H_2(g)$

Predict signs for $\Delta_r H°$ and $\Delta_r S°$:

The reaction of sodium with water gives off heat, and the heat frequently ignites the hydrogen gas that is evolved. $\Delta_r H° = -$.

Regarding entropy, the system changes from one with a solid (low entropy) and a liquid (higher entropy) to a solution (*frequently* higher entropy than liquid) and a gas (high entropy). So, you would predict that the entropy would increase, i.e., $\Delta_r S° = +$.

Now for the calculation:

$\Delta_r H° = [1 \cdot \Delta_f H° \, NaOH(aq) + ½ \cdot \Delta_f H° \, H_2(g)] - [1 \cdot \Delta_f H° \, Na(s) + 1 \cdot \Delta_f H° \, H_2O(\ell)]$

$= [(1 \text{ mol})(-469.15 \text{ kJ/mol}) + (½ \text{ mol})(0)] - [(1 \text{ mol})(0) + (1 \text{ mol})(-285.83 \text{ kJ/mol})]$

$= -183.32 \text{ kJ/mol-rxn}$

$\Delta_r S° = [1 \cdot S° \, NaOH(aq) + ½ \cdot S° \, H_2(g)] - [1 \cdot S° \, Na(s) + 1 \cdot S° \, H_2O(\ell)]$

$= [(1 \text{ mol})(48.1 \text{ J/K·mol}) + (½ \text{ mol})(130.7 \text{ J/K·mol})]$

$- [(1 \text{ mol})(51.21 \text{ J/K·mol}) + (1 \text{ mol})(69.95 \text{ J/K·mol})]$

$= -7.7 \text{ J/K·mol-rxn}$

As expected, the $\Delta_rH°$ for the reaction is negative. A surprise comes in the calculation for $\Delta_rS°$. While you anticipate the sign to be positive, you find a slightly negative number—reflecting the order (hence a decrease in entropy) that can occur as solutions occur.

18.66. For the fermentation of glucose:

$\Delta_rH° = 2 \cdot \Delta_fH°[C_2H_5OH(\ell)] + 2 \cdot \Delta_fH°[CO_2(g)] - 1 \cdot \Delta_fH°[C_6H_{12}O_6(aq)]$

$\Delta_rH° = 2$ mol $(-277.0$ kJ/mol$) + 2$ mol $(-393.509$ kJ/mol$) - 1$ mol $(-1260.0$ kJ/mol$)$

$ = -81.0$ kJ/mol-rxn

$\Delta_rS° = 2 \cdot S°[C_2H_5OH(\ell)] + 2 \cdot S°[CO_2(g)] - 1 \cdot S°[C_6H_{12}O_6(aq)]$

$\Delta_rS° = 2$ mol $(160.7$ J/K·mol$) + 2$ mol $(213.74$ J/K·mol$) - 1$ mol $(289$ J/K·mol$)$

$ = 460.$ J/K·mol-rxn

$\Delta_rG° = \Delta_rH° - T\Delta_rS° = -81.0$ kJ/mol-rxn $- (298$ K$)(460.$ J/K·mol-rxn$)(1$ kJ/10^3 J$)$

$ = -218.1$ kJ/mol-rxn

The reaction is product-favored at equilibrium.

18.67. For the reaction: $BCl_3(g) + 3/2$ $H_2(g) \rightarrow B(s) + 3$ $HCl(g)$

	$BCl_3(g)$	$H_2(g)$	$B(s)$	$HCl(g)$
$S°$ (J/K·mol)	290.17	130.7	5.86	186.2
$\Delta_fH°$ (kJ/mol)	–402.96	0	0	–92.31

$\Delta_rH° = [3 \cdot \Delta_f H° \text{ HCl}(g) + 1 \cdot \Delta_f H° \text{ B}(s)] - [1 \cdot \Delta_fH° \text{ BCl}_3(g) + 3/2\ \Delta_fH° \text{ H}_2(g)]$

$\Delta_rH° = [(3$ mol$)(-92.31$ kJ/mol$) + (1$ mol$)(0)] - [(1$ mol$)(-402.96$ kJ/mol$) + (3/2$ mol$)(0)]$

$ = 126.03$ kJ/mol-rxn

$\Delta_rS° = [3 \cdot S° \text{ HCl}(g) + 1 \cdot S° \text{ B}(s)] - [1 \cdot S° \text{ BCl}_3(g) + 3/2\ S° \text{ H}_2(g)]$

$\Delta_rS° = [(3$ mol$)(186.2$ J/K·mol$) + (1$ mol$)(5.86$ J/K·mol$)]$

$ -[(1\text{mol})(290.17$ J/K·mol$) + (3/2$ mol$)(130.7$ J/K·mol$)]$

$ = 78.2$ J/K·mol-rxn

$\Delta_rG° = \Delta_rH° - T\Delta_rS° = 126.03$ kJ $- (298.15$ K$)(78.2$ J/K$)(1.000$ kJ/1000 J$) = 103$ kJ/mol-rxn

The reaction is not product-favored at equilibrium.

18.68. Estimate vapor pressure of ethanol at 37 °C:

You can use the Clausius-Clapeyron equation. At the normal boiling point of ethanol, 78 °C, the vapor pressure of ethanol is 1.0 atm and the enthalpy of vaporization is 39.3 kJ/mol.

$$\ln\left(\frac{P_2}{P_1}\right) = \frac{\Delta_{vap}H}{R}\left(\frac{1}{T_1} - \frac{1}{T_2}\right)$$

$$\ln\left(\frac{P_2}{760 \text{ mm Hg}}\right) = \frac{39.3 \text{ kJ/mol}}{0.0083145 \text{ kJ/K}\cdot\text{mol}}\left(\frac{1}{351 \text{ K}} - \frac{1}{310. \text{ K}}\right)$$

$P_2 = 128$ mm Hg

18.69. Calculate $\Delta_rG°$ for conversion of N_2O_4 to NO_2:

$\Delta_rG° = -RT \ln K = -(8.3145 \times 10^{-3}$ kJ/K·mol)(298 K) ln 0.14 = 4.87 kJ

Compare with the calculated $\Delta_fG°$ values:

$\Delta_rG° = 2\, \Delta_fG°[NO_2(g)] - \Delta_fG°[N_2O_4(g)] = (2 \cdot 51.23$ kJ/mol) $- (1 \cdot 97.73$ kJ/mol) $= 4.73$ kJ

18.70. Estimate the boiling point of water at 630 mm Hg:

$H_2O(\ell) \rightleftarrows H_2O(g)$

$\Delta_rH° = \Delta_fH°[H_2O(g)] - \Delta_fH°[H_2O(\ell)] = 1$ mol $(-241.83$ kJ/mol$) - 1$ mol $(-285.83$ kJ/mol$)$

$\Delta_rH° = 44.00$ kJ/mol-rxn

(The enthalpy of vaporization above is at 298 K. The enthalpy of vaporization at 373 K is 40.7 kJ/mol-rxn. Either value is adequate for this estimation.)

$$\ln\left(\frac{P_2}{P_1}\right) = \frac{\Delta_{vap}H}{R}\left(\frac{1}{T_1} - \frac{1}{T_2}\right)$$

$$\ln\left(\frac{630 \text{ mm Hg}}{760 \text{ mm Hg}}\right) = \frac{44.00 \text{ kJ/mol}}{0.0083145 \text{ kJ/K}\cdot\text{mol}}\left(\frac{1}{373 \text{ K}} - \frac{1}{T_2}\right)$$

$T_2 = 368$ K $= 95$ °C

18.71. Calculate $\Delta_rG°$ for conversion of butane to isobutane, given $K = 2.50$:

$\Delta_rG° = -RT \ln K = -(8.3145 \times 10^{-3}$ kJ/K·mol)(298.15 K) ln 2.50 $= -2.27$ kJ/mol-rxn

18.72. Concerning the reaction of coal with steam:

(a) $\Delta_rG° = \Delta_fG°[CO(g)] - \Delta_fG°[H_2O(g)] = 1$ mol $(-137.168$ kJ/mol$) - 1$ mol $(-228.59$ kJ/mol$)$

$\Delta_rG° = 91.42$ kJ/mol-rxn

(b) $\Delta_rG° = 91.4$ kJ/mol $= -RT\ln K_p = -(8.3145 \times 10^{-3}$ kJ/K·mol$)(298$ K$)\ln K_p$
$\ln K_p = -36.9$ and $K_p = 1 \times 10^{-16}$ (1 sf)

(c) The reaction is not product-favored at equilibrium at 25 °C.
$\Delta_rH° = \Delta_fH°[CO(g)] - \Delta_fH°[H_2O(g)]$
$\quad = 1$ mol $(-110.525$ kJ/mol$) - 1$ mol $(-241.83$ kJ/mol$) = 131.31$ kJ/mol-rxn
$\Delta_rS° = S°[CO(g)] + S°[H_2(g)] - \{S°[C(s)] + S°[H_2O(g)]\}$
$\Delta_rS° = 1$ mol $(197.674$ J/K·mol$) + 1$ mol $(130.7$ J/K·mol$)$
$\quad\quad\quad - [1$ mol $(5.6$ J/K·mol$) + 1$ mol $(188.84$ J/K·mol$)]$
$\Delta_rS° = 133.9$ J/K·mol-rxn

$$T = \frac{\Delta_rH°}{\Delta_rS°} = \frac{131.31 \times 10^3 \text{ J/mol-rxn}}{133.9 \text{ J/K·mol-rxn}} = 980.4 \text{ K} = 707.3 \text{ °C}$$

18.73. For the reaction: $2 SO_3(g) \rightleftarrows 2 SO_2(g) + O_2(g)$

$\Delta_rH° = [2 \cdot \Delta_fH° SO_2(g) + 1 \cdot \Delta_fH° O_2(g)] - [2 \cdot \Delta_fH° SO_3(g)]$
$\quad = [(2$ mol$)(-296.84$ kJ/mol$) + 0] - [(2$ mol$)(-395.77$ kJ/mol$)] = 197.86$ kJ
$\Delta_rS° = [2 \cdot S° SO_2(g) + 1 \cdot S° O_2(g)] - [2 \cdot S° SO_3(g)]$
$\quad = [(2$ mol$)(248.21$ J/K·mol$) + (1$ mol$)(205.07$ J/K·mol$)] - [(2$ mol$)(256.77$ J/K·mol$)]$
$\quad = 187.95$ J/K·mol-rxn

(a) Is the reaction product-favored at equilibrium at 25 °C?

$\Delta_rG° = \Delta_rH° - T\Delta_rS° = 197.86$ kJ $- (298.15$ K$)(187.95$ J/K$)(1.000$ kJ/1000 J$)$
$\quad = 141.82$ kJ/mol-rxn

The reaction is not product-favored at equilibrium.

(b) The reaction can become product-favored at equilibrium if there is a temperature at which $\Delta_rG° < 0$. To see if such a temperature is feasible, set $\Delta_rG° = 0$ and solve for T.
$\Delta_rG° = \Delta_rH° - T\Delta_rS°$
$0 = 197.86$ kJ $- T(0.18795$ kJ/K$)$
$$T = \frac{197.86 \text{ kJ}}{0.18795 \text{ kJ/K}} = 1052.7 \text{ K or } (1052.7 - 273.1) = 779.6 \text{ °C}$$

(c) The equilibrium constant for the reaction at 1500 °C. Since you know that $\Delta_rG° = \Delta_rH° - T\Delta_rS° = -RT\ln K$, you can solve for K if you know $\Delta_rG°$ at 1500 °C

$\Delta_rG° = \Delta_rH° - T\Delta_rS°$
$\quad = 197.86$ kJ $- (1773$ K$)(187.95$ J/K$)(1.000$ kJ/1000 J$) = -135.4$ kJ

Substitute into the equation ($\Delta_rG° = -RT\ln K$):

-135.4 kJ $= -(8.314$ J/K·mol$)(1$ kJ/1000 J$)(1773$ K$)\ln K$ and
$K = 9.7 \times 10^3$ or 1×10^4 (1 sf)

18.74. Regarding the oxidation of methane with O_2:

(a) $\Delta S°$(system) = $S°[CH_3OH(\ell)] - \{S°[CH_4(g)] + {}^1\!/_2\, S°[O_2(g)]\}$

$\Delta S°$(system) = 1 mol (127.19 J/K·mol)

$\qquad\qquad\qquad\qquad$ −[1 mol (186.26 J/K·mol) + $^1\!/_2$ mol (205.07 J/K·mol)]

$\Delta S°$(system) = −161.61 J/K·mol-rxn

$\Delta_r H° = \Delta_f H°[CH_3OH(\ell)] - \Delta_f H°[CH_4(g)]$

$\qquad$ = 1 mol (−238.4 kJ/mol) − 1 mol (−74.87 kJ/mol) = −163.5 kJ/mol-rxn

$\Delta S°$(surroundings) = $-\Delta_r H°/T$ = −[(−163.5 kJ/mol-rxn)(10³ J/1 kJ)/298 K]

$\qquad$ = 548.7 J/K·mol-rxn

ΔS(universe) = $\Delta S°$(system) + $\Delta S°$(surroundings)

$\qquad$ = −161.61 J/K·mol-rxn + 548.7 J/K·mol-rxn = 387.1 J/K·mol-rxn

(b) $\Delta_r G° = \Delta_r H° - T\Delta_r S°$ = −163.5 kJ/mol-rxn − (298 K)(−161.61 J/K·mol-rxn)(1 kJ/10³ J)
$\Delta_r G°$ = −115.3 kJ/mol-rxn
The reaction is product-favored at equilibrium at 25 °C.

18.75. Reaction: $H_2S(g) + 2\,O_2(g) \rightarrow H_2SO_4(\ell)$

	$H_2S(g)$	$O_2(g)$	$H_2SO_4(\ell)$
$\Delta_f H°$ (kJ/mol)	−20.63	0	−814
$S°$ (J/K·mol)	205.79	205.07	156.9

$\Delta_r H°$ = [(1 mol)(−814 kJ/mol)] − [(1 mol)(−20.63 kJ/mol) + 0] = −793 kJ/mol-rxn

$\Delta_r S°$ = [(1 mol)(156.9 J/K·mol)] −[(1 mol)(205.79 J/K·mol) + (2 mol)(205.07 J/K·mol)]

$\qquad$ = −459.0 J/K·mol-rxn

$\Delta_r G° = \Delta_f H° - T\Delta_r S°$ = −793 kJ − (298.15 K)(− 459.0 J/K)(1 kJ/10³ J) = −657 kJ/mol-rxn

The reaction is product-favored at equilibrium at 25 °C ($\Delta G° < 0$) and enthalpy-driven ($\Delta_r H° < 0$).

18.76. Regarding the reaction of $CaCO_3$ with SO_2:

(a) $CaCO_3(s) + SO_2(g) + {}^1\!/_2\, H_2O(\ell) \rightleftarrows CaSO_3\cdot{}^1\!/_2\, H_2O(s) + CO_2(g)$

$\Delta_r H° = \Delta_f H°\,[CaSO_3\cdot{}^1\!/_2\, H_2O(s)] + \Delta_f H°[CO_2(g)]$

$\qquad\qquad\qquad\qquad - \{\Delta_f H°[CaCO_3(s)] + \Delta_f H°[SO_2(g)] + {}^1\!/_2\,\Delta_f H°[H_2O(\ell)]\}$

$\Delta_r H° = 1 \text{ mol } (-1311.7 \text{ kJ/mol}) + 1 \text{ mol } (-393.509 \text{ kJ/mol})$
$\quad - [1 \text{ mol } (-1207.6 \text{ kJ/mol}) + 1 \text{ mol } (-296.84 \text{ kJ/mol}) + \frac{1}{2} \text{ mol } (-285.83 \text{ kJ/mol})]$
$\Delta_r H° = -57.9 \text{ kJ/mol-rxn}$
$\Delta_r S° = S°[CaSO_3 \cdot \frac{1}{2}H_2O(s)] + S°[CO_2(g)]$
$\quad - \{S°[CaCO_3(s)] + S°[SO_2(g)] + \frac{1}{2} S°[H_2O(\ell)]\}$
$\Delta_r S° = 1 \text{ mol } (121.3 \text{ J/K·mol}) + 1 \text{ mol } (213.74 \text{ J/K·mol})$
$\quad - [1 \text{ mol } (91.7 \text{ J/K·mol}) + 1 \text{ mol } (248.21 \text{ J/K·mol}) + \frac{1}{2} \text{ mol } (69.95 \text{ J/K·mol})]$
$\Delta_r S° = -39.8 \text{ J/K·mol-rxn}$
$\Delta_r G° = \Delta_r H° - T\Delta_r S° = -57.9 \text{ kJ/mol-rxn} - (298 \text{ K})(-39.8 \text{ J/K·mol-rxn})(1 \text{ kJ}/10^3 \text{ J})$
$\quad = -46.0 \text{ kJ/mol-rxn}$

For the 2nd rxn: $CaCO_3(s) + SO_2(g) + \frac{1}{2} H_2O(\ell) + \frac{1}{2} O_2(g) \rightleftarrows CaSO_4 \cdot \frac{1}{2}H_2O(s) + CO_2(g)$

$\Delta_r H° = \Delta_f H°[CaSO_4 \cdot \frac{1}{2}H_2O(s)] + \Delta_f H°[CO_2(g)]$
$\quad - \{\Delta_f H°[CaCO_3(s)] + \Delta_f H°[SO_2(g)] + \frac{1}{2} \Delta_f H°[H_2O(\ell)]\}$

$\Delta_r H° = 1 \text{ mol } (-1574.65 \text{ kJ/mol}) + 1 \text{ mol } (-393.509 \text{ kJ/mol})$
$\quad - [1 \text{ mol } (-1207.6 \text{ kJ/mol}) + 1 \text{ mol } (-296.84 \text{ kJ/mol}) + \frac{1}{2} \text{ mol } (-285.83 \text{ kJ/mol})]$

$\Delta_r H° = -320.8 \text{ kJ/mol-rxn}$
$\Delta_r S° = S°[CaSO_4 \cdot \frac{1}{2}H_2O(s)] + S°[CO_2(g)]$
$\quad - \{S°[CaCO_3(s)] + S°[SO_2(g)] + \frac{1}{2} S°[H_2O(\ell)] + \frac{1}{2} S°[O_2(g)]\}$

$\Delta_r S° = 1 \text{ mol } (134.8 \text{ J/K·mol}) + 1 \text{ mol } (213.74 \text{ J/K·mol})$
$\quad - [1 \text{ mol } (91.7 \text{ J/K·mol}) + 1 \text{ mol } (248.21 \text{ J/K·mol}) + \frac{1}{2} \text{ mol } (69.95 \text{ J/K·mol})$
$\quad\quad + \frac{1}{2} \text{ mol } (205.07 \text{ J/K·mol})]$

$\Delta_r S° = -128.9 \text{ J/K·mol-rxn}$
$\Delta_r G° = \Delta_r H° - T\Delta_r S° = -320.8 \text{ kJ/mol-rxn} - (298 \text{ K})(-128.9 \text{ J/K·mol-rxn})(1 \text{ kJ}/10^3 \text{ J})$
$\Delta_r G° = -282.4 \text{ kJ/mol-rxn}$. The second reaction is more product-favored at equilibrium.

(b) $CaSO_3 \cdot \frac{1}{2}H_2O(s) + CO_2(g) \rightleftarrows CaCO_3(s) + SO_2(g) + \frac{1}{2} H_2O(\ell)$

$CaCO_3(s) + SO_2(g) + \frac{1}{2} H_2O(\ell) + \frac{1}{2} O_2(g) \rightleftarrows CaSO_4 \cdot \frac{1}{2}H_2O(s) + CO_2(g)$

$CaSO_3 \cdot \frac{1}{2}H_2O(s) + \frac{1}{2} O_2(g) \rightleftarrows CaSO_4 \cdot \frac{1}{2}H_2O(s)$: Net equation

$\Delta_r G° = -(-46.0 \text{ kJ/mol-rxn}) + (-282.4 \text{ kJ/mol-rxn}) = -236.4 \text{ kJ/mol-rxn}$

The reaction is product-favored at equilibrium.

18.77. Calculate the $\Delta_r G°$ for the transition of S_8 (rhombic) $\rightarrow S_8$ (monoclinic):

(a) At 80 °C: $\Delta_r G° = \Delta_r H° - T\Delta_r S°$

$\Delta_r G° = 3.213 \text{ kJ} - (353 \text{ K})(0.0087 \text{ kJ/K}) = 0.14 \text{ kJ/mol-rxn}$

At 110 °C: $\Delta_r G° = \Delta_r H° - T\Delta_r S°$

$\Delta_r G° = 3.213 \text{ kJ} - (383 \text{ K})(0.0087 \text{ kJ/K}) = -0.12 \text{ kJ/mol-rxn}$

The rhombic form of sulfur is the more stable at lower temperature, while the monoclinic form is the more stable at higher temperature. The transition to monoclinic form is product-favored at equilibrium at temperatures above 110 degrees C.

(b) The temperature at which $\Delta_rG° = 0$:

$\Delta_rG° = 3.213$ kJ $− (T)(0.0087$ kJ/K) substituting: $0 = 3.213$ kJ $− (T)(0.0087$ kJ/K)

$T = \dfrac{3.213 \text{ kJ}}{0.0087 \text{ kJ/K}} = 370$ K or 96°C

96 °C is the temperature at which the phase transition begins.

18.78. Calculate the entropy change for: HCl(g) → HCl(aq)
$\Delta_rS° = S°[\text{HCl(aq)}] − S°[\text{HCl(g)}]$

= 1 mol (56.5 J/K·mol) − 1 mol (186.2 J/K·mol) = −129.7 J/K·mol-rxn

Yes, the negative value indicates a decrease in entropy, which is expected when going from a gas to a solution.

In the Laboratory

18.79. Is decomposition of silver(I) oxide product-favored at equilibrium at 25 °C ?

Calculate: $\Delta_rH°$ and $\Delta_rS°$:

$\Delta_rH° = ([4 · \Delta_fH° \text{ Ag(s)}] + [1 · \Delta_fH° \text{ O}_2\text{(g)}]) − [2 · \Delta_fH° [\text{Ag}_2\text{O(s)}]$

$\Delta_rH° = 0$ kJ $− [2$ mol · −31.1 kJ/mol$] = 62.2$ kJ/mol-rxn

and for $\Delta_rS°$:

$\Delta_rS° = ([4 · S° \text{ Ag(s)}] + [1 · S° \text{ O}_2\text{(g)}]) − [2 · S° [\text{Ag}_2\text{O(s)}]$

$\Delta_rS° = ([4\text{mol} · 42.55 \text{ J/K·mol}] + [1\text{mol} · 205.07 \text{ J/K·mol}]) − [2\text{mol} · 121.3 \text{ J/K·mol}]$

= [170.2 J/K + 205.07 J/K] − [242.6 J/K] = +132.7 J/K

While enthalpic considerations do **not** favor product formation, entropic considerations **do**.

The Gibbs Free Energy change would be:

$\Delta_rG° = ([4 · \Delta_fG° \text{ Ag(s)}] + [1 · \Delta_fG° \text{ O}_2\text{(g)}]) − [2 · \Delta_fG° \text{ Ag}_2\text{O(s)}]$

= (0 kJ) − (2 mol · −11.32 kJ/mol) = 22.64 kJ, so this change **does not favor** product formation. The signs of $\Delta_rH°$ and $\Delta_rS°$ indicate that there **may be** some temperature at which the reaction is product-favored at equilibrium. So, calculate the temperature at which $\Delta_rG° = 0$:

$\Delta_rG° = 62.2$ kJ $− (T)(0.1327$ kJ/K) Note the conversion of ΔS units to kJ!

$0 = 62.2$ kJ $− (T)(0.1327$ kJ/K) and solving for T:

$T = \dfrac{62.2 \text{ kJ}}{0.1327 \text{ kJ/K}} = 469$ K or 196°C

At temperatures greater than 196 °C, the reaction would be product-favored at equilibrium.

18.80. For the reaction of copper(II) oxide with hydrogen:

$\Delta_r G° = \Delta_f G°[H_2O(g)] – \Delta_f G°[CuO(s)]$

= 1 mol (–228.59 kJ/mol) – 1 mol (–128.3 kJ/mol) = –100.3 kJ/mol-rxn

The reaction is product-favored at equilibrium.

18.81. Calculate $\Delta_f G°$ for HI(g) at 350 °C, given the following equilibrium partial pressures:

P(H₂) = 0.132 bar, P(I₂) = 0.295 bar, and P(HI) =1.61 bar. At 350 °C, 1 bar, I₂ is a gas.

For the equation: ½ H₂(g) + ½ I₂(g) ⇌ HI(g)

Calculate $K_p = \dfrac{P_{HI}}{P_{H_2}^{1/2} \cdot P_{I_2}^{1/2}} = \dfrac{(1.61)}{(0.363)(0.543)} = 8.16$

Knowing that $\Delta_r G° = –RT \ln K$, you can solve:

$\Delta_r G° = –RT \ln K = – (8.3145\ \text{J/K·mol})(623.15\ \text{K}) \ln 8.16 = –10{,}873\ \text{J}$ or –10.9 kJ/mol

18.82. Equilibrium constant for formation of NiO at 1627 °C:

$\Delta_f G° = –72.1$ kJ/mol = $–(8.3145 \times 10^{–3}$ kJ/K·mol$)(1900.\ \text{K}) \ln K_p$ and $K_p = 96$

When $P_{O_2} = 1.00\ \text{mm Hg} \cdot \dfrac{1\ \text{atm}}{760\ \text{mm Hg}} = 0.00131$ atm

$Q = \dfrac{1}{P_{O_2}^{1/2}} = \dfrac{1}{(0.00131)^{1/2}} = 27.6$ as $Q < K_p$ the reaction will proceed in the forward direction when P_{O_2} is less than 1.00 mm Hg.

18.83. Regarding the conversion of titanium(IV) oxide to titanium carbide:

(a) Calculate $\Delta_r G°$ and K for the reaction at 727 °C:

$\Delta_r G° = ([2 \cdot \Delta_f G°\ CO(g)] + [1 \cdot \Delta_f G°\ TiC(s)]) – ([1 \cdot \Delta_f G°\ [TiO_2(s)] + [3 \cdot \Delta_f G°\ [C(s)])$

$\Delta_r G° = ([2\ \text{mol} \cdot –200.2\ \text{kJ/mol}] + [1\ \text{mol} \cdot –162.6\ \text{kJ/mol}]) – ([1\ \text{mol} \cdot –757.8\ \text{kJ/mol}] + [0])$

= (–400.4 kJ + –162.6 kJ) – (–757.8 kJ) = –563.0 kJ + 757.8 kJ = 194.8 kJ/mol-rxn

K would equal:

$\Delta_r G° = –RT \ln K$; 194.8×10^3 J = $–(8.3145$ J/K·mol$)(1000\ \text{K}) \ln K$

[Note the conversion of the energy units of $\Delta G°$ to accommodate J in the value of R.]

$\ln K = \dfrac{194.8 \times 10^3\ \text{J}}{–(8.3145\ \text{J/K·mol})(1000\ \text{K})} = –23.43$ and $K = 6.7 \times 10^{–11}$

(b) The value of *K* indicates that the reaction is **not product-favored at equilibrium at this temperature**.

(c) Three of the four substances in the equilibrium are solids, hence do not appear in the *K* expression. The *K* expression would have the composition: $K = P^2(CO)$.

According to Le Chatelier's principle, reducing the concentration (and the pressure) of CO would tend to shift the equilibrium to the right, favoring product formation.

18.84. What is *K* for the process: *cis*-Pt(NH$_3$)$_2$Cl$_2$ ⇌ *trans*-Pt(NH$_3$)$_2$Cl$_2$

$\Delta_rG° = \Delta_fG°[\text{trans-Pt(NH}_3)_2\text{Cl}_2] - \Delta_fG°[\text{cis-Pt(NH}_3)_2\text{Cl}_2]$
$\Delta_rG° = 1 \text{ mol }(-222.8 \text{ kJ/mol}) - 1 \text{ mol }(-228.7 \text{ kJ/mol}) = 5.9 \text{ kJ/mol-rxn}$
$\Delta_rG° = -RT\ln K$; 5.9 kJ/mol = $-(8.3145 \times 10^{-3}$ kJ/K·mol$)(298$ K$)\ln K$ and solving for *K*:
K = 0.09 so with $\Delta_rG° > 1$ the *cis* isomer is more thermodynamically stable.

Summary and Conceptual Questions

18.85. An examination of the equation Hg(ℓ) ⇌ Hg(g) shows that the equilibrium constant expression would be $K_p = P_{Hg(g)}$. So, to find the temperature at which K_p = 1.00 bar and 1/760 bar, you need only to find the temperature at which the vapor pressure of mercury is 1.00 bar and 1/760 bar, respectively.

First, calculate the *T* for K_p = 1.00 bar. At the equilibrium point, you can calculate *T* at which K_p = 1.00 bar if you know $\Delta G°$. Since at equilibrium, $\Delta G° = 0$, you can rewrite the equation: $\Delta G° = \Delta H° - T\Delta_rS°$ to read: $\Delta H°/\Delta_rS° = T$

$\Delta_rH° = (1 \text{ mol})(61.38 \text{ kJ/mol}) - (1 \text{ mol})(0) = 61.38$ kJ

For entropy: $\Delta_rS° = (1 \text{ mol})(174.97 \text{ J/K·mol}) - (1 \text{ mol})(76.02 \text{ J/K·mol}) = 98.95$ J/K

Substituting into the equation:

$$T = \frac{\Delta_rH°}{\Delta_rS} = \frac{\left(61.38 \text{ kJ} \cdot \frac{1000 \text{ J}}{1 \text{ kJ}}\right)}{98.95 \text{ J/K}} = 620.3 \text{ K or } 347.2 \text{ °C}$$

Temperature at which K_p = 1/760: Using the Clausius-Clapeyron equation:

$$\ln\left(\frac{P_2}{P_1}\right) = \frac{\Delta H}{R}\left(\frac{1}{T_1} - \frac{1}{T_2}\right) \text{ and } \ln\left(\frac{1}{760}\right) = \frac{61.38 \times 10^3 \text{ J/mol}}{8.3145 \text{ J/K·mol}}\left(\frac{1}{620.3 \text{ K}} - \frac{1}{T_2}\right)$$

$$-6.6333 = \frac{61.38 \times 10^3 \text{ J/mol}}{8.3145 \text{ J/K·mol}}\left(\frac{1}{620.3 \text{ K}} - \frac{1}{T_2}\right); \frac{-6.6333 \cdot 8.3145 \text{ J/K·mol}}{61.38 \times 10^3 \text{ J/mol}} = \left(\frac{1}{620.3 \text{ K}} - \frac{1}{T_2}\right)$$

so -8.98×10^{-4} K^{-1} = $\left(\frac{1}{620.3 \text{ K}} - \frac{1}{T_2}\right)$ and solving for T_2: T_2 = 398.3 K or 125.2 °C.

In summary, K_p = 1 at 347.2 °C and is 1/760 at 125.2 °C.

18.86. Provide corrections to the statements:

(a) The entropy **of the universe** increases in all spontaneous reactions.

(b) Reactions with a negative **standard** free energy change ($\Delta G° < 0$) are product-favored **at equilibrium** and can **occur at any rate, not necessarily a fast rate.**

(c) While many spontaneous processes are exothermic, **some are not spontaneous if their entropic contribution is large enough**.

(d) Endothermic processes **can be spontaneous at high temperatures if the standard entropy change for the system for the process is positive ($\Delta S° > 0$)**.

18.87. Following statements false or true?

(a) The entropy of a substance increases on going from the liquid to the vapor state at any temperature. **True**. For a given substance, the entropy of the vapor state of that substance is greater than for the liquid state.

(b) An exothermic reaction will always be spontaneous. **False**. While exothermic reactions are *frequently spontaneous* entropy does play a role. Should the entropy decrease enough, that may cause the reaction to be non-spontaneous (i.e., reactant-favored).

(c) Reactions with a + $\Delta_r H°$ and a +$\Delta_r S°$ can never be product-favored at equilibrium. **False**. At high temperatures, such reactions can be product-favored at equilibrium ($\Delta_r G° < 0$).

(d) If $\Delta_r G°$ is < 0, the reaction will have an equilibrium constant greater than 1. **True**. Since $\Delta_r G°$ and K are related by the expression: $\Delta_r G° = -RT\ln K$, if $\Delta G° < 0$, then mathematically K will be greater than 1.

18.88. The entropy of a pure crystal is zero at 0 K. A substance cannot have $S = 0$ J/K·mol at standard conditions (25 °C, 1 bar). All substances have positive entropy values at temperatures above 0 K. Based on the third law of thermodynamics, negative values of entropy cannot occur. The only exception to this is the entropy of the solvation process. When water molecules are constrained to a more ordered arrangement in a solution than in pure water, a higher degree of order results and entropy is negative.

18.89. If you dissolve a solid (e.g., table salt), the process is product-favored at equilibrium. ($\Delta_r G° < 0$).

If $\Delta_r H° = 0$, you can write: $\Delta_r G° = \Delta_r H° - T\Delta_r S°$ and $(-) = 0 - (+)\Delta_r S°$.

The only mathematical condition for which this equation is true is if $\Delta_r S° = +$, hence the process is entropy driven.

18.90. Regarding the formation of NO from its elements:

(a) $\Delta_r H° = 2\,\Delta_f H°[NO(g)] = 2\text{ mol }(90.29\text{ kJ/mol}) = 180.58\text{ kJ/mol-rxn}$
$\Delta_r S° = 2\,S°[NO(g)] - \{S°[N_2(g)] + S°[O_2(g)]\}$
$\Delta_r S° = 2\text{ mol }(210.76\text{ J/K·mol}) - \{1\text{ mol }(191.56\text{ J/K·mol}) + 1\text{ mol }(205.07\text{ J/K·mol})\}$

$= 24.89\text{ J/K·mol-rxn}$

$\Delta_r G° = \Delta_r H° - T\Delta_r S° = 180.58\text{ kJ/mol-rxn} - (298\text{ K})(24.89\text{ J/K·mol-rxn})(1\text{ kJ}/10^3\text{ J})$

$= 173.16\text{ kJ/mol-rxn}$

$\Delta_r G° = -RT\ln K$; $173.16\text{ kJ/mol} = -(8.3145 \times 10^{-3}\text{ kJ/K·mol})(298\text{ K})\ln K_p$

$K_p = 4.4 \times 10^{-31}$

The reaction is not product-favored at equilibrium at this temperature.

(b) $\Delta G° = 180.58\text{ kJ/mol-rxn} - (973\text{ K})(24.89\text{ J/K·mol-rxn})(1\text{ kJ}/10^3\text{ J}) = 156.4\text{ kJ/mol-rxn}$
$\Delta_r G° = -RT\ln K$; $156.4\text{ kJ} = -(8.3145 \times 10^{-3}\text{ kJ/K·mol})(973\text{ K})\ln K_p$ and solving for K_p:
$K_p = 4 \times 10^{-9}$

The reaction is not product-favored at equilibrium at this temperature.

(c) Calculate the equilibrium partial pressures:

$$K = \frac{P_{NO}^2}{P_{N_2} \cdot P_{O_2}} = 4 \times 10^{-9} = \frac{(2x)^2}{(1.00-x)^2} \approx \frac{(2x)^2}{(1.00)^2}$$ and solving for x: $x = 3 \times 10^{-5}$ bar

$P_{NO} = 2x = 6 \times 10^{-5}$ bar and $P_{N_2} = P_{O_2} = 1.00$ bar

18.91. For the reaction: $2\text{ C}_2\text{H}_6(g) + 7\text{ O}_2(g) \rightarrow 4\text{ CO}_2(g) + 6\text{ H}_2\text{O}(g)$

(a) Predict whether signs of $\Delta_r S°$(system), $\Delta S°$(surroundings), $\Delta S°$(universe) are greater than, equal to, or less than 0.

$\Delta_r S°$(system) will be > 0, since 9 mol of gas form 10 mol of gas as the reaction proceeds.

$\Delta S°$(surroundings) will be > 0, since the reaction liberates heat, and would increase the entropy of the surroundings. With both $\Delta_r S°$(system) and $\Delta S°$(surroundings) increasing, $\Delta S°$(universe) would also increase.

(b) Predict signs of $\Delta_r H°$, and $\Delta_r G°$: Since the reaction is exothermic, $\Delta_r H°$ would be "–". With a negative $\Delta_r H°$ and an increasing entropy, $\Delta_r G°$ would be "–" as well.

(c) Will value of K_p be very large, very small, or nearly 1? With the relatively large number of moles of carbon dioxide and water being formed, $\Delta_r H°$ will be *large and negative*, and with the increasing entropy, $\Delta_r G°$ will also be relatively *large and negative*. Since $\Delta_r G° = -RT\ln K$, you anticipate K_p being very large.

Will K_p be larger or smaller at temperatures greater than 298 K?
Rearrange the expression:

$\dfrac{\Delta G}{RT} = -\ln K$. As temperature increases, the term on the left will decrease, resulting in a larger value of K (–ln K decreases, so K increases).

18.92. For the melting of benzene(s), what are the signs for:

(a) $\Delta_r H°$: positive (endothermic process)
(b) $\Delta_r S°$: positive (solid → liquid)
(c) $\Delta_r G°$ at 5.5 °C: zero (equilibrium)
(d) $\Delta_r G°$ at 0.0 °C: positive (reactant-favored)
(e) $\Delta_r G°$ at 25.0 °C: negative (product-favored)

18.93. Calculate the $\Delta_r S°$ for:

(1) C(s) + 2 H$_2$(g) → CH$_4$(g)

$\Delta_r S_1°$ = (1 mol)(+186.26 J/K·mol) – [(1mol)(+5.6 J/K·mol) + (2mol)(+130.7 J/K·mol)]

= –80.7 J/K·mol-rxn

(2) CH$_4$(g) + ½ O$_2$(g) → CH$_3$OH(ℓ)

$\Delta_r S_2°$ = (1 mol)(+127.19 J/K·mol)

– [(1mol)(+186.26 J/K·mol) + (½ mol)(+205.07 J/K·mol)]

= –161.60 J/K·mol-rxn

(3) C(s) + 2 H$_2$(g) + ½ O$_2$(g) → CH$_3$OH(ℓ)

$\Delta_r S_3°$ = (1 mol)(+127.19 J/K·mol)

– [(1mol)(+5.6 J/K·mol) + (2 mol)(+130.7 J/K·mol) + (½ mol)(+205.07 J/K·mol)]

= –242.3 J/K·mol-rxn

So, $\Delta_r S_1°$ + $\Delta_r S_2°$ = (–80.7 J/K) + (–161.60 J/K) = –242.3 J/K·mol-rxn.

Entropy values for reactions are additive.

18.94. Predict the algebraic signs for each:

	$\Delta H°$	$\Delta S°$	$\Delta G°$
(a) Decomposition of H$_2$O(ℓ)	+	+	+
(b) Decomposition of nitroglycerine	–	+	–
(c) Combustion of gasoline	–	+	–

18.95. For the reaction of Mg with H₂O:

(a) Confirm that Mg(s) + 2 H₂O(ℓ) → Mg(OH)₂(s) + H₂(g) is a product-favored reaction.

	Mg(s)	H₂O(ℓ)	Mg(OH)₂(s)	H₂(g)
$\Delta_fH°$ (kJ/mol)	0	–285.83	–924.54	0
$S°$ (J/K·mol)	32.67	69.95	63.18	130.7

$\Delta_rH°$ = [(1 mol)(–924.54 kJ/mol) + (1mol)(0)] – [(1mol)0 + (2 mol)(–285.83 kJ/mol)]
   = –352.88 kJ/mol-rxn

$\Delta_rS°$ = [(1 mol)(63.18 J/K·mol) + (1mol)(130.7 J/K·mol)] –
   [(1mol)(32.67 J/K·mol) + (2 mol)(69.95 J/K·mol)]
   = 21.31 J/K·mol-rxn

$\Delta_rG° = \Delta_fH° – T\Delta_rS°$
   = –352.88 kJ – (298.15 K)(21.31 J/K)(1.000 kJ/1000 J) = –359.23 kJ/mol-rxn

With a negative $\Delta_rG°$, you anticipate the reaction to be product-favored at equilibrium.

(b) Mass of Mg to produce sufficient energy to heat 225 mL of water (density = 0.995 g/mL) from 25 °C to the boiling point (100 °C)? [100 – 25 = 75 °C or 75 K]

Heat required: $225 \text{ mL} \left(\dfrac{0.995 \text{ g}}{1 \text{ mL}}\right)\left(\dfrac{4.184 \text{ J}}{\text{g·K}}\right) 75 \text{ K} = 70251.98 \text{ J or } 70.3 \text{ kJ}$

The $\Delta_rH°$ = –352.88 kJ for 1 mol of Mg.

$70.3 \text{ kJ}\left(\dfrac{1 \text{ mol Mg}}{352.88 \text{ kJ}}\right) = 0.20$ mol Mg or 24.3 g Mg/mol · 0.20 mol Mg = 4.8 g Mg

18.96. Calculate the equilibrium vapor pressure:

The process: I₂(s) ⇌ I₂(g) $\Delta G°$ = (19.327 kJ/mol – 0 kJ/mol) = 19.327 kJ/mol

$\Delta_fG°$ = 19.327 kJ/mol = –RTln K = – (0.0083145 kJ/K · mol)(298 K) ln K

solving for K: K = 4.10 × 10⁻⁴ and K = P_{I_2} = 4.10 × 10⁻⁴ bar

18.97. Regarding the reaction of hydrazine and oxygen:

(a) Equation for the reaction: $N_2H_4(\ell) + O_2(g) \rightarrow 2\ H_2O(\ell) + N_2(g)$

Oxygen is the **oxidizing agent,** and hydrazine is the **reducing agent.** There are several ways to assess this. Note that the oxidation state for O_2 is 0 (as reactant) and –2 (as product)—it has been reduced (by hydrazine). Note that hydrazine **loses H**—in going from reactant to product—a definition for being oxidized.

(b) Calculate $\Delta_rH°$, $\Delta_rS°$, and $\Delta_rG°$:

$\Delta_rH° = [2 \cdot \Delta_fH°\ H_2O(\ell) + 1 \cdot \Delta_fH°\ N_2(g)] - [1 \cdot \Delta_fH°\ N_2H_4(\ell) + 1 \cdot \Delta_fH°\ O_2(g)]$

$= [(2\ mol)(-285.830\ kJ/mol) + 0] - [(1\ mol)(50.63\ kJ/mol) + 0] = -622.29\ kJ/mol\text{-rxn}$

$\Delta_rS° = [2 \cdot S°\ H_2O(\ell) + 1 \cdot S°\ N_2(g)] - [1 \cdot S°\ N_2H_4(\ell) + 1 \cdot S°\ O_2(g)]$

$= [(2\ mol)(69.95\ J/K\cdot mol) + (1\ mol)(191.56\ J/K\cdot mol)]$
$- [(1\ mol)(121.52\ J/K\cdot mol) + (1\ mol)(205.07\ J/K\cdot mol)] = 4.87\ J/K\cdot mol\text{-rxn}$

$\Delta_rG° = \Delta_rH° - T\Delta_rS°$

$= -622.29\ kJ - (298\ K)(4.87\ J/K)(1.000\ kJ/1000.\ J) = -623.74\ kJ/mol\text{-rxn}$

(c) Temperature change of 5.5×10^4 L of water (assuming 1 mole of N_2H_4 reacts):

1 mol of hydrazine releases –622.29 kJ,

Heat = $m \cdot c \cdot \Delta T$ [Assume denity of water = 0.996 g/mL]

$622.29 \times 10^3\ J = 5.5 \times 10^4\ L \left(\dfrac{996\ g}{1\ L}\right)\left(\dfrac{4.184\ J}{g \cdot K}\right)\Delta T$ Solving for ΔT:

$\dfrac{622.29 \times 10^3\ J}{5.5 \times 10^4\ L \left(\dfrac{996\ g}{1\ L}\right)\left(\dfrac{4.184\ J}{g \cdot K}\right)} = \Delta T$ Solving for ΔT gives: 2.7×10^{-3} K.

(d) Solubility of O_2 = 0.000434 g O_2/100g water.

$5.5 \times 10^4\ L \left(\dfrac{996\ g}{1\ L}\right)\left(\dfrac{4.34 \times 10^{-4}\ g\ O_2}{100\ g\ H_2O}\right)\left(\dfrac{1\ mol\ O_2}{32.00\ g\ O_2}\right) = 7.5\ mol\ O_2$ (approximately 240 g)

(e) If hydrazine is present in 5% solution, what mass of hydrazine solution is needed to consume the O_2 present?

$7.5\ mol\ O_2 \left(\dfrac{1\ mol\ N_2H_4}{1\ mol\ O_2}\right)\left(\dfrac{32.05\ g\ N_2H_4}{1\ mol\ N_2H_4}\right)\left(\dfrac{100\ g\ solution}{5.00\ g\ N_2H_4}\right) = 4.8 \times 10^3$ g solution (2 sf)

(f) Assuming N_2 escapes as gas, calculate V of N_2 at STP:

The balanced equation tells us that 7.5 mol of O_2 will liberate 7.5 mol of N_2.
At STP, 7.5 mol of this gas will occupy (7.5 mol × 22.4 L/mol) or 170 L. (2 sf)

18.98. Considering the formation of diamond from graphite:

C(graphite) → C(diamond)

(a) $\Delta_rS° = S°[C(diamond)] – S°[C(graphite)]$

$\Delta_rS° = 1$ mol (2.377 J/K·mol) – 1 mol (5.6 J/K·mol) = –3.2 J/K·mol-rxn

$\Delta_rH° = \Delta_fH°[C(diamond)] = 1$ mol (1.8 kJ/mol) = 1.8 kJ/mol-rxn

$\Delta_rG° = \Delta_rH° – T\Delta_rS° = 1.8$ kJ/mol-rxn – (298 K)(–3.2 J/K·mol-rxn)(1 kJ/10³ J)

= 2.8 kJ/mol-rxn

(b) Nonstandard conditions of extremely high pressure and temperature must be used to "force" the carbon atoms closer to one another, overcoming the unfavorable thermodynamics and allowing the conversion of graphite to diamond.

18.99. The key phrase needed to answer the question: "What is the sign......" is "Iodine dissolves readily....". This phrase tells us that $\Delta_rG°$ is negative.

Enthalpy-driven processes are exothermic. Since $\Delta H° = 0$ kJ/mol for this reaction tells us that the process is NOT enthalpy-driven. Since the iodine goes from the solid state to the "solution" state, you anticipate an increase in entropy and conclude that the process is entropy-driven.

18.100. The equation for the reaction of $Fe_2O_3(s)$ with C: $Fe_2O_3(s) + 3\ C(s) \rightarrow 2\ Fe(s) + 3\ CO(g)$

$\Delta_rS° = (2$ mol$)(S°[Fe(s)]) + (3$ mol$)(S°[CO(g)]) – (1$ mol$)(S°[Fe_2O_3(s)]) – (3$ mol$)(S°[C(s)])$

= (2 mol)(27.78 J/mol·K) + (3 mol)(197.64 J/mol·K) – (1 mol)(87.40 J/mol·K)

– (3 mol)(5.6 J/mol·K) = 544.3 J/K

$\Delta_rH° = (3$ mol$)(\Delta_fH°[CO(g)]) – (1$ mol$)(\Delta_fH°[Fe_2O_3(s)])$

= (3 mol)(–110.525 kJ/mol) – (1 mol)(–825.5 kJ/mol) = 493.9 kJ

$\Delta_rG° = 493.9$ kJ – T(0.5443 kJ/K)

As $\Delta_rS°$ is positive, $\Delta_rG°$ decreases as temperature increases.

To determine the T at which the reaction is product-favored, solve for $\Delta_rG° = 0$.

$\dfrac{\Delta H}{\Delta S} = \dfrac{493.9 \text{ kJ}}{0.5443 \text{ kJ/K}} = 907.4$ K thus the reaction is spontaneous above 907.4 K (634.3 °C).

18.101. Equation for decomposition of 1 mol of $CH_3OH(g)$ to elements:

$CH_3OH(g) \rightarrow C(s) + 2\ H_2(g) + ½\ O_2(g)$

(a) According to Appendix L, the $\Delta_fH°$ formation for methanol is –201.0 kJ/mol. The equation above is the reverse of that process, so you anticipate that the $\Delta_rH°$ for this process will be positive (endothermic). Additionally, the decomposition has a positive

$\Delta_r S°$, since the number of moles of gas increase during the process. So, the product-favored nature increases as T increases.

(b) Determine $\Delta_r S°$ for the reaction, then solve for the temperature at which the reaction become product-favored at equilibrium.

$\Delta_r S° = (1 \text{ mol})(S°[C(s)]) + (2 \text{ mol})(S°[H_2(g)]) + (1/2 \text{ mol})(S°[O_2(g)])$

$- (1 \text{ mol})(S°[CH_3OH(g)])$

$\Delta_r S° = (1 \text{ mol})(5.6 \text{ J/K·mol}) + (2 \text{ mol})(130.7 \text{ J/K·mol}) + (1/2 \text{ mol})(205.07 \text{ J/K·mol})$

$- (1 \text{ mol})(239.7 \text{ J/K·mol})$

$\Delta_r S° = 129.8 \text{ J/K}$

$\Delta_r G° = \Delta_r H° - T\Delta_r S°$, if $\Delta_r G° = 0$, then $T = \Delta_r H°/\Delta_r S° = (201.0 \text{ kJ/mol})/(0.1298 \text{ kJ/K})$, $T = 1550$ K.

The reaction becomes product-favored at equilibrium at temperatures above 1550 K. There is no temperature between 400 K and 1000 K at which the reaction is product-favored at equilibrium.

18.102. For the reaction of NO with Cl_2 to form NOCl:

(a) $\Delta_r S°$ for $2 \text{ NO}(g) + Cl_2(g) \rightarrow 2 \text{ NOCl}(g)$

$\Delta_r S° = (2 \text{ mol})(S°[NOCl(g)]) - (2 \text{ mol})(S°[NO(g)]) - (1 \text{ mol})(S°[Cl_2(g)])$

$\Delta_r S° = (2 \text{ mol})(261.8 \text{ J/mol·K}) - (2 \text{ mol})(210.76 \text{ J/mol·K}) - (1 \text{ mol})(223.08 \text{ J/mol·K})$

$= -121.0 \text{ J/K}$

For the reaction of one mole of NO to give 1 mole of NOCl, $\Delta_r S° = -60.5 \text{ J/K}$

(b) $\Delta S°$(system) change with T: **yes**

(c) $\Delta S°$(surroundings) change with T: **yes**

(d) $\Delta S°$(universe) change with T: **yes**

(e) Exothermic reactions lead to $\Delta S°$ (universe) > 0; **No**. Exothermic reactions lead to positive $\Delta S°$(surroundings) values. The sign of $\Delta S°$(universe) depends on both $\Delta S°$(surroundings) and $\Delta S°$(system).

(f) Reaction spontaneous at 298 K? 700 K?

$\Delta_r H° = (2 \text{ mol})(\Delta_f H°[NOCl(g)]) - (2 \text{ mol})(\Delta_f H°[NO(g)])$

$= (2 \text{ mol})(51.71 \text{ kJ/mol}) - (2 \text{ mol})(90.29 \text{ kJ/mol}) = -77.16 \text{ kJ}$

$\Delta_r G° = -77.16 \text{ kJ} - T(-0.1210 \text{ kJ/K})$

The reaction is spontaneous at 298 K ($\Delta_r G° = -41.1 \text{ kJ}$), but not spontaneous at 700 K ($\Delta_r G° = 7.5 \text{ kJ}$).

18.103. (a) Calculate $\Delta_rG°$ for the two reactions:

$$2\ H_2O(\ell) \rightarrow 2\ H_2(g) + O_2(g)$$

$\Delta_rG° = [2 \cdot \Delta_fG°\ H_2(g) + \Delta_fG°\ O_2(g)] - [2 \cdot \Delta_fG°\ H_2O(\ell)\ ]$

$\Delta_rG° = (2\ mol \cdot 0\ kJ/mol + 0\ kJ/mol) - (2\ mol \cdot -237.15\ kJ/mol)$

$= +474.3\ kJ$

and for: $CH_4(g) + H_2O(g) \rightarrow 3\ H_2(g) + CO(g)$

$\Delta_rG° = [3 \cdot \Delta_fG°\ H_2(g) + 1 \cdot \Delta_fG°\ CO(g)] - [1 \cdot \Delta_fG°\ CH_4(g) + 1 \cdot \Delta_fG°\ H_2O(g)\ ]$

$\Delta_rG° = [(3\ mol \cdot 0\ kJ/mol) + (1\ mol \cdot -137.168\ kJ/mol)]$
$\qquad - [(1mol \cdot -50.8\ kJ/mol) + (1mol \cdot -228.59\ kJ/mol)\ ]$

$\Delta_rG° = (-137.168\ kJ) - (-279.39\ kJ) = +142.2\ kJ$

(b) On a per-mol of H_2 basis:

For the electrolysis of water: $+474.30/2$ mol H_2 or 237.2 kJ/mol H_2.

For the methane reaction: $+142.2$ kJ/3 mol H_2 or 47.4 kJ/mol H_2.

(c) The $\Delta_rG°$ for the methane reaction is less than that for the electrolysis of water.

18.104. Consider the combustion of H_2 and CH_4:

(a) For the reaction: $H_2(g) + ½\ O_2(g) \rightarrow H_2O(\ell)$

$\Delta_rG° = 1 \cdot \Delta_fG°\ H_2O(\ell) - [1 \cdot \Delta_fG°\ H_2(g) + ½ \cdot \Delta_fG°\ O_2(g)]$

$= [1\ mol \cdot -237.15\ kJ/mol] - [0 + 0] = -237.15\ kJ/mol\text{-rxn}$

For the reaction: $CH_4(g) + 2\ O_2(g) \rightarrow CO_2(g) + 2\ H_2O(\ell)$

$\Delta_rG° = [1 \cdot \Delta_fG°\ CO_2(g) + 2 \cdot \Delta_fG°\ H_2O(\ell)] - [1 \cdot \Delta_fG°\ CH_4(g) + 2 \cdot \Delta_fG°\ O_2(g)]$

$= [1\ mol \cdot -394.359\ kJ/mol + 2\ mol \cdot -237.15\ kJ/mol] - [1\ mol \cdot -50.8\ kJ/mol + 0]$

$= -817.9\ kJ/mol\text{-rxn}$

(b) $\left(\dfrac{-237.15\ kJ}{mol-rxn}\right)\left(\dfrac{1\ mol-rxn}{1\ mol\ H_2}\right)\left(\dfrac{1\ mol\ H_2}{2.0159\ g}\right) = -117.64\ kJ/g\ H_2$

$\left(\dfrac{-817.9\ kJ}{mol-rxn}\right)\left(\dfrac{1\ mol-rxn}{1\ mol\ CH_4}\right)\left(\dfrac{1\ mol\ CH_4}{16.043\ g\ CH_4}\right) = -50.98\ kJ/g\ CH_4$

(c) Based solely on these numbers, the combustion of hydrogen would be better as this generates more free energy per gram.

18.105. For the formation of NH₃ from N₂ and H₂:

(a) Calculate $\Delta_r G°$ at 298 K, 800. K, and 1300. K for: N₂(g) + 3 H₂(g) ⇌ 2 NH₃(g)

At 298 K: $\Delta_r H° $ = (2 mol · –45.90 kJ/mol)– (0 + 0) = –91.80 kJ

and $\Delta_r S°$ = (2 mol · 192.77 J/K·mol) – (1mol · 191.56 J/K·mol + 3mol · 130.7 J/K·mol)

= –198.12 J/K and

$\Delta_r G° = \Delta_f H° - T\Delta_r S°$ = –91.80 kJ – (298 K)(–0.19812 kJ/K) = –32.8 kJ/mol-rxn

At 800. K:

$\Delta_r G° = \Delta_f H° - T\Delta_r S°$ = –107.4 kJ – (800. K)(–0.2254 kJ//K) = 73 kJ/mol-rxn

At 1300. K:

$\Delta_r G° = \Delta_f H° - T\Delta_r S°$ = –112.4 kJ – (1300. K)(–0.2280 kJ/K) = 184.0 kJ/mol-rxn

A quick examination of the values of $\Delta_r G°$ indicates that the free energy change becomes more positive as *T* increases.

(b) Calculate *K* for the reaction at 298 K, 800. K, 1300. K:

At 298 K:

$\ln K = \dfrac{-32.8 \times 10^3 \text{ J}}{-(8.3145 \text{ J/K·mol})(298 \text{ K})} = 13.21$ and $K = 5.5 \times 10^5$

At 800. K:

$\ln K = \dfrac{72.9 \times 10^3 \text{ J}}{-(8.3145 \text{ J/K·mol})(800. \text{ K})} = -10.96$ and $K = 1.7 \times 10^{-5}$

At 1300. K:

$\ln K = \dfrac{184.0 \times 10^3 \text{ J}}{-(8.3145 \text{ J/K·mol})(1300. \text{ K})} = -17.02$ and $K = 4.0 \times 10^{-8}$

(c) For which *T* will the mole fraction of NH₃ be largest?
The partial pressure of ammonia (and hence the mol fraction) is greatest for the temperature at which *K* is greatest (298 K in this case).

18.106. For the reaction of pyruvate to lactose:

(a) $\Delta_r G°' = \Delta_r G° + RT \ln \dfrac{[\text{lactate}][\text{NAD}^+]}{[\text{pyruvate}][\text{NADH}][\text{H}^+]}$

$\Delta_r G° = -25.1 \text{ kJ/mol} - (8.3145 \times 10^{-3} \text{ kJ/K·mol})(298 \text{ K})\ln \dfrac{[1][1]}{[1][1][1\times 10^{-7}]} = -65.0 \text{ kJ/mol}$

(b) Calculate *K'*:

$\Delta_r G°' = -RT \ln K'$
–25.1 kJ/mol = – (8.3145 × 10⁻³ kJ/K·mol)(298) ln *K'* and *K'* = 2.5 × 10⁴

(c) $\Delta_r G' = \Delta_r G^{o\prime} + RT \ln Q'$

$\Delta_r G' = -25.1 \text{ kJ/mol} + (8.3145 \times 10^{-3} \text{ kJ/K} \cdot \text{mol})(298 \text{ K})\ln\dfrac{[3700 \text{ μmol/L}][540 \text{ μmol/L}]}{[380 \text{ μmol/L}][50 \text{ μmol/L}]} = -13.5 \text{ kJ/mol}$

18.107. Following statements true or false:

(a) False; Stoichiometry will determine the product formed in the greatest amount.

(b) True; Lower activation energies favor product formation.

(c) False; Thermodynamics indicates feasibility of a process. Kinetics determines the rates.

(d) False; Phase diagram for carbon (Figure 1) indicates areas in which diamond is not more stable than graphite.

Solution and Answer Guide

Kotz Treichel Townsend Treichel, Chemistry and Chemical Reactivity 11e, 978-0-357-85140-1, Chapter 19: Principles of Chemical Reactivity: Electron Transfer Reactions

TABLE OF CONTENTS

Applying Chemical Principles 713
Practicing Skills 716
General Questions 739
In the Laboratory 758
Summary and Conceptual Questions 761

Applying Chemical Principles

Electric Batteries versus Gasoline

19.1.1. Energy released by a 3.6 V Li battery:

(a) Energy released per mol of Li:

Energy = $\Delta G° = -nFE°$

$$\Delta G° = -3.6 \text{ V} \left(\frac{1 \text{ mol e}^-}{1 \text{ mol Li}}\right)\left(\frac{96485 \text{ C}}{1 \text{ mol e}^-}\right)\left(\frac{1 \text{ J}}{1 \text{ V}\cdot\text{C}}\right)\left(\frac{1 \text{ kJ}}{1000 \text{ J}}\right) = -347 \text{ kJ/mol Li} = -350 \text{ kJ/mol Li}$$

(b) Energy released per kg of Li:

$$\left(\frac{-347 \text{ kJ}}{\text{mol Li}}\right)\left(\frac{1 \text{ mol Li}}{6.94 \text{ g Li}}\right)\left(\frac{1000 \text{ g Li}}{1 \text{ kg Li}}\right) = -5.00 \times 10^4 \text{ kJ}$$

19.1.2. Energy contained in 15 gallons of gasoline:

$$15 \text{ gallon}\left(\frac{4 \text{ qts}}{1 \text{ gallon}}\right)\left(\frac{1 \text{ L}}{1.057 \text{ qts}}\right)\left(\frac{0.70 \text{ kg gasoline}}{1 \text{ L}}\right)\left(\frac{46 \text{ MJ}}{1 \text{ kg gasoline}}\right)\left(\frac{1000 \text{ kJ}}{1 \text{ MJ}}\right) = 1.8 \times 10^6 \text{ kJ}$$

19.1.3. Mass of Li batteries to produce 1.8×10^6 kJ:

$$1.8 \times 10^6 \text{ kJ}\left(\frac{1 \text{ kg Li}}{5.00 \times 10^4 \text{ kJ}}\right) = 37 \text{ kg Li}$$

© 2024 Cengage Learning, Inc. All Rights Reserved. May not be scanned, copied or duplicated, or posted to a publicly accessible website, in whole or in part.

Sacrifice!

19.2.1. Explain why an insulator negates the effect of a corrosion protector:

To be effective as a sacrificial anode, the zinc and copper must be in contact. An insulator will prevent the flow of electrons from the zinc to copper, preventing zinc from keeping the copper reduced.

19.2.2. Metals that could serve as a sacrificial anode:

The reduction potential for copper is: $Cu^{2+} + 2\ e^- \rightarrow Cu$; $E° = 0.337$ V

The reduction potentials for the metals in question are:

$Cr^{3+} + 3\ e^- \rightarrow Cr \qquad E° = -0.74$ V

$Fe^{2+} + 2\ e^- \rightarrow Fe \qquad E° = -0.44$ V

$Ni^{2+} + 2\ e^- \rightarrow Ni \qquad E° = -0.25$ V

$Sn^{2+} + 2\ e^- \rightarrow Sn \qquad E° = -0.14$ V

$Ag^+ + e^- \rightarrow Ag \qquad E° = +0.799$ V

For a metal to be a sacrificial anode, it would have to provide a + voltage when Cu is the cathode, i.e. $E° = 0.337 - X$ would need to be positive (where X is the reduction potential of the other metal). Hence (a) tin, (c) iron, (d) nickel, (e) chromium

19.2.3. Which of the following metals could serve as a sacrificial anode on a steel hull, assuming that the reduction potential for steel is equivalent to that of iron ($E° = -0.44$ V)?

The reduction potentials are:

$Sn^{2+} + 2\ e^- \rightarrow Sn \qquad E° = -0.14$ V

$Ag^+ + e^- \rightarrow Ag \qquad E° = +0.799$ V

$Fe^{2+} + 2\ e^- \rightarrow Fe \qquad E° = -0.44$ V

$Ni^{2+} + 2\ e^- \rightarrow Ni \qquad E° = -0.25$ V

$Cr^{3+} + 3\ e^- \rightarrow Cr \qquad E° = -0.74$ V

The only reduction potential that is more negative than that of iron is that belonging to chromium, making chromium the only metal in this list capable of replacing Zn as a sacrificial anode. Compare the reduction potential for zinc ($Zn^{2+} + 2\ e^- \rightarrow Zn$ $E° = -0.763$ V).

19.2.4. For the production of copper(II) hydroxide from Cu in oxygenated water:

(a) Oxidation half–reaction: $Cu(s) \rightarrow Cu^{2+}(aq) + 2\ e^-$

Reduction half–reaction: $½\ O_2(g) + H_2O(\ell) + 2\ e^- \rightarrow 2\ OH^-(aq)$

(b) Associated equations:

$Cu^{2+}(aq) + 2\ e^- \rightarrow Cu(s)$ $E° = 0.337$ V

$\Delta G° = -nFE° = -6.503 \times 10^4$ J/mol

½ $O_2(g) + H_2O(\ell) + 2\ e^- \rightarrow 2\ OH^-(aq)$ $E° = 0.40$ V $\Delta G° = -nFE° = -7.72 \times 10^4$ J/mol

$Cu(OH)_2(s) \rightarrow Cu^{2+}(aq) + 2\ OH^-(aq)$ $K_{sp} = 2.2 \times 10^{-20}$ and writing this equation in reverse:

$Cu^{2+}(aq) + 2\ OH^-(aq) \rightarrow Cu(OH)_2(s)$ $1/K_{sp} = 4.55 \times 10^{19}$

$\Delta G° = -RT\ln K = -1.121 \times 10^5$ J/mol

Adding the equations:

$Cu(s) \rightarrow Cu^{2+}(aq) + 2\ e^-$ $\Delta G° = 6.503 \times 10^4$ J/mol (Reverse of reaction requires change of sign of $\Delta G°$)

½ $O_2(g) + H_2O(\ell) + 2\ e^- \rightarrow 2\ OH^-(aq)$ $\Delta G° = -7.72 \times 10^4$ J/mol

$Cu^{2+}(aq) + 2\ OH^-(aq) \rightarrow Cu(OH)_2(s)$ $\Delta G° = -1.121 \times 10^5$ J/mol

$Cu(s) + ½\ O_2(g) + H_2O(\ell) \rightarrow Cu(OH)_2(s)$ $\Delta G° = -1.243 \times 10^5$ J/mol NET REACTION

$\Delta G° = -RT\ln K = -1.243 \times 10^5$ J/mol $= -(8.3145$ J/K·mol$)(298$ K$)\ln K$
and $K = 6.1 \times 10^{21}$

19.2.5. For the cell: $Zn\ |\ Zn(OH)_2(s)\ |\ OH^-(aq)\ ||\ Cu(OH)_2(s)\ |\ Cu(s)$

(a) The balanced equation for the cathode reaction: $Cu(OH)_2(s) + 2\ e^- \rightarrow Cu(s) + 2\ OH^-(aq)$

(b) The balanced equation for the anode reaction: $Zn(s) + 2\ OH^-(aq) \rightarrow Zn(OH)_2(s) + 2\ e^-$

(c) The balanced equation for the overall reaction:
$Cu(OH)_2(s) + Zn(s) \rightarrow Zn(OH)_2(s) + Cu(s)$

(d) What is the potential (in volts) of the cell at pH = 7.90 and 25 °C?

This problem simplifies as all reactants and products are solids, so in the "ln" term of the Nernst equation, all those quantities would have a value of 1, and the ln(1) = 0.
NOTE: The reduction potentials are those *in base*:

$Cu(OH)_2(s) + 2\ e^- \rightarrow Cu(s) + 2\ OH^-(aq)$ $E° = -0.36$ V

$Zn(OH)_2(s) + 2\ e^- \rightarrow Zn(s) + 2\ OH^-(aq)$ $E° = -1.245$ V

$E°_{cell} = E°_{cathode} - E°_{anode} = (-0.36\ V) - (-1.245\ V) = 0.89$ V and $E_{cell} = E°_{cell} = 0.89$ V

Practicing Skills

Balancing Equations for Oxidation-Reduction Reactions

19.1. Balance the following: overall process is

(a) $Cr(s) \rightarrow Cr^{3+}(aq) + 3\ e^-$ oxidation

(b) $AsH_3(g) \rightarrow As(s) + 3\ H^+(aq) + 3\ e^-$ oxidation

(c) $VO_3^-(aq) + 6\ H^+(aq) + 3\ e^- \rightarrow V^{2+}(aq) + 3\ H_2O(\ell)$ reduction

(d) $2\ Ag(s) + 2\ OH^-(aq) \rightarrow Ag_2O(s) + H_2O(\ell) + 2\ e^-$ oxidation

Note: e^- are used to balance charge; H^+ balances only H atoms; H_2O (or OH^- in base) balances both H and O atoms.

19.2. Balance the following: overall process is

(a) $H_2O_2(aq) \rightarrow O_2(g) + 2\ H^+(aq) + 2\ e^-$ oxidation

(b) $H_2C_2O_4(aq) \rightarrow 2\ CO_2(g) + 2\ H^+(aq) + 2\ e^-$ oxidation

(c) $NO_3^-(aq) + 4\ H^+(aq) + 3\ e^- \rightarrow NO(g) + 2\ H_2O(\ell)$ reduction

(d) $MnO_4^-(aq) + 2\ H_2O(\ell) + 3\ e^- \rightarrow MnO_2(s) + 4\ OH^-(aq)$ reduction

19.3. Balance the equations (in acidic solutions):

Balancing redox equations in neutral or acidic solutions may be accomplished by:

1. Separating the equation into two equations which represent reduction and oxidation
2. Balancing mass of elements (other than H or O)
3. Balancing mass of O by adding H_2O
4. Balancing mass of H by adding H^+
5. Balancing charge by adding electrons
6. Balancing electron gain (in the reduction half-equation) with electron loss (in the oxidation half-equation)
7. Combining the two half equations

For the parts of this problem, each step is identified with a number corresponding to the list above. In addition, the physical states of all species will be omitted in all but the final step. While this omission is <u>not generally recommended</u>, it should increase the clarity of the steps involved here. In addition, when a step leaves a half equation unchanged from the previous step, we have omitted the half equation.

(a) Ag(s) + NO$_3^-$(aq) → NO$_2$(g) + Ag$^+$(aq)

Oxidation half-equation	Reduction half-equation	Step
Ag → Ag$^+$	NO$_3^-$ → NO$_2$	1 & 2
	NO$_3^-$ → NO$_2$ + H$_2$O	3
	2 H$^+$ + NO$_3^-$ → NO$_2$ + H$_2$O	4
Ag → Ag$^+$ + e$^-$	2 H$^+$ + NO$_3^-$ + e$^-$ → NO$_2$ + H$_2$O	5 & 6
2 H$^+$(aq) + NO$_3^-$(aq) + Ag(s) → Ag$^+$(aq) + NO$_2$(g) + H$_2$O(ℓ)		7

(b) MnO$_4^-$(aq) + HSO$_3^-$(aq) → Mn^{2+}(aq) + SO$_4^{2-}$(aq)

Oxidation half-equation	Reduction half-equation	Step
HSO$_3^-$ → SO$_4^{2-}$	MnO$_4^-$ → Mn^{2+}	1 & 2
H$_2$O + HSO$_3^-$ → SO$_4^{2-}$	MnO$_4^-$ → Mn^{2+} + 4 H$_2$O	3
H$_2$O + HSO$_3^-$ → SO$_4^{2-}$ + 3 H$^+$	8 H$^+$ + MnO$_4^-$ → Mn^{2+} + 4 H$_2$O	4
H$_2$O + HSO$_3^-$ → SO$_4^{2-}$ + 3 H$^+$ + 2 e$^-$	8 H$^+$ + MnO$_4^-$ + 5 e$^-$ → Mn^{2+} + H$_2$O	5
5 H$_2$O + 5 HSO$_3^-$ → 5 SO$_4^{2-}$ + 15 H$^+$ + 10 e$^-$	16 H$^+$ + 2 MnO$_4^-$ + 10 e$^-$ → 2 Mn^{2+} + 8 H$_2$O	6
5 HSO$_3^-$(aq) + H$^+$(aq) + 2 MnO$_4^-$(aq) → 5 SO$_4^{2-}$(aq) + 2 Mn^{2+}(aq) + 3 H$_2$O(ℓ)		7

(c) Zn(s) + NO$_3^-$(aq) → Zn^{2+}(aq) + N$_2$O(g)

Oxidation half-equation	Reduction half-equation	Step
Zn → Zn^{2+}	NO$_3^-$ → N$_2$O	1
	2 NO$_3^-$ → N$_2$O	2
	2 NO$_3^-$ → N$_2$O + 5 H$_2$O	3
	10 H$^+$ + 2 NO$_3^-$ → N$_2$O + 5 H$_2$O	4
Zn → Zn^{2+} + 2 e$^-$	10 H$^+$ + 2 NO$_3^-$ + 8 e$^-$ → N$_2$O + 5 H$_2$O	5
4 Zn → 4 Zn^{2+} + 8 e$^-$		6
10 H$^+$(aq) + 2 NO$_3^-$(aq) + 4 Zn(s) → 4 Zn^{2+}(aq) + N$_2$O(g) + 5 H$_2$O(ℓ)		7

(d) $Cr(s) + NO_3^-(aq) \rightarrow Cr^{3+}(aq) + NO(g)$

Oxidation half-equation	Reduction half-equation	Step
$Cr \rightarrow Cr^{3+}$	$NO_3^- \rightarrow NO$	1 & 2
	$NO_3^- \rightarrow NO + 2\ H_2O$	3
	$4\ H^+ + NO_3^- \rightarrow NO + 2\ H_2O$	4
$Cr \rightarrow Cr^{3+} + 3\ e^-$	$4\ H^+ + NO_3^- + 3\ e^- \rightarrow NO + 2\ H_2O$	5 & 6
$Cr(s) + 4\ H^+(aq) + NO_3^-(aq) \rightarrow NO(g) + 2\ H_2O(\ell) + Cr^{3+}(aq)$		7

19.4. Balance the following redox reactions, all of which occur in acid solution.

(a) $Sn(s) + H^+(aq) \rightarrow Sn^{2+}(aq) + H_2(g)$

Oxidation half-equation	Reduction half-equation	Step
$Sn(s) \rightarrow Sn^{2+}(aq)$	$H^+(aq) \rightarrow H_2(g)$	1 & 2
		3
	$2\ H^+(aq) \rightarrow H_2(g)$	4
$Sn(s) \rightarrow Sn^{2+}(aq) + 2\ e^-$	$2\ H^+(aq) + 2\ e^- \rightarrow H_2(g)$	5 & 6
$Sn(s) + 2\ H^+(aq) \rightarrow Sn^{2+}(aq) + H_2(g)$		7

(b) $Cr_2O_7^{2-}(aq) + Fe^{2+}(aq) \rightarrow Cr^{3+}(aq) + Fe^{3+}(aq)$

Oxidation half-equation	Reduction half-equation	Step
$Fe^{2+} \rightarrow Fe^{3+}$	$Cr_2O_7^{2-} \rightarrow 2\ Cr^{3+}$	1 & 2
	$Cr_2O_7^{2-} \rightarrow 2\ Cr^{3+} + 7\ H_2O$	3
	$Cr_2O_7^{2-} + 14\ H^+ \rightarrow 2\ Cr^{3+} + 7\ H_2O$	4
$6\ Fe^{2+} \rightarrow 6\ Fe^{3+} + 6\ e^-$	$Cr_2O_7^{2-} + 14\ H^+ + 6\ e^- \rightarrow 2\ Cr^{3+} + 7\ H_2O$	5 & 6
$Cr_2O_7^{2-}(aq) + 14\ H^+(aq) + 6\ Fe^{2+}(aq) \rightarrow 2\ Cr^{3+}(aq) + 7\ H_2O(\ell) + 6\ Fe^{3+}(aq)$		7

(c) MnO_4^- (aq) + Cl^-(aq) → Mn^{2+}(aq) + Cl_2(g)

Oxidation half-equation	Reduction half-equation	Step
$Cl^- → Cl_2$	$MnO_4^- → Mn^{2+}$	1 & 2
	MnO_4^- (aq) → Mn^{2+} + 4 H_2O	3
	MnO_4^- (aq) + 8 H^+ → Mn^{2+} + 4 H_2O	4
5(2 Cl^-(aq) → Cl_2(g) + 2 e^-)	2(MnO_4^- (aq) + 8 H^+(aq) + 5 e^- → Mn^{2+}(aq) + 4 $H_2O(\ell)$)	5 & 6
2 MnO_4^-(s) + 16 H^+(aq) + 10 Cl^-(aq) → 2 Mn^{2+}(aq) + 8 $H_2O(\ell)$ + 5 Cl_2(g)		7

(d) CH_2O + Ag^+ → HCO_2H + Ag

Oxidation half-equation	Reduction half-equation	Step
$CH_2O → HCO_2H$	$Ag^+ → Ag$	1 & 2
$CH_2O + H_2O → HCO_2H$		3
$CH_2O + H_2O → HCO_2H + 2 H^+$		4
$CH_2O + H_2O → HCO_2H + 2 H^+ + 2 e^-$	2 Ag^+(aq) + 2 e^- → 2 Ag(s)	5 & 6
CH_2O(aq) + $H_2O(\ell)$ + 2 Ag^+(aq) → HCO_2H(aq) + 2 H^+(aq) + 2 Ag(s)		7

19.5. Balancing redox equations in basic solutions may be accomplished in several steps.

There is only a *slight change* from the "acidic solution" procedure (*italicized below*).

1. Separating the equation into two equations which represent reduction and oxidation
2. Balancing mass of elements (other than H or O)
3. Balancing mass of O by adding H_2O
4. Balancing mass of H by adding H^+
5. Balancing charge by adding electrons
6. Balancing electron gain (in the reduction half-equation) with electron loss (in the oxidation half-equation)
7. Combine the two half equations, removing any redundancies.
8. *Add as many OH^- to both sides of the equation as there are H^+ ions, to form water.*
9. *Remove any redundancies in H_2O molecules.*

As before, each step is identified with a number corresponding to the list above, and physical states of all species will be omitted in all but the final step.

(a) Al(s) + H$_2$O(ℓ) → Al(OH)$_4^-$(aq) + H$_2$(g)

Oxidation half-equation	Reduction half-equation	Step
Al → Al(OH)$_4^-$	H$_2$O → H$_2$	1 & 2
Al + 4 H$_2$O → Al(OH)$_4^-$	H$_2$O → H$_2$ + H$_2$O	3
Al + 4 H$_2$O → Al(OH)$_4^-$ + 4 H$^+$	2 H$^+$ + H$_2$O → H$_2$ + H$_2$O	4
Al + 4 H$_2$O → Al(OH)$_4^-$ + 4 H$^+$ + 3 e$^-$	2 H$^+$ + H$_2$O + 2 e$^-$ → H$_2$ + H$_2$O	5
2Al + 8 H$_2$O → 2 Al(OH)$_4^-$ + 8 H$^+$ + 6 e$^-$	6 H$^+$ + 3 H$_2$O + 6 e$^-$ → 3 H$_2$ + 3 H$_2$O	6
2 Al + 8 H$_2$O → 2 Al(OH)$_4^-$ + 3 H$_2$ + 2 H$^+$		7
2 Al + 8 H$_2$O + 2 OH$^-$ → 2 Al(OH)$_4^-$ + 3 H$_2$ + 2 H$^+$ + 2 OH$^-$		8
2 Al(s) + 6 H$_2$O(ℓ) + 2 OH$^-$(aq) → 2 Al(OH)$_4^-$(aq) + 3 H$_2$(g)		9

(b) CrO$_4^{2-}$(aq) + SO$_3^{2-}$(aq) → Cr(OH)$_3$(s) + SO$_4^{2-}$(aq)

Oxidation half-equation	Reduction half-equation	Step
SO$_3^{2-}$ → SO$_4^{2-}$	CrO$_4^{2-}$ → Cr(OH)$_3$	1 & 2
H$_2$O + SO$_3^{2-}$ → SO$_4^{2-}$	CrO$_4^{2-}$ → Cr(OH)$_3$ + H$_2$O	3
H$_2$O + SO$_3^{2-}$ → SO$_4^{2-}$ + 2 H$^+$	5 H$^+$ + CrO$_4^{2-}$ → Cr(OH)$_3$ + H$_2$O	4
H$_2$O + SO$_3^{2-}$ → SO$_4^{2-}$ + 2 H$^+$ + 2 e$^-$	5 H$^+$ + CrO$_4^{2-}$ + 3 e$^-$ → Cr(OH)$_3$ + H$_2$O	5
3 H$_2$O + 3 SO$_3^{2-}$ → 3 SO$_4^{2-}$ + 6 H$^+$ + 6 e$^-$	10 H$^+$ + CrO$_4^{2-}$ + 6 e$^-$ → 2Cr(OH)$_3$ + 2H$_2$O	6
H$_2$O + 3 SO$_3^{2-}$ + 4 H$^+$ + 2 CrO$_4^{2-}$ → 2 Cr(OH)$_3$ + 3 SO$_4^{2-}$		7
H$_2$O + 3 SO$_3^{2-}$ + 4 H$^+$ + 4 OH$^-$ + 2 CrO$_4^{2-}$ → 2 Cr(OH)$_3$ + 3 SO$_4^{2-}$ + 4 OH$^-$		8
5 H$_2$O(ℓ) + 3 SO$_3^{2-}$(aq) + 2 CrO$_4^{2-}$(aq) → 2 Cr(OH)$_3$(s) + 3 SO$_4^{2-}$(aq) + 4 OH$^-$(aq)		9

(c) $Zn(s) + Cu(OH)_2(s) \rightarrow Zn(OH)_4^{2-}(aq) + Cu(s)$

Oxidation half-equation	Reduction half-equation	Step
$Zn \rightarrow Zn(OH)_4^{2-}$	$Cu(OH)_2 \rightarrow Cu$	1 & 2
$4\ H_2O + Zn \rightarrow Zn(OH)_4^{2-}$	$Cu(OH)_2 \rightarrow Cu + 2\ H_2O$	3
$4\ H_2O + Zn \rightarrow Zn(OH)_4^{2-} + 4\ H^+$	$2\ H^+ + Cu(OH)_2 \rightarrow Cu + 2\ H_2O$	4
$4\ H_2O + Zn \rightarrow Zn(OH)_4^{2-} + 4\ H^+ + 2\ e^-$	$2\ H^+ + Cu(OH)_2 + 2\ e^- \rightarrow Cu + 2\ H_2O$	5 & 6
$2\ H_2O + Zn + Cu(OH)_2 \rightarrow Zn(OH)_4^{2-} + 2\ H^+ + Cu$		7
$2\ H_2O + Zn + Cu(OH)_2 + 2\ OH^- \rightarrow Zn(OH)_4^{2-} + 2\ H^+ + 2\ OH^- + Cu$		8
$Zn(s) + Cu(OH)_2(s) + 2\ OH^-(aq) \rightarrow Zn(OH)_4^{2-}(aq) + Cu(s)$		9

(d) $HS^-(aq) + ClO_3^-(aq) \rightarrow S(s) + Cl^-(aq)$

Oxidation half-equation	Reduction half-equation	Step
$HS^- \rightarrow S$	$ClO_3^- \rightarrow Cl^-$	1 & 2
	$ClO_3^- \rightarrow Cl^- + 3\ H_2O$	3
$HS^- \rightarrow S + H^+$	$6\ H^+ + ClO_3^- \rightarrow Cl^- + 3\ H_2O$	4
$HS^- \rightarrow S + H^+ + 2\ e^-$	$6\ e^- + 6\ H^+ + ClO_3^- \rightarrow Cl^- + 3\ H_2O$	5
$3\ HS^- \rightarrow 3\ S + 3\ H^+ + 6\ e^-$		6
$3\ HS^- + 3\ H^+ + ClO_3^- \rightarrow Cl^- + 3\ H_2O + 3\ S$		7
$3\ HS^- + 3\ H^+ + 3\ OH^- + ClO_3^- \rightarrow Cl^- + 3\ H_2O + 3\ S + 3\ OH^-$		8
$3\ HS^-(aq) + ClO_3^-(aq) \rightarrow Cl^-(aq) + 3\ S(s) + 3\ OH^-(aq)$		9

19.6. Balancing the following redox equations in basic solutions:

(a) $MnO_4^-(aq) + I^-(aq) \rightarrow MnO_2(s) + IO_3^-(aq)$

Oxidation half-equation	Reduction half-equation	Step
$I^- \rightarrow IO_3^-$	$MnO_4^- \rightarrow MnO_2$	1 & 2
$I^- + 3\ H_2O \rightarrow IO_3^-$	$MnO_4^- \rightarrow MnO_2 + 2\ H_2O$	3
$I^- + 3\ H_2O \rightarrow IO_3^- + 6\ H^+$	$MnO_4^- + 4\ H^+ \rightarrow MnO_2 + 2\ H_2O$	4
$I^- + 3\ H_2O \rightarrow IO_3^- + 6\ H^+ + 6\ e^-$	$2(MnO_4^- + 4\ H^+ + 3\ e^- \rightarrow MnO_2 + 2\ H_2O)$	5 & 6
$I^- + 3\ H_2O + 2\ MnO_4^- + 8\ H^+ + 6\ e^- \rightarrow IO_3^- + 6\ H^+ + 6\ e^- + 2\ MnO_2 + 4\ H_2O$ $I^- + 2\ MnO_4^- + 8\ H^+ + 6\ e^- \rightarrow IO_3^- + 6\ H^+ + 6\ e^- + 2\ MnO_2 + H_2O$ $I^- + 2\ MnO_4^- + 2\ H^+ + 6\ e^- \rightarrow IO_3^- + 6\ e^- + 2\ MnO_2 + H_2O$		7
$I^- + 2\ MnO_4^- + 2\ H_2O + 6\ e^- \rightarrow IO_3^- + 2\ OH^- + 6\ e^- + 2\ MnO_2 + H_2O$ $I^- + 2\ MnO_4^- + H_2O + 6\ e^- \rightarrow IO_3^- + 2\ OH^- + 6\ e^- + 2\ MnO_2$		8
$I^-(aq) + 2\ MnO_4^-(aq) + H_2O(\ell) \rightarrow IO_3^-(aq) + 2\ OH^-(aq) + 2\ MnO_2(s)$		9

(b) $NiO_2(s) + Zn(s) \rightarrow Ni(OH)_2(s) + Zn(OH)_2(s)$

Oxidation half-equation	Reduction half-equation	Step
$Zn \rightarrow Zn(OH)_2$	$NiO_2 \rightarrow Ni(OH)_2$	1 & 2
$Zn + 2\ H_2O \rightarrow Zn(OH)_2$		3
$Zn + 2\ H_2O \rightarrow Zn(OH)_2 + 2\ H^+$	$NiO_2 + 2\ H^+ \rightarrow Ni(OH)_2$	4
$Zn + 2\ H_2O \rightarrow Zn(OH)_2 + 2\ H^+ + 2\ e^-$	$NiO_2 + 2\ H^+ + 2\ e^- \rightarrow Ni(OH)_2$	5 & 6
$Zn(s) + 2\ H_2O + NiO_2 \rightarrow Zn(OH)_2 + Ni(OH)_2$		7
		8
$NiO_2(s) + 2\ H_2O(\ell) + Zn(s) \rightarrow Ni(OH)_2(s) + Zn(OH)_2(s)$		9

(c) Fe(OH)$_2$(s) + CrO$_4^{2-}$(aq) → Fe(OH)$_3$(aq) + [Cr(OH)$_4$]$^-$(aq)

Oxidation half-equation	Reduction half-equation	Step
Fe(OH)$_2$ → Fe(OH)$_3$	CrO$_4^{2-}$ → [Cr(OH)$_4$]$^-$	1 & 2
Fe(OH)$_2$ + H$_2$O → Fe(OH)$_3$		3
Fe(OH)$_2$ + H$_2$O → Fe(OH)$_3$ + H$^+$	CrO$_4^{2-}$ + 4 H$^+$ → [Cr(OH)$_4$]$^-$	4
3 Fe(OH)$_2$ + 3 H$_2$O → 3 Fe(OH)$_3$ + 3 H$^+$ + 3 e$^-$	CrO$_4^{2-}$ + 4 H$^+$ + 3 e$^-$ → [Cr(OH)$_4$]$^-$	5 & 6
3 Fe(OH)$_2$ + 3 H$_2$O + CrO$_4^{2-}$ + 4 H$^+$ → 3 Fe(OH)$_3$ + 3 H$^+$ + [Cr(OH)$_4$]$^-$		7
3 Fe(OH)$_2$ + 3 H$_2$O + CrO$_4^{2-}$ + H$^+$ + OH$^-$ → 3 Fe(OH)$_3$ + [Cr(OH)$_4$]$^-$ + OH$^-$		8
3 Fe(OH)$_2$(s) + CrO$_4^{2-}$(aq) + 4 H$_2$O(ℓ) → 3 Fe(OH)$_3$(s) + [Cr(OH)$_4$]$^-$(aq) + OH$^-$(aq)		9

(d) N$_2$H$_4$(aq) + Ag$_2$O(s) → N$_2$(g) + Ag(s)

Oxidation half-equation	Reduction half-equation	Step
N$_2$H$_4$ → N$_2$	Ag$_2$O → 2Ag	1 & 2
	Ag$_2$O → 2 Ag + H$_2$O	3
N$_2$H$_4$ → N$_2$ + 4 H$^+$	Ag$_2$O + 2 H$^+$ → 2 Ag + H$_2$O	4
N$_2$H$_4$ → N$_2$ + 4 H$^+$ + 4 e$^-$	2 Ag$_2$O + 4 H$^+$ + 4 e$^-$ → 4 Ag + 2 H$_2$O	5 & 6
N$_2$H$_4$ + 2 Ag$_2$O → N$_2$ + 4 Ag + 2 H$_2$O		7
		8
N$_2$H$_4$(aq) + 2 Ag$_2$O(s) → N$_2$(g) + 2 H$_2$O(ℓ) + 4 Ag(s)		9

Constructing Voltaic Cells

19.7. For the reaction: 2 Cr(s) + 3 Fe^{2+}(aq) → 2 Cr^{3+}(aq) + 3 Fe(s):

Electrons in the external circuit flow from the Cr electrode to the Fe electrode. Negative ions move in the salt bridge from the iron half-cell to the chromium half-cell. The half-reaction at the anode is Cr(s) → Cr^{3+}(aq) + 3 e$^-$ and that at the cathode is Fe^{2+}(aq) + 2 e$^-$ → Fe(s). Note that the reaction shows that Cr is being oxidized (to Cr^{3+}) by Fe^{2+} that is being reduced (to Fe). The electrons leave the anode and head to the cathode via the external circuit (wire). With reduction occurring in the cathode half-cell, a net deficit of positive ions accumulates, necessitating the assistance of + ions from the salt bridge.

19.8. For the voltaic cell:

(a) Oxidation (anode): Mg(s) → Mg²⁺(aq) + 2 e⁻
Reduction (cathode): 2 H⁺(aq) + 2 e⁻ → H₂(g)

(b) Oxidation occurs in the Mg/Mg²⁺ anode compartment and reduction occurs in the H⁺/H₂ cathode compartment.

(c) Electrons in the external circuit flow from the Mg electrode to the positive (site of H⁺ reduction) electrode. Negative ions move in the salt bridge from the H⁺/H₂ half-cell to the Mg/Mg²⁺ half-cell. The half-reaction at the anode and the cathode are shown in (a).

19.9. Like SQ19.7, you can complete the paragraph in part (c), by deciding on the spontaneous or product-favored reaction between the iron and oxygen half-cells.

The reduction potentials are:

$$O_2(g) + 4\,H^+(aq) + 4\,e^- \rightarrow 2\,H_2O(\ell) \qquad E° = 1.229\text{ V}$$

$$Fe^{2+}(aq) + 2\,e^- \rightarrow Fe(s) \qquad E° = -0.44\text{ V}$$

(a) Oxidation half-reaction: 2 Fe(s) → 2 Fe²⁺(aq) + 4 e⁻ (with balanced # electrons)

Reduction half-reaction: O₂(g) + 4 H⁺(aq) + 4 e⁻ → 2 H₂O(ℓ)

Net cell reaction: 2 Fe(s) + O₂(g) + 4 H⁺(aq) → 2 H₂O(ℓ) + 2 Fe²⁺(aq)

How to decide which half-reaction occurs as oxidation? An examination of Table 19.1 (or Appendix M) shows O₂ as a stronger oxidizing agent than Fe²⁺.

Alternatively, Fe is a stronger reducing agent than H₂O. Either of these conclusions points to the direction of reaction, which is product-favored at equilibrium.

(b) The anode half-reaction is the oxidation half-reaction: 2 Fe(s) → 2 Fe²⁺(aq) + 4 e⁻

At the cathode, the reduction half-reaction: O₂(g) + 4 H⁺(aq) + 4 e⁻ → 2 H₂O(ℓ)

(c) Electrons in the external circuit flow from the Fe electrode to the positive cathode (site of the O₂ half-reaction). Negative ions move in the salt bridge from the oxygen half-cell to the iron half-cell.

19.10. For the voltaic cell consisting of tin and chlorine:
(a) Oxidation: Sn(s) → Sn²⁺(aq) + 2 e⁻
Reduction: Cl₂(g) + 2 e⁻ → 2 Cl⁻(aq)
Overall: Sn(s) + Cl₂(g) → Sn²⁺(aq) + 2 Cl⁻(aq)
(b) Oxidation occurs in the anode compartment and reduction occurs in the cathode compartment.
(c) Electrons in the external circuit flow from the Sn electrode to the positive (site of Cl₂ reduction) electrode. Negative ions move in the salt bridge from the Cl₂/Cl⁻ half-cell to the Sn/Sn²⁺ half-cell.

19.11. Equations for the oxidation and reduction half-cell reactions and the overall equation:

(a) oxidation: $Cu(s) \rightarrow Cu^{2+}(aq) + 2\ e^-$

reduction: $Fe^{3+}(aq) + e^- \rightarrow Fe^{2+}(aq)$

balanced overall: $2\ Fe^{3+}(aq) + Cu(s) \rightarrow 2\ Fe^{2+}(aq) + Cu^{2+}(aq)$

(b) oxidation: $Pb(s) + SO_4^{2-}(aq) \rightarrow PbSO_4(s) + 2\ e^-$

reduction: $Fe^{3+}(aq) + e^- \rightarrow Fe^{2+}(aq)$

balanced overall: $Pb(s) + 2\ Fe^{3+}(aq) + SO_4^{2-}(aq) \rightarrow 2\ Fe^{2+}(aq) + PbSO_4(s)$

19.12. Equations for the oxidation and reduction half-cell reactions and the overall equation:

(a) oxidation: $Pb(s) \rightarrow Pb^{2+}(aq) + 2\ e^-$

reduction: $Fe^{3+}(aq) + e^- \rightarrow Fe^{2+}(aq)$

balanced overall: $Pb(s) + 2\ Fe^{3+}(aq) \rightarrow Pb^{2+}(aq) + 2\ Fe^{2+}(aq)$

(b) oxidation: $2\ Hg(\ell) + 2\ Cl^-(aq) \rightarrow Hg_2Cl_2(s) + 2\ e^-$

reduction: $Ag^+(aq) + e^- \rightarrow Ag(s)$

balanced overall: $2\ Hg(\ell) + 2\ Cl^-(aq) + 2\ Ag^+(aq) \rightarrow Hg_2Cl_2(s) + 2\ Ag(s)$

19.13. Cell notation for the reaction: $Cu(s) + Cl_2(g) \rightarrow 2\ Cl^-(aq) + Cu^{2+}(aq)$

$Cu(s)\ |\ Cu^{2+}(aq)\ ||\ Cl^-(aq)\ |\ Cl_2(g)\ |\ Pt(s)$

Anode half equation is written first: $(Cu(s)\ |\ Cu^{2+}(aq))$; followed by cathode $(Cl^-(aq)\ |\ Cl_2(g)\ |\ Pt(s))$. A salt bridge ($||$) separates the half-cells.

19.14. Cell notation for the reaction: $Fe(s) + AgCl(s) \rightarrow Fe^{2+}(aq) + Ag(s) + Cl^-(aq)$

oxidation: $Fe(s) \rightarrow Fe^{2+}(aq) + 2\ e^-$

reduction: $AgCl(s) + e^- \rightarrow Cl^-(aq) + Ag(s)$

$Fe(s)\ |\ Fe^{2+}(s)\ ||\ Cl^-(aq)\ ||\ AgCl(s)\ |\ Ag(s)$

Commercial Electrochemical Cells

19.15. Similarities and differences between dry cells, alkaline batteries, and Ni-cad batteries:

The first two types of cells are non-rechargeable batteries—also called primary batteries. They also share the common anode, zinc. Ni-cad batteries are rechargeable. Alkaline and Ni-cad batteries are in a basic environment, whereas dry cells are in an acidic environment.

19.16. Reactions during recharge of a lead storage battery:
Lead(II) sulfate is oxidized to lead(IV) oxide and reduced to elemental lead.

Standard Electrochemical Potentials

19.17. Calculate $E°$ for each of the following, and decided if it is product-favored at equilibrium as written:

$$E°_{cell} = E°_{cathode} − E°_{anode}$$

(a) $2\ I^−(aq) + Zn^{2+}(aq) \to I_2(s) + Zn(s)$

Cathode reaction: $Zn^{2+}(aq) + 2\ e^− \to Zn(s)$ $E° = −0.763$ V

Anode reaction: $2\ I^−(aq) \to I_2(s) + 2\ e^−$ $\underline{E° = +0.535\ V}$

             Cell voltage: $E° = −1.298$ V (reactant-favored)

(b) $Zn^{2+}(aq) + Ni(s) \to Zn(s) + Ni^{2+}(aq)$

Cathode reaction: $Zn^{2+}(aq) + 2\ e^− \to Zn(s)$ $E° = −0.763$ V

Anode reaction: $Ni(s) \to Ni^{2+}(aq) + 2\ e^−$ $\underline{E° = −0.25\ V}$

             Cell voltage: $E° = −0.51$ V (reactant-favored)

(c) $2\ Cl^−(aq) + Cu^{2+}(aq) \to Cu(s) + Cl_2(g)$

Cathode reaction: $Cu^{2+}(aq) + 2\ e^− \to Cu(s)$ $E° = +0.337$ V

Anode reaction: $2\ Cl^−(aq) \to Cl_2(g) + 2\ e^−$ $\underline{E° = +1.36\ V}$

             Cell voltage: $E° = −1.02$ V (reactant-favored)

(d) $Fe^{2+}(aq) + Ag^+(aq) \to Fe^{3+}(aq) + Ag(s)$

Cathode reaction: $Ag^+(aq) + e^− \to Ag(s)$ $E° = +0.799$ V

Anode reaction: $Fe^{2+}(aq) \to Fe^{3+}(aq) + e^−$ $\underline{E° = +0.771\ V}$

             Cell voltage: $E° = 0.028$ V (product-favored)

19.18. Calculate $E°$ for each of the following, and decided if it is product-favored at equilibrium as written:

$$E°_{cell} = E°_{cathode} − E°_{anode}$$

(a) $I_2(s) + Mg(s) \to Mg^{2+}(aq) + 2\ I^−(aq)$

Cathode reaction: $I_2(s) + 2\ e^− \to 2\ I^−(aq)$ $E° = +0.535$ V

Anode reaction: $Mg(s) + 2\ e^− \to Mg^{2+}(s)$ $\underline{E° = −2.37\ V}$

             Cell voltage: $E° = 2.91$ V (product-favored)

(b) $Zn^{2+}(aq) + Ni(s) \to Zn(s) + Ni^{2+}(aq)$

Cathode reaction: $Zn^{2+}(aq) + 2\ e^− \to Zn(s)$ $E° = −0.763$ V

Anode reaction: Ni(s) → Ni^{2+}(aq) + 2 e$^-$ $\quad\quad$ $E° = –0.25$ V

$\quad\quad\quad\quad\quad\quad\quad$ Cell voltage: $\quad\quad$ $E° = –0.51$ V (reactant-favored)

(c) Sn^{2+}(aq) + 2 Ag$^+$(aq) → Sn^{4+}(aq) + 2 Ag(s)

$\quad\quad$ Cathode reaction: 2 Ag$^+$(aq) + 2 e$^-$ → 2 Ag(s) $\quad\quad$ $E° = +0.799$ V

$\quad\quad$ Anode reaction: Sn^{2+}(aq) → Sn^{4+}(aq) + 2 e$^-$ $\quad\quad$ $E° = +0.15$ V

$\quad\quad\quad\quad\quad\quad\quad$ Cell voltage: $\quad\quad$ $E° = 0.65$ V (product-favored)

(d) 2 Zn(s) + O$_2$(g) + 2 H$_2$O(ℓ) + 4 OH$^-$(aq) → 2 [Zn(OH)$_4$]$^{2-}$(aq)

$\quad\quad$ Cathode reaction: O$_2$(g) + 2 H$_2$O(ℓ) + 4 e$^-$ → 4 OH$^-$(aq) $\quad\quad$ $E° = +0.40$ V

$\quad\quad$ Anode reaction: 2 Zn(s) + 8 OH$^-$(aq) → 2 [Zn(OH)$_4$]$^{2-}$(aq) + 2 e$^-$ $\quad\quad$ $E° = –1.22$ V

$\quad\quad\quad\quad\quad\quad\quad$ Cell voltage: (product favored) $\quad\quad$ $E° = 1.62$ V

19.19. Balance the equations, calculate $E°$, and decide if they are product-favored at equilibrium:
(a) Sn^{2+}(aq) + 2 Ag(s) → Sn(s) + 2 Ag$^+$(aq)

$\quad\quad$ Cathode reaction: Sn^{2+}(aq) + 2 e$^-$ → Sn(s) $\quad\quad$ $E° = –0.14$ V
$\quad\quad$ Anode reaction: 2 Ag(s) → 2 Ag$^+$(aq) + 2 e$^-$ $\quad\quad$ $E° = +0.799$ V

$\quad\quad\quad\quad\quad\quad\quad$ Cell voltage: $\quad\quad$ $E° = –0.94$ V (reactant-favored)

(b) 2 Al(s) + 3 Sn^{4+}(aq) → 3 Sn^{2+}(aq) + 2 Al^{3+}(aq)

$\quad\quad$ Cathode reaction: 3 Sn^{4+}(aq) + 6 e$^-$ → 3 Sn^{2+}(aq) $\quad\quad$ $E° = +0.15$ V

$\quad\quad$ Anode reaction: 2 Al(s) → 2 Al^{3+}(aq) + 6 e$^-$ $\quad\quad$ $E° = –1.66$ V
$\quad\quad\quad\quad\quad\quad\quad$ Cell voltage: $\quad\quad$ $E° = 1.81$ V (product-favored)

(c) ClO$_3^-$(aq) + 5 Ce^{3+}(aq) + 6 H$^+$(aq) → 1/2 Cl$_2$(g) + 5 Ce^{4+}(aq) + 3 H$_2$O(ℓ)

$\quad\quad$ Cathode reaction: ClO$_3^-$(aq) + 5 e$^-$ + 6 H$^+$(aq) →
$\quad\quad\quad\quad$ 1/2 Cl$_2$(g) + 3 H$_2$O(ℓ) $\quad\quad$ $E° = +1.47$ V

$\quad\quad$ Anode reaction: 5 Ce^{3+}(aq) → 5 Ce^{4+}(aq) + 5 e$^-$ $\quad\quad$ $E° = +1.61$ V
$\quad\quad\quad\quad\quad\quad\quad$ Cell voltage: $\quad\quad$ $E° = –0.14$ V (reactant-favored)

(d) 3 Cu(s) + 2 NO$_3^-$(aq) + 8 H$^+$(aq) →
$\quad\quad\quad$ 3 Cu^{2+}(aq) + 2 NO(g) + 4 H$_2$O(ℓ)

$\quad\quad$ Cathode reaction: 2 NO$_3^-$(aq) + 8 H$^+$(aq) + 6 e$^-$ →
$\quad\quad\quad\quad$ 2 NO (g) + 4 H$_2$O(ℓ) $\quad\quad$ $E° = +0.96$ V

$\quad\quad$ Anode reaction: 3 Cu(s) → 3 Cu^{2+}(aq) + 6 e$^-$ $\quad\quad$ $E° = +0.337$ V

$\quad\quad\quad\quad\quad\quad\quad$ Cell voltage: $\quad\quad$ $E° = 0.62$ V (product-favored)

19.20. Balance the equations, calculate $E°$, and decide if they are product-favored at equilibrium:

(a) $I_2(s) + 2\ Br^-(aq) \rightarrow 2\ I^-(aq) + Br_2(\ell)$

Cathode reaction: $I_2(s) + 2\ e^- \rightarrow 2\ I^-(aq)$		$E° = +0.535$ V
Anode reaction: $2\ Br^-(aq) \rightarrow Br_2(\ell) + 2\ e^-$		$E° = +1.08$ V
	Cell voltage:	$E° = –0.55$ V (reactant-favored)

(b) $2\ Fe^{2+}(aq) + Cu^{2+}(aq) \rightarrow 2\ Fe^{3+}(aq) + Cu(s)$

Cathode reaction: $Cu^{2+}(aq) + 2\ e^- \rightarrow Cu(s)$		$E° = +0.337$ V
Anode reaction: $2\ Fe^{2+}(aq) \rightarrow 2\ Fe^{3+}(aq) + 2\ e^-$		$E° = +0.771$ V
	Cell voltage:	$E° = –0.434$ V (reactant-favored)

(c) $6\ Fe^{2+}(aq) + Cr_2O_7^{2-}(aq) + 14\ H^+(aq) \rightarrow 6\ Fe^{3+}(aq) + 7\ H_2O(\ell) + 2\ Cr^{3+}(aq)$

Cathode reaction: $Cr_2O_7^{2-}(aq) + 14\ H^+(aq) + 6\ e^- \rightarrow$

$\qquad\qquad 2\ Cr^{3+}(aq) + 7\ H_2O(\ell)\qquad E° = +1.33$ V

Anode reaction: $6\ Fe^{2+}(aq) \rightarrow 6\ Fe^{3+}(aq) + 6\ e^-\qquad E° = +0.771$ V

$\qquad\qquad$ Cell voltage: $\qquad\qquad E° = +0.56$ V (product-favored)

(d) $2\ MnO_4^-(aq) + H^+(aq) + 5\ HNO_2(aq) \rightarrow 2\ Mn^{2+}(aq) + 5\ NO_3^-(aq) + 3\ H_2O(\ell)$

Cathode reaction: $2\ MnO_4^-(aq) + 16\ H^+(aq) + 10\ e^- \rightarrow$

$\qquad\qquad 2\ Mn^{2+}(aq) + 8\ H_2O(\ell)\qquad E° = +1.51$ V

Anode reaction: $5\ HNO_2(aq) + 5\ H_2O(\ell) \rightarrow$

$\qquad\qquad 5\ NO_3^-(aq) + 15\ H^+(aq) + 10\ e^-\quad E° = +0.94$ V

$\qquad\qquad$ Cell voltage: $\qquad\qquad E° = +0.57$ V (product-favored)

Ranking Oxidizing and Reducing Agents

19.21. From the following half-reactions:

(a) The metal most easily oxidized:

From the list **Al** is the most easily oxidized metal. Having the most negative reduction potential, Al is the strongest reducing agent of the group, and reducing agents are oxidized as they perform their task.

(b) Metals on the list capable of reducing Fe^{2+} to Fe:

Zn and **Al** have more negative reduction potentials than Fe, hence are stronger reducing agents, and can reduce Fe^{2+} to Fe.

(c) A balanced equation for the reaction of Fe^{2+} with Sn. Is the reaction product-favored at equilibrium?

$Fe^{2+}(aq) + Sn(s) \rightarrow Fe(s) + Sn^{2+}(aq)$; Since Fe is a stronger reducing agent than Sn, this reaction would have an $E° < 0$, and the reaction would be reactant-favored at equilibrium.

(d) A balanced equation for the reaction of Zn^{2+} with Sn. Is the reaction product-favored at equilibrium?

$Zn^{2+}(aq) + Sn(s) \rightarrow Zn(s) + Sn^{2+}(aq)$; Zn is a stronger reducing agent than Sn, this reaction would have an $E° < 0$, and the reaction would be reactant-favored at equilibrium.

19.22. From the following half-reactions:

(a) Strongest and weakest oxidizing agent:

MnO_4^- is the strongest oxidizing agent and SO_4^{2-} is the weakest oxidizing agent.

(b) Oxidizing agent capable of oxidizing Cr^{3+} to $Cr_2O_7^{2-}$:

Both BrO_3^- and MnO_4^- can oxidize Cr^{3+} to $Cr_2O_7^{2-}$

(c) Balanced equation for NO_3^- reacting with SO_2:

$NO_3^-(aq) + 4\ H^+(aq) + 3\ e^- \rightarrow NO(g) + 2\ H_2O(\ell)$ Cathode reaction

$3\ SO_2(g) + 6\ H_2O(\ell) \rightarrow 3\ SO_4^{2-}(aq) + 12\ H^+(aq) + 6\ e^-$ Anode reaction

$2\ NO_3^-(aq) + 2\ H_2O(\ell) + 3\ SO_2(g) \rightarrow 2\ NO(g) + 4\ H^+(aq) + 3\ SO_4^{2-}(aq)$

$E°_{cell} = E°_{cathode} - E°_{anode} = (+0.96\ V) - (+0.20\ V) = +0.76\ V$ (product-favored)

(d) Balanced equation for $Cr_2O_7^{2-}$ reacting with Mn^{2+}:

$5\ Cr_2O_7^{2-}(aq) + 70\ H^+(aq) + 30\ e^- \rightarrow 10\ Cr^{3+}(aq) + 35\ H_2O(\ell)$

$6\ Mn^{2+}(aq) + 24\ H_2O(\ell) \rightarrow 6\ MnO_4^-(aq) + 48\ H^+(aq) + 30\ e^-$

$5\ Cr_2O_7^{2-}(aq) + 22\ H^+(aq) + 6\ Mn^{2+}(aq) \rightarrow 10\ Cr^{3+}(aq)$

$+ 11\ H_2O(\ell) + 6\ MnO_4^-(aq)$

$E°_{cell} = E°_{cathode} - E°_{anode} = (+1.33\ V) - (+1.51\ V) = -0.18\ V$ (reactant-favored)

19.23. Element from the group that is the best reducing agent:

Cr has the most negative standard reduction potential of the group. Recall that the more negative the reduction potential, the stronger a substance is as a reducing agent.

19.24. Elements easier to oxidize than $H_2(g)$: (b) Zn, (c) Fe, and (e) Cr

19.25. Ion from the group that is most easily reduced:

The species with the most positive reduction potential is the strongest oxidizing agent of the list, and with that role, becomes the most easily reduced. Ag^+ fits that role from this list.

19.26. Ions that are more easily reduced than H^+: (a) Cu^{2+}(aq) and (d) Ag^+(aq)

19.27. Regarding the halogens:

(a) The halogen most easily reduced is the one with the most positive reduction potential. F_2 has the most reduction potential of the halogens.

(b) MnO_2 has a reduction potential of 1.23V. Both F_2 and Cl_2 have more positive reduction potentials than MnO_2 and are better oxidizing agents than MnO_2.

19.28. Identify the most easily oxidized ion in each case:
(a) Ion most easily oxidized to the elemental form: I^-(aq)
(b) Halide ions more easily oxidized than H_2O: Br^-(aq) and I^-(aq)

Electrochemical Cells Under Nonstandard Conditions

19.29. The voltage of a cell that has dissolved species at 0.025 M:

Calculate the standard voltage of the cell:

(1) $2\ H_2O(\ell) + 2\ e^- \rightarrow H_2(g) + 2\ OH^-(aq)$ $E° = -0.8277$ V

(2) $[Zn(OH)_4]^{2-}(aq) + 2\ e^- \rightarrow Zn(s) + 4\ OH^-(aq)$ $E° = -1.22$ V

The net equation is: $Zn(s) + 2\ H_2O(\ell) + 2\ OH^-(aq) \rightarrow [Zn(OH)_4]^{2-}(aq) + H_2(g)$

The equilibrium expression (Q) would be: $\dfrac{[Zn(OH)_4^-]P_{H_2}}{[OH^-]^2}$

The hydrogen pressure is 1.0 bar, and you know the concentrations of the other terms: 0.025 M. Note that n (in the Nernst equation) corresponds to 2, since the balanced overall equation indicates that 2 moles of electrons are lost and 2 moles of electrons are gained.

Calculate $E°_{cell} = E°_{cathode} - E°_{anode} = -0.8277 - (-1.22) = 0.39$ V

The Nernst equation: $E_{cell} = E°_{cell} - \dfrac{0.0257}{n} \ln \dfrac{[Zn(OH)_4^-]P_{H_2}}{[OH^-]^2}$

$$E_{cell} = 0.39 - \frac{0.0257}{2} \ln \frac{[0.025] \cdot 1}{[0.025]^2} = 0.39 - \frac{(0.0257 \cdot 3.69)}{2} = 0.34 \text{ V}$$

19.30. The voltage of a cell that has dissolved species at 0.20 M:

Anode reaction: $2 \text{ Fe}^{2+}(aq) \rightarrow 2 \text{ Fe}^{3+}(aq) + 2 \text{ e}^-$

Cathode reaction: $\text{H}_2\text{O}_2(aq) + 2 \text{ H}^+(aq) + 2 \text{ e}^- \rightarrow 2 \text{ H}_2\text{O}(\ell)$

$E°_{cell} = E°_{cathode} - E°_{anode} = (1.77 \text{ V}) - (0.771 \text{ V}) = +1.00 \text{ V}$

$$E_{cell} = E°_{cell} - \frac{0.0257}{n} \ln \frac{[\text{Fe}^{3+}]^2}{[\text{Fe}^{2+}]^2[\text{H}_2\text{O}_2][\text{H}^+]^2} = +1.00 \text{ V} - \frac{0.0257}{2} \ln \frac{(0.20)^2}{(0.20)^2(0.20)(0.20)^2}$$

$E_{cell} = 0.94 \text{ V}$

19.31. The voltage of a cell that has Ag in a 0.25 M solution of Ag$^+$ and Zn electrode in 0.010 M Zn^{2+}: Calculate the standard voltage of the cell:

(1) $\text{Zn}^{2+}(aq) + 2\text{e}^- \rightarrow \text{Zn}(s)$ $E° = -0.763 \text{ V}$

(2) $\text{Ag}^+(aq) + \text{e}^- \rightarrow \text{Ag}(s)$ $E° = +0.7994 \text{ V}$

The net equation is given as: $2 \text{ Ag}^+(aq) + \text{Zn}(s) \rightarrow \text{Zn}^{2+}(aq) + 2 \text{ Ag}(s)$.

The equilibrium expression (Q) would be: $\dfrac{[\text{Zn}^{2+}]}{[\text{Ag}^+]^2}$

The cell will run in the direction that is product favored, so you can calculate

$E°_{cell} = E°_{cathode} - E°_{anode} = +0.799 - (-0.763) = 1.5624 \text{ V}$

Also, the balanced equation shows that 2 moles (n) of electrons are transferred.

Using the Nernst equation $E_{cell} = E°_{cell} - \dfrac{0.0257}{n} \ln \dfrac{[\text{Zn}^{2+}]}{[\text{Ag}^+]^2}$

$$E_{cell} = 1.5624 - \frac{0.0257}{2} \ln \frac{[0.010]}{[0.25]^2} = 1.5624 - \frac{0.0257}{2}(-1.83) = 1.5624 + 0.0235 = 1.59 \text{ V}$$

19.32. Calculate the cell potential for the non-standard cell described:

$\text{Zn}(s) + \text{Cu}^{2+}(aq) \rightarrow \text{Zn}^{2+}(aq) + \text{Cu}(s)$

$E°_{cell} = E°_{cathode} - E°_{anode} = (0.337 \text{ V}) - (-0.763 \text{ V}) = +1.100 \text{ V}$

$$E_{cell} = E°_{cell} - \frac{0.0257}{n} \ln \frac{[\text{Zn}^{2+}]}{[\text{Cu}^{2+}]} = +1.100 \text{ V} - \frac{0.0257}{2} \ln \frac{0.40}{4.8 \times 10^{-3}} = 1.043 \text{ V}$$

19.33. Determine the [Ag$^+$] in a cell having a voltage of 1.48 V and composed of Ag wire electrode and Zn metal placed in 1.0 M Zn(NO$_3$)$_2$:

Calculate the standard voltage of the cell:

(1) Zn^{2+}(aq) + 2 e$^-$ → Zn(s) $\qquad$ $E°$ = –0.763 V

(2) Ag$^+$(aq) + e$^-$ → Ag(s) $\qquad$ $E°$ = +0.799 V

The net equation is given as: 2 Ag$^+$(aq) + Zn(s) → Zn^{2+}(aq) + 2 Ag(s).

The equilibrium expression (Q) would be: $\dfrac{[\text{Zn}^{2+}]}{[\text{Ag}^+]^2}$ The cell will run in the direction that is product favored, so you can calculate $E°_{cell} = E°_{cathode} - E°_{anode}$ = +0.799 – (–0.763) = 1.562 V

Also, the balanced equation shows that 2 moles(n) of electrons are transferred.

Using the Nernst equation: $E_{cell} = E°_{cell} - \dfrac{0.0257}{n} \ln \dfrac{[\text{Zn}^{2+}]}{[\text{Ag}^+]^2}$

Given the E_{cell} = 1.48 V, you should be able to calculate the value of the "ln term". Recall that you know the concentration of [Zn^{2+}], but don't know the [Ag$^+$].

$1.48 = 1.562 - \dfrac{0.0257}{2} \ln \dfrac{[\text{Zn}^{2+}]}{[\text{Ag}^+]^2}$ or $1.48 = 1.562 - \dfrac{0.0257}{2} \ln \dfrac{[\text{Zn}^{2+}]}{[\text{Ag}^+]^2}$

$\dfrac{-(1.48 - 1.562)2}{0.0257} = \ln \dfrac{[\text{Zn}^{2+}]}{[\text{Ag}^+]^2}$ or $6.381 = \ln \dfrac{[\text{Zn}^{2+}]}{[\text{Ag}^+]^2}$ giving $\dfrac{[\text{Zn}^{2+}]}{[\text{Ag}^+]^2} = 590.7$

So $\dfrac{1.0}{590.7}$ = [Ag$^+$]^2 and [Ag$^+$] = 0.040 M

19.34. Calculate the cell potential for the non-standard cell described:

Overall equation: Fe(s) + 2 H$^+$(aq) → Fe^{2+}(aq) + H$_2$(g)

$E°_{cell} = E°_{cathode} - E°_{anode}$ = (0.00 V) – (–0.44 V) = +0.44 V

E_{cell} = 0.49 V = 0.44 V – $\dfrac{0.0257}{2} \ln \dfrac{[\text{Fe}^{2+}]1.0}{(1.0)^2}$ and [Fe^{2+}] = 0.020 M

Electrochemistry, Thermodynamics, and Equilibrium

19.35. Calculate $\Delta_rG°$ and K for the reactions:

(a) $2\ Fe^{3+}(aq) + 2\ I^-(aq) \rightleftarrows 2\ Fe^{2+}(aq) + I_2(aq)$

Using the potentials: $Fe^{3+}(aq) + e^- \rightarrow 2\ Fe^{2+}$ $E° = 0.771\ V$
$2\ I^-(aq) \rightarrow I_2(aq) + 2\ e^-$ $E° = 0.621\ V$

$E°_{cell} = (0.771\ V - 0.621\ V) = 0.150\ V$

The relationship between $\Delta G°$ and $E°$ is: $\Delta G° = -nFE°$:

$\Delta_rG° = -(2\ mol\ e^-)(96,485\ C/mol\ e^-)(0.150\ V)$ and $1V = 1J/C$ so

$\Delta_rG° = -(2\ mol\ e^-)(96,485\ C/mol\ e^-)(0.150\ J/C) = -28940\ J$ or $-28.9\ kJ$

$\Delta_rG° = -RT\ln K$ so $-\Delta G°/RT = \ln K$

$-(-28940\ J)/[(8.3145\ J/mol·K)(298.15\ K)] = \ln K$ so $11.68 = \ln K$ and $K = 1 \times 10^5$

(b) $I_2(aq) + 2\ Br^-(aq) \rightarrow 2\ I^-(aq) + Br_2(\ell)$

Using the potentials: $Br_2(\ell) + 2\ e^- \rightarrow 2\ Br^-(aq)$ $E° = 1.08\ V$

$I_2(aq) + 2\ e^- \rightarrow 2\ I^-(aq)$ $E° = 0.621\ V$

You calculate an $E°_{cell}$, noting that in the cell reaction given, molecular iodine is being reduced (the cathode) and bromide ion is being oxidized (the anode).

$E°_{cell} = (0.621\ V - 1.08\ V) = -0.459\ V$

The relationship between $\Delta_rG°$ and $E°$ is: $\Delta_rG° = -nFE°$ so

$\Delta_rG° = -(2\ mol\ e^-)(96,485\ C/mol\ e^-)(-0.459\ V)$ and $1\ V = 1\ J/C$ so

$\Delta_rG° = -(2\ mol\ e^-)(96,485\ C/mol\ e^-)(-0.459\ J/C) = +88,573\ J$ or $+88.6\ kJ$ (3 sf)

$\Delta_rG° = -RT\ln K$ so $-\Delta G°/RT = \ln K$

$-(88573\ J)/[(8.3145\ J/mol·K)(298.15\ K)] = \ln K$ so $-35.73 = \ln K$ and $K = 3 \times 10^{-16}$

19.36. Calculate $\Delta_rG°$ and K:

(a) Cathode reaction: $Zn^{2+}(aq) + 2\ e^- \rightarrow Zn(s)$

Anode reaction: $2\ Ag(s) + 2\ Br^-(aq) \rightarrow 2\ AgBr(aq) + 2\ e^-$
$E°_{cell} = E°_{cathode} - E°_{anode} = (-0.763\ V) - (0.0713\ V) = -0.8343\ V$
$\Delta_rG° = -nFE° = -(2\ mol\ e^-)(96,485\ C/mol\ e^-)(-0.8343\ V)(1\ J/1\ C·V)(1\ kJ/10^3\ J)$

$= 161.0\ kJ = 161\ kJ$ (3 sf)

$\Delta_rG° = -RT\ln K$ so $-\Delta G°/RT = \ln K$

$-(1.610 \times 10^5\ J)/[(8.3145\ J/mol·K)(298.15\ K)] = \ln K$ so $-6.49 = \ln K$ and $K = 6.2 \times 10^{-29}$

(b) Anode reaction: Ni(s) → Ni²⁺(aq) + 2 e⁻
Cathode reaction: 2 Ag⁺(aq) + 2 e⁻ → 2 Ag(s)
$E°_{cell} = E°_{cathode} - E°_{anode}$ = (0.799 V) – (–0.25 V) = 1.049 V
$\Delta_r G° = -nFE°$ = –(2 mol e⁻)(96,485 C/mol e⁻)(1.049 V)(1 J/1 C·V)(1 kJ/10³ J)
= –202.4 kJ = –202 kJ (3 sf)

$\Delta_r G° = -RT \ln K$ so $-\Delta G°/RT = \ln K$

–(–2.024 × 10⁵ J)/[(8.3145 J/mol·K)(298.15 K)] = ln K so 81.7 = ln K and $K = 3 \times 10^{25}$

19.37. Calculate K_{sp} for AgBr using the following reactions:

(1) AgBr(s) + 1 e⁻ → Ag(s) + Br⁻(aq)   E° = 0.0713 V

(2) Ag⁺(aq) + 1e⁻ → Ag(s)   E° = 0.7994 V

Write the K_{sp} expression for AgBr: AgBr(s) ⇌ Ag⁺(aq) + Br⁻(aq)

Note that you can accomplish this as an overall reaction, by reversing equation (2) and adding that to equation (1). Equation (1) is presently written as a reduction (naturally) and the *reverse* of Equation (2) would be an oxidation—so the roles for the "cell" are defined.

$E°_{cell} = E°_{cathode} - E°_{anode}$ = 0.0713 V – 0.7994 V = –0.7281 V

Using the $\Delta_r G°$ relationships: $\Delta_r G° = -nFE°$ = –(1 mol e⁻)(96485 C/mol e⁻)(–0.7281 V)
= 7.025 × 10⁴ J

$\Delta_r G° = -RT \ln K$ so $-\Delta G°/RT = \ln K$

–(7.025 × 10⁴ J)/[(8.3145 J/mol·K)(298.15 K)] = ln K so –28.34 = ln K and $K = 4.9 \times 10^{-13}$

19.38. Calculate K_{sp} for Hg₂Cl₂:

Cathode reaction: Hg₂Cl₂(s) + 2 e⁻ → 2 Hg(ℓ) + 2 Cl⁻(aq)

Anode reaction: 2 Hg(ℓ) → Hg₂²⁺(aq) + 2 e⁻

Write the K_{sp} expression: Hg₂Cl₂(s) ⇌ Hg₂²⁺(aq) + 2 Cl⁻(aq)

$E°_{cell} = E°_{cathode} - E°_{anode}$ = (0.27 V) – (0.789 V) = –0.519 V

Using the $\Delta_r G°$ relationships: $\Delta_r G° = -nFE°$ = –(2 mol e⁻)(96485 C/mol e⁻)(–0.519 V)
= 1.00 × 10⁵ J

$\Delta_r G° = -RT \ln K$ so $-\Delta G°/RT = \ln K$

–(1.00 × 10⁵ J)/[(8.3145 J/mol·K)(298.15 K)] = ln K so –40.4 = ln K and $K = 3 \times 10^{-18}$

19.39. Calculate the $K_{formation}$ for AuCl₄⁻(aq)

(1) AuCl₄⁻(aq) + 3 e⁻ → Au(s) + 4 Cl⁻(aq)   E° = 1.00 V

(2) Au³⁺(aq) + 3 e⁻ → Au(s)   E° = 1.50 V

The formation reaction for the complex is: Au³⁺(aq) + 4 Cl⁻(aq) ⇌ AuCl₄⁻(aq)

To achieve this reaction as a net reaction, you need to reverse equation (1) and add it to equation (2).

The $E°_{cell}$ for that process is: $E°_{cathode} - E°_{anode} = 1.50 - 1.00 = 0.50$ V

Using the $\Delta_rG°$ relationships: $\Delta_rG° = -nFE° = -(3 \text{ mol e}^-)(96485 \text{ C/mol e}^-)(0.50 \text{ V})$
$$= -1.45 \times 10^5 \text{ J}$$

$\Delta_rG° = -RT\ln K$ so $-\Delta G°/RT = \ln K$

$-(-1.45 \times 10^5 \text{ J})/[(8.3145 \text{ J/mol·K})(298.15 \text{ K})] = \ln K$ so $58.4 = \ln K$ and $K = 2 \times 10^{25}$

19.40. Calculate the $K_{formation}$ for $[Zn(OH)_4]^{2-}$:

Anode reaction: $Zn(s) + 4 \text{ OH}^-(aq) \rightarrow [Zn(OH)_4]^{2-}(aq) + 2 \text{ e}^-$
Cathode reaction: $Zn^{2+}(aq) + 2 \text{ e}^- \rightarrow Zn(s)$

Net reaction: $Zn^{2+}(aq) + 4 \text{ OH}^-(aq) \rightleftarrows [Zn(OH)_4]^{2-}(aq)$

$E°_{cell} = E°_{cathode} - E°_{anode} = (-0.763 \text{ V}) - (-1.22 \text{ V}) = 0.457$ V

Using the $\Delta_rG°$ relationships: $\Delta_rG° = -nFE° = -(2 \text{ mol e}^-)(96485 \text{ C/mol e}^-)(0.457 \text{ V})$
$$= -8.82 \times 10^4 \text{ J}$$

$\Delta_rG° = -RT\ln K$ so $-\Delta G°/RT = \ln K$

$-(-8.82 \times 10^4 \text{ J})/[(8.3145 \text{ J/mol·K})(298.15 \text{ K})] = \ln K$ so $35.6 = \ln K$ and $K = 3 \times 10^{15}$

Electrolysis

19.41. Diagram of an electrolysis apparatus for molten NaCl:

Reactions occurring:

$2 \text{ Na}^+(\ell) + \text{e}^- \rightarrow \text{Na}(\ell)$

(cathode, reduction)

$2 \text{ Cl}^-(\ell) \rightarrow \text{Cl}_2(g) + 2 \text{ e}^-$

(anode, oxidation)

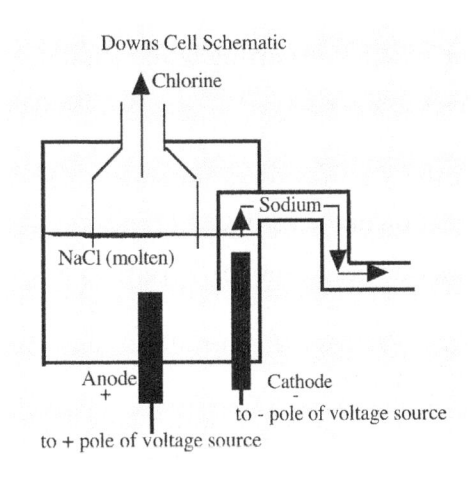

19.42. Apparatus used to electrolyze $CuCl_2(aq)$:

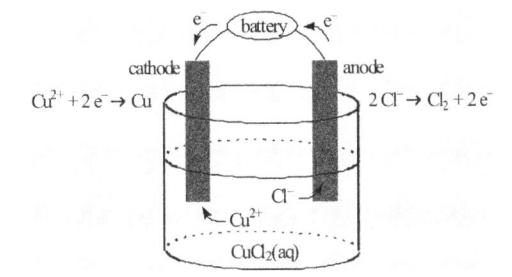

19.43. For the electrolysis of a solution of KF(aq), what product is expected at

the anode: O_2 or F_2: Figure 19.21 provides some help with this concept. The **bottom line** is that the process that occurs in an electrolysis is the one requiring the smaller applied potential. For electrolysis: $E°_{cell} = E°_{cathode} - E°_{anode}$. For aqueous KF, those voltages are:

2 F⁻ (aq) → F_2(g) + 2 e⁻ $\qquad E° = +2.87$ V

2 H_2O(ℓ) → O_2(g) + 4 H⁺(aq) + 4 e⁻ $\qquad E° = +1.23$ V

At the cathode: 2 H_2O(ℓ) + 2 e⁻ → H_2(g) + 2 OH⁻(aq) $\quad E° = –0.83$ V

[K has such a large negative reduction potential, that it *will not be reduced*.]

So, the two choices are: $E°_{cell} = E°_{cathode} - E°_{anode}$

For fluorine oxidation: $E°_{cell} = –0.83 – (+2.87) = –3.7$ V

For oxygen oxidation: $E°_{cell} = –0.83 – (+1.23) = –2.06$ V

Oxygen oxidation will require the lower applied potential and will be produced at the anode.

19.44. More likely product to appear at the cathode in the electrolysis of $CaCl_2$:
Ca^{2+} is much more difficult to reduce than water, so H_2 is more likely to be formed at the cathode.

19.45. For the electrolysis of KBr (aq):

(a) The reaction occurring at the cathode: 2 H_2O(ℓ) + 2 e⁻ → H_2(g) + 2 OH⁻(aq)

(b) The reaction occurring at the anode: 2 Br⁻(aq) → Br_2(ℓ) + 2 e⁻

19.46. For the electrolysis of aqueous Na_2S:

(a) Reaction at the cathode: 2 H_2O(ℓ) + 2 e⁻ → H_2(g) + 2 OH⁻(aq)

(b) Reaction at the anode: S^{2-}(aq) → S(s) + 2 e⁻

Counting Electrons

19.47. Solutions to problems of this sort are best solved by beginning with a factor containing the desired units. Connecting this factor to data provided usually gives a direct path to the answer.

units desired
↓

$12.2 \text{ min} \left(\dfrac{60 \text{ s}}{1 \text{ min}}\right)\left(\dfrac{0.150 \text{ C}}{1 \text{ s}}\right)\left(\dfrac{1 \text{ mol e}^-}{96485 \text{ C}}\right)\left(\dfrac{1 \text{ mol Ni}}{2 \text{ mol e}^-}\right)\left(\dfrac{58.693 \text{ g}}{1 \text{ mol Ni}}\right) = 0.0334$ g Ni

The second factor ($\frac{1 \text{ mol Ni}}{2 \text{ mol e}^-}$) is arrived at from the reduction half-reaction:

$Ni^{2+}(aq) + 2 e^- \rightarrow Ni(s)$.

All other factors are either data or common unity factors (e.g., $\frac{60 \text{ s}}{1 \text{ min}}$).

19.48. Mass of Ag that forms in 3.50 hours with a current of 1.50 amperes:
Charge = current × time = (1.50 A)(3.50 h)(60.0 min/h)(60.0 s/min) = 1.89×10^4 C

$\text{mol e}^- = (1.89 \times 10^4 \text{ C}) \left(\frac{1 \text{ mol e}^-}{96,500 \text{ C}} \right) = 0.196 \text{ mol e}^-$

$\text{mass of Ag} = 0.196 \text{ mol e}^- \left(\frac{1 \text{ mol Ag}}{1 \text{ mol e}^-} \right) \left(\frac{107.87 \text{ g Ag}}{1 \text{ mol Ag}} \right) = 21.1 \text{ g Ag}$

19.49. This is like SQ 19.47, except the problem is solved in reverse.

$0.50 \text{ g Cu} \left(\frac{1 \text{ mol Cu}}{63.546 \text{ g}} \right) \left(\frac{2 \text{ mol e}^-}{1 \text{ mol Cu}} \right) \left(\frac{96485 \text{ C}}{1 \text{ mol e}^-} \right) \left(\frac{1 \text{ s}}{0.66 \text{ C}} \right) = 2300 \text{ s}$

19.50. Time to prepare 2.5 g Zn through electrolysis:

$\text{mol e}^- = (2.5 \text{ g Zn}) \frac{1 \text{ mol Zn}}{65.39 \text{ g Zn}} \cdot \frac{2 \text{ mol e}^-}{1 \text{ mol Zn}} = 0.076 \text{ mol e}^-$

$\text{charge} = (0.076 \text{ mol e}^-) \frac{96500 \text{ C}}{1 \text{ mol e}^-} = 7.4 \times 10^3 \text{ C}$

time = 7.4×10^3 C/2.12 amperes = 3500 s (or 58 min)

19.51. Voltaic cell made of aluminum and oxygen from air:

$55 \text{ g Al} \left(\frac{1 \text{ mol Al}}{26.982 \text{ g/mol}} \right) \left(\frac{3 \text{ mol e}^-}{1 \text{ mol Al}} \right) \left(\frac{96500 \text{ C}}{1 \text{ mol e}^-} \right) = 5.9 \times 10^5 \text{ C}$

19.52. Mass of Cd to produce 0.25 amperes in 1.00 hour:

charge = current × time =

$(0.25 \text{ ampere})(1.00 \text{ hr}) \left(\frac{60.0 \text{ min}}{1 \text{ hr}} \right) \left(\frac{60.0 \text{ s}}{1 \text{ min}} \right) \left(\frac{1 \text{ C}}{1 \text{ amp} \cdot \text{s}} \right) = 9.0 \times 10^2 \text{ C}$

$$\text{mass of Cd} = \left(9.0\times 10^2 \text{ C}\right)\left(\frac{1 \text{ mol e}^-}{96500 \text{ C}}\right)\left(\frac{1 \text{ mol Cd}}{2 \text{ mol e}^-}\right)\left(\frac{112.41 \text{ g Cd}}{1 \text{ mol Cd}}\right) = 0.52 \text{ g Cd}$$

Corrosion: Redox Reactions in the Environment

19.53. Which of the following metals, coated on Fe, will provide cathodic protection to corrosion:

For the metals given below, you seek metals that are more readily oxidized than iron.

$Fe^{2+} + 2 \text{ e}^- \rightarrow Fe(s)$ $E° = -0.44$ V

$Mg^{2+} + 2 \text{ e}^- \rightarrow Mg(s)$ $E° = -2.37$ V

$Ni^{2+} + 2 \text{ e}^- \rightarrow Ni(s)$ $E° = -0.25$ V

$Sn^{2+} + 2 \text{ e}^- \rightarrow Sn(s)$ $E° = -0.14$ V

$Cu^{2+} + 2 \text{ e}^- \rightarrow Cu(s)$ $E° = +0.337$ V

From this list, only Mg can provide cathodic protection to Fe.

19.54. Which of the following metals, coated on Ni, will provide cathodic protection to corrosion:

For the metals given below, you seek metals that are more readily oxidized than nickel.

$Ni^{2+} + 2 \text{ e}^- \rightarrow Ni(s)$ $E° = -0.25$ V

$Mg^{2+} + 2 \text{ e}^- \rightarrow Mg(s)$ $E° = -2.37$ V

$Cr^{2+} + 2 \text{ e}^- \rightarrow Cr(s)$ $E° = -0.91$ V

$Zn^{2+} + 2 \text{ e}^- \rightarrow Zn(s)$ $E° = -0.763$ V

$Cu^{2+} + 2 \text{ e}^- \rightarrow Cu(s)$ $E° = +0.337$ V

All the metals except Cu will provide cathodic protection to corrosion to Ni.

19.55. Calculate $E°_{cell}$ for the corrosion of Fe, and decide if it is product-favored at equilibrium:

$Fe^{2+} + 2 \text{ e}^- \rightarrow Fe(s)$ $E° = -0.44$ V

$O_2(g) + 2 H_2O(\ell) + 4 \text{ e}^- \rightarrow 4 \text{ OH}^-(aq)$ $E° = 0.40$ V

$E°_{cell} = E°_{cathode} - E°_{anode} = (0.40 \text{ V}) - (-0.44 \text{ V}) = +0.84$ V

The + value for $E°_{cell}$ indicates that the reaction is product-favored at equilibrium.

Decreasing the pH (lowering the concentration of OH⁻, a product) will favor the reaction even more, so it will make the reaction *more* product-favored at equilibrium. Le Chatelier's principle in action.

19.56. In the presence of O_2 and acid, calculate $E°_{cell}$ for the corrosion of Fe, and decide if it is product-favored at equilibrium:

$Fe^{2+} + 2\ e^- \rightarrow Fe(s)$ $E° = -0.44$ V

(Note: For this process, the reverse reaction will occur.)

$O_2(g) + 4\ H^+(aq) + 4\ e^- \rightarrow 2\ H_2O(\ell)$ $E° = +1.229$ V

$E°_{cell} = E°_{cathode} - E°_{anode} = (+1.229\ V) - (-0.44\ V) = +1.669$ V

The + value for $E°_{cell}$ indicates that the reaction is product-favored at equilibrium.

Decreasing the pH (increasing the concentration of H^+, a reactant) will favor the reaction even more, so it will make the reaction *more* product-favored at equilibrium.

General Questions

19.57. Balanced equations for the following half-reactions:

(a) $UO_2^+(aq) \rightarrow U^{4+}(aq)$ (acid solution)

Reduction half-equation	
$UO_2^+(aq) \rightarrow U^{4+}(aq)$	Balance all non H, O
$UO_2^+(aq) \rightarrow U^{4+}(aq) + 2\ H_2O(\ell)$	Balance O with H_2O
$4\ H^+(aq) + UO_2^+(aq) \rightarrow U^{4+}(aq) + 2\ H_2O(\ell)$	Balance H with H^+
$4\ H^+(aq) + UO_2^+(aq) + 1\ e^- \rightarrow U^{4+}(aq) + 2\ H_2O(\ell)$	Balance charge with e^-

(b) $ClO_3^-(aq) \rightarrow Cl^-(aq)$ (acid solution)

Reduction half-equation	
$ClO_3^-(aq) \rightarrow Cl^-(aq)$	Balance all non H, O
$ClO_3^-(aq) \rightarrow Cl^-(aq) + 3\ H_2O(\ell)$	Balance O with H_2O
$6\ H^+(aq) + ClO_3^-(aq) \rightarrow Cl^-(aq) + 3\ H_2O(\ell)$	Balance H with H^+
$6\ H^+(aq) + ClO_3^-(aq) + 6\ e^- \rightarrow Cl^-(aq) + 3\ H_2O(\ell)$	Balance charge with e^-

(c) $N_2H_4(aq) \rightarrow N_2(g)$ (basic solution)

Oxidation half-equation	
$N_2H_4(aq) \rightarrow N_2(g)$	Balance all non H, O
$N_2H_4(aq) \rightarrow N_2(g) + 2\ H_2O(\ell)$	Balance H with H_2O
$4\ OH^-(aq) + N_2H_4(aq) \rightarrow N_2(g) + 4\ H_2O(\ell)$	Balance O and H with OH^-
$4\ OH^-(aq) + N_2H_4(aq) \rightarrow N_2(g) + 4\ H_2O(\ell) + 4\ e^-$	Balance charge with e^-

(d) $ClO^-(aq) \rightarrow Cl^-(aq)$ (basic solution)

Reduction half-equation	
$ClO^-(aq) \rightarrow Cl^-(aq)$	Balance all non H, O
$ClO^-(aq) \rightarrow Cl^-(aq) + OH^-(aq)$	Balance O with OH^-
$H_2O(\ell) + ClO^-(aq) \rightarrow Cl^-(aq) + OH^-(aq) + OH^-(aq)$	Balance H with H_2O rebalance H with OH^-
$H_2O(\ell) + ClO^-(aq) + 2\ e^- \rightarrow Cl^-(aq) + 2\ OH^-(aq)$	Balance charge with e^-

19.58. Balance the following equations:

These equations are balanced with the steps noted explicitly in SQ19.3 and SQ19.5. In the cases below, the oxidation is given first, and the reduction second. Numbers of electrons needed to balance electron gain and loss are designated in front of the appropriate half-reaction (e.g. 2[ ). The overall balanced equation is given in the third line of each equation.

(a) $Zn(s) \rightarrow Zn^{2+}(aq) + 2\ e^-$

   $2[VO^{2+}(aq) + 2\ H^+(aq) + e^- \rightarrow V^{3+}(aq) + H_2O(\ell)]$

   $Zn(s) + 2\ VO^{2+}(aq) + 4\ H^+(aq) \rightarrow Zn^{2+}(aq) + 2\ V^{3+}(aq) + 2\ H_2O(\ell)$

(b) $3[Zn(s) \rightarrow Zn^{2+}(aq) + 2\ e^-]$

   $2[VO_3^-(aq) + 6\ H^+(aq) + 3\ e^- \rightarrow V^{2+}(aq) + 3\ H_2O(\ell)]$

   $3\ Zn(s) + 2\ VO_3^-(aq) + 12\ H^+(aq) \rightarrow 3\ Zn^{2+}(aq) + 2\ V^{3+}(aq) + 6\ H_2O(\ell)$

(c) Zn(s) + 2 OH⁻(aq) → Zn(OH)₂(s) + 2 e⁻

ClO⁻(aq) + H₂O(ℓ) + 2 e⁻ → Cl⁻(aq) + 2 OH⁻(aq)

Zn(s) + OCl⁻(aq) + H₂O(ℓ) → Zn(OH)₂(s) + Cl⁻(aq)

(d) 3[ClO⁻(aq) + H₂O(ℓ) + 2 e⁻ → Cl⁻(aq) + 2 OH⁻(aq)]

2[[Cr(OH)₄]⁻(aq) + 4 OH⁻(aq) → CrO₄²⁻(aq) + 4 H₂O(ℓ) + 3 e⁻]

3 ClO⁻(aq) + 2 [Cr(OH)₄]⁻(aq) + 2 OH⁻(aq) → 3 Cl⁻(aq) + 2 CrO₄²⁻(aq) + 5 H₂O(ℓ)

19.59. For the electrochemical cell involving Mg and Ag:

(a) Parts of the cell:

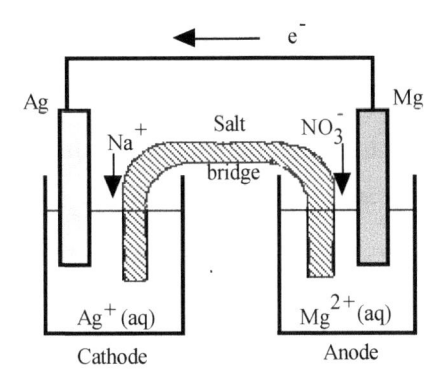

(b) Anode (oxidation):

Mg → Mg²⁺ + 2 e⁻

Cathode (reduction):

2 Ag⁺ + 2 e⁻ → 2 Ag

Net reaction:

2 Ag⁺ + Mg → Mg²⁺ + 2 Ag

(c) The flow of electrons in the outer circuit, as described on the diagram above is from the Mg electrode to the Ag electrode. The ion flow in the salt bridge is also shown on the diagram above. The salt bridge is necessary to negate the charge differential that would grow as the cell operates. For example, in the Ag half-cell, the compartment that originally contained equal amounts of + and – charges (from Ag⁺ and NO₃⁻ ions respectively) would accrue a net "−" charge, as the "+" silver ions are reduced. The Na⁺ ions from the salt bridge would flow into the Ag compartment, neutralizing that growing negative charge.

19.60. Construct a voltaic cell in which overall voltage is ~1.5 V and ~ 0.50 V:

The half-reaction for the reduction of Zn:

$$Zn^{2+}(aq) + 2\ e^- \rightarrow Zn(s) \qquad E° = -0.763\ V$$

(a) Zn as a cathode: $E°_{anode} = E°_{cathode} - E°_{cell} = -0.763\ V - (1.5\ V) = -2.3\ V$

Magnesium (–2.37 V) would be an appropriate choice

$E°_{cell} = E°_{cathode} - E°_{anode} = -0.763\ V - (-2.37\ V) = 1.61\ V$

Zn as an anode: $E°_{cathode} = E°_{cell} + E°_{anode} = 1.5\ V + (-0.763\ V) = 0.74\ V$

SbCl$_6^-$ ($E° = 0.75$ V) would be an appropriate choice.

$E°_{cell} = E°_{cathode} - E°_{anode} = 0.75\ V - (-0.763\ V) = 1.51\ V$

(b) Zn as a cathode: $E°_{anode} = E°_{cathode} - E°_{cell} = -0.763\ V - 0.5\ V = -1.3\ V$

CdS(s) ($E° = -1.21$ V) is an appropriate choice.

$E°_{cell} = E°_{cathode} - E°_{anode} = -1.21\ V + (-0.763\ V) = +0.45\ V$ for Ni

Zn as an anode: $E°_{cathode} = E°_{cell} + E°_{anode} = 0.5\ V + (-0.763\ V) = -0.26\ V$

Nickel ($E° = -0.25$) or vanadium ($E° = -0.255$ V) are appropriate choices.

$E°_{cell} = E°_{cathode} - E°_{anode} = -0.25\ V + (+0.763\ V) = +0.51\ V$ for Ni

$E°_{cell} = E°_{cathode} - E°_{anode} = -0.255\ V + (+0.763\ V) = +0.508\ V$ for V

19.61. Electrochemical cells with specific voltages:

(a) Half-cells that might be used in conjunction with the Ag$^+$/Ag half-cell to produce a cell with voltage close to 1.7 V. Consider cells in which the silver half-cell could function either as cathode or anode. $E°$ Ag$^+$ = 0.799 V

Recall that $E°_{cell} = E°_{cathode} - E°_{anode}$.

With the desired $E°$ cell to be 1.7 V, you can substitute the Ag/Ag$^+$ half-cell first as cathode, then as anode.

$E°_{cell} = E°_{cathode} - E°_{anode}$
1.7 V = 0.799 V $- E°_{anode}$, and $E°_{anode} = 0.799 - 1.7 = -0.90$ V

The chromium half-cell, with $E° = -0.91$ V, is an appropriate half-cell, and the $E°_{cell}$ would be: $E°_{cell} = 0.799\ V - (-0.91\ V) = +1.71\ V$

Now for a half-cell in which Ag is the anode:
$E°_{cell} = E°_{cathode} - E°_{anode}$

1.7 V = $E°_{cathode} - 0.799$ V and $E°_{cathode} = 0.799 + 1.7 = 2.5$ V
The textbook does not contain any standard reduction potentials in the vicinity of 2.5 V.

(b) Half-cells that might be used in conjunction with the Ag$^+$/Ag half-cell to produce a cell with voltage close to 0.5 V. Consider cells in which the silver half-cell could function either as cathode or anode. $E°$ Ag$^+$ = 0.799 V. As in (a) above, you can substitute the Ag/Ag$^+$ half-cell first as cathode, then as anode, for a $E°$ cell = 0.5 V

(i) $E°_{cell} = E°_{cathode} - E°_{anode}$ so 0.5 V = 0.799 V − $E°_{anode}$, and

$E°_{anode}$ = 0.799 − 0.5 = 0.3 V

The Cu couple half-cell with E° = 0.337 V is a possible half-cell. The $E°_{cell}$ for that is: $E°_{cell}$ = 0.799 − 0.337 = 0.462 V

Another possibility involving silver as the cathode is to oxidize Hg(ℓ) (E° = 0.27 V) to Hg_2Cl_2(s) at the anode ($E°_{cell}$ = 0.53 V).

(ii) Substituting with the Ag couple as anode:

$E°_{cell}$ = 0.5 V = $E°_{cathode}$ − 0.799 V, and $E°_{cathode}$ = 0.5 + 0.799 = 1.3 V.
One possibility involving silver as the anode is to reduced $N_2H_5^+$(aq) (E° = 1.24 V) to NH_4^+(aq) at the cathode ($E°_{cell}$ = 0.44 V).

Another possibility involving silver as the anode is to reduced $Cr_2O_7^{2-}$(aq) (E° = 1.33 V) to Cr^{3+}(aq) at the cathode ($E°_{cell}$ = 0.53 V).

19.62. Which of the reactions are product-favored at equilibrium:
(a) Cu(s) + I_2(s) → Cu^{2+}(aq) + 2 I^-(aq); $E°_{cell}$ = 0.621 V − 0.337 V = 0.284 V
$E°_{cell}$ > 0 product-favored
(b) 2 Cl^-(aq) + I_2(s) → Cl_2(g) + 2 I^-(aq); $E°_{cell}$ = 0.621 V − 1.36 V = −0.739 V
$E°_{cell}$ < 0 reactant-favored
(c) 2 K^+(aq) + 2 Cl^-(aq) → Cl_2(g) + 2 K(s); $E°_{cell}$ = −2.925 V − 1.36 V = −4.29 V
$E°_{cell}$ < 0 reactant-favored
(d) 2 K(s) + 2 H_2O(ℓ) → 2 K^+(aq) + H_2(g) + 2 OH^-(aq); $E°_{cell}$ = −0.8277 V − (−2.925) V
= 2.097 V

$E°_{cell}$ > 0 product-favored

19.63. Examine the reduction potentials for:

Au^+(aq) + 1 e^- → Au(s)   E° = +1.68 V

Ag^+(aq) + 1 e^- → Ag(s)   E° = +0.799 V

Cu^{2+}(aq) + 2 e^- → Cu(s)   E° = +0.337 V

Sn^{2+}(aq) + 2 e^- → Sn(s)   E° = −0.14 V

Co^{2+}(aq) + 2 e^- → Co(s)   E° = −0.28 V

Zn^{2+}(aq) + 2 e^- → Zn(s)   E° = −0.763 V

To clarify the trends, values are listed in the descending reduction potential typical of Reduction Potential Charts, as in Table 19.1 and Appendix M.

(a) Zn^{2+} is the weakest oxidizing agent.

Strength of oxidizing agents increases with *increasing* + E°.

(b) Au⁺ is the strongest oxidizing agent.

(c) Zn(s) is the strongest reducing agent.

Since its oxidized partner is the weakest oxidizing agent, Zn metal is the strongest reducing agent.

(d) Au(s) is the weakest reducing agent. See (c).

(e) Will Sn(s) reduce Cu^{2+} to Cu(s)? Yes. Use the "NW to SE" rule to see that this will give a positive $E°_{cell}$.

(f) Will Ag(s) reduce Co^{2+}(aq) to Co(s)? No. See part (e)

(g) Which metal ions from the list can be reduced by Sn(s)?
Using the logic as in part (e) Au⁺, Ag⁺, and Cu^{2+} can be reduced by Sn.

(h) Metals that can be oxidized by Ag⁺(aq) –Any metal "below" Ag: Cu, Sn, Co, Zn

19.64. Examine the reduction potentials for:

$F_2(g) + 2 e^- \rightarrow 2 F^-(aq)$ $E° = +2.87$ V

$Cl_2(g) + 2 e^- \rightarrow 2 Cl^-(aq)$ $E° = +1.36$ V

$O_2(g) + 4 H^+(aq) + 4 e^- \rightarrow 2 H_2O(\ell)$ $E° = +1.229$ V

$Br_2(\ell) + 2 e^- \rightarrow 2 Br^-(aq)$ $E° = +1.08$ V

$I_2(s) + 2 e^- \rightarrow 2 I^-(aq)$ $E° = +0.535$ V

$S(s) + 2 H^+(aq) + 2 e^- \rightarrow H_2S(aq)$ $E° = +0.14$ V

$Se(s) + 2 H^+(aq) + 2 e^- \rightarrow H_2Se(aq)$ $E° = -0.40$ V

To clarify the trends, values are listed in the descending reduction potential typical of Reduction Potential Charts, as in Table 19.1 and Appendix M.

(a) Se(s) is the weakest oxidizing agent. Strength of oxidizing agents
increases with the *increasing* +$E°$

(b) F⁻(aq) is the weakest reducing agent. Strongest oxidizing agent has reduced form that is the weakest reducing agent.

(c) $Cl_2(g)$ and $F_2(g)$ can oxidize H_2O to O_2.

(d) $F_2(g)$, $Cl_2(g)$, $O_2(g)$, $Br_2(\ell)$, and $I_2(s)$ can oxidize H_2S to S.

(e) O_2 capable of oxidizing I⁻? Yes See (a).

(f) S capable of oxidizing I⁻? No See (a).

(g) Is $H_2S(aq) + Se(s) \rightarrow 2 H_2Se(aq) + S(s)$ product-favored at equilibrium? No
As written half-equations produces cell with $E° < 0$.

(h) Is $H_2S(aq) + I_2(s) \rightarrow 2 H^+(aq) + 2 I^-(aq) + S(s)$ product-favored at equilibrium? Yes As written half-equations produces cell with $E° > 0$.

19.65. Examine the following reductions $E°$

$Cu^{2+}(aq) + 2\ e^- \rightarrow Cu(s)$ +0.337 V

$Fe^{2+}(aq) + 2\ e^- \rightarrow Fe(s)$ –0.44 V

$Cr^{3+}(aq) + 3\ e^- \rightarrow Cr(s)$ –0.74 V

$Mg^{2+}(aq) + 2\ e^- \rightarrow Mg(s)$ –2.37 V

(a) In which of the voltaic cells would the S.H.E. be the cathode?

$E°_{cell}$ must be positive, and the $E°$ S.H.E.= 0.00 V, so for $E°_{cell}$ = 0.00 V – ($E°_{anode}$) to be positive, the $E°_{anode}$ would have to be *negative*, so the iron, chromium, and magnesium half-cells fit this description.

(b) Voltaic cell with the highest and lowest potentials?

For $E°_{cell}$ = 0.00 V – ($E°_{anode}$) to have the largest positive value the $E°_{anode}$ would have to be the most negative (magnesium).

For $E°_{cell}$ = 0.00 V – ($E°_{anode}$) to have the smallest value the $E°_{anode}$ would have to be the least negative (iron). While Cu^{+2} is a tempting choice, note that the Cu couple, when paired with the S.H.E. would give a negative voltage, unless the SHE is the anode, in which case Cu^{2+} is the half-cell that produces the lowest potential.

19.66. Concerning the combination of the half-cells to make a voltaic cell:
(a) Copper serves as cathode: Cu-Zn and Cu-Ni
Nickel serves as anode: Ag-Ni and Cu-Ni
(b) Ag-Zn: $E°_{cell}$ = (0.799 V) – (–0.763 V) = 1.562 V
Cu-Ni: $E°_{cell}$ = (–0.337 V) – (0.25 V) = –0.587 V

19.67. Mass of Al metal produced from electrolysis of Al^{3+} salt with 5.0 V and 1.0×10^5 amperes in 24 hr.

$$1.0 \times 10^5\ \text{amp} \left(\frac{3600\ \text{s}}{1\ \text{hr}}\right)\left(\frac{24\ \text{hr}}{1\ \text{day}}\right)\left(\frac{26.982\ \text{g Al}}{1\ \text{mol Al}}\right)\left(\frac{1\ \text{mol Al}}{3\ \text{mol e}^-}\right)\left(\frac{1\ \text{mol e}^-}{96500\ \text{C}}\right)\left(\frac{1\ \text{C}}{1\ \text{amp} \times \text{s}}\right) = 8.1 \times 10^5\ \text{g Al/day}$$

19.68. For the cell described:

(a) reaction observed at the cathode: $Ag^+(aq) + e^- \rightarrow Ag(s)$

reaction observed at the anode: $2\ Ag(s) + SO_4^{2-}(aq) \rightarrow Ag_2SO_4(s) + 2\ e^-$

net reaction: $2\ Ag^+(aq) + SO_4^{2-}(aq) \rightarrow Ag_2SO_4(s)$

cell voltage: $E°_{cell} = E°_{cathode} - E°_{anode}$ = 0.799 V – 0.653 V = 0.146 V

(b) Calculate the K_{sp} for $Ag_2SO_4(s)$:
$Ag_2SO_4(s) \rightleftarrows 2\ Ag^+(aq) + SO_4^{2-}(aq)$ $E°_{cell} = -0.146$ V

$$\ln K = \frac{nE^0}{0.0257\ \text{V}} = \frac{(2)(-0.146\ \text{V})}{0.0257\ \text{V}} = -11.4 \text{ and solving for } K: K_{sp} = 1 \times 10^{-5}$$

19.69. For the cell described:
(a) $E°_{cell} = 0.142$ V $= E°_{cathode} - E°_{anode}$, so 0.142 V $= -0.126 - E°_{anode}$
[Note that the value, –0.126 V, came from the Reduction Potential Table]
Rearranging to solve: $E°_{anode} = -0.126 - 0.142 = -0.268$ V
(b) Given these data, estimate K_{sp} for PbCl$_2$.

Using Equation 19.3, you calculate the equilibrium constant (Q):
$$E_{cell} = E°_{cell} - \frac{0.0257}{n} \ln Q$$
If the system is at equilibrium, $E_{cell} = 0$, and $Q = K_{sp}$.
$$0.0 = -0.142 - \frac{0.0257}{2} \ln K_{sp} \text{ and rearranging: } \frac{-(0.142)2}{0.0257} = \ln K_{sp}$$
$\ln K_{sp} = -11.05$ and solving for K: $K_{sp} = 1.59 \times 10^{-5} = 2 \times 10^{-5}$ (1 sf)

19.70. What is the value of $E°$ for the half-reaction:
Identify two half-reactions that, when added, give the desired reaction:
The oxidation half-reaction: $2[Ag(s) \rightarrow Ag^+(aq) + e^-]$
The reduction half-reaction: $Ag_2CrO_4(s) + 2 e^- \rightarrow 2 Ag(s) + CrO_4^{2-}(aq)$
The net equation: $Ag_2CrO_4(s) \rightarrow 2 Ag^+(aq) + CrO_4^{2-}(aq)$
$$E° = \frac{0.0257 \text{ V}}{n} \ln K_{sp} = \frac{0.0257 \text{ V}}{2} \ln(1.1 \times 10^{-12}) = -0.35 \text{ V}$$
$E°_{cathode} = E°_{cell} + E°_{anode} = -0.35$ V $+ 0.799$ V $= 0.45$ V

19.71. What is $\Delta_r G°$ for the reaction, given that $E°_{cell} = +2.12$V
Note that the # of electrons (n) would be 2 (Zn $\rightarrow$ Zn^{2+})

$$\Delta_r G° = -nFE° = -2 \text{ mole e}^- \left(\frac{96500 \text{ C}}{1 \text{ mol e}^-}\right)(+2.12 \text{ V})\left(\frac{1 \text{ J}}{1 \text{ V} \cdot \text{C}}\right)\left(\frac{1 \text{ kJ}}{1000 \text{ J}}\right) = -409 \text{ kJ}$$

19.72. Number of kwh needed to produce 1 metric ton of Al:

$$1.0 \times 10^3 \text{ kg}\left(\frac{10^3 \text{ g}}{1 \text{ kg}}\right)\left(\frac{1 \text{ mol Al}}{26.982 \text{ g}}\right)\left(\frac{3 \text{ mol e}^-}{1 \text{ mol Al}}\right)\left(\frac{96,500 \text{ C}}{1 \text{ mol e}^-}\right)(5.0 \text{ V})\left(\frac{1 \text{ J}}{1 \text{ V} \cdot \text{C}}\right)\left(\frac{1 \text{ kWh}}{3.6 \times 10^6 \text{ J}}\right)$$
$$= 1.5 \times 10^4 \text{ kWh}$$

19.73. Mass of Cl$_2$ and Na produced by 7.0 V with a current of 4.0×10^4 amp, flowing for 1 day:

Mass of chlorine:

$$\left(\frac{70.90 \text{ g Cl}_2}{1 \text{ mol Cl}_2}\right)\left(\frac{1 \text{ mol Cl}_2}{2 \text{ mol e}^-}\right)\left(\frac{1 \text{ mol e}^-}{96500 \text{ C}}\right)\left(\frac{1 \text{ C}}{1 \text{ amp} \cdot \text{s}}\right)(4.0 \times 10^4 \text{ amp})\left(\frac{3600 \text{ s}}{1 \text{ hr}}\right)\left(\frac{24 \text{ hr}}{1 \text{ day}}\right) = 1.3 \times 10^6 \text{ g Cl}_2$$

and for sodium:

$$\left(\frac{22.990 \text{ g Na}}{1 \text{ mol Na}}\right)\left(\frac{1 \text{ mol Na}}{1 \text{ mol e}^-}\right)\left(\frac{1 \text{ mol e}^-}{96500 \text{ C}}\right)\left(\frac{1 \text{ C}}{1 \text{ amp} \cdot \text{s}}\right)(4.0 \times 10^4 \text{ amp})\left(\frac{3600 \text{ s}}{1 \text{ hr}}\right)\left(\frac{24 \text{ hr}}{1 \text{ day}}\right) = 8.2 \times 10^5 \text{ g Na}$$

The energy consumed:

$$\left(\frac{1 \text{ kwh}}{3.6 \times 10^6 \text{ J}}\right)\left(\frac{1 \text{ J}}{1 \text{ V} \cdot \text{C}}\right)(7.0 \text{ V})\left(\frac{1 \text{ C}}{1 \text{ amp} \cdot \text{s}}\right)(4.0 \times 10^4 \text{ amp})\left(\frac{3600 \text{ s}}{1 \text{ hr}}\right)\left(\frac{24 \text{ hr}}{1 \text{ day}}\right) = 6700 \text{ kwh (2sf)}$$

19.74. Charge on the rhodium ion & formula for rhodium sulfate:

$$0.038 \text{ g}\left(\frac{1 \text{ mol Rh}}{102.91 \text{ g}}\right) = 3.7 \times 10^{-4} \text{ mol Rh}$$

$$3.00 \text{ hr}\left(\frac{60.0 \text{ min}}{1 \text{ hr}}\right)\left(\frac{60 \text{ s}}{1 \text{ min}}\right)(0.0100 \text{ amp})\left(\frac{1 \text{ C}}{1 \text{ amp} \cdot \text{s}}\right)\left(\frac{1 \text{ mol e}^-}{96500 \text{ C}}\right) = 0.00112 \text{ mol e}^-$$

$$\frac{0.00112 \text{ mol e}^-}{3.7 \times 10^{-4} \text{ mol Rh}} = 3 \text{ mol e}^-/\text{mol Rh} \quad \text{So the charge on the ion is: } Rh^{3+}.$$

The formula for the sulfate would then be: $Rh_2(SO_4)_3$.

19.75. To calculate the charge on the Ru^{n+} ion, you need to know two things:

1. How many moles of elemental ruthenium are reduced?
2. How many moles of electrons caused that reduction?

Moles of ruthenium: $0.345 \text{ g Ru}\left(\frac{1 \text{ mol Ru}}{101.07 \text{ g Ru}}\right) = 3.41 \times 10^{-3} \text{ mol Ru}$

Moles of electrons: $0.44 \text{ amp}\left(\frac{1 \text{ C}}{1 \text{ amp} \times \text{s}}\right)\left(\frac{60 \text{ s}}{1 \text{ min}}\right)(25 \text{ min})\left(\frac{1 \text{ mol e}^-}{96500 \text{ C}}\right) = 6.8 \times 10^{-3} \text{ mol e}^-$

Recall that our general reduction reactions are written $M^{+x} + x \text{ e}^- \rightarrow M$.
If you know the number of $\frac{\text{moles of electrons}}{\text{mol of metal}}$, you know the charge on the cation. For the Ru^{n+} ion you have $\frac{6.8 \times 10^{-3} \text{ mol e}^-}{3.41 \times 10^{-3} \text{ mol Ru}} = 2.0 \text{ mol e}^{-1}/\text{mol Ru}$.
The ion is therefore the Ru^{2+} ion! The formula for the nitrate salt would be $Ru(NO_3)_2$.

19.76. Mass of Zn consumed when 35 amp-hours are drawn from a cell:

$$35 \text{ amp-hours}\left(\frac{1 \text{ C}}{1 \text{ amp} \cdot \text{s}}\right)\left(\frac{3600 \text{ s}}{1 \text{ hr}}\right)\left(\frac{1 \text{ mol e}^-}{96,500 \text{ C}}\right)\left(\frac{1 \text{ mol Zn}}{2 \text{ mol e}^-}\right)\left(\frac{65.38 \text{ g}}{1 \text{ mol Zn}}\right) = 43 \text{ g Zn}$$

19.77. Mass of Cl₂ produced by electrolysis with 3.0 × 10⁵ amperes at 4.6 V in a 24-hr day:

$$\left(\frac{70.90 \text{ g Cl}_2}{1 \text{ mol Cl}_2}\right)\left(\frac{1 \text{ mol Cl}_2}{2 \text{ mol e}^-}\right)\left(\frac{1 \text{ mol e}^-}{96500 \text{ C}}\right)\left(\frac{1 \text{ C}}{1 \text{ amp} \cdot \text{s}}\right)(3.0 \times 10^5 \text{ amp})\left(\frac{3600 \text{ s}}{1 \text{ hr}}\right)(24 \text{ hr}) = 9.5 \times 10^6 \text{ g Cl}_2$$

19.78. Equations for reactions at cathode and anode in KBr:
Molten KBr:
   Cathode: K⁺(ℓ) + e⁻ → K(s)
   Anode: 2 Br⁻(ℓ) → Br₂(ℓ) + 2 e⁻
   Products: K(s) and Br₂(ℓ)
Aqueous KBr:
   Cathode: 2 H₂O(ℓ) + 2 e⁻ → H₂(g) + 2 OH⁻(aq)
   Anode: 2 Br⁻(aq) → Br₂(ℓ) + 2 e⁻
   Products: H₂(g), OH⁻(aq), and Br₂(ℓ)

19.79. The products formed in the electrolysis of aqueous CuSO₄ are Cu(s) and O₂(g).

Write equations for the anode and cathode reactions.

   Anode: 2 H₂O(ℓ) → O₂(g) + 4 H⁺(aq) + 4 e⁻

   Cathode: Cu²⁺(aq) + 2 e⁻ → Cu(s)

19.80. Products formed in electrolysis of aqueous CdSO₄:

From a thermodynamic perspective,
   Cathode reaction: 2 e⁻ + SO₄²⁻(aq) + 4 H⁺ → SO₂(g) + 2 H₂O(ℓ)
   Anode reaction: 2 H₂O(ℓ) → O₂(g) + 4 H⁺(aq) + 4 e⁻
   Products: O₂(g), H⁺(aq), SO₂(g), and H₂O
However, the reaction at the anode is likely to be kinetically controlled so that
Cathode reaction: Cd²⁺(aq) + 2 e⁻ → Cd(s) and products would be O₂(g), H⁺(aq), and Cd(s)

19.81. The fact that H₂ is formed at the cathode rather than NO reflects the relative **rates** of the two reactions. If water is reduced faster than nitric acid, you would expect hydrogen gas to be preferentially produced. At the anode, you anticipate the formation of oxygen gas, according to the equation: 2 H₂O(ℓ) → O₂(g) + 4 H⁺(aq) + 4 e⁻

19.82. For the electrolysis of Al₂O₃ in cryolite:

Predicted product at anode: O₂

Reaction at cathode: $Al^{3+}(\ell) + 3\ e^- \rightarrow Al(s)$

Reaction at anode: $2\ O^{2-}(\ell) + 4\ e^- \rightarrow O_2(g)$

19.83. The half-cells: Pt | Fe³⁺(aq), (0.50 M) and Fe²⁺(aq), (1.0 × 10⁻⁵ M) and Hg²⁺(0.020 M) | Hg. Which electrode is the anode?

Examine the two half-reactions:

$Fe^{3+}(aq) + e^- \rightarrow Fe^{2+}(aq)$ $E° = +0.771$ V

$Hg^{2+}(aq) + 2\ e^- \rightarrow Hg(\ell)$ $E° = +0.855$ V

Under standard conditions, the mercury electrode is the cathode and the iron electrode is the anode.

The cell reaction is: $Hg^{2+}(aq) + 2\ Fe^{2+}(aq) \rightarrow Hg(\ell) + 2\ Fe^{3+}(aq)$

$E°_{cell} = E°_{cathode} - E°_{anode} = (+0.855\ V) - (+0.771\ V) = 0.084$ V

What is the cell potential?

Using the Nernst equation: $E_{cell} = E°_{cell} - \dfrac{0.0257}{n} \ln \dfrac{[Fe^{3+}]^2}{[Hg^{2+}][Fe^{2+}]^2}$

Recall that the "ln" term has "right-hand species over left-hand species"!

$E_{cell} = 0.084 - \dfrac{0.0257}{2} \ln \dfrac{[0.50]^2}{[0.020][1.0 \times 10^{-5}]^2} = 0.084 - \left(\dfrac{0.0257}{2}(25.55)\right) = -0.244$ V

Voltaic cells *do not* operate at "negative" voltages. Consequently, the E_{cell} will be +0.244 V. This also tells you that the reaction *will not occur* as predicted. So the spontaneous process is $Hg(\ell) + 2\ Fe^{3+}(aq) \rightarrow Hg^{2+}(aq) + 2\ Fe^{2+}(aq)$, with mercury functioning as the anode.

19.84. Will voltage of cell be higher/lower/the same as that predicted?

As [Ag⁺] < 1.0 M, the reaction will shift to the left (reactant-favored); $E_{cell} < E°_{cell}$

$E_{cell} = E°_{cell} - \dfrac{0.0257}{2} \ln \dfrac{[1.0]}{[1.0 \times 10^{-3}]^2} = 0.45 - 0.18 = 0.27$ V

19.85. Cell potential for: Pt |H₂(P = 1 bar) | H⁺(aq, 1.0 M) ‖ Fe³⁺(aq, 1.0M), Fe²⁺(aq, 1.0M) | Pt

The cell convention indicates that the spontaneous reaction is:

H₂(g) + 2 Fe³⁺(aq) → 2 H⁺(aq) + 2 Fe²⁺(aq). Note that you can simply read the convention "from left to right". Note also that the equation is NOT balanced automatically, so you check the number of electrons transferred, 2. With the conditions noted, the cell is in its "standard" state. Calculate $E°_{cell}$: $E°_{cell} = E°(Fe^{3+}) - E°(H_2) = (+0.771\ V) - (+0.0\ V) = +0.771\ V$

Note in the equation above that hydrogen ions are *produced* as the reaction occurs. A higher pH [a lower concentration of H⁺] favors the "right side" of this equation and the reaction would be more favored.

To demonstrate the change in the reaction with pH quantitatively, calculate the cell potential for a reaction in which [H⁺(aq)] is 1.0×10^{-7} M, pH = 7.

Using the Nernst equation: $E_{cell} = E°_{cell} - \dfrac{0.0257}{n} \ln \dfrac{[H^+]^2 [Fe^{2+}]^2}{[Fe^{3+}]^2}$

$E_{cell} = +0.771 - \dfrac{0.0257}{2} \ln \dfrac{[1.0 \times 10^{-7}]^2 [1.0]^2}{[1.0]^2}$

$E_{cell} = +0.771 - (0.0128)(-32.236) = 0.771 + 0.414 = 1.185\ V$

So, the reaction is more favorable at a higher pH.

19.86. As the cell is discharged, the concentration of Cu²⁺ will increase and the concentration of Ag⁺ will decrease until $E_{cell} = 0$ and the reaction is at equilibrium. The cell potential will then be constant.

19.87. Two Ag⁺(aq)|Ag half-cells are constructed. The first has [Ag⁺] = 1.0 M, the second has [Ag⁺] = 1.0 × 10⁻⁵ M. When linked together with a salt bridge and external circuit, a cell potential is observed. (This kind of voltaic cell is referred to as a concentration cell.)

(a) Draw a picture of this cell consisting of one half-cell with [Ag⁺] = 1.0 M and the second with [Ag⁺] = 1.0 × 10⁻⁵ M.

See Figure 19.5. In this case, both electrodes are made of silver metal. On one side, the solution contains 1.0 × 10⁻⁵ M Ag⁺ and on the other the solution contains 1.0 M Ag⁺. The cathode is the electrode on the side that has the 1.0 M Ag⁺ solution [half-reaction: Ag⁺(aq) + e⁻ → Ag(s)], and the anode is the electrode on the side that has the 1.0 × 10⁻⁵ M Ag⁺ solution [half-reaction: Ag(s) → Ag⁺(aq) + e⁻]. Connecting the two electrodes is a wire. Electrons flow through the wire from the anode to the cathode. A salt bridge also connects the two compartments.

(b) Calculate the cell potential:

Since the silver ion concentrations are *not* 1.0 M, you need to use the Nernst equation.

$E_{cell} = E°_{cell} - \dfrac{0.0257}{1} \ln \dfrac{[Ag^+]}{[Ag^+]}$ As *both* half-cells are Ag, the $E°_{cell} = 0.0\ V$.

Substituting the concentrations into the Nernst equation:

$$E_{cell} = 0.00 - \frac{0.0257}{1} \ln \frac{[1.0 \times 10^{-5}]}{[1.0]} = -0.0257(-11.51) = 0.30 \text{ V}$$

19.88. Calculate K_{eq} for the following:

(a) $Co(s) + Ni^{2+}(aq) \rightleftarrows Co^{2+}(aq) + Ni(s)$

Note that Co(s) is the species oxidized.
$E°_{cell} = E°_{cathode} - E°_{anode} = -0.25 \text{ V} - (-0.28 \text{ V}) = +0.03 \text{ V}$
The reaction is product-favored at equilibrium.

As $n = 1$, $\ln K = \dfrac{nE°}{0.0257 \text{ V}} = \dfrac{(1)(0.03 \text{ V})}{0.0257 \text{ V}} = 1.167$ and $K = 3$ (to 1 sf)

(b) $Fe^{3+}(aq) + Cr^{2+}(aq) \rightleftarrows Cr^{3+}(aq) + Fe^{2+}(aq)$

Note that $Cr^{2+}(aq)$ is the species oxidized.
$E°_{cell} = E°_{cathode} - E°_{anode} = +0.771 \text{ V} - (-0.41 \text{ V}) = +1.18 \text{ V}$

The reaction is product favored at equilibrium.

As $n = 1$, $\ln K = \dfrac{nE°}{0.0257 \text{ V}} = \dfrac{1(+1.18)}{0.0257} = 45.9$ and $K = 9 \times 10^{19}$

19.89. Equilibrium constants for the following reactions. Is the equilibrium, as written, reactant- or product-favored at equilibrium?

(a) $2 \text{ Cl}^-(aq) + Br_2(\ell) \rightleftarrows Cl_2(aq) + 2 \text{ Br}^-(aq)$

$E°_{cell} = E°_{cathode} - E°_{anode} = (+ 1.08 \text{ V}) - (+ 1.36 \text{ V}) = - 0.28 \text{ V}$

Using Equation 19.7, $\ln K = \dfrac{nE^0}{0.0257}$ so $\ln K = \dfrac{2(-0.28)}{0.0257} = -21.79$ and $K = 3.4 \times 10^{-10}$, and the equilibrium, as written, is reactant-favored at equilibrium.

(b) $Fe^{2+}(aq) + Ag^+(aq) \rightleftarrows Fe^{3+}(aq) + Ag(s)$

$E°_{cell} = E°_{cathode} - E°_{anode} = (+ 0.799 \text{ V}) - (+ 0.771 \text{ V}) = + 0.028 \text{ V}$

and $\ln K = \dfrac{nE^0}{0.0257}$ so $\ln K = \dfrac{1(+0.028)}{0.0257} = +1.09$ and $K = 3.0$, and the equilibrium, as written, is product-favored at equilibrium.

19.90. Calculate $\Delta_r G°$ for the following:

(a) $ClO_3^-(aq) + 5 \text{ Cl}^-(aq) + 6 \text{ H}^+(aq) \rightleftarrows 3 \text{ Cl}_2(g) + 3 \text{ H}_2O(\ell)$

Note that $Cl^-(aq)$ is the species oxidized.
$E°_{cell} = E°_{cathode} - E°_{anode} = +1.47 \text{ V} - (+1.36 \text{ V}) = +0.11 \text{ V}$

$\Delta_rG° = -nFE = -(5 \text{ mol e}^-)(96,500 \text{ C/mol e}^-)(+0.11 \text{ V})(1 \text{ J}/1 \text{ C·V})(1 \text{ kJ}/10^3 \text{ J})$

$= -5.3 \times 10^1 \text{ kJ/mol rxn}$

(b) AgCl(s) + Br⁻(aq) ⇌ AgBr(s) + Cl⁻(aq)

The half reactions for this reaction are

AgCl(s) + e⁻ → Ag(s) + Cl⁻(aq) and Ag(s) + Br⁻(aq) → AgBr(s) + e⁻

$E°_{cell} = E°_{cathode} - E°_{anode} = +0.222 \text{ V} - (+0.0713 \text{ V}) = +0.151 \text{ V}$

$\Delta_rG° = -nFE = -(1 \text{ mol e}^-)(96,500 \text{ C/mol e}^-)(+0.151 \text{ V})(1 \text{ J}/1 \text{ C·V})(1 \text{ kJ}/10^3 \text{ J})$

$= -14.6 \text{ kJ/mol rxn}$

19.91. Calculate $\Delta_rG°$ for the following reactions:

(a) 3 Cu(s) + 2 NO₃⁻(aq) + 8 H⁺(aq) → 3 Cu²⁺(aq) + 2 NO(g) + 4 H₂O(ℓ)

Using Equation 19.6, $\Delta G° = -nFE°_{cell}$ and

$E°_{cell} = E°_{cathode} - E°_{anode} = (+0.96 \text{ V}) - (+0.337 \text{ V}) = +0.623 \text{ V}$

$\Delta G° = -nFE°_{cell} = -(6 \text{ mol e}^-)(96,500 \text{ C/mol e}^-)(+0.623 \text{ V}) = -3.6 \times 10^5 \text{ VC}$

and since 1 VC = 1 J, $\Delta G° = -3.6 \times 10^5$ J or -360 kJ

(b) H₂O₂(aq) + 2 Cl⁻(aq) + 2 H⁺(aq) → Cl₂(g) + 2 H₂O(ℓ)

$E°_{cell} = E°_{cathode} - E°_{anode} = (+1.77 \text{ V}) - (+1.36 \text{ V}) = +0.41 \text{ V}$

$\Delta G° = -nFE°_{cell} = -(2 \text{ mol e}^-)(96,500 \text{ C/mol e}^-)(+0.41 \text{ V}) = -7.9 \times 10^4$ J and

$\Delta G° = -79$ kJ

19.92. Balanced equations for the reduction half-reactions: See SQ19.3 for the explicit steps in balancing a reduction half-reaction.
(a) HCO₂H + 2 H⁺ + 2 e⁻ → HCHO + H₂O
(b) C₆H₅CO₂H + 6 H⁺ + 6 e⁻ → C₆H₅CH₃ + 2 H₂O
(c) CH₃CH₂CHO + 2 H⁺ + 2 e⁻ → CH₃CH₂CH₂OH
(d) CH₃OH + 2 H⁺ + 2 e⁻ → CH₄ + H₂O

19.93. Balance the equations:

(a) Separating the equation into an oxidation and reduction half-reaction you get:

Ag⁺ + e⁻ → Ag   (reduction)

C₆H₅CHO → C₆H₅CO₂H   (oxidation)

Using the method outlined earlier in this chapter, you balance the oxidation half-equation:

(Numbers in parentheses correspond to the steps listed in SQ19.3)

C₆H₅CHO + H₂O → C₆H₅CO₂H   (3)

$C_6H_5CHO + H_2O \rightarrow C_6H_5CO_2H + 2\ H^+$ (4)

$C_6H_5CHO + H_2O \rightarrow C_6H_5CO_2H + 2\ H^+ + 2\ e^-$ (5)

Note that the reduction half equation gains 1 electron/ silver ion. To balance electron gain with electron loss, you multiply the reduction equation by 2.

2 Ag$^+$ + 2̶e̶$^-$ → 2 Ag

$C_6H_5CHO + H_2O \rightarrow C_6H_5CO_2H + 2\ H^+ +$ 2̶e̶$^-$
_____

2 Ag$^+$ + $C_6H_5CHO + H_2O \rightarrow C_6H_5CO_2H + 2\ H^+ + 2$ Ag

(b) Separating the equation into an oxidation and reduction half-equation you get:

$Cr_2O_7^{2-} \rightarrow Cr^{3+}$ (reduction)

$C_2H_5OH \rightarrow CH_3CO_2H$ (oxidation)

Performing the "steps" on each half-equation

$Cr_2O_7^{2-} \rightarrow Cr^{3+}$	$C_2H_5OH \rightarrow CH_3CO_2H$	(1)
$Cr_2O_7^{2-} \rightarrow 2\ Cr^{3+}$	$C_2H_5OH \rightarrow CH_3CO_2H$	(2)
$Cr_2O_7^{2-} \rightarrow 2\ Cr^{3+} + 7\ H_2O$	$C_2H_5OH + H_2O \rightarrow CH_3CO_2H$	(3)
$14\ H^+ + Cr_2O_7^{2-} \rightarrow 2\ Cr^{3+} + 7\ H_2O$	$C_2H_5OH + H_2O \rightarrow CH_3CO_2H + 4H^+$	(4)

$14\ H^+ + Cr_2O_7^{2-} + 6\ e^- \rightarrow 2\ Cr^{3+} + 7\ H_2O$

$C_2H_5OH + H_2O \rightarrow CH_3CO_2H + 4\ H^+ + 4\ e^-$ (5)

Multiplying the reduction half-equation by 2, and the oxidation half-equation by 3 will equalize electron gain with electron loss.

$28\ H^+ + 2\ Cr_2O_7^{2-} + 12\ e^- \rightarrow 4\ Cr^{3+} + 14\ H_2O$

$3\ C_2H_5OH + 3\ H_2O \rightarrow CH_3CO_2H + 12\ H^+ + 12\ e^-$

Adding the two equations, and removing duplications:

$28\ H^+ + 2\ Cr_2O_7^{2-} +$ 1̶2̶e̶$^- \rightarrow 4\ Cr^{3+} + 14\ H_2O$

$3\ C_2H_5OH + 3\ H_2O \rightarrow 3\ CH_3CO_2H + 12\ H^+ +$ 1̶2̶e̶$^-$
_____

$16\ H^+ + 2\ Cr_2O_7^{2-} + 3\ C_2H_5OH \rightarrow 3\ CH_3CO_2H + 11\ H_2O + 4\ Cr^{3+}$

19.94. For the cell described:
(a) Calculate cell potential: $E°_{cell} = E°_{cathode} - E°_{anode} = (0.799\ V) - (0.771\ V) = 0.028\ V$
(b) Net ionic equation for the cell reaction: $Ag^+(aq) + Fe^{2+}(aq) \rightarrow Ag(s) + Fe^{3+}(aq)$
(c) Reduction takes place at the silver cathode and oxidation occurs at the platinum electrode in the Fe^{2+}/Fe^{3+} solution.

(d) $E_{cell} = E°_{cell} - \dfrac{0.0257}{n} \ln \dfrac{[Fe^{3+}]}{[Ag^+][Fe^{2+}]} = 0.028 \text{ V} - \dfrac{0.0257}{1} \ln \dfrac{1.0}{(0.10)(1.0)} = -0.031 \text{ V}$

The net cell reaction is now the reverse: $Ag(s) + Fe^{3+}(aq) \rightarrow Ag^+(aq) + Fe^{2+}(aq)$

19.95. Comparing the silver/zinc battery with the lead storage battery:

(a) The stoichiometry of the silver/zinc battery indicates a reaction of one mole each of silver oxide, zinc, and water. The mass of one mole of each:

1 mol Ag₂O	231.7 g
1 mol Zn	65.4 g
1 mol H₂O	18.0 g
	315.1 g

The energy associated with the battery is:

$1.59 \text{ V} \left(\dfrac{96500 \text{ C}}{1 \text{ mol e}^-}\right)\left(\dfrac{1 \text{ J}}{1 \text{ V}\cdot\text{C}}\right)\left(\dfrac{2 \text{ mol e}^-}{1 \text{ mol reactant}}\right)\left(\dfrac{1 \text{ mol reactant}}{315.1 \text{ g}}\right) = 973.88$ J/g or 0.974 kJ/g

(b) Performing the same calculations for the lead storage battery, using a stoichiometric amount for the overall battery reaction:

1 mol Pb	207.2 g
1 mol PbO₂	239.2 g
2 mol H₂SO₄	196.2 g
	642.6 g

$2.0 \text{ V} \left(\dfrac{96500 \text{ C}}{1 \text{ mol e}^-}\right)\left(\dfrac{1 \text{ J}}{1 \text{ V}\cdot\text{C}}\right)\left(\dfrac{2 \text{ mol e}^-}{1 \text{ mol reactant}}\right)\left(\dfrac{1 \text{ mol reactant}}{642.6 \text{ g}}\right) = 600.$ J/g or 0.60 kJ/g

(c) The silver/zinc battery produces more energy/gram.

19.96. Regarding a lead storage battery:

(a) Coulombs flowing in 15 hr: $15 \text{ hr} \left(\dfrac{3600 \text{ s}}{1 \text{ hr}}\right)\left(\dfrac{1 \text{ C}}{1 \text{ amp}\cdot\text{s}}\right) 1.5 \text{ amp} = 8.1 \times 10^4 \text{ C}$

$(8.1 \times 10^4 \text{ C})\left(\dfrac{1 \text{ mol e}^-}{96,500 \text{ C}}\right)\left(\dfrac{1 \text{ mol Pb}}{2 \text{ mol e}^-}\right)\left(\dfrac{207.2 \text{ g}}{1 \text{ mol Pb}}\right) = 87$ g Pb

(b) Mass of PbO₂:

$(8.1 \times 10^4 \text{ C})\left(\dfrac{1 \text{ mol e}^-}{96,500 \text{ C}}\right)\left(\dfrac{1 \text{ mol PbO}_2}{2 \text{ mol e}^-}\right)\left(\dfrac{239.2 \text{ g}}{1 \text{ mol PbO}_2}\right) = 1.0 \times 10^2$ g PbO₂

(c) Molarity of H₂SO₄ needed:

$(8.1 \times 10^4 \text{ C})\left(\dfrac{1 \text{ mol e}^-}{96,500 \text{ C}}\right)\left(\dfrac{2 \text{ mol H}_2\text{SO}_4}{2 \text{ mol e}^-}\right)\left(\dfrac{1}{0.50 \text{ L}}\right) = 1.7$ M H₂SO₄

19.97. Using the procedure outlined in SQ19.3:

(a1) $Mn^{2+} + NO_3^- \rightarrow NO + MnO_2$

Oxidation half-equation	Reduction half-equation	Step
$Mn^{2+} \rightarrow MnO_2$	$NO_3^- \rightarrow NO$	1 & 2
$2 H_2O + Mn^{2+} \rightarrow MnO_2$	$NO_3^- \rightarrow NO + 2 H_2O$	3
$2 H_2O + Mn^{2+} \rightarrow MnO_2 + 4 H^+$	$4 H^+ + NO_3^- \rightarrow NO + 2 H_2O$	4
$2 H_2O + Mn^{2+} \rightarrow MnO_2 + 4 H^+ + 2 e^-$	$4 H^+ + NO_3^- + 3 e^- \rightarrow NO + 2 H_2O$	5
$6 H_2O + 3 Mn^{2+} \rightarrow 3 MnO_2 + 12 H^+ + 6 e^-$	$8 H^+ + 2 NO_3^- + 6 e^- \rightarrow 2 NO + 4 H_2O$	6
$2 H_2O(\ell) + 3 Mn^{2+}(aq) + 2 NO_3^-(aq) \rightarrow 2 NO(g) + 3 MnO_2(s) + 4 H^+(aq)$		7

(a2) $NH_4^+ + MnO_2 \rightarrow N_2 + Mn^{2+}$

Oxidation half-equation	Reduction half-equation	Step
$2 NH_4^+ \rightarrow N_2$	$MnO_2 \rightarrow Mn^{2+}$	1&2
	$MnO_2 \rightarrow Mn^{2+} + 2 H_2O$	3
$2 NH_4^+ \rightarrow N_2 + 8 H^+$	$4 H^+ + MnO_2 \rightarrow Mn^{2+} + 2 H_2O$	4
$2 NH_4^+ \rightarrow N_2 + 8 H^+ + 6 e^-$	$4 H^+ + MnO_2 + 2 e^- \rightarrow Mn^{2+} + 2 H_2O$	5
	$12 H^+ + 3 MnO_2 + 6 e^- \rightarrow 3 Mn^{2+} + 6 H_2O$	6
$2 NH_4^+(aq) + 4 H^+(aq) + 3 MnO_2(s) \rightarrow N_2(g) + 3 Mn^{2+}(aq) + 6 H_2O(\ell)$		7

(b) $E°_{cell}$ for the two reactions:

For a1: $E°_{cell} = E°_{cathode} - E°_{anode} = 0.96 \text{ V} - 1.23 \text{ V} = -0.27 \text{ V}$

For a2: $E°_{cell} = E°_{cathode} - E°_{anode} = 1.23 \text{ V} - (-0.272 \text{ V}) = 1.50 \text{ V}$

19.98. Plate a cylindrical object:

(a) anode: Ni(s, impure) → Ni^{2+}(aq) + 2 e$^-$
cathode: Ni^{2+}(aq) + 2 e$^-$ → Ni(s, pure)

(b) Volume of Ni added = (cylinder volume with Ni surface) – (original cylinder volume)
= [π(2.50 + 0.40 cm)^2(20.00 + 0.80 cm)] – [π(2.5 cm)^2(20.00 cm)]
= 157 cm^3 Ni

$$(157 \text{ cm}^3)\left(\frac{8.90 \text{ g}}{1 \text{ cm}^3}\right)\left(\frac{1 \text{ mol Ni}}{58.693 \text{ g}}\right)\left(\frac{2 \text{ mol e}^-}{1 \text{ mol Ni}}\right)\left(\frac{96,500 \text{ C}}{1 \text{ mol e}^-}\right)(2.50 \text{ V})\left(\frac{1 \text{ J}}{1 \text{ C·V}}\right)\left(\frac{1 \text{ kWh}}{3.6 \times 10^6 \text{ J}}\right)$$

= 3.2 kWh

19.99. For the disproportionation of iron(II):

(a) The half-reactions that make up the disproportionation reaction:
Fe^{2+}(aq) + 2 e$^-$ → Fe(s) $E°$ = –0.44 V (cathode)
2 Fe^{2+}(aq) → 2 Fe^{3+}(aq) + 2 e$^-$ $E°$ = 0.771 V (anode)

3 Fe^{2+}(aq) → 2 Fe^{3+}(aq) + Fe(s) $E°$ = –0.44 V – 0.771 V = –1.21 V

(b) The disproportionation reaction is **not product-favored** at equilibrium.

(c) $\ln K = \dfrac{(2 \text{ mol e}^-)\left(96,500 \dfrac{\text{C}}{\text{mol e}^-}\right)\left(-1.21 \dfrac{\text{J}}{\text{C}}\right)}{\left(8.314 \dfrac{\text{J}}{\text{K·mol}}\right)(298 \text{ K})}$ = –94.26 and $K_{formation}$ = 1 × 10^{-41}

19.100. Regarding the disproportionation of Cu(I) in solution:

(a) Half-equations: Cu$^+$(aq) + e$^-$ → Cu(s) and Cu$^+$(aq) → Cu^{2+}(aq) + e$^-$

(b) $E°_{cell} = E°_{cathode} - E°_{anode}$ = (0.521 V) – (0.153 V) = 0.368 V

The reaction is product-favored at equilibrium.

(c) K_{eq} for the reaction: $\ln K = \dfrac{nE°}{0.0257 \text{ V}} = \dfrac{(1)(0.368 \text{ V})}{0.0257 \text{ V}}$ = 14.3 and K = 1.66 × 10^6

= 2 × 10^6 (1 sf)

The equilibrium expression is:

2 Cu$^+$(aq) ⇌ Cu^{2+}(aq) + Cu(s) K = 1.66 × 10^6

The equilibrium constant is very large, so assume that nearly all Cu$^+$ is converted to Cu and Cu^{2+}. At equilibrium, [Cu$^+$] = x M and [Cu^{2+}] = 0.050 M. Solve the equilibrium expression for x.

$$K = \dfrac{[\text{Cu}^{2+}]}{[\text{Cu}^+]^2} = \dfrac{(0.050)}{x^2} = 1.66 \times 10^6; x = [\text{Cu}^+] = 1.7 \times 10^{-4} \text{ M} = 2 \times 10^{-4}$$

19.101. Regarding a lithium-ion battery:
   (a) Oxidation numbers for Co in the battery substances:
   Li(on C)(s) + CoO$_2$(s) → 6 C(s) + LiCoO$_2$(s)
   In CoO$_2$, $x + (2 \times -2) = 0$, so $x = +4$; in LiCoO$_2$, $+1 + x + (2 \times -2) = 0$, so $x = +3$
   (b) cathode reaction: CoO$_2$(s) + Li$^+$(solv) + e$^-$ → LiCoO$_2$(s) as Co is reduced from +4 to +3
   anode reaction: Li(on C)(s) → Li$^+$(solv) + e$^-$ as Li is oxidized from 0 to +1
   (c) Elemental lithium reacts with water, although not as actively as other alkali metals, so the electrolyte **cannot** be dissolved in water.

19.102. Regarding a lithium-ion camera battery:
   (a) Moles of electrons in one hour:

   $$7.5 \text{ amp} \left(\frac{1 \text{ C}}{\text{amp}\times\text{s}}\right)\left(\frac{3600 \text{ s}}{1 \text{ hr}}\right)\left(\frac{1 \text{ mol e}^-}{96500 \text{ C}}\right) = 0.28 \text{ mol e}^-$$

   (b) Mass of lithium oxidized in 1.0 hours:

   $$0.28 \text{ mol e}^- \left(\frac{1 \text{ mol Li}}{1 \text{ mol e}^-}\right)\left(\frac{6.94 \text{ g Li}}{1 \text{ mol Li}}\right) = 1.9 \text{ g Li}$$

19.103. Can sodium or potassium be used to protect a ship's hull?

   Neither Na or K could be used to protect the hull of a ship for an important reason. Both Na and K react with water vigorously.

19.104. Balanced equation for the reaction occurring between copper and steel:

   Iron in steel pipes reacts with copper plumbing in a corrosive redox reaction.

   Cu^{2+}(aq) + Fe(s) → Fe^{2+}(aq) + Cu(s)

   The cell potential is:

   $E°_{cell} = E°_{cathode} - E°_{anode} = +0.337 - (-0.44) = 0.78$ V

   To stop this corrosion reaction galvanized materials steel pipes or nails are coated in zinc changing the reaction from Cu and Fe to the Zn and Cu. The balanced equation:

   Cu^{2+}(aq) + Zn(s) → Zn^{2+}(aq) + Cu(s)

   The cell potential is:

   $E°_{cell} = E°_{cathode} - E°_{anode} = +0.337 - (-0.763) = 1.10$ V

   Notice that the Zn-Cu reaction, having a larger positive cell potential, is the preferred reaction. In other words, Zn becomes a sacrificial anode to keep the steel pipes from corroding.

In the Laboratory

19.105. Consider the electrochemical cell at right:

(a) Cell diagram:

(b) As shown in part (a), Cd serves as the anode, so it must be oxidized to the 2+ cation. Likewise, Ni ions would be reduced to elemental Ni. The balanced equation is:
$Ni^{2+}(aq) + Cd(s) \rightarrow Cd^{2+}(aq) + Ni(s)$

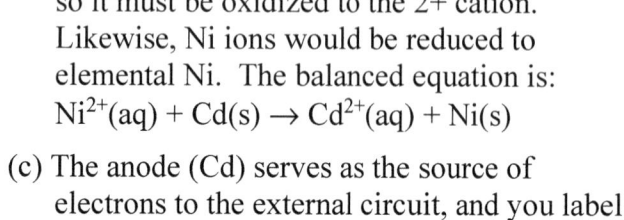

(c) The anode (Cd) serves as the source of electrons to the external circuit, and you label it "−". The cathode is labeled "+".

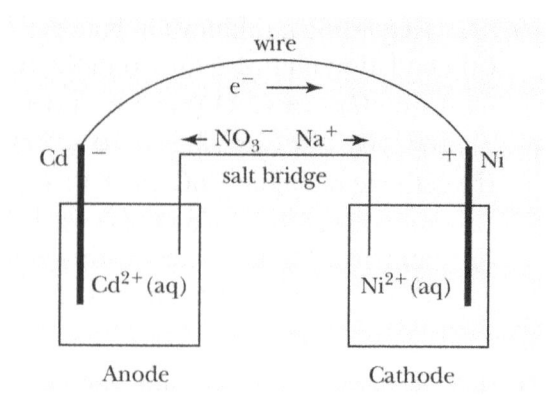

(d) The $E°_{cell} = E°_{cathode} − E°_{anode} = (−0.25) − (−0.403) = +0.15$ V

(e) As shown on the diagram, electrons flow from Cd to Ni compartments.

(f) The direction of travel for the sodium and nitrate ions are shown on the diagram.

(g) The K for the reaction:
$\Delta_r G° = −nFE°_{cell} = −(2 \text{ mol e}^−)(96500 \text{ C/mol e}^−)(+0.15 \text{ J/C}) = 28950$ J

$\Delta_r G° = −RT \ln K$ and $\ln K = \dfrac{nFE^0}{RT}$

$\ln K = \dfrac{(2 \text{ mol e}^-)\left(\dfrac{96{,}500 \text{ C}}{1 \text{ mol e}^-}\right)\left(\dfrac{0.15 \text{ J}}{1 \text{ C}}\right)}{\left(8.314 \dfrac{\text{J}}{\text{K} \cdot \text{mol}}\right)(298 \text{ K})} = 11.68$ and $K = 1 \times 10^5$

(h) If $[Cd^{2+}] = 0.010$ M and $[Ni^{2+}] = 1.0$ M what is the value for E_{cell}?

Using the Nernst equation $E_{cell} = E°_{cell} - \dfrac{0.0257}{n} \ln \dfrac{[Cd^{2+}]}{[Ni^{2+}]}$

$E_{cell} = 0.15 - \dfrac{0.0257}{2} \ln \dfrac{[0.010]}{[1.0]} = 0.15 - (0.0257/2) \cdot (-4.605) = 0.15 + 0.0592 = 0.21$ V

The net reaction is still that given in part(b). These concentration changes increase the cell potential enhancing the reaction.

(i) Lifetime battery use:

You have 1.0 L of each solution. You should determine the limiting reagent if there is one.

The spontaneous reaction reduces Ni^{2+} and oxidizes Cd.

The nickel solution contains: 1.0 L(1.0 M) = 1.0 mol Ni^{2+}. The cadmium electrode weighs 50.0 g, so you have: 50.0 g Cd(1 mol Cd/112.41 g Cd) = 0.445 mol Cd, so Cd is the limiting reagent.

Now you can calculate:

$$0.445 \text{ mol Cd} \left(\frac{2 \text{ mol e}^-}{1 \text{ mol Cd}}\right)\left(\frac{96500 \text{ C}}{1 \text{ mol e}^-}\right)\left(\frac{1 \text{ amp} \cdot \text{s}}{1 \text{ C}}\right)\left(\frac{1}{0.050 \text{ amp}}\right) = 1.7 \times 10^6 \text{ s or 480 hr}$$

19.106. Current flowing in the circuit:

$$0.052 \text{ g Ag} \left(\frac{1 \text{ mol Ag}}{107.87 \text{ g}}\right)\left(\frac{1 \text{ mol e}^-}{1 \text{ mol Ag}}\right)\left(\frac{96,500 \text{ C}}{1 \text{ mol e}^-}\right) = 47 \text{ C}$$

Current = charge/time = (47 C)/(450 s) = 0.10 amperes

19.107. What amount of Au will be deposited by the current that deposits 0.089g Ag in 10 minutes?

$$0.089 \text{ g Ag} \left(\frac{1 \text{ mol Ag}}{107.87 \text{ g}}\right)\left(\frac{1 \text{ mol e}^-}{1 \text{ mol Ag}}\right) = 8.3 \times 10^{-4} \text{ mol e}^-$$

$$8.3 \times 10^{-4} \text{ mol e}^- \left(\frac{1 \text{ mol Au}}{3 \text{ mol e}^-}\right)\left(\frac{196.97 \text{ g Au}}{1 \text{ mol Au}}\right) = 0.054 \text{ g Au}$$

19.108. Arrange A, B, C, D in order of increasing strength as reducing agents:

(a) Reducing agent strength: H_2 < A and C

(b) C is stronger than B, D, and A

(c) D is stronger than B
   B < D < H_2 < A < C

19.109. Explain the reaction of Cu^{2+} with I^-:

The appropriate reduction potentials:

$Br_2 + 2 \text{ e}^- \rightarrow 2 \text{ Br}^-$ $E° = 1.08$ V

$Cl_2 + 2 \text{ e}^- \rightarrow Cl^-$ $E° = 1.36$ V

$I_2 + 2 \text{ e}^- \rightarrow 2I^-$ $E° = 0.535$ V

The reduction potentials indicate that I^- is the strongest reducing agent which can reduce the Cu(II) ion to Cu(I). The Cu^+ ion then reacts with I^- to form insoluble CuI. The reaction equation is:

2 Cu^{2+}(aq) + 4 I^-(aq) → 2 CuI(s) + I_2(aq)

19.110. Determine the amount of O₂ in a sample of water:

(a) Balance in acid, then convert to a basic solution.

2[2 H₂O(ℓ) + Mn²⁺(aq) → MnO₂(s) + 4 H⁺(aq) + 2 e⁻]

O₂(g) + 4 H⁺(aq) + 4 e⁻ → 2 H₂O(ℓ)
_____

2 Mn²⁺(aq) + O₂(g) + 2 H₂O(ℓ) → 2 MnO₂(s) + 4 H⁺(aq)

basic solution: 2 Mn²⁺(aq) + O₂(g) + 4 OH⁻(aq) → 2 MnO₂(s) + 2 H₂O(ℓ)

(b) MnO₂(s) + 4 H⁺(aq) + 2 e⁻ → 2 H₂O(ℓ) + Mn²⁺(aq)

2 I⁻(aq) → I₂(aq) + 2 e⁻
_____

MnO₂(s) + 4 H⁺(aq) + 2 I⁻(aq) → 2 H₂O(ℓ) + Mn²⁺(aq) + I₂(aq)

(c) 2 S₂O₃²⁻(aq) → S₄O₆²⁻(aq) + 2 e⁻
I₂(aq) + 2 e⁻ → 2 I⁻(aq)
_____

2 S₂O₃²⁻(aq) + I₂(aq) → 2 I⁻(aq) + S₄O₆²⁻(aq)

(d) $0.00245 \text{ L} \left(\dfrac{0.0112 \text{ mol S}_2\text{O}_3^{2-}}{1 \text{ L}} \right) \left(\dfrac{1 \text{ mol I}_2}{2 \text{ mol S}_2\text{O}_3^{2-}} \right) \left(\dfrac{1 \text{ mol MnO}_2}{1 \text{ mol I}_2} \right) \left(\dfrac{1 \text{ mol O}_2}{2 \text{ mol MnO}_2} \right)$

$= 6.86 \times 10^{-6}$ mol O₂ and [O₂] = $\dfrac{6.86 \times 10^{-6} \text{ mol O}_2}{0.025 \text{ L}} = 2.7 \times 10^{-4}$ M

Summary and Conceptual Questions

19.111. In the electrolysis of 150 g of CH₃SO₂F:

(a) The mass of HF required to electrolyze 150 g of CH₃SO₂F:

$150 \text{ g CH}_3\text{SO}_2\text{F} \left(\dfrac{1 \text{ mol CH}_3\text{SO}_2\text{F}}{98.09 \text{ g CH}_3\text{SO}_2\text{F}} \right) \left(\dfrac{3 \text{ mol HF}}{1 \text{ mol CH}_3\text{SO}_2\text{F}} \right) \left(\dfrac{20.01 \text{ g HF}}{1 \text{ mol HF}} \right) = 92$ g HF

$150 \text{ g CH}_3\text{SO}_2\text{F} \left(\dfrac{1 \text{ mol CH}_3\text{SO}_2\text{F}}{98.09 \text{ g CH}_3\text{SO}_2\text{F}} \right) \left(\dfrac{1 \text{ mol CF}_3\text{SO}_2\text{F}}{1 \text{ mol CH}_3\text{SO}_2\text{F}} \right) \left(\dfrac{152.06 \text{ g CF}_3\text{SO}_2\text{F}}{1 \text{ mol CF}_3\text{SO}_2\text{F}} \right)$

$= 230$ g CF₃SO₂F

$150 \text{ g CH}_3\text{SO}_2\text{F} \left(\dfrac{1 \text{ mol CH}_3\text{SO}_2\text{F}}{98.09 \text{ g CH}_3\text{SO}_2\text{F}} \right) \left(\dfrac{3 \text{ mol H}_2}{1 \text{ mol CH}_3\text{SO}_2\text{F}} \right) \left(\dfrac{2.02 \text{ g H}_2}{1 \text{ mol H}_2} \right) = 9.3$ g H₂

(b) H₂ produced at the anode or cathode? Since H is being reduced (from +1 to 0), it will be produced at the cathode.

(c) Energy consumed:
$$\left(\frac{1 \text{ kWh}}{3.60 \times 10^6 \text{ J}}\right)\left(\frac{1 \text{ J}}{1 \text{ V} \cdot \text{C}}\right)(8.0 \text{ V})\left(\frac{250 \text{ C}}{1 \text{ s}}\right)(24 \text{ hr})\left(\frac{3600 \text{ s}}{1 \text{ hr}}\right) = 48 \text{ kWh}$$

19.112. Considering the hydrogen-oxygen fuel cell:

(a) Efficiency = $\dfrac{\Delta_f G°[H_2O(\ell)]}{\Delta_f H°[H_2O(\ell)]} \times 100\% = \dfrac{-237.15 \text{ kJ/mol}}{-285.83 \text{ kJ/mol}} \times 100\% = 82.969\%$

(b) Efficiency = $\dfrac{\Delta_f G°[H_2O(g)]}{\Delta_f H°[H_2O(g)]} \times 100\% = \dfrac{-228.59 \text{ kJ/mol}}{-241.83 \text{ kJ/mol}} \times 100\% = 94.525\%$

(c) The efficiency is greater for the gaseous product, possibly due to energy loss when converting gaseous water to the liquid phase.

19.113. Since the reaction depends on the oxidation of elemental hydrogen to water (2 mol e⁻ per mol H₂), you must determine the amount of H₂ present:

$$n = \frac{(200. \text{ atm})(1.0 \text{ L})}{(0.0821 \frac{\text{L} \cdot \text{atm}}{\text{K} \cdot \text{mol}})(298 \text{ K})} = 8.2 \text{ mol H}_2$$

The amount of time this cell can produce current:

$$8.2 \text{ mol H}_2 \left(\frac{2 \text{ mol e}^-}{1 \text{ mol H}_2}\right)\left(\frac{96{,}500 \text{ C}}{1 \text{ mol e}^-}\right)\left(\frac{1 \text{ A} \cdot \text{s}}{1 \text{ C}}\right)\left(\frac{1}{1.5 \text{ A}}\right) = 1.1 \times 10^6 \text{ s } (290 \text{ hrs})$$

19.114. For the reduction of water:
(a) It is much more advantageous to reduce water in an acidic solution. The calculations in (b) will show this.
(b) pH = 7:
$$E_{cell} = E°_{cell} - \frac{0.0257}{n} \ln [OH^-]^2 P_{H_2} = -0.83 \text{ V} - \frac{0.0257}{2} \ln(1.0 \times 10^{-7})^2(1) = -0.42 \text{ V}$$
pH = 1:
$$E_{cell} = -0.83 \text{ V} - \frac{0.0257}{2} \ln(1.0 \times 10^{-13})^2(1) = -0.06 \text{ V}$$

19.115. Regarding the oxidation of glucose:

(a) Amount of glucose(mole) needed to furnish 2400 kcal per 24 hours:

$$2400 \text{ kcal}\left(\frac{4.184 \text{ kJ}}{1 \text{ kcal}}\right)\left(\frac{1 \text{ mol glucose}}{2800 \text{ kJ}}\right) = 3.6 \text{ mol glucose}$$

Amount of oxygen consumed in the process:

From the balanced equation, note that 1 mol glucose requires 6 mol oxygen, so

3.6 mol glucose (6 mol O$_2$/1 mol glucose = 22 mol O$_2$

(b) Moles of electrons to reduce 22 mol O$_2$:
From Appendix M, note that 1 mol of oxygen requires 4 mol of electrons, so
22 mol oxygen (21.5 mol to 3 sf) would require 4 × 21.5 or 86 mol electrons.

(c) Current from the combustion of 3.6 mol glucose:

$$\frac{3.6 \text{ mol glucose}}{24 \text{ hr}} \left(\frac{86 \text{ mol e}^-}{3.6 \text{ mol glucose}}\right)\left(\frac{1 \text{ hr}}{3600 \text{ s}}\right)\left(\frac{96500 \text{ C}}{1 \text{ mol e}^-}\right)\left(\frac{1 \text{ ampere}\cdot\text{s}}{1 \text{ C}}\right) = 96 \text{ amperes}$$

(d) Watts expended for 1.0 V:

$$96 \text{ amperes}\left(\frac{1 \text{ watt}\cdot\text{s}}{1 \text{ J}}\right)\left(\frac{1 \text{ J}}{1 \text{ V}\cdot\text{C}}\right)(1.0 \text{ V})\left(\frac{1 \text{ C}}{1 \text{ ampere}\cdot\text{s}}\right) = 96 \text{ watts}$$

Solution and Answer Guide

Kotz Treichel Townsend Treichel, Chemistry and Chemical Reactivity 11e, 978-0-357-85140-1, Chapter 20: Nuclear Chemistry

TABLE OF CONTENTS

Applying Chemical Principles 763
Practicing Skills..768
General Questions...781
In the Laboratory...783
Summary and Conceptual Questions...786

Applying Chemical Principles

A Primordial Nuclear Reactor

20.1.1. Relative abundance of ^{235}U 2.0×10^9 years ago:

Need rate constants for both isotopes:

$$k_{238} = \frac{\ln(2)}{4.468 \times 10^9 \text{ y}} = 1.551 \times 10^{-10} \text{ y}^{-1}$$

$$k_{235} = \frac{\ln(2)}{7.038 \times 10^8 \text{ y}} = 9.849 \times 10^{-10} \text{ y}^{-1}$$

Now using the integrated rate equation for first-order processes: $\ln\left(\dfrac{N_t}{N_0}\right) = -kt$

For U-238:

$$\ln\left(\frac{99.274}{N_0}\right) = -(1.551 \times 10^{-10} \text{ y}^{-1})(2.0 \times 10^9 \text{ y})$$

$$\ln\left(\frac{99.274}{N_0}\right) = -0.31$$

$\ln(99.274) - \ln(N_0) = -0.31$

$4.59788 - \ln(N_0) = -0.31$

$\ln(N_0) = 0.31 + 4.59788 = 4.91$

$N_0 = e^{4.91} = 1.4 \times 10^2$

For U-235:

$$\ln\left(\frac{0.720}{N_0}\right) = -(9.849 \times 10^{-10} \text{ y}^{-1})(2.0 \times 10^9 \text{ y})$$

$$\ln\left(\frac{0.720}{N_0}\right) = -1.97$$

$$\ln(0.720) - \ln(N_0) = -1.97$$

$$-0.329 - \ln(N_0) = -1.97$$

$$\ln(N_0) = 1.97 - 0.329 = 1.64$$

$$N_0 = e^{1.64} = 5.2$$

The relative abundance of ^{235}U was: $\frac{5.2}{135 + 5.2} \times 100\% = 3.7\%$

20.1.2. Ratio of rates of decomposition for ^{238}U /^{235}U:

The decompositions are proportional to their rate constants (determined in 20.1.1):

Ratio: $\frac{\text{rate for }^{235}U}{\text{rate for }^{238}U} = \frac{9.849 \times 10^{-10} \text{ y}^{-1}}{1.551 \times 10^{-10} \text{ y}^{-1}} = 6.348$

^{235}U decomposes 6.348 times more quickly than ^{238}U.

20.1.3. Equation for the decay of ^{235}U by α emission: $^{235}_{92}U \rightarrow\ ^{231}_{90}Th + ^{4}_{2}\alpha$

20.1.4. Atomic weight of Uranium:

	% Abundance	Fractional Abundance (= % Abundance/100)	Mass	Contribution
^{235}U	0.720	0.00720	235.0439	1.69
^{238}U	99.274	0.99274	238.0508	236.32
			Sum	238.01

The calculation is done by calculating the product of (fractional abundance × mass) for each isotope and adding the resulting products.

Technetium-99m and Medical Imaging

20.2.1. β-decay of ^{99}Mo to ^{99}Tc:

$$^{99}_{42}\text{Mo} \rightarrow {}^{99}_{43}\text{Tc} + {}^{0}_{-1}\beta$$

20.2.2. Oxidation number of Tc in TcO$_4^-$:

Sum of oxidation numbers = –1

1(Tc) + 4(O) = –1; 1(Tc) + 4(–2) = –1; Tc = +7

Electron configuration of Tc^{+7}: The atom has 43 electrons, so a +7 oxidation state would have 36 electrons and be isoelectronic with Kr.

Having a filled outer shell, the ion would have 0 unpaired electrons, and be diamagnetic.

20.2.3. Amount (moles) of Na^{99m}TcO$_4$ in 1.0 µg of the salt? Mass of ^{99m}Tc?

Molar Mass = 22.990 + 99 + 4(15.999) = 186 g/mol

Amount of Na^{99m}TcO$_4$

$$= 1.0 \times 10^{-6} \text{ g Na}^{99m}\text{TcO}_4 \left(\frac{1 \text{ mol Na}^{99m}\text{TcO}_4}{186 \text{ g Na}^{99m}\text{TcO}_4} \right) = 5.4 \times 10^{-9} \text{ mol Na}^{99m}\text{TcO}_4$$

Mass of Tc: $1.0 \times 10^{-6} \text{ g Na}^{99m}\text{TcO}_4 \left(\frac{99 \text{ g }^{99m}\text{Tc}}{186 \text{ g Na}^{99m}\text{TcO}_4} \right) = 5.3 \times 10^{-7} \text{ g }^{99m}\text{Tc}$

20.2.4. Mass of 1.0 µg ^{99m}Tc remaining after 24 hours: $t_{1/2}$ = 6.01 hr

With 24 hours representing 4 half-lives, (1.0 mg)(1/2)^4 = 0.063 µg ^{99m}Tc remain after 24 hr.

20.2.5. Particle produced in the decay of 99-Tc to 99-Ru:

The mass number of ^{99}Ru is the same as that of ^{99}Tc. The emitted particles must have a mass number of 0. The atomic number of ^{99}Ru is one more than the atomic number of ^{99}Tc, so the emitted particle must have an atomic charge of –1. The emitted particle is a beta particle.

The nuclear equation for this decay is ${}^{99}_{43}\text{Tc} \rightarrow {}^{99}_{44}\text{Ru} + {}^{0}_{-1}\beta$.

20.2.6. Explain relative binding of TcO$_4^-$ to Al$_2$O$_3$ versus MoO$_4^{2-}$:

Given that the attraction for an anion to a column is related to its charge, the more highly charged MoO$_4^{2-}$ anion is more tightly bound to the column than the TcO$_4^-$ ion.

The Age of Meteorites

20.3.1. A balanced equation for the radioactive decomposition of ^{87}Rb: $^{87}_{37}\text{Rb} \rightarrow {}^{87}_{38}\text{Sr} + {}^{0}_{-1}\beta$

With the mass number constant (87), the mass number of the second product must be 0. As the atomic number *increases* from 37 to 38, the second product must have an atomic charge of −1, so the beta particle fits the requirement.

20.3.2. The process by which Rb-87 decays is: (b) β-emission. See 20.3.1.

20.3.3. What is the rate constant for the decay of the Rb-87, given the half-life is 4.965×10^{10} y?

$$k = \frac{\ln(2)}{4.965 \times 10^{10} \text{ y}} = \frac{0.6931}{4.965 \times 10^{10} \text{ y}} = 1.396 \times 10^{-11} \text{ y}^{-1}$$

20.3.4. Fraction of Rb-87 decayed in 4.5 billion years (4.5×10^9 y):

The rate equation for the decay is: $\ln\left(\frac{N_t}{N_0}\right) = -kt$

And substituting:

$$\ln\left(\frac{N_t}{N_0}\right) = -(1.396 \times 10^{-11} \text{ y}^{-1})(4.5 \times 10^9 \text{ y})$$

$$\ln\left(\frac{N_t}{N_0}\right) = -0.063$$

$$\frac{N_t}{N_0} = e^{-0.063} = 0.939$$

This is the fraction remaining after 4.5×10^9 years. The fraction decayed after this time is 1 − 0.939 = 0.061 or 6.1 %.

20.3.5. Data from the samples of a meteorite:

Sample	Sr-86	Sr-87	Rb-87	Sr-87/Sr-86	Rb-87/Sr-86
1	1.000	0.819	0.839	0.819	0.839
2	1.063	0.855	0.506	0.804	0.476
3	0.950	0.824	1.929	0.867	2.03
4	1.011	0.809	0.379	0.800	0.375

A strontium-rubidium isochron plot of the data provided:

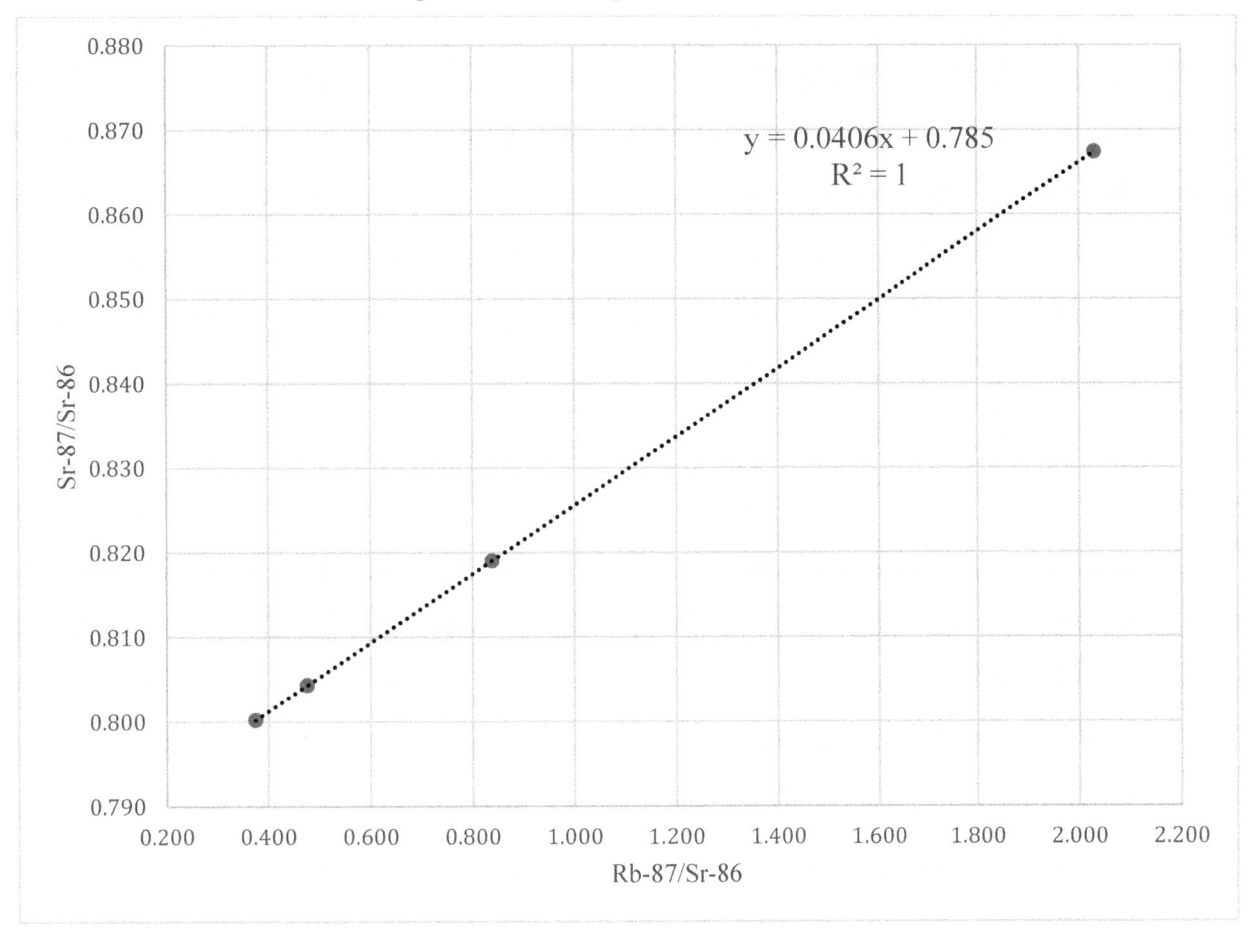

[Plot: Sr-87/Sr-86 vs Rb-87/Sr-86, with linear fit y = 0.0406x + 0.785, R² = 1]

The slope is 0.0406 = $e^{kt} - 1$ thus 1.0406 = e^{kt} and kt = 0.03980. In question 20.3.3 above, k was found to be 1.396 × 10⁻¹¹ y⁻¹. Dividing 0.03980 by k gives t = 2.85 × 10⁹ yr, the age of the meteorite.

20.3.6. Derive the equation:

Start with $\ln\left(\dfrac{^{87}\text{Rb}_t}{^{87}\text{Rb}_0}\right) = -kt$

Raising e to both sides of this equation gives

$\dfrac{^{87}\text{Rb}_t}{^{87}\text{Rb}_0} = e^{-kt}$ and so $^{87}\text{Rb}_0 = e^{kt} \times {}^{87}\text{Rb}_t$ (eq. 1)

Now, the concentration of Sr-87 will be equal to the amount initially present and the amount formed by the decomposition of ^{87}Rb:

$^{87}\text{Sr}_t = {}^{87}\text{Sr}_0 + {}^{87}\text{Sr}$ formed from the decomposition of ^{87}Rb

$^{87}Sr_t = {}^{87}Sr_0 + ({}^{87}Rb_0 - {}^{87}Rb_t)$

This equation can be solved for $^{87}Rb_0$:

$^{87}Rb_0 = {}^{87}Sr_t + {}^{87}Rb_t - {}^{87}Sr_0$ (eq. 2)

Next, set eq.1 = eq.2; they have $^{87}Rb_0$ in common.

$e^{kt} \times {}^{87}Rb_t = {}^{87}Sr_t + {}^{87}Rb_t - {}^{87}Sr_0$

and solving for $^{87}Sr_t$:

$^{87}Sr_t = (e^{kt} \times {}^{87}Rb_t) - {}^{87}Rb_t + {}^{87}Sr_0$

$^{87}Sr_t = {}^{87}Rb_t(e^{kt} - 1) + {}^{87}Sr_0$

As you want the ratio of $^{87}Sr_t$ to $^{86}Sr_t$, divide this equation by $^{86}Sr_t$ to obtain the desired equation: $\dfrac{{}^{87}Sr_t}{{}^{86}Sr_t} = \dfrac{{}^{87}Rb_t}{{}^{86}Sr_t}(e^{kt} - 1) + \dfrac{{}^{87}Sr_0}{{}^{86}Sr_t}$

Practicing Skills

Important Concepts

20.1. Rank α, β, γ in terms of:

(a) Increasing mass: $\gamma < \beta < \alpha$. Gamma is the least massive. α is the nucleus of an He atom and the most massive.

(b) Increasing penetrating power: $\alpha < \beta < \gamma$

20.2. Information used to identify α, β particles:

Information on mass and charge, including experiments such as that shown in Figure 2 on page 73 of the text, helped identify these particles.

20.3. Data for graph of binding energy/nucleon: Once the mass defect of an isotope is determined, Einstein's equation allows one to calculate the binding energy associated with that mass defect. Dividing the binding energy by the number of nucleons provides the desired data.

20.4. How does one use Figure 20.3 to predict decomposition type?
Isotopes that fall to the left of the band of stability (high n/p ratio) undergo beta emission, and isotopes that fall to the right of the band of stability (low n/p ratio) undergo positron emission or beta capture to become more stable. Isotopes beyond $Z = 83$ are unstable and undergo alpha emission.

20.5. Nuclear reactions are carried out by bombarding one nucleus with, typically, another particle—neutron, alpha particle, and so forth. The resulting products are then characterized. For example, one can bombard U-238 with a neutron. As the products emit β particles, atoms of neptunium-239 and plutonium-239 can be formed. Neutrons are effective because they are neutrally charged and not repelled by the positive nuclear charge. This makes it easier for them to penetrate a nucleus and change the nuclear constitution.

20.6. Neutron bombardment of ^{59}Co: $^{59}_{27}\text{Co} + ^{1}_{0}\text{n} \rightarrow ^{60}_{27}\text{Co}$

20.7. Carbon-14 can be used for dating old objects by using the ratio of ^{14}C to ^{12}C in the object under examination. Organisms exchange (through respiration) ^{14}C and ^{12}C with the atmosphere, so that a living organism contains a ratio of ^{14}C: ^{12}C that is the same as in the atmosphere around it. When the organism dies, the exchange ceases, and the ^{14}C in the dead object decays. An assumption is that the amount of ^{14}C in the atmosphere remains constant.

20.8. A limitation of radiocarbon dating is a result of the half-life for ^{14}C (5730 years). Objects that aren't very old have very small changes in the amount of ^{14}C from current living species, and estimation by this technique is subject to large error. At the other end of the timescale, after 60,000 years, more than 10 half-lives have elapsed, almost all of the ^{14}C has decayed; less than 0.1% [= 100%(1/2)^{10}] of the initial amount remains.

20.9. Number of α and β decays between U-238 and Pb-206:

$^{238}_{92}\text{U} \xrightarrow{\alpha} ^{234}_{90}\text{Th} \xrightarrow{\beta} ^{234}_{91}\text{Pa} \xrightarrow{\beta} ^{234}_{92}\text{U} \xrightarrow{\alpha} ^{230}_{90}\text{Th}$

$^{230}_{90}\text{Th} \xrightarrow{\alpha} ^{226}_{88}\text{Ra} \xrightarrow{\alpha} ^{222}_{86}\text{Rn} \xrightarrow{\alpha} ^{218}_{84}\text{Po} \xrightarrow{\alpha} ^{214}_{82}\text{Pb}$

$^{214}_{82}\text{Pb} \xrightarrow{\beta} ^{214}_{83}\text{Bi} \xrightarrow{\beta} ^{214}_{84}\text{Po} \xrightarrow{\alpha} ^{210}_{82}\text{Pb} \xrightarrow{\beta} ^{210}_{83}\text{Bi}$

$^{210}_{83}\text{Bi} \xrightarrow{\beta} ^{210}_{84}\text{Po} \xrightarrow{\alpha} ^{206}_{82}\text{Pb}$

Counting decays yields 8 α and 6 β decays. Another approach is to recognize that each alpha decay results in a decrease of 4 in mass number and 2 in atomic number, whereas each beta decay does not change the mass number but increases the atomic number by 1. The overall change in mass number in going from ^{238}U to ^{206}Pb is 238 – 206 = 32. This means that 32/4 = 8 alpha decays occurred. These would result in a decrease of in atomic number by 16. The atomic number, however, only decreases by 92 83 = 10, so 6 beta decays must have occurred.

20.10. Identify the product of U-235 decay involving 7 α and 4 β decays:

Regardless of the sequence, 7 α particles will result in a decrease in mass number of 28 and a decrease of 14 in atomic number. The additional loss of 4 beta particles will ADD 4 protons to the atomic number with 0 added to the number of neutrons. Thus, the final mass number is 235 – 28 = 207 and the final atomic number is 92 – 14 + 4 = 82. The final product is $^{207}_{82}\text{Pb}$.

20.11. Describe initiation, propagation, and termination with U-235:

Initiation: $^{235}_{92}U + ^{1}_{0}n \longrightarrow ^{236}_{92}U$ (neutron bombardment)

Propagation: $^{236}_{92}U \rightarrow ^{92}_{36}Kr + ^{141}_{56}Ba + 3\ ^{1}_{0}n$ + energy (other products are also possible)

Note the production of multiple neutrons that can result in additional U-235 nuclei splitting.

Termination: occurs when a moderator (such as graphite in a reactor) absorbs the neutrons, or when all the U-235 is consumed.

20.12. Mass converted to energy when U-235 releases 2×10^{10} kJ/mol:

Einstein's famous equation connecting energy and mass: $E = mc^2$.

Note that 1 J = 1 kg-m²/s²: Multiply by 1000 to convert to g-m²/s²

$$m = \frac{E}{c^2} = \frac{2 \times 10^{13}\ J}{(2.9979 \times 10^8\ m/s)^2} = \frac{2 \times 10^{16}\ \frac{g \cdot m^2}{s^2}}{8.9874 \times 10^{16}\ \frac{m^2}{s^2}} = 0.2\ g\ (1\ sf)$$

20.13. Describe a moderator and its function:

In a nuclear reactor, a moderator is a substance (such as graphite or water) that absorbs neutrons and thus reduces the amount of nuclear reaction occurring at a given time.

20.14. The element generated in the reaction is Kr-93: $^{235}_{92}U + ^{1}_{0}n \rightarrow ^{141}_{56}Ba + 2\ ^{1}_{0}n + ^{93}_{36}Kr$

20.15. Units associated with:

(a) curie: a curie is 3.7×10^{10} disintegrations/second

(b) rad: a measure of radiation dose; denotes the absorption of 0.01 J/kg tissue

20.16. Hazards of radiation and the use of radiation in medicine:
Radiation is hazardous to plants and animals (including humans) insofar as it damages tissue. This damage may result in changes to the DNA, leading to cancerous tissue. Diagnostic procedures using nuclear chemistry are beneficial and essential in medical imaging, which entails the creation of images of specific parts of the body. Technetium-99m is used in more than 85% of the diagnostic scans done in hospitals each year. Another medical imaging technique based on nuclear chemistry is *positron emission tomography* (PET). To treat most cancers, it is necessary to use radiation that can penetrate the body to the location of the tumor. Gamma radiation from a cobalt-60 source is commonly used.

20.17. Oxygen-15 used in medical imaging:

(1) $^{14}_{7}N + ^{2}_{1}H \rightarrow ^{15}_{8}O + ^{1}_{0}n$

(2) $^{15}_{8}O \rightarrow ^{14}_{7}N + ^{0}_{+1}\beta$

(3) $^{0}_{+1}\beta + ^{0}_{-1}e \rightarrow 2\gamma$

20.18. Preparation of Tc-99m from Mo-98:

Tc-99m is prepared by the beta decay of Mo-99. Mo-99 is prepared by neutron bombardment of Mo-98 according to the equation: $^{98}_{42}Mo + ^{1}_{0}n \rightarrow ^{99}_{42}Mo \rightarrow ^{99m}_{43}Tc + ^{0}_{-1}\beta$

Nuclear Reactions

20.19. Balance the following nuclear equations, supplying the missing particle.

(a) $^{54}_{26}Fe + ^{4}_{2}He \rightarrow 2\,^{1}_{0}n + ^{56}_{28}Ni$

(b) $^{27}_{13}Al + ^{4}_{2}He \rightarrow ^{1}_{0}n + ^{30}_{15}P$

(c) $^{32}_{16}S + ^{1}_{0}n \rightarrow ^{1}_{1}H + ^{32}_{15}P$

(d) $^{96}_{42}Mo + ^{2}_{1}H \rightarrow ^{1}_{0}n + ^{97}_{43}Tc$

(e) $^{98}_{42}Mo + ^{1}_{0}n \rightarrow ^{99}_{43}Tc + ^{0}_{-1}\beta$

(f) $^{18}_{9}F \rightarrow ^{18}_{8}O + ^{0}_{+1}\beta$

20.20. Balance the following nuclear equations, supplying the missing particle.

(a) $^{9}_{4}Be + ^{1}_{1}H \rightarrow ^{6}_{3}Li + ^{4}_{2}He$

(b) $^{27}_{13}Al + ^{1}_{0}n \rightarrow ^{24}_{11}Na + ^{4}_{2}He$

(c) $^{40}_{20}Ca + ^{1}_{0}n \rightarrow ^{40}_{19}K + ^{1}_{1}H$

(d) $^{241}_{95}Am + ^{4}_{2}He \longrightarrow ^{243}_{97}Bk + 2\,^{1}_{0}n$

(e) $^{246}_{96}Cm + ^{12}_{6}C \longrightarrow 4\,^{1}_{0}n + ^{254}_{102}No$

(f) $^{238}_{92}U + ^{16}_{8}O \rightarrow ^{249}_{100}Fm + 5\,^{1}_{0}n$

20.21. Balance the following nuclear equations, supplying the missing particle.

(a) $^{111}_{47}Ag \rightarrow ^{111}_{48}Cd + ^{0}_{-1}\beta$

(b) $^{87}_{36}Kr \rightarrow ^{0}_{-1}\beta + ^{87}_{37}Rb$

(c) $^{231}_{91}Pa \rightarrow ^{227}_{89}Ac + ^{4}_{2}\alpha$

(d) $^{230}_{90}Th \rightarrow ^{4}_{2}He + ^{226}_{88}Ra$

(e) $^{82}_{35}Br \rightarrow ^{82}_{36}Kr + ^{0}_{-1}\beta$

(f) $^{24}_{11}Na \rightarrow ^{24}_{12}Mg + ^{0}_{-1}\beta$

20.22. Complete the nuclear equations, writing the mass number, atomic number, and symbol for the remaining particle:

(a) $^{19}_{10}\text{Ne} \rightarrow\ ^{0}_{+1}\beta + ^{19}_{9}\text{F}$

(b) $^{59}_{26}\text{Fe} \rightarrow\ ^{0}_{-1}\beta + ^{59}_{27}\text{Co}$

(c) $^{40}_{19}\text{K} \rightarrow\ ^{0}_{-1}\beta + ^{40}_{20}\text{Ca}$

(d) $^{37}_{18}\text{Ar} + ^{0}_{-1}e \rightarrow\ ^{37}_{17}\text{Cl}$

(e) $^{55}_{26}\text{Fe} + ^{0}_{-1}e \rightarrow\ ^{55}_{25}\text{Mn}$

(f) $^{26}_{13}\text{Al} \rightarrow\ ^{25}_{12}\text{Mg} + ^{1}_{1}\text{H}$

20.23. Decay series for U-235:

$^{235}_{92}\text{U} \xrightarrow{\alpha}\ ^{231}_{90}\text{Th} \xrightarrow{\beta}\ ^{231}_{91}\text{Pa} \xrightarrow{\alpha}\ ^{227}_{89}\text{Ac} \xrightarrow{\beta}\ ^{227}_{90}\text{Th}$

$\quad\quad + ^{4}_{2}\text{He} \quad\quad + ^{0}_{-1}e \quad\quad + ^{4}_{2}\text{He} \quad\quad + ^{0}_{-1}e$

$^{227}_{90}\text{Th} \xrightarrow{\alpha}\ ^{223}_{88}\text{Ra} \xrightarrow{\alpha}\ ^{219}_{86}\text{Rn} \xrightarrow{\alpha}\ ^{215}_{84}\text{Po} \xrightarrow{\alpha}\ ^{211}_{82}\text{Pb}$

$\quad\quad + ^{4}_{2}\text{He} \quad\quad + ^{4}_{2}\text{He} \quad\quad + ^{4}_{2}\text{He} \quad\quad + ^{4}_{2}\text{He}$

$^{211}_{82}\text{Pb} \xrightarrow{\beta}\ ^{211}_{83}\text{Bi} \xrightarrow{\beta}\ ^{211}_{84}\text{Po} \xrightarrow{\alpha}\ ^{207}_{82}\text{Pb}$

$\quad\quad + ^{0}_{-1}e \quad\quad + ^{0}_{-1}e \quad\quad + ^{4}_{2}\text{He}$

Decay particles are listed in the line *underneath* the listing for the main particle.

For example, the first decay is $^{235}_{92}\text{U} \xrightarrow{\alpha}\ ^{231}_{90}\text{Th} + ^{4}_{2}\text{He}$.

20.24. Decay series for Th-232:

$^{232}_{90}\text{Th} \xrightarrow{\alpha}\ ^{228}_{88}\text{Ra} \xrightarrow{\beta}\ ^{228}_{89}\text{Ac} \xrightarrow{\beta}\ ^{228}_{90}\text{Th} \xrightarrow{\alpha}\ ^{224}_{88}\text{Ra}$

$\quad\quad + ^{4}_{2}\text{He} \quad\quad + ^{0}_{-1}e \quad\quad + ^{0}_{-1}e \quad\quad + ^{4}_{2}\text{He}$

$^{224}_{88}\text{Ra} \xrightarrow{\alpha}\ ^{220}_{86}\text{Rn} \xrightarrow{\alpha}\ ^{216}_{84}\text{Po} \xrightarrow{\alpha}\ ^{212}_{82}\text{Pb} \xrightarrow{\beta}\ ^{212}_{83}\text{Bi}$

$\quad\quad + ^{4}_{2}\text{He} \quad\quad + ^{4}_{2}\text{He} \quad\quad + ^{4}_{2}\text{He} \quad\quad + ^{0}_{-1}e$

$^{212}_{83}\text{Bi} \xrightarrow{\beta}\ ^{212}_{84}\text{Po} \xrightarrow{\alpha}\ ^{208}_{82}\text{Pb}$

$\quad\quad + ^{0}_{-1}e \quad\quad + ^{4}_{2}\text{He}$

Decay particles are listed in the line *underneath* the listing for the main particle.

For example, the first decay is. $^{232}_{90}\text{Th} \xrightarrow{\alpha}\ ^{228}_{88}\text{Ra} + ^{4}_{2}\text{He}$

Nuclear Stability and Nuclear Decay

20.25. The particle emitted in the following reactions:

(a) $^{198}_{79}\text{Au} \rightarrow {}^{198}_{80}\text{Hg} + {}^{0}_{-1}\beta$ (b) $^{222}_{86}\text{Rn} \rightarrow {}^{218}_{84}\text{Po} + {}^{4}_{2}\alpha$

(c) $^{137}_{55}\text{Cs} \rightarrow {}^{137}_{56}\text{Ba} + {}^{0}_{-1}\beta$ (d) $^{110}_{49}\text{In} \rightarrow {}^{110}_{48}\text{Cd} + {}^{0}_{+1}\beta$

20.26. The particle emitted in the following reactions:

(a) $^{67}_{31}\text{Ga} + {}^{0}_{-1}e \rightarrow {}^{67}_{30}\text{Zn}$ (b) $^{38}_{19}\text{K} \rightarrow {}^{38}_{18}\text{Ar} + {}^{0}_{+1}\beta$

(c) $^{99m}_{43}\text{Tc} \rightarrow {}^{99}_{43}\text{Tc} + \gamma$ (d) $^{56}_{25}\text{Mn} \rightarrow {}^{56}_{26}\text{Fe} + {}^{0}_{-1}\beta$

20.27. Predict the probable mode of decay for each of the following:

(a) $^{80}_{35}\text{Br}$ (large neutron/proton ratio - beta emission) $^{80}_{35}\text{Br} \rightarrow {}^{80}_{36}\text{Kr} + {}^{0}_{-1}\beta$

(b) $^{240}_{98}\text{Cf}$ (large isotope - alpha emission) $^{240}_{98}\text{Cf} \rightarrow {}^{236}_{96}\text{Cm} + {}^{4}_{2}\alpha$

(c) $^{61}_{27}\text{Co}$ (large neutron/proton ratio - beta emission) $^{61}_{27}\text{Co} \rightarrow {}^{61}_{28}\text{Ni} + {}^{0}_{-1}\beta$

(d) $^{11}_{6}\text{C}$ (large proton/neutron ratio - positron emission or electron capture)

$^{11}_{6}\text{C} \rightarrow {}^{11}_{5}\text{B} + {}^{0}_{+1}\beta$ or $^{11}_{6}\text{C} + {}^{0}_{-1}e \rightarrow {}^{11}_{5}\text{B}$

20.28. Predict the probable mode of decay for each of the following:

(a) $^{54}_{25}\text{Mn}$ (large neutron/proton ratio - beta emission) $^{54}_{25}\text{Mn} \rightarrow {}^{54}_{26}\text{Fe} + {}^{0}_{-1}\beta$

(b) $^{241}_{95}\text{Am}$ (large isotope - alpha emission) $^{241}_{95}\text{Am} \rightarrow {}^{237}_{93}\text{Np} + {}^{4}_{2}\alpha$

(c) $^{110}_{47}\text{Ag}$ (large neutron/proton ratio - beta emission) $^{110}_{47}\text{Ag} \rightarrow {}^{110}_{48}\text{Cd} + {}^{0}_{-1}\beta$

(d) $^{197m}_{80}\text{Hg}$ (metastable isotopes frequently decay with gamma emission)

$^{99m}_{80}\text{Hg} \rightarrow {}^{99}_{80}\text{Hg} + \gamma$

20.29. Beta particle and positron emission:

(a) Beta particle emission occurs (usually) when the *ratio of neutrons/protons is high*:

Hydrogen-3 has 1 proton and 2 neutrons—**beta particle emission** (forms $^{3}_{2}\text{He}$)

Fluorine-20 has 9 protons and 11 neutrons-- **beta particle emission** (forms $^{20}_{10}\text{Ne}$)

(b) Positron emission occurs when *the ratio of neutron/proton is too low*:

Sodium-22 has 11 protons and 11 neutrons—**positron emission** (forms $^{22}_{10}\text{Ne}$)

20.30. Beta particle and positron emission:

(a) ^{32}P has a high n/p ratio (17/15) and will likely decay by beta emission.

(b) ^{38}K has a low n/p ratio (19/19) and will likely decay by positron emission.

20.31. The change in mass (Δm) for ^{10}B is:

$\Delta m = [5(1.00783) + 5(1.008665)] - 10.01294 = 10.08248 - 10.01294 = 0.06954$ g/mol nuclei

Binding energy is:

$$\Delta mc^2 = (6.954 \times 10^{-5} \text{ kg/mol nuclei})(2.9979 \times 10^8 \text{ m/s})^2 \left(\frac{1 \text{ J}}{1 \text{ kg} \cdot \text{m}^2 \cdot \text{s}^{-2}}\right)$$

$$= 6.249 \times 10^{12} \text{ J/mol}$$

The **binding energy per mol nucleon**:

$$\frac{6.249 \times 10^9 \text{ kJ/mol nuclei}}{10 \text{ mol nucleons/mol nuclei}} = 6.249 \times 10^8 \text{ kJ/mol nucleon}$$

The mass change for ^{11}B is:

$\Delta m = [5(1.00783) + 6(1.008665)] - 11.00931 = 11.09114 - 11.00931 = 0.08183$ g/mol nuclei

Binding energy is:

$$\Delta mc^2 = (8.183 \times 10^{-5} \text{ kg/mol nuclei})(2.9979 \times 10^8 \text{ m/s})^2 \left(\frac{1 \text{ J}}{1 \text{ kg} \cdot \text{m}^2 \cdot \text{s}^{-2}}\right)$$

$$= 7.354 \times 10^{12} \text{ J/mol nuclei} = 7.354 \times 10^9 \text{ kJ/mol nuclei}$$

The **binding energy per mol nucleon**:

$$\frac{7.354 \times 10^9 \text{ kJ/mol nuclei}}{11 \text{ mol nucleons/mol nuclei}} = 6.686 \times 10^8 \text{ kJ/mol nucleon}$$

20.32. Calculate the binding energy for ^{30}P and ^{31}P:

For ^{30}P: $^{30}_{15}\text{P} \rightarrow 15\ ^{1}_{1}\text{H} + 15\ ^{1}_{0}\text{n}$

mass change: $\Delta m = [(15 \times 1.00783) + (15 \times 1.008665)] - 29.97832 = 0.26911$ g/mol nuclei

Binding energy is:

$$\Delta mc^2 = (2.6911 \times 10^{-4} \text{ kg/mol nuclei})(2.9979 \times 10^8 \text{ m/s})^2 \left(\frac{1 \text{ J}}{1 \text{ kg} \cdot \text{m}^2 \cdot \text{s}^{-2}}\right)$$

$$= 2.4186 \times 10^{13} \text{ J/mol nuclei (or } 2.4186 \times 10^{10} \text{ kJ/mol nuclei)}$$

The **binding energy per mol nucleon**:

$$\frac{2.4186 \times 10^{10} \text{ kJ/mol nuclei}}{30 \text{ mol nucleons/mol nuclei}} = 8.0619 \times 10^8 \text{ kJ/mol nucleon}$$

For ^{31}P: $^{31}_{15}$P $\rightarrow$ 15 $^{1}_{1}$H + 16 $^{1}_{0}$n

mass change: Δm = [(15 × 1.00783) + (16 × 1.008665)] − 30.97376 = 0.28233 g/mol nuclei

Binding energy is: Δmc^2 = $(2.8233 \times 10^{-4}$ kg/mol nuclei$)(2.9979 \times 10^8$ m/s$)^2 \left(\dfrac{1 \text{ J}}{1 \text{ kg} \cdot \text{m}^2 \cdot \text{s}^{-2}}\right)$

$= 2.5374 \times 10^{13}$ J/mol nuclei (or 2.5374×10^{10} kJ/mol nuclei)

The **binding energy per mol nucleon**:
$\dfrac{2.5374 \times 10^{10} \text{ kJ/mol nuclei}}{31 \text{ mol nucleons/mol nuclei}} = 8.1852 \times 10^8$ kJ/mol nucleon

20.33. The binding energy per mol of nucleon for calcium-40:
The change in mass (Δm) for ^{40}Ca is:

mass change: Δm = [(20 × 1.00783) + (20 × 1.008665)] − 39.96259 = 0.36731 g/mol nuclei

Binding energy is: Δmc^2 = $(3.6731 \times 10^{-4}$ kg/mol nuclei$)(2.9979 \times 10^8$ m/s$)^2 \left(\dfrac{1 \text{ J}}{1 \text{ kg} \cdot \text{m}^2 \cdot \text{s}^{-2}}\right)$

$= 3.3012 \times 10^{13}$ J/mol nuclei (or 3.3012×10^{10} kJ/mol nuclei)

The **binding energy per mol nucleon**: $\dfrac{3.3012 \times 10^{10} \text{ kJ/mol nuclei}}{40 \text{ mol nucleons/mol nuclei}} = \dfrac{8.2529 \times 10^8 \text{ kJ}}{\text{mol nucleon}}$

This value matches well with the value shown in Figure 20.4.

20.34. The binding energy per mol of nucleon for ^{56}Fe is:

$^{56}_{26}$Fe $\rightarrow$ 26 $^{1}_{1}$H + 30 $^{1}_{0}$n

Δm = [(26 × 1.00783) + (30 × 1.008665)] − 55.9349 = 0.52863 g/mol

Binding energy is: Δmc^2 = $(5.2863 \times 10^{-4}$ kg/mol nuclei$)(2.9979 \times 10^8$ m/s$)^2 \left(\dfrac{1 \text{ J}}{1 \text{ kg} \cdot \text{m}^2 \cdot s^{-2}}\right)$

$= 4.7510 \times 10^{13}$ J/mol nuclei or 4.7510×10^{10} kJ/mol nuclei

The **binding energy per mol nucleon**: $\dfrac{4.7510 \times 10^{10} \text{ kJ/mol nuclei}}{56 \text{ mol nucleons/mol nuclei}} = \dfrac{8.4839 \times 10^8 \text{ kJ}}{\text{mol nucleon}}$

This value matches well with the value shown in Figure 20.4.

20.35. The binding energy per mol of nucleon for ^{16}O:

$^{16}_{8}$O $\rightarrow$ 8 $^{1}_{1}$H + 8 $^{1}_{0}$n

Δm = [(8 × 1.00783) + (8 × 1.008665)] − 15.99491 = 0.13705 g/mol nuclei

Binding energy is: $\Delta mc^2 = (1.3705 \times 10^{-4} \text{ kg/mol nuclei})(2.9979 \times 10^8 \text{ m/s})^2 \left(\dfrac{1 \text{ J}}{1 \text{ kg} \cdot \text{m}^2 \cdot \text{s}^{-2}}\right)$

$= 1.2317 \times 10^{13}$ J/mol nuclei or 1.2317×10^{10} kJ/mol nuclei

The **binding energy per mol nucleon**: $\dfrac{1.2317 \times 10^{10} \text{ kJ/mol nuclei}}{16 \text{ mol nucleons/mol nuclei}} = \dfrac{7.6983 \times 10^{8} \text{ kJ}}{\text{mol nucleon}}$

20.36. Binding energy per mol nucleon for nitrogen-14:

$^{14}_{7}\text{N} \rightarrow 7\, ^{1}_{1}\text{H} + 7\, ^{1}_{0}\text{n}$, so, the change in mass (Δm) is:

$\Delta m = [(7 \times 1.00783) + (7 \times 1.008665)] - 14.003074 = 0.11239$ g/mol nuclei

Binding energy is:

$\Delta mc^2 = (1.1239 \times 10^{-4} \text{ kg/mol nuclei})(2.9979 \times 10^8 \text{ m/s})^2 \left(\dfrac{1 \text{ J}}{1 \text{ kg} \cdot \text{m}^2 \cdot \text{s}^{-2}}\right)$

$= 1.0101 \times 10^{13}$ J/mol nuclei or 1.0101×10^{10} kJ/mol nuclei

The **binding energy per mol nucleon**:

$\dfrac{1.0101 \times 10^{10} \text{ kJ/mol nuclei}}{14 \text{ mol nucleons/mol nuclei}} = 7.2150 \times 10^{8}$ kJ/mol nucleon

Nucleosynthesis of the Elements

20.37. Nuclear reactions to produce light elements:

(a) $^{3}_{2}\text{He} + ^{4}_{2}\text{He} \rightarrow ^{7}_{4}\text{Be} + \gamma$

(b) $^{7}_{4}\text{Be} + ^{0}_{-1}\text{e} \rightarrow ^{7}_{3}\text{Li}$

20.38. Nuclear reactions in triple alpha process:

(a) $^{4}_{2}\text{He} + ^{4}_{2}\text{He} \rightarrow ^{8}_{4}\text{Be}$

(b) $^{8}_{4}\text{Be} + ^{4}_{2}\text{He} \rightarrow ^{12}_{6}\text{C}$

(c) $^{12}_{6}\text{C} + ^{4}_{2}\text{He} \rightarrow ^{16}_{8}\text{O} + \gamma$

Rates of Radioactive Decay

20.39. For ^{64}Cu, $t_{1/2} = 12.7$ hr

The fraction remaining as ^{64}Cu following n half-lives is equal to $\left(\dfrac{1}{2}\right)^n$.

Note that 63.5 hours corresponds to **exactly five** half-lives.

The fraction remaining as ^{64}Cu is $\left(\dfrac{1}{2}\right)^5 = \dfrac{1}{32} = 0.03125$.

The mass remaining is: $(0.03125)(25.0\ \mu g) = 0.781\ \mu g$.

20.40. For ^{198}Au, $t_{1/2} = 2.69$ days
So, 10.8 days is: 10.8 days/2.69 days = 4 half-lives.
The mass remaining = $2.8\ \mu g \times \left(\dfrac{1}{2}\right)^4 = 0.18\ \mu g$

20.41. For ^{131}I, $t_{1/2} = 8.04$ days

(a) The equation for β–decay of ^{131}I is: $^{131}_{53}\text{I} \rightarrow\ ^{0}_{-1}\beta + ^{131}_{54}\text{Xe}$

(b) The amount of ^{131}I remaining after 40.2 days:

For ^{131}I, $t_{1/2}$ is 8.04 days, so 40.2 days/8.04 days = 5 half-lives:

The fraction of ^{131}I remaining is $\left(\dfrac{1}{2}\right)^5 = \dfrac{1}{32} = 0.03125$.

Mass of the original 2.4 μg remaining: $(0.03125)(2.4\ \mu g) = 0.075\ \mu g$

20.42. For ^{32}P, $t_{1/2} = 14.3$ days:

(a) The equation for β–decay of ^{32}P is: $^{32}_{15}\text{P} \rightarrow\ ^{0}_{-1}\beta + ^{32}_{16}\text{S}$

(b) Mass of ^{32}P remaining after 28.6 days:

28.6 days/14.3 days = 2 half-lives so $4.8\ \mu g \times \left(\dfrac{1}{2}\right)^2 = 1.2\ \mu g$

20.43. To determine the mass of gallium-67 left after 13 days, determine the number of half-lives corresponding to 13 days.

$13\ \text{days}\left(\dfrac{24\ \text{hr}}{1\ \text{day}}\right) = 312$ hours

The rate constant is $k = \dfrac{\ln(2)}{t_{1/2}} = \dfrac{\ln(2)}{78.27\ \text{h}} = 0.008856\ \text{h}^{-1}$

The fraction remaining: $\ln(x) = -(0.008856\ \text{h}^{-1})(312\ \text{h})$ and solving for x yields 0.063 (where x represents the fraction of Ga-67 remaining.)

The amount of gallium-67 remaining is then $(0.063)(0.015\ \text{mg}) = 9.5 \times 10^{-4}$ mg

20.44. Regarding ^{131}I:

(a) Equation for the decomposition: $^{131}_{53}\text{I} \rightarrow {}^{0}_{-1}\beta + {}^{131}_{54}\text{Xe}$

(b) Time required for the sample to decrease to 35.0 % of its original activity:

$$k = \frac{0.693}{8.02 \text{ days}} = 0.0864 \text{ days}^{-1}$$

$$\ln\left(\frac{N_t}{N_0}\right) = -kt$$

$$\ln\left(\frac{35.0}{100}\right) = -(0.0864 \text{ days}^{-1})t$$

$t = 12.1$ days

20.45. For the decomposition of Radon-222:

(a) The balanced equation for the decomposition of Rn-222 with α particle emission.

$^{222}_{86}\text{Rn} \rightarrow {}^{4}_{2}\alpha + {}^{218}_{84}\text{Po}$

(b) Time required for the sample to decrease to 20.0% of its original activity:

Since this decay follows 1st order kinetics, you can calculate a rate constant:

$$k = \frac{0.693}{t_{1/2}} = \frac{0.693}{3.82 \text{ days}} = 0.181 \text{ days}^{-1}$$

With this rate constant, using the 1st order integrated rate equation, calculate the time required:

$$\ln\left(\frac{20.0}{100}\right) = -(0.181 \text{ days}^{-1})t$$

$$t = \frac{-1.609}{-0.181 \text{ days}^{-1}} = 8.87 \text{ days}$$

20.46. Regarding ^{90}Sr:

(a) $\ln\left(\frac{975 \text{ dpm}}{1.0 \times 10^3 \text{ dpm}}\right) = -k(1.0 \text{ y})$

$k = 0.025 \text{ y}^{-1}$

$$t_{1/2} = \frac{0.693}{k} = \frac{0.693}{0.025 \text{ y}^{-1}} = 27 \text{ y}$$

(b) Now calculate the time required to drop to 1.0% of original value:

$$\ln\left(\frac{1.0}{100}\right) = -(0.025 \text{ y}^{-1})t$$

$$t = 1.8 \times 10^2 \text{ y}$$

20.47. For the decay of cobalt-60, $t_{1/2}$ is 5.27 y:

(a) Time for Co-60 to decrease to 1/8 of its original activity:

Following the methodology of previous questions, determine the rate constant:

$$k = \frac{0.693}{t_{1/2}} = \frac{0.693}{5.27 \text{ y}} = 0.131 \text{ y}^{-1}$$

Substituting into the equation:

$$\ln\left(\frac{1}{8}\right) = -(0.131 \text{ y}^{-1})t$$

$$\ln(0.125) = -(0.131 \text{ y}^{-1})t$$

$$t = 15.8 \text{ y}$$

A "short-cut" is available here if you notice that 1/8 corresponds to $\left(\frac{1}{2}\right)^3$.

Said another way, one-eighth of the Co-60 will remain after **three half-lives** have passed, so 3(5.27 y) = 15.8 y!!

(b) Fraction of Co-60 remaining after 1.0 year:

Now solve for the fraction on the "left-hand side" of the rate equation:

$$\ln(\text{fraction remaining}) = -kt = -0.131 \text{ y}^{-1}(1.0 \text{ y})$$

$\ln(\text{fraction remaining}) = -0.13$ and $e^{-0.13} = $ fraction remaining

fraction remaining = 0.88, so 88% remains after 1.0 year.

20.48. Graph showing disintegrations of scandium-46:

The following table and graph were constructed. Note that both go out further in time than the requested 1.0 year.

Days	DPM	# half-lives
0	70000	0
83.8	35000	1
167.6	17500	2
251.4	8750	3
335.2	4375	4
419	2187.5	5
502.8	1093.75	6

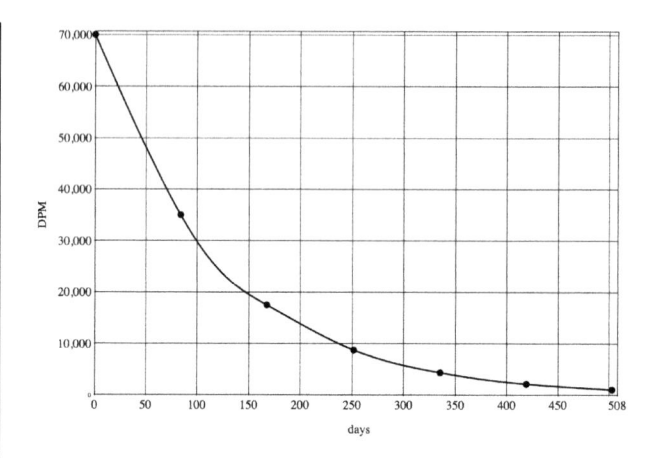

Nuclear Reactions

20.49. For the decay of plutonium-239: $^{239}_{94}\text{Pu} + ^{4}_{2}\alpha \rightarrow ^{240}_{95}\text{Am} + ^{1}_{1}\text{H} + 2\,^{1}_{0}\text{n}$

20.50. Absorption of neutrons by plutonium-239:
Two neutron absorption: $^{239}_{94}\text{Pu} + 2\,^{1}_{0}\text{n} \rightarrow ^{241}_{94}\text{Pu}$
Beta-particle emission: $^{241}_{94}\text{Pu} \rightarrow ^{0}_{-1}\beta + ^{241}_{95}\text{Am}$

20.51. Synthesis of ^{287}Fl: $^{242}_{94}\text{Pu} + ^{48}_{20}\text{Ca} \rightarrow 3\,^{1}_{0}\text{n} + ^{287}_{114}\text{Fl}$

20.52. Synthesis of ^{246}Cf: $^{238}_{92}\text{U} + ^{12}_{6}\text{C} \rightarrow ^{246}_{98}\text{Cf} + 4\,^{1}_{0}\text{n}$

20.53. Complete the following equations using deuterium bombardment:

(a) $^{114}_{48}\text{Cd} + ^{2}_{1}\text{H} \rightarrow ^{115}_{48}\text{Cd} + ^{1}_{1}\text{H}$

(b) $^{6}_{3}\text{Li} + ^{2}_{1}\text{H} \rightarrow ^{7}_{4}\text{Be} + ^{1}_{0}\text{n}$

(c) $^{40}_{20}\text{Ca} + ^{2}_{1}\text{H} \rightarrow ^{38}_{19}\text{K} + ^{4}_{2}\alpha$

(d) $^{63}_{29}\text{Cu} + ^{2}_{1}\text{H} \rightarrow ^{65}_{30}\text{Zn} + \gamma$

20.54. Discovery, contributors, significance:

(a) 1896, discovery of radioactivity: H. Becquerel discovered that a covered photographic plate darkened when exposed to uranium. This indicated the complexity of the atom, and spurred research for the fundamental particles.

(b) 1898, discovery of Ra and Po: Pierre and Marie Curie identified Ra and Po as trace components of pitchblende, linking Ra and Po as members of the decay series of uranium.

(c) 1919, first artificial nuclear reaction: Patrick Blackett, while a colleague of Ernest Rutherford, explained experiments performed by Rutherford and colleagues in which N atoms were bombarded with alpha particles and protons were ejected. Blackett proposed that an artificial nuclear reaction had occurred with the formation of atoms of O. This led to many subsequent nuclear reactions and the synthesis of new elements.

20.55. The equation for the bombardment of Boron-10 with a neutron, and the subsequent release of an alpha particle: $^{10}_{5}B + ^{1}_{0}n \rightarrow ^{4}_{2}\alpha + ^{7}_{3}Li$

20.56. Balance the reaction, and identify the unknown species:

(a) $^{14}_{7}N + ^{4}_{2}He \rightarrow ^{17}_{8}O + ^{1}_{1}H$

(b) $^{9}_{4}Be + ^{4}_{2}He \rightarrow ^{12}_{6}C + ^{1}_{0}n$

(c) $^{27}_{13}Al + ^{4}_{2}He \rightarrow ^{30}_{15}P + ^{1}_{0}n$

(d) $^{239}_{94}Pu + ^{4}_{2}He \rightarrow ^{242}_{96}Cm + ^{1}_{0}n$

General Questions

20.57. The rate constant, $k = \dfrac{\ln(2)}{t_{1/2}} = \dfrac{0.69315}{4.965 \times 10^{10} \text{ y}} = 1.396 \times 10^{-11} \text{ y}^{-1}$

At some time, t, you have 1.8×10^{-3} mol ^{87}Rb. You also have 1.6×10^{-3} mol ^{87}Sr, which resulted from the decay of an equal amount of ^{87}Rb. So, the initial amount of ^{87}Rb = (1.6 + 1.8) × 10^{-3} mol.

Substituting into the first-order equation you get:

$\ln\left(\dfrac{1.8 \times 10^{-3} \text{ mol}}{3.4 \times 10^{-3} \text{ mol}}\right) = -\left(1.396 \times 10^{-11} \text{ y}^{-1}\right)t$

$t = \dfrac{\ln(0.53)}{-1.396 \times 10^{-11} \text{ y}^{-1}} = \dfrac{-0.64}{-1.396 \times 10^{-11} \text{ y}^{-1}} = 4.6 \times 10^{10} \text{ y}$

20.58. Reaction of Li with a neutron: $^{6}_{3}\text{Li} + ^{1}_{0}\text{n} \rightarrow ^{3}_{1}\text{H} + ^{4}_{2}\text{He}$

20.59. Graph for P-31 of disintegrations per minute as a function of time for a period of 1 year:

Calculate rate constant:

$$k = \frac{\ln(2)}{t_{1/2}} = \frac{0.69315}{14.28 \text{ days}} = 0.04854 \text{ day}^{-1}$$

$$\ln\left(\frac{x}{3.2 \times 10^{6} \text{ dpm}}\right) = -(0.04854 \text{ day}^{-1})t$$

The results of the calculation are plotted on the graph. Multiple substitutions of 14.28 days for t gives data points corresponding to a half-life.

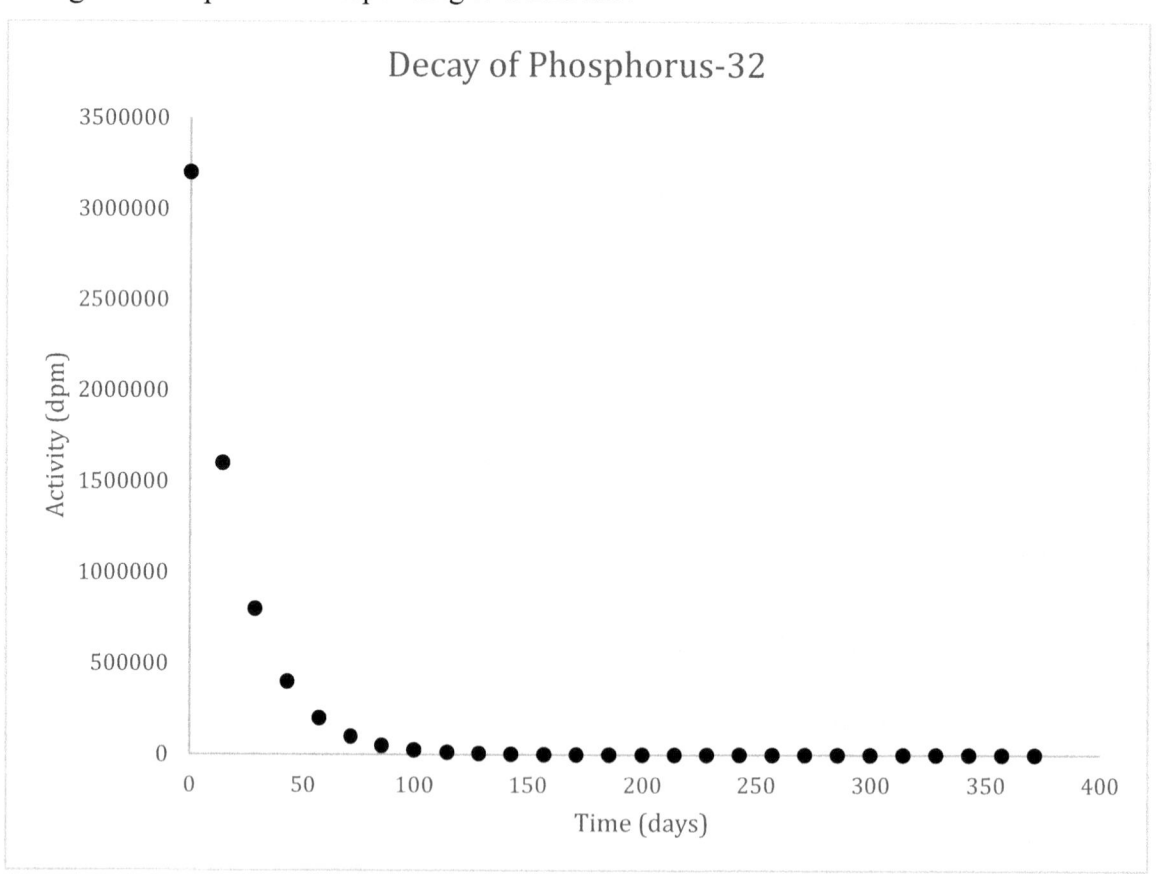

Decay of Phosphorus-32

(Scatter plot: Activity (dpm) vs Time (days); activity starts at ~3,200,000 dpm and decays exponentially to near zero by ~200 days, extending to ~400 days.)

20.60. For the decomposition:

Calculate the rate constant: $k = \dfrac{\ln(2)}{t_{1/2}} = \dfrac{0.693}{7.04 \times 10^{8} \text{ y}} = 9.84 \times 10^{-10} \text{ y}^{-1}$

Calculate elapsed time to reduce percentage from 3.0% to 0.72%:

$$\ln\left(\frac{0.72}{3.0}\right) = -(9.84 \times 10^{-10} \text{ y}^{-1})t$$

$$t = 1.4 \times 10^9 \text{ y}$$

20.61. The decay of Uranium-238 to produce plutonium-239:

(a) $^{238}_{92}\text{U} + ^{1}_{0}\text{n} \rightarrow ^{239}_{92}\text{U} + \gamma$

(b) $^{238}_{92}\text{U} \rightarrow ^{239}_{93}\text{Np} + ^{0}_{-1}\beta$

(c) $^{239}_{93}\text{Np} \rightarrow ^{239}_{94}\text{Pu} + ^{0}_{-1}\beta$

(d) $^{239}_{94}\text{Pu} + ^{1}_{0}\text{n} \rightarrow 2\ ^{1}_{0}\text{n} + \text{other nuclei} + \text{energy}$

20.62. Calculate energy associated with neutron bombardment of lithium-6:

(a) $\Delta m = 7.01600 - [6.01512 + 1.008665] = -0.00779$ g/mol

$\Delta E = \Delta mc^2 = (-7.79 \times 10^{-6} \text{ kg/mol})(2.9979 \times 10^8 \text{ m/s})^2 = -7.00 \times 10^{11}$ J/mol

$\left(\dfrac{-7.00 \times 10^{11} \text{ J}}{1 \text{ mol nucleons}}\right)\left(\dfrac{1 \text{ mol nucleons}}{6.022 \times 10^{23} \text{ atoms}}\right) = -1.16 \times 10^{-12}$ J/atom

(b) $\lambda = \dfrac{hc}{E} = \dfrac{(6.626 \times 10^{-34} \text{ J} \cdot \text{s})(2.998 \times 10^8 \text{ m/s})}{1.16 \times 10^{-12} \text{ J}} = 1.71 \times 10^{-13}$ m = 0.171 pm

20.63. Regarding the synthesis of Livermorium:

(a) $? + ^{248}_{96}\text{Cm} \rightarrow ^{296}_{116}\text{Lv} : ^{48}_{20}\text{Ca}$

(b) $^{296}_{116}\text{Lv} \rightarrow ? + ^{293}_{116}\text{Lv} : 3\ ^{1}_{0}\text{n}$

20.64. Energy involved in fusion of deuterium and tritium to form helium-4:

$^{2}_{1}\text{H} + ^{3}_{1}\text{H} \rightarrow ^{4}_{2}\text{He} + ^{1}_{0}\text{n}$

$\Delta m = [4.00260 + 1.008665] - [2.01410 + 3.01605] = -0.01889$ g/mol

$E = (\Delta m)c^2 = (-1.889 \times 10^{-5} \text{ kg/mol})(2.9979 \times 10^8 \text{ m/s})^2 = -1.697 \times 10^{12}$ J/mol

$= -1.697 \times 10^9$ kJ/mol

In the Laboratory

20.65. The age of the fragment can be determined if: (a) you calculate the rate constant and, (b) use the 1st order integrated rate equation.

(a) $k = \dfrac{0.693}{t_{1/2}} = \dfrac{0.693}{5730 \text{ y}} = 1.21 \times 10^{-4} \text{ y}^{-1}$

(b) Now calculate the time required for the carbon-14 /carbon-12 to decay to 72% of that ratio in living organisms.

$\ln\left(\dfrac{72}{100}\right) = -(1.21 \times 10^{-4} \text{ y}^{-1})t$

$t = \dfrac{-0.33}{-1.21 \times 10^{-4} \text{ y}^{-1}} = 2.7 \times 10^{3} \text{ y}$

20.66. Regarding the wood from a Thracian chariot:

Determine the rate constant:

$k = \dfrac{0.693}{t_{1/2}} = \dfrac{0.693}{5730 \text{ y}} = 1.21 \times 10^{-4} \text{ y}^{-1}$

Use the integrated rate law to determine t:

$\ln\left(\dfrac{11.2}{14.0}\right) = -(1.21 \times 10{-4} \text{ y}^{-1})t$

$t = 1845 \text{ y} = 1850 \text{ y (3sf)}$

The chariot was made 1850 years before the analysis was performed. If the ^{14}C activity is 14.0 dpm/g in 2023, the chariot was made in 2023 − 1845 = 178 AD. The value is only significant to the tens place, so the best that can be said is that the chariot was made around 180 AD.

20.67. To determine the half-life of polonium-210, plot ln (dpm) vs time.

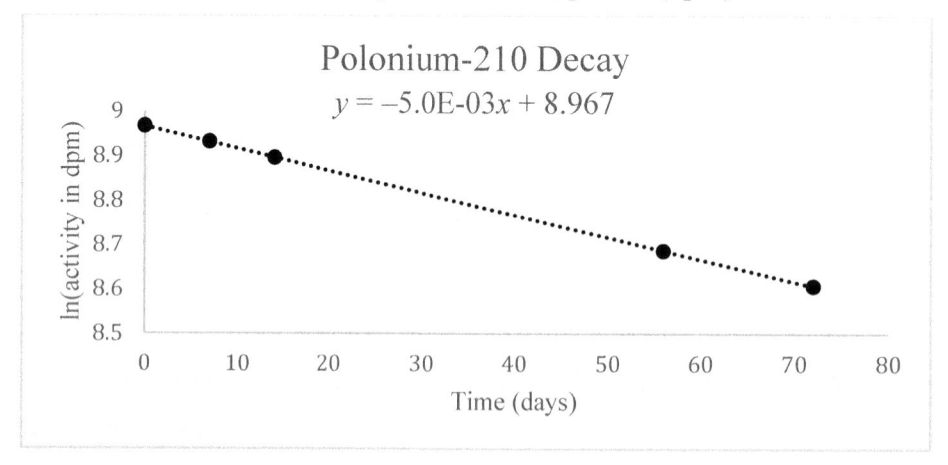

Polonium-210 Decay
y = −5.0E-03x + 8.967

The slope of the line is −5.0 × 10⁻³ day⁻¹. This is equal to −k, so $k = 5.0 \times 10^{-3} \text{ day}^{-1}$.

$t_{1/2} = \dfrac{0.693}{k} = \dfrac{0.693}{5.0 \times 10^{-3} \text{ day}^{-1}} = 140 \text{ days}$

20.68. Regarding the formation and decay of sodium-24:

(a) $^{23}_{11}\text{Na} + ^{1}_{0}\text{n} \rightarrow ^{24}_{11}\text{Na}$ and $^{24}_{11}\text{Na} \rightarrow ^{0}_{-1}\beta + ^{24}_{12}\text{Mg}$

(b) Determine k and the half-life:

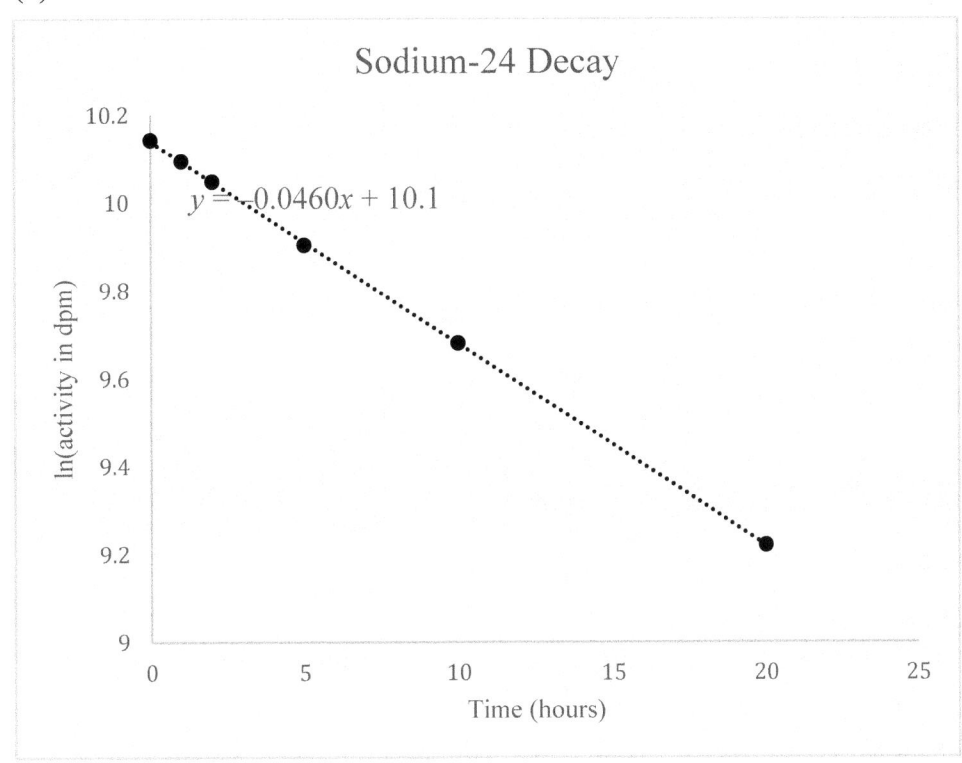

Sodium-24 Decay

$y = -0.0460x + 10.1$

The slope of the line is –0.0460 hour^{-1}. This is equal to –k, so $k = 0.0460$ hour^{-1}.

$$t_{1/2} = \frac{0.693}{0.0460 \text{ hour}^{-1}} = 15.1 \text{ hours}$$

20.69. If the ratio of $\frac{\text{Pb-206}}{\text{U-238}}$ is 0.33, a ratio of 0.25 of Pb-206 to 0.75 of U-238 is equivalent and indicates that 0.75 (75%) of the U-238 initially present in the rock has not decayed. Based on that measurement calculate a rate constant and use the 1st order rate equation to calculate the time required for this to occur:

$$k = \frac{0.693}{4.5 \times 10^9 \text{ y}} = 1.5 \times 10^{-10} \text{ y}^{-1}$$

$$\ln\left(\frac{0.75}{1.00}\right) = -(1.5 \times 10^{-10} \text{ y}^{-1})t$$

$$t = \frac{-0.29}{-1.5 \times 10^{-10} \text{ y}^{-1}} = 1.9 \times 10^9 \text{ y}$$

20.70. Volume of the circulatory system:

$(2.0 \times 10^6 \text{ dps})(1.0 \text{ mL}) = (1.5 \times 10^4 \text{ dps})(x)$
x = volume of circulatory system = 130 mL

Summary and Conceptual Questions

20.71. Tons of coal needed to be equivalent to one pound of ^{235}U:

$1 \text{ lb } ^{235}\text{U} \left(\dfrac{453.59 \text{ g } ^{235}\text{U}}{1 \text{ lb } ^{235}\text{U}} \right) \left(\dfrac{1 \text{ mol } ^{235}\text{U}}{235 \text{ g } ^{235}\text{U}} \right) \left(\dfrac{2.1 \times 10^{10} \text{ kJ}}{1 \text{ mol } ^{235}\text{U}} \right) = 4.05 \times 10^{10} \text{ kJ}$

Now compare this amount of energy to the energy yielded per ton of coal.

$\dfrac{4.05 \times 10^{10} \text{ kJ}}{2.6 \times 10^7 \text{ kJ/ton coal}} = 1600$ tons of coal

20.72. Regarding collision of an electron and a positron:

The equation for this process: $^{0}_{-1}\text{e} + ^{0}_{+1}\text{e} \rightarrow 2\gamma$

(a) mass of electron = mass of positron
$\Delta E = \Delta mc^2 = [2(9.109 \times 10^{-28} \text{ g})](1 \text{ kg}/10^3 \text{ g})(2.9979 \times 10^8 \text{ m/s})^2 = 1.637 \times 10^{-13}$ J
$= 1.637 \times 10^{-16}$ kJ

(b) Frequency of γ rays:

$\nu = \dfrac{E}{h} = \dfrac{1.637 \times 10^{-13} \text{ J}/2 \text{ }\gamma\text{-rays}}{6.626 \times 10^{-34} \text{ J} \cdot \text{s}} = 1.236 \times 10^{20} \text{ s}^{-1}$

20.73. If you assume that the catch represents a homogeneous sample of the tagged fish, the problem is straight forward. The percentage of tagged fish in the sample is $\dfrac{27}{5250} = 0.00514$ or 0.51%. If the 1000 tagged fish represent 0.51% of the fish in the lake, the number is approximately: $\dfrac{1000}{0.00514} = 190{,}000$ fish

20.74. Use radioisotope O-15 to explain the source of an O atom in a reaction product.

Assume that the O atom of the alcohol is "tagged" with radioactive oxygen (^{15}O). If the O in the water comes from the —OH of the acid, the water is free of the radioactive ^{15}O isotope.

H₃C—C(=O)—O—H + H—^{15}O—CH₃ ⟶ H₃C—C(=O)—^{15}O—CH₃ + H—O—H

If the O in the water comes from the alcohol, the water will contain radioactive oxygen.

$$H_3C-\overset{\overset{O}{\|}}{C}-O-H \; + \; H-{}^{15}O-CH_3 \longrightarrow H_3C-\overset{\overset{O}{\|}}{C}-O-CH_3 \; + \; H-{}^{15}O-H$$

(If this experiment is conducted, it turns out that the ^{15}O appears in the methyl acetate and not in the water.)

20.75. Production and decay of ^{210}Po:

(a) Synthesis reaction steps:

$${}^{209}_{83}Bi + {}^{1}_{0}n \rightarrow {}^{210}_{83}Bi$$

$${}^{210}_{83}Bi \rightarrow {}^{210}_{84}Po + {}^{0}_{-1}\beta$$

$${}^{210}_{84}Po \rightarrow {}^{206}_{82}Pb + {}^{4}_{2}\alpha$$

(b) Amount of ^{210}Po remaining after 3 weeks:

$$k = \frac{0.693}{t_{1/2}} = \frac{0.693}{138 \text{ d}} = 5.02 \times 10^{-3} \text{ d}^{-1}$$

3 weeks = 21 days

$$\ln\left(\frac{N_t}{N_0}\right) = -kt$$

$$\ln(N_t) - \ln(N_0) = -kt \text{ thus } \ln(N_t) = -kt + \ln(N_0)$$

$$\ln(N_t) = -(5.02 \times 10^{-3} \text{ d}^{-1})(21 \text{ d}) + \ln(10 \mu g)$$

$$\ln(N_t) = -0.105 + 2.30$$

$$\ln(N_t) = 2.20$$

$$N_t = e^{2.20} = 9.0 \text{ μg still remain}$$

20.76. Sequence from Th-232 to Th-228:

$${}^{232}_{90}Th \rightarrow {}^{228}_{88}Ra + {}^{4}_{2}\alpha$$

$${}^{228}_{88}Ra \rightarrow {}^{228}_{89}Ac + {}^{0}_{-1}\beta$$

$${}^{228}_{89}Ac \rightarrow {}^{228}_{90}Th + {}^{0}_{-1}\beta$$

20.77. For protactinium:

(a) The series containing protactinium-231 is the U-235 series, corresponding to the (4n + 3) series. (A Closer Look: Radioactive Decay Series, page 997.)

(b) A series of reactions to produce Pa-231:

First, alpha decay of U-235: $^{235}_{92}\text{U} \rightarrow {}^{231}_{90}\text{Th} + {}^{4}_{2}\alpha$

followed by the beta decay of Th-231: $^{231}_{90}\text{Th} \rightarrow {}^{231}_{91}\text{Pa} + {}^{0}_{-1}\beta$.

(c) Quantity of ore to provide 1.0 g of Pa-231 assuming 100% yield:

If the ore is 1 part per million, and you want 1.0 g, then you need 1,000,000 g of ore.

(d) Decay for Pa-231: $^{231}_{91}\text{Pa} \rightarrow {}^{227}_{89}\text{Ac} + {}^{4}_{2}\alpha$

20.78. A mathematical equation for the rate of delay:

$$\frac{\Delta N}{\Delta t} = -kN$$

$$N = (1.0 \times 10^{-3} \text{ g } {}^{238}\text{U})\left(\frac{1 \text{ mol } {}^{238}\text{U}}{238 \text{ g } {}^{238}\text{U}}\right)\left(\frac{6.022 \times 10^{23} \text{ nuclei}}{1 \text{ mol } {}^{238}\text{U}}\right) = 2.5 \times 10^{18} \text{ nuclei}$$

12 dps = $k(2.5 \times 10^{18}$ nuclei) and solving for k: $k = 4.7 \times 10^{-18}$ s^{-1}

$$t_{1/2} = \frac{0.693}{k} = \frac{0.693}{4.7 \times 10^{-18} \text{ s}^{-1}} = 1.5 \times 10^{17} \text{ s}$$

$$1.5 \times 10^{17} \text{ s}\left(\frac{1 \text{ h}}{3600 \text{ s}}\right)\left(\frac{1 \text{ day}}{24 \text{ h}}\right)\left(\frac{1 \text{ year}}{365 \text{ days}}\right) = 4.6 \times 10^{9} \text{ years}$$

This is very close to the literature value of 4.5×10^{9} years.

20.79. Which of the isotopes are anticipated in uranium ore:

Since both radium and polonium are decay products of uranium, they must belong to either the 4n + 2 (U-238) or 4n + 3 (U-235) decay series. The 4n + 3 series would give rise to isotopes with mass numbers:
235 → 231 → 227 → 223 → 219 → 215 → 211;
The 4n + 2 series would give rise to isotopes with mass numbers:
238 → 234 → 230 → 226 → 222 → 218 → 214 → 210.
Of the 5 isotopes given, Ra-226, and Po-210 are both members of the 4n + 2 decay series and have a sufficient half-life to be detected.

20.80. Nuclear equation to produce ^{293}Ts:

$^{249}_{97}\text{Bk} + {}^{48}_{20}\text{Ca} \rightarrow {}^{293}_{117}\text{Ts} + 4\,{}^{1}_{0}\text{n}$

Solution and Answer Guide

Kotz Treichel Townsend Treichel, Chemistry and Chemical Reactivity 11e, 978-0-357-85140-1, Chapter 21: The Chemistry of the Main Group Elements

TABLE OF CONTENTS

Applying Chemical Principles 789
Practicing Skills...790
General Questions..811
In the Laboratory..821
Summary and Conceptual Questions...825

Applying Chemical Principles

Lead in the Environment

21.1.1. Atoms of Pb in 1.0 L of 50-ppb blood: (Assume d of blood = 1.0 g/mL)

$$1.0\,L\left(\frac{50.\,g\,Pb}{1\times10^9\,g\,blood}\right)\left(\frac{1\,g\,blood}{1\,mL\,blood}\right)\left(\frac{1000\,mL\,blood}{1.0\,L\,blood}\right)\left(\frac{1\,mol\,Pb}{207.2\,g\,Pb}\right)\left(\frac{6.022\times10^{23}\,atoms\,Pb}{1\,mol\,Pb}\right)$$

$$= 1.5 \times 10^{17}\,Pb\,atoms$$

21.1.2. Mass of Pb in 750 mL of wine containing 2000 ppm Pb: (d of wine = 1.0 g/mL)

$$750\,mL\,wine\left(\frac{2000\,g\,Pb}{1\times10^6\,g\,wine}\right)\left(\frac{1\,g\,wine}{1\,mL\,wine}\right) = 1.5\,g\,Pb\,\text{or}\,2\,g\,Pb\,\text{(to 1 sf)}$$

21.1.3. Concentration of Pb in 4.7 L of blood from digesting a 0.15 g paint chip:

$$\frac{0.15\,g\,paint}{4.7\,L}\left(\frac{12\,g\,Pb}{100\,g\,paint}\right)\left(\frac{1\,mol\,Pb}{207.2\,g\,Pb}\right) = 1.8\times10^{-5}\,M\,Pb$$

Hydrogen Storage

21.2.1. Lewis structure for ammonia borane. Formal charges on atoms:

Formal charges:
$N = 5 - ½(8) = +1$
$B = 3 - ½(8) = -1$
$H = 1 - ½(2) = 0$

© 2024 Cengage Learning, Inc. All Rights Reserved. May not be scanned, copied or duplicated, or posted to a publicly accessible website, in whole or in part.

21.2.2. Mass of H in 1 kg of H₃NBH₃:

$$1.00 \times 10^3 \text{ g H}_3\text{NBH}_3 \left(\frac{6.048 \text{ g H}}{30.87 \text{ g H}_3\text{NBH}_3} \right) = 196 \text{ g H}$$

21.2.3. The H₂ density (g/L) in ammonia borane:

Volume of the same mass of H₃NBH₃ (1.00 kg):

$$1.00 \times 10^3 \text{ g H}_3\text{NBH}_3 \left(\frac{1 \text{ cm}^3 \text{ H}_3\text{NBH}_3}{0.780 \text{ g H}_3\text{NBH}_3} \right) = 1280 \text{ cm}^3 \text{ H}_3\text{NBH}_3 \text{ or } 1.28 \text{ L H}_3\text{NBH}_3$$

From 21.2.2, you know that this mass of ammonia borane contains 196 g H.

The hydrogen density is: $\dfrac{196 \text{ g H}}{1.28 \text{ L H}_3\text{NBH}_3} = 153 \text{ g/L}$

21.2.4. Proportion of H₂ released: (140 g H₂/153 g H₂) × 100% = 92%

21.2.5. Lewis structure for borazine; Is the compound isoelectronic with benzene:

With 3 B (9 valence electrons) and 3 N (15 valence electrons), there is a total of 24 valence electrons, the same as the number of valence electrons in benzene (C₆H₆). In short, borazine is isoelectronic with benzene.

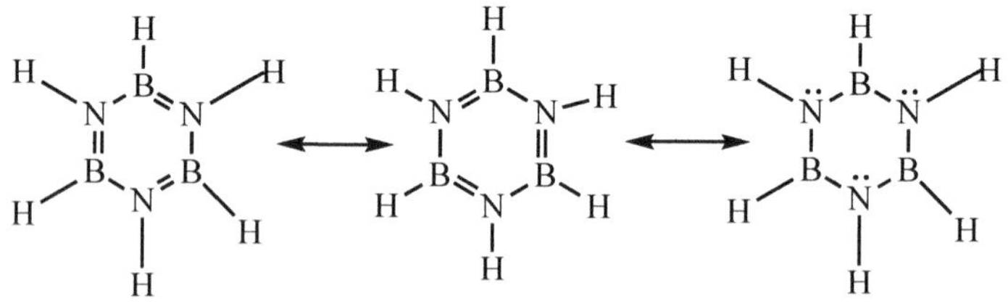

Practicing Skills

Properties of the Elements

21.1. Identify the incorrect formula:

Incorrect formula is Ca₂O₃. As a member of Group 2A (2), Ca forms a 2+ ion.

Oxygen forms a 2– ion, so a proper formula of calcium oxide would be CaO.

21.2. The name of the compound P₄O₁₀:

(d) Tetraphosphorus decaoxide is the correct name. Covalently bonded compounds are named by giving the prefix indicating the number of atoms, hence tetra- and deca-.

21.3. Least likely to be an oxidation state for Se in its compounds: +3.

21.4. Highest oxidation state for antimony in its compounds:
Antimony, located in Group 5A (15), would have +5 as its highest oxidation state.

21.5. Examples of two basic oxides, and equations showing the oxide formation from its elements:

4 Li(s) + O$_2$(g) → 2 Li$_2$O(s)

Lithium oxide is basic owing to the reaction: Li$_2$O(s) + H$_2$O(ℓ) → 2 LiOH(aq)

2 Ca(s) + O$_2$(g) → 2 CaO(s)

CaO is basic owing to the reaction: CaO(s) + H$_2$O(ℓ) → Ca(OH)$_2$(s)

21.6. Two acidic oxides:

N$_2$(g) + 2 O$_2$(g) → 2 NO$_2$(g)   2 NO$_2$(g) + H$_2$O(ℓ) → HNO$_3$(aq) + HNO$_2$(aq)

2 S(s) + 3 O$_2$(g) → 2 SO$_3$(g)   SO$_3$(g) + H$_2$O(ℓ) → H$_2$SO$_4$(aq)

21.7. Name and symbol with a valence configuration ns^2np^1:

The configuration shown is characteristic of the elements in Group 3A (13).

Symbol	B	Al	Ga	In	Tl
Name	Boron	Aluminum	Gallium	Indium	Thallium

Of these, technically only boron and aluminum have the configuration [noble gas]ns^2np^1 because the others in this group also have inner shell d electrons beyond the previous noble gas electron configuration.

21.8. Symbols and names for four monatomic ions that are isoelectronic with Ar:

S^{2-}, sulfide ion; Cl$^-$, chloride ion; K$^+$, potassium ion; Ca^{2+}, calcium ion (others are possible).

21.9. Select an alkali metal and write the balanced equation with Cl$_2$:

2 Na(s) + Cl$_2$(g) → 2 NaCl(g)

Reactions of alkali metals with halogens are exothermic given the excellent ability of these metals to function as reducing agents, and the ability of Cl$_2$ to function as an oxidizing agent.

Typically compounds formed from elements of greatly differing electronegativities (a metal and a nonmetal) are ionic in nature.

21.10. An alkaline earth metal and the equation for its reaction with oxygen:

2 Mg(s) + O$_2$(g) → 2 MgO(s)
The reaction is likely to be exothermic and the product is ionic.

21.11. Predict color, state of matter, water solubility of compound in Study Question 21.9:

Such compounds are typically colorless, high melting solids, and water soluble.

21.12. Predict color, state of matter, water solubility of compound in Study Question 21.10:

Such compounds are typically colorless or white, high melting solids, with water solubility that decreases with the mass of the metal ion.

21.13. Calcium is **not** expected to be found free in the Earth's crust. Its great reactivity with oxygen and water would dispose the metal to exist as an oxide or hydroxide.

21.14. Of the first 10 elements in the periodic table, which are found as free elements in Earth's crust?

Only C is found as the free element in Earth's crust. H, Li, Be, B, N, O, and F are found in compounds. He and Ne occur as free elements as gases in the atmosphere, as opposed to the crust.

21.15. The oxides listed in order of increasing basicity:
Metal oxides are basic, and nonmetal oxides are acidic. The three listed range from a nonmetal oxide (CO$_2$) to a metal oxide, SnO$_2$. So, the order is: CO$_2$ < SiO$_2$ < SnO$_2$.

21.16. The oxides listed in order of increasing basicity: SO$_3$ < SiO$_2$ < Al$_2$O$_3$ < Na$_2$O

21.17. Balanced equations for the following reactions:

(a) 2 Na(s) + Br$_2$(ℓ) → 2 NaBr(s)

(b) 2 Mg(s) + O$_2$(g) → 2 MgO(s)

(c) 2 Al(s) + 3 F$_2$(g) → 2 AlF$_3$(s)

(d) C(s) + O$_2$(g) → CO$_2$(g)

21.18. Balanced equations for the following reactions:

(a) 2 K(s) + I$_2$(g) → 2 KI(s)

(b) 2 Ba(s) + O$_2$(g) → 2 BaO(s)

(c) 16 Al(s) + 3 S$_8$(s) → 8 Al$_2$S$_3$(s)

(d) Si(s) + 2 Cl$_2$(g) → SiCl$_4$(ℓ)

Hydrogen

21.19. Element not reacting with hydrogen: (a) neon. Noble gases are relatively inert!

21.20. Method most suitable to prepare large quantities of H$_2$:

(c) The reaction of methane and water at high temperature.

21.21. Balanced chemical equation for hydrogen gas reacting with oxygen, chlorine, and nitrogen:

2 H$_2$(g) + O$_2$(g) → 2 H$_2$O(g)

H$_2$(g) + Cl$_2$(g) → 2 HCl(g)

3 H$_2$(g) + N$_2$(g) → 2 NH$_3$(g)

21.22. Equation for the reaction of potassium and hydrogen:

2 K(s) + H$_2$(g) → 2 KH(s)

The compound produced in this reaction, potassium hydride, is ionic, a solid, has a high melting point, and reacts vigorously with water.

21.23. The reaction: CH$_4$(g) + H$_2$O(g) → CO(g) + 3 H$_2$(g). Calculate $\Delta_rH°$, $\Delta_rS°$, and $\Delta_rG°$:

Data from Appendix L:

	CH$_4$(g)	H$_2$O(g)	CO(g)	H$_2$(g)
$\Delta_fH°$ (kJ/mol)	−74.87	−241.83	−110.525	0
$S°$ (J/K·mol)	+186.26	+188.84	+197.674	+130.7
$\Delta_fG°$ (kJ/mol)	−50.8	−228.59	−137.168	0

$\Delta_r H° = [(1\text{mol/mol-rxn})(–110.525 \text{ kJ/mol}) + (3 \text{ mol/mol-rxn})(0 \text{ kJ/mol})] –$

$[(1\text{mol/mol-rxn})(–74.87 \text{ kJ/mol}) + (1\text{mol/mol-rxn})(–241.83 \text{ kJ/mol})]$

$= 206.18 \text{ kJ/mol-rxn}$

$\Delta_r S° = [(1\text{mol/mol-rxn})(+197.674 \text{ J/K·mol}) + (3 \text{ mol/mol-rxn})(+130.7 \text{ J/K·mol})] –$

$[(1\text{mol/mol-rxn})(+186.26 \text{ J/K·mol}) + (1\text{mol/mol-rxn})(+188.84 \text{ J/K·mol})]$

$= +214.7 \text{ J/K · mol-rxn}$

$\Delta_r G° = [(1\text{mol/mol-rxn})(–137.168 \text{ kJ/mol}) + (3 \text{ mol/mol-rxn})(0 \text{ kJ/mol})] –$

$[(1\text{mol/mol-rxn})(–50.8 \text{ kJ/mol}) + (1\text{mol/mol-rxn})(–228.59 \text{ kJ/mol})]$

$= 142.2 \text{ kJ/mol-rxn}$

21.24. For the reaction of C(s) and H₂O to give CO(g) + H₂(g), calculate $\Delta_f H°$, $\Delta_r S°$, and $\Delta_r G°$: Reaction is $C(s) + H_2O(g) \rightarrow CO(g) + H_2(g)$. Data from Appendix L:

$\Delta_r H° = [(1 \text{ mol/mol-rxn})\Delta_f H°(CO(g)] – [(1 \text{ mol/mol-rxn})\Delta_f H°(H_2O(g)]$

$= (1 \text{ mol/mol-rxn})(–110.525 \text{ kJ/mol}) – (1 \text{ mol/mol-rxn})(–241.83 \text{ kJ/mol})$

$= 131.31 \text{ kJ/mol-rxn}$

$\Delta_r S° = [(1 \text{ mol/mol-rxn})S°(CO(g)) + (1 \text{ mol/mol-rxn})S°(H_2(g))] –$

$[(1 \text{ mol/mol-rxn})S°(C(s)) + (1 \text{ mol/mol-rxn})S°(H_2O(g))]$

$= (1 \text{ mol/mol-rxn})(197.674 \text{ J/K·mol}) + (1 \text{ mol/mol-rxn})(130.7 \text{ J/K·mol})$

$– [(1 \text{ mol/mol-rxn})(5.6 \text{ J/K·mol}) + (1 \text{ mol/mol-rxn})(188.84 \text{ J/K·mol})]$

$= 133.9 \text{ J/K · mol-rxn}$

$\Delta_r G° = \Delta_r H° – T\Delta_r S° = 131.31 \text{ kJ/mol-rxn} – (298 \text{ K})(133.9 \text{ J/K·mol-rxn})(1 \text{ kJ}/10^3 \text{ J})$

$= 91.4 \text{ kJ/mol-rxn}$

21.25. Prepare a balanced equation for each of the 3 steps and show that the sum is the decomposition of water to form hydrogen and oxygen.

$\cancel{2 \text{ SO}_2(g)} + \cancel{A} \; 2 \text{ H}_2\text{O}(\ell) + \cancel{2 \text{ I}_2(s)} \rightarrow \cancel{2 \text{ H}_2\text{SO}_4(\ell)} + \cancel{4 \text{ HI}(g)}$

$\cancel{2 \text{ H}_2\text{SO}_4(\ell)} \rightarrow \cancel{2 \text{ H}_2\text{O}(\ell)} + \cancel{2 \text{ SO}_2(g)} + \text{O}_2(g)$

$\cancel{4 \text{ HI}(g)} \rightarrow 2 \text{ H}_2(g) + \cancel{2 \text{ I}_2(s)}$

$2 \text{ H}_2\text{O}(\ell) \rightarrow 2 \text{ H}_2(g) + \text{O}_2(g)$

21.26. Mass of H₂ produced by reaction of steam with CH₄, petroleum, and coal:

Methane: $CH_4(g) + H_2O(g) \rightarrow 3 H_2(g) + CO(g)$

$$1 \text{ mol CH}_4 \left(\frac{3 \text{ mol H}_2}{1 \text{ mol CH}_4} \right) \left(\frac{2.02 \text{ g H}_2}{1 \text{ mol H}_2} \right) = 6.06 \text{ g H}_2$$

Petroleum: $CH_2(\ell) + H_2O(g) \rightarrow 2 H_2(g) + CO(g)$

$$1 \text{ mol CH}_2 \left(\frac{2 \text{ mol H}_2}{1 \text{ mol CH}_2} \right) \left(\frac{2.02 \text{ g H}_2}{1 \text{ mol H}_2} \right) = 4.04 \text{ g H}_2$$

Coal: $2 CH(s) + 2 H_2O(g) \rightarrow 3 H_2(g) + 2 CO(g)$

$$1 \text{ mol CH} \left(\frac{3 \text{ mol H}_2}{2 \text{ mol CH}} \right) \left(\frac{2.02 \text{ g H}_2}{1 \text{ mol H}_2} \right) = 3.03 \text{ g H}_2$$

Alkali Metals

21.27. Not a property of Na: (b) Has a high melting point (> 400 °C)
Na has a melting point of 97.8 °C.

21.28. Components of Na₂O₂: (c) two Na⁺ ions and one O₂²⁻ ion

21.29. Equations for the reaction of sodium with the halogens:

$2 Na(s) + F_2(g) \rightarrow 2 NaF(s)$

$2 Na(s) + Cl_2(g) \rightarrow 2 NaCl(s)$

$2 Na(s) + Br_2(\ell) \rightarrow 2 NaBr(s)$

$2 Na(s) + I_2(s) \rightarrow 2 NaI(s)$

Physical properties of the alkali metal halides:
(1) ionic solids (2) high melting and boiling points (3) white color (4) water soluble

21.30. Equations for the reaction of Li, Na, and K with O₂:

$4 Li(s) + O_2(g) \rightarrow 2 Li_2O(s)$ lithium oxide
$2 Na(s) + O_2(g) \rightarrow Na_2O_2(s)$ sodium peroxide
$K(s) + O_2(g) \rightarrow KO_2(s)$ potassium superoxide

21.31. In the electrolysis of aqueous NaCl:

(a) The balanced equation for the process:
$2 NaCl(aq) + 2 H_2O(\ell) \rightarrow 2 NaOH(aq) + Cl_2(g) + H_2(g)$

(b) Anticipated mass ratios:

$$\frac{1 \text{ mol Cl}_2}{2 \text{ mol NaOH}} = \frac{70.9 \text{ g Cl}_2}{80.0 \text{ g NaOH}} = 0.886 \text{ g Cl}_2/\text{g NaOH}$$

Actual: $\dfrac{1.14 \times 10^{10} \text{ kg Cl}_2}{1.19 \times 10^{10} \text{ kg NaOH}} = 0.958 \dfrac{\text{kg Cl}_2}{\text{kg NaOH}}$

The difference in ratios means that alternative methods of producing chlorine are used. One of these is the Kel-Chlor process, which produces Cl₂ from the oxidation of HCl. (The other product is water.)

21.32. Regarding the electrolysis of KCl and CsI:

(a) Anode: 2 Cl⁻(aq) → Cl₂(g) + 2 e⁻

Cathode: 2 H₂O(ℓ) + 2 e⁻ → 2 OH⁻(aq) + H₂(g)

Chloride ion is oxidized and water is reduced.

(b) Anode: 2 I⁻(aq) → I₂(s) + 2 e⁻

Cathode: 2 H₂O(ℓ) + 2 e⁻ → 2 OH⁻(aq) + H₂(g)

Iodide ion is oxidized and water is reduced.

Alkaline Earth Elements

21.33. Calcium compound that does not dissolve in HCl:

(c) gypsum would not dissolve in HCl. The other compounds (either carbonates or hydroxides) would react with HCl—and dissolve.

21.34. Not an industrial process for calcium minerals: (c) converting slaked lime Ca(OH)₂ to lime. The reverse of this process is an important one.

21.35. Balanced equations for the reaction of magnesium with nitrogen and oxygen:

3 Mg(s) + N₂(g) → Mg₃N₂(s)

2 Mg(s) + O₂(g) → 2 MgO(s)

21.36. Regarding the reaction of calcium with hydrogen gas:

(a) Ca(s) + H₂(g) → CaH₂(s)
(b) CaH₂(s) + 2 H₂O(ℓ) → Ca(OH)₂(s) + 2 H₂(g)

21.37. Uses of limestone:

Agricultural: to furnish Ca²⁺ to plants and neutralize acidic soils

Building: lime (CaO) is used in mortar and absorbs CO₂ to form CaCO₃

Steel-making: CaCO₃ furnishes lime (CaO) in the basic oxygen process. The lime reacts with gangue (SiO₂) to form calcium silicate.

The balanced equation for the reaction of limestone with carbon dioxide in water:

$$CaCO_3(s) + H_2O(\ell) + CO_2(g) \rightarrow Ca^{2+}(aq) + 2\ HCO_3^-(aq)$$

This reaction is important in the formation of "hard water" (not particularly a great happening for plumbing) and stalagmites and stalactites (aesthetically pleasing in caves). (See photos of the reactions on page 145 of *Chemistry & Chemical Reactivity*, Figure 3.5.)

21.38. Hard water:

Hard water contains metal ions such as Ca^{2+} and Mg^{2+}. Hard water forms as ground water flows through mineral beds having slightly soluble salts such as carbonates. Hard water ions react with soap to form insoluble soap scum. The ions decrease the lathering ability of the soap. Hard water leads to deposits of boiler scale in hot water heaters.

21.39. The amount of SO₂ that could be removed by 1200 kg of CaO by the reaction:

$$CaO(s) + SO_2(g) \rightarrow CaSO_3(s):$$

$$1.2 \times 10^6\ g\ CaO \left(\frac{1\ mol\ CaO}{56.077\ g\ CaO}\right)\left(\frac{1\ mol\ SO_2}{1\ mol\ CaO}\right)\left(\frac{64.06\ g\ SO_2}{1\ mol\ SO_2}\right) = 1.4 \times 10^6\ g\ SO_2$$

21.40. Calculate *K* for the reaction between calcium hydroxide and magnesium ions:

$Ca(OH)_2(s) \rightleftharpoons Ca^{2+}(aq) + 2\ OH^-(aq)$	$K_{sp} = 5.5 \times 10^{-6}$
$Mg^{2+}(aq) + 2\ OH^-(aq) \rightleftharpoons Mg(OH)_2(s)$	$1/K_{sp} = 1/(5.6 \times 10^{-12})$
$Ca(OH)_2(s) + Mg^{2+}(aq) \rightleftharpoons Mg(OH)_2(s) + Ca^{2+}(aq)$	$K = 9.8 \times 10^5$

Adding Ca(OH)₂ to sea water will lead to the precipitation of Mg(OH)₂, which can then be further processed to ultimately yield Mg metal.

Boron and Aluminum

21.41. Ranking of Al in the Earth's crust: (c) third—behind O and Si.

21.42. Gallium and aluminum hydroxides: (d) dissolve in acid and in base—as is typical of amphoteric hydroxides.

21.43. Structure for the cyclic anion in the salt K₃B₃O₆, and the chain anion in Ca₂B₂O₅

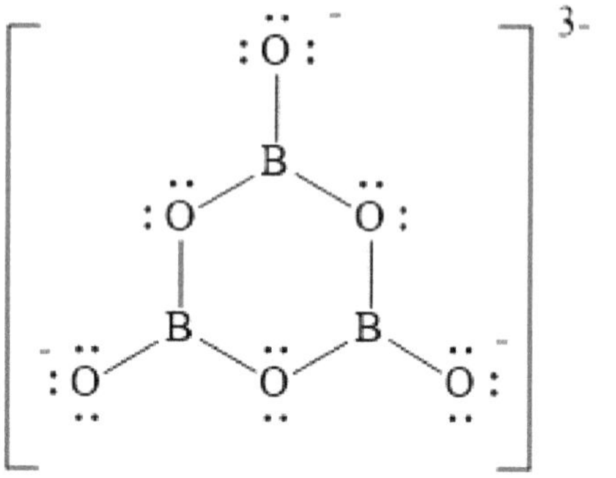

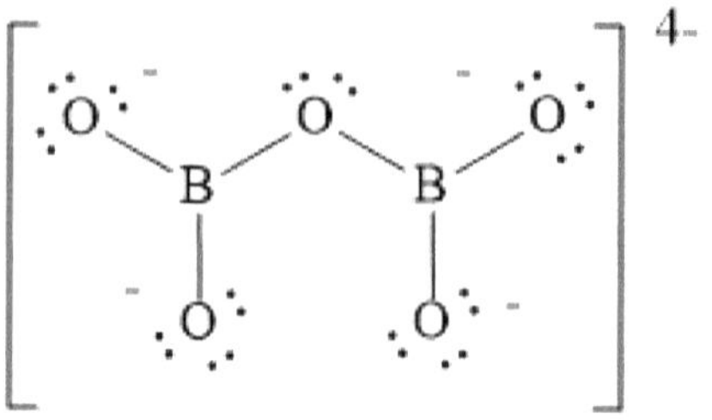

21.44. Regarding the reactions of BCl₃ with water:

(a) BCl₃(g) + 3 H₂O(ℓ) → B(OH)₃(s) + 3 HCl(aq)

(b) Enthalpy change for the hydrolysis of BCl₃:

Δ_rH° = [Δ_fH° B(OH)₃(s) + 3 Δ_fH° HCl(aq)] − [Δ_fH° BCl₃(g) + 3 Δ_fH° H₂O(ℓ)]

Δ_rH° = [(1 mol/mol-rxn)(−1094 kJ/mol) + (3 mol/mol-rxn)(−167.159 kJ/mol)]

− [(1 mol/mol-rxn)(−403 kJ/mol) + (3 mol/mol-rxn)(−285.83 kJ/mol)]

Δ_rH° = −335 kJ/mol-rxn

21.45. For the reactions of boron hydrides:

(a) Balanced Equation: 2 B₅H₉(g) + 12 O₂(g) → 5 B₂O₃(s) + 9 H₂O(ℓ)

(b) Enthalpy of combustion for B₅H₉(g):

Δ_combH° = [9 Δ_fH° H₂O(ℓ) + 5 Δ_fH° B₂O₃(s)] − [2 Δ_fH° B₅H₉(g) + 12 Δ_fH° O₂(g)]

Δ_combH° = [(9 mol/mol-rxn)(−285.83 kJ/mol) + (5 mol/mol-rxn)(−1271.9 kJ/mol)]

− [(2 mol/mol-rxn)(73.2 kJ/mol)) + (12 mol/mol-rxn)(0 kJ/mol)]

= −9078.4 kJ/mol-rxn

In kJ/mol: −9078.4 kJ /2 mol B₅H₉(g) = −4539.2 kJ/mol

Enthalpy of combustion for B₂H₆ (from the text) is −2170.4 kJ/mol, so the enthalpy of combustion for B₅H₉ is more than twice as great as the value for diborane.

(c) Compare Enthalpies of Combustion for C₂H₆(g) with B₂H₆(g):

For diborane: B₂H₆(g) + 3 O₂(g) → B₂O₃(s) + 3 H₂O(ℓ)   Δ_combH° = −2170.4 kJ/mol

Energy per gram = (−2170.4 kJ/mol) × (1 mol/27.67 g B₂H₆) = −78.44 kJ/g B₂H₆.

For C: 2 C₂H₆(g) + 7 O₂(g) → 4 CO₂(g) + 6 H₂O(ℓ)

$\Delta_{comb}H° = [(6 \text{ mol/mol-rxn})(-285.83 \text{ kJ/mol}) + (4 \text{ mol/mol-rxn})(-393.509 \text{ kJ/mol})]$

$- [(2 \text{ mol/mol-rxn})(-83.85 \text{ kJ/mol}) + (7 \text{ mol/mol-rxn})(0 \text{ kJ/mol})]$

$\Delta_{comb}H° = -3121.32$ kJ or -1560.66 kJ/mol C_2H_6. Therefore, the energy evolved per gram of $C_2H_6 = -51.90$ kJ/g.

Diborane transfers more energy per gram than ethene.

21.46. For the preparation of diborane (B_2H_6):

The reaction is: $2 \text{ NaBH}_4(s) + I_2(s) \rightarrow B_2H_6(g) + 2 \text{ NaI}(s) + H_2(g)$

$NaBH_4$ is oxidized and I_2 is reduced.

21.47. The equations for the reaction of aluminum with HCl, Cl_2 and O_2:
$2 \text{ Al}(s) + 6 \text{ HCl}(aq) \rightarrow 2 \text{ Al}^{3+}(aq) + 6 \text{ Cl}^-(aq) + 3 H_2(g)$

$2 \text{ Al}(s) + 3 Cl_2(g) \rightarrow 2 \text{ AlCl}_3(s)$

$4 \text{ Al}(s) + 3 O_2(g) \rightarrow 2 \text{ Al}_2O_3(s)$

21.48. For the reaction of Al with H_2O:

(a) $2 \text{ Al}(s) + 3 H_2O(\ell) \rightarrow 3 H_2(g) + Al_2O_3(s)$

(b) $\Delta_rH° = \Delta_fH° \text{ Al}_2O_3(s) - 3 \Delta_fH° H_2O(\ell)$

$= (1 \text{ mol/mol-rxn})(-1675.7 \text{ kJ/mol}) - (3 \text{ mol/mol-rxn})(-285.83 \text{ kJ/mol})$

$= -818.2$ kJ/mol-rxn

$\Delta_rS° = [S° \text{ Al}_2O_3(s) + 3 S° H_2(g)] - [2 S° \text{ Al}(s) + 3 S° H_2O(\ell)]$

$\Delta_rS° = [(1 \text{ mol/mol-rxn})(50.92 \text{ J/K·mol}) + (3 \text{ mol/mol-rxn})(130.7 \text{ J/K·mol})]$

$- [(2 \text{ mol/mol-rxn})(28.3 \text{ J/K·mol}) + (3 \text{ mol/mol-rxn})(69.95 \text{ J/K·mol})]$

$= 176.6$ J/K · mol-rxn

$\Delta_rG° = [\Delta_fG° \text{ Al}_2O_3(s) - 3 \Delta_fG° H_2O(\ell)]$

$= (1 \text{ mol/mol-rxn})(-1582.3 \text{ kJ/mol}) - (3 \text{ mol/mol-rxn})(-237.15 \text{ kJ/mol})$

$= -870.9$ kJ

The negative free energy change for the reaction indicates that the reaction is product-favored at equilibrium. There is also both a negative enthalpy change for the reaction as well as a positive entropy change, so the reaction is favored by both enthalpy and entropy.

(c) A thin film of Al_2O_3 on the surface of the metal is slow to react with water and other substances, preventing further reaction.

21.49. The equation for the reaction of aluminum dissolving in aqueous NaOH:

2 Al(s) + 2 NaOH(aq) + 6 H$_2$O(ℓ) → 2 NaAl(OH)$_4$(aq) + 3 H$_2$(g)

Volume of H$_2$ (in mL) produced when 13.2 g of Al react:

$$13.2 \text{ g Al}\left(\frac{1 \text{ mol Al}}{26.982 \text{ g Al}}\right)\left(\frac{3 \text{ mol H}_2}{2 \text{ mol Al}}\right) = 0.734 \text{ mol H}_2$$

$$V = \frac{(0.734 \text{ mol H}_2)\left(0.082057 \frac{\text{L} \cdot \text{atm}}{\text{K} \cdot \text{mol}}\right)}{735 \text{ mm Hg}\left(\frac{1 \text{ atm}}{760 \text{ mm Hg}}\right)} = 18.4 \text{ L}$$

21.50. Regarding reactions of Al$_2$O$_3$:

(a) Al$_2$O$_3$(s) + 3 SiO$_2$(s) → Al$_2$(SiO$_3$)$_3$(s)

(b) Al$_2$O$_3$(s) + CaO(s) → Ca(AlO$_2$)$_2$(s)

21.51. The equation for the reaction of aluminum oxide with sulfuric acid:
Al$_2$O$_3$(s) + 3 H$_2$SO$_4$(aq) → Al$_2$(SO$_4$)$_3$(s) + 3 H$_2$O(ℓ)

Mass of Al$_2$O$_3$ to make 1 kg of Al$_2$(SO$_4$)$_3$:

$$1.00 \times 10^3 \text{ g Al}_2(\text{SO}_4)_3\left(\frac{1 \text{ mol Al}_2(\text{SO}_4)_3}{342.1 \text{ g Al}_2(\text{SO}_4)_3}\right)\left(\frac{1 \text{ mol Al}_2\text{O}_3}{1 \text{ mol Al}_2(\text{SO}_4)_3}\right)$$

$$\left(\frac{102.0 \text{ g Al}_2\text{O}_3}{1 \text{ mol Al}_2\text{O}_3}\right)\left(\frac{1 \text{ kg}}{1 \times 10^3 \text{ g}}\right) = 0.298 \text{ kg Al}_2\text{O}_3$$

Mass of H$_2$SO$_4$ to make 1 kg of Al$_2$(SO$_4$)$_3$:

$$1.00 \times 10^3 \text{ g Al}_2(\text{SO}_4)_3\left(\frac{1 \text{ mol Al}_2(\text{SO}_4)_3}{342.1 \text{ g Al}_2(\text{SO}_4)_3}\right)\left(\frac{3 \text{ mol H}_2\text{SO}_4}{1 \text{ mol Al}_2(\text{SO}_4)_3}\right)$$

$$\left(\frac{98.07 \text{ g H}_2\text{SO}_4}{1 \text{ mol H}_2\text{SO}_4}\right)\left(\frac{1 \text{ kg}}{1 \times 10^3 \text{ g}}\right) = 0.860 \text{ kg H}_2\text{SO}_4$$

21.52. For the manufacture of aerated concrete blocks:

$$0.56 \text{ g Al}\left(\frac{1 \text{ mol Al}}{26.982 \text{ g Al}}\right)\left(\frac{3 \text{ mol H}_2}{2 \text{ mol Al}}\right) = 0.031 \text{ mol H}_2$$

$$V = \frac{nRT}{P} = \frac{(0.031 \text{ mol H}_2)(0.082057 \text{ L} \cdot \text{atm/K} \cdot \text{mol})(299 \text{ K})}{745 \text{ mm Hg}\left(\frac{1 \text{ atm}}{760 \text{ mm Hg}}\right)} = 0.78 \text{ L}$$

Silicon

21.53. Silicon reaction puzzle:

Compound A: SiO$_2$ (oxidation of Si)

Compound B: SiO$_2$ + 2 Na$_2$CO$_3$(ℓ) → Na$_4$SiO$_4$ + 2 CO$_2$ (B is Na$_4$SiO$_4$)

Compound C: Na$_4$SiO$_4$ + 4 HCl(aq) → 4 NaCl + 2 H$_2$O + SiO$_2$ (C is SiO$_2$)

21.54. Oxidation state of Si in [Si$_3$O$_9$]$^{6-}$:
(a) Bookkeeping of electrons tells us: 3(Si) + 9(O) = 6 –. With O having a –2 oxidation state, 3(Si) = +12, and so Si = +4.
(b) Structure of anion: With each atom in the ring having a tetrahedral electron-pair geometry, it is not planar.

21.55. In a pyroxene each tetrahedral silicon atom is surrounded by four oxygen atoms, as shown in the structure below. The SiO$_4$ tetrahedra are linked because each silicon atom shares a bridging O atom with a neighboring Si atom. In addition, each Si atom has two "terminal" oxygen atoms. The ratio of O atoms to Si atoms is 3:1 in the anion SiO$_3^{2-}$. Common minerals have, among others, Mg^{2+} or Fe^{2+} cations as counterions.

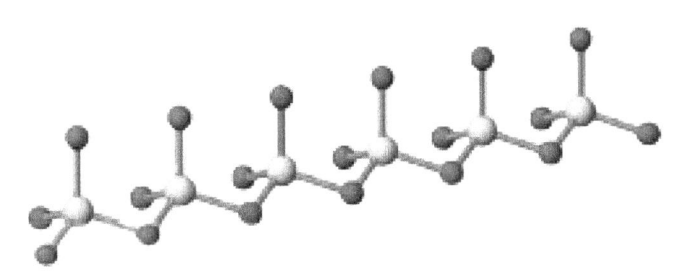

21.56. Production of ultrapure Si from sand:

Sand (SiO$_2$) is reduced to Si by reaction with coke (C) in an electric furnace at 3000 °C. The Si obtained this way is reacted with chlorine to form SiCl$_4$(ℓ), which is purified by distillation. The SiCl$_4$(ℓ) is reduced to Si by reaction with very pure magnesium or zinc. This "pure" silicon is made ultrapure by a process known as zone refining.

21.57. The structure for the silicate anion is shown below:

Like the pyroxenes (Study Question 21.55), each silicon atom is associated with a net of 3 O atoms. There are six SiO_3^{2-} "units" linked through bridging O atoms to give the $[Si_6O_{18}]^{12-}$ ring. The **net charge** on the anion is 12–.

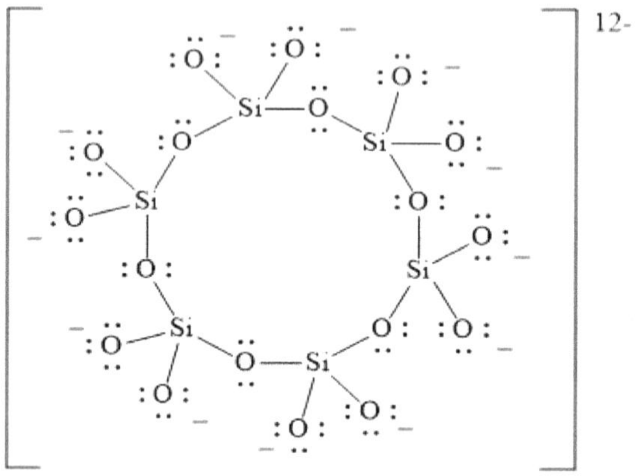

21.58. Ribbon structure for silicate:

The $Si_2O_5^{2-}$ repeating unit results in a ribbon structure consisting of two parallel chains. Each tetrahedral SiO_4 unit shares three oxygen atoms with adjacent SiO_4 tetrahedra.

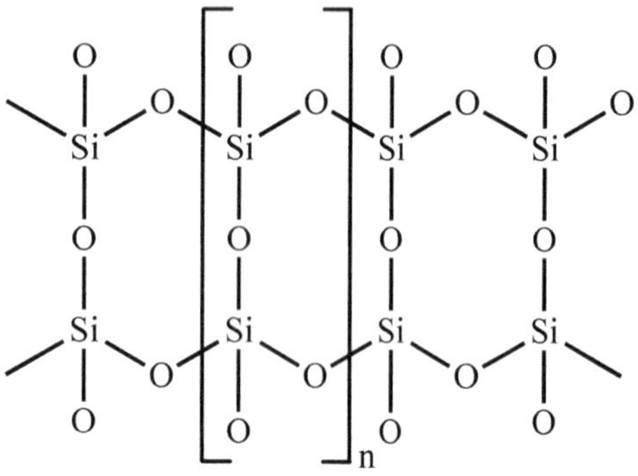

Nitrogen and Phosphorus

21.59. Lewis structures for N_2O:

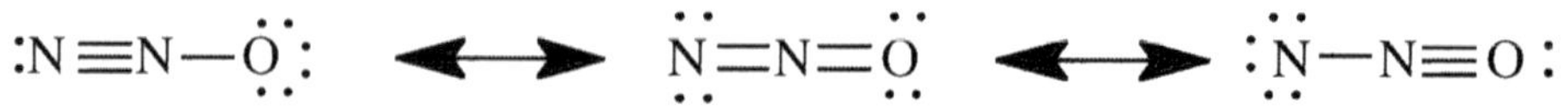

Analysis of formal charges shows the leftmost structure with the O atom (more electronegative than N) bearing a formal charge of –1. In the central resonance structure the O atom has a formal charge of 0 (zero). The rightmost structure would have a formal charge of +1 on the very electronegative O atom, an undesirable charge distribution. Therefore, the more probable forms of N_2O are the leftmost and center forms with an "average" N–N bond order of between 2 and 3.

21.60. Statement about NH₃ that is incorrect:

(b) aqueous solutions of NH₃ are acidic (Aqueous ammonia is a weak base.)

21.61. The $\Delta_f G°$ data from Appendix L are shown below:

compound:	NO(g)	NO₂(g)	N₂O(g)	N₂O₄(g)
$\Delta_f G°$ (kJ/mol)	+86.58	+51.23	+104.20	+97.73

To ask if the oxide is stable with respect to decomposition is to ask about $\Delta_r G°$ for the process: $N_xO_y \rightarrow (x/2) N_2(g) + (y/2) O_2(g)$

This reaction is *the reverse* of the free energy of formation, $\Delta_f G°$, for each of these oxides. $\Delta_r G°$ for such a process has an *opposite sign* from the data given above. Since that sign is negative, the process is product-favored at equilibrium. Hence, **all** the oxides shown above are unstable with respect to decomposition.

21.62. Calculate enthalpy and free energy change for 2 NO₂ → N₂O₄ reaction.

$\Delta_r H° = [\Delta_f H° \text{ N}_2\text{O}_4(g) - 2 \Delta_f H° \text{ NO}_2(g)]$

= (1 mol/mol-rxn)(9.08 kJ/mol) – (2 mol/mol-rxn)(33.1 kJ/mol)

= –57.1 kJ/mol-rxn

$\Delta_r G° = [\Delta_f G° \text{ N}_2\text{O}_4(g) - 2 \Delta_f G° \text{ NO}_2(g)]$

= (1 mol/mol-rxn)(97.73 kJ/mol) – (2 mol/mol-rxn)(51.23 kJ/mol)

= –4.73 kJ/mol-rxn

The reaction is exothermic, and the free energy change is negative indicating that the reaction is product-favored at equilibrium.

21.63. Calculate $\Delta_r H°$ and $\Delta_r G°$ for the reaction: 2 NO(g) + O₂(g) → 2 NO₂(g)

	NO(g)	O₂(g)	NO₂(g)
$\Delta_f H°$ (kJ/mol)	+90.29	0	+33.1
$\Delta_f G°$ (kJ/mol)	+86.58	0	+51.23

$\Delta_r H° = [(2 \text{ mol/mol-rxn})(+33.1 \text{ kJ/mol})] - [(2 \text{ mol/mol-rxn})(+90.29 \text{ kJ/mol})]$

= –114.4 kJ/mol-rxn

$\Delta_r G° = [(2 \text{ mol/mol-rxn})(+51.23 \text{ kJ/mol})] - [(2 \text{ mol/mol-rxn})(+86.58 \text{ kJ/mol})]$

= –70.7 kJ/mol-rxn

The reaction is exothermic (negative $\Delta_r H^\circ$) and product-favored at equilibrium (negative $\Delta_r G^\circ$).

21.64. For the industrial synthesis of nitric acid:

(a) $\Delta_r G^\circ = [\Delta_f G^\circ \text{ HNO}_3\text{(aq)} + \Delta_f G^\circ \text{ H}_2\text{O}(\ell)] - [\Delta_f G^\circ \text{ NH}_3\text{(g)}]$

$$\Delta_r G^\circ = [(1 \text{ mol/mol-rxn})(-111.25 \text{ kJ/mol}) + (1 \text{ mol/mol-rxn})(-237.15 \text{ kJ/mol})] - [(1 \text{ mol/mol-rxn})(-16.37 \text{ kJ/mol})] = -332.03 \text{ kJ/mol-rxn}$$

(b) $\Delta_r G^\circ = -332.03 \text{ kJ/mol} = -RT \ln K = -(8.3145 \times 10^{-3} \text{ kJ/K·mol})(298 \text{ K}) \ln K$

and $K = 2 \times 10^{58}$

21.65. Regarding the use of hydrazine in steam boilers:

(a) The reaction of hydrazine with dissolved oxygen:

$$N_2H_4\text{(aq)} + O_2\text{(g)} \rightarrow N_2\text{(g)} + 2\ H_2O(\ell)$$

(b) Mass of hydrazine to consume the oxygen in 3.00×10^4 L of water:

$$3.00\times10^4 \text{ L H}_2\text{O}\left(\frac{0.0044 \text{ g O}_2}{0.100 \text{ L H}_2\text{O}}\right)\left(\frac{1 \text{ mol O}_2}{32.0 \text{ g O}_2}\right)\left(\frac{1 \text{ mol N}_2\text{H}_4}{1 \text{ mol O}_2}\right)\left(\frac{32.05 \text{ g N}_2\text{H}_4}{1 \text{ mol N}_2\text{H}_4}\right)$$

$$= 1.32\times10^3 \text{ g N}_2\text{H}_4$$

21.66. Regarding the use of Na_2SO_3 in steam boilers:

$$3.00\times10^4 \text{ L H}_2\text{O}\left(\frac{0.0044 \text{ g O}_2}{0.100 \text{ L H}_2\text{O}}\right)\left(\frac{1 \text{ mol O}_2}{32.0 \text{ g O}_2}\right)\left(\frac{2 \text{ mol Na}_2\text{SO}_3}{1 \text{ mol O}_2}\right)\left(\frac{126.0 \text{ g}}{1 \text{ mol Na}_2\text{SO}_3}\right)$$

$$= 1.04\times10^4 \text{ g Na}_2\text{SO}_3$$

21.67. For diphosphorous acid, $H_4P_2O_5$: The structure of diphosphorous acid, $H_4P_2O_5$, is shown below. The maximum number of protons that can be ionized in this acid is 2 per formula unit. The "acidic" H atoms are those attached to the O atoms, and NOT those directly bonded to the P atoms.

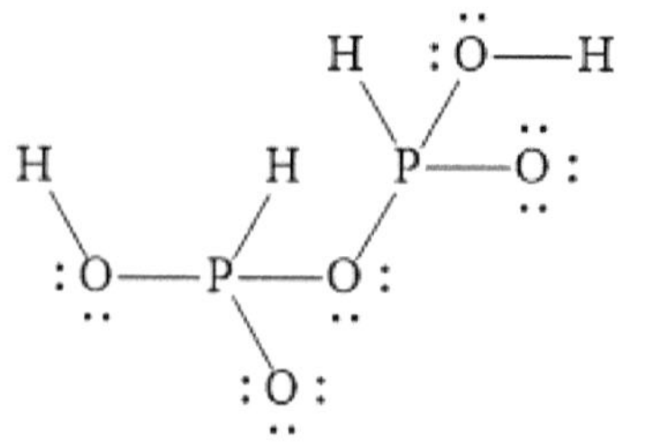

21.68. Lewis structure of the azide ion: The azide ion has three resonance structures:

$$\left[\ddot{\text{N}}=\text{N}=\ddot{\text{N}} \right]^{-} \longleftrightarrow \left[:\ddot{\text{N}}-\text{N}\equiv\text{N}: \right]^{-} \longleftrightarrow \left[:\text{N}\equiv\text{N}-\ddot{\text{N}}: \right]^{-}$$

The central nitrogen has two groups attached (two nitrogen atoms and zero lone pairs) so the ion is linear.

Oxygen and Sulfur

21.69. Not a common oxidation number for S in compounds: (c) +3

21.70. Statement about O_2 that is not true: (d) all electrons in O_2 are paired.
If liquid O_2 is poured between the poles of a strong magnet the oxygen is attracted to the magnet because of the unpaired electrons. See the photo on page 458 of *Chemistry & Chemical Reactivity*.

21.71. Regarding the reactions of S with O_2:

(a) Allowable release of SO_2: (0.30%)

$$1.80 \times 10^6 \text{ kg H}_2\text{SO}_4 \left(\frac{1000 \text{ g}}{1 \text{ kg}} \right) \left(\frac{1 \text{ mol H}_2\text{SO}_4}{98.07 \text{ g H}_2\text{SO}_4} \right) \left(\frac{1 \text{ mol SO}_2}{1 \text{ mol H}_2\text{SO}_4} \right)$$

$$\left(\frac{64.06 \text{ g SO}_2}{1 \text{ mol SO}_2} \right) \left(\frac{0.30 \text{ g SO}_2 \text{ vented}}{100 \text{ g SO}_2} \right) \left(\frac{1 \text{ kg}}{1000 \text{ g}} \right)$$

$= 3.53 \times 10^3$ kg SO_2 = 3.5×10^3 kg SO_2 (2 sf)

(b) Mass of $Ca(OH)_2$ to remove 3.53×10^3 kg SO_2:

$$3.53 \times 10^3 \text{ kg SO}_2 \left(\frac{1000 \text{ g}}{1 \text{ kg}} \right) \left(\frac{1 \text{ mol SO}_2}{64.06 \text{ g SO}_2} \right) \left(\frac{1 \text{ mol Ca(OH)}_2}{1 \text{ mol SO}_2} \right)$$

$$\left(\frac{74.09 \text{ g Ca(OH)}_2}{1 \text{ mol Ca(OH)}_2} \right) \left(\frac{1 \text{ kg}}{1000 \text{ g}} \right)$$

$= 4.1 \times 10^3$ kg $Ca(OH)_2$

21.72. In the production of sulfuric acid:

(1) $S(s) + O_2(g) \rightarrow SO_2(g)$ $\Delta_r H° = \Delta_f H° \text{ SO}_2 = -296.84$ kJ/mol-rxn

(2) $SO_2(g) + \frac{1}{2} O_2(g) \rightarrow SO_3(g)$ $\Delta_r H° = \Delta_f H° \text{ SO}_3 - \Delta_f H° \text{ SO}_2 = -98.93$ kJ/mol-rxn

(3) $SO_3(g) + H_2O(\text{in H}_2\text{SO}_4) \rightarrow H_2SO_4(\ell)$ $\Delta_r H° = -130$ kJ/mol-rxn

---

$S(s) + \frac{3}{2} O_2(g) + H_2O(\text{in H}_2\text{SO}_4) \rightarrow H_2SO_4(\ell)$

$\Delta_r H° = \Delta_r H°(1) + \Delta_r H°(2) + \Delta_r H°(3) = -526$ kJ/mol-rxn = -530 kJ/mol-rxn (2 sf)

$$\Delta H° = 1.00 \times 10^3 \text{ kg H}_2\text{SO}_4 \left(\frac{10^3 \text{ g}}{1 \text{ kg}}\right)\left(\frac{1 \text{ mol H}_2\text{SO}_4}{98.07 \text{ g H}_2\text{SO}_4}\right)\left(\frac{1 \text{ mol-rxn}}{1 \text{ mol H}_2\text{SO}_4}\right)\left(\frac{-526 \text{ kJ}}{1 \text{ mol-rxn}}\right)$$

$$= -5.4 \times 10^6 \text{ kJ (2 sf)}$$

The negative sign tells you that 5.4×10^6 kJ of energy is released per metric ton of sulfuric acid produced.

21.73. The disulfide ion S_2^{2-} Lewis dot structure:

$$\left[:\ddot{\underset{..}{S}}-\ddot{\underset{..}{S}}: \right]^{2-}$$

21.74. Oxidation number for F in the named compounds:

S_2F_2 :F̈—S̈—S̈—F̈:   S oxidation number is +1.

SF_2 :F̈—S̈—F̈:   S oxidation number is +2

SF_4 (see-saw structure with 4 F atoms and 1 lone pair on S)   S oxidation number is +4 (lone pairs on F not shown)

SF_6 (octahedral structure with 6 F atoms on S)   S oxidation number is +6 (lone pairs on F not shown)

S_2F_{10} (two octahedral S atoms each with 5 F atoms, joined by S—S bond)

S oxidation number is +5 (lone pairs on F not shown)

The Halogens

21.75. Halogen with the highest bond dissociation energy:

Cl–Cl has the highest bond dissociation energy. See the Table of Bond Dissociation Energies in the chapter.

21.76. Statement that is not correct:

(c) F_2 is prepared industrially by the electrolysis of aqueous NaF. The preparation involves the electrolysis of KF in anhydrous HF.

21.77. Calculate the equivalent net cell potential for the oxidation of Mn^{2+} by BrO_3^-:

Cathode, reduction: $2\ BrO_3^-(aq) + 12\ H^+(aq) + 10\ e^- \rightarrow Br_2(aq) + 6\ H_2O(\ell)$

Anode, oxidation: $2[Mn^{2+}(aq) + 4\ H_2O(\ell) \rightarrow MnO_4^-(aq) + 8\ H^+(aq) + 5\ e^-]$

$2\ Mn^{2+}(aq) + 2\ H_2O(\ell) + 2\ BrO_3^-(aq) \rightleftarrows 2\ MnO_4^-(aq) + 4\ H^+(aq) + Br_2(aq)$

$E°_{cell} = E°_{cathode} - E°_{anode} = 1.44\ V - 1.51\ V = -0.07\ V$

The half-reaction for the reduction of the bromate anion to bromine is combined with the half-reaction for the oxidation of the Mn(II) ion to permanganate ion to give the overall reaction. The overall reaction has a negative net cell potential indicating that it is not product-favored at equilibrium.

21.78. Calculate the concentration of HClO in a 0.10 M NaOCl solution.

The equation for hydrolysis of the hypochlorite ion is:

$OCl^-(aq) + H_2O(\ell) \rightleftarrows HClO(aq) + OH^-(aq)$

The ionization constant for this reaction (K_b) is obtained from the ionization constant ($K_a = 3.5 \times 10^{-8}$) for HClO and K_w.

$K_b = \dfrac{K_w}{K_a} = 2.9 \times 10^{-7} = \dfrac{[HClO][OH^-]}{[OCl^-]} = \dfrac{(x)(x)}{0.10 - x} \approx \dfrac{x^2}{0.10}$

$x = [HClO] = [OH^-] = 1.7 \times 10^{-4}\ M$

21.79. The balanced equation for the reaction of Cl_2 with Br^-

Cathode, reduction: $Cl_2(g) + 2\ e^- \rightarrow 2\ Cl^-$

Anode, oxidation: $2\ Br^-(aq) \rightarrow Br_2(\ell) + 2\ e^-$

$Cl_2(g) + 2\ Br^-(aq) \rightarrow 2\ Cl^-(aq) + Br_2(\ell)$

$E°_{cell} = E°_{cathode} - E°_{anode} = 1.36\ V - 1.08\ V = 0.28\ V$

Bromide ions are losing electrons (and donating them to chlorine), causing chlorine to be reduced, Therefore, bromide ion is the reducing agent and chlorine is the oxidizing agent. The cell potential is positive, making this process a product-favored reaction at equilibrium.

21.80. Oxidizing agents to prepare Cl_2 from Cl^-: The reduction potential for Cl_2 is +1.36 V. The convert Cl^- ions to Cl_2 will require an oxidizing agent with a cell potential more positive than +1.36 V. A few of the possibilities are:

BrO_3^-: $BrO_3^-(aq) + 6\ H^+(aq) + 6\ Cl^-(aq) \rightarrow Br^-(aq) + 3\ H_2O(\ell) + 3\ Cl_2(g)$

MnO_4^-: $2\ MnO_4^-(aq) + 10\ Cl^-(aq) + 16\ H^+(aq) \rightarrow 2\ Mn^{2+}(aq) + 5\ Cl_2(g) + 8\ H_2O(\ell)$

Ce^{4+}: $2\ Ce^{4+}(aq) + 2\ Cl^-(aq) \rightarrow 2\ Ce^{3+}(aq) + Cl_2(g)$

21.81. Mass of F_2 that can be produced per 24 hr by a current of 5.00×10^3 amps (at 10.0 V). Recall that 1 amp = 1 C/s. The half-reaction taking place is $2\ F^-(aq) \rightarrow F_2(g) + 2\ e^-$.

$$24\ h \left(\frac{3600\ s}{1\ h}\right)\left(\frac{5.00 \times 10^3\ C}{1\ s}\right)\left(\frac{1\ mol\ e^-}{96{,}485\ C}\right)\left(\frac{1\ mol\ F_2}{2\ mol\ e^-}\right)\left(\frac{38.00\ g\ F_2}{1\ mol\ F_2}\right) = 8.51 \times 10^4\ g\ F_2$$

21.82. Molecular structure for BrF_3 shown below:

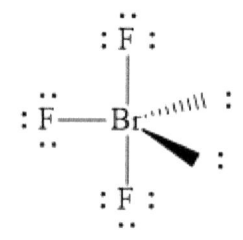

There are five groups around the central bromine (three bond pairs and two lone pairs). The electron-pair geometry is trigonal-bipyramidal and the molecular geometry is T-shaped. Two of the fluorine atoms are in axial positions and one is in an equatorial position. The ideal bond angle would be 90°. The bond angles in BrF_3 will be less than the ideal 90° because of the lone pair electrons on the Br atom. The lone pairs in the equator of the molecule are somewhat larger than bond pairs and push the bond pairs away, narrowing what is otherwise a 90° F–Br–F angle.

The Noble Gases

21.83. Species in which Xe is in the +4 oxidation state: (c) XeF_3^+

(Xe) + 3(F) = +1 so if F = –1, then (Xe) + –3 = +1, and Xe = +4

21.84. Electron pair geometry for Xe in XeOF₄:

Electron-pair geometry is (d), octahedral.

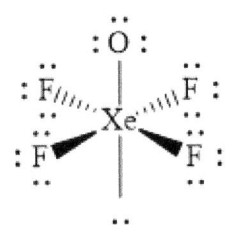

21.85. The reaction: Xe(g) + 2 F₂(g) → XeF₄(g) has $\Delta_f H° = -218$ kJ/mol

The bond-dissociation enthalpy (*D*) for F–F = +155 kJ/mol. What is *D* for Xe–F?

$\Delta_f H° = D$(bonds broken) – *D*(bonds formed)

–218 kJ/mol-rxn = (2 mol F–F/mol-rxn)(+155 kJ/mol F–F) –

(4 mol Xe–F/mol-rxn)(*D*(Xe–F))

–218 kJ/mol-rxn = +310. kJ/mol-rxn – (4 mol Xe–F/mol-rxn)(*D*(Xe–F))

–218 kJ/mol-rxn – 310. kJ/mol-rxn = – (4 mol Xe–F/1-mol-rxn)(*D*(Xe-F))

–528 kJ/mol-rxn = –(4 mol Xe–F)(*D* (Xe–F))

D(Xe–F) = +132 kJ/mol Xe–F

(Note that this solution to the problem assumes that XeF₄ is in the gas state and that the enthalpy of formation reported for the compound is for the gaseous compound. In reality, the reported enthalpy of formation is for the solid. A more complete solution would therefore involve a step involving the transformation of the solid to the gas state and require knowing the enthalpy for this process.)

21.86. The molecular and electron pair geometries for XeO₃F₂ are both trigonal bipyramidal.

21.87. Quantity of Ar present in 1.00 L of air at 298 K and 1.00 atm pressure:

$\text{argon} = 1.00 \text{ L dry air} \left(\dfrac{0.93 \text{ L Ar}}{100 \text{ L dry air}} \right) = 0.0093 \text{ L Ar}$

The ideal gas law will tell you the amount of argon:
$$n = \frac{PV}{RT} = \frac{(1.00 \text{ atm})(0.0093 \text{ L})}{\left(0.082057 \frac{\text{L} \cdot \text{atm}}{\text{K} \cdot \text{mol}}\right)(298 \text{ K})} = 3.8 \times 10^{-4} \text{ mol Ar}$$

The mass is:
$$3.8 \times 10^{-4} \text{ mol Ar}\left(\frac{39.95 \text{ g Ar}}{1 \text{ mol Ar}}\right) = 0.015 \text{ g Ar in 1.00 L of air}$$

To isolate 1.00 mol of Ar, you would need
$$1.00 \text{ mol Ar}\left(\frac{1.00 \text{ L air}}{3.8 \times 10^{-4} \text{ mol Ar}}\right) = 2.6 \times 10^{3} \text{ L air}$$

21.88. For the reaction of XeF$_6$ with water:

(a) The reaction: XeF$_6$(s) + 3 H$_2$O(ℓ) → XeO$_3$(s) + 6 HF(aq)

(b) Molecular geometry of XeO$_3$: Trigonal pyramidal

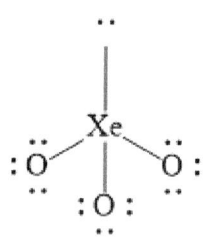

(c) Xe(g) + 3/2 O$_2$(g) → XeO$_3$(g)

Δ_rH = D(bonds broken) – D(bonds formed)

Δ_rH = 3/2 D(O=O) – 3 D(Xe–O)

+402 kJ/mol-rxn = (3/2 mol/mol-rxn)(498 kJ/mol) – (3 mol/mol-rxn)D(Xe–O)

D(Xe–O) = 115 kJ/mol

(Note that this solution to the problem assumes that XeO$_3$ is in the gas state and that the enthalpy of formation reported for the compound is for the gaseous compound. In reality, the reported enthalpy of formation is for the solid. A more complete solution would therefore involve a step involving the transformation of the solid to the gas state and require knowing the enthalpy for this process.)

General Questions

21.89. Describe the third-period elements:

Atomic No.	Element	(a) Type	(b) Color	(c) State
11	Sodium	metal	silvery	solid
12	Magnesium	metal	silvery	solid
13	Aluminum	metal	silvery	solid
14	Silicon	metalloid	black, shiny	solid
15	Phosphorus	nonmetal	red, white, and black allotropes	solid
16	Sulfur	nonmetal	yellow	solid
17	Chlorine	nonmetal	pale green	gas
18	Argon	nonmetal	colorless	gas

21.90. Structure of B_2Cl_4:
(a) Lewis dot structure of B_2Cl_4:

$$\begin{array}{c} :\ddot{\text{Cl}}: \quad\quad :\ddot{\text{Cl}}: \\ \backslash \quad\quad\quad / \\ \text{B}\!-\!\text{B} \\ / \quad\quad\quad \backslash \\ :\ddot{\text{Cl}}: \quad\quad :\ddot{\text{Cl}}: \end{array}$$

(b) B atoms are sp^2 hybridized and have trigonal planar geometry

21.91. Complete and balance the equations:

(a) $2\ KClO_3(s) + \text{heat} \rightarrow 2\ KCl(s) + 3\ O_2(g)$

(b) $2\ H_2S(g) + 3\ O_2(g) \rightarrow 2\ H_2O(g) + 2\ SO_2(g)$

(c) $2\ Na(s) + O_2(g) \rightarrow Na_2O_2(s)$

(d) $P_4(s) + 3\ KOH(aq) + 3\ H_2O(\ell) \rightarrow PH_3(g) + 3\ KH_2PO_2(aq)$

(e) $NH_4NO_3(s) + \text{heat} \rightarrow N_2O(g) + 2\ H_2O(g)$

(f) $2\ In(s) + 3\ Br_2(\ell) \rightarrow 2\ InBr_3(s)$

(g) $SnCl_4(\ell) + 2\ H_2O(\ell) \rightarrow 4\ HCl(aq) + SnO_2(s)$

21.92. Regarding barium oxide and barium peroxide:

(a) Equation for heating BaO in pure oxygen: 2 BaO(s) + O$_2$(g) → 2 BaO$_2$(s)

(b) Equation for heating BaO$_2$ in Fe: 3 BaO$_2$(s) + 2 Fe(s) → Fe$_2$O$_3$(s) + 3 BaO(s)

21.93. Mass of silicon sand to produce silicon carbide:

$$1.0 \times 10^5 \text{ metric tons SiC} \left(\frac{100 \text{ metric tons sand}}{70 \text{ metric tons SiC}} \right) = 1.4 \times 10^5 \text{ metric tons sand}$$

21.94. Volume required to store H$_2$ at 25 °C and 1.0 atm pressure:

Amount of H$_2$: $1.0 \times 10^3 \text{ g H}_2 \left(\dfrac{1 \text{ mol H}_2}{2.02 \text{ g}} \right) = 5.0 \times 10^2 \text{ mol H}_2$

$$V = \frac{nRT}{P} = \frac{(5.0 \times 10^2 \text{ mol})(0.082057 \text{ L·atm/K·mol})(298 \text{ K})}{1.0 \text{ atm}} = 1.2 \times 10^4 \text{ L}$$

21.95. Calculate Δ$_r$G° for the decomposition of the metal carbonates for Mg, Ca, and Ba

Identity of Metal (M)	Δ$_f$G° MCO$_3$(s)	Δ$_f$G° MO(s)	Δ$_f$G° CO$_2$(g)	Δ$_f$G° M
Mg	−1028.2	−568.93	−394.359	0
Ca	−1129.16	−603.42	−394.359	0
Ba	−1134.41	−520.38	−394.359	0

The units in this table for all Δ$_f$G° values are kJ/mol.

The reaction: MCO$_3$(s) → MO(s) + CO$_2$(g)

Δ$_r$G° = Δ$_f$G° MO + Δ$_f$G° CO$_2$ − Δ$_f$G° MCO$_3$

Δ$_r$G° MgCO$_3$ = [(−568.93 kJ/mol)(1 mol/mol-rxn) + (−394.359 kJ/mol)(1 mol/mol-rxn)]
  − (−1028.2 kJ/mol)(1 mol/mol-rxn) = 64.9 kJ/mol-rxn

Δ$_r$G° CaCO$_3$ = [(−603.42 kJ/mol)(1 mol/mol-rxn) + (−394.359 kJ/mol)(1 mol/mol-rxn)]
  − (−1129.16 kJ/mol)(1 mol/mol-rxn) = 131.38 kJ/mol-rxn

Δ$_r$G° BaCO$_3$ = [(−520.38 kJ/mol)(1 mol/mol-rxn) + (−394.359 kJ/mol)(1 mol/mol-rxn)]
  − (−1134.41 kJ/mol)(1 mol/mol-rxn) = 219.67 kJ/mol-rxn

The relative tendency for decomposition is then MgCO$_3$ > CaCO$_3$ > BaCO$_3$.

21.96. Regarding the use of ammonium perchlorate:

The equation for the reaction: $2\ NH_4ClO_4(s) \rightarrow N_2(g) + Cl_2(g) + 2\ O_2(g) + 4\ H_2O(g)$

(a) Mass of water and oxygen:

$$6.35 \times 10^5\ kg\ NH_4ClO_4 \left(\frac{1000\ g}{1\ kg}\right)\left(\frac{1\ mol\ NH_4ClO_4}{117.5\ g\ NH_4ClO_4}\right)\left(\frac{4\ mol\ H_2O}{2\ mol\ NH_4ClO_4}\right)$$

$$\left(\frac{18.02\ g\ H_2O}{1\ mol\ H_2O}\right)\left(\frac{1\ kg}{1000\ g}\right) = 1.95 \times 10^5\ kg\ H_2O$$

$$6.35 \times 10^5\ kg\ NH_4ClO_4 \left(\frac{1000\ g}{1\ kg}\right)\left(\frac{1\ mol\ NH_4ClO_4}{117.5\ g\ NH_4ClO_4}\right)\left(\frac{2\ mol\ O_2}{2\ mol\ NH_4ClO_4}\right)$$

$$\left(\frac{32.00\ g\ O_2}{1\ mol\ O_2}\right)\left(\frac{1\ kg}{1000\ g}\right) = 1.73 \times 10^5\ kg\ O_2$$

(b) Mass of Al required to use the oxygen: $4\ Al(s) + 3\ O_2(g) \rightarrow 2\ Al_2O_3(s)$

$$1.73 \times 10^5\ kg\ O_2 \left(\frac{1000\ g}{1\ kg}\right)\left(\frac{1\ mol\ O_2}{32.00\ g\ O_2}\right)\left(\frac{4\ mol\ Al}{3\ mol\ O_2}\right)$$

$$\left(\frac{26.982\ g\ Al}{1\ mol\ Al}\right)\left(\frac{1\ kg}{1000\ g}\right) = 1.94 \times 10^5\ kg\ Al$$

(c) Mass of Al_2O_3 produced:

$$1.73 \times 10^5\ kg\ O_2 \left(\frac{1000\ g}{1\ kg}\right)\left(\frac{1\ mol\ O_2}{32.00\ g\ O_2}\right)\left(\frac{2\ mol\ Al_2O_3}{3\ mol\ O_2}\right)$$

$$\left(\frac{102.0\ g\ Al_2O_3}{1\ mol\ Al_2O_3}\right)\left(\frac{1\ kg}{1000\ g}\right) = 3.67 \times 10^5\ kg\ Al_2O_3$$

21.97. For the reaction of metals with hydrogen halides:

(a) Since $\Delta_rG° < 0$ for the reaction to be product-favored at equilibrium, calculate the value for

$\Delta_fG°MX$ that will make the $\Delta_rG° = 0$.

You know the free energy change for the reaction, $\Delta_rG°$, is equal to $\Delta_fG° (MX_n) - n\Delta_fG°(HX)$. When HX = HCl, then $\Delta_rG° = \Delta_fG° (MX_n) - n(-95.1\ kJ/mol)$.

To be product-favored at equilibrium, $\Delta_rG° < 0$, and this will be true when

$n(-95.1\ kJ) > \Delta_fG°(MX_n)$.

(b) Examine $\Delta_fG°$ MX values for each metal.

metal	Ba	Pb	Hg	Ti
$\Delta_fG°$ MX (kJ/mol)	−810.4	−314.10	−178.6	−737.2
n	2	2	2	4
$n(−95.1)$	−190.2	−190.2	−190.2	−380.4

For barium, lead, and titanium, $n(−95.1 \text{ kJ}) > \Delta_fG°(\text{MX})$ and so these reactions are expected to be product-favored at equilibrium.

21.98. Lewis dot structures for the polyhalides:

(a) I_3^- is linear

(b) $BrCl_2^-$ is linear

(c) ClF_2^+ is bent

(d) I_5^- ion: The ion has 5 × 7 for five I atoms + 1 for the negative charge = 36 valence electrons (18 pairs) The I atoms are placed in a row and connected by single bonds. After completing the octets of all the I atoms except the central one, 28 valence electrons (14 pairs) have been used. The remaing eight valence electrons (four pairs) are also placed on the central I atom, completing and expanding its octet. The Lewis structure is

The central I atom has six electron pairs around it (two bonding pairs and four lone pairs). This indicates an octahedral electron-pair geometry around this atom. There are two different arrangements of the atoms possible in this geometry, one where the two iodine-containing groups occupy positions across from each other and one where they occupy adjacent positions around the central atom.

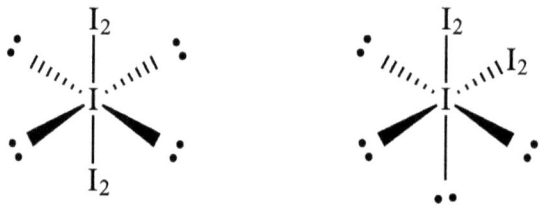

The structure on the left predicts a linear structure for the ion, and the one on the right predicts a bent structure. Because the text indicates that the ion is not linear, the structure on the right is correct (though there is some evidence that the linear structure does exist under some conditions).

21.99. The average O–F bond energy in OF$_2$, given that the $\Delta_f H° = +24.5$ kJ/mol:

The reaction is represented as: O$_2$(g) + 2 F$_2$(g) → 2 OF$_2$(g)

$\Delta H = \Sigma D$(bonds broken) – ΣD(bonds formed)

Bonds broken: 1 mol O=O = (1 mol/mol-rxn) · 498 kJ/mol = 498 kJ/mol-rxn

2 mol F–F = (2 mol/mol-rxn) · 155 kJ/mol = 310 kJ/mol-rxn

Total input = 808 kJ/mol-rxn

Bonds formed: 4 mol O–F = (4 mol/mol-rxn)x (where x = O–F bond energy)

Total input = (4 mol/mol-rxn)x

$\Delta H = \Sigma D$(bonds broken) – ΣD(bonds formed)

+49.0 kJ/mol-rxn = 808 kJ/mol-rxn – 4x,

+49.0 kJ/mol-rxn – 808 kJ/mol-rxn = –(4 mol/mol-rxn)x

–759 kJ/mol-rxn = –(4 mol/mol-rxn)x

x = 190 kJ/mol

The O–F bond energy is 190 kJ/mol

21.100. Mass of CaF$_2$ required to obtain [F$^-$] = 2.0 × 10^{-5} M in 1.0 × 10^6 L of water. (Assume adding the CaF$_2$ does not significantly affect the volume.)

$$1.0 \times 10^6 \text{ L} \left(\frac{2.0 \times 10^{-5} \text{ mol F}^-}{\text{L}} \right) \left(\frac{1 \text{ mol CaF}_2}{2 \text{ mol F}^-} \right) \left(\frac{78.07 \text{ g CaF}_2}{1 \text{ mol CaF}_2} \right) = 780 \text{ g CaF}_2 \text{ (2 sf)}$$

(It is worthwhile to check that this concentration of F$^-$ can be achieved by dissolution of CaF$_2$ in water. The concentrations of the two ions are 2.0 × 10^{-5} M F$^-$, and 1.0 × 10^{-5} M Ca^{2+} (half the F$^-$ concentration).

CaF$_2$ is CaF$_2$(s) ⇌ Ca^{2+}(aq) + 2 F$^-$(aq) K_{sp} – 5.3 × 10^{-11}

Q = [Ca^{2+}][F$^-$]^2 = (1.0 × 10^{-5})(2.0 × 10^{-5})^2 = 4.0 × 10^{-15}

$Q < K_{sp}$, so the solution is unsaturated. Therefore, dissolving this mass of CaF$_2$ in water should be fine since its solubility has not been exceeded.)

21.101. For the equation: $H_2NN(CH_3)_2(\ell) + 2\ N_2O_4(\ell) \rightarrow 3\ N_2(g) + 4\ H_2O(g) + 2\ CO_2(g)$

(a) The oxidizing and reducing agents: Noting that $N_2O_4(\ell)$ loses O as it reacts, the oxide is reduced, which makes $N_2O_4(\ell)$ the **oxidizing agent.** This means that $H_2NN(CH_3)_2(\ell)$ serves as the **reducing agent**.

(b) What mass of $N_2O_4(\ell)$ is consumed if 4100 kg of $H_2NN(CH_3)_2(\ell)$ reacts?

$$4100\text{ kg H}_2\text{NN(CH}_3)_2 \left(\frac{1000\text{ g}}{1\text{ kg}}\right)\left(\frac{1\text{ mol H}_2\text{NN(CH}_3)_2}{60.10\text{ g H}_2\text{NN(CH}_3)_2}\right) = 6.82 \times 10^4\text{ mol H}_2\text{NN(CH}_3)_2$$

$$= 6.8 \times 10^4\text{ mol H}_2\text{NN(CH}_3)_2$$

$$6.82 \times 10^4\text{ mol H}_2\text{NN(CH}_3)_2 \left(\frac{2\text{ mol N}_2\text{O}_4}{1\text{ mol H}_2\text{NN(CH}_3)_2}\right)\left(\frac{92.01\text{ g N}_2\text{O}_4}{1\text{ mol N}_2\text{O}_4}\right)\left(\frac{1\text{ kg}}{1000\text{ g}}\right)$$

$$= 1.3 \times 10^4\text{ kg N}_2\text{O}_4\text{ consumed (2 sf)}$$

What mass of N_2, H_2O, CO_2 will be formed knowing that 6.82×10^4 mol of $H_2NN(CH_3)_2$ is consumed? (Answers are given to 2 significant figures.)

$$6.82 \times 10^4\text{ mol H}_2\text{NN(CH}_3)_2 \left(\frac{3\text{ mol N}_2}{1\text{ mol H}_2\text{NN(CH}_3)_2}\right)\left(\frac{28.01\text{ g N}_2}{1\text{ mol N}_2}\right)\left(\frac{1\text{ kg}}{1000\text{ g}}\right)$$

$$= 5.7 \times 10^3\text{ kg N}_2$$

$$6.82 \times 10^4\text{ mol H}_2\text{NN(CH}_3)_2 \left(\frac{4\text{ mol H}_2\text{O}}{1\text{ mol H}_2\text{NN(CH}_3)_2}\right)\left(\frac{18.02\text{ g H}_2\text{O}}{1\text{ mol H}_2\text{O}}\right)\left(\frac{1\text{ kg}}{1000\text{ g}}\right)$$

$$= 4.9 \times 10^3\text{ kg H}_2\text{O}$$

$$6.82 \times 10^4\text{ mol H}_2\text{NN(CH}_3)_2 \left(\frac{2\text{ mol CO}_2}{1\text{ mol H}_2\text{NN(CH}_3)_2}\right)\left(\frac{44.01\text{ g CO}_2}{1\text{ mol CO}_2}\right)\left(\frac{1\text{ kg}}{1000\text{ g}}\right)$$

$$= 6.0 \times 10^3\text{ kg CO}_2$$

21.102. Regarding liquid HCN:

(a) Structure for the trimer:

(b) Energy of the trimerization:

3 HCN(ℓ) → (HCN)₃(ℓ)

Three C—H bonds do not change during the reaction.

$\Delta_r H° = $ (3 mol/mol-rxn) · D(C≡N) –

[(3 mol/mol-rxn) · D(C—N) + (3 mol/mol-rxn) · D(C=N)]

$\Delta_r H° = $ (3 mol/mol-rxn)(887 kJ/mol) –

[(3 mol/mol-rxn)(305 kJ/mol) + (3 mol/mol-rxn)(615 kJ/mol)]

$\Delta_r H° = $ –99 kJ/mol-rxn

21.103. Examine the enthalpy change for the reaction:

2 N₂(g) + 5 O₂(g) + 2 H₂O(ℓ) → 4 HNO₃(aq)

$\Delta_r H° = $ [(4 mol/mol-rxn)(–207.36 kJ/mol)] – [(2 mol/mol-rxn)(0 kJ/mol) +

(5 mol/mol-rxn)(0 kJ/mol) + (2 mol/mol-rxn)(–285.83 kJ/mol)] = –257.78 kJ/mol-rxn

The reaction is **exothermic** so it is a reasonable "first-guess" that this might be a way to "fix" nitrogen. The only way to be certain is to calculate free energy change, $\Delta_r G°$.

$\Delta_r G° = $ [(4 mol/mol-rxn)(–111.25 kJ/mol)] – [(2 mol/mol-rxn)(0 kJ/mol) +

(5 mol/mol-rxn)(0 kJ/mol) + (2 mol/mol-rxn)(–237.15 kJ/mol)] = 29.30 kJ/mol-rxn

The positive value for $\Delta_r G°$ indicates that the **reaction is not product-favored at equilibrium at 25 °C**. The decrease in the number of moles of gas (an entropy decrease) is a factor that works against this process at 25 °C. The favorable $\Delta_r H°$ does indicate that a lower temperature might be feasible, and below 268 K the $\Delta_r G$ is favorable. However, at this temperature water is a solid.

21.104. Concerning oxoanions of phosphorus:

(a) The structure of [P₄O₁₃]ⁿ⁻ is given below. The formal charges are shown for each atom with a nonzero formal charge is shown. Each terminal oxygen carries a single negative formal charge for a total negative charge of 10–. Each P atom has a formal charge of +1, for a total positive charge of 4+. The overall charge is therefore 6–. A completely protonated acid has 6 ionizable protons.

(b) P₄O₁₂⁸⁻ has a cyclic structure, shown below. The formal charges for each atom with a nonzero formal charge is shown. The net charge is 4–. A completely protonated anion has 4 ionizable protons.

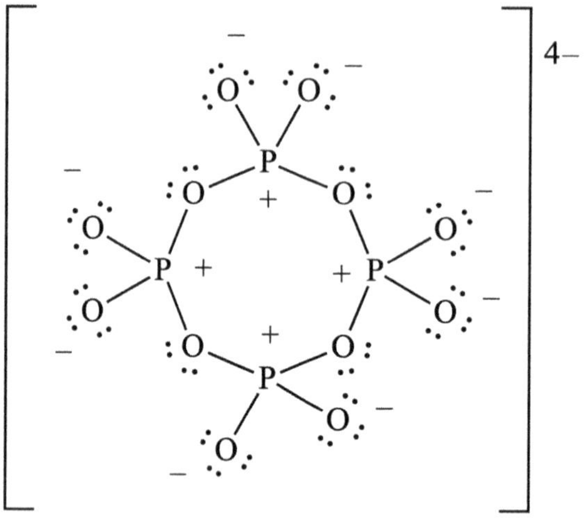

21.105. Derive empirical and molecular formulas for compounds A-E:

	A	B	C	D	E
%B	78.3	81.2	83.1	85.7	88.5
%H	21.7	18.8	16.9	14.3	11.5
mol B	7.24	7.51	7.69	7.93	8.19
mol H	21.5	18.7	16.8	14.2	11.4
ratio H/B	2.97 ≈ 3	2.48 ≈ 2.5	2.18 ≈ 2.2	1.78 ≈ 1.8	1.39 ≈ 1.4
	A	B	C	D	E
#B atoms	1	2	5	5	5
#H atoms	3	5	11	9	7
MM calc	13.83	26.66	65.14	63.12	61.11
MM known	27.7	53.3	65.1	63.1	122.2
Emp.formula/ Molec.formula	2.00	2.00	0.999	1.00	2.000
Molecular Formula	B₂H₆	B₄H₁₀	B₅H₁₁	B₅H₉	B₁₀H₁₄

This problem requires one to calculate empirical formulas for the boron hydrides. Using the %B (and %H) as masses, calculate the number of moles of each element. Next, determine the ratio of hydrogen/boron atoms to get an empirical formula. Now, using the number of B and H atoms, with their respective atomic masses, calculate molar mass (in the row entitled MM calc). These values are compared with the known molar masses (in the row labeled MM known). Using the molar masses provided with the data (MM known) divided by the calculated "empirical mass", the number of empirical formulas per molecular formula is obtained. Multiplying that ratio by the number of B and H atoms in the empirical formulas gives the molecular formula (final row).

The rows (#B) and (#H) were obtained by expressing the data in the row (Ratio of H/B atoms) as a fraction. For example, 2.97 is 3, 2.5 becomes 2+1/2, 2.2 is 2+1/5, 1.8 is 1+4/5, 1.4 is 1+2/5. Converting these fractions to integers gives the numbers in each #B atoms and #H atoms row.

21.106. Finding the formula of a compound of cesium and oxygen:

(a) Assume 100.00 g of compound and calculate the amount (moles) of each element:

$$19.39 \text{ g O}\left(\frac{1 \text{ mol O}}{15.999 \text{ g}}\right) = 1.212 \text{ mol O}$$

$$80.61 \text{ g Cs}\left(\frac{1 \text{ mol Cs}}{132.91 \text{ g}}\right) = 0.6065 \text{ mol Cs}$$

$$\frac{1.212 \text{ mol O}}{0.6065 \text{ mol Cs}} = \frac{2 \text{ mol O}}{1 \text{ mol Cs}}, \text{ so, the empirical formula is CsO}_2\text{, cesium superoxide.}$$

(b) The reaction of cesium superoxide and hydrogen peroxide:

$$2 \text{ CsO}_2(s) + \text{H}_2\text{O}_2(\ell) \rightarrow 2 \text{ CsOH(aq)} + 2 \text{ O}_2(g)$$

21.107. For the PdH$_x$ compound formed:

Amount (moles) of Pd involved: 0.192 g Pd/106.42 g/mol = 1.804×10^{-3} mol Pd

The amount of H$_2$ absorbed can be determined by the *difference* in pressures exerted by the gas in the 2.25 L flask. $P = 113$ mm Hg $-$ 108 mm Hg $= 5$ mm Hg.

$$n = \frac{PV}{RT} = \frac{\left(\frac{5 \text{ mm Hg}}{760 \text{ mm Hg/atm}}\right)(2.25 \text{ L})}{\left(0.082057 \frac{\text{L} \cdot \text{atm}}{\text{K} \cdot \text{mol}}\right)(296.2 \text{ K})} = 6.1 \times 10^{-4} \text{ mol H}_2 \text{ absorbed}$$

$$6.1 \times 10^{-4} \text{ mol H}_2 \left(\frac{2 \text{ mol H}}{1 \text{ mol H}_2}\right) = 1.2 \times 10^{-3} \text{ mol H}$$

The x in PdH$_x$ is the ratio of the amount of H to the amount of Pd:

$$\frac{1.2 \times 10^{-3} \text{ mol H}}{1.80 \times 10^{-3} \text{ mol Pd}} = \frac{0.67 \text{ mol H}}{\text{mol Pd}}$$

Thus $x = 0.67$. This indicates a 2 H:3 Pd ratio, Pd$_3$H$_2$.

21.108. Concerning the chemistry of gallium:

(a) Equations for reacting gallium hydroxide with HCl and NaOH:

Ga(OH)$_3$(s) + 3 HCl(aq) + 3 H$_2$O(ℓ) → [Ga(H$_2$O)$_6$]Cl$_3$(aq)

Ga(OH)$_3$(s) + NaOH(aq) → Na[Ga(OH)$_4$](aq)

(b) K_a for Al^{3+} = 7.9 × 10^{-6} < K_a for Ga^{3+} = 1.2 × 10^{-3}; Al^{3+} is a weaker acid than Ga^{3+}.

21.109. Is the following reaction product- or reactant-favored?

2 ZnS(s) + 3 O$_2$(g) → 2 ZnO(s) + 2 SO$_2$(g)

Calculate the $\Delta_r G°$ for the reaction:

$\Delta_r G° = [2\Delta_f G°$ ZnO + 2 $\Delta_f G°$ SO$_2$] – 2$\Delta_f G°$ ZnS

$\Delta_r G° = $ [(2 mol/mol-rxn)(–318.30 kJ/mol) + (2 mol/mol-rxn)(–300.13 kJ/mol)] –

(2 mol/mol-rxn)(–201.29 kJ/mol)

= (–636.60 kJ/mol-rxn + –600.26 kJ/mol-rxn) – (–402.58 kJ/mol-rxn)

= –1236.86 kJ/mol-rxn + 402.58 kJ/mol-rxn = –834.28 kJ/mol-rxn

With a negative $\Delta_r G°$, the reaction is product-favored at equilibrium. A negative standard entropy change for the reaction (owing to a decrease in the number of moles of gas in the reaction) makes the reaction less product-favored at a higher temperature.

21.110. Calculate $\Delta_r G°$ for the reaction of Ag with the hydrogen halides:

Notice that the values in the table given in the problem are for –$\Delta_f G°$ of each compound. Thus, each value must be adjusted to be used in the equations below.

$\Delta_r G° = $ (1 mol/mol-rxn)($\Delta_f G°$ AgX(s)) – (1 mol/mol-rxn)($\Delta_f G°$ HX(g))

For X = F$^-$: $\Delta_r G° = $ (1 mol/mol-rxn)(–193.8 kJ/mol) – (1 mol/mol-rxn)(–273.2 kJ/mol)

= +79.4 kJ/mol-rxn

This reaction is reactant-favored at equilibrium.

For X = Cl$^-$: $\Delta_r G° = $ (1 mol/mol-rxn)(–109.76 kJ/mol) – (1 mol/mol-rxn)(–95.09 kJ/mol)

= –14.67 kJ/mol-rxn

This reaction is product-favored at equilibrium.

For X = Br⁻: $\Delta_r G° =$ (1 mol/mol-rxn)(–96.90 kJ/mol) – (1 mol/mol-rxn)(–53.45 kJ/mol)

$$= -43.45 \text{ kJ/mol-rxn}$$

This reaction is product-favored at equilibrium.

For X = I⁻: $\Delta_r G° =$ (1 mol/mol-rxn)(–66.19 kJ/mol) – (1 mol/mol-rxn)(–1.56 kJ/mol)

$$= -64.63 \text{ kJ/mol-rxn}$$

This reaction is product-favored at equilibrium.

Recall that HF is a weak acid whereas HCl, HBr, and HI are strong acids, and that acid strength increases as you move down Group 7A (17). As might be expected, the reactions become more product favored as the strength of the acid used increases.

In the Laboratory

21.111. In the synthesis of dichlorodimethylsilane:

(a) The balanced equation: Si(s) + 2 CH₃Cl(g) → (CH₃)₂SiCl₂(ℓ)

(b) Stoichiometric amount of CH₃Cl to react with 2.65 g silicon:

$$2.65 \text{ g Si} \left(\frac{1 \text{ mol Si}}{28.085 \text{ g Si}} \right) \left(\frac{2 \text{ mol CH}_3\text{Cl}}{1 \text{ mol Si}} \right) = 0.189 \text{ mol CH}_3\text{Cl}$$

$$P = \frac{(0.189 \text{ mol CH}_3\text{Cl})\left(0.082057 \frac{\text{L} \cdot \text{atm}}{\text{K} \cdot \text{mol}}\right)(297.7 \text{ K})}{5.60 \text{ L}} = 0.823 \text{ atm}$$

(c) Mass of (CH₃)₂SiCl₂ produced assuming 100% yield:

$$2.65 \text{ g Si} \left(\frac{1 \text{ mol Si}}{28.085 \text{ g Si}} \right) \left(\frac{1 \text{ mol (CH}_3)_2\text{SiCl}_2}{1 \text{ mol Si}} \right) \left(\frac{129.1 \text{ g (CH}_3)_2\text{SiCl}_2}{1 \text{ mol (CH}_3)_2\text{SiCl}_2} \right)$$

$$= 12.2 \text{ g (CH}_3)_2\text{SiCl}_2$$

21.112. Concerning the reaction of NaBH₄ with AgNO₃:

(a) 2 NaBH₄(s) + 2 AgNO₃(aq) + 6 H₂O(ℓ)

$$\rightarrow 2 \text{ Ag(s)} + 7 \text{ H}_2\text{(g)} + 2 \text{ B(OH)}_3\text{(aq)} + 2 \text{ NaNO}_3\text{(aq)}$$

(b) Mass of Ag produced: Determine the limiting reagent to determine mass.

From AgNO₃: $0.575 \text{ L} \left(\frac{0.011 \text{ mol AgNO}_3}{1 \text{ L}} \right) \left(\frac{2 \text{ mol Ag}}{2 \text{ mol AgNO}_3} \right) \left(\frac{107.87 \text{ g}}{1 \text{ mol Ag}} \right) = 0.68 \text{ g Ag}$

From NaBH₄: $13.0 \text{ g NaBH}_4 \left(\frac{1 \text{ mol NaBH}_4}{37.83 \text{ g NaBH}_4} \right) \left(\frac{2 \text{ mol Ag}}{2 \text{ mol NaBH}_4} \right) \left(\frac{107.87 \text{ g}}{1 \text{ mol Ag}} \right) = 37.1 \text{ g Ag}$

As AgNO₃ is the limiting reagent; 0.68 g Ag is produced.

21.113. Identify the species A-E: Begin by calculating the amount of the gas evolved:

$$n = \frac{PV}{RT} = \frac{\left(\frac{209 \text{ mm Hg}}{760 \text{ mm Hg/atm}}\right)(0.450 \text{ L})}{\left(0.082057 \frac{\text{L}\cdot\text{atm}}{\text{K}\cdot\text{mol}}\right)(298 \text{ K})} = 5.06 \times 10^{-3} \text{ mol gas}$$

So, **A** → **B** + gas. This is a typical decomposition for carbonate salts.

Gas + Ca(OH)$_2$ → **C** (a solid precipitate, indicative of a carbonate salt), so **C** is probably CaCO$_3$, and the gas is probably CO$_2$.

Assuming the gas is CO$_2$, **B** is mostly likely a metal oxide. When **B** reacts with H$_2$O, a basic solution results. (Metal oxides behave this way!)

When **D** is heated in flame, a green color results, indicating presence of Ba^{2+}, so **D** is a barium salt. If D is a barium salt, then so are **A**, **B**, and **E**. So, it is likely **B** is BaO and **A** is BaCO$_3$. What about **D**?

Assuming that the gas is CO$_2$, calculate the mass of CO$_2$.

$$5.06 \times 10^{-3} \text{ mol CO}_2 \left(\frac{44.01 \text{ g CO}_2}{1 \text{ mol CO}_2}\right) = 0.223 \text{ g CO}_2$$

Subtracting this mass from the 1.000 g of **A** gives 0.777 g of **B**.

As **B** is assumed to be BaO, calculate the mass and amount of **B**:

$$0.777 \text{ g BaO} \left(\frac{137.33 \text{ g Ba}}{153.33 \text{ g BaO}}\right) = 0.696 \text{ g Ba or } 5.07 \times 10^{-3} \text{ mol Ba}$$

Note the 1:1 mol correspondence between mol Ba and mol CO$_2$ (the gas). This makes sense, as **B** and CO$_2$ should be in a 1:1 mole ratio if **A** is BaCO$_3$.

Subtract the mass of Ba from the mass of the BaCl$_x$ salt.

1.055 g BaCl$_x$ – 0.696 g Ba = 0.359 g of Cl, which corresponds to:

$$0.359 \text{ g Cl}\left(\frac{1 \text{ mol Cl}}{35.45 \text{ g Cl}}\right) = 0.0101 \text{ mol Cl}$$

Dividing the amount of Cl by the amount of Ba: 0.0101 mol Cl/0.00507 mol Ba = 2.00 mol Cl/mol Ba. So **D** is BaCl$_2$.

What about **E**? If **B** is BaO (as shown), then the reaction with H$_2$SO$_4$ will form BaSO$_4$ (compound **E**).

In summary: **A** is BaCO$_3$; **B** is BaO; **C** is CaCO$_3$; **D** is BaCl$_2$; **E** is BaSO$_4$

21.114. The half-reaction equations are (note that H$^+$ is used rather than H$_3$O$^+$ in these equations):

Oxidation, anode: N$_2$H$_5^+$(aq) → N$_2$(g) + 5 H$^+$(aq) + 4 e$^-$

Reduction, cathode: IO₃⁻(aq) + 6 H⁺(aq) + 5 e⁻ → 1/2 I₂(aq) + 3 H₂O(ℓ)

are combined according to the procedures in Chapter 19 to give:

5 N₂H₅⁺(aq) + 4 IO₃⁻(aq) → 5 N₂(g) + 2 I₂(aq) + 12 H₂O(ℓ) + H⁺(aq)

$E°_{cell} = E°_{cathode} − E°_{anode} = 1.195$ V − (−0.23 V) = 1.43 V

21.115. BrO₂ decomposes on heating to two other oxides:

A: less volatile oxide, golden yellow

B: more volatile oxide, brown. Now known to be Br₂O.

What is the formula of A, Br$_x$O$_y$?

A sample of **A** was treated with sodium iodide, which released I₂. What is the chemistry that occurred? This is a redox reaction.

The oxidation half-reaction involves I⁻ being oxidized to I₂. Its half-reaction is

Oxidation: 2 I⁻ → I₂ + 2e⁻

The reduction half-reaction involves Br$_x$O$_y$ being reduced to Br⁻. The exact half-reaction can't be written at this point because the values of x and y are not known, but using these variables and assigning n as the number of electrons involved in the half-reaction and balancing in the usual manner shown in Chapter 19 gives

Reduction: n e⁻ + 2y H⁺ + Br$_x$O$_y$ → x Br⁻ + y H₂O

The electrons lost in the oxidation and the reduction must the same, so the oxidation half-reaction must occur n times for every two times the reduction half-reaction occurs, resulting in a total of 2n electrons being transferred and the overall balanced equation

Overall Reaction: 4y H⁺ + 2 Br$_x$O$_y$ + 2n I⁻ → n I₂ + 2x Br⁻ + 2y H₂O

This reaction was carried out twice using the same mass of compound A. Two titrations were then performed, one for each sample. The first was the titration of the I₂ formed with sodium thiosulfate, which required 17.7 mL of 0.065 M Na₂S₂O₃.

I₂(aq) + 2 S₂O₃²⁻(aq) → 2 I⁻(aq) + S₄O₆²⁻(aq)

$17.7 \text{ mL} \left(\dfrac{1 \text{ L}}{1000 \text{ mL}} \right) \left(\dfrac{0.065 \text{ mol S}_2\text{O}_3^{2-}}{1 \text{ L}} \right) \left(\dfrac{1 \text{ mol I}_2}{2 \text{ mol S}_2\text{O}_3^{2-}} \right) = 5.8 \times 10^{-4}$ mol I₂

This can be related to the number of electrons transferred in the original reaction with Br$_x$O$_y$:

5.8×10^{-4} mol I₂ $\left(\dfrac{2 \text{ mol e}^-}{1 \text{ mol I}_2} \right) = 1.2 \times 10^{-3}$ mol e⁻

The second titration was the titration of the Br⁻ formed with AgNO₃, which required 14.4 mL of 0.020 M AgNO₃.

Br⁻(aq) + Ag⁺(aq) → AgBr (s)⁻

$$14.4 \text{ mL}\left(\frac{1 \text{ L}}{1000 \text{ mL}}\right)\left(\frac{0.020 \text{ mol Ag}^+}{1 \text{ L}}\right)\left(\frac{1 \text{ mol Br}^-}{1 \text{ mol Ag}^+}\right) = 2.9 \times 10^{-4} \text{ mol Br}^-$$

This is equal to the amount of Br in the original Br$_x$O$_y$ in the original reaction, so the amount of Br is 2.9 × 10^{-4} mol. You can now determine the ratio of the amount of electrons transferred to the amount of Br.

$$\frac{1.2 \times 10^{-3} \text{ mol e}^-}{2.9 \times 10^{-4} \text{ mol Br}} = \frac{4.0 \text{ mol e}^-}{1 \text{ mol Br}}$$

This means that each Br atom gained 4 e–. The final oxidation number of Br in the overall reaction is –1, so the initial oxidation number must have been +3. The oxidation number of O does not change in the reaction and is –2. To have a neutral compound, the formula of compound **A** must be Br$_2$O$_3$.

Structures of Br$_2$O and Br$_2$O$_3$ are shown below. Br$_2$O has a bent or angular molecular geometry. Br$_2$O$_3$ has a Br atom attached to one of the O atoms in a pyramidal structure.

21.116. Give empirical and molecular formulas for **A** and **B**:

Compounds **A** and **B** have the same empirical formula.
Reaction: PCl$_5$ (ℓ) + NH$_4$Cl(s) → HCl(g) + **A** + **B** + **C**.

When reacted in equal molar amounts, the product is a compound of P, N, and Cl plus 5.14 L of HCl at STP.

5.14 L of HCl at STP represent (5.14 L/22.4 L/mol) or 0.229 mol HCl

$$\text{Amount of PCl}_5 = 12.41 \text{ g PCl}_5\left(\frac{1 \text{ mol PCl}_5}{208.22 \text{ g PCl}_5}\right) = 0.05960 \text{ mol PCl}_5$$

$$\frac{\text{Amount of HCl}}{\text{amount of PCl}_5} = \frac{0.229 \text{ mol HCl}}{0.005960 \text{ mol PCl}_5} = \frac{3.85 \text{ mol HCl}}{1 \text{ mol PCl}_5}$$

That is, the reaction produces approximately 4 mol of HCl per mole of PCl$_5$.

Therefore, the reaction can be written:

PCl$_5$(ℓ) + NH$_4$Cl(s) → [PNCl$_2$](s) + 4 HCl(g)

This leads to an empirical formula of PNCl$_2$.

Knowing that **A** + **B** have the same empirical formula, PNCl$_2$, determine what a "PNCl$_2$" species would weigh.

P + N + 2 Cl = 115.9 g per unit.

If **A** has a molar mass of 347.7 g/mol, divide the mass of the PNCl₂ unit into it.

$$\frac{347.7 \text{ g A/mol A}}{115.9 \text{ g/mol empirical formula}} = 3.000 \text{ empirical formula units per molecule of A}$$

Thus, **A** has the molecular formula (PNCl₂)₃ or P₃N₃Cl₆.

Knowing that **B** has the molar mass of 463.5 g/mol, the number of "empirical formula units" in a mole of **B** can be calculated.

$$\frac{463.5 \text{ g B/mol B}}{115.9 \text{ g/mol empirical formula}} = 4.000 \text{ empirical formula units per molecule of B}$$

So, **B** has the molecular formula (PNCl₂)₄ or P₄N₄Cl₈.

A reasonable structure for **A** is shown below:

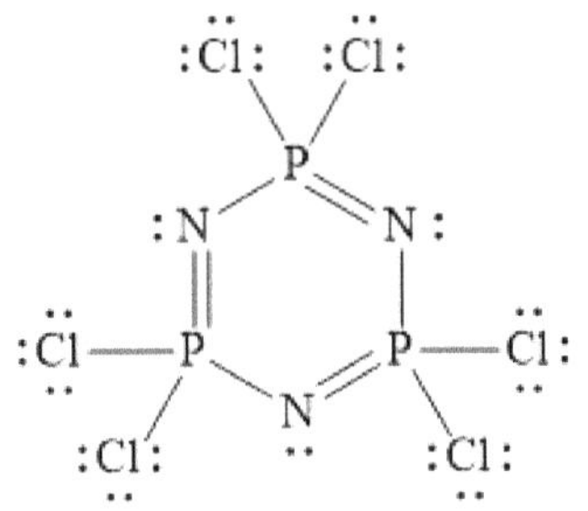

Summary and Conceptual Questions

21.117. For N₂O₃:

(a) Explain differences in bond lengths for the three N-O bonds:

The Lewis dot picture indicates that the "left" N has a N=O double bond, whereas the "right" N has two oxygens, each connected with a N–O "one and a half" bond, owing to resonance. The N-O bonds on the right" are expected to be equal in length (as observed, 121 pm) and longer than the N=O double bond (114.2 pm).

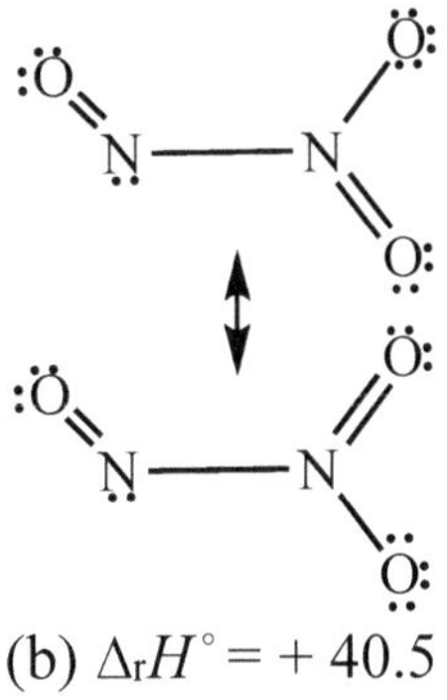

(b) $\Delta_rH° = + 40.5$ kJ/mol N₂O₃ and $\Delta_rG° = -1.59$ kJ/mol. What are $\Delta_rS°$ and K for the reaction at 298 K?

Given that $\Delta_rG° = \Delta_rH° - T\Delta_rS°$, $\Delta_rS° = -(\Delta_rG° - \Delta_rH°)/T = (\Delta_rH° - \Delta_rG°)/T$

$\Delta_r S° = (+40.5 \text{ kJ/mol} +1.59 \text{ kJ/mol})/298 \text{ K} = [(42.09 \times 10^3 \text{ J/mol})/298 \text{ K}]$

$= 141 \text{ J/K} \cdot \text{mol}$

K can be calculated from the relationship $\Delta_r G° = -RT \ln K$.

$-1.59 \text{ kJ/mol} = -(8.3145 \text{ J/K} \cdot \text{mol})(298 \text{ K})\ln K)$

$\dfrac{-1.59 \times 10^3 \text{ J/mol}}{-(8.3145 \text{ J/K} \cdot \text{mol})(298 \text{ K})} = \ln K = 6.42 \times 10^{-1}$ and $K = 1.90$

(c) Calculate $\Delta_f H°$ for $N_2O_3(g)$

The reaction: $N_2O_3(g) \rightarrow NO(g) + NO_2(g)$ has $\Delta_r H = +40.5 \text{ kJ/mol } N_2O_3$.

$\Delta_r H = [\Delta_f H° NO(g) + \Delta_f H° NO_2(g)] - \Delta_f H° N_2O_3(g)$

$+40.5 \text{ kJ/mol-rxn} = [(+90.29 \text{ kJ/mol})(1 \text{ mol/mol-rxn}) + (+33.1 \text{ kJ/mol})(1 \text{ mol/mol-rxn})]$

$- (1 \text{ mol/mol-rxn})(\Delta_f H° N_2O_3(g))$

$+40.5 \text{ kJ/mol-rxn} = 123.39 \text{ kJ/mol-rxn} - (1 \text{ mol/mol-rxn})\Delta_f H° N_2O_3(g)$

Rearranging: $123.39 \text{ kJ/mol} - 40.5 \text{ kJ/mol} = \Delta_f H° N_2O_3(g) = +82.9 \text{ kJ/mol}$

21.118. Estimate the radius of a lead atom:

Unit cell volume:

$\dfrac{4 \text{ Pb atoms}}{\text{unit cell}} \left(\dfrac{207.2 \text{ g}}{1 \text{ mol Pb}}\right)\left(\dfrac{1 \text{ cm}^3}{11.350 \text{ g}}\right)\left(\dfrac{1 \text{ mol Pb}}{6.0221 \times 10^{23} \text{ atoms}}\right) = \dfrac{1.213 \times 10^{-22} \text{ cm}^3}{\text{unit cell}}$

Unit cell edge length:

$V = 1.231 \times 10^{-22} \text{ cm}^3 = (\text{edge length})^3$

edge length $= \sqrt[3]{1.213 \times 10^{-22} \text{ cm}^3} = 4.950 \times 10^{-8} \text{ cm}$

face diagonal $= 4(\text{radius}) = \sqrt{2}(\text{edge length})$

radius $= \dfrac{\sqrt{2}(4.950 \times 10^{-8} \text{ cm})}{4} = 1.750 \times 10^{-8} \text{ cm} = 175.0 \text{ pm}$

21.119. To extinguish a Na fire, addition of water is NOT ADVISABLE, since the reaction of the water with the Na would produce HEAT and HYDROGEN GAS—which would also burn!! The optimal solution would be to smother the fire—perhaps with sand.

21.120. Regarding tin(IV) oxide:

(a) Number of tin and oxide ions present in a unit cell:

8 corner Ti × 1/8 = 1 Ti + 1 internal Ti = 2 Ti total

4 face O × 1/2 = 2 O + 2 internal O = 4 O total

(b) Possible to convert SnO₂(s) into SnCl₄(ℓ):

The reaction: SnO₂(s) + 4 HCl(g) → SnCl₄(ℓ) + 2 H₂O(g)

Δ_rG° = [Δ_fG° SnCl₄(ℓ) + 2 Δ_fG° H₂O(g)] − [Δ_fG° SnO₂(s) + 4 Δ_fG° HCl(g)]

Δ_rG° = [(1 mol/mol-rxn)(−440.15 kJ/mol) + (2 mol/mol-rxn)(−228.59 kJ/mol)]

   − [(1 mol/mol-rxn)(−515.88 kJ/mol) + (4 mol/mol-rxn)(−95.09 kJ/mol)]

Δ_rG° = −1.09 kJ/mol-rxn

Δ_rG° = −RT ln K

−1.09 kJ/mol-rxn = −(8.3145 × 10⁻³ kJ/K·mol)(298 K)(ln K)

ln K = 0.440 and K = 1.55

Δ_rG° is less than zero and K is greater than 1, but only slightly so. While both indicate that the reaction will be product-favored at equilibrium, significant amounts of reactants will be present at equilibrium.

21.121. Stoppered flask contains H₂, N₂, or O₂. What experiment would identify the gas?

Several possibilities exist. Chilling a small sample of the gas with liquid N₂ (bp −196°C), would cause O₂ to liquefy (bp −183 °C)—but not H₂, and N₂ only slowly. If the gas liquefies—it's O₂! If the gas doesn't liquefy at all, the gas is either H₂ (bp −253°C) or N₂ (bp −196°C). Allowing a small sample of the gas to escape through a narrow opening (perhaps the tip of a plastic medicine dropper) while holding a smoldering splint at the tip of the opening would confirm the presence of H₂ (the escaping gas would burn), or N₂ (the escaping gas would **not** burn). *CAUTION: Burning H₂ in glass vessels is TO BE AVOIDED. Serious physical damage frequently results!*

21.122. Regarding nitric acid:

(a) You can rationalize the equivalence of the two N—O bond lengths by writing resonance structures for this molecule.

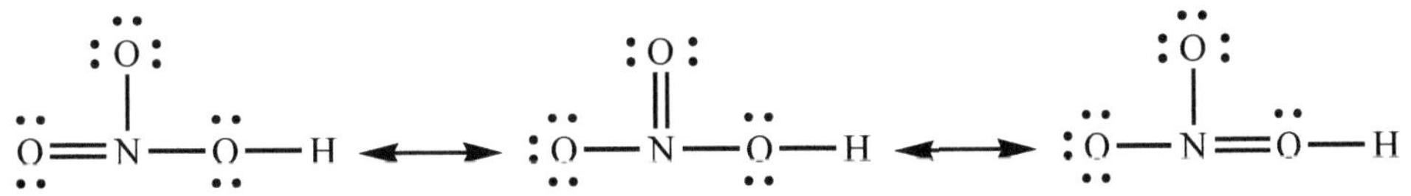

The structure on the right is a minor contributor to the overall resonance hybrid due to the positive formal charge that would be on the O (a very electronegative atom) bonded to H. The other two resonance structures are dominant and equivalent to each other. The greater bond order (approximately 1.5) of the N–O bonds involving the oxygens with no hydrogen attached explains the shorter bond length when compared to the approximately N–O single bond involving the O with the hydrogen attached.

(b) The central atom, nitrogen, is surrounded by three sets of bonding electrons, and VSEPR predicts that this atom is trigonal-planar. Oxygen, in the —OH group, is surrounded by four electron pairs (in the two most important resonance structures), Two are bonding pairs and two are lone pairs, leading to the bent molecular geometry.

(c) The hybridization of the central N is sp^2. There is an empty p orbital on N that is perpendicular to the plane of the molecule; this can overlap with p orbitals on the two terminal oxygen atoms to form a π bond.

21.123. The change to more positive values in **reduction potentials** as one descends the group (from Al to Tl) indicates a diminishing ability for the metals to act as a reducing agent. Another noticeable trend is the large **differences** between the reduction potentials of Al and Ga, and between In and Tl.

Thus, the oxidation of these metals to the 3+ ion becomes less favored on moving down Group 3A (13). In fact, thallium is more stable in the +1 oxidation state rather than the +3 oxidation state. This same tendency for elements further down a group to be more stable with lower oxidation numbers is also seen in Group 4A (14) for Ge and Pb and Group 5A (15) for Bi.

21.124. Regarding obtaining Mg from seawater:

(a) Volume of seawater to obtain 1 kg of magnesium:

$$1.00 \times 10^3 \text{ g} \left(\frac{1 \text{ mol Mg}}{24.305 \text{ g}}\right)\left(\frac{1 \text{ L}}{0.050 \text{ mol Mg}^{2+}}\right) = 820 \text{ L seawater}$$

Mass of lime needed to precipitate the Mg in that sample:

The reaction: $CaO(s) + H_2O(\ell) + Mg^{2+}(aq) \rightarrow Mg(OH)_2(s) + Ca^{2+}(aq)$

$$1.00 \text{ kg Mg}\left(\frac{1000 \text{ g}}{1 \text{ kg}}\right)\left(\frac{1 \text{ mol Mg}}{24.305 \text{ g Mg}}\right)\left(\frac{1 \text{ mol CaO}}{1 \text{ mol Mg}^{2+}}\right)\left(\frac{56.08 \text{ g CaO}}{1 \text{ mol CaO}}\right)\left(\frac{1 \text{ kg}}{1000 \text{ g}}\right)$$

$$= 2.31 \text{ kg CaO}$$

(b) Mass of Mg produced at the cathode:

$MgCl_2(\ell) \rightarrow Mg(s) + Cl_2(g)$

$$1.2 \times 10^3 \text{ kg MgCl}_2 \left(\frac{10^3 \text{ g}}{1 \text{ kg}}\right)\left(\frac{1 \text{ mol MgCl}_2}{95.21 \text{ g MgCl}_2}\right)\left(\frac{1 \text{ mol Mg}}{1 \text{ mol MgCl}_2}\right)$$

$$\left(\frac{24.305 \text{ g Mg}}{1 \text{ mol Mg}}\right)\left(\frac{1 \text{ kg}}{10^3 \text{ g}}\right) = 3.1 \times 10^2 \text{ kg Mg}$$

Cl₂ gas is produced at the anode.

Mass of Cl₂ produced:

$$1.2 \times 10^3 \text{ kg MgCl}_2 \left(\frac{10^3 \text{ g}}{1 \text{ kg}}\right)\left(\frac{1 \text{ mol MgCl}_2}{95.21 \text{ g MgCl}_2}\right)\left(\frac{1 \text{ mol Cl}_2}{1 \text{ mol MgCl}_2}\right)$$

$$\left(\frac{70.90 \text{ g Cl}_2}{1 \text{ mol Cl}_2}\right)\left(\frac{1 \text{ kg}}{10^3 \text{ g}}\right) = 8.9 \times 10^2 \text{ kg Cl}_2$$

Faradays consumed:

$$1.2 \times 10^3 \text{ kg MgCl}_2 \left(\frac{10^3 \text{ g}}{1 \text{ kg}}\right)\left(\frac{1 \text{ mol MgCl}_2}{95.21 \text{ g MgCl}_2}\right)\left(\frac{2 \text{ mol e}^-}{1 \text{ mol MgCl}_2}\right)\left(\frac{1 \text{ Faraday}}{1 \text{ mol e}^-}\right)$$

$$= 2.5 \times 10^4 \text{ Faradays}$$

(c) Joules required per mole:

$$\frac{18.5 \text{ kWh}}{1 \text{ kg Mg}}\left(\frac{1 \text{ kg}}{1000 \text{ g}}\right)\left(\frac{3.6 \times 10^6 \text{ J}}{1 \text{ kWh}}\right)\left(\frac{24.305 \text{ g Mg}}{1 \text{ mol Mg}}\right) = 1.6 \times 10^6 \text{ J/mol Mg}$$

Energy consumption compared with decomposition:

The decomposition reaction is the reverse of the formation of the compound, so

$-(\Delta_f H° \text{ MgCl}_2(s)) = -(641.62 \text{ kJ/mol}) = 641.62 \text{ kJ/mol} = 6.4162 \times 10^5 \text{ J/mol}$

Decomposing MgCl₂(s) therefore should require less energy per mole of Mg produced than the industrial process described. However, this does not consider all the energy required for the electrolytic process in the earlier parts of this question because the electrolytic process requires that the MgCl₂ must be molten. An additional input of energy is needed to do this.

21.125. Comparing the chemistry of C and Si:

(a) Reactions of CH₄ and SiH₄ with H₂O:

CH₄(g) + 2 H₂O(ℓ) → CO₂(g) + 4 H₂(g)

SiH₄(g) + 2 H₂O(ℓ) → SiO₂(s) + 4 H₂(g)

(b) Calculate $\Delta_r G°$ for reactions above:

For the reaction involving CH₄:

$\Delta_r G°$ = [(1 mol/mol-rxn)($\Delta_f G°$ CO₂(g)) + (4 mol/mol-rxn)($\Delta_f G°$ H₂(g))]

− [(1 mol/mol-rxn)($\Delta_f G°$ CH₄(g)) + (2 mol/mol-rxn)($\Delta_f G°$ H₂O(ℓ))]

$\Delta_r G°$ = [(1 mol/mol-rxn)(−394.359 kJ/mol) + (4 mol/mol-rxn)(0 kJ/mol)]

− [(1 mol/mol-rxn)(−50.8 kJ/mol) + (2 mol/mol-rxn)(−237.15 kJ/mol)]

= +130.7 kJ/mol-rxn

For the reaction involving SiH₄:

$\Delta_r G° = [(1 \text{ mol/mol-rxn})(\Delta_f G° \text{ SiO}_2(s)) + (4 \text{ mol/mol-rxn})(\Delta_f G° \text{ H}_2(g))]$

$\qquad - [(1 \text{ mol/mol-rxn})(\Delta_f G° \text{ SiH}_4(g)) + (2 \text{ mol/mol-rxn})(\Delta_f G° \text{ H}_2\text{O}(\ell)]$

$\Delta_r G° = [(1 \text{ mol/mol-rxn})(-856.97 \text{ kJ/mol}) + (4 \text{ mol/mol-rxn})(0 \text{ kJ/mol})]$

$\qquad - [(1 \text{ mol/mol-rxn})(56.84 \text{ kJ/mol}) + (2 \text{ mol/mol-rxn})(-237.15 \text{ kJ/mol})]$

$= -439.51 \text{ kJ/mol-rxn}$

The formation of SiO₂ is product favored at equilibrium.

(c) Regarding the polarity of C–H and Si–H bonds: Carbon is more electronegative than hydrogen, whereas hydrogen is more electronegative than silicon. The C–H bond is polar with the partial negative charge on C and the partial positive charge on H, while the Si—H bond is polar with the partial positive charge on silicon and the partial negative charge on hydrogen.

(d) Lewis structures for

(CH₃)₂CO and (CH₃)₂SiO:

Does a silicon compound similar to ethene exist? No, silicon prefers to form four single bonds rather than form a double bond, therefore a polymer of [SiH₂]ₙ is expected.

21.126. B–N bond length in boron nitride:

Atoms in a unit cell:

8 N on corners, 8 × 1/8 = 1; 6 N on faces, 6 × ½ = 3; this gives 4 N total

4 B on inside; 4 × 1 = 4; 4 B total, so the unit cell contains 4 BN formula units

$\dfrac{24.82 \text{ g BN}}{1 \text{ mol BN}} \left(\dfrac{1 \text{ mol BN}}{6.0221 \times 10^{23} \text{ BN units}} \right) \left(\dfrac{4 \text{ BN units}}{1 \text{ unit cell}} \right) = 1.649 \times 10^{-22}$ g/unit cell

Volume = $\dfrac{1.649 \times 10^{-22} \text{ g}}{1 \text{ unit cell}} \left(\dfrac{1 \text{ cm}^3}{3.45 \text{ g}} \right) = 4.78 \times 10^{-23}$ cm³/unit cell

Edge length = $(4.78 \times 10^{-23} \text{ cm}^3)^{1/3} = 3.63 \times 10^{-8}$ cm = 363 pm

Diagonal of face = $\sqrt{2}$ × cell edge = $\sqrt{2}$ × 363 pm = 513 pm

The N–N distance along the face diagonal is one-half this distance: ½(513 pm) = 257 pm.

Now consider the angle formed between an interior boron atom and two nitrogen atoms in corners. The angle is the tetrahedral angle, 109.5°. If this angle is bisected, a right triangle is formed that has the hypotenuse being the desired B–N distance and having one angle equal to ½(109.5°). The side opposite this angle has a length equal to half the N–N distance, or ¼ of the face diagonal.

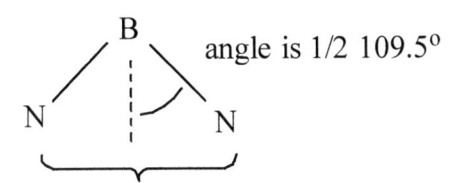

Recall from trigonometry that the sine of an angle is equal to the length of the opposite side divided by the length of the hypotenuse. Thus,

sin(109.5°/2) = ¼ face diagonal/B–N distance

sin(54.75°) = [¼(513 pm)]/B–N distance

sin(54.75°) = (128 pm)/B–N distance

sin(54.75°) = 0.8166, so 0.8166 = (128 pm)/B–N distance, and B–N distance = 157 pm

21.127. Concerning the reactions of the xenate anion, $HXeO_4^-$, with base:

$2\ HXeO_4^-(aq) + 2\ OH^-(aq) \rightarrow XeO_6^{4-}(aq) + Xe(g) + O_2(g) + 2\ H_2O(\ell)$

Oxidation numbers in $HXeO_4^-$: Assume H and O have their usual oxidation numbers in compounds, +1 and –2, respectively.

(+1) + Xe + 4(–2) = –1

Xe oxidation state = +6

Oxidation numbers in XeO_6^{4-}: Assume O has it usual oxidation number of –2, in this ion.

Xe + 6(–2) = –4

Xe oxidation state = +8

In atomic Xe the oxidation state is 0.

For $HXeO_4^- \rightarrow XeO_6^{4-}$, Xe is changing oxidation state from +6 to +8; it is oxidized.

For $HXeO_4^- \rightarrow Xe$, Xe is changing oxidation state from +6 to 0; it is reduced.

The balanced equation also indicates that another oxidation is taking place.

For $OH^- \rightarrow O_2$, O is changing oxidation state from –2 to 0; it is oxidized.

Three half-reactions can be written:

$6\ e^- + 3\ H_2O(\ell) + HXeO_4^-(aq) \rightarrow Xe(g) + 7\ OH^-(aq)$

$5\ OH^-(aq) + HXeO_4^-(aq) \rightarrow XeO_6^{4-}(aq) + 3\ H_2O(\ell) + 2\ e^-$

$4\ OH^-(aq) \rightarrow O_2(g) + 2\ H_2O(\ell) + 4\ e^-$

The sum of these three half-reactions gives the net reaction shown. Notice that the reduction half-reaction shows 6 electrons being gained. The two oxidation half-reactions show a total of 2 + 4 = 6 electrons being lost. Thus, the extents of oxidation and reduction are equal.

This process is a *disproportionation reaction* as the same element (Xe) is being **both** oxidized and reduced. It goes from +6 in $HXeO_4^-$ to +8 in XeO_6^{4-} and to 0 in Xe. These changes are accompanied by the oxidation of oxygen from –2 to 0. Finally, the $HXeO_4^-$ and OH^- ions are reducing agents and together supply 6 electrons to reduce the $HXeO_4^-$ ion to Xe.

Solution and Answer Guide

Kotz Treichel Townsend Treichel, Chemistry and Chemical Reactivity 11e, 978-0-357-85140-1, Chapter 22: The Chemistry of the Transition Elements

TABLE OF CONTENTS

Applying Chemical Principles 833
Practicing Skills 835
General Questions 848
In the Laboratory 857
Summary and Conceptual Questions 859

Applying Chemical Principles

Blue!

22.1.1. Preparation and formula of blue:

$$4.5392 \text{ g Y}_2\text{O}_3 \left(\frac{1 \text{ mol Y}_2\text{O}_3}{225.809 \text{ g Y}_2\text{O}_3} \right)\left(\frac{2 \text{ mol Y}}{1 \text{ mol Y}_2\text{O}_3} \right) = 0.040204 \text{ mol Y}$$

$$5.302 \text{ g In}_2\text{O}_3 \left(\frac{1 \text{ mol In}_2\text{O}_3}{277.637 \text{ g In}_2\text{O}_3} \right)\left(\frac{2 \text{ mol In}}{1 \text{ mol In}_2\text{O}_3} \right) = 0.038194 \text{ mol In}$$

$$0.1585 \text{ g MnO}_2 \left(\frac{1 \text{ mol MnO}_2}{86.936 \text{ g MnO}_2} \right)\left(\frac{1 \text{ mol Mn}}{1 \text{ mol MnO}_2} \right) = 0.001823 \text{ mol Mn}$$

$$\frac{\text{mol In}}{\text{mol Y}} = \frac{0.038194 \text{ mol}}{0.040204 \text{ mol}} = 0.95$$

$1 - x = 0.95$ thus $x = 0.05$

22.1.2. (a) Place electrons in appropriate d orbitals in the diagram:

Mn^{2+} has 5 d electrons. One electron goes in each of the five orbitals.

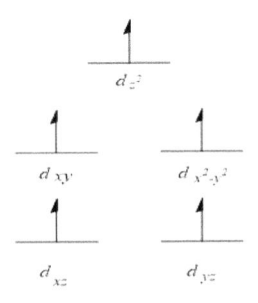

(b) The d_{z^2} orbital is highest in energy as it points directly at the oxide ions on the z axis. The d_{xz} and d_{yz} orbitals do not lie in the plane of the three ligands in the trigonal plane and so are less affected and are the lowest energy orbitals.

22.1.3. In Prussian blue a CN⁻ ligand bridges an iron(II) and an iron(III) ion.

Cis-platin: Accidental discovery of a Chemotherapy Agent

22.2.1 Quantity of *cis*-platin remaining after 24 hours:

With a half-life of 2.5 hours, 24 hours represents 9.6 half-lives.

We can calculate the rate constant for the decomposition: $k = \dfrac{0.693}{t_{1/2}} = \dfrac{0.693}{2.5 \text{ hr}} = 0.2772 \text{ hr}^{-1}$

Now the integrated rate equation can be used to determine the amount remaining:

$\ln\left(\dfrac{x}{10.0 \text{ mg}}\right) = -(0.2772 \text{ hr}^{-1})(24 \text{ hr}) = -6.653$ and $x = 0.0129$ mg or 0.013 mg (2 sf)

22.2.2. Systematic name for *cis*-platin: *cis*-diamminedichloroplatinum(II)

22.2.3. Distribution of *d* electrons in Pt(II) in *d*-orbital splitting diagram:

Having lost two electrons, the electron configuration of Pt is [Xe] $4f^{14}5d^8$. The four pairs of electrons then occupy the lowest energy orbitals. The *d*-orbitals split into four groups, in order of increasing energy: $d_{xz} = d_{yz} < d_{z^2} < d_{xy} < d_{x^2-y^2}$.

The Rare Earths

22.3.1. Electron configurations for Nd and Eu using the spectroscopic notation:

Nd: Atomic number 60: [Xe]$4f^46s^2$

Eu: Atomic number 63: [Xe]$4f^76s^2$

22.3.2. Electron configurations for Ce^{3+} and Nd^{3+}:

Ce^{3+}: With 55 electrons— [Xe]$4f^1$—having lost the 5*d* and 6*s* electrons.

Nd^{3+}: With 57 electrons—[Xe]$4f^3$—having lost the 6*s* electrons and 1*f* electron (from Nd, see 22.3.1)

22.3.3. Concentration of Ce salt when titrated with 0.181 g Fe^{2+}

$(NH_4)_2Ce(NO_3)_6$ has Ce in a +4 oxidation state. When it reacts with Fe^{2+}, the Ce is reduced to Ce^{3+}. This means that 1 mol of electrons is transferred from the Fe^{2+} ion to the Ce^{4+} ion, reducing the latter to the +3 oxidation state.

The stoichiometry is:

$$0.181 \text{ g Fe} \left(\frac{1 \text{ mol Fe}}{55.845 \text{ g Fe}} \right) \left(\frac{1 \text{ mol Ce}}{1 \text{ mol Fe}} \right) \left(\frac{1}{0.03133 \text{ L}} \right) = 0.103 \text{ M}$$

22.3.4. Balance equations for reactions involving Ce(IV) and CO:

Balanced by inspection: $2 \text{ CeO}_2(s) + \text{CO}(g) \rightarrow \text{Ce}_2\text{O}_3(s) + \text{CO}_2(g)$

Balanced by inspection: $2 \text{ Ce}_2\text{O}_3(s) + \text{O}_2(g) \rightarrow 4 \text{ CeO}_2(s)$

22.3.5
If there are N^{3-} ions in the faces, there are (6 faces)(1/2 ion in each face) = net of 3 N^{3-} ions. This accounts for a total of –9 in ion charges. The lanthanide ions at the 8 corners must be La^{3+} ions. There are 8 corners and 1/8 of each corner ion is within the cell [(8 corners)(1/8 ion for each corner) = net of one La^{3+} ion. The tungsten ion in the center of the cell is a +6 ion. Sum of ion charges: (–9 for N^{3-} ions in the cube faces) + 1 La^{3+} ion + 1 W^{6+} ion = overall charge = 0.

Practicing Skills

Properties of Transition Elements

22.1. Of the elements, Co, Cm, Cd, Ce, Cf, the actinides are Cm and Cf.

22.2. Of the elements, Ta, Tc, Ti, Th, Tm, the only actinide is Th.

22.3. Elements with unfilled 4d orbitals:

Rh: [Kr] 4$d^8$5s^1 Given the electron configurations here, only Rh and Ru

Re: [Xe] 4f^{14}5$d^5$6s^2 have unfilled 4d orbitals.

Ru: [Kr] 4$d^7$5s^1

Rf: [Rn] 5f^{14}6$d^2$7s^2

Ra: [Rn]7s^2

22.4. Elements with unfilled 3d orbitals:

Cd: [Kr] $4d^{10}5s^2$

Ce: [Xe] $4f^15d^16s^2$

Co: [Ar] $3d^74s^2$

Cr: [Ar] $3d^54s^1$

Cu: [Ar]$3d^{10}4s^1$

Given the electron configurations here, Co and Cr have unfilled 3d orbitals.

22.5. Identify the following as chemical or physical properties:

(a) can be oxidized — chemical

(b) have unpaired electrons — physical

(c) solids at 25 °C — physical

(d) metallic luster — physical

(e) compounds often colored — physical

22.6. Common chemical and physical properties of iron:

chemical: (a) can be oxidized to +2 or +3 oxidation state; (b) forms several oxides (e.g., rust).

physical: (a) solid at 25 °C; (b) compounds are colored; (c) iron has a metallic luster.

22.7 Among the transition elements:

(a) The densest element is osmium, Os.

(b) The metal with the lowest melting point is mercury, Hg, a liquid at room T.

(c) A radioactive *d*-block element used in lung scans is Tc.

(d) Essential elements in the human body are iron, cobalt, and molybdenum (Fe, Co, Mo).

22.8. Among the transition elements:

(a) The metal with the highest melting point is tungsten, W, with a melting point of 3686 K.

(b) A commonly used metal in construction is iron, Fe.

(c) A radioactive *f*-block element, radium, Ra, is useful in X-rays, etc.

(d) Two of the metals in nitrogenase are molybdenum and iron (Mo and Fe).

22.9. Give electron configurations and determine if paramagnetic.

(a) Cr^{3+} [Ar]$3d^3$ ↑ ↑ ↑ (3d) ☐ (4s) paragmagnetic

(b) V^{2+} [Ar]$3d^3$ ↑ ↑ ↑ (3d) ☐ (4s) paragmagnetic

(c) Ni^{2+} [Ar]3d^8 [↕|↕|↕|↑|↑] [ ] paramagnetic

(d) Cu$^+$ [Ar]3d^{10} [↕|↕|↕|↕|↕] [ ] diamagnetic

22.10. Identify transition metal cations with the following electron configurations:

(a) [Ar]3d^6: Fe^{2+}, Co^{3+}

(b) [Ar]3d^{10}: Cu$^+$, Zn^{2+}

(c) [Ar]3d^5: Cr$^+$, Mn^{2+}

(d) [Ar]3d^8: Ni^{2+}

22.11. A cation from the first series transition metal that is isoelectronic with:

(a) Mn^{2+}: Fe^{3+} has 5 3d electrons so it is isoelectronic with Mn^{2+}

(b) Zn^{2+}: Cu^{1+} has 10 3d electrons

(c) Fe^{2+}: Co^{3+} has 6 3d electrons

(d) Cr^{3+}: V^{2+} has 3 3d electrons

22.12. Match the isoelectronic ions from the list: Cu^{1+}, Mn^{2+}, Fe^{2+}, Co^{3+}, Fe^{3+}, Zn^{2+}, Ti^{2+}, V^{3+}.
Ti^{2+} and V^{3+} are isoelectronic with 20 electrons.
Mn^{2+} and Fe^{3+} are isoelectronic with 23 electrons.
Fe^{2+} and Co^{3+} are isoelectronic with 24 electrons.

Formulas of Coordination Compounds

22.13. The lanthanide contraction is given as an explanation for the fact that the 6th period transition metals have (b) atomic radii like the 5th period transition elements.

22.14. Describe how atomic radii of the transition elements change across a period.
The atomic radii of transition elements is dependent on the outermost ns electrons. As you move across a period (e.g., the 4th period), the 4s electron pair is in place, but electrons are placed in 3d orbitals. As the nuclear charge increases, 3d and 4s electrons are more strongly attracted to the nucleus, and the radius decreases. Toward the end of the d-block, however, electron-electron repulsions overcome the increased nuclear charge, and atomic sizes increase somewhat. See Figure 7.9 in *Chemistry & Chemical Reactivity*.

22.15. Most common form of iron found in Earth's crust: (b) iron oxide.

22.16. Consider the reaction: CuFeS$_2$ + 3 CuCl$_2$ → 4 CuCl + FeCl$_2$ + 2 S:
S is in a −2 oxidation state in CuFeS$_2$, and in a 0 oxidation state in S, so S is oxidized.
Cu is in a +2 oxidation state in CuFeS$_2$, and in a +1 oxidation state in CuCl, so Cu is reduced.

22.17. In the pyrometallurgy of iron, C and CO serve as reducing agents.

22.18. The function of CaO in a blast furnace is as a base. The basic oxide CaO reacts with SiO_2 to give $CaSiO_3$. Recall that metal oxides function as bases and nonmetal oxides as acids.

22.19. Classify each of the following ligands as monodentate or polydentate:
- (a) CH_3NH_2 — monodentate (lone pair on N)
- (b) CH_3CN — monodentate (lone pair on N)
- (c) N_3^- — monodentate (lone pair on a N atom)
- (d) ethylenediamine — bidentate (lone pairs on 2 terminal N atoms)
- (e) Br^- — monodentate (lone pair on Br)
- (f) phenanthroline — bidentate (lone pairs on 2 N atoms)

22.20. NH_4^+ is incapable of serving as a ligand because it does not have any unshared electron pairs.

22.21. Name or structural formula for ligands:

	Name	Formula
(a)	ethylenediamine	$H_2NCH_2CH_2NH_2$
(b)	oxalate ion	$C_2O_4^{2-}$
(c)	ammine	NH_3
(d)	thiocyanate ion	SCN^-

22.22. Name or structural formula for ligands:

	Name	Formula
(a)	acetylacetonate	$CH_3COCHCOCH_3^-$
(b)	nitrite ion	NO_2^-
(c)	aqua	H_2O
(d)	nitro	NO_2^-

22.23. Oxidation number of the metal in the following complexes:

Compound	Metal	Oxidation Number
(a) [Mn(NH$_3$)$_6$]SO$_4$	Mn	+2

Ammonia is a neutral ligand. The presence of the binegative anion SO$_4^{2-}$ means that Mn has a +2 oxidation number.

(b) K$_3$[Co(CN)$_6$]	Co	+3

The 3 K$^+$ ions mean that the complex ion has a –3 charge. Each CN$^-$ has a –1 charge, so Co has a +3 oxidation number.

(c) [Co(NH$_3$)$_4$Cl$_2$]Cl	Co	+3

The chloride anion means that the complex ion must have a net +1 charge. While the ammonia ligand is neutral, each Cl has a –1 charge, so Co has a +3 oxidation number.

(d) Cr(en)$_2$Cl$_2$	Cr	+2

The ethylenediamine ligand is neutral. Each each Cl has a –1 charge, so Cr has a +2 oxidation number.

22.24. Oxidation number of the metal in the following complexes:
(a) [Fe(NH$_3$)$_6$]$^{2+}$, because ammonia is a neutral ligand, Fe has a +2 oxidation state.
(b) [Zn(CN)$_4$]$^{2-}$, with CN as a −1 ligand, Zn has a +2 oxidation state.
(c) [Co(NH$_3$)$_5$(NO$_2$)]$^+$, with ammonia being a neutral ligand, and NO$_2^-$ having a −1 charge, Co must has a +2 oxidation state.
(d) [Cu(en)$_2$]$^{2+}$, with en being neutral, Cu has a +2 oxidation state.

22.25. Formula for the complex ion: [Ni(en)(NH$_3$)$_3$(H$_2$O)]$^{2+}$.
The complex has to have a 2+ charge, since each of the three types of ligands is neutral.

22.26. Formula for the complex ion: [Cr(NH$_3$)$_2$(en)$_2$]$^{3+}$
The complex has to have a 3+ charge, since each of the attached ligands is neutral.

Naming Coordination Compounds

22.27. Formulas for the following compounds or ions:

(a) dichlorobis(ethylenediamine)nickel(II)	Ni(en)$_2$Cl$_2$
(b) potassium tetrachloroplatinate(II)	K$_2$[PtCl$_4$]
(c) potassium dicyanocuprate(I)	K[Cu(CN)$_2$]
(d) tetraamminediaquairon(II)	[Fe(NH$_3$)$_4$(H$_2$O)$_2$]$^{2+}$

22.28. Formulas for the following compounds or ions:

(a) diamminetriaquahydroxochromium(II) nitrate [Cr(NH$_3$)$_2$(H$_2$O)$_3$(OH)]NO$_3$

(b) hexaammineiron(III) nitrate [Fe(NH$_3$)$_6$](NO$_3$)$_3$

(c) pentacarbonyliron(0) [Fe(CO)$_5$]

(d) ammonium tetrachlorocuprate(II) (NH$_4$)$_2$[CuCl$_4$]

22.29. Names for the following compounds or ions:

(a) [Ni(C$_2$O$_4$)$_2$(H$_2$O)$_2$]$^{2-}$ diaquabis(oxalato)nickelate(II) ion

(b) [Co(en)$_2$Br$_2$]$^+$ dibromobis(ethylenediamine)cobalt(III) ion

(c) [Co(en)$_2$(NH$_3$)Cl]$^{2+}$ amminechlorobis(ethylenediamine)cobalt(III) ion

(d) Pt(NH$_3$)$_2$(C$_2$O$_4$) diammineoxalatoplatinum(II)

22.30. Names for the following compounds or ions:

(a) [Co(H$_2$O)$_4$Cl$_2$]$^+$ tetraaquadichlorocobalt(III) ion

(b) Co(H$_2$O)$_3$F$_3$ triaquatrifluorocobalt(III)

(c) [Pt(NH$_3$)Br$_3$]$^-$ amminetribromoplatinate(II) ion

(d) [Co(en)(NH$_3$)$_3$Cl]$^{2+}$ triamminechloroethylenediaminecobalt(III) ion

22.31. The name or formula for the ions or compounds shown below:

(a) [Fe(H$_2$O)$_5$(OH)]$^{2+}$ pentaaquahydroxoiron(III) ion

(b) K$_2$[Ni(CN)$_4$] potassium tetracyanonickelate(II)

(c) K[Cr(C$_2$O$_4$)$_2$(H$_2$O)$_2$] potassium diaquabis(oxalato)chromate(III)

(d) (NH$_4$)$_2$[PtCl$_4$] ammonium tetrachloroplatinate(IV)

22.32. The name or formula for the ions or compounds shown below:

(a) tetraaquadichlorochromium(III) chloride [CrCl$_2$(H$_2$O)$_4$]Cl

(b) pentaamminesulfatochromium(III) chloride [Cr(NH$_3$)$_5$SO$_4$]Cl

(c) sodium tetrachlorocobaltate(II) Na$_2$[CoCl$_4$]

(d) tris(oxalato)ferrate(III) ion [Fe(C$_2$O$_4$)$_3$]$^{3-}$

Isomerism

22.33. Isomerism possible for the specified complex:

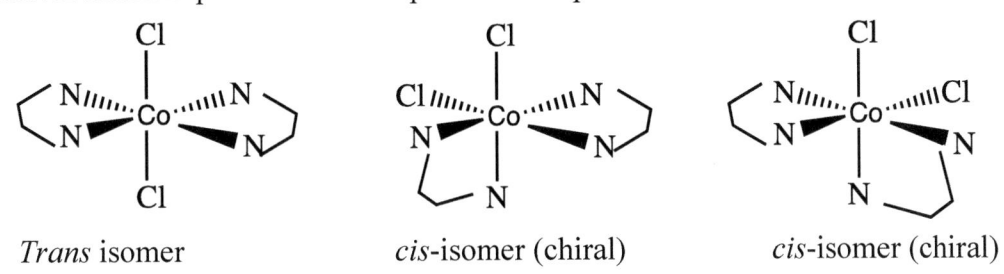

Trans isomer

cis-isomer (chiral)

cis-isomer (chiral)

22.34. Isomerism possible for the specified complex:

geometric isomers

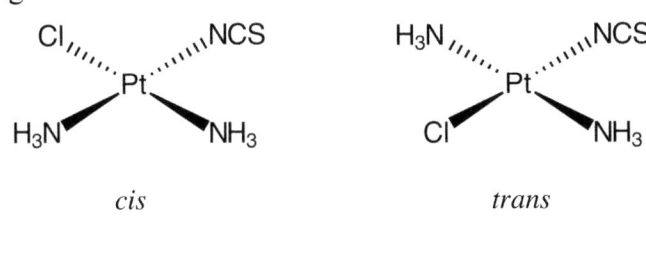

cis

trans

linkage isomers

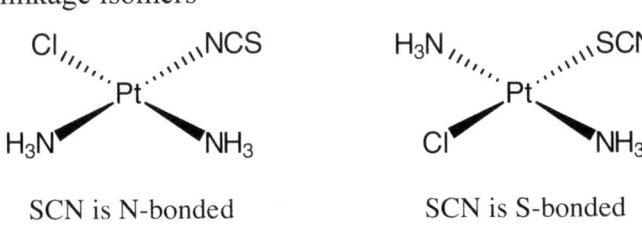

SCN is N-bonded

SCN is S-bonded

22.35. Geometric isomers of:

(a) $Fe(NH_3)_4Cl_2$

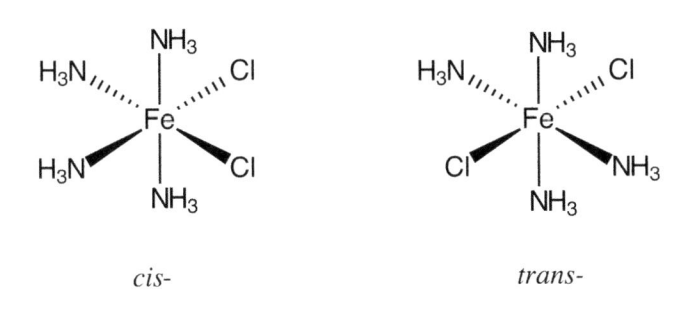

cis-

trans-

(b) Pt(NH$_3$)$_2$(SCN)(Br)

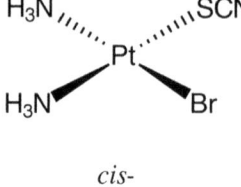

cis-   trans-

(c) Co(NH$_3$)$_3$(NO$_2$)$_3$

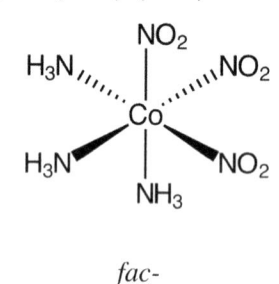

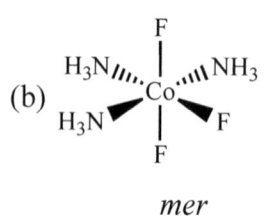

fac-   mer-

(d) [Co(en)Cl$_4$]$^-$

22.36. Geometric isomers possible?

(a) trans   cis

(b) mer   fac

(c) No geometrical isomers possible.

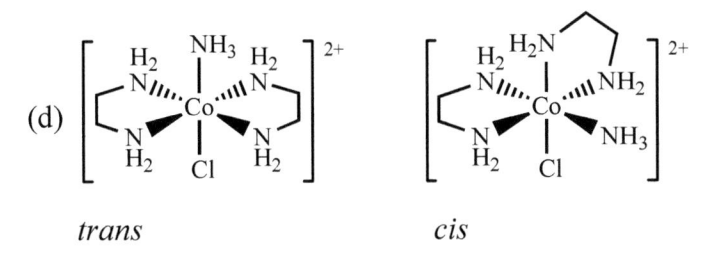

(d) [structure: trans-[Co(en)₂(NH₃)(Cl)]²⁺]   [structure: cis-[Co(en)₂(NH₃)(Cl)]²⁺]

   trans   *cis*

22.37. Which of the following species has a chiral center?

(a) [Fe(en)₃]²⁺   Yes

[structures: two mirror-image Fe(en)₃ complexes]

The two mirror images shown above are not superimposable, and therefore possess a chiral center.

(b) *trans*-[Co(en)₂Br₂]⁺   No.

[structure: trans-[Co(en)₂Br₂]⁺]

(c) *fac*-[Co(en)(H₂O)Cl₃]   No

[structures: two mirror-image fac-[Co(en)(H₂O)Cl₃] complexes]

As you can see by the two structures above, a 180° rotation along the Cl–Co–H₂O axis, would make these two mirror images superimposable. The *mer* complex would also be – superimposable, and therefore possesses no chiral center.

(d) Pt(NH₃)(H₂O)Cl(NO₂)

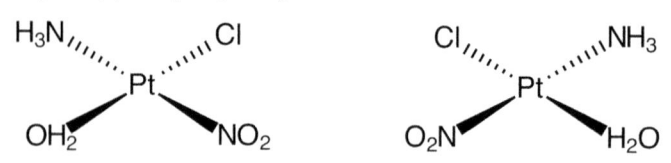

Above are mirror images of the complex. Flip the complex over like a pancake and it will overlap with the second image. There are no nonsuperimposable isomers.

22.38. Four isomers for the complex:

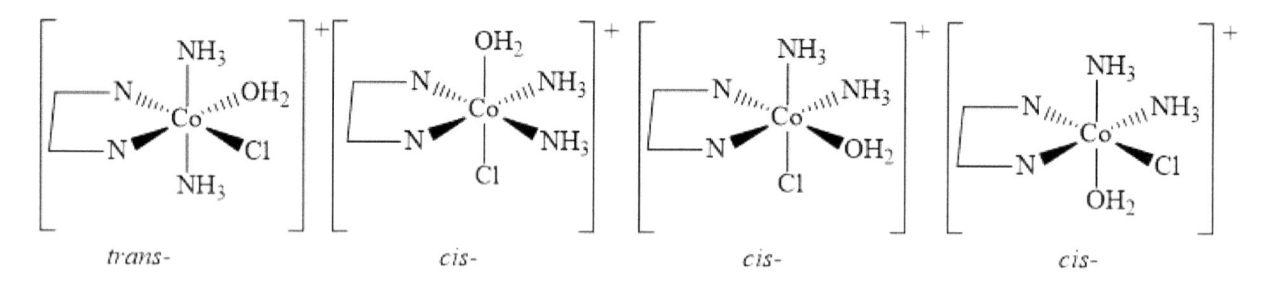

trans-   *cis-*   *cis-*   *cis-*

Note bonding of NH₃ groups to Co.

Note bonding of H₂O and Cl groups to Co. Both of these *cis-* isomers are chiral. Vertical axis is axis of asymmetry.

Magnetic Properties of Complexes

22.39. Force of attraction between metal and ligand in coordination complex:

Ligand field theory describes the attraction as electrostatic, with the positively charged metal ion and negatively charged (or polar molecule) ligands.

22.40. Correct statements about ligand field theory:

(c) Coordination of the ligands to the metal causes splitting in the *d* orbital energy levels.

22.41. The lower energy *d* orbitals in an octahedral complex d_{xy}, d_{yz}, d_{xz}.

22.42. Lowest energy and highest energy *d* orbitals in square planar complex:

Lowest energy: d_{xy}, d_{yz}

Highest energy: $d_{x^2-y^2}$

22.43. The counterions in these examples have been omitted. The counterions do not affect the magnetic behavior of the complex ion.

(a) $[Mn(CN)_6]^{4-}$

Mn^{2+} has a d^5 configuration.

low spin
strong field ligand

1 unpaired electron
paramagnetic

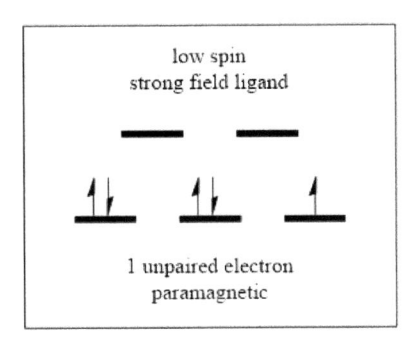

(b) $[Co(NH_3)_6]^{3+}$
Co^{3+} has a d^6 configuration

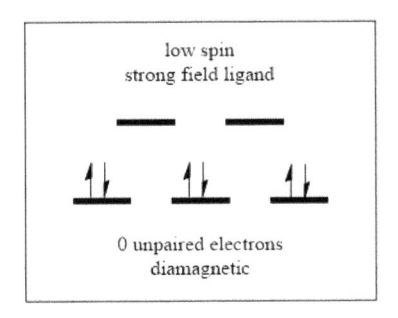

low spin
strong field ligand

0 unpaired electrons
diamagnetic

(c) $[Fe(H_2O)_6]^{3+}$
Fe^{3+} has a d^5 configuration (like Mn^{2+}) and 1 unpaired electron

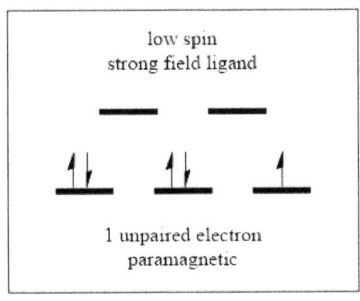

low spin
strong field ligand

1 unpaired electron
paramagnetic

(d) $[Cr(en)_3]^{2+}$

Cr^{2+} has the d^4 configuration

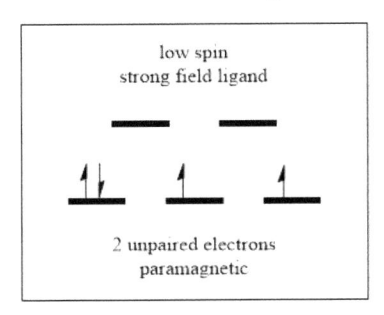

low spin
strong field ligand

2 unpaired electrons
paramagnetic

22.44. Electron configuration of the metal and number of unpaired electrons, if any:

(a) Fe^{2+}, d^6, 4 unpaired electrons	(c) Cr^{2+}, d^4, 4 unpaired electrons
↑ ↑ ↑↓ ↑ ↑	↑ ___ ↑ ↑ ↑
(b) Mn^{2+}, d^5, 5 unpaired electrons	(d) Fe^{3+}, d^5, 5 unpaired electrons
↑ ↑ ↑ ↑ ↑	↑ ↑ ↑ ↑ ↑

22.45. For the following (high spin) tetrahedral complexes, determine the number of unpaired electrons:

(a) $[FeCl_4]^{2-}$ (d^6, paramagnetic, 4 unpaired)	(c) $[MnCl_4]^{2-}$ (d^5, paramagnetic, 5 unpaired)
↑ ↑ ↑ ↑↓ ↑	↑ ↑ ↑ ↑ ↑
(b) $[CoCl_4]^{2-}$ (d^7, paramagnetic, 3 unpaired)	(d) $[ZnCl_4]^{2-}$ (d^{10}, diamagnetic, 0 unpaired)
↑ ↑ ↑ ↑↓ ↑↓	↑↓ ↑↓ ↑↓ ↑↓ ↑↓

22.46. For the following (high spin) tetrahedral complexes, determine the number of unpaired electrons:

(a) Zn^{2+}, d^{10}, no unpaired electrons	(c) Mn^{2+}, d^5, 5 unpaired electrons
↑↓ ↑↓ ↑↓ ↑↓ ↑↓	↑ ↑ ↑ ↑ ↑
(b) V^{5+}, d^0, no d electrons, no unpaired electrons	(d) Cu^{2+}, d^9, 1 unpaired electron
	↑↓ ↑↓ ↑ ↑↓ ↑↓

22.47. For [Fe(H$_2$O)$_6$]$^{2+}$:

(a) The coordination number of iron is 6. Six monodentate ligands (H$_2$O) are attached.

(b) The coordination geometry is octahedral. (Six groups attached to the central metal ion).

(c) The oxidation state of iron is +2. The charge on the complex is 2+. (Water is a neutral ligand.)

(d) Fe^{2+} is a d^6 case. Water is a weak-field ligand (leading to a high-spin complex); there are 4 unpaired electrons.

(e) The complex is paramagnetic.

22.48. For the complex [Co(en)(NH$_3$)$_2$Cl$_2$]ClO$_4$:

(a) The coordination number of cobalt is 6. One bidentate and 4 monodentate ligands are attached.

(b) The coordination geometry of the complex ion is octahedral. (Six groups attached to the central metal ion).

(c) The oxidation state of cobalt is +3. The charge on the complex ion is 1+. The perchlorate anion has a 1– charge, so the overall charge on the complex ion is 1+.

(d) No unpaired electrons on the central cobalt(III) ion ([Ar]3d^6, low spin).

(e) With no unpaired electrons, the complex is diamagnetic.

22.49. The weak-field ligand, Cl$^-$, permits either tetrahedral or square-planar geometries. With the d^8 Ni^{2+} ion, such weak-field splitting leads to 2 unpaired electrons in the tetrahedral NiCl$_4^{2-}$ ion. The strong-field ligand, CN$^-$, results in a square-planar complex with no unpaired electrons. The complex ion would be diamagnetic.

tetrahedral NiCl$_4^{2-}$	square planar Ni(CN)$_4^{2-}$
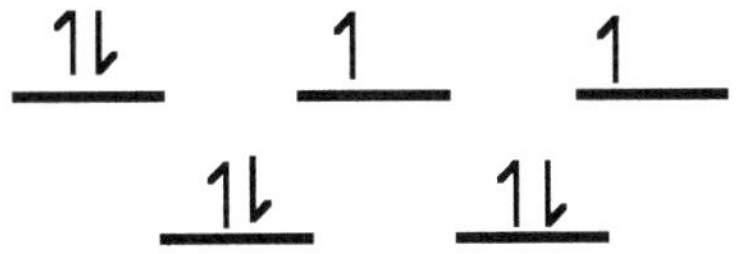 ↑↓ ↑ ↑ ↑↓ ↑↓	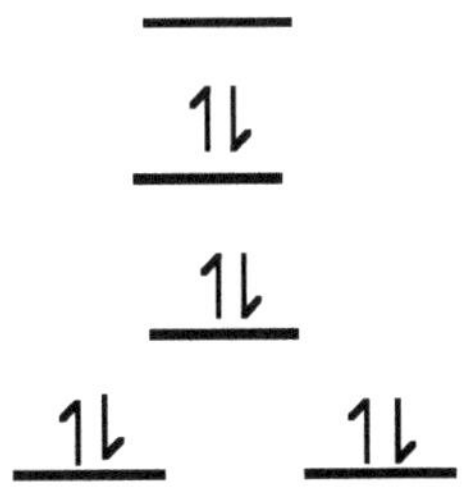 ___ ↑↓ ↑↓ ↑↓ ↑↓

22.50. Explain the change in magnetic property for aqueous iron(II) sulfate:

When FeSO$_4$ dissolves in water it forms [Fe(H$_2$O)$_6$]$^{2+}$; addition of ammonia converts this to [Fe(NH$_3$)$_6$]$^{2+}$. The hexaaqua complex is high spin (d^6, paramagnetic, 4 unpaired electrons), but the ammonia complex is low spin (diamagnetic, no unpaired electrons). Ammonia is a stronger field ligand than water.

Spectroscopy of Complexes

22.51. Color of light absorbed by the Ti(III) ion:

500 nm light is in the green region of the spectrum. The transmitted light, and the color of the solution, is magenta.

22.52. Color of the solution when absorbing 700 nm light:
The light absorbed (700 nm) is in the red region of the spectrum. Therefore, the light transmitted (the color of the solution) is blue.

General Questions

22.53. Describe an experiment to determine if:

Nickel in $K_2[NiCl_4]$ is in a square planar or tetrahedral environment.

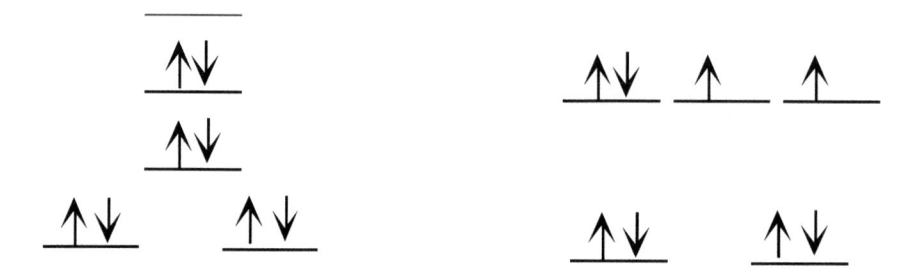

d- orbitals for square-planar complex   d- orbitals for tetrahedral complex

Square-planar nickel(II) is diamagnetic whereas tetrahedral nickel(II) is paramagnetic. Measuring the magnetic moment would discriminate between the two.

22.54. Greatest number of unpaired electrons:

(a) Cr^{3+}, d^3, 3 unpaired electrons   (c) Fe^{2+}, d^6, 4 unpaired electrons
(b) Mn^{2+}, d^5, 5 unpaired electron   (d) Ni^{2+}, d^8, 2 unpaired electrons

22.55. Number of unpaired electrons for high-spin and low-spin complexes of Fe^{2+}:

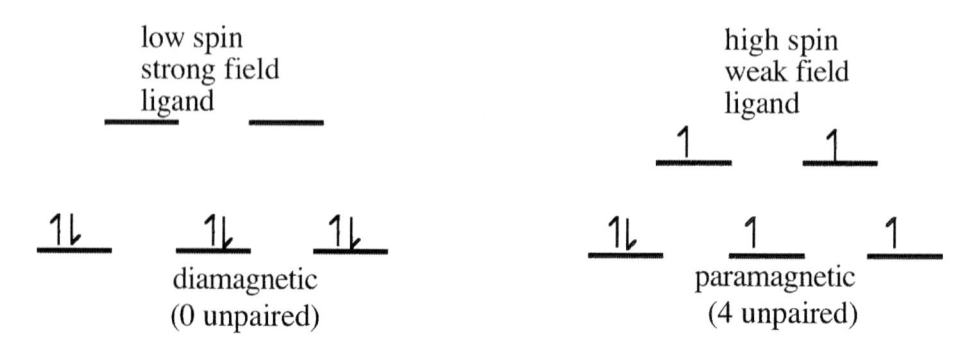

low spin
strong field
ligand

high spin
weak field
ligand

diamagnetic
(0 unpaired)

paramagnetic
(4 unpaired)

22.56. Moles of AgCl expected to precipitate:

Only one chloride is not coordinated directly to the metal, so 1.0 mole of AgCl will precipitate.

22.57. Which of the following complexes are square planar?

Of the four complexes, $[Ni(CN)_4]^{2-}$ and $[Pt(CN)_4]^{2-}$ are square planar, since such complexes have the geometry assumed by d^8 metal ions. The other complexes are tetrahedral.

22.58. Identify chiral complexes: Only (b) cis-$[Fe(C_2O_4)_2Cl_2]^{2-}$ has a nonsuperimposable mirror image.

22.59. Geometric isomers of $Pt(NH_3)(CN)Cl_2$: There are two.

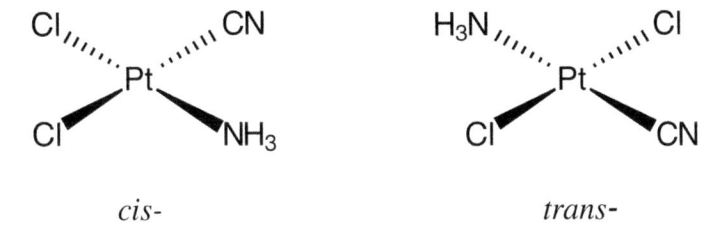

cis-   *trans-*

22.60. For the coordination compound $[Fe(en)_2Cl_2]Cl$:

(a) Oxidation number of iron: Fe^{3+}; Overall charge is +1, 2 Cl ligands (each −1).

(b) The coordination number for iron is 6.

(c) The coordination geometry is octahedral.

(d) 1 unpaired electron ($[Ar]3d^5$, low spin)

(e) Paramagnetic

(f) Two possible geometric isomers, *cis* and *trans*

22.61. For the high-spin complex $Mn(NH_3)_4Cl_2$ determine:

(a) The oxidation number of manganese is +2. With an overall neutral complex, and 2 chloride ions (each with a −1 charge) and 4 neutral ammonia ligands, manganese has to be +2.

(b) With six monodentate ligands, the coordination number for manganese is 6.

(c) With six monodentate ligands, the coordination geometry for manganese is octahedral.

(d) Number of unpaired electrons: 5. The ligands attached are weak-field ligands resulting in five unpaired electrons.

(e) With unpaired electrons, the complex is paramagnetic.

(f) *Cis* and *trans* geometric isomers are possible.

22.62. Cation and anion in Magnus's green salt:

cation: $[Pt(NH_3)_4]^{2+}$ tetraammineplatinum(II) ion

anion: $[PtCl_4]^{2-}$ tetrachloroplatinate(II) ion

22.63. Structures for cis- and trans- isomers of $CoCl_3 \cdot 4\,NH_3$:

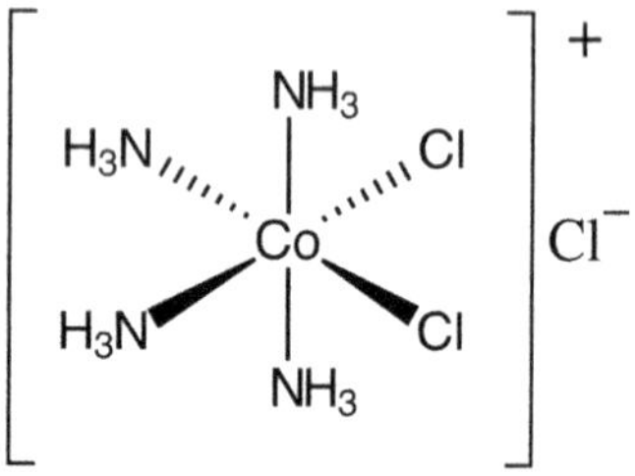

cis-tetraamminedichlorocobalt(III) chloride

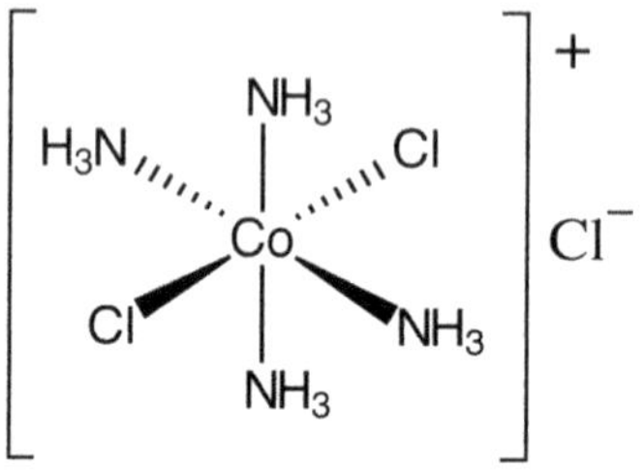

trans-tetraamminedichlorocobalt(III) chloride

22.64. Formula and name of square planar complex of Pt: $[Pt(NO_2)Cl(NH_3)_2]$, diamminechloronitroplatinum(II)

Two geometric isomers

cis

trans

22.65. Formula of a complex containing a Co^{3+} ion, two ethylenediamine molecules, one water molecule, and one chloride ion: $[Co(en)_2(H_2O)Cl]^{2+}$

Both ethylenediamine and water are neutral, so with a (3+) and (1−) charge, the net charge on the ion is 2+.

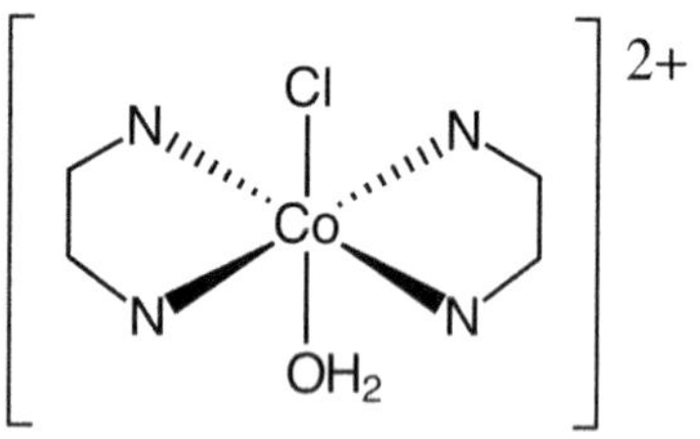

22.66. Two geometric isomers of [Cr(dmen)₃]³⁺ can exist.

fac   *mer*

22.67. Regarding diethylenetriamine:

(a) Structures for the *fac-* and *mer-* isomers of Cr(dien)Cl₃. In these diagrams the curved lines represent the tridentate ligand H₂N-CH₂CH₂-NH-CH₂CH₂-NH₂ with attachments to the metal ion through the electron pairs on three N atoms.

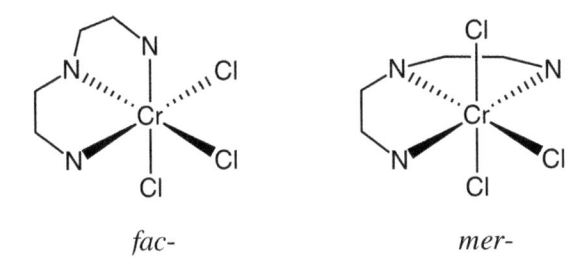

fac-   *mer-*

(b) Two different isomers of Cr(dien)BrCl₂:

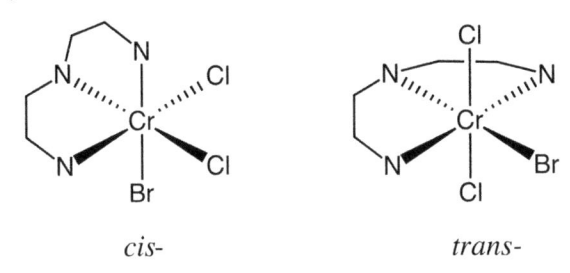

cis-   *trans-*

(c) The geometric isomers for isomers of [Cr(dien)₂]³⁺

fac-   *mer-*

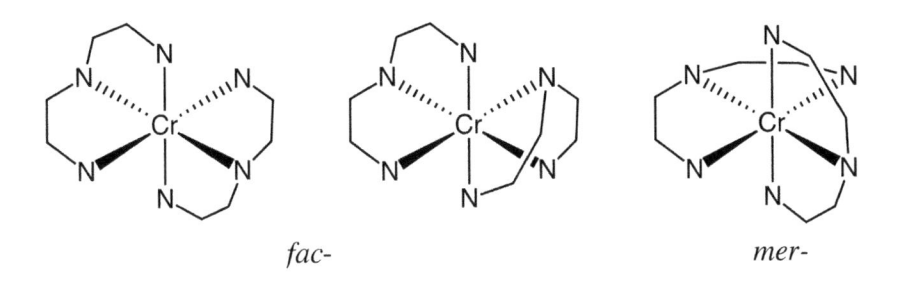

22.68 Write electron configuration of the $[CoF_6]^{3-}$ and $[Co(NH_3)_6]^{3+}$:

$[CoF_6]^{3-}$ (paramagnetic) $[Co(NH_3)_6]^{3+}$ (diamagnetic)

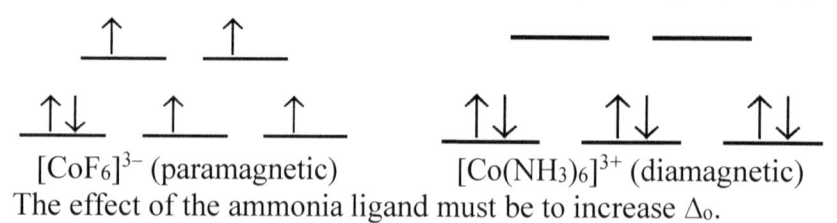

The effect of the ammonia ligand must be to increase Δ_o.

22.69. Geometric isomers for $[Co(en)(NH_3)(H_2O)_2]^{3+}$:

Chiral

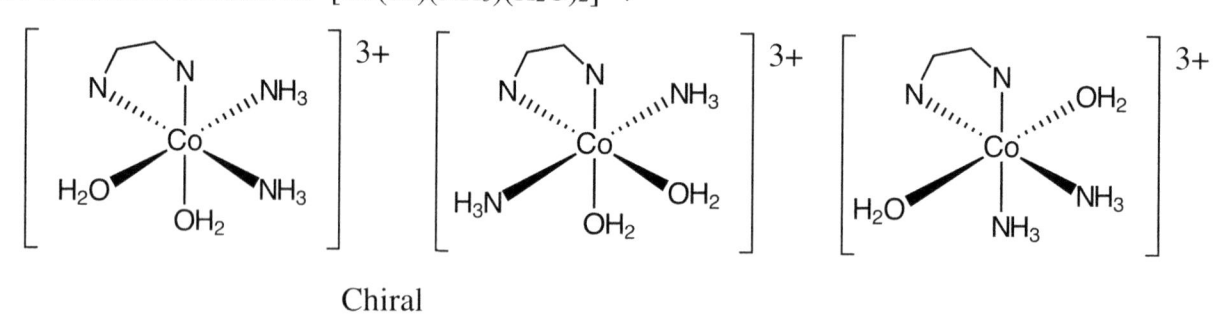

22.70. No *trans* isomer for Pt(en)Cl₂: The bond angles and bond lengths are such that the nitrogen atoms cannot span the diagonal of the Pt complex and accomplish reasonable overlap with the Pt orbitals to form the bonds.

22.71. Effects of CN⁻ and H₂O ligands on magnitude of Δ_o.

For the two Mn^{2+} complexes:

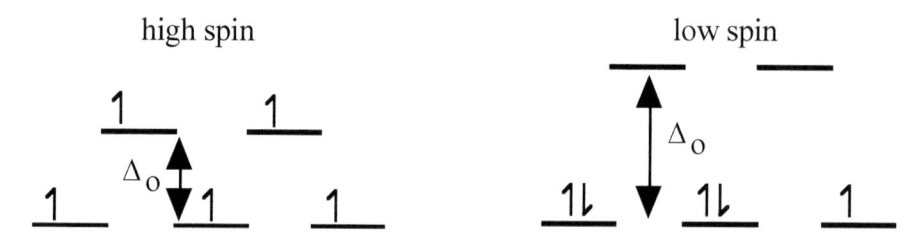

high spin   low spin

Weak field $[Mn(H_2O)_6]^{2+}$ Strong field $[Mn(CN)_6]^{4-}$

The cyano ligand is a *strong field* ligand, so the difference in energies of the doubly versus triply degenerate *d* orbitals is greater than with the *weak field* aqua ligand. The difference in these energy levels is referred to as Δ_o, as shown in the diagrams above.

22.72. Account for the magnetism of the two compounds:

In K$_4$[Cr(CN)$_6$] and K$_4$[Cr(SCN)$_6$], Cr has an oxidation number of +2 ([Ar]3d^4).

$$\underline{\uparrow\downarrow}\ \underline{\uparrow}\ \underline{\uparrow}\quad\underline{\ \ }\ \underline{\ \ }$$
$$\text{Cr}^{2+}$$
paramagnetic, 2 unpaired e⁻

$$\underline{\uparrow}\ \underline{\ \ }\quad\underline{\uparrow}\ \underline{\uparrow}\ \underline{\uparrow}$$
$$\text{Cr}^{2+}$$
paramagnetic, 4 unpaired e⁻

The SCN⁻ ligand is a weaker field ligand than the CN⁻ ligand. The SCN⁻ ligand occurs to the left (lower) in the spectrochemical series relative to CN⁻.

22.73. Provide systematic names or formulas:

(a) (NH$_4$)$_2$[CuCl$_4$]   ammonium tetrachlorocuprate(II)

(b) [Cr(H$_2$O)$_4$Cl$_2$]Cl   tetraaquadichlorochromium(III) chloride

(c) [Co(H$_2$O)(H$_2$NCH$_2$CH$_2$NH$_2$)$_2$(SCN)](NO$_3$)$_2$

   aquabis(ethylenediamine)thiocyanatocobalt(III) nitrate

22.74. Complex absorbs 425-nm light, the blue-violet end of the visible spectrum. So red and green are transmitted, and the complex appears to be yellow.

22.75. For the complex ion [Co(CO$_3$)$_3$]$^{3-}$:

(a) Color of the complex? The light absorbed is in the orange region of the spectrum. Therefore, the light transmitted (the color of the solution) is blue or cyan.

(b) Carbonate, a strong or weak-field ligand? According to Table 22.3, CO$_3^{2-}$ belongs between F⁻ and C$_2$O$_4^{2-}$, with a relatively small d orbital splitting.

(c) Should [Co(CO$_3$)$_3$]$^{3-}$ be para- or diamagnetic? Δ$_o$ should be small, and therefore the complex should be high spin and paramagnetic.

22.76. For Cu(H$_2$NCH$_2$CO$_2$)$_2$(H$_2$O)$_2$:

(a) Oxidation state of Cu? With 2 glycinate ligands (each with a 1– charge), the oxidation state of copper is +2.

(b) Coordination number of copper? Six pairs of electrons are shared. The coordination number = 6.

(c) Number of unpaired electrons? The Cu^{2+} ion configuration is [Ar]3d^9, so there is one unpaired electron.

(d) Is the complex dia- or paramagnetic? With an unpaired electron, complex is paramagnetic.

22.77. Structures for the five geometric isomers of Cu(H₂NCH₂CO₂)₂(H₂O)₂:

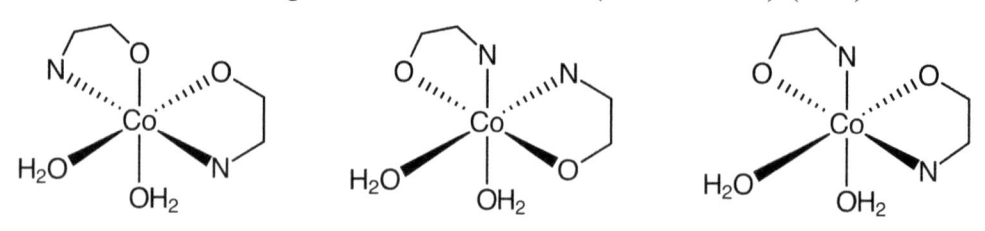

The three isomers above each have a nonsuperimposable mirror image—that is they contain a chiral center. The two isomers shown below have no chiral center.

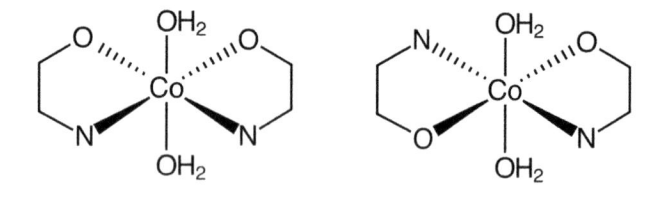

22.78. Empirical formula for the compound Mn(CO)ₓ(CH₃)ᵧ:

Determine moles of CH₃:

$$0.0290 \text{ g H}_2\text{O}\left(\frac{1 \text{ mol H}_2\text{O}}{18.02 \text{ g H}_2\text{O}}\right)\left(\frac{2 \text{ mol H}}{1 \text{ mol H}_2\text{O}}\right)\left(\frac{1 \text{ mol CH}_3}{3 \text{ mol H}}\right) = 1.07 \times 10^{-3} \text{ mol CH}_3$$

Determine moles of C overall:

$$0.283 \text{ g CO}_2\left(\frac{1 \text{ mol CO}_2}{44.01 \text{ g CO}_2}\right)\left(\frac{1 \text{ mol C}}{1 \text{ mol CO}_2}\right) = 6.43 \times 10^{-3} \text{ mol C total}$$

Determine moles of C in CO (and not in CH₃):

$$6.43 \times 10^{-3} \text{ mol C total} - \left(\frac{1 \text{ mol C}}{1 \text{ mol CH}_3}\right) 1.07 \times 10^{-3} \text{ mol CH}_3 = 5.36 \times 10^{-3} \text{ mol C}$$

$$\frac{5.36 \times 10^{-3} \text{ mol C in CO}}{1.07 \times 10^{-3} \text{ mol CH}_3} = \frac{5 \text{ mol CO}}{1 \text{ mol CH}_3}$$

The empirical formula would be [Mn(CO)₅(CH₃)] so $x = 5$ and $y = 1$.

22.79. Explain the magnetic properties of the Ni(II) and Pd(II) complexes of M(PR₃)₂Cl₂.

For Ni²⁺ and Pd²⁺, the electron configuration is d^8, with each losing 2 s electrons in the formation of the ion. (See the solution to Study Question 22.53 as you consider this solution.)

(a) The diagram below shows eight electrons distributed in the *d* orbitals of a square planar or tetrahedral complex.

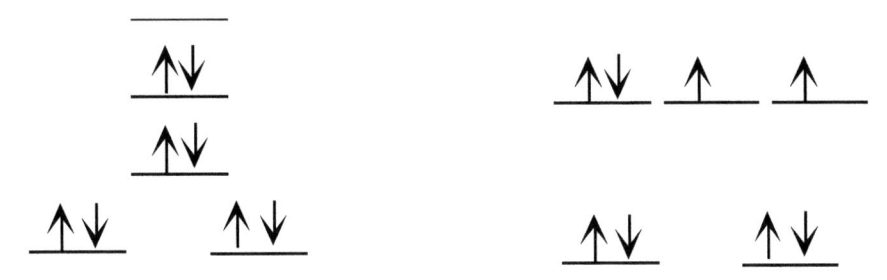

d- orbitals for square-planar complex   d- orbitals for tetrahedral complex

The Ni(II) complex is reported to be paramagnetic, which means it is tetrahedral. On the other hand, the Pd(II) complex is diamagnetic, so it is the square planar complex.

(b) The tetrahedral complex will have no isomers. In contrast, the square planar Pd(II) complex will have two isomers.

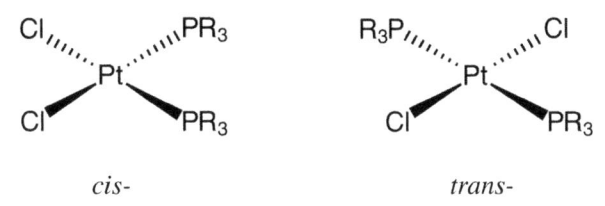

cis-   trans-

22.80. Regarding Ni(CO)₄:

(a) For the process: Ni(s) + 4 CO(g) → Ni(CO)₄(g)

$\Delta_r H° = \Delta_f H°[Ni(CO)_4(g)] - 4\,\Delta_f H°[CO(g)]$

$\Delta_r H° = (1\text{ mol})(-602.9\text{ kJ/mol}) - (4\text{ mol})(-110.525\text{ kJ/mol}) = -160.8\text{ kJ}$

$\Delta_r S° = [S°\,Ni(CO)_4(g)] - [S°\,Ni(s) + 4\,S°\,CO(g)]$

$\Delta_r S° = [(1\text{ mol})(410.6\text{ J/K·mol})]$

$\qquad - [(1\text{ mol})(29.87\text{ J/K·mol}) + (4\text{ mol})(197.674\text{ J/K·mol})] = -410.0\text{ J/K}$

$\Delta_r G° = \Delta_r H° - T\Delta_r S° = -160.8\text{ kJ} - (298\text{ K})(-410.0\text{ J/K})(1\text{ kJ}/10^3\text{ J}) = -38.6\text{ kJ}$

Now solving for K: $\Delta_r G° = -RT \ln K$

$-38.6\text{ kJ} = -(8.3145 \times 10^{-3}\text{ kJ/K})(298\text{ K}) \ln K$

$\ln K = 15.6$ and $K = 6 \times 10^6$

(b) Product-favored as $\Delta_r G° < 0$.

(c) To find out if Ni(CO)₄ could be decomposed to form pure Ni, calculate the temperature at which the formation of Ni(CO)₄ could be reversed and the decomposition of the compound (to form Ni metal) becomes favorable.

$\Delta_r G° = 0 = \Delta_r H° - T\Delta_r S° = -160.8 \text{ kJ} - T(-410.0 \text{ J/K})(1 \text{ kJ}/10^3 \text{ J})$

T = 392 K or greater. Beyond the temperature Ni(CO)$_4$ is unstable with respect to decomposition to Ni and CO.

Reacting impure nickel with excess CO gas to form Ni(CO)$_4$(g) is a product-favored reaction. However, heating the carbonyl complex above 392 K will reverse the formation reaction, isolating pure nickel metal.

22.81. Regarding cerium ions:
(a) Electron configurations for:
Ce: [Xe]$6s^2 5d^1 4f^1$
Ce^{3+}: [Xe]$4f^1$
Ce^{4+}: [Xe]

(b) Ce^{3+} is paramagnetic, with 1 unpaired electron. The 4+ ion has a closed, filled outer shell with no unpaired electrons, so it is diamagnetic.

(c) The figure indicates that the Ce^{4+} ions occupy the corners and faces of a face-centered cubic lattice giving **four** Ce^{4+} ions per unit cell. O^{2-} ions occupy the tetrahedral sites in this lattice for a total of eight O^{2-} ions. This yields the formula Ce$_4$O$_8$ or more simply CeO$_2$.

22.82. Neodymium and its various oxidation states:
(a) Electron configurations for:

Nd   [Xe]$6s^2 4f^4$

Nd^{2+} [Xe]$4f^4$

Nd^{3+} [Xe]$4f^3$

Nd^{4+} [Xe]$4f^2$

(b) All neodymium ions are paramagnetic.

(c) Empirical formula for Nd/Fe/B magnets: Assume 1.0000 g of the material.

Mol Nd: (0.2668 g Nd)(1 mol/144.24 g) = 0.001850 mol Nd

Mol Fe: (0.7232 g Fe)(1 mol/55.845 g) = 0.01295 mol Fe

Mol B: (1.0000 − 0.2668 g − 0.7232 g) = 0.0100 g B

(0.0100 g B)(1 mol/10.81 g) = 0.000925 mol B

Mol: Mol ratios:

0.001850 mol Nd/0.000925 mol B = 2.00 mol Nd/mol B

0.01295 mol Fe/0.000925 mol B = 14.0 mol Fe/mol B

and an empirical formula of: Nd$_2$Fe$_{14}$B

In the Laboratory

22.83. Suggested structures for compounds A and B:

A + BaCl₂ → ppt (BaSO₄)   implies that A = [Co(NH₃)₅Br]SO₄

B + BaCl₂ → no ppt   implies that B = [Co(NH₃)₅SO₄]Br

Complex A has the sulfate ion in the "outer sphere" (not the coordination sphere) of the transition metal compound. As such, this compound (like many ionic compounds) dissolves in water—and dissociates, liberating sulfate ions. Barium ions react with the sulfate ions to produce the precipitate.

Complex B has the sulfate ion as a part of the inner sphere of the compound. This ion is bound tightly to the cobalt ion, and not available to the barium ions. Hence, no precipitate of BaSO₄ forms.

The reaction between A and BaCl₂:
[Co(NH₃)₅Br]SO₄(aq) + BaCl₂(aq) → BaSO₄(s) + [Co(NH₃)₅Br]²⁺(aq) + 2 Cl⁻(aq)

22.84. For the compound of chromium(III) with water and chloride ion:

Determine the empirical formula:

$19.51 \text{ g} \left(\dfrac{1 \text{ mol Cr}}{51.996 \text{ g Cr}} \right) = 0.3752 \text{ mol Cr} = 1 \text{ mol Cr}$

$39.92 \text{ g} \left(\dfrac{1 \text{ mol Cl}}{35.45 \text{ g Cl}} \right) = 1.126 \text{ mol Cl} = 3 \text{ mol Cl}$

$40.57 \text{ g} \left(\dfrac{1 \text{ mol H}_2\text{O}}{18.015 \text{ g H}_2\text{O}} \right) = 2.252 \text{ mol H}_2\text{O} = 6 \text{ mol H}_2\text{O}$

So the formula is [Cr(H₂O)₆]Cl₃ and its name is hexaaquachromium(III) chloride. Structure of the complex ion:

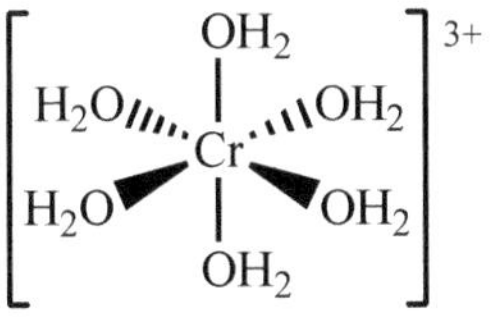

$$\begin{bmatrix} & \text{OH}_2 & \\ \text{H}_2\text{O} \cdots & | & \cdots \text{OH}_2 \\ & \text{Cr} & \\ \text{H}_2\text{O} & | & \text{OH}_2 \\ & \text{OH}_2 & \end{bmatrix}^{3+}$$

Reaction of the compound with AgNO₃: Cl⁻(aq) + Ag⁺(aq) → AgCl(s)

22.85. A 0.213 g sample of UO₂(NO₃)₂ contains 5.41 × 10⁻⁴ mol of UO₂(NO₃)₂.

$0.213 \text{ g UO}_2(\text{NO}_3)_2 \left(\dfrac{1 \text{ mol g UO}_2(\text{NO}_3)_2}{394.036 \text{ g UO}_2(\text{NO}_3)_2} \right) = 5.41 \times 10^{-4} \text{ mol UO}_2(\text{NO}_3)_2$

(a) If MnO₄⁻ is reduced to Mn²⁺, each mol of MnO₄⁻ gains 5 mol of electrons (Mn⁷⁺ is reduced to Mn²⁺). The number of mol of MnO₄⁻:

$$0.01247 \text{ L} \left(\frac{0.0173 \text{ mol MnO}_4^-}{1 \text{ L}} \right) = 2.16 \times 10^{-4} \text{ mol MnO}_4^-$$

With this information, we know that (5 mol e⁻/mol MnO₄⁻)(2.16 × 10⁻⁴ mol MnO₄⁻) = 1.08 ×10⁻³ mol e⁻.

Knowing that we have 5.41 × 10⁻⁴ mol of UO₂(NO₃)₂ and 1.08 ×10⁻³ mol e⁻, calculate the number of electrons per mol of uranium salt.

$$\frac{1.08 \times 10^{-3} \text{ mol electrons}}{5.41 \times 10^{-4} \text{ mol UO}_2(\text{NO}_3)_2} = 2 \frac{\text{mol e}^-}{\text{mol UO}_2(\text{NO}_3)_2}$$

With U having an oxidation state of +6 (in UO₂²⁺), and each mol of U gaining 2 mol of electrons, the U would be reduced to +4. So, $n = 4$.

(b) Balanced net ionic equation for the reduction of UO₂²⁺ by zinc:

Zn will be oxidized to Zn²⁺, its most common ion: Zn → Zn²⁺ + 2e⁻

The half-reaction for the reduction of the uranyl ion is

UO₂²⁺ + 4 H⁺ + 2 e⁻ → U⁴⁺ + 2 H₂O

and so the overall equation is

UO₂²⁺(aq) + 4 H⁺(aq) + Zn(s) → U⁴⁺(aq) + 2 H₂O(ℓ) + Zn²⁺(aq)

(c) The net ionic equation for the oxidation of U⁴⁺ to UO₂²⁺ by MnO₄⁻:

The oxidation process is: U⁴⁺ → UO₂²⁺. The reduction process is: MnO₄⁻ → Mn²⁺

In acid, the oxidation equation is: U⁴⁺ + 2 H₂O → UO₂²⁺ + 4 H⁺ + 2 e⁻.

The reduction equation (in acid) is: MnO₄⁻ + 8 H⁺ + 5 e⁻ → Mn²⁺ + 4 H₂O

To equalize electron gain and loss, multiply the oxidation equation by 5, and the reduction equation by 2:

5 U⁴⁺ + 10 H₂O → 5 UO₂²⁺ + 20 H⁺ + 10 e⁻

2 MnO₄⁻ + 16 H⁺ + 10 e⁻ → 2 Mn²⁺ + 8 H₂O

Adding the two equations, and removing redundancies (8 H₂O; 16 H⁺; 10 e⁻):

2 MnO₄⁻(aq) + 5 U⁴⁺(aq) + 2 H₂O(ℓ) → 5 UO₂²⁺(aq) + 4 H⁺(aq) + 2 Mn²⁺(aq).

22.86. Weight percent of KClO₃ in the original sample:

Mol Fe²⁺ used: (0.0500 L)(0.0960 mol Fe²⁺/L) = 0.00480 mol Fe²⁺

Unreacted Fe²⁺: (0.01299 L)(0.08362 M mol/L) = 0.001086 mol Fe²⁺

Net Fe²⁺: 0.00480 mol – 0.001086 mol = 0.00371 mol Fe²⁺

Amount of KClO₃ that reacts with that amount of Fe²⁺:

$$0.00371 \text{ mol Fe}^{2+}\left(\frac{1 \text{ mol KClO}_3}{6 \text{ mol Fe}^{2+}}\right)\left(\frac{122.55 \text{ g KClO}_3}{1 \text{ mol KClO}_3}\right) = 0.07577 \text{ g KClO}_3$$

Percent of KClO₃ in firework: $\left(\dfrac{0.07577 \text{ g KClO}_3}{0.1342 \text{ g firework}}\right)100 = 56.47\%$

Summary And Conceptual Questions

22.87. Regarding the austenite unit cell and the nitinol alloy:

(a) Dimensions of the austenite unit cell, assuming Ti and Ni touch along the unit cell diagonal. [Atomic radii: Ti = 145 pm; Ni = 125 pm] The body diagonal consists of the diameter of a Ni atom and the radii of two Ti atoms: (250 pm + 290 pm) = 540 pm

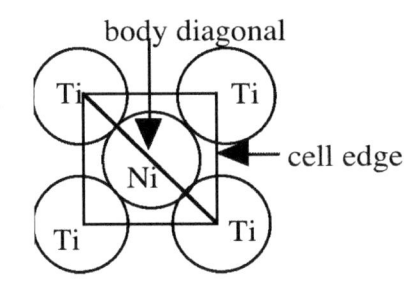

The diagonal = $\sqrt{3}$(cell edge)

540 pm = $\sqrt{3}$(cell edge) and $\dfrac{540 \text{ pm}}{\sqrt{3}}$ = cell edge = 311.8 pm (312 to 3 sf)

The volume of the cell is (311.8 pm)³ or (3.118 × 10⁻⁸ cm)³ = 3.03 × 10⁻²³ cm³.

(b) Calculate density of the alloy based on this unit cell.

The unit cell contains 8(1/8) Ti atoms and 1 Ni atom, for a total of 1 atom of EACH metal.

The mass of 1 Ti atom = $\dfrac{47.867 \text{ g Ti}}{6.022 \times 10^{23} \text{ Ti atoms}}$ = 7.95 × 10⁻²³ g Ti

The mass of 1 Ni atom = $\dfrac{58.693 \text{ g Ni}}{6.022 \times 10^{23} \text{ Ni atoms}}$ = 9.75 × 10⁻²³ g Ni

So, the mass of the two atoms = 1.77 × 10⁻²² g (or the sums of the two masses above).

Given that the volume of the unit cell is 3.03 × 10⁻²³ cm³, the calculated density is 1.77 × 10⁻²² g/3.03 × 10⁻²³ cm³ or 5.84 g/cm³. The agreement with the literature value (6.5 g/cm³) isn't great! Given the larger literature value, one can only assume that the atoms don't pack as well as we might anticipate.

(c) Are Ti and Ni atoms paramagnetic or diamagnetic?
Ti: [Ar]3d²4s². With 2 electrons in the 3d sublevel, Ti is paramagnetic (2 unpaired electrons)
Ni: [Ar]3d⁸4s². With 8 electrons in the 3d sublevel, Ni is paramagnetic (2 unpaired electrons)

22.88. Regarding the chelate effect for the nickel(II) complexes:

Substituting 10^8 and 10^{18} into the expression ($-RT \ln K$) produces $\Delta_r G°$ values of -45.6 kJ (ammine) and -102.7 kJ (en) respectively.

Ammine complex: $\Delta_r G° = -(8.3145 \times 10^{-3}$ kJ/K · mol-rxn$)(298.15$ K$)\ln(10^8)$

$= -45.6$ kJ/mol-rxn

Ethylenediamine complex: $\Delta_r G° = -(8.3145 \times 10^{-3}$ kJ/K · mol-rxn$)(298.15$ K$)\ln(10^{18})$

$= -102.7$ kJ/mol-rxn

The difference in $\Delta_r H$ values is much less than this (~8 kJ), so entropy must play a role. While there are fewer molecules in the reaction of ethylenediamine (3 molecules versus 6 ligand molecules for the NH_3 reaction), the change in entropy (as the much larger bidentate ethylenediamine ligands form the complex) is greater.

Solution and Answer Guide

Kotz Treichel Townsend Treichel, Chemistry and Chemical Reactivity 11e, 978-0-357-85140-1, Chapter 23: Carbon: Not Just Another Element

TABLE OF CONTENTS

Applying Chemical Principles 861
Practicing Skills...863
General Questions..885
In the Laboratory..894
Summary and Conceptual Questions: 897

Applying Chemical Principles

An Awakening with L-DOPA

23.1.1. L-DOPA is chiral.
What is the center of chirality in the molecule?
In the figure, the chiral C is indicated.

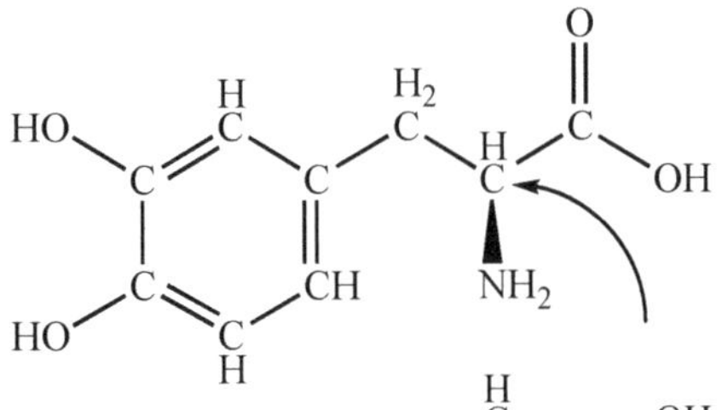

23.1.2. Is dopamine or epinephrine chiral?
What is the center of chirality in the molecule?
While dopamine does not possess a chiral C, epinephrine does. The chiral C of epinephrine is indicated in the figure.

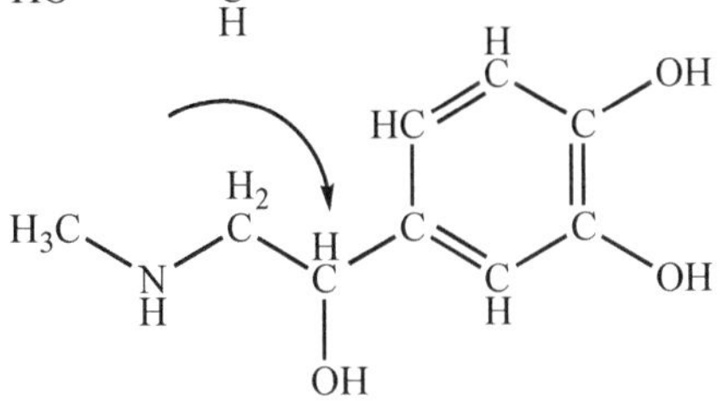

23.1.3. What number of moles corresponds to 5.0 g of L-DOPA?

With a molecular formula of $C_9H_{11}NO_4$, the molar mass is 197.190.

$$5.0 \text{ g L-DOPA}\left(\frac{1 \text{ mol L-DOPA}}{197.190 \text{ g L-DOPA}}\right) = 0.025 \text{ mol L-DOPA}$$

Green Adhesives

23.2.1. Structures of phenol, urea, and formaldehyde.

phenol	urea	formaldehyde
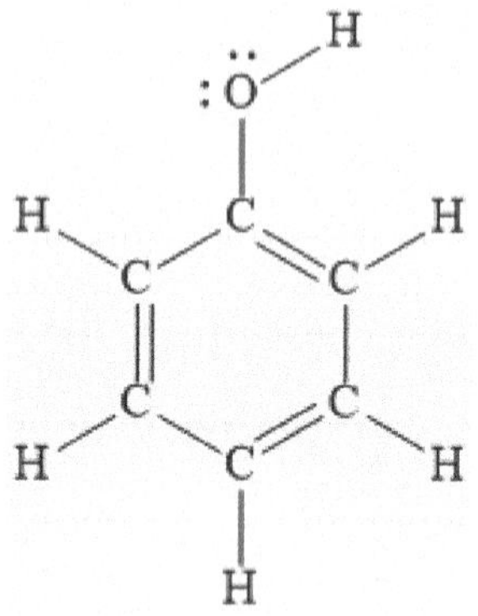 (structure shown)	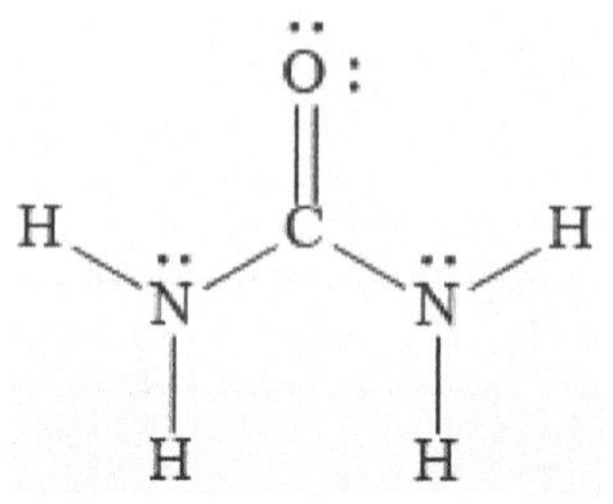 (structure shown)	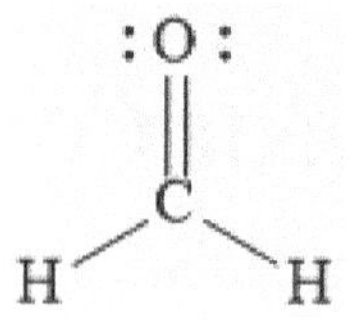 (structure shown)

23.2.2. Bonding in formaldehyde:

The electron-pair and molecular geometries are both trigonal-planar. The C–H σ bonds are each formed by the overlap of an sp^2 hybrid orbital on the C atom with the $1s$ orbital of the H atom. The σ bond between the C and O is formed by the overlap of an sp^2 hybrid orbital on the C with an sp^2 hybrid orbital on the O. The π bond between the C and O is formed by the overlap of a $2p$ orbital on C with a $2p$ orbital on O.

23.2.3. Compare structures of nylon-6,6 with a protein.

Similarities:

Both the nylon and the protein are polyamides, containing a C(=O)-N linkage.

Differences:

In proteins there is one direction for the amide linkage: CONH. In nylon-6,6 two orientations are present: CONH and NHCO.

The nylon contains multiple C atoms between the amide linkages while in proteins, there is only one.

Proteins are chiral, while nylon isn't.

Proteins have numerous types of side chains that can be connected to the central backbone, while nylon has only CH linkages.

Bisphenol A

23.3.1. Atom economy for the reaction of acetone with phenol to make BPA:

Add the molar masses of $C_3H_6O + 2(C_6H_6O) = 246.31$

Calculate the molar mass of BPA. $C_{15}H_{16}O_2$: 228.29

Atom economy = 228.29/246.31 = 0.927 or 92.7%

23.3.2. Describe polymers of polycarbonate and epoxy resins as addition or condensation: The graphic that accompanies this ACP section indicates the polymerization between Bisphenol A, and either phosgene or epichlorohydrin. In each case, as the polymerization continues, small molecules are also formed as a result—indicative of a condensation polymerization.

23.3.3. Amount of BPA ingested by an infant weighing 15 lbs in one day:

Estimate of ingestion = 13 µg/kg:

$$15 \text{ lb infant} \left(\frac{0.454 \text{ kg}}{1 \text{ lb}}\right)\left(\frac{13 \text{ µg BPA}}{1 \text{ kg}}\right) = 89 \text{ µg BPA}$$

so the infant would ingest more than 50 micrograms/day.

23.3.4. With a weight of 156 lbs, how much BPA does one ingest per day:

$$156 \text{ lb person} \left(\frac{0.454 \text{ kg}}{1 \text{ lb}}\right)\left(\frac{1.5 \text{ µg BPA}}{\text{kg}}\right) = 1.1 \times 10^2 \text{ µg BPA}$$

23.3.5. Mass of 0.050 M NaOH to react with 300 mg of BPA:

$$\left(\frac{300 \times 10^{-3} \text{ g}}{1.00 \text{ L}}\right)\left(\frac{1 \text{ mol BPA}}{228.29 \text{ g BPA}}\right)\left(\frac{2 \text{ mol NaOH}}{1 \text{ mol BPA}}\right)\left(\frac{1.00 \text{ L solution}}{0.050 \text{ mol NaOH}}\right) = 0.053 \text{ L or 53 mL}$$

Practicing Skills

Why Carbon?

23.1 & 23.2. Structures for C with various bonding configurations:

	(a)	(b)	(c)	(d)
Structure	(tetrahedral C)	(trigonal C)	≡C—	=C=
Example	H—C(H)(H)—H (CH₄)	H₂C=CH₂	H—C≡C—H	H₂C=C=CH₂
Hybridization	sp^3	sp^2	sp	sp
Bond angle	109.5°	120°	180°	180°

23.3. Does light of wavelength 400 nm possess energy to break a 346 kJ/mol bond?

$$E = \frac{hc}{\lambda} = \frac{(6.626 \times 10^{-34} \text{ J} \cdot \text{s})(2.998 \times 10^8 \text{ m/s})}{400 \times 10^{-9} \text{ m}} = 4.97 \times 10^{-19} \text{ J/photon}$$

$$4.97 \times 10^{-19} \text{ J/photon} \left(\frac{6.022 \times 10^{23} \text{ photons}}{1 \text{ mol photons}}\right)\left(\frac{1 \text{ kJ}}{1000 \text{ J}}\right) = 299 \text{ kJ/mol photons}$$

Violet light does not possess sufficient energy to break a 346 kJ/mol bond.

23.4. The decomposition and/or polymerization reactions of ethene both require a certain amount of energy to accomplish the reaction, the "energy of activation". Room temperature does not possess sufficient energy to provide the energy of activation.

23.5. Compounds with identical elemental compositions, but different atom-to-atom connections are called (b) structural isomers.

23.6. Compounds with the same attachment of atoms, but different orientations in space are called (a) stereoisomers, (c) *cis-trans* isomers, or (d) optical isomers.

23.7. Structural and geometric isomers for compounds with formula C_4H_{10}:

There are two structural isomers:

Butane: $CH_3-CH_2-CH_2-CH_3$ and methylpropane: $(CH_3)_2CHCH_3$

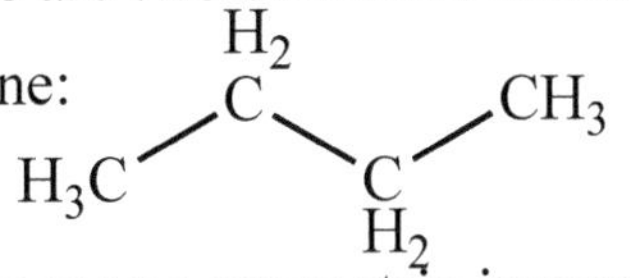

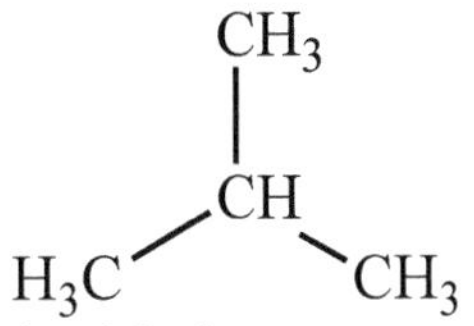

There are no geometric isomers, and neither compound is chiral.

23.8. Non-cyclical structural and geometric isomers for compounds with formula C_4H_8:

The three isomers are:

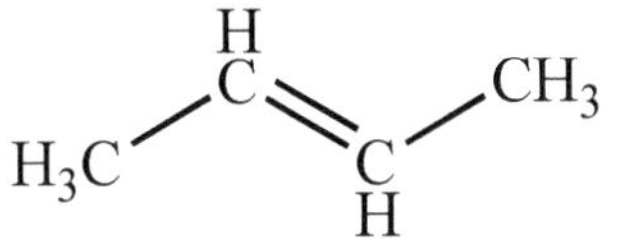

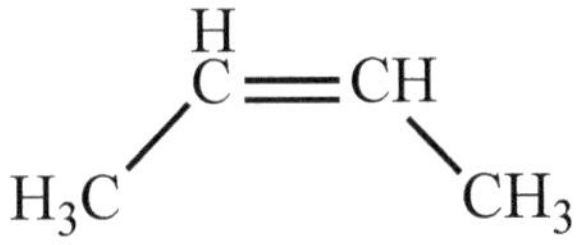

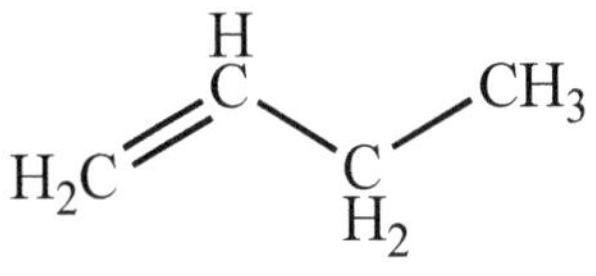

The first two compounds are geometric isomers: *trans*-2-butene and *cis*-2-butene. The third is a structural isomer: 1-butene.

None are chiral.

Alkanes and Cycloalkanes

23.9. The straight chain alkane with the formula C_7H_{16} is heptane. The prefix "hept" tells us that there are seven carbons in the chain.

23.10. Molecular formula for a 12-carbon alkane:
The general formula for an alkane is C_nH_{2n+2}. The molecular formula for an alkane with 12 carbons is $C_{12}H_{26}$.

23.11. Of the formulas given, which represents an alkane?
Alkanes have saturated carbon atoms—with 4 bonds. Since bridging C atoms have two bonds to C, then the other 2 bonds are to H atoms. Terminal C atoms on the end of the chains have only 1 bond to the chain, leaving 3 bonds to H. The net result is that alkanes have the general formula: C_nH_{2n+2}. Both (b) and (c) have this general formula.

23.12. Which can be a cycloalkane:
Of the substances shown, only C_5H_{10} can be a cycloalkane. Cycloalkanes have the general formula: C_nH_{2n}.

23.13. Structure for 3-ethyl-2-methylhexane:

If we number the "main chain" from left to right, placing a CH_3 group on C-2, and a CH_3CH_2 group on C-3 completes the structure. For a 5-carbon chain, one can envision the following:

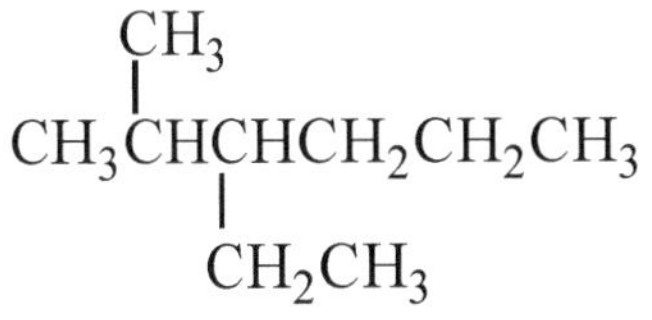

$CH_3CHCHCH_2CH_2CH_3$ with CH_3 and CH_2CH_3 substituents

This structure would be named: 3,3-diethylpentane.

There are other isomeric structures.

$$\text{CH}_3\text{CH}_2\underset{\underset{\text{CH}_2\text{CH}_3}{|}}{\overset{\overset{\text{CH}_2\text{CH}_3}{|}}{\text{C}}}\text{CH}_2\text{CH}_3$$

23.14. Structure for 2,2,4-trimethylpentane:

$$\text{H}_3\text{C}-\underset{\underset{\text{CH}_3}{|}}{\overset{\overset{\text{CH}_3}{|}}{\text{C}}}-\text{CH}_2-\underset{}{\overset{\overset{\text{CH}_3}{|}}{\text{CH}}}-\text{CH}_3$$

Two other possible isomers:

2,3,4-trimethylpentane	2,2,3-trimethylpentane
$\text{H}_3\text{C}-\overset{\overset{\text{CH}_3}{|}}{\text{CH}}-\overset{\overset{\text{CH}_3}{|}}{\text{CH}}-\overset{\overset{\text{CH}_3}{|}}{\text{CH}}-\text{CH}_3$	$\text{H}_3\text{C}-\underset{\underset{\text{CH}_3}{|}}{\overset{\overset{\text{CH}_3}{|}}{\text{C}}}-\overset{\overset{\text{CH}_3}{|}}{\text{CH}}-\text{CH}_2-\text{CH}_3$

23.15. Systematic name for the alkane: Numbering the longest C chain from one end to the other indicates that the longest C chain has 4 carbons, the root name is therefore *butane*. On the 2nd and 3rd C there is a 1-C chain. Since we truncate the –ane ending and change it to –yl, these are *methyl groups*. Indicating that the methyl groups are on the 2nd and 3rd carbon atoms, we get 2,3-dimethylbutane.

23.16. Systematic name for the given alkane:
2,5-dimethylheptane

One possible isomer is 2,2,4,4-tetramethylpentane:

$$\text{H}_3\text{C}-\underset{\underset{\text{CH}_3}{|}}{\overset{\overset{\text{CH}_3}{|}}{\text{C}}}-\text{CH}_2-\underset{\underset{\text{CH}_3}{|}}{\overset{\overset{\text{CH}_3}{|}}{\text{C}}}-\text{CH}_3$$

23.17. Structures for the given compounds:

(a) 2,3-dimethylhexane

$$\text{CH}_3\overset{\overset{\text{CH}_3}{|}}{\text{CH}}\underset{\underset{\text{CH}_3}{|}}{\text{CH}}\text{CH}_2\text{CH}_2\text{CH}_3$$

(b) 2,3-dimethyloctane

$$\text{CH}_3\overset{\overset{\text{CH}_3}{|}}{\text{CH}}\underset{\underset{\text{CH}_3}{|}}{\text{CH}}\text{CH}_2\text{CH}_2\text{CH}_2\text{CH}_3$$

(c) 3-ethylheptane

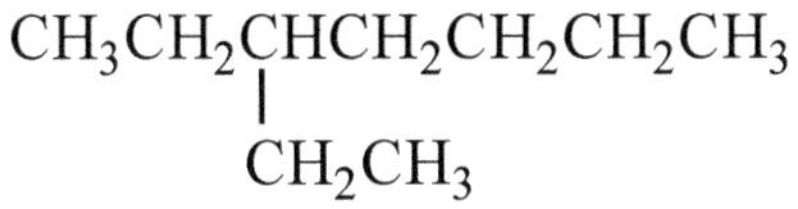

CH₃CH₂CHCH₂CH₂CH₂CH₃
         |
         CH₂CH₃

(d) 3-ethyl-2-methylhexane

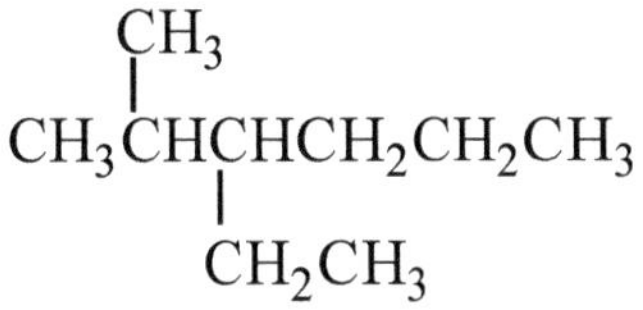

         CH₃
         |
CH₃CHCHCH₂CH₂CH₃
       |
       CH₂CH₃

23.18. Structures for the given compounds:

(a) 3-ethylpentane	(b) 2,2-dimethylpentane
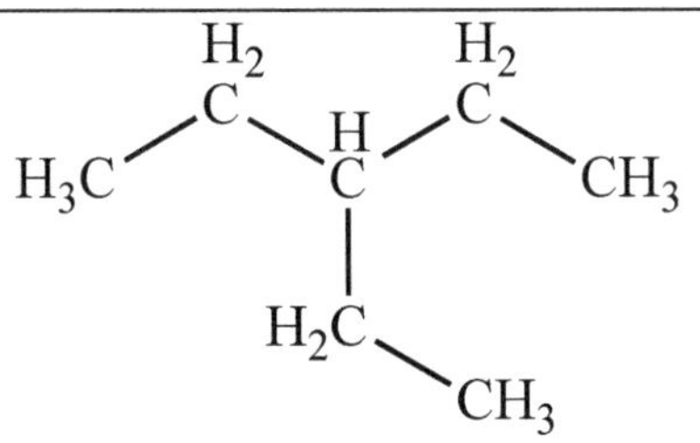 *structure shown*	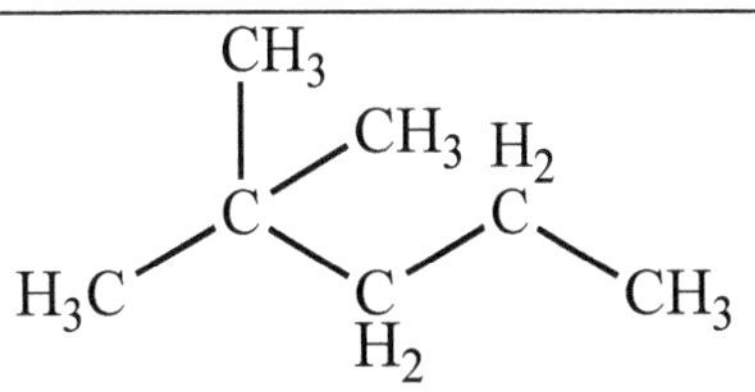 *structure shown*
(c) 2,3-dimethylpentane	(d) 2,4-dimethylpentane
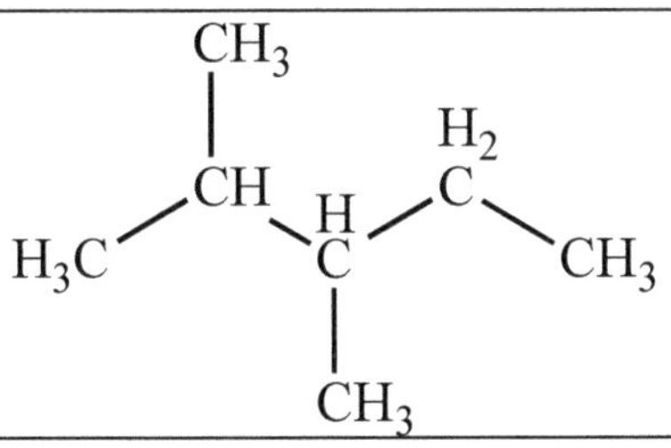 *structure shown*	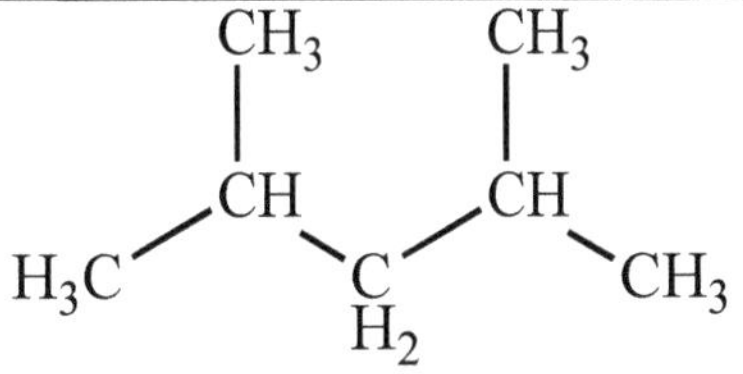 *structure shown*

23.19. Structures for all compounds with a seven-carbon chain and one methyl substituent.

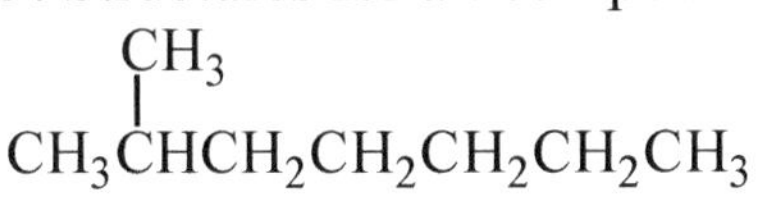

           CH₃
           |
CH₃CHCH₂CH₂CH₂CH₂CH₃

2-methylheptane

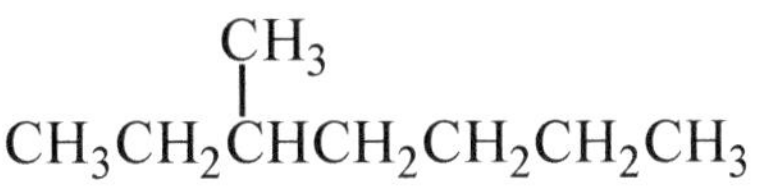

              CH₃
              |
CH₃CH₂CHCH₂CH₂CH₂CH₃

3-methylheptane

                 CH₃
                 |
CH₃CH₂CH₂CHCH₂CH₂CH₃

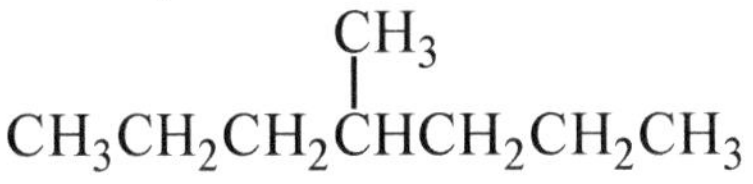

4-methylheptane

There are 3 structures. *3-methylheptane has a chiral carbon*, since on C-3, there are four different groups: a methyl group (CH₃) an ethyl group (CH₃CH₂), a H, and a butyl group (CH₂CH₂CH₂CH₃).

23.20. Structures for the following:

(a) 2,2-dimethylhexane	(b) 2,3-dimethylhexane (chiral center on C-3)
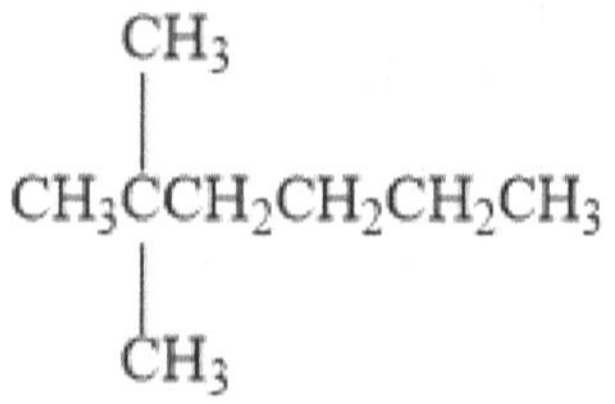 CH₃ | CH₃CCH₂CH₂CH₂CH₃ | CH₃	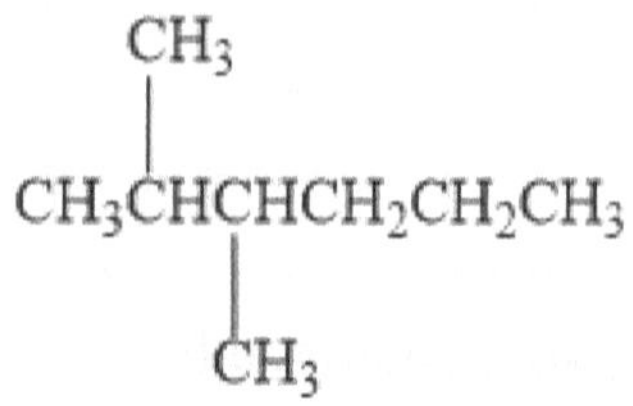 CH₃ | CH₃CHCHCH₂CH₂CH₃ | CH₃
(c) 2,4-dimethylhexane (chiral center on C-4)	(d) 2,5-dimethylhexane
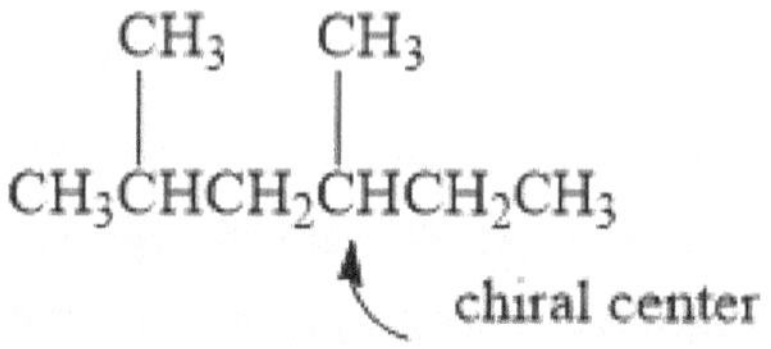 CH₃ CH₃ | | CH₃CHCH₂CHCH₂CH₃ ↑ chiral center	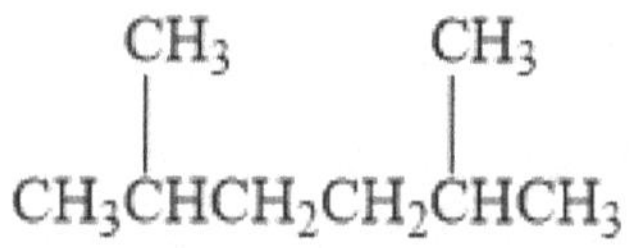 CH₃ CH₃ | | CH₃CHCH₂CH₂CHCH₃

23.21. Chair form of cyclohexane, with axial and equatorial H's identified:

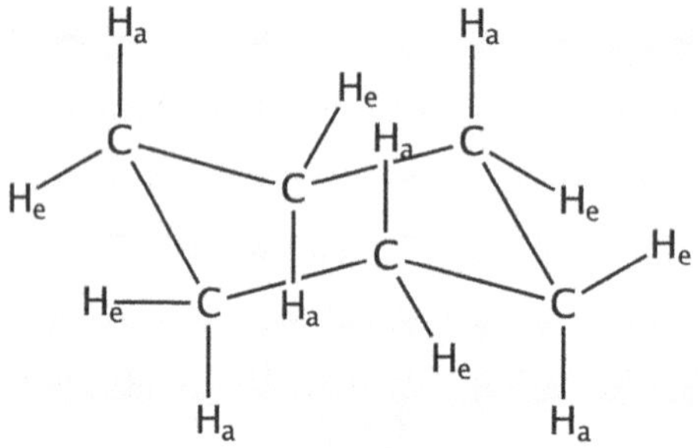

In the structure, the axial hydrogen atoms are labeled with a subscript "a", and the equatorial hydrogen atoms are labeled with a subscript "e."

23.22. Cycloheptane; Is the ring planar?

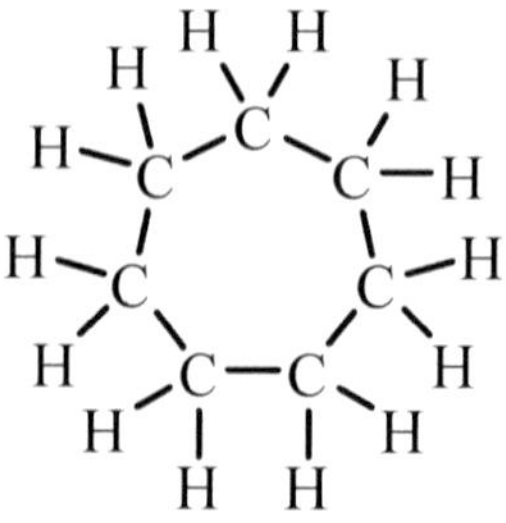

The ring is not planar. The geometry around each carbon atom is tetrahedral.

23.23. Structures and names of the two ethylheptanes:

$$\underset{\text{3-ethylheptane}}{CH_3CH_2\overset{\overset{\displaystyle CH_2CH_3}{|}}{C}HCH_2CH_2CH_2CH_3} \qquad \underset{\text{4-ethylheptane}}{CH_3CH_2CH_2\overset{\overset{\displaystyle CH_2CH_3}{|}}{C}HCH_2CH_2CH_3}$$

Neither of the isomers has a chiral carbon.

23.24. Two isomers with formula C_8H_{18}, each having a methyl and ethyl substituent group:

$$\underset{\text{3-ethyl-2-methylpentane}}{CH_3\overset{\overset{\displaystyle CH_3}{|}}{C}H\underset{\underset{\displaystyle CH_2CH_3}{|}}{C}HCH_2CH_3} \qquad \underset{\text{3-ethyl-3-methylpentane}}{CH_3CH_2\overset{\overset{\displaystyle CH_3}{|}}{\underset{\underset{\displaystyle CH_2CH_3}{|}}{C}}CH_2CH_3}$$

23.25. Physical properties of C_4H_{10}: Assuming that we're talking about butane, the following properties are pertinent: mp = –138.4 °C, bp = –0.5°C (it's a colorless gas at room *T*). According to the *CRC Handbook of Chemistry and Physics*, it is slightly soluble in water, but more so in less polar ether, chloroform, and ethanol.

Predicted properties for $C_{12}H_{26}$: One would guess that the material is colorless (no features that would invoke color). Given the much longer chain, one could guess that it might be a liquid (It is! mp = –9.6°C and bp = 216°C). We would also guess that it would be less water soluble than butane--(and it is indeed reported *not* to be water soluble), but soluble in alcohol and ether.

23.26. Balanced equations for:

(a) Reaction of methane with Cl_2: $CH_4(g) + 4\ Cl_2(g) \rightarrow CCl_4(\ell) + 4\ HCl(g)$

(b) Oxidation of cyclohexane with O_2: $C_6H_{12}(\ell) + 9\ O_2(g) \rightarrow 6\ CO_2(g) + 6\ H_2O(g)$

Alkenes and Alkynes

23.27. Molecular formulas of five carbon atom alkane, alkene, and alkyne:

alkane – C_5H_{12} alkene – C_5H_{10} alkyne – C_5H_8

23.28. Molecular formulas of six carbon cyclic alkane and alkene:

cyclic alkane – C_6H_{12} alkene - C_6H_{12}

Notice that these two molecular formulas are the same.

23.29. *Cis*- and *trans*- isomers of 4-methyl-2-hexene:

cis-   trans-

23.30. Structural requirement for alkene to have cis- and trans- isomers:

To have *cis* and *trans* isomers, the alkene must have different atoms or groups on the carbons that are double bonded:
That is, A ≠ B and X ≠ Y.

$$\begin{array}{c} A \\ \diagdown \\ B \end{array} C=C \begin{array}{c} X \\ \diagup \\ Y \end{array}$$

Alkanes do not have *cis* and *trans* isomerism because free rotation occurs around carbon–carbon single bonds. Alkynes do not have *cis* and *trans* isomerism because the *sp* hybridization of the carbon atoms produces a linear molecule.

23.31. For a hydrocarbon with the formula C_5H_{10}:

(a) Structures and names for alkenes with formula C_5H_{10}:

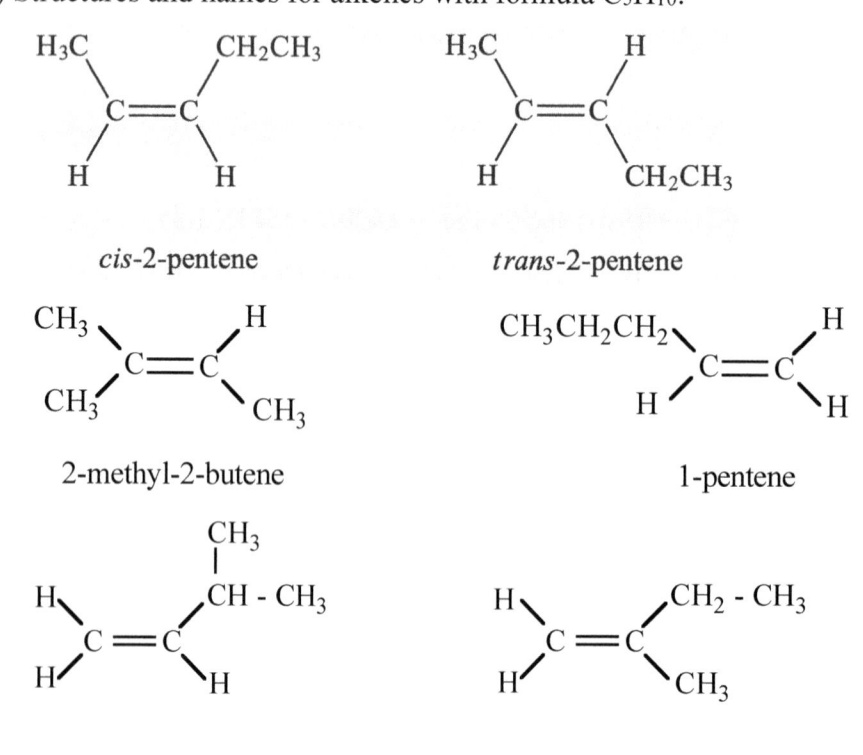

cis-2-pentene   trans-2-pentene

2-methyl-2-butene   1-pentene

3-methyl-1-butene   2-methyl-1-butene

(b) The cycloalkane with the formula C$_5$H$_{10}$:

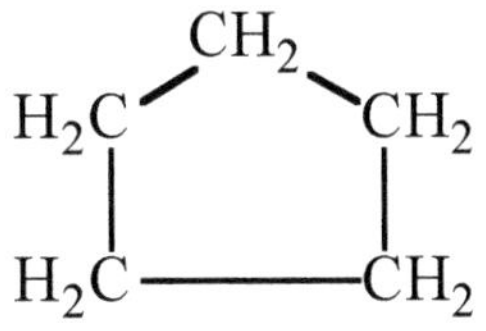

[Cyclopentane structure: five CH$_2$ groups in a ring]

23.32. Alkenes with formula C$_7$H$_{14}$ and a seven-carbon chain:

1-heptene: 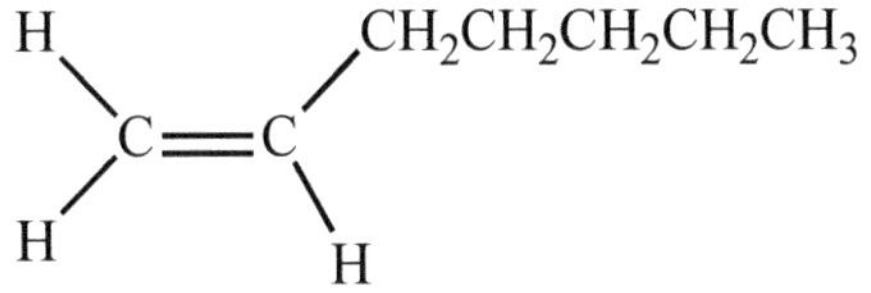 H$_2$C=CH–CH$_2$CH$_2$CH$_2$CH$_2$CH$_3$	
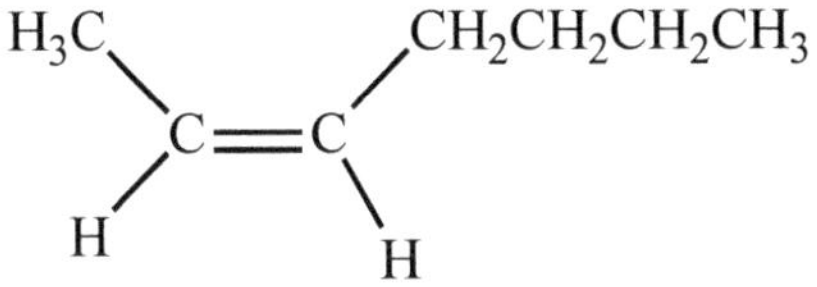 *cis*-2-heptene (H$_3$C and CH$_2$CH$_2$CH$_2$CH$_3$ on same side)	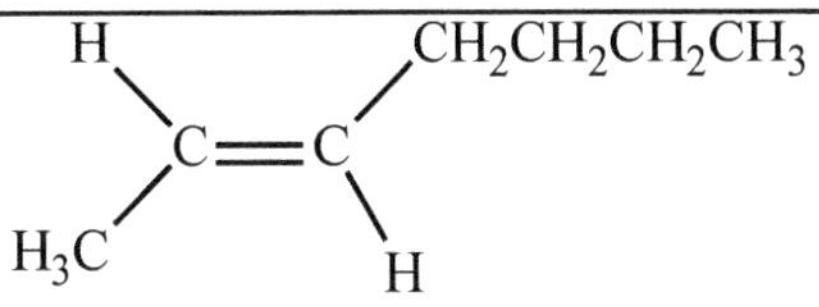 *trans*-2-heptene (H$_3$C and CH$_2$CH$_2$CH$_2$CH$_3$ on opposite sides)
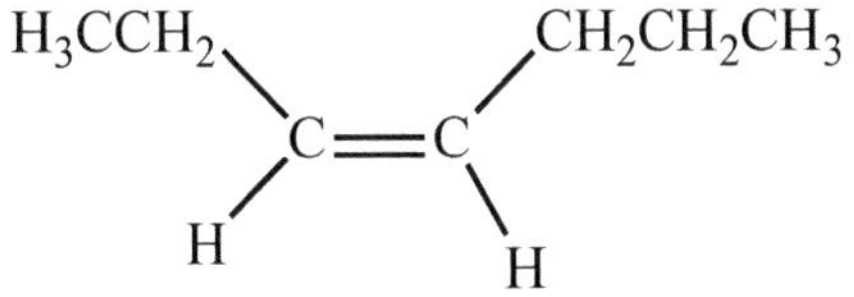 *cis*-3-heptene (H$_3$CCH$_2$ and CH$_2$CH$_2$CH$_3$ on same side)	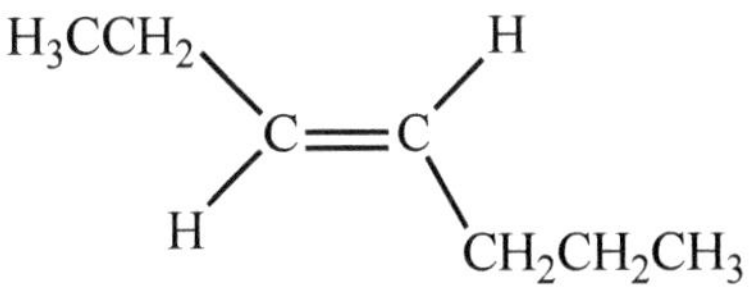 *trans*-3-heptene (H$_3$CCH$_2$ and CH$_2$CH$_2$CH$_3$ on opposite sides)

23.33. Structures and names of products of the following reactions:

(a) bromination of propene: Br atoms add to the C1 and C2 atoms, forming 1,2-dibromopropane

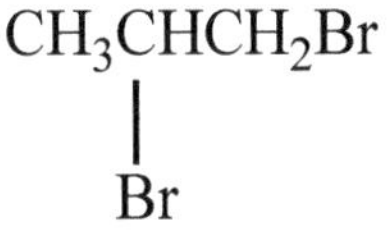

CH$_3$CHCH$_2$Br
  |
  Br

(b) hydrogenation of 2-pentene: H atoms add across the double bond, resulting in the formation of an alkane, specifically pentane.

CH$_3$CH$_2$CH$_2$CH$_2$CH$_3$

23.34. Structures and names of products of the following reactions:

(a) hydrogenation of 2-pentene: forms 2-methylpentane

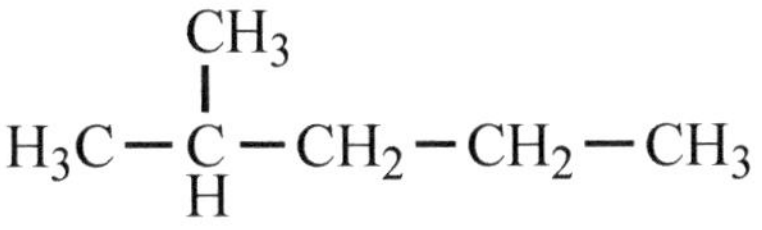

H$_3$C–CH(CH$_3$)–CH$_2$–CH$_2$–CH$_3$

(b) bromination of 2-pentyne: forms 2,2,3,3-tetrabromopentane

$$H_3C-\underset{\underset{Br}{|}}{\overset{\overset{Br}{|}}{C}}-\underset{\underset{Br}{|}}{\overset{\overset{Br}{|}}{C}}-CH_2-CH_3$$

23.35. The alkenes, which upon addition of HBr yields CH₃CH₂CHBrCH₃ is:

1-butene   trans-2-butene   cis-2-butene

The addition of HBr is accomplished by adding (H) on one "side" of the double bond and (Br) on the other. The equation for the reaction with 1-butene is:

CH₃CH₂CH=CH₂ + HBr → CH₃CH₂CHBrCH₃

23.36. Alkene, which upon addition of Br₂ yields 2,3-dibromo-2-methylhexane:

$$H_3C-\overset{\overset{CH_3}{|}}{C}=CHCH_2CH_2CH_3 + Br_2 \rightarrow H_3C-\underset{\underset{Br}{|}}{\overset{\overset{CH_3}{|}}{C}}-\underset{\underset{Br}{|}}{C}HCH_2CH_2CH_3$$

2-methyl-2-hexene

23.37. Four alkenes with the formula C₃H₅Cl:

trans-1-chloropropene   cis-1-chloropropene   2-chloropropene   3-chloro-1-propene

23.38. Seven dichloropropene isomers:

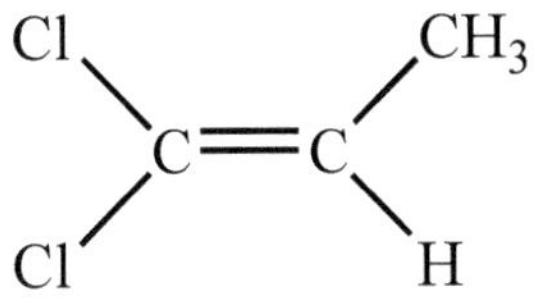

$$\underset{\text{1,1-dichloropropene}}{\begin{array}{c}Cl\\ \diagdown\\ Cl\end{array}C=C\begin{array}{c}CH_3\\ \diagup\\ H\end{array}}$$ $$\underset{\text{2,3-dichloropropene}}{\begin{array}{c}H\\ \diagdown\\ H\end{array}C=C\begin{array}{c}CH_2Cl\\ \diagup\\ Cl\end{array}}$$ $$\underset{\text{3,3-dichloropropene}}{\begin{array}{c}H\\ \diagdown\\ H\end{array}C=C\begin{array}{c}CHCl_2\\ \diagup\\ H\end{array}}$$

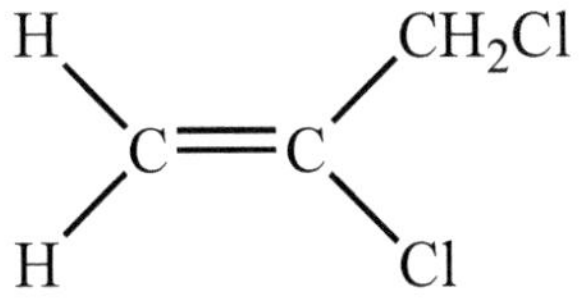

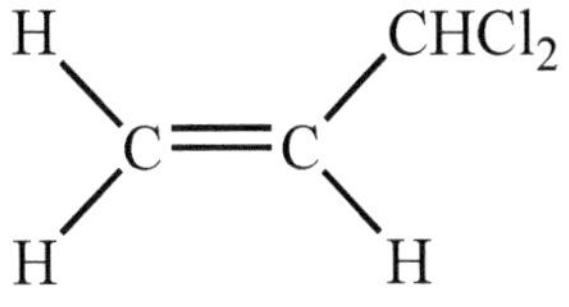

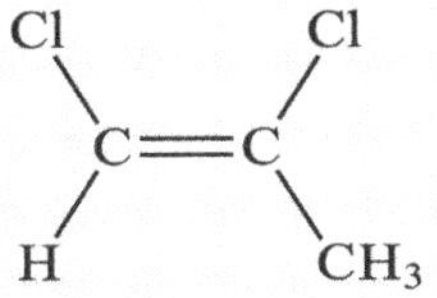

$$\underset{cis\text{-1,2-dichloropropene}}{\begin{array}{c}Cl\\ \diagdown\\ H\end{array}C=C\begin{array}{c}Cl\\ \diagup\\ CH_3\end{array}}$$ $$\underset{trans\text{-1,2-dichloropropene}}{\begin{array}{c}Cl\\ \diagdown\\ H\end{array}C=C\begin{array}{c}CH_3\\ \diagup\\ Cl\end{array}}$$

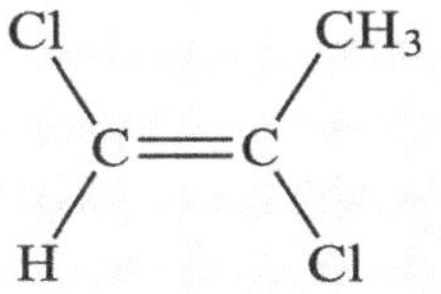

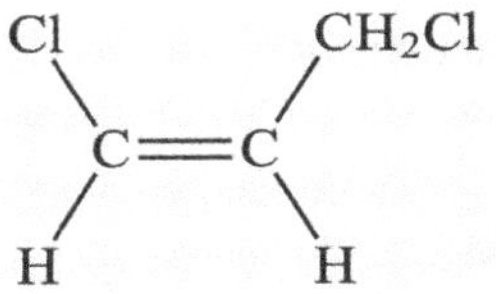

$$\underset{cis\text{-1,3-dichloropropene}}{\begin{array}{c}Cl\\ \diagdown\\ H\end{array}C=C\begin{array}{c}CH_2Cl\\ \diagup\\ H\end{array}}$$ $$\underset{trans\text{-1,3-dichloropropene}}{\begin{array}{c}Cl\\ \diagdown\\ H\end{array}C=C\begin{array}{c}H\\ \diagup\\ CH_2Cl\end{array}}$$

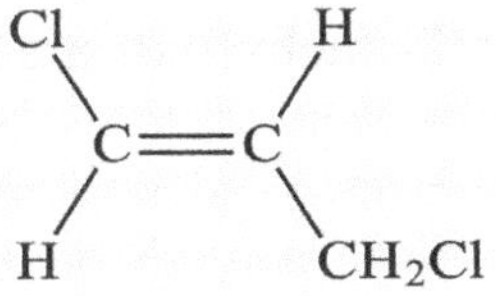

23.39. The hydrogenation of 1-hexene proceeds, frequently catalyzed by Pd:

$CH_3CH_2CH_2CH_2CH=CH_2 + H_2 \rightarrow CH_3CH_2CH_2CH_2CH_2CH_3$

Hydrogenation is used in the food industry to convert liquid oils to solids and to make them less susceptible to spoilage.

23.40. Distinguish between cycloalkane and an alkene:

If the compound is an alkene it will react with bromine to form a substituted alkane. The alkane would not react with bromine unless exposed to light.

23.41. Structures and names of two products of HBr reaction with 1-pentene:

1-bromopentane

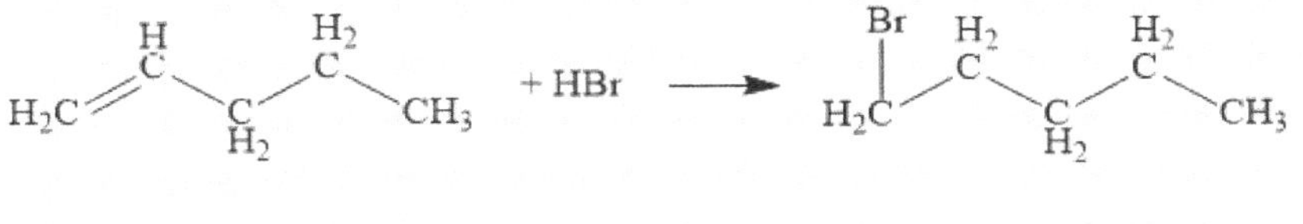

2-bromopentane

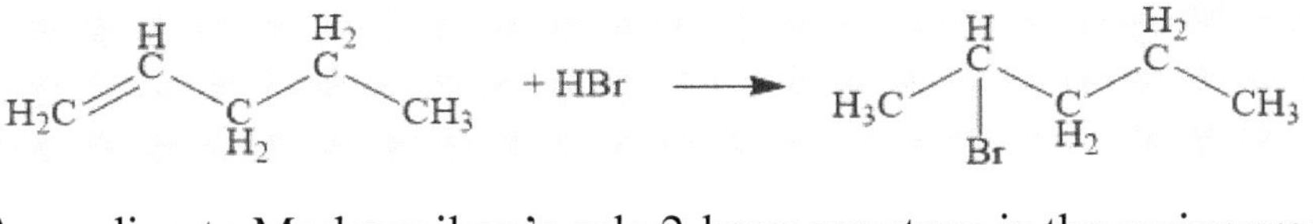

According to Markovnikov's rule 2-bromopentane is the major product.

23.42. Structures and names of two products of HCl reaction with 2-methyl-2hexene:

[Reaction 1: 2-methyl-2-hexene + HCl → 2-chloro-2-methyl-hexane]

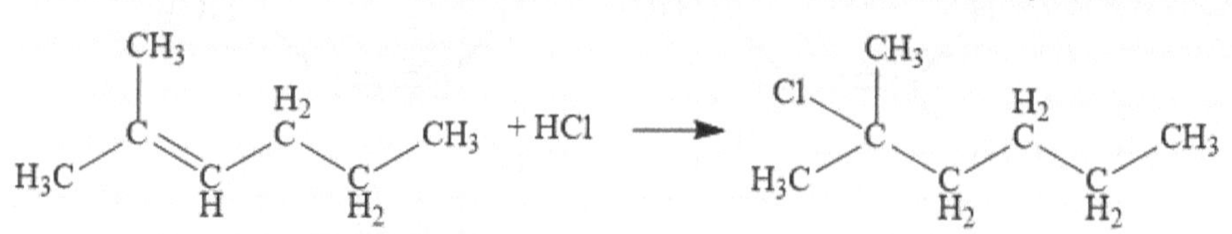

[Reaction 2: 2-methyl-2-hexene + HCl → 3-chloro-2-methyl-hexane]

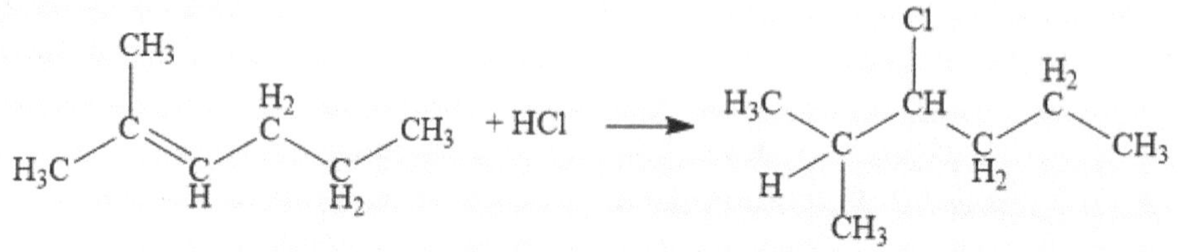

According to Markovnikov's rule 2-chloro-2-methyl-hexane is the major product.

Aromatic Compounds

23.43. Structural formulas for:

(a) *m*-dichlorobenzene

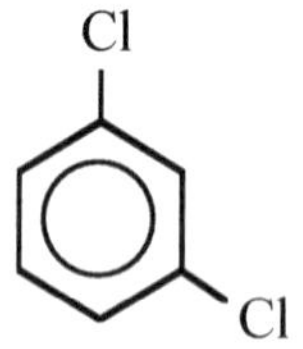

(b) *p*-bromotoluene

23.44. Structural formulas for:

(a) 1,2-dinitrobenzene

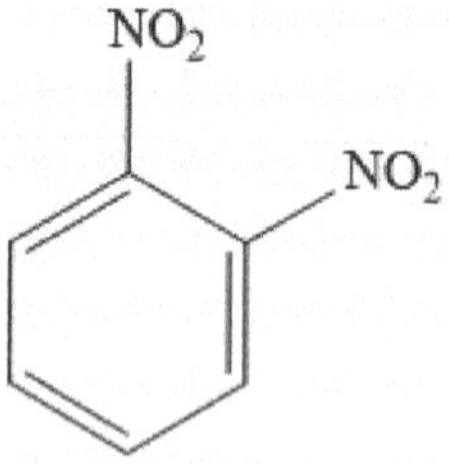

(b) 4-chlorophenol

23.45. Systematic names for each of the compounds:

(a) 1,2-dichlorobenzene

(b) 1,3,5-trinitrobenzene

(c) 1,2,4-tribromobenzene

23.46. Systematic names for each of the compounds:

(a) 1-chloro-2-nitrobenzene

(b) 1,4-dinitrobenzene

(c) 1-chloro-2-ethylbenzene

23.47. The alkylated product of *p*-xylene:

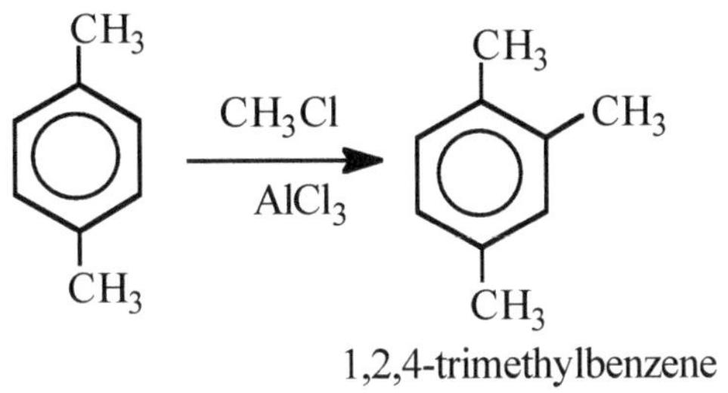
CH3
CH3Cl
AlCl3
CH3
CH3
CH3
CH3
1,2,4-trimethylbenzene

23.48. Equation for preparation of hexylbenzene:

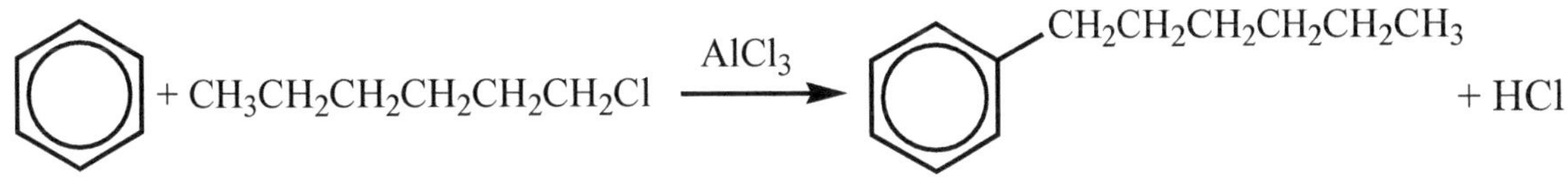
+ CH3CH2CH2CH2CH2CH2Cl
AlCl3
CH2CH2CH2CH2CH2CH3
+ HCl

23.49. Structures and names of the two isomers following nitration of 1,2-dimethylbenzene.

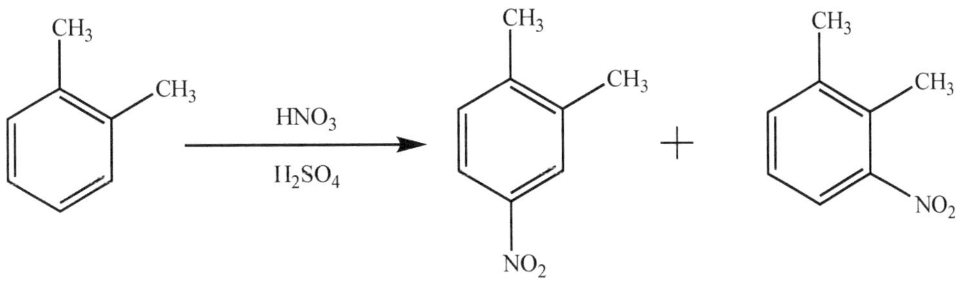
CH3
CH3
HNO3
H2SO4
CH3
CH3
NO2
+
CH3
CH3
NO2

1,2-dimethyl-
4-nitrobenzene

1,2-dimethyl-
3-nitrobenzene

23.50. Nitration products of toluene:

Alcohols, Ethers, and Amines

23.51. Systematic names of the alcohols:

    (a) 1-propanol   primary alcohol (OH group on C connected to 1 other C "group"

    (b) 1-butanol   primary alcohol (OH group on C connected to 1 other C "group"

    (c) 2-methyl-2-propanol   tertiary alcohol (OH group on C connected to 3 other C "groups"

    (d) 2-methyl-2-butanol   tertiary alcohol (OH group on C connected to 3 other C "groups"

23.52. Structural formulas and designation as primary, secondary, or tertiary alcohol:

(a) 1-pentanol — primary alcohol

(b) 2-pentanol — secondary alcohol

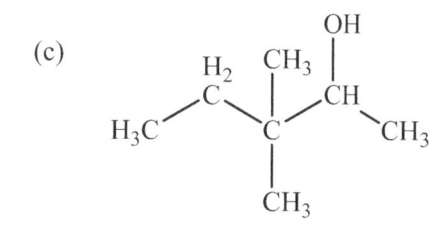

(c) 3,3-dimethyl-2-pentanol — secondary alcohol

(d) 3,3-dimethyl-1-pentanol — primary alcohol

23.53. Formulas and structures of amines:

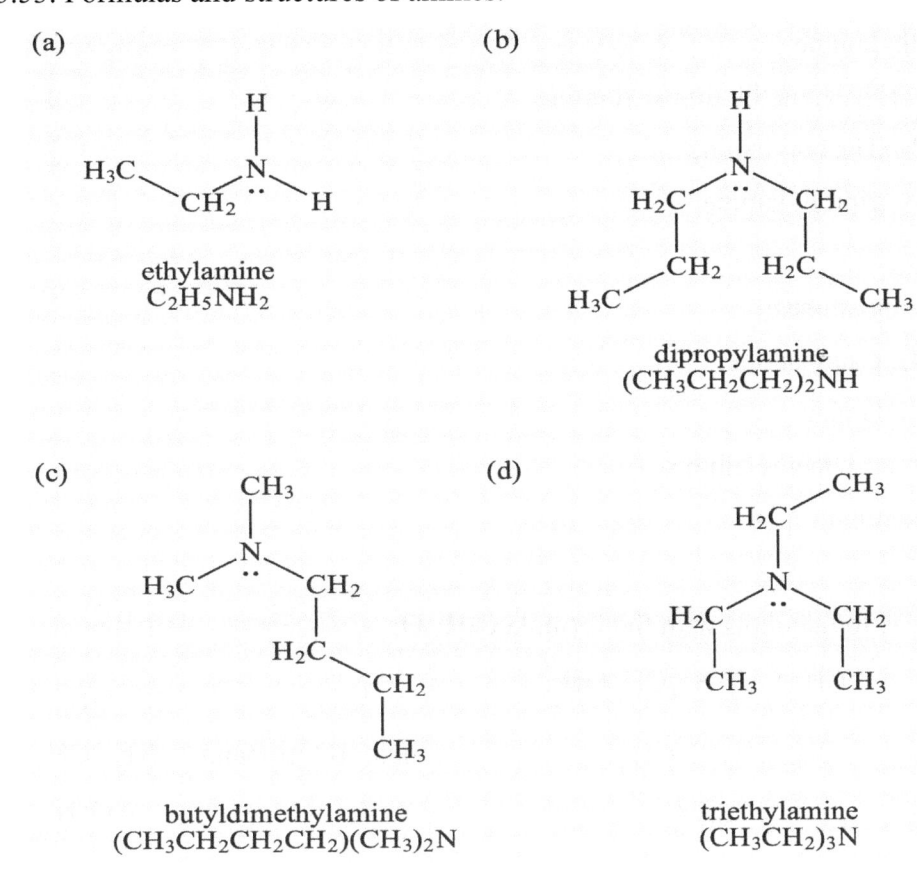

(a) ethylamine $C_2H_5NH_2$

(b) dipropylamine $(CH_3CH_2CH_2)_2NH$

(c) butyldimethylamine $(CH_3CH_2CH_2CH_2)(CH_3)_2N$

(d) triethylamine $(CH_3CH_2)_3N$

23.54. Formulas and structures of amines:

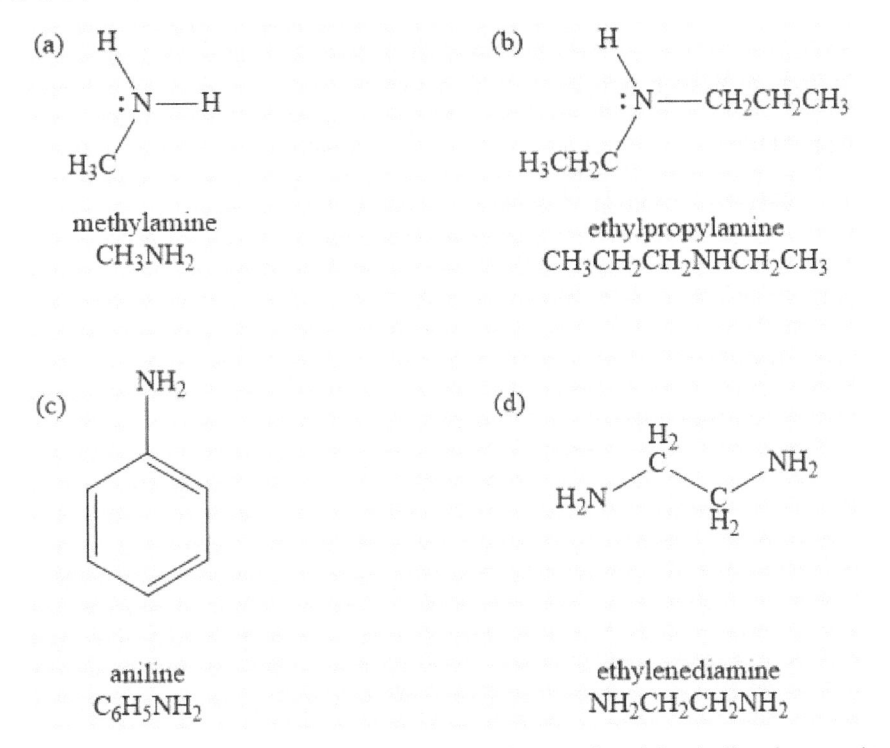

(a) methylamine CH_3NH_2

(b) ethylpropylamine $CH_3CH_2CH_2NHCH_2CH_3$

(c) aniline $C_6H_5NH_2$

(d) ethylenediamine $NH_2CH_2CH_2NH_2$

23.55. Name the following amines:

(a) (C₂H₅)₂NH: diethylamine

(b) (CH₃)(CH₃CH₂CH₂)₂N: methyldipropylamine

(c) CH₃CH₂CH₂CH₂NH₂: butylamine

(d) (CH₃CH₂CH₂)₃N: tripropylamine

23.56. Name the following amines:

(a) CH₃CH₂CH₂NH₂: propylamine

(b) (CH₃)₃N: trimethylamine

(c) (CH₃)(C₂H₅)NH: ethylmethylamine

(d) C₆H₁₃NH₂: hexylamine

23.57. Structural formulas for alcohols with the formula C₄H₁₀O:

1-butanol   CH₃CH₂CH₂CH₂OH

2-butanol   CH₃CH₂CHCH₃
                         |
                        OH

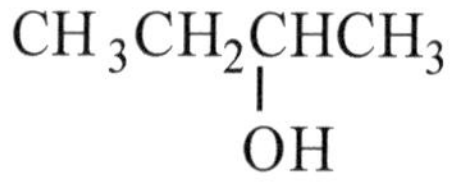

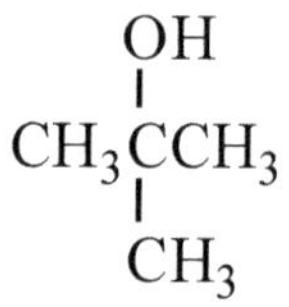

2-methyl-2-propanol   CH₃CCH₃
                         | |
                        OH CH₃

2-methyl-1-propanol   CH₃CHCH₂OH
                         |
                        CH₃

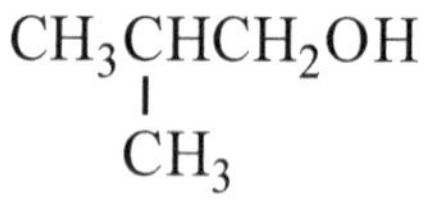

23.58. Primary amines with the formula C₄H₉NH₂:

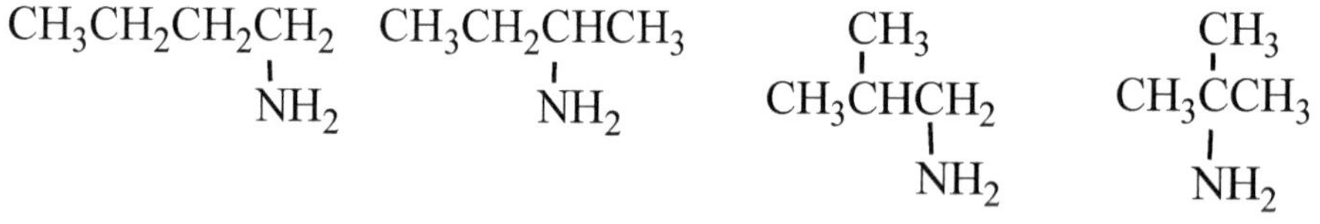

CH₃CH₂CH₂CH₂   CH₃CH₂CHCH₃   CH₃   CH₃
     | | | |
    NH₂   NH₂   CH₃CHCH₂   CH₃CCH₃
                                          | |
                                         NH₂   NH₂

23.59. Amines treated with acid:

(a) C₆H₅NH₂(ℓ) + HCl(aq) → (C₆H₅NH₃)⁺Cl⁻(aq)

(b) (CH₃)₃N(aq) + H₂SO₄(aq) → [(CH₃)₃NH]⁺HSO₄⁻(aq)

23.60. Predict the reaction of dopamine with HCl(aq):

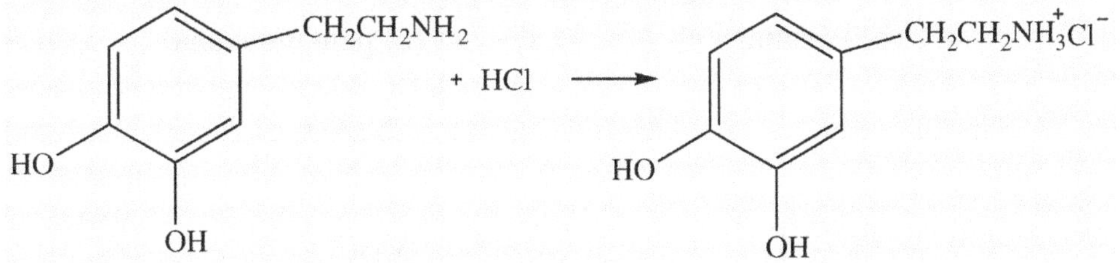

23.61. Oxidation of the following alcohols—in the case of the primary alcohols (OH on the end) we form a carboxylic acid. For the secondary alcohol (b), we form a ketone. Note that the amine (d), is not oxidized:

(a) 2-methyl-1-pentanol

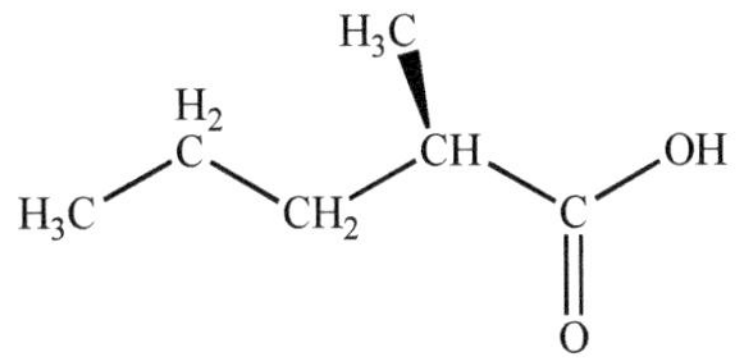

(b) 3-methyl-2-pentanol

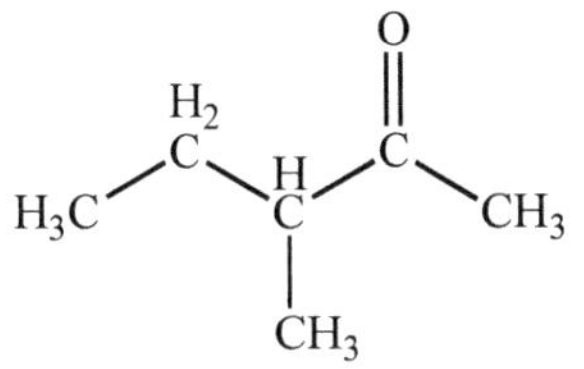

(c) HOCH₂CH₂CH₂CH₂OH

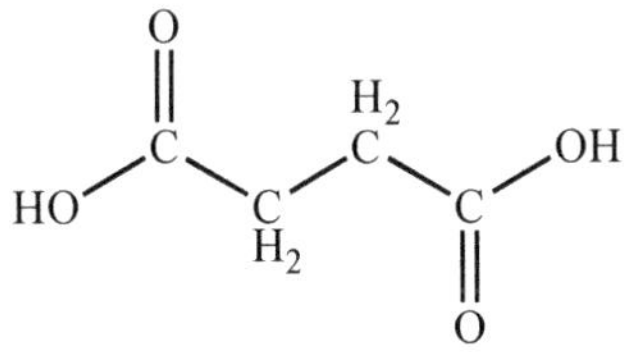

(d) H₂NCH₂CH₂CH₂OH

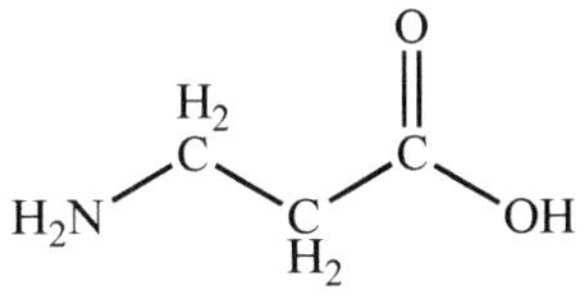

23.62. Name and formula for the requested alcohols:

Compounds with a Carbonyl Group

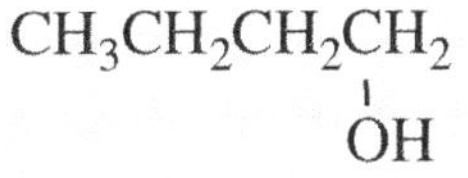

CH₃CH₂CH₂CH₂
          |
          OH
(a) 1-butanol

CH₃CH₂CH₂CH₂CHCH₃
                |
                OH
(b) 2-hexanol

23.63. Structural formulas for:

(a) 2-pentanone	(b) hexanal	(c) pentanoic acid
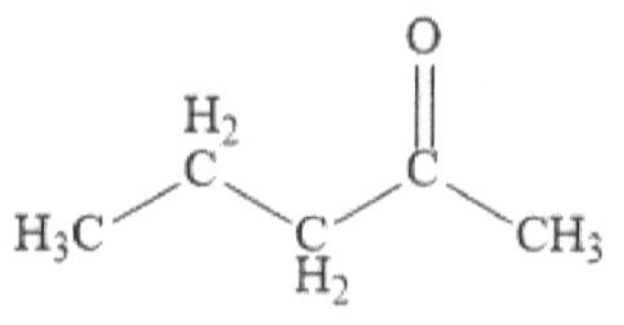 CH₃-CH₂-CH₂-C(=O)-CH₃	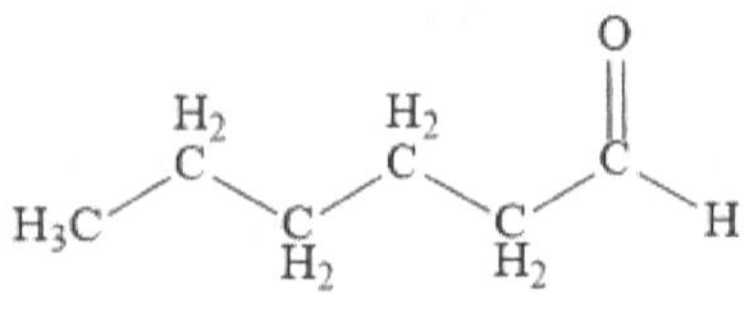 CH₃-CH₂-CH₂-CH₂-CH₂-CHO	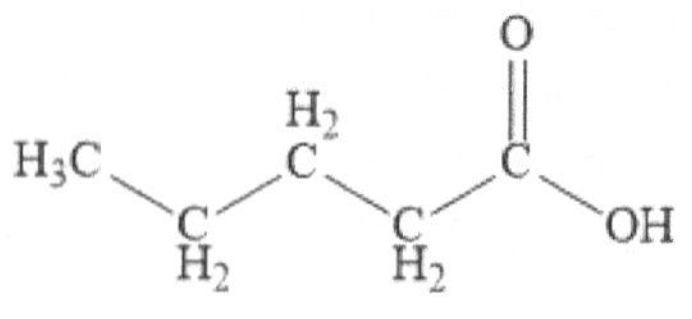 CH₃-CH₂-CH₂-CH₂-COOH

23.64. Structural formulas for acids and esters:

(a) 2-methylhexanoic acid	CH₃CH₂CH₂CH₂CH(CH₃)C(=O)-OH
(b) pentyl butanoate	CH₃CH₂CH₂C(=O)OCH₂CH₂CH₂CH₂CH₃
(c) octyl acetate	CH₃C(=O)OCH₂CH₂CH₂CH₂CH₂CH₂CH₂CH₃

23.65. Name the following compounds:

(a) carboxylic acid; 3-methylpentanoic acid

(b) ester; methyl propanoate

(c) ester; butyl acetate (or butyl ethanoate)

(d) carboxylic acid; *p*-bromobenzoic acid

23.66. Identify the class of compound, and provide a systematic name:

(a) ketone; propanone

(b) aldehyde; butanal

(c) ketone; 2-pentanone

23.67. Formula and systematic name for organic product of each of the following reactions:

(a) The oxidation product is pentanoic acid: $CH_3CH_2CH_2CH_2\underset{\underset{O}{\|}}{C}-OH$

(b) The reduction product is 1-pentanol: $CH_3CH_2CH_2CH_2CH_2OH$

(c) The reduction yields 2-octanol: $CH_3CH_2CH_2CH_2CH_2CH_2\underset{\underset{OH}{|}}{C}HCH_3$

(d) No reaction occurs.

23.68. Formula and systematic name for organic product of each of the following reactions:

(a) reduction of butanal by LiAlH₄: $CH_3CH_2CH_2CH_2OH$; 1-butanol

(b) oxidation of 1-butanol by KMnO₄: $CH_3CH_2CH_2\underset{\underset{O}{\|}}{C}-OH$; butanoic acid

(c) oxidation of butanal: [structure: HO-C(=O)-CH₂-CH₂-CH₃]; butanoic acid

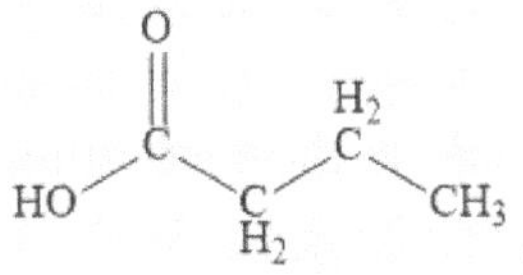

23.69. The following equations show the preparation of propyl propanoate:

$$CH_3CH_2CH_2OH \xrightarrow[H^+]{KMnO_4} CH_3CH_2CO_2H$$

$$CH_3CH_2CO_2H + HOCH_2CH_2CH_3 \xrightarrow{H^+} CH_3CH_2CO_2CH_2CH_2CH_3$$

23.70. Name and structure of reaction product:
Using the rules given in the chapter, the name of this compound would be predicted to be 2-propyl benzoate, but this is not its correct name and can be confused with another compound that is properly called this in which there is a propyl group on carbon 2 of the benzene ring and in which the carboxylic acid group of benzoic acid has been deprotonated. A more systematic name of the desired compound is 1-methylethyl benzoate, and a common name is isopropyl benzoate.

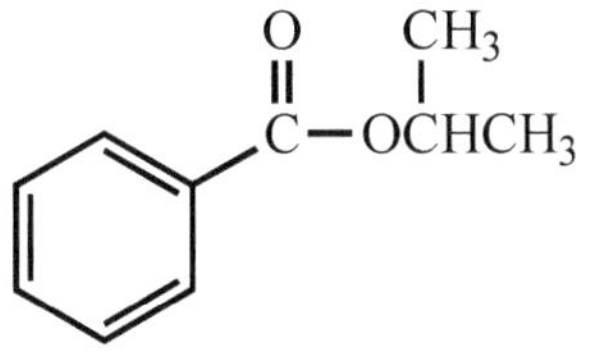

[structure: phenyl-C(=O)-OCH(CH₃)CH₃]

23.71. The products of the hydrolysis of the ester are 1-butanol and sodium acetate:

[Structure of 1-butanol] and [structure of sodium acetate]

23.72. The products of the hydrolysis of the ester are 2-propanol (isopropyl alcohol) and sodium benzoate:

[Structure of 2-propanol] and [structure of sodium benzoate]

23.73. Regarding phenylalanine:

(a) Carbon 3 has three groups attached, and a trigonal-planar geometry.

(b) The O–C–O bond angle would be 120°.

(c) Carbon 2 has four different groups attached, so the molecule is chiral.

(d) The H attached to the carboxylic acid group (carbon 3) is acidic.

23.74. Regarding Vitamin C:

(a) Approximately value for O–C–O bond angle: 120°
(b) All of the C—O—H bond angles should be approximately 109°
(c) The molecule is chiral, there are two chiral carbon atoms in the molecule
(d) Shortest bond: C=O
(e) Functional groups present: alcohol, alkene, ester

23.75. The reaction of a carboxylic acid with an amine is a condensation reaction, forming the amide, so important in proteins!

$$CH_3CH_2CH_2COH + H_2NCH_3 \longrightarrow CH_3CH_2CH_2CNHCH_3 + H_2O$$

23.76. Equation for reaction forming acetaminophen:

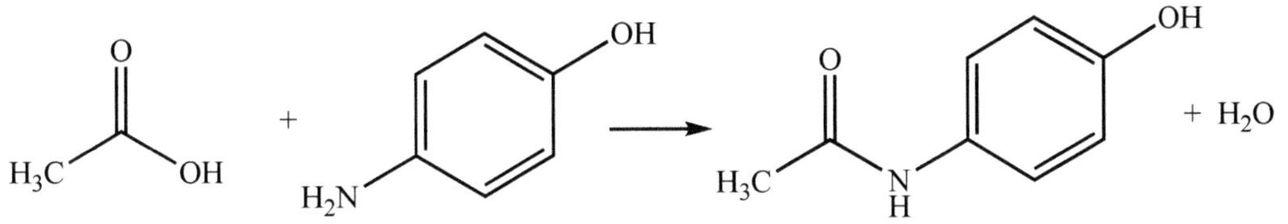

[Structural equation: acetic acid + 4-aminophenol → acetaminophen + H₂O]

Functional Groups

23.77. Functional groups present:

(a) the OH group makes this molecule an *alcohol*.

(b) the carbonyl group (C=O) adjacent to the N–H makes this molecule an *amide*.

(c) the carbonyl group (C=O) adjacent to the O–H makes this molecule a *carboxylic acid*.

(d) the carbonyl group (C=O) with an attached O–C makes this molecule an *ester*.

23.78. Regarding the reactions involved:

(a) The reduction of the ketone results in the formation of the alcohol, 2-butanol.

(b) The resulting product is an ester.

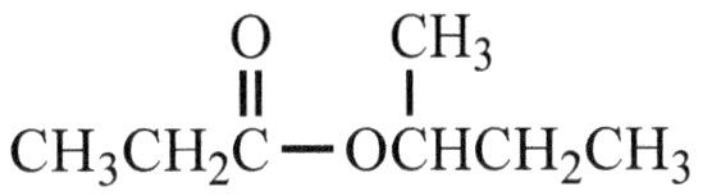

$$CH_3CH_2\overset{O}{\underset{\|}{C}}-O\overset{CH_3}{\underset{|}{C}}HCH_2CH_3$$

(c) The addition of hydrogen across the double bond, results in the formation of the alcohol, 1-propanol.

(d) The addition of NaOH to the acid results in the formation of the salt, sodium propanoate, $CH_3CH_2CO_2^-Na^+$.

Polymers

23.79. Regarding polyvinyl acetate:

(a) An equation for the formation of polyvinyl acetate from vinyl acetate:

   n CH₂CHOCOCH₃ → [–CH₂CH(OCOCH₃)–]$_n$

(b) The structure for polyvinylacetate:

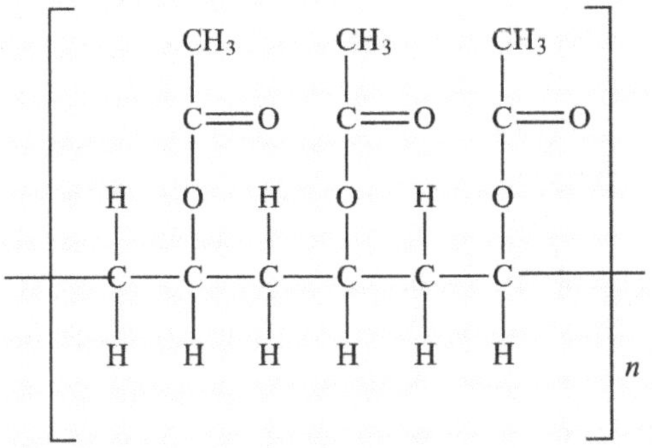

[Structure of polyvinyl acetate polymer showing repeating units with CH₃, C=O, O groups attached to the carbon backbone]

(c) Prepare polyvinyl alcohol from polyvinyl acetate:

Polyvinyl acetate is an ester. Hydrolysis of the ester (structure shown above) with NaOH will produce the sodium salt, NaCH₃CO₂, and polyvinyl alcohol. Acidification with a strong acid (e.g. HCl) will produce polyvinyl alcohol (structure below).

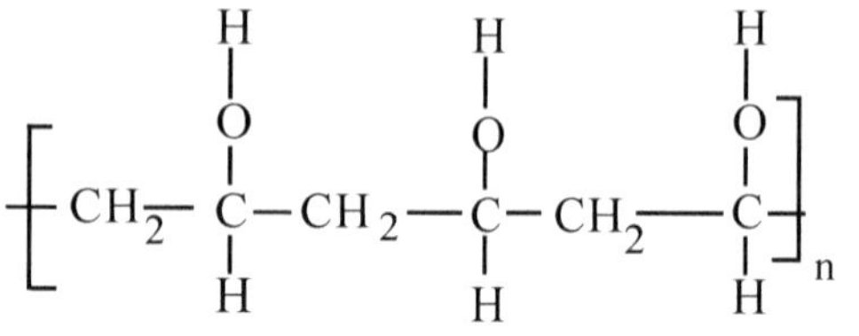

23.80. Regarding neoprene:

(a) n H₂C=C(Cl)−C(H)=CH₂ → [−C(H)(H)−C(Cl)=C(H)−C(H)(H)−]$_n$

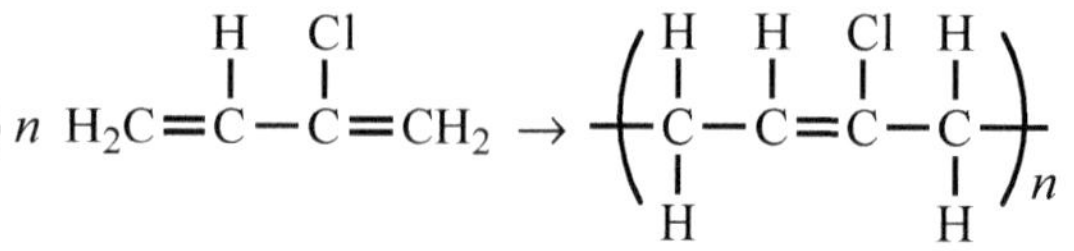

(b) [−C(H)(H)−C(Cl)=C(H)−C(H)(H)−C(H)(H)−C(Cl)=C(H)−C(H)(H)−C(H)(H)−C(Cl)=C(H)−C(H)(H)−]$_n$

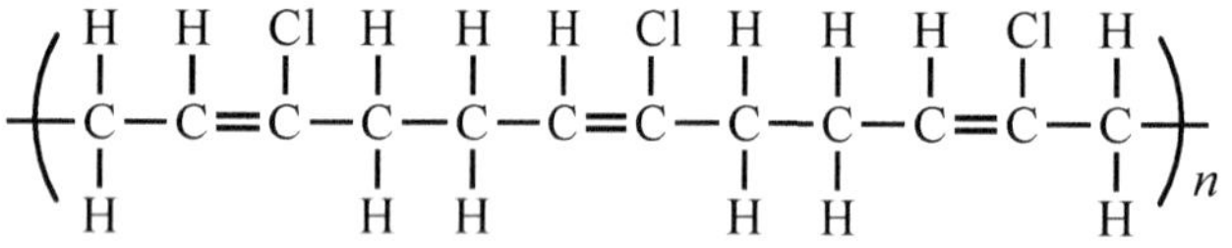

23.81. Polymerize acrylonitrile:

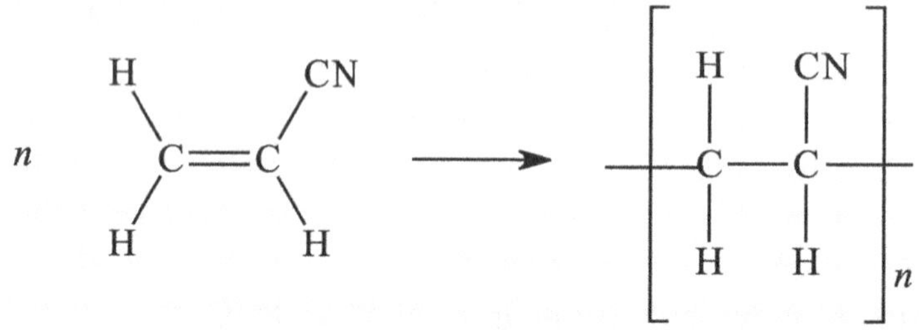

23.82. Regarding methyl methacrylate:

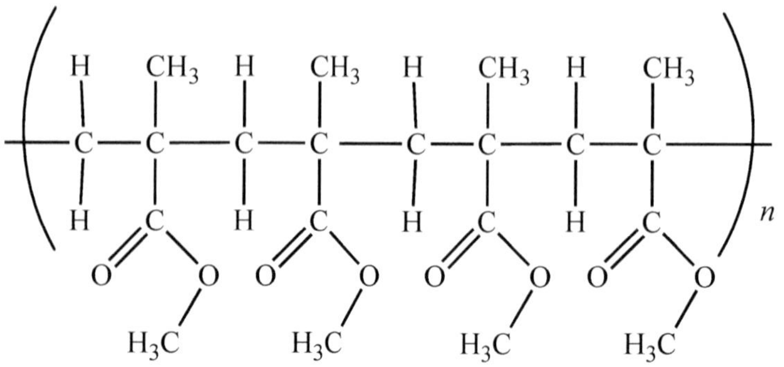

General Questions

23.83. Regarding the three compounds of formula $C_2H_2Cl_2$:

(a) geometric isomers of $C_2H_2Cl_2$:

$$\underset{cis \text{ isomer}}{\overset{ClCl}{\underset{HH}{C=C}}} \qquad \underset{trans \text{ isomer}}{\overset{ClH}{\underset{HCl}{C=C}}}$$

(b) structural isomer of $C_2H_2Cl_2$:

$$\overset{ClH}{\underset{ClH}{C=C}}$$

23.84. Regarding 2-butanol: Structure and mirror image:

The chiral C atom is the atom to which the OH group is attached. Mirror images of the molecule are not superimposable.

23.85. Isomers of C_6H_{12} with a 6-C chain.

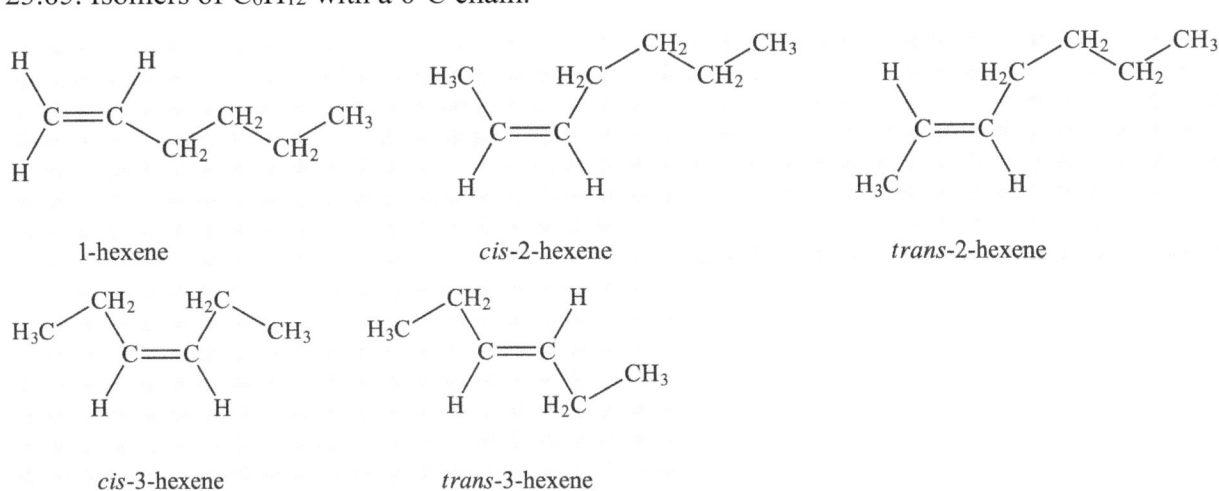

1-hexene

cis-2-hexene

trans-2-hexene

cis-3-hexene

trans-3-hexene

23.86. Structures and names for alkanes and cycloalkanes with formula C₄H₈:

cis-2-butene   trans-2-butene   2-methylpropene   1-butene

methylcyclopropane   cyclobutane

23.87. Reactions of cis-2-butene:

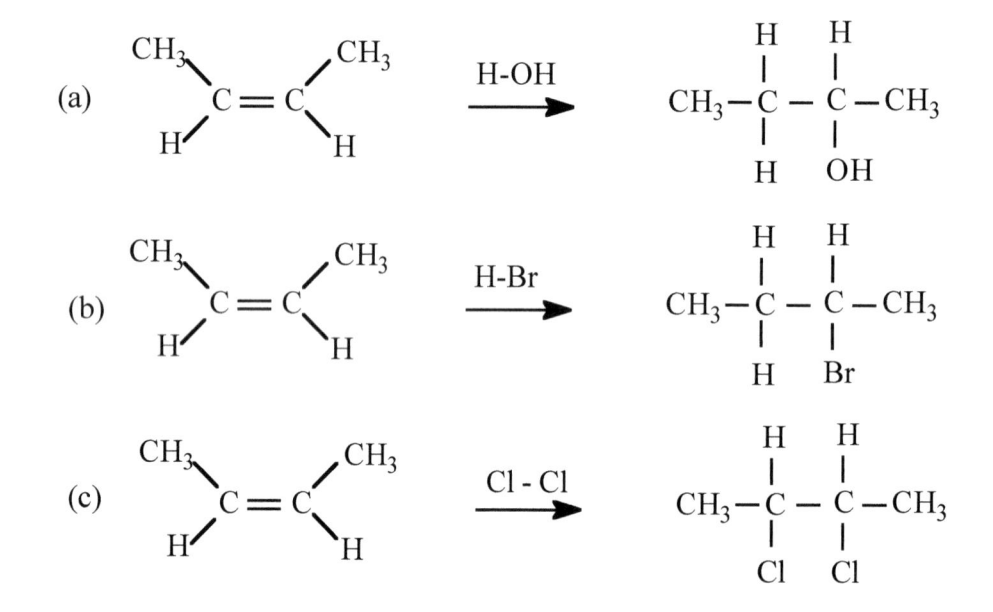

(a) cis-2-butene + H–OH → CH₃–CH(H)–CH(OH)–CH₃

(b) cis-2-butene + H–Br → CH₃–CH(H)–CH(Br)–CH₃

(c) cis-2-butene + Cl–Cl → CH₃–CH(Cl)–CH(Cl)–CH₃

23.88. Products of oxidation of the alcohols:

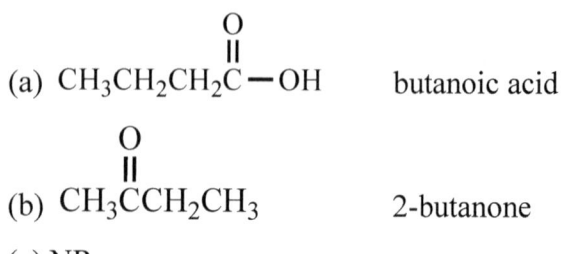

(a) CH₃CH₂CH₂C(=O)–OH   butanoic acid

(b) CH₃C(=O)CH₂CH₃   2-butanone

(c) NR

(d) 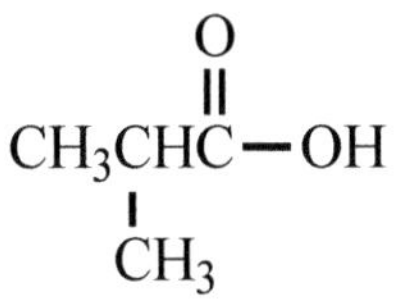 $\underset{\underset{CH_3}{|}}{CH_3CH}\overset{\overset{O}{\|}}{C}-OH$ 2-methylpropanoic acid

23.89. In (a) acetic acid reacts with NaOH (a base) to form a salt (an ionic compound). In (b) there is also an acid-base reaction. The amine (base) reacts with HCl (acid) to form a salt.

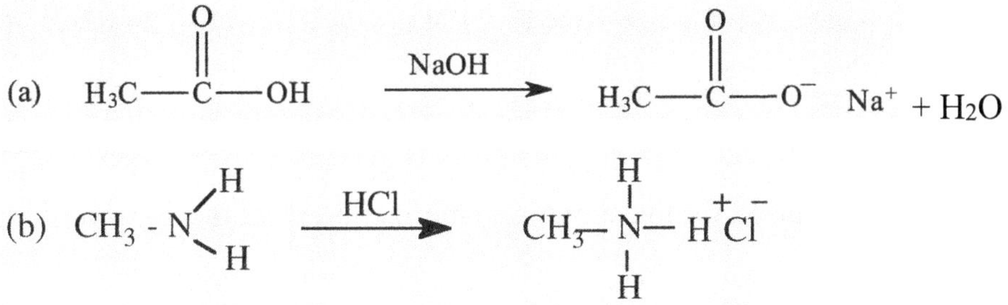

(a) $H_3C-\overset{\overset{O}{\|}}{C}-OH \xrightarrow{NaOH} H_3C-\overset{\overset{O}{\|}}{C}-O^-\ Na^+ + H_2O$

(b) $CH_3-N\overset{H}{\underset{H}{\diagdown}} \xrightarrow{HCl} CH_3-\overset{\overset{H}{|}}{\underset{\underset{H}{|}}{N}}-H^+\ Cl^-$

23.90. Equations for the following reactions:

(a) $H_3C\overset{\overset{O}{\|}}{C}OH + HOCH_2CH_3 \longrightarrow H_3C\overset{\overset{O}{\|}}{C}OCH_2CH_3$

(b) $\begin{array}{l}H_2C-O\overset{\overset{O}{\|}}{C}(CH_2)_{16}CH_3\\ |\\ HC-O\overset{\overset{O}{\|}}{C}(CH_2)_{16}CH_3\\ |\\ H_2C-O\overset{\overset{O}{\|}}{C}(CH_2)_{16}CH_3\end{array} + 3\ H_2O \longrightarrow \begin{array}{l}H_2COH\\ |\\ HCOH\\ |\\ H_2COH\end{array} + 3\ CH_3(CH_2)_{16}\overset{\overset{O}{\|}}{C}-OH$

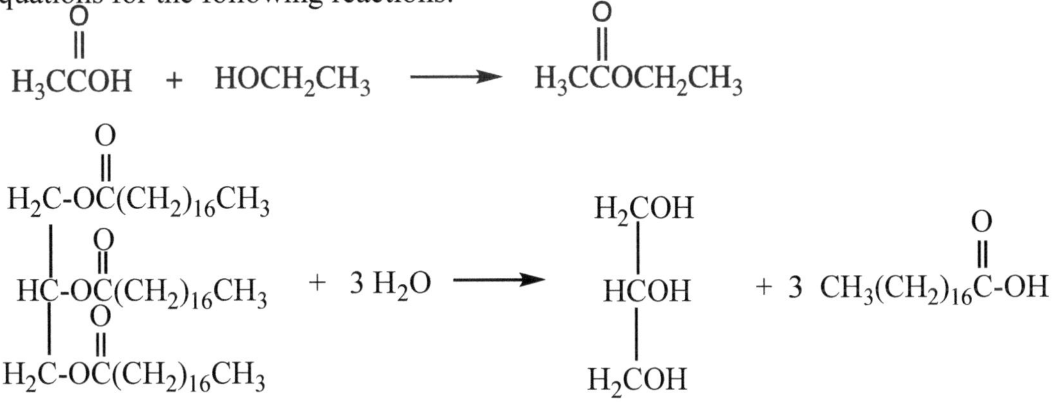

Catalytic amounts of acid are required for these two hydrolysis reactions to occur in a reasonable time-frame.

23.91. Equation for the formation of polymers:

(a) n CH₂=CH(C₆H₅) → [-CH(C₆H₅)-CH₂-]ₙ

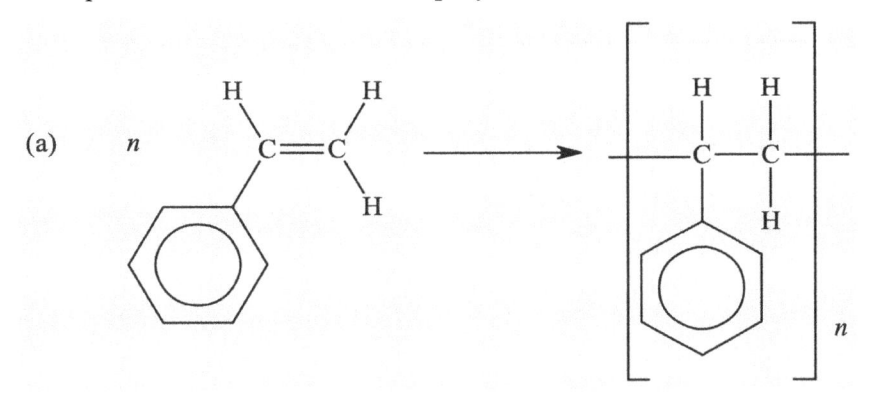

(b) 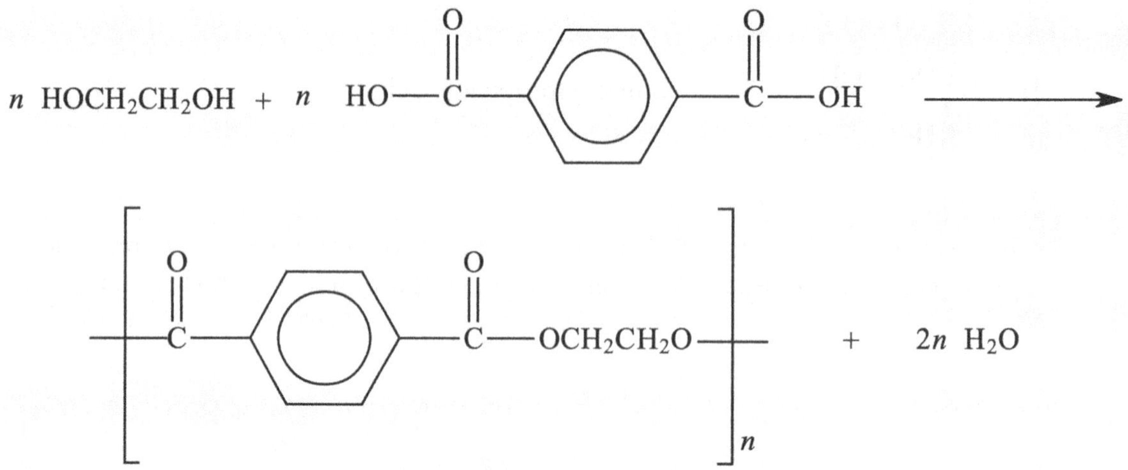 n HOCH$_2$CH$_2$OH + n HO−C(=O)−C$_6$H$_4$−C(=O)−OH ⟶

$\left[-C(=O)-C_6H_4-C(=O)-OCH_2CH_2O- \right]_n$ + $2n$ H$_2$O

Catalytic amounts of acid are required for these two polymerization reactions to occur in a reasonable time-frame.

23.92. Equations for the requested reactions:

(a) 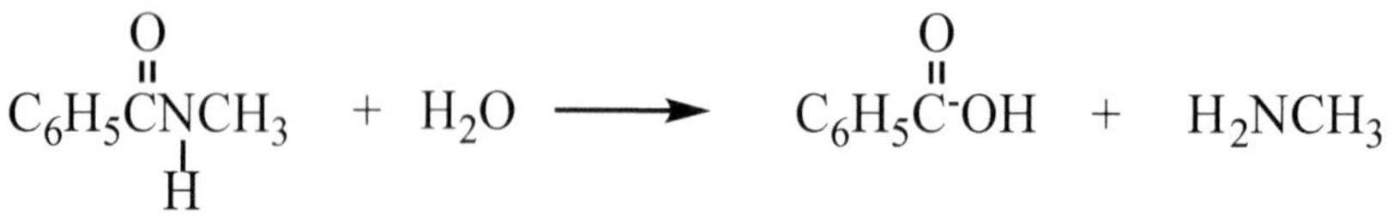 C$_6$H$_5$C(=O)N(H)CH$_3$ + H$_2$O ⟶ C$_6$H$_5$C(=O)-OH + H$_2$NCH$_3$

(b) 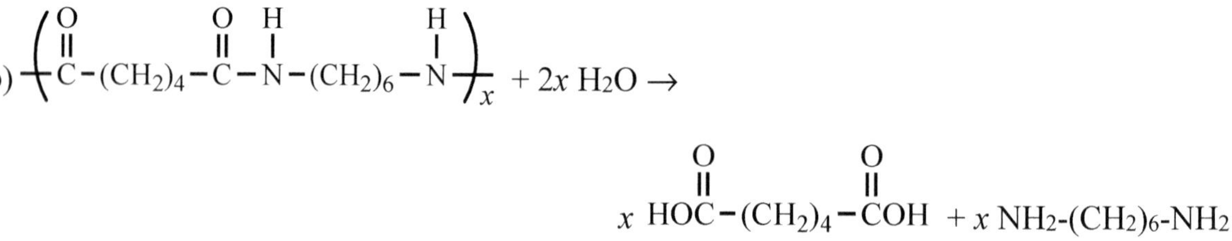 $\left[-C(=O)-(CH_2)_4-C(=O)-N(H)-(CH_2)_6-N(H)- \right]_x$ + $2x$ H$_2$O →

x HOC(=O)−(CH$_2$)$_4$−C(=O)OH + x NH$_2$-(CH$_2$)$_6$-NH$_2$

Catalytic amounts of acid are required for these two hydrolysis reactions to occur in a reasonable time-frame.

23.93. Structures for:

(a) 2,2-dimethylpentane

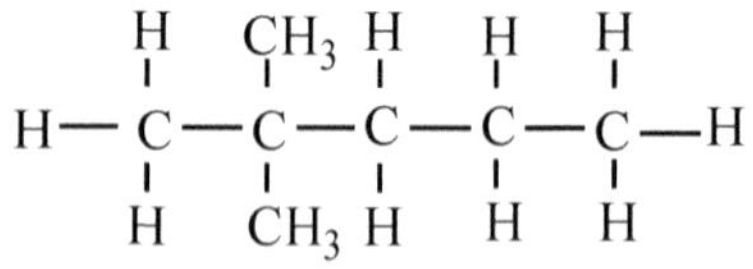

H−C(H)(H)−C(CH$_3$)(CH$_3$)−C(H)(H)−C(H)(H)−C(H)(H)−H

(b) 3,3-diethylpentane

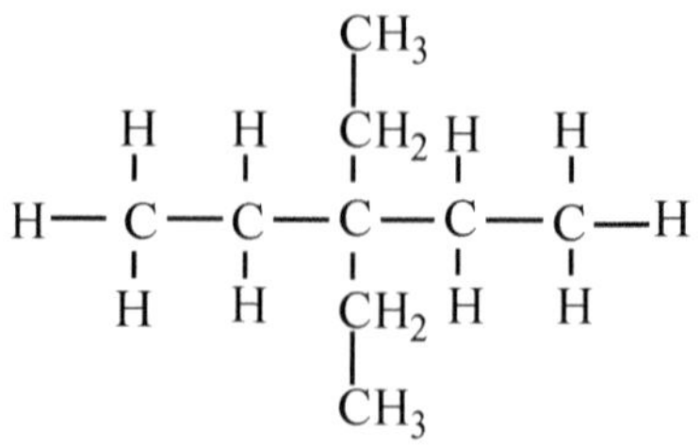

H−C(H)(H)−C(H)(H)−C(CH$_2$CH$_3$)(CH$_2$CH$_3$)−C(H)(H)−C(H)(H)−H

(c) 3-ethyl-2-methylpentane

$$H-\underset{\underset{H}{|}}{\overset{\overset{H}{|}}{C}}-\underset{\underset{H}{|}}{\overset{\overset{CH_3}{|}}{C}}-\underset{\underset{\underset{CH_3}{|}}{CH_2}}{\overset{\overset{H}{|}}{C}}-\underset{\underset{H}{|}}{\overset{\overset{H}{|}}{C}}-\underset{\underset{H}{|}}{\overset{\overset{H}{|}}{C}}-H$$

(d) 3-ethylhexane

$$H-\underset{\underset{H}{|}}{\overset{\overset{H}{|}}{C}}-\underset{\underset{H}{|}}{\overset{\overset{H}{|}}{C}}-\underset{\underset{\underset{CH_3}{|}}{CH_2}}{\overset{\overset{H}{|}}{C}}-\underset{\underset{H}{|}}{\overset{\overset{H}{|}}{C}}-\underset{\underset{H}{|}}{\overset{\overset{H}{|}}{C}}-CH_3$$

23.94. Structural isomers:

(a) $CH_3CH_2CH_2-OH$ 1-propanol   primary alcohol

$CH_3\overset{\overset{OH}{|}}{C}HCH_3$ 2-propanol   secondary alcohol

$H_3C-O-CH_2CH_3$ ethylmethylether   ether

(b)

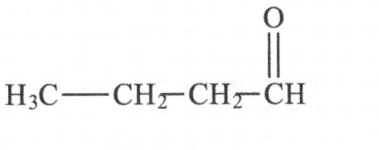

$H_3C-CH_2-CH_2-\overset{\overset{O}{\|}}{C}H$ $H_3C-\overset{\overset{O}{\|}}{C}-CH_2-CH_3$

aldehyde
butanal

ketone
butanone

23.95. Structural isomers for $C_3H_6Cl_2$:

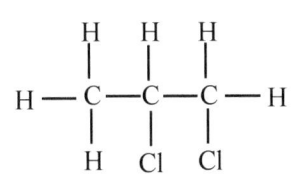

$$H-\underset{\underset{H}{|}}{\overset{\overset{H}{|}}{C}}-\underset{\underset{Cl}{|}}{\overset{\overset{H}{|}}{C}}-\underset{\underset{Cl}{|}}{\overset{\overset{H}{|}}{C}}-H$$

1,2-dichloropropane

$$H-\underset{\underset{H}{|}}{\overset{\overset{H}{|}}{C}}-\underset{\underset{Cl}{|}}{\overset{\overset{Cl}{|}}{C}}-\underset{\underset{H}{|}}{\overset{\overset{H}{|}}{C}}-H$$

2,2-dichloropropane

$$H-\underset{\underset{Cl}{|}}{\overset{\overset{Cl}{|}}{C}}-\underset{\underset{H}{|}}{\overset{\overset{H}{|}}{C}}-\underset{\underset{H}{|}}{\overset{\overset{H}{|}}{C}}-H$$

1,1-dichloropropane

$$H-\underset{\underset{Cl}{|}}{\overset{\overset{H}{|}}{C}}-\underset{\underset{H}{|}}{\overset{\overset{H}{|}}{C}}-\underset{\underset{Cl}{|}}{\overset{\overset{H}{|}}{C}}-H$$

1,3-dichloropropane

23.96. Structural formulas for compounds with formula C_4H_6BrCl:

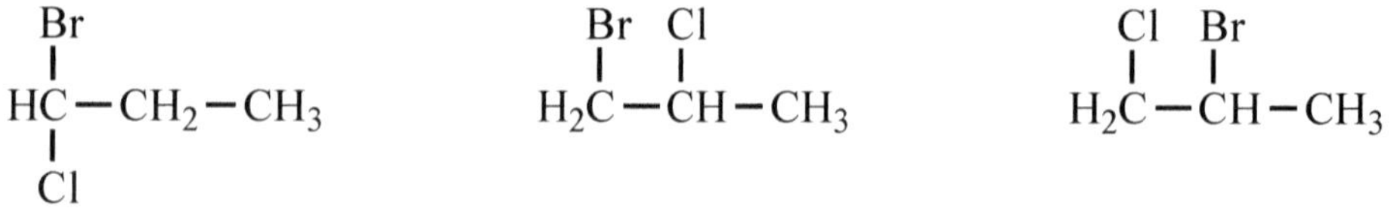

Br | HC—CH$_2$—CH$_3$ | Cl	Br Cl | | H$_2$C—CH—CH$_3$	Cl Br | | H$_2$C—CH—CH$_3$
1-bromo-1-chloropropane	1-bromo-2-chloropropane	2-bromo-1-chloropropane

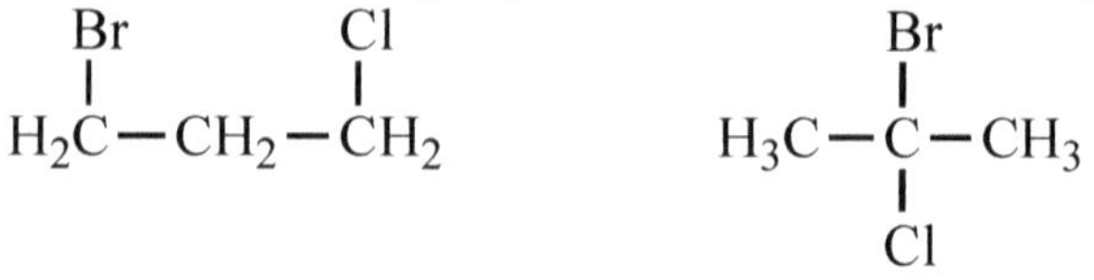

Br Cl | | H$_2$C—CH$_2$—CH$_2$	Br | H$_3$C—C—CH$_3$ | Cl
1-bromo-3-chloropropane	2-bromo-2-chloropropane

23.97. Structural isomers for trimethylbenzene:

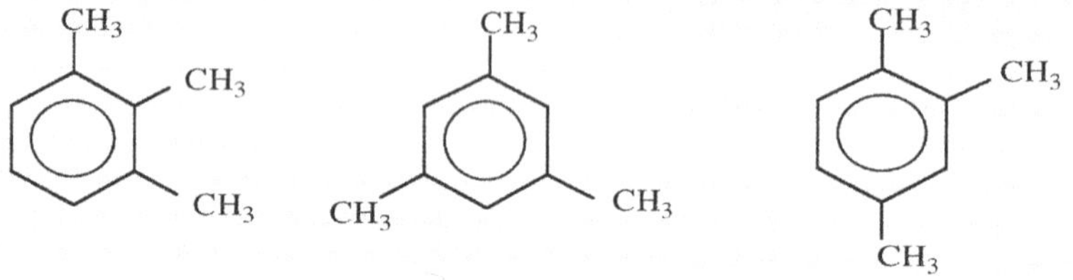

1,2,3-trimethylbenzene   1,3,5-trimethylbenzene   1,2,4-trimethylbenzene

23.98. Formulas and names for isomers of dichlorobenzene:

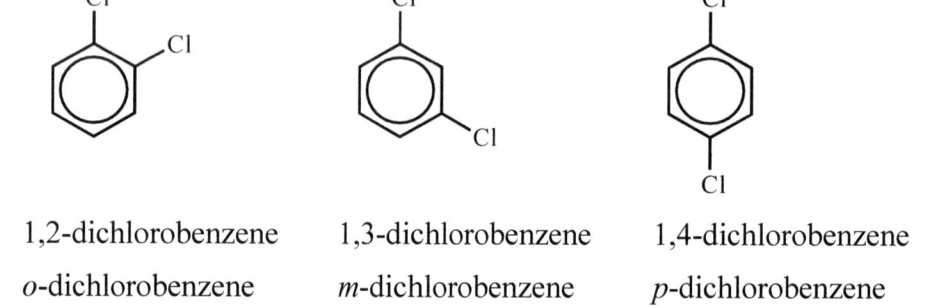

1,2-dichlorobenzene   1,3-dichlorobenzene   1,4-dichlorobenzene

o-dichlorobenzene   *m*-dichlorobenzene   *p*-dichlorobenzene

23.99. Atoms in lysine needed to be replaced to make cadaverine:

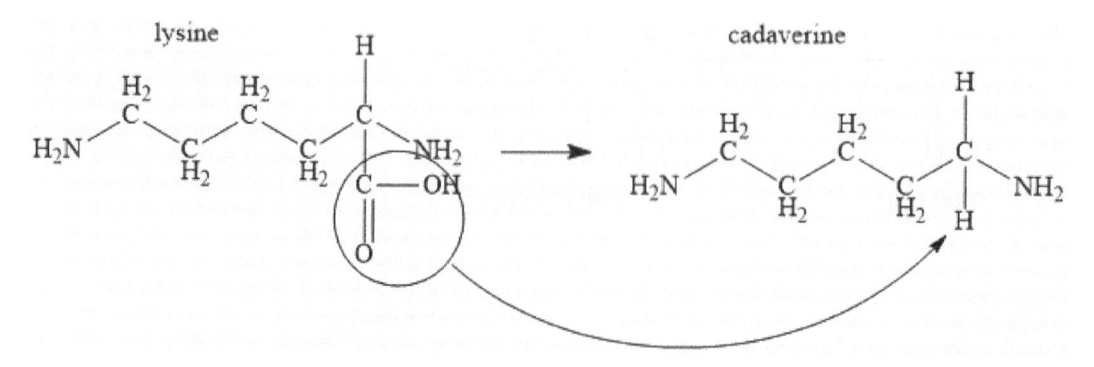

23.100. Regarding hippuric acid:

Hippuric acid is an acid because it contains a carboxylic acid group.

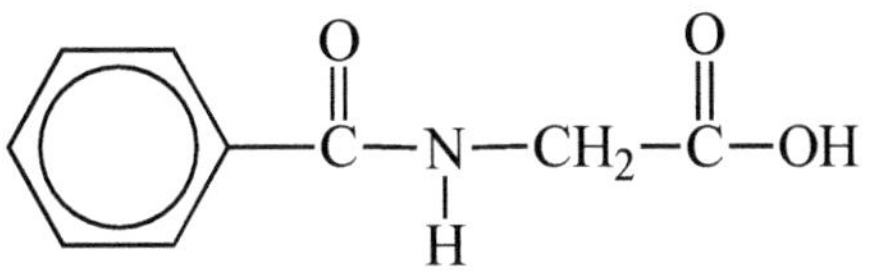

23.101. Reaction of cis-2-butene:

(a)

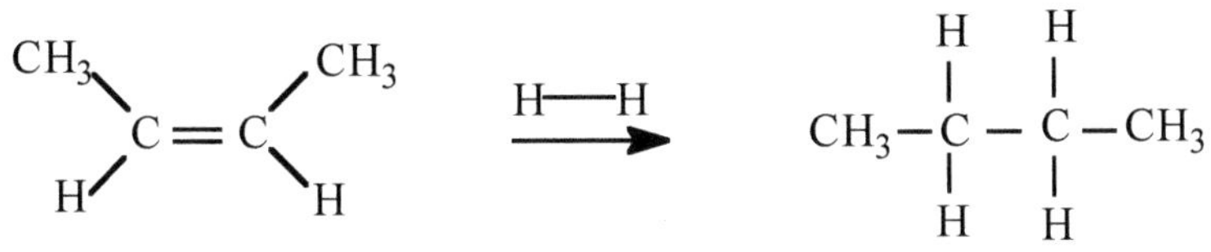

butane, not chiral

(b) an isomer of butane

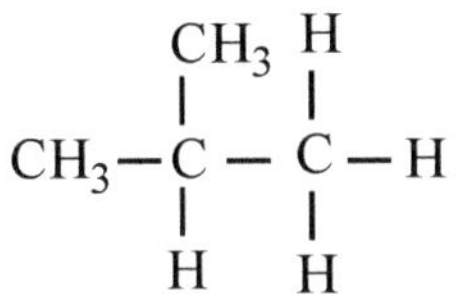

23.102. Name the following compounds and identify the functional groups contained therein:

(a) 2-pentanol   alcohol (–OH group attached to the carbon chain)

(b) 2-pentanone   ketone (C=O group sandwiched between carbon atoms)

(c) 2-methylpropanal aldehyde (C=O group attached to a terminal carbon atom)

(d) butanoic acid carboxylic acid (COOH group attached to the carbon chain)

23.103. Regarding glyceryl trilaurate:

(a) Equation for the saponification of glyceryl trilaurate.

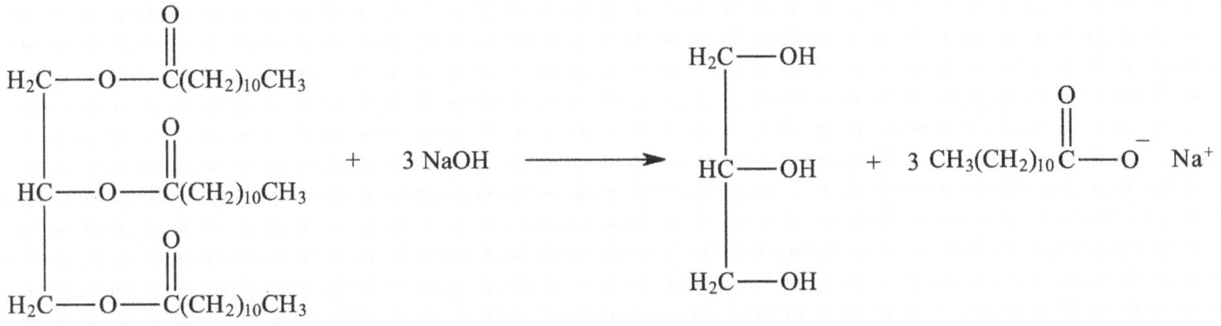

glyceryl trilaurate + 3 NaOH → glycerol + 3 sodium laurate

(b) Prepare biodiesel fuel from the fat:

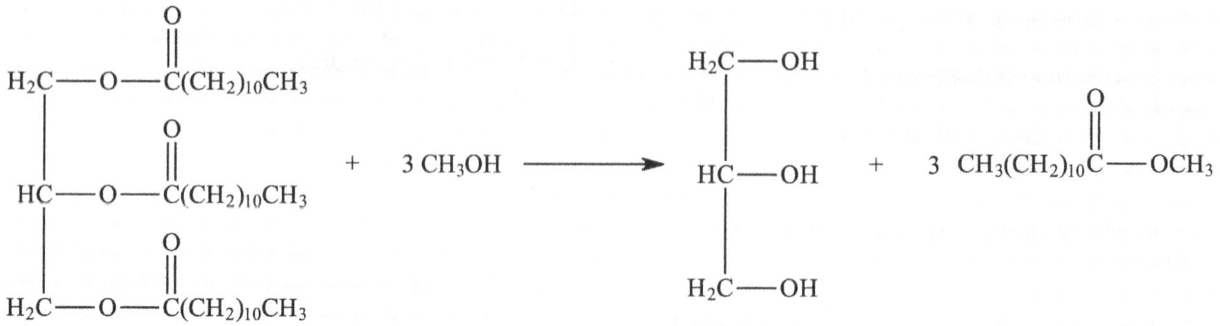

$$\text{glyceryl trilaurate} + 3\,CH_3OH \longrightarrow \text{glycerol} + 3\,\text{methyl laurate}$$

where glyceryl trilaurate is $H_2C{-}O{-}C(O)(CH_2)_{10}CH_3$, $HC{-}O{-}C(O)(CH_2)_{10}CH_3$, $H_2C{-}O{-}C(O)(CH_2)_{10}CH_3$; glycerol is $H_2C(OH){-}CH(OH){-}CH_2(OH)$; and methyl laurate is $CH_3(CH_2)_{10}C(O){-}OCH_3$.

23.104. Reaction of PET with methanol:

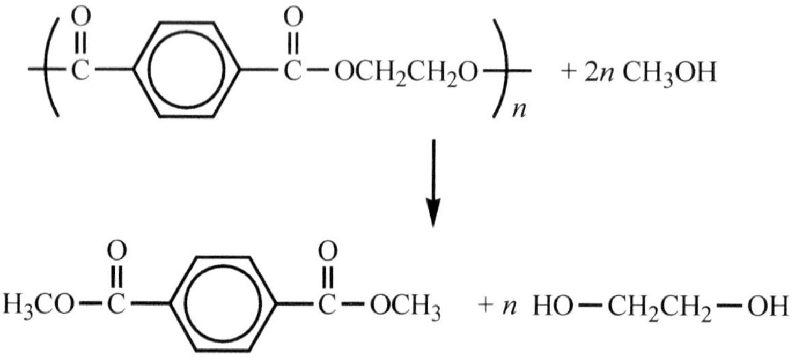

$$\left[{-}C(O){-}C_6H_4{-}C(O){-}OCH_2CH_2O{-} \right]_n + 2n\,CH_3OH$$

$$\longrightarrow n\,H_3CO{-}C(O){-}C_6H_4{-}C(O){-}OCH_3 + n\,HO{-}CH_2CH_2{-}OH$$

23.105. Show products of the following reactions of $CH_2{=}CHCH_2OH$:

(a) Hydrogenation of the compound:

$$CH_2{=}CHCH_2OH \longrightarrow CH_3CH_2CH_2OH$$

(b) Oxidation of the compound:

$$CH_2{=}CHCH_2OH \longrightarrow CH_2{=}CHCO_2H$$

(c) Addition polymerization:

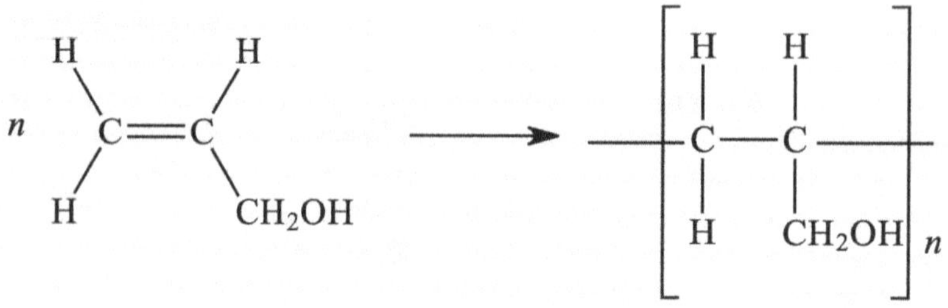

$$n\;\underset{H\quad CH_2OH}{\overset{H\quad\quad H}{C{=}C}} \longrightarrow \left[\underset{H\quad CH_2OH}{\overset{H\quad\;H}{-C{-}C-}} \right]_n$$

(d) Ester formation with acetic acid:

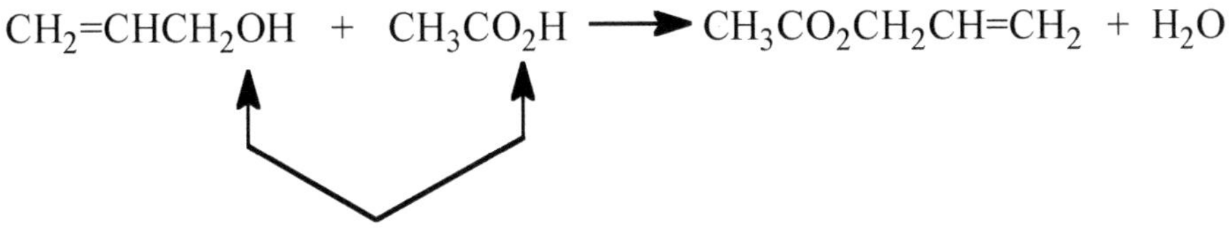

$$CH_2=CHCH_2OH + CH_3CO_2H \longrightarrow CH_3CO_2CH_2CH=CH_2 + H_2O$$

23.106. Reaction between glycerol and stearic acid:

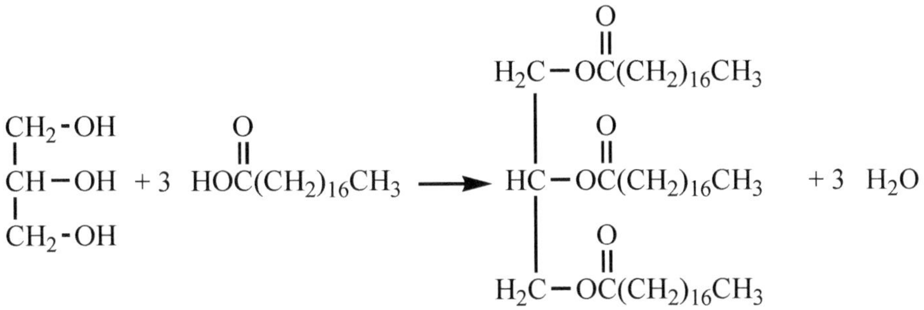

$$\begin{array}{c}CH_2\text{-}OH \\ | \\ CH\text{-}OH \\ | \\ CH_2\text{-}OH\end{array} + 3\ HOC(CH_2)_{16}CH_3 \longrightarrow \begin{array}{c}H_2C-OC(CH_2)_{16}CH_3 \\ | \\ HC-OC(CH_2)_{16}CH_3 \\ | \\ H_2C-OC(CH_2)_{16}CH_3\end{array} + 3\ H_2O$$

23.107. For the reactions in question:

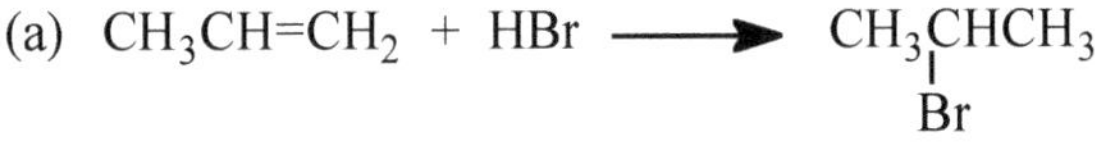

(a) $CH_3CH=CH_2 + HBr \longrightarrow CH_3\underset{Br}{\overset{|}{C}}HCH_3$

2-bromopropane

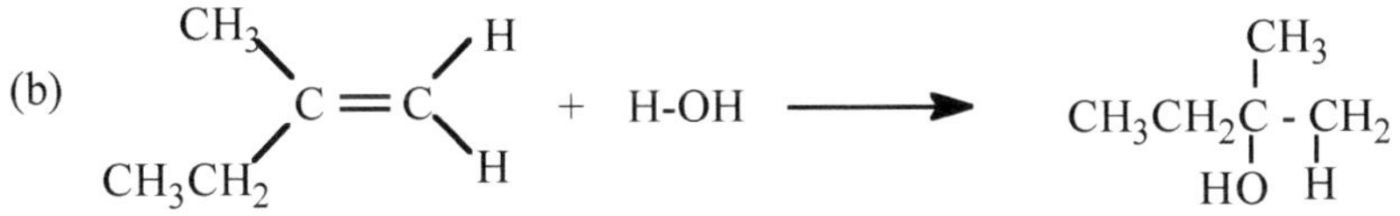

(b)
$$\underset{CH_3CH_2}{\overset{CH_3}{\diagdown}}C=C\underset{H}{\overset{H}{\diagup}} + H\text{-}OH \longrightarrow CH_3CH_2\underset{HO}{\overset{CH_3}{\underset{|}{\overset{|}{C}}}}\text{-}\underset{H}{\overset{}{CH_2}}$$

2-methyl-2-butanol

(c)
$$\underset{CH_3}{\overset{H}{\diagdown}}C=C\underset{CH_3}{\overset{CH_3}{\diagup}} + H\text{-}OH \longrightarrow CH_3\underset{H}{\overset{}{\underset{|}{\overset{|}{C}}H}}\text{-}\underset{OH}{\overset{CH_3}{\underset{|}{\overset{|}{C}}CH_3}}$$

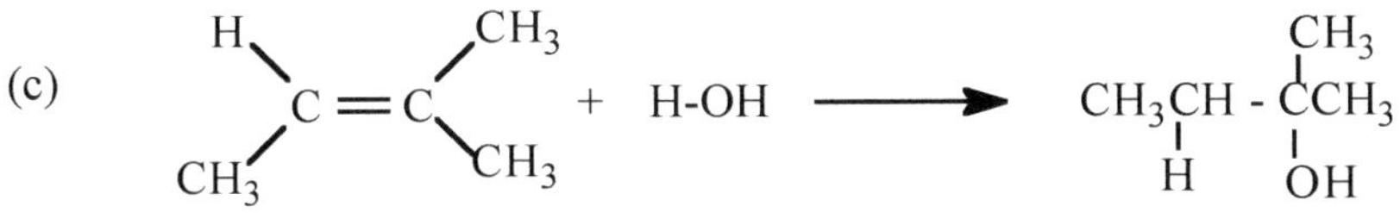

2-methyl-2-butanol

Given the identical names for the product formed in (b) and (c), the products of these two reactions are the same.

23.108. Ethers with formula $C_4H_{10}O$:

$CH_3CH_2OCH_2CH_3 \qquad CH_3OCH_2CH_2CH_3 \qquad CH_3O\underset{}{\overset{CH_3}{\underset{}{\overset{|}{C}}H}}CH_3$

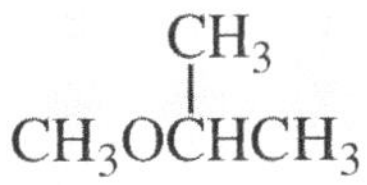

23.109. Regarding theobromine and caffeine:

(a) Structures for theobromine and caffeine:

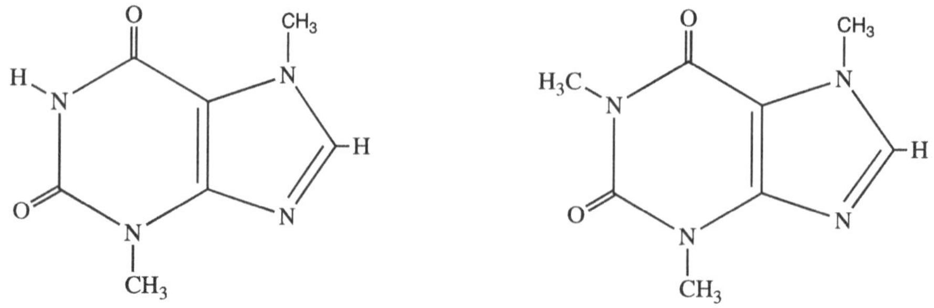

Theobromine   Caffeine

The difference between the two is the methyl group on the 6-member ring at left.

(b) Mass of theobromine in 5.00 g sample of cocoa containing 2.16% theobromine:

$$5.00 \text{ g cocoa}\left(\frac{2.16 \text{ g theobromine}}{100 \text{ g cocoa}}\right) = 0.108 \text{ g theobromine}$$

23.110. Equation for the formation of Nylon-6:

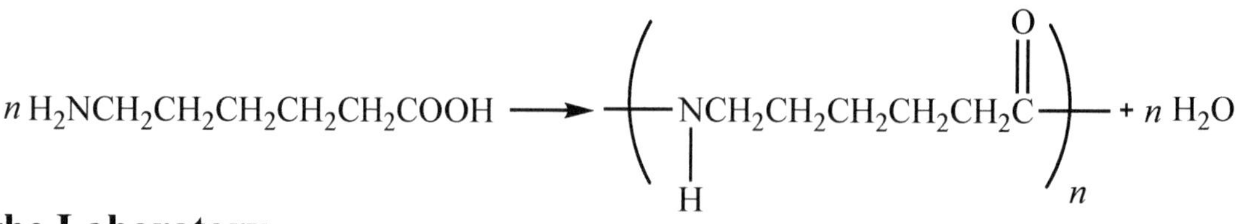

$$n\ H_2NCH_2CH_2CH_2CH_2CH_2COOH \longrightarrow \left(-NCH_2CH_2CH_2CH_2CH_2\overset{O}{\underset{\|}{C}}-\right)_n + n\ H_2O$$

with H on the N.

In the Laboratory

23.111. Which of the following produce acetic acid when reacted with KMnO₄:

Both (b) and (c) produce acetic acid. The alcohol (c) is oxidized to the aldehyde (b) which subsequently is oxidized to acetic acid.

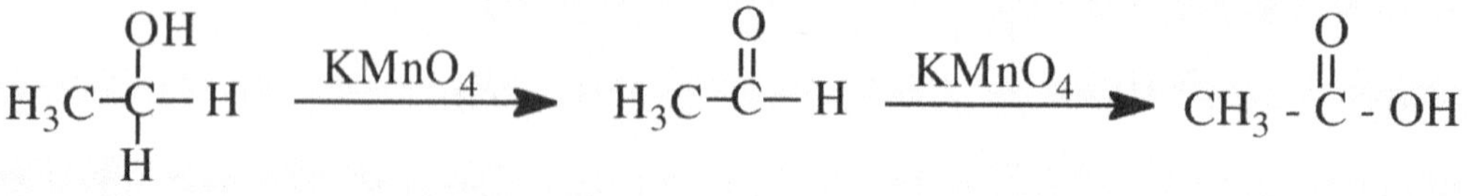

$$H_3C-\underset{H}{\overset{OH}{\underset{|}{C}}}-H \xrightarrow{KMnO_4} H_3C-\overset{O}{\underset{\|}{C}}-H \xrightarrow{KMnO_4} CH_3-\overset{O}{\underset{\|}{C}}-OH$$

(c)   (b)

23.112. Consider the reactions of C₃H₇OH:

(a) The reactant: 1-propanol
(b) A structural isomer of the reactant: CH₃CHCH₃ with OH, known as: 2-propanol
(c) Product of reaction A: propene
(d) Product of reaction B: propyl ethanoate

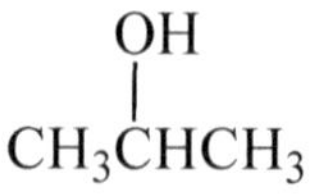

23.113. The reaction between bromine and cyclohexene:

The double bond in cyclohexene adds elemental bromine across the double bond so that the adjacent carbon atoms (originally participating in the double bond) each have one Br atom bound to them. Benzene does not add Br₂ under these conditions.

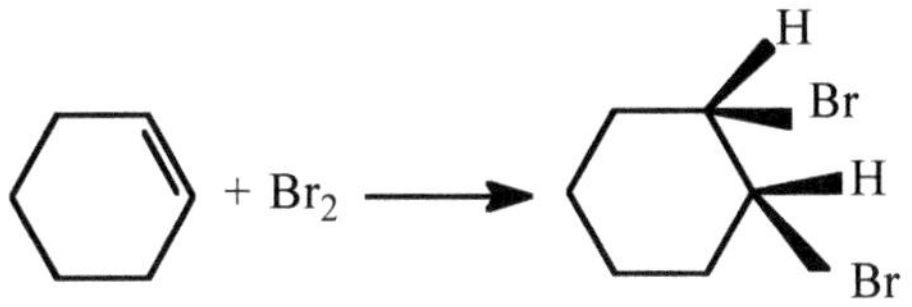

23.114. Provide names and structures for alcohol A and acid B:

$$\underset{\text{2-methyl-1-propanol}}{CH_3CH(CH_3)CH_2-OH} \qquad \underset{\text{2-methylpropanoic acid}}{CH_3CH(CH_3)COOH}$$

23.115. Identify an unknown colorless liquid with formula C₃H₆O:

(a) Two structures with formula C₃H₆O:

$$\underset{\text{propanone (acetone)}}{CH_3COCH_3} \qquad \underset{\text{propanal}}{CH_3CH_2CHO}$$

(b) Oxidation of the compound gives an acidic solution indicating that the unknown is propanal.

(c) Oxidation of propanal gives propanoic acid: CH_3CH_2COOH

23.116. A test to distinguish between 1-pentene and cyclopentane:

Cyclopentane will not react with bromine in the dark, while 1-pentene will.

23.117. Test to distinguish between 2-propanol and methyl ethyl ether:
Methyl ethyl ether will not react with an oxidizing agent like KMnO₄. The alcohol will react with KMnO₄, forming a ketone.

23.118. Identify the ester of formula C₄H₈O₂, and write the equation for the hydrolysis:

$$CH_3CH_2C(O)-O-CH_3 + H_2O \longrightarrow CH_3CH_2C(O)-O-H + CH_3OH$$

23.119. Addition of water to an alkene, X, gives an alcohol, Y. Oxidation of Y gives 3,3-dimethyl-2-pentanone. The structure for X and Y:

$$\underset{\text{3,3-dimethyl-2-pentanone}}{CH_3\overset{O}{\overset{\|}{C}}\overset{CH_3}{\underset{CH_3}{\overset{|}{C}}}CH_2CH_3} \longleftarrow \underset{\underset{\text{3,3-dimethyl-2-pentanol}}{Y}}{CH_3\overset{HO}{\overset{|}{C}H}\overset{CH_3}{\underset{CH_3}{\overset{|}{C}}}CH_2CH_3} \longleftarrow \underset{\underset{\text{3,3-dimethyl-1-pentene}}{X}}{CH_2=CH\overset{CH_3}{\underset{CH_3}{\overset{|}{C}}}CH_2CH_3}$$

23.120. Regarding 2-aminobenzoic acid:

(a) The reagents react in a 1:1:1 mol ratio. Whichever is present in the smallest amount (moles) is the limiting reactant.

$$4.0 \text{ g } C_6H_4(CO_2H)NH_2 \left(\frac{1 \text{ mol } C_6H_4(CO_2H)NH_2}{137.1 \text{ g } C_6H_4(CO_2H)NH_2}\right) = 0.029 \text{ mol } C_6H_4(CO_2H)NH_2$$

$$2.2 \text{ g NaNO}_2 \left(\frac{1 \text{ mol NaNO}_2}{69.0 \text{ g NaNO}_2}\right) = 0.032 \text{ mol NaNO}_2$$

$$5.3 \text{ g KI} \left(\frac{1 \text{ mol KI}}{166.0 \text{ g KI}}\right) = 0.032 \text{ mol KI}$$

$$0.029 \text{ mol } C_6H_4(CO_2H)NH_2 \left(\frac{1 \text{ mol } C_6H_4(CO_2H)I}{1 \text{ mol } C_6H_4(CO_2H)NH_2}\right)\left(\frac{248.0 \text{ g } C_6H_4(CO_2H)I}{1 \text{ mol } C_6H_4(CO_2H)I}\right)$$
$$= 7.2 \text{ g } C_6H_4(CO_2H)I$$

(b) Yes. 3-iodobenzoic acid and 4-iodobenzoic acid are possible isomers.

(c) $C_6H_4(CO_2H)I + NaOH \rightarrow NaC_6H_4(CO_2)I + H_2O$

$$0.01562 \text{ L}\left(\frac{0.101 \text{ mol NaOH}}{1 \text{ L}}\right)\left(\frac{1 \text{ mol } C_6H_4(CO_2H)I}{1 \text{ mol NaOH}}\right) = 0.00158 \text{ mol } C_6H_4(CO_2H)I$$

$$\frac{0.399 \text{ g } C_6H_4(CO_2H)I}{0.00158 \text{ mol } C_6H_4(CO_2H)I} = 253 \text{ g/mol } C_6H_4(CO_2H)I$$

(actual molar mass 248 g/mol)

23.121. Transesterification between PET and CH₃OH:

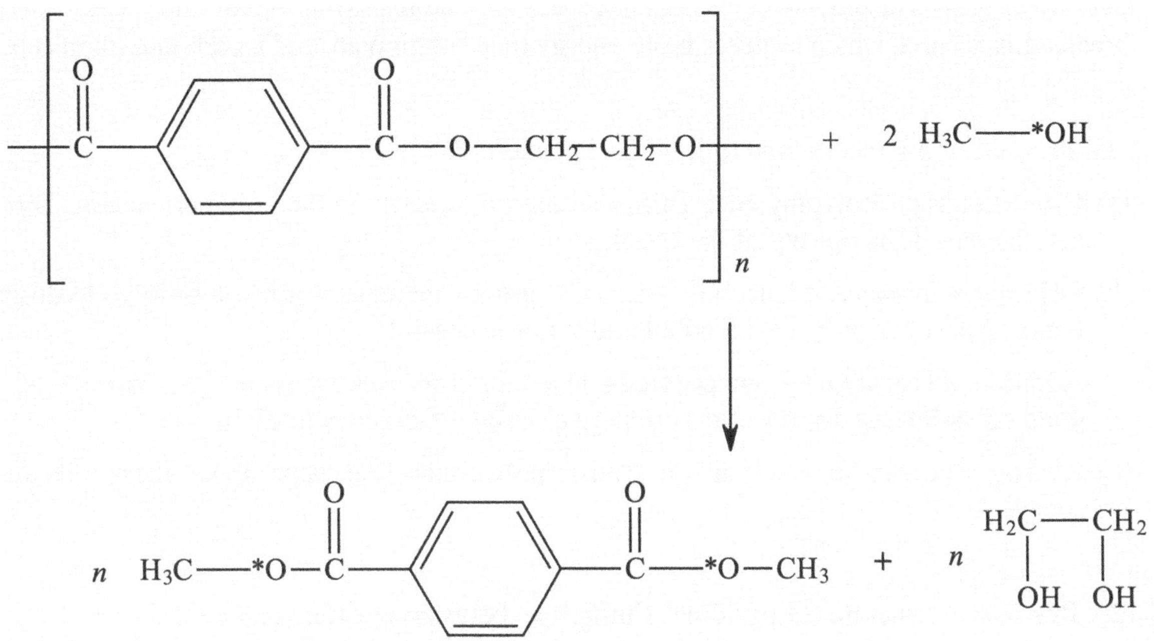

[PET polymer structure] + 2 H₃C—*OH

↓

n H₃C—*O—C(=O)—C₆H₄—C(=O)—*O—CH₃ + n HOCH₂—CH₂OH

The * indicates the O-18 labeled atom in the reactants and products.

23.122. Structures resulting from hydrolysis of Vitamin B-5:

Hydrolysis results in the cleavage of the amide bond, the addition of H₂O into that cleaved bond (box below), and the formation of a carboxylic acid and an amine.

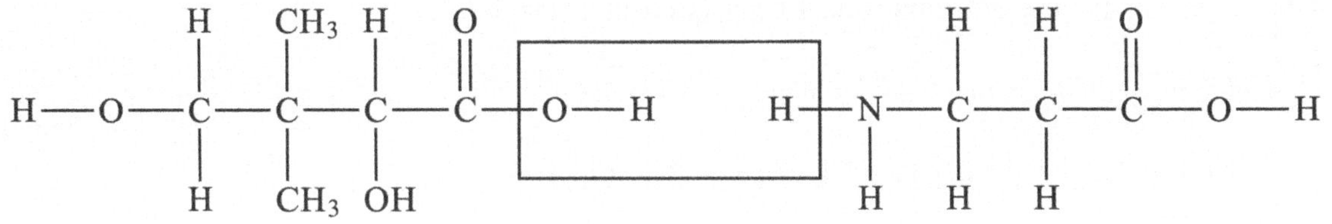

H—O—C(H)(H)—C(CH₃)(CH₃)—C(H)(OH)—C(=O)—[O—H   H—N(H)]—C(H)(H)—C(H)(H)—C(=O)—O—H

Summary and Conceptual Questions:

23.123. Modes by which C can achieve an octet of electrons: Since C has 4 valence electrons (it is in group 4A [14]), it needs to acquire an octet by gaining 4 electrons.

This can be done by forming 4 single bonds to other atoms (e.g., 4 H atoms in methane, CH₄). It can also participate in a double bond and two single bonds (e.g. ethene—or ethylene, C₂H₄), two double bonds (as in allene, C₃H₄), OR a triple bond and one single bond (e.g. ethyne—or acetylene, C₂H₂).

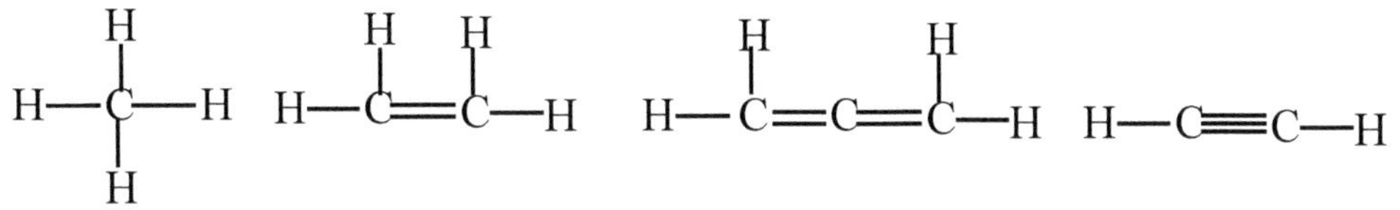

H—CH₃   H₂C=CH₂   H₂C=C=CH₂   H—C≡C—H

23.124. The restricted rotation around C=C double bonds is due to the π bond formed from overlap of unhybridized *p* orbitals on adjacent carbon atoms. Rotation around a C=C bond breaks this π bond, which requires more energy than rotation around a carbon-carbon σ bond.

23.125. Properties imparted by the following characteristics:

(a) Cross-linking in polyethylene: Brings structural integrity to the polymeric chain. This also increases the rigidity of the chain.

(b) OH groups in polyvinyl alcohol: These OH groups increase water solubility as hydrogen-bonding between polyvinyl alcohol and water is possible.

Additionally these OH group provide a locus for cross-linking agents (e.g. borax is a good cross-linking agent in the commercial polymer called Slime™).

(c) Hydrogen bonding in a polyamide: Causes polyamides (e.g. peptides) to form coils and sheets.

23.126. Resonance structure for pyridine; similarities between pyridine and benzene:

Pyridine is isoelectronic with benzene. A C—H group in benzene has been replaced by an N atom in pyridine. Both benzene and pyridine have multiple resonance structures.

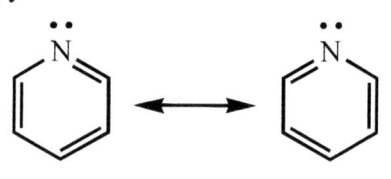

23.127. Equations for the combustion of $C_2H_6(g)$ and $C_2H_5OH(\ell)$:

$CH_3CH_3(g) + 7/2\ O_2(g) \rightarrow 2\ CO_2(g) + 3\ H_2O(\ell)$

$CH_3CH_2OH(\ell) + 3\ O_2(g) \rightarrow 2\ CO_2(g) + 3\ H_2O(\ell)$

(a) Enthalpy of combustion of ethane and ethanol:

$\Delta_rH°\ C_2H_6 = [2\ \Delta_fH°\ CO_2 + 3\ \Delta_fH°\ H_2O] - [\Delta_fH°\ CH_3CH_3 + 7/2\ \Delta_fH°\ O_2]$

$= [(2\ mol/mol\text{-}rxn)(-393.509\ kJ/mol) + (3\ mol/mol\text{-}rxn)(-285.83\ kJ/mol)]$

$- [(1\ mol/mol\text{-}rxn)(-83.85\ kJ/mol) + (7/2\ mol/mol\text{-}rxn)(0\ kJ/mol)]$

$= -1560.66\ kJ/mol\text{-}rxn$

$\Delta_rH°\ C_2H_5OH = [2\ \Delta_fH°\ CO_2 + 3\ \Delta_fH°\ H_2O] - [1\ \Delta_fH°\ CH_3CH_3 + 3\ \Delta_fH°\ O_2]$

$= [(2\ mol/mol\text{-}rxn)(-393.509\ kJ/mol) + (3\ mol/mol\text{-}rxn)(-285.83\ kJ/mol)]$

$- [(1\ mol/mol\text{-}rxn)(-277.0\ kJ/mol) + (3\ mol/mol\text{-}rxn)(0\ kJ/mol)]$

$= -1367.5\ kJ/mol\text{-}rxn$

On a per gram basis:

–1560.66 kJ/mol ethane · 1 mol/30.070 g or approximately –51.901 kJ/g of ethane and

–1367.5 kJ/mol ethanol · 1 mol/46.069 g or approximately –29.684 kJ/g ethanol

(b) If ethanol is partially oxidized ethane, the ΔH of reaction is **less negative**. So partially oxidized ethane has less energy to release during the combustion process.

23.128. Regarding plastics and their recycling codes:

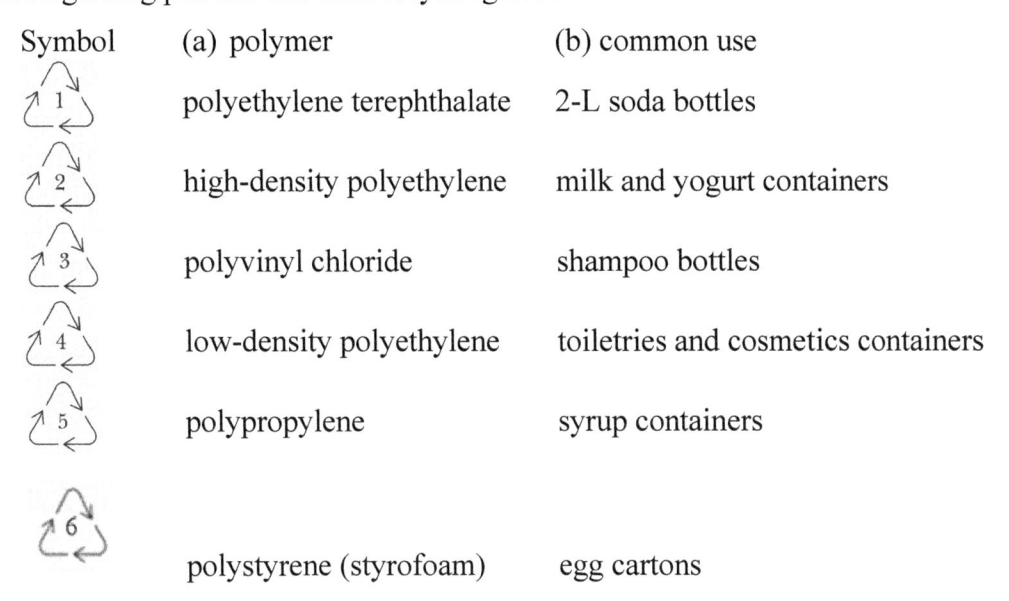

Symbol	(a) polymer	(b) common use
1	polyethylene terephthalate	2-L soda bottles
2	high-density polyethylene	milk and yogurt containers
3	polyvinyl chloride	shampoo bottles
4	low-density polyethylene	toiletries and cosmetics containers
5	polypropylene	syrup containers
6	polystyrene (styrofoam)	egg cartons

(c) Assuming the plastic containers are crushed so they will not retain air, the PET plastics should sink in a water bath. HDPE and PP will float and can be skimmed off. By heating this mixture to approximately 140 °C, the HDPE will melt, allowing separation.

23.129. Regarding maleic acid:

(a) Combustion of 0.125 g (125 mg) of maleic acid gives 0.190 g (190. mg) of CO_2 and .0388 g (38.8 mg) of H_2O. The empirical formula of maleic acid:

The combustion of maleic acid (unbalanced) can be represented as: $C_xH_yO_z + O_2 \rightarrow CO_2 + H_2O$

It's important to note that while **all** the C in CO_2 originated in the maleic acid and **all** the H in H_2O originated in the maleic acid—**not all** of the oxygen originates in the maleic acid. So, we begin by calculating the masses of C and H, and subtracting those masses from the 0.125 g of compound to determine the mass of O present in maleic acid.

$$0.190 \text{ g } CO_2 \left(\frac{12.01 \text{ g C}}{44.01 \text{ g } CO_2} \right) = 0.0518 \text{ g C}$$

$$0.0388 \text{ g H}_2\text{O}\left(\frac{2.016 \text{ g H}}{18.02 \text{ g H}_2\text{O}}\right) = 0.00434 \text{ g H}$$

Mass O = 0.125 g compound – (0.0518 g C + 0.00434 g H) = 0.069 g O

$$0.0518 \text{ g C}\left(\frac{1 \text{ mol C}}{12.01 \text{ g C}}\right) = 0.00432 \text{ mol C}$$

Now we can calculate the # of moles of each of these atoms:

$$0.00434 \text{ g H}\left(\frac{1 \text{ mol H}}{1.008 \text{ g H}}\right) = 0.00431 \text{ mol H}$$

$$0.069 \text{ g O}\left(\frac{1 \text{ mol O}}{16.00 \text{ g O}}\right) = 0.0043 \text{ mol O}$$

Note that the **ratio** of C, H, and O present in maleic acid indicates equal numbers of C, H, and O atoms. The empirical formula is then $C_1H_1O_1$—or CHO.

(b) 0.261 g of maleic acid requires 34.60 mL of 0.130 M NaOH. What is the molecular formula of maleic acid:

The text indicates that there are 2 carboxylic acid groups per molecule of maleic acid.

The titration indicates: $0.03460 \text{ L}\left(\dfrac{0.130 \text{ mol NaOH}}{1 \text{ L}}\right) = 0.004498 \text{ mol NaOH}$

Since 1 mol of NaOH reacts with 1 mol of an acidic group (a –COOH group), we know that there are 0.002249 mol of maleic acid present.

The molecular weight of maleic acid is: 0.261 g maleic acid/0.002249 mol maleic acid or 116 g/mol. The empirical formula (CHO) would have a formula weight of (12.0 + 1.01 + 16.0) = 29.0 g. The molecular formula is:

$$\frac{116 \text{ g/mol molecular formula}}{29.0 \text{ g/mol empirical formula}} = 4.00 \text{ empirical formulas/molecular formula}$$

or a molecular formula of $C_4H_4O_4$.

(c) A Lewis structure for maleic acid:

$$\text{H–}\ddot{\text{O}}\text{–C–C=C–C–}\ddot{\text{O}}\text{–H}$$

(with =O above each terminal C, and H attached to each middle C)

More experiments would need to be performed to determine that the *cis-* configuration is correct for maleic acid.

(d) Hybridization used by C atoms: All four C atoms are attached to three other groups (including a double bond), making the hybridization of all sp^2.

(e) Bond angles around each carbon are 120°.

Solution and Answer Guide

Kotz Treichel Townsend Treichel, Chemistry and Chemical Reactivity 11e, 978-0-357-85140-1, Chapter 24: Biochemistry

TABLE OF CONTENTS

Applying Chemical Principles	901
Practicing Skills	902
General Questions	913
In the Laboratory	915
Summary and Conceptual Questions	916

Applying Chemical Principles

Polymerase Chain Reaction

24.1.1. # of copies of double-stranded DNA would be present after 20 cycles of PCR?

As noted in the Figure after each cycle, the number of copies doubles:

DNA original → 2 copies (after 1 cycle)

2 copies DNA → 4 copies (after 2 cycles)

4 copies DNA → 8 copies (after 3 cycles)

The # of copies following n cycles is then: 2^n. After 20 cycles, there are 2^{20} copies of DNA or 1,048,576 copies.

24.1.2. Explain why forces holding DNA molecules are not covalent bonds:

Covalent bonds would need more energy to break than the thermal energy present at 95 °C. The weaker hydrogen bonds are easily broken at this temperature.

24.1.3. Regarding primers to use in PCR:
(a) Primers that have high numbers of cytosines and guanines?

Primers initially bond to the strand of DNA to be replicated. Cytosine and guanine have 3 possible sites for bonding (H-bonding) to the complementary bases on DNA. Uracil, adenine, and thymine have but 2 possible sites. Use of cytosine and guanine provide for stronger bonding. [There are also additional reasons for stronger binding between G and C than between A and T (or U) than just the number of hydrogen bonds per base pair, but these go beyond the level of this course.]

(b) Since C bonds to G, a long primer sequence of C's followed by G's would potentially fold over and have the C's and G's of the primer bonded to itself—instead of bonding to the DNA.

(c) A large excess of primers favors the bonding of the primers to the target (DNA) strands instead of the target strands bonding to each other.

Practicing Skills

Proteins

24.1. Lewis structures for:

(a) Isoleucine with amino and carboxyl groups in un-ionized form:

(b) Isoleucine in zwitterionic form:

(c) At physiological pH, the zwitterion predominates.

24.2. Concerning tryptophan:

(a) structure with un-ionized amino and carboxylic acid groups:

(b) zwitterionic form:

(c) At physiological pH, the zwitterion predominates.

24.3. Classify the amino acids as having a polar or nonpolar R group:
See Figure 24.2 for structures of the specific amino acids:
Polar R groups: serine, lysine, aspartic acid
Nonpolar R groups: alanine, leucine, phenylalanine,

24.4. Classify the amino acids as having a polar or nonpolar R group:
Polar: arginine, cysteine, glutamine, threonine
Nonpolar: isoleucine, tryptophan

24.5. Two different ways in which glycine and alanine may be combined:

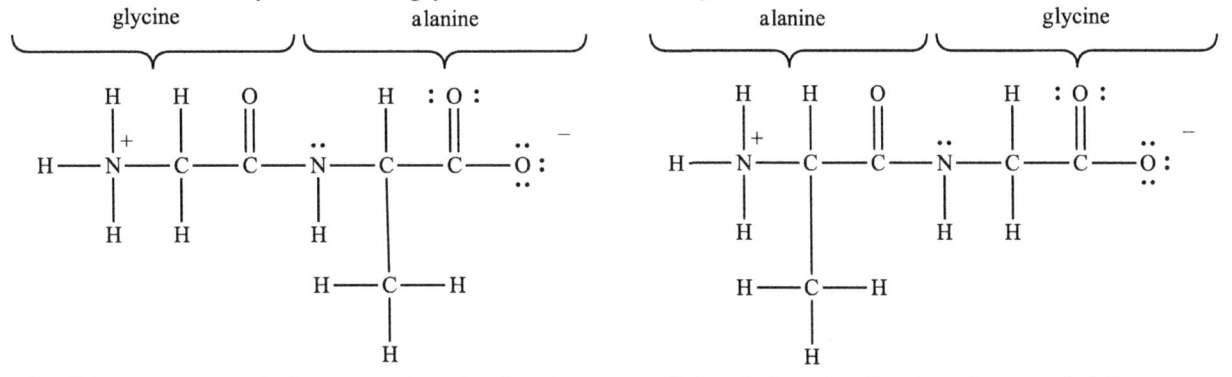

[Structural diagrams of glycine-alanine and alanine-glycine dipeptides]

The differences result from varying the "amino group" bonded to the "carboxyl group". That is, in the first molecule the carboxyl group of glycine is bonded to the amino group of alanine, whereas in the second the carboxyl group of alanine is bonded to the amino group of glycine.

24.6. Do the two sequences represent the same compound:
No, for valine-asparagine, the carboxyl group of valine and the amino group of asparagine are used to make the amide bond between the two peptides. For asparagine-valine, the amino group of valine and the carboxyl group of asparagine are used to make the amide bond between the two peptides. This results in two different molecules.

24.7. Lewis structure for the tripeptide: serine-leucine-valine:

[Structural diagram of serine-leucine-valine tripeptide]

24.8. Lewis structure for the tripeptide: tyrosine-histidine-glycine:

tyrosine histidine glycine

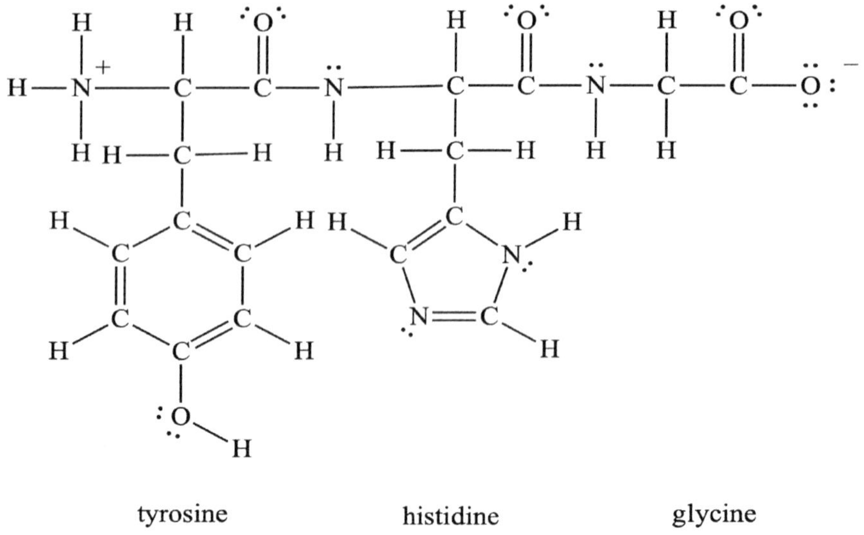

24.9. Structure corresponding to:

(a) Type of structure is the amino acid sequence: primary, since the primary structure shows the structural composition of the amino acid chain.

(b) Arrangement of different peptide chains with respect to each other: quaternary, since the quaternary structure concerns how different chains interact.

(c) Type of structure that refers to how the chain is folded: tertiary

(d) Structure arising from H-bonding within a chain: secondary, since secondary structure refers to the H-bonding network formed between amide linkages in the protein backbone.

24.10. Describe the type of structure shown in the figure:
The structure shown in the figure is secondary protein structure.

Carbohydrates

24.11. Structural formula for α-D glucose and β-D glucose:

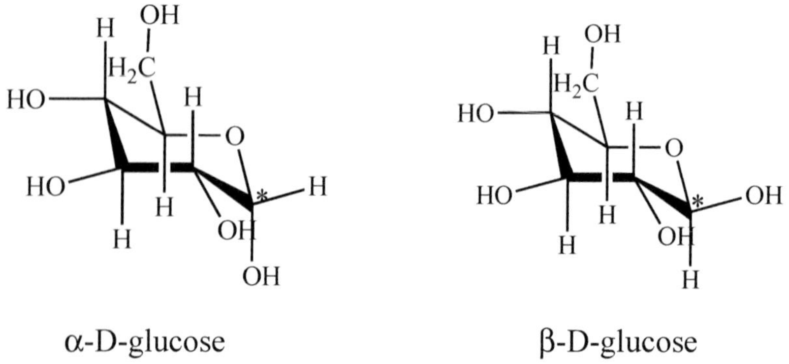

α-D-glucose   β-D-glucose

When the ring closes, if the OH is "up" we have the beta form, and if the OH is "down" we have the alpha form. The asterisk indicates the C at which the ring closes.

24.12. Glucose molecules joined by β-1,4 linkage:

(This disaccharide is called cellobiose.)

24.13. Two simplest monosaccharides:

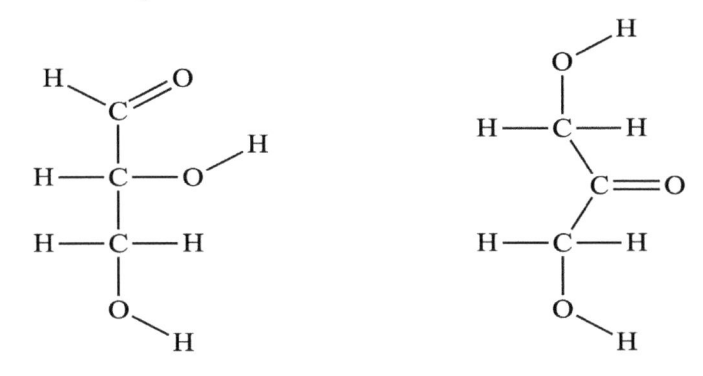

D-glyceraldehyde   dihydroxyacetone

24.14. Disaccharide components and linkages:

Sucrose is made of glucose and fructose joined by an α,β-1,2 glycosidic bond.

Lactose is made of galactose and glucose joined by a β-1,4 glycosidic bond.

Maltose is made of two glucose units joined by an α-1,4 glycosidic bond.

24.15. Difference between amylose starch and cellulose:
Amylose is a human digestible starch made of linear chains of glucose units connected by an α-1,4 glycosidic bonds.

Cellulose is not digestible by humans because it is made of linear chains of glucose units connected by β-1,4 glycosidic bonds.

The reason cellulose is not digestible is that humans do not have enzymes capable of breaking the β-1,4 glycosidic bonds in cellulose.

24.16. Difference between amylopectin starch and glycogen:

Both amylopectin starch and glycogen are polysaccharides made of glucose units connected by α-1,4 glycosidic bonds with branches connected to the main chain with α-1,6 glycosidic bonds. The major difference is that the branches in glycogen occur every 10 or so glucose units. Branches in amylopectin starch occur every 30 or so glucose units. Amylopectin starch is found in plants and glycogen is present in animals.

Nucleic Acids

24.17. For the sugar β-D-ribose:

(a) The structural formula for the sugar β-D-ribose:

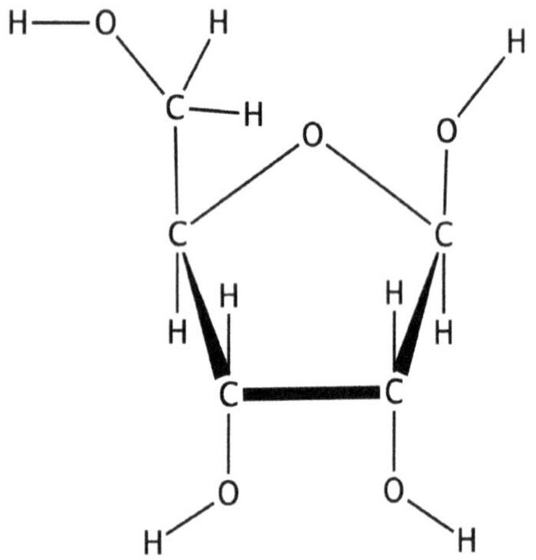

(b) Structural formula for adenosine:

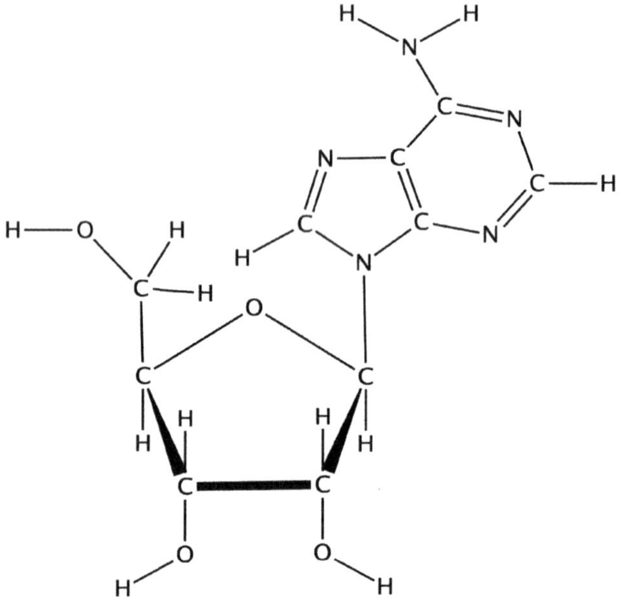

Note that for the nucleoside the adenine (base) is attached to the ribose ring. The attachment occurs via the formation of water (H- from the adenine moiety, and OH from the ribose molecule.)

(c) Structural formula for nucleotide: adenosine 5′-monophosphate.

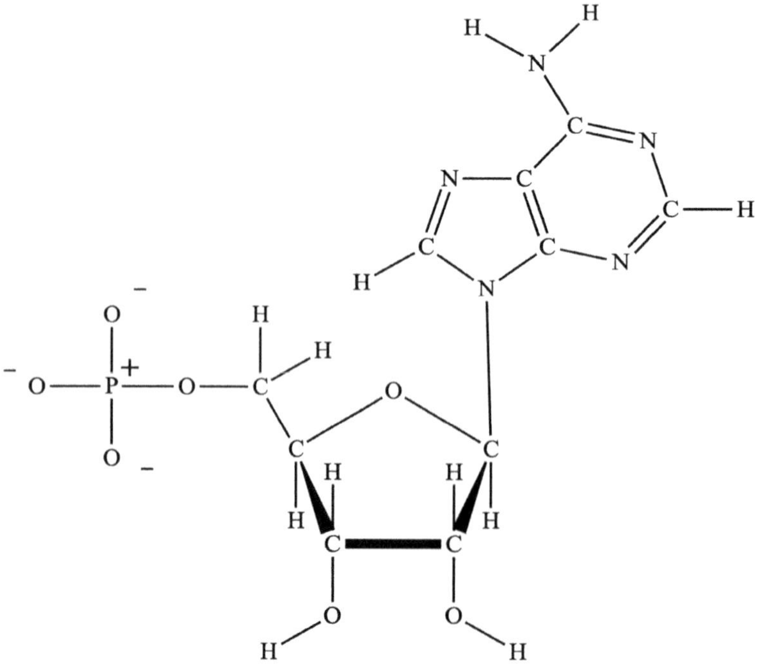

Note that the phosphate group is attached to the 5′ C.

24.18. For the sugar β-D-2-deoxyribose:

(a) sugar β-D-2-deoxyribose:

(b) structure for nucleoside deoxyadenosine:

(c) structure for nucleotide deoxyadenosine-5′-monophosphate

24.19. Structural formula for the tetranucleotide AUGC:

24.20. Structural formula for the tetradeoxynucleotide CGTA:

24.21. For the nucleotide sequence in DNA: 5′-ACGCGATTC-3′:
(a) The sequence of the complementary strand would be found by noting that every A has a T as a complement; every C a G. Recall that the 5′ end of a DNA strand pairs with the 3′ end of its complement. So if we write the complement to the sequence above, we'd get: 3′-TGCGCTAAG-5′. Since the two strands of DNA have 5′ ends paired with 3′ ends, we write the complement as the reverse of the listing above, namely: 5′-GAATCGCGT-3′.
(b) Write the sequence 5′ to 3′ for a strand of m-RNA to complement the original strand of DNA. For RNA, thymine (in DNA) is replaced with uracil (in RNA). Hence taking the complementary strand from part (a), we replace every T with U: 5′-GAAUCGCGU-3′
(c) Beginning with the m-RNA strand shown in part (b), we now code each "letter" with its complement, so 5′-GAA-3′ has the complement 3′-CUU-5′, 5′-UCG-3′ has the complement 3′-AGC-5′, and 5′-CGU-3′ has the complement 3′-GCA-5. Rewriting the anticodons in order from the 5′-ends to the 3′-ends:
5′-UUC-3′; 5′-CGA-3′; 5′-ACG 3′
(d) The sequence of amino acids coded by the three codons on mRNA (part(b)):
GAA codes for Glu (glutamic acid)
UCG codes for Ser (serine)
CGU codes for Arg (arginine)

24.22. Regarding the nucleotide structure in DNA:
(a) Complementary strand: 5′-ATCCTACGA-3′
(b) mRNA strand complementary to the original strand: 5′-AUCCUACGA-3′
(c) sequences for the anticodons: 5′-GAU-3′, 5′-UAG-3′, 5′-UCG-3′
(d) sequence of amino acids selected by this mRNA: isoleucine, leucine, arginine

Lipids and Cell Membranes

24.23 Phospholipid bilayer:

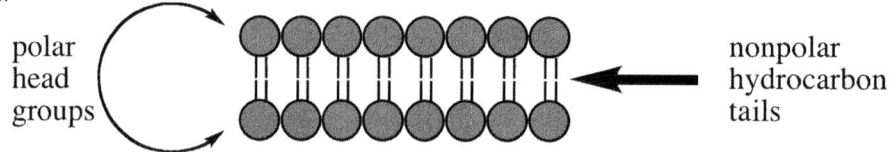

polar head groups — nonpolar hydrocarbon tails

24.24. Regarding oleic acid:
(a) Lewis structure for oleic acid:

(b) nonpolar / polar

(c) The polar end of the molecule points into the water, with the nonpolar end pointing out of the water.

24.25. The structure that all steroids have in common is:

There are three six-membered rings and a five-member ring. As is fairly common in larger organic molecules, these geometric structures are assumed to have C atoms at each vertex.

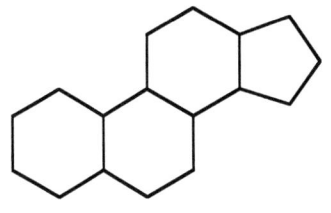

24.26. Regarding the structures of the sex hormones:

(a) Compare structures: Testosterone has an additional methyl group on carbon 10 of the ring structure that makes ring A no longer aromatic. Due to this change, the two carbon atoms to the left of this also each have two hydrogen atoms attached in testosterone instead of one as in estrogen. Testosterone also has a carbonyl group on ring A, whereas estrogen has a hydroxyl group at this point.

(b) Both are steroids as they possess the basic steroid ring structure.

Metabolism

24.27. Calculate the $\Delta_r H°$ for the oxidation of glucose:

$C_6H_{12}O_6(s) + 6\ O_2(g) \rightarrow 6\ CO_2(g) + 6\ H_2O(\ell)$

$\Delta_r H° = [6\ \Delta_f H°\ CO_2(g) + 6\ \Delta_f H°\ H_2O(\ell)] - [1\ \Delta_f H°\ C_6H_{12}O_6(s) + 6\ \Delta_f H°\ O_2(g)]$

$= [(6\ \text{mol}\ CO_2/\text{mol-rxn})(-393.509\ \text{kJ/mol}\ CO_2) + (6\ \text{mol}\ H_2O/\text{mol-rxn})(-285.83\ \text{kJ/mol}\ H_2O)] - [(1\ \text{mol}\ C_6H_{12}O_6/\text{mol-rxn})(-1273.3\ \text{kJ/mol}\ C_6H_{12}O_6) + (6\ \text{mol}\ O_2/\text{mol-rxn})(0\ \text{kJ/mol}\ O_2)]$

$\Delta_r H° = [-2361.054\ \text{kJ/mol-rxn} + -1714.98\ \text{kJ/mol-rxn}] - [-1273.3\ \text{kJ/mol-rxn}]$
$= -4076.03\ \text{kJ/mol-rxn} - (-1273.3\ \text{kJ/mol-rxn}) = -2802.7\ \text{kJ/mol-rxn}$. The reported value is $-2803\ \text{kJ/mol-rxn}$.

24.28. Calculate $\Delta_r H°$ for the conversion of glucose into ethanol:

$\Delta_f H°[C_6H_{12}O_6(s)] = -1273.3$ from problem 24.27

$\Delta_r H° = [2\ \Delta_f H°\ C_2H_5OH(\ell) + 2\ \Delta_f H°\ CO_2(g)] - \Delta_f H°\ C_6H_{12}O_6(s)$

$\Delta_r H° = [(2\ \text{mol}\ C_2H_5OH/\text{mol-rxn})(-277.0\ \text{kJ/mol}\ C_2H_5OH) + (2\ \text{mol}\ CO_2/\text{mol-rxn})(-393.509\ \text{kJ/mol}\ CO_2)] - [(1\ \text{mol}\ C_6H_{12}O_6/\text{mol-rxn})(-1273.3\ \text{kJ/mol}\ C_6H_{12}O_6)]$

$\Delta_r H° = -67.7\ \text{kJ/mol}$

24.29. For the reaction: $NADH + H^+ + ½\ O_2 \rightarrow NAD^+ + H_2O$

(a) Species undergoing oxidation: The loss of H is one metric for defining oxidation, so NADH undergoes oxidation. (From Figure 24.26, you can see that the oxidation of NADH involves the loss of two electrons.)

(b) Species undergoing reduction: O_2 is gaining H during the process, so O_2 undergoes reduction. (The oxidation number of each oxygen changes from 0 in O_2 to -2 in H_2O, indicating a gain of two electrons.)

(c) Species functioning as oxidizing agent: As O₂ is removing the H from NADH, O₂ is functioning as the oxidizing agent.

(d) Species functioning as reducing agent: As NADH is donating the H to oxygen, NADH is functioning as the reducing agent.

24.30. For the conversation of ethanol to acetaldehyde:

CH₃CH₂OH is oxidized—as it loses H (in addition, the average oxidation number of each carbon atom in ethanol (C₂H₆O) is –2, whereas it is –1 in acetaldehyde (C₂H₄O), indicating a loss of one electron per carbon atom and thus oxidation occurring).

NAD⁺ is reduced—as it gains H (the conversion of NAD⁺ to NADH involves the gain of two electrons, reduction).

24.31. Moles of ATP formed from one mole of glucose:

If the process were 100% efficient all the free energy from glucose oxidation would form ATP. Thus, the maximum number of moles of ATP formed is $\frac{2850 \text{ kJ/mol-rxn}}{30.5 \text{ kJ/mol-rxn}} = 93.4 \approx 93$ moles of ATP formed. Since only about 32 moles of ATP are formed per glucose oxidation, the process is less than 100% efficient. It is actually $\left(\frac{32}{93}\right)100 = 34\%$ efficient.

24.32. (a) Overall balanced equation and cell potential for reaction involving half-cell reactions:

The half-reaction with the greater reduction potential is predicted to occur as the reduction, and that with the smaller reduction potential is predicted to occur as the oxidation in the reaction that is product-favored at equilibrium.

$1/2\ O_2 + 2H^+ + 2e^- \rightarrow H_2O$

$QH_2 \rightarrow Q + 2H^+ + 2e^-$

$1/2\ O_2 + QH_2 \rightarrow Q + H_2O$

(b) $E°'_{cell} = E°'(\text{cathode}) - E°'(\text{anode}) = 0.815 \text{ V} - 0.415 \text{ V} = 0.770 \text{ V}$

(c) Value of $\Delta_r G°'$ for this reaction:

$\Delta_r G°' = -nFE°'_{cell}$

$= -\left(\frac{2 \text{ mol e}^-}{1 \text{ mol-rxn}}\right)(96{,}485 \text{ C/mol e}^-)(0.770 \text{ V})$

$= -1.49 \times 10^5 \frac{\text{C} \cdot \text{V}}{\text{mol-rxn}} = -1.49 \times 10^5 \frac{\text{J}}{\text{mol-rxn}} = -149 \frac{\text{kJ}}{\text{mol-rxn}}$

(d) Positive value of $E°'_{cell}$ and negative value of $\Delta_r G°'$ both confirm that the reaction is product-favored at equilibrium.

General Questions

24.33. Two Lewis structures for the dipeptide alanine-isoleucine that show the resonance structures of the amide linkage. Resonance structures are a result of shifting pairs of electrons, in this case, the lone pair of electrons from N to create the C=N bond.

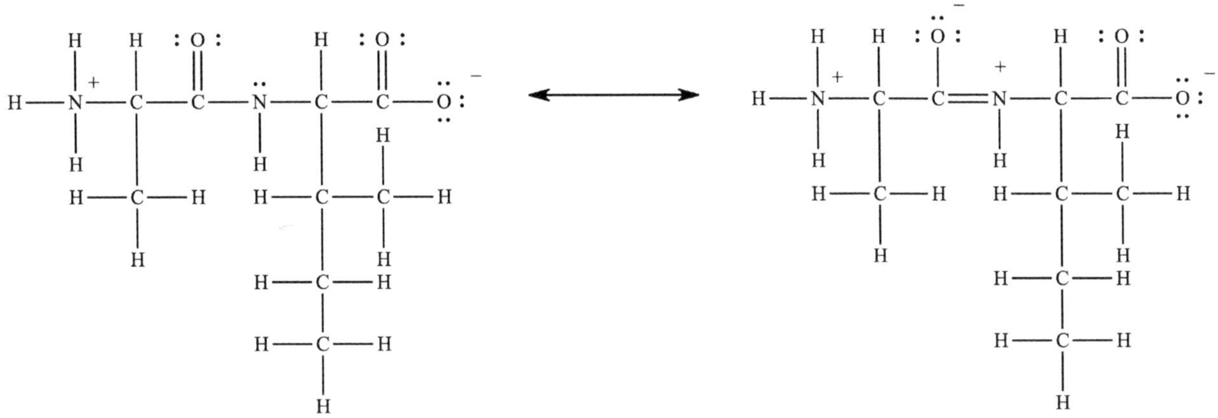

The lone pair of electrons from N is shifted to the C atom, and the C=O electrons are shifted to a "lone pair" position on O. The resulting O bears a "–" charge, while the N bears a "+" charge.

24.34. The equilibrium constant for the conversion of α-D-glucose to β-D-glucose:

K = [β-D-glucose]/[α-D-glucose] = 0.63/0.37 = 1.7

24.35. Regarding the production of glucose by photosynthesis:

(a) The enthalpy change for the production of one mole of glucose at 25°C:

The process may be represented: $6\ CO_2(g) + 6\ H_2O(\ell) \rightarrow C_6H_{12}O_6(s) + 6\ O_2(g)$

The enthalpy change is:

$\Delta_rH°$ = [$\Delta_fH°$ $C_6H_{12}O_6(s)$ + 6·$\Delta_fH°$ $O_2(g)$] − [6·$\Delta_fH°$ $CO_2(g)$ + 6·$\Delta_fH°$ $H_2O(\ell)$]

= [(1 mol $C_6H_{12}O_6$/mol-rxn)(−1273.3 kJ/mol $C_6H_{12}O_6$) + (6 mol O_2/mol-rxn)·(0 kJ/mol O_2))] − [(6 mol CO_2/mol-rxn)(−393.509 kJ/mol CO_2) + (6 mol H_2O/mol-rxn)(−285.83 kJ/mol H_2O)]

$\Delta_rH°$ = 2803 kJ for 1 mol of glucose.

(b) The enthalpy change for 1 molecule of glucose is found by dividing the energy change found in (a) by Avogadro's number:

$\dfrac{2803\ kJ}{1\ mol\ glucose}\left(\dfrac{1\ mol\ glucose}{6.022\times 10^{23}\ molecules\ glucose}\right)$ = 4.654×10^{-21} kJ/molecule = 4.654 × 10^{-18} J/molecule

(c) What is the energy of a photon of light with a wavelength of 650 nm?

To calculate the energy of one photon of light with 650 nm wavelength, first calculate the frequency of the radiation:

$$\nu = \frac{c}{\lambda} = \frac{2.9979 \times 10^8 \text{ m/s}}{650 \text{ nm}}\left(\frac{1 \times 10^9 \text{ nm}}{1 \text{ m}}\right) = 4.61 \times 10^{14} \text{ s}^{-1}$$

$$E = h\nu = \left(6.626 \times 10^{-34} \frac{\text{J} \cdot \text{s}}{\text{photon}}\right)(4.61 \times 10^{14} \text{ s}^{-1}) = 3.1 \times 10^{-19} \frac{\text{J}}{\text{photon}}$$

(d) One molecule of glucose requires 4.654×10^{-18} J. One photon with wavelength 650 nm has 3.1×10^{-19} J of energy, one photon is **insufficient** to cause the production of one molecule of glucose.

24.36. Regarding amino acids selected:

(a) Amino acid selected by the mRNA codon GAA: glutamic acid

(b) Sequence in original DNA that led to the codon being present: 5′-TTC-3′

(c) The DNA sequence would now be 5′-TGC-3′. This would lead to an mRNA codon of 5′-GCA-3′. This codes for the amino acid alanine.

24.37. Number of possible codons: There are 64, resulting from there being 4 bases grouped in units of 3 (the codons), so we have (4 bases) · (4 bases) · (4 bases) = 4^3 = 64.

An example may help answer the remaining part of this question. Suppose the triplet codon we have is UUU. That codon ALWAYS means that phenylalanine is the amino acid that is associated with that grouping of bases. The codon UUC also ALWAYS means that phenylalanine is the amino acid that is associated with that group of bases. So MORE THAN ONE CODON is associated with a SPECIFIC AMINO ACID (here UUC and UUU are both associated with phenylalanine). However, each codon ALWAYS HAS a specific amino acid. Another way to look at this is that there are more possible codons than are required to code for 20 amino acids, so more than one codon may be used for a given amino acid.

24.38. Nucleotide length to provide one sequence for each person on earth:

$4^x = 8 \times 10^9$

Taking the log of each side: $x(\log 4) = \log(8 \times 10^9)$; $x = 16.4$, so 17 nucleotides needed

24.39. For the reaction associated with respiration:

$C_6H_{12}O_6(s) + 6\ O_2(g) \rightarrow 6\ CO_2(g) + 6\ H_2O(\ell)$, we see that O_2 is gaining H atoms, so O_2 is being reduced. $C_6H_{12}O_6$ is losing H atoms to O, and is therefore oxidized. Another way to look at this is that the O atoms are going from an oxidation number of 0 in O_2 to –2 in H_2O. This is a reduction in oxidation number–a reduction. The average oxidation number of each carbon atom goes from 0 in glucose to +4 in CO_2. This is an increase in oxidation number– an oxidation.

24.40. Sum of the two equations, and the $\Delta_rG^{o\prime}$ for the coupled reactions:

Glucose + P$_i$ → Glucose-6-phosphate + H$_2$O $\Delta_rG^{o\prime}$ = +13.8 kJ/mol-rxn

ATP + H$_2$O → ADP + P$_i$ $\Delta_rG^{o\prime}$ = −30.5 kJ/mol-rxn

Overall Glucose + ATP → Glucose-6-phosphate + ADP $\Delta_rG^{o\prime}$ = −16.7 kJ/mol-rxn

The process is product-favored at equilibrium as $\Delta_rG^{o\prime} < 0$.

In the Laboratory

24.41. The table containing the data and the necessary manipulations of that data is below:

[S] mol/L × 10^5	Rate mol/L·min × 10^5	1/[S]	1/Rate
1.0	0.63	1.0	1.587
1.4	0.88	0.71	1.136
2.0	1.2	0.50	0.833
5.0	2.6	0.20	0.385
6.7	3.1	0.15	0.323

The graph of 1/[S] vs 1/Rate is below:

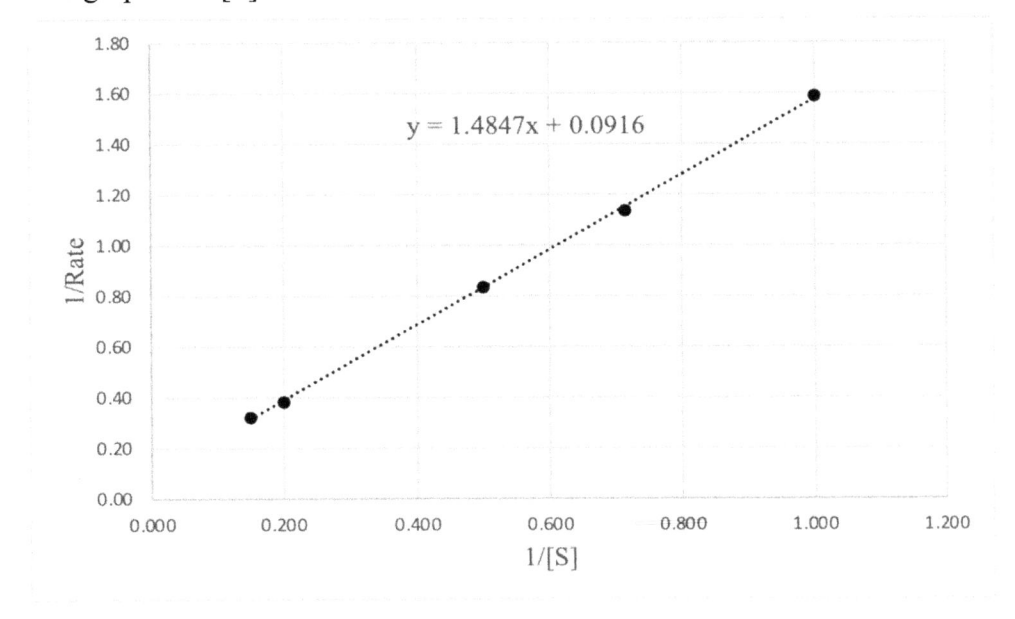

y = 1.4847x + 0.0916

The equation for the straight line indicates an intercept (1/[S]= 0) of 0.0916. For simplicity of graphing, we multiplied the Rate by the factor 10^5. So the intercept = 0.0916×10^5. Since this corresponds to the 1/Rate, then Rate = 1/916 = 1.1×10^{-4} mol/L · min.

24.42. Molar mass of insulin:

Express the pressure in units of mm Hg: $21.8 \text{ mm H}_2\text{O} \left(\dfrac{1 \text{ mm Hg}}{13.6 \text{ mm H}_2\text{O}} \right) = 1.60$ mm Hg

Convert pressure to units of atmospheres: $1.60 \text{ mm Hg} \left(\dfrac{1 \text{ atm}}{760 \text{ mm Hg}} \right) = 0.00211$ atm

Osmotic Pressure Equation: $\Pi = cRT$; solving for c: $c = \dfrac{\Pi}{RT}$

$c = \dfrac{\Pi}{RT} = \dfrac{0.00211 \text{ atm}}{\left(0.082057 \dfrac{\text{L} \cdot \text{atm}}{\text{K} \cdot \text{mol}} \right)(298 \text{ K})} = 8.63 \times 10^{-5}$ M

$8.63 \times 10^{-5} \dfrac{\text{mol}}{\text{L}} = \dfrac{\dfrac{50.0 \times 10^{-3} \text{ g}}{\text{molar mass}}}{0.100 \text{ L}}$; and molar mass = 5.80×10^3 g/mol

Summary and Conceptual Questions

24.43. Amino acid that isn't chiral: As glycine has two H atoms attached to the alpha carbon, glycine isn't chiral. Recall that chiral carbons MUST HAVE 4 DIFFERENT groups attached—and with 2 H atoms, glycine doesn't qualify.

24.44. Regarding DNA's double-helical strand:

(a) Hydrogen bonds hold the strand together.

(b) In order for replication to take place, the two strands must be able to separate from each other. If they were joined by covalent bonds, it would take much more energy to separate them. Normal conditions would not provide this much energy, and other bonds in the molecule might also break with this input of energy.

24.45. Do DNA sequences ATGC and CGTA represent the same molecule?

As one can see, while the individual pieces are the same, the assembly does not result in the same molecular structure—the two are **different**.

24.46. Regarding the reliability of DNA replication:

(a) Fraction of molecules produced with correct nucleotide sequence: $0.95^{10} = 0.60$

(b) Fraction of molecules in 10-nucleotide sequence produced with correct nucleotide sequence with greater accuracy: $0.99999999^{10} = 0.99999990$

24.47. Describe the transcription and translation processes:

(a) Describe what occurs in the process of transcription:

The information contained in DNA is "transcribed" by the process of transcription. The process is NOT a straight copy, but "an exchange", in which complementary nitrogen bases appear in the "product"---mRNA. So when DNA has an "A", that "A" is transcribed into mRNA as U, a "C" is transcribed as "G", a "G" as "C: So a sequence in DNA that is 5′-CGCAA-3′ is transcribed into mRNA as 3′-GCGUU-5′.

(b) Describe what occurs in the process of translation: The information that mRNA contains is "decoded" by tRNA. That decoding results in the formation of an amino acid sequence (much as we did in 21(d) above).

24.48. Which of the statements are true:

(a) Breaking P–O bond in ATP is an exothermic process: false

(b) Making bond between P and OH of water is exothermic process: true

(c) Breaking bonds is an endothermic process: true

(d) Energy released in hydrolysis of ATP may be used to run endothermic cell reactions: true

Solution and Answer Guide

Kotz Treichel Townsend Treichel, Chemistry and Chemical Reactivity 11e, 978-0-357-85140-1, Chapter 25: Environmental Chemistry Earth's Environment, Energy, and Sustainability

TABLE OF CONTENTS

Applying Chemical Principles	919
Practicing Skills	921
General Questions	934
In the Laboratory	937
Summary and Conceptual Questions	938

Applying Chemical Principles

Chlorination of Water Supplies

25.1.1. For the reaction of Cl_2 in H_2O to form HCl and HOCl: $Cl_2 + H_2O \rightarrow HCl + HOCl$

(a) The reaction is a disproportionation since Cl_2 changes oxidation number from 0 to –1 (in HCl) and from 0 to +1 (in HOCl)

(b) The $E°_{cell}$ and K for the reaction:

Equations of interest: (1) $2 HClO + 2 H^+ + 2 e^- \rightarrow Cl_2 + 2 H_2O \quad E° = +1.63$ V

(2) $Cl_2 + 2 e^- \rightarrow 2 Cl^- \quad E° = +1.36$ V

Reverse eq(1): $Cl_2 + 2 H_2O \rightarrow 2 HClO + 2 H^+ + 2 e^-$

Add eq(2) $\quad Cl_2 + 2 e^- \rightarrow 2 Cl^-$

Net equation: $2 Cl_2 + 2 H_2O \rightarrow 2 HClO + 2 H^+ + 2 Cl^-$ for which

$E°_{cell} = E°_{cathode} - E°_{anode} = (+1.36 V - +1.63 V) = -0.27$ V

And using the Gibbs Free Energy equation: $\Delta G = \Delta G° - RT \ln K$ and $\Delta G° = -nFE°$

Solving for $\ln K$:

$$\ln K = \frac{-nFE°_{cell}}{-RT} = \frac{-2 \text{ mol e}^- (96{,}500 \text{ C/mol e}^-)(-0.27 \text{ V})}{(-8.3145 \text{ J/mol} \cdot \text{K})(298.15 \text{ K})} = -21$$

$K = e^{-21} = 7.4 \times 10^{-10}$

(c) How is E affected by changing pH?

As H^+ is a product of the reaction, increasing pH would favor the products more and decreasing pH would favor the reactants more!

25.1.2. The overall K is calculated in (b) above: 7.4×10^{-10}.

25.1.3. For a solution that is 0.010 M OCl⁻, what is the concentration of HOCl and what is the pH:

For OCl⁻ acting as a base the reaction is: $OCl^- + H_2O \rightarrow HOCl + OH^-$

The K_b for OCl⁻ is:

$$K_b = \frac{1.0 \times 10^{-14}}{3.5 \times 10^{-8}} = 2.86 \times 10^{-7}$$

The equilibrium is:

$$\frac{[HOCl][OH^-]}{[OCl^-]} = \frac{x^2}{0.010 - x} = 2.86 \times 10^{-7}$$

With $K_b = 2.86 \times 10^{-7}$, the value of x is small compared to 0.010, and the denominator can be approximated as 0.010.

$$\frac{x^2}{0.010} = 2.9 \times 10^{-7}; \; x^2 = 2.9 \times 10^{-9} \text{ and } x = 5.4 \times 10^{-5};$$

so [HOCl] = [OH⁻] = 5.4×10^{-5}; pOH = 4.27, and pH = 9.73

25.1.4. Reaction between H_2O_2 and ClO⁻ to give Cl⁻ and O_2:

$H_2O_2(aq) + ClO^-(aq) \rightarrow Cl^-(aq) + O_2(g) + H_2O(\ell)$

25.1.5. Warning about mixing NH_3 and HOCl warranted?

Ammonia and hypochlorous acid react to form chloramine according to the equation:

$HClO(aq) + NH_3(aq) \rightarrow NH_2Cl(g) + H_2O(\ell)$. Chloramine gas ($NH_2Cl$) is reported to be toxic.

Hard Water

25.2.1. A solution contains 50 ppm Mg^{2+} and 150 ppm Ca^{2+}. What mass of solid should result if CaO is added to precipitate $MgCO_3$ and $CaCO_3$?

Amount of CaO needed to precipitate $CaCO_3$ and $MgCO_3$:

$$1.0 \text{ L} \left(\frac{50. \text{ mg Mg}^{2+}}{1.0 \text{ L}} \right) \left(\frac{1.00 \text{ g Mg}^{2+}}{1000 \text{ mg Mg}^{2+}} \right) \left(\frac{1 \text{ mol Mg}^{2+}}{24.305 \text{ g Mg}^{2+}} \right) \left(\frac{1 \text{ mol CaO}}{1 \text{ mol Mg}^{2+}} \right) \left(\frac{56.08 \text{ g CaO}}{1 \text{ mol CaO}} \right) = 0.115 \text{ g CaO}$$

$$1.0 \text{ L} \left(\frac{150. \text{ mg Ca}^{2+}}{1.0 \text{ L}} \right) \left(\frac{1.00 \text{ g Ca}^{2+}}{1000 \text{ mg Ca}^{2+}} \right) \left(\frac{1 \text{ mol Ca}^{2+}}{40.078 \text{ g Ca}^{2+}} \right) \left(\frac{1 \text{ mol CaO}}{1 \text{ mol Ca}^{2+}} \right) \left(\frac{56.08 \text{ g CaO}}{1 \text{ mol CaO}} \right) = 0.210 \text{ g CaO}$$

Total mass of oxides: 0.115 + 0.210 = 0.325 g CaO or to 2 sf, 330 mg CaO.

We get 2 moles of CaCO₃ per mole of Ca²⁺, 1 mol of CaCO₃ and 1 mol of MgCO₃ per mole of Mg²⁺.

CaCO₃ from Mg²⁺ reaction:

$$1.0 \text{ L} \left(\frac{50. \text{ mg Mg}^{2+}}{1.0 \text{ L}} \right) \left(\frac{1.00 \text{ g Mg}^{2+}}{1000 \text{ mg Mg}^{2+}} \right) \left(\frac{1 \text{ mol Mg}^{2+}}{24.305 \text{ g Mg}^{2+}} \right) \left(\frac{1 \text{ mol CaCO}_3}{1 \text{ mol Mg}^{2+}} \right) \left(\frac{100.1 \text{ g CaCO}_3}{1 \text{ mol CaCO}_3} \right) = 0.206 \text{ g CaCO}_3$$

CaCO₃ from Ca²⁺ reaction:

$$1.0 \text{ L} \left(\frac{150. \text{ mg Ca}^{2+}}{1.0 \text{ L}} \right) \left(\frac{1.00 \text{ g Ca}^{2+}}{1000 \text{ mg Ca}^{2+}} \right) \left(\frac{1 \text{ mol Ca}^{2+}}{40.078 \text{ g Ca}^{2+}} \right) \left(\frac{2 \text{ mol CaCO}_3}{1 \text{ mol Ca}^{2+}} \right) \left(\frac{100.1 \text{ g CaCO}_3}{1 \text{ mol CaCO}_3} \right) = 0.749 \text{ g CaCO}_3$$

MgCO₃ from Mg²⁺ reaction:

$$1.0 \text{ L} \left(\frac{50. \text{ mg Mg}^{2+}}{1.0 \text{ L}} \right) \left(\frac{1.00 \text{ g Mg}^{2+}}{1000 \text{ mg Mg}^{2+}} \right) \left(\frac{1 \text{ mol Mg}^{2+}}{24.305 \text{ g Mg}^{2+}} \right) \left(\frac{1 \text{ mol MgCO}_3}{1 \text{ mol Mg}^{2+}} \right) \left(\frac{84.31 \text{ g MgCO}_3}{1 \text{ mol MgCO}_3} \right) = 0.173 \text{ g MgCO}_3$$

Total mass of solids = 0.206 g + 0.749 g + 0.173 g = 1.128 g = 1.13 g solids (2 sf).

25.2.2. The reaction of CaCO₃ with vinegar:

CaCO₃(s) + 2 CH₃CO₂H(aq) → Ca(CH₃CO₂)₂(aq) + CO₂(g) + H₂O(ℓ).

This reaction would be classified as a gas-forming reaction.

Practicing Skills

The Atmosphere

25.1 Atmospheric gas present in highest concentration in dry air?

According to Table 25.1, the gases in question are present in the following concentrations(ppm):

CH₄	O₃	N₂O	CO
1.8	0.4	0.3	0.1

CH₄ is present in the highest quantity.

25.2. Bonds requiring most energy to be broken:

The double bond in O₂ would require the most energy to be broken. Ozone's O-O bonds would be less strong owing to the existence of the "1.5" bonds between the oxygen atoms.

25.3. Concentration of water vapor = 40,000 ppm. Table 25.1 indicates that the concentrations (in ppm) refer to the relative numbers of particles and hence to the mole fractions. Assuming that the total pressure is 760 torr, the pressure that the water vapor exerts is:
$$\frac{40,000}{1,000,000}(760 \text{ mm Hg}) = 30.4 \text{ mm Hg}$$

Assuming that this is 100% humidity, we can see, from Appendix G, that the temperature at which the vapor pressure of water is approximately 30 torr is 29 °C.

25.4. There is a high concentration of CO₂ sources found in a typical city (human respiration, combustion engines, power plants) so the CO₂ level is likely higher than the average concentration.

25.5. Regarding nitrous acid:

(a) Lewis dot structure for nitrous acid (HONO):   H—Ö—N=Ö

Electron pair geometries: N—3 groups attached, so trigonal planar.

O between H and N—4 groups attached, so tetrahedral.

Terminal O—1 bond to N, so linear.

(b) Using N-O bond dissociation enthalpy to calculate the energy of light needed to break this bond:
According to Table 8.9, the bond dissociation enthalpy for N-O is 201 kJ/mol.

Converting to J/bond: $\dfrac{201 \text{ kJ}}{1 \text{ mol bonds}} \cdot \dfrac{1 \text{ mol bonds}}{6.022 \times 10^{23} \text{ bonds}} \cdot \dfrac{1000 \text{ J}}{1 \text{ kJ}} = 3.34 \times 10^{-19} \text{ J/bond}$

$E = h\nu = \dfrac{hc}{\lambda}$, so $\lambda = \dfrac{hc}{E} = \dfrac{(6.626 \times 10^{-34} \text{ J·s})(2.9979 \times 10^{8} \text{ m/s})}{3.34 \times 10^{-19} \text{ J}} = 5.95 \times 10^{-7} \text{ m}$

corresponding to 595 nm.

25.6. [CO] = 10. ppm (by volume) at P = 1.00 atm and 25 °C.

$\dfrac{10. \text{ L CO}}{10^6 \text{ L air}} = 10.$ ppm by volume so let's ask "What mass of CO is present?"

Using the Ideal Gas Law: mass = $\dfrac{(28\frac{\text{g CO}}{\text{mol}})(1\text{ atm})(10\text{ L})}{(0.082057\frac{\text{L}\cdot\text{atm}}{\text{K}\cdot\text{mol}})(298\text{K})} = 11.45$ g CO

Next question:" What is the mass of 10^6 L of air?"

We need the molar mass of air, which is reported to be 28.96 g air/mol.

mass = $\dfrac{(28.96\frac{\text{g air}}{\text{mol}})(1\text{ atm})(10^6\text{L})}{(0.082057\frac{\text{L}\cdot\text{atm}}{\text{K}\cdot\text{mol}})(298\text{K})} = 1.18 \times 10^6$ g air

So, the concentration is $\dfrac{11.45 \text{ g CO}}{1.18 \times 10^6 \text{ g air}} = 9.66 \times 10^{-6}$ g CO/g air

and 9.66 g CO/10^6 g air or 9.7 ppm (by mass) (2 sf)

As to the concentration in mg/L:
We have 11.45 g CO/10^6 L = 11.45×10^{-6} g/L and converting mass to units of mg:

$\dfrac{11.45 \times 10^{-6}\text{ g}}{1 \text{ L}} \cdot \dfrac{10^3 \text{ mg}}{1 \text{ g}} = 0.011$ mg/L (2 sf)

The Aqua Sphere

25.7. Of the substances listed, $Al_2(SO_4)_3$ is not listed as a pathogen antagonist. Aluminum sulfate is used in the purification process, but not as a substance to kill pathogens.

25.8. (b) Melting of the polar ice cap will result in a rise of sea levels.

Ice at the polar ice cap is already floating in the ocean so it will not increase sea levels as it melts.

25.9. To establish the ratios of a conjugate pair in solution, one can use the Henderson-Hasselbalch equation:

$\text{pH} = \text{p}K_a + \log\dfrac{[\text{conjugate base}]}{[\text{acid}]}$

$8.10 = 10.32 + \log\dfrac{[\text{conjugate base}]}{[\text{acid}]}$

The 10.32 value is the –log of the K_a for HCO_3^- (4.8×10^{-11}).

Solving for the ratio:

$$-2.22 = \log \frac{[CO_3^{2-}]}{[HCO_3^-]} \text{ or } 6.03 \times 10^{-3} = \frac{[CO_3^{2-}]}{[HCO_3^-]}$$

The reciprocal of this ratio is: $\frac{[HCO_3^-]}{[CO_3^{2-}]} = 167 = 170 \text{ (2 sf)}$

25.10. Estimate minimum mass of solid following evaporation of 1.0 L seawater:

Calculate the mass associated with each ion (assume carbonate is present as HCO3⁻)

$0.55 \text{ mol Cl}^- \left(\frac{35.5 \text{ g}}{1 \text{ mol Cl}^-} \right) = 20.0 \text{ g Cl}^-$

$0.46 \text{ mol Na}^+ \left(\frac{23.0 \text{ g}}{1 \text{ mol Na}^+} \right) = 11.0 \text{ g Na}^+$

$0.052 \text{ mol Mg}^{2+} \left(\frac{24.3 \text{ g}}{1 \text{ mol Mg}^{2+}} \right) = 1.3 \text{ g Mg}^{2+}$

$0.01 \text{ mol Ca}^{2+} \left(\frac{40.1 \text{ g}}{1 \text{ mol Ca}^{2+}} \right) = 0.4 \text{ g Ca}^{2+}$

$0.01 \text{ mol K}^+ \left(\frac{39.1 \text{ g}}{1 \text{ mol K}^+} \right) = 0.4 \text{ g K}^+$

$0.03 \text{ mol HCO}_3^- \left(\frac{61.0 \text{ g}}{1 \text{ mol HCO}_3^-} \right) = 2.0 \text{ g HCO}_3^-$

$0.001 \text{ mol HPO}_4^{2-} \left(\frac{96.0 \text{ g}}{1 \text{ mol HPO}_4^{2-}} \right) = 0.1 \text{ g HPO}_4^{2-}$

Minimum mass = 35 g

25.11. Mass of NaCl obtainable from 1.0 L of seawater (concentration of Na⁺ = 460 mmol/L from Table 25.2)

$\frac{460 \text{ mmol Na}^+}{1 \text{ L}} \left(\frac{1 \text{ mmol NaCl}}{1 \text{ mmol Na}^+} \right) \left(\frac{58.44 \text{ mg NaCl}}{1 \text{ mmol NaCl}} \right) \left(\frac{1 \text{ g NaCl}}{1000 \text{ mg NaCl}} \right) = 27 \text{ g NaCl}/\text{L}$

25.12. Using data from Table 25.2, estimate charge balance in seawater:

Assuming a 1.0 L sample,

total positive charge = 0.46 mol(1+/ion) + 0.052 mol(2+/ion) + 0.01 mol(2+/ion)

+ 0.01 mol(1+/ion) = ~0.6

total negative charge = 0.55 mol(1–/ion) + 0.03 mol(1.5–/ion) + 0.001 mol(1–/ion) = ~0.6

25.13. Balanced net ionic equations:

$Ca(OH)_2(s) + Mg^{2+}(aq) \rightarrow Mg(OH)_2(s) + Ca^{2+}(aq)$

$Mg(OH)_2(s) + 2\ H_3O^+(aq) \rightarrow Mg^{2+}(aq) + 4\ H_2O(\ell)$

$MgCl_2(\ell) \rightarrow Mg(s) + Cl_2(g)$

25.14. Mass of Mg available from 1.0 L of seawater:

$\dfrac{1.0\ L}{1} \cdot \dfrac{0.052\ mol\ Mg^{2+}}{1\ L} \cdot \dfrac{1\ mol\ Mg}{1\ mol\ Mg^{2+}} \cdot \dfrac{24.3\ g\ Mg}{1\ mol\ Mg} = 1.3\ g\ Mg$ (2 sf)

Volume of seawater to provide 100. kg Mg:

$\dfrac{100.\ kg\ Mg}{1} \cdot \dfrac{10^3\ g}{1\ kg} \cdot \dfrac{1\ L}{1.3\ g\ Mg} = 7.9 \times 10^4$ L seawater

25.15. Balance the indicated equations:

Step numbers indicated are those used in Chapter 19.

(i) $NH_4^+(aq) + NO_2^-(aq) \rightarrow N_2(g) + H_2O(\ell)$

Oxidation half-equation	Reduction half-equation	Step
$2\ NH_4^+(aq) \rightarrow N_2(g)$	$2\ NO_2^-(aq) \rightarrow N_2(g)$	1 & 2
	$2\ NO_2^-(aq) \rightarrow N_2(g) + 4\ H_2O(\ell)$	3
$2\ NH_4^+(aq) \rightarrow N_2(g) + 8\ H^+(aq)$	$2\ NO_2^-(aq) + 8\ H^+(aq) \rightarrow N_2(g) + 4\ H_2O(\ell)$	4
$2\ NH_4^+(aq) \rightarrow N_2(g) + 8\ H^+(aq) + 6\ e^-$	$2\ NO_2^-(aq) + 8\ H^+(aq) + 6\ e^- \rightarrow N_2(g) + 4\ H_2O(\ell)$	5 & 6
$2\ NH_4^+(aq) + 2\ NO_2^-(aq) \rightarrow 2\ N_2(g) + 4\ H_2O(\ell)$		7
$NH_4^+(aq) + NO_2^-(aq) \rightarrow N_2(g) + 2\ H_2O(\ell)$		

(ii) $NH_4^+(aq) + O_2(g) \rightarrow NO_2^-(aq) + H_2O(\ell)$

Oxidation half-equation	Reduction half-equation	Step
$NH_4^+(aq) \rightarrow NO_2^-(aq)$	$O_2(g) \rightarrow H_2O(\ell)$	1 & 2
$NH_4^+(aq) + 2 H_2O(\ell) \rightarrow NO_2^-(aq)$	$O_2(g) \rightarrow 2 H_2O(\ell)$	3
$NH_4^+(aq) + 2 H_2O(\ell) \rightarrow NO_2^-(aq) + 8 H^+(aq)$	$4 H^+(aq) + O_2(g) \rightarrow 2 H_2O(\ell)$	4
$NH_4^+(aq) + 2 H_2O(\ell) \rightarrow NO_2^-(aq) + 8 H^+(aq) + 6 e^-$	$4 H^+(aq) + O_2(g) + 4 e^- \rightarrow 2 H_2O(\ell)$	5
$2 NH_4^+(aq) + 4 H_2O(\ell) \rightarrow 2 NO_2^-(aq) + 16 H^+(aq) + 12 e^-$	$12 H^+(aq) + 3 O_2(g) + 12 e^- \rightarrow 6 H_2O(\ell)$	6
$2 NH_4^+(aq) + 3 O_2(g) \rightarrow 2 NO_2^-(aq) + 4 H^+(aq) + 2 H_2O(\ell)$		7

25.16. Regarding Ag^+ in water:

(a) Ag_2S solubility should increase because hydrolysis of the sulfide ion reduces sulfide ion concentration on the product side of the Ag_2S equilibrium shifting the equilibrium to the product, dissolved, side of the equilibrium.

(b) From Appendix H, Table H.1 you know that

$HS^-(aq) \rightleftarrows H^+(aq) + S^{2-}(aq) \qquad K_{2a} = 1 \times 10^{-19}$

Reverse the equilibrium equation and add it to the one for water autoionization:

$H^+(aq) + S^{2-}(aq) \rightleftarrows HS^-(aq) \qquad 1/K_{2a} = 1 \times 10^{19}$

$H_2O(\ell) \rightleftarrows H^+(aq) + OH^-(aq) \qquad K_w = 1 \times 10^{-14}$

$S^{2-}(aq) + H_2O(\ell) \rightleftarrows HS^-(aq) + OH^-(aq) \qquad K_{b1} = 1 \times 10^5$

Note that for a diprotic acid, $K_{b1} = K_w/K_{a2}$.

(c) Add the equilibrium equation for K_{sp} to the equation for K_{b1}:

$Ag_2S(s) \rightleftarrows 2 Ag^+(aq) + S^{2-}(aq) \qquad K_{sp} = 8 \times 10^{-51}$

$S^{2-}(aq) + H_2O(\ell) \rightleftarrows HS^-(aq) + OH^-(aq) \qquad K_{b1} = 1 \times 10^5$

$Ag_2S(s) + H_2O(\ell) \rightleftarrows 2 Ag^+(aq) + HS^-(aq) + OH^-(aq) \qquad K = 8 \times 10^{-46}$

(d) The new value of K is larger than the K_{sp} for Ag_2S, so the Ag_2S solubility increases as predicted by Le Chatelier's principle. However, the equilibrium constant is still very small and, thus, the solubility of Ag_2S is almost negligible.

Energy, Fossil Fuels

25.17. Of the sources listed, hydroelectric power is a renewable energy source.

25.18. Second largest source of energy:

According to Figure 25.14, the second largest is **renewables**.

25.19. Mass of H_2 expected from the reaction of steam with 100. g of methane (CH_4), petroleum (CH_2) and coal (C):

$$CH_4(g) + H_2O(g) \rightarrow CO(g) + 3\ H_2(g)$$
$$CH_2(g) + H_2O(g) \rightarrow CO(g) + 2\ H_2(g)$$
$$C(s) + H_2O(g) \rightarrow CO(g) + H_2(g)$$

Based on the equations above, we convert the mass to moles of the C containing specie, use the stoichiometric ratio indicated by the balanced equation, and convert to mass of H_2, using the molar mass of H_2.

For methane: $CH_4(g) + H_2O(g) \rightarrow 3\ H_2(g) + CO(g)$

$$100.\ g\ CH_4 \left(\frac{1\ mol\ CH_4}{16.04\ g\ CH_4}\right)\left(\frac{3\ mol\ H_2}{1\ mol\ CH_4}\right)\left(\frac{2.02\ g\ H_2}{1\ mol\ H_2}\right) = 37.7\ g\ H_2$$

For petroleum: $CH_2(\ell) + 2\ H_2O(g) \rightarrow 2\ H_2(g) + CO(g)$

$$100.\ g\ CH_2 \left(\frac{1\ mol\ CH_2}{14.03\ g\ CH_2}\right)\left(\frac{2\ mol\ H_2}{1\ mol\ CH_2}\right)\left(\frac{2.02\ g\ H_2}{1\ mol\ H_2}\right) = 28.8\ g\ H_2$$

For coal: $C(s) + H_2O(g) \rightarrow H_2(g) + CO(g)$

$$100.\ g\ C \left(\frac{1\ mol\ C}{12.01\ g\ C}\right)\left(\frac{1\ mol\ H_2}{1\ mol\ C}\right)\left(\frac{2.02\ g\ H_2}{1\ mol\ H_2}\right) = 16.8\ g\ H_2$$

25.20. Estimate carbon mass percentage in gasoline:

Burning gasoline releases 47 kJ/g, burning C releases 32.8 kJ/g, and burning H_2 releases 119.9 kJ/g.
Fraction of C in gasoline + Fraction of H in gasoline = 1
47 kJ/g = 32.8 kJ/g (fraction of C in gasoline) + 119.9 kJ/g (fraction of H in gasoline)
47 kJ/g = 32.8 kJ/g (x) + 119.9 kJ/g (1 − x)
x = 0.84, so we estimate that gasoline is approximately 84% C and 16% H

25.21. Calculate the energy evolved (in kJ) when 70. lb of coal is burned. Assuming that coal can be represented as carbon (C).

$$\frac{70.\text{ lb C}}{1} \cdot \frac{454 \text{ g C}}{1 \text{ lb C}} \cdot \frac{33 \text{ kJ}}{1 \text{ g C}} = 1.0 \times 10^6 \text{ kJ}$$

25.22. Calculate the enthalpy change per gram and per liter of isooctane:

$$\frac{5.45 \times 10^3 \text{ kJ}}{1 \text{ mol C}_8\text{H}_{18}} \cdot \frac{1 \text{ mol C}_8\text{H}_{18}}{114.2 \text{ g}} = 47.7 \text{ kJ/g}$$

$$\frac{5.45 \times 10^3 \text{ kJ}}{1 \text{ mol C}_8\text{H}_{18}} \cdot \frac{1 \text{ mol C}_8\text{H}_{18}}{114.2 \text{ g}} \cdot \frac{0.688 \text{ g}}{1 \text{ mL}} \cdot \frac{1000 \text{ mL}}{1 \text{ L}} = 3.28 \times 10^4 \text{ kJ/L}$$

25.23. Energy consumption in U.S. is: 7.0 gallons of oil (or 70. lb coal) per person per day.

Compare the energy produced by 7.0 gallons of oil with that of 70. lb of coal (1.0×10^6 kJ — see problem 25.21). From Table 25.3, the energy released per gram of crude petroleum (43 kJ/g) is found.

$$7.0 \text{ gal oil} \left(\frac{4 \text{ qts}}{1 \text{ gal}}\right)\left(\frac{1000 \text{ mL}}{1.0567 \text{ qt}}\right)\left(\frac{0.8 \text{ g oil}}{1 \text{ mL oil}}\right)\left(\frac{43 \text{ kJ}}{1 \text{ g oil}}\right) = 9.1 \times 10^5 \text{ kJ} \approx 9 \times 10^5 \text{ kJ (1 sf)}$$

The energy released is slightly less for petroleum than for coal but close to the same amount.

25.24. Energy to power a 100 W light bulb for 24 hrs:

$$\frac{100 \text{ W}}{1} \cdot \frac{1 \text{ J/s}}{1 \text{ W}} \cdot \frac{1 \text{ kJ}}{1000 \text{ J}} \cdot \frac{3600 \text{ s}}{1 \text{ hr}} \cdot \frac{24 \text{ hr}}{1 \text{ day}} = 8640 \text{ kJ/day} \quad \text{and} \quad \frac{8640 \text{ kJ}}{1 \text{ day}} \cdot \frac{1 \text{ g coal}}{33 \text{ kJ}} = 260 \text{ g coal}$$

25.25. Confirm that oxidation of 1.0 L of CH₃OH to form CO₂(g) and H₂O(ℓ) in a fuel cell provides at least 5.0 kW-h of energy (Density of CH₃OH = 0.787 g/mL).

We need to convert 1.0 L of methanol into mass and moles:

$$\frac{1.0 \text{ L CH}_3\text{OH}}{1} \cdot \frac{787 \text{ g CH}_3\text{OH}}{1.0 \text{ L}} \cdot \frac{1 \text{ mol CH}_3\text{OH}}{32.04 \text{ g CH}_3\text{OH}} = 25 \text{ mol CH}_3\text{OH}$$

Now the question is what energy change occurs when methanol burns.

That question is answered by using thermodynamic data (from Appendix L)

2 CH₃OH(ℓ) + 3 O₂(g) → 2 CO₂(g) + 4 H₂O(ℓ)

Δ_rH° = [2 Δ_fH° CO₂(g) + 4 Δ_fH° H₂O(ℓ)] – [2 Δ_fH° CH₃OH(ℓ) + 3 Δ_fH° O₂(g)]

$\Delta_rH° = [2 \text{ mol} \cdot (-393.509 \text{ kJ/mol}) + 4 \text{ mol} \cdot (-285.83 \text{ kJ/mol})]$

$\qquad - [2 \text{mol} \cdot (-238.4 \text{ kJ/mol}) + 3 \text{ mol} \cdot 0)]$

$= -1453.5 \text{ kJ or } -726.8 \text{ kJ/mol}$

This energy change is associated with 2 mol of CH_3OH (balanced equation).

The energy change per mol is then multiplied by the number of moles of methanol associated with 1.0 L of CH_3OH

$$\frac{-1453.5 \text{ kJ}}{2 \text{ mol } CH_3OH} \cdot \frac{25 \text{ mol } CH_3OH}{1} = -1.8 \times 10^4 \text{ kJ} \quad (2 \text{ sf})$$

Now converting that energy change to units of kwh:

$$\frac{1.8 \times 10^4 \text{ kJ}}{1} \cdot \frac{1000 \text{ J}}{1 \text{ kJ}} \cdot \frac{1 \text{ watt}}{1 \text{ J/s}} \cdot \frac{1 \text{ hr}}{3600 \text{ s}} \cdot \frac{1 \text{ kw}}{1000 \text{ watt}} = 5.0 \text{ kwh}$$

25.26. Regarding energy released per gram in combustion:

For each of the four substances, we need to solve for the energy released during the reaction: $X + O_2(g) \rightarrow XO_2 (+ H_2O)$. The data for isooctane is obtained from SQ25.22.

Then convert kJ/mol to kJ/g by dividing by the appropriate atomic/molecular weight:

C_8H_{18}: $(5.45 \times 10^3 \text{ kJ/mol})(1 \text{ mol}/114.23 \text{ g}) = 47.7 \text{ kJ/g}$

CH_4: $(802 \text{ kJ/mol})(1 \text{ mol}/16.04 \text{ g}) = 50.0 \text{ kJ/g}$

$C(s)$: $(393.5 \text{ kJ/mol})(1 \text{ mol}/12.011 \text{ g}) = 32.8 \text{ kJ/g}$

H_2: $(241.83 \text{ kJ/mol})(1 \text{ mol}/2.01594 \text{ g}) = 119.96 \text{ kJ/g}$

Listed by decreasing energy/g: $H_2 > CH_4 > C_8H_{18} > C(s)$; hydrogen is best per gram

25.27. Regarding the energy striking a parking lot:

(a) Energy striking the area of the parking lot:

The parking lot is 325 m long and 50.0 m wide, or a surface area of 16250 m²

If the solar radiation is 2.6×10^7 J/m² (per day), the parking lot would receive: 2.6×10^7 J/m² $\cdot$ 16250 m² = 4.2×10^{11} J (per day).

(b) Mass of coal to supply that energy:

$$4.2 \times 10^{11} \text{ J} \cdot \frac{1 \text{ kJ}}{1 \times 10^3 \text{ J}} \cdot \frac{1 \text{ g coal}}{33 \text{ kJ}} \cdot \frac{1 \text{ kg coal}}{1 \times 10^3 \text{ g coal}} = 1.3 \times 10^4 \text{ kg coal}$$

25.28. Quantity of energy lost per day:

$$\frac{1 \text{ day}}{1} \cdot \frac{24 \text{ hr}}{1 \text{ day}} \cdot \frac{1.0 \times 10^6 \text{ J}}{1 \text{ hr}} \cdot \frac{1 \text{ kJ}}{1 \times 10^3 \text{ J}} = 2.4 \times 10^4 \text{ kJ}$$

kWh lost through door: $\dfrac{2.4 \times 10^4 \text{ kJ}}{1} \cdot \dfrac{1000 \text{ J}}{1 \text{ kJ}} \cdot \dfrac{1 \text{ kwh}}{3.60 \times 10^6 \text{ J}} = 6.7$ kwh (per day)

25.29. Energy consumed to drive 1.00 mile by a car rated at 55.0 mpg. The density of gasoline is 0.737 g/cm³; gasoline produces 48.0 kJ/g.

$$\frac{48.0 \text{ kJ}}{1 \text{ g gasoline}} \cdot \frac{0.737 \text{ g gasoline}}{1 \text{ cm}^3 \text{ gasoline}} \cdot \frac{1000 \text{ cm}^3}{1 \text{ L}} \cdot \frac{3.785 \text{ L}}{1 \text{ gal}} \cdot \frac{1 \text{ gal}}{55.0 \text{ mile}} \cdot \frac{1.00 \text{ mile}}{1} = 2.43 \times 10^3 \text{ kJ}$$

25.30. Calculate efficiency of microwave oven:

Assume the density of water is 1.00 g/mL
q_{water} = (225 g)(4.184 J/g·K)(340. K – 293 K) = 4.4×10^4 J

$$\frac{1100 \text{ W}}{1} \cdot \frac{90 \text{ s}}{1} \cdot \frac{1 \text{ J/s}}{1 \text{ W}} = 9.9 \times 10^4 \text{ J}$$

Efficiency = $\dfrac{4.4 \times 10^4 \text{ J}}{9.9 \times 10^4 \text{ J}} \cdot 100\% = 44\%$ efficient

25.31. Regarding methane hydrate:

(a) water molecules in the cage: With each point in the lattice (red spheres) representing an oxygen atom, there are 20 O atoms corresponding to 20 water molecules

(b) # of hydrogen bonds involved. Each edge contains a H-bond, so we need to determine the number of edges. The diagram at right shows that the front half of the cage has 10 edges common to those 6 pentagons. The back half would also

have 10 edges common to the "back 6 pentagons" and the unnumbered edges (10 of them) would connect the pentagons in the front half to the pentagons in the back half—for a total of 30 edges, and therefore 30 hydrogen bonds along the edges.

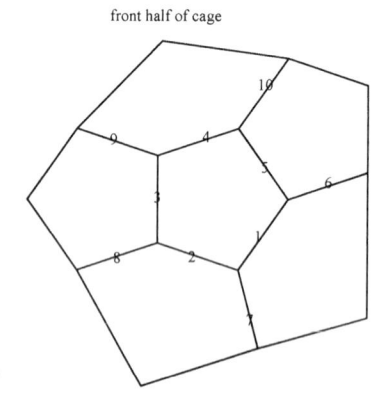

(c) # of faces in the cage: The "front" half of the picture shown has 5 faces around the periphery and 1 in front—for a total of 6 faces. The "back" half would have the same number, and therefore there are 12 faces in the cage.

25.32. Energy from 1.00 m³ of methane hydrate:

There are 1000 L to 1 m³, so the volume is 1.64×10^5 L of CH4. Use the ideal gas law (at STP) to determine the moles of methane.

$$n = \frac{PV}{RT} = \frac{(1.00 \text{ atm})(1.64 \times 10^5 \text{ L})}{(0.082057 \text{ L} \cdot \text{atm/K} \cdot \text{mol})(273.15 \text{ K})} = 7.317 \times 10^3 \text{ mol CH}_4$$

(802 kJ/mol)(7.317 × 10³ mol) = 5.87×10^6 kJ

[SQ 25.26 is the source of 802 kJ/mol.]

Environmental Impact of Fossil Fuels

25.33. Of the species given, O₂ is not a greenhouse gas.

25.34. Correlation between atmospheric CO₂ and ocean pH?

(a) As CO₂ increases, the pH decreases.

25.35. Regarding dinitrogen monoxide:

(a) Molecular geometry is linear. Lewis dot and formal charges:

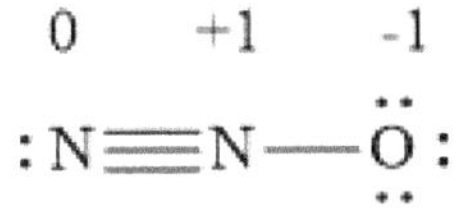

$$\overset{0}{:}N\equiv\overset{+1}{N}-\overset{-1}{\underset{..}{\overset{..}{O}}}:$$

(b) Why is N-O-N not likely? More electronegative O should not get positive formal charge; also, formal charges are greater than in structure in part (a).

Formal charges for an N-O-N structure:

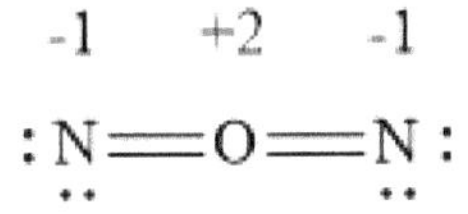

$$\overset{-1}{:}N=\overset{+2}{O}=\overset{-1}{N}:$$

(c) Mass of NO₂ per liter:

(800 × 10⁻⁹ mol/L)(44.0128 g/mol) = 4×10^{-5} g/L

25.36. Mass of coal to produce 1.74×10^6 tons of SO₂:

$$1.74 \times 10^6 \text{ tons SO}_2 \left(\frac{2000 \text{ pounds}}{1 \text{ ton}}\right)\left(\frac{454 \text{ g}}{1 \text{ pound}}\right) = 1.58 \times 10^{12} \text{ g SO}_2$$

$0.70(1.58 \times 10^{12} \text{ g SO}_2) = 1.11 \times 10^{12}$ g SO₂ from coal-fired plants

Mass of (2% S) coal: $0.020 \left(\dfrac{64.06 \text{ g/mol SO}_2}{32.06 \text{ g/mol S}} \right) x = 1.11 \times 10^{12}$ g

And solving for x: $x = 2.8 \times 10^{13}$ g coal

Mass of coal expressed in metric tons:

2.8×10^{13} g coal $\left(\dfrac{1 \text{ kg coal}}{1000 \text{ g coal}} \right) \left(\dfrac{1 \text{ metric ton coal}}{1000 \text{ kg coal}} \right) = 2.8 \times 10^{7}$ metric tons coal

Alternative Fuels

25.37. Choice that is *not* a limitation for H₂ as a fuel:
(c) H₂ is explosive.

25.38. Which of the following statements is true?
(c) Petroleum is a mix of hydrocarbons.

25.39. Regarding the use of methyl myristate as a fuel:
(a) Balanced equation for the combustion of methyl myristate:

$C_{13}H_{27}CO_2CH_3(\ell) + O_2(g) \rightarrow CO_2(g) + H_2O(g)$ (unbalanced)

$2\ C_{13}H_{27}CO_2CH_3(\ell) + 43\ O_2(g) \rightarrow 30\ CO_2(g) + 30\ H_2O(g)$

(b) Standard enthalpy change for the oxidation of 1.00 mol of methyl myristate

$2\ C_{13}H_{27}CO_2CH_3(\ell) + 43\ O_2(g) \rightarrow 30\ CO_2(g) + 30\ H_2O(g)$

$\Delta_fH°$(kJ/mol)   −771.0   0   −393.509   −241.83

$\Delta_rH° = \sum\Delta_fH°$ products − $\sum\Delta_fH°$ reactants
= [(30 mol)(−393.509 kJ/mol) + (30 mol)(−241.83 kJ/mol)]
− [(2 mol)(−771.0 kJ/mol) + (0 kJ/mol)]

$\Delta_rH° = (−19060.17$ kJ$) - (−1542.0$ kJ$) = −17518.2$ kJ for 2 mol of methyl myristate!
So, for **1 mole** of the compound: –8759.1 kJ/mol

(c) For hexadecane, the calculation:

The balanced equation: $2\ C_{16}H_{34}(\ell) + 49\ O_2(g) \rightarrow 32\ CO_2(g) + 34\ H_2O(g)$

$\Delta_rH° = \sum\Delta_fH°$ products − $\sum\Delta_fH°$ reactants

$\Delta_rH°$ = [(32 mol)(−393.509 kJ/mol) + (34 mol)(−241.83 kJ/mol)]
− [(2 mol)(−456.1 kJ/mol) + (49 mol)(0 kJ/mol)]

$\Delta_rH° = -20814.5$ kJ $- (-912.2$ kJ$) = -19902.3$ kJ—for 2 mol of $C_{16}H_{34}$ (ℓ), so the energy change **per mole** is −9951.2 kJ. The **greater energy per mole** is realized with hexadecane.

To determine the greater energy per liter, we need to convert the molar masses of each into mass (g) and calculate the volume associated with 1 mol of each fuel. The molar mass of methyl myristate is 242.4 g/mol and that of hexadecane is 226.4 g/mol.

For methyl myristate:

$$\frac{-8759.1 \text{ kJ}}{1 \text{ mol}} \cdot \frac{1 \text{ mol}}{242.4 \text{ g}} \cdot \frac{0.86 \text{ g}}{1 \text{ mL}} \cdot \frac{1000 \text{ mL}}{1 \text{ L}} = -31075.436 \text{ kJ/L or } -3.1 \times 10^4 \text{ kJ/L (2 sf)}$$

For hexadecane:

$$\frac{-9951.2 \text{ kJ}}{1 \text{ mol}} \cdot \frac{1 \text{ mol}}{226.4 \text{ g}} \cdot \frac{0.77 \text{ g}}{1 \text{ mL}} \cdot \frac{1000 \text{ mL}}{1 \text{ L}} = -33837.451 \text{ kJ/L or } -3.4 \times 10^4 \text{ kJ/L (2 sf)}$$

25.40. Equilibrium constant for the decomposition reaction:

$\Delta_r H° = \Sigma \Delta_f H°$ products $- \Sigma \Delta_f H°$ reactants

$\Delta_r H° = (1\text{mol})(\Delta_f H°[CO_2(g)]) + (1 \text{ mol})(\Delta_f H°[H_2(g)]) - (1 \text{ mol})(\Delta_f H°[HCO_2H(g)])$

$= -393.509 \text{ kJ} + 0 \text{ kJ} - (-378.6 \text{ kJ}) = -14.9 \text{ kJ/mol-rxn}$

$\Delta S° = (1\text{mol})(\Delta S°[CO_2(g)]) + (1 \text{ mol})(\Delta S°[H_2(g)]) - (1 \text{ mol})(\Delta S°[HCO_2H(g)])$

$= 213.74 \text{ J/mol·K} + 130.7 \text{ J/mol·K} - 248.70 \text{ J/mol·K} = 95.7 \text{ J/mol·K}$

$\Delta G° = \Delta H° - T\Delta S° = -14.9 \text{ kJ/mol} - (298 \text{ K})(1 \text{ kJ}/1000 \text{ J})(95.7 \text{ J/mol·K}) = -43.4 \text{ kJ/mol}$, product-favored

Next, calculate K: $\Delta G° = -RT \ln K$ or $\dfrac{\Delta G°}{-RT} = \ln K$

$$\ln K = \frac{-43.4 \frac{\text{kJ}}{\text{mol}}}{-(8.314472 \frac{\text{J}}{\text{K} \cdot \text{mol}})(\frac{1 \text{ kJ}}{1000 \text{ J}})(298 \text{ K})} = 17.5 \text{ and } K = 4 \times 10^7$$

Green Chemistry

25.41. Regarding reactions associated with acetonitrile:

(a) Lewis structures:

In these structures, bonding pairs of electrons are indicated with lines. Only lone pairs are indicated with dots.

Ethanol:
Both carbons have 4 atoms attached, and the molecular geometry is tetrahedral. The O atom is bent (having two bonding pairs and two lone pairs)

Ethylamine:

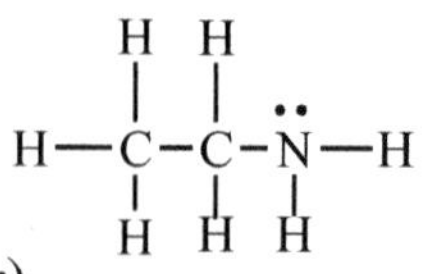

Both carbons have 4 atoms attached, and the molecular geometry is tetrahedral. The N atom is pyramidal (having three bonding pairs and one lone pair)

Acetonitrile:

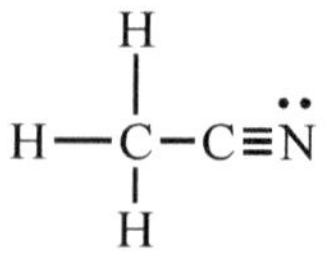

The methyl carbon has 4 atoms attached, and the molecular geometry is tetrahedral.

The C attached to the N has only two atoms attached, so the molecular geometry is linear. The N atom is attached to one atom and one lone pair, and the molecular geometry is linear.

Acrylonitrile:

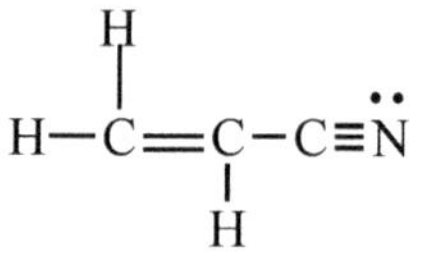

The carbon atoms associated with the double bond have 3 atoms attached so the molecular geometry is trigonal planar. The C attached to the N has only two atoms attached, so the molecular geometry is linear. The N atom is attached to one atom and one lone pair, and the molecular geometry is linear.

(b) Balanced equation for the steps in the synthesis.

Step 1: $CH_3CH_2OH + NH_3 \rightarrow CH_3CH_2NH_2 + H_2O$

Step 2: $CH_3CH_2NH_2 + O_2 \rightarrow CH_3CN + 2\ H_2O$

(c) Atom economy of the synthesis of acetonitrile from ethanol.

The reactants are ethanol, ammonia, and oxygen, with a total of 2 C, 9 H, 3 O + 1 N with a combined "molar mass" of 95.1. The molar mass of acrylonitrile is 41.1. So the atom economy is 41.1/95.1 or 0.432 and can be expressed as 43%.

25.42. Regarding trifluoromethanesulfonic acid:

The geometry around the C is tetrahedral, and the geometry around the S is tetrahedral. Their hybridization is sp^3.

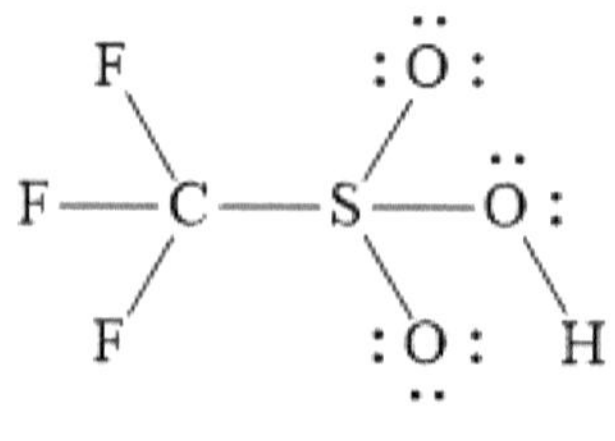

General Questions

25.43. Calculate the wavelength of light with sufficient energy to break the C-Cl bond:

According to Table 8.8, the C-Cl bond has a bond energy = 339 kJ/mol.

Convert this energy to J/bond:

$$\frac{339 \text{ kJ}}{1 \text{ mol bonds}} \cdot \frac{1 \text{ mol bonds}}{6.022 \times 10^{23} \text{ bonds}} \cdot \frac{10^3 \text{ Joules}}{1 \text{ kJ}} = 5.63 \times 10^{-19} \text{ J}$$

Since $E = h\nu = h \cdot \dfrac{c}{\lambda}$ and $\lambda = \dfrac{hc}{E}$

$$\frac{6.626 \times 10^{-34} \text{ J·s/photon} \cdot 2.998 \times 10^8 \text{m/s}}{5.63 \times 10^{-19} \text{ J}} = 3.53 \times 10^{-7} \frac{\text{m}}{\text{photon}}$$

or 353 nanometers. This wavelength is in the UV region.

25.44. Regarding the formation of NO from N_2 and O_2:

$\Delta_r H° = (1 \text{ mol})(\Delta_f H°[NO(g)]) - (½ \text{ mol})(\Delta_f H°[N_2(g)]) - (½ \text{ mol})(\Delta_f H°[O_2(g)])$

$\qquad = 90.29 \text{ kJ} - 0 - 0 = 90.29 \text{ kJ/mol}$

$\Delta S° = (1 \text{mol})(\Delta S°[NO(g)]) - (½ \text{ mol})(\Delta S°[N_2(g)]) - (½ \text{ mol})(\Delta S°[O_2(g)])$

$\qquad = 210.76 \text{ J/mol·K} - ½(191.56 \text{ J/mol·K}) - ½(205.07 \text{ J/mol·K}) = 12.45 \text{ J/mol·K}$

At 298 K, $\Delta G° = \Delta H° - T\Delta S°$

$\qquad = 90.29 \text{ kJ/mol} - (298 \text{ K})(1 \text{ kJ/1000 J})(12.45 \text{ J/mol·K}) = 86.58 \text{ kJ/mol}$

(value matches $\Delta_f G°$ at 298 in Appendix L)

At 1000 K, $\Delta G° = \Delta H° - T°\Delta S°$

$\qquad = 90.29 \text{ kJ/mol} - (1000 \text{ K})(1 \text{ kJ/1000 J})(12.45 \text{ J/mol·K}) = 77.84 \text{ kJ/mol}$

Now calculate the values for K at the two temperatures:

At 298 K, $\Delta G° = -RT \ln K$

$\qquad = -(8.314472 \text{ J/mol·K})(1 \text{ kJ/1000 J})(298 \text{ K})\ln K = 86.58 \text{ kJ/mol}$

and solving for K: $K = 6.7 \times 10^{-16}$

At 1000 K, $\Delta G° = -RT \ln K$

$\qquad = -(8.314472 \text{ J/mol·K})(1 \text{ kJ/1000 J})(1000 \text{ K})\ln K = 77.84$

and solving for K: $K = 8.6 \times 10^{-5}$

25.45. Balanced half-equation for (i) oxidation of ammonium ion to nitrite ion, and (ii) subsequent oxidation of nitrite ion to nitrate ion:

The steps noted in Chapter 19 were used to balance both equations

(i) $NH_4^+(aq) + 2 H_2O(\ell) \rightarrow NO_2^-(aq) + 8 H^+(aq) + 6 e^-$

(ii) $NO_2^-(aq) + H_2O(\ell) \rightarrow NO_3^-(aq) + 2 H^+(aq) + 2 e^-$

25.46. Methods of synthesizing each compound:

Na_3PO_4 – react H_3PO_4 with excess NaOH

Na_2HPO_4 – react H_3PO_4 with limited NaOH

NaH_2PO_4 – react H_3PO_4 with limited NaOH

K_3PO_4 – react H_3PO_4 with excess KOH

$(NH_4)_2HPO_4$ – react H_3PO_4 with $NH_3(\ell)$ or $NH_3(aq)$

$CaHPO_4$ – react H_3PO_4 with excess $Ca(OH)_2$

$Ca(H_2PO_4)_2$ – react H_3PO_4 with limited $Ca(OH)_2$

25.47 Compare ethanol and isooctane as fuels:

(a) Enthalpy change for combustion of 1.00 kg of ethanol and of 1.00 kg of isooctane:

$$C_2H_5OH\,(\ell) + 3\,O_2(g) \rightarrow 2\,CO_2(g) + 3\,H_2O(\ell)$$

$\Delta_fH°$(kJ/mol)   –277.0   0   –393.509   –285.83

$\Delta_rH° = \sum\Delta_fH°$ products $- \sum\Delta_fH°$ reactants

$\Delta_rH° = [(2\,\text{mol})(-393.509\,\text{kJ/mol}) + (3\,\text{mol})(-285.83\,\text{kJ/mol})]$

$- [(1\,\text{mol})(-277.0\,\text{kJ/mol}) + (3\,\text{mol})(0\,\text{kJ/mol})]$

$= (-1644.508\,\text{kJ}) - (-277.0\,\text{kJ}) = -1367.5$ kJ/mol ethanol.

$\Delta_rH°$ on a *per gram basis*:

$$\frac{-1.3675 \times 10^3 \text{ kJ}}{1 \text{ mol } C_2H_5OH}\left(\frac{1 \text{ mol } C_2H_5OH}{46.069 \text{ g } C_2H_5OH}\right) = -29.684 \text{ kJ/g } C_2H_5OH$$

for 1.00 kg of ethanol, the enthalpy change is -29.684×10^3 kJ or -2.9684×10^4 kJ

For isooctane: $2\,C_8H_{18}(\ell) + 25\,O_2(g) \rightarrow 16\,CO_2(g) + 18\,H_2O(\ell)$

$\Delta_fH°$(kJ/mol)   –259.3   0   –393.509   –285.83

$\Delta_rH° = \sum\Delta_fH°$ products $- \sum\Delta_fH°$ reactants

$= [(16\,\text{mol})(-393.509\,\text{kJ/mol}) + (18\,\text{mol})(-285.83\,\text{kJ/mol})]$

$- [(2\,\text{mol})(-259.3\,\text{kJ/mol}) + (0\,\text{kJ/mol})]$

$\Delta_rH° = (-11441.084\,\text{kJ}) - (-518.6\,\text{kJ}) = -10922.484$ kJ for 2 mol of isooctane.

$= -5461.2$ kJ/mol isooctane

On a *per gram basis*: $\dfrac{-5.4612 \times 10^3 \text{ kJ}}{1 \text{ mol } C_8H_{18}}\left(\dfrac{1 \text{ mol } C_8H_{18}}{114.232 \text{ g } C_8H_{18}}\right) = -47.808$ kJ/g C_8H_{18}

for 1.00 kg of isooctane, the enthalpy change is -47.808×10^3 kJ or -4.7808×10^4 kJ

Comparing the two fuels, *isooctane delivers more energy than ethanol.*

(b) Which fuel produces more CO_2 per kg?

Consider the balanced equations for the two fuels. From the balanced equations, 1 mole of ethanol produces 2 moles of CO_2, while 1 mole of isooctane produces 8.
1 mol of ethanol has a mass of 46.1 grams, and 1 mol of isooctane a mass of 114.2 grams. Divide the mass of each into the # of moles of CO_2:

$$\frac{2 \text{ mol } CO_2}{46.1 \text{ g } C_2H_5OH} \cdot \frac{1000 \text{ g } C_2H_5OH}{1 \text{ kg } C_2H_5OH} = \frac{43.4 \text{ mol } CO_2}{\text{kg } C_2H_5OH} \text{ and for isooctane:}$$

$$\frac{8 \text{ mol } CO_2}{114.2 \text{ g } C_8H_{18}} \cdot \frac{1000 \text{ g } C_8H_{18}}{1 \text{ kg } C_8H_{18}} = \frac{70.0 \text{ mol } CO_2}{\text{kg } C_8H_{18}}$$

Note that the amount of CO_2
(as well as the mass) is greater per kg for isooctane.

(c) Which is better fuel in terms of energy production and greenhouse gases?

Isooctane produces more energy per kg of fuel than ethanol, but also produces more greenhouse gas (carbon dioxide) per kg.

25.48. Using the graph provided:

(a) pH at which $[H_2CO_3] = [HCO_3^-]$; the graph indicates a pH of ~6.4.

(b) pH at which $[HCO_3^-] = [CO_3^{2-}]$; the graph indicates a pH of ~10.3

(c) predominant species at pH = 8: HCO_3^-

(d) species present at pH = 7: HCO_3^- and H_2CO_3

In the Laboratory

25.49. Graduated cylinder contains 100. mL of liquid water at 0 °C. An ice cube of volume 25 cm³ is dropped into the cylinder.

(a) If 92% of the ice is below water level, then the corresponding volume of water is displaced by the ice, resulting in an apparent gain of volume. The volume increase is 0.92 × 25 cm³ or 23 cm³. So, the volume in the graduated cylinder will increase to 123 mL.

(b) When the ice melts, the volume will be equal to the initial volume of the water (100. mL) **plus** the volume of the ice, which is now liquid water. So, the question is, "What mass of water is contained in the original 25 cm³ of ice? With a density of 0.92, the mass of the ice will be 23 g. This mass plus the mass of the original water (now **both** with a density of 1.0 g/cm³) will be 123 g and with a volume of 123 mL.

25.50. Percentage of As₂O₃ in the mineral sample:

$$0.0457 \text{ L}\left(\frac{0.0480 \text{ mol I}_3^-}{1 \text{ L}}\right)\left(\frac{1 \text{ mol H}_3\text{AsO}_3}{1 \text{ mol I}_3^-}\right)\left(\frac{1 \text{ mol As}_2\text{O}_3}{2 \text{ mol H}_3\text{AsO}_3}\right)\left(\frac{197.84 \text{ g As}_2\text{O}_3}{1 \text{ mol As2O}_3}\right) = 0.2170 \text{ g As}_2\text{O}_3$$

$$\% \text{ As}_2\text{O}_3 = \frac{0.2170 \text{ g As}_2\text{O}_3}{0.562 \text{ g mineral}} \times 100\% = 38.6\%$$

Summary and Conceptual Questions

25.51. Renewable resources are defined as those that can be concurrently replenished. Nonrenewable resources are defined as those for which the energy is used and not replenished. Of the energy resources listed, those that **are** renewable: solar, geothermal, and wind.

25.52. Three most abundant gases in the atmosphere:

N₂, O₂, Ar (in dry air); none are greenhouse gases. For moist air, the third most abundant gas can be H₂O, which is a greenhouse gas.

25.53. Identify sources of mercury, lead, and arsenic pollution:

Arsenic: Wood preservatives, pesticides, and electronic devices are major areas. Other problems arise from leaching from old waste dumps or mines—contaminating water.

Mercury: Coal-fired power plants are the primary sources of mercury pollution, as mercury is exhausted via smoke stacks. Other industrial processes provide additional areas of concern. The well-known case of Minamata disease in the 1950s was a result of the improper disposal of mercury from chlor-alkali plants in Japan.

Lead: Lead-based paints, batteries, pipes, solder, and lead-glazed pottery are sources. The removal of lead from paints has decreased that source as a major area of concern, but lead-acid batteries (still common in automobiles) can be problematic, when these batteries are recycled. Soil containing residues of emissions from lead-based gasoline is also a source.

25.54. Regarding the rate of increase of atmospheric CO₂:
Atmospheric CO₂ levels increased from about 338 ppm in 1980 to 421 ppm in 2022, a rate of 2 ppm/year. For the 30 years from 2020 to 2050, this would result in an increase of ~60 ppm to concentration of about 466 ppm.

25.55. Sulfur compounds in the atmosphere:

(i) Sulfur compounds that are atmospheric pollutants:

Sulfur dioxide and sulfur trioxide are the more common sulfur compounds that provide pollution. The oxidation of these substances into sulfate particles contributes to harm external surfaces. The reaction of these gases with water is partly responsible for the increased acidic levels of rain. Additionally, SF_6, is used in industrial processes for its electrical insulating abilities. Regrettably the "greenhouse-factor" of this compound is greater than that of CO_2, so levels of this substance must be carefully monitored.

(ii) Origin of those substances:

The oxides of sulfur mostly result from combustion of coal. Coal-fired power plants are major sources of SO_2, as sulfur is a component of coal, with SO_3 and sulfate being produced via further oxidation in the atmosphere. SF_6 is a result of having been produced via synthetic chemical processes.

(iii) Steps to prevent the entry of substances into the atmosphere?
Removal of sulfur from fuels (whether it be petroleum or coal) before combustion is one major avenue of removal of sulfur. Scrubbers are devices that remove sulfur oxides via a chemical reaction to produce useful products—e.g., calcium sulfate which can be used as a fertilizer or a component of drywall products (gypsum).